HANDBOOK OF PAINT AND COATING RAW MATERIALS

HANDBOOK OF PAINT AND COATING RAW MATERIALS

Volume 2
Chemical Products
with Trade Name Cross-Reference

Compiled by

Michael and Irene Ash
Synapse Information Resources

Gower

Published by
Gower Publishing Limited
Gower House
Croft Road
Aldershot
Hampshire GU11 3HR
England

Gower
Old Post Road
Brookfield
Vermont 05036
U.S.A.

British Library Cataloguing-in-Publication Data
Ash, Michael
Handbook of paint and coating raw materials: an international guide to 11,000 products by trade name, chemical, function, and manufacturer
1. Paint materials 2. Protective coatings
I. Title II. Ash, Irene
667.6

ISBN Volume 1 0-566-07824-4, ISBN Volume 2 0-566-07823-6, ISBN Volume set 0-566-07787-6

Library of Congress Cataloguing-in-Publication Data
Ash, Michael
Handbook of paint and coating raw materials: an international guide to 11,000 products by trade name, chemical, function, and manufacturer/compiled by Michael and Irene Ash.
p. cm
Includes index
ISBN Volume 1 0-566-07824-4 (cloth), ISBN Volume 2 0-566-07823-6, ISBN Volume set 0-566-07787-6
1. Paint—Handbooks, manuals, etc. 2. Coatings—Handbooks, manuals, etc. I. Ash, Irene. II. Title
TP935.A85 1996
667′,9′ 0212—dc20 96-704
CIP

The text has been typeset in Arial Narrow by Synapse Information Resources, Inc.

Printed and bound in Great Britain by
Hartnolls Limited, Bodmin, Cornwall

Contents

Preface

These two volumes of the ***Handbook of Paint and Coating Raw Materials*** reference work describe more than 11,000 trade name and generic chemical ingredients that are used in the formulation of paint and coating products for industry and commerce. Extensive information, gathered from more than 1300 worldwide manufacturers, distributors, trade and research journals, is presented for each product profiled in this publication.

The $13 billion paints and coatings market is experiencing profound changes that significantly impact the composition of paint formulations. Mounting environmental concerns have created a race to find material alternatives that reduce VOC (volatile organic compound) content and HAPs (hazardous air pollutants).

Government regulations have placed pressure on manufacturers to use low VOC, non-HAP solvents and the industry is attempting to respond by developing alternative paint and coating systems that do not compromise quality. Chlorinated solvents are under heavy attack. The switch from hydrocarbon systems to waterborne systems creates problems that impact foam, wetting, water sensitivity, flow and leveling, coverage of surface area, which in turn increase the dependence on other additives for the overall quality of the coating system . Reducing solvent content, in an attempt to reduce VOC, pushes resin manufacturers to increase the production of high-solid systems. However, raising solid content increases viscosity of the coating system and thereby decreases surface coverage efficiency. These issues, combined with price escalation of raw materials, have created a highly competitive environment for formulators as well as all key personnel involved in the paint and coating industry.

Thus, with a worldwide focus on alternative materials for the development of efficient coatings, it is essential for those associated with this industry to have immediate access to current, accurate and comprehensive information on these ingredients so that material selection for end formulations can be based on informed decision making. This reference provides such a resource.

This volume covers Chemical Products and incorporates a trade name cross-reference. It is divided into:

Part I—Chemical Product Reference contains an alphabetical listing of approximately 2000 paint and coating chemicals. Each entry includes: CAS (Chemical Abstract Service) numbers, EINECS (European Inventory of Existing Commercial Chemical Substances) numbers, synonyms, molecular and empirical formulas, definition, classification, general properties, uses, regulatory, toxicological, and precaution information wherever possible. In addition, each chemical entry lists the trade name products that are equivalent to the chemical compound or contains that chemical compound as the trade name product's major chemical constituent. Synonyms are thoroughly cross-referenced back to the main entry.

Part II—Chemical Product Functional Cross-Reference contains an alphabetical listing of major coating ingredient functional categories. More than 70 categories are included, e.g., carriers/vehicles; corrosion inhibitors; defoamers; diluents; fillers and extenders; flame/fire retardants; flatting/matte agents; flow control agents; gloss aids; mar resistance aids; pigments and dyes; plasticizers; preservatives; slip agents; solvents; stabilizers; suspending agents; tackifiers; thickeners; uv absorbers/stabilizers; water repellents; wetting agents, etc. Each functional category entry is followed by an alphabetical listing of the generic chemicals that possess that functional attribute.

Part III—Manufacturers Directory contains detailed contact information for the manufacturers of both trade name products and generic chemicals that are referenced in the two volumes of this handbook. Wherever possible telephone, telefax, and telex numbers, toll-free 800 numbers, and complete mailing addresses are included for each manufacturer.

The ***Appendices*** contain the following cross-references:

CAS Number to Chemical Cross-Reference orders chemical compounds found in Part I by CAS numbers.

EINECS Number to Chemical Cross-Reference orders chemical compounds found in Part I by EINECS numbers.

The *Glossary* contains important terms associated with paint and coating technology.

The companion volume covers Trade Name Products and is divided into:

Part I—Trade Name Reference contains approximately 9000 alphabetical entries of trade name paint and coating raw materials. Each entry references its manufacturer, chemical composition, associated CAS and EINECS identifying numbers, general properties, applications and functions, toxicology, and compliance and regulatory information as provided by the manufacturer and other sources.

Part II—Trade Name Functional Cross-Reference contains an alphabetical listing of major paint and coating ingredient functional categories. More than 70 categories are included, e.g., carriers/vehicles; coalescing agents; defoamers; diluents; fillers and extenders; flame/fire retardants, flatting/matte agents; flow control agents; gloss aids; mar resistance aids; pigments and dyes; plasticizers; preservatives; slip agents; solvents; stabilizers; surface modifiers; suspending agents; tackifiers; thickeners; uv absorbers/stabilizers; water repellents; wetting agents, etc. Each functional category entry is followed by an alphabetical listing of the trade name products that possess that functional attribute.

Part III—Manufacturers Directorycontains detailed contact information for the manufacturers of the trade name products and generic chemicals that are referenced in the two volumes of this handbook. Wherever possible, telephone, telefax, and telex numbers, toll-free 800 numbers, and complete mailing addresses are included for each manufacturer.

The ***Appendices*** contain the following cross-references:

*CAS Number to Trade Name Cross-Reference*orders many trade names found in Part I by identifying CAS numbers; it should be noted that trade names contain more than one chemical component and the associated CAS numbers in this section refer to each trade name product's primary chemical component.

*EINECS Number to Trade Name Cross-Reference*orders many trade names found in Part I by identifying EINECS numbers that refer to each trade name product's primary chemical component.

These two volumes are the culmination of many months of research, investigation of product sources, and sorting through a variety of technical data sheets and brochures acquired through personal contacts and correspondences with major chemical manufacturers worldwide as well as trade journals. We are especially grateful to Roberta Dakan for her skills in chemical information database management. Her tireless efforts have been instrumental in the production of this reference.

M. & I. Ash

NOTE:

The information contained in this reference is accurate to the best of our knowledge; however, no liability will be assumed by the publisher or the authors for the correctness or comprehensiveness of such information. The determination of the suitability of these products for prospective use is the responsibility of the user. It is herewith recommended that those who plan to use any of the products referenced seek the manufacturers instructions for the handling of that chemical.

Regulations and standards are complicated and are dependent on both the chemistry and the application involved. Variation in laws and standards exist from country to country. The ultimate decision on compliance must be made by the user with full understanding of the application as well as working with the producer or distributor of the ingredients used in the formulation.

Abbreviations

abs.	absolute
ABS	acrylonitrile-butadiene-styrene
absorp.	absorption
ACGIH	American Conference of Governmental Industrial Hygienists
ACN	acrylonitrile
act.	active
adsorp.	adsorption
AEB	average extent of burning
agric.	agricultural
alc.	alcohol
Am., Amer.	American
amts.	amounts
anhyd.	anhydrous
APHA	American Public Health Association
applic(s).	application(s)
aq.	aqueous
ASA	acrylic-styrene-acrylonitrile
ASTM	American Society for Testing and Materials
ATB	average time of burning
ATH	alumina trihydrate
at.wt.	atomic weight
aux.	auxiliary
avail.	available
avg.	average
a.w.	atomic weight
BGA	Federal Republic of Germany Health Dept. certification
BHA	butylated hydroxyanisole
BHT	butylated hydroxytoluene
biochem.	biochemical
biodeg.	biodegradable
blk.	black
BMC	bulk molding compound
BOD	biological oxygen demand
b.p.	boiling point
BP	British Pharmacopeia
BR	butadiene rubbers, polybutadienes
B&R	Ball & Ring
br., brn.	brown
brnsh.	brownish
B/S	butadiene/styrene
BSS	British Standard Sieve
C	degrees Centigrade
ca.	circa
CAB	cellulose acetate butyrate
CAS	Chemical Abstracts Service
CC	closed cup
cc	cubic centimeter(s)
CCl_4	carbon tetrachloride
CEL	corporate exposure limit
CFR	Code of Federal Regulations (U.S.)
char.	characteristic, characterized
CHDM	cyclohexanedimethanol
chel.	chelation
chem(s).	chemical(s)
CI	Color Index
CIP	cure-in-place
cks	centistoke(s)
cl.	clear
CL	ceiling concentration
cm	centimeter(s)
cm^3	cubic centimeter(s)

CMC	carboxymethylcellulose
CMC	critical Micelle concentration
CNS	central nervous system
CO	carbon monoxide
COC	Cleveland Open Cup
COD	chemical oxygen demand
coeff.	coefficient
COF	coefficient of friction
compat.	compatible
compd(s).	compound(s)
compr.	compression
conc(s).	concentrated, concentration
conduct.	conductive
const.	constant
contg.	containing
cosolv.	cosolvent
cp	centipoise(s)
CPE	chlorinated polyethylene
cps	centipoise(s)
CPVC	chlorinated polyvinyl chloride
CR	chloroprene rubber, polychloroprene
cryst.	crystalline, crystallization
cs, cSt	centistoke(s)
CTFA	Cosmetic, Toiletry and Fragrance Association
ctks	centistoke(s)
DAB	Deutsche. Arzneibuch
DBNPG	Dibromoneopentyl glycol
DBP	dibutyl phthalate
DBU	Diazabicycloundecene
dec.	decomposes
decomp.	decomposition
DEG	diethylene glycol
deliq.	deliquescent
dens.	density
deriv(s).	derivative(s)
descrip.	description
dg	decigram(s)
DGE	diglycidyl ether
DI	deionized
diam.	diameter
DIBK	Diisobutyl ketone
DIDA	diisodecyl adipate
dielec.	dielectric
dil.	dilute
DIOP	diisooctyl phthalate
disp.	dispersible, dispersion
dissip.	dissipation
dist.	distilled
distrib.	distributor
distort.	distortion
DIY	do-it-yourself
dk.	dark
DMF	dimethyl formamide
DOI	distinction of image (paints)
DOP	dioctyl phthalate
DOT	U.S. Department of Transportation
DPM	PPG-2 methyl ether
DW	distilled water, deionized water
EAA	ethylene acrylic acid
eb, EB	electron beam
EC	European Community
ECN	epoxy cresol novolac
ECTFE	ethylene/chlorotrifluoroethylene copolymer
EEW	epoxide equivalent weight

e.g.	for example
EINECS	European Inventory of Existing Commercial Chemical Substances
elec.	electrical
elong.	elongation
EMA, E/MA	ethylene-methyl acrylate
EMI	electromagnetic interference
EO	ethylene oxide
EOEOEA	2-(2-ethoxyethoxy) ethyl acrylate
EOTMPTA	ethoxylated trimethylolpropane triacrylate
EP	European Pharmacopoeia; extreme pressure
EPA	U.S. Environmental Protection Agency
EPDM	ethylene-propylene-diene rubber, ethylene-propylene terpolymer
EPM	ethylene-propylene (copolymer) rubber
EPR	ethylene-propylene rubber
EPS	expandable polystyrene
equip.	equipment
equiv.	equivalent
ES	electrostatic
ESD	electrostatic discharge
esp.	especially
ESP	electrostatic protection
EVA	ethylene vinyl acetate
exc.	excellent
F	degrees Fahrenheit
FAO	Food and Agriculture Organization (United Nations)
FCC	Food Chemicals Codex
FDA	Food and Drug Administration (U.S.)
FEP	fluorinated ethylene propylene
FKM	fluoroelastomer
flamm.	flammable, flammability
flex.	flexural
f.p.	freezing point
FRP	fiberglass-reinforced plastics
ft	foot, feet
F-T	Fischer-Tropsch
g	gram(s)
G	giga
gal	gallon(s)
G-H	Gardner-Holdt
GI	gastro-intestinal
glac.	glacial
GPTA	glyceryl propoxy triacrylate
gr.	gravity
gran.	granules, granular
GRAS	generally regarded as safe
grn.	green
GRP	glass-reinforced plastics, glass-reinforced polyester
GVS	Gardner varnish scale (color)
h	hour(s)
HALS	hindered amine light stabilizer
HAP	hazardous air pollutant
HC	hydrocarbon
HCl	hydrochloride, hydrochloric acid
HDDA	hexanediol diacrylate
HDPE	high-density polyethylene
HDT	heat distortion (deflection) temp.
HF	hydrofluoric acid
Hg	mercury
HIPS	high-impact polystyrene
HLB	hydrophilic lipophilic balance
hr	hour(s)
HTV	high temperature vulcanizing
hyd.	hydroxyl
hydrog.	hydrogenated

Hz	hertz
IBOA	Isobornyl acrylate
i.b.p.	initial boiling point
IC	integrated circuit
I&I	industrial and institutional
IM	intramuscular
in.	inch(es)
inc.	increases, increased
INCI	International Nomenclature Cosmetic Ingredient
incl.	including
incompat.	incompatible
ind.	industries, industrial
ing.	ingestion
ingred(s).	ingredient(s)
inh.	inhalation
inj.	injection
inorg.	inorganic
insol.	insoluble
IntÕl.	International
IP	intraperitoneal
IPA	isopropyl alcohol
IPDI	isophorone diisocyanate
IPP	isopropyl palmitate
IR	infrared
IV	intravenous
J	joule
JECFA	Joint Expert Committee on Food Additives
JHOSPA	Japan Hygienic Olefine and Styrene Plastics Association
JHPA	Japan Hygienic PVC Association
JP	Japanese Pharmacopoeia
JSFA	Japan Standards for Food Additives
k	kilo
KB	Kauri-Butanol
kg	kilogram(s)
KU	Krebs units
l	liter(s)
lb	pound(s)
LC50	lethal concentration 50%
LD50	lethal dose 50%
LDLo	lowest published lethal dose
LDPE	low-density polyethylene
lel	lower explosive level
liq.	liquid
LLDPE	linear low-density polyethylene
lt.	light
Ltd.	Limited
m	milli or meter(s)
m-	meta
M	mega
M	mole
MAK	maximum allowable conc. (German Research Society)
manuf.	manufacturer
max.	maximum
MD	machine direction, mold direction
MDA	methylene dianiline
MDI	methylene diphenylene isocyanate
MDPE	medium density polyethylene
mech.	mechanial
med.	medium
MEHQ	hydroquinone monomethyl ether
MEK	methyl ethyl ketone
mfg.	manufacture
mg	milligram(s)
MIBK	methyl isobutyl ketone

microcryst.	microcrystalline
microgran.	microgranules, microgranular
mil	1/1000th inch
min	minute(s)
min.	mineral, minimum
misc.	miscible, miscellaneous
mixt(s).	mixture(s)
ml	milliliter(s)
mm	millimeter(s)
mo, mos	month(s)
mod.	moderately
monocl.	monoclinic
m.p.	melting point
mPa•s	millipascal-second(s)
mppcf	million particles per cubic foot
mus	mouse
MVTR	moisture vapor transmission rate
m.w.	molecular weight
N	normal
n(20/D)	index of refraction for the sodium D line @ 20 C
nat.	natural
NC	nitrocellulose
NCO	isocyanate radical
need.	needles
neut.	neutral, neutralized
NF	National Formulary
NMP	N-methyl pyrrolidone
no.	number
nonaq.	nonaqueous
nonbiodeg.	nonbiodegradable
nonflamm.	nonflammable
nonyel.	nonyellowing
NPCA	National Paint & Coatings Association
NPG	neopentyl glycol
NR	natural rubber, isoprene rubber (natural)
NSF	National Sanitation Foundation
NV	nonvolatiles
o-	ortho
OC	open cup
ODC	ozone-depleting compound
ODP	ozone-depletion potential
OEM	original equipment manufacturer
OMS	odorless mineral spirits
OPP	oriented polypropylene
OPV	overprint varnishes
org.	organic
OSHA	Occupational Safety and Health Administration
OTC	over-the-counter
o/w	oil-in-water
oz	ounce
p-	para
Pa	Pascal
PAN	polyacrylonitrile
PBT	polybutylene terephthalate
pbw	parts by weight
PC	polycarbonate
PCA	2-pyrrolidone-5-carboxylic acid
PE	polyethylene
PEA	2-phenoxyethyl acrylate
PEG	polyethylene glycol
PEL	permissible exposure level
percut.	percutaneous
PES	polyether sulfone
PET	polyethylene terephthalate

petrol.	petroleum
pH	hydrogen-ion concentration
phr	parts per hundred of rubber or resin
pkg.	packaging
PM, P-M	Pensky-Martens
PMCC	Pensky-Martens closed cup
PMMA	polymethyl methacrylate
PO	propylene oxide
POE	polyoxyethylene, polyoxyethylated
POM	polyoxymethylene
PO-NPGDA	propoxylated neopentyl glycol diacrylate
POP	polyoxypropylene, polyoxypropylated
powd.	powder
PP	polypropylene
ppb	parts per billion
PPG	polypropylene glycol
pph	parts per hundred (percent)
ppm	parts per million
PPO	polyphenylene oxide
PPS	polyphenylene sulfide
pract.	practically
prep(s).	preparation(s)
prod(s).	product(s), production
props.	properties
ps	poise
PS	polystyrene
psi	pounds per square inch
psia	pounds per square inch absolute
pt.	point
Pt-Co	platinum-cobalt
PTFE	polytetrafluoroethylene
PTMEG	polytetramethylene ether glycol
PU	polyurethane
PUR	polyurethane
PVA	polyvinyl alcohol
PVAc	polyvinyl acetate
PVAL	polyvinyl alcohol
PVB	polyvinyl butyral
PVC	polyvinyl chloride
PVC	pigment volume concentration
PVDC	polyvinylidene chloride
PVDF	polyvinylidene fluoride
PVM/MA	polyvinyl methyl ether/maleic anhydride
PVP	polyvinylpyrrolidone
quat.	quaternary
R&B	Ring & Ball
rbt	rabbit
R&D	research and development
RCRA	Resource Conservation and Recovery Act (40CFR §261)
rdsh.	reddish
rec.	recommended
ref.	refractive
regs.	regulations
rep.	represents
resist.	resistance, resistant, resistivity
resp.	respectively
RFI	radio frequency interference
r.h.	relative humidity
rhomb.	rhombic
RIM	reaction injection molded/molding
rpm	revolutions per minute
R.T.	room temperature
RTM	resin transfer molding
RTV	room temperature vulcanizing

s	second(s), secondary
SAN	styrene-acrylonitrile
sapon.	saponification
SARA	Superfund Amendments and Reauthorization Act
sat.	saturated
S/B	styrene/butadiene
SBR	styrene/butadiene rubber
SD	specially denatured
SDA	specially denatured alcohol
SE	self-emulsifying
sec.	secondary
sl.	slight, slightly
sm.	small
SMC	sheet molding compound
soften.	softening
sol.	soluble, solubility
sol'n.	solution
solid.	solidification
solv(s).	solvent(s)
sp.	specific
spec.	specification, specialty
SR	styrene rubber
SSU	Saybolt Universal Seconds
std.	standard
STEL	short term exposure limit
Stod.	Stoddard solvent
str.	strength
subcut.	subcutaneous
subl.	sublimes
surf.	surface
SUS	Saybolt Universal Seconds
susp.	suspension
syn.	synthetic
t	tertiary
TAPPI	Technical Association of the Pulp & Paper Industry
TCC	Tag closed cup
TDI	toluene diisocyanate
tech.	technical
temp.	temperature
tens.	tensile, tension
tert	tertiary
TFE	tetrafluoroethylene
TGIC	triglycidyl isocyanurate
THF	tetrahydrofuran
thru	through
TKPP	tetrapotassium pyrophosphate
TLV	Threshold Limit Value
TMA	trimellitic anhydride
TMC	thick molding compound
TMPEOTA	trimethylolpropane ethoxy triacrylate
TMPTA	trimethylolpropane triacrylate
TOC	Tag open cup
TOFA	tall oil fatty acid
TPE	thermoplastic elastomer
TPGDA	tripropylene glycol diacrylate
TPO	thermoplastic polyolefin
TPR	thermoplastic rubber
TPU	thermoplastic polyurethane
TSCA	Toxic Substances Control Act
TWA	time weighted average
typ.	typical
uel	upper explosive limits
UF	urea formaldehyde
UHMWPE	ultra high molecular weight polyethylene

UN no.	United Nations Substance Identification Number (for transport purposes)
unsat.	unsaturated
USDA	U.S. Department of Agriculture
USP	Unites States Pharmacopeia
uv, UV	ultraviolet
V	volt
VA	vinyl acetate
VdC, VDC	vinylidene chloride
veg.	vegetable
visc.	viscous, viscosity
VM&P	Varnish Makers and Painters
VOC	volatile organic compounds
vol.	volume
v/v	volume by volume
wh.	white
WHO	World Health Organization (United Nations)
wks	weeks
w/o	water-in-oil
wt.	weight
w/v	weight by volume
w/w	weight by weight
XLPE	crosslinked polyethylene
yel.	yellow
ylsh.	yellowish
yr	year
#	number
%	percent
<	less than
>	greater than
≤	less than or equal to
≥	greater than or equal to
@	at
α	alpha
β	beta
δ, Δ	delta
ε	epsilon
γ	gamma
ω	omega
μ	micron, micrometer
μg	microgram
≈	approximately equal to

Part I
Chemical Product Reference

Chemical Product Reference

Abietic acid
CAS 514-10-3; EINECS 208-178-3
Synonyms: Abietinic acid; Sylvic acid
Definition: Active ingredient of rosin, occurring with other acids
Empirical: $C_{19}H_{29}COOH$
Properties: Yellowish powd.; sol. in alcohol, ether, chloroform, benzene; insol. in water; m.w. 302.44; m.p. 171-175 C
Precaution: Combustible
Toxicology: Poison by intravenous route; heated to dec., emits acrid smoke and irritating fumes
Uses: Varnish driers; mfg. of ester gums, soaps; fermentation industry
Manuf./Distrib.: Aldrich; Arizona; Penta Mfg.; Sigma

Abietic acid methyl ester. *See* Methyl abietate
Abietinic acid. See Abietic acid
ABS. See Acrylonitrile-butadiene-styrene
Absolute alcohol. See Alcohol
Aceite de Algodon. See Cottonseed oil
Aceite de ricino. See Castor oil

Acetal copolymer
Synonyms: POM; Polyoxymethylene
Uses: Thermoplastic engineering resin for injection molding, extrusion, blow molding; for automotive, industrial, appliance, seasonal outdoor, etc.
Manuf./Distrib.: Ashland; Berje; Hoechst Celanese; Hüls AG; LNP; Penta Mfg.; Westlake Plastics
Trade names: Celcon® M25; Celcon® M25-01

Acetaldehyde
CAS 75-07-0; EINECS 200-836-8
Synonyms: Acetic aldehyde; Ethyl aldehyde; Ethanal
Classification: Aldehyde
Empirical: C_2H_4O
Formula: CH_3CHO
Properties: Colorless fuming liq., pungent fruity odor; misc. in water, alcohol, ether; m.w. 44.06; dens. 0.788 (16/4 C); m.p. -123.5 C; b.p. 20.8 C; flash pt. (CC) -38 C; ref. index 1.3316 (20 C)
Precaution: Flamm. liq. (DOT); can react violently with acid anhydrides, alcohols, ketones, phenols, NH_3, halogens, etc.; reaction with oxygen may lead to detonation; keep cold
Toxicology: LD50 (oral, rat) 1930 mg/kg; poison by intratracheal/IV routes; human systemic irritant by inh.; narcotic; human mutagenic data; experimental tumorigen, teratogen; skin and severe eye irritant; heated to decomp., emits acrid smoke and fumes
Uses: Manufacture of acetic acid, acetic anhydride, n-butanol, peracetic acid, pentaerythritol, pyridines, 1,3-butylene glycol, trimethylolpropane; intermediate for paints; synthetic flavors; in nonalcoholic beverages, ice cream, ices, candy, baked goods
Regulatory: FDA 21CFR §177.2410, 182.60, 27CFR §21.93, GRAS; FEMA GRAS
Manuf./Distrib.: Aldrich; Allchem Ind.; J.T. Baker; BP Chems. Ltd; Eastman; Fluka; Hoechst Celanese; Hüls UK; Mitsui Petrochem. Ind.; Penta Mfg.; Spectrum Chem. Mfg.; Sigma; Wacker-Chemie GmbH

Acetate/ethylene copolymer
Uses: Material for latex emulsions
Manuf./Distrib.: Air Prods.; Ashland; Nat'l. Starch & Chem.; Reichhold

(Acetato)phenylmercury. *See* Phenylmercuric acetate
Acetic acid barium salt. See Barium acetate
Acetic acid, butyl ester. See n-Butyl acetate
Acetic acid, sec-butyl ester. See s-Butyl acetate
Acetic acid, ethenyl ester. See Vinyl acetate
Acetic acid, ethenyl ester, homopolymer. See Polyvinyl acetate (homopolymer)
Acetic acid, ethenyl ester, polymer with ethene. See Ethylene/VA copolymer
Acetic acid, ethyl ester. See Ethyl acetate
Acetic acid 2-ethylhexyl ester. See 2-Ethylhexyl acetate
Acetic acid hexyl ester. See Hexyl acetate
Acetic acid isobutyl ester. See Isobutyl acetate
Acetic acid, lead salt. See Lead acetate
Acetic acid manganese (II) salt (2:1). See Manganese acetate
Acetic acid manganese (2+) salt. See Manganese acetate
Acetic acid methyl ester. See Methyl acetate
Acetic acid 1-methylethyl ester. See Isopropyl acetate
Acetic acid 1-methylpropyl ester. See s-Butyl acetate
Acetic acid 2-methylpropyl ester. See Isobutyl acetate
Acetic acid pentyl ester. See Amyl acetate
Acetic acid n-propyl ester. See Propyl acetate
Acetic acid, vinyl ester. See Vinyl acetate
Acetic acid vinyl ester polymers. See Polyvinyl acetate (homopolymer)
Acetic aldehyde. See Acetaldehyde
Acetic ether. See Ethyl acetate
Acetin. See Triacetin
Acetoacetic acid ethyl ester. See Ethylacetoacetate
Acetoacetic ester. See Ethylacetoacetate
Acetoacetone. See Acetylacetone

2-(Acetoacetoxy) ethyl acrylate
CAS 21282-96-2
Synonyms: Butanoic acid, 3-oxo-, 2-[(1-oxo-2-propenyl)oxy]ethyl ester
Empirical: $C_9H_{12}O_5$
Properties: M.w. 200.19
Toxicology: LD50 (oral, rat) 1132 mg/kg, (dermal, rat) > 2000 mg/kg; mod. skin irritant
Uses: Visc. reducer in polymerization (varnishes, adhesives, polyesters)
Trade names: Lonzamon® AAEA

2-(Acetoacetoxy) ethyl methacrylate
CAS 21282-97-3
Synonyms: 2 (Methacryloyloxy) ethyl acetoacetate
Empirical: $C_{10}H_{10}O_5$
Formula: $CH_3COCH_2CO_2CH_2CH_2O_2CC(CH_3)=CH_2$
Properties: Liq.; m.w. 214.22; sp.gr. 1.12 (20/20 C); b.p. 100 C; flash pt. 134 C
Toxicology: LD50 (oral, rat) > 5000 mg/kg; pract. nonirritating to skin
Uses: Visc. reducer in polymerization (varnishes, adhesives, polyesters); resin intermediate
Manuf./Distrib.: Aldrich
Trade names: Lonzamon® AAEMA

Acetone
CAS 67-64-1; EINECS 200-662-2
Synonyms: Dimethylketone; 2-Propanone; β-Ketopropane; Pyroacetic ether
Classification: Aliphatic ketone
Empirical: C_3H_6O
Formula: CH_3COCH_3
Properties: Colorless volatile transparent liq., sweetish odor, pungent sweetish taste; misc. with water, alcohol, chloroform, ether, most volatile oils; m.w. 58.09; dens. 0.792 (20/20 C); m.p. -94.3 C; b.p. 56.2 C; flash pt. (OC) 15 F
Precaution: DOT: flamm. liq.; dangerous fire risk; explosive limit in air 2.6-12.8%
Toxicology: LD50 (oral, mouse) 3000 mg/kg; TLV 750 ppm in air; narcotic in high conc.; moderately toxic by ingestion and inhalation; lg. doses may cause narcosis; peeling and splitting of nails, skin rashes; inh. irritates lungs
Uses: Solvent for paints, varnishes and lacquers; for cleaning and drying precision equip.; delustrant for

cellulose acetate fibers; extraction solvent for food use; mfg. of acetate rayon, photographic film, smokeless powd., paint remvoers, glues, cements, plastics, pharmaceuticals
Regulatory: FDA 21CFR §73.1, 73.30, 73.345, 73.615, 173.210, 175.105, 175.320, 176.300, 182.99; 40CFR §180.1001; 30 ppm tolerance in spice oleoresins; FEMA GRAS; Japan approved with restrictions; USP/NF, BP compliance
Manuf./Distrib.: Aldrich; Allchem Ind.; AlliedSignal; Aristech; Ashland; BASF; J.T. Baker; Baychem; BP Chems. Ltd; R.E. Carroll; Chemcentral; Dow; Eastman; Elf Atochem N. Am.; Exxon; Fluka; C.P. Hall; Harcros; Mitsui Petrochem.; Montedipe SpA; Samson; Shell; Sigma; Spectrum Chem. Mfg.; Sunnyside; Texaco; Union Carbide; Van Waters & Rogers
Trade names containing: Araldite® EPN 1138 A-85; D.E.N. 438-A85; D.E.R. 661-A80; Tactix 741

Acetone oxime
CAS 127-06-0; EINECS 204-820-1
Synonyms: Acetoxime; Acetonoxime; 2-Propanone oxime; β-Isonitrosopropane
Empirical: C_3H_7NO
Formula: $(CH_3)_2CNOH$
Properties: Colorless cryst., chloral-like odor; sol. in alcohols, ethers, water; m.w. 73.09; m.p. 61 C; b.p. 136.3 C
Precaution: Combustible
Toxicology: LD50 (ipr, mouse) 4000 mg/kg; mod. toxic by intraperitoneal route; emits toxic NO_x fumes when heated to decomp.
Uses: Used in sealants and coatings; intermediate; primer for diesel fuels
Manuf./Distrib.: Aldrich; AlliedSignal; Fluka; Sigma

Acetonic acid. *See* Lactic acid
Acetonoxime. See Acetone oxime

Acetophenone
CAS 98-86-2; EINECS 202-708-7
Synonyms: Ketone methyl phenyl; Methyl phenyl ketone; Acetylbenzene; 1-Phenylethanone; Hypnone
Classification: Aromatic organic compd.
Empirical: C_8H_8O
Formula: $C_6H_5COCH_3$
Properties: Colorless liq., sweet pungent odor and taste; sol. in alcohol, chloroform, ether, fatty oils, glycerol; sl. sol. in water; m.w. 120.15; dens. 1.030 (20/20 C); m.p. 20.5 C; b.p. 201.7 C; flash pt. 82.2 C; ref. index 1.5339 (20 C)
Precaution: DOT: Combustible liq.
Toxicology: LD50 (oral, rat) 815 mg/kg; narcotic in high concs.; poison by intraperitoneal route; moderate toxicity by ingestion; skin and eye irritant; may cause allergic reaction
Uses: Perfumery, solvent, intermediate for pharmaceuticals, resins, paints, etc.; flavoring agent; polymerization catalyst, organic synthesis
Regulatory: FDA 21CFR §172.515; FEMA GRAS; Japan approved for flavoring
Manuf./Distrib.: Aldrich; Alemark; Amber Syn.; ARCO; BP Chems. Ltd; EniChem Am.; Fluka; R.W. Greeff; Janssen Chimica; Mitsui Petrochem. Ind.; Mitsui Toatsu; Montedipe SpA; Penta Mfg.; Rhone-Poulenc; Sigma; Spectrum Chem. Mfg.

Acetoxime. *See* Acetone oxime
1-Acetoxy-2-butoxyethane. *See* Butoxyethanol acetate
1-Acetoxy-2-methoxyethane. *See* Methoxyethanol acetate
Acetoxyphenylmercury. *See* Phenylmercuric acetate
α-Acetoxytoluene. *See* Benzyl acetate

Acetylacetonate
Uses: Catalyst for esterification and olefin polymerization; used in coatings; resin crosslinking agent for automotive prods., coatings, elastomers, films/paints, graphic arts, plastics
Manuf./Distrib.: Monomer-Polymer & Dajac Labs; Penta Mfg.; Union Carbide; Wacker Chems.
Trade names: Tyzor AA

Acetylacetone
CAS 123-54-6; EINECS 204-634-0
Synonyms: 2,4-Pentanedione; Pentanedione-2,4; Diacetylmethane; Acetoacetone
Empirical: $C_5H_8O_2$
Formula: $CH_3COCH_2COCH_3$

Acetylated palm kernel glycerides

Properties: Colorless to sl. yel. liq., pleasant odor; misc. with alcohol, benzene, ether, chloroform, acetone, glac. acetic acid, propylene glycol; insol. in water; m.w. 100.13; dens. 0.952-0.962; m.p. -23.2 C; b.p. 139 C (746 mm); flash pt. (OC) 105 F
Precaution: Flamm. liq. exposed to heat or flame; incompat. with oxidizing materials; light-sensitive; refrigerate
Toxicology: LD50 (oral, rat) 1000 mg/kg; mod. toxic by ingestion, IP, inh. routes; skin and severe eye irritant; narcotic in high doses
Uses: Solvent for cellulose acetate; intermediate; chelating agent for metals, paint driers, lubricant additives, pesticides.
Regulatory: FDA 21CFR §172.515; BP compliance
Manuf./Distrib.: Aceto; Aldrich; Fluka; Monomer-Polymer & Dajac Labs; Penta Mfg.; Sigma; Spectrum Chem. Mfg.; Union Carbide; Wacker Chems.

Acetylated palm kernel glycerides
Synonyms: Glycerides, palm kernel oil mono-, di-, and tri-, acetates
Classification: Acetyl ester
Uses: Lubricant and plasticizer for plastics and coatings; cosolv. for polar additives

Acetylbenzene. *See* Acetophenone
Acetylcellulose. *See* Cellulose acetate
Acetylene black. *See* Carbon black
Acetylenic glycol. *See* Tetramethyl decynediol
2-(Acetyloxy)-1,2,3-propanetricarboxylic acid, tributyl ester. *See* Acetyl tributyl citrate
2-(Acetyloxy)-1,2,3-propanetricarboxylic acid, triethyl ester. *See* Acetyl triethyl citrate

Acetyl tributyl citrate
CAS 77-90-7; EINECS 201-067-0
Synonyms: ATBC; 2-(Acetyloxy)-1,2,3-propanetricarboxylic acid, tributyl ester; Tributyl acetyl citrate
Classification: Aliphatic ester
Definition: Ester of citric acid
Empirical: $C_{20}H_{34}O_8$
Formula: $CH_3COOC_3H_4(COOC_4H_9)_3$
Properties: Colorless sl. visc. liq., sweet herbaceous odor; sol. in alcohol; insol. in water; m.w. 402.49; dens. 1.14; b.p. > 300 C; flash pt. 204 C; ref. index 1.4408
Toxicology: Heated to decomp., emits acrid smoke and irritating fumes
Uses: Plasticizer for vinyl resins
Regulatory: FDA 21CFR §172.515, 175.105, 175.300, 175.320, 178.3910, 181.22, 181.27; FEMA GRAS
Manuf./Distrib.: AC Ind.; Aldrich; Allchem Ind.; Morflex; Pfizer Int'l.; Reilly Chems. SA; Unitex
Trade names: Citroflex® A-4; Citroflex® A-4 Special; Estaflex® ATC; Uniplex 84

Acetyl triethyl citrate
CAS 77-89-4; EINECS 201-066-5
Synonyms: ATEC; 2-(Acetyloxy)-1,2,3-propanetricarboxylic acid, triethyl ester; Tricarballylic acid-β-acetoxytributyl ester; Triethyl acetylcitrate
Classification: Aliphatic ester
Empirical: $C_{14}H_{22}O_8$
Formula: $CH_3COOC_3H_4(COOC_2H_5)_3$
Properties: Colorless, odorless liq.; sl. sol. in water; m.w. 318.36; dens. 1.135 (25 C); flash pt. 187 C
Precaution: Combustible
Toxicology: Moderate toxicity by intraperitoneal route; mild toxicity by ingestion
Uses: Plasticizer for cellulosics, primarily ethyl cellulose
Regulatory: FDA 21CFR §175.105, 175.300, 175.320, 178.3910, 181.22, 181.27
Manuf./Distrib.: Aldrich; Morflex; Unitex
Trade names: Citroflex® A-2

Acid butyl phosphate. *See* Butyl acid phosphate
Acids, lanolin. *See* Lanolin acid
Acids, soy. *See* Soy acid
Acids, tall oil. *See* Tall oil acid
Acroleic acid. *See* Acrylic acid

Acrylamide monomer
CAS 79-06-1
Formula: $CH_2CHCONH_2$

Properties: M.w. 71.08; m.p. 83-86 C
Precaution: Store refrigerated
Toxicology: Poisonous material
Uses: Monomer for prod. of paints
Manuf./Distrib.: AMRESCO; Bio-Rad Labs; Cytec Ind.; Jarchem Ind.; Research Organics; Spectrum Chem. Mfg.

2-Acrylamido-2-methylpropane sulfonate. *See* 2-Acrylamido-2-methylpropanesulfonic acid

2-Acrylamido-2-methylpropanesulfonic acid
CAS 15214-89-8; EINECS 239-268-0
Synonyms: 2-AMPS; 2-Acrylamido-2-methylpropane sulfonate; 2-Acryloylamido-2-methylpropanesulfonic acid monomer
Formula: $H_2C{=}CHCONHC(CH_3)_2CH_2SO_3H$
Properties: M.w. 207.25; m.p. 195 C
Precaution: Corrosive
Toxicology: Cancer suspect agent
Manuf./Distrib.: Aldrich; Fluka; Sigma
Trade names: AMPS® 2401; AMPS® 2404; Lubrizol® 2401 Monomer; Lubrizol® 2404 Monomer

2-Acrylamido-2-methylpropane sulfonic acid sodium salt. *See* Sodium 2-acrylamido-2-methylpropanesulfonate

Acrylate monomer
Definition: Any of several monomers used in the mfg. of thermosetting acrylic surface coating resins

Acrylates copolymer
Synonyms: Acrylic/acrylate copolymer; Acrylic copolymer
Definition: Polymers of two or more monomers consisting of acrylic acid, methacrylic acid, or their simple esters
Uses: Surface coatings, emulsions, paints, paper and leather finishes
Regulatory: FDA 21CFR §175.105, 175.210, 175.300, 175.320, 176.170, 176.180, 177.1010, 178.3790
Manuf./Distrib.: Air Prods.; Ashland; Chemcentral; GCA; General Latex & Chem.; BFGoodrich; Hampshire; Henkel; S.C. Johnson Polymer; Lenape; D.N. Lukens; Nat'l. Starch & Chem.; Reichhold; Rhone-Poulenc; Rohm & Haas; Samson; Scott Bader; StanChem; Synthron; Tamms Ind.; Thibaut & Walker; Tri-Iso; Union Carbide; Zeneca Resins; Zinchem
Trade names: Acrysol® ASE-95; Acrysol® TT-615; Airflex® 600-BP Series; Carboset® 515; Carboset® CR710; Carboset® CR712; Carboset® CR714; Carboset® CR780; Carboset® CR785; Carboset® EI255; Carboset® EL484; Carboset® GA1086; Carboset® GA1324; Carboset® GA1604; Carboset® PL958; Colloid™ 226/30; Colloid™ 226/35; Esi-Cryl 246; Esi-Cryl 263; Esi-Cryl 460; Esi-Cryl 595; Esi-Cryl 604; Esi-Cryl Beyond; Esi-Cryl Respond I; Esi-Graph 743; Flexbond® 972; Flow Agent WR-100; Hycar® 26237; Hycar® 26256; HyStretch V-43; Modaflow® 2100; Modaflow® Powder 2000; Modaflow® Powder III; Modaflow® Resin; Mor-Flo™ 1001; NeoCryl® A-601; NeoCryl® A-621; NeoCryl® A-630; NeoCryl® B-700; NeoCryl® B-723; NeoCryl® B-725; NeoCryl® B-1019; NeoCryl® B-1135; NeoCryl® BT-8; NeoCryl® BT-20; NeoCryl® BT-175; NeoCryl® S-1004; Rheolate® 1; SCX™-835; SCX™-839; Texicryl® 13-002; Texicryl® 13-011; Texicryl® 13-300, 13-302; Texicryl® 13-666
Trade names containing: Acryloid® B-66; Acryloid® B-72; Acryloid® F-10, 40%; Acryloid® NAD-10; Forbest VP S7; Multiflow® Resin

Acrylic acid
CAS 79-10-7; EINECS 201-177-9
Synonyms: Acroleic acid; Acrylic acid, glacial; 2-Propenoic acid; Propene acid; Vinylformic acid; Ethylene carboxylic acid
Empirical: $C_3H_4O_2$
Formula: $H_2C{:}CHCOOH$
Properties: Colorless liq., acrid odor; misc. with water, alcohol, ether, acetone, benzene; m.w. 72.07; dens. 1.052 (20/20 C); b.p. 140.9 C; m.p. 12.1 C; flash pt. (OC) 54.5 C; ref. index 1.4224 (20 C); polymerizes readily
Precaution: May polymerize explosively; flamm. liq.; corrosive material; fire hazard exposed to heat or flame
Toxicology: LD50 (oral, rat) 250 mg/kg; severe skin and eye irritant; toxic by inhalation; TLV 10 ppm in air; experimental teratogen
Uses: Monomer for polyacrylic and polymethacrylic acids, other acrylic acids, acrylic polymers, paints
Manuf./Distrib.: Aldrich; Allchem Ind.; Ashland; BASF; Elf Atochem N. Am.; Fluka; Hoechst Celanese; Hüls AG; Penta Mfg.; Rohm & Haas; Union Carbide

Acrylic acid butyl ester. *See* Butyl acrylate
Acrylic acid, dicyclopentadienyl ester. *See* Dicyclopentenyl acrylate
Acrylic acid 2,3-epoxypropyl ester. *See* Glycidyl acrylate
Acrylic acid 2-ethoxyethanol diester. *See* Diethylene glycol diacrylate
Acrylic acid 2-ethylhexyl ester. *See* Octyl acrylate
Acrylic acid, glacial. *See* Acrylic acid
Acrylic acid, hexyl ester. *See* n-Hexyl acrylate
Acrylic acid homopolymer. *See* Polyacrylic acid
Acrylic acid isodecyl ester. *See* Isodecyl acrylate
Acrylic acid methyl ester. *See* Methyl acrylate (monomer)
Acrylic acid-1-methyltrimethylene ester. *See* 1,3-Butylene glycol diacrylate
Acrylic acid polymer. *See* Polyacrylic acid
Acrylic acid, polymers. *See* Acrylic resin
Acrylic acid, propylenebis-(oxypropylene) ester. *See* PPG-3 diacrylate
Acrylic acid resin. *See* Acrylic resin
Acrylic acid resin. *See* Polyacrylic acid
Acrylic/acrylate copolymer. *See* Acrylates copolymer

Acrylic/acrylonitrile copolymer
Uses: Latex for use as a release coating; provides high gloss, block-resist. coatings with exc. release against pressure-sensitive adhesives used in many tape and label applics.
Trade names: Darex® 409L

Acrylic/alkyd copolymer
Uses: Industrial finishes, machinery enamels, primers, maintenance paints, force-cured coatings, bake applics.
Trade names: Chempol 10-1300

Acrylic copolymer. *See* Acrylates copolymer

Acrylic elastomer
Synonyms: Polyacrylic rubber
Uses: Low temp.-resistant rubber used for textile, nonwoven, adhesive bindings, coatings, finishes, seals and gaskets
Trade names: HyStretch T-35; HyStretch V-29; HyStretch V-43HX

Acrylic fiber. *See* Acrylic resin

Acrylic monomer
Uses: Intermediate for paints
Manuf./Distrib.: Albright & Wilson Am.; Ashland; BASF; Cyro Ind.; Cytec Ind.; GCA; Henkel; Morton Int'l.; Nat'l. Starch & Chem.; Rohm & Haas; San Esters; Sartomer; Seegott; Stanchem; 3M; Tri-Iso; UCB Radcure; Union Carbide; United Min. & Chem.; Van Waters & Rogers

Acrylic polymer. *See* Acrylic resin; Polyacrylic acid

Acrylic resin
CAS 9003-01-4
Synonyms: Acrylic polymer; Acrylic acid, polymers; Acrylic acid resin; Acrylic fiber; Polyacrylate; Poly(acrylic acid); 2-Propenoic acid homopolymer
Definition: Thermoplastic polymer or copolymer of acrylic acid, methacrylic acid, esters of these acids, or acrylonitrile
Empirical: $(C_3H_4O_2)_4$
Formula: $[-CH_2CH(CO_2H)-]$
Properties: Varies from hard to brittle solids to fibrous elastomeric structures, to viscous liqs.; able to transmit light for sheet and rod forms
Toxicology: Heated to decomp., emits acrid smoke and fumes
Uses: Hard, shatterproof transparent or colored material, glass substitute in medical, dental, and industrial applics.; impact modifier, processing aid for PVC
Manuf./Distrib.: Air Prods.; Akzo Nobel; Albright & Wilson Am.; Alco; Aldrich; Bayer; Cytec Ind.; DSM UK; DuPont; BFGoodrich; Goodyear; Henkel/Coatings & Inks; Hoechst Celanese; S.C. Johnson Polymer; Lenape Ind.; Lomas Int'l.; D.N. Lukens; Morton Int'l.; Nat'l. Starch & Chem.; Reichhold; Rhone-Poulenc;

Rohm & Haas; Seegott; Shell; Sigma; Sybron; Synthron; Thibaut & Walker; UCB SA; Union Carbide; Zeneca Resins

Trade names: Acrysol® TT-935; Acrysol® WS-50; BYK®-356; Carboflow® 20; Carboset® 511; Carboset® 519; Carboset® 525; Carboset® 526; Carboset® 527; Carboset® 537; Carboset® 560; Carboset® GA33; Carboset® GA1594; Carboset® GA1914; Carboset® XL-11; Carboset® XL-19; Disparlon® L-1980; Disparlon® L-1983; Disparlon® L-1984; Disperse-Ayd 9100; DT 100NA; DT 211NA; Elvacite® 2008, 2009, 2010, 2021, 2041, 6010; Elvacite® 2042, 2043; Esi-Cryl 301; Esi-Cryl 405; Esi-Cryl 900; Esi-Graph 752; Esi-Graph 1055; Fulatex® PD-442; Fulatex® PD-444; Fulatex® PD-449; Fulatex® PD-450; Fulatex® PD-461; Fulatex® PD-723; Fulatex® PD-3496-F; G-Cryl® 130; G-Cryl® 250; G-Cryl® 345; G-Cryl® 535; G-Cryl® 5000; G-Cryl® 5002; G-Cryl® 5003; G-Cure® 105; G-Cure® 105P70; G-Cure® 106P70; G-Cure® 107A70; G-Cure® 108A70; G-Cure® 109HE75; G-Cure® 806; G-Cure® 867; G-Cure® 868; G-Cure® 869; Hycar® 2671; Hycar® 2679; Hycar® 26084; Hycar® 26092; Hycar® 26106; Hycar® 26120; Hycar® 26138; Hycar® 26322; Hycar® 26345; Joncryl® 67; Joncryl® 95; Joncryl® 142; Joncryl® 530; Joncryl® 537; Joncryl® 538; Joncryl® 540; Joncryl® 586; Joncryl® 587; Joncryl® 611; Joncryl® 678; Joncryl® 690; Joncryl® 1907; Joncryl® 1908; Joncryl® 1954; Maincote® AE-58; Maincote® HG-54; Maincote® Hydur™ 30; Maincote® PR-71; Modifier QM-1007; Mor-Flo™ 33; Mor-Flo™ 155; Mor-Flo™ 604; Mor-Flo™ 614; Mor-Lok™ 1035; Mor-Lok™ 1043; NeoCryl® A-612; NeoCryl® A-614; NeoCryl® A-622; NeoCryl® A-623; NeoCryl® A-625; NeoCryl® A-634; NeoCryl® A-645; NeoCryl® A-650; NeoCryl® A-655; NeoCryl® A-5044; NeoCryl® A-5045; NeoCryl® A-5090; NeoCryl® A-6015; NeoCryl® A-6044; NeoCryl® A-6051; NeoCryl® A-6069; NeoCryl® A-6074; NeoCryl® A-6075; NeoCryl® A-6092; NeoCryl® B-723LV; NeoCryl® B-735; NeoCryl® B-736; NeoCryl® BT-24; Parcryl® 200, 300; Parcryl® 250; Parcryl® 311; Parcryl® 400I; Parcryl® 450I; Parcryl® 500I; Parcryl® 900; Rheolate® 101; Rhoplex® AC-33; Rhoplex® B-15J; Rhoplex® Multilobe™ 200; Rhoplex® WL-51; Rhoplex® WL-71; Rhoplex® WL-81; Rhoplex® WL-91; Rhoplex® WL-92; Rhoplex® WL-96; Rohatol® D 362; SCX™-815; SCX™-817; SCX™-819; SCX™-1520; SCX™-1959; SCX™-2500; SCX™-2560; Synthemul® 40-412; Synthemul® 40-423; Synthemul® 40-425; Synthemul® 40-428; Tint-Ayd® NV Series; Tint-Ayd® PC Series; Ucar® Acrylic 503; Ucar® Acrylic 505; Ucar® Acrylic 515; Ucar® Acrylic 516; Ucar® Acrylic 518; Ucar® Acrylic 522; Ucar® Acrylic 525; Ucar® Latex 123; Ucar® Latex 163; Ucar® Latex 413; Ucar® Latex 429; Ucar® Latex 430; Ucar® Latex 439; Ucar® Latex 446; Ucar® Latex 516; Ucar® Latex 518; Ucar® Latex 522; Ucar® Latex 525; Ucar® Latex 603; Ucar® Latex 604; Ucar® Latex 624; Ucar® Latex 625; Ucar® Latex 651; Ucar® Vehicle 435; Ucar® Vehicle 441; Ucar® Vehicle 443; Ucar® Vehicle 452; Ucar® Vehicle 455; Ucar® Vehicle 462; Ucar® Vehicle 4550; Versaflex® 1; Vultacryl

Trade names containing: Acryloid® AT-400; Acryloid® AT-410; Acryloid® AT-954; Acryloid® AU-608B; Acryloid® AU-608S; Acryloid® AU-608X; Acryloid® AU-946; Acryloid® AU-1003; Acryloid® AU-1004; Acryloid® AU-1033; Acrysol® WS-68; Arolon 559-G4-70; Carboflow® 21S; Carboset® 514A; Carboset® 514H; Carboset® 531; Disparlon® 1970; Disparlon® AP-20; Disparlon® AP-30; Disparlon® AQ-200; Disparlon® L-1985-50; Disparlon® LC-955; Disparlon® OX-60; Disparlon® OX-70; Disperse-Ayd 15; ECO-CRYL™ 9790; Edaplan LA 402; HC-913; Joncryl® 56; Joncryl® 62; Joncryl® 500; Joncryl® 502; Joncryl® 504; Joncryl®-510; Joncryl®-588; Konform® AR 2000; Mearlite® Radiant Pearl STL; Mearlite® Ultra Bright UTL; Mearlite® Ultra Bright UWA; Microspersion™ 114; Microspersion™ 250; Modarez MFP Powd.; OPPalyte® 250 ASW, 350 ASW; Resin QR-1166; Resin QR-1269; Resin QR-1281; SCX™-507; SCX™-514; SCX™-901; SCX™-902; SCX™-906; SCX™-907; SCX™-915; Sobral 586; Sobral 587; Sobral 588; Synthemul® 40-426; Synthemul® 40-427; TEGO® Flow 300; TEGO® Flow ZFS 460; Tint-Ayd® AP Series; Tint-Ayd® ST Series; XU-19044.01

See also Polyacrylic acid

Acrylic/styrene

Uses: High gloss polymer for flexo/gravure ink and coating applics., paper/paperboard applics.; binder

Trade names: Esi-Graph 745

Acrylic, thermosetting

Uses: Thermosetting resin for general prod. finishing, exterior coil coating

Trade names containing: Acryloid® AT-9LO; Acryloid® AT-51; Acryloid® AT-63; Acryloid® AT-70IBA; Acryloid® AT-76; Acryloid® AT-81; Acryloid® AT-85; Acryloid® WR-97

Acrylic-urethane

Trade names: Mor-Flo™ 645

Acrylonitrile

CAS 107-13-1; EINECS 203-466-5

Synonyms: Propenenitrile; 2-Propenenitrile; Acrylonitrile monomer; Vinyl cyanide; Cyanoethylene

Empirical: C_3H_3N

Formula: $H_2C{:}CHCN$

Acrylonitrile-butadiene copolymer

Properties: Colorless, mobile liq., mild odor; sol. in org. solvs., partially misc. in water; m.w. 53.06; dens. 0.8004 (25 C); f.p. -83 C; b.p. 77.3-77.4 C; flash pt. (TOC) 32 F
Toxicology: LD50 (oral, rat) 78 mg/kg; TLV: 2 ppm; toxic by inhalation and skin absorp.
Uses: Monomer for acrylic and modacrylid fibers, paints; in prod. of ABS and acrylonitrile styrene copolymers
Manuf./Distrib.: Aldrich; Allchem Ind.; Asahi Chem. Ind.; BP Chems. Ltd; Cytec Ind.; DSM; DuPont; Fluka; Mitsui Toatsu; Monomer-Polymer & Dajac Labs; Monsanto

Acrylonitrile-butadiene copolymer. *See* Acrylonitrile-butadiene rubber

Acrylonitrile-butadiene rubber
Synonyms: NBR; Acrylonitrile rubber; Acrylonitrile-butadiene copolymer; Nitrile rubber; Nitrile elastomer
Classification: Syn. rubber
Formula: $[CH_2CH{=}CHCH_2CH_2CH(CN)]_x$
Properties: Dens. 0.98; tens. str. 1000-3000 psi; elong. 100-700%; combustible
Uses: Oil well parts, general-purpose oil-resistant applics., gaskets, grommets, o-rings; modifier for PVC and ABS resins
Trade names: Heveasyn Nitrile Latex; Hycar® 1552; Hycar® 1561; Hycar® 1562; Hycar® 1571; Hycar® 1572; Hycar® 1572X64; Nipol® DN-601; Perbunan N Latex 2818

Acrylonitrile-butadiene-styrene
CAS 9003-56-9
Synonyms: ABS
Definition: Thermoplastic resin grafted from the 3 monomers from which its name is derived
Properties: Dens. ≈ 1.04; tens. str. ≈ 6500 psi; flex. str. ≈ 10,000 psi
Precaution: Combustible
Uses: Automotive body parts, telephones, pkg., shower stalls, etc.; impact modifier for PVC; latex used for waterproof and greaseproof coatings, leather finishes, paints, paper coatings, and nonwoven binders and saturation
Regulatory: FDA 21CFR §181.32
Manuf./Distrib.: Aiscondel SA; Aldrich; Allchem Ind.; Bamberger Polymers; BASF; GE Plastics; LNP; Mitsui Toatsu; Monsanto; Reichhold/Emulsion Polymers; Shuman Plastics
Trade names: Hycar® 1577; Hycar® 1578X1

Acrylonitrile copolymer
Synonyms: Propenenitrile copolymer; Vinyl cyanide copolymer
Uses: Specialty elastomer for automotive and industrial applics., e.g., seals, gaskets, membranes, cable jackets, extruded profiles, rubber linings, hoses, bellows, sleeves, oil field parts; textile coatings
Trade names: Tylac® 68073-00; Tylac® 68074-00; Tylac® 68076-00; Tylac® 69719-00; Tylac® 97696-00; Tylac® 97942-00

Acrylonitrile monomer. *See* Acrylonitrile
Acrylonitrile rubber. *See* Acrylonitrile-butadiene rubber
2-Acryloylamido-2-methylpropanesulfonic acid monomer. *See* 2-Acrylamido-2-methylpropanesulfonic acid
3-Acryloyloxypropionic acid. *See* β-Carboxyethyl acrylate
Activated alumina. *See* Alumina
Activated aluminum oxide. *See* Alumina
Activated attapulgite. *See* Attapulgite
ADA. *See* Azodicarbonamide
Adeps lanae. *See* Lanolin

Adipic acid
CAS 124-04-9; EINECS 204-673-3
Synonyms: Dicarboxylic acid C_6; Hexanedioic acid; 1,6-Hexanedioic acid; 1,4-Butanedicarboxylic acid
Classification: Organic dicarboxylic acid
Empirical: $C_6H_{10}O_4$
Formula: $HOOC(CH_2)_4COOH$
Properties: Wh. monoclinic prisms, pract. odorless; very sol. in alcohol; sol. in acetone; sl. sol. in water; m.w. 146.16; dens. 1.360 (25/4 C); m.p. 152 C; b.p. 337.5 C; flash pt. (CC) 385 F
Precaution: Combustible
Toxicology: LD50 (oral, mouse) 1900 mg/kg; no known human toxicity; may cause occupational asthma
Uses: Manufacture of nylon and polyurethane foams; intermediate for paints; preparation of esters for use as plasticizers and lubricants; food additive (acidulant); baking powders; buffer, neutralizing agent; adhesives
Regulatory: FDA 21CFR §131.44, 172.515, 184.1009, GRAS; FEMA GRAS; USDA 9CFR §318.7; Japan

approved; Europe listed (ADI 0-5 mg/kg, free acid basis); UK approved
Manuf./Distrib.: Aldrich; Allchem Ind.; AlliedSignal; Asahi Chem. Ind.; Ashland; BASF AG; Chemical; Coyne; DuPont; Fluka; C.P. Hall; Monsanto; Penta Mfg.; Rhone-Poulenc; Samson; Sigma; Spectrum Chem. Mfg.; UCB SA; United Min. & Chem.; Van Waters & Rogers
Trade names: Adi-pure®

Adipic acid bis (2-ethylhexyl) ester. *See* Dioctyl adipate
Adipic acid dimethyl ester. *See* Dimethyl adipate
AEP. *See* Aminoethylpiperazine
Agalmatolite. *See* Pyrophyllite
AGE. *See* Allyl glycidyl ether
Agricultural limestone. *See* Limestone
Alabaster. *See* Calcium sulfate dihydrate
Albacol. *See* n-Propyl alcohol

Alcohol
CAS 64-17-5; EINECS 200-578-6
Synonyms: EtOH; Ethyl alcohol; Ethyl alcohol, undenatured; Ethanol; Ethanol, undenatured; Distilled spirits; Absolute alcohol
Definition: Undenatured ethyl alcohol
Empirical: C_2H_6O
Formula: CH_3CH_2OH
Properties: Colorless limpid, volatile liq., vinous odor, pungent taste; misc. with water, methanol, ether, chloroform, 95% acetone; m.w. 46.08; dens. 0.816 (15.5 C); b.p. 78.3 C; f.p. -117.3 C; flash pt. (CC) 12.7 C; ref. index 1.365 (15 C)
Precaution: DOT: Flamm. liq.; dangerous fire and explosion hazard; can react vigorously with oxidizers; reacts violently with many chemicals
Toxicology: TLV 1000 ppm in air; CNS depressant; moderately toxic by ingestion; eye and mucous membrane irritant; experimental tumorigen, teratogen
Uses: Solvent; extraction medium; mfg. of acetaldehyde, denatured alcohol, paints, waterborne coatings, pharmaceuticals (tonics, colognes), perfumery, organic synthesis; octane booster in gasoline; topical antiseptic; active ingred. in OTC drug prods.
Regulatory: FDA 21CFR §169.3, 169.175, 169.176, 169.177, 169.178, 169.180, 169.181, 172.340, 172.560, 175.105, 176.200, 176.210, 177.1440, 184.1293, GRAS; 27CFR §2.5, 2.12; FEMA GRAS; USP/NF, BP, Ph.Eur. compliance; Japan restricted
Manuf./Distrib.: Aldrich; Ashland; J.T. Baker; Baychem; BP Chems. Inc; Chemcentral; Coyne; Eastman; Fluka; Georgia-Pacific Resins; Gist-brocades SpA; Grain Processing; Great Western; C.P. Hall; Harcros; Quantum; Samson; Sigma; Sunnyside; Union Carbide; Van Waters & Rogers; Vista
Trade names containing: Beckamine 21-511; Ethoquad® T/12; Filmex® A-2; Filmex® B; Filmex® C; Filmex® D-1; Filmex® D-2; Gantrez® ES-225; Gantrez® ES-425; Geropon® SS-O-75; GP-197 Resin Sol'n.; GPRI™ 4000; GPRI™ BKS-2600; GPRI™ BKS-2700; GPRI™ BKS-2900; GPRI™ BKS-2901; Monawet MO-70E; Monawet MO-75E; Resimene® 797; Resimene® 2040; Silbond® H-6C

Alcohol C-3. *See* n-Propyl alcohol
Alcohol C-5. *See* Amyl alcohol
Alcohol C-6. *See* Hexyl alcohol
Alcohol C-8. *See* Caprylic alcohol
Alcohol C_8. *See* 2-Ethylhexanol; Isooctyl alcohol
Alcohol C-10. *See* n-Decyl alcohol
Alcohol C16. *See* Cetyl alcohol

Alcohol denatured
CAS 64-17-5; EINECS 200-578-6
Synonyms: Denatured alcohol; Denatured ethyl alcohol; Denatured ethanol; Denatured spirits
Definition: Alcohol denatured with one or more denaturing agents
Empirical: C_2H_6O
Formula: CH_3CH_2OH
Properties: M.w. 46.08
Precaution: Fire hazard; mod. explosion hazard
Toxicology: TLV: 1000 ppm in air; potentially poisonous by ingestion
Uses: Mfg. of acetaldehyde and other chemicals, solvents, antifreeze, brake fluids, fuels; solv. used in protective coatings, printing inks, paint and varnish removers

Alcohols, lanolin

Manuf./Distrib.: Aldrich; Ashland; BP Chems.; Fluka; Midwest Grain Prods.; Sigma; Spectrum Chem. Mfg.
Trade names: Eastman® CDA-19

Alcohols, lanolin. *See* Lanolin alcohol
Aldehyde C-1. *See* Formaldehyde
Aldehyde C-3. *See* Propionaldehyde
Algaroba. *See* Locust bean gum

Algin
CAS 9005-38-3
Synonyms: Sodium alginate; Sodium polymannuronate; Alginic acid, sodium salt
Classification: Hydrophilic polysaccharide
Definition: Purified carbohydrate prod. extracted from brown seaweeds; sodium salt of alginic acid
Empirical: $(C_6H_7O_6Na)_n$
Properties: Cream-colored powd., pract. odorless and tasteless; sol. in water; insol. in alcohol, ether, chloroform; m.w. 198.11
Toxicology: LD50 (IV, rat) 1000 mg/kg; poison by IV and intraperitoneal routes; heated to decomp., emits toxic fumes of Na_2O
Uses: Stabilizer in mfg. of ice cream; emulsifier; firming agent; flavor enhancer; formulation aid; processing aid; surfactant; texturizer; thickener; for hard candy, confections, frostings, fruit juice, gelatins, sauces, toppings, etc.
Regulatory: FDA 21CFR §133.133, 133.134, 133.162, 133.178. 133.179, 150.141, 150.161, 173.310, 184.1724, GRAS; FEMA GRAS; Japan approved; Europe listed; UK approved; FDA approved for orals; USP/NF, BP, Ph.Eur. compliance
Manuf./Distrib.: Aldrich; Allchem Ind.; Colony Ind.; Fluka; Int'l. Chem. Inc.; Kelco; Penta Mfg.; Sigma; San Yuan; Spectrum Chem. Mfg.
Trade names: Colloid 488T

Alginic acid, ester with 1,2-propanediol. *See* Propylene glycol alginate
Alginic acid, sodium salt. *See* Algin
Alkane C-4. *See* Butane
Alkane C-6. *See* Hexane
Alkane C_7. *See* Heptane
Alkanes, C7-8-iso-. *See* C7-8 isoparaffin
Alkanes, C8-9-iso-. *See* C8-9 isoparaffin
Alkanes, C10-11-iso-. *See* C10-11 isoparaffin
Alkanes, C11-12-iso-. *See* C11-12 isoparaffin
Alkanes, C11-13-iso-. *See* C11-13 isoparaffin
Alkanes, C13-14-iso-. *See* C13-14 isoparaffin
Alkyd. *See* Alkyd resin

Alkyd resin
CAS 63148-69-6; 68459-31-4
Synonyms: Alkyd
Classification: Thermosetting coating polymer
Uses: Vehicles in exterior house paints, marine paints, and baking enamels; elec. components, encapsulation
Manuf./Distrib.: Akzo Nobel; Arakawa USA; Bayer AG; Chemical; Croda Resins Ltd; Degen; DSM Resins UK; GFI; Hoechst Celanese; PPG Ind.; Ranbar Tech.; Reichhold Chemie AG; Scott Bader; Seegott; Thibaut & Walker; U.S. Polymers
Trade names: Aquachem® 896; Arolon 580-W-42; Beckosol® 10-560; Cellokyd 492; Chempol 11-3071; Chempol 13-1905; Chempol 13-2211; G-4495-100; G-4549-100; G-4587-100; G-4709-100; G-4726-100; Silmar® S802; Sobral 923-68; Sobral AD-002; Super Alkyd® 574-75TK; Super Alkyd® 800-80TK; Super Alkyd® 801-75TK; Super Alkyd® 840-90; Super Alkyd® 845-90; Super Alkyd® 846-90; Tint-Ayd® High Strength Tinting Black; Ultramoll® TGN
Trade names containing: Aroflat 3113-P-30; Arolon 921-G4-70; Aroplaz 111-M-70; Aroplaz 310-V-50; Aroplaz 1247-T-70; Aroplaz 3667-Z-80; Aroplaz 6056-MX-90; Aroplaz 6440-A4-85; Arothix 4000-P-40; Beckosol® 10-027; Beckosol® 10-539; Beckosol® 11-035; Beckosol® 12-514; Beckosol® 13-400; Cellokyd 2708; Cellokyd 2829 PMA/T; Cellokyd 3378 HFN; Flat-Ayd® FA-3B; Formulator; G-4067-M-50; G-4070-X-70; G-4412-M-50; G-4412-M-50HV; G-4412-V-50; G-4561-X-70; G-4594-OMS-35; G-4602-SD-50; G-4617-M-35; G-4635-M-60; G-4664-T-60; G-4689-HM-30; G-4696-M-50; G-4699-V-50; G-4714-P-75; G-4732-VBT-50; G-4744-B-75; G-4825-PnB-80; Kelpol 835-M-50; Kelpol 3755-X-80; Kelsol 3902-BG4-75; Kelsol 3905-B2G-75; Kelsol 3906-B2G-75; Kelsol 3922-G-80; Kelsol 3950-B2G-70; Kelsol 3970-G4-75; Kelsol 3990-B2G-75; Keltrol 1001-M-60; Kelvar 3758-M-85; Mearlite® Ultra Bright UDQ; Mearlite® Ultra Bright

UMT; Mearlite® Ultra Fine OFS; Mearlite® Ultra Fine OFW; Sicoflush® A; Sobral 12-101; Sobral 71NX60; Sobral 81X; Sobral 91NX; Sobral 663X50; Sobral 1308; Sobral 1341; Sobral 2750; Sobral 9249; Sobral 9255; Sobral 9257; Sobral AA-536; Sobral AD-640; Sobral AN-001; Sobral AR-530; Sobral AS-531; Sobral P470; Styresol 13-031; Styresol 13-040; Tint-Ayd® AL Series; Wallkyd 11-024

Alkylbenzyldimethylammonium chloride. *See* Benzalkonium chloride
Alkyl dimethyl benzyl ammonium chloride. *See* Benzalkonium chloride

Alkyl monomer
Uses: Monomer for prod. of paints
Manuf./Distrib.: ARCO

Allomaleic acid. *See* Fumaric acid
Allyl-(2,3-dihydroxypropane) ether. *See* Glyceryl-1-allyl ether
Allyl-2,3-epoxypropyl ether. *See* Allyl glycidyl ether

Allyl glycidyl ether
CAS 106-92-3; EINECS 203-442-4
Synonyms: AGE; [(2-Propenyloxy)methyl]oxirane; Allyl-2,3-epoxypropyl ether; 1-Allyloxy-2,3-epoxypropane
Classification: Glycidyl ether monomer
Empirical: $C_6H_{10}O_2$
Formula: $CH_2CHCH_2OCH_2=CH_2CHO$
Properties: Colorless liq., pleasant odor; sol. in methanol, toluene; partly sol. in water; m.w. 114.15; dens. 0.9698 (20 C); f.p. -100 C; b.p. 153.9 C; flash pt. (OC) 135 F; ref. index 1.434
Precaution: Incompat. with strong acids, bases, oxidizing agents, amines; hazardous decomp. prods.: CO, hydrocarbons
Toxicology: LD50 (oral, rat) 922 mg/kg, (dermal, rabbit) 2550 mg/kg; poison by ingestion; moderately toxic by inhalation and skin contact; strong skin and eye irritant; TLV 5 ppm; may cause cancer
Uses: Modifier for elastomers, adhesives, fibers, reactive diluent for epoxies, reactive intermediate for coatings, sizing/finishing agent for fiberglass, silane intermediate in elec. coatings; stabilizer of chlorinated compds., vinyl resins, rubber; defoamer
Manuf./Distrib.: Aldrich; Alemark; Amber Syn.; CPS; Fluka; Monomer-Polymer & Dajac Labs; Raschig; Spectrum Chem. Mfg.
Trade names: Ageflex AGE

Allyl methacrylate
CAS 96-05-9; EINECS 202-473-0
Empirical: $C_7H_{10}O_2$
Formula: $CH_2:C(CH_3)COOCH_2CH:CH_2$
Properties: M.w. 126.16; b.p. 59-60 C (35 mm); flash pt. 33 C; ref. index 1.4360 (20 C)
Uses: Silane monomer intermediate, crosslinker, polymer modifier for high impact plastics, adhesives, acrylic elastomers, optical polymers
Manuf./Distrib.: Aldrich; CPS; Elf Atochem N. Am.; Fluka; Monomer-Polymer & Dajac Labs; Polysciences; Rhone-Poulenc Surf. & Spec.; Richman; Rohm Tech; San Esters
Trade names containing: Sipomer® AM; SR-201

3-Allyloxy-2,2-bis-(allyloxymethyl)-1-propanol. *See* Pentaerythrityl triallyl ether
1-Allyloxy-2,3-epoxypropane. *See* Allyl glycidyl ether
1-Allyloxy-2,3-propanediol. *See* Glyceryl-1-allyl ether
Allyl 2,4,6-tribromophenyl ether. *See* Tribromophenyl allyl ether

Alumina
CAS 1344-28-1; EINECS 215-691-6
Synonyms: Aluminum oxide; Aluminum oxide (2:3); Tabular alumina; Alumina, tabular; Aluminium oxydes C; Aluminium oxide alumite; Calcined alumina; Alumina, calcined; Activated alumina; Alumina, activatedi; Alumite; Alundum; Activated aluminum oxide
Classification: Inorganic compd.
Empirical: Al_2O_3
Properties: Wh. cryst. powd., balls, or lumps, odorless, tasteless; pract. insol. in water; very sl. sol. in min. acids; m.w. 101.96; dens. 3.5-4; m.p. 2050 C; b.p. 2977 C
Precaution: Noncombustible; incompat. with hot chlorinated rubber; exothermic reaction above 200 C with halocarbon vapors produces toxic HCl and phosgene
Toxicology: Toxic by inhalation of dust; inh. may cause Shaver's disease; TLV:TWA 10 mg/m^3 (dust); may be

irritating to respiratory tract; eye irritant by mech. abrasion; skin drying, peeling; experimental tumorigen and neoplastigen by implant

Uses: Prod. of aluminum, polyethylene, abrasives, refractories, ceramics, elec. insulators, catalysts and catalyst supports, paper, spark plugs, crucibles and lab ware, adsorbent for gases/water vapors, chromatographic analysis, heat-resist. fibers; filler for paints and varnishes; adsorbent; food additives

Regulatory: FDA 21CFR §73.1010, 177.1460; exempt from certification; BP compliance (anhyd., basic)

Manuf./Distrib.: Air Prods.; Alcan; Alcoa; Aldrich; Atomergic Chemetals; BA Chem. Ltd; Degussa; Ferro/Transelco; Fluka; Hüls Am.; Lonza Sarl; Nissan Chem. Ind.; Norton Chem. Process Prods.; Rhone-Poulenc; San Yuan; Sigma; Spectrum Chem. Mfg.; Sumitomo Chem.; Zircar Prods.

Trade names: Alcoa® A-10; Alcoa® A-12; Alcoa® A-14; Alcoa® A-16; Alcoa® A-17; Alcoa® RG-300; Alumalux® 39; Premalox FG; Realox® A-152; Realox® A-1000; Realox® A-1520; Realox® A-2750; Realox® A-3000FL; Realox® A-3500; Realox® A-4000; Realox® RG-100; Versal™ 150; Versal™ 850; Versal™ 900; Versal™ GH

Trade names containing: Tronox® CR-800; Tronox® CR-822; Tronox® CR-828

Alumina, activated. *See* Alumina
Alumina, calcined. *See* Alumina
Alumina hydrate. *See* Alumina trihydrate; Aluminum hydroxide
Alumina hydrated. *See* Aluminum hydroxide

Alumina silicate

Uses: Filler for plastics, coatings, foams

Alumina, tabular. *See* Alumina

Alumina trihydrate

CAS 21645-51-2

Synonyms: Alumina hydrate; Aluminum hydrate; Hydrated alumina; Aluminum hydroxide; Hydrated aluminum oxide; Aluminum trihydroxide

Classification: Inorganic compd.

Empirical: $Al_2O_3 \cdot 3H_2O$

Properties: White powd., balls, or lumps; sol. in min. acids, caustic soda; insol. in water; m.w. 156.01; dens. 2.42; m.p. 2030 C; releases water on heating

Toxicology: Irritant

Uses: Prod. of aluminum and salts, glass, ceramics, organic lakes, flame retardants, mattrress batting; extender in paints; flame retardant for plastics, rubber reinforcing agent, paper coating, filler, cosmetics

Regulatory: BP compliance

Manuf./Distrib.: Alcan; Alcoa; Aldrich; Atomergic Chemetals; R.E. Carroll; Climax Perf. Materials; Croxton & Garry; Fluka; Franklin Ind. Mins.; J.M. Huber; D.N. Lukens; Nyco Mins.; Polyad; Reheis; Sigma; Whittaker, Clark & Daniels

Trade names: Alcan BW53, BW103, BW153; Alcan FRF 5; Alcan FRF 10; Alcan FRF 20; Alcan FRF 30; Alcan FRF 40; Alcan FRF 60; Alcan FRF 80; Alcan FRF 85; Alcan FRF LV2; Alcan FRF LV4; Alcan FRF LV5; Alcan FRF LV6; Alcan FRF LV7; Alcan FRF LV8; Alcan FRF LV9; Alcan Superfine 4; Alcan Superfine 7; Alcan Superfine 11; Alcan Ultrafine UF7; Alcan Ultrafine UF11; Alcan Ultrafine UF15; Alcan Ultrafine UF25; Alcan Ultrafine UF35; Alcoa® Grade C-231; Alcoa® Grade C-331/333; Baco DH101; Haltex™ 304; Haltex™ 310; Hydral® 705; Hydral® PGA; Micral® 532; Micral® 632; Micral® 916; Micral® 932; Micral® 1000; Micral® 1500; SB-331; SB-332; SB-335; SB-336; SB-632; SB-805; SB-932; SpaceRite™ S-3; SpaceRite™ S-11; SpaceRite™ S-23; WH-31

See also Aluminum hydroxide

Aluminic acid. *See* Aluminum hydroxide
Aluminium oxide alumite. *See* Alumina
Aluminium oxydes C. *See* Alumina
Aluminophosphoric acid. *See* Aluminum orthophosphate

Aluminosilicate

CAS 1327-36-2

Definition: A compd. of aluminum silicate with metal oxides or other radicals

Uses: Catalyst in refining petroleum, to soften water, in detergents

Trade names: Extendospheres® XOL-50; Extendospheres® XOL-200

Aluminosilicic acid, magnesium salt. *See* Magnesium aluminum silicate

Aluminum
CAS 7429-90-5; EINECS 231-072-3
Synonyms: CI 77000; Aluminum powder; Aluminum bronze
Classification: Metallic element
Empirical: Al
Properties: Silvery wh. cryst. solid; sol. in HCl, sulfuric acid; insol. in water, alcohol; m.w. 26.98; dens. 2.708; m.p. 660 C; b.p. 2450 C
Precaution: Flamm. in air; explosive; no stable isotopes; moisture-sensitive; dangerous when wet
Toxicology: TLV 10 mg/m^3 of air; (sol. salt) 2 mg/m^3 of air; (welding fumes) 5 mg/m^3 of air
Uses: Building and construction, corrosion-resistant chemical equip. (desalination plants), die-cast auto parts, elec. industry (power transmission lines), photoengraving plates, permanent magnets, cryogenic technology, machinery, tubes for ointments
Regulatory: FDA 21CFR §73.1645, 73.2645, 175.105, 175.300, 177.1460; BP compliance; Japan approved
Manuf./Distrib.: Aarbor Int'l.; Aldrich; Canbro; Fluka; Harcros; D.N. Lukens; Punda Mercantile; Reade Advanced Materials; Royale Pigments & Chems.; Seegott; U.S. Aluminum; Vanguard Chem. Int'l.; Whittaker, Clark & Daniels
Trade names: Standart® Chromal I; Standart® Chromal IV; Standart® Chromal X; Standart® Chromalux IV; Standart® Chromalux X; Standart® Lack/Lac AT; Standart® Lack/Lac CT; Standart® Lack/Lac E 900; Standart® Lack/Lac GTT; Standart® Lack/Lac K 900; Standart® Lack/Lac L 900; Standart® Lack/Lac LT; Standart® Lack/Lac OT; Standart® Resist AT; Standart® Resist LT; Standart® Resist Rotoflex Brilliant; Standart® Rubal CT; Standart® Rubal RT; Standart® Special PC 20; Standart® Super Lining GGT

Aluminum bronze. *See* Aluminum
Aluminum Dextran. *See* Aluminum stearate
Aluminum, dihydroxy (octadecanoato-o-). *See* Aluminum stearate

Aluminum distearate
CAS 300-92-5; EINECS 206-101-8
Definition: Aluminum salt of stearic acid
Empirical: $C_{36}H_{71}O_5Al$
Formula: $[CH_3(CH_2)_{16}COO]_2Al(OH)$
Properties: White powd.; insol. in water, alcohol, ether; forms gel w/ aliphatic and aromatic hydrocarbons; m.w. 556.51; dens. 1.009; m.p. 145 C
Uses: Thickener in paints, inks, greases; water repellent; lubricant in plastics and cordage, in cement prod.
Regulatory: FDA 21CFR §172.863, 173.340, 175.105, 175.300, 175.320, 176.170, 176.200, 176.210, 177.1200, 177.1460, 177.2260, 178.3910, 179.45, 181.22, 181.29
Trade names: Synpro® Aluminum Distearate 303; Witco® Aluminum Stearate 22

Aluminum 2-ethyl hexanate
Uses: Gelling agent for printing inks, varnishes
Trade names: Forbest 5724

Aluminum hydrate. *See* Aluminum hydroxide; Alumina trihydrate

Aluminum hydroxide
CAS 21645-51-2; EINECS 244-492-7
Synonyms: Alumina hydrate; Alumina hydrated; Alumina trihydrate; Aluminic acid; Aluminum oxide hydrate; Aluminum hydrate ; Aluminum trihydroxide; Aluminum trihydrate; Hydrated alumina; Trihydrated alumina
Classification: Inorganic compd.
Empirical: AlH_3O_3
Formula: $Al(OH)_3$
Properties: White cryst. powd. or gran.; insol. in water; sol. in min. acids, caustic soda; m.w. 78.01; dens. 2.42; m.p. loses water @ 300 C
Precaution: Incompat. with chlorinated rubber
Toxicology: TLV:TWA 2 mg (Al)/m^3; poison by intraperitoneal route; human systemic effects by ing.; no known skin toxicity; mutagenic data
Uses: Adsorbent, emulsifier, filtering medium, dyes, paints, printing inks, waterproofing fabrics, textile finishing; flame retardant; mordant in dyeing
Regulatory: FDA 21CFR §175.300, 177.1200, 177.2600, 182.90; BP compliance (dried, gel, mixt., oral susp.)
Manuf./Distrib.: Alcan; Alcoa; Aldrich; Atomergic Chemetals; BA Chem. Ltd; Fluka; J.M. Huber/Engineered Mins.; Lohmann; Nyco Mins.; Reheis; Rhone-Poulenc; Seimi Chem.; Sigma; Vista; Whittaker, Clark & Daniels

Aluminum-magnesium hydroxy carbonate

Trade names: H-100; H-105; H-109; H-600; H-800; H-900; H-990; Martinal® OL-104; Martinal® ON-310; Trikup®
See also Alumina trihydrate

Aluminum-magnesium hydroxy carbonate
Uses: Acid neutralizing agent used in prod. of polyolefin resins
Trade names: L-55R®

Aluminum magnesium silicate. *See* Magnesium aluminum silicate
Aluminum monobasic stearate. *See* Aluminum stearate
Aluminum monostearate. *See* Aluminum stearate

Aluminum naphthenate
Properties: Rubbery yel. substance
Uses: Drier for paints; detergent in lube oils; high thickening power
Manuf./Distrib.: D.N. Lukens; Vanguard Chem. Int'l.

Aluminum nitride
CAS 24304-00-5; EINECS 246-140-8
Formula: AlN
Properties: Cryst. solid; dec. by water into $Al(OH)_3$ and NH_3; m.w. 40.99; dens. 3.10; m.p. 2150 C; b.p. dec. @ 2773 K; Mohs hardness 9+
Precaution: incompat. with water and water vapor; high humidity or exposure to moisture causes slow hydrolysis and formation of ammonia
Toxicology: ACGIH TLV 10 mg/m^3; overexposure may cause skin irritation, burning of eyes; chronic inh. of dust may cause lung disorders/disease; ingestion may release ammonia gas
Uses: As semiconductor in electronics, nitriding of steel; filler for elastomeric, epoxy, polymeric systems, in coatings
Manuf./Distrib.: Advanced Refractory Tech.; Aldrich; Atomergic Chemetals; Carborundum; Dow; Mandoval Ltd; Noah; Reade Advanced Materials
Trade names: AlNel™ A-100; AlNel™ A-100S; AlNel™ A-100 WR, A-200WR; AlNel™ A-200; AlNel™ A-200SD

Aluminum octoate
Uses: Drier for paints
Manuf./Distrib.: Archway Sales; Baychem; Lenape Ind.; D.N. Lukens; OM Group; Vanguard Chem. Int'l.; Witco/Oleo-Surf.

Aluminum orthophosphate
CAS 7784-30-7
Synonyms: Aluminum phosphate; Aluminophosphoric acid; Phosphoric acid, aluminum salt (1:1)
Empirical: $AlPO_4$
Properties: Wh. cryst.; insol. in water and alcohol; sl. sol. in HCl, nitric acid; m.w. 121.95; dens. 2.566; m.p. 1500 C; isomorphous with quartz
Toxicology: Sol'ns. are corrosive to tissue
Uses: Flux for ceramics, dental cements, cosmetics, paints and varnishes, pharmaceuticals, pulp and paper, antacid
Manuf./Distrib.: Aldrich; Fluka; Sigma

Aluminum oxide. *See* Alumina
Aluminum oxide (2:3). *See* Alumina
Aluminum oxide hydrate. *See* Aluminum hydroxide
Aluminum phosphate. *See* Aluminum orthophosphate
Aluminum powder. *See* Aluminum

Aluminum silicate
CAS 1327-36-2; 12141-46-7; 14504-95-1; EINECS 215-475-1; 235-253-8
Synonyms: Pyrophyllite; CI 77004
Definition: Complex inorganic salt with 1 mole alumina, 1-3 moles silica
Formula: Al_2O_5Si
Properties: Varying proportions of Al_2O_3 and SiO_2; wh. mass, crystals, or whiskers; high str.; m.w. 162.05; insol. in water
Toxicology: Essentially harmless when given orally or applied to skin
Uses: Clay, extender used in reinforced plastics, dental cements, glass industry, paint filler, mfg. of precious

stones, enamels
Regulatory: FDA 21CFR §175.300, 177.1200, 177.1460, 177.2600, 184.1155
Manuf./Distrib.: Boehle; Burgess Pigment; R.E. Carroll; CE Mins.; Cimbar Perf. Mins.; Dry Branch Kaolin; ECC Int'l.; Engelhard; C.P. Hall; J.M. Huber; Kaopolite; Kyowa Chem. Ind.; Lenape Ind.; D.N. Lukens; Solvay SA; Southern Clay Prods.; Takeda Chem. Ind.; Tamms Ind.; Thiele Kaolin; Tomita Pharmaceutical; U.S. Silica; R.T. Vanderbilt; Whittaker, Clark & Daniels; Jesse S. Young
Trade names: Akrodip™ Dry; Aluminum Silicate P820; Bentolite® D; Iceberg®; Icecap® K; Optiwhite®; Optiwhite® MX; Optiwhite® P; SnowTex® 45; Suspengel Elite; Suspengel Micro; Suspengel Ultra; Tisyn®
See also Pyrophyllite

Aluminum silicate, calcined
Uses: Reinforcing extender, pigment for rubber and polymer systems, paints, wire and cable insulation
Trade names: Altowhite LL; Satintone® 5; Satintone® 5HB; Satintone® OP; Satintone® Plus; Satintone® Special; Satintone® W; Translink® 37; Translink® 77; Translink® 445; Translink® 555; Translink® HF-900
See also Kaolin

Aluminum silicate, hydrous
Uses: Reinforcing extender, pigment in rubber and polymer systems, paints; blocking agent for adhesives
Trade names: ASP® 072; ASP® 100; ASP® 102; ASP® 170; ASP® 172; ASP® 200; ASP® 400; ASP® 400P; ASP® 600; ASP® 602; ASP® 672; ASP® 900; ASP® 902; ASP® NC2; ASP® NCF; ASP® NCS; ASP® RO; ASP® Superfine; Buca®; Catalpo®
Trade names containing: ASP® 101

Aluminum sodium silicate. *See* Sodium silicoaluminate

Aluminum stearate
CAS 7047-84-9; EINECS 230-325-5
Synonyms: Aluminum, dihydroxy (octadecanoato-o-); Octadecanoic acid aluminum salt; Aluminum Dextran; Stearic acid aluminum dihydroxide salt; Aluminum monobasic stearate; Aluminum monostearate
Classification: Aliphatic organic compd.
Definition: Aluminum salt of stearic acid
Empirical: $C_{18}H_{37}AlO_4$
Formula: $CH_3(CH_2)_{16}COOAl(OH)_2$
Properties: Wh. to ylsh. fine powd., faint char. odor; insol. in water, alcohol, ether; sol. in alkali, petrol., turpentine oil; m.w. 344.48; dens. 1.070; m.p. 115 C
Toxicology: TWA 10 mg/m^3 (ACGIH); TLV:TWA 2 mg(Al)/m^3; no known toxicity; questionable carcinogen; heated to decomp., emits acrid smoke and irritating fumes
Uses: Paint, varnish drier, greases, waterproofing agent, cement additive, plastics lubricants, cutting compds., flatting agent, cosmetics, pharmaceuticals, and defoaming agent
Regulatory: FDA §121.1099, 172.863, 173.340, 181.29; USP/NF compliance
Manuf./Distrib.: AC Ind.; Ashland; Elf Atochem N. Am.; Ferro/Grant; Fluka; Harcros; Lohmann; Magnesia GmbH; Norac; Spectrum Chem. Mfg.; Syn. Prods.; Witco/Oleo-Surf.; Witco/Polymer Addit.
Trade names: Cecavon AL 11; Cecavon AL 12; Synpro® Aluminum Monostearate NF

Aluminum trihydrate. *See* Aluminum hydroxide
Aluminum trihydroxide. *See* Alumina trihydrate; Aluminum hydroxide

Aluminum tristearate
CAS 637-12-7; EINECS 211-279-5
Synonyms: Tribasic aluminum stearate
Definition: Aluminum salt of stearic acid
Empirical: $C_{54}H_{105}O_6 \cdot Al$
Formula: $[CH_3(CH_2)_{16}COO]_3Al$
Properties: Hard material; sol. in alcohol, benzene, oil turpentine, min. oils; pract. insol. in water; m.w. 877.35; m.p. 117-120 C
Uses: Paint and varnish drier, greases, waterproofinga gent, cement additive, lubricants, cutting compds., flatting agent, cosmetics and pharmaceuticals, defoaming agent in beet sugar and yeast processing
Regulatory: FDA 21CFR §172.863, 173.340, 175.105, 175.300, 175.320, 176.170, 176.200, 176.210, 177.1200, 177.1460, 177.260, 178.3910, 179.45, 181.22, 181.29

Aluminum-zinc phosphate hydrate
Uses: Wh., nontoxic, anticorrosive pigments for coatings; versatile use in a wide range of vehicles
Trade names: Heucophos™ ZPA

Alumite. *See* Alumina
Alundum. *See* Alumina
Amber. *See* Succinic acid
Amber acid. *See* Succinic acid
Amide C-18. *See* Stearamide
Amides, tallow, hydrogenated. *See* Hydrogenated tallow amide

Amidoamine
Uses: Hydrophobic component for formulating fabric finishes; softener, lubricant; epoxy curing agent
Manuf./Distrib.: Ashland; Chemron; Croda; Scher; Witco/Oleo-Surf.
Trade names: Ancamide 2396; Epi-Cure® 3010; Epi-Cure® 3015; Epi-Cure® 3025; Epi-Cure® 3060; Epotuf Hardener 37-620; Genamid® 151; Genamid® 235; Genamid® 250; Genamid® 490; Genamid® 491; Genamid® 747; Genamid® 2000; HY 955; HY 3640; Uni-Rez® 2800

Amine dinonylnaphthalene sulfonate
Trade names containing: Nacorr® 1754

Amine dodecylbenzene sulfonate
Uses: Emulsifier, wetting agent, dispersant, solubilizer, degreaser; for cosmetic, industrial, agric., metal, fuel oil additive, cleaning applics.
Trade names: Rhodacal® 2283

Amines, coco alkyl, acetates. *See* Cocamine acetate
Amines, coco alkyl dimethyl. *See* Dimethyl cocamine
Amines, dicoco alkyl. *See* Dicocamine
Aminic acid. *See* Formic acid
3-Aminoaniline. *See* m-Phenylenediamine
m-Aminoaniline. *See* m-Phenylenediamine
Aminobenzene. *See* Aniline
2-Aminobenzoic acid. *See* Anthranilic acid
o-Aminobenzoic acid. *See* Anthranilic acid
3-Aminobutanoic acid, n-coco alkyl derivatives. *See* Cocaminobutyric acid
Aminocaproic lactam. *See* Caprolactam
2-Aminoethanol. *See* Ethanolamine
2-Aminoethyl alcohol. *See* Ethanolamine
β-Aminoethyl alcohol. *See* Ethanolamine

N-2-Aminoethyl-3-aminopropyl trimethoxysilane
CAS 1760-24-3; EINECS 212-164-2
Synonyms: N-β-(-Aminoethyl) γ-aminopropyltrimethoxysilane; N-[3-(Trimethoxysilyl) propyl] ethylenediamine
Classification: Diamino silane
Empirical: $C_8H_{22}N_2O_3Si$
Formula: $NH_2CH_2CH_2NH(CH_2)_3Si(OCH_3)_3$
Properties: Liq.; m.w. 222.41; dens. 1.028 (20/4 C); b.p. 261-263 C; ref. index 1.442 (25 C); flash pt. 121 C
Precaution: Corrosive; moisture sensitive
Toxicology: Poison by intravenous route, mild toxicity by ingestion and skin contact; severe eye irritant
Uses: Coupling agent for epoxies, phenolics, melamines, nylons, PVC, acrylics, urethanes, nitrile rubbers
Manuf./Distrib.: Aldrich; Fluka
Trade names: Dow Corning® 21; Silquest® A-1120

N-β-(-Aminoethyl) γ-aminopropyltrimethoxysilane. *See* N-2-Aminoethyl-3-aminopropyl trimethoxysilane
Aminoethylethandiamine. *See* Diethylenetriamine

Aminoethylpiperazine
CAS 140-31-8; EINECS 205-411-0
Synonyms: AEP; 1-(2-Aminoethyl) piperazine; 2-Piperazinoethylamione
Classification: Amine
Empirical: $C_6H_{15}N_3$
Formula: $H_2NC_2H_4NCH_2CH_2NHCH_2CH_2$
Properties: Colorless or pale yel liq.; sol. in water; m.w. 129.24; dens. 0.9837; b.p. 222.0 C; f.p. 17.6 C; flash pt. 200 F; ref. index 1.499
Precaution: Combustible; corrosive

Toxicology: Strong irritant to tissue; poison by intraperitoneal route, toxic by ingestion and skin contact
Uses: Epoxy curing agent, intermediate for pharmaceuticals, anthelmintics, surface-active agents, synthetic fibers, corrosion inhibitors, asphalt additives, emulsion breakers, emulsifiers, textile chems.; used for civil engineering, coatings, adhesives, grouts, casting and elec. encapsulation; yields systems with improved impact and shorter pot life
Manuf./Distrib.: Akzo Nobel; Aldrich; BASF AG; Diamine & Chems. Ltd; Dow; Fabrichem; Fluka; Texaco; Tosoh; Union Carbide
Trade names: D.E.H. 39; Epi-Cure® 3200

1-(2-Aminoethyl) piperazine. *See* Aminoethylpiperazine
Aminoethyl tall oil imidazoline. *See* Tall oil aminoethyl imidazoline
Aminoform. *See* Hexamethylenetetramine
2-Amino-2-(hydroxymethyl)-1,3-propanediol. *See* Tris (hydroxymethyl) aminomethane
1-Amino-3-methoxypropane. *See* Methoxypropylamine
Aminomethyl propanol. *See* 2-Amino-2-methyl-1-propanol

2-Amino-2-methyl-1-propanol
CAS 124-68-5; EINECS 204-709-8
Synonyms: AMP; Isobutanolamine; Aminomethyl propanol; Isobutanol-2 amine
Classification: Substituted aliphatic alcohol
Empirical: $C_4H_{11}NO$
Formula: $CH_3(CH_3)(NH_2)CH_2OH$
Properties: Solid or visc. liq.; misc. with water; sol. in alcohol; m.w. 89.1; dens. 0.93 (20/4 C); m.p. 30 C; b.p. 165 C (760 mm); flash pt. (TOC) 153 F
Precaution: Flamm. exposed to heat or flame
Toxicology: Moderately toxic by ingestion
Uses: Multifunctional additive for latex paint; pigment codispersant; solubilizer for resins; emulsifier; neutralizer for boiler water treatment; corrosion inhibitor; catalyst (as acid salt); absorbent for acidic gases, organic synthesis, cosmetics
Regulatory: FDA 21CFR §175.105, 176.170
Manuf./Distrib.: Aldrich; Allchem Ind.; ANGUS; Ashland; Fluka; Janssen Chimica; Sigma
Trade names: AMP; AMP-95®

N-3-Aminopropyl-N-tallow alkyl trimethylene diamine. *See* Tallow dipropylene triamine

Aminopropyltriethoxysilane
CAS 919-30-2; EINECS 213-048-4
Synonyms: γ-Aminopropyl triethoxysilane; 3-(Triethoxysilyl)-1-propanamine
Formula: $NH_2(CH_2)_3Si(OC_2H_5)_3$
Properties: Liq.; m.w. 221.37; dens. 0.942; b.p. 217 C; flash pt. 104 C
Toxicology: Poison by intraperitoneal route; moderate toxicity by ingestion and skin contact; skin and eye irritant
Uses: Sizing of glass fibers for making laminates; filler for thermosets, thermoplastics; as primer or additive, esp. in epoxy, phenolic, PU, vinyl plastisols and imide systems
Manuf./Distrib.: Aldrich; Fluka; Gelest; Hüls Am.; PCR; Sigma
Trade names: Dow Corning® Z-6011; Silquest® A-1100; Silquest® A-1101; Silquest® A-1102

γ-Aminopropyl triethoxysilane. *See* Aminopropyltriethoxysilane

Aminopropyltrimethoxysilane
CAS 13822-56-5; EINECS 237-511-5
Synonyms: 3-(Trimethoxysilyl)propylamine; 3-Aminopropyltrimethoxysilane
Formula: $NH_2(CH_2)_3Si(OCH_3)_3$
Properties: Water-white liq.; m.w. 179.29; dens. 1.01; b.p. 91-92 C (15 mm); flash pt. 83 C; ref. index 1.42
Precaution: Corrosive; moisture-sensitive
Uses: Class fabric sizing, binder, adhesion promoter, release agent, lubricant; adhesion promoter for coatings; features active hydrogen reaction
Manuf./Distrib.: Aldrich
Trade names: Silquest® A-1110

3-Aminopropyltrimethoxysilane. *See* Aminopropyltrimethoxysilane
5-Amino-1,3,3-trimethylcyclohexylmethanamine. *See* Isophorone diamine
Aminotrimethylolmethane. *See* Tris (hydroxymethyl) aminomethane

1,2-Aminozophenylene. *See* 1H-Benzotriazole

Ammonia
CAS 7664-41-7; EINECS 231-635-3
Classification: Inorganic compd.
Empirical: H_3N
Formula: NH_3
Properties: Colorless gas or liq., sharp, intensely irritating odor; easily liquefied by pressure; sol. in water, alcohol, ether; m.w. 17.03; f.p. -77 C; b.p. -33.5 C
Precaution: Moderate fire risk; caustic
Toxicology: Corrosive; inh. of conc. fumes may be fatal; irritating to eyes and mucous membranes; shown to produce cancer of the skin in human doses of 1000 mg/kg of body wt.
Uses: Fertilizers, refrigerant, nitriding of steel, condensation catalyst, neutralizing agent, petroleum industry, latex preservative, explosives
Regulatory: FDA 27CFR §21.95; Japan approved; BP compliance
Manuf./Distrib.: Air Prods.; Aldrich; AlliedSignal; Asahi Chem. Ind.; Chemical; Chevron; Coyne; Cytec Ind.; Fluka; General Chem.; LaRoche Ind.; Mitsui Toatsu; Molycorp; Monsanto; Nissan Chem. Ind.; Norsk Hydro AS; Occidental; Olin; PPG Ind.; Veckridge
Trade names containing: SMA® 1440H; Snowtex N

Ammonia aqueous. *See* Ammonium hydroxide
Ammonia solution, strong. *See* Ammonium hydroxide
Ammonia water. *See* Ammonium hydroxide
Ammonioformaldehyde. *See* Hexamethylenetetramine

Ammonium acrylamidomethyl propane sulfonate
Trade names: AMPS® 2411 Monomer

Ammonium acrylate
Uses: Low-foaming dispersant for water-based coatings, slurries, adhesives, etc.; best for water sensitivity
Trade names: Busperse® 2143; Troythix™ LLBA

Ammonium, alkyldimethyl (phenylmethyl)-, chloride. *See* Benzalkonium chloride

Ammonium C9-10 perfluoroalkyl sulfonate
Trade names containing: Fluorad® FC-120

Ammonium heptamolybdate. *See* Ammonium molybdate

Ammonium hydroxide
CAS 1336-21-6; EINECS 215-647-6
Synonyms: Ammonia solution, strong; Strong ammonia solution; Ammonia water; Aqueous ammonia; Ammonia aqueous; Aqua ammonium; Spirit of Hartshorn
Classification: Inorganic base
Empirical: H_5NO
Formula: NH_4OH
Properties: Clear colorless liq., very pungent odor, acrid taste; sol. in water; m.w. 35.06; dens. 0.90; m.p. -77 C; flash pt. none; pH 13.6
Precaution: DOT: Corrosive material; vapor will ignite, but not readily
Toxicology: LD50 (oral, rat) 350 mg/kg; poison by ingestion; inhalation irritant; severe eye irritant; liq. can inflict burns; heated to decomp., emits NH_3 and NO_x
Uses: Alkali, leavening agent, pH control agent for baked goods, caramel, cheese, processed fruits, puddings; boiler water additive
Regulatory: FDA 21CFR §163.110, 182.90, 184.1139, GRAS; Europe listed; UK approved; USP/NF compliance
Manuf./Distrib.: Aldrich; AlliedSignal; Chemical; Coyne; LaRoche Ind.; Ruger; Sigma; Spectrum Chem. Mfg.
Trade names containing: Carboset® 514H; Carboset® GA1926; Carboset® GA1931; Zinc Oxide #1

Ammonium-L-2-hydroxy propionate. *See* Ammonium lactate

Ammonium lactate
CAS 515-98-0
Synonyms: Lactic acid, ammonium salt; Ammonium-L-2-hydroxy propionate

Empirical: $C_3H_9NO_3$
Formula: $CH_3CHOHCOONH_4$
Properties: Colorless to yel. syrupy liq.; sol. in water, alcohol; m.w. 107.08; dens. 1.19-1.21 (15 C); m.p. 91-94 C
Uses: Catalyst for esterification and olefin polymerization; used in coatings; resin crosslinking agent for automotive prods., coatings, elastomers, films/paints, graphic arts, plastics
Manuf./Distrib.: AMRESCO; Kraft Chem.; Lohmann; Magnablend; Penta Mfg.; Purac Am.; Spectrum Chem. Mfg.; Universal Preserv-A-Chem
Trade names: Tyzor LA

Ammonium laureth sulfate
CAS 32612-48-9 (generic); 67762-19-0
Synonyms: Ammonium lauryl ether sulfate
Definition: Ammonium salt of ethoxylated lauryl sulfate
Formula: $(C_2H_4O)_n \cdot C_{12}H_{26}O_4S \cdot H_3N$, n = 1-4
Toxicology: Moderate toxicity by ingestion; skin and eye irritant
Uses: Surfactant, emulsifier for emulsion polymerization (acrylics, styrene-acrylic, vinyl acrylics, S/B, paints, coatings), hair and skin detergents; breaks up and holds oils and soil
Regulatory: FDA 21CFR §175.105
Manuf./Distrib.: Allchem Ind.; Aquatec Quimica SA; Ashland; Clariant; Great Western; Lonza; Pilot; Rhone-Poulenc Surf. & Spec.; Sea-Land; Stepan; Witco/Oleo-Surf.
Trade names: Polystep® B-11

Ammonium laureth-12 sulfate
CAS 32612-48-9 (generic)
Definition: Ammonium salt of ethoxylated lauryl sulfate
Formula: $(C_2H_4O)_n \cdot C_{12}H_{26}O_4S \cdot H_3N$, n = 12
Uses: Surfactant, emulsifier for acrylic copolymers, emulsion polymerization, paints and coatings
Regulatory: FDA 21CFR §175.105
Trade names: Polystep® B-22

Ammonium laureth-30 sulfate
CAS 32612-48-9 (generic)
Formula: $(C_2H_4O)_n \cdot C_{12}H_{26}O_4S \cdot H_3N$, n = 30
Uses: Emulsifier for emulsion polymerization (acrylics, styrene-acrylics, vinyl acrylics), floor finish latexes, paints and coatings
Trade names: Polystep® B-20

Ammonium lauryl ether sulfate. *See* Ammonium laureth sulfate

Ammonium lauryl sulfate
CAS 2235-54-3; 68081-96-9; EINECS 218-793-9
Synonyms: Sulfuric acid, monododecyl ester, ammonium salt
Definition: Ammonium salt of lauryl sulfate
Formula: $C_{12}H_{26}O_4S \cdot H_3N$
Properties: M.w. 283.48
Toxicology: Skin and eye irritant
Uses: Detergent, emulsifier, foaming agent, dispersant, wetting agent; for personal care prods., coatings, carpet shampoos, firefighting, dry wall mfg., dyes, chemical specialties, emulsion polymerization
Regulatory: FDA 21CFR §175.105, 175.210, 176.170, 177.1200
Manuf./Distrib.: Allchem Ind.; Aquatec Quimica SA; Ashland; Clariant; Lonza; Pilot; Rhone-Poulenc Surf. & Spec.; Sea-Land; Stepan; Witco/Oleo-Surf.
Trade names: Polystep® B-7; Rhodapon® L-22; Ufarol Am 30; Ufarol Am 70

Ammonium manganese pyrophosphate. *See* Manganese violet

Ammonium molybdate
CAS 12027-67-7; 13106-76-8; EINECS 234-722-4
Synonyms: Ammonium paramolybdate; Ammonium heptamolybdate
Formula: $(NH_4)_6Mo_7O_{24} \cdot 4H_2O$
Properties: Wh. crystals; sol. in water; insol. in alcohol, acetone; m.w. 1235.89; sp.gr. 2.50; pH 5.0-5.5 (5%)
Toxicology: LD50 (oral, rat) 333 mg/kg; poison by ingestion and intraperitoneal routes; an irritant
Uses: In soil additives, enamel bonding agents, protective and decorative metal coatings, iron and steel alloys,

lubricants, petrol. refining catalysts, pigments, corrosion inhibitors, smoke suppressants, prod. of molybdenum metal
Manuf./Distrib.: AAA Molybdenum Prods.; Aldrich; Am. Biorganics; Climax Molybdenum BV; Climax Perf. Materials; GFS; Noah; Research Organics; San Yuan; Spectrum Chem. Mfg.

Ammonium naphthalene sulfonate
Uses: Dispersant for pigments, carbon black, dyestuffs, in ceramics, paints; viscosity depressant; emulsion polymerization
Trade names: Lomar® PWA; Lomar® PWA Liq.

Ammonium nonoxynol-4 sulfate
CAS 9051-57-4; 31691-97-1 (generic); 63351-73-5
Definition: Ammonium salt of sulfated nonoxynol-4
Formula: $(C_2H_4O)_xC_{15}H_{24}O_4S \cdot H_3N$
Uses: High foaming surfactant, wetting agent, dispersant, emulsifier for shampoos, skin cleansers, lt. duty cleaners, emulsion polymerization (vinyl acetate, acrylates, methacrylates, styrene, butadiene, latex); for adhesives, paints, paper coatings, textile, and industrial and architectural coatings
Regulatory: FDA 21CFR §178.3400
Trade names: Abex® EP-100; Polystep® B-1

Ammonium nonoxynol-9 sulfate
Uses: Emulsfier, stabilizer for emulsion polymerization; for adhesives, paints, paper, textile, and industrial coatings
Trade names: Abex® EP-110

Ammonium nonoxynol-30 sulfate
Uses: Emulsifier for emulsion polymerization; for adhesives, paints, paper, textile, and industrial coatings
Trade names: Abex® EP-120

Ammonium paramolybdate. *See* Ammonium molybdate

Ammonium polyacrylate
CAS 9003-03-6
Synonyms: Poly(acrylic acid), ammonium salt; 2-Propenoic acid, homopolymer, ammonium salt
Definition: Ammonium salt of polyacrylic acid
Empirical: $(C_3H_4O_2)_x \cdot xH_3N$
Uses: Dispersant for paints and coatings; thickening and stabilizing agent for syn. latices; used in coatings, adhesives, dipped, cast, and molded goods, cements for rug backing, spraying, spreading, brushing, and extruding compds.
Regulatory: FDA 21CFR §175.105
Trade names: BYK®-156; Daxad® 37LA7-35; Ser-AD FX 504

Ammonium polymethacrylate
Uses: Stabilizer and thickener in latex rubber processing; dispersant for pigments and fillers commonly used in latex paint systems; also for polymerization and clay coating
Trade names: Daxad® 34A9

Ammonium polyphosphate sol'n.
CAS 14728-39-3
Uses: Flame retardant for plastics, adhesives, elastomers, paints
Manuf./Distrib.: Albright & Wilson Am.; Fabrichem; Hoechst Celanese; LaRoche Ind.; Monsanto
Trade names: Amgard® MC; Exolit® 422; Hostaflam® 422; Hostaflam® AP 422; Hostaflam® AP 750
Trade names containing: Hostaflam® AP 462

Ammonium/sodium acrylate
Uses: Dispersant for water-based coatings, slurries, adhesives, etc.
Trade names: Busperse® 2144

Ammonium xylenesulfonate
CAS 26447-10-9; EINECS 247-710-9
Synonyms: Benzenesulfonic acid, dimethyl-, ammonium salt
Definition: Ammonium salt of ring sulfonated mixed xylene isomers
Formula: $C_8H_{10}O_3S \cdot H_3N$

Properties: Sol. in water; m.w. 203.26
Precaution: Flamm.
Toxicology: Chronic toxicity or skin effects are not known; narcotic in high doses
Uses: Hydrotrope, stabilizer, solubilizer used in formulating detergents, inks, electroplating baths, dyestuffs, polymers; lacquer solv. used in nail polishes
Manuf./Distrib.: Witco/Oleo-Surf.

Ammonium zirconium carbonate
CAS 68309-95-5
Synonyms: AZC; Zirconate (2-), bis[carbonate (2-)-0] dihydroxy-diammonium
Uses: Binder insolubilizer for paper and board coatings, nonwoven fabrics; deposit control agent in wet-end paper processes; str. additive for alkaline-sized paper; component of waterborne inks, overprint varnishes, adhesives; textile, leather, and paper water-repellent formulations; antimicrobial treatment for textiles and timber; coagulant in mfg. of syn. leather cloth; metal pretreatment coatings; precursor to zirconia in catalyst systems
Trade names: Bacote™ 20

Amodimethicone
Definition: Silicone polymer end blocked with amino functional groups
Formula: $HO[Si(CH_3)_2O][SiOH(CH_2)_3NHCH_2CH_2NH_{2-}O]H$
Uses: Silicone polymer curing to a durable, detergent-resistant film for mold release, particle treatment, textile finishes, and polish applics.
Trade names: GP-4 Silicone Fluid; SM 2059

Amorphous silica. *See* Silica, amorphous
AMP. *See* 2-Amino-2-methyl-1-propanol
2-AMPS. *See* 2-Acrylamido-2-methylpropanesulfonic acid

Amyl acetate
CAS 628-63-7; EINECS 211-047-3
Synonyms: Amylacetic ester; Banana oil; Pear oil; Acetic acid pentyl ester; Pentyl acetate
Definition: Ester of amyl alcohol and acetic acid
Empirical: $C_7H_{12}O_2$
Formula: $CH_3COOC_5H_{11}$
Properties: Liq.; m.w. 130.19; dens. 0.876; m.p. -100; f.p. 75 F; b.p. 142 C; flash pt. 75 F; ref. index 1.4013
Precaution: Flamm.
Toxicology: Irritant; TLV 100 ppm
Uses: Solvent for lacquers and paints, extraction of penicillin, photographic film, leather and nail polishes, warning odor, flavoring agent, printing and finishing fabrics, solvent for phosphors in fluorescent lamps
Regulatory: FDA 21CFR §175.105; FEMA GRAS; BP compliance
Manuf./Distrib.: Aldrich; Ashland; BP Chems. Ltd; Chemcentral; Coyne; Fluka; Penta Mfg.; Sigma; Union Carbide; Van Waters & Rogers

Amylacetic ester. *See* Amyl acetate

Amyl alcohol
CAS 71-41-0 (n-); 75-85-4 (t-); EINECS 200-752-1; 227-907-6
Synonyms: Amyl hydrate; Alcohol C-5; Pentyl alcohol; 2-Methyl-1-butanol; n-Butyl carbinol; 1-Pentanol (n-); 2-Pentanol (t-)
Empirical: $C_5H_{12}O$
Formula: $CH_3(CH_2)_4OH$, eight isomers possible
Properties: Clear liq., mild char. odor; sol. in water; misc. with alcohol, ether; m.w. 88.15; dens. 0.8168 (20/20 C); m.p. -79 C; b.p. 137.8 C; flash pt. (CC) 91 F; ref. index 1.409
Toxicology: LD50 (oral, rat) 3030 mg/kg; moderately toxic by ingestion and skin contact; eye irritant and upper respiratory irritant by inhalation; severe skin and eye irritant; narcotic
Uses: Solv. for paints; n-: Raw material for pharmaceutical preparations, organic synthesis solvent; t-: Solvent, flotation agent, organic synthesis, medicine (sedative)
Regulatory: FDA §121.1164; FEMA GRAS
Manuf./Distrib.: Aldrich; Allchem Ind.; Ashland; BASF; Fluka; Hoechst Celanese; MTM Spec. Chems.; Sigma; Union Carbide; Vista

Amylase
CAS 9000-92-4; EINECS 232-567-7

Amylcarbinol

Synonyms: Mylase 100; 1,4-D-Glucan glucanohydrolase
Classification: Enzyme
Definition: Starch-degrading enzyme
Properties: Off-white powd. or suspension
Toxicology: May produce hypersensitivity reactions
Uses: Enzyme for textile desizing, conversion of starch-glucose sugars in syrups, baking, dry cleaning, pharmaceuticals; prep. of paper coatings and sizes in paper industry
Manuf./Distrib.: Fluka
Trade names: Amizyme DTX-11; Amizyme TX-8; BAN 120 L; BAN 240 L; BAN 360 S; BAN 1000 S

Amylcarbinol. *See* Hexyl alcohol
Amyl hydrate. *See* Amyl alcohol
n-Amyl methyl ketone. *See* Methyl n-amyl ketone
Analine oil. *See* Aniline
Anchoic acid. *See* Azelaic acid
Anhydrite (natural form). *See* Calcium sulfate
1,4-Anhydro-D-glucitol, 6-hexadecanoate. *See* Sorbitan palmitate
Anhydrosorbitol dioleate. *See* Sorbitan dioleate
Anhydrosorbitol monolaurate. *See* Sorbitan laurate
Anhydrosorbitol monooleate. *See* Sorbitan oleate
Anhydrosorbitol monostearate. *See* Sorbitan stearate
Anhydrosorbitol trioleate. *See* Sorbitan trioleate
Anhydrosorbitol tristearate. *See* Sorbitan tristearate
Anhydrotrimellitic acid. *See* Trimellitic anhydride
Anhydrous gypsum. *See* Calcium sulfate
Anhydrous lanolin. *See* Lanolin

Aniline
CAS 62-53-3; EINECS 200-539-3
Synonyms: Aminobenzene; Analine oil; Phynlamine; Kyanol
Empirical: C_6H_7N
Formula: $C_6H_5NH_2$
Properties: Colorless oily liq., char. odor; m.w. 93.13; dens. 1.021 (20/4 C); f.p. -6.2 C; b.p. 184 C; flash pt. 158 F; ref. index 1.586 (20 C)
Precaution: Photosensitive
Toxicology: LD50 (oral, rat) 440 mg/kg; severe eye irritant; mild skin irritant
Uses: Intermediate for paints
Manuf./Distrib.: Aldrich; Allchem Ind.; AlliedSignal; Aristech; BASF AG; Bayer; First Chem.; Fluka; Sigma; Spectrum Chem. Mfg.; Uniroyal; United Min. & Chem.

Aniline, N,N-dimethyl. *See* n,n-Dimethylaniline
Animal starch. *See* Glycogen
Annulene. *See* Benzene

Anthranilic acid
CAS 118-92-3; EINECS 204-287-5
Synonyms: o-Aminobenzoic acid; 2-Aminobenzoic acid; o-Carboxyaniline; 2-Carboxyaniline; Vitamin L
Classification: Aromatic organic compd.
Empirical: $C_7H_7NO_2$
Formula: $C_6H_4(NH_2)(CO_2H)$
Properties: Yellowish crystals, sweetish taste; sol. in hot water, alcohol, ether; m.w. 137.14; m.p. 144-146 C, sublimes
Precaution: Combustible
Toxicology: Experimental tumorigen and reproductive effects; moderate toxicity by intraperitoneal route; irritant
Uses: Intermediate for dyes, paints, drugs, perfumes, pharmaceuticals
Regulatory: BP compliance
Manuf./Distrib.: Aldrich; BASF; BASF AG; Fluka; PMC Specialties; Sigma; Spectrum Chem. Mfg.; Triple Crown Am.
Trade names: AA

Antimonic acid. *See* Antimony pentoxide
Antimonic anhydride. *See* Antimony pentoxide

Antimonic oxide. *See* Antimony pentoxide
Antimonious oxide. *See* Antimony trioxide
Antimonious sulfide. *See* Antimony trisulfide
Antimony bloom. *See* Antimony trioxide
Antimony (III) oxide. *See* Antimony trioxide
Antimony (III) sulfide. *See* Antimony trisulfide
Antimony orange. *See* Antimony trisulfide
Antimony oxide. *See* Antimony trioxide

Antimony pentoxide
CAS 1314-60-9; EINECS 215-237-7
Synonyms: Antimonic anhydride; Antimonic oxide; Antimonic acid; Diantimony pentoxide; Stibic anhydride
Empirical: O_5Sb_2
Formula: Sb_2O_5
Properties: Yellowish powd., cube; sl. sol. in water; pract. insol. in HNO_3; sl. sol. in warm KOH or HCl; m.w. 323.52
Toxicology: LD50 (rats) 4 g/kg; toxic by intraperitoneal route
Uses: Flame/fire retardant for clothing, latex, resins
Manuf./Distrib.: Aldrich; All Chemie Ltd; Atomergic Chemetals; Cerac; Great Western; Laurel Ind.; Noah; Nyacol Prods.; PQ
Trade names containing: Nyacol® AB40

Antimony peroxide. *See* Antimony trioxide

Antimony trioxide
CAS 1309-64-4; EINECS 215-175-0
Synonyms: Antimony white; Flowers of antimony; Antimony bloom; Antimonious oxide; Antimony (III) oxide; Antimony oxide; Antimony peroxide
Empirical: Sb_2O_3
Formula: O_3Sb_2
Properties: White crystalline powd., odorless; very sl. sol. in water; sol. in KOH, HCl and sulfuric acids, strong alkalies; m.w. 291.52; dens. 5.67; m.p. 655 C; b.p. 1425 C
Toxicology: LD50 (rats) > 20 g/kg; TLV:TWA 500 μ g/m^3; poison by intravenous and subcutaneous routes; confirmed carcinogen, moderate toxicity otherwise
Uses: Flameproofing of textiles, paper, plastics; as paint pigment; ceramic opacifier; catalyst; intermediate staining iron and copper; phosphorus; mordant; glass decolorizer
Manuf./Distrib.: Aldrich; Amspec; Asarco; Ashland; Atomergic Chemetals; Chemisphere Ltd; Chemson Ltd; Elf Atochem N. Am.; Fluka; Hoechst Celanese; Holtrachem; Laurel Ind.; D.N. Lukens; Miljac; Nihon Kagaku Sangyo; Noah; Punda Mercantile; Reade Advanced Materials; Revelli; H.M. Royal; Sigma; Sino-Am. Pigment Systems; Spectrum Chem. Mfg.; United Min. & Chem.
Trade names: FireShield® H; FireShield® HPM; FireShield® HPM-UF; FireShield® L; Monykup®; Octoguard FR-10; Thermoguard® S; Thermoguard® UF
Trade names containing: Octoguard FR-11; Octoguard FR-12; Octoguard FR-13; Octoguard FR-15

Antimony trisulfide
CAS 1345-44-6; EINECS 215-713-4
Synonyms: Antimony (III) sulfide; Antimonious sulfide; Antimony orange
Empirical: S_3Sb_2
Formula: Sb_2S_3
Properties: Red to blk. cryst.; sol. in H_2SO_4; m.w. 339.68; m.p. 546 C; b.p. ≈ 1150 C
Precaution: Flamm. when exposed to hea or flame
Toxicology: LD50 (ipr, mouse) 209 mg/kg
Uses: Vermilion or yellow pigment, antimony salts, pyrotechnics, matches, percussion caps, camouflage paints, ruby glass
Manuf./Distrib.: Atomergic Chemetals; Barium & Chems.; BASF; Great Western; Noah; Spectrum Chem. Mfg.

Antimony white. *See* Antimony trioxide
Apple acid. *See* N-Hydroxysuccinic acid
Aqua ammonium. *See* Ammonium hydroxide
Aqueous ammonia. *See* Ammonium hydroxide
Arachis oil. *See* Peanut oil

Aromatic 100
Properties: M.w. 120; sp.gr. 0.873 (60/60 F); b.p. 313-343 F; flash pt. (TCC) 108 F; ref. index 1.4993 (20 C)
Uses: Solvent, diluent for coatings
Trade names containing: Acryloid® F-10, 40%; Cellokyd 3378 HFN; G-4602-SD-50; Joncryl® 502; Luxate® HT2090; Slip-Ayd® SL-551; Tint-Ayd® ST Series

Aromatic 150
Uses: Solvent
Trade names containing: G-4602-SD-50; Uformite® 21-806

Arsenic pentasulfide
Empirical: As_2S_5
Properties: Yel. to orange powd.; sol. in nitric acid and alkalies; insol. in water; m.w. 310.12
Uses: Paint pigments, light filters, other arsenic compounds
Manuf./Distrib.: Atomergic Chemetals; Great Western

Artificial almond oil. *See* Benzaldehyde
Artificial ant oil. *See* Furfural
Artificial oil of ants. *See* Furfural
as-Trimethylbenzene. *See* 1,2,4-Trimethylbenzene
asym-Dichloroethylene. *See* Vinylidene chloride monomer
Asymmetrical trimethyl benzene. *See* 1,2,4-Trimethylbenzene
ATBC. *See* Acetyl tributyl citrate
ATEC. *See* Acetyl triethyl citrate

Attapulgite
CAS 1337-76-4
Synonyms: Fuller's earth; Activated attapulgite; Colloidal activated attapulgite; Palygorskite; Dioctrahedral smectite
Definition: A hydrated aluminum-magnesium silicate, chief ingredient in Fuller's earth
Toxicology: Nuisance dust when < 1% cryst. silica is present (PEL 5.00 mg/m^3, TLV/TWA 10 mg/m^3 total dust, 5 mg/m^3 respirable); on decomp. by heating, emits acrid smoke and fumes
Uses: Drilling fluids, decolorizing oils, filter medium, absorbent, adsorbent; thickener, stabilizer, suspending and flatting agent in paints; suspending, bodying, and sag control agent in adhesives, sealants, and mastics
Regulatory: FDA GRAS; USP/NF, BP compliance
Trade names: Attagel 40; Attagel 50; Min-U-Gel® 400; Vistrol; Vistrol 1265
Trade names containing: M-P-A® 14

Azacyclotridecane-2-one, homopolymer. *See* Nylon-12
Azacyclotridecane-2-one polyamide. *See* Nylon-12

1-Aza-3,7-dioxo-bicyclo-2,8-diisopropyl-5-ethyl(3.3.0)-octane
CAS 79185-77-6
Manuf./Distrib.: Aldrich
Trade names: Zoldine® RD-20

AZC. *See* Ammonium zirconium carbonate

Azelaic acid
CAS 123-99-9; EINECS 204-669-1
Synonyms: Nonanedoic acid; Dicarboxylic acid C9; 1,7-Heptanedicarboxylic acid; Anchoic acid
Empirical: $C_9H_{16}O_4$
Formula: $HOOC(CH_2)_7COOH$
Properties: Yel. to wh. cryst. powd.; sol. in hot water, alcohol, org. solvs.; m.w. 188.23; dens. 1.029 (20/4 C); m.p. 106 C; b.p. 365 C (dec.); flash pt. 210 C
Precaution: Combustible when exposed to heat or flame
Toxicology: Skin and eye irritant
Uses: Organic synthesis, lacquers, alkyd resins, polyamides, polyester, adhesives, low temp. plasticizers, urethane elastomers
Manuf./Distrib.: Aldrich; Allchem Ind.; Charkit; Chemical; Fluka; Henkel/Emery; Sigma

Azimidobenzene. *See* 1H-Benzotriazole
Azobisformamide. *See* Azodicarbonamide

1,1´-Azobisformamide. *See* Azodicarbonamide

Azodicarbonamide

CAS 123-77-3; EINECS 204-650-8

Synonyms: ADA; 1,1´-Azobisformamide; Azoformamide; Azobisformamide; Diazenedicarboxamide; Azodicarboxamide

Empirical: $C_2H_4N_4O_2$

Formula: $H_2NCON=NCONH_2$

Properties: Orange-red crystals; sol. in hot water; insol. in cold water, alcohol; m.w. 116.08; m.p. 225 C (dec.)

Precaution: Flamm.

Toxicology: Heated to decomp., emits toxic fumes of NO_x

Uses: Blowing and foaming agent for plastics and wire coatings; bleaching agent in cereal flour, maturing agent for flour

Regulatory: FDA 21CFR §172.806, 178.3010; Europe listed; UK approved

Manuf./Distrib.: Aldrich; Allchem Ind.; Charkit; Elf Atochem SA; Fabrichem; Fluka; Gist-brocades Food Ingreds.; Olin; Plastics & Chems.; Quimicel; Rit-Chem; Uniroyal; Witco/Polymer Addit.

Trade names: Celogen® AZ 120

Azodicarboxamide. *See* Azodicarbonamide
Azoformamide. *See* Azodicarbonamide

Baking soda. *See* Sodium bicarbonate
Banana oil. *See* Amyl acetate
Barite. *See* Barium sulfate

Barium acetate

CAS 543-80-6; EINECS 208-849-0

Synonyms: Acetic acid barium salt; Barium diacetate

Empirical: $C_4H_6BaO_4$

Formula: $(CH_3COO)_2Ba$

Properties: Wh. cryst.; water-sol.; m.w. 255.44

Toxicology: LD50 (oral, rat) 921 mg/kg; poisonous by ingestion, subcutaneous, and intravenous routes

Uses: Chemical reagent, acetates, textile mordant, catalyst manufacture, paint and varnish driers

Manuf./Distrib.: Aldrich; Allan; Barium & Chems.; Bernardy Chimie SA; Fluka; Hoechst Celanese; Lohmann; Mallinckrodt; Noah; Sigma; Spectrum Chem. Mfg.

Barium-cadmium complex

Uses: Stabilizer for flexible and rigid PVC compds., calendering, extrusion, inj. molding, powd. molding, slush and rotational molding, hose and profile, filled systems, film and sheet; stabilizer for external coatings, roofing sheets, canvas material, tents, portable inflatable structures, coated metal surfaces

Trade names: Bärostab® XBC 5230; Mark® TT; Mark® WS

Barium chromate

CAS 10294-40-3; EINECS 233-660-5

Synonyms: Lemon chrome; Ultramarine yellow; Baryta yellow

Formula: $BaCrO_4$

Properties: Yel. heavy cryst. powd.; sol. in acids; insol. in water

Precaution: Combustible

Toxicology: Poison, carcinogen

Uses: Safety matches, corrosion inhibitor in metal-joining compounds, rust-inhibitive pigment for paints, ceramics, fuses, pyrotechnics, metal primers, ignition control devices

Manuf./Distrib.: Aldrich; Atomergic Chemetals; Barium & Chems.; BASF AG; Davis Colors; Fluka; Great Western; Nat'l. Chem.; Noah; Punda Mercantile; Sigma

Barium diacetate. *See* Barium acetate

Barium dinonylnaphthalene sulfonate

CAS 25619-56-1

Formula: $C_{28}H_{44}O_3S \cdot {}^1/_2Ba$

Properties: M.w. 529.39

Uses: Rust and corrosion inhibitor, acid neutralizer, stabilizer used in metal working lubricants, greases, lubricating oils, protective coatings, and specialty prods.

Barium metaborate
CAS 13701-59-2
Synonyms: Boric acid barium salt
Empirical: $BaB_2O_4 \cdot H_2O$
Formula: $BaOB_2O_3 \cdot H_2O$
Properties: Wh. powd.; sol. 3-4 g/l water; m.w. 176.54; m.p. 1000 C
Uses: Flame retardant for plastics; rust-inhibitive pigment, extender for paints
Manuf./Distrib.: Atomergic Chemetals; Buckman Labs; CasChem; D.N. Lukens; Seegott; Sino-Am. Pigment Systems; Joseph Storey
Trade names: Bubond® 70; Bubond® 390; Bubond® 391; Busan® 11-M1; Busan® 11-M2; Butrol® 11-M3; Butrol® 22; Butrol® 23

Barium molybdate
Formula: $BaMoO_4$
Properties: White powd.; sl. sol. in acids and water; dens. 4.7 g/cc; m.p. Å 1600 C
Toxicology: TLV:TWA 0.5 mg/m^3
Uses: Electronic and optical equip., pigment in paints and other protective coatings
Manuf./Distrib.: AAA Molybdenum Prods.

Barium naphthenate
Uses: Drier for paints
Manuf./Distrib.: Ferro/Bedford; Ideas; D.N. Lukens

Barium phenate
Uses: Heat stabilizer for PVC coatings, films, and fabricated materials

Barium sulfate
CAS 7727-43-7; EINECS 231-784-4
Synonyms: Barytes (natural); Sulfuric acid barium salt (1:1); Barium sulfate (1:1); Barite; Heavy sparr; Tiff; Blanc fixe (artificial, precipitated); Precipitated barium sulfate; Basofor
Classification: Inorganic salt
Empirical: BaO_4S
Formula: $BaSO_4$
Properties: White or yellowish fine powd. free from grittiness, odorless, tasteless; sol. in conc. sulfuric acid; pract. insol. in water, dilute acids, alcohol; m.w. 233.40; dens. 4.25-4.5; m.p. 1580 C
Toxicology: TLV:TWA 10/mg^3 (total dust); poisonous when ingested; frequently causes skin reactions when applied
Uses: Extender, weighting mud in oil drilling, paper coating, printing inks, paints; filler and delustrant for textiles, rubber, plastics, and lithographic inks; radiation shield; x-ray contrast media; storage batteries; extender pigment
Regulatory: FDA approved for use in intrauterine prods.; BP compliance
Manuf./Distrib.: 20 Microns Ltd.; Aldrich; Am. Biorganics; Archway Sales; J.T. Baker; Barium & Chems.; Barker Ind.; Baychem; R.E. Carroll; Cimbar Perf. Mins.; EM Ind.; Fluka; J.M. Huber; Lenape Ind.; Lomas Int'l.; Mallinckrodt; Mitsubishi Chem.; Ore & Chem.; Rit-Chem; H.M. Royal; Sachtleben Chemie GmbH; San Yuan; Sigma; Sino-Am. Pigment Systems; Spectrum Chem. Mfg.
Trade names: Bariace™ B-30; Bariace™ B-34; Barifine™ BF-1; Barifine™ BF-10; Barifine™ BF-20; Barifine™ BF-21; Barimite™; Barimite™ 200; Barimite™ UF; Barimite™ XF; Bartex® 10; Bartex® 65; Bartex® 80; Bartex® OWT; Cimbar™ 325; Cimbar™ CF; Cimbar™ UF; Cimbar™ XF; G-50 Barytes™; Huberbrite® 1; Huberbrite® 3; Huberbrite® 7; Huberbrite® 10; Huberbrite® 12; No. 22 Barytes™; Polywate 200; Polywate 325; Sparwhite W-5HB; Sparwhite W-10; Sparwhite W-10HB; Sparwhite W-20; Sparwhite W-20HB; Sparwhite W-44; Sparwhite W-44HB

Barium sulfate (1:1). *See* Barium sulfate

Barium-zinc complex
Uses: Heat and light stabilizer for flexible and rigid PVC, plastisols, topcoats, flooring, sheet and film; stabilizer/activator for chemically blown PVC
Trade names: Baerostab UBZ 6370 TS; Bärostab® UBZ 614; Bärostab® UBZ 630 TS; Bärostab® UBZ 632 TS; Bärostab® UBZ 671 TS; Bärostab® UBZ 776 X

Baryta yellow. *See* Barium chromate

Barytes (natural). *See* Barium sulfate
Basic bismuth chloride. *See* Bismuth oxychloride
Basic lead silica chromate. *See* Lead silicochromate
Basic lead white silicate. *See* Lead silicate
Basofor. *See* Barium sulfate
BBOT. *See* 2,2′-(2,5-Thiophenediyl)bis[5-t-butylbenzoxazole]
BBP. *See* Butyl benzyl phthalate
BDMA. *See* N-Benzyldimethylamine
BDO. *See* 1,4-Butanediol

Beeswax
CAS 8006-40-4 (white); 8012-89-3 (yellow); EINECS 232-383-7
Synonyms: Cera alba; White wax; White beeswax; Yellow wax; Yellow beeswax
Definition: Purified wax from the honeycomb of the bee, *Apis mellifera*; commonly called white wax when bleached, yellow wax when not bleached
Properties: Brown or white (bleached) solid with faint odor; insol. in water; sl. sol. in alcohol; sol. in chloroform, ether, and oils; dens. 0.95; m.p. 62-65 C; acid no. 17-24; sapon. no. 84
Precaution: Combustible when heated
Toxicology: Essentially nontoxic; mild allergen; may cause contact dermatitis, human intolerance reaction
Uses: Food additive, flavoring agent, candy glaze/polish, furniture and floor waxes, shoe polishes, leather dressings, anatomical specimens, artificial fruit, textile sizes and finishes, church candles, cosmetic creams, adhesive compds.; pharmaceutic aid; wood and paper finishes
Regulatory: FDA 21CFR §184.1973, GRAS; FEMA GRAS (white); Japan approved; Europe listed; UK approved; FDA approved for orals, topicals; USP/NF, BP, Ph.Eur. compliance
Manuf./Distrib.: Aldrich; British Wax Refining; CC Pollen; Dussek Campbell Inc; Fluka; ICI Spec. Chems.; Koster Keunen; Maruzen Fine Chems.; Penta Mfg.; Frank B. Ross; Spectrum Chem. Mfg.; Strahl & Pitsch
Trade names: Koster Keunen Beeswax; Ross Beeswax

Beeswax, synthetic. *See* Synthetic beeswax

Behenamide
CAS 3061-75-4; EINECS 221-304-1
Synonyms: Behenic acid amide; Docosanamide
Classification: Aliphatic amide
Empirical: $C_{22}H_{45}NO$
Formula: $CH_3(CH_2)_{20}CONH_2$
Properties: M.w. 339.60; m.p. 111-112 C
Uses: Antifoam in mfg. of detergents, in floor polishes, dripless candles; lubricant, slip, antiblock, and mold release in plastics
Regulatory: FDA 21CFR 117.1200
Manuf./Distrib.: Croda Universal Ltd.; Witco/Oleo-Surf.
Trade names: Kemamide® B

Behenic acid amide. *See* Behenamide
Behenic imidazoline. *See* Behenyl hydroxyethyl imidazoline

Behenyl acrylate
Uses: As a flexibilizer for coatings and inks
Trade names: Photomer® 4822

Behenyl hydroxyethyl imidazoline
Synonyms: Behenyl imidazoline; 1H-Imidazole-1-ethanol, 4,5-dihydro-2-docosanyl-; Behenic imidazoline
Classification: Heterocyclic compd.
Empirical: $C_{26}H_{52}N_2O$
Properties: Tan solid; m.w. 383; m.p. 45-49 C
Uses: Dispersant, wetting agent, emulsifier, microbide; acid cleaning, polishes, surface treatment, textile processing, paints, metal processing, agric., cosmetic intermediate
Regulatory: FDA 21CFR §175.300, 176.210, 177.2800

Behenyl imidazoline. *See* Behenyl hydroxyethyl imidazoline

Bentonite
CAS 1302-78-9; EINECS 215-108-5

Synonyms: Soap clay; Mineral soap; Wilkinite; CI 77004
Definition: Native hydrated colloidal aluminum silicate clay
Formula: $Al_2O_3 \cdot 4SiO_2 \cdot nH_2O$
Properties: Light to cream-colored impalpable powd., odorless, sl. earthy taste; hygroscopic; forms colloidal suspension in water, thixotropic properties; insol. in water and org. solvs.; pH 9.5-10.5
Toxicology: LD50 (IV, rat) 35 mg/kg; poison by intravenous route causing blood clotting; inert and generally nontoxic
Uses: Oil-well drilling fluids, cement slurries for oil-well casings, thickener, fireproofing, cosmetics, decolorizing agent, filler in ceramics, emulsifier for oils, suspending agent in pharmaceuticals, base for plasters; extender for paints; food additive, colorant, pigment, stabilizer for wine
Regulatory: FDA 21CFR §175.105, 175.300, 177.1460, 184.1155, GRAS; Japan restricted (0.5% max. residual); Europe listed; UK approved; USP/NF, BP, Ph.Eur. compliance; FDA approved for orals, topicals; USP/NF compliance
Manuf./Distrib.: Akzo Nobel; Aldrich; Allchem Ind.; Am. Colloid; Asbury Graphite Mills; R.E. Carroll; Cimbar Perf. Mins.; Dry Branch Kaolin; Fluka; Lomas Int'l.; D.N. Lukens; Norsk Hydro AS; Punda Mercantile; H.M. Royal; L.A. Salomon; Seegott; Sigma; Southern Clay Prods.; Spectrum Chem. Mfg.; Süd-Chemie AG; Tamms Ind.; U.S. Silica; R.T. Vanderbilt; Whittaker, Clark & Daniels; Wyo-Ben
Trade names: Aquamont® 600; Aquamont® 610; Bentolite® H; Bentolite® L; Bentolite® WH; Bentonite® 789; Bentonite® HS; Bentonite® L-3; Bentonite® L-10; Bentonite® NF; Korthix; Korthix H

Benzaldehyde
CAS 100-52-7; EINECS 202-860-4
Synonyms: Benzoic aldehyde; Benzic aldehyde; Benzene methylal; Benzene carbonal; Artificial almond oil; Benzenecarbaldehyde
Classification: Aromatic org. compd.
Empirical: C_7H_6O
Formula: C_6H_5CHO
Properties: Colorless liq., bitter almond odor, burning taste; sl. sol. in water; misc. in alcohol, ether, oils; m.w. 106.13; sp.gr. 1.041; m.p. -26 C; b.p 179 C; flash pt. 148 F; ref. index 1.544
Precaution: DOT: Combustible liq.; light-sensitive; strong reducing agent; acts violently with oxidizers
Toxicology: LD50 (oral, rat) 1300 mg/kg; highly toxic; poison by ingestion and intraperitoneal routes; allergen; feeble local anesthethic; skin irritant; CNS depressant; 1 g/kg may be fatal in humans
Uses: Chemical intermediate for dyes, flavoring agent, perfumes, aromatic alcohols, paints; solvent for oils, resins, cellulose acetate and nitrate; manufacture of cinnamic acid, benzoic acid; pharmaceuticals; photographic chemicals
Regulatory: FDA 21CFR §182.60, 27CFR §21.65, 21.151, GRAS; FEMA GRAS; Japan approved as flavoring; USP/NF, BP compliance
Manuf./Distrib.: Aceto; Aldrich; Allchem Ind.; J.T. Baker; Bayer; Chemical; Chemisphere; DSM; Elf Atochem SA; Fluka; R.W. Greeff; Haarmann & Reimer; Hoechst Celanese; Janssen Chimica; Kalama; Mitsubishi Chem.; Penta Mfg.; Punda Mercantile; Rit-Chem; Schweizerhall; Sigma; SNIA UK; Spectrum Chem. Mfg.; Van Waters & Rogers

Benzalkonium chloride
CAS 8001-54-5; 61789-71-7; 63449-41-2; 68391-01-5; 68424-85-1; 68989-00-4; 85409-22-9; EINECS 204-479-9; 263-080-8; 264-151-6; 269-919-4; 270-325-2; 273-544-1; 287-089-1
Synonyms: Alkyl dimethyl benzyl ammonium chloride; Octyl-octadecyl dimethyl ethylbenzyl ammonium chlorides; Alkylbenzyldimethylammonium chloride; Ammonium, alkyldimethyl (phenylmethyl)-, chloride
Classification: Quaternary ammonium salt
Definition: Mixt. of alkylbenzyldimethylammonium chlorides
Formula: $C_6H_5CH_2N(CH_3)_2RCl$, $R = C_8H_{17}$ to $C_{18}H_{37}$
Properties: Wh. or ylsh.-wh. amorphous powd., aromatic odor, bitter; sol. in water, alcohol, acetone; insol. in ether; m.p. 34-37 C
Precaution: DOT: Corrosive material; heated to decomp., emits toxic fumes of Cl^- and NO_x
Toxicology: LD50 (rats, oral) 400 mg/kg; highly toxic; poison by parenteral, ingestion, intraperitoneal, intravenous routes; large systemic doses can cause nausea, vomiting, muscle paralysis, CNS depression, local tissue damage; eye irritant
Uses: Cationic detergent; surface antiseptic, bactericide, fungicide, preservative in pharmaceuticals; latex coagulation, flotation, electrostatic paints, demulsification of hydrocarbons; emulsifier for emulsion polymerization of acrylic monomer
Usage level: 0.1-0.3%
Regulatory: USA not restricted; FDA 21CFR §175.105, 178.1010; FDA approved for ophthalmics, injectables, otics, topicals; Japan, Europe listed; USP/NF, BP, Ph.Eur. compliance

Manuf./Distrib.: Akzo Nobel; Aldrich; Allchem Ind.; Chemron; EM Ind.; Fluka; Gresco Mfg.; Lonza Ltd.; Mason; Rhone-Poulenc France; Sigma; Spectrum Chem. Mfg.; Stepan; Stepan Europe; Witco/Oleo-Surf.
Trade names: Noramium DA.50

Benzene
CAS 71-43-2; EINECS 200-753-7
Synonyms: Benzol; Cyclohexatriene; Annulene
Empirical: C_6H_6
Properties: Colorless to light-yel. mobile, nonpolar liq., aromatic odor; misc. with alcohol, ether, chloroform, CCl_4, acetone, oils; m.w. 78.11; dens. 0.8790 (20/4 C); b.p. 80.1 C; f.p. 5.5 C; flash pt. (CC) -11 C; ref. index 1.50110 (20 C)
Precaution: Highly flamm.; fire risk
Toxicology: LD50 (rats, oral) 3.8 ml/kg; toxic; carcinogen; TLV 10 ppm in air; poison by inhalation; absorbed thru the skin causing systemic effects; chronic exposure can cause bone marrow damage, leukemia
Uses: Mfg. of ethylbenzene, dodecylbenzene, cyclohexane, phenol, nitrobenzene, chlorobenzene; dyes, medicinals, artificial leather, airplane dopes, lacquers; solv. for waxes, resins, oils, paints
Manuf./Distrib.: Aldrich; ARCO; Ashland; J.T. Baker; BP Chems. Ltd; Exxon; Fluka; Janssen Chimica; Marathon Oil; Mitsubishi Petrochem.; Mitsui Petrochem. Ind.; Mobil; Occidental; Shell; Sigma; Spectrum Chem. Mfg.

Benzenecarbaldehyde. *See* Benzaldehyde
Benzene carbonal. *See* Benzaldehyde
Benzenecarboxylic acid. *See* Benzoic acid
Benzene chloride. *See* Chlorobenzene
Benzene, 1-chloro-4 (trifluoromethyl). *See* p-Chlorobenzotrifluoride
1,3-Benzenediamine. *See* m-Phenylenediamine
m-Benzenediamine. *See* m-Phenylenediamine
Benzene-1,3-dicarboxylic acid. *See* Isophthalic acid
Benzene-1,4-dicarboxylic acid. *See* Terephthalic acid
1,4-Benzenedicarboxylic acid. *See* Terephthalic acid
m-Benzenedicarboxylic acid. *See* Isophthalic acid
p-Benzenedicarboxylic acid. *See* Terephthalic acid
1,2-Benzenedicarboxylic acid, 2-butoxy-2-oxoethyl, butyl ester. *See* n-Butyl phthalyl-n-butyl glycolate
1,2-Benzenedicarboxylic acid, butyl cyclohexyl ester. *See* Butyl cyclohexyl phthalate
1,2-Benzenedicarboxylic acid, butyl octyl ester. *See* Butyl octyl phthalate
1,2-Benzenedicarboxylic acid, butyl phenylmethyl ester. *See* Butyl benzyl phthalate
1,2-Benzenedicarboxylic acid, dibutyl ester. *See* Dibutyl phthalate
1,2-Benzenedicarboxylic acid, dicyclohexyl ester. *See* Dicyclohexyl phthalate
1,2-Benzenedicarboxylic acid, diisodecyl ester. *See* Diisodecyl phthalate
1,2-Benzenedicarboxylic acid, diisononyl ester. *See* Diisononyl phthalate
1,2-Benzenedicarboxylic acid, diisotridecyl ester. *See* Diisotridecyl phthalate
1,2-Benzenedicarboxylic acid dimethyl ester. *See* Dimethyl phthalate
1,2-Benzenedicarboxylic acid dioctyl ester. *See* Dioctyl phthalate
1,4-Benzenedicarboxylic acid, dioctyl ester. *See* Dioctyl terephthalate
1,2-Benzenedicarboxylic acid, lead (2+) salt, basic. *See* Lead phthalate, basic
1,2-Benzenedicarboxylic acid, 3,4,5,6-tetrabromo-, disodium salt. *See* Disodium tetrabromophthalate
Benzene, 1,2-dichloro-4-(trifluoromethyl)-. *See* 3,4-Dichlorobenzotrifluoride
Benzene, 1,2-dimethyl. *See* o-Xylene
1,3-Benzenediol. *See* Resorcinol
1,4-Benzenediol. *See* Hydroquinone
Benzene, ethenyl-, homopolymer. *See* Polystyrene
Benzene formic acid. *See* Benzoic acid
Benzene methylal. *See* Benzaldehyde
Benzenenitrile. *See* Benzonitrile
Benzenesulfonic acid butyl amide. *See* N,N-Butyl benzene sulfonamide
Benzenesulfonic acid, dimethyl-. *See* Xylene sulfonic acid
Benzenesulfonic acid, dimethyl-, ammonium salt. *See* Ammonium xylenesulfonate
1,2,3,4,5-Benzenetetracarboxylic acid. *See* Pyromellitic acid
Benzene-1,2,4,5-tetracarboxylic acid. *See* Pyromellitic acid
Benzene-1,2,4,5-tetracarboxylic dianhydride. *See* Pyromellitic dianhydride
1,2,4,5-Benzenetetracarboxylic dianhydride. *See* Pyromellitic dianhydride
1,2,4,5-Benzenetetracarboxylic 1,2:4,5 dianhydride. *See* Pyromellitic dianhydride

1,2,4-Benzenetricarboxylic acid anhydride. *See* Trimellitic anhydride
Benzenol. *See* Phenol
Benzic aldehyde. *See* Benzaldehyde

Benzil dimethyl ketal
Uses: UV photoinitiator for UV curable coatings and inks for paper, plastics, wood, and elec. applics.
Manuf./Distrib.: Morre-Tec Ind.
Trade names: Photomer® 0051

4-(2-Benzimidazolyl) thiazol. *See* Thiabendazole
Benzin. *See* Naphtha
Benzine. *See* Naphtha

1,2-Benzisothiazolin-3-one
CAS 2634-33-5; EINECS 220-120-9
Synonyms: BIT
Empirical: C_7H_5NOS
Properties: M.w. 151.18
Uses: Microbiostat preservative for aq. formulations, latex emulsions, metalworking fluids, aq. paints, adhesives, and polishes
Manuf./Distrib.: Zeneca Biocides
Trade names: Proxel® BD 20
Trade names containing: Biochek 410; Proxel® GXL

Benzisotriazole. *See* 1H-Benzotriazole
Benzoate of soda. *See* Sodium benzoate
Benzoate sodium. *See* Sodium benzoate
1H,3H-Benzo(1,2-c:4,5-c′)difuran-1,3,5,7-tetrone. *See* Pyromellitic dianhydride
Benzochinamide. *See* Benzoguanamine

Benzoguanamine
CAS 91-76-9; 23844-24-8
Synonyms: 6-Phenyl-1,3,5-triazine-2,4-diamine; 2,4-Diamino-6-phenyl-s-triazine; Benzochinamide; Benzquinamide; Quantril
Empirical: $C_9H_9N_5$
Formula: $C_6H_5C_3N_3(NH_2)_2$
Properties: Cryst.; sol. in alcohol; insol. in water, benzene, ether; pract. insol. in chloroform; m.w. 187.20
Toxicology: LD50 (oral, rat) 1050 mg/kg; poison by ingestion, intraperitoneal, and intravenous routes
Uses: Intermediate for paints; intermediate for pesticides, pharmaceuticals and dyestuffs; resin modifier
Manuf./Distrib.: Allchem Ind.; Chemical; Monsanto; Punda Mercantile; SKW Chems.; Triple Crown Am.; United Min. & Chem.

Benzoguanamine formaldehyde resin
Trade names containing: Beckamine 27-809

Benzoic acid
CAS 65-85-0; EINECS 200-618-2
Synonyms: Benzenecarboxylic acid; Benzene formic acid; Carboxybenzene; Phenylformic acid; Phenylcarboxylic acid; Dracylic acid
Classification: Aromatic acid
Empirical: $C_7H_6O_2$
Formula: C_6H_5COOH
Properties: White scales, needles, crystals, benzoin odor; sol. in alcohol, ether, chloroform, benzene, carbon disulfide; sl. sol. in water; m.w. 122.13; dens. 1.2659; m.p. 121.25 C; b.p. 249.2 C; subl. at 100 C; flash pt. 121.1 C
Precaution: Combustible when exposed to heat or flame; reactive with oxidizing materials
Toxicology: LD50 (oral, rat) 2530 mg/kg; mod. toxic by ingestion, IP routes; poison by subcut. route; severe eye/skin irritant; may cause human intolerance reaction, asthma, hyperactivity in children; heated to decomp., emits acrid smoke and irritating fumes
Uses: Sodium and butyl benzoates, plasticizers, benzoyl chlorides, food preservatives, antimicrobial agent, flavors, perfumes, antifungal agent , retarder for rubbers and latexes
Regulatory: FDA 21CFR §150.141, 150.161, 166.40, 166.110, 175.300, 184.1021, GRAS 0.1% max. in foods; USA EPA registered; Japan 0.2% max.; Europe listed 0.5% max.; FEMA GRAS; cleared by MID to retard

flavor reversion in oleomargarine at 0.1%; Japan approved with limitations; Europe listed; UK approved; approved for orals, rectals, topicals; USP/NF, BP, Ph.Eur. compliance
Manuf./Distrib.: Aceto; Aldrich; Ashland; J.T. Baker; Baychem; Browning; Chemical; Jan Dekker BV; Elf Atochem SA; Fluka; R.W. Greeff; Int'l. Sourcing; Mallinckrodt; E. Merck; Mitsubishi Chem.; Napp Tech.; Penta Mfg.; Punda Mercantile; Schweizerhall; Sigma; Van Waters & Rogers; Velsicol

Benzoic acid, diester with diethylene glycol. *See* Diethylene glycol dibenzoate
Benzoic acid nitrile. *See* Benzonitrile
Benzoic acid sodium salt. *See* Sodium benzoate
Benzoic aldehyde. *See* Benzaldehyde

Benzoin ether
Uses: Initiator used in UV light curing of polyester resins; UV sensitizer for lacquers and FRP laminates
Trade names: Trigonal® 14

Benzoin isobutyl ether. *See* 2-(2-Methylpropoxy)-1,2-diphenylethanone
Benzol. *See* Benzene

Benzonitrile
CAS 100-47-0; EINECS 202-855-7
Synonyms: Phenyl cyanide; Cyanobenzene; Benzenenitrile; Benzoic acid nitrile
Empirical: C_7H_5N
Formula: C_6H_5CN
Properties: Colorless liq., almond-like odor; sol. in alcohol; sl. sol. in cold water; m.w. 103.13; dens. 1.246 (20/4 C); m.p. -13 C; b.p. 190.7 C
Precaution: Combustible
Toxicology: LD50 (oral, mouse) 971 mg/kg; high toxicity; absorbed by skin; poison by IP and subcut. routes; skin irritant; heated to decomp., emits toxic fumes of CN^- and NO_x
Uses: Manufacture of benzoquanamine; intermediate for rubber chemicals; solvent for nitrile rubber, specialty lacquers, resins and polymers, anhydrous metallic salts
Manuf./Distrib.: Aldrich; Alemark; Amber Syn.; BASF; J.T. Baker; Fluka; Penta Mfg.; PMC Specialties; Sigma; SKW Chems.; Spectrum Chem. Mfg.; Triple Crown Am.

Benzophenone
CAS 119-61-9; EINECS 204-337-6
Synonyms: Benzoylbenzene; Diphenyl ketone; Diphenylmethanone
Classification: Organic compd.
Empirical: $C_{13}H_{10}O$
Formula: $C_6H_5COC_6H_5$
Properties: Wh. rhombic cryst., persistent rose-like odor; sol. in fixed oils; sl. sol. in propylene glycol; m.w. 182.23; sp.gr. 1.0976 (α, 50/50 C), 1.108 (β, 23/40 C); m.p. 49 C (α), 26 C (β), 47 C (γ); b.p. 305 C
Precaution: Combustible when heated; incompat. with oxidizers
Toxicology: LD50 (oral, mouse) 2895 mg/kg; moderately toxic by ingestion and intraperitoneal routes; heated to decomp., emits acrid and irritating fumes
Uses: UV absorber for sunscreens; uv protective agents in polymers, petrol. waxes, etc.; photosensitizers in printing inks, wood and metal finishes; intermediate for mfg. of antihistamines, insecticides; flavoring agent
Regulatory: FDA 21CFR §172.515; FEMA GRAS; BP compliance
Manuf./Distrib.: Aldrich; Allchem Ind.; Berje; Chemisphere; Elf Atochem N. Am.; EM Ind.; Fluka; GCA; R.W. Greeff; Monomer-Polymer & Dajac Labs; Penta Mfg.; Plastics & Chems.; Reedy Int'l.; Schweizerhall; Sigma; Spectrum Chem. Mfg.; 3V; Velsicol
Trade names: BP; Photomer® 0081; Photomer® BP; Velsicure® BTF

Benzophenone-1
CAS 131-56-6; EINECS 205-029-4
Synonyms: 2,4-Dihydroxybenzophenone; Benzoresorcinol; 4-Benzoyl resorcinol
Classification: Organic benzophenone deriv.
Formula: $C_6H_5COC_6H_3(OH)_2$
Properties: Lt. yel. crystalline solid; insol. in water; sol. in ethanol, methanol, MEK, ethyl acetate; m.p. 142 C; b.p. 194 C (1 mm)
Uses: UV absorber in polymers and in outdoor paints and coatings, varnishes
Manuf./Distrib.: Aceto; Aldrich; EM Ind.; Ferro/Bedford; Fluka; R.W. Greeff; Haarmann & Reimer; Hoechst Celanese; Quest Int'l.; Sartomer; Sigma
Trade names: Uvinul® 400; Uvinul® 3000

Benzophenone-3
CAS 131-57-7; EINECS 205-031-5
Synonyms: 2-Hydroxy-4-methoxybenzophenone; (2-Hydroxy-4-methoxyphenyl) phenylmethanone; Oxybenzone
Classification: Organic benzophenone deriv.
Empirical: $C_{14}H_{12}O_3$
Properties: Yellowish cryst., rose-like odor; sol. in min. oil, peanut oil, ethanol, PEG-8, oleyl alcohol, castor oil; insol. in water; m.w. 228.26; m.p. 62-63.5 C
Toxicology: Poison by intraperitoneal route; mild toxicity by ingestion; LD50 (rat, oral) > 10 g/kg; LD50 (rabbit, dermal) > 16 g/kg
Uses: Sunscreen; uv absorber for plastics (ABS, cellulosics, polyester, PS, flexible and rigid PVC, VDC); protects clear varnishes and lacquers, linseed oil-based alkyds and phenolic coatings intended for use on uv-sensitive surfaces
Regulatory: FDA 21CFR §177.1010
Manuf./Distrib.: Aceto; Aldrich; EM Ind.; Ferro/Bedford; Fluka; R.W. Greeff; Haarmann & Reimer; Hoechst Celanese; Inchema; Quest Int'l.; Sartomer; Sigma; Spectrum Chem. Mfg.
Trade names: Cyasorb® UV 9; Syntase® 62; UV-Absorber 325; Uvinul® 3040; Uvinul® M-40

Benzophenone-6
CAS 131-54-4; EINECS 205-027-3
Synonyms: 2,2´-Dihydroxy-4,4´-dimethoxybenzophenone; Bis(2-hydroxy-4-methoxyphenyl) methanone
Classification: Organic benzophenone deriv.
Empirical: $C_{15}H_{14}O_5$
Properties: Crystals; m.w. 274.26; m.p. 139-140 C
Uses: UV light absorber, esp. in paints, plastics (epoxy, polyester)
Manuf./Distrib.: Aceto; Aldrich; EM Ind.; Ferro/Bedford; R.W. Greeff; Haarmann & Reimer; Hoechst Celanese; Quest Int'l.; Sartomer
Trade names: Uvinul® 3049

Benzophenone-8
CAS 131-53-3; EINECS 205-026-8
Synonyms: 2,2´-Dihydroxy-4-methoxybenzophenone; Dioxybenzone
Classification: Organic benzophenone deriv.
Empirical: $C_{14}H_{12}O_4$
Properties: Powd.; m.p. 68 C
Toxicology: Practically nontoxic in single oral doses; LD50 (rat, oral) > 10 g/kg; nonirritating to rabbit eye or skin
Uses: UV absorber, light stabilizer for flexible PVC; lt. stabilizer and uv absorber for coatings and plastics, e.g., alkyds, phenolics, PU coatings; stabilizer for polyester film
Manuf./Distrib.: Aldrich; EM Ind.; Ferro/Bedford; R.W. Greeff; Haarmann & Reimer; Hoechst Celanese; Quest Int'l.; Sartomer; Sigma
Trade names: Cyasorb® UV 24

Benzophenone-9
CAS 3121-60-6; EINECS 221-498-8
Synonyms: Disodium 2,2´-dihydroxy-4,4´-dimethoxy-5,5´-disulfobenzophenone
Classification: Organic benzophenone deriv.
Empirical: $C_{15}H_{14}O_8S \cdot Na$
Uses: UV absorber in cosmetic formulations, textiles and water-based paints
Manuf./Distrib.: EM Ind.; Ferro/Bedford; R.W. Greeff; Haarmann & Reimer; Hoechst Celanese; Quest Int'l.; Sartomer
Trade names: Uvinul® DS 49

Benzophenone-11
CAS 1341-54-4
Classification: Organic benzophenone deriv.
Definition: Mixture of benzophenone-6 and -2 and other tetra-substituted benzophenone materials
Uses: UV absorber used in NC lacquer, fluorescent paint, inks, and for protecting furniture woods, colored liq. toiletries and cleaning agents, isocyanate systems, and butyrate metal lacquers
Manuf./Distrib.: EM Ind.; Ferro/Bedford; R.W. Greeff; Haarmann & Reimer; Hoechst Celanese; Quest Int'l.; Sartomer
Trade names: Uvinul® 3093

Benzophenone-12
CAS 1843-05-6; EINECS 217-421-2
Synonyms: 2-Hydroxy-4-n-octoxybenzophenone; 2-Hydroxy-4-(octyloxy) benzophenone; [2-Hydroxy-4-(octyloxy)phenyl]phenylmethanone; Octabenzone
Classification: Organic benzophenone deriv.
Empirical: $C_{21}H_{26}O_3$
Properties: Crystals; m.w. 326.42; m.p. 45-46 C
Toxicology: Very low acute toxicity for animals; LD50 (rat, oral) > 10 g/kg; nonirritating to rabbit skin or eyes
Uses: UV and lt. stabilizer and for plastics and coatings, e.g., polyethylene, PP, PVC, and EVA; uses incl. pipe, storage tanks, and auto, marine, garden prods., auto refinish and industrial coatings, adhesives and sealants
Regulatory: FDA 21CFR §178.2010
Manuf./Distrib.: Aldrich; EM Ind.; Enterprise; Ferro/Bedford; Great Lakes; R.W. Greeff; Haarmann & Reimer; Hoechst Celanese; Quest Int'l.; Sartomer; 3V
Trade names: Cyasorb® UV 531; UV-Chek® AM-300; UV-Chek® AM-301

3,3´,4,4´-Benzophenone tetracarboxylic dianhydride
CAS 2421-28-5; EINECS 219-348-1
Synonyms: BTDA; 5,5´-Carbonylbis 1,3-isobenzofurandione; Benzophenone-3,3´-4,4´-tetracarboxylic dianhydride; 4,4´-Carbonylbis(phthalic anhydride); 4,4´-Carbonyldiphthalic anhydride; 4,4´-Carbonyldiphthalic acid anhydride
Classification: Aromatic carboxylic anhydride
Empirical: $C_{17}H_6O_7$
Properties: Amber to lt. tan powd. or flakes, odorless, tasteless; sol. 0.26 g/100 g water; m.w. 322.22; sp.gr. 1.57 (30 C); m.p. 221-225 C; b.p. 380-400 C; pH 2.4 (2.1 g/L)
Precaution: Incompat. with alkali metal hydroxides, amines, ammonia, alcohols, thiols, strong oxidizing/reducing agents
Toxicology: LD50 (oral, rats) > 12,800 mg/kg (as tetra-acid); fumes from molten BTDA are highly toxic; solid considered nontoxic; may cause respiratory irritation/allergic response on overexposure
Uses: Epoxy curing agent; monomer in unsat. and saturated polyester resins; ester derivs. as intermediates in prod. of polyimides, lubricants, plasticizers; intermediate for paints, alkyd, PU resins, polypyrrones; end-uses incl. electronic transfer molding compds., wire coatings, aerospace adhesives, advanced structural composites, uv-cured inks, insulating coatings and foams, flame retardant prods., release agents, molded engineering parts, refinery catalysts
Manuf./Distrib.: Aldrich; Allco; Chemie Linz N. Am.; Fluka
Trade names: Allco BTDA

Benzophenone-3,3´-4,4´-tetracarboxylic dianhydride. *See* 3,3´,4,4´-Benzophenone tetracarboxylic dianhydride
Benzoresorcinol. *See* Benzophenone-1
2(3H)-Benzothiazolethione, zinc salt. *See* Zinc 2-mercaptobenzothiazole
2-(Benzothiazolylthio) methyl ester. *See* 2-Thiocyanomethylthiobenzothiazole

1H-Benzotriazole
CAS 95-14-7; EINECS 202-394-1
Synonyms: Benzisotriazole; 1,2-Aminozophenylene; Azimidobenzene
Empirical: $C_6H_5N_3$
Properties: Needle-like crystals; sparingly sol. in water; sol. in alcohol, benzene, toluene, chloroform, DMF; m.w. 119.12; m.p. 98.5 C; b.p. 204 (15 mm); may explode during vacuum distillation
Toxicology: Poison by intravenous route; moderately toxic by ingestion, intraperitoneal routes
Uses: Chelating agent and sesquestrant for copper ions; corrosion inhibitor for copper, brass, bronze; used in antifreeze, cleaners, coatings, detergents, functional fluids, metalworking fluids, pkg. materials, polishes
Manuf./Distrib.: Aldrich; Atomergic Chemetals; Bayer; Clariant; Fluka; PMC Specialties; Sigma
Trade names: Cobratec® 99

2-(2H-Benzotriazol-2-yl)-4,6-bis(1,1-dimethylpropyl)phenol. *See* 2-(2´-Hydroxy-3,5´-di-t-amylphenyl) benzotriazole
2-(2H-Benzotriazol-2-yl)-4-methylphenol. *See* Drometrizole
2-(2H-Benzotriazol-2-yl)-4-(1,1,3,3-tetramethylbutyl)phenol. *See* Octrizole
Benzoylbenzene. *See* Benzophenone
Benzoyloxytributylstannane. *See* Tributyltin benzoate
o-Benzoylphenol. *See* 2-Hydroxybenzophenone
4-Benzoyl resorcinol. *See* Benzophenone-1

Benzquinamide. *See* Benzoguanamine

Benzyl acetate
CAS 140-11-4; EINECS 205-399-7
Synonyms: Phenylmethyl acetate; α-Acetoxytoluene; Benzyl ethanoate
Classification: Aromatic org. compd.
Definition: Ester of benzyl alcohol and acetic acid
Empirical: $C_9H_{10}O_2$
Formula: $CH_3COOCH_2C_6H_5$
Properties: Colorless liq., sweet floral fruity odor; sol. in alcohol, most fixed oils, propylene glycol; insol. in water, glycerin; m.w. 150.19; sp.gr. 1.06; m.p. -51.5 C; b.p. 213.5 C; flash pt. (CC) 216 F; ref. index 1.501
Precaution: Combustible liq.
Toxicology: LD50 (oral, rat) 2490 mg/kg; moderately toxic by ingestion and subcutaneous routes; poison by inhalation; antipsychotic; heated to decomp., emits irritating fumes
Uses: Artificial jasmine and other perfumes; soap perfume; flavoring agent; solvent and high boiler for cellulose acetate and nitrate, natural and synthetic resins; oils; lacquers; polishes; printing inks; varnish removers
Regulatory: FDA 21CFR §172.515; FEMA GRAS; Japan approved as flavoring
Manuf./Distrib.: Aldrich; Allan; Berje; Chemisphere; Fluka; Haarmann & Reimer; Janssen Chimica; Kalama; Koyo; MTM Spec. Chems.; Penta Mfg.; Quest Int'l.; Rhone-Poulenc N. Am.

Benzyl butyl phthalate. *See* Butyl benzyl phthalate

N-Benzyldimethylamine
CAS 103-83-3; EINECS 203-149-1
Synonyms: DMBA; BDMA; N-(Phenylmethyl) dimethylamine; N,N-Dimethylbenzylamine
Empirical: $C_9H_{13}N$
Formula: $C_6H_5CH_2N(CH_3)_2$
Properties: Colorless to lt. yel. liq.; m.w. 135.23; dens. 0.894 (27 C); b.p. 180-182 C
Toxicology: Poison by ingestion; moderately toxic by inhalation, skin contact; corrosive severe skin and eye irritant
Uses: Intermediate, dehydrohalogenating catalyst, corrosion inhibitor, acid neutralizer, potting compounds, adhesives, paints, cellulose modifier; catalyst for rigid PU foam
Manuf./Distrib.: Aceto; Aldrich; Anhydrides & Chems.; Fluka; R.W. Greeff; Penta Mfg.; PMC Specialties; Schweizerhall; Sigma; Synthron; Zeeland

Benzyldimethyl ketal
CAS 24650-42-8
Uses: Photoinitiator for wh. coatings, adhesives, inks, photopolymers, electronic photoresists, polyester-styrene wood filler composites
Manuf./Distrib.: Aldrich; Alemark; Amber Syn.; Fabrichem; Fluka
Trade names: Esacure® KB1; Esacure® KB60

Benzyl ethanoate. *See* Benzyl acetate

Benzyl phthalate
Synonyms: Dibenzyl phthalate
Formula: $C_6H_4(COOCH_2C_6H_5)_2$
Properties: M.w. 346.14; m.p. 42-44 C; b.p. 277 C
Uses: Plasticizer for PVC, paints and acrylic coatings, caulks and sealants based on chlorinated rubber, butyl rubber, polysulfides, or PU
Trade names: Santicizer 278

Benzytol. *See* Chloroxylenol
BGE. *See* Butyl glycidyl ether

BHA
CAS 25013-16-5; EINECS 204-442-7; 246-563-8
Synonyms: Butylated hydroxyanisole; t-Butyl-4-methoxyphenol; (1,1-Dimethylethyl)-4-methoxyphenol; 3-t-Butyl-4-hydroxyanisole
Definition: Mixture of isomers of tertiary butyl-substituted 4-methoxyphenols
Empirical: $C_{11}H_{16}O_2$
Properties: Wh. or sl. yel. waxy solid, faint char. odor; insol. in water; sol. in petrol. ether, 50% or higher alcohol, propylene glycol, chloroform, fats, oils; m.w. 180.27; m.p. 48-55 C; b.p. 264-270 C (733 mm)

Precaution: Combustible
Toxicology: LD50 (oral, mouse) 2000 mg/kg; suspected carcinogen; moderate toxicity by ingestion, intraperitoneal routes; may cause rashes, hyperactivity; heated to decomp., emits acrid and irritating fumes
Uses: Antioxidant, preservative for foods, cosmetics, plastics, etc.
Usage level: 0.02% max. (preservation of fixed oils, fats, vitamin oil concs.)
Regulatory: FDA 21CFR §166.110, 172.110, 172.515, 172.615, 173.340, 175.105, 175.125, 175.300, 175.380, 175.390, 176.170, 176.210, 177.1010, 177.1210, 177.1350, 178.3120, 178.3570, 179.45,; 181.22, 181.24 (0.005% migrating from food pkg.), 182.3169 (0.02% max. of fat or oil), GRAS; FEMA GRAS; USDA 9CFR 318.7, 381.147; Japan approved 0.2-1 g/kg; Europe approved; UK approved; approved for orals, rectals, topicals; USP/NF, BP compliance
Manuf./Distrib.: Aceto; Aldrich; Allchem Ind.; Eastman; Fluka; Int'l. Chem. Inc.; Penta Mfg.; Sigma; Spectrum Chem. Mfg.; UOP
Trade names containing: Centrophil® M

BHT
CAS 128-37-0; EINECS 204-881-4
Synonyms: DBPC; Butylated hydroxytoluene; 2,6-Di-t-butyl-4-methylphenol; 2,6-Di-t-butyl-p-cresol; 2,6-Bis (1,1-dimethylethyl)-4-methylphenol
Classification: Substituted toluene
Empirical: $C_{15}H_{24}O$
Formula: $[C(CH_3)_3]_2CH_3C_6H_2OH$
Properties: Wh. cryst. solid, faint char. odor; insol. in water, propylene glycol; sol. in toluene, alcohols, MEK, acetone, Cellosolve, petrol. ether, chloroform, benzene, most HC solvs.; m.w. 220.39; sp.gr. 1.048 (20/4 C); m.p. 68 C; b.p. 265 C; flash pt. (TOC) 260 F
Precaution: Combustible exposed to heat or flame; reactive with oxidizing materials
Toxicology: TLV: 10 mg/m^3; LD50 (oral, rat) 890 mg/kg; moderately toxic by ingestion; poison by IP, IV routes; suspected carcinogen; human skin irritant; eye irritant; may cause rashes, hyperactivity; heated to decomp., emits acrid smoke and fumes
Uses: Antioxidant for foods, animal feed, petrol. prods., syn. rubbers, plastics, soaps; antiskinning agent in paints and inks
Usage level: 0.02% max. (preservation of fixed oils, fats, vitamin oil concs.)
Regulatory: FDA 21CFR §137.350, 166.110, 172.115, 172.615 (0.1% max.), 173.340 (0.1% of defoamer), 175.105, 175.125, 175.300, 175.380, 175.390, 176.170, 176.210, 177.1010, 177.1210, 177.1350, 177.2260, 177.2600, 178.3120, 178.3570, 179.45, 181.22; 181.24 (0.005% migrating from food pkg.), 182.3173 (0.02% max. of fat/oil), GRAS; USDA 9CFR §318.7, 381.147; Japan 0.2-1 g/kg; Europe, UK; USP/NF, BP, Ph.Eur. compliance
Manuf./Distrib.: Aceto; Aldrich; Allchem Ind.; Ashland; Fluka; Great Lakes; Int'l. Chem. Inc.; Penta Mfg.; PMC Specialties; Raschig; Rhone-Poulenc N. Am.; Sigma; Uniroyal
Trade names: Akrochem® Antioxidant BHT; CAO®-3

Bicarbonate of soda. *See* Sodium bicarbonate
Biphenyl, isopropyl. *See* 1,1´-Biphenyl (1-methylethyl)

1,1´-Biphenyl (1-methylethyl)
CAS 25640-78-2
Synonyms: Biphenyl, isopropyl; Isopropyl diphenyl
Empirical: $C_{15}H_{16}$
Properties: M.w. 196.31
Trade names: Nusolv™ ABP-62

Bis (acetylacetonato) titanium oxide. *See* Titanium acetylacetonate
Bis(2-aminoethyl) amine. *See* Diethylenetriamine
N,N´-Bis(2-aminoethyl)-1,2-ethanediamine. *See* Triethylenetetramine
Bis(4-aminophenyl) sulfone. *See* 4,4´-Diaminodiphenyl sulfone
Bis(2-benzothiazolylthio)zinc. *See* Zinc 2-mercaptobenzothiazole
Bis (2-butoxyethyl) ether. *See* Diethylene glycol dibutyl ether
2,5-Bis (5-t-butyl-2-benzoxazolyl) thiophene. *See* 2,2´-(2,5-Thiophenediyl)bis[5-t-butylbenzoxazole]
2,5-Bis (5-t-butyl-2-benzoxazol-2-yl) thiophene. *See* 2,2´-(2,5-Thiophenediyl)bis[5-t-butylbenzoxazole]
2,2´-Bis(6-t-butyl-p-cresyl)methane. *See* 2,2´-Methylenebis (6-t-butyl-4-methylphenol)
Bis (3-t-butyl-4-hydroxy-6-methylphenyl) sulfide. *See* 4,4´-Thiobis-6-(t-butyl-m-cresol)

Biscumylphenyl trimellitate
Uses: Plasticizer, process aid for extrusion of PVC, urethanes; lubricant and process aid for filled PS, PC; reactive diluent and flow promoter for epoxy powd. coatings

Bis (2,4-di-t-butylphenyl) pentaerythritol diphosphite
CAS 26741-53-7; EINECS 247-952-5
Synonyms: Bis (2,4-di-t-butylphenyl) pentaerythrityl diphosphite
Formula: $C_5H_{18}[O_2POC_6H_3(C[CH_3]_3)2]_2$
Properties: M.w. 604
Uses: Antioxidant; impact modifier for PVC
Manuf./Distrib.: Aldrich
Trade names containing: Ultranox® 626; Ultranox® 626A; Ultranox® 627A

Bis (2,4-di-t-butylphenyl) pentaerythrityl diphosphite. *See* Bis (2,4-di-t-butylphenyl) pentaerythritol diphosphite
Bis(dimethylcarbamodithioato,S,S´) zinc. *See* Zinc dimethyldithiocarbamate
2,6-Bis (1,1-dimethylethyl)-4-methylphenol. *See* BHT
Bis (1,1-dimethylethyl) peroxide. *See* Di-t-butyl peroxide
1,3-Bis(2,3-epoxypropoxy)-2,2-dimethyl propane. *See* Neopentyl glycol diglycidyl ether
Bis (2-ethoxyethyl) ether. *See* Diethylene glycol diethyl ether
Bis (2-ethylhexyl) fumarate. *See* Dioctyl fumarate
Bis(2-ethylhexyl) hexanedioate. *See* Dioctyl adipate
Bis (2-ethylhexyl) maleate. *See* Dioctyl maleate
Bis (2-ethylhexyl) phthalate. *See* Dioctyl phthalate
Bis (2-ethylhexyl)-S-sodium sulfosuccinate. *See* Dioctyl sodium sulfosuccinate
Bis (4-hydroxy-5-t-butyl-2-methylphenyl) sulfide. *See* 4,4´-Thiobis-6-(t-butyl-m-cresol)
Bis(2-hydroxyethyl)amine. *See* Diethanolamine
N,N-Bis(2-hydroxyethyl)aniline. *See* Phenyldiethanolamine
1,3-Bis(2-hydroxyethyl)-5,5-dimethyl-2,4-imidazolidinedione. *See* DEDM hydantoin
Bis (2-hydroxyethyl) tallow amine. *See* PEG-2 tallowamine
Bis (hydroxylamine) sulfate. *See* Hydroxylamine sulfate
Bis(2-hydroxy-4-methoxyphenyl) methanone. *See* Benzophenone-6
1,4-Bis(hydroxymethyl)cyclohexane (cis and trans). *See* 1,4-Cyclohexanedimethanol

Bishydroxymethyl dimethyl hydantoin
Trade names containing: Biochek 240

2,2-Bis(hydroxymethyl)-1,3-propanediol. *See* Pentaerythritol
2,2-Bis(hydroxymethyl)propionic acid. *See* Dimethylolpropionic acid
2,2-Bis (4-hydroxyphenol) propane. *See* Bisphenol A
Bis (hydroxypropyl) ether. *See* Dipropylene glycol
Bis[1-hydroxy-2(1H)-pyridinethinato-O,S]-(T-4) zinc. *See* Zinc pyrithione
Bis (1,4-isocyanatophenyl) methane. *See* MDI
Bis (4-isocyanatophenyl) methane. *See* MDI
Bis(p-isocyanatophenyl) methane. *See* MDI
Bis(mercaptobenzothiazolato)zinc. *See* Zinc 2-mercaptobenzothiazole
1,2-Bis(methacryloyoxy) ethane. *See* Ethylene glycol dimethacrylate
Bis (2-methoxyethyl) ether. *See* Diethylene glycol dimethyl ether
Bis(methoxypropyl) ether. *See* Dipropylene glycol dimethyl ether

Bis(methylethyl)-1,1´-biphenyl
CAS 69009-90-1
Synonyms: Diisopropyl biphenyl
Empirical: $C_{18}H_{22}$
Properties: M.w. 238.37
Uses: Solvent possessing exc. solvency, chem. and thermal stability, nonvolatility; for aerosols, adhesives, electronic parts mfg. and cleaning, industrial cleaning, inks and paper coatings, metal cleaning, paints, plastics, sealants, tile mfg.
Trade names: Nusolv™ ABP-103

Bis(2-methylpropyl) hexanedioate. *See* Diisobutyl adipate
1,4-Bis(2-methylpropyl) sulfobutanedioate, sodium salt. *See* Diisobutyl sodium sulfosuccinate
Bismuth carbonate basic. *See* Bismuth subcarbonate

Bismuth chloride oxide. *See* Bismuth oxychloride
Bismuth (III) carbonate basic. *See* Bismuth subcarbonate

Bismuth nitrate
CAS 10361-46-3; EINECS 233-792-3
Synonyms: Bismuthoxy nitrate; Bismuthyl nitrate; Bismuth trinitrate; Bismuth ternitrate
Empirical: $Bi(NO_3)_3$
Properties: Colorless cryst.; sl. hygroscopic; sol. in dil. nitric acid, alcohol, acetone; m.w. 395.01; m.p. 30 C; b.p. 75-80 C
Precaution: Oxidizing material; potential fire risk near organic materials
Toxicology: Poison by intravenous route
Uses: Preparation of other bismuth salts, bismuth luster on tin, luminous paints and enamels, precipitation of alkaloids
Manuf./Distrib.: AC Ind.; Advance Research Chems.; Aldrich; Am. Int'l.; Atomergic Chemetals; Fluka; Mallinckrodt; Nihon Kagaku Sangyo; Noah; Shepherd; Sigma; Spectrum Chem. Mfg.

Bismuth nitrate, basic. *See* Bismuth subnitrate

Bismuth octoate
Uses: Drier for coatings
Manuf./Distrib.: AC Ind.; OM Group; Shepherd
Trade names: Nuodex Octoate Bismuth® 24%

Bismuth oxycarbonate. *See* Bismuth subcarbonate

Bismuth oxychloride
CAS 7787-59-9; EINECS 232-122-7
Synonyms: Basic bismuth chloride; Bismuth chloride oxide; Bismuth subchloride; CI 77163; Chlorooxobismuthine; Synthetic pearl; Pearl white; Pigment white 14
Classification: Inorganic pigment
Empirical: BiClO
Formula: BiOCl
Properties: Wh. cryst. powd.; sol. in acids; insol. in water; m.w. 260.48; dens. 7.717
Toxicology: Irritant; toxic if ingested
Uses: Cosmetics, pigment, dry cell cathodes, artificial pearls
Regulatory: FDA 21CFR §73.1162, 73.2162; permanently listed
Manuf./Distrib.: Aldrich; Am. Int'l.; Atomergic Chemetals; Fluka; Great Western; ISP Van Dyk; Mallinckrodt; Mearl; Spectrum Chem. Mfg.
Trade names containing: Mearlite® GPN; Mearlite® Radiant Pearl STL; Mearlite® Ultra Bright UDQ; Mearlite® Ultra Bright UMS; Mearlite® Ultra Bright UMT; Mearlite® Ultra Bright USD; Mearlite® Ultra Bright USS; Mearlite® Ultra Bright UTL; Mearlite® Ultra Bright UWA; Mearlite® Ultra Fine OFS; Mearlite® Ultra Fine OFW

Bismuthoxy nitrate. *See* Bismuth nitrate
Bismuth oxynitrate. *See* Bismuth subnitrate

Bismuth subcarbonate
CAS 5892-10-4; EINECS 227-567-9
Synonyms: Bismuth oxycarbonate; Bismuth carbonate basic; Bismuth (III) carbonate basic
Empirical: Bi_2CO_5
Formula: $(BiO)_2CO_3$
Properties: Wh. powd., odorless; sol. in nitric acid; insol. in water, alcohol; m.w. 510.01; dens. 6.860; stable in air but affected slowly by light
Precaution: Light-sensitive
Uses: Bismuth compounds, cosmetics, opacifier in x-ray diagnosis, enamel fluxes, ceramic glazes
Regulatory: FDA approved for orals; BP, Ph.Eur. compliance
Manuf./Distrib.: Advance Research Chems.; Aldrich; Am. Int'l.; Atomergic Chemetals; Fluka; Mallinckrodt; Metalspecialties; Shepherd; Spectrum Chem. Mfg.

Bismuth subchloride. *See* Bismuth oxychloride

Bismuth subnitrate
CAS 1304-85-4; EINECS 215-136-8

Bismuth ternitrate

Synonyms: Bismuth nitrate, basic; Bismuth oxynitrate
Classification: Inorganic salt
Empirical: $Bi_5H_9N_4O_{22}$
Formula: $Bi_5(OH)_9(NO_3)_4O$
Properties: Hygroscopic; m.w. 1461.99
Uses: Cosmetics, ceramic glazes, enamel fluxes
Manuf./Distrib.: Am. Int'l.; Atomergic Chemetals; Celtic Chem. Ltd; R.W. Greeff; Mallinckrodt; Metalspecialties; Sigma; Spectrum Chem. Mfg.; Summer Labs

Bismuth ternitrate. *See* Bismuth nitrate
Bismuth trinitrate. *See* Bismuth nitrate
Bismuthyl nitrate. *See* Bismuth nitrate

Bis-(1-octyloxy-2,2,6,6,tetramethyl-4-piperidinyl) sebacate
CAS 129757-67-1
Uses: Light stabilizer for PP
Trade names: Tinuvin® 123

Bis (1,2,2,6,6-pentamethyl-4-piperidinyl) (3,5-di-t-butyl-4-hydroxybenzyl) butyl propanedioate
CAS 63843-89-0
Synonyms: Propanedioic acid, ((3,5-bis-(1,1-dimethylethyl)-4-hydroxyphenyl)methyl)-butyl, bis (1,2,2,6,6-pentamethyl-4-piperidinyl) ester
Empirical: $C_{42}H_{72}N_2O_5$
Uses: UV stabilizer used in coatings systems
Trade names: Tinuvin® 144

Bis (2,4-pentanedionato) titanium oxide. *See* Titanium acetylacetonate

Bisphenol A
CAS 80-05-7; EINECS 201-245-8
Synonyms: 4,4´-Isopropylidenediphenol (CTFA); p,p-Isopropylidenediphenol; 2,2-Bis (4-hydroxyphenol) propane; 4,4´-(1-Methylethylidene) bisphenol
Empirical: $C_{15}H_{16}O_2$
Formula: $(CH_3)_2C(C_6H_4OH)_2$
Properties: White flakes, phenolic odor; insol. in water; sol. in alcohol; sl. sol. in CCl_4; m.w. 228.31; dens. 1.195 (25/25 C); m.p. 153 C; b.p. 220 C (4 mm); flash pt. 79.4 C
Toxicology: Poison by intraperitoneal route; moderate toxicity by ingestion and skin contact
Uses: Intermediate in mfg. of epoxy, polycarbonate, phenoxy, polysulfone, polyester resins, flame retardant, rubber chemicals, fungicide; hard resin for finished prods. incl. laminates, powd. coatings, bondable coatings, molding powds.
Manuf./Distrib.: Akzo Nobel; Aldrich; Aristech; Ashland; Fluka; ICC Ind.; ICI Surf. Am.; Mitsui Petrochem. Ind.; Mitsui Toatsu; Punda Mercantile; Rhone-Poulenc N. Am.; Shell; Sigma; Van Waters & Rogers

Bisphenol A dicyanate monomer/prepolymer
Uses: Co-reacts with and cures epoxy resins; for pultrusion, filament winding applics., for alloying with thermoplastic tougheners, in laminates, powd. coatings, molding powders, printed wiring boards, fiber-reinforcing prepregs, adhesives
Trade names: AroCy® B-50

Bisphenol A epoxy acrylate
Uses: For use in screen ink vehicles, clear coatings for paper, wood and metal decorating, and laminating adhesives
Trade names: Ebecryl® 3500; Ebecryl® 3605; Ebecryl® 3701; Ebecryl® 3703
Trade names containing: Ebecryl® 3701-20T

Bisphenol A epoxy diacrylate
Uses: For use in screen ink vehicles, clear coatings for paper, wood and metal decorating, and laminating adhesives
Trade names: Ebecryl® 600; Ebecryl® 3200; Ebecryl® 3600; Ebecryl® 3700; Photomer® 3015; Photomer® 3016
Trade names containing: Ebecryl® 605; Ebecryl® 1608; Ebecryl® 3502; Ebecryl® 3700-20H; Ebecryl® 3700-20T; Ebecryl® 3700-25R

Bisphenol A, hydrogenated
Empirical: $C_{16}H_{28}O_2$
Properties: M.w. 236.30; m.p. 150 C
Uses: In prep. of alkyd, polyester, and epoxy resins where good color stability and improved weatherability are important; for casting, laminating, coatings and fiber prod.
Manuf./Distrib.: Allchem Ind.; Milliken; Rhone-Poulenc N. Am.
Trade names: Millad® HBPA

Bis(phenoxarsin-10-yl)ether. *See* 10,10′-Oxybisphenoxyarsine
Bis(10-phenoxyarsinyl)oxide. *See* 10,10′-Oxybisphenoxyarsine
N,N-Bis-stearoylethylenediamide. *See* Ethylene distearamide

Bis (tribromophenoxy) ethane
Properties: Wh. cryst. powd.
Uses: Flame retardant for many thermoplastic and thermoset systems; used in adhesives, coatings, textiles, lt.-stable applics.
Manuf./Distrib.: Great Lakes
Trade names: Great Lakes FF-680™

Bis(tributyltin) oxide. *See* Tributyltin oxide
Bis (tri-n-butyltin) oxide. *See* Tributyltin oxide

Bis (trichloromethyl) sulfone
CAS 3064-70-8
Synonyms: Chlorosulfona; Methane, sulfonylbis trichloro-; Sulfonylbis (trichloromethane)
Empirical: $C_2Cl_6O_2S$
Properties: Off-wh. cryst. powd., pungent odor; m.w. 300.78; m.p. 36-38 C
Toxicology: LD50 (oral, rat) 691 mg/kg, (IV, mouse) 18 mg/kg; poison by intravenous route; moderately toxic by ingestion; heated to decomp., emits very toxic fumes of Cl^- and SO_x
Uses: Microbiocide for paint, adhesives, latex, ink, mineral slurries, metalworking fluids
Trade names: Amerstat® 294

Bis-(γ-trimethoxysilylpropyl) amine
CAS 82985-35-1; EINECS 280-084-5
Formula: $[(CH_3O)_3Si(CH_2)_3]_2$
Properties: Clear liq.; m.w. 341.5; sp.gr. 1.040; b.p. 152 C (4 mm); flash pt. 113 C
Precaution: Moisture-sensitive
Toxicology: Irritant
Uses: Coupling agent, adhesion promoter providing improved wet adhesion, corrosion resist., weatherability, pigment disp. to adhesives, coatings, inks, sealants; min./resin coupling for improved composite str. in rubber and elastomers; moisture-cure crosslinking in polymer modification; filler treatment in thermoset and thermoplastic resins; also in fiberglass, textiles, foundry, crude oil extraction
Manuf./Distrib.: Aldrich; Fluka
Trade names: Silquest® A-1170

Bis (trimethylsilyl) amine. *See* Hexamethyldisilazane
BIT. *See* 1,2-Benzisothiazolin-3-one
BKF. *See* 2,2′-Methylenebis (6-t-butyl-4-methylphenol)
Black lead. *See* Graphite
Blanc fixe (artificial, precipitated). *See* Barium sulfate
Bleached shellac. *See* Shellac
BLO. *See* Butyrolactone
Blue copperas. *See* Cupric sulfate pentahydrate
Blue lead. *See* Lead sulfate, blue basic
Blue stone. *See* Cupric sulfate pentahydrate
Blue vitriol. *See* Cupric sulfate pentahydrate
BNPD. *See* 2-Bromo-2-nitropropane-1,3-diol
Boletic acid. *See* Fumaric acid
Bolus alba. *See* Kaolin

Bone black
CAS 8021-99-6
Synonyms: CI Pigment black 9; Bone charcoal; Bone char

Bone char

Definition: Black pigment produced from carbonization of bones
Precaution: Flamm. as suspended dust; nonflamm. in bulk
Uses: Pigment in paints, varnishes; mfg. activated carbon; decolorizing agent and filter medium; cementation agent; adsorptive medium in gas masks; clarifying shellac; water purification
Manuf./Distrib.: Allchem Ind.; Ebonex; Landers-Segal Color
Trade names: Cosmic Black 3D; Cosmic Black #6; Cosmic Black #7; Cosmic Black 8-S; Cosmic Black 300; Cosmic Black 500 WB; Cosmic Black #700; Cosmic Black 2172; Cosmic Black AP; Cosmic Black BS; Cosmic Black CEM; Cosmic Black D-2; Cosmic Black DB; Cosmic Black D Special; Cosmic Black FCG-1; Cosmic Black MCW; Cosmic Black VBB; Ebonex 3D; Ebonex SC-5

Bone char. *See* Bone black
Bone charcoal. *See* Bone black
BOP. *See* Butyl octyl phthalate
Boracic acid. *See* Boric acid
Borax, fused. *See* Sodium borate

Boric acid
CAS 10043-35-3; EINECS 233-139-2
Synonyms: Boracic acid; Orthoboric acid
Classification: Inorganic acid
Empirical: BH_3O_3
Formula: H_3BO_3
Properties: Wh. or colorless cryst. powd., gran., odorless, almost tasteless or sl. acidic taste; sol. in water, alcohol, glycerol; insol. in ether, benzene; m.w. 61.83; dens. 1.435; m.p. 171 C
Precaution: Hygroscopic
Toxicology: Sl. to mod. toxic; acute poisoning causes digestive upsets in man, CNS stimulation, depression, renal damage, circulatory collapse; irritant; contraindicated with perforated eardrum or broken skin
Uses: Heat-resistant glass, porcelain enamels, boron chemicals, metallurgy, flame retardant in cellulosic insulation, mattress batting and cotton textiles, fungus control on citrus fruits, ointment and eyewash, nickel electroplating baths, preservative; medicine (protective disinfectant), antiseptic, cosmetics, photography, leather finishing
Usage level: 0.01-1.0%
Regulatory: FDA 21CFR §175.105, 176.180, 181.22, 181.30; USA not restricted; Europe listed; permitted in Switzerland and Sweden as preservative in some processed seafoods; FDA approved for ophthalmics, otics, topicals; USP/NF, BP, Ph.Eur. compliance
Manuf./Distrib.: Advance Research Chems.; Aldrich; Allchem Ind.; Am. Int'l.; Coyne; Dragoco; Fluka; Janssen Chimica; Nippon Denko; Occidental; Research Organics; San Yuan; Seeler Ind.; Sigma; Spectrum Chem. Mfg.; U.S. Borax; Veckridge

Boric acid barium salt. *See* Barium metaborate
Boric acid, zinc salt. *See* Zinc borate
Boric anhydride. *See* Boron oxide
Boric oxide. *See* Boron oxide
2-Bornanone. *See* Camphor
Bornan-2-one. *See* Camphor

Boron carbide
CAS 12069-32-8
Empirical: B_4C
Properties: Hard blk. cryst.; insol. in water; m.w. 55.26; m.p. 2350 C; b.p. > 3500 C
Toxicology: Avoid inhalation of dust or particles
Uses: Raw material for paint mfg.; alloying agent; abrasive; neutron absorber; reinforcing agent in composites for military aircraft
Manuf./Distrib.: Advanced Refractory Tech.; Atlantic Equip. Engrs.; Atomergic Chemetals; Electro Abrasives; Noah; Reade Advanced Materials

Boron fluoride. *See* Boron trifluoride

Boron oxide
CAS 1303-86-2; EINECS 215-125-8
Synonyms: Boron trioxide; Boric anhydride; Boric oxide; Fused boric acid
Empirical: B_2O_3
Formula: B_2O_3

Properties: Vitreous colorless cryst.; sl. bitter taste; hygroscopic; sol. in hot water, alcohol; m.w. 69.62; m.p. 450 C; b.p. 1860 C
Toxicology: LD50 (oral, mouse) 3163 mg/kg; eye and skin irritant; TLV: 10 mg/m^3 in air
Uses: Production of boron, heat-resistant glassware, fire-resistant additive for paints, electronics, liquid encapsulation techniques, herbicide
Manuf./Distrib.: Advance Research Chems.; Aldrich; Atlantic Equip. Engrs.; Atomergic Chemetals; Carborundum; Fluka; Noah; Reade Advanced Materials; Sigma; Spectrum Chem. Mfg.; U.S. Borax

Boron trifluoride
CAS 7637-07-2; EINECS 231-569-5
Synonyms: Boron fluoride
Empirical: BF_3
Properties: Colorless pungent gas in a dry atm.; m.w. 67.81; m.p. -127 C; b.p. -100 C
Precaution: Nonflamm. gas
Toxicology: Highly toxic; corrosive to skin; poisonous by inhalation and tissue
Uses: Catalyst in organic synthesis, production of diborane, instruments for measuring neutron intensity, soldering fluxes, gas brazing; fumigant; used in electronic potting compds., elec. varnishes, adhesives, filament winding, fiberglass composites, prepregs
Manuf./Distrib.: Advance Research Chems.; Air Prods.; Akzo Nobel; Aldrich; AlliedSignal; Atomergic Chemetals; Eagle-Picher; Elf Atochem N. Am.; Leepoxy Plastics; Voltaix
Trade names: Leecure B-110; Leecure B-550; Leecure B-610; Leecure B-612; Leecure B-614; Leecure B-950; Leecure B-1310; Leecure B-1550; Leecure B-1600; Leecure B-1700

Boron trifluoride-amine complex
Uses: Epoxy curing agent for hot-dip coatings of electronic components, low-temp. cure prepregs, lacquers
Trade names: Anchor® 1115; Anchor® 1171; Anchor® 1222

Boron trioxide. *See* Boron oxide

Borosilicate glass
CAS 65997-17-3
Definition: A soda-lime glass contg. approx. 5% boric oxide; heat-resistant glass
Uses: Extender/filler for plastics, etc.; improves flow and leveling, provides hiding and corrosion protection, reduces VOCs in paint and coating systems, incl. primers, alkyd interior enamels, PU acrylic topcoats, epoxy polyamide topcoats; extends wh. pigments
Manuf./Distrib.: Aldrich; Sigma
Trade names: Sphericel® 110P8

BR. *See* Polybutadiene
Brassica campestris oil. *See* Rapeseed oil
Brazil wax. *See* Carnauba
2-Bromo-2-(bromomethyl) glutaronitrile. *See* Methyldibromo glutaronitrile
2-Bromo-2-(bromomethyl) pentanedinitrile. *See* Methyldibromo glutaronitrile
Bromoethylene. *See* Vinyl bromide

2-Bromo-4´-hydroxyacetophenone
CAS 2491-38-5
Synonyms: 1-(4-Hydroxyphenyl)-2-bromoethanone
Empirical: $C_8H_7BrO_2$
Properties: M.w. 215.05
Uses: Bactericide for water-based paints, adhesives, waxes, polishes
Trade names: Busan® 90
Trade names containing: Busan® 1130

2-Bromo-2-nitropropane-1,3-diol
CAS 52-51-7; EINECS 200-143-0
Synonyms: BNPD; 1,3-Propanediol, 2-bromo-2-nitro; Bronopol; 2-Bromo-2-nitro-1,3-propanediol
Classification: Substituted aliphatic diol
Empirical: $C_3H_6BrNO_4$
Properties: Cryst., odorless; sol. in water, alcohol; sl. sol. in chloroform, acetone, ether; m.w. 200.01; m.p. 130-133 C
Toxicology: Poison by ingestion, subcutaneous, intravenous, intraperitoneal routes; moderately toxic by skin contact; eye and skin irritant

2-Bromo-2-nitro-1,3-propanediol

Uses: Broad spectrum antimicrobial agent; for adhesives, coatings, paints, starch, pigment and extender slurries, latex, antifoam emulsions, and inks
Usage level: 0.01-0.1%
Regulatory: FDA 21CFR §175.105; USA CIR approved to 0.1%, EPA registered; Europe listed; BP compliance
Manuf./Distrib.: Aldrich; ANGUS; Boots Co plc; CM Chem. Prods.; Fluka; Girindus Chemie; K3; Sigma
Trade names: Canguard® 409; Myacide® AS Plus
Trade names containing: Canguard® 409-40

2-Bromo-2-nitro-1,3-propanediol. *See* 2-Bromo-2-nitropropane-1,3-diol
Bromphthal. *See* Tetrabromophthalic anhydride
Bronopol. *See* 2-Bromo-2-nitropropane-1,3-diol
Bronze powder. *See* Copper
Brown copper oxide. *See* Copper oxide (ous)
BTDA. *See* 3,3′,4,4′-Benzophenone tetracarboxylic dianhydride

Bumetrizole
CAS 3896-11-5; EINECS 223-445-4
Synonyms: 2-(3′-t-Butyl-2′-hydroxy-5′-methylphenyl)-5-chlorobenzotriazole; UV Absorber 6; 2-(2′-Hydroxy-3′-t-butyl-t′-methylphenyl)-5-chlorobenzotriazole; 2-t-Butyl-6-(5-chloro-2H-benzotriazol-2-yl)-p-cresol
Classification: Benzotriazole
Empirical: $C_{17}H_{18}ClN_3O$
Properties: Sl. yel. powd.; m.p. 138 C
Uses: UV stabilizer/absorber for polyolefins, polyester resins, coatings
Manuf./Distrib.: Aldrich
Trade names: Lowilite® 26; Tinuvin® 326

Burnt lime. *See* Calcium oxide

Butadiene/acrylonitrile copolymer
CAS 9003-18-3
Synonyms: 1,3-Butadiene, polymer with 2-propenenitrile; 2-Propenenitrile, polymer with 1,3-butadiene
Empirical: $(C_4H_6 \cdot C_3H_3N)_x$
Formula: $-[CH_2CH=CHCH_2]_x.[CHCNCH_2]_y$
Properties: Dens. 0.980
Toxicology: Suspected carcinogen
Uses: Elastomer for low temp. oil well specialties, belt covers, idler rolls, o-rings, seals, gaskets, hydraulic hose, footwear, packings, fuel hose, printer's blankets and roll covers, etc.; in plasticizer masterbatches, as impact modifier for thermoplastics, as visc. modifier; used in coatings and adhesives
Regulatory: FDA 21CFR §175.125, 175.300, 175.320, 177.1200, 177.2600
Manuf./Distrib.: Aldrich; Sigma
Trade names: Nipol® 1432
Trade names containing: Nipol® 1452X8

1,3-Butadiene-2-methyl polymer with 2-methyl-1-propene. *See* Isobutylene/isoprene copolymer

Butadiene monomer
Uses: Monomer for prod. of paints
Manuf./Distrib.: Amoco; ARCO; BP Chems.; Elf Atochem N. Am.; Huntsman; Shell; Wiley Organics

1,3-Butadiene, polymer with 2-propenenitrile. *See* Butadiene/acrylonitrile copolymer
Butadiene rubber. *See* Polybutadiene
1,3-Butadiene-styrene copolymer. *See* Polybutadiene-styrene copolymer; Styrene-butadiene polymer
Butadiene-styrene polymer. *See* Polybutadiene-styrene copolymer
Butadiene-styrene resin. *See* Polybutadiene-styrene copolymer; Styrene-butadiene polymer
Butal. *See* n-Butyraldehyde
Butaldehyde. *See* n-Butyraldehyde
Butalyde. *See* n-Butyraldehyde
Butanal. *See* n-Butyraldehyde
n-Butanal. *See* n-Butyraldehyde
Butanal oxime. *See* m-Butyraldehyde oxime

Butane
CAS 106-97-8; EINECS 203-448-7

Synonyms: n-Butane; Alkane C-4
Classification: Hydrocarbon
Empirical: C_4H_{10}
Formula: $CH_3CH_2CH_2CH_3$
Properties: Colorless gas, faint disagreeable odor; easily liquefied under pressure @ R.T.; one vol. water dissolves 0.15 vol.; 1 vol. alcohol dissolves 18 vol.s (17 C, 770 mm); m.w. 58.12; dens. 0.599; vapor pressure 1620 mm Hg (17 psig, 21 C); f.p. -138 C; b.p. -0.5 C; flash pt. (CC) -76 F
Precaution: Flamm. gas; very dangerous fire hazard exposed to heat, flame, oxidizers; highly explosive; explosive limits 1.9-8.5%
Toxicology: TLV:TWA 800 ppm; mildly toxic by inh.; causes drowsiness; asphyxiant; narcotic in high concs.; heated to decomp., emits acrid smoke and fumes
Uses: Hydrocarbon propellant; paint aerosols
Regulatory: FDA 21CFR §173.350, 184.1165, GRAS; approved for topicals; USP/NF compliance
Manuf./Distrib.: Air Prods.; Aldrich; Electrochem Ltd; Phillips; Stanchem

n-Butane. *See* Butane
Butanecarboxylic acid. *See* n-Valeric acid
1-Butanecarboxylic acid. *See* n-Valeric acid
1,4-Butanedicarboxylic acid. *See* Adipic acid
1,4-Butanedioic acid. *See* Succinic acid
Butanedioic acid, sulfo-, 1,4-dihexyl ester, sodium salt. *See* Dihexyl sodium sulfosuccinate
Butanedioic anhydride. *See* Succinic anhydride
Butanediol. *See* 1,4-Butanediol

1,4-Butanediol
CAS 110-63-4; EINECS 203-786-5
Synonyms: BDO; 1,4-Butylene glycol; 1,4-Dihydroxybutane; Tetramethylene glycol; Butanediol
Formula: $HO(CH_2)_4OH$
Properties: Sol. in alcohol; sl. sol. in ether; m.w. 90.12; dens. 1.017; m.p. 16 C; b.p. 230, 120-122 C (10 mm); flash pt. > 110 C
Precaution: Combustible
Toxicology: LD50 (oral, rat) 1525 mg/kg; toxic by ingestion
Uses: Intermediate; used in paints, polyurethane formulation in the hard segment as a curative; plasticizer
Manuf./Distrib.: Aldrich; Allchem Ind.; ARCO; Ashland; BASF; BASF AG; Browning; Chemcentral; DuPont; Fluka; Hüls UK; ISP; Tri-Iso; Van Waters & Rogers

1,3-Butanediol diacrylate. *See* 1,3-Butylene glycol diacrylate

Butanetriol
Uses: Intermediate for paints
Manuf./Distrib.: BASF

Butanoic acid, 3-amino-, N-coco alkyl derivatives. *See* Cocaminobutyric acid
Butanoic acid, 3-oxo-, 2-[(1-oxo-2-propenyl)oxy]ethyl ester. *See* 2-(Acetoacetoxy) ethyl acrylate
1-Butanol. *See* Butyl alcohol
Butan-1-ol. *See* Butyl alcohol

2-Butanol
CAS 78-92-2; EINECS 240-029-8
Synonyms: SBA; 2-Butyl alcohol; s-Butyl alcohol; sec-Butyl alcohol; Secondary butyl alcohol; Methyl ethyl carbinol; Ethylmethyl carbinol; sec-Butanol; Butan-2-ol; Butylene hydrate; 1-Methyl propanol; 2-Hydroxybutane
Empirical: $C_4H_{10}O$
Formula: $CH_3CH_2CH(OH)CH_3$
Properties: Colorless liq.; m.w. 74.12; dens. 0.807 (20/4 C); m.p. -89 C; b.p. 99-100 C; flash pt. 23 C; ref. index 1.3972 (20 C)
Precaution: Flamm.
Toxicology: LD50 (oral, rat) 6480 mg/kg; poison by intravenous and intraperitoneal routes; mildly toxic by ingestion; skin and eye irritant
Uses: Solvent for coatings, paint removers, industrial cleaners, prep. of methyl ethyl ketone
Regulatory: FDA 21CFR §172.515
Manuf./Distrib.: Ashland; Chemcentral; Fluka; Shell
Trade names containing: Resimene® 717

Butan-2-ol. *See* 2-Butanol
n-Butanol. *See* Butyl alcohol
sec-Butanol. *See* 2-Butanol
4-Butanolide. *See* Butyrolactone
2-Butanone. *See* Methyl ethyl ketone
2-Butanone oxime. *See* Methyl ethyl ketoxime
2-Butanone peroxide. *See* Methyl ethyl ketone peroxide
2-Butenedioic acid. *See* Fumaric acid; Maleic acid
cis-Butenedioic acid. *See* Maleic acid
trans-Butenedioic acid. *See* Fumaric acid
2-Butenedioic acid bis (2-ethylhexyl) ester. *See* Dioctyl fumarate
2-Butenedioic acid, dibutyl ester. *See* Dibutyl maleate
2-Butenedioic acid, polymer with methoxyethene, butyl ester. *See* PVM/MA copolymer, butyl ester
2-Butenedioic acid, polymer with methoxyethene, 1-methylethyl ester. *See* PVM/MA copolymer, isopropyl ester
cis-Butenedioic anhydride. *See* Maleic anhydride
1-Butene, homopolymer. *See* Polybutene
2-Butenoic acid. *See* Crotonic acid

Butoxydiglycol
CAS 112-34-5; EINECS 203-961-6
Synonyms: PEG-2 butyl ether; Butyl 'Carbitol'; Butyldiglycol; Diethylene glycol butyl ether; Diethylene glycol monobutyl ether; 2-(2-Butoxyethoxy) ethanol
Empirical: $C_8H_{18}O_3$
Formula: $C_4H_9OCH_2CH_2OCH_2CH_2OH$
Properties: Colorless liq., faint butyl odor; sol. in oils and water; m.w. 162.26; dens. 0.9553 (20/4 C); m.p. -68.1 C; b.p. 230.6 C; flash pt. 100 C; ref. index 1.4316 (20 C)
Precaution: Combustible; heated to dec., emits acrid smoke and irritating fumes
Toxicology: LD50 (oral, rat) 6.56 g/kg; moderately toxic by ingestion, intraperitoneal; mildly toxic by skin contact; severe eye irritant
Uses: Solvent for nitrocellulose, coatings, oils, dyes, gums, soaps, polymers; plasticizer intermediate
Regulatory: FDA 21CFR §175.105, 176.180
Manuf./Distrib.: Aldrich; Allchem Ind.; ARCO; Ashland; Brown; Eastman; Fluka; Occidental; Shell; Union Carbide
Trade names: Butyl Carbitol®; Butyl Di-Icinol; Butyl Dioxitol; Dowanol® DB; Eastman® DB

2-Butoxy-1,2-diphenylethanone
Trade names containing: Esacure® EB3

Butoxyethanol
CAS 111-76-2; EINECS 203-905-0
Synonyms: 2-Butoxyethanol; Ethylene glycol monobutyl ether; Glycol butyl ether; Ethylene glycol butyl ether; Butyl glycol
Classification: Ether alcohol
Empirical: $C_6H_{14}O_2$
Formula: $HOCH_2CH_2OC_4H_9$
Properties: Colorless liq.; mild pleasant odor; m.w. 118.20; dens. 0.9019 (20/20 C); m.p. -74.8 C; b.p. 171.2 C; flash pt. (COC) 160 F
Precaution: Flamm. liq. exposed to heat or flame; incompat. with oxidizers, heat, flame
Toxicology: TLV 25 ppm in air; LD50 (IP, rat) 20 mg/kg; poison by ing., skin contact, IP, IV routes; mod. toxic by inh., subcut.; human systemic effects; experimental reproductive effects; skin/eye irritant; heated to dec., emits acrid smoke, irritating fumes
Uses: Solvent for nitrocellulose resins, spray lacquers, cosolvent, gas chromatography
Regulatory: FDA 21CFR §173.315, 175.105, 176.210, 177.1650, 178.1010, 178.3297, 178.3570
Manuf./Distrib.: Aldrich; Allchem Ind.; ARCO; Ashland; Brown; Eastman; Fluka; Occidental; Shell; Sigma; Union Carbide
Trade names: Butyl Cellosolve®; Butyl Icinol; Butyl Oxitol; Dowanol® EB; Eastman® EB
Trade names containing: Acryloid® AT-9LO; Acryloid® AT-76; Acryloid® AT-81; Acryloid® AT-85; Acryloid® WR-97; Arolon 559-G4-70; Arolon 921-G4-70; Beckosol® 13-400; Disparlon® AQ-200; Edaplan LA 402; Fluorad® FC-120; GPRI™ BKUA-2370; Kelsol 3902-BG4-75; Kelsol 3905-B2G-75; Kelsol 3906-B2G-75; Kelsol 3950-B2G-70; Kelsol 3970-G4-75; Kelsol 3990-B2G-75; K-Sperse® 132; K-Sperse® 152; 8521 MX60; Nacorr® 1352; Nacorr® 1652; Nacorr® 1754; Nacure® 1051; Nacure® 1557; Perenol® S5;

Perenol® S500; Slip-Ayd® SL-018; Slip-Ayd® SL-295A; Slip-Ayd® SL-530; Sobral 586; Sobral 588; Surfynol® 104BC; Synthemul® 40-426; Synthemul® 40-427

2-Butoxyethanol. *See* Butoxyethanol

Butoxyethanol acetate
CAS 112-07-2; EINECS 203-933-3; 203-934-9
Synonyms: Ethylene glycol butyl ether acetate; Ethylene glycol monobutyl ether acetate; 1-Acetoxy-2-butoxyethane; 2-Butoxyethanol acetate; 2-Butoxyethyl acetate; Ethanol, 2-butoxy-, acetate
Empirical: $C_8H_{16}O_3$
Formula: $CH_3CH_2CH_2CH_2OCH_2CH_2OCOCH_3$
Properties: Colorless liq., mild odor; sol. 1.1% in water; m.w. 160.2; dens. 0.941 (20/20 C); b.p. -64 C; b.p. 186 C (760 mm); ref. index 1.4142 (20 C)
Precaution: Combustible liq. and vapor; peroxide former
Toxicology: May cause blood disorders, kidney damage; harmful if inhaled or absorbed thru skin
Uses: Solvent for high-solids coatings (NC, butyrate, ethyl cellulose acrylic); coalescing aid for water-disp. coatings
Manuf./Distrib.: Aldrich; Allchem Ind.; ARCO; Ashland; Eastman; Fluka; Occidental; Oxiteno
Trade names: Butyl Cellosolve® Acetate; Eastman® EB Acetate

2-Butoxyethanol acetate. *See* Butoxyethanol acetate
2-Butoxyethanol phosphate. *See* Tributoxyethyl phosphate
2-(2-Butoxyethoxy) ethanol. *See* Butoxydiglycol
2-(2-Butoxyethoxy) ethyl acetate. *See* Diethylene glycol butyl ether acetate
2-Butoxyethyl acetate. *See* Butoxyethanol acetate
2-Butoxyethyl ether. *See* Diethylene glycol dibutyl ether
Butoxyl. *See* Methoxybutyl acetate
Butoxymethyl oxirane. *See* Butyl glycidyl ether

Butoxypropanol
CAS 5131-66-8; EINECS 225-878-4
Synonyms: 1-Butoxy-2-propanol; Propylene glycol n-butyl ether; Propylene glycol butyl ether
Classification: Ether alcohol
Empirical: $C_7H_{16}O$
Formula: $CH_3CH_2CH_2CH_2OCH_2CHOHCH_3$
Properties: M.w. 132.23; sp.gr. 0.884 (20/20 C); f.p. -148 F; b.p. 170.2 C; flash pt. (TCC) 138 F; ref.index 1.4074 (20 C)
Toxicology: LD50 (oral, rat) 2200 mg/kg; mod. toxic by ingestion and by skin contact; heated to decomp., emits acrid smoke and irritating fumes
Uses: Solvent for coatings, cleaners, electronic, and ink applics.; strong coupling ability; used in lt.-duty and hard-surf. cleaners, water-reducible polyester and alkyd resin prod.
Trade names: Arcosolv® PNB; Dowanol® PnB
Trade names containing: G-4825-PnB-80

1-Butoxy-2-propanol. *See* Butoxypropanol
1-t-Butoxy-2-propanol. *See* Propylene glycol t-butyl ether

Butoxytriglycol
CAS 143-22-6; EINECS 205-592-6
Synonyms: Triethylene glycol butyl ether; Triethylene glycol monobutyl ether; PEG-3 butyl ether
Empirical: $C_{10}H_{22}O_4$
Formula: $C_4H_9O(C_2H_4O)_3H$
Properties: Liq.; misc. in water; m.w. 206.32; dens. 1.0021 (20/20 C); b.p. dec.; f.p. -47.6 C; flash pt. 143 C
Toxicology: Moderately toxic by skin contact; mildly toxic by ingestion; skin and severe eye irritant
Uses: Plasticizer, intermediate
Manuf./Distrib.: Ashland; Fluka; Oxiteno
Trade names containing: Icinol BE33

Button lac. *See* Shellac
Butyl acetate. *See* n-Butyl acetate

n-Butyl acetate
CAS 123-86-4; EINECS 204-658-1

s-Butyl acetate

Synonyms: Acetic acid, butyl ester; Butyl acetate
Definition: Ester of butyl alcohol and acetic acid
Empirical: $C_6H_{12}O_2$
Formula: $CH_3COOC_4H_9$
Properties: Colorless liq., fruity odor; sol. in alcohol, ether, hydrocarbons; sl. sol. in water; m.w. 116.18; dens. 0.8826 (20/20 C); b.p. 126.3 C; f.p. -75 C; flash pt. (TOC) 36.6 C); ref. index 1.2951 (20 C)
Precaution: Flamm. liq.; mod. explosive when exposed to flame
Toxicology: LD50 (oral, rat) 14.13 g/kg; mildly toxic by inhalation, ingestion; moderately toxic by intraperitoneal route; skin irritant; severe eye irritant; mild allergen; TLV 150 ppm in air; heated to decomp., emits acrid and irritating fumes
Uses: Solv. for paints; mfg. of lacquers, artificial leather, photographic films, plastics, safety glass, flavoring agent
Regulatory: FDA 21CFR §172.515, 175.105, 175.320, 177.1200; FEMA GRAS; Japan approved as flavoring
Manuf./Distrib.: Aldrich; Allchem Ind.; Ashland; J.T. Baker; BASF; Baychem; BP Chems. Ltd; Chemcentral; Chisso Am.; Coyne; Eastman; Fluka; General Chem.; Great Western; Harcros; Hoechst AG; Hüls AG; Janssen Chimica; Penta Mfg.; Samson; Sigma; Spectrum Chem. Mfg.; Sunnyside; Union Carbide; Van Waters & Rogers
Trade names containing: Acryloid® AU-608B; Aroplaz 6440-A4-85; Cellokyd 2708; Joncryl®-588; Konform® AR 2000; Luxate® HB9075; Luxate® HT2090; Resin QR-1166; Resin QR-1269; SCX™-507; SCX™-902; SCX™-907; SCX™-915; XU-19044.01

s-Butyl acetate
CAS 105-46-4; EINECS 203-300-1
Synonyms: Acetic acid, sec-butyl ester; Acetic acid 1-methylpropyl ester
Definition: Ester of butyl alcohol and acetic acid
Formula: $CH_3COOCH(CH_3)(C_2H_5)$
Properties: Colorless liq.; misc. with alcohol, ether; insol. in water; dens. 0.870 (20/4 C); b.p. 112.2 C; flash pt. (OC)31 C; ref. index 1.389 (20 C); flamm.
Toxicology: TLV 200 ppm in air
Uses: Solv. for nitrocellulose lacquers, thinners, nail enamels, leather finishes
Manuf./Distrib.: Aldrich; Schweizerhall

Butylacetic acid. *See* Caproic acid
Butyl 12-(acetyloxy)-9-octadecenoate. *See* Butyl acetyl ricinoleate

Butyl acetyl ricinoleate
CAS 140-04-5; EINECS 205-393-4
Synonyms: Butyl 12-(acetyloxy)-9-octadecenoate
Definition: Butyl ester of acetyl ricinoleic acid
Empirical: $C_{24}H_{44}O_4$
Properties: Yel. oily liq., mild odor; misc. with org. solvs.; pract. insol. in water; dens. 0.940 (20/20 C); f.p. cloudy -32 C; solid. pt. -65 C; flash pt. 110 C; ref. index 1.4614 (20 C)
Precaution: Combustible
Uses: Plasticizer for vinyls, emulsifier, lubricant, detergent, protective coatings, special cleaning compds., quick-breaking emulsions
Regulatory: FDA 21 CFR §175.105, 177.2600, 177.2800
Manuf./Distrib.: CasChem
Trade names: Flexricin® P-6

Butyl acid phosphate
CAS 12788-93-1
Synonyms: Acid butyl phosphate; n-Butyl acid phosphate; Butyl phosphoric acid
Empirical: $C_4H_{10}O_4P$
Properties: Water-wh. liq.; sol. in alcohol, acetone, toluene; insol. in water, petrol., naphtha; m.w. 153.1; dens. 1.120-1.125 (25/40 C); flash pt. (COC) 230 F
Precaution: DOT: Corrosive material; combustible when exposed to heat or flame
Toxicology: Toxic and corrosive; heated to decomp., emits highly toxic fumes of PO_x
Uses: High reactivity catalyst used in coatings and inks; for appliance, automotive, floor, furniture, paper, and other finishes; compat. with alkyds, cellulosics, epoxies, melamine, urethane, B/S; also for antistatic agents, corrosion inhibitors, solder fluxes, lubricants, and tanning chems.
Manuf./Distrib.: Akzo Nobel; Occidental; Witco/Oleo-Surf.
Trade names: Albrite® Butyl Acid Phosphate

n-Butyl acid phosphate. *See* Butyl acid phosphate

Butyl acrylate
CAS 141-32-2; EINECS 205-480-7
Synonyms: Butyl-2-propenoate; n-Butyl acrylate; Acrylic acid butyl ester
Empirical: $C_7H_{12}O_2$
Formula: $CH_2:CHCOOC_4H_9$
Properties: Water-wh., extremely reactive monomer; insol. in water; m.w. 128.19; dens. 0.89 (25/25 C); b.p. 69 C (50 mm); f.p. -64.6 C; flash pt. (OC) 120 F
Precaution: Flamm.
Toxicology: LD50 (oral, rat) 900 mg/kg; TLV:TWA 10 ppm (air); moderately toxic by ingestion, inhalation, skin contact, intraperitoneal route; skin and eye irritant
Uses: Intermediate in organic synthesis, polymers and copolymers for solvent coatings, adhesives, paints, binders, emulsifiers
Manuf./Distrib.: Aldrich; Allchem Ind.; Ashland; BASF; Fluka; Hoechst Celanese; ICD Group; Monomer-Polymer & Dajac Labs; Union Carbide

n-Butyl acrylate. *See* Butyl acrylate

Butyl alcohol
CAS 71-36-3; EINECS 200-751-6
Synonyms: n-Butyl alcohol; 1-Butanol; n-Butanol; Butan-1-ol; Propyl carbinol
Classification: Aliphatic alcohol
Empirical: $C_4H_{10}O$
Formula: $CH_3(CH_2)_2CH_2OH$
Properties: Colorless clear mobile liq., char. penetrating vinous odor; sol. in water; misc. with alcohol, ether, many org. solvs.; m.w. 74.14; dens. 0.8109 (20/20 C); m.p. -90 C; f.p. -89.0 C; b.p. 117.7 C; flash pt. 35 C; ref. index 1.3993 (20 C)
Precaution: DOT: Flamm. liq.; mod. explosive exposed to flame; incompat. with Al, oxidizing materials
Toxicology: TLV:CL 50 ppm in air; LD50 (oral, rat) 790 mg/kg; poison by IV route; mod. toxic by skin contact, ingestion, subcutaneous, IP routes; skin and severe eye irritant; may be narcotic in high doses; heated to decomp., emits acrid smoke and fumes
Uses: Preparation of esters, solv. for resins, plasticizers, paints; flavoring agent
Regulatory: FDA 21CFR §73.1, 172.515, 172.560, 175.105, 175.320, 176.200, 177.1200, 177.1440, 177.1650; 27CFR §21.99; FEMA GRAS; FDA approved for orals; USP/NF, BP compliance
Manuf./Distrib.: Aldrich; Allchem Ind.; Ashland; J.T. Baker; BASF; Baychem; BP Chems. Ltd; R.E. Carroll; Chemcentral; Chisso Am.; Coyne; Eastman; Fluka; Harcros; Hoechst Celanese; Hüls Am.; Penta Mfg.; Punda Mercantile; Shell; Sigma; Spectrum Chem. Mfg.; Sunnyside; Union Carbide; Van Waters & Rogers; Vista
Trade names: Isanol®
Trade names containing: Acryloid® A-21; Acryloid® A-21LV; Acryloid® AT-51; Ageflex NB-50; Albrite® PA-75; Ancamide 400-BX-60; Ancamide 700-B-75; Beckamine 21-500; Beckamine 21-511; Beckamine 27-556; Beckosol® 12-514; G-4732-VBT-50; G-4744-B-75; GP-197 Resin Sol'n.; GPRI™ 4000; GPRI™ 7550; GPRI™ 7557; GPRI™ 7570; GPRI™ 7590; GPRI™ 7597; GPRI™ BKS-2640; GPRI™ CKS-3892; Kelsol 3902-BG4-75; Nacorr® 1754; Nacure® 1557; Resimene® 750; Resimene® 881; Resimene® 901; Resimene® 918; Resimene® 920; Resimene® 1406; Resimene® 7512; Resimene® BM-5901; Resimene® CE-4514; Resimene® CE-6517; Slip-Ayd® SL-280; Slip-Ayd® SL-551; Surfynol® SM-745; Suspend-Ayd® 1; Suspend-Ayd® 2; Tint-Ayd® EP Series; Uformite® 21-806; Uformite® 27-803; Uformite® 27-804; Uformite® 27-805

2-Butyl alcohol. *See* 2-Butanol
n-Butyl alcohol. *See* Butyl alcohol
s-Butyl alcohol. *See* 2-Butanol
sec-Butyl alcohol. *See* 2-Butanol

t-Butyl alcohol
CAS 75-65-0; EINECS 200-889-7
Synonyms: TBA; 2-Methyl-2-propanol; Trimethyl carbinol; 2-Propanol, 2-methyl-
Empirical: $C_4H_{10}O$
Formula: $(CH_3)_3COH$
Properties: Colorless liq., unpleasant odor; m.w. 74.12; dens. 0.775; m.p. 25-26 C; b.p. 83 C; flash pt. 11 C; ref. index 1.3870 (20 C)
Precaution: DOT: Flamm. liq.

n-Butylaldehyde

Toxicology: Irritant to mucous membranes; inh. of 25 ppm causes pulmonary problems in man; ingestion can cause headache, dizziness, drowsiness; skin contact can cause contact dermatitis
Uses: Solvent, coupling agent, processing aid for pharmaceuticals, personal care prods., aq. coatings and adhesives, agric. formulations, polymer processing, cleaners/disinfectants
Regulatory: FDA 21CFR §176.200, 178.3910, 27CFR §21.100
Manuf./Distrib.: AC Ind.; Aldrich; Allchem Ind.; ARCO; Ashland; Fluka; Hüls Am.; Sigma; Spectrum Chem. Mfg.
Trade names: Tebol™; Tebol™ 99

n-Butylaldehyde. *See* n-Butyraldehyde

t-Butylaminoethyl methacrylate
CAS 3775-90-4
Classification: Amine monomer
Formula: $CH_2=C(CH_3)COOCH_2CH_2NHC(CH_3)_3$
Properties: Liq.; m.w. 185.27; dens. 0.914 (25 C); b.p. 100-105 C (12 mm); flash pt. (COC) 96.1 C
Precaution: Combustible; reactive monomer
Toxicology: Severe skin and eye irritant; harmful if swallowed
Uses: Automotive dip tanks, coating applics., industrial/consumer adhesives, dye and lube oil additives, intermediate for water treatment chemicals, textile chemicals, dispersant, antistat, stabilizer, ion exchange resins, emulsifier, precipitating agent
Manuf./Distrib.: Aldrich; CPS; Monomer-Polymer & Dajac Labs; Rohm Tech
Trade names: Ageflex FM-4
Trade names containing: Sipomer® TBM

Butylated hydroxyanisole. *See* BHA
Butylated hydroxytoluene. *See* BHT
Butylated polyvinylpyrrolidone. *See* Butylated PVP

Butylated PVP
Synonyms: Butylated polyvinylpyrrolidone
Definition: Polymer of butylated vinylpyrrolidone
Uses: Moisture barrier, adhesive, protective colloid, and microencapsulating resin in cosmetics; pigment dispersant; as solubilizer for dyes; in petroleum industry as sludge and detergent dispersant; protective colloid in coatings; suspending aid in polymerization; dyeing assistant; antiredeposition agent in drycleaning; esp. as dispersant in aq. agric. chemicals or pigmented skin care prods.
Trade names: Ganex® P-904

N,N-Butyl benzene sulfonamide
CAS 3622-84-2; EINECS 222-823-6
Synonyms: Benzenesulfonic acid butyl amide
Empirical: $C_{20}H_{19}NO_2S$
Formula: $C_6H_5SO_2NHC_3H_9$
Properties: Amber to straw liq., pleasant odor; m.w. 337.46; dens. 1.148; b.p. 189-190 C (4.5 mm); ref. index 1.5235
Toxicology: Toxic by ingestion
Uses: Plasticizer for some cellulose compds., emulsion adhesives, pkg., caulk, printing ink, surf. coatings; synthesis of dyes, pharmaceuticals, other organic chemicals; in resin mfg.
Manuf./Distrib.: Advance Coatings; Aldrich; Allchem Ind.; Nipa Hardwicke; Unitex
Trade names: Dellatol® BBS; Plasthall® BSA

Butyl benzyl phthalate
CAS 85-68-7; EINECS 201-622-7
Synonyms: BBP; 1,2-Benzenedicarboxylic acid, butyl phenylmethyl ester; Benzyl butyl phthalate
Classification: Aromatic ester
Formula: $C_4H_9OOCC_6H_4COOC_7H_7$
Properties: Clear oily liq., sl. odor; m.w. 312.39; dens. 1.116 (25/25 C); m.p. < -35 C; b.p. 370 C; flash pt. 390 F; ref. index 1.535-1.540
Toxicology: LD50 (oral, rat) 20,400 mg/kg; moderate toxicity by ingestion, intraperitoneal routes
Uses: High-solvating plasticizer for polyvinyl and cellulosic resins incl. NC lacquers and films, PVAc, acrylic coatings, chlorinated rubber, cellulose propionate, polyamide; organic intermediate
Regulatory: FDA 21CFR §175.105, 176.170, 176.180, 177.2420, 178.3740
Manuf./Distrib.: Aldrich; Allchem Ind.; Ashland; Bayer; Monsanto

Trade names: Santicizer 160; Unimoll® BB
Trade names containing: Vinyzene® BP-5-5 160

n-Butyl carbinol. *See* Amyl alcohol
Butyl 'Carbitol'. *See* Butoxydiglycol
Butyl carbobutoxymethyl phthalate. *See* n-Butyl phthalyl-n-butyl glycolate
2-t-Butyl-6-(5-chloro-2H-benzotriazol-2-yl)-p-cresol. *See* Bumetrizole
Butyl citrate. *See* Tributyl citrate

Butyl cyclohexyl phthalate
CAS 84-64-0; EINECS 201-548-5
Synonyms: 1,2-Benzenedicarboxylic acid, butyl cyclohexyl ester
Classification: Ester
Formula: $C_4H_9OOCC_6H_4COOC_6H_{11}$
Properties: Clear liq., very mild odor; misc. with most org. solvs.; m.w. 304.39; dens. 1.078
Precaution: Combustible
Uses: Plasticizer for PVC, plastisols, automotive undercoatings, plastics, flooring, processing aid for molded ethylcellulose prods.
Manuf./Distrib.: CPS; Hüls; Unitex

Butyldiglycol. *See* Butoxydiglycol
Butyldiglycol acetate. *See* Diethylene glycol butyl ether acetate
1,3-Butylene diacrylate. *See* 1,3-Butylene glycol diacrylate
1,4-Butylene glycol. *See* 1,4-Butanediol

1,3-Butylene glycol diacrylate
CAS 19485-03-1
Synonyms: Acrylic acid-1-methyltrimethylene ester; 1,3-Butylene diacrylate; 2-Propenoic acid-1-methyl-13-propanediyl ester; 1,3-Butanediol diacrylate
Empirical: $C_{10}H_{14}O_4$
Properties: M.w. 198.24
Toxicology: LD50 (oral, rat) 3540 mg/kg, (skin, rabbit) 450 mg/kg; mod. toxic by ingestion and skin contact; heated to decomp., emits acrid smoke and irritating fumes
Uses: Curing agent; polymerizes to hard, insol., infusible, thermoset resin
Manuf./Distrib.: Aldrich; Monomer-Polymer & Dajac Labs
Trade names containing: SR-212

1,3-Butylene glycol dimethacrylate
CAS 11890-98-8
Properties: M.w. 226.26
Uses: Crosslinker for plastisols, hard rubber rolls, cast acrylic sheet and rods, coagent for rubber compding., impregnant for metal and wood composites, adhesives, and glass-reinforced plastics
Manuf./Distrib.: CPS; Monomer-Polymer & Dajac Labs
Trade names containing: SR-297

Butylene hydrate. *See* 2-Butanol
Butylene oxide. *See* Tetrahydrofuran
Butyl ester of PVM/MA copolymer (INCI). *See* PVM/MA copolymer, butyl ester
Butyl ether. *See* Dibutyl ether
Butylethylacetic acid. *See* 2-Ethylhexoic acid
Butyl ethylhexyl phthalate. *See* Butyl octyl phthalate

Butyl glycidyl ether
CAS 2426-08-6; EINECS 219-376-4
Synonyms: BGE; Glycidyl butyl ether; Butoxymethyl oxirane
Classification: Glycidyl ether monomer
Empirical: $C_7H_{14}O_2$
Properties: M.w. 130.19; dens. 0.910; b.p. 164-166 C; flash pt. 55 C
Toxicology: TLV:TWA 25 ppm; moderately toxic by ingestion, skin contact, intraperitoneal route; mildly toxic by inhalation
Uses: Reactive diluent in epoxy resins, laminating, flooring, elec. casting and encapsulants and highly filled coatings
Manuf./Distrib.: Aldrich; Alemark; Amber Syn.; Ashland; CPS; Fluka; Monomer-Polymer & Dajac Labs; Sigma

Butyl glycol

Trade names: Epodil® 741; Heloxy® 61
Trade names containing: EPON® Resin 815; Epotuf Resin 37-130

Butyl glycol. *See* Butoxyethanol

t-Butyl hydroperoxide
CAS 75-91-2; EINECS 200-915-7
Synonyms: TBHP; 1,1-Dimethylethylhydroperoxide; 2-Hydroperoxy-2-methylpropane
Classification: Organic peroxide
Empirical: $C_4H_{10}O_2$
Formula: $(CH_3)_3COOH$
Properties: Water-white liq.; sl. sol. in water; sol. in org. solvs.; very sol. in esters and alcohols; m.w. 90.12; dens. 0.896 (20/4 C); m.p. -8 C; b.p. 35 C; flash pt. 62 C; ref. index 1.4007
Precaution: Flamm. oxidizing liq.; very dangerous fire hazard when exposed to heat or flame or by spontaneous chem. reaction; mod. explosive; heated to decomp., emits acrid smoke and fumes
Toxicology: LD50 (oral, rat) 560 mg/kg, (dermal, rat) 790 mg/kg; poison by ingestion and inhalation; severe skin and eye irritant; mutagenic data; may cause severe depression, cyanosis, death at highest dosage levels
Uses: Catalyst in polymerization reactions; oxidative membrane in RBC suspensions; introduces peroxy group into organic molecules; raw material for paints
Manuf./Distrib.: Akzo Nobel; Aldrich; Allchem Ind.; ARCO; Elf Atochem N. Am.; Fluka; SAF Bulk Chems.; Sigma; Witco/Polymer Addit.
Trade names: T-Hydro® Sol'n.

3-t-Butyl-4-hydroxyanisole. *See* BHA
2-(3′-t-Butyl-2′-hydroxy-5′-methylphenyl)-5-chlorobenzotriazole. *See* Bumetrizole
Butyl 2-hydroxypropanoate. *See* Butyl lactate
Butyl α-hydroxypropionate. *See* Butyl lactate
n-Butyl-S(-)-2-hydroxypropionate. *See* Butyl lactate

4,4′-Butylidenebis (6-t-butyl-m-cresol)
CAS 85-60-9
Synonyms: 4,4′-Butylidenebis (3-methyl-6-t-butylphenol)
Formula: $C_{26}H_{38}O_2$
Properties: Wh. powd.; m.w. 382.64; m.p. 209 C (min.)
Toxicology: Mildly toxic by ingestion
Uses: Antioxidant for NR, synthetics, latexes; polymer stabilizer
Trade names: Santowhite® Powd.

4,4′-Butylidenebis (3-methyl-6-t-butylphenol). *See* 4,4′-Butylidenebis (6-t-butyl-m-cresol)
Butyl-3-iodo-2-propynylcarbamate. *See* Iodopropynyl butylcarbamate

Butyl lactate
CAS 138-22-7
Synonyms: n-Butyl lactate; n-Butyl-S(-)-2-hydroxypropionate; Butyl 2-hydroxypropanoate; Butyl α-hydroxypropionate
Classification: Ester
Empirical: $C_7H_{14}O_3$
Formula: $CH_3CHOHCOOC_4H_9$
Properties: Water-wh. liq., mild odor; misc. with many lacquer solvs., diluents, oils; sl. sol. in water; hydrolyzed in acids and alkalies; m.w. 146.19; dens. 0.974-0.984 (20/20 C); m.p. -43 C; flash pt. (TOC) 75.5 C
Precaution: Combustible
Toxicology: TLV 5 ppm in air
Uses: Dipped latex prods., coatings, specialty finishes, inks, diluents, adhesives, intermediates, photoresist solvs., screen printing of electronic parts, flavors and fragrances, solvs. for nitrocellulose, antiskinning agent, dry cleaning fluids
Regulatory: FDA 21CFR §172.515; FEMA GRAS
Manuf./Distrib.: Aldrich; CPS; Penta Mfg.; Purac Biochem; Van Waters & Rogers
Trade names: Purasolv® BL

n-Butyl lactate. *See* Butyl lactate

Butyl methacrylate
CAS 97-88-1; EINECS 202-615-1
Synonyms: Butyl 2-methyl-2-propenoate
Definition: Ester of n-butyl alcohol and methacrylic acid
Empirical: $C_8H_{14}O_2$
Formula: $CH2:C(CH_3)COO(CH_2)_3CH_3$
Properties: Colorless liq., ester odor; m.w. 142.22; dens. 0.895 (20/4 C); b.p. 163 C; flash pt. (TOC) 126 F
Toxicology: Poison by intraperitoneal route; mildly toxic by ingestion, inhalation, skin contact; skin irritant
Uses: Monomer for resins, solvent coatings, adhesives, oil additives; emulsions for textiles, leather, paper finishing
Regulatory: FDA 21CFR §175.300, 177.2420
Manuf./Distrib.: Aldrich; Degussa; Fluka; ICI Spec. Chems.; Janssen Chimica; Mitsubishi Gas; Monomer-Polymer & Dajac Labs; Rohm & Haas; Rohm Tech; Sigma

t-Butyl-4-methoxyphenol. *See* BHA
t-Butyl methyl ether. *See* Methyl t-butyl ether
Butyl 2-methyl-2-propenoate. *See* Butyl methacrylate
Butyl octadecanoate. *See* Butyl stearate
n-Butyl octadecanoate. *See* Butyl stearate

Butyl octyl phthalate
CAS 84-78-6; EINECS 201-562-1
Synonyms: BOP; 1,2-Benzenedicarboxylic acid, butyl octyl ester; Butyl ethylhexyl phthalate; Phthalic acid, butyl octyl ester
Empirical: $C_{20}H_{30}O_4$
Formula: $C_4H_9OOCC_6H_4COOC_8H_{17}$
Properties: Water-wh. liq., mild char. odor; m.w. 334.46; dens. 0.988-0.996 (20/20 C); f.p. -48 C; b.p. 350 C (760 mm); sapon. no. 298-308; flash pt. (COC) 208 C; ref. index 1.4868
Precaution: Combustible
Uses: Plasticizer for PVC, PS, PVB, PVAc, cellulosics, and rotational molding and dip coating
Manuf./Distrib.: Aristech; Ashland; BASF; BP Oil; Eastman; Harwick; Unitex
Trade names: Kodaflex® HS-3

t-Butyl peroxide. *See* Di-t-butyl peroxide

4-t-Butylphenol
CAS 98-54-4; EINECS 202-679-0
Synonyms: PTBP; p-t-Butylphenol
Empirical: $C_{10}H_{14}O$
Formula: $(CH_3)_3CC_6H_4OH$
Properties: Wh. cryst.; sol. in alcohol, ether; insol. in water; m.w. 150.24; dens. 1.03; f.p. -7 C; m.p. 98-99 C; b.p. 239 C; flash pt. (OC) 230 F
Precaution: Combustible
Toxicology: Irritant to skin and eyes
Uses: Plasticizer for cellulose acetate; intermediate for antioxidants, starches, phenolic resins, paints; pour pt. depressor and emulsion breaker for petrol. oils, some plastics; syn. lubricants; insecticides; motor-oil additives
Manuf./Distrib.: Aceto; AC Ind.; Aldrich; Fabrichem; Fluka; Hüls Am.; Janssen Chimica; Jarchem Ind.; PMC Specialties; Punda Mercantile; Schenectady

p-t-Butylphenol. *See* 4-t-Butylphenol
4-t-Butylphenyl-2,3-epoxypropyl ether. *See* p-t-Butyl phenyl glycidyl ether

p-t-Butyl phenyl glycidyl ether
CAS 3101-60-8
Synonyms: 4-t-Butylphenyl-2,3-epoxypropyl ether
Properties: M.w. 206.29; dens. 1.038; b.p. 165-170 (14 mm); flash pt. 101 C
Uses: Reactive epoxy diluent for tooling, elec. applics., flooring, casting, laminating
Manuf./Distrib.: Aldrich
Trade names containing: Epotuf Resin 37-128; Epotuf Resin 37-135

Butyl phosphate
Properties: Liq.; sl. sol. in water; sp.gr. 1.13; acid no. 430; flash pt. (TCC) 265 F

Butyl phosphoric acid

Uses: Acid catalyst in resin curing, chemical intermediate in formulation of rust preventatives, antistats, textile lubricants, oil additives, heavy metal extractants
Manuf./Distrib.: Albright & Wilson
Trade names: Findet DD

Butyl phosphoric acid. *See* Butyl acid phosphate
Butyl phthalate. *See* Dibutyl phthalate

n-Butyl phthalyl-n-butyl glycolate
CAS 85-70-1; EINECS 201-624-8
Synonyms: 1,2-Benzenedicarboxylic acid, 2-butoxy-2-oxoethyl, butyl ester; Butyl carbobutoxymethyl phthalate; Dibutyl-o-carboxybenzoyloxyacetate; Dibutyl-o-(o-carboxybenzoyl) glycolate
Classification: Aromatic ester
Empirical: $C_{18}H_{24}O_6$
Formula: $C_4H_9OOCC_6H_4COOCH_2COOC_4H_9$
Properties: Colorless liq., odorless; insol. in water; m.w. 336.42; dens. 1.093-1.103 (25/25 C); b.p. 219 C (5 mm); flash pt. 390 F; ref. index 1.4880; extremely light-stable
Precaution: Combustible
Toxicology: LD50 (oral, rat) 7 g/kg; mildly toxic by IP; experimental teratogen, reproductive effects; mutagenic data; eye irritant; heated to decomp., emits acrid and irritating fumes
Uses: Plasticizer for PVC
Regulatory: FDA 21CFR §175.105, 175.300, 175.320, 181.22, 181.27
Manuf./Distrib.: Aldrich; Fabrichem; Morflex
Trade names containing: Morflex® 190

Butyl-2-propenoate. *See* Butyl acrylate
Butyl propionate. *See* n-Butyl propionate

n-Butyl propionate
CAS 590-01-2
Synonyms: Propanoic acid butyl ester; Butyl propionate
Empirical: $C_7H_{14}O_2$
Formula: $CH_3CH_2COOC_4H_9$
Properties: Colorless liq., earthy faintly sweet odor, apricot-like taste; very sol. in alcohol, ether; very sl. sol. in water; m.w. 130.19; dens. 0.8754 (20/4 C); m.p. -89 C; f.p. 32 C; b.p. 145-146 C; ref. index 1.401 (20 C)
Precaution: Flamm.
Toxicology: Skin and eye irritant
Uses: Solvent for nitrocellulose, paints, retarder in lacquer thinner, ingredient of perfumes, flavors
Regulatory: FDA 21CFR §172.515; FEMA GRAS
Manuf./Distrib.: Aldrich; Penta Mfg.; Union Carbide
Trade names: Ucar® n-Butyl Propionate

Butyl rubber. *See* Isobutylene/isoprene copolymer

Butyl stearate
CAS 123-95-5; EINECS 204-666-5
Synonyms: Butyl octadecanoate; n-Butyl octadecanoate; Octadecanoic acid butyl ester
Definition: Ester of butyl alcohol and stearic acid
Empirical: $C_{22}H_{44}O_2$
Formula: $CH_3(CH_2)_{16}COO(CH_2)_3CH_3$
Properties: Crystals; sl. sol. in water; sol. in alcohol, ether; m.w. 340.60; dens. 0.86 (20/4 C); m.p. 17-22 C; b.p. 343 C; flash pt. (CC) 160 C; ref. index 1.4430 (20 C)
Toxicology: No known toxicity; heated to decomp., emits acrid smoke and irritating fumes
Uses: Solv., spreading and softening agent in plastics, textiles, cosmetics, rubbers; flavoring agent; plasticizer for food pkg. materials; lubricant and slip agent for coatings and inks
Regulatory: FDA 21CFR §172.515, 173.340, 175.105, 175.300, 175.320, 176.200, 176.210, 177.2600, 177.2800, 178.3910, 181.22, 181.27; FDA approved for topicals; FEMA GRAS
Manuf./Distrib.: Aldrich; Amerchol; Anar; Aquatec Quimica SA; Ashland; Fluka; C.P. Hall; Henkel/Organic Prods.; Inolex; Kenrich Petrochems.; Mosselman NV; Penta Mfg.; Sea-Land; Sigma; Spectrum Chem. Mfg.; Stepan; Union Camp; Unitex; Witco/Oleo-Surf.
Trade names: Lexolube® NBS

Butyltin mercaptide
Uses: Heat stabilizer for PVC; catalyst for polyurethane coatings
Trade names: Dabco® T-120; Okstan M 69 S

Butyl titanate. *See* Tetrabutyl titanate
Butyral. *See* n-Butyraldehyde

n-Butyraldehyde
CAS 123-72-8; EINECS 204-646-6
Synonyms: Butal; Butaldehyde; Butalyde; Butanal; n-Butanal; n-Butylaldehyde; Butyric aldehyde; Butyral; Butyric aldehyde
Classification: Aldehyde
Empirical: C_4H_8O
Formula: $CH_3CH_2CH_2CHO$
Properties: Colorless liq.; sol. in water; misc. with ether @ 75 C; m.w. 72.10; sp.gr. 0.800; m.p. -99 C; f.p. 12 F; b.p. 75-76 C; flash pt. (CC) 20 F; ref. index 1.3843
Precaution: DOT: Flamm. liq.; dangerous fire hazard when exposed to heat and flame; incompat. with oxidizing materials; reacts vigorously with chlorosulfonic acid
Toxicology: LD50 (oral, rat) 2490 mg/kg; mod. toxic by ingestion, inh., skin contact, intraperitoneal, and subcutaneous routes; severe skin and eye irritant; suspected carcinogen; heated to decomp., emits acrid smoke and fumes
Uses: Plasticizers, rubber accelerators, solvents, high polymers; intermediate for paints; flavoring agent
Regulatory: FDA 21CFR §172.515; FEMA GRAS
Manuf./Distrib.: Aldrich; BASF AG; Chisso Am.; Eastman; Fluka; Hoechst Celanese; Neste UK; Penta Mfg.; Union Carbide; Van Waters & Rogers

Butyraldehyde oxime. *See* m-Butyraldehyde oxime

m-Butyraldehyde oxime
CAS 110-69-0; EINECS 203-792-8
Synonyms: Butyraldoxime; n-Butyraldoxime; Butanal oxime; Butyraldehyde oxime
Empirical: C_4H_9NO
Properties: Liq.; m.w. 87.14; dens. 0.923; m.p. -29.5 C; b.p. 152 C; flash pt. (CC) 136 F
Precaution: Flamm. when exposed to heat or flame; highly explosive; incompat. with oxidizing materials, metallic impurities
Toxicology: Poison by intraperitoneal route; heated to decomp., emits toxic fumes of NO_x
Uses: Antiskinning agent used in coating vehicles
Manuf./Distrib.: AlliedSignal; OM Group

Butyraldoxime. *See* m-Butyraldehyde oxime
n-Butyraldoxime. *See* m-Butyraldehyde oxime
Butyric aldehyde. *See* n-Butyraldehyde
Butyrolactam. *See* 2-Pyrrolidone

Butyrolactone
CAS 96-48-0; EINECS 202-509-5
Synonyms: BLO; Dihydro-2(3H)-furanone; 4-Hydroxybutanoic acid lactone; γ-Butyrolactone; 4-Butanolide
Classification: Lactone
Empirical: $C_4H_6O_2$
Properties: Colorless oily liq., mild caramel odor; misc. with water; sol. in methanol, ethanol, acetone, ether, benzene; m.w. 86.09; dens. 1.120; m.p. -45 C; b.p. 204-205 C; flash pt. 98 C; ref. index 1.4348
Precaution: Combustible when exposed to heat or flame; reactive with oxidizing materials
Toxicology: LD50 (oral, rat) 1800 mg/kg; mod. toxic by ingestion, IV, intraperitoneal routes; suspected tumorigen; heated to decomp., emits acrid and irritating fumes
Uses: Intermediate for synthesis of polyvinylpyrrolidone, piperidine, phenylbutyric acid, etc.; solv. for polyacrylonitrile, cellulose acetate, methyl methacrylate polymers, polystyrene, paints; constituent of paint removers, textile aids, drilling oils; flavoring agent for candy, soy milk; solv. for PAN, PS, fluorinated hydrocarbons, cellulose triacetate, shellac; in paint removers, petrol. processing, inks
Regulatory: FEMA GRAS
Manuf./Distrib.: Aldrich; Allchem Ind.; BASF; Fabrichem; Fluka; Gelest; Great Western; ISP; Janssen Chimica; Schweizerhall; Sigma; Spectrum Chem. Mfg.; UCB SA
Trade names: Agrisynth® BLO; Agsol Ex BLO; BLO®; Dynasolve 699

γ-Butyrolactone. *See* Butyrolactone

C. *See* Carbon
CA. *See* Cellulose acetate
CAB. *See* Cellulose acetate butyrate

C7 acid triglyceride
Uses: Lubricant for machinery for confectionery and food industries; mold release agent in processing of plastics; additive to cutting oils, lacquer and varnish systems; antiblocking agent for artificial skins

C14 acid triglyceride
Uses: Compressing aid in tableting technical substances; hydrophobing agent against moisture; lubricant for machinery, lacquer and varnish formulations; additive to liq. lubricating systems
Trade names: Softenol® 3114

C18 acid triglyceride
Definition: Triester of glycerin and C18-36 acid
Formula: $CH_2OCO(CH_2)_xCH_3{\cdot}HCOCO(CH_2)_xCH_3{\cdot}CH_2OCO(CH_2)_xCH_3$, avg. x = 16-34
Uses: Compressing aid in tableting technical substances; hydrophobing agent against moisture; lubricant for machinery, lacquer and varnish formulations; additive to liq. lubricating systems
Trade names: Softenol® 3118

Cadmium
CAS 7440-43-9; EINECS 231-152-8
Formula: Cd
Properties: Blue-white, soft hexagonal crystals, malleable metal or grayish-white powd.; sol. in acids; at.wt.112.41; dens. 8.642; m.p. 320.9 C; b.p. 767 C; ref. index 1.13
Precaution: Flamm. in powd. form
Toxicology: LD50 (oral, rat) 225 mg/kg; TLV 0.05 mg/m^3 in air; TLV:TWA 0.01mg (Cd)/m^3 (dust); highly toxic; poison by inhalation; probable human carcinogen; can cause blood, kidney, calcium disorders, bone problems
Uses: In alloys; electrodeposited and dipped coatings on metals; pigment in paints
Manuf./Distrib.: Aldrich; Asarco; Atlantic Equip. Engrs.; Atomergic Chemetals; Belmont Metals; Fluka; Pasminco Europe; Spectrum Chem. Mfg.; Zinc Corp. of Am.

Cadmium naphthenate
Uses: Drier for paints
Manuf./Distrib.: Archway Sales; Baychem; Lomas Int'l.

Cajeputene. *See* dl-Limonene
Calcined alumina. *See* Alumina
Calcined litharge. *See* Lead (II) oxide
Calcined magnesia. *See* Magnesium oxide
Calcium acetylide. *See* Calcium carbide
Calcium alkylbenzenesulfonate. *See* Calcium dodecylbenzene sulfonate

Calcium carbide
CAS 75-20-7; EINECS 200-848-3
Synonyms: Calcium acetylide
Formula: CaC_2
Properties: Gray rhombic cryst., garlic-like odor; m.w. 64.10; dens. 2.222; m.p. ≈ 2300 C
Precaution: Flamm. solid; dangerous when wet; contact with moisture yields explosive acetylene gas; corrosive
Uses: Generation of acetylene gas for welding, chloroethylenes, vinyl acetate monomer, acetylene chemicals, reducing agent; mfg. of graphite; extender for paints
Manuf./Distrib.: Aldrich; Elkem Metals; San Yuan; Sigma; Spectrum Chem. Mfg.

Calcium carbonate
CAS 471-34-1; 1317-65-3; EINECS 207-439-9
Synonyms: Carbonic acid calcium salt (1:1); Calcium carbonate (1:1); CI 77220; Precipitated calcium carbonate; Precipitated chalk

Classification: Inorganic salt
Empirical: $CaCO_3$
Properties: White powd. or colorless crystals, odorless, tasteless; hygroscopic; very sl. sol. in water; sol. in acids with CO_2; insol. in alcohol; m.w. 100.09; dens. 2.7-2.95; m.p. 825 C (dec.); stable in air
Precaution: Ignites on contact with F_2; incompat. with acids, alum, ammonium salts
Toxicology: TLV 5 mg/m^3 of air; LD50 (oral, rat) 6450 mg/kg; no known toxicity; severe eye and moderate skin irritant
Uses: Filler; alkali; source of lime; neutralizing agent; opacifying agent in paper; putty; tooth powds.; antacid; whitewash; portland cement; paint; rubber; plastics; insecticides; in chemical analysis
Regulatory: FDA 21CFR §73.1070, 137.105, 137.155, 137.160, 137.165, 137.170, 137.175, 137.180, 137.185, 137.350, 169.115, 175.300, 177.1460, 181.22, 181.29, 182.5191, 184.1191, 184.1409, GRAS; BATF 27CFR §240.1051, limitation 30 lb/1000 gal of wine; Japan approved (1-2%); Europe listed; UK approved; FDA approved for implants, orals, otics; USP/NF, BP, Ph.Eur. compliance
Manuf./Distrib.: 20 Microns Ltd.; Aldrich; AluChem; Am. Ingreds.; Archway Sales; J.T. Baker; BASF; R.E. Carroll; Cerac; ECC Int'l.; EM Ind.; Engelhard; Fluka; Genstar Stone Prods.; Georgia Marble; C.P. Hall; J.M. Huber; Landers-Segal Color; Lohmann; Mallinckrodt; Nichia Kagaku Kogyo; Pfizer Int'l.; Punda Mercantile; H.M. Royal; Seegott; Sigma; Tamms Ind.; Whittaker, Clark & Daniels; Jesse S. Young; Zeneca Resins
Trade names: Albacar® 5970; Albafil®; Albaglos®; Amical® SC3; Amical® SC7; Atomite®; Calwhite®; Calwhite® II; Camel-CAL®; Camel-CAL® ST; Camel-CAL® Slurry; Camel-CARB®; Camel-TEX®; Camel-WITE®; Camel-WITE® Slurry; Camel-WITE® ST; Carbital® 35; Carbital® 50; Carbital® 75; Carbital® 90; Carbokup®; CC™-103; Drikalite®; Duramite®; Gamaco®; Gamaco® II; Gama-Fil® D-2; Gama-Plas®; Gama-Sperse® 5; Gama-Sperse® 80; Gama-Sperse® 140; Gama-Sperse® 255; Gama-Sperse® 6451; Gama-Sperse® 6532; G-White; Hallcote® 573; HO-M60 Milled; HO-M60 Stabilized; Hubercarb® Optifil; Hubercarb® Optifil T; Hubercarb® Q 1; Hubercarb® Q 2; Hubercarb® Q 3; Hubercarb® Q 4; Hubercarb® Q 6; Hubercarb® Q 6-20; Hubercarb® Q 20-60; Hubercarb® Q 40-200; Hubercarb® Q 60; Hubercarb® Q 100; Hubercarb® Q 200; Hubercarb® Q 325; Hubercarb® W 2; Hubercarb® W 3; Hubercarb® W3N; Hubercarb® W 4; H-White; Macromite; Magnum Gloss; Micro-White® 07 Slurry; Micro-White® 10; Micro-White® 10 Slurry; Micro-White® 15; Micro-White® 15 Slurry; Micro-White® 25; Micro-White® 25 Slurry; Micro-White® 40; Micro-White® 100; Mikhart® 1; Mikhart® 1T; Multifex-MM®; No. 1 White™; No. 10 White; No. 12 White; Opacimite™; SC-53; Snowflake White™; Supermite®; Super-Pflex® 100; Troykyd® Anti-Float Powd.; Troyso™ AFP; Ultramite; Ultra-Pflex®; Winnofil S, SP; XO White
Trade names containing: Dualite® M6001AE; Hubercarb® Q 1T; Hubercarb® Q 2T; Hubercarb® Q 3T; Hubercarb® Q 200T; Hubercarb® W 2T; Hubercarb® W 3T
See also Limestone

Calcium carbonate (1:1). *See* Calcium carbonate
Calcium carbonate, natural. *See* Limestone

Calcium chromate
Uses: Rust-inhibitive pigment for paints
Manuf./Distrib.: Atomergic Chemetals; Engelhard; Nat'l. Chem.; Noah; Van Waters & Rogers

Calcium difluoride. *See* Calcium fluoride

Calcium dinonylnaphthalene sulfonate
CAS 57855-77-3
Synonyms: Dinonylnaphthalene sulfonic acid, calcium salt
Empirical: $C_{28}H_{44}O_3S \cdot {}^1/_2Ca$
Properties: M.w. 480.80
Uses: Rust inhibitor; demulsifier in oils; grease penetrant
Trade names containing: Nacorr® 1352

Calcium dodecylbenzene sulfonate
CAS 26264-06-2; 68953-96-8; EINECS 247-557-8
Synonyms: Dodecylbenzene sulfonic acid calcium salt; Calcium alkylbenzenesulfonate
Empirical: $C_{18}H_{30}O_3S \cdot {}^1/_2Ca$
Properties: M.w. 346.54
Uses: Emulsifier for herbicides/pesticides, polymerization, industrial use; dispersant for textiles, coatings, leather applics.
Trade names: Nansa® EVM50; Rhodacal® 70/B; Rhodacal® CA
Trade names containing: Nansa® EVM70/B; Nansa® EVM70/E

Calcium drier
Uses: Drier used in org. coatings, inks, polyesters
Trade names: Troymax™ Drier Calcium 4%; Troymax™ Drier Calcium 5%; Troymax™ Drier Calcium 6%; Troymax™ Drier Calcium 8%; Troymax™ Drier Calcium 10%

Calcium fluoride
CAS 7789-75-5; EINECS 232-188-7
Synonyms: Calcium difluoride; Fluorite; Fluorspar
Empirical: CaF_2
Properties: Color or wh. cryst. powd.; hygroscopic; luminous when heated; insol. in water; sol. in ammonium salts; m.w. 78.08; dens. 3.180; m.p. 1360 C; b.p. ≈ 2500 C
Toxicology: LD50 (oral, rat) 4250 mg/kg; mildly toxic by ingestion; irritant
Uses: Source of fluorine, flux, ceramics, phosphors, paint pigment, catalyst in wood preservative, spectroscopy, electronics, lasers, high-temperature dry-film lubricants
Manuf./Distrib.: Aldrich; Cerac; Fluka; GE; Noah; Sigma; Solvay GmbH; Spectrum Chem. Mfg.

Calcium hydrogen orthophosphate. *See* Calcium phosphate dibasic
Calcium hydrogen phosphate. *See* Calcium phosphate dibasic
Calcium hydrogen phosphate anhydrous. *See* Calcium phosphate dibasic

Calcium linoleate
Formula: $Ca(C_{18}H_{31}O_2)_2$
Properties: Wh. amorphous powd.; sol. in alcohol; insol. in water; m.w. 598.56
Uses: Drier for paints
Manuf./Distrib.: Akzo Nobel; Troy

Calcium metaborate
Uses: Rust-inhibitive pigment for paints
Manuf./Distrib.: Buckman Labs
Trade names: Butrol® 9100

Calcium metasilicate
CAS 1344-95-2
Empirical: $CaSiO_3$
Properties: Wh. powd.; insol. in water; dens. 2.9
Toxicology: TLV: 10 mg/m^3; irritating dust
Uses: Extender for paints; absorbent, antacid, filler for paper, paper coatings, cosmetics; anticaking additive for foods; mfg. of glass and Portland cement
Manuf./Distrib.: Am. Colloid; Nyco Mins.; R.T. Vanderbilt
See also Calcium silicate; Wollastonite

Calcium monohydrogen phosphate. *See* Calcium phosphate dibasic

Calcium naphthalene sulfonate
Uses: Dispersant for pigments; stabilizer for dispersions and emulsions; plasticizer for concrete, mortar, and gypsum
Trade names: Tamol® NHC 3001; Tamol® NHC 9301

Calcium naphthenate
Definition: Calcium deriv. of cycloparaffin hydrocarbon
Properties: Light sticky tenacious mass; sol. in CCl_4, gasoline, benzene, ether; insol. in water
Precaution: Combustible
Uses: Drier for paints; adhesives; waterproofing; wood filler; varnishes; hardening agent in plastic compds.
Manuf./Distrib.: Akzo Nobel; Archway Sales; Baychem; Boehle; Dussek Campbell Ltd; Hüls Am.; Lomas Int'l.; D.N. Lukens; OM Group; Troy
Trade names: Nuodex Napthenate® Calcium 4%; Nuodex Napthenate® Calcium 5%

Calcium octadecanoate. *See* Calcium stearate

Calcium octoate
Uses: Drier for paints

Manuf./Distrib.: Aceto; Archway Sales; Baychem; Boehle; Dussek Campbell Ltd; Hüls Am.; Lomas Int'l.; D.N. Lukens; OM Group; Troy; Witco/Oleo-Surf.

Calcium oxide

CAS 1305-78-8; EINECS 215-138-9

Synonyms: Lime; Unslaked lime; Fluxing lime; Quicklime; Calx; Burnt lime

Classification: Inorganic oxide

Empirical: CaO

Properties: Wh. or gray cryst. or powd., odorless; sol. in acids, glycerol, sugar sol'n.; sol. in water forming $Ca(OH)_2$ and generating heat; pract. insol. in alcohol; m.w. 56.08; dens. 3.40; m.p. 2570 C; b.p. 2850 C

Precaution: Noncombustible; powd. may react explosively with water; mixts. with ethanol may ignite if heated

Toxicology: TLV 2 mg/m^3; strong caustic; may cause severe irritation to skin, mucous membranes, eyes

Uses: Refractory, sewage treatment, insecticides, fungicides, mfg. of steel and aluminum; flotation of nonferrous ores; mfg. of glass, paper, Ca salts; in drilling fluids, lubricants; laboratory; desiccant for extruded rubber goods; alkali; nutrient, dietary supplement, dough conditioner, yeast food, hog/poultry scald agent

Regulatory: FDA 21CFR §182.1210, 182.5210, 184.1210, GRAS; USDA 9CFR §318.7, 381.147; Europe listed; UK approved

Manuf./Distrib.: Aldrich; AluChem; Ash Grove Cement; Atlantic Equip. Engrs.; Cerac; Fluka; GE; Hüls Am.; Mallinckrodt; Mins. Tech.; Mississippi Lime; Pfizer Int'l.; Reade Advanced Materials; Sigma; Specialty Mins.; Smith Lime Flour; U.S. Gypsum

Trade names containing: BYK®-2615

Calcium phosphate dibasic

CAS 7757-93-9; EINECS 231-826-1

Synonyms: DCP-0; Dicalcium phosphate; Dicalcium orthophosphate anhyd.; Calcium monohydrogen phosphate; Calcium hydrogen phosphate; Calcium hydrogen phosphate anhydrous; Calcium hydrogen orthophosphate; Phosphoric acid calcium salt (1:1)

Classification: Inorganic salt

Empirical: $CaHPO_4$

Properties: Wh. cryst. powd., odorless, tasteless; deliq.; sol. in dilute HCl, nitric, and acetic acids; insol. in alcohol; sl. sol. in water; m.w. 136.07; dens. 2.306; loses water at 109 C; stable in air

Toxicology: No known toxicity; skin and eye irritant; nuisance dust

Uses: Foods, pharmaceuticals, medicine, glass, fertilizer, stabilizer for plastics, dough conditioner, yeast food, nutrient, dietary supplement; dispersant in tableting; pigment in paints; secondary abrasive in dentifrices

Regulatory: FDA 21CFR §181.29, 182.1217, 182.5217, 182.8217, GRAS; Japan approved (1% max. as calcium); Europe listed; UK approved; FDA approved for dentals, orals; USP/NF, BP, Ph.Eur. compliance

Manuf./Distrib.: AC Ind.; Albright & Wilson Am.; Aldrich; Brown; Coyne; EM Ind.; Fluka; FMC; GE; Int'l. Chem. Inc.; Janssen Chimica; Lohmann; Mallinckrodt; Mendell; Occidental; Rhone-Poulenc Basic; Sigma; San Yuan; Spectrum Chem. Mfg.

Trade names: Albrite® Dicalcium Phosphate Anhyd.

Calcium resinate

CAS 9007-13-0

Synonyms: Limed rosin

Formula: $Ca(C_{44}H_{62}O_4)_2$

Properties: Ylsh.-wh. powd., rosin odor; sol. in acid; insol. in water; m.w. 1349.5

Precaution: Flamm. solid; reactive with oxidizing materials

Toxicology: Heated to decomp., emits acrid smoke and fumes

Uses: Waterproofing, paint driers, porcelains, perfumes, cosmetics, enamels; color diluent for egg shells

Regulatory: FDA 21CFR §73.1

Manuf./Distrib.: Arakawa USA; Barium & Chems.

Calcium silicate

CAS 1344-95-2; 10101-39-0; EINECS 215-710-8

Synonyms: Silicic acid, calcium salt; Okenite

Definition: Hydrous or anhydrous silicate with varying proportions of calcium oxide and silica

Empirical: Common forms: $CaSiO_3$, Ca_2SiO_4, Ca_3SiO_5

Properties: Wh. or cream-colored powd.; pract. insol. in water; forms a gel with min. acids; dens. 2.10; bulk dens. 15-16 lb/ft^3; absorp. power 600% (water); surf. area 95-175 m^2/g; pH 8.4-10.2 (5% aq. susp.)

Toxicology: Pract. nontoxic orally, but inh. may cause respiratory tract irritation; nuisance dust

Uses: Constituent of lime glass, portland cement; reinforcing filler in elastomers and plastics; absorbent for liqs., gases, vapors; suspending agent; pigment and pigment extender; binder for refractories; in chromatography, road building, paints; anticaking agent, filter aid for foods (baking powd., table salt)

Calcium stearate

Regulatory: FDA 21CFR §172.410, 175.300, 177.1460, 182.2227, GRAS (limitation 2% in table salt, 5% in baking powd.); Europe listed; UK approved; approved for orals; USP/NF compliance
Manuf./Distrib.: Aldrich; Celite; Chem-Materials; Crosfield; Degussa; Great Western; J.M. Huber; Kraft Chem.; D.N. Lukens; H.M. Royal; Seegott; R.T. Vanderbilt
Trade names: Micro-Cel® T-38
See also Calcium metasilicate

Calcium stearate
CAS 1592-23-0; EINECS 216-472-8
Synonyms: Calcium octadecanoate; Stearic acid calcium salt; Octadecanoic acid calcium salt
Classification: Aliphatic organic compd.
Definition: Calcium salt of stearic acid
Empirical: $C_{18}H_{35}O_2 \cdot ½Ca$
Formula: $Ca(C_{18}H_{35}O_2)_2$
Properties: Wh. impalpable powd., sl. char. odor; insol. in water, alcohol, ether; sl. sol. in hot alcohol, hot veg. and min. oils; m.w. 707.00; bulk dens. 20 lb/ft^3; m.p. 149 C
Toxicology: Nontoxic; heated to decomp., emits acrid smoke and irritating fumes
Uses: Water repellent; flatting agent in paints, emulsions; release agent for plastic molding powds.; stabilizer for PVC resins; lubricant; in pencils and crayons; food grade as conditioning agent in foods and pharmaceuticals, anticaking agent, binder, emulsifier, lubricant, release agent, stabilizer, thickener
Regulatory: FDA 21CFR §169.179, 169.182, 172.863, 173.340; must conform to FDA specs for fats or fatty acids derived from edible oils, 175.105, 175.300, 175.320, 176.170, 176.200, 176.210, 177.1200, 177.2260, 177.2410, 177.2600, 178.2010, 179.45; 181.22, 181.29, 184.1229, GRAS; FDA approved for implants, orals, rectals, topicals; USP/NF, BP, Ph.Eur. compliance
Manuf./Distrib.: Adeka Fine Chem.; Allchem Ind.; Chemisphere; Cometals; Eka Nobel Ltd; Elf Atochem N. Am./Wire Mill Prods.; Ferro/Grant; Harwick; Henkel/Organic Prods.; In-Cide Tech.; Lohmann; Mallinckrodt; Miljac; Norac; PPG Ind.; Spectrum Chem. Mfg.; R.T. Vanderbilt; Witco/Oleo-Surf.; Witco/Polymer Addit.
Trade names: Akrodip™ CS-400; Hallcote® CaSt 50; Hallcote® CSD; Interstab® CA-18-1; Lubracal® 48; Lubracal® 53; Lubracal® 60; Nopco® 1097A; Petrac® Calcium Stearate CP-11; Petrac® Calcium Stearate CP-11 LS; Petrac® Calcium Stearate CP-11 LSG; Petrac® Calcium Stearate Disp. CW-1250; Quikote™ C; Quikote™ C-LD; Quikote™ C-LD-A; Quikote™ C-LM; Quikote™ C-LMLD; Synpro® Calcium Stearate 15
Trade names containing: Estane® 5707 F1; Estane® 5715

Calcium sulfate
CAS 7778-18-9 (anhyd.); EINECS 231-900-3
Synonyms: Anhydrite (natural form); Calcium sulfate (1:1); Calcium sulfonate; Gypsum; Anhydrous gypsum; Plaster of Paris
Classification: Inorganic salt
Empirical: CaO_4S
Formula: $Ca \cdot H_2O_4S$
Properties: Wh. to sl. yel.-wh. powd. or crystals, odorless, tasteless; sl. sol. in water; sol. in 3 N HCl; m.w. 136.14; dens. 2.964; m.p. 1450 C
Precaution: Reacts violently with aluminum when heated; mixts. with phosphorus ignite at high temps.
Toxicology: Ingestion may result in intestinal obstruction because it absorbs water and hardens; no known toxicity on the skin; irritant; heated to decomp., emits toxic fumes of SO_x
Uses: Extender in paints; Insol. anhydride: in cement formulations, as paper filler; Sol. anhydride: drying agent, desiccant; nutrient, dietary supplement, yeast food, dough conditioner/strengthener, firming agent, sequestrant, anticaking agent, coloring agent, leavening agent, pH control agent, processing aid, stabilizer, texturizer, thickener
Regulatory: FDA 21CFR §133, 133.102, 133.106, 133.111, 133.141, 133.165, 133.181, 133.195, 137.105, 137.155, 137.160, 137.165, 137.170, 137.175, 137.180, 137.185, 150.141, 150.161, 155.200, 175.300, 177.1460, 184.1230, GRAS; BATF 27CFR §240.1051, limitation 16.69 lb/1000 gal; Japan approved with restrictions (1% max.); Europe listed; UK approved; FDA approved for orals; USP/NF compliance
Manuf./Distrib.: Advance Research Chems.; Aldrich; AMRESCO; Am. Ingreds.; J.T. Baker; R.E. Carroll; EM Ind.; Fluka; Franklin Ind. Mins.; Kemira Kemi AB; Lohmann; H.M. Royal; Seegott; U.S. Gypsum; Van Waters & Rogers; Whittaker, Clark & Daniels

Calcium sulfate (1:1). *See* Calcium sulfate

Calcium sulfate dihydrate
CAS 10101-41-4; EINECS 231-900-3
Synonyms: Native calcium sulfate; Precipitated calcium sulfate; Gypsum; Alabaster

Empirical: $CaO_4S \cdot 2H_2O$
Formula: $CaSO_4 \cdot 2H_2O$
Properties: Lumps or powd.; hygroscopic; sol. in water; very slowly sol. in glycerol; pract. insol. in most org. solvs.; m.w. 172.10; dens. 2.32
Toxicology: Irritant
Uses: Mfg. of portland cement, plaster of Paris, artificial marble, sulfuric acid; as white pigment and filler in paints, pharmaceuticals, paper, insecticide dusts, yeast, polishing powds.; soil and water treatment; plaster casts (pharmaceutical); flame retardant for plastics
Regulatory: FDA approved for orals; USP/NF, BP, JP compliance
Manuf./Distrib.: Aldrich; R.E. Carroll; EM Ind.; Fluka; Franklin Ind. Mins.; Sigma

Calcium sulfonate
CAS 61789-86-4
Uses: Rust preventive, detergent for diesels; pigment dispersant for thermoplastic and epoxy color concs.; visc. stabilizer
Manuf./Distrib.: Ashland; King Ind.; Lubrizol; Stepan; Witco/Petroleum Spec.
Trade names: Ircogel® 900; Lubrizol® 2064; Lubrizol® 2152; SACI® 200HM; SACI® 445a; SACI® 450W; SACI® 4192; SACI® 4194; SACI® 4215; SACI® 4253
Trade names containing: Ircogel® 905; Ircogel® 906; Ircogel® 907; K-Sperse® 131; K-Sperse® 132; SACI® 100a; SACI® 200; SACI® 200a; SACI® 214; SACI® 230; SACI® 230a; SACI® 300; SACI® 300a; SACI® 350; SACI® 350a; SACI® 400 EPW; SACI® 440 EOW; SACI® 500; SACI® 500a; SACI® 552a; SACI® 560; SACI® 560a; SACI® 570; SACI® 700; SACI® 760; SACI® 2452; SACI® 8100; SACI® 8200
See also Calcium sulfate

Calcium tallate
Uses: Drier for paints
Manuf./Distrib.: Akzo Nobel; Dussek Campbell Ltd

Calcium-zinc complex
Uses: Heat stabilizer for rigid and plasticized PVC, profiles, bottles, pipe, plastisols; stabilizer for flooring base coats, artificial leather, wallpaper, conveyor belts; compat. with isocyanate adhesion promoter
Trade names: Bärostab® CT 927; Bärostab® CT 9141X-TS; Synpron 1522

Calcium-zinc-molybdenum complex
Uses: Corrosion inhibiting pigment for coatings
Trade names: Moly-White® 212

C-8 alcohols. *See* Caprylic alcohol

C20-40 alcohols
Definition: Mixture of syn. fatty alcohols with 20-40 carbons in the alkyl chain
Uses: Antioxidant, heat stabilizer, uv stabilizer, visc. depressant; promotes emollient protective films for skin care prods.; chemical intermediate for oxidation, ethoxylation, sulfation, amination, esterification; coemulsifier and direct additive in coatings; plastics additive (processing aid, lubricant, dispersant)
Trade names: Unilin® 350 Alcohol; Unilin® 425 Alcohol

C30-50 alcohols
Definition: Mixture of syn. fatty alcohols with 30-50 carbons in the alkyl chain
Uses: Surfactants; chemical intermediate; in coatings; plastics additive (processing aid, lubricant, dispersant); cosmetics
Trade names: Unilin® 550 Alcohol

C40-60 alcohols
Definition: Mixture of syn. fatty alcohols with 30-50 carbons in the alkyl chain
Uses: Functional polymer for modification of PP, PVC, polyethylene, PS, and high-performance engineering resins; acts as antioxidant, heat stabilizer, uv stabilizer, or visc. depressant; promotes emollient protective films onto skin; chemical intermediate for oxidation, ethoxylation, sulfation, amination, esterification; coemulsifier and direct additive in coatings; plastics additive (processing aid, lubricant, dispersant)
Trade names: Unilin® 700 Alcohol

C6 alkyl acetate. *See* Oxo-hexyl acetate
C7 alkyl acetate. *See* Oxo-heptyl acetate

C8 alkyl acetate
CAS 108419-32-5
Properties: Sp.gr. 0.87; flash pt. (TCC) 171 F
Uses: Solv. for high solids, maintenance, marine, and industrial waterborne coatings, acrylic resin polymerization media, metalworking fluids, urethane foam and industrial cleaners and degreasers, paint strippers, pigment dispersions; pesticide emulsifiable concs., textile scouring aids, NC or screen printing inks, penetrating and lubricating oils, vinyl plastisols; good solvency, low surf. tens. and interfacial tens., good wetting, lubricity; visc. modifier; latex coalescer
Trade names: Exxate® 800

C9 alkyl acetate
CAS 108419-33-6
Properties: Sp.gr. 0.87; flash pt. (TCC) 194 F
Uses: Solv. for high solids, maintenance, marine, and industrial waterborne coatings, acrylic resin polymerization media, metalworking fluids, urethane foam and industrial cleaners and degreasers, paint strippers, pigment dispersions; pesticide emulsifiable concs., textile scouring aids, NC or screen printing inks, penetrating and lubricating oils, vinyl plastisols; good solvency, low surf. tens. and interfacial tens., good wetting, lubricity; visc. modifier; latex coalescer
Trade names: Exxate® 900

C10 alkyl acetate
CAS 108419-34-7
Properties: Sp.gr. 0.87; flash pt. (TCC) 212 F
Uses: Solv. for high solids, maintenance, marine, and industrial waterborne coatings, acrylic resin polymerization media, metalworking fluids, urethane foam and industrial cleaners and degreasers, paint strippers, pigment dispersions; pesticide emulsifiable concs., textile scouring aids, NC or screen printing inks, penetrating and lubricating oils, vinyl plastisols; good solvency, low surf. tens. and interfacial tens., good wetting, lubricity; visc. modifier; latex coalescer
Trade names: Exxate® 1000

C13 alkyl acetate
CAS 108419-35-8
Properties: Sp.gr. 0.87; flash pt. (TCC) 261 F
Uses: Solv. for high solids, maintenance, marine, and industrial waterborne coatings, acrylic resin polymerization media, metalworking fluids, urethane foam and industrial cleaners and degreasers, paint strippers, pigment dispersions; pesticide emulsifiable concs., textile scouring aids, NC or screen printing inks, penetrating and lubricating oils, vinyl plastisols; good solvency, low surf. tens. and interfacial tens., good wetting, lubricity; visc. modifier; latex coalescer
Trade names: Exxate® 1300

Calx. *See* Calcium oxide
2-Camphanone. *See* Camphor

Camphor
CAS 76-22-2 (DL); 464-49-3 (D-+); EINECS 207-355-2
Synonyms: 1,7,7-Trimethylbicyclo[2.2.1] heptan-2-one; Gum camphor; Camphora; 2-Camphanone; 2-Bornanone; Bornan-2-one
Classification: Aliphatic cyclohexyl compd.
Definition: Ketone derived from wood of the camphor tree, *Cinnamomum camphora* or prepared synthetically
Empirical: $C_{10}H_{16}O$
Properties: Colorless or wh. translucent cryst. mass, char. fragrant penetrating odor, sl. bitter cooling taste; sol. in alc., chloroform, ether, fixed/volatile oils; sl. sol. in water; m.w. 152.24; dens. 0.992 (25/4 C); m.p. 179.7 C; b.p. 204 C; subl. @ ambient temp./pressure; flash pt. 150 F
Precaution: Combustible; keep tightly closed away from heat; incompat. with potassium permanganate
Toxicology: LD50 (IP, mouse) 3000 mg/kg; ingestion by humans may cause nausea, vomiting, vertigo, mental confusion, delirium, coma, respiratory failure, death; chronic exposure may cause transient hepatic and renal damage
Uses: Medicine, plasticizer for cellulose nitrate, other explosives and lacquers, insecticides, moth and mildew proofing, tooth powders, flavoring, embalming, pyrotechnics, intermediate; active OTC drug prod. ingred.
Regulatory: FDA 21CFR §172.510, 172.515, 175.105, 27CFR §21.65, 21.151; FEMA GRAS; Japan approved; banned in U.S. for internal use
Manuf./Distrib.: Aldrich; Berje; Buckton Scott Ltd; Fluka; R.W. Greeff; Lonza; Penta Mfg.; Quest Int'l.; San Yuan; Schweizerhall; Sigma; Spectrum Chem. Mfg.

Camphora. *See* Camphor
Candelilla. *See* Candelilla wax

Candelilla synthetic
CAS 136097-95-5
Uses: Gellant, thickener, stabilizer, and moisturizer cosmetics, leather dressings, furniture and other polish; cements; varnishes; sealing wax; elec. insulating compositions; phonograph records; paper sizing; rubber; lubricants; adhesives; waterproofing
Trade names: Koster Keunen Synthetic Candelilla

Candelilla wax
CAS 8006-44-8; EINECS 232-347-0
Synonyms: Candelilla
Definition: Wax from various *Euphorbiaceae* species
Properties: Yel.-brown to translucent solid; pract. insol. in water; sparingly sol. in alcohol; sol. in acetone, benzene, carbon disulfide, hot petrol. ether, gasoline, oils, turpentine, CCl_4; dens. 0.983; m.p. 67-68 C; acid no. 10-20; sapon. no. 50-65; ref. index 1.4555
Precaution: Combustible
Toxicology: No known toxicity; heated to decomp., emits acrid smoke and irritating fumes
Uses: Mfg. of cosmetics, rubber substitutes, polishes, candles, sealing waxes, varnishes, leather, creams; for waterproofing; elec. insulation; inks; molding compositions; sizing paper; hardening other waxes; protective coating for citrus; masticatory substance in chewing gum base
Regulatory: FDA 21CFR §175.105, 175.320, 176.180, 184.1976, GRAS; FEMA GRAS; Japan approved; approved for orals, topicals
Manuf./Distrib.: Aldrich; Koster Keunen; Penta Mfg.; Frank B. Ross; Spectrum Chem. Mfg.; Stevenson Cooper; Strahl & Pitsch
Trade names containing: Ross Beeswax Substitute No. 628/5

Canola oil. *See* Rapeseed oil
CAP. *See* Cellulose acetate propionate

Caproic acid
CAS 142-62-1; EINECS 205-550-7
Synonyms: Carboxylic acid C-6; Hexanoic acid; Hexoic acid; Butylacetic acid; Pentylformic acid
Empirical: $C_6H_{12}O_2$
Formula: $CH_3(CH_2)_4COOH$
Properties: Colorless oily liq., odor of Limburger cheese; very sol. in ether, fixed oils; sl. sol. in water; m.w. 116.18; dens. 0.9295 (20/20 C); f.p. -3.4 C; b.p. 205 C; flash pt. (COC) 215 F; ref. index 1.415-1.418
Precaution: DOT: Corrosive material; combustible
Toxicology: No known toxicity
Uses: Analytical chemistry, flavors, manufacture of rubber chemicals, varnish driers, resins, pharmaceuticals
Regulatory: FDA 21CFR §172.515, 173.315; FEMA GRAS; Japan approved as flavoring
Manuf./Distrib.: Agri-Pharm; Aldrich; Allchem Ind.; Chisso Am.; Condor; Fluka; Janssen Chimica; Penta Mfg.; Schweizerhall; Sigma; Spectrum Chem. Mfg.

Caprolactam
CAS 105-60-2
Synonyms: ε-Caprolactam; Aminocaproic lactam; 2-Oxohexamethylenimine
Empirical: $C_6H_{11}NO$
Formula: $(CH_2)_5NH \cdot CO$
Properties: Wh. flakes; sol. in water and most org. solvs.
Toxicology: LD50 (oral, rat) 1210 mg/kg; mod. toxic skin and eye irritant
Uses: Intermediate for paints
Manuf./Distrib.: AlliedSignal; BASF; United Min. & Chem.

ε-Caprolactam. *See* Caprolactam

Caprolactone acrylate monomer
Formula: $CH_2HOCOCH_2CH_2OCOCH_2CH_2CH_2CH_2CH_2OH$
Trade names: SR-495

ε-Caprolactone homopolymer
Trade names: Tone® Polymer P-767

6-Caprolactone monomer. *See* ε-Caprolactone monomer

ε-Caprolactone monomer
CAS 502-44-3; EINECS 207-938-1
Synonyms: 6-Hexanalactone, 2-oxepanone; 6-Caprolactone monomer
Empirical: $C_6H_{10}O_2$
Properties: M.w. 114.14; dens. 1.030; b.p. 96-97.5 C (10 mm), 98-99 (2 mm); flash pt. 109 C
Uses: Intermediate in adhesives, urethane coatings, elastomers; solvent; diluent for epoxy resins; synthetic fibers; organic synthesis
Manuf./Distrib.: Aldrich; Fluka; Union Carbide

Capryl alcohol. *See* Caprylic alcohol

Capryl hydroxyethyl imidazoline
CAS 37478-68-5; EINECS 253-521-2
Synonyms: Caprylic imidazoline; 2-Nonyl-4,5-dihydro-1H-imidazole-1-ethanol; 1H-Imidazole-1-ethanol, 4,5-dihydro-2-nonyl-
Classification: Heterocyclic compd.
Empirical: $C_{14}H_{28}N_2O$
Uses: Wetting agent, emulsifier for nonpolar liq., detergent, thickener, corrosion inhibitor, antistat, softener, bactericide used in paint and textile industries
Trade names: Monazoline CY

Caprylic alcohol
CAS 111-87-5; EINECS 203-917-6
Synonyms: n-Octyl alcohol; pri-Octyl alcohhol; Heptyl carbinol; 1-Octanol; n-Octanol; C-8 alcohols; Alcohol C-8; Capryl alcohol
Classification: Fatty alcohol
Empirical: $C_8H_{18}O$
Formula: $CH_3(CH_2)_6CH_2OH$
Properties: Colorless liq., fresh orange-rose odor, sl. herbaceous taste; sol. in water; misc. in alcohol, ether, and chloroform; m.w. 130.26; dens. 0.827; m.p. -16.7 C; b.p. 194.5 C; flash pt. 178 F; ref. index 1.429 (20 C)
Precaution: Combustible liq. when exposed to heat or flame; can react with oxidizers
Toxicology: LD50 (oral, mouse) 1790 mg/kg; poison by intravenous route; moderately toxic by ingestion; skin irritant
Uses: Wetting agent, emulsifier, surfactant intermediate, raw material for plasticizers; emulsion polymerization; tobacco-offshoots controller; detergent intermediate; solv. for paints; flavoring agent, intermediate, solv.
Regulatory: FDA 21CFR §172.230, 172.515 (only for encapsulating lemon, dist. lime, orange, peppermint, and spearmint oils), 172.864, 173.280, 175.105, 175.300, 176.210, 177.1010, 177.1200, 177.2800, 178.3480; FEMA GRAS
Manuf./Distrib.: Albemarle; Aldrich; Fluka; Henkel/Emery; M. Michel; Penta Mfg.; Schweizerhall; Sigma; Spectrum Chem. Mfg.; Union Camp; Van Waters & Rogers; Vista

Caprylic/capric acid
CAS 67762-36-1
Uses: Used in alkyd resins, rubber compding., water repellents, polishes, soaps, abrasives, cutting oils, candles, crayons, emulsifiers; food grades as lubricant, release agent, binder, defoamer in foods, intermediate for food emulsifiers
Trade names: Prifrac 2912

Caprylic imidazoline. *See* Capryl hydroxyethyl imidazoline

Caprylyl pyrrolidone
Synonyms: N-Octyl-2-pyrrolidone
Classification: Substituted heterocyclic compd.
Trade names: Surfadone® LP-100

Captan
CAS 133-06-2; EINECS 205-087-0
Synonyms: cis-N-[(Trichloromethyl) thio]-4-cyclohexene-1,2-dicarboximide; N-Trichloromethylthio-4-cyclohexene-1,2-dicarboximide; N-Trichloromethylthiotetrahydrophthalimide

Classification: Organic compd.
Empirical: $C_9H_8Cl_3NO_2S$
Properties: Crystal, odorless; pract. insol. in water; partly sol. in benzene, chloroform, tetrachloroethane; m.w. 300.59; dens. 1.74; m.p. 178 C
Toxicology: Moderately toxic by inhalation; irritant; TLV 5 mg/m^3 of air; LD50 (rat, oral) 9000 mg/kg
Uses: Seed treatment, fungicide, mildewcidte, and bacteriostat in plants, plastics, solv-based coatings, leather, fabrics, fruit preservation; gas odorant
Regulatory: FDA 21CFR §176.170
Manuf./Distrib.: Industrias Quimicas del Valles SA; R.T. Vanderbilt
Trade names: Fungitrol® C; Vancide® 89

Carbamic acid, butyl-3-iodo-2-propynyl ester. *See* Iodopropynyl butylcarbamate
Carbamic acid, homopolymer (9C1), octadecyl-ethenyl ester. *See* Polyvinyl octadecyl carbamate
Carbinol. *See* Methyl alcohol
'Carbitol'. *See* Ethoxydiglycol
'Carbolic acid'. *See* Phenol

Carbomer
CAS 9007-16-3; 9003-01-4; 9007-17-4; 9007-20-9; 76050-42-5
Synonyms: Carboxypolymethylene
Definition: Homopolymer of acrylic acid crosslinked with an allyl ether of pentaerythritol or an allyl ether of sucrose
Properties: Wh. fluffy powd., odorless; hygroscopic; dens. 1.41
Toxicology: Essentially nontoxic
Uses: Emulsifier, thickener, stabilizer, suspending agent; for lubricating, quenching and silicone emulsions; graphite, polyethylene, fiber and paper suspensions, textile back coatings, pharmaceuticals, cosmetics
Usage level: 0.4% (suspending agent), 0.5-5% (aq. ointment base)
Regulatory: FDA approved for orals, rectals, topicals; BP compliance
Manuf./Distrib.: Aldrich; Sigma
Trade names: Carbopol® 910; Carbopol® 940

Carbon
CAS 7440-44-0; EINECS 231-153-3
Synonyms: C
Classification: Nonmetallic element
Empirical: C
Properties: Blk. cryst., powd. or diamond form; at.wt. 12.01; dens. 1.8-2.1 (amorphous); m.p. 3652-3697 C; b.p. ≈ 4200 C
Toxicology: Moderately toxic by intravenous route
Uses: Reducing agent, purifying metals, electronics; in rubber compds., coatings, and plastics
Manuf./Distrib.: Aldrich; Fluka; Sigma
Trade names: Synthetic Carbon 910-44M
Trade names containing: Xytrex® 451; Xytrex® 455; Xytrex® 642

Carbon black
CAS 1333-86-4; EINECS 215-609-9
Synonyms: Thermal black; Charcoal; Vegetable carbon; Channel black; Furnace black; Acetylene black; Lamp black; CI 77266
Definition: Finely divided particles of elemental carbon obtained by incomplete combustion of hydrocarbons (channel or impingement process)
Empirical: C
Properties: Insol. in water, ethanol, veg. oil
Toxicology: TLV 3.5 mg/m^3; low toxicity by ingestion, inhalation, and skin contact; nuisance dust in high concs.; suspected carcinogen
Uses: Inorganic colorant, filler, reinforcement; tire treads, belt covers, other abrasion-resistant rubber prods.; UV light absorber; colorant for paints, printing inks; color additive, purification agent in food processing
Regulatory: Banned by U.S. FDA; may be used in foods for the European community; UK approved
Manuf./Distrib.: Akrochem; Akzo Nobel; Archway Sales; BASF; Cabot; Chem-Materials; Columbian Chems.; Cookson Pigments; Daniel Prods.; Degussa; Engelhard; Exxon; Heucotech Ltd; Hilton Davis; J.M. Huber; Landers-Segal Color; D.N. Lukens; Reade Advanced Materials; Royale Pigments & Chems.; San Yuan; Spectrum Chem. Mfg.; Tamms Ind.; Tiarco; R.T. Vanderbilt; Witco/Concarb; Zeochem
Trade names: Black Pearls® 120; Black Pearls® 130; Black Pearls® 160; Black Pearls® 170; Black Pearls®

280; Black Pearls® 430; Black Pearls® 450; Black Pearls® 460; Black Pearls® 470; Black Pearls® 480; Black Pearls® 490; Black Pearls® 700; Black Pearls® 800; Black Pearls® 880; Black Pearls® 900; Black Pearls® 1000; Black Pearls® 1100; Black Pearls® 1300; Black Pearls® 1400; Black Pearls® 2000; Black Pearls® 3700; Black Pearls® L; Continex® N326; Dapro High Jetness Masstone Blacks; Elftex® 005; Elftex® 008; Elftex® 012; FW 200; Mitsubishi Carbon Black; Monarch® 120; Monarch® 280; Monarch® 700; Monarch® 800; Monarch® 880; Monarch® 900; Monarch® 1000; Monarch® 1100; Monarch® 1300; Monarch® 1400; Octojet 104; Octojet 494; Octojet 495; Octojet 588; Octojet 619; Octojet 645; Octojet 868; Printex® U; Raven® 850; Raven® 890; Raven® 890H; Raven® 1000; Raven® 1020; Raven® 1035; Raven® 1040; Raven® 1060 Ultra; Raven® 1080 Ultra; Raven® 1170; Raven® 1200; Raven® 1250; Raven® 1255; Raven® 1500; Raven® 2000; Raven® 2500 Ultra; Raven® 3500; Raven® 5000 Ultra II; Raven® 7000; Regal® 300R; Regal® 330; Regal® 330R; Regal® 400; Regal® 400R; Regal® 660; Regal® 660R; Special Black 4; Special Black 4A; Special Black 5; Special Black 6; Special Black 100 Powd.; Tint-Ayd® Blue Tone Tinting Blacks; Vulcan® XC72R

Trade names containing: Ampacet 19153; Ampacet 19153-S; Colloids PE 48/61/30; MDI EC-940; Plasblak® PE 4135

Carbonic acid calcium salt (1:1). *See* Calcium carbonate
Carbonic acid monosodium salt. *See* Sodium bicarbonate
Carbonic acid, 1,2-propylene glycol ester. *See* Propylene carbonate
Carbon silicide. *See* Silicon carbide
5,5´-Carbonylbis 1,3-isobenzofurandione. *See* 3,3´,4,4´-Benzophenone tetracarboxylic dianhydride
4,4´-Carbonylbis(phthalic anhydride). *See* 3,3´,4,4´-Benzophenone tetracarboxylic dianhydride
1,3-Carbonyl dioxypropane. *See* Propylene carbonate
4,4´-Carbonyldiphthalic acid anhydride. *See* 3,3´,4,4´-Benzophenone tetracarboxylic dianhydride
4,4´-Carbonyldiphthalic anhydride. *See* 3,3´,4,4´-Benzophenone tetracarboxylic dianhydride
Carborundeum. *See* Silicon carbide
Carborundum. *See* Silicon carbide
2-Carboxyaniline. *See* Anthranilic acid
o-Carboxyaniline. *See* Anthranilic acid
Carboxybenzene. *See* Benzoic acid

β-Carboxyethyl acrylate
CAS 24615-84-7
Synonyms: 2-Propenoic acid, 2-carboxyethyl ester; 3-Acryloyloxypropionic acid
Empirical: $C_6H_8O_4$
Properties: M.w. 144.13
Uses: Reactive monomer for emulsion polymerization, adhesion promoter; latexes; adhesives
Manuf./Distrib.: Rhone-Poulenc Surf. & Spec.; UCB Radcure
Trade names containing: Sipomer® β-CEA

Carboxylated styrene butadiene
Uses: Used in interior vapor barrier primer sealers and flat wall paint
Trade names: Rovene™ 4002; Rovene™ 4076; Rovene™ 4106; Rovene™ 4112; Rovene™ 4170

Carboxylic acid C_5. *See* n-Valeric acid
Carboxylic acid C-6. *See* Caproic acid
Carboxylic acid C_9. *See* Nonanoic acid
Carboxylic acid C_{18}. *See* Stearic acid

Carboxymethylcellulose
CAS 9000-11-7
Synonyms: Cellulose carboxymethyl ether
Classification: Synthetic gum
Toxicology: Shown to cause cancer in animals when ingested; toxicity on skin not known; may cause immune responses
Uses: Resin for paints
Regulatory: FDA approved for orals; FEMA GRAS; BP compliance
Manuf./Distrib.: Aqualon; Chemcentral; Colony Ind.; Hercules; Hoechst Celanese; Int'l. Chem. Inc.; Punda Mercantile; Sigma; Van Waters & Rogers

Carboxymethylcellulose sodium
CAS 9004-32-4; EINECS 265-995-8; 201-178-4
Synonyms: CMC; Cellulose gum; Cellulose carboxymethyl ether sodium salt; Carmellose sodium; Sodium carboxymethylcellulose; Sodium CMC

carboxymethylcellulose; Sodium CMC

Definition: Sodium salt of the carboxylic acid R-O-CH_2COONa

Formula: $[(C_6H_7O_2(OH)_2OCH_2COOH]_n$

Properties: Colorless or wh. powd. or gran., odorless; hygroscopic; water sol. depends on degree of substitution; insol. in org. liqs.; m.w. 21,000-500,000; visc. various; m.p. > 300 C; pH 6.5-8.5 (1%)

Toxicology: Nontoxic; LD50 (oral, rat) 27,000 mg/kg

Uses: In drilling muds; in detergents as soil-suspending agent; in emulsion paints, adhesives, inks, textile sizes; as protective colloid; cosmetics; in pharmaceuticals as suspending agent, excipient, viscosity modifier; food stabilizer, binder, thickener, extender, boiler water additive; in rubber industry

Regulatory: FDA 21CFR §133.134, 133.178, 133.179, 150.141, 150.161, 173.310, 175.105, 175.300, 182.70, 182.1745, GRAS; USDA 9CFR 318.7, limitation 1.5%, must be added dry; 9CFR 381.147; Japan restricted (2% max.); Europe listed; UK approved; approved for dentals, injectables, orals, topicals; USP/NF, BP, Ph.Eur. compliance

Manuf./Distrib.: Aldrich; Aqualon; Courtaulds Water Soluble Polymers; J.W.S. Delavau; Fabrichem; FMC; Fluka; Hercules/Aqualon; Sigma

Trade names: Aqualon® CMC-T; CMC Daicel 1160; Tylose® CBR Grades; Tylose® CR

Carboxypolymethylene. *See* Carbomer
Carmellose sodium. *See* Carboxymethylcellulose sodium

Carnauba

CAS 8015-86-9; EINECS 232-399-4

Synonyms: Brazil wax; Carnauba wax; Cera carnauba

Definition: Exudate from leaves of Brazilian wax palm tree *Copernicia prunifera*

Properties: Yel. greenish brown lumps, solid, char. odor; sol. in ether, alkalis, warm benzene, warm chloroform, toluene; sl. sol. in boiling alcohol; insol. in water; dens. 0.995 (15/15 C); m.p. 82-85.5 C; acid no. 2-7; sapon. no. 78-89; ref. index 1.4500

Precaution: Protect from light

Toxicology: Essentially nontoxic; heated to decomp., emits acrid smoke and irritating fumes

Uses: Shoe polishes, leather finishes, varnishes, waterproofing, furniture and floor polishes; hardening candles; substitute for beeswax; plasticizer for dental compds.; purified form in cosmetics, pharmaceuticals; anticaking agent, candy glaze/polish, formulation aid, lubricant, release agent, surface-finishing agent

Regulatory: FDA 21CFR §175.320, 184.1978, GRAS; Japan approved; Europe listed (permitted only in chocolate prods.); UK approved; approved for orals, topicals; USP/NF, BP, Ph.Eur., JP compliance

Manuf./Distrib.: Aldrich; Alfa Chem; Allchem Ind.; Koster Keunen; Penta Mfg.; Frank B. Ross; Shamrock Tech.; Sigma; Specialty Prods.; Spectrum Chem. Mfg.; Stevenson Cooper; Strahl & Pitsch

Trade names: Drewax™ E-7030; Drewax™ E-7920; Emulsion C-340; Michem® Emulsion 32020; Michem Lube® 156; Michem Lube® 160; Shamrock S-Nauba #1-80; Shamrock S-Nauba #1-120; Shamrock S-Nauba #1 Flake; Shamrock S-Nauba #3-80; Shamrock S-Nauba #3-120; Shamrock S-Nauba #3 Flake; Shamrock S-Nauba 5021 N5; Slip-Ayd® SL-535E

Trade names containing: Michem® Experimental Emulsion 87140; Michem Lube® 110; Michem Lube® 155; Michem Lube® 162; Michem Lube® 180; Michem Lube® 182; Michem Lube® 188; Michem Lube® 209; Slip-Ayd® SL-506; Slip-Ayd® SL-508; Slip-Ayd® SL-511

Carnauba synthetic

CAS 136097-96-6

Uses: Stabilizer used for lipsticks, creams; shoe, furniture, and car polishes; lacquers, varnishes; phonograph records; hardener for candles; leather finishes; elec.-insulating composition; waterproofing textiles, wood, inks

Trade names: Koster Keunen Synthetic Carnauba

Carnauba wax. *See* Carnauba
Carob bean gum. *See* Locust bean gum
Carob flour. *See* Locust bean gum
Carthanus tinctorious oil. *See* Safflower oil

Casein

CAS 9000-71-9; EINECS 232-555-1

Synonyms: Milk protein, casein

Definition: Mixt. of phosphoproteins obtained from cow's milk

Properties: Hygroscopic

Toxicology: Mild sensitive reactions in persons allergic to cow's milk

Cashew nutshell liq.

Uses: Dietary supplements; frozen desserts; raw material for paints
Regulatory: FDA 21CFR §166.110, 182.90, GRAS; Japan restricted; BP compliance
Manuf./Distrib.: Aabbitt Adhesives; Aldrich; Allchem Ind.; Am. Biorganics; Am. Casein; Browning; Chemical; Fluka; Int'l. Casein; Nat'l. Casein; Research Organics; Sigma; Spectrum Chem. Mfg.; Trugman Nash; Ultra Additives
See also Milk protein

Cashew nutshell liq. *See* Cashew nut shell oil

Cashew nut shell oil
CAS 8007-24-7; EINECS 232-355-4
Synonyms: Cashew nutshell liq.; Cashew shell liq.
Definition: Mixt. of mono- and dihydric phenols contg. cardanol and cardol with long alkyl unsat. chain substitutes
Formula: Contains 90% anacardic acid, $C_{22}H_{32}O_3$
Properties: Sp.gr. 0.943-0.968; visc. 600 cps
Toxicology: Strong irritant
Uses: Resin; chemical intermediate for phenolic resin synthesis and modification, varnishes and impregnating materials, plasticizers, germicides, insecticides, coloring material, indelible inks, lubricants, preservatives
Manuf./Distrib.: Arista Ind.

Cashew shell liq. *See* Cashew nut shell oil

Castor oil
CAS 1323-38-2; 8001-79-4; EINECS 232-293-8
Synonyms: Ricinus oil; Aceite de ricino; Huile de ricini; Oleum ricini; Ricini oleum; Oil of Palma Christi; Tangantangan oil
Definition: Fixed oil obtained from seeds of *Ricinus communis*
Properties: Colorless to pale yel. viscous liq., characteristic odor; sol. in alcohol; misc. with glac. acetic acid, chloroform, ether; dens. 0.961; m.p. -12 C; b.p. 313 C; flash pt. (CC) 445 F; iodine no. 83-88; sapon. no. 176-187; hyd. no. 160-168; ref. index 1.478
Precaution: Combustible when exposed to heat; spontaneous heating may occur
Toxicology: Moderately toxic by ingestion; allergen; eye irritant; purgative, laxative in large doses
Uses: Plasticizer in lacquers and nitrocellulose; polyurethane coatings; drying oil; hydraulic fluids; industrial lubricants; used in mfg. of Turkey red oil, cosmetics, leather treatment; antisticking agent, release agent for foods; component of protective coatings; drying oil
Usage level: 0.08-23 mg (solid oral dosage forms); 5-12.5% (topicals)
Regulatory: FDA 21CFR §73.1, 172.510, 172.876, 175.300, 176.210, 177.2600, 177.2800, 178.3120, 178.3570, 178.3910, 181.22, 181.26, 181.28; FEMA GRAS; FDA approved for injectables, orals, topicals; USP/NF, BP, Ph.Eur., JP compliance
Manuf./Distrib.: Air Prods.; Aldrich; Amber Syn.; Am. Oil & Supply; Arista Ind.; Ashland; CasChem; Climax Perf. Materials; Degen; Fanning; Fluka; Harcros; ICI Surf. Am.; Lipo; Norman, Fox; Penta Mfg.; Sigma; Spectrum Chem. Mfg.; Süd-Chemie Rheologicals; Van Waters & Rogers; Welch, Holme & Clark; Zeochem
Trade names: Castor Oil No. 1; #1 Oil; Pale Pressed Castor Oil; York #1 Castor Oil; York #3 Castor Oil; Z-1 Blown Castor Oil; Z-6 Blown Castor Oil

Castor oil, acetylated. *See* Glyceryl triacetyl ricinoleate
Castor oil, acetylated and hydrogenated. *See* Glyceryl (triacetoxystearate)
Castor oil acid. *See* Ricinoleic acid

Castor oil, dehydrated
EINECS 264-705-7
Synonyms: DCO
Definition: Castor oil with 5% water removed
Uses: Drying oil for protective coatings and alkyd resins
Regulatory: FDA 21CFR §181.26
Manuf./Distrib.: Alnor Oil; Arista Ind.; Welch, Holme & Clark
Trade names: Castung 103 G-H; Copolymer 186; York Dehydrated Castor Oil

Castor oil, hydrogenated. *See* Glyceryl tris-12-hydroxystearate; Hydrogenated castor oil

Castor oil, polymerized
Definition: Rubber-like polymer resulting from combination of castor oil with sulfur or diisocyanates

Uses: Pigment wetting/dispersing; plasticizer, processing aid for resins, gums, polymers, rubber; lubricant, penetrant, coupling solv.; adhesion promoter; for lacquers, inks, adhesives, lubricants, polishes, hydraulic fluids
Trade names: #15 Oil; Pale 4; Pale 16; Pale 170; Pale 1000; Vorite 105; Vorite 110; Vorite 115; Vorite 120; Vorite 125

Castor oil sulfated. *See* Sulfated castor oil
Castorwax. *See* Hydrogenated castor oil
Caustic soda. *See* Sodium hydroxide

C12 dibasic acid
CAS 693-23-2; EINECS 211-746-3
Synonyms: DDDA; Dodecanedioic acid; Decamethylenedicarboxylic acid; Dicarboxylic acid C_{12}; 1,12-Dodecanedioic acid; Decane-1,10-dicarboxylic acid; 1,10-Decanedicarboxylic acid; 1,10-Dicarboxydecane
Empirical: $C_{12}H_{22}O_4$
Formula: $HOOC(CH_2)_{10}COOH$
Properties: M.p. 127-129 C; flash pt. 128 C
Uses: Corrosion inhibitor raw material when reacted with amines; curative for powd. coatings; used in adhesives, syn. lubricants, fibers, plasticizers, polyester coatings, epoxy and polyamide resins
Manuf./Distrib.: Aldrich; Fluka; Sigma
Trade names: DDDA

'Cellosolve'. *See* Ethoxyethanol

Cellulase
CAS 9012-54-8; EINECS 232-734-4
Classification: Enzyme complex
Definition: Derived from *Aspergillus niger*
Properties: Off-wh. powd.; m.w. ≈ 31,000
Precaution: Refrigerate
Uses: Digestive aid in medicine and brewing industry; aids bacteria in the hydrolysis of cellulose; paper size and coatings, textile desizing, starch adhesives, cleaning starch processing equip.
Regulatory: FDA 21C FR §173.120, GRAS; UK, Japan approved
Manuf./Distrib.: Enzyme Development; Fluka; Schweizerhall; Sigma; Solvay Enzymes; Spectrum Chem. Mfg.
Trade names: Cellulase TRL

Celluloid. *See* Nitrocellulose

Cellulose
CAS 9004-34-6; 65996-61-4; EINECS 232-674-9
Synonyms: Wood pulp, bleached; Crystalline cellulose; α-Cellulose; Cotton fiber; Cellulose powder; Powdered cellulose; Cellulose gel
Definition: Natural polysaccharide derived from plant fibers
Properties: Colorless to wh. solid, odorless; sl. sol. in sodium hdyroxide sol'n.; insol. in water, dil. acids, and org. solvs.; m.w. 160,000-560,000; dens. ≈ 1.5
Toxicology: No known toxicity; cannot be digested by humans; nuisance dust; heated to decomp., emits acrid smoke and irritating fumes
Uses: Fibers: reinforcing filler, visc./sag control for textile, paper, rubber, plastics, coatings, grouts; for mfg. of nitrocellulose, cellulose acetate, etc.; in chromatography; as ion exchange material; microcrystalline: as binder-disintegrants in table; colloidal: stabilizer, emulsifier, thickener, dispersant, anticaking agent, binding agent in foods
Usage level: 5-20% (tablet binders, disintegrants), 30% (capsule diluent)
Regulatory: FDA 21CFR §177.2260; GRAS; Europe listed; use in baby foods not permitted in UK; FDA approved for buccals, dentals, orals; powdered USP/NF, BP, Ph.Eur., JP compliance
Manuf./Distrib.: Aldrich; Am. Fillers & Abrasives; Degussa; Eastman; Fluka; FMC; Hercules Ltd; Mendell; NB Entrepreneurs; Reed; Sigma; Spectrum Chem. Mfg.
Trade names: CF 31,000C Coarse; CF 32,500L; CF 42,500T; CF 42,500T Medium; CF 70,000WDK, Ex. Superfine; CF 72,500L; CF S-40605T; Fibra-Cel® BH-200; Fibra-Cel® CBR-40
See also Microcrystalline cellulose

α-**Cellulose.** *See* Cellulose

Cellulose acetate
CAS 9004-35-7
Synonyms: CA; Cellulose acetate ester; Acetylcellulose
Classification: Cellulosics
Properties: Triacetate insol. in water, alcohol, ether, sol. in glacial acetic acid; tetraacetate insol. in water, alcohol, ether, glacial acetic acid, methanol; pentaacetate insol. in water, sol. in alcohol; m.w. ≈ 37,000; dens. 1.300
Uses: Mfg. of rubber and celluloid substitutes, nonflamm. photographic and cinema films, airplane dopes, varnishes and lacquers; waterproofing and sizing for textiles; coating skins; insulating elec. wires
Regulatory: FDA 21CFR §182.90, GRAS; USP/NF, BP, Ph.Eur. compliance
Manuf./Distrib.: Aldrich; Courtaulds Chems.; Eastman; Fluka; FMC; Rotuba Extruders; Scientific Adsorbents; Sigma
Trade names: CA-394-60S; CA-398-3; CA-398-6; CA-398-10; CA-398-30

Cellulose, acetate butanoate. *See* Cellulose acetate butyrate

Cellulose acetate butyrate
CAS 9004-36-8
Synonyms: CAB; Cellulose acetobutyrate; Cellulose, acetate butanoate
Definition: Butyric acid ester of a partially acetylated cellulose
Properties: Dens. 1.250
Uses: Thermoplastic for automotive, tool, building, furniture industries, domestic appliances, lighting, elec., radio, and TV industries, optical, photographic, stationery, toy, toiletries, pkg. applics.; coatings, printing inks, lacquers
Regulatory: FDA 21CFR §175.105, 175.230, 175.300, 177.1200
Manuf./Distrib.: Aldrich; Allchem Ind.; Eastman; FMC; Sigma; Van Waters & Rogers; Whitfield Chem. Ltd
Trade names: CAB-171-15S; CAB-321-0.1; CAB-381-0.1; CAB-381-0.5; CAB-381-2; CAB-381-2BP; CAB-381-20; CAB-381-20BP; CAB-500-5; CAB-531-1; CAB-551-0.01; CAB-551-0.2; CAB-553-0.4

Cellulose acetate ester. *See* Cellulose acetate
Cellulose, acetate propanoate. *See* Cellulose acetate propionate

Cellulose acetate propionate
CAS 9004-39-1
Synonyms: CAP; Cellulose propionate; Cellulose acetate propionate ester; Cellulose, acetate propanoate
Definition: Propionic acid ester of a partially acetylated cellulose
Uses: Thermoplastic for automotive, tool, building, furniture industries, domestic appliances, lighting, elec., radio, and TV industries, optical, photographic, stationery, toy, toiletries, and pkg. applics.; paints, printing inks, paper and cloth coatings
Regulatory: FDA 21CFR §175.105, 175.230, 175.300, 177.1200
Manuf./Distrib.: Aldrich; Allchem Ind.; Eastman; SAF Bulk Chems.
Trade names: CAP-482-0.5; CAP-482-20; CAP-504-0.2

Cellulose acetate propionate ester. *See* Cellulose acetate propionate
Cellulose acetobutyrate. *See* Cellulose acetate butyrate
Cellulose carboxymethyl ether. *See* Carboxymethylcellulose
Cellulose carboxymethyl ether sodium salt. *See* Carboxymethylcellulose sodium
Cellulose ethyl ether. *See* Ethylcellulose
Cellulose gel. *See* Cellulose; Microcrystalline cellulose
Cellulose gum. *See* Carboxymethylcellulose sodium
Cellulose, 2-hydroxyethyl ether. *See* Hydroxyethylcellulose
Cellulose, 2-hydroxypropyl ether. *See* Hydroxypropylcellulose
Cellulose hydroxypropyl methyl ether. *See* Hydroxypropyl methylcellulose
Cellulose 2-hydroxypropyl methyl ether. *See* Hydroxypropyl methylcellulose
Cellulose methyl ether. *See* Methylcellulose
Cellulose nitrate. *See* Nitrocellulose
Cellulose powder. *See* Cellulose
Cellulose propionate. *See* Cellulose acetate propionate
Cellulose tetranitrate. *See* Nitrocellulose

Cellulose triacetate
CAS 9012-09-3
Synonyms: Triacetyl cellulose

Formula: $C_6H_7O_2(OOCCH_3)_3$
Properties: Ylsh. amorphous; insol. in water, alcohol, ether; sol. in glacial acetic acid, tetrachloroethane; m.w. approx. 72,000-74,000
Uses: Protective coatings resistant to most solvents; textile fibers; base for magnetic tape
Manuf./Distrib.: Aldrich; Courtaulds Chems.; Eastman; Fluka; FMC
See also Cellulose acetate

Cera alba. *See* Beeswax
Cera carnauba. *See* Carnauba

Ceresin
CAS 8001-75-0; EINECS 232-290-1
Synonyms: White ozokerite wax; Cerin; Ceresine; Ceresin wax; Earth wax; Mineral wax
Definition: Waxy mixture of hydrocarbons obtained by purification of ozokerite
Properties: Wh. or yel. waxy cake; sol. in alcohol, benzene, chloroform, naphtha, hot oils; insol. in water; dens. 0.92-0.94; m.p. 68-72 C
Precaution: Combustible
Toxicology: May cause allergic reactions
Uses: Candles, sizing; bottles for hydrofluoric acid; shoe and leather polishes; antifouling paints; cosmetics; waterproofing textiles; substitute for beeswax; dental wax compds.; rubber compounding
Regulatory: FDA 21CFR §175.105
Manuf./Distrib.: M. Argueso; Astor Corp.; Jonk BV; Koster Keunen; Frank B. Ross; Scheel; Spectrum Chem. Mfg.; Stevenson Cooper; Strahl & Pitsch
Trade names: Koster Keunen Ceresine; Ross Ceresine Wax
See also Ozokerite

Ceresine. *See* Ceresin
Ceresin wax. *See* Ceresin
Cerin. *See* Ceresin

Cerium naphthenate
Properties: Rubbery; nearly insol. without org. stabilizers; difficult to dry
Uses: Drier for paints
Manuf./Distrib.: Archway Sales; Baychem; Boehle; Hüls Am.; OM Group; Rhone-Poulenc N. Am.
Trade names: Nuodex Napthenate® Cerium 6%

Ceteareth-6
CAS 68439-49-6 (generic)
Synonyms: PEG-6 cetyl/stearyl ether; POE (6) cetyl/stearyl ether
Definition: PEG ether of cetearyl alcohol
Formula: $R(OCH_2CH_2)_nOH$, R = blend of cetyl and stearyl radicals, avg. n = 6
Uses: Emulsifier, lubricant, emollient, and conditioner for cosmetics, textiles, metal cleaners, industrial, institutional and household cleaners, emulsion polymerization; dyeing assistant, leveling agent; coating material for foam suppressant, enzymes, etc.; dyeing auxiliaries
Trade names: Marlipal® 1618/6

Ceteareth-8
CAS 68439-49-6 (generic)
Synonyms: PEG-8 cetyl/stearyl ether; PEG 400 cetyl/stearyl ether; POE (8) cetyl/stearyl ether
Definition: PEG ether of cetearyl alcohol
Formula: $R(OCH_2CH_2)_nOH$, R rep. alkyl groups from cetyl/stearyl alcohol and avg. n =8
Uses: Detergent, emulsifier, leveling agent, intermediate, used for cosmetics, household formulations, silicone emulsification, textile processing; coating material for foam suppressant, enzymes, etc.; dyeing auxiliaries
Regulatory: FDA 21CFR §177.2800
Trade names: Marlipal® 1618/8

Ceteareth-10
CAS 68439-49-6 (generic)
Synonyms: PEG-10 cetyl/stearyl ether; POE (10) cetyl/stearyl ether
Definition: PEG ether of cetearyl alcohol
Formula: $R(OCH_2CH_2)_nOH$, R = blend of cetyl and stearyl radicals, avg. n = 10
Uses: Rubber emulsifier, lubricant, dye and fulling assistant; conditioner and emollient for personal care

prods.; emulsion polymerization; floor waxes, paper finishes, pharmaceuticals
Regulatory: FDA 21CFR §177.2800

Ceteareth-11
CAS 68439-49-6 (generic)
Synonyms: PEG-11 cetyl/stearyl ether; POE (11) cetyl/stearyl ether
Definition: PEG ether of cetearyl alcohol
Formula: $R(OCH_2CH_2)_nOH$, R = blend of cetyl and stearyl radicals, avg. n = 11
Uses: Solubilizer, emulsifier for cosmetic and pharmaceutical preparations; coating material for foam suppressant, enzymes, etc.; dyeing auxiliaries
Regulatory: FDA 21CFR §177.2800
Trade names: Marlipal® 1618/11

Ceteareth-12
CAS 68439-49-6 (generic)
Synonyms: PEG-12 cetyl/stearyl ether; POE (12) cetyl/stearyl ether
Definition: PEG ether of cetearyl alcohol
Formula: $R(OCH_2CH_2)_nOH$, R = blend of cetyl and stearyl radicals, avg. n = 12
Uses: Emulsifier, solubilizer for personal care and household prods., essential oils; wetting agent and dispersant for paint systems
Regulatory: FDA 21CFR §177.2800

Ceteareth-14
CAS 68439-49-6 (generic)
Synonyms: PEG-14 cetyl/stearyl ether; POE (14) cetyl/stearyl ether
Definition: PEG ether of cetearyl alcohol
Formula: $R(OCH_2CH_2)_nOH$, R = blend of cetyl and stearyl radicals, avg. n = 14
Uses: Rubber emulsifier, lubricant, dye and fulling assistant; conditioner and emollient for personal care prods.; emulsion polymerization; floor waxes, paper finishes, pharmaceuticals; textile lubricant

Ceteareth-18
CAS 68439-49-6 (generic)
Synonyms: PEG-18 cetyl/stearyl ether; POE (18) cetyl/stearyl ether
Definition: PEG ether of cetearyl alcohol
Formula: $R(OCH_2CH_2)_nOH$, R = blend of cetyl and stearyl radicals, avg. n = 18
Uses: Dispersant; prod. of powd. detergents; binding agent and base material for solid cleaning agents (toilet sticks); coating material for foam suppressant, enzymes, etc.; dyeing auxiliaries
Regulatory: FDA 21CFR §177.2800
Trade names: Marlipal® 1618/18

Ceteareth-27
CAS 68439-49-6 (generic)
Synonyms: PEG-27 cetyl/stearyl ether; POE (27) cetyl/stearyl ether
Definition: PEG ether of cetearyl alcohol
Formula: $R(OCH_2CH_2)_nOH$, R = blend of cetyl and stearyl radicals, avg. n = 27
Uses: Detergent, dispersant, wetting agent, emulsifier for commerical and home cleaners, heavy-duty detergents, metal cleaners, detergent tablets, electrolytic cleaning, latexes
Regulatory: FDA 21CFR §177.2800
Trade names: Plurafac® A-38

Ceteareth-30
CAS 68439-49-6 (generic)
Synonyms: PEG-30 cetyl/stearyl ether; POE (30) cetyl/stearyl ether
Definition: PEG ether of cetearyl alcohol
Formula: $R(OCH_2CH_2)_nOH$, R = blend of cetyl and stearyl radicals, avg. n = 30
Uses: Wetting agent, dispersant, detergent, emulsifier, leveling agent, intermediate, used for cosmetics, household formulations, silicone emulsification, textile processing, paints, rubber, floor waxes
Regulatory: FDA 21CFR §177.2800; approved for topicals
Trade names: Rhodasurf® E-15

Ceteareth-55
CAS 68439-49-6 (generic)
Synonyms: PEG-55 cetyl/stearyl ether; POE (55) cetyl/stearyl ether

Definition: PEG ether of cetearyl alcohol
Formula: $R(OCH_2CH_2)_nOH$, R = blend of cetyl and stearyl radicals, avg. n = 55
Uses: Detergent, dispersant, wetting agent, emulsifier for commercial and home cleaning formulations, heavy-duty detergents, metal cleaners, detergent tablets, electrolytic cleaning, latexes
Regulatory: FDA 21CFR §177.2800
Trade names: Plurafac® A-39

Ceteareth-80
Uses: Dispersant; prod. of powd. detergents; binding agent and base material for solid cleaning agents (toilet sticks); coating material for foam suppressant, enzymes, etc.; dyeing auxiliaries
Trade names: Marlipal® 1618/80

Cetearyl methacrylate
Synonyms: Cetyl-stearyl methacrylate
Uses: Monomer for mfg. of polymers used as oil additives, visc. index improvers, pour pt. depressants; internal plasticizer for adhesives, uv-curable resins, coatings
Trade names containing: Empicryl® 6030

Cetethyl morpholinium ethosulfate
CAS 78-21-7; EINECS 201-094-8
Synonyms: Cetyl ethyl morpholinium ethosulfate; Quaternium-25; Morpholinium, 4-ethyl-4-hexadecyl, ethyl sulfate
Classification: Quaternary salt
Empirical: $C_{22}H_{46}NO \cdot C_2H_5O_4S$
Properties: M.w. 465.73
Uses: Antistatic agent; textile specialties; antistatic coating for cellulose acetate
Trade names: Barquat® CME-35

Cetin. *See* Cetyl palmitate

Cetoleth-5
Uses: Emulsifier for emulsion polymerization; surface-active dispersant, wetting agent, and compatibility aid for enhancing color acceptance in high solids coatings (alkyds, epoxies, uv and eb-cured coatings)
Trade names: Disponil® O 5

Cetyl alcohol
CAS 124-29-8; 36653-82-4; EINECS 253-149-0
Synonyms: Palmityl alcohol; C16 linear primary alcohol; Alcohol C16; 1-Hexadecanol; Hexadecyl alcohol
Classification: Fatty alcohol
Empirical: $C_{16}H_{34}O$
Formula: $CH_3(CH_2)_{14}CH_2OH$
Properties: Wh. waxy solid; partially sol. in alcohol and ether; insol. in water; m.w. 242.27; dens. 0.8176 (49.5 C); m.p. 49.3 C; b.p. 344 C; acid no. ≤ 2; iodine no. ≤ 5; hyd. no. 218-238; flash pt. > 110 C; ref. index 1.4283; surface-active dispersant, wetting agent, and compatibility aid for enhancing color acceptance in high solids coatings (alkyds, epoxies, uv and eb-cured coatings); emulsifier for emulsion polymerization
Precaution: Flamm. when exposed to heat or flame; can react with oxidizing materials
Toxicology: LD50 (oral, rat) 6400 mg/kg; mod. toxic by ingestion, intraperitoneal routes; eye and human skin irritant; can cause hives; heated to decomp., emits acrid smoke and fumes
Uses: Perfumery; emulsifier, emollient, coupling agent, surface-active dispersant, wetting agent; foam stabilizer in detergents; opacifier; thickener; chemical intermediate; cosmetics, pharmaceuticals; external mold release for acrylics; emulsion polymerization; color enhancer in high solids coatings; flavoring agent, color diluent, intermediate for confectionery, food supplements in tablet form, gum
Regulatory: FDA 21CFR §73.1, 73.1001, 172.515, 172.864, 175.105, 175.300, 176.200, 177.1010, 177.1200, 177.2800, 178.3480, 178.3910; FEMA GRAS; FDA approved for ophthalmics, orals, otics, rectals, topicals; USP/NF, BP, Ph.Eur. compliance
Manuf./Distrib.: Aarhus Oliefabrik A/S; AC Ind.; Albemarle; Aldrich; Allchem Ind.; Amerchol; Brown; Chemron; Croda; Fluka; R.W. Greeff; Henkel/Emery; Lipo; Lonza; M. Michel; Norman, Fox; Penta Mfg.; Procter & Gamble; Rhone-Poulenc Surf. & Spec.; Sea-Land; Sigma; Spectrum Chem. Mfg.; Stepan; Vista
Trade names containing: Ross Beeswax Substitute No. 628/5

Cetyl esters
CAS 8002-23-1; 17661-50-6; 136097-97-7; EINECS 241-640-2
Synonyms: Synthetic spermaceti; Synthetic spermaceti wax; Cetyl esters wax

Cetyl esters wax

Classification: Synthetic wax
Definition: A mixt. of sat. fatty alcohols (C14 to C18) and sat. fatty acids (C14 to C18)
Properties: Wh. to off. wh. translucent flakes, faint odor, bland mild taste; sol. in boiling alcohol, ether, chloroform; insol. in water; dens. 0.820; m.p. 43-47 C; acid no. 5 max.; iodine no. 1 max.; sapon. no. 109-120
Uses: Synthetic spermaceti NF; emollient and visc. builder for cosmetic and pharmaceutical preps.; processing aid and lubricant for plastics; gloss/slip aid for varnish
Regulatory: FDA approved for orals, topicals; USP/NF compliance
Manuf./Distrib.: Koster Keunen; Robeco; Sigma; Werner G. Smith; Spectrum Chem. Mfg.; Witco/Oleo-Surf.
Trade names: Ross Spermaceti Wax Substitute #573; W.G.S. Synaceti 116 NF/USP

Cetyl esters wax. *See* Cetyl esters
Cetyl ethyl morpholinium ethosulfate. *See* Cetethyl morpholinium ethosulfate
Cetylic acid. *See* Palmitic acid

Cetyl palmitate
CAS 540-10-3; EINECS 208-736-6
Synonyms: Hexadecanoic acid, hexadecyl ester; Palmitic acid, hexadecyl ester; Cetin
Definition: Ester of cetyl alcohol and palmitic acid
Formula: $C_{15}H_{31}COOC_{16}H_{33}$
Properties: Wh. crystalline wax-like substance; sol. in abs. alcohol, ether; insol. in water; m.w. 480.50; dens. 0.832; m.p. 50 C; b.p. 360 C; acid no. max.; iodine no. 5 max.; sapon. no. 110-130; ref. index 1.4398
Precaution: Combustible
Toxicology: Nontoxic
Uses: Base for ointments; mfg. of candles and soaps; emollient additive; internal lubricant and binder in pressed powds.; in metalworking lubricant coatings
Regulatory: FDA approved for topicals
Manuf./Distrib.: Aldrich; Croda; Koster Keunen; Sigma; Werner G. Smith; Spectrum Chem. Mfg.; Stepan; Witco/Oleo-Surf.
Trade names: Waxenol® 815; Waxenol® 816

Cetyl-stearyl methacrylate. *See* Cetearyl methacrylate
Channel black. *See* Carbon black
Charcoal. *See* Carbon black
CHDA. *See* 1,4-Cyclohexanedicarboxylic acid
CHDM. *See* 1,4-Cyclohexanedimethanol
Chile salpeter. *See* Sodium nitrate
Chile saltpeter. *See* Sodium nitrate
China clay. *See* Kaolin
Chinawood oil. *See* Tung oil
Chinese bean oil. *See* Soybean oil
Chinese tung oil. *See* Tung oil
Chinese white. *See* Zinc oxide
Chlorallyl methenamine chloride. *See* Quaternium-15
Chlorcosane. *See* Paraffin, chlorinated
Chlorinated paraffin. *See* Paraffin, chlorinated
N-(3-Chloroallyl)hexaminium chloride. *See* Quaternium-15
1-(3-Chloroallyl)-3,5,7-triaza-1-azoniaadamantane chloride. *See* Quaternium-15
Chloroalonil. *See* Tetrachloroisophthalonitrile

Chlorobenzene
CAS 108-90-7; EINECS 203-628-5
Synonyms: Monochlorobenzene; Phenyl chloride; Benzene chloride
Empirical: C_6H_5Cl
Properties: Clear volatile liq., faint almond-like not unpleasant odor; insol. in water; sol. in alcohol, benzene, chloroform, ether; misc. with most org. solvs.; m.w. 112.50; dens. 1.105 (25/25 C); m.p. -45 C; b.p. 131.6 C; flash pt. (CC) 29.4 C; ref. index 1.5
Precaution: Moderate fire risk; explosive limits 1.8-9.6% in air
Toxicology: TLV 75 ppm in air; avoid inhalation and skin contact
Uses: Pesticide intermediate; mfg. of phenol, aniline, DDT; solvent carrier for methylene diisocyanate; solvent for paints; heat transfer medium
Manuf./Distrib.: Aldrich; Allchem Ind.; Bayer/Fibers, Orgs., Rubber; Elf Atochem SA; Esprit; Fluka; Janssen Chimica; Monsanto; PPG Ind.; Sigma

4-Chlorobenzotrifluoride. *See* p-Chlorobenzotrifluoride

p-Chlorobenzotrifluoride
CAS 98-56-6
Synonyms: PCBTF; Parachlorobenzotrifluoride; Toluene, p-chloro-α,α,α,-trifluoro; Benzene, 1-chloro-4 (trifluoromethyl); 4-Chlorobenzotrifluoride
Empirical: $C_7H_4F_3Cl$
Properties: Water-wh. liq.; aromatic odor; m.w. 180.56; f.p. -36 C; m.p. -33.2 C; b.p. 138.7 C; flash pt. (CC) 116 F
Precaution: Flamm. liq.
Toxicology: LD50 (oral, rat) 13 g/kg, (oral, mouse) 11,500 mg/kg; mildly toxic by inh. or ingestion
Uses: Non-ozone depleting solv. for industrial solv. cleaning, aerosols, adhesives, coatings, inks, electronic applics.; 1,1,1-trichloroethane alternative
Manuf./Distrib.: Aldrich; Fluka; Occidental; Sigma
Trade names: Oxsol® 100
Trade names containing: Oxsol® 253; Oxsol® 325; Oxsol® 550

Chlorobutadiene. *See* Polychloroprene
2-Chlorobutadiene-1,3. *See* Polychloroprene
2-Chloro-1,3-butadiene. *See* Polychloroprene
Chlorocosane. *See* Paraffin, chlorinated
Chlorocresol. *See* p-Chloro-m-cresol
4-Chloro-m-cresol. *See* p-Chloro-m-cresol

p-Chloro-m-cresol
CAS 59-50-7; EINECS 200-431-6
Synonyms: PCMC; 4-Chloro-3-methyl phenol; 4-Chloro-m-cresol; Parachlorometacresol; Chlorocresol; 2-Chloro-5-hydroxytoluene
Empirical: C_7ClH_7O
Properties: Colorless dimorphous cryst., phenolic odor; very sol. in alcohol; sol. in ether, fixed oils, hot water; m.w. 142.59; m.p. 65-68 C; b.p. 235 C
Toxicology: Weak irritant; may produce digestive disturbances, nervous disorders, fainting, dizziness, mental changes, skin eruptions; eye irritant
Uses: Preservative for latex coatings, emulsions, and adhesives; esp. useful for starch and casein-based systems; less active in highly alkaline systems; may cause yellowing
Usage level: 0.1-0.15% (injectables), 0.075-0.12% (topicals)
Regulatory: FDA 21CFR §175.105, 176.200, 176.210, 178.3120; FDA approved for topicals; USP/NF, BP compliance
Manuf./Distrib.: Aldrich; Esprit; Fabrichem; Fluka
Trade names: Nuosept® PCMC

4-Chloro-3,5-dimethylphenol. *See* Chloroxylenol
1-Chloro-2,3-epoxypropane. *See* Epichlorohydrin
2-Chloroethanol phosphate. *See* Trichloroethyl phosphate
Chloroethene. *See* Trichloroethane; Vinyl chloride
Chloroethene homopolymer. *See* Polyvinyl chloride
Chloroethylene. *See* Vinyl chloride
Chloroethylene polymer. *See* Polyvinyl chloride
2-Chloro-5-hydroxytoluene. *See* p-Chloro-m-cresol

Chloromethoxy propyl mercuric acetate
Uses: Preservative for water-based paints

Chloromethylisothiazolinone. *See* Methylchloroisothiazolinone
5-Chloro-2-methyl-4-isothiazolin-3-one. *See* Methylchloroisothiazolinone
4-Chloro-3-methyl phenol. *See* p-Chloro-m-cresol
Chlorooxobismuthine. *See* Bismuth oxychloride
Chloroprene. *See* Polychloroprene
β-Chloroprene. *See* Polychloroprene
Chloroprene rubber. *See* Polychloroprene
Chloropropylene oxide. *See* Epichlorohydrin
3-Chloro-1,2-propylene oxide. *See* Epichlorohydrin
Chlorosulfona. *See* Bis (trichloromethyl) sulfone

Chlorothalonil. *See* Tetrachloroisophthalonitrile

2-Chlorothioxanthone
CAS 86-39-5
Properties: M.w. 246.72; m.p. 152.5-153.5 C
Uses: Photoinitiator used to decrease drying time of uv-cured inks and coatings
Manuf./Distrib.: Aldrich; Eastern Chem.; PMC Specialties; Sigma; Triple Crown Am.
Trade names: Ultra-Cure I-100

Chlorotrifluoroethylene/vinylidenefluoride copolymer
Uses: Used in O-rings, oil and shaft seals, gaskets, hose, diaphragms, mechanical goods, propellants, fiber optic components, elec. coatings
Trade names: Kel-F Brand 800 Resin

Chloroxylenol
CAS 88-04-0; EINECS 201-793-8
Synonyms: PCMX; p-Chloro-m-xylenol; Benzytol; 4-Chloro-3,5-xylenol; 4-Chloro-3,5-dimethylphenol; Parachlorometaxylenol
Classification: Organic compd.
Empirical: C_8H_9OCl
Formula: $C_6H_2(CH_3)_2OHCl$
Properties: Wh. cryst. or cryst. powd., char. phenolic odor; sol. in 1 part of 95% alcohol, ether, benzene, terpenes, fixed oils; very sl. sol. in water; m.w. 156.61; m.p. 114-116 C; b.p. 246 C
Toxicology: Toxic by ingestion; strong irritant absorbed by skin
Uses: Active ingredient in germicides, antiseptics; mildew preventive; chemical intermediate; preservative in polymer emulsions, adhesives, latex paints, metalworking cutting fluids
Usage level: 0.2-0.8%; 0.1-0.15% (otics, topicals)
Regulatory: USA EPA registered; Japan approved; Europe listed; FDA approved for otics, topicals; BP compliance
Manuf./Distrib.: Aldrich; Fluka; Nipa Hardwicke; Sigma; Thomas Swan
Trade names: Nipacide® MX; Ottasept®

4-Chloro-3,5-xylenol. *See* Chloroxylenol
p-Chloro-m-xylenol. *See* Chloroxylenol

Chrome antimony titanate
Uses: Pigment for paints
Manuf./Distrib.: Ferro/Color; Shepherd Color

Chrome titanate
Uses: Pigment for paints
Manuf./Distrib.: BASF; Bayer; Engelhard; Ferro/Color; Heucotech Ltd

Chromia. *See* Chromium oxide (ic)

Chromic acid
CAS 7738-94-5; EINECS 231-801-5
Synonyms: Chromic (VI) acid; Chromic anhydride
Formula: CrH_2O_4
Properties: Dk. purplish-red cryst.; deliq.; sol. in water, alcohol, min. acids; m.w. 118.02; dens. 1.67-12.82; m.p. 196 C
Precaution: Powerful oxidizing agent; may explode on contact with reducing agents; may ignite in contact with org. materials
Toxicology: ACGIH TLV 0.05 mg/m^3 (air); poison; corrosive to skin; human carcinogen; heated to decomp., emits acrid smoke and irritating fumes
Uses: Chemicals (chromates, oxidizing agents, catalysts), medicine, process engraving, anodizing, ceramic glazes, colored glass, metal cleaning, inks, tanning, paints, textile mordant, etchant for plastics
Manuf./Distrib.: Aldrich; Allchem Ind.; British Chrome & Chems.; Coyne; Elf Atochem N. Am.; McGean-Rohco; Occidental; Olin; Rit-Chem; San Yuan; Spectrum Chem. Mfg.; Veckridge

Chromic acid strontium salt. *See* Strontium chromate
Chromic acid strontium salt (1:1). *See* Strontium chromate
Chromic acid zinc salt (1:1). *See* Zinc chromate

Chromic anhydride. *See* Chromic acid
Chromic oxide. *See* Chromium oxide (ic)
Chromic phosphate. *See* Chromium phosphate
Chromic (VI) acid. *See* Chromic acid

Chromium
CAS 7440-47-3; EINECS 231-157-5
Classification: Metallic element
Empirical: Cr
Properties: M.w. 52
Precaution: Powd. will explode spontaneously in air; ignites and is potentially explosive in CO_2 atm.; other violent reactions possible; incompat. with oxidants
Toxicology: Metal dust is an irritant; hexavalent chromium compds. are suspect human carcinogens; human poison by ing.; may cause GI disturbances, kidney damage, and circulatory shock
Uses: Alloying and plating element for corrosion resistance, stainless steels, protective coatings, nuclear and high-temperature research, constituent of inorganic pigments
Manuf./Distrib.: Aldrich; Atlantic Equip. Engrs.; Atomergic Chemetals; Atramet; Cerac; Elkem Metals; Fluka; Noah; Powmet; Spectrum Chem. Mfg.

Chromium (III) oxide. *See* Chromium oxide (ic)
Chromium orthophosphate. *See* Chromium phosphate

Chromium oxide hydrated
Uses: Grn. pigment for paints
Manuf./Distrib.: D.N. Lukens; Whittaker, Clark & Daniels

Chromium oxide (ic)
CAS 1308-38-0; EINECS 215-160-9
Synonyms: Chromic oxide; Chromia; Chromium (III) oxide; Green cinnabar; Chromium sesquioxide
Empirical: Cr_2O_3
Uses: Metallurgy, green paint pigment, ceramics, catalyst in organic synthesis, green granules in asphalt roofing, component of refractory brick, abrasive
Manuf./Distrib.: Am. Colors; Atlantic Equip. Engrs.; Atomergic Chemetals; John K. Bice; British Chrome & Chems.; Cleveland Pigment & Color; Engelhard; Ferro/Color; Great Western; Harcros; Hoover Color; Landers-Segal Color; D.N. Lukens; Min. Pigments; Noah; Reade Advanced Materials; Seegott; Spartan Color; Tamms Ind.; Universal Color Disp.; Whittaker, Clark & Daniels

Chromium phosphate
CAS 7789-04-0
Synonyms: Chromic phosphate; Chromium orthophosphate
Empirical: $CrPO_4 \cdot 4H_2O$ or $CrPO_4 \cdot 6H_2O$
Properties: Tetrahydrate: Grn. cryst.; sol. in acids; insol. in water; Hexahydrate: Violet cryst.; dens. 2.12 (14 C)
Toxicology: Carcinogen
Uses: Green rust-inhibitive pigment for paints; catalyst
Manuf./Distrib.: Atomergic Chemetals; Nat'l. Chem.

Chromium sesquioxide. *See* Chromium oxide (ic)

C5 hydrocarbon resin
Uses: Tackifier, processing aid for adhesives, caulks, sealants, elastomers, polymers
Trade names: Eastotac™ H-100; Eastotac™ H-100E; Eastotac™ H-100L; Eastotac™ H-100R; Eastotac™ H-100W; Eastotac™ H-115E; Eastotac™ H-115L; Eastotac™ H-115R; Eastotac™ H-115W; Eastotac™ H-130E; Eastotac™ H-130L; Eastotac™ H-130R; Eastotac™ H-130W; Eastotac™ H-142R; Wingtack® 95; Wingtack® Plus

CI 77000. *See* Aluminum
CI 77004. *See* Aluminum silicate
CI 77004. *See* Bentonite
CI 77007. *See* Ultramarine blue
CI 77007. *See* Ultramarine violet
CI 77019. *See* Mica
CI 77019. *See* Talc

CI 77163. *See* Bismuth oxychloride
CI 77220. *See* Calcium carbonate
CI 77266. *See* Carbon black
CI 77322. *See* Cobalt oxide (ous)
CI 77400. *See* Copper powder
CI 77402. *See* Copper oxide (ous)
CI 77577. *See* Lead (II) oxide
CI 77578. *See* Lead oxide, red
CI 77742. *See* Manganese violet
CI 77788. *See* Pigment yellow 53
CI 77891. *See* Titanium dioxide
CI 77947. *See* Zinc oxide

CI Fluorescent Brightener 236
CAS 3333-62-8
Classification: Coumarin fluorescent brightener
Empirical: $C_{25}H_{15}N_3O_2$
Properties: Green-yel. powd.; insol. in water; m.p. 482-486 F
Precaution: Dusts may be explosive with spark or flame initiation; incompat. with strong oxidizing agents; thermal decomp. may produce CO_x and NO_x
Toxicology: LD50 (oral, rat) > 5000 mg/kg; virtually nontoxic; nonirritating to skin and eyes; nonsensitizing
Uses: Fluorescent whitener/brightener for laundry detergents esp. for synthetics and wool, whitening soap, polymers and plastics, aq. inks and coatings, powd. coatings
Regulatory: FDA 21CFR §177.1520, 177.2800
Trade names: Leucopure EGM Powd.

Cinene. *See* dl-Limonene
Cinnamene. *See* Styrene
Cinnamenol. *See* Styrene
Cinnamol. *See* Styrene
CI Pigment black 9. *See* Bone black
CI Pigment black 13. *See* Cobalt oxide (ous)
CI Pigment red 105. *See* Lead oxide, red
CI Pigment yellow 32. *See* Strontium chromate
CI Pigment Yellow 46. *See* Lead (II) oxide

C7-8 isoparaffin
CAS 64742-48-9; EINECS 274-273-1
Synonyms: Alkanes, C7-8-iso-
Definition: Mixt. of branched chain aliphatic hydrocarbons with 7 or 8 carbons in the alkyl chain
Uses: Solvent for coatings; minimizes VOC; virtually HAPS free; extremely low surf. tens. to improve flow and wetting; effective thinner and cleaner; suitable for odorless alkyd and acrylic paints
Regulatory: FDA 21CFR §172.882, 173.340
Trade names: Isopar® C

C8-9 isoparaffin
CAS 64742-48-9
Synonyms: Alkanes, C8-9-iso-
Definition: Mixt. of branched chain aliphatic hydrocarbons with 8 or 9 carbons in alkyl chain
Uses: Solvent for coatings
Regulatory: FDA 21CFR §172.882, 173.340
Trade names: Isopar® E

C10-11 isoparaffin
CAS 64742-48-9
Synonyms: Alkanes, C10-11-iso-
Definition: Mixt. of branched chain aliphatic hydrocarbons with 9-11 carbons in the alkyl chain
Uses: Solvent, diluent, carrier for personal care prods. for sprays; solvent for coatings; minimizes VOC; virtually HAPS free; extremely low surf. tens. to improve flow and wetting; effective thinner and cleaner; suitable for odorless alkyd/acrylic paints
Regulatory: FDA 21CFR §172.882, 173.340
Trade names: Isopar® G

C11-12 isoparaffin
CAS 64742-48-9
Synonyms: Alkanes, C11-12-iso-
Definition: Mixt. of branched chain aliphatic hydrocarbons with 11 or 12 carbons in the alkyl chain
Uses: Solvent, diluent, carrier for personal care prods. for sprays; solvent for coatings; minimizes VOC; virtually HAPS free; extremely low surf. tens. to improve flow and wetting; effective thinner and cleaner; suitable for odorless alkyd/acrylic paints
Regulatory: FDA 21CFR §172.882, 173.340
Trade names: Isopar® H; Isopar® K

C11-13 isoparaffin
CAS 64742-48-9
Synonyms: Alkanes, C11-13-iso-
Definition: Mixture of branched chain aliphatic hydrocarbons with 11 to 13 carbons in the alkyl chain
Uses: Solvent for skin care formulations; solvent for coatings; minimizes VOC; virtually HAPS free; extremely low surf. tens. to improve flow and wetting; effective thinner and cleaner; suitable for odorless alkyd/acrylic paints
Regulatory: FDA 21CFR §172.882, 173.340
Trade names: Isopar® L

C13-14 isoparaffin
CAS 64742-48-9
Synonyms: Alkanes, C13-14-iso-
Definition: Mixture of branched chain aliphatic hydrocarbons with 13 or 14 carbons in the alkyl chain
Uses: Solvent for hair prods.; solvent for coatings; minimizes VOC; virtually HAPS free; extremely low surf. tens. to improve flow and wetting; effective thinner and cleaner; suitable for odorless alkyd/acrylic paints
Regulatory: FDA 21CFR §172.882, 173.340
Trade names: Isopar® M

Citric acid
CAS 77-92-9 (anhyd.); EINECS 201-069-1
Synonyms: 2-Hydroxy-1,2,3-propanetricarboxylic acid; β-Hydroxytricarballylic acid
Classification: Organic acid
Empirical: $C_6H_8O_7$
Formula: $HOC(COOH)(CH_2COOH)_2$
Properties: Colorless translucent crystals or powd., odorless, strongly acidic tart taste; very sol. in water and alcohol; very sl. sol. in ether; m.w. 192.43; dens. 1.542; m.p. 153 C
Precaution: Combustible; potentially explosive reaction with metal nitrates
Toxicology: LD50 (oral, rat) 6730 mg/kg; poison by IV; mod. toxic by subcutaneous and intraperitoneal routes; mildly toxic by ingestion; severe eye, mod. skin irritant; some allergenic props.; erodes tooth enamel; heated to decomp., emits acrid smoke, fumes
Uses: Preparation of citrates, acidifier, flavoring extracts, confections, soft drinks; antioxidant in foods; sequestering agent; dispersant; detergent builder; metal cleaner; curing accelerator; intermediate for paints
Regulatory: FDA 21CFR §131.111, 131.112, 131.136, 131.138, 131.144, 131.146, 133, 145.131, 145.145, 146.187, 150.141, 150.161, 155.130, 161.190, 166.40, 166.110, 169.115, 169.140, 169.150, 172.755, 173.160, 173.165, 173.280, 182.1033, 182.6033, GRAS; USDA 9CFR §318.7, 381.147; BATF 27CFR §240.1051, limitation 5.8 lb/1000 gal; FEMA GRAS; Japan approved; Europe listed; UK approved; FDA approved for injectables, buccals, inhalants, nasals, ophthalmics, topicals, orals, otics; USP/NF, BP, Ph.Eur.
Manuf./Distrib.: Aldrich; Ashland; J.T. Baker; Baychem; Bayer/Fibers, Orgs., Rubber; Browning; Cargill; Chemical; Fluka; R.W. Greeff; Haarmann & Reimer; Harcros; Hoffmann-LaRoche; Int'l. Chem. Inc.; Lohmann; Penta Mfg.; PMC Specialties; Poly Research; Research Organics; San Yuan; Schweizerhall; Sigma; Spectrum Chem. Mfg.; TR-AMC; U.S. Petrochem. Ind.; Van Waters & Rogers
Trade names containing: Centrophil® M

Citric acid esters of mono- and diglycerides of fatty acids
CAS 68990-05-6; 91744-38-6; 97593-31-2
Synonyms: Mono- and diglycerides citrates; Citroglycerides
Properties: Sol. in hot water, lipids; HLB 10-12
Uses: Emulsifier, surfactant
Regulatory: FDA 21CFR §172.832; Europe listed; UK approved
Trade names: Acidan

Citroglycerides. *See* Citric acid esters of mono- and diglycerides of fatty acids
Citrucel. *See* Methylcellulose
C13 linear primary alcohol. *See* Tridecyl alcohol
C16 linear primary alcohol. *See* Cetyl alcohol
CM. *See* Polyethylene elastomer, chlorinated
CMC. *See* Carboxymethylcellulose sodium
Coal tar naphtha. *See* Naphtha

Cobalt
CAS 7440-48-4; EINECS 231-158-0
Synonyms: Super cobalt
Classification: Metallic element
Empirical: Co
Properties: Steel gray, shiny, hard, somewhat malleable metal; powd., cryst.; ferromagnetic; at.wt. 58.93; dens. 8.9; m.p. 1493 C; b.p. 3100 C
Toxicology: Toxic by inhalation (powd.); ingestion may cause nausea, vomiting, and diarrhea
Uses: Oxidizing agent, lamp filaments, in mfg. of cobalt steel; in porcelain, glass, pottery, enamels; pigment in paints; Japan approved; not permitted in certain foods
Manuf./Distrib.: Aldrich; Atlantic Equip. Engrs.; Atomergic Chemetals; Cerac; Fluka; Noah; Pentad Group; Powmet

Cobalt acetate (ous)
CAS 6147-53-1; EINECS 200-755-8
Synonyms: Cobaltous acetate; Cobaltous acetate tetrahydrate; Cobalt diacetate tetrahydrate
Empirical: $C_4H_6CoO_4 \cdot 4H_2O$
Formula: $Co(C_2H_3O_2)_2 \cdot 4H_2O$
Properties: Reddish-violet cryst.; hygroscopic, deliq.; sol. in water, acids, alcohol; m.w. 249.11; dens. 1.7043; m.p. loses water @ 140 C
Toxicology: LD50 (oral, rat) 708 mg/kg; mod. toxic by ingestion; skin and eye irritant; ACGIH TLV:TWA 0.05 mg (Co)/m^3; human mutagenic data; heated to decomp., emits acrid smoke and irritating fumes
Uses: Inks, paint and varnish driers, catalyst, anodizing, mineral supplement in feed additives, foam stabilizer
Manuf./Distrib.: AC Ind.; Aldrich; Alemark; Atomergic Chemetals; Barker Ind.; Celtic Chem. Ltd; Fluka; Mallinckrodt; Min. R&D; Nihon Kagaku Sangyo; Noah; OM Group; Shepherd; Sigma; Spectrum Chem. Mfg.

Cobalt black. *See* Cobalt oxide (ous)
Cobalt diacetate tetrahydrate. *See* Cobalt acetate (ous)

Cobalt drier
Uses: Used in org. coatings, inks, polyesters
Trade names: Troymax™ Drier Cobalt 6%; Troymax™ Drier Cobalt 12%

Cobalt-2-ethylhexoate. *See* Cobalt octoate
Cobaltic oxide monohydrate. *See* Cobalt oxide (ic)

Cobalt linoleate
Synonyms: Cobaltous linoleate
Definition: Obtained by boiling a cobalt salt and sodium linoleate
Formula: $Co(C_{18}H_{31}O_2)_2$
Properties: Brown amorphous powd.; sol. in alcohol, ether, acids; insol. in water; m.w. 617.42
Precaution: Combustible
Toxicology: Heated to decomp., emits acrid smoke and irritating fumes
Uses: Drier for paints
Regulatory: FDA 21CFR §181.25
Manuf./Distrib.: Shepherd; Troy

Cobalt monoxide. *See* Cobalt oxide (ous)

Cobalt naphthenate
CAS 61789-51-3
Synonyms: Naphthenic acid cobalt salt; Cobaltous naphthenate
Definition: Obtained by treating cobaltous hydroxide with naphthenic acid
Properties: Brn. amorphous powd. or bluish-red solid; sol. in oil, alcohol, ether; insol. in water; dens. 0.95; flash pt. 120 F; autoignition temp. 529 F; contains 6% Co

Precaution: Flamm. when exposed to heat or flame
Toxicology: TWA 0.1 mg(Co)/m^3 (fume, dust); LD50 (oral, rat) 3900 mg/kg; mod. toxic by ingestion; heated to decomp., emits acrid smoke and irritating fumes
Uses: Paint and varnish drier, bonding rubber to steel and other metals
Regulatory: FDA 21CFR §181.25
Manuf./Distrib.: Akzo Nobel; Archway Sales; Baychem; Boehle; Dussek Campbell Ltd; Fluka; Hüls Am.; Lomas Int'l.; D.N. Lukens; Nuodex Espanola SA; OM Group; Shepherd; Sigma; Troy
Trade names: Nuodex Napthenate® Cobalt 6%

Cobalt octoate
CAS 136-52-7
Synonyms: Cobalt-2-ethylhexoate
Formula: $C_4H_9CH(C_2H_5)COOH$
Properties: Blue liq.; dens. 1.013 (25 C)
Uses: Accelerator, paint drier, whitener, catalyst
Manuf./Distrib.: Aceto; Aldrich; Archway Sales; Baychem; Boehle; Dussek Campbell Ltd; Hüls Am.; Lomas Int'l.; D.N. Lukens; OM Group; Shepherd; Thor; Troy; Witco/Oleo-Surf.
Trade names: Nuodex Octoate® Cobalt 6%; Nuodex Octoate® Cobalt 12%

Cobaltous acetate. *See* Cobalt acetate (ous)
Cobaltous acetate tetrahydrate. *See* Cobalt acetate (ous)
Cobaltous linoleate. *See* Cobalt linoleate
Cobaltous naphthenate. *See* Cobalt naphthenate
Cobaltous oxide. *See* Cobalt oxide (ous)
Cobaltous sulfate. *See* Cobalt sulfate (ous)
Cobalt oxide. *See* Cobalt oxide (ous)
Cobalt (II) oxide. *See* Cobalt oxide (ous)
Cobalt (III) oxide. *See* Cobalt oxide (ic)

Cobalt oxide (ic)
CAS 1308-04-9; EINECS 215-156-7
Synonyms: Cobaltic oxide monohydrate; Cobalt (III) oxide
Empirical: Co_2O_3
Formula: Co_2O_3
Properties: M.w. 165.86
Precaution: Reacts violently with hydrogen peroxide
Toxicology: ACGIH TLV:TWA 0.05 mg(Co)/m^3
Uses: Pigment, coloring enamels, glazing pottery
Manuf./Distrib.: AC Ind.; Aldrich; Alfa Aesar Johnson Matthey; Atlantic Equip. Engrs.; Atomergic Chemetals; Carbochem; Cerac; Great Western; Noah; OM Group; Reade Advanced Materials; San Yuan; Spectrum Chem. Mfg.

Cobalt oxide (ous)
CAS 1307-96-6; EINECS 215-154-6
Synonyms: Cobaltous oxide; Cobalt oxide; CI 77322; CI Pigment black 13; Cobalt black; Cobalt monoxide; Monocobalt oxide; Cobalt (II) oxide
Formula: CoO
Properties: Grayish powd. or cryst.; insol. in water; sol. in acids or alkalies; m.w. 74.93; dens. 5.7-6.7; m.p. ≈ 1935 C
Precaution: Reacts violently with hydrogen peroxide
Toxicology: LD50 (rat, oral) 202 mg/kg; ACGIH TLV:TWA 0.05 mg (Co)/m^3; poison by ingestion, subcut. routes; experimental carcinogen and tumorigen
Uses: Pigments for ceramics and paints; oxidation catalyst for drying oils; preparation of cobalt-metal catalysts, cobalt salts, coloring glass; feed additive
Manuf./Distrib.: Aldrich; Atomergic Chemetals; Brandeis; Cerac; Great Western; The Hall Chem. Co; Noah; OM Group; Reade Advanced Materials

Cobalt sulfate. *See* Cobalt sulfate (ous)
Cobalt sulfate (1:1). *See* Cobalt sulfate (ous)
Cobalt (II) sulfate. *See* Cobalt sulfate (ous)
Cobalt (II) sulfate (1:1). *See* Cobalt sulfate (ous)

Cobalt sulfate (ous)
CAS 10124-43-3
Synonyms: Cobalt sulfate; Cobalt sulfate (1:1); Cobaltous sulfate; Cobalt (II) sulfate; Cobalt (II) sulfate (1:1)
Definition: Obtained by action of sulfuric acid on cobaltous oxide
Empirical: $O_4S \cdot Co$
Formula: $CoSO_4$
Properties: Red to lavender dimorphic, orthorhombic cryst.; dissolves slowly in boiling water; m.w. 154.99; dens. 3.71; stable to 708 C
Toxicology: TLV:TWA 0.05 mg(Co)/m^3; LD50 (oral, rat) 424 mg/kg; poison by IV, IP routes; mod. toxic by ingestion; heated to decomp., emits toxic fumes of SO_x
Uses: Ceramics, pigments, glazes, in plating baths for cobalt, additive to soils, catalyst, paint and ink drier, storage batteries
Regulatory: FDA 21CFR §173.310 (catalyst in boiler water), 189.120 (prohibited from direct addition or use in human food)
Manuf./Distrib.: AC Ind.; Atomergic Chemetals; Barker Ind.; Brandeis; Elf Atochem N. Am.; IMC Group; Johnson Matthey plc; Mallinckrodt; McGean-Rohco; Nihon Kagaku Sangyo; OM Group; San Yuan; Shepherd; Spectrum Chem. Mfg.

Cobalt tallate
Definition: Cobalt deriv. of refined tall oil, of varying composition
Precaution: Combustible
Toxicology: Heated to decomp., emits acrid smoke and irritating fumes
Uses: Drier for paints
Regulatory: FDA 21CFR §181.25
Manuf./Distrib.: Akzo Nobel; Baychem; Dussek Campbell; OM Group; Shepherd

Cocamidopropylamine oxide
CAS 68155-09-9; EINECS 268-938-5
Synonyms: Cocamidopropyl dimethylamine oxide; Coco amides, N-[3-(dimethylamino)propyl], N-oxide; N-[3-(Dimethylamino)propyl]coco amides-N-oxide
Classification: Tertiary amine oxide
Formula: $RCO\text{-}NH(CH_2)_3N(CH_3)_3O$, RCO- represents the coconut fatty acids
Uses: Detergent, wetting agent, emulsifier, softener, conditioner, foam booster/stabilizer for rug shampoos, laundry detergents, dishwashing, shampoos, cleaners, antistatic softeners; foam stabilizer in foam rubber, electroplating, paper coatings
Trade names: Rhodamox® CAPO

Cocamidopropyl dimethylamine oxide. *See* Cocamidopropylamine oxide

Cocamine
CAS 61788-46-3; EINECS 262-977-1
Synonyms: Coconut amine
Classification: Primary aliphatic amine
Formula: RNH_2, R represents the coconut radical
Uses: Emulsifier, flotation agent, corrosion inhibitor, stripping agent for paints; mold release for rubber and plastics
Trade names: Armeen® C; Armeen® CD

Cocamine acetate
CAS 61790-57-6; EINECS 263-147-1
Synonyms: Amines, coco alkyl, acetates
Uses: Surface coating agent for pigments, anticaking agent for fertilizer; emulsifier, dispersant, and softening agent for textiles; min. flotation reagent; emulsifier, bactericide, corrosion inhibitor, flocculant

Cocaminobutyric acid
CAS 68649-05-8; EINECS 272-021-5
Synonyms: Butanoic acid, 3-amino-, N-coco alkyl derivatives; 3-Aminobutanoic acid, n-coco alkyl derivatives
Classification: Substituted amino acid
Formula: $R\text{-}NH\text{-}CHCH_2COOHCH_3$, R represents the coconut radical
Uses: Pigment softening, dispersing agent; antifogging agent, foam booster, stabilizer, wetting agent in alkaline paint strippers, latex emulsions, latex rubber reclamation, inks, plastic films, cosmetics; cooling tower corrosion inhibitor
Trade names: Armeen® Z

Cocoalkonium chloride
Synonyms: Coco dimonium chloride; Coco dimethyl benzyl ammonium chloride; Cocoalkyl dimethyl benzyl ammonium chloride
Classification: Quaternary ammonium salt
Uses: Bactericide, fungicide, sanitizer, disinfectant, swimming pool treatment, demulsification of hydrocarbons, cosmetics, latex coagulation, flotation, electrostatic paints

Cocoalkyl dimethyl benzyl ammonium chloride. *See* Cocoalkonium chloride
Coco amides, N-[3-(dimethylamino)propyl], N-oxide. *See* Cocamidopropylamine oxide
Coco dimethyl amine. *See* Dimethyl cocamine
Coco dimethyl benzyl ammonium chloride. *See* Cocoalkonium chloride
Coco dimonium chloride. *See* Cocoalkonium chloride
Coconut amine. *See* Cocamine
Coconut butter. *See* Coconut oil
Coconut hydroxyethyl imidazoline. *See* Cocoyl hydroxyethyl imidazoline

Coconut oil
CAS 8001-31-8; EINECS 232-282-8
Synonyms: Copra oil; Coconut butter; Coconut palm oil
Classification: Saturated fat
Definition: Fixed oil obtained from kernels of seeds of *Cocos nucifera*
Properties: Wh. fatty solid or liq., sweet nutty taste; very sol. in chloroform, ether, CS_2; pract. insol. in water; dens. 0.903 (0/4 C); m.p. 21-27 C; acid no. < 6; iodine no. 8-9.5; sapon. no. 255-258; ref. index 1.4485-1.4495; surf. tens. 33.4 dynes/cm
Precaution: Flamm. solid when exposed to heat or flame; may spontaneously heat and ignite if stored wet and hot
Toxicology: May cause allergic skin reaction
Uses: Emollient, superfatting agent, clouding agent, detergent, wetting agent, emulsifier used in cosmetic, and pharmaceutical industries; drying oil for paints; coating agent, emulsifier, formulation aid, texturizer in baked goods, candy, desserts, margarine
Regulatory: FDA 21CFR §175.105, 175.300, 176.200, 176.210, 177.2800, 182.70; GRAS; FDA approved for orals, topicals; BP compliance
Manuf./Distrib.: ABITEC; Akzo Nobel; Aldrich; Alnor Oil; Amber Syn.; Arista Ind.; Lomas Int'l.; Rhone-Poulenc Surf. & Spec.; Sigma; Spectrum Chem. Mfg.; Tri-K Ind.; Welch, Holme & Clark
Trade names containing: Centrophil® M

Coconut palm oil. *See* Coconut oil

Cocoyl hydroxyethyl imidazoline
CAS 61791-38-6; EINECS 263-170-7
Synonyms: 1H-Imidazole-1-ethanol, 4,5-dihydro-2-norcocoyl-; Cocoyl imidazoline; Coconut hydroxyethyl imidazoline
Classification: Heterocyclic compd.
Formula: $R=N-N-(CH_2)_2OH$, R is derived from coconut fatty radical
Uses: Emulsifier, antistat, corrosion inhibitor, softener for textiles, plastics, asphalt, tar emulsion breakers, paints, printing inks; water repellent treatment of cement, concrete, and plaster; fungicide
Trade names: Monazoline C; Schercozoline C; Varine C

Cocoyl imidazoline. *See* Cocoyl hydroxyethyl imidazoline
Collodion. *See* Nitrocellulose
Collodion cotton. *See* Nitrocellulose
Collodion wool. *See* Nitrocellulose
Colloidal activated attapulgite. *See* Attapulgite
Colloidal silica. *See* Silica, colloidal
Cologel. *See* Methylcellulose
Colophane. *See* Rosin
Colophonium. *See* Rosin
Colophony. *See* Rosin
Columbian spirits. *See* Methyl alcohol
Colza oil. *See* Rapeseed oil

Copper
CAS 7440-50-8; EINECS 231-159-6
Synonyms: Bronze powder; Copper bronze; Gold bronze
Formula: Cu
Properties: Reddish metal; at.wt. 63.54; dens. 8.96; m.p. 1083 C; b.p. 2595 C
Precaution: Fine particles may burn in air; reacts violently with ammonium nitrate, bromates, iodates, chlorates, hydrogen peroxide, sodium peroxide, sulfuric acid
Toxicology: TLV 0.2 mg/m^3 (fume), 1 mg/m^3 (dusts and mists); poison to humans by ingestion, inhalation; may cause damage to liver, kidneys, spleen, and blood
Uses: Electric wiring, switches, plumbing, heating, roofing; chemical and pharmaceutical machinery; alloys; coatings; cooking utensils; dietary supplement; herbicide for potable water
Regulatory: FDA 21CFR §193.90, herbicides residue tolerance 1 ppm in potable water; GRAS for use in dietary supplements; Japan restricted; BP compliance
Manuf./Distrib.: Aldrich; Alfa Aesar Johnson Matthey; Am. Chemet; Asarco; Atotech USA; Belmont Metals; Fluka; Fry's Metals Ltd; Noah

Copperas. *See* Ferrous sulfate heptahydrate
Copper bronze. *See* Copper

Copper linoleate
CAS 7721-15-5
Synonyms: (Z,Z)-9,12-Octadecadienoic acid, copper salt
Empirical: $C_{18}H_{32}O_2 \cdot xCu$
Uses: Drier for paints
Manuf./Distrib.: Troy

Copper metallic powder. *See* Copper powder

Copper naphthenate
CAS 1338-02-9
Synonyms: Naphthenic acid copper salt
Definition: Copper salt of petroleum naphthenic acids
Empirical: $C_{13}H_{25}CuO_2$
Properties: Green-blue solid; sol. in gasoline, benzene, and min. oil distillates; m.w. 276.89; dens. 1.055; flash pt. 100 F
Precaution: Fire hazard when exposed to heat or flame
Toxicology: Poison by ingestion and inhalation
Uses: Wood, canvas and rope preservative; drier and antifouling in paints; insecticide, fungicide
Manuf./Distrib.: Akzo Nobel; Archway Sales; Baychem; Boehle; Dussek Campbell Ltd; Hüls Am.; ISK Biosciences; KMZ Chem. Ltd; Lomas Int'l.; D.N. Lukens; OM Group; Troy
Trade names: Fungitrol® Copper 8%

Copper (III) nitrate. *See* Copper nitrate (ic)

Copper nitrate (ic)
CAS 10031-43-3; EINECS 221-838-5
Synonyms: Cupric nitrate trihydrate; Cupric nitrate hexahydrate; Cupric nitrate; Copper (III) nitrate
Empirical: $CuN_2O_6 \cdot 3H_2O$ or $CuN_2O_6 \cdot 6H_2O$
Formula: $Cu(NO_3)_2 \cdot 3H_2O$ or $Cu(NO_3)_2 \cdot 6H_2O$
Properties: Trihydrate: Dk. blue cryst., deliq.; sol. in water and ethanol; m.w. 241.60; Hexahydrate: Blue cryst. flakes, deliq.; sol. in water, ethanol; m.w. 295.66; sp.gr. 2.0
Uses: Light-sensitive papers, analytical reagent, textile dyeing mordant, nitrating agent, insecticide, coloring copper black, electroplating, paints, varnishes, enamels, pharmaceuticals, catalyst, copper salts, fuel additives
Manuf./Distrib.: Blythe, William Ltd; Fluka; Mallinckrodt; Sigma; Spectrum Chem. Mfg.

Copper octoate
Uses: Drier for paints
Manuf./Distrib.: Archway Sales; Baychem; Hüls Am.; Lomas Int'l.; D.N. Lukens

Copper (I) oxide. *See* Copper oxide (ous)
Copper (II) oxide. *See* Copper oxide (ic)

Copper oxide (ic)
CAS 1317-38-0; EINECS 215-269-1
Synonyms: Cupric oxide; Copper (II) oxide
Empirical: CuO
Formula: CuO
Properties: Blk. fine powd.; m.w. 79.54; dens. 6.4; b.p. dec. @ 1026 C
Precaution: Incompat. with metals and reductants; explodes when heated with powd. aluminum, hydrogen, magnesium, etc.; ignites on contact with hydrogen trisulfide; incandescent reactions
Uses: Ceramic colorant, reagent in analytical chemistry, fungicide, insecticide, catalyst, purification of hydrogen, batteries and electrodes, electroplating, solvent, desulfurizing oils, rayon, metallurgical and welding fluxes, antifouling paints; phosphors; trace mineral in animal feeds
Manuf./Distrib.: Aldrich; Am. Chemet; Carbochem; Cerac; Chemisphere Ltd; Fluka; Goldschmidt Ind. Chems.; Nihon Kagaku Sangyo; Noah; SCM Metal Prods.; Shance; Sigma; Spectrum Chem. Mfg.

Copper oxide (ous)
CAS 1317-39-1
Synonyms: Copper suboxide; Copper protoxide; Cuprous oxide; CI 77402; Dicopper monoxide; Red copper oxide; Red cuprous oxide; Brown copper oxide; Yellow cuprocide; Copper (I) oxide
Empirical: Cu_2O
Properties: Red cubic cryst.; insol. in water; sol. in acids and ammonia; m.w. 143.08; dens. 6.0; m.p. 1235 C; b.p. loses O_2 @ 1800 C
Precaution: Violent, potentially explosive reaction with conc. peroxyformic acid; violent reaction when heated with aluminum
Toxicology: ACGIH TLV:TWA 1 mg(Cu)/m^3; LD50 (oral, rat) 470 mg/kg; mod. toxic by ingestion; experimental reproductive effects
Uses: Red pigment for paints; antifouling paints; fungicide; porcelain red glaze; red glass; electroplating
Manuf./Distrib.: Am. Chemet; Goldschmidt Ind. Chems.; Noah; Punda Mercantile; Revelli; SCM Metal Prods.; Spectrum Chem. Mfg.; Van Waters & Rogers

Copper powder
CAS 7440-50-8; EINECS 231-159-6
Synonyms: CI 77400; Copper metallic powder
Definition: Color additive consisting of powdered metallic copper
Empirical: Cu
Properties: Powd.; a.w. 63.54
Precaution: DOT: Flamm. solid
Toxicology: TLV 0.2 mg/m^3 (fume), 1 mg/m^3 (dusts and mists); poison to humans by ingestion
Uses: Color additive; metallic paint pigment
Regulatory: FDA 21CFR §73.1647, 73.2647
Manuf./Distrib.: Aarbor Int'l.; Aldrich; Am. Chemet; Atlantic Equip. Engrs.; Atomergic Chemetals; BASF; Canbro; Crescent Bronze Powd.; Fluka; Lenape; Noah; Obron Atlantic; Punda Mercantile; Reade Advanced Materials; SCM Metal Prods.; Sheffield Bronze Paint; Spectrum Chem. Mfg.; U.S. Bronze Powds.; Zinc Corp. of Am.

Copper protoxide. *See* Copper oxide (ous)
Copper suboxide. *See* Copper oxide (ous)
Copper sulfate. *See* Cupric sulfate anhyd.
Copper (II) sulfate. *See* Cupric sulfate anhyd.
Copper sulfate (ic). *See* Cupric sulfate pentahydrate
Copper sulfate pentahydrate. *See* Cupric sulfate pentahydrate
Copper sulfate tribasic. *See* Cupric sulfate anhyd.

Copper tallate
CAS 61789-22-8
Synonyms: Tall oil, copper salt
Classification: Metallic soap
Uses: Drier for paints
Manuf./Distrib.: Dussek Campbell Ltd

Copra oil. *See* Coconut oil

Corn oil
CAS 8001-30-7; EINECS 232-281-2
Synonyms: Maize oil; Zea mays oil

Corn starch

Definition: Refined fixed oil obtained from wet milling of corn, *Zea mays*
Properties: Pale yel. oily liq., faint char. odor and taste; insol. in water; sol. in ether, chloroform, amyl acetate, benzene, CS_2; sl. sol. in alcohol; dens. 0.914-0.921; m.p. -10 C; acid no. 2-6; iodine no. 109-133; sapon. no. 187-193; flash pt. 321 C; ref. index 1.470-1.474
Precaution: Combustible liq. when exposed to heat or flame; dangerous spontaneous heating may occur; light-sensitive
Toxicology: Nontoxic; human skin irritant; experimental teratogen; may be an allergen
Uses: Used in preparing foodstuffs, coating agent, emulsifier, formulation aid, texturizer, in bakery prods., margarine, salad oil, dietary supplement; soap, lubricants, leather finishing, hair dressing; solvent; paints
Regulatory: FDA 21CFR §175.105, 175.300, 176.200, 176.210, GRAS; FDA approved for injectables, orals, topicals; USP/NF, BP, JP compliance
Manuf./Distrib.: ABITEC; Alba Int'l.; Arista Ind.; Ashland; Fluka; Grain Processing; Penta Mfg.; Sigma; A.E. Staley Mfg.; Tri-K Ind.; Welch, Holme & Clark

Corn starch
CAS 9005-25-8; EINECS 232-679-6
Synonyms: Starch, corn; Maize starch
Definition: Granules obtained from mature grains of corn, *Zea mays*; carbohydrate polymer consisting primarily of amylose and amylopectin
Properties: Wh. powd. or spheroidal gran.
Toxicology: No ill effects unless massive doses are given; use of powd. in rubber gloves may cause contact urticaria
Uses: Source of glucose; filler in baking powder; thickening agent in food prods.; dietary supplement; adhesives, coatings; additive in plastics
Regulatory: FDA 21CFR §175.105, 178.3520, 182.70, 182.90; GRAS for use in dietary supplements; BP, Ph.Eur. compliance
Manuf./Distrib.: ADM Corn Processing; Aldrich; Am. Maize Prods.; Cerestar UK; Grain Processing; Nalco; Nat'l. Starch & Chem.; Penford Prods.; Sigma; A.E. Staley Mfg.

Corn sugar gum. *See* Xanthan gum
Cosmetic talc. *See* Talc
Cotton fiber. *See* Cellulose
Cotton oil. *See* Cottonseed oil

Cottonseed oil
CAS 8001-29-4; EINECS 232-280-7
Synonyms: Deodorized winterized cottonseed oil; Cotton oil; Oleum Gossypii seminis; Aceite de Algodon
Definition: Refined fixed oil from seeds of *Gossypium hirsutum*
Properties: Pale yel. oily liq., nearly odorless; sol. in ether, benzene, chloroform; sl. sol. in alcohol; dens. 0.915-0.921; f.p. 0-5 C; solid. pt. 31-33 C; iodine no. 109-120; sapon. no. 190-198; flash pt. (CC) 486 F
Precaution: Combustible liq. when exposed to heat of flame; may be dangerous hazard due to spontaneous heating; light-sensitive
Toxicology: Experimental tumorigen and teratogen; allergen
Uses: Coating agent, emulsifier, formulation aid, texturizer for use in cooking oil, margarine, salad oil, shortening; drying oil for paints
Usage level: 56-92% (intramuscular preps.), 0.002-402 mg (solid oral dosage forms)
Regulatory: FDA 21CFR §175.105, 175.300, 176.200, 176.210, 177.2800, GRAS; FDA approved for intramuscular injectables, orals; USP/NF compliance
Manuf./Distrib.: ABITEC; Aldrich; Alnor Oil; Amber Syn.; Arista Ind.; Ruger; Sigma; Tri-K Ind.; Welch, Holme & Clark
Trade names containing: Centrophil® M

Coumarone-indene resin
Classification: Thermosetting resin
Properties: Soft and sticky at R.T.; hardens on heating to solid; soften. pt. 126 C; ref. index 1.63-1.64
Toxicology: Heated to decomp., emits acrid smoke and irritating fumes
Uses: Extender in epoxy systems, adhesives, printing inks, floor tile binder; friction tape; paints; varnishes, enamels; protective coating on grapefruit, lemons, limes, oranges, tangerines; in chewing gum
Regulatory: FDA 21CFR §172.215
Manuf./Distrib.: Allchem Ind.; Natrochem; Neville
Trade names: Cumar® LX-509; Cumar® P-10; Cumar® P-25; Cumar® R-1; Cumar® R-3 (W-1); Cumar® R-5 (W-1$^1/_2$); Cumar® R-6 (W-2, 2$^1/_2$); Cumar® R-7; Cumar® R-9; Cumar® R-10 (V-1); Cumar® R-11 (V-1$^1/_2$); Cumar® R-12 (V-2, V-2$^1/_2$); Cumar® R-12A (V-3); Cumar® R-13; Cumar® R-14 (MH-1); Cumar® R-15

(MH-1½); Cumar® R-16 (MH-2, MH-2½); Cumar® R-16A (MH-3); Cumar® R-17 (RH-17); Cumar® R-19; Cumar® R-21; Cumar® R-27; Cumar® R-28; Cumar® R-29, 10°; Cumar® R-29, 25°; Epodil® VFT-V6

C11-15 pareth-7
CAS 68131-40-8 (generic)
Synonyms: Pareth-15-7
Definition: PEG ether of a mixture of syn. C11-15 fatty alcohols with avg. 7 moles ethylene oxide
Uses: Detergent, emulsifier, wetting agent, stabilizer used in textile processing applics.; textile lubricant; dyeing aid, dye leveling agent for leather and textile processing; household and industrial detergents; paper, paints, agric., metal cleaners; oilfield chems., water treatment
Regulatory: FDA 21CFR §178.3400
Manuf./Distrib.: Sigma
Trade names: Tergitol® 15-S-7

C11-15 pareth-9
CAS 68131-40-8 (generic)
Synonyms: Pareth-15-9
Definition: PEG ether of a mixture of syn. C11-15 fatty alcohols with avg. 9 moles of ethylene oxide
Uses: Detergent, emulsifier, wetting agent, stabilizer used in textile processing applics.; textile lubricant; dyeing aid, dye leveling agent for leather and textile processing; household and industrial detergents; paper, paints, agric., metal cleaners; oilfield chems., water treatment
Regulatory: FDA 21CFR §178.3400
Manuf./Distrib.: Sigma
Trade names: Tergitol® 15-S-9

C11-15 pareth-12
CAS 6813-14-0 (generic)
Synonyms: Pareth-15-12
Definition: PEG ether of a mixture of syn. C11-15 fatty alcohols with avg. 12 moles of ethylene oxide
Uses: Detergent, emulsifier, wetting agent, stabilizer used in textile processing applics.; textile lubricant; dyeing aid, dye leveling agent for leather and textile processing; household and industrial detergents; paper, paints, agric., metal cleaners; oilfield chems., water treatment
Regulatory: FDA 21CFR §178.3400
Trade names: Tergitol® 15-S-12

C12-15 pareth-7
CAS 68131-39-5 (generic)
Synonyms: Pareth-25-7; PEG-7 C12-15 fatty alcohol ether
Definition: PEG ether of a mixture of syn. C12-15 fatty alcohols with avg. 7 moles of ethylene oxide
Uses: Detergent, emulsifier, wetting agent, dispersant for sanitizers, metal cleaners, hard surface cleaners; detergent intermediate; for paper, paint, leather; coupler and solubilizer for perfumes and org. additives
Trade names: Rhodasurf® 25-7

C12-15 pareth-9
CAS 68131-39-5 (generic)
Synonyms: Pareth-25-9
Definition: PEG ether of a mixture of syn. C12-15 fatty alcohols with avg. 9 moles of ethylene oxide
Uses: Detergent intermediate used in preparation of sulfates for high-foaming liq. detergents; textile processing, metal cleaners, paper, paint, leather applics.
Trade names: Rhodasurf® 25-9

C20-40 pareth-3
Definition: PEG ether of a mixture of syn. C20-40 alcohols with avg. 3 moles ethylene oxide
Uses: Emulsifier for cosmetics; vehicle; coatings, metalworking fluids, pulp/paper processing, textiles; mold release for plastics
Trade names: Unithox® 420

C20-40 pareth-10
Definition: PEG ether of a mixture of syn. C20-40 alcohols with avg. 10 moles ethylene oxide
Uses: Component for mold release agents; metalworking additive; coatings, pulp/paper processing, textiles
Trade names: Unithox® 450; Unithox® D-100 Disp.
Trade names containing: Petrolite® 01 Disp.

C20-40 pareth-40
Definition: PEG ether of a mixture of syn. C20-40 alcohols with avg. 40 moles ethylene oxide
Uses: Emulsifier for cosmetics; vehicle; coatings, metalworking fluids, pulp/paper processing, textiles; mold release for plastics
Trade names: Unithox® 480

C30-50 pareth-3
Definition: PEG ether of a mixture of syn. C30-50 alcohols with avg. 3 moles ethylene oxide
Uses: O/w emulsifier for cosmetics, vehicle for inert, difficult-to-disperse colorants, oils, and waxes; also for coatings, metalworking fluids, pulp/paper processing, textiles, mold releases
Trade names: Unithox® 520

C30-50 pareth-10
Definition: PEG ether of a mixture of syn. C30-50 alcohols with avg. 10 moles ethylene oxide
Uses: Emulsifier for cosmetics; vehicle; coatings, metalworking fluids, pulp/paper processing, textiles; mold release for plastics
Trade names: Unithox® 550; Unithox® D-150 Disp.

C40-60 pareth-3
Definition: PEG ether of a mixture of syn. C40-60 alcohols with avg. 3 moles of ethylene oxide
Uses: O/w emulsifier for cosmetics, vehicle for inert, difficult-to-disperse colorants, oils, and waxes; provides silky lubricating feel, superior film-forming chars.; also for coatings, metalworking fluids, pulp/paper processing, textiles, mold releases
Trade names: Unithox® 720

C40-60 pareth-10
Definition: PEG ether of a mixture of syn. C40-60 alcohols with avg. 10 moles ethylene oxide
Uses: Emulsifier for cosmetics; vehicle; coatings, metalworking fluids, pulp/paper processing, textiles; mold release for plastics
Trade names: Unithox® 750; Unithox® D-300 Disp.
Trade names containing: Petrolite® 07 Disp.; Petrolite® 29 Disp.; Petrolite® 37 Disp.

CR. *See* Polychloroprene
Cresol. *See* Cresylic acid
p-Cresol dicyclopentadiene butylated polymer. *See* p-Cresol/dicyclopentadiene butylated reaction product

p-Cresol/dicyclopentadiene butylated reaction product
CAS 68610-51-5; EINECS 271-867-2
Synonyms: p-Cresol dicyclopentadiene butylated polymer
Properties: Powd.; sol. in acetone, ethyl acetate, ethanol, methylene chloride, liq. phosphites; insol. in water; m.w. 600-700; dens. 1.10; flash pt. (COC) 215 C
Uses: Antioxidant for rubber goods esp. latex applics.; also for EVA hot melts, thermoplastic rubber, styrenics
Trade names: Ralox® LC

Cresyl diphenyl phosphate. *See* Diphenylcresyl phosphate

o-Cresyl glycidyl ether
CAS 2210-79-9
Synonyms: Glycidyl 2-methylphenyl ether
Empirical: $C_{10}H_{12}O_2$
Properties: M.w. 164.20
Toxicology: Irritant
Uses: Reactive epoxy diluent for tooling, elec. applics, coatings, flooring, casting, laminating, and decoupage
Manuf./Distrib.: Aldrich; Ashland; Raschig
Trade names: Epodil® 742
Trade names containing: EPON® Resin 813

Cresylic acid
CAS 1319-77-3; EINECS 215-293-2
Synonyms: Cresol; Tricresol
Definition: Commercial mixtures of phenolic materials boiling above the cresol range
Empirical: C_7H_8O
Properties: Ylsh. to brn.-yel. or pink liq.; phenolic odor; m.w. 108.15; dens. 1.030-1.038; m.p. 10.9-33.5 C; b.p.

191-203 C; flash pt. 178 F
Precaution: Flamm. when exposed to heat or flame
Toxicology: LD50 (oral, rat) 1454 mg/kg; poison; mod. toxic by ingestion; corrosive to skin
Uses: Phosphate esters, phenolic resins, wire enamel solvent, plasticizers, gasoline additives, laminates, coating for magnet wire, disinfectants, metal cleaning, flotation agents, surfactants, intermediates, oil additives, solvent refining, pesticides; exc. electrical insulators
Manuf./Distrib.: Allchem Ind.; Crowley Tar Prods.; PMC Specialties; Spectar Ltd

Cresyl phosphate. *See* Tricresyl phosphate

Crotonic acid
CAS 3724-65-0
Synonyms: β-Methacrylic acid; 2-Butenoic acid
Classification: Aliphatic organic compd.
Empirical: $C_4H_6O_2$
Formula: $CH_3CH{:}CHCOOH$
Properties: Colorless needle-like cryst.; m.w. 86.09; dens. 1.018 (15/4 C); m.p. 72 C; b.p. 185 C; flash pt. (COC) 190 F
Precaution: DOT: Corrosive material; flamm. when exposed to heat and flames
Toxicology: LD50 (oral, rat) 1000 mg/kg; poison by intraperitoneal route; powerful irritant
Uses: Synthesis of resins, polymers, plasticizers, paints, drugs
Manuf./Distrib.: Aldrich; Allchem Ind.; Atomergic Chemetals; Chisso Am.; Eastman; Hoechst Celanese; Penta Mfg.; Spectrum Chem. Mfg.

Crystalline cellulose. *See* Cellulose
CSM. *See* Polyethylene elastomer, chlorosulfonated
CSP. *See* Cupric sulfate pentahydrate
Cubic niter. *See* Sodium nitrate

Cumene hydroperoxide
CAS 80-15-9
Synonyms: α,α-Dimethylbenzyl hydroperoxide
Classification: Organic peroxide
Empirical: $C_9H_{12}O_2$
Formula: $C_6H_5C(CH_3)_2OOH$
Properties: Colorless to pale yel. liq.; sl. sol. in water; sol. in alcohol, acetone, esters, hydrocarbons; m.w. 152.2; flash pt. 175 F
Precaution: Combustible; avoid contact with strong min. acids, other oxidizers, reducing agents, accelerators
Toxicology: Toxic by inhalation and skin absorption
Uses: Prod. of acetone and phenol; polymerization catalyst; curing agent for polyester resins; used in gel coats
Manuf./Distrib.: AC Ind.; Aldrich; Aristech; Elf Atochem N. Am.; Fluka; JLM Marketing; Monomer-Polymer & Dajac Labs; Sigma; Witco/Polymer Addit.
Trade names: CHP-5

Cumylphenyl acetate
Properties: Sp.gr. 1.03; m.p. < 0 C; flash pt. (COC) 250 C
Uses: Plasticizer for urethanes; high flash pt. reactive diluent for epoxy, for polyamide-cured epoxy floorings and coal tar pipe coatings; comonomer and impact modifier for phenolics
Manuf./Distrib.: Kenrich Petrochems.
Trade names: Kenplast® ES-2HP

Cupric nitrate. *See* Copper nitrate (ic)
Cupric nitrate hexahydrate. *See* Copper nitrate (ic)
Cupric nitrate trihydrate. *See* Copper nitrate (ic)
Cupric oxide. *See* Copper oxide (ic)
Cupric sulfate. *See* Cupric sulfate pentahydrate

Cupric sulfate anhyd.
CAS 7758-98-7; EINECS 231-847-6
Synonyms: Copper sulfate; Copper (II) sulfate; Copper sulfate tribasic
Empirical: CuO_4S
Formula: $CuSO_4$
Properties: Blue cryst. or cryst. gran. or powd., nauseous metallic taste; sol. in water; insol. in alcohol; m.w.

159.60; dens. 3.6

Precaution: Reacts violently with hydroxylamine, magnesium

Toxicology: LD50 (oral, rat) 300 mg/kg; strong irritant; experimental tumorigen; human poison, systemic effects by ingestion: gastritis, diarrhea, nausea, vomiting, hemolysis; no known skin toxicity; mutagenic data; heated, emits toxic fumes of SO_x

Uses: Soil additive, pesticides, feed additive, germicides, textile mordant, leather, pigments, batteries, electroplated coatings, copper salts, analytical reagent, medicine, wood preservative, lithography, ore flotation, petroleum, rubber, steel mfg.

Regulatory: FDA 21CFR §184.1261, 582.80; Japan approved (0.6 mg/L as copper in milk); FDA approved for orals

Manuf./Distrib.: Aldrich; Allchem Ind.; Farleyway Chem. Ltd; Fluka; Sigma; Spectrum Chem. Mfg.

Cupric sulfate pentahydrate

CAS 7758-99-8; EINECS 231-847-6

Synonyms: CSP; Copper sulfate pentahydrate; Copper sulfate (ic); Cupric sulfate; Blue vitriol; Blue stone; Blue copperas

Classification: Inorganic salt

Empirical: $CuO_4S \cdot 5H_2O$

Formula: $CuSO_4 \cdot 5H_2O$

Properties: Blue cryst. or cryst. gran. or powd., nauseous metallic taste; very sol. in water; sol. in methanol, glycerin; sl. sol. in ethanol; m.w. 249.70; dens. 2.286 (15.6/4 C)

Toxicology: TLV:TWA 1 mg (Cu)/m^3; LD50 (oral, rat) 960 mg/kg; human poison by unspecified routes; moderately toxic by ingestion; heated to dec., emits toxic fumes of SO_x

Uses: Soil additive, pesticides, feed additive, germicides, textile mordant, leather, pigments, batteries, electroplated coatings, copper salts, analytical reagent, medicine, wood preservative, lithography, ore flotation, petroleum, rubber, steel mfg.

Regulatory: FDA 21CFR §184.1261, GRAS

Manuf./Distrib.: Aldrich; Allchem Ind.; Farleyway Chem. Ltd; Fluka; Sigma; Spectrum Chem. Mfg.

Cuprous oxide. *See* Copper oxide (ous)
Cyamopsis gum. *See* Guar gum
Cyanobenzene. *See* Benzonitrile
Cyanoethylene. *See* Acrylonitrile
Cyanoguanidine. *See* Dicyandiamide
Cyanourotriamine. *See* Melamine
Cyanuramide. *See* Melamine
Cyanurotriamide. *See* Melamine

Cyclodextrin

CAS 7585-39-9 (β); 10016-20-3 (α); 17465-86-0 (γ); EINECS 231-493-2 (β); 233-007-4 (α); 241-482-4 (γ)

Synonyms: α: α-Cyclodextrin; Cyclomaltohexaose; Schardinger α-dextrin; β: β-Cyclodextrin; Cyclomaltoheptaose; γ: γ-Cyclodextrin; Cyclomaltooctaose

Definition: Cyclic polysaccharide comprised of six to eight glucopyranose units

Properties: M.w. 972.86; m.p. 278 (dec.)

Uses: Complex hosting guest molecules; increases the sol. and bioavailability of other substances; masks flavor, odor, or coloration; stabilizes against light, oxidation, heat, and hydrolysis; turns liqs. or volatiles into stable solid powds.; for use in pharmaceuticals, cosmetics, toiletries, foods, tobacco, pesticides, textiles, paints, plastics, synthesis, polymers; Japan approved; not permitted in certain foods

Manuf./Distrib.: Aldrich; Am. Maize Prods.; Fluka; Janssen Chimica; Pfanstiehl Labs; Sigma; U.S. Biochemical; Wacker Chems.

Trade names: Beta W7; Gamma W8

α-Cyclodextrin. *See* Cyclodextrin

β-Cyclodextrin

CAS 7585-39-9; EINECS 231-493-2

Synonyms: Cyclomaltoheptaose; Schardinger β-dextrin

Definition: Cyclic polysaccharide comprised of 7 glucopyranosyl units

Empirical: $C_{42}H_{70}O_{35}$

Formula: $(C_6H_{10}O_5)_7$

Properties: Wh. fine cryst. powd., pract. odorless, sl. sweet taste; sparingly sol. in water; m.p. 290-300 C (dec.)

Uses: Complex hosting guest molecules; increases the sol. and bioavailability of other substances; masks

flavor, odor, or coloration; stabilizes against light, oxidation, heat, and hydrolysis; turns liqs. or volatiles into stable solid powds.; for use in pharmaceuticals, cosmetics, toiletries, foods, tobacco, pesticides, textiles, paints, plastics, synthesis, polymers
Regulatory: USP/NF compliance
Manuf./Distrib.: Am. Maize Prods.; Fluka; Sigma; Wacker Chems.
See also Cyclodextrin

γ-Cyclodextrin. *See* Cyclodextrin

Cyclo (dioctyl) pyrophosphato dioctyl titanate
Trade names containing: Ken-React® KR OPPR

Cyclohexane
CAS 110-82-7; EINECS 203-806-2
Synonyms: Hexahydrobenzene; Hexamethylene; Hexanaphthene
Classification: Aliphatic organic compd.
Empirical: C_6H_{12}
Properties: Colorless mobile liq., pungent odor; insol. in water; sol. in alcohol, acetone, benzene; m.w. 84.16; dens. 0.779 (20/4 C); m.p. 6.5 C; b.p. 80.7 C; f.p. 6.3 C; flash pt. (CC) -18.3 C; ref. index 1.4264
Precaution: DOT: Flamm. liq.; dangerous fire hazard exposed to heat or flame; reactive with oxidizers; mod. explosion hazard as vapor exposed to flame; explosive mixed hot with liq. dinitrogen tetraoxide
Toxicology: TLV 300 ppm in air; LD50 (oral, rat) 29,820 mg/kg; poison by IV; mod. toxic by ingestion, inhalation, skin contact; systemic and skin irritant; high concs. may act as narcotic; mutagenic data; heated to decomp., emits acrid smoke, irritating fumes
Uses: Mfg. of nylon; solvent for cellulose ethers, fats, oils, waxes, coatings; paint and varnish remover; glass substitutes; in analytic chemistry; chemical intermediate; in fungicidal formulations
Regulatory: FDA approved for orals; BP compliance
Manuf./Distrib.: Aldrich; Ashland; Exxon Europe; Fluka; Huntsman; Phillips; Sigma; Spectrum Chem. Mfg.; Texaco; Triple Crown Am.

1,2-Cyclohexanediamine. *See* 1,2-Diaminocyclohexane

1,4-Cyclohexanedicarboxylic acid
CAS 1076-97-7
Synonyms: CHDA
Uses: Monomer for resin processing and coatings applics.
Manuf./Distrib.: Aldrich
Trade names: 1,4-CHDA

1,2-Cyclohexanedicarboxylic anhydride. *See* Hexahydrophthalic anhydride

1,4-Cyclohexanedimethanol
CAS 105-08-8; EINECS 203-268-9
Synonyms: CHDM; 1,4-Bis(hydroxymethyl)cyclohexane (cis and trans)
Empirical: $C_8H_{16}O_2$
Formula: $C_6H_{10}(CH_2OH)_2$
Properties: M.w. 144.21; b.p. 283 C; flash pt. 161 C
Precaution: Combustible
Uses: Monomer for polymer applics., paints
Manuf./Distrib.: Aldrich; Eastman; Fluka; Hüls Am.

1,4-Cyclohexanedimethanol dibenzoate
CAS 35541-81-2
Synonyms: Hexahydro terephthalic acid
Formula: $C_6H_{10}(COOH)_2$
Properties: M.w. 172.09
Uses: Plasticizer, modifier for hot-melt adhesives (EVA, block copolymer, polyester, polyamide, PU), delayed tack latex and hot-melt coatings
Manuf./Distrib.: Aldrich; Velsicol
Trade names: Benzoflex® 352

Cyclohexanol
CAS 108-93-0; EINECS 203-630-6

Cyclohexanone

Synonyms: Hexahydrophenol; Hexalin
Empirical: $C_6H_{11}OH$
Formula: $CH_2(CH_2)_4CHOH$
Properties: Colorless oily liq.; camphor-like odor; hygroscopic; sl. sol. in water; misc. with org. solvs.; m.w. 100.09; dens. 0.937 (37.4 C); m.p. 23 C; b.p. 160.9 C; flash pt. 67.7 C; ref. index 1.4656 (20 C)
Precaution: Combustible
Toxicology: LD50 (rat, oral) 2.06 g/kg; TLV 50 ppm
Uses: Soap making to incorporate solvents and phenolic insecticides; source of adipic acid for nylon, textile finishing; solvent for alkyd and phenolic resins, paints, rubber
Manuf./Distrib.: Aldrich; AlliedSignal; Ashland; J.T. Baker; BASF; BASF AG; Chemcentral; Fluka; Harcros; Hüls Am.; Spectrum Chem. Mfg.; Sunnyside; UCB SA; Van Waters & Rogers

Cyclohexanone
CAS 108-94-1; EINECS 203-631-1
Synonyms: Ketohexamethylene; Pimelic ketone
Empirical: $C_6H_{10}O$
Formula: $CO(CH_2)_4CH_2$
Properties: Oily liq., peppermint-acetone odor; sl. sol. in water; sol. in alcohol, ether, common org. solvs.; m.w. 98.08; dens. 0.947 (20/4 C); m.p. -32.1 C; b.p. 156.7 C; flash pt. (CC) 63 C; ref. index 1.4507
Precaution: DOT: Flamm. liq.
Toxicology: LD50 (rat, oral) 1.62 ml/kg
Uses: Used as paint and varnish remover; solvent for cellulose acetate, nitrocellulose, natural resins, vinyl resins, paints, rubber, waxes, fats; in prod. of adipic acid for nylon, cyclohexanone resins
Manuf./Distrib.: Aldrich; Allchem Ind.; AlliedSignal; Ashland; J.T. Baker; BASF; Chemcentral; Coyne; Fluka; Harcros; Sigma; Spectrum Chem. Mfg.; Stanchem; Sunnyside; Union Carbide; Van Waters & Rogers

Cyclohexatriene. *See* Benzene
Cyclohexene. *See* Polydipentene

Cyclohexyl acrylate
CAS 3066-71-5
Formula: $CH_2:CHCOOC_6H_{11}$
Properties: Colorless liq.; sol. in alcohol and ether; insol. in water; m.w. 154.20; dens. 0.975 (20 C, 4 mm); b.p. 88 C
Uses: Curing agent
Manuf./Distrib.: Monomer-Polymer & Dajac Labs
Trade names containing: SR-220

Cyclohexyl methacrylate
CAS 101-43-9; EINECS 202-943-5
Classification: Monomer
Empirical: $C_{10}H_{16}O_2$
Formula: $H_2C:C(CH_3)COOC_6H_{11}$
Properties: Colorless monomeric liq.; insol. in water; m.w. 168.24; dens. 0.9626 (20/20 C); b.p. 210 C
Precaution: Combustible
Uses: Polymer modifier for optical lens systems, adhesives, floor polishes, vinyl polymerization, anaerobic adhesives; dental resins; encapsulation of electronic assemblies
Manuf./Distrib.: Aldrich; CPS; Fluka; Monomer-Polymer & Dajac Labs; Polysciences; Rohm Tech; Sigma
Trade names containing: Ageflex CHMA; Sipomer® CHM; SR-208

1-Cyclohexyl-2-pyrrolidone. *See* N-Cyclohexyl pyrrolidone

N-Cyclohexyl pyrrolidone
CAS 6837-24-7; EINECS 229-919-7
Synonyms: 1-Cyclohexyl-2-pyrrolidone
Empirical: $C_{10}H_{17}NO$
Properties: Liq.; m.w. 167.25; sp.gr. 1.026; f.p. 12 C; b.p. 284 C; flash pt. (CC) 293 F; ref. index 1.4950
Toxicology: LD50 (oral, rat) 0.37 g/kg; primary skin irritant; may cause permanent eye damage
Uses: Solv. for plating engineering plastics, esp. EMS applics.; dyeing of aromatic amide fibers; photo resists; gas separation; elec. insulating coatings; chem. synthesis; absorp. of sulfur dioxide and nitric acid; aromatic extraction; high temp. polyimide processing aid
Manuf./Distrib.: Aldrich; Fluka
Trade names: CHP®

Cyclomaltoheptaose. *See* Cyclodextrin; β-Cyclodextrin
Cyclomaltohexaose. *See* Cyclodextrin
Cyclomaltooctaose. *See* Cyclodextrin

Cyclopentadiene
CAS 542-92-7
Synonyms: 1,3-Cyclopentadiene
Empirical: C_5H_6
Properties: Colorless liq., sweet char. odor; sol. in most org. solvs.; insol. in water; m.w. 66.10; dens. 0.805; m.p. -85 C; b.p. 42.5 C; flash pt. 77 F
Precaution: Fire hazard when exposed to heat or flame
Uses: Mfg. of paints
Manuf./Distrib.: Polysat; Reichhold

1,3-Cyclopentadiene. *See* Cyclopentadiene
Cyclopentadiene dimer. *See* Dicyclopentadiene

DACH. *See* 1,2-Diaminocyclohexane
DAMP. *See* 1,3-Pentanediamine
DAP. *See* Diallyl phthalate
Dapsone. *See* Sulfolane
DBM. *See* Dibutyl maleate
DBNPA. *See* 2,2-Dibromo-3-nitrilopropionamide
DBP. *See* Dibutyl phthalate
DBPC. *See* BHT
DBTL. *See* Dibutyltin dilaurate
DBU. *See* Diazabicycloundecene
DCBTF. *See* 3,4-Dichlorobenzotrifluoride
1,2-DCE. *See* Ethylene dichloride
DCHP. *See* Dicyclohexyl phthalate
DCM. *See* Methylene chloride
DCO. *See* Castor oil, dehydrated
DCP-0. *See* Calcium phosphate dibasic
DDAO. *See* Lauramine oxide
DDBSA. *See* Dodecylbenzene sulfonic acid
DDDA. *See* C12 dibasic acid
4,4′-DDS. *See* 4,4′-Diaminodiphenyl sulfone
DDSA. *See* Dodecenyl succinic anhydride
DEA. *See* Diethanolamine
Deadburned magnesite. *See* Magnesium oxide
Deanol. *See* Dimethylethanolamine
DECA. *See* Decabromodiphenyl oxide
Decabromobiphenyl ether. *See* Decabromodiphenyl oxide
Decabromobiphenyl oxide. *See* Decabromodiphenyl oxide

Decabromodiphenyl oxide
CAS 1163-19-5; EINECS 214-604-9
Synonyms: DECA; Decabromobiphenyl ether; Decabromobiphenyl oxide; Decabromophenyl ether; Pentabromophenyl ether; 1,1′-Oxybis (2,3,4,5,6-pentabromobenzene)
Empirical: $C_{12}Br_{10}O$
Properties: Wh. powd., odorless; m.w. 959.22; m.p. 300-310 C
Toxicology: Extremely hazardous substance; experimental neoplastigen, teratogenic and reproductive effects; heated to decomp., emits toxic fumes of Br^-
Uses: Flame retarant for HIPS, thermoset and thermoplastic polyesters, polypropylene, crosslinked polyethylene, elastomers, wire and cable insulation, adhesives, coatings, textile coatings
Manuf./Distrib.: Albemarle; Aldrich; Allchem Ind.; Elf Atochem N. Am.; Fluka; Great Lakes
Trade names: FR-1210; Great Lakes DE-83™; Great Lakes DE-83R™; Octoguard FR-01; Saytex® 102E
Trade names containing: Octoguard FR-11; Octoguard FR-12; Octoguard FR-13; Octoguard FR-15

Decabromophenyl ether. *See* Decabromodiphenyl oxide
Decaglycerol tetraoleate. *See* Polyglyceryl-10 tetraoleate

Decaglyceryl tetraoleate. *See* Polyglyceryl-10 tetraoleate

Decahydronaphthalene
CAS 91-17-8; EINECS 202-046-9
Synonyms: Perhydronaphthalene; Dekalin; Naphthalane
Definition: Mixture of cis and trans
Empirical: $C_{10}H_{18}$
Properties: Insol. in water; sol. in alcohol; m.w. 138.14; dens. 0.8963; m.p. -43.26 C; b.p. 185.5 C; flash pt. 58 C; ref. index 1.474
Precaution: Flamm.
Uses: Solvent and stabilizer for shoe creams and floor waxes; solv. in paint and lacquers, oils, resins, rubber, and asphalt
Manuf./Distrib.: Aldrich; Fluka; Sigma
Trade names: Decalin

Decamethylenedicarboxylic acid. *See* C12 dibasic acid
Decane-1,10-dicarboxylic acid. *See* C12 dibasic acid
1,10-Decanedicarboxylic acid. *See* C12 dibasic acid
Decanedioic acid. *See* Sebacic acid
1-Decanol. *See* n-Decyl alcohol

n-Decyl alcohol
CAS 112-30-1; 68526-85-2; EINECS 203-956-9
Synonyms: Alcohol C-10; Noncarbinol; Nonylcarbinol; 1-Decanol; Decylic alcohol
Classification: Fatty alcohol
Empirical: $C_{10}H_{22}O$
Formula: $CH_3(CH_2)_8CH_2OH$
Properties: Mod. visc. liq., sweet odor; sol. in alcohol, ether; insol. in water; m.w. 158.32; dens. 0.8297 (20/4 C); m.p. 7 C; b.p. 232.9 C; flash pt. (OC) 180 F; ref. index 1.43587
Precaution: Flamm. when exposed to heat or flame
Toxicology: LD50 (oral, rat) 4720 mg/kg; moderately toxic by skin contact; irritating to eyes, skin, respiratory system; heated to decomp., emits acrid smoke and irritating fumes
Uses: In mfg. of plasticizers, lubricants, petrol. additives, herbicides, surfactants, solvents, coatings; emulsion polymerization; moderate antifoaming capacity; flavoring agent
Regulatory: FDA 21CFR §172.515, 172.864, 175.300, 176.170, 178.3480, 178.3910; FEMA GRAS; Japan approved as flavoring
Manuf./Distrib.: Albemarle; Aldrich; Brown; Fluka; Henkel/Emery; M. Michel; Penta Mfg.; Schweizerhall; Sigma; Vista
Trade names: Exxal® 10

Decylic alcohol. *See* n-Decyl alcohol

DEDM hydantoin
CAS 26850-24-8; EINECS 248-052-5
Synonyms: Di-(2-hydroxyethyl)-5,5-dimethyl hydantoin; 1,3-Bis(2-hydroxyethyl)-5,5-dimethyl-2,4-Imidazolidinedione; Diethylol dimethyl hydantoin
Classification: Organic compd.
Empirical: $C_9H_{16}N_2O_4$
Uses: Intermediate for epoxies, urethane resins, and antistatic lubricants for the textile and plastics industries; crosslinker; for high-temp. resist. polyester or imide-amide wire enamels, acrylic, PU spec. coatings
Trade names: Dantocol® DHE

DEG. *See* Diethylene glycol
DEHP. *See* Dioctyl phthalate
Dekalin. *See* Decahydronaphthalene
Denatured alcohol. *See* Alcohol denatured
Denatured ethanol. *See* Alcohol denatured
Denatured ethyl alcohol. *See* Alcohol denatured
Denatured spirits. *See* Alcohol denatured
Deodorized winterized cottonseed oil. *See* Cottonseed oil
DETA. *See* Diethylenetriamine
DETDA. *See* Diethyl toluene diamine

Diacetone alcohol
CAS 123-42-2; EINECS 204-626-7
Synonyms: 4-Hydroxy-4-methyl-2-pentanone
Classification: Ketone
Empirical: $C_6H_{12}O_2$
Formula: $(CH_3)_2C(OH)CH_2COCH_3$
Properties: Sol. in water; m.w. 116.16; sp.gr. 0.940 (20/20 C); f.p. -47 F; b.p. 145.2-172 C; flash pt. (TCC) 126 F; ref. index 1.4234
Toxicology: TLV 50 ppm
Uses: Solvent for nitrocellulose, cellulose acetate, oils, resins, waxes, fats, dyes, tars, lacquers, dopes, coatings, wood preservatives, rayon, artificial leather, metal cleaning; laboratory reagent; hydraulic fluids; textile stripping agent
Regulatory: FDA 21CFR §175.105
Manuf./Distrib.: Aldrich; Allchem Ind.; Ashland; Baychem; BP Chems. Ltd; Browning; R.E. Carroll; Chemcentral; Coyne; Elf Atochem N. Am.; Fabrichem; Fluka; Great Western; Harcros; Hoechst Celanese; Rhone-Poulenc Surf. & Spec.; Shell; Sunnyside; Union Carbide; Van Waters & Rogers

1,2-Diacetoxyethane. *See* Ethylene glycol diacetate
Diacetyl ether. *See* Ethylacetoacetate
Diacetyl manganese. *See* Manganese acetate
Diacetylmethane. *See* Acetylacetone

Dialkyl dimethyl ammonium chloride
CAS 68514-95-4
Synonyms: Quaternium 31; Dimethyl dialkyl ammonium chloride
Toxicology: Heated to decomp., emits acrid smoke and irritating fumes
Uses: Softener
Regulatory: FDA 21CFR §173.400
Trade names containing: Adogen® 432

Diallyl dimethyl ammonium chloride. *See* Dimethyl diallyl ammonium chloride

Diallyl fumarate
Uses: Allyl monomer; crosslinking agent
Manuf./Distrib.: Monomer-Polymer & Dajac Labs
Trade names containing: SR-204

Diallyl monomer
Uses: Monomer for prod. of paints
Manuf./Distrib.: Ashland; FMC; GCA; C.P. Hall; Nat'l. Starch & Chem.; PPG Ind.; Punda Mercantile

Diallyl phthalate
CAS 131-17-9; EINECS 205-016-3
Synonyms: DAP
Empirical: $C_{14}H_{14}O_4$
Formula: $C_6H_4(COOCH_2CH:CH_2)_2$
Properties: Colorless oily liq.; insol. in water; m.w. 246.27; dens. 1.120 (20/20 C); b.p. 158-165 (4 mm); f.p. -70 C; flash pt. 165.5 C
Precaution: Combustible
Toxicology: Toxic by ingestion
Uses: Primary plasticizer; intermediate; mfg. of paints
Manuf./Distrib.: Aldrich; Allchem Ind.; ARCO; Ashland; BP Chems. Ltd; Chemcentral; Fluka; FMC; C.P. Hall; Monomer-Polymer & Dajac Labs; Occidental; Punda Mercantile; Rogers; Samson; Sigma; Van Waters & Rogers
Trade names: Arcal™ DAP; DAP; Ftalidap®

1,3-Diaminobenzene. *See* m-Phenylenediamine
m-Diaminobenzene. *See* m-Phenylenediamine

1,2-Diaminocyclohexane
CAS 694-83-7; EINECS 211-776-7
Synonyms: DACH; 1,2-Cyclohexanediamine
Empirical: $C_6H_4N_2$

p,p´-Diaminodiphenylmethane

Properties: Misc. with water; m.w. 114.19; sp.gr. 0.94; b.p. 183 C (760 mm)
Precaution: Corrosive liq.
Uses: Epoxy curing agent; chelating agent for oilfield, textile, water treatment, detergent fields; herbicide intermediate; polyamide resins for adhesives, films, plastics, inks; PU for extenders; catalysts; scale/corrosion inhibitors; coatings
Manuf./Distrib.: Air Prods.; Aldrich; Alfa Aesar Johnson Matthey; DuPont; Fluka; Johnson Matthey; Milliken
Trade names: DCH-99

p,p´-Diaminodiphenylmethane. *See* 4,4´-Methylene dianiline

4,4´-Diaminodiphenyl sulfone
CAS 80-08-0; EINECS 201-248-4
Synonyms: 4,4´-DDS; 4,4´-Sulfonyldianiline; Bis(4-aminophenyl) sulfone
Empirical: $C_{12}H_{12}N_2O_2S$
Formula: $(NH_2C_6H_4)_2SO_2$
Properties: Cryst.; nearly insol. in water; sol. in acetone, alcohol; m.w. 248.32; m.p. 176 C
Toxicology: Poison by ingestion, intraperitoneal, subcutaneous routes; moderately toxic unspecified
Uses: Curing agent; hardener for adhesives, PWB laminates, prepregs, composites, powd. coatings; exc. high temp. and chem. resist.
Manuf./Distrib.: Aldrich; Crown Metro; Fluka; R S A ; Sigma
Trade names: HT 976

1,2-Diaminoethane. *See* Ethylenediamine
1,3-Diaminopentane. *See* 1,3-Pentanediamine
2,4-Diamino-6-phenyl-s-triazine. *See* Benzoguanamine
Diantimony pentoxide. *See* Antimony pentoxide

Diaryl iodonium hexafluoroantimonate
Uses: Photoinitiator for use in coatings (metal, paper, wood, plastic) and inks; enhances adhesion to metallic substrates; offers fast cure speeds in epoxy, vinyl ether, and other cationically cured resin systems
Trade names: SarCat™ CD-1012

Diatomaceous earth
CAS 6067-86-0; 7631-86-9; 68855-54-9; EINECS 231-545-4
Synonyms: Kieselguhr; Diatomite; Diatomaceous silica; Infusorial earth; Siliceous earth
Definition: Mineral material consisting chiefly of the siliceous frustules and fragments of various species of diatoms
Properties: Wh. to pale buff soft bulky solid; insol. in acids except HF; sol. in strong alkalies; dens. 1.9-2.35; bulk dens. 8-15 lb/ft^3; oil absorp. 135-185%; 88% silica
Precaution: Noncombustible
Toxicology: TLV:TWA 10 mg/m^3 (dust); poison by inhalation and ingestion; dust may cause fibrosis of the lungs
Uses: Filtration, clarifying, decolorizing; insulation absorbent; mild abrasive; catalyst carrier; anticaking agents in fertilizer; conditioning agent for dusts, soil; filler, processing aid for rubber; extender for paints
Regulatory: FDA 21CFR §73.1, 133.146, 160.105, 160.185, 172.230, 172.480, 173.340, 175.300, 177.1460, 177.2410, 182.90, 193.135, 561.145, 573.940; USDA 9CFR §318.7; Japan restricted (0.5% max. residual); USP/NF, BP compliance
Manuf./Distrib.: Aldrich; Ashland; Celite; Coyne; CR Mins.; Eagle-Picher; Great Western; Grefco; Lomas Int'l.; D.N. Lukens; L.A. Salomon; Seefast (Europe) Ltd; Seegott; Sigma; Spectrum Chem. Mfg.; Tamms Ind.; Van Waters & Rogers; Whittaker, Clark & Daniels
Trade names: Celite® 110; Celite® 266; Celite® 281; Celite® 289; Celite® 499; Celite® Super Fine Super Floss; Celite® Super Floss; Celite® White Mist

Diatomaceous earth, amorphous
CAS 61790-53-2
Definition: Diatomaceous earth contg. < 1% cryst. silica
Uses: Filler, flatting agent, rheology control agent in agric., asphalt, roof coatings, catalysts, cement, dental, paints, paper, plastics, polishes, cleaners, caulks, adhesives, sealants, encapsulants
Manuf./Distrib.: Aldrich; Sigma
Trade names: DiaFil® 10; DiaFil® 110; DiaFil® 190; DiaFil® 520; DiaFil® 525; DiaFil® 530; DiaFil® 570; DiaFil® 590

Diatomaceous silica. *See* Diatomaceous earth
Diatomite. *See* Diatomaceous earth

Diazabicycloundecene
CAS 6674-22-2; EINECS 229-713-7
Synonyms: DBU; 1,8-Diazabicyclo[5.4.0]undec-7-ene (1,5-5); [2,3,4,6,7,8,9,10-Octapyrimidol[1,2-a]azepine]
Empirical: $C_9H_{16}N_2$
Properties: M.w. 152.24; dens. 1.022 (20/4 C); b.p. 115 C (11 mm); ref. index 1.521 (20 C)
Uses: Accelerator for phenolic novolac and other epoxy cures; highly active catalyst for PU elastomers, shoe soles, coatings
Manuf./Distrib.: Air Prods.; Aldrich; BASF; BASF AG; Fluka; SAF Bulk Chems.; Schweizerhall; Sigma
Trade names: Polycat® DBU; Polycat® SA-1; Polycat® SA-102; Polycat® SA-610/50

1,8-Diazabicyclo[5.4.0]undec-7-ene (1,5-5). *See* Diazabicycloundecene

Diazabicycloundecene, 2-ethylhexanoic acid salt
Uses: Epoxy curing agent
Trade names: Amicure® SA-102

4,5-Diazaphenanthrene. *See* 1,10-Phenanthroline
Diazenedicarboxamide. *See* Azodicarbonamide
Diazobicyclooctane. *See* Triethylene diamine
1,4-Diazobicyclo [2.2.2] octane. *See* Triethylene diamine
DIBA. *See* Diisobutyl adipate
Dibasic lead carbonate. *See* Lead carbonate (basic)
Dibenzoyldiethyleneglycol ester. *See* Diethylene glycol dibenzoate
Dibenzoyl dipropylene glycol ester. *See* Dipropylene glycol dibenzoate
Dibenzyl phthalate. *See* Benzyl phthalate
DIBK. *See* Diisobutyl ketone
DIBP. *See* Diisobutyl phthalate
Dibromocyanoacetamide. *See* 2,2-Dibromo-3-nitrilopropionamide
α,α-Dibromo-α-cyanoacetamide. *See* 2,2-Dibromo-3-nitrilopropionamide
1,2-Dibromo-2,4-dicyanobutane. *See* Methyldibromo glutaronitrile

Dibromoethyldibromocyclohexane
CAS 3322-93-8; EINECS 222-036-8
Empirical: $C_8H_{12}Br_4$
Properties: M.w. 427.84
Uses: Flame retardant for expandable PS, crystal and high impact PS, SAN, adhesives, coatings, textile treatment, polyurethanes
Manuf./Distrib.: Albemarle
Trade names: Saytex® BCL-462

2,2-Dibromo-3-nitrilopropionamide
CAS 10222-01-2
Synonyms: DBNPA; α,α-Dibromo-α-cyanoacetamide; Dibromocyanoacetamide
Empirical: $C_3H_2Br_2N_2O$
Properties: M.w. 241.89
Toxicology: LD50 (oral, mammal) 118 mg/kg, (IV, mouse) 10 mg/kg; poison by ingestion and IV routes; severe skin and eye irritant; heated to decomp., emits very toxic fumes of Br^- and NO_x
Uses: Microbiocide, paper mill slimicide, antimicrobial agent for enhanced oil recovery systems; preservative for metalworking fluids containing water; also used in paints, latex, adhesives, mineral slurries, inks
Regulatory: FDA 21CFR §173.320 (limitation 2-10 ppm)
Trade names: Amerstat® 300; Biochek 20

Dibromostyrene
Uses: Reactive flame retardant for plastics, adhesives, coatings
Manuf./Distrib.: Great Lakes
Trade names: Great Lakes DBS™

2,2′-Dibutoxyethyl ether. *See* Diethylene glycol dibutyl ether
Dibutyl-1,2-benzene dicarboxylate. *See* Dibutyl phthalate
Dibutyl 'Carbitol'. *See* Diethylene glycol dibutyl ether
Dibutyl-o-(o-carboxybenzoyl) glycolate. *See* n-Butyl phthalyl-n-butyl glycolate
Dibutyl-o-carboxybenzoyloxyacetate. *See* n-Butyl phthalyl-n-butyl glycolate

2,6-Di-t-butyl-p-cresol. *See* BHT
Di-n-butyldiacetoxystannane. *See* Dibutyltin diacetate
Dibutyldiglycol. *See* Diethylene glycol dibutyl ether

Dibutyl ether
CAS 142-96-1
Synonyms: Butyl ether; n-Dibutyl ether
Empirical: $C_8H_{18}O$
Formula: $CH_3(CH_2)_3O(CH_2)_3CH_3$
Properties: Colorless liq.; mild odor; stable; sol. in water, alcohol; m.w. 130.26; m.p. -95.2 C; b.p. 142 C; flash pt. 77 F
Precaution: Flamm. liq.
Toxicology: LD50 (oral, rat) 7400 mg/kg; mildly toxic by inhalation, ingestion, and skin contact
Uses: Solv. for paints
Manuf./Distrib.: Focus

n-Dibutyl ether. *See* Dibutyl ether

Dibutyl fumarate
CAS 105-76-9
Synonyms: Fumaric acid, dibutyl ester
Formula: $(C_4H_9)OOCCH:CHCOO(C_4H_9)$
Properties: Colorless liq.; insol. in water; dens. 0.9873 (20 C); b.p. 285.2 C; f.p. -15.6 C
Precaution: Combustible
Uses: Monomeric plasticizer for PVC, PVAc, paints; copolymers, intermediate
Manuf./Distrib.: AC Ind.; Lenape; Merrand Int'l.; Monomer-Polymer & Dajac Labs; Penta Mfg.; Unitex
Trade names: Staflex DBF

2-(3´,5´-Di-t-butyl-2´-hydroxyphenyl)-5-chlorobenzotriazole
CAS 3864-99-1; EINECS 223-383-8
Synonyms: Phenol, 2-(5-chloro-2H-benzo-triazol-2-yl)-4,6-bis (1,1-dimethylethyl)-
Empirical: $C_{20}H_{24}ClN_3O$
Properties: M.w. 357.86
Uses: UV lt. absorber, stabilizer for plastics, coatings; for polyolefins, EPDM, SAN, acrylic, flexible PVC, cold-cured polyesters, PU, dyes
Manuf./Distrib.: Aldrich
Trade names: Tinuvin® 327

Dibutyl maleate
CAS 105-76-0; EINECS 203-328-4
Synonyms: DBM; 2-Butenedioic acid, dibutyl ester; Di-n-butyl maleate
Empirical: $C_{12}H_{20}O_4$
Formula: $C_4H_9OOCCH:CHCOOC_4H_9$
Properties: Colorless oily liq.; insol. in water; m.w. 228.29; dens. 0.9964 (20/20 C); b.p. 274-277; flash pt. (OC) 285 F
Precaution: Combustible
Uses: Copolymers, plasticizer for PVAc, paints, intermediate
Manuf./Distrib.: AC Ind.; Aldrich; Allchem Ind.; Aristech; Ashland; Ferro/Bedford; Fluka; GFI; Hoechst AG; Lenape; Merrand Int'l.; Monomer-Polymer & Dajac Labs; Pentagon Chems. Ltd; Penta Mfg.; Tiarco; Unitex

Di-n-butyl maleate. *See* Dibutyl maleate
2,6-Di-t-butyl-4-methylphenol. *See* BHT

Di (butyl, methyl pyrophosphato) ethylene titanate di (dioctyl, hydrogen phosphite)
Uses: Coupling agent, adhesion promoter, antioxidant, antistat, antifoam, accelerator, blowing agent activator, catalyst, curative, corrosion inhibitor, dispersion aid, emulsifier, flame retardant, foaming agent, grinding and process aid
Trade names containing: Ken-React® KR 262ES

Di-t-butyl peroxide
CAS 110-05-4; EINECS 203-733-6
Synonyms: DTBP; t-Butyl peroxide; Bis (1,1-dimethylethyl) peroxide
Classification: Organic peroxide

Empirical: $C_8H_{18}O_2$
Formula: $(CH_3)_3COOC(CH_3)_3$
Properties: Clear water-white liq.; sol. in styrene, ketones, most aliphatic and aromatic hydrocarbons; insol. in water; m.w. 146.23; dens. 0.791 (25/25 C); f.p. -40 C; b.p. 111 C; flash pt. (CC) 65 F
Precaution: Flamm. oxidizing liq.; dangerous fire hazard
Toxicology: LD50 (rat, oral) > 25,000 mg/kg
Uses: Polymerization catalyst for resins (e.g., olefins, styrene, styrenated alkyds, silicones); ignition accelerator for diesel fuel; organic synthesis; intermediate; raw material for paints
Manuf./Distrib.: Akzo Nobel; Aldrich; Amber Syn.; Elf Atochem N. Am.; Fluka; Monomer-Polymer & Dajac Labs; Witco/Polymer Addit.

N,N′-Di-s-butyl-p-phenylenediamine
CAS 101-96-2
Empirical: $C_{14}H_{24}N_2$
Formula: $CH_3CH_2CH(CH_3)NHC_6H_4NHCH(CH_3)CH_2CH_3$
Properties: Amber to red liq.; sol. in gasoline, abs. alcohol, benzene; insol. in water or caustic sol'ns.; dens. 0.94 (15.5/15.5 C); f.p. 20 C; flash pt. (COC) 290 F
Precaution: Combustible
Toxicology: Toxic by ingestion, inhalation, and skin absorption; strong irritant to tissue; causes skin burns; heated to dec., emits toxic fumes of NO_x
Uses: Oxidation inhibitor and stabilizer in gasoline; metal deactivator for crude oil distillates
Trade names containing: V-Pyrol

Dibutyl phthalate
CAS 84-74-2; EINECS 201-557-4
Synonyms: DBP; 1,2-Benzenedicarboxylic acid, dibutyl ester; Dibutyl-1,2-benzene dicarboxylate; Butyl phthalate; Di-n-butyl phthalate
Definition: Aromatic diester of butyl alcohol and phthalic acid
Empirical: $C_{16}H_{22}O_4$
Formula: $C_6H_4(COOC_4H_9)_2$
Properties: Colorless stable oily liq., odorless; misc. with common org. solvs.; insol. in water; m.w. 278.17; dens. 1.0484 (20/20 C); b.p. 340 C; f.p. -35 C; flash pt. (COC) 340 F; ref. index 1.4920
Precaution: Combustible; reacts exposively with chlorine
Toxicology: Toxic; TLV 5 mg/m^3 of air; irritant; if ingested can cause gastrointestinal upset; vapor irritating to eyes and mucous membranes
Uses: Plasticizer in nitrocellulose, lacquers, elastomers, explosives, nail polishes; solvent for perfumes, oils; perfume fixative; textile lubricating agent; as insect repellent
Regulatory: FDA 21CFR §175.105, 175.300, 176.170, 176.300, 177.1200, 177-2420, 177.2600; FDA approved for orals; BP, Ph.Eur. compliance
Manuf./Distrib.: Aldrich; Allchem Ind.; Aristech; Ashland; BP Chems. Ltd; Chisso Am.; Coyne; Daihachi Chem. Ind.; Eastman; Fluka; Great Western; C.P. Hall; Mitsubishi Gas; Sigma; Spectrum Chem. Mfg.; Unitex
Trade names: Diplast® A; Eastman® DBP; Kodaflex® DBP; Palatinol® C; Palatinol® DBP; Unimoll® DB; Uniplex 150
Trade names containing: Araldite® GY 502; Bonding Agent 2001; Epotuf Resin 37-134; Kodaflex® HS-4; Mearlite® Ultra Bright UDQ; Mearlite® Ultra Bright UMS; Mearlite® Ultra Bright UMT; Morflex® 190

Di-n-butyl phthalate. *See* Dibutyl phthalate

Dibutyltin (bis) mercaptide
Uses: Catalyst for PU coatings
Trade names: Dabco® T-131

Dibutyltin diacetate
CAS 1067-33-0; EINECS 213-928-8
Synonyms: Di-n-butyldiacetoxystannane
Empirical: $C_{12}H_{24}O_4Sn$
Formula: $(C_4H_9)_2Sn(C_2H_3O_2)_2$
Properties: Clear yel. liq.; sol. in water and most org. solvs.; m.w. 351.01; f.p. < 12 C; b.p. 130 C (2 mm); flash pt. 143 C; ref. index 1.471
Precaution: Combustible
Toxicology: TLV 0.1 mg/m^3 of air; toxic by skin absorption
Uses: Stabilizer for chlorinated organics; catalyst for PU coatings, condensation reactions

Dibutyltin dilaurate

Manuf./Distrib.: Aldrich; Fluka
Trade names: Dabco® T-1; Metacure® T-1

Dibutyltin dilaurate
CAS 77-58-7; EINECS 201-039-8
Synonyms: DBTL
Empirical: $C_{32}H_{64}O_4Sn$
Formula: $(C_4H_9)_2Sn(OCOC_{10}H_{20}CH_3)_2$
Properties: Clear pale yel. liq.; sol. in acetone and benzene; insol. in water; m.w. 631.56; dens. 1.066; m.p. 23 C; f.p. 8 C; flash pt. 440 F
Precaution: Combustible
Toxicology: TLV 0.1 mg/m^3 of air; LD50 (oral, rat), 3954 mg/kg (sl. toxic), (dermal, rabbit) > 2000 mg/kg (sl. toxic)
Uses: Stabilizer for vinyl resins, lacquers, elastomers; catalyst for urethane and silicones
Manuf./Distrib.: Air Prods.; Aldrich; Alfa Aesar Johnson Matthey; Cardinal Stabilizers; Elf Atochem N. Am.; Ferro/Bedford; Fluka; Johnson Matthey; KMZ Chem. Ltd; OM Group; Witco/Oleo-Surf.
Trade names: Dabco® T-12; Metacure® T-12

Dibutyltin disulfide
Uses: Catalyst for PU coatings
Trade names: Dabco® T-5

Dicalcium orthophosphate anhyd. *See* Calcium phosphate dibasic
Dicalcium phosphate. *See* Calcium phosphate dibasic
1,10-Dicarboxydecane. *See* C12 dibasic acid
Dicarboxylic acid C_6. *See* Adipic acid
Dicarboxylic acid C9. *See* Azelaic acid
Dicarboxylic acid C_{12}. *See* C12 dibasic acid

3,4-Dichlorobenzotrifluoride
CAS 328-84-7
Synonyms: DCBTF; Benzene, 1,2-dichloro-4-(trifluoromethyl)-
Empirical: $C_7H_3F_3Cl_2$
Properties: M.w. 215.00; dens. 1.478; m.p. -13 to -12 C; b.p. 173-174 C; flash pt. 65 C
Toxicology: Irritant
Manuf./Distrib.: Aldrich; Occidental
Trade names: Oxsol® 1000

Dichlorodimethylsilane. *See* Dimethyldichlorosilane
1,2-Dichloroethane. *See* Ethylene dichloride
1,1-Dichloroethene. *See* Vinylidene chloride monomer
Dichloroethylene. *See* Ethylene dichloride
1,1-Dichloroethylene. *See* Vinylidene chloride monomer
Dichloromethane. *See* Methylene chloride

Dicocamine
CAS 61789-76-2; EINECS 263-086-0
Synonyms: Dicoco alkyl amine; Amines, dicoco alkyl
Definition: Secondary aliphatic amine derived from coconut acid
Formula: R-2NH, R represents the coconut radical
Uses: Synthesis intermediate, anticaking agent, flotation, antistripping for road making, soil stabilization; auxs. for fuel additives, rust inhibition, paint; chemical intermediate for quats., betaines, amine oxides; emulsifier, corrosion inhibitor
Trade names: Noram 2C

Dicoco alkyl amine. *See* Dicocamine
Dicoco dimethyl ammonium chloride. *See* Dicocodimonium chloride

Dicocodimonium chloride
CAS 61789-77-3; EINECS 263-087-6
Synonyms: Dicoco dimethyl ammonium chloride; Quaternium-34
Classification: Quaternary ammonium salt
Formula: $[RN(CH_3)_2R]^+Cl^-$ where R rep. the coconut radical

Uses: Emulsifier, corrosion inhibitor, coupling agent, bactericide, textile softener, asphalt emulsifier, hair conditioner; for car spray waxes, dust control oil, spot removal, petrol. processing, detergents, textiles, fabric softeners
Regulatory: FDA 21CFR §172.710, 177.1200
Trade names containing: Noramium M2C

Dicoco nitrite
CAS 71487-01-9
Trade names containing: Arquad® 2C-70 Nitrite

Dicopper monoxide. *See* Copper oxide (ous)

Dicyandiamide
CAS 461-58-5; EINECS 207-312-8
Synonyms: Cyanoguanidine
Classification: Aliphatic organic compd.
Empirical: $C_2H_4N_4$
Formula: $NH_2C(NH)(NHCN)$
Properties: Pure white crystals; sol. in liq. ammonia; partly sol. in hot water; m.w. 84.08; dens. 1.4 (25 C); m.p. 207-209 C; stable when dry
Precaution: Nonflamm.
Uses: Fertilizers; nitrocellulose stabilizer; organic synthesis, esp. of melamine, barbituric acid, and guanidine salts; explosives; curing agent, catalyst for epoxy resin, powd. coatings, structural laminates, adhesives, film adhesives; pharmaceuticals; fireproofing compds.; stabilizer in detergents; modifier for starch prod.
Regulatory: BP compliance
Manuf./Distrib.: Aldrich; Allchem Ind.; Andrulex Trading Ltd; Chemical; CVC Spec. Chem.; Fluka; Monomer-Polymer & Dajac Labs; San Yuan; Sigma; SKW Chems.
Trade names: Amicure® CG-1200; Amicure® CG-1400; Amicure® CG-NA

1,3-Dicyanotetrachlorobenzene. *See* Tetrachloroisophthalonitrile

Dicyclo (dioctyl) pyrophosphato titanate
Uses: Coupling agent, adhesion promoter, antioxidant, antistat, antifoam, accelerator, blowing agent activator, catalyst, curative, corrosion inhibitor, dispersion aid, emulsifier, flame retardant, foaming agent, grinding and process aid
Trade names containing: Ken-React® KR OPP2

Dicyclohexyl phthalate
CAS 84-61-7; EINECS 201-545-9
Synonyms: DCHP; 1,2-Benzenedicarboxylic acid, dicyclohexyl ester
Formula: $C_6H_4(COOC_6H_{11})_2$
Properties: Wh. gran. solid; mildly aromatic odor; sol. in most org. solvs.; insol. in water; m.w. 330.43; dens. 1.20 (25/25 C); m.p. 62-65 C; flash pt. 405 F; nonvolatile
Precaution: Combustible
Uses: Plasticizer for foil lacquers, cellophane lacquers, nitrocellulose, ethylcellulose, chlorinated rubber, PVAc, PVC, and other polymers
Manuf./Distrib.: Aldrich; Bayer; Morflex; Schweizerhall; Unitex
Trade names: Edenol DCHP; Morflex® 150; Unimoll® 66; Unimoll® 66 M

Dicyclopentadiene
CAS 77-73-6; EINECS 201-052-9
Synonyms: 4,7-Methano-3a,4,7,7a-tetrahydroindene; Cyclopentadiene dimer
Definition: Cyclopentadiene dimer
Empirical: $C_{10}H_{12}$
Properties: Liq.; sol. in alcohol; insol. in water; m.w. 132.2; dens. 0.979 (20/20 C); m.p. 33.6 C; b.p. 172 C; flash pt. 26 C; flamm.; moderate fire risk
Toxicology: Toxic by ingestion; TLV 5 ppm in air
Uses: Chemical intermediate for insecticides, EPDM elastomers, metallocenes, paints and varnishes, flame retardants for plastics
Manuf./Distrib.: Aldrich; Allchem Ind.; ARCO; Crowley Chem.; Exxon; Fluka; Monomer-Polymer & Dajac Labs; Shell

Dicyclopentenyl acrylate
CAS 12542-30-2; 50976-02-8
Synonyms: Acrylic acid, dicyclopentadienyl ester
Empirical: $C_{13}H_{14}O_2$
Properties: M.w. 202.27
Toxicology: LD50 (oral, rat) 11 g/kg; skin irritant
Uses: Monomer for air-drying adhesives, coatings, caulks, sealants, elastomers; crosslinks acrylics, unsat. polyesters and alkyd
Manuf./Distrib.: Rhone-Poulenc Surf. & Spec.
Trade names: Sipomer® DCPA

Dicyclopentenyl methacrylate
CAS 51178-59-7; 31621-69-9
Properties: M.w. 218.30; dens. 1.050; b.p. 137 C (13 mm); flash pt. > 110 C
Precaution: Light-sensitive
Toxicology: Irritant
Uses: Monomer for concrete resurfacing, air-drying adhesives and coatings, elastomers; improves cohesive str. through crosslinking
Manuf./Distrib.: Aldrich; Rhone-Poulenc Surf. & Spec.
Trade names: Sipomer® DCPM

DIDA. *See* Diisodecyl adipate

Di (dioctylphosphato) ethylene titanate
Uses: Coupling agent, adhesion promoter, antioxidant, antistat, antifoam, accelerator, blowing agent activator, catalyst, curative, corrosion inhibitor, dispersion aid, emulsifier, flame retardant, foaming agent, grinding and process aid
Trade names containing: Ken-React® KR 212

Di (dioctylpyrophosphato) ethylene titanate
Uses: Coupling agent, adhesion promoter, antioxidant, antistat, antifoam, accelerator, blowing agent activator, catalyst, curative, corrosion inhibitor, dispersion aid, emulsifier, flame retardant, foaming agent, grinding and process aid
Trade names containing: Ken-React® KR 238S

DIDP. *See* Diisodecyl phthalate

Diethanolamine
CAS 111-42-2; EINECS 203-868-0
Synonyms: DEA; 2,2´-Iminobisethanol; Bis(2-hydroxyethyl)amine; Di(2-hydroxyethyl) amine; 2,2´-Iminodiethanol
Classification: Aliphatic amine
Empirical: $C_4H_{11}NO_2$
Formula: $(HOCH_2CH_2)_2NH$
Properties: Colorless cryst. or liq., mild ammoniacal odor; very sol. in water and alcohol; misc. with acetone, chloroform, glycerin; insol. in ether, benzene; m.w. 105.09; dens. 1.0881 (30/4 C); m.p. 28 C; b.p. 268 C; flash pt. 300 F; ref. index 1.4770
Precaution: Combustible; DOT: corrosive material; light-sensitive
Toxicology: LD50 (rat, oral) 12.76 g/kg; TLV 3 ppm in air; may be irritating to skin and mucous membranes
Uses: Gas scrubbing; rubber chemicals intermediate; mfg. of surfactants for textiles, herbicides, petrol. demulsifier; emulsifier and dispersant in coatings, agric., cosmetics, pharmaceuticals; textile lubricants; humectant; softening agent; in organic synthesis; crosslinker for flexible PU foam slabstock
Regulatory: FDA 21CFR §175.105, 176.170, 176.180, 176.210, 177.2600; FDA approved for injectables (IV), ophthalmics; USP/NF, BP compliance
Manuf./Distrib.: Akzo Nobel; Aldrich; Allchem Ind.; Ashland; BASF AG; Coyne; Fluka; Great Western; Hüls AG; Occidental; Oxiteno; Sigma; Union Carbide
Trade names: DEA Commercial Grade; DEA Low Freeze Grade

Diethanolaminobenzene. *See* Phenyldiethanolamine
Diethanolaniline. *See* Phenyldiethanolamine
N,N-Diethanolaniline. *See* Phenyldiethanolamine

2,2-Diethoxyacetophenone
CAS 6175-45-7; EINECS 228-220-4
Synonyms: α,α-Diethoxyacetophenone; Phenylglyoxal 2-diethyl acetal
Empirical: $C_{12}H_{16}O_3$
Formula: $C_6H_5COCH(OC_2H_5)_2$
Properties: M.w. 208.26; dens. 1.034 (20/4 C); b.p. 131-134 C (10 mm); flash pt. > 110 C; ref. index 1.499 (20 C)
Toxicology: Toxic by inh., skin contact, ingestion; irritant; suspected carcinogen
Uses: Photoinitiator for uv-curable systems, coatings, adhesives, inks
Manuf./Distrib.: Aldrich; First Chem.; Fluka; Sigma
Trade names: Firstcure™ DEAP

α,α-Diethoxyacetophenone. *See* 2,2-Diethoxyacetophenone
1,2-Diethoxyethane. *See* Ethylene glycol diethyl ether

Diethylamine
CAS 109-89-7; EINECS 203-716-3
Synonyms: N-Ethylethanamine
Classification: Aliphatic organic compd.
Empirical: $C_4H_{11}N$
Formula: $(C_2H_5)_2NH$
Properties: Liq.; misc. with water, alcohol; m.w. 73.14; dens. 0.707; m.p. -50 C; b.p. 55 C; flash pt. -28 C; ref. index 1.3850 (20 C)
Precaution: DOT: Flamm. liq.
Toxicology: Corrosive
Uses: Rubber chemicals, textile specialties, selective solvent, dyes, flotation agents, resins, pesticides, polymerization inhibitor, pharmaceuticals, petroleum chemicals, electroplating, corrosion inhibitors
Regulatory: FDA approved for injectables (IM, IV); BP compliance
Manuf./Distrib.: AC Ind.; Air Prods.; Aldrich; Allchem Ind.; Ashland; BASF; BASF AG; Coyne; Elf Atochem N. Am.; Fluka; Sigma; Spectrum Chem. Mfg.; Union Carbide
Trade names containing: Virco-Pet® 40; Virco-Pet® 50

N,N-Diethylaminobenzene. *See* Diethyl aniline
(Diethylamino) ethane. *See* Triethylamine

Diethylaminoethyl acrylate
CAS 2426-54-2
Classification: Amine monomer
Formula: $CH_2=CHCOOCH_2CH_2N(C_2H_5)_2$
Properties: M.w. 171.24; dens. 0.925 (20/4 C); b.p. 202 C
Toxicology: Corrosive; severe skin and eye irritant; harmful if swallowed or inhaled
Uses: Industrial and automotive coatings, dye and lube oil additives, cationic quat. monomers, intermediate for water treatment chemicals, etc.
Manuf./Distrib.: Aldrich; CPS; Monomer-Polymer & Dajac Labs; Rhone-Poulenc Spec.
Trade names: Ageflex FA-2

Diethylaminoethyl acrylate dimethyl sulfate quat.
CAS 21810-39-9
Classification: Quat. ammonium monomer
Formula: $[CH_2=CHCOOCH_2CH_2N(C_2H_5)_2CH_3]^+CH_3OSO_3^-$
Properties: M.w. 297.37
Uses: Antistatic finish for polyester fibers, flocculant and coagulant polymers for water treatment, min. recovery, ion exchange resins, adhesives, acid dye receptivity, retention aid for paper mfg.
Trade names containing: Ageflex FA-2Q50DMS

Diethylaminoethyl methacrylate
CAS 105-16-8
Classification: Amine monomer
Formula: $CH_2=C(CH_3)COOCH_2CH_2N(C_2H_5)_2$
Properties: M.w. 185.29
Precaution: Combustible
Toxicology: Severe eye and skin irritant; harmful if swallowed
Uses: Industrial and automotive coatings, dye additives, intermediate for water treatment and oil field

chemicals, stabilizer for fuel oils
Manuf./Distrib.: Aldrich; CPS; Monomer-Polymer & Dajac Labs; Polysciences; Rhone-Poulenc Spec.
Trade names: Ageflex FM-2; Sipomer® 2E1M

Diethyl aniline
CAS 91-66-7; EINECS 202-088-8
Synonyms: N,N-Diethylaniline; N,N-Diethylaminobenzene; Diethylphenylamine
Empirical: $C_{10}H_{15}N$
Formula: $(C_2H_5)_2NC_6H_5$
Properties: Colorless to yel. liq.; sl. sol. in alcohol, chloroform, ether; m.w. 149.26; dens. 0.9302 (25/4 C); m.p. -38 C; b.p. 215-216 C; ref. index 1.5394 (24 C)
Precaution: Photosensitive
Toxicology: LD50 (oral, rat) 782 mg/kg; mod. toxic by ingestion, intraperitoneal, skin contact; heated to decomp., emits toxic fumes of NO_x
Uses: Intermediate for paints
Manuf./Distrib.: Aceto; Albemarle; Allchem Ind.; BASF; Bayer; Buffalo Color; Fabrichem; Fluka; Hoechst Celanese; Zeneca

N,N-Diethylaniline. *See* Diethyl aniline
Diethyl 'Carbitol'. *See* Diethylene glycol diethyl ether

3,5-Diethyl-2,4-diaminotoluene
Uses: Chain extender for PU; hardener for epoxy resins
Manuf./Distrib.: Lonza Ltd
Trade names containing: DETDA 80

3,5-Diethyl-2,6-diaminotoluene
Uses: Chain extender for PU; hardener for epoxy resins
Manuf./Distrib.: Lonza Ltd
Trade names containing: DETDA 80

Diethyldiglycol. *See* Diethylene glycol diethyl ether
Diethylene dioxide. *See* 1,4-Dioxane
1,4-Diethylene dioxide. *See* 1,4-Dioxane
Diethylene ether. *See* 1,4-Dioxane

Diethylene glycol
CAS 111-46-6; EINECS 203-872-2
Synonyms: DEG; Dihydroxydiethyl ether; Diglycol; 2,2′-Oxybisethanol
Classification: Aliphatic diol
Empirical: $C_4H_{10}O_3$
Formula: $CH_2OHCH_2OCH_2CH_2OH$
Properties: Colorless syrupy liq., pract. odorless, sweetish taste; misc. with water, ethanol, acetone, ether, ethylene glycol; m.w. 106.12; dens. 1.1184 (20/20 C); b.p. 245 C; f.p. -80 C; flash pt. 255 F; extremely hygroscopic; lowers f.p. of water
Precaution: Combustible
Toxicology: Hazardous for household use in conc. of 10% or more; may cause acidosis and renal failure
Uses: Prod. of polyurethane and unsat. polyester resins, triethylene glycol, paints; textile softener; solvent for nitrocellulose, dyes and oils; dehydration of natural gas, elasticizers, and surfactants; humectant for tobacco, casein, and synthetic sponge
Regulatory: FDA 21CFR §175.105, 177.2420
Manuf./Distrib.: Aldrich; Allchem Ind.; Ashland; BASF; Baychem; BP Chems. Ltd; Chemcentral; Coyne; DuPont; Eastman; Ferro/Grant; Fluka; C.P. Hall; Harcros; Hoechst Celanese; Itochu Spec.; Mitsui Petrochem. Ind.; Mobil; Occidental; Olin; Samson; Shell; Sigma; Spectrum Chem. Mfg.; Sunnyside; Texaco; Union Carbide; Van Waters & Rogers

Diethylene glycol butyl ether. *See* Butoxydiglycol

Diethylene glycol butyl ether acetate
CAS 124-17-4; EINECS 204-685-9
Synonyms: 2-(2-Butoxyethoxy) ethyl acetate; Diethylene glycol monobutyl ether acetate; Butyldiglycol acetate
Empirical: $C_{10}H_{20}O_4$
Formula: $CH_3COOCH_2CH_2OCH_2CH_2O(CH_2)_3CH_3$

Properties: Colorless liq.; mod. water sol.; m.w. 204.3; dens. 0.977 (20/4 C); i.b.p. 235 C; flash pt. 102 C; ref. index 1.43 (20 C)
Precaution: Peroxide former
Toxicology: Prolonged/repeated skin contact may cause drying, cracking, or irritation
Uses: Solvent for printing inks, coatings, separation processes; coalescing aid
Manuf./Distrib.: Aldrich; Allchem Ind.; ARCO; Ashland; Eastman; Fluka; Occidental; Spectrum Chem. Mfg.; Union Carbide
Trade names: Butyl Carbitol® Acetate; Eastman® DB Acetate

Diethylene glycol diacrylate
CAS 4074-88-8
Synonyms: Oxydiethylene diacrylate; Acrylic acid 2-ethoxyethanol diester
Empirical: $C_{10}H_{14}O_5$
Toxicology: LD50 (oral, rat) 770 mg/kg; mod. toxic by ingestion; poison by skin contact
Trade names containing: SR-230

Diethylene glycol dibenzoate
CAS 120-55-8; EINECS 204-407-6
Synonyms: PEG 100 dibenzoate; PEG-2 dibenzoate; POE (2) dibenzoate; Benzoic acid, diester with diethylene glycol; Dibenzoyldiethyleneglycol ester
Definition: PEG diester of benzoic acid
Properties: Sp.gr. 1.1765-1.178 (20 C); m.p. 16 C; flash pt. (PMCC) 232 C; ref. index 1.5424-1.5449 (20 C)
Uses: Plasticizer for polymers, adhesives, latex coatings
Regulatory: FDA 21CFR §175.105, 176.170
Manuf./Distrib.: Aldrich; Allchem Ind.; Hüls; Kalama; Unitex; Velsicol
Trade names: Benzoflex® 2-45
Trade names containing: Benzoflex® 50

Diethylene glycol dibutyl ether
CAS 112-73-2; EINECS 204-001-9
Synonyms: 2,2′-Dibutoxyethyl ether; Bis (2-butoxyethyl) ether; 2-Butoxyethyl ether; Dibutyl 'Carbitol'; Dibutyldiglycol
Empirical: $C_{12}H_{26}O_3$
Formula: $C_4H_9O(C_2H_4O)_2C_4H_9$
Properties: Almost colorless liq.; sl. sol. in water; m.w. 218.38; dens. 0.8853 (20/20 C); f.p. -60.2 C; b.p. 256 C; flash pt. 118 C; ref. index 1.423 (20 C)
Precaution: Combustible
Toxicology: Moderately toxic by ingestion; mildly toxic by skin contact; skin and eye irritant; heated to dec., emits acrid smoke and irritating fumes
Uses: Solvent used in electrochemistry, polymer and boron chemistry; physical processes such as gas absorption, extraction, stabilization; coatings and inks; diluent in VC dispersions
Manuf./Distrib.: Aldrich; Brand-Nu Labs; Ferro/Grant; Fluka; Great Western; Hoechst AG; Sigma

Diethylene glycol diethyl ether
CAS 112-36-7; EINECS 203-963-7
Synonyms: Diglycol diethyl ether; Bis (2-ethoxyethyl) ether; Diethyldiglycol; Diethyl 'Carbitol'; Ethoxyethyl ether; 2-Ethoxyethyl ether
Empirical: $C_8H_{18}O_3$
Formula: $C_2H_5O(C_2H_4O)_2C_2H_5$
Properties: Colorless liq.; sol. in hydrocarbons and water; m.w. 162.93; dens. 0.9082 (20/20 C); b.p. 189 C; f.p. -44.3 C; flash pt. 180 F; ref. index 1.412 (20 C); very stable
Precaution: Combustible
Uses: Solvent for nitrocellulose, resins, lacquers; organic synthesis
Manuf./Distrib.: Aldrich; Ferro/Grant; Fluka; Sigma

Diethylene glycol dimethacrylate
CAS 2358-84-1
Synonyms: Oxydiethylene methacrylate
Formula: $[H_2C=C(CH_3)CO_2CH_2CH_2]_2O$
Properties: M.w. 242.27; dens. 1.082; b.p. 134 C (2 mm); flash pt. > 230 F
Uses: Crosslinker for rubber vulcanization, moisture barrier films and coatings, photopolymer printing plates and letterpress inks, conversion coatings and adhesives

Diethylene glycol dimethyl ether

Manuf./Distrib.: Aldrich; CPS; Monomer-Polymer & Dajac Labs; Polysciences; Rohm Tech; Sartomer
Trade names containing: Ageflex DEGDMA; SR-231

Diethylene glycol dimethyl ether
CAS 111-96-6; EINECS 203-924-4
Synonyms: Diglycol methyl ether; Dimethyldiglycol; Diglyme; Dimethoxydiglycol (INCI); 2-Methoxyethyl ether; Bis (2-methoxyethyl) ether
Empirical: $C_6H_{14}O_3$
Formula: $CH_3(OCH_2CH_2)_2OCH_3$
Properties: Colorless liq., mild odor; misc. with water and hydrocarbons; m.w. 134.18; dens. 0.9451 (20/20 C); f.p. -58 C; b.p. 162 C; ref. index 1.408 (20 C)
Precaution: Combustible
Uses: Solvent, anhydrous reaction medium for organo-metallic synthesis; reaction solv. in reduction, isomerization, alkylation condensation, polymerization; extract and separation solv.; paints
Manuf./Distrib.: Aldrich; Brand-Nu Labs; Ferro/Grant; Fluka; Great Western; Hoechst AG; Sigma
Trade names: Diglyme
Trade names containing: SILIKOFTAL® Nonstick 50; Silikophen® Non-stick 50

Diethylene glycol dodecyl ether. *See* Laureth-2
Diethylene glycol ethyl ether. *See* Ethoxydiglycol
Diethylene glycol ethyl ether acetate. *See* Ethoxydiglycol acetate

Diethylene glycol 2-ethylhexyl ether
CAS 1559-36-0
Synonyms: Diethylene glycol mono-2-ethylhexyl ether
Trade names containing: Eastman® EEH

Diethylene glycol n-hexyl ether
CAS 112-59-4; EINECS 203-988-3
Synonyms: 3,6-Dioxadodecanol-1; Diethylene glycol monohexyl ether; n-Hexoxyethoxyethanol; 2-(2-Hexyloxyethoxy)ethanol; Hexyl Carbitol; Hexyldiglycol
Empirical: $C_{10}H_{22}O_3$
Formula: $C_6H_{13}O[C_2H_4O]_2H$
Properties: Water-wh. liq.; sol. 3% in water; m.w. 190.28; sp.gr. 0.9346 (20/20 C); f.p. -33 C; b.p. 259.1 C; flash pt. 140.5 C; ref. index 1.438 (20 C)
Precaution: Combustible
Uses: High-boling solv. for coatings
Manuf./Distrib.: Aldrich; Fluka; Union Carbide
Trade names: Hexyl Carbitol®

Diethylene glycol laurate. *See* PEG-2 laurate
Diethylene glycol methyl ether. *See* Methoxydiglycol
Diethylene glycol monobutyl ether. *See* Butoxydiglycol
Diethylene glycol monobutyl ether acetate. *See* Diethylene glycol butyl ether acetate
Diethylene glycol monoethyl ether. *See* Ethoxydiglycol
Diethylene glycol monoethyl ether acetate. *See* Ethoxydiglycol acetate
Diethylene glycol mono-2-ethylhexyl ether. *See* Diethylene glycol 2-ethylhexyl ether
Diethylene glycol monohexyl ether. *See* Diethylene glycol n-hexyl ether
Diethylene glycol monolaurate self-emulsifying. *See* PEG-2 laurate SE
Diethylene glycol monomethyl ether. *See* Methoxydiglycol
Diethylene glycol monooleate. *See* PEG-2 oleate
Diethylene glycol monooleate self-emulsifying. *See* PEG-2 oleate SE
Diethylene glycol monopropyl ether. *See* Diethylene glycol propyl ether
Diethylene glycol monostearate self-emulsifying. *See* PEG-2 stearate SE

Diethylene glycol phenyl ether
CAS 104-68-7
Synonyms: Ethanol, 2-(2-phenoxyethoxy); Phenyl 'Carbitol'
Empirical: $C_{10}H_{14}O_3$
Properties: Liq.; m.w. 182.24; dens. 1.1158 (20/20 C); f.p. -50 C; b.p. 207 C (55 mm)
Toxicology: LD50 (oral, rat) 2140 mg/kg; mod. toxic; skin and eye irritant
Manuf./Distrib.: Kenrich Petrochems.
Trade names containing: Ken-React® NZ 66A

Diethylene glycol propyl ether
CAS 6881-94-3; EINECS 229-985-7
Synonyms: Diethylene glycol monopropyl ether; 2-(2-Propoxyethoxy) ethanol
Empirical: $C_7H_{16}O_3$
Formula: $CH_2OHCH_2OCH_2CH_2OCH_2CH_2CH_3$
Properties: Liq.; sol. in water; m.w. 148.2; dens. 0.963 (20 C); b.p. 202 C (760 mm); f.p. < -90 C; flash pt. (TCC) 93 C
Precaution: Potential peroxide former
Toxicology: Causes eye irritation
Uses: Coupling solv. for sol'n. and water-dilutable coatings; active solv. for NC, acrylic copolymers, epoxy resins, chlorinated rubber, alkyd resins; coalescing aid in architectural latex coatings, floor polishes; hydraulic brake fluids; household and general industrial cleaners
Manuf./Distrib.: Ashland; Eastman; Great Western
Trade names: Eastman® DP

Diethylene glycol stearate. *See* PEG-2 stearate
Diethylene oxide. *See* 1,4-Dioxane; Tetrahydrofuran

Diethylenetriamine
CAS 111-40-0; EINECS 203-865-4
Synonyms: DETA; Aminoethylethandiamine; 2,2′-Iminodiethylamine; Bis(2-aminoethyl) amine
Empirical: $C_4H_{13}N_3$
Formula: $NH_2C_2H_4NHC_2H_4NH_2$
Properties: Yel. liq., ammoniacal odor; sol. in water and hydrocarbons; m.w. 103.17; dens. 0.9542 (20/20 C); b.p. 206.7 C; f.p. -39 C; flash pt. 215 F; strongly alkaline; hygroscopic; corrosive to copper and its alloys
Toxicology: Toxic by ingestion, inhalation and skin absorption; strong irritant to eyes and skin; TLV 1 ppm in air
Uses: Solvent for sulfur, acid gases, various resins, dyes; saponification agent for acidic materials; fuel component; epoxy curing agent; corrosion inhibitors; surfactants; intermediate for textile finishes; curing agent for two-pkg. protective coatings
Manuf./Distrib.: Akzo Nobel; Aldrich; Allchem Ind.; Ashland; BASF AG; Coyne; Diamine & Chems. Ltd; Fabrichem; Fluka; Janssen Chimica; Nayler Chem. Ltd; Rhone-Poulenc Surf. & Spec.; Sigma; Tosoh; Union Carbide
Trade names: D.E.H. 20; DETA; Epi-Cure® 3223

N,N-Diethylethanamine. *See* Triethylamine
Diethyl ether. *See* Ethyl ether
Di(2-ethylhexyl) adipate. *See* Dioctyl adipate
Di (2-ethylhexyl) fumarate. *See* Dioctyl fumarate
Di-(2-ethylhexyl) maleate. *See* Dioctyl maleate
Di-2-ethylhexyl phosphate. *See* Dioctyl phosphate
Di-2-ethylhexyl phosphoric acid. *See* Dioctyl phosphate
Di(2-ethylhexyl) phthalate. *See* Dioctyl phthalate
Di-2-ethylhexyl terephthalate. *See* Dioctyl terephthalate

Diethyl ketone
CAS 96-22-0; EINECS 202-490-3
Synonyms: 3-Pentanone; Dimethylacetone; Propione; Methacetone
Empirical: $C_5H_{10}O$
Formula: $CH_3CH_2COCH_2CH_3$
Properties: Liq., acetone odor; misc. with alcohol, ether; sol. in 25 parts water; m.w. 86.14; dens. 0.814; m.p. -42 C; b.p. 99-103 C; flash pt. 55 F; ref. index 1.392
Precaution: Flamm.; dangerous fire hazard
Toxicology: LD50 (oral, rat) 2.1 g/kg
Uses: Flavorings, medicine, organic synthesis; solv. for paints
Manuf./Distrib.: Aldrich; BASF; Fluka; Janssen Chimica; Penta Mfg.; Sigma; Union Carbide

Diethylol dimethyl hydantoin. *See* DEDM hydantoin
Diethylphenylamine. *See* Diethyl aniline

Diethyl toluene diamine
CAS 68479-98-1
Synonyms: DETDA

1,1-Difluoroethene homopolymer

Classification: Orthoalkylated aromatic
Uses: Intermediate for mfg. of pesticides, miticides, dyestuffs, antioxidants, pharmaceuticals, synthetic resins, fragrances; curing agent for polyurethane and epoxy resins; used in filament winding, elec. encapsulation, prepregs, tooling, potting and casting, laminating, coating, molding, and adhesive applics.
Trade names: Ethacure® 100

1,1-Difluoroethene homopolymer. *See* Polyvinylidene fluoride
Diglycidyl ether of neopentyl glycol. *See* Neopentyl glycol diglycidyl ether
Diglycol. *See* Diethylene glycol

Diglycol/CHDM/isophthalates/SIP copolymer
Synonyms: Diglycol/cyclohexanedimethanol/isophthalates/sulfoisophthalates copolymer
Definition: Copolymer of diethylene glycol, 1,4-cyclohexanedimethanol, and the simple esters of isophthalic acid and sulfoisophthalic acid
Uses: Dispersant providing adhesion, bonding, dust suppression, and asbestos mitigation to household adhesives, dyes, hard surf. cleaners, protective coatings, shoe polish, pesticides/herbicides; flocculant for wastewater; cosmetic film-former (hair spray, mousse, moisturizer, nail polish); ingred. in oil-free cosmetics, hand creams, sun screens
Trade names: Eastman® AQ 38S; Eastman® AQ 55S

Diglycol/cyclohexanedimethanol/isophthalates/sulfoisophthalates copolymer. *See* Diglycol/CHDM/isophthalates/SIP copolymer
Diglycol diethyl ether. *See* Diethylene glycol diethyl ether
Diglycol laurate. *See* PEG-2 laurate
Diglycol methyl ether. *See* Diethylene glycol dimethyl ether
Diglycol oleate. *See* PEG-2 oleate
Diglycol stearate. *See* PEG-2 stearate
Diglyme. *See* Diethylene glycol dimethyl ether

Di-n-heptyl phthalate
CAS 3648-21-3
Properties: M.w. 362; sp.gr. 0.991-0.993
Uses: Plasticizer
Manuf./Distrib.: Chisso Am.; Frinton Labs
Trade names: Diplast® E

Dihexyl sodium sulfosuccinate
CAS 3006-15-3; 2373-38-8; 6001-97-4; EINECS 221-109-1
Synonyms: Sodium dihexyl sulfosuccinate; Butanedioic acid, sulfo-, 1,4-dihexyl ester, sodium salt
Definition: Sodium salt of the diester of 1-methylamyl alcohol and sulfosuccinic acid
Formula: $C_{16}H_{30}O_7S \cdot Na$
Properties: Clear visc. liq.; m.w. 389.51
Precaution: Flamm. when exposed to heat, flame, oxidizers
Uses: Dispersant, textile wetting agent, emulsifier, solubilizer; used for emulsion polymerization, battery separators, electroplating, ore leaching; wetting agent/emulsifier for waterborne industrial coatings (PVAc, acrylics), printing inks
Regulatory: FDA 21CFR §175.105, 176.170, 177.1210, 178.3400
Manuf./Distrib.: Aquatec Quimica SA; Sigma
Trade names: Hydropalat® 680
Trade names containing: Monawet MM-80

Dihydroabietyl alcohol. *See* Hydroabietyl alcohol
Dihydro-2,5-furandione. *See* Succinic anhydride
2,5-Dihydrofuran-2,5-dione. *See* Maleic anhydride
Dihydro-2(3H)-furanone. *See* Butyrolactone

Dihydrogenated tallow benzylmonium hectorite
Definition: Reaction prod. of hectorite and a dihydrog. tallow benzyl monomethyl quaternary ammonium salt
Uses: Thickener, sag/level control, antisettling agent, and suspension agent for aromatic systems, acid-catalyzed systems, acrylic coatings, alkyds, antifouling paints, architectural paints, automotive OEM, refinish, and primers; baking systems, bituminous paints, butyl caulks, epoxies, plastisols/organosols, polyester coatings, two-part PU, varnishes, vinyl paints
Trade names: Bentone® SD-3

Dihydrogenated tallow dimethyl ammonium chloride. *See* Quaternium-18

Dihydrogenated tallow methylamine
CAS 61788-63-4; 67700-99-6; EINECS 262-991-8
Classification: Tertiary aliphatic amine
Formula: $R_2N\text{-}CH_3$, R represents tallow radical
Uses: Chemical intermediate for quat. ammonium derivs., betaines, amine oxides; acid scavenger in petrol. prods.; epoxy hardener, catalyst in mfg. of flexible PU foams; auxs. for fuel additives, rust inhibition, paint
Trade names: Noram M2SH

4,5-Dihydro-7-nortall oil-1H-imidazole-1-ethanol. *See* Tall oil hydroxyethyl imidazoline
1,2-Dihydro-2,2,4-trimethylquinoline homopolymer. *See* 2,2,4-Trimethyl-1,2-dihydroquinoline polymer
1,2-Dihydro-2,2,4-trimethylquinoline polymer. *See* 2,2,4-Trimethyl-1,2-dihydroquinoline polymer
1,4-Dihydroxybenzene. *See* Hydroquinone
m-Dihydroxybenzene. *See* Resorcinol
p-Dihydroxybenzene. *See* Hydroquinone
2,4-Dihydroxybenzophenone. *See* Benzophenone-1
1,4-Dihydroxybutane. *See* 1,4-Butanediol
Dihydroxydiethyl ether. *See* Diethylene glycol
2,2´-Dihydroxy-4,4´-dimethoxybenzophenone. *See* Benzophenone-6
2,2´-Dihydroxydipropyl ether. *See* Dipropylene glycol
Di(2-hydroxyethyl) amine. *See* Diethanolamine
Dihydroxyethylaniline. *See* Phenyldiethanolamine
N,N-Di(2-hydroxyethyl)aniline. *See* Phenyldiethanolamine
N,N-Di(β-hydroxyethyl)aniline. *See* Phenyldiethanolamine
Di-(2-hydroxyethyl)-5,5-dimethyl hydantoin. *See* DEDM hydantoin
1,6-Dihydroxyhexane. *See* Hexamethylene glycol
2,2´-Dihydroxyisopropyl ether. *See* Dipropylene glycol
2,2´-Dihydroxy-4-methoxybenzophenone. *See* Benzophenone-8
1,2-Dihydroxypropane. *See* Propylene glycol
2,3-Dihydroxypropyl octadecanoate. *See* Glyceryl stearate

Diiodomethyl p-tolyl sulfone
CAS 20018-09-1; EINECS 243-468-3
Uses: Mildewcide, fungicide, preservative for adhesives and sealants, paints, and in lumber, construction, home improvement, textile, leather, and automotive industries
Trade names: Amical® 48; Amical® Flowable; Amical® Wettable Powd.
Trade names containing: Amical® 50

Diisoamyl ketone
Uses: Solv. for paints
Manuf./Distrib.: Eastman

Diisobutyl adipate
CAS 141-04-8; EINECS 205-450-3
Synonyms: DIBA; Bis(2-methylpropyl) hexanedioate; Diisobutyl hexanedioate; Hexanedioic acid diisobutyl ester
Definition: Diester of isobutyl alcohol and adipic acid
Empirical: $C_{14}H_{26}O_4$
Formula: $[C_2H_4COOCH_2CH(CH_3)_2]_2$
Properties: Colorless liq., odorless; sol. in most org. solvs.; insol. in water; m.w. 258.40; dens. 0.950 (25 C); b.p. 278-280 C; f.p. -20 C
Precaution: Combustible
Toxicology: LD50 (IP, rat) 5950 mg/kg; mod. toxic by IP route; mildly toxic by ingestion; experimental teratogen, reproductive effects; heated to decomp., emits acrid smoke and irritating fumes
Uses: Plasticizer for some cellulosics, PVAc, PVC, VCA
Regulatory: FDA 21CFR §175.105, 181.22, 181.27
Manuf./Distrib.: Aceto; Aldrich
Trade names containing: DBE-IB

Diisobutyl glutarate
Trade names containing: DBE-IB

Diisobutyl hexanedioate. *See* Diisobutyl adipate

Diisobutyl ketone
CAS 108-83-8; EINECS 203-620-1
Synonyms: DIBK; 2,6-Dimethyl-4-heptanone; Isovalerone
Empirical: $C_9H_{18}O$
Formula: $(CH_3)_2CHCH_2COCH_2CH(CH_3)_2$
Properties: M.w. 142.24; sp.gr. 0.8076; b.p. 325 F; flash pt. (TCC) 120 F; ref. index 1.4130 (20 C)
Precaution: Flamm.
Toxicology: LD50 (oral, rat) 4300 mg/kg; ACGIH TLV 100 ppm, STEL 150 ppm; OSHA PEL 100 ppm; irritant to respiratory system
Uses: Solvent for nitrocellulose, rubber, synthetic resins; lacquers, coatings, organic synthesis, roll-coating inks, stains
Regulatory: FEMA GRAS
Manuf./Distrib.: Aldrich; Allchem Ind.; Ashland; Chemcentral; Coyne; Eastman; Fluka; Great Western; Harcros; Hüls AG; Hüls Am.; Samson; Stanchem; Sunnyside; Union Carbide; Van Waters & Rogers
Trade names containing: BYK®-S 706

Diisobutyl nonyl phenol
CAS 4306-88-1
Uses: Primary antioxidant; for syn. rug-backing, latex paints, rosin, ester gums, in gasoline and aviation fuels, insulating oils, paraffin wax
Trade names: Uvi-Nox® 1494

Diisobutyl phthalate
CAS 84-69-5
Synonyms: DIBP; Phthalic acid diisobutyl ester
Empirical: $C_{16}H_{22}O_4$
Properties: Sp.gr. 1.040 (20 C); f.p. -50 C; m.p. -64 C; flash pt. (PMCC) 185 C; ref. index 1.490
Toxicology: LD50 (oral, rat) 15 g/kg, (oral, mouse) 13 g/kg; mod. toxic by intraperitoneal route; mildly toxic by ingestion and skin contact
Uses: Plasticizer for PVC, PVB, PS, cellulosics, paints, as vehicles for pigment dispersions
Manuf./Distrib.: Aldrich; Allchem Ind.; Hüls; Unitex
Trade names: Diplast® B; Nuoplaz® DIBP

Diisobutyl sodium sulfosuccinate
CAS 127-39-9; EINECS 204-839-5
Synonyms: Sodium diisobutyl sulfosuccinate; 1,4-Bis(2-methylpropyl) sulfobutanedioate, sodium salt; Sulfosuccinic acid diisobutyl ester sodium salt
Definition: Sodium salt of the diester of isobutyl alcohol and sulfosuccinic acid
Empirical: $C_{12}H_{22}NaO_7S$
Properties: Wh. powd.; sol. 760 g/l in water (25 C); sol. in glycerol, pine oil, oleic acid; m.w. 332.35
Toxicology: Irritating to eyes, mucous membranes
Uses: Wetting agent; emulsion polymerization of styrene, butadiene and copolymers; aq. paints
Regulatory: FDA 21CFR §178.3400
Trade names: Geropon® CYA/DEP; Monawet MB-45

Diisobutyl succinate
Trade names containing: DBE-IB

4,4´-Diisocyanatodiphenylmethane. *See* MDI
1,6-Diisocyanatohexane. *See* Hexamethylene diisocyanate
2,4-Diisocyanatotoluene. *See* Toluene diisocyanate

Diisodecyl adipate
CAS 27178-16-1
Synonyms: DIDA; Hexanedioic acid, diisodecyl ester
Definition: Diester of branched chain decyl alcohols and adipic acid
Empirical: $C_{26}H_{50}O_4$
Formula: $C_{10}H_{21}OOC(CH_2)_4COOC_{10}H_{21}$
Properties: Lt. colored oily liq., mild odor; insol. in glycerols, glycols, and some amines; sol. in most other org. solvs.; m.w. 426; dens. 0.918 (20/20 C); m.p. -70 C; b.p. 239-246 C (4 mm); f.p. -71 C; flash pt. 225 F
Precaution: Combustible

Uses: Primary plasticizer for polymers; lacquer marring resistance of low temp. plasticizer
Regulatory: FDA 21CFR §175.105, 177.2600
Manuf./Distrib.: C.P. Hall; Hatco; Henkel; Hüls; ICI Am.; Inolex; Werner G. Smith
Trade names: Staflex DIDA

Diisodecyl phthalate
CAS 68515-49-1; 26761-40-0; EINECS 247-977-1
Synonyms: DIDP; 1,2-Benzenedicarboxylic acid, diisodecyl ester
Formula: $C_6H_4(COOC_{10}H_{21})_2$
Properties: Clear liq., mild odor; insol. in glycerol, glycols, some amines; sol. in most other organics; m.w. 746; dens. 0.966 (20/20 C); f.p. -50 C; b.p. 250-257 C (4 mm); visc. 108 cp (20 C); flash pt. 450 F
Precaution: Combustible
Toxicology: Irritant
Uses: Plasticizer for PVC, PS, cellulosics, paints, as vehicles for pigment dispersions
Manuf./Distrib.: Allchem Ind.; Aristech; Ashland; BASF; Coyne; Exxon; C.P. Hall; Harwick; Hatco; Hoechst AG; Hüls; ICC Ind.; Occidental; Velsicol
Trade names: Diplast® R; Jayflex® DIDP
Trade names containing: Vinyzene® BP-5-2 DIDP; Vinyzene® BP-5-5 DIDP; Vinyzene® IT-3000 DIDP

Diisononyl phthalate
CAS 14103-61-8; 68515-48-0; EINECS 249-079-5
Synonyms: DINP; 1,2-Benzenedicarboxylic acid, diisononyl ester
Empirical: $C_{26}H_{42}O_4$
Properties: M.w. 418.62; sp.gr. 0.973; flash pt. (TCC) 415 F
Uses: Plasticizer for PVC, PS, cellulosics, paints, as vehicles for pigment dispersions
Manuf./Distrib.: Aldrich; Allchem Ind.; Aristech; Ashland; BASF; Chemisphere Ltd; Chisso Am.; Exxon; C.P. Hall; Harwick; Henkel; ICC Ind.; Unitex
Trade names: Diplast® N; Jayflex® DINP
See also Dinonyl phthalate

Diisopropanolamine
CAS 110-97-4; EINECS 203-820-9
Synonyms: DIPA; 1,1′-Iminobis-2-propenol; 1,1′-Iminobis(propan-2-ol)
Classification: Aliphatic amine
Empirical: $C_6H_{15}NO_2$
Formula: $HN(CH_2CHOHCH_3)_2$
Properties: Wh. cryst. solid; misc. with water; m.w. 133.19; dens. 1.004; m.p. 44.5-45.5 C; b.p. 249-250 C (745 mm); flash pt. 126 C
Precaution: Combustible; protect from light
Toxicology: LD50 (oral, rat) 6720 mg/kg; heated to decomp., emits toxic fumes of NO_x
Uses: Emulsifying agents for polishes, textile specialties, leather compounds, insecticides, cutting oils, aq. paints; neutralizing agent for cosmetics and toiletries
Regulatory: FDA 21CFR § 175.105, 176.210; FDA approved for topicals; USP/NF
Manuf./Distrib.: Aldrich; Ashland; BASF; BASF AG; Coyne; Fluka; Ruger; Van Waters & Rogers
Trade names: DIPA Commercial Grade; DIPA Low Freeze Grade 85; DIPA Low Freeze Grade 90; DIPA NF Grade

Diisopropyl biphenyl. *See* Bis(methylethyl)-1,1′-biphenyl
Diisopropyl oxide. *See* Isopropyl ether

Diisotridecyl phthalate
EINECS 248-368-3
Synonyms: 1,2-Benzenedicarboxylic acid, diisotridecyl ester
Uses: Plasticizer for NC, vinyl, and other coatings

2,5-Diketotetrahydrofuran. *See* Succinic anhydride

Dilinoleic acid
CAS 6144-28-1; 61788-89-4
Synonyms: 9,12-Octadecadienoic acid, dimer; Dimer acid
Definition: 36-Carbon dicarboxylic acid formed by the catalytic dimerization of linoleic acid
Empirical: $C_{36}H_{64}O_4$
Properties: Gardner 6-9 liq.

Dimelamine phosphate

Precaution: Combustible
Uses: Lubricant, corrosion inhibitor, mildness additive in household detergents, plastics, and protective coatings
Regulatory: FDA 21CFR §176.200, 176.210, 178.3910
Manuf./Distrib.: Allchem Ind.; Arizona; Henkel/Emery; Unichema N. Am.; Union Camp; Welch, Holme & Clark; Witco/Oleo-Surf.

Dimelamine phosphate
Uses: Flame retardant for polymers, intumescent coatings and paints, plastics (polyesters, polymethyl methacrylate, polyolefins, PS, PU foams), and paper
Manuf./Distrib.: DSM Melamine Am.
Trade names: Amgard® ND

Dimer acid
CAS 68783-41-5
Definition: Dibasic acid usually contg. 36 carbons
Properties: Visc. liq.
Uses: Polymer building block; polymerizes with alcohols and polyols to yield plasticizers, lubricating oils, hydraulic fluids; intermediate for paints
Manuf./Distrib.: Aldrich; Allchem Ind.; Arizona; Henkel/Emery; Rit-Chem; Unichema N. Am.; Union Camp; United Min. & Chem.; Welch, Holme & Clark; Witco/Oleo-Surf.
Trade names: Pripol 1009; Pripol 1013; Pripol 1022
See also Dilinoleic acid

Dimethicone
CAS 9006-65-9; 9016-00-6; 63148-62-9; 68037-74-1 (branched)
Synonyms: PDMS; Dimethylpolysiloxane; Dimethyl silicone; Polydimethylsiloxane
Definition: Silicone oil consisting of dimethylsiloxane polymers
Empirical: $(C_2H_6OSi)_xC_4H_{12}Si$
Formula: $(CH_3)_3SiO[Si(CH_3)_2O]_nSi(CH_3)_3$
Properties: Colorless clear visc. oily liq., odorless; sol. in hydrocarbon solvs., benzene, toluene, xylene, hexane, petrol. spirits, amyl acetate; misc. in chloroform, ether; insol. in water, methanol, acetone, alcohol; m.w. 340-250,000; dens. 0.96-0.97; visc. 20-12,500 cst; ref. index 1.400-1.404
Toxicology: Very low toxicity; ingestion of large amts. may cause stomach upset; suspected neoplastigen; heated to dec., emits acrid smoke and irritating fumes
Uses: Ointment and topical drug vehicle, skin protectant, foam preventative, surface active agent, release material, antifoaming agent; used in adhesive, ink, latex, soap, starch, and paint mfg. and other aq. industrial systems
Regulatory: FDA 21CFR §145.180, 145.181, 146.185, 173.340, 175.105, 175.300, 176.170, 176.200, 176.210, 177.2260, 177.2600, 177.2800, 178.3570, 178.3910, 181.22, 181.28; USDA 9CFR §318.7, 381.147; Europe listed; UK approved; FDA approved for topicals; USP/NF, BP, Ph.Eur. compliance
Manuf./Distrib.: Aldrich; Dow Corning; Sigma
Trade names: 200® Fluid, 50 cs; AF 60; AF 9020; BYK®-331; BYK®-336; BYK®-341; BYK®-344; Dehydran® 1293; Dimethicone L-45 Series; Dow Corning® 18; Dow Corning® 51; Dow Corning® 175; Dow Corning® 200 Fluid; Ecco Additive DS; Ecco Additive PI; Foamkill® 830F; GP-7101 Silicone Wax; GP-7102 Silicone Wax; SF69; SF81-50; SF96®; SF96® (100 cst); SF96® (350 cst); SF96® (500 cst); SF96® (1000 cst); SF99; Silquest® Y-11343; SM 70 Glass Coating Emulsion; TEGO® Flow ATF; Union Carbide® LE-467; Viscasil®; Viscasil® 60,000
Trade names containing: BYK®-LP W 6246; Cab-O-Sil® TS-720; Foamaster® DS; SWS-290; SWS-295; SWS-299; TEGO® Foamex 835; TEGO® Foamex N

Dimethicone copolyol
CAS 63148-55-0; 64365-23-7; 67762-96-3
Synonyms: Dimethylsiloxane-glycol copolymer
Classification: Polymer
Toxicology: Very low toxicity
Uses: Silicone surfactant, dispersant, emulsifier, leveling and flow control agent, antifogging agent, lubricant, antiblock, slip additive for adhesives, agric., automotive, coatings, printing inks, textiles, household specialties, cutting fluids; in flexible slabstock PU foam
Regulatory: FDA approved for injectables (percutaneous)
Trade names: Alkasil® NE 58-50; Dow Corning® 190 Surfactant; GP-217 Silicone Polyol Copolymer; GP-226 Silicone Polyol Copolymer; SF1188; Silwet® L-722; Silwet® L-7001; Silwet® L-7002; Silwet® L-7500; Silwet® L-7600; Silwet® L-7602; Silwet® L-7604; Silwet® L-7605; TEGO® Foamex 1488; TEGO®

Foamex 3062; TEGO® Foamex 7447; TEGO® Glide 100; TEGO® Glide 405; TEGO® Glide 410; TEGO® Glide 470; TEGO® Glide B 1484; TEGO® Glide ZG 400; TEGO® Wet KL 245
Trade names containing: TEGO® Glide 406
See also Polysiloxane polyether copolymer

Dimethicone/mercaptopropyl methicone copolymer
Classification: Silicone polymer
Uses: Release agent for plastics and rubber; internal lubricant and release agent for sulfur and peroxide cure rubber; coreactant in vinyl polymerization; synthesis of org./silicone copolymers; heat stabilizer for dimethyl silicone fluids; in corrosion inhibitor coatings, uv and eb-cured inks and coatings; cosmetic and personal care applics. incl. hair care
Trade names: GP-71-SS Mercapto Modified Silicone Fluid

Dimethiconol
CAS 31692-79-2
Synonyms: Poly[oxy(dimethylsilylene)], α-hydro-ω-hydroxy-
Definition: Dimethyl silicone terminated with hydroxyl groups
Uses: Silicone surfactant, textile softener, etc.; raw material for silicone RTV systems, textile/paper coatings; plasticizer/processing aid for silicone elastomers, in water repelllents
Manuf./Distrib.: Fluka
Trade names: Masil® SFR 70; Masil® SFR 100; Masil® SFR 750; Masil® SFR 2000; Masil® SFR 3500; Masil® SFR 18,000; Masil® SFR 50,000

Dimethoxydiglycol (INCI). *See* Diethylene glycol dimethyl ether
1,2-Dimethoxyethane. *See* Ethylene glycol dimethyl ether

2,2-Dimethoxy-2-phenylacetophenone
CAS 24650-42-8; EINECS 246-386-6
Synonyms: Ethanone, 2,2-dimethoxy-1,2-didphenyl-(9CI); 1,2-Diphenylethane-1,2-dione, dimethyl ketal
Empirical: $C_{16}H_{16}O_3$
Properties: Wh. cryst. solid; sol. in acrylates, dichlormethane, DOP; m.w. 256.30; m.p. 63-68 C
Precaution: Light-sensitive
Toxicology: May cause skin and eye irritation; may be harmful by inhalation, ingestion, or skin absorption
Uses: Photoinitiator for uv-curable systems incl. coatings, adhesives, inks, printing plates
Manuf./Distrib.: Aldrich; First Chem.; Fluka
Trade names: Firstcure™ BDK

Dimethylacetic acid. *See* Isobutyric acid
Dimethylacetone. *See* Diethyl ketone

Dimethyl adipate
CAS 627-93-0; EINECS 211-020-6
Synonyms: Dimethyl hexanedioate; Methyl adipate; Adipic acid dimethyl ester
Definition: Diester of methyl alcohol and adipic acid
Empirical: $C_8H_{14}O_4$
Formula: $CH_3O_2C(CH_2)_4CO_2CH_3$
Properties: Colorless liq.; m.w. 174.22; dens. 1.062 (20/4 C); m.p. 9-11 C; b.p. 115C (13 mm); flash pt. 107 C
Toxicology: Moderately toxic by intraperitoneal route; suspected teratogen and reproductive effects; heated to dec., emits acrid smoke and irritating fumes
Uses: Solv. for coatings, cleaners, inks, textile lubricants, urethane prod.; plasticizer for flexible thermoset polyester; polymer intermediate for polyester polyols for urethanes, wet-str. paper resins, polyester resins; specialty chemical intermediate
Manuf./Distrib.: Aldrich; Ashland; DuPont; Eastern Chem.; Fluka; Morflex; Sigma; UCB SA
Trade names: DBE-6; Santosol™ DMA
Trade names containing: DBE; DBE-2, -2SPG; DBE-3; DBE-9; Santosol™ DME; Santosol™ DME-2; Santosol™ DME-3

(Dimethylamino) benzene. *See* n,n-Dimethylaniline
2-Dimethylaminoethanol. *See* Dimethylethanolamine
N,N-Dimethylaminoethanol. *See* Dimethylethanolamine

Dimethylaminoethyl acrylate
CAS 2439-35-2; EINECS 219-460-0
Empirical: $C_7H_{13}O_2N$
Formula: $CH_2=CHCOOCH_2CH_2N(CH_3)_2$
Properties: Colorless to sl. brn. liq.; sol. in water, alcohol, ethyl acetate, benzene; m.w. 143.19; dens. 0.943; f.p. -75 C; m.p. -55 C; b.p. 64 C (12 mm); flash pt. 58 C; ref. index 1.4380 (20 C)
Toxicology: Highly toxic; corrosive; severe skin and eye irritant; harmful if swallowed or inhaled
Uses: Adhesion promoter in UV and EB cured coatings for metal, plastic, paper, and wood; catalyst; intermediate for water treatment chemicals, etc.; textile finishing, adhesives
Manuf./Distrib.: Aldrich; CPS; Elf Atochem N. Am.; Monomer-Polymer & Dajac Labs
Trade names: Ageflex FA-1

Dimethylaminoethyl acrylate dimethyl sulfate
CAS 13106-44-0
Classification: Quat. ammonium monomer
Uses: Antistatic finish for polyester fibers, flocculant and coagulant polymers for water treatment, min. recovery, ion exchange resins, adhesives, acid dye receptivity, retention aid for paper mfg.
Manuf./Distrib.: Aldrich
Trade names containing: Ageflex FA-1Q80DMS

Dimethylaminoethyl acrylate methyl chloride quat.
CAS 44992-01-0
Empirical: $C_8H_{16}O_2NCl$
Formula: $[CH_2=CHCOOCH_2CH_2N(CH_3)_3]^+Cl^-$
Properties: M.w. 193.68
Uses: Antistatic finish for polyester fibers, flocculant and coagulant polymers for water treatment, min. recovery, ion exchange resins, adhesives, acid dye receptivity, retention aid for paper mfg.
Trade names: Ageflex FA-1Q75MC
Trade names containing: Ageflex FA-1Q80MC

Dimethylaminoethyl methacrylate
CAS 2867-47-2; EINECS 220-688-8
Synonyms: 2-(Dimethylamino)ethyl 2-methyl-2-propenoate
Classification: Amine monomer; aliphatic ester amine
Empirical: $C_8H_{15}NO_2$
Formula: $CH_2=C(CH_3)COOCH_2CH_2N(CH_3)_2$
Properties: Liq.; sol. in water and org. solvs.; m.w. 157.21; dens. 0.933 (25 C); b.p. 182-190 C; flash pt. 73.9 C
Precaution: Combustible; flamm. when exposed to sparks, heat, open flames, or oxidizers
Toxicology: LD50 (oral, rat) 1751 mg/kg; poisonous; harmful if swallowed; causes severe eye burns, skin and mucous membrane irritation; strong lachrymator
Uses: Visc. index improvers, automotive and industrial coatings, dye additives, lube oil additives, detergent and sludge dispersant in lubricant oils, flocculant for waste water treatment, retention aid for paper making, acrylic floor polish; resin and rubber modifier; acid scavenger in PU foams; corrosion inhibitor
Regulatory: FDA 21CFR §178.3520
Manuf./Distrib.: Aldrich; CPS; Elf Atochem N. Am.; Fluka; Monomer-Polymer & Dajac Labs; Polysciences; Rhone-Poulenc Spec. Chem.; Rohm Tech; Sigma
Trade names: Ageflex FM-1; Sipomer® 2M1M

Dimethylaminoethyl methacrylate dimethyl sulfate quat.
CAS 6891-44-7
Classification: Quat. ammonium monomer
Formula: $[CH_2=C(CH_3)COOCH_2CH_2N(CH_3)_3]^+CH_3OSO_3^-$
Properties: M.w. 283.35
Uses: Antistatic finish for polyester fibers, flocculant and coagulant polymers for water treatment, min. recovery, ion exchange resins, adhesives, acid dye receptivity, retention aid for paper mfg.
Manuf./Distrib.: Aldrich
Trade names containing: Ageflex FM-1Q80DMS

Dimethylaminoethyl methacrylate methyl chloride quat.
CAS 5039-78-1
Classification: Quat. ammonium monomer
Formula: $[CH_2=CH(CH_3)COOCH_2CH_2N(CH_3)_3]^+Cl^-$

Properties: M.w. 207.70
Uses: Antistatic finish for polyester fibers, flocculant and coagulant polymers for water treatment, min. recovery, ion exchange resins, adhesives, acid dye receptivity, retention aid for paper mfg.
Manuf./Distrib.: Aldrich
Trade names containing: Ageflex FM-1Q75MC

2-(Dimethylamino)ethyl 2-methyl-2-propenoate. *See* Dimethylaminoethyl methacrylate

Dimethylaminomethyl phenol
CAS 25338-55-0
Synonyms: m-Hydroxy-N,N-dimethyl aniline
Formula: $C_6H_4OHCH_2N(CH_3)_2$
Properties: Dk. red liq., phenolic odor; sol. in org. solvs.; moderately sol. in water; m.w. 137.09; dens. 1.010 (25/25 C); m.p. 85 C; b.p. 265-268 C
Toxicology: Toxic by ingestion, inhalation, and skin absorption
Uses: Antioxidants, stabilizers, catalysts, intermediates; R.T. epoxy curing agent; activator for epoxy resins cured with wide variety of hardener types incl. polyamide, amidoamines, polymercaptans, and anhydrides; for coatings, adhesives, castings; potting, encapsulation
Manuf./Distrib.: Sigma
Trade names: Ancamine® 1110

2-Dimethylaminomethyl propanol. *See* 2-Dimethylamino-2-methyl-1-propanol

2-Dimethylamino-2-methyl-1-propanol
CAS 7005-47-2; EINECS 230-279-6
Synonyms: 2-Dimethylaminomethyl propanol
Classification: Amino alcohol
Empirical: $C_6H_{15}NO$
Formula: $(CH_3)_2C(CH_2OH)N(CH_3)_2$
Properties: M.w. 117.22; dens. 0.910 (20/4 C); b.p. 158-160 C
Toxicology: Poison by intraperitoneal route; heated to dec., emits toxic NO_x
Uses: Solubilizer for resins in aq. coatings; emulsifier for waxes; vapor-phase corrosion inhibitor; urethane catalyst; titanate solubilizer; raw material for synthesis
Manuf./Distrib.: Aldrich; ANGUS; Fluka
Trade names: DMAMP-80

N-[3-(Dimethylamino)propyl]coco amides-N-oxide. *See* Cocamidopropylamine oxide

Dimethylaminopropylmethacrylamide
CAS 5205-93-6; EINECS 226-002-3
Synonyms: DMAPMA
Formula: $H_2C{=}C(CH_2)COOCH_2N(CH_3)_3$
Properties: Pale yel. liq.; bulk dens. 7.8 lb/gal; f.p. 140 C; ref. index 1.4763
Uses: Reactive cationic monomer, incorporation of amine group into polymers, paints, coating agent, textile finishing
Manuf./Distrib.: Aldrich; Kenrich Petrochems.; Monomer-Polymer & Dajac Labs; Rohm Tech
Trade names containing: Ken-React® NZ 38J

4-Dimethylaminotoluene. *See* N,N-Dimethyl-p-toluidine

n,n-Dimethylaniline
CAS 121-69-7; EINECS 204-493-5
Synonyms: Aniline, N,N-dimethyl; (Dimethylamino) benzene; N,N-Dimethylphenylamine; N,N-Dimethylbenzeneamine; Dimethylphenylamine
Empirical: $C_8H_{11}N$
Formula: $C_6H_5N(CH_3)_2$
Properties: Ylsh. to brnsh. oily liq.; sol. in alcohol, ether; insol. in water; m.w. 121.18; dens. 0.956 (20/4 C); m.p. 2-4 C; b.p. 76-78 C (10 mm); flash pt. $\approx$ 48 C; ref. index 1.558
Precaution: Photosensitive; combustible; explodes on contact with benzoyl peroxide
Toxicology: LD50 (oral, rat) 1410 mg/kg; toxic by inhalation, skin contact, ingestion; CNS depressant; suspected carcinogen; TLV: 5 ppm in air; heated to decomp., emits highly toxic fumes of aniline and NO_x
Uses: Dyes, intermediates, solvent, manufacture of vanillin, stabilizer (acid acceptor), reagent; intermediate for paints

Dimethylbenzene

Manuf./Distrib.: Aceto; Aldrich; Allchem Ind.; BASF; BASF AG; Buffalo Color; DuPont; Fabrichem; Fluka; Hoechst Celanese; ICI Am.; Monomer-Polymer & Dajac Labs; Sigma; Zeneca

Dimethylbenzene. *See* Xylene
1,2-Dimethylbenzene. *See* o-Xylene
o-Dimethylbenzene. *See* o-Xylene
N,N-Dimethylbenzeneamine. *See* n,n-Dimethylaniline
Dimethyl 1,2-benzenedicarboxylate. *See* Dimethyl phthalate
Dimethylbenzenesulfonic acid. *See* Xylene sulfonic acid
N,N-Dimethylbenzylamine. *See* N-Benzyldimethylamine
α,α-Dimethylbenzyl hydroperoxide. *See* Cumene hydroperoxide
Dimethyl butanedioate. *See* Dimethyl succinate

Dimethyl caprylamide-capramide
Uses: Mutual solvent for polar and nonpolar ingred. in cosmetics, cleaners, corrosion inhibitor; modifier for polymer coatings
Trade names: Hallcomid® M-8-10

Dimethyl carbinol. *See* Isopropyl alcohol

Dimethyl cocamine
CAS 61788-93-0; EINECS 263-020-0
Synonyms: Coco dimethyl amine; Amines, coco alkyl dimethyl; Dimethyl coconut amine
Classification: Tertiary aliphatic amine
Formula: $R\text{-}N(CH_3)_2$, R represents the coconut radical
Uses: Chemical intermediate, raw material for surfactants
Trade names: Noram DMC

Dimethyl coconut amine. *See* Dimethyl cocamine
Dimethyl dialkyl ammonium chloride. *See* Dialkyl dimethyl ammonium chloride

Dimethyl diallyl ammonium chloride
CAS 7398-69-8; EINECS 230-993-8
Synonyms: Diallyl dimethyl ammonium chloride
Classification: Quat. ammonium monomer
Empirical: $C_8H_{16}ClN$
Formula: $(CH_3)_2N(CH_2CH{=}CH_2)_2^+Cl^-$
Properties: M.w. 161.80
Precaution: Nonhazardous (DOT); protect from oxiders, strong alkali; mildly corrosive
Uses: Antistatic finish for polyester fibers, flocculant and coagulant polymers for water treatment, min. recovery, ion exchange resins, adhesives, acid dye receptivity, retention aid for paper mfg.; demulsifier for petrol. recovery, elec. conductive paper and coatings, wet and dry str. resins, antistatic additives and coatings
Manuf./Distrib.: Aldrich; CPS; Fluka; Monomer-Polymer & Dajac Labs
Trade names: Ageflex mDMDAC
Trade names containing: Ageflex NB-50

Dimethyldichlorosilane
CAS 75-78-5; 68611-44-9; EINECS 200-901-0
Synonyms: Dichlorodimethylsilane
Empirical: $C_2H_6Cl_2Si$
Formula: $(CH_3)_2SiCl_2$
Properties: Colorless liq.; sol. in benzene and ether; m.w. 129.0; dens. 1.062 (20 C); m.p. -76 C; b.p. 70 C; f.p.-86 C; flash pt. -8.9 C; ref. index 1.4023 (25 C)
Precaution: Highly flamm.; dangerous fire and explosion risk; keep away from ignition sources
Toxicology: Irritating to eyes, respiratory system, and skin; causes burns; PEL 6 mg/m^3
Uses: Intermediate for silicone prods.
Manuf./Distrib.: Aldrich; Dow Corning; Fluka; Hüls AG; Hüls Am.; PCR; SAF Bulk Chems.; Sigma
Trade names containing: Cab-O-Sil® TS-610

Dimethyldiglycol. *See* Diethylene glycol dimethyl ether
Dimethyl di(hydrogenated tallow)ammonium chloride. *See* Quaternium-18
Dimethyldithiocarbamic acid sodium salt. *See* Sodium dimethyldithiocarbamate

N,N-Dimethyl-1-dodecanamine-N-oxide. *See* Lauramine oxide
N,N-Dimethyldodecylamine-N-oxide. *See* Lauramine oxide
Dimethylenediamine. *See* Ethylenediamine

Dimethylethanolamine
CAS 108-01-0; EINECS 203-542-8
Synonyms: DMAE; 2-Dimethylaminoethanol; N,N-Dimethylaminoethanol; Deanol; N,N-Dimethyl ethanolamine
Empirical: $C_4H_{11}NO$
Formula: $(CH_3)_2NCH_2CH_2OH$
Properties: Colorless liq., amine odor; sol. in water, alcohol, ether; m.w. 89.14; dens. 0.8866 (20/4 C); visc. 3.8 cps (20 C); f.p. -64 C; b.p. 133 C; flash pt. (OC) 105 F; ref. index 1.430
Precaution: Flamm.
Toxicology: LD50 (oral, rat) 2 g/kg; moderately toxic by ingestion, inhalation, skin contact, intraperitoneal, subcutaneous routes; skin and eye irritant; central nervous system stimulant
Uses: Synthesis of dyestuffs, paints, pharmaceuticals, textile auxiliaries; catalyst for PU foam and slabstock; epoxy curing agents; polymer aggregation agent
Manuf./Distrib.: Air Prods.; Aldrich; Allchem Ind.; Ashland; BASF; BASF AG; Elf Atochem N. Am.; Fluka; Great Western; Nippon Nyukazai; Pelron; Sigma; Texaco; Union Carbide

N,N-Dimethyl ethanolamine. *See* Dimethylethanolamine

Dimethyl ether
CAS 115-10-6; EINECS 204-065-8
Synonyms: Methane, oxybis-; Oxybismethane; Methyl ether
Classification: Organic compd.
Empirical: C_2H_6O
Formula: CH_3OCH_3
Properties: Colorless compressed gas or liq.; sol. in water, alcohol; f.p. -141.4 C; b.p. -24.5 C; flash pt. -41 C
Precaution: Highly flamm.; dangerous fire and explosion hazard
Toxicology: Mildly toxic by inhalation
Uses: Aerosol propellant; paint aerosols; refrigerant, solvent, extraction agent
Manuf./Distrib.: Air Prods.; Aldrich; DuPont; Stanchem

1,1-Dimethylethylhydroperoxide. *See* t-Butyl hydroperoxide
(1,1-Dimethylethyl)-4-methoxyphenol. *See* BHA

Dimethyl formamide
CAS 68-12-2; EINECS 200-679-5
Synonyms: DMF; DMFA
Empirical: C_3H_7NO
Formula: $CHCON(CH_3)_2$
Properties: Water-wh. liq.; a dipolar aprotic solv.; misc. with water, most org. solvs.; m.w. 73.10; dens. 0.953-0.954 (15.6/15.6 C); b.p. 152.8 C; f.p. -61 C; flash pt. 57.7 C
Precaution: Combustible; moderate fire risk
Toxicology: LD50 (oral, rat) 2800 mg/kg; toxic by skin absorption; strong irritant to skin and tissue; TLV 10 ppm in air; on decomp., emits toxic fumes of NO_x
Uses: Solvent in vinyl resins and acetylene, butadiene, acid gases, paints; polyacrylic fibers; catalyst in carboxylation reactions, organic synthesis, carrier for gases
Manuf./Distrib.: Aceto; Air Prods.; Aldrich; Allchem Ind.; Ashland; J.T. Baker; BASF; Baychem; Brown; Browning; Chemcentral; Coyne; DuPont; Fluka; Great Western; Harcros; ICI Spec. Chems.; Mallinckrodt; Mitsubishi Gas; Monomer-Polymer & Dajac Labs; Nissan Chem. Ind.; Sigma; Spectrum Chem. Mfg.; UCB SA; Van Waters & Rogers

Dimethylformocarbothialdine. *See* 3,5-Dimethyl tetrahydro-2-H,1,3,5-thiadiazone-2-thione

Dimethyl glutarate
CAS 1119-40-0; EINECS 214-277-2
Synonyms: Dimethyl pentanedioate
Empirical: $C_7H_{12}O_4$
Formula: $CH_3OCO(CH_2)_3COOCH_3$
Properties: M.w. 160.17; dens. 1.087 (20/4 C); b.p. 96-103 C; flash pt. 97 C; ref. index 1.424 (20 C)
Uses: Solv. for coatings, cleaners, inks, textile lubricants, urethane prod.; plasticizer for flexible thermoset

polyester; polymer intermediate for polyester polyols for urethanes, wet-str. paper resins, polyester resins; specialty chemical intermediate
Manuf./Distrib.: Aldrich; Ashland; Fluka; Sigma
Trade names: DBE-5; Santosol™ DMG
Trade names containing: DBE; DBE-2, -2SPG; DBE-3; DBE-9; Santosol™ DME; Santosol™ DME-2; Santosol™ DME-3

Dimethylglycol. *See* Ethylene glycol dimethyl ether
2,6-Dimethyl-4-heptanone. *See* Diisobutyl ketone
Dimethyl hexanedioate. *See* Dimethyl adipate

Dimethyl hexynediol
CAS 142-30-3
Synonyms: 2,5-Dimethyl-3-hexyne-2,5-diol
Definition: Di-tertiary acetylenic diol
Empirical: $C_8H_{14}O_2$
Formula: $(CH_3)_2COHCCCOH(CH_3)_2$
Properties: Wh. cryst.; sol. in water; sl. sol. in benzene, CCl_4, naphtha; very sol. in acetone, alcohol, ethyl acetate; dens. 0.949 (20/20 C); m.p. 94-95 C; b.p. 205-206 C
Uses: As intermediate in synthesis of flavor and fragrance compds. and organic peroxides; antifoaming agent; coupling agent in resin coatings; wire-drawing lubricant
Manuf./Distrib.: Aldrich; Air Prods.

2,5-Dimethyl-3-hexyne-2,5-diol. *See* Dimethyl hexynediol
3,5-Dimethyl-1-hexyne-3-ol. *See* Dimethyl hexynol

Dimethyl hexynol
CAS 107-54-0; EINECS 203-500-9
Synonyms: 3,5-Dimethyl-1-hexyne-3-ol
Empirical: $C_8H_{14}O$
Formula: $HC{\equiv}CCOH(CH_3)CH_2CH(CH_3)_2$
Properties: Colorless liq., camphor-like odor; sl. sol. in water; dens. 0.8545 (20/20 C); b.p. 150-151 C; flash pt. 56.6 C
Precaution: Moderate fire risk
Uses: Stabilizer for chlorinated organic compds., surface active agents, intermediate, solvent lubricant; glass cleaner additive; surfactant, wetting agent used for paper coatings, inks, floor polishes, and glass cleaning formulations; cleaner in silicon wafer industry
Manuf./Distrib.: Aldrich
Trade names: Surfynol® 61

Dimethylhydroxybenzene. *See* Xylenol
Dimethylketone. *See* Acetone

Dimethyl maleate
CAS 624-48-6; EINECS 210-848-5
Synonyms: Methyl maleate
Empirical: $C_6H_8O_4$
Formula: $CH_3OCOCH{:}CHCOOCH_3$
Properties: Liq., mild char. odor; m.w. 144.13; sp.gr. 1.15 (20 C); b.p. 205 C; acid no. 1; flash pt. (OC) 235 F; ref. index 1.441
Precaution: Combustible when exposed to heat or flame; can react with oxidizing agents
Toxicology: LD50 (skin, rabbit) 530 mg/kg; mod. toxic by ingestion and skin contact; eye irritant; may cause skin irritation
Uses: Monomer for coatings; internal modifier for PS and PVC systems
Manuf./Distrib.: Albright & Wilson Am.; Allchem Ind.; Chemie Linz N. Am.; Fluka; Monomer-Polymer & Dajac Labs; Rhone-Poulenc
Trade names: Sipomer® DMM

Dimethylmethane. *See* Propane

Dimethyl octynediol
CAS 1321-87-5; 78-66-0; EINECS 201-131-8
Synonyms: 3,6-Dimethyl-4-octyne-3,6-diol

Classification: Aliphatic alcohol
Formula: $C_2H_5(CH_3)COHC:CCOH(CH_3)C_2H_5$
Properties: Wh. cryst.; very sol. in acetone, alcohol, benzene; mod. sol. in water; sl. sol. in kerosene; .w. 170.35; m.p. 55 C; b.p. 222 C; f.p. > 110 C
Precaution: Combustible
Uses: Defoamer, wetting agent used in pesticide concs.; solubilizer and clarifier in shampoos; developer compds.; electroplating baths
Regulatory: FDA 21CFR §175.105
Manuf./Distrib.: Aldrich
Trade names containing: Surfynol® 82S

3,6-Dimethyl-4-octyne-3,6-diol. *See* Dimethyl octynediol

Dimethyl oleamide
Uses: Solubilizer, solvent, dispersant, wetting agent; modifier for coatings
Trade names: Hallcomid® M-18-OL

Dimethylolpropane. *See* Neopentyl glycol
Dimethylolpropane diacrylate. *See* Neopentyl glycol diacrylate

Dimethylolpropionic acid
CAS 4767-03-7; EINECS 225-306-3
Synonyms: DMPA; 2,2-Bis(hydroxymethyl)propionic acid
Empirical: $C_5H_{10}O_4$
Formula: $(HOCH_2)_2C(CH_3)COOH$
Properties: Off-wh. cryst. solid; hygroscopic; sol. in water and methanol; sl. sol. in acetone; insol. in benzene; m.w. 134; m.p. 192-194 C
Toxicology: Essentially nontoxic; LD50 (mouse, oral) > 5000 mg/kg; sl. irritating to abraded skin; moderately irritating to eyes
Uses: In prep. of water-sol. alkyd resins, polyester resins, surfactants, chemical intermediates, syn. lubricants, plasticizers, pharmaceuticals, cosmetics; produces coatings with outstanding thermal, hydrolytic, and color stability
Manuf./Distrib.: Aldrich; Allchem Ind.; Fabrichem; Fluka; Hoechst AG
Trade names: DMPA®

Dimethyl oxazolidine
CAS 51200-87-4; EINECS 257-048-2
Synonyms: 4,4-Dimethyloxazolidine; Oxazolidine A
Classification: Heterocyclic compd.
Empirical: $C_5H_{11}NO$
Properties: Sol. in water and oil; m.w. 101.17; dens. 0.942; flash pt. 48 C; ref. index 1.4320
Toxicology: LD50 (oral, rat) 950 mg/kg, (dermal, rabbit) 1400mg/kg; irritant; moderately toxic by ingestion and skin contact; mildly toxic by inhalation; heated to decomp., emits toxic fumes of NO_x
Uses: Preservative, antimicrobial for cosmetics, paints and coatings
Regulatory: USA not restricted; Euorpe listed
Manuf./Distrib.: Aldrich
Trade names: Canguard® 327; Nuosept® 101

4,4-Dimethyloxazolidine. *See* Dimethyl oxazolidine
Dimethyl pentanedioate. *See* Dimethyl glutarate
Dimethylphenol. *See* Xylenol
Dimethylphenylamine. *See* n,n-Dimethylaniline
N,N-Dimethylphenylamine. *See* n,n-Dimethylaniline

Dimethyl phthalate
CAS 131-11-3; EINECS 205-011-6
Synonyms: DMP; Dimethyl 1,2-benzenedicarboxylate; Methyl phthalate; 1,2-Benzenedicarboxylic acid dimethyl ester; Phthalic acid dimethyl ester
Definition: Diester of methyl alcohol and phthalic acid
Empirical: $C_{10}H_{10}O_4$
Formula: $C_6H_4(CO_2CH_3)_2$
Properties: Colorless oily liq., sl. aromatic odor; misc. with alcohol, ether; insol. in water, paraffinic hydrocarbons; sl. sol. in min. oil; m.w. 194.08; dens. 1.189 (25/25 C); visc. 17.2 cps; m.p. 5.5 C; b.p. 282.0

C; flash pt. 149 C; ref. index 1.5138 (25 C)
Precaution: Combustible
Toxicology: LD50 (oral, rat) 6.9 ml/kg; TLV 5 mg/m^3 of air (TWA); moderately toxic by ingestion and intraperitoneal routes; midly toxic by inhalation; heated to decomp., emits acrid smoke and irritating fumes
Uses: Solvent for resin, plasticizer for cellulose acetate and nitrocellulose lacquers; plastics, rubber; coating agents; safety glass
Regulatory: FDA 21CFR §175.105, 177.1010, 177.2420
Manuf./Distrib.: Aldrich; Allan; Allchem Ind.; Ashland; BASF; BP Chems. Inc; Daihachi Chem. Ind.; Eastman; Fluka; Great Western; Hüls Am.; Morflex; UCB SA; Unitex
Trade names: Palatinol® M

Dimethylpolysiloxane. *See* Dimethicone
2,2-Dimethyl-1,3-propanediol. *See* Neopentyl glycol
2,2-Dimethyl-1,3-propanediol diacrylate. *See* Neopentyl glycol diacrylate
2,2-Dimethyl-1,3-propanediol dibenzoate. *See* Neopentyl glycol dibenzoate
2,2′-((2,2-Dimethyl-1,3-propanediyl)bis(oxymethylene))bisoxirane. *See* Neopentyl glycol diglycidyl ether
2,2-Dimethylpropanoic acid. *See* Neopentanoic acid
2,2-Dimethylpropionic acid. *See* Neopentanoic acid
α,α-Dimethylpropionic acid. *See* Neopentanoic acid
Dimethyl silicone. *See* Dimethicone
Dimethylsiloxane-glycol copolymer. *See* Dimethicone copolyol

Dimethyl succinate
CAS 106-65-0; EINECS 203-419-9
Synonyms: Dimethyl butanedioate; Methyl succinate
Empirical: $C_6H_{10}O_4$
Formula: $CH_3OCOCH_2CH_2COOCH_3$
Properties: Colorless liq., ethereal winey odor; m.w. 146.14; dens. 1.119 (20/4 C); m.p. 16-19 C; b.p. 190-193 C; flash pt. 90 C; ref. index 1.419
Uses: Light and heat stabilizer for polyolefins, ABS polymer systems, flexible PVC, food pkg.; solv. for coatings, cleaners, inks, textile lubricants, urethane prod.; plasticizer for flexible thermoset polyester; polymer intermediate for polyester polyols for urethanes, wet-str. paper resins, polyester resins; specialty chemical intermediate
Regulatory: FDA 21CFR §172.515; FEMA GRAS
Manuf./Distrib.: Aldrich; Ashland; Chemie Linz N. Am.; DuPont; Fluka; Penta Mfg.; Schweizerhall; Sigma
Trade names: DBE-4; Santosol™ DMS
Trade names containing: DBE; DBE-2, -2SPG; DBE-3; DBE-9; Santosol™ DME; Santosol™ DME-2; Santosol™ DME-3

3,5-Dimethyl tetrahydro-2-H,1,3,5-thiadiazone-2-thione
CAS 533-74-4
Synonyms: DMTT; Tetrahydro-3,5-dimethyl-2H-1,3,5-thiadiazine-2-thione; 3,5-Dimethyltetrahydro-1,3,5,2H-thiadiazine-2-thione; Dimethylformocarbothialdine
Empirical: $C_5H_{10}N_2S_2$
Properties: Wh. cryst., nearly odorless; sl. sol. in water, alcohol; sol. in acetone; m.w. 162; m.p. 100 C
Toxicology: LD50 (oral, rat) 363 mg/kg, (IP, rat) 87 mg/kg; TLV 10 mg/m^3; toxic by ingestion, inhalation, IP routes; skin and severe eye irritant; heated to decomp., emits very toxic fumes of NO_x and SO_x
Uses: Fungicide, bactericide, microbiocide for paint, adhesives, leather, latex, mineral slurries, inks, glue casein, starch, paper, and metal working fluids
Manuf./Distrib.: BASF
Trade names: Amerstat® 233

3,5-Dimethyltetrahydro-1,3,5,2H-thiadiazine-2-thione. *See* 3,5-Dimethyl tetrahydro-2-H,1,3,5-thiadiazone-2-thione

N,N-Dimethyl-p-toluidine
CAS 99-97-8; EINECS 202-805-4
Synonyms: 4-Dimethylaminotoluene
Empirical: $C_9H_{13}N$
Formula: $CH_3(C_6H_4N(CH_3)_2$
Properties: M.w. 135.23; dens. 0.936 (20/4 C); b.p. 210-211 C (760 mm); flash pt. 7 C; ref. index 1.546 (20 C)
Precaution: Combustible; vapor is flamm. at elevated temps.; photosensitive

Toxicology: Toxic by inhalation, skin contact, ingestion; may cause skin and eye irritation
Uses: Accelerator for unsat. polyester and acrylate polymerizations
Manuf./Distrib.: Aceto; Aldrich; Fabrichem; First Chem.; Fluka; Monomer-Polymer & Dajac Labs; Polysciences; R S A; Sigma
Trade names: Firstcure™ DMPT

2,2-Dimethyltrimethylene acrylate. *See* Neopentyl glycol diacrylate
2,2-Dimethyltrimethylene ester acrylic acid. *See* Neopentyl glycol diacrylate
Dimethyl trimethylene glycol. *See* Neopentyl glycol
Dinonyl 1,2-benzene dicarboxylate. *See* Dinonyl phthalate

Dinonylnaphthalene disulfonic acid
Trade names containing: Nacure® 155; Nacure® 3327; Nacure® 3525; Nacure® X49-110; Nacure® XP-383

Dinonylnaphthalene monosulfonic acid. *See* Dinonylnaphthalene sulfonic acid

Dinonylnaphthalene sulfonic acid
CAS 25322-17-2
Synonyms: Dinonylnaphthalene monosulfonic acid
Empirical: $C_{28}H_{44}O_3S$
Properties: M.w. 460.72
Trade names containing: Nacure® 1051; Nacure® 1323; Nacure® 1419; Nacure® 1557

Dinonylnaphthalene sulfonic acid, calcium salt. *See* Calcium dinonylnaphthalene sulfonate

Dinonyl phthalate
CAS 28553-12-0; EINECS 237-954-4
Synonyms: Dinonyl 1,2-benzene dicarboxylate; Di-n-nonyl phthalate; Diisononyl phthalate
Empirical: $C_{26}H_{42}O_4$
Properties: Colorless liq.; m.w. 418.68; sp.gr. 0.974 (20/4 C); b.p. 205-220 C; flash pt. ≈ 200 C; ref. index 1.486
Precaution: Combustible
Uses: General-purpose low-volatility plasticizer for vinyl resins; pure grade as stationary liq. phase in chromatography
Manuf./Distrib.: Fluka; Sigma

Di-n-nonyl phthalate. *See* Dinonyl phthalate
DINP. *See* Diisononyl phthalate
Dioctadecyl dimethyl ammonium chloride. *See* Distearyldimonium chloride
Dioctrahedral smectite. *See* Attapulgite

Dioctyl adipate
CAS 103-23-1; EINECS 203-090-1
Synonyms: DOA; Bis(2-ethylhexyl) hexanedioate; Adipic acid bis (2-ethylhexyl) ester; Di(2-ethylhexyl) adipate; Hexanedioic acid, bis (2-ethylhexyl) ester
Classification: Aliphatic organic compd.
Definition: Diester of 2-ethylhexyl alcohol and adipic acid
Empirical: $C_{22}H_{42}O_4$
Formula: $[CH_2CH_2COOCH_2CH(C_2H_5)C_4H_9]_2$
Properties: Lt.-colored oily liq.; insol. in water; m.w. 370.64; dens. 0.9268 (20/20 C); b.p. 417 C; flash pt. 196 C; ref. index 1.4472
Precaution: Combustible
Toxicology: LD50 (oral, rat) 9110 mg/kg); suspected carcinogen and tetratogen; moderately toxic by IV route; mildly toxic by ingestion and skin contact; mutagenic data; eye and skin irritant; on decomp., emits acrid smoke and irritating fumes
Uses: Plasticizer, commonly blended with general purpose plasticizers (DOP, DIOP); solvent, aircraft lubricants; paints, as vehicles for pigment dispersions
Regulatory: FDA 21CFR §175.105, 175.300, 177.1200, 177.1210, 177.1400, 177.2600, 178.3740; FDA approved for injectables
Manuf./Distrib.: Aldrich; Allchem Ind.; Ashland; BASF; Chisso Am.; Coyne; Eastman; Esprit; Fluka; Hüls AG; Inolex; Monsanto; Sigma
Trade names: Diplast® DOA; Palatinol® DOA; Uniplex 125-A

Dioctyl ammonium sulfosuccinate
Trade names: Octowet 70A

Dioctyl dodecanedioate dioate
Uses: Lubricant additive; as textile surf. finishes, softeners, thread lubricants and/or antistats; plasticizer used in PVC, NC, and rubber; food contact applics.; plastisols; dip coating formulations
Trade names: Plasthall® DODD

Dioctyl fumarate
CAS 141-02-6; EINECS 220-835-6
Synonyms: DOF; Di (2-ethylhexyl) fumarate; Bis (2-ethylhexyl) fumarate; 2-Butenedioic acid bis (2-ethylhexyl) ester; 2-Ethylhexyl fumarate
Empirical: $C_{20}H_{36}O_4$
Formula: $C_8H_{17}OOCCH:CHCOOC_8H_{17}$
Properties: Clear mobile liq., mild odor; m.w. 340.56; dens. 0.942 (20/20 C); b.p. 211-220 C; flash pt. (COC) 365 F
Precaution: Combustible exposed to heat and flame; can react with oxidizing material
Toxicology: LD50 (IP, mouse) 250 mg/kg; poison by intraperitoneal route; eye and severe skin irritant
Uses: Monomer for polymerization and copolymerization; internal plasticizer for copolymerization with vinyl acetate, vinyl chloride, acrylates, and styrene; exterior paint formulations
Manuf./Distrib.: Monomer-Polymer & Dajac Labs
Trade names: Staflex DOF

Dioctyl maleate
CAS 142-16-5; 2915-53-9; EINECS 205-524-5
Synonyms: Bis (2-ethylhexyl) maleate; Di-N-octyl maleate; Di-(2-ethylhexyl) maleate
Definition: Diester of 2-ethylhexyl alcohol and maleic acid
Empirical: $C_{20}H_{38}O_5$
Formula: $C_8H_{17}OCOCH=CHCOOC_8H_{17}$
Properties: Liq., char. mild odor; m.w. 358.52; sp.gr. 0.960-0.970; m.p. -50 to -85 C; flash pt. (COC) 182 C; ref. index 1.4480-1.4500
Toxicology: LD50 (oral, rat) 14,200 mg/kg; mildly toxic by ingestion; heated to decomp., emits acrid smoke and fumes
Uses: Emollient, fragrance coupler, solubilizer for benzophenone-3, antitackifier in antisperspirants, conditioner for hair prods.; in copolymerization of PVC and vinyl acetates; plasticizer for vinyl resins; also in latex paints
Manuf./Distrib.: Allchem Ind.; Aristech; BASF; Finetex; Hoechst AG; Hüls; Monomer-Polymer & Dajac Labs; Unitex
Trade names: Nuoplaz® DOM; Staflex DOM

Di-N-octyl maleate. *See* Dioctyl maleate

Dioctylphenyl hydrogen phosphate diethylamine salt
CAS 64051-35-0
Manuf./Distrib.: Albright & Wilson Am.
Trade names containing: Virco-Pet® 50

Dioctyl phosphate
Synonyms: Di-2-ethylhexyl phosphate; Di-2-ethylhexyl phosphoric acid
Uses: Complexing agent; coupling and surf. treatment agent for fillers, flame retardants, and reinforcing agents used in thermoplastic and thermoset polymers; improves elong. and impact resist. in filled resins; coupling agent which improves bond between fillers/pigments and polymers; suitable for coated prods.
Manuf./Distrib.: Albright & Wilson Am.
Trade names: DEHPA®

Dioctyl phthalate
CAS 117-81-7; 117-84-0; EINECS 204-211-0
Synonyms: DOP; DEHP; Di(2-ethylhexyl) phthalate; Bis (2-ethylhexyl) phthalate; Di-s-octyl phthalate; 1,2-Benzenedicarboxylic acid dioctyl ester
Definition: Diester of 2-ethylhexyl alcohol and phthalic acid
Empirical: $C_{24}H_{38}O_4$
Formula: $C_6H_4[COOCH_2CH(C_2H_5)C_4H_9]_2$
Properties: Lt.-colored liq., odorless, bitter taste; insol. in water; misc. with min. oil; m.w. 390.62; dens. 0.9861

(20/20 C); b.p. 231 C (5 mm); flash pt. 218 C; ref. index 1.4836
Precaution: Combustible
Toxicology: TLV 5 mg/m^3; STEL 10 mg/m^3; LD50 (oral, rat) 30,600 mg/kg; poison by IV; mildly toxic by ingestion; skin and severe eye irritant; suspected human carcinogen; experimental teratogen; affects human GI tract; heated to decomp., emits acrid smoke
Uses: Plasticizer for many resins and elastomers
Regulatory: FDA 21CFR §175.105, 175.300, 175.310, 715.380, 175.390, 176.170, 176.210, 176.1210, 177.1010, 177.1200, 177.1210, 177.1400, 177.2600, 178.3120, 178.3910, 181.22, 181.27; FDA approved for ophthalmics, injectables; BP compliance
Manuf./Distrib.: Aldrich; Allchem Ind.; Aristech; BASF; Chemisphere Ltd; Chisso Am.; Coyne; Daihachi Chem. Ind.; Eastman; Fluka; Great Western; C.P. Hall; Hoechst AG; Hüls AG; Mitsubishi Gas; Sigma; Spectrum Chem. Mfg.; UCB SA
Trade names: Diplast® L8; Diplast® O; Kodaflex® DOP; Staflex DOP
Trade names containing: BYK®-2615; Kodaflex® HS-4; Vinyzene® BP-5-2 DOP; Vinyzene® BP-5-5 DOP

Di-s-octyl phthalate. *See* Dioctyl phthalate

Dioctyl sodium sulfosuccinate
CAS 577-11-7; 1369-66-3; EINECS 209-406-4
Synonyms: DSS; Sodium dioctyl sulfosuccinate; Sodium 1,4-bis(2-ethylhexyl) sulfosuccinate; Dioctyl sulfosodiumsuccinate; Dioctyl sulfosuccinate sodium salt; Sodium di(2-ethylhexyl) sulfosuccinate; Bis (2-ethylhexyl)-S-sodium sulfosuccinate; 2-Ethylhexyl sulfosuccinate sodium; Docusate sodium
Definition: Sodium salt of the diester of 2-ethylhexyl alcohol and sulfosuccinic acid
Empirical: $C_{20}H_{38}O_7S \cdot Na$
Formula: $C_8H_{17}OOCCH_2CH(SO_3Na)COOC_8H_{17}$
Properties: Wh. wax-like solid, char. octyl alcohol odor; slowly sol. in water; freely sol. in alcohol, glycerol, CCl_4, acetone, xylene, hexane; m.w. 445.63; m.p. 173-179 C
Precaution: Hygroscopic
Toxicology: LD50 (oral, rat) 1900 mg/kg; moderately toxic by ingestion, intraperitoneal routes; poison by intravenous route; skin, severe eye irritant; heated to decomp., emits toxic fumes of SO_x and Na_2O
Uses: Food additive, emulsifier, wetting agent (processing aid in sugar industry); stabilizer for hydrophilic colloids; wetting agent, dispersant, emulsifier in cosmetic, pharmaceutical, emulsion polymerization, and industrial applics.; adjuvant in tablet; emulsifier wax for polish, firefighting, germicide, metal cleaner, mold release agent; dispersant in paints and inks, paper, photography, process aid, rust preventative, soldering flux, wallpaper removal
Regulatory: FDA 21CFR §73.1, with cocoa, 131.130, 131.132, 133.124, 133.133, 133.134, 133.162, 133.178, 133.179, 163.114, 163.117, 169.115, 169.150, 172.520, 172.808, 172.810, 175.105, 175.300, 175.320, 176.170, 176.210, 177.1200, 177.2800, 178.1010, 178.3400; USDA 9 CFR §318.7, 381.147; FDA approved for injectables (IM), orals, topicals; USP/NF, BP compliance
Manuf./Distrib.: Alco; Aldrich; Aquatec Quimica SA; Brotherton Ltd; Calgene; Cytec Ind.; Eastern Color & Chem.; EM Ind.; Finetex; Fluka; Hart Prods.; Henkel/Organic Prods.; Hickson Spec.; McIntyre; Mona Ind.; Rhone-Poulenc Surf. & Spec.; Sigma; Spectrum Chem. Mfg.; Witco/Oleo-Surf.
Trade names: Aerosol® OT-75%; Aerosol® OT-100%; Aerosol® OT-S; Calgene DOSS-70; Calgene DOSS-75; Chemax DOSS/70E; Chemax DOSS/70HFP; Chemax DOSS-75E; Drewfax® 0007; Empimin® OP70; Empimin® OT75; Geropon® SDS; Marlinat® DF 8; Monawet MO-65-150; Monawet MO-70-150; Nopco® 1186A; Octowet 60-I; Octowet 70; Octowet 70PG; Octowet 75; Octowet 75E; Rewopol® SBDO 70; Secosol® DOS 70; Triton® GR-5M; Triton® GR-7M
Trade names containing: Geropon® 99; Geropon® SS-O-70PG; Geropon® SS-O-75; Monawet MO-70; Monawet MO-70E; Monawet MO-70R; Monawet MO-70S; Monawet MO-75E; Monawet MO-84R2W; Monawet MO-85P

Dioctyl sulfosodiumsuccinate. *See* Dioctyl sodium sulfosuccinate
Dioctyl sulfosuccinate sodium salt. *See* Dioctyl sodium sulfosuccinate

Dioctyl terephthalate
CAS 422-86-2; EINECS 225-091-6
Synonyms: DOTP; Di-2-ethylhexyl terephthalate; 1,4-Benzenedicarboxylic acid, dioctyl ester
Empirical: $C_{24}H_{38}O_4$
Formula: $C_6H_4(COOCH_2CH[C_2H_5]C_4H_9)_2$
Properties: Liq.; m.w. 390.57; sp.gr. 0.984; b.p. 400 C; f.p. -48 C; flash pt. (COC) 238 C; ref. index 1.489
Precaution: Combustible
Uses: Plasticizer for PVC, PS, cellulosics, rubber; applics. incl. wire coatings, automotive and furniture upholstery; compatible with acrylics, CAB, cellulose nitrate, polyvinyl butyral, styrene, oxidizing alkyds,

nitrile rubber
Manuf./Distrib.: Ashland; Eastman; C.P. Hall; Harwick
Trade names: Kodaflex® DOTP

N,N´-Dioleoylethylenediamine. *See* Ethylene dioleamide
3,6-Dioxadodecanol-1. *See* Diethylene glycol n-hexyl ether
1,4-Dioxan. *See* 1,4-Dioxane

1,4-Dioxane
CAS 123-91-1; EINECS 204-661-8
Synonyms: Diethylene ether; p-Dioxane; 1,4-Dioxan; 1,4-Diethylene dioxide; Diethylene dioxide; Tetrahydro-1,4-dioxin; Glycol ethylene ether; Diethylene oxide
Classification: Aliphatic organic compd.; ether
Empirical: $C_4H_8O_2$
Formula: $OCH_2CH_2OCH_2CH_2$
Properties: Colorless liq., ethereal odor; stable; misc. with water, most org. solvs.; m.w. 88.11; dens. 1.0356 (20/20 C); b.p. 101.3 C; f.p. 10-12 C; flash pt. 18.3 C; ref. index 1.4220
Precaution: DOT: Flamm. liq.; dangerous fire risk
Toxicology: LD50 (oral, mouse) 5700 mg/kg, (skin, rabbit) 7600 mg/kg; carcinogen; toxic by inhalation, absorbed by skin; TLV 25 ppm in air; OSHA 100 ppm min. air; emits acrid smoke and fumes
Uses: Stabilizer for chlorinated hydrocarbons; solv. for adhesives, dyes, cellulose, lacquer, paints, wax, pharmaceuticals, coatings, natural and syn. rubbers, dry cleaning, metal surf. finishes, chem. reaction and extraction, transistors
Regulatory: BP compliance
Manuf./Distrib.: AC Ind.; Aldrich; Alemark; Allchem Ind.; Amber Syn.; Ashland; J.T. Baker; BASF; CPS; Ferro/Grant; Fluka; Mallinckrodt; Osaka Org. Chem. Ind.; Sigma; Spectrum Chem. Mfg.; Toho Chem. Ind.; Union Carbide; Van Waters & Rogers

p-Dioxane. *See* 1,4-Dioxane
1,3-Dioxolan-2-one, 4-methyl. *See* Propylene carbonate
Dioxybenzone. *See* Benzophenone-8
N,N-Dioxyethylaniline. *See* Phenyldiethanolamine
DIPA. *See* Diisopropanolamine

Dipentaerythritol
CAS 126-58-9
Formula: $(CH_2OH)_3CCH_2OCH_2C(CH_2OH)_3$
Properties: Off-white free-flowing powd.; m.w. 254.28; dens. 1.33 (25/4 C); m.p. 212-220 C
Uses: Used as intermediate in mfg. of alkyds and drying oils, paints and coatings
Manuf./Distrib.: Aldrich; Allchem Ind.; Amber Syn.; Chemical; Honeywill & Stein
Trade names: Hercules® Tech. Di-PE

Dipentaerythrityl pentaacrylate
Uses: Highly reactive monomer exhibiting exc. chem. and abrasion resist.; for pigmented systems to improve cure and wetting characteristics
Manuf./Distrib.: Monomer-Polymer & Dajac Labs
Trade names: Photomer® 4399
Trade names containing: SR-399

Dipentene (INCI). *See* dl-Limonene
Dipentene polymer. *See* Polydipentene
Diphenyl cresol phosphate. *See* Diphenylcresyl phosphate

Diphenylcresyl phosphate
CAS 26444-49-5
Synonyms: Cresyl diphenyl phosphate; Phosphoric acid methylphenyl diphenyl ester (9CI); Tolyl diphenyl phosphate; Diphenyl cresol phosphate
Formula: $C_{19}H_{17}O_4P$
Properties: M.w. 340.33; sp.gr. 1.204-1.208; f.p. -38 C; flash pt. (COC) 233-237 C; ref. index 1.560
Toxicology: LD50 (oral, rat) 6400 mg/kg; mildly toxic by ingestion; heated to dec., emits PO_x
Uses: Plasticizer for PVC, PVB, PS, cellulosics, coatings
Manuf./Distrib.: Bayer; FMC; Velsicol
Trade names: Disflamoll® DPK

1,2-Diphenylethane-1,2-dione, dimethyl ketal. *See* 2,2-Dimethoxy-2-phenylacetophenone
Diphenyl-2-ethylhexyl phosphate. *See* Diphenyl octyl phosphate

Diphenyl isodecyl phosphite
CAS 26544-23-0; EINECS 247-777-4
Synonyms: DPDP
Empirical: $C_{22}H_{31}O_3P$
Formula: $(C_6H_5O)_2POC_{10}H_{21}$
Properties: Liq.; m.w. 374; dens. 1.022-1.032; b.p. 190 C (5 mm); flash pt. (PMCC) 154 C; ref. index 1.5160-1.5190
Uses: Chelating agent with metal carboxylates as polymer additives esp. for chlorinated polymers; color, heat, and light stabilizer for PC, polyurethanes, ABS polymers, coatings; sec. stabilizer for PVC
Manuf./Distrib.: Akzo Nobel; Dover
Trade names: Weston® DPDP

Diphenyl isooctyl phosphite
CAS 26401-27-4; EINECS 247-658-7
Synonyms: DPIOP
Formula: $(C_6H_5O)_2POC_8H_{17}$
Properties: Liq.; m.w. 346; dens. 1.040-1.047; b.p. 190 C (5 mm); ref. index 1.5210-1.5230; flash pt. (PMCC) 182 C
Uses: Color and processing stabilizer for ABS, PC, polyurethane, coatings, PET fiber; sec. stabilizer for PVC
Manuf./Distrib.: Aldrich
Trade names: Weston® ODPP

Diphenyl ketone. *See* Benzophenone

Di (phenyl mercury) dodecenyl succinate
Uses: Preservative and mildew inhibitor for aq. latex paints

Diphenylmethane-4,4´-diisocyanate. *See* MDI
4,4´-Diphenylmethane diisocyanate. *See* MDI
p,p-Diphenylmethane diisocyanate. *See* MDI
Diphenylmethanone. *See* Benzophenone

Diphenyl octyl phosphate
CAS 1241-94-7
Synonyms: Diphenyl-2-ethylhexyl phosphate; 2-Ethylhexyl diphenyl ester phosphoric acid; 2-Ethyl-1-hexanol ester with diphenyl phosphate; 2-Ethylhexyl diphenyl phosphate
Empirical: $C_{20}H_{27}O_4P$
Properties: M.w. 362.44; sp.gr. 1.088-1.093; flash pt. (COC) 224 C; ref. index 1.507-1.510
Toxicology: LDLo (IV, rabbit) 272 mg/kg; poison by IV route; heated to decomp., emits toxic fumes of PO_x
Uses: Flame-retardant low-temp. plasticizer for PVC, cellulose nitrate, CAB, ethyl cellulose, polymethyl methacrylate, PS, buna N rubber; used in vinyl film and sheeting, textile coatings, plastisols, organosols, adhesives, and pkg. materials; dip, rotationally, extruded and inj. molded parts, mechanical foam
Regulatory: FDA 21CFR §181.27
Manuf./Distrib.: Akzo Nobel; Ashland; Bayer; Harwick; Monsanto
Trade names: Santicizer 141

Diphenyl-sulfon-3,3´-disulfohydrazide
Uses: Blowing agent used in coatings for textiles
Trade names: Porofor® D33

Diphosphoric acid, ammonium manganese (3+) salt (1:1:1). *See* Manganese violet
Diphosphoric acid tetrapotassium salt. *See* Tetrapotassium pyrophosphate
Diphosphoric acid tetrasodium salt. *See* Tetrasodium pyrophosphate
Dipropanediol dibenzoate. *See* Dipropylene glycol dibenzoate

Dipropylene glycol
CAS 110-98-5; 25265-71-8; EINECS 203-821-4; 246-770-3
Synonyms: DPG; Di-1,2-propylene glycol; Bis (hydroxypropyl) ether; 2,2´-Dihydroxydipropyl ether; 1,1´-Oxybis-2-propanol; 1,1´-Oxydi-2-propanol; 2,2´-Dihydroxyisopropyl ether; Methyl-2(methyl-2) oxybispropanol

Di-1,2-propylene glycol

Classification: Mixture of diols
Empirical: $C_6H_{14}O_3$
Formula: $CH_3CHOHCH_2OCH_2CHOHCH_3$
Properties: Colorless, sl. visc. liq., nearly odorless; hygroscopic; sol. in toluene, water; m.w. 134.18; dens. 1.023; b.p. 233 C; flash pt. (OC) 280 F; ref. index 1.4410
Precaution: Combustible when exposed to heat and flame; can react vigorously with oxidizing materials
Toxicology: LD50 (oral, rat) 14,850 mg/kg; mildly toxic by ingestion; skin and eye irritant; mutagenic data; heated to decomp., emits acrid smoke and irritating fumes
Uses: Solvent in hydraulic brake fluids, cutting oils, textile lubricants, inks; polyester and alkyd resins, reinforced plastics, plasticizers, solvents, fuel additives, in paints, cosmetics
Regulatory: FDA 21CFR §175.105, 176.170, 176.200, 178.3910
Manuf./Distrib.: Aldrich; Allchem Ind.; ARCO; Ashland; Berje; Brown; Chemcentral; Coyne; Fluka; Great Western; C.P. Hall; Olin; PMC Specialties; Samson; Stanchem; Sunnyside; Texaco; Union Carbide; Van Waters & Rogers
Trade names: Adeka Dipropylene Glycol
Trade names containing: Amicure® 33-LV; Dabco® 33-LV®; Proxel® GXL; Surfynol® DF-110D

Di-1,2-propylene glycol. *See* Dipropylene glycol

Dipropylene glycol butyl ether
CAS 29911-28-2
Empirical: $C_{10}H_{22}O_3$
Properties: M.w. 190.32
Toxicology: LDLo (oral, rat) 2000 mg/kg; mod. toxic by ingestion; heated to decomp., emits acrid smoke and irritating fumes
Uses: Slow evaporating solv. with good solvency for coating resins, cleaners, cosmetics, agric., electronics, ink, textile and adhesives prods.
Manuf./Distrib.: Aldrich
Trade names: Arcosolv® DPNB; Dowanol® DPnB

Dipropylene glycol t-butyl ether
CAS 132739-31-2
Trade names: Arcosolv® DPTB

Dipropylene glycol dibenzoate
CAS 94-51-9; 27138-31-4; EINECS 202-340-7
Synonyms: 3,3´-Oxydyl-1-propanol dibenzoate; 3,3´-Oxydi-1-propanol dibenzoate; Dibenzoyl dipropylene glycol ester; Dipropanediol dibenzoate
Empirical: $C_{20}H_{22}O_5$
Properties: Lt.-colored liq.; insol. in water; m.w. 342.42; dens. 1.1271 (20/20 C); m.p. 200 C; b.p. 250 C (10 mm)
Precaution: Combustible
Toxicology: LD50 (oral, rat) 9800 mg/kg; mildly toxic by ingestion; heated to decomp., emits acrid smoke and fumes
Uses: Plasticizer for cellulosics, PVC, PS, PVB, PVAc, VCA; latex and lacquer coating applics.; enhances film formation and surf. wetting in PVAc homopolymer emulsion adhesives
Regulatory: FDA 21CFR §175.105, 176.170
Manuf./Distrib.: Aldrich; Allchem Ind.; Ashland; Kalama; Unitex; Velsicol
Trade names: Benzoflex® 9-88
Trade names containing: Benzoflex® 50

Dipropylene glycol dimethyl ether
CAS 111109-77-4; EINECS 404-640-5
Synonyms: Bis(methoxypropyl) ether
Empirical: $C_8H_{18}O_3$
Formula: $CH_3OCH_2CH(CH_3)OCH_2CH(CH_3)OCH_3$
Properties: Colorless liq., low odor; sol. 53 g/100 g water; m.w. 162.23; sp.gr. 0.898; f.p. $<$ -71 C; b.p. 175 C; flash pt. 65 C; ref. index 1.407
Uses: Solvent used in coatings incl. PU coatings, agric. formulations, resin prod., cleaning prods.
Manuf./Distrib.: Dow; Fluka

Dipropylene glycol methyl ether. *See* PPG-2 methyl ether
Dipropylene glycol methyl ether acetate. *See* PPG-2 methyl ether acetate

Dipropylene glycol monomethyl ether. *See* PPG-2 methyl ether
Dipropylene glycol monopropyl ether. *See* Dipropylene glycol n-propyl ether
Dipropylene glycol propyl ether. *See* Dipropylene glycol n-propyl ether

Dipropylene glycol n-propyl ether
CAS 29911-27-1
Synonyms: Dipropylene glycol propyl ether; Dipropylene glycol monopropyl ether
Formula: $C_3H_7O[CH_2(CH)CH_3O]_2H$
Properties: Colorless liq., low odor; mod. sol. in water; m.w. 176.2; sp.gr. 0.922; b.p. 212 C; surf. tens. 27.6 dynes/cm
Uses: Coalescent, solv. for water- or solv.-borne coatings
Manuf./Distrib.: Ashland; Dow
Trade names: Arcosolv® DPNP; Dowanol® DPnP

Dipropyl methane. *See* Heptane
Disodium 2,2´-dihydroxy-4,4´-dimethoxy-5,5´-disulfobenzophenone. *See* Benzophenone-9
Disodium dithionite. *See* Sodium hydrosulfite
Disodium ethylene bisdithiocarbamate. *See* Nabam
Disodium ethylene-1,2-bisdithiocarbamate. *See* Nabam

Disodium ethylene bisdithiocarbamate ethylene diamine
Trade names containing: Amerstat® 274

Disodium maleic anhydride/diisobutylene copolymer
Uses: Pigment dispersant and stabilizer for water based paints
Trade names: Empicryl® APD; Empicryl® APD/B

Disodium nonoxynol-10 sulfosuccinate
CAS 67999-57-9 (generic); 9040-38-4
Synonyms: Sulfobutanedioic acid, nonoxynol-10 ester, disodium salt
Definition: Disodium salt of the half ester of nonoxynol-10 and sulfosuccinic acid
Uses: Emulsifier, solubilizer, wetting agent, surfactant, dispersant, foamer, foam stabilizer, surf. tens. depressant; used in PVAc acrylic emulsions, paints; textile emulsions, pad-bath additive, textile wetting
Regulatory: FDA 21CFR §175.105
Trade names: Monawet 1240

Disodium tallow sulfosuccinamate
CAS 90268-48-7; EINECS 290-850-0
Synonyms: Sulfobutanedioic acid, tallow ester, disodium salt
Classification: Organic compd.
Formula: $RNHCOCH_2CHSO_3NaCOONa$, R = tallow alkyl groups
Uses: Foaming and antigelling agent for latex foam backings and coatings; emulsion polymerization; flotation agent
Trade names: Rewopol® B 1003

Disodium tetrabromophthalate
EINECS 246-890-6
Synonyms: 1,2-Benzenedicarboxylic acid, 3,4,5,6-tetrabromo-, disodium salt
Uses: Flame retardant for adhesives, coatings, and textiles
Trade names: Great Lakes FR-756™

Distearyl dimethyl ammonium chloride. *See* Distearyldimonium chloride

Distearyldimonium chloride
CAS 107-64-2; EINECS 203-508-2
Synonyms: Distearyl dimethyl ammonium chloride; Quaternium-5; Dioctadecyl dimethyl ammonium chloride
Classification: Quaternary ammonium salt
Empirical: $C_{38}H_{80}N \cdot Cl$
Uses: Fabric softener, conditioner, antistat for commercial and institutional laundries; conditioner, antistat for hair prods.; foam and visc. builder in personal care prods.; antistatic coating for ABS, acrylic, cellulosics, nylon, polyacetal, PP, PS, PVC
Regulatory: FDA 21CFR §172.712, 177.1200

Distearyl pentaerythritol diphosphite

Manuf./Distrib.: Fluka
Trade names: Arosurf® TA-100

Distearyl pentaerythritol diphosphite
CAS 3806-34-6; EINECS 223-276-6
Empirical: $C_{41}H_{82}O_6P_2$
Properties: Solid; m.w. 732; dens. 0.920-0.935; m.p. 40-70 C; ref. index 1.4560-1.4590; flash pt. (PMCC) 185 C
Uses: Color and m.w. stabilizer, melt flow aid for polymer processing (polyolefins, polyesters, elastomers, styrenics, engineering thermoplastics), adhesives, coatings
Manuf./Distrib.: Aldrich
Trade names: Weston® 618F
Trade names containing: Weston® 619F

Distilled spirits. *See* Alcohol
Ditallowalkonium chloride. *See* Quaternium-18

Ditridecyl sodium sulfosuccinate
CAS 2673-22-5; EINECS 220-219-7
Synonyms: Sodium bistridecyl sulfosuccinate; Sodium ditridecyl sulfosuccinate; Sulfobutanedioic acid, 1,4-ditridecyl ester, sodium salt
Definition: Sodium salt of the diester of tridecyl alcohol and sulfosuccinic acid
Empirical: $C_{30}H_{58}O_7S \cdot Na$
Uses: Emulsifier, surfactant, detergent, foam modifier, wetting agent, dispersant; emulsion polymerization; dispersant, processing aid for resins, pigments, polymers, paints, and dyes
Regulatory: FDA 21CFR §175.105, 176.180, 178.3400
Trade names: Calgene DTSS-70; Monawet MT-70E
Trade names containing: Monawet MT-70

Ditrimethylolpropane tetraacrylate
CAS 94108-97-1
Formula: $(H_2C{=}CHCO_2CH_2)_2C(C_2H_5)(CH_2)_2O$
Properties: M.w. 466.53; dens. 1.101; flash pt. > 110 C; ref. index 1.4790
Uses: Curing agent
Manuf./Distrib.: Aldrich; Sartomer
Trade names containing: SR-355

Ditrimethylolpropane triacrylate
Uses: Highly reactive monomer exhibiting exc. chem. and abrasion resist.; for pigmented systems to improve cure and wetting characteristics
Trade names: Photomer® 4355

Divinylene oxide. *See* Furan polymer
DMAE. *See* Dimethylethanolamine
DMAPMA. *See* Dimethylaminopropylmethacrylamide
DMBA. *See* N-Benzyldimethylamine
DME. *See* Ethylene glycol dimethyl ether
DMF. *See* Dimethyl formamide
DMFA. *See* Dimethyl formamide
DMP. *See* Dimethyl phthalate; 2,4,6-Tris (dimethylaminomethyl) phenol
DMPA. *See* Dimethylolpropionic acid
DMTT. *See* Tetrahydro-3,5-dimethyl-2H-1,3,5-thiadiazine-2-thione
DOA. *See* Dioctyl adipate
Docosanamide. *See* Behenamide
13-Docosenamide. *See* Erucamide
cis 13-Docosenamide. *See* Erucamide
Docusate sodium. *See* Dioctyl sodium sulfosuccinate
Dodecafluoropentane. *See* Perfluoropentane
Dodecahydro-1,4a-dimethyl-7-(1-methylethyl)-1-phenanthrenemethanol. *See* Hydroabietyl alcohol
Dodecanedioic acid. *See* C12 dibasic acid
1,12-Dodecanedioic acid. *See* C12 dibasic acid
Dodecanoic acid 1,2-ethanediyl ester. *See* Glycol dilaurate
Dodecanoic acid, 2-hydroxypropyl ester. *See* Propylene glycol laurate

Dodecanoic acid, monoester with 1,2-propanediol. *See* Propylene glycol laurate
Dodecanoic acid, zinc salt. *See* Zinc laurate

Dodecenyl succinic anhydride
CAS 25377-73-5; 26544-38-7; EINECS 246-917-1
Synonyms: DDSA; 2-Dodecenylsuccinic anhydride; 2-Dodecen-1-ylsuccinic anhydride
Empirical: $C_{16}H_{26}O_3$
Properties: Clear lt. yel. visc. liq.; m.w. 266.38; dens. 1.003-1.008 (60 F); b.p. 150 C (3 mm); flash pt. 343-347 F
Precaution: Combustible when exposed to open flame
Toxicology: Toxic by inhalation,skin contact; poison by intraperitoneal route; irritant and sensitizer; may emit toxic fume
Uses: Epoxy curing agent; corrosion inhibitor for nonaq. lubricating oils, intermediate for prep. of alkyd or unsat. polyester resins, platicizers, paints, intermediate in chem. reactions
Manuf./Distrib.: Aldrich; Allchem Ind.; Anhydrides & Chems.; Buffalo Color; Cambridge Ind. Co. of Am.; Dixie; Fluka; Humphrey; Lubrizol; Milliken
Trade names: Milldride® DDSA

2-Dodecenylsuccinic anhydride. *See* Dodecenyl succinic anhydride
2-Dodecen-1-ylsuccinic anhydride. *See* Dodecenyl succinic anhydride
Dodecyl acrylate. *See* Lauryl acrylate

Dodecylbenzene
CAS 123-01-3; EINECS 204-591-8
Synonyms: 1-Phenyldodecane
Empirical: $C_{18}H_{30}$
Formula: $C_{12}H_{25}C_6H_5$
Properties: M.w. 246.44; dens. 0.856 (20/4 C); b.p. 331 C; flash pt. 109 C; ref. index 1.482 (20 C)
Toxicology: Toxic by ingestion
Uses: Intermediate for detergents of ABS or LAS type; in specialty coatings and other industrial applics.
Manuf./Distrib.: Aldrich; Fluka; Sigma
Trade names: Naxel™ DDB 500

Dodecylbenzene sodium sulfonate. *See* Sodium dodecylbenzenesulfonate

Dodecylbenzene sulfonic acid
CAS 27176-87-0; 68411-32-5; 68584-22-5; 68608-88-8; 85536-14-7; EINECS 248-289-4
Synonyms: DDBSA
Classification: Substituted aromatic acid
Empirical: $C_{18}H_{30}O_3S$
Properties: Wh. to lt. yel. flakes, granules, and powd.; biodegradable
Uses: Anionic detergent raw material used in emulsifiers, heavy and lt. duty detergents, hand cleaning gels, machine degreasers, tank cleaners; emulsion polymerization; catalyst; metalworking; in paints and coatings; catalyst in acid catalyzed reactions
Regulatory: FDA 21CFR §176.210, 178.1010
Manuf./Distrib.: Allchem Ind.; Ashland; Biddle Sawyer; Pilot; Rhone-Poulenc Surf. & Spec.; Stepan; Tradig; Witco/Oleo-Surf.
Trade names: Polystep® A-13; Polystep® A-17
Trade names containing: Nacure® 5076; Nacure® 5225; Nacure® 5414; Nacure® 5528; Nacure® 5925

Dodecylbenzene sulfonic acid calcium salt. *See* Calcium dodecylbenzene sulfonate
Dodecylbenzenesulfonic acid, compd. with 2,2′,2′′-nitrilotris[ethanol] (1:1). *See* TEA-dodecylbenzene-sulfonate
Dodecylbenzenesulfonic acid, comp. with 2-propanamine (1:1). *See* Isopropylamine dodecylbenzene-sulfonate
Dodecylbenzenesulfonic acid, potassium salt. *See* Potassium dodecylbenzene sulfonate
Dodecylbenzenesulfonic acid sodium salt. *See* Sodium dodecylbenzenesulfonate
Dodecyl methacrylate. *See* Lauryl methacrylate
Dodecyl 2-methyl-2-propenoate. *See* Lauryl methacrylate
2-[2-(Dodecyloxy)ethoxy]ethanol. *See* Laureth-2
N-Dodecyl-2-pyrrolidone. *See* Lauryl pyrrolidone
Dodecyl sodium sulfate. *See* Sodium lauryl sulfate
Dodecylsulfate sodlum salt. *See* Sodium lauryl sulfate

Dodoxynol-10
CAS 9014-92-0 (generic); 26401-47-8 (generic)
Synonyms: PEG-10 dodecyl phenyl ether; POE (10) dodecyl phenyl ether
Definition: Ethoxylated alkyl phenol
Formula: $C_{12}H_{25}C_6H_4(OCH_2CH_2)_nOH$, avg. n = 10
Uses: Detergent, wetting agent, emulsifier, dispersant, penetrant, stabilizer, coemulsifier for agric., household/industrial cleaners; dedusting agent; paint pigment dispersant
Trade names: T-Det® DD-10

DOF. *See* Dioctyl fumarate
DOP. *See* Dioctyl phthalate
DOTP. *See* Dioctyl terephthalate
DPDP. *See* Diphenyl isodecyl phosphite
DPG. *See* Dipropylene glycol
DPIOP. *See* Diphenyl isooctyl phosphite
DPM. *See* PPG-2 methyl ether
Dracylic acid. *See* Benzoic acid

Drometrizole
CAS 2440-22-4; EINECS 219-470-5
Synonyms: 2-(2′-Hydroxy-5′-methylphenyl) benzotriazole; 2-(2H-Benzotriazol-2-yl)-4-methylphenol
Classification: Benzotriazole deriv.
Empirical: $C_{13}H_{11}N_3O$
Properties: Powd.; m.w. 225.27; m.p. 128 C
Toxicology: LD50 (oral, mouse) 6500 mg/kg; mildly toxic by ingestion; eye irritant; heated to decomp., emits toxic fumes of NO_x
Uses: UV light stabilizer for polymers (ABS, cellulosics, epoxy, polyester, PS, flexible and rigid PVC, VDC), coatings
Regulatory: FDA 21CFR §178.2010
Manuf./Distrib.: 3V
Trade names: Tinuvin® P; Topanex 100BT

DSE. *See* Nabam
DSS. *See* Dioctyl sodium sulfosuccinate
DTBP. *See* Di-t-butyl peroxide
Dutch oil. *See* Ethylene dichloride

Earthnut oil. *See* Peanut oil
Earth wax. *See* Ceresin
EC. *See* Ethylcellulose
ECTFE. *See* Ethylene-chlorotrifluoroethylene copolymer
EDA. *See* Ethylenediamine
EE. *See* Ethoxyethanol
EEA. *See* Ethoxyethanol acetate; Ethylacetoacetate
EEP. *See* Ethyl 3-ethoxypropionate
EGDS. *See* Glycol distearate
EGMS. *See* Glycol stearate
2-EH. *See* 2-Ethylhexanol
EHEC. *See* Ethyl hydroxyethyl cellulose
EHGE. *See* 2-Ethylhexyl glycidyl ether
Elainic acid. *See* Oleic acid
Enzactin. *See* Triacetin
EOEOEA. *See* 2-(2-Ethoxyethoxy) ethyl acrylate
EPDM. *See* EPDM rubber

EPDM rubber
Synonyms: EPT; EPDM; Ethylene-propylene-diene terpolymer; Ethylene propylene terpolymer
Uses: Rubber for hose, molded and extruded goods, sponge, roofing, automotive parts, gaskets, tire tubes, coated fabrics, footwear, wire/cable coating; modifier for polyolefins and other thermoplastic resins
Trade names: Royaltherm® 1421; Royaltherm® 1721; Vistalon 6505

Epichlorohydrin
CAS 106-89-8; EINECS 203-439-8
Synonyms: Chloropropylene oxide; 1-Chloro-2,3-epoxypropane; 3-Chloro-1,2-propylene oxide
Empirical: C_3H_5ClO
Properties: Highly volatile unstable liq., chloroform-like odor; sl. sol. in water; m.w. 92.53; f.p. -25 C; b.p. 115.2 C; flash pt. 33.9 C; ref. index 1.4358
Precaution: Flamm.; mod. fire risk
Toxicology: Toxic by inhalation, ingestion, and skin absorp.
Uses: Intermediate for paints; major raw material for epoxy and phenoxy resins; mfg. of glycerol; solv. for cellulose sters and ethers; high wet-str. resins for paper industry
Manuf./Distrib.: BASF; Jarchem Ind.; Rit-Chem; Shell; United Min. & Chem.

EPM. *See* EPM rubber

EPM rubber
CAS 9010-79-1
Synonyms: EPR; EPM; Ethylene/propylene copolymer (INCI); EPR rubber; Ethene, polymer with 1-propene
Properties: Dens. 0.860
Uses: Rubber for inj. molded and extruded goods (elec. components, wire insulation, o-rings, brake components); modifying PP and other plastics
Regulatory: FDA 21CFR §177.1210
Manuf./Distrib.: Aldrich; Hüls AG
Trade names: Petrolite® CP-7

Epoxidized flaxseed oil. *See* Epoxidized linseed oil

Epoxidized linseed oil
Synonyms: Epoxidized flaxseed oil
Properties: Clear pale yel. liq., low odor; sol. < 0.1% in water; sp.gr. 1.03; dec. 550 F; flash pt. (CC) 435 F
Precaution: Avoid oxidizing agents, strong acids, bases and amines; may produce CO on burning
Toxicology: LD50 (oral, rat) 30 gma/kg; nontoxic orally; no eye or skin irritation
Uses: Plasticizer/stabilizer for epoxy, PVC, paints; food pkg.
Manuf./Distrib.: Elf Atochem N. Am.; Ferro/Bedford; Union Carbide; Witco/Polymer Addit.

Epoxidized linseed oil acrylate
Uses: Oligomer; forms flexible films readily; pigment wetter/coupling agent; especially useful in wet offset litho inks and overprints where intercoat adhesion and delamination occur
Trade names: Photomer® 3082

Epoxidized 1,2-polybutadiene
CAS 129288-65-9
Uses: As sole resin in an epoxy system or as modifiers for epoxy systems; in flexible and impact-resist. coatings and potting compds.
Manuf./Distrib.: Elf Atochem N. Am.
Trade names: ADK CIZER BF-1000; Poly bd® 600; Poly bd® 605

Epoxidized soybean oil
CAS 8013-07-8; EINECS 232-391-0
Synonyms: Soybean oil, epoxidized
Definition: Modified oil obtained from soybean oil by epoxidation
Properties: Clear pale yel. liq., low odor; sol. < 0.1% in water; sp.gr. 0.99; dec. 550 F; m.p. 25 F; iodine no. 6 max.; flash pt. (CC) 430 F
Precaution: Avoid oxidizing agents, strong acids, bases and amines
Toxicology: Nonhazardous; LD50 (oral, rat) 30 gm/kg; no eye or skin irritation (rabbit); heated to decomp., emits acrid smoke and irritating fumes
Uses: Plasticizer, stabilizer for PVC, epoxy, chlorinated rubber, coatings, inks; acid scavenger; food pkg.
Regulatory: FDA 21CFR §175.105, 177.1650, 178.3910, 181.22, 181.27
Manuf./Distrib.: Ashland; Elf Atochem N. Am.; FMC; Ferro/Bedford; C.P. Hall; Henkel; Hüls; Merrand Int'l.; Union Carbide; Witco/Polymer Addit.
Trade names: ADK CIZER O-130P; Estabex® 2307; Estabex® 2307 DEOD; Paraplex® G-60; Paraplex® G-62; Plasthall® ESO
Trade names containing: Vinyzene® BP-5-5

Epoxidized soybean oil acrylate

Uses: Offers good flexibility, pigment wetting; for use in uv and eb curing compositions incl. overprint varnishes, lithographic inks

Trade names: CN 111; Craynor 111; Ebecryl® 860; Photomer® 3005

Epoxy acrylate

Uses: Curing agent; for paper clear coatings, wood top coatings, screen inks, litho inks, polyethylene coatings, metal decorative coatings, adhesive papers, wood fillers, solder masks and photoresists

Manuf./Distrib.: Monomer-Polymer & Dajac Labs

Trade names: CN 103; CN 104; CN 114; CN 120; Craynor 104; Craynor 114; Ebecryl® 3201; Ebecryl® 3702

Trade names containing: CN 104 A80; CN 104 B80; CN 104 C75; CN 104 D80; CN 104 F50; CN 114 A80; CN 114 E80; CN 120 A60; CN 120 A75; CN 120 B80; CN 120 C60; CN 120 C80; CN 120 D80; CN 120 E50; CN 120 S80; Craynor 104 A80; Craynor 104 B80; Craynor 104 C75; Craynor 104 D80; Craynor 114 A80; Craynor 114 D75; Craynor 114 E80; Ebecryl® 3604; Photomer® 3016-20T; Photomer® 3016-40R; Photomer® 3016-40T; Photomer® 3038

Epoxy, bisphenol A

CAS 25036-25-3; 25068-38-6

Uses: Heat stabilizer for PVC; exc. halogen capture for flame retardant resins; surf. coatings

Manuf./Distrib.: Archway Sales; Ashland; Baychem; Cardolite; Ciba-Geigy; GCA; General Polymers; Lenape Ind.; Lomas Int'l.; McWhorter; Reichhold; Samson; Shell; United Min. & Chem.

Trade names: Araldite® GT 6060; Araldite® GT 6063; Araldite® GT 6084; Araldite® GT 6097; Araldite® GT 6099; Araldite® GT 6243; Araldite® GT 6248; Araldite® GT 6259; Araldite® GT 6450; Araldite® GT 7013; Araldite® GT 7014; Araldite® GT 7071; Araldite® GT 7072; Araldite® GT 7074; Araldite® GT 7097; Araldite® GT 7099; Araldite® GT 7220; Araldite® GT 7226; Araldite® GT 7255; Araldite® GT 9013; Araldite® GT 9496; Araldite® GT 9516; Araldite® GT 9545; Araldite® GY 2600; Araldite® GY 6008; Araldite® GY 6010; Araldite® GY 6020; Araldite® GY 9513; Araldite® GY 9613; Araldite® GY 9667; Araldite® PY 258; Araldite® XU 0248; D.E.R. 330; D.E.R. 331; D.E.R. 337; D.E.R. 361; D.E.R. 362; D.E.R. 383; D.E.R. 661; D.E.R. 662; D.E.R. 663U; D.E.R. 664; D.E.R. 664U; D.E.R. 665; D.E.R. 667; D.E.R. 668; D.E.R. 669; D.E.R. 6225; Epi-Rez® 3515-W-60; Epi-Rez® 3519-W-50; Epi-Rez® 3522-W-60; Epi-Rez® WD-510; EPON® Resin 825; EPON® Resin 826; EPON® Resin 828; EPON® Resin 1001-F; EPON® Resin 1002F; EPON® Resin 1004F; EPON® Resin 1007F; EPON® Resin 1009F; GT 6097; GT 7004; GT 7013; GT 7014; GT 7072; GT 7097; GT 7220; GT 7226; GT 7255; GT 9013; GT 9496; HZ 0949U; RDX 61010; XU GT 0259; XU GT 0273; XU HZ 0365; XUS 19000.01

Trade names containing: Araldite® GZ 471 X-75; Araldite® GZ 488 N-40; Araldite® GZ 488 PMA-32; Araldite® GZ 540 X-90; Araldite® GZ 571 KX-75; Araldite® GZ 571 T-75; Araldite® GZ 597 KT-55; Araldite® GZ 6097 PM-55; Araldite® GZ 7071 N-80; Araldite® GZ 7071 OX-65; Araldite® GZ 7071 PM-75; Araldite® GZ 7071 T-65; Araldite® GZ 7097 PM-55; Araldite® GZ 7488 N-50; Araldite® GZ 7488 PMA-40; Araldite® GZ 9749 OX-65; D.E.R. 324; D.E.R. 325; D.E.R. 337-X90; D.E.R. 353; D.E.R. 642U; D.E.R. 660-PA80; D.E.R. 661-A80; D.E.R. 667-PMT55; D.E.R. 671-MAK75; D.E.R. 671-MK75; D.E.R. 671-PM75; D.E.R. 671-PMA75; D.E.R. 671-PMK75; D.E.R. 671-PMT70; D.E.R. 671-PMX75; D.E.R. 671-T75; D.E.R. 671-X75; D.E.R. 671-XM75; D.E.R. 672U; D.E.R. 684-EK40; EPON® Resin 813; EPON® Resin 815; EPON® Resin 2002-FC-10; EPON® Resin Custom Solution CS-241

Epoxy, bisphenol F

Uses: For adhesives, tank linings, flooring, filament winding, casting, pultrusion, RTM, surf. coating applics

Trade names: Araldite® GY 281; Araldite® GY 282; Araldite® GY 285; Araldite® GY 308; Araldite® LY 9703; Araldite® PT 810; Araldite® PY 306; Araldite® XD 4955; D.E.R. 354

Trade names containing: D.E.R. 353; D.E.R. 642U; D.E.R. 672U

Epoxy cresol novolac

CAS 29690-82-2

Uses: For high temp. adhesives, coatings, electrical and laminating product areas

Trade names: Araldite® ECN 1235; Araldite® ECN 1273; Araldite® ECN 1280; Araldite® ECN 1299; Araldite® ECN 9495; Araldite® ECN 9511; Araldite® ECN 9699; EPON® Resin 164

2-(3,4-Epoxycyclohexyl) ethyltrimethoxysilane

CAS 3388-04-3; EINECS 222-217-1

Synonyms: β-(3,4-Epoxycyclohexyl) ethyltrimethoxysilane

Empirical: $C_{11}H_{22}O_4Si$

Properties: Liq.; m.w. 246.42; dens. 1.07; b.p. 310 C; ref. index 1.449 (20 C); flash pt. 146 C

Toxicology: LD50 (oral, rat) 12,300 mg/kg, (dermal, rabbit) 6300 mg/kg; mildly toxic by ingestion and skin contact; skin irritant; heated to decomp., emits acrid smoke and fumes

Uses: Coupling agent, chem. intermediate, blocking agent, release agent, lubricant, primer, reducing agent; adhesion promoter for coatings
Manuf./Distrib.: Aldrich; Fluka
Trade names: Silquest® A-186

β-(3,4-Epoxycyclohexyl) ethyltrimethoxysilane. *See* 2-(3,4-Epoxycyclohexyl) ethyltrimethoxysilane

Epoxy novolac
Uses: Resin for elec. potting, encapsulation, and casting, high-pressure laminating, filament winding, coatings, adhesives
Trade names: D.E.N. 431; D.E.N. 438; D.E.N. 439; D.E.N. 444; Epi-Rez® 5003-W-55; EPON® Resin 1031; Epotuf Resin 37-170
Trade names containing: D.E.N. 438-A85; D.E.N. 438-EK85; D.E.N. 438-MAK80; D.E.N. 438-MK75; D.E.N. 439-EK85; D.E.N. 444-MAK75

Epoxy novolac acrylate
Uses: Provides heat and solv. resist. to solder resists, marking inks, adhesion on metalized substrates, low shrinkage coatings, heat-resist. applics.
Trade names: Ebecryl® 629
Trade names containing: CN 112 C60; Craynor 112 C60; Ebecryl® 639; Ebecryl® 3603

Epoxy phenol novolac
Uses: Performance polymer for electronic, aerospace, surf. coating applics.
Manuf./Distrib.: Aceto; Ashland; Cardolite; Chemcentral; Ciba-Geigy; Focus; Georgia-Pacific Resins; Lomas Int'l.; Reichhold
Trade names: Araldite® EPN 1138; Araldite® EPN 1139; Araldite® EPN 1179; Araldite® PY 307-1

1,2-Epoxy-3-phenoxypropane. *See* Phenyl glycidyl ether
2,3-Epoxy-1-propanol acrylate. *See* Glycidyl acrylate
2,3-Epoxypropyl ester acrylic acid. *See* Glycidyl acrylate
2,3-Epoxypropyl methacrylate. *See* Glycidyl methacrylate
2,3-Epoxypropyl phenyl ether. *See* Phenyl glycidyl ether
3-(2,3-Epoxypropyl)-propyltrimethoxysilane. *See* 3-Glycidoxypropyltrimethoxysilane

Epoxy resin
CAS 25928-94-3
Definition: Derived from epichlorohydrin and diethylene glycol
Formula: $-OCH_2CHOCH_2$
Toxicology: LD50 (oral, rat) 2200 mg/kg; strong skin irritant in uncured state; poison by inhalation; moderately toxic by ingestion; heated to decomp., emits toxic fumes of Cl^-
Uses: Plasticizer for surface coatings, adhesive; casting metal-forming tools and dies; encapsulation of elec. parts; stabilizer, modifier for other resins
Manuf./Distrib.: Archway Sales; Asahi Chem. Ind.; Ashland; Cardolite; Chemcentral; Ciba-Geigy; Conap; Dow Plastics; Ferro/Bedford; GCA; Hardman; Henkel; Hoechst Celanese; Key Polymer; McWhorter; Monomer-Polymer & Dajac Labs; Monsanto; Morton Int'l.; Reichhold; Rhone-Poulenc/Perf. Resins & Coatings; Samson; Sartomer; Seegott; Shell; Union Carbide; United Min. & Chem.; Witco/Polymer Addit.
Trade names: Adeka Optomer KR Series; Cardolite® NC-513LC; Cardolite® NC-514; D.E.R. 662UH; D.E.R. 669E; D.E.R. 732; D.E.R. 736; D.E.R. 755; D.E.R. 6224; Epi-Rez® 5520-W-60; EPON® HPT® Resin 1077; EPON® Resin 832; EPON® Resin 2002; EPON® Resin 2003; EPON® Resin 2004; EPON® Resin 2005; EPON® Resin 2012; EPON® Resin 2014; EPON® Resin 2022; EPON® Resin 2042; EPON® Resin 3001; EPON® Resin 3002; EPON® Resin 8101; EPON® Resin 8111; EPON® Resin 8131; EPON® Resin 8161; EPON® Resin 58134; EPON® Resin SU-8; Epotuf Resin 37-001; Epotuf Resin 37-007; Epotuf Resin 37-140; Epotuf Resin 37-141; Epotuf Resin 37-151; Master Bond EP3FL; Master Bond EP11EET; Master Bond EP19; Master Bond EP21CE; Master Bond EP36; Master Bond EP39M; Master Bond Supreme 10LV; PY 284; RDX 68654; Scotchkote® 213, 214; Scotchkote® 306; Tactix 740; Waterpoxy™ 1401; Witcobond® XW
Trade names containing: Araldite® EPN 1138 A-85; Araldite® EPN 1138 MAK-80; Araldite® EPN 1138 X-85; Araldite® GY 502; Aroflint 303-X-90; Epi-Rez® 3520-WY-55; Epi-Rez® 3540-WY-55; Epi-Rez® 5522-WY-55; Epi-Tex® 199; Epi-Tex® 199-E; Epi-Tex® 1662; Epi-Tex® 1663; EPON® Resin 834-X-90; EPON® Resin 875-O-70; EPON® Resin 876-QX-50; EPON® Resin 1001-CX-75; EPON® Resin 1001-T-75; EPON® Resin 1001-X-75; EPON® Resin 1001-Y-75; EPON® Resin 1004-QX-55; EPON® Resin 1007-Q-40; EPON® Resin 1007-Q-45; EPON® Resin 2024; EPON® Resin 8132; Epotuf Resin 37-127; Epotuf Resin 37-128; Epotuf Resin 37-130; Epotuf Resin 37-134; Epotuf Resin 37-135; Epotuf Resin 38-509;

Epoxy resin, brominated

Epotuf Resin 38-515; Epotuf Resin 38-519; 8521 MX60; Sobral EE-632; Tactix 741; Thiokol® FEC-2232; Tint-Ayd® EP Series

Epoxy resin, brominated
CAS 3072-84-2
Uses: Flame-retardant resin for elec. potting, encapsulation, and casting, high-pressure laminating, filament winding, and coatings; flame retardant for thermoplastics
Trade names: D.E.R. 542; EPON® Resin 1183; EPON® Resin 5183; Epotuf Resin 37-200; F-2200; Thermoguard® 220

EPR. *See* EPM rubber
EPR rubber. *See* EPM rubber
EPT. *See* EPDM rubber

Erucamide
CAS 112-84-5; EINECS 204-009-2
Synonyms: Erucic acid amide; Erucylamide; 13-Docosenamide; cis 13-Docosenamide
Classification: Aliphatic amide
Empirical: $C_{22}H_{43}ON$
Formula: $CH_3(CH_2)_7CH{=}CH(CH_2)_{11}CONH_2$
Properties: Solid; sol. in isopropanol; sl. sol. in alcohol, acetone; dens. 0.888; m.p. 75-80 C
Precaution: Combustible
Uses: Foam stabilizer; solvent for waxes and resins, emulsions; slip/antiblock agent for polyethylene; lubricant, mold release for rubber and plastics; lamination of polyethylene to cellophane and in polyethylene extrusion coatings; food contact applics.
Regulatory: FDA 21CFR §175.105, 177.1200, 177.1210, 178.3860
Manuf./Distrib.: Akzo Nobel; Aldrich; Chemax; Cookson Spec. Additives; Croda Universal Ltd.; Sigma; Syn. Prods.; Witco/Oleo-Surf.
Trade names: Petrac® Eramide®

Erucic acid amide. *See* Erucamide
Erucylamide. *See* Erucamide
ETFE. *See* Ethylene/tetrafluoroethylene copolymer
Ethanal. *See* Acetaldehyde

Ethane diamide, n-(2-ethoxyphenyl)-n´-(4-ethylphenyl)
Uses: UV light absorber for coatings
Trade names: Sanduvor® VSU Disp.

Ethanediamide, n-(2-ethoxyphenyl)-n´-(4-ethylphenyl). *See* 2-Ethyl, 2´-ethoxy-oxalanilide

Ethane diamide, n-(2-ethoxyphenyl)-n´-(4-isododecyl phenyl)
Uses: UV light absorber for coatings
Trade names: Sanduvor® 3206

1,2-Ethanediamine. *See* Ethylenediamine
Ethanediamine, N-(2-ethoxyphenyl)-N´-(2-ethylphenyl)-. *See* 2-Ethyl, 2´-ethoxy-oxalanilide
1,2-Ethanedicarboxylic acid. *See* Succinic acid
1,2-Ethanediol. *See* Glycol
Ethane-1,2-diol. *See* Glycol
1,2-Ethanediol dimethacrylate. *See* Ethylene glycol dimethacrylate
1,2-Ethanediylbiscarbamodithioic acid disodium salt. *See* Nabam
N,N´-1,2-Ethanediylbisoctadecanamide. *See* Ethylene distearamide
N,N´-1,2-Ethanediylbis-9-octadecenamide. *See* Ethylene dioleamide
Ethane, 1,1,1-trichloro-. *See* Trichloroethane
Ethanol. *See* Alcohol

Ethanolamine
CAS 141-43-5; EINECS 205-483-3
Synonyms: MEA; 2-Aminoethanol; β-Ethanolamine; Ethylolamine; β-Hydroxyethylamine; 2-Aminoethyl alcohol; β-Aminoethyl alcohol; Monoethanolamine; Glycinol; 2-Hydroxyethylamine
Classification: Monoamine
Empirical: C_2H_7NO

Formula: $NH_2CH_2CH_2OH$
Properties: Colorless clear mod. visc. liq., ammoniacal odor; hygroscopic; misc. with water, alcohol, acetone, glycerin; sol. in chloroform; sl. sol. in benzene; m.w. 61.10; dens. 1.012; m.p. 10.5 C; b.p. 170 C; flash pt. 93 C; ref. index 1.4540
Precaution: DOT: Corrosive material; flamm. exposed to heat or flame; powerful reactive base
Toxicology: ACGIH TLV:TWA 3 ppm; LD50 (oral, rat) 2140 mg/kg, (skin, rat) 1500 mg/kg; poison by IP route; mod. toxic by ingestion, skin contact, subcut., IV routes; corrosive irritant to eyes, skin, mucous membranes; heated to decomp., emits toxic fumes of NO_x
Uses: Scrubbing acid gases, esp. in synthesis of ammonia; nonionic detergents for dry cleaning wool treatment, emulsion paints, polishes, agricultural sprays; chemical intermediate; pharmaceuticals; corrosion inhibitor; chem. intermediate for mfg. rubber, vulcanization accelerators
Regulatory: FDA 21CFR §173.315, 175.105, 176.210, 176.300, 178.3120; not permitted for use in foods intended for babies and young infants in UK; USP/NF, BP compliance
Manuf./Distrib.: Aldrich; Allchem Ind.; BP Chems. Ltd; Fluka; Hüls AG; Int'l. Chem. Inc.; Occidental; Oxiteno; Research Organics; Sigma; Spectrum Chem. Mfg.; Texaco; Union Carbide
Trade names: MEA Commercial Grade; MEA Low Freeze Grade; MEA Low Iron Grade; MEA Low Iron-Low Freeze Grade; MEA NF Grade

β-Ethanolamine. *See* Ethanolamine
Ethanol, 2-butoxy-, acetate. *See* Butoxyethanol acetate
Ethanol, 2-butoxy-, phosphate (3:1). *See* Tributoxyethyl phosphate
Ethanol, 2-(2-ethoxyethoxy). *See* Ethoxydiglycol
Ethanol, 2-(hydroxymethylamino)-. *See* 2[(-Hydroxymethyl) amino] ethanol
Ethanol, 2-(2-methoxyethoxy)-. *See* Methoxydiglycol
Ethanol, 2-(2-phenoxyethoxy). *See* Diethylene glycol phenyl ether
Ethanol, 2-propoxy. *See* Ethylene glycol propyl ether
Ethanol,2,2′,2′′-(propyldyne tris(methyleneoxy))tri-, triacrylate. *See* Trimethylolpropane ethoxy triacrylate
1-Ethanol-2-thiol. *See* 2-Mercaptoethanol
Ethanol, undenatured. *See* Alcohol
Ethanone, 2,2-dimethoxy-1,2-didphenyl-(9Cl). *See* 2,2-Dimethoxy-2-phenylacetophenone
Ethene, homopolymer. *See* Polyethylene
Ethene, homopolymer, oxidized. *See* Polyethylene, oxidized
Ethene polymer. *See* Polyethylene
Ethene, polymer with 1-propene. *See* EPM rubber
Ethenol homopolymer. *See* Polyvinyl alcohol
Ethenyl acetate. *See* Vinyl acetate
Ethenyl acetate, homopolymer. *See* Polyvinyl acetate (homopolymer)
Ethenylbenzene. *See* Styrene
Ethenylbenzene, homopolymer. *See* Polystyrene
Ethenylbenzene polymer with 1,3-butadiene. *See* Polybutadiene-styrene copolymer
6-Ethenyl-6-(methoxyethoxy)-2,5,7,10-tetraoxa-6-silaundecane. *See* Vinyltris(2-methoxyethoxy) silane
1-Ethenyl-2-pyrrolidinone. *See* N-Vinyl-2-pyrrolidone
1-Ethenyl-2-pyrrolidinone homopolymer. *See* PVP
1-Ethenyl-2-pyrrolidinone, polymer with acetic acid ethenyl ester. *See* PVP/VA copolymer
1-Ethenyl-2-pyrrolidinone, polymer with ethenylbenzene. *See* Styrene/PVP copolymer
Ether. *See* Ethyl ether
Ethocel. *See* Ethylcellulose

N-(p-Ethoxycarbonylphenyl)-N′-ethyl-N′-phenylformamidine
CAS 65816-20-8
Empirical: $C_{18}H_{20}N_2O_2$
Properties: Wh. to pale yel. powd.; sol. > 50 g/100 g in ethanol, methanol, IPA, butyl acetate; insol. in water; m.w. 296.4; dens. 1.077 (65 C); b.p. 215 C (2 mm); m.p. 60-65 C
Uses: UV absorber for cellulosics; used in surf. coatings, polishes, dyestuffs, carpet treatments
Manuf./Distrib.: Aldrich
Trade names: Givsorb® UV-2

Ethoxydiglycol
CAS 111-90-0; EINECS 203-919-7
Synonyms: Diethylene glycol monoethyl ether; Diethylene glycol ethyl ether; 'Carbitol'; Ethanol, 2-(2-ethoxyethoxy); 2-(2-Ethoxyethoxy) ethanol
Classification: Ether alcohol
Empirical: $C_6H_{14}O_3$

Formula: $CH_2OHCH_2OCH_2CH_2OC_2H_5$

Properties: Colorless liq., mild pleasant odor; hygroscopic; misc. with water, common org. solvs.; m.w. 134.20; dens. 1.0272 (20/20 C); b.p. 195-202 C; flash pt. 96.1 C; ref. index 1.425 (25 C)

Precaution: Combustible liq. and vapor; peroxide former

Toxicology: LD50 (oral, rat) 5500 mg/kg; moderately toxic by ingestion and other routes; skin and eye irritant; heated to decomp., emits acrid smoke

Uses: Solvent for dyes, nitrocellulose, resins; textiles, textile printing, soaps; lacquers, quick-dry varnishes, enamels, wood stains; organic synthesis; brake fluid diluent; mfg. of plasticizers; solv., thinner in nail enamels

Regulatory: FDA 21CFR §175.105, 176.180

Manuf./Distrib.: Aldrich; Allchem Ind.; Ashland; Eastman; Fluka; Great Western; Occidental; Oxiteno; Sigma; Spectrum Chem. Mfg.; Union Carbide

Trade names: Carbitol® Low Gravity; Dioxitol-Low Gravity; Eastman® DE; Ethyl Di-Icinol; Poly-Solv® DE (Low Gravity)

Trade names containing: Dioxitol-High Gravity; Icinol EE22; Poly-Solv® DE (High Gravity)

Ethoxydiglycol acetate

CAS 112-15-2; EINECS 203-940-1

Synonyms: Diethylene glycol monoethyl ether acetate; 2-(2-Ethoxyethoxy)ethanol acetate; Diethylene glycol ethyl ether acetate

Empirical: $C_8H_{16}O_4$

Formula: $CH_3OCH_2CH_2OCH_2CH_2OAc$

Properties: Colorless liq.; sol. in water; m.w. 176.2; sp.gr. 1.011 (20 C); m.p. -25 C; b.p. 214 C; ref. index 1.42 (20 C)

Toxicology: Irritant

Uses: Solvent, coalescing aid for latex paints

Regulatory: FDA 21CFR §175.105

Manuf./Distrib.: Aldrich; ARCO; Eastman; Fluka

Trade names: Eastman® DE Acetate

Ethoxyethanol

CAS 110-80-5; EINECS 203-804-1

Synonyms: EE; Ethyl glycol; Ethylene glycol ethyl ether; Ethylene glycol monoethyl ether; 2-Ethoxyethanol; 'Cellosolve'

Classification: Ether alcohol

Empirical: $C_4H_{10}O_2$

Formula: $CH_3CH_2OCH_2CH_2OH$

Properties: Colorless liq., pract. odorless; misc. with hydrocarbons, alcohol, ether, acetone, liq. esters, water; m.w. 90.14; dens. 0.9311 (20/20 C); m.p. -70 C; b.p. 135.6 C; flash pt. 48.9 C; ref. index 1.4060 (25 C)

Precaution: Flamm.; reacts with oxidizing materials; mod. explosion hazard in form of vapor when exposed to flames

Toxicology: TLV 5 ppm (skin); LD50 (oral, rat) 3 g/kg; toxic by skin absorption; moderately toxic by ingestion, IV, IP routes; mildly toxic by inhalation, subcut. routes

Uses: Solvent for nitrocellulose, alkyd resins, lacquer, lacquer thinners, dyeing and printing textiles, varnish removers, cleaning solutions

Regulatory: FDA 21CFR §73.1, 175.105, 177.2600

Manuf./Distrib.: Aldrich; Allchem Ind.; ARCO; Ashland; Brown; Fluka; Great Western; Occidental; Oxiteno; Sigma; Union Carbide

Trade names: Cellosolve®; Ethyl Icinol; Oxitol; Poly-Solv® EE

2-Ethoxyethanol. *See* Ethoxyethanol

Ethoxyethanol acetate

CAS 111-15-9; EINECS 203-839-2

Synonyms: EEA; Ethylene glycol ethyl ether acetate; Ethylene glycol monoethyl ether acetate; 2-Ethoxyethanol acetate; 2-Ethoxyethyl acetate

Classification: Ester

Definition: Ester of ethoxyethanol and acetic acid

Formula: $CH_3COOC_2H_4OC_2H_5$

Properties: Colorless liq., pleasant odor; sol. in $\approx$ 6 parts water; m.w. 132.16; dens. 0.975 (20/20 C); b.p. 156 C; flash pt. (OC) 56 C; ref. index 1.4030

Precaution: Combustible; reacts with oxidizing materials

Toxicology: LD50 (rat, oral) 5.1 g/kg; toxic by ingestion and skin absorp.; heated to decomp., emits acrid smoke

and irritating fumes
Uses: Solvent for coatings (NC, cellulose acetate, ethyl cellulose, butyrate, acrylic, urethane), automobile lacquers; retards evaporation and imparts high gloss
Regulatory: FDA 21CFR §175.105
Manuf./Distrib.: Aldrich; Allchem Ind.; ARCO; Ashland; Fluka; Occidental; Union Carbide
Trade names: Cellosolve® Acetate
Trade names containing: GPRI™ CKSB-2001

2-Ethoxyethanol acetate. *See* Ethoxyethanol acetate
2-(2-Ethoxyethoxy) ethanol. *See* Ethoxydiglycol
2-(2-Ethoxyethoxy)ethanol acetate. *See* Ethoxydiglycol acetate
2-[2-(2-Ethoxyethoxy) ethoxy] ethanol. *See* Ethoxytriglycol
Ethoxyethoxyethyl acrylate. *See* 2-(2-Ethoxyethoxy) ethyl acrylate

2-(2-Ethoxyethoxy) ethyl acrylate
Synonyms: EOEOEA; Ethoxyethoxyethyl acrylate
Uses: Monomer for radiation-curable ink and coating formulations
Manuf./Distrib.: Monomer-Polymer & Dajac Labs; Morton Int'l.
Trade names containing: CN 960 H90; CN 961 H81; CN 961 H90; CN 962 H90; CN 963 H90; CN 964 H90; CN 965 H90; CN 966 H90; CN 970 H75; CN 971 H90; CN 972 H90; CN 973 H85; Ebecryl® 8800; Ebecryl® 8800-20R; SR-256

2-Ethoxyethyl acetate. *See* Ethoxyethanol acetate
Ethoxyethyl ether. *See* Diethylene glycol diethyl ether
2-Ethoxyethyl ether. *See* Diethylene glycol diethyl ether
N-(2-Ethoxyphenyl)-N′-(2-ethylphenyl) ethanediamide. *See* 2-Ethyl, 2′-ethoxy-oxalanilide
1-Ethoxy-2-propanol. *See* Propylene glycol ethyl ether

Ethoxypropanol acetate
Trade names containing: BYK®-A 525

Ethoxypropionic acid ethyl ester. *See* Ethyl 3-ethoxypropionate
3-Ethoxypropionic acid ethyl ester. *See* Ethyl 3-ethoxypropionate

Ethoxytriglycol
CAS 112-50-5; EINECS 203-978-9
Synonyms: Triethylene glycol monoethyl ether; Triethylene glycol ethyl ether; PEG-3 ethyl ether; Triglycol monoethyl ether; 2-[2-(2-Ethoxyethoxy) ethoxy] ethanol
Empirical: $C_8H_{18}O_4$
Formula: $C_2H_5O(C_2H_4O)_3H$
Properties: Colorless liq.; sol. in water; m.w. 178.26; dens. 1.0208 (20/20 C); b.p. 255.4 C; f.p. -18.7 C; flash pt. (OC)135 C
Precaution: Combustible; can react with oxidizing materials
Toxicology: LD50 (oral, rat) 10,610 mg/kg, (dermal, rabbit) 8 g/kg; mildly toxic by ingestion and skin contact; eye irritant; heated to decomp., emits acrid smoke and fumes
Uses: Chemical intermediate
Manuf./Distrib.: Ashland; Fluka; Oxiteno
Trade names containing: Icinol EE22; Icinol ME33

Ethyl acetate
CAS 141-78-6; EINECS 205-500-4
Synonyms: Acetic ether; Acetic acid, ethyl ester; Ethyl ethanoate; Vinegar naphtha
Classification: Aliphatic organic compd.
Definition: Ester of ethyl alcohol and acetic acid
Empirical: $C_4H_8O_2$
Formula: $CH_3COOC_2H_5$
Properties: Colorless liq., fragrant fruity odor, acetous burning taste; sol. in chloroform, alcohol, ether; sl. sol. in water; m.w. 88.12; dens. 0.902 (20/4 C); bulk dens. 0.8945 g/ml (25 C); b.p. 77 C; f.p. -83.6 C; flash pt. -4.4 C; ref. index 1.3723
Precaution: DOT: Flamm. liq.; very dangerous fire hazard exposed to heat or flame; can react vigorously with oxidizers
Toxicology: TLV 400 ppm in air; LD50 (oral, rat) 5620 mg/kg; poison by inhalation; mildly toxic by ingestion; irritant to eyes, skin, mucous membranes; mutagenic data; mildly narcotic; CNS depressant; heated to

decomp., emits acrid smoke and irritating fumes

Uses: General solvent in coatings and plastics, organic synthesis, smokeless powders, artificial leather, photographic films and plates, pharmaceuticals, synthetic fruit essences, flavoring agent; cleaning textiles; nail polish removers; color diluent, flavoring agent, solv. for food use

Regulatory: FDA 21CFR §73.1, 172.560, 173.228, 175.320, 177.1200, 182.60, GRAS; 27CFR §21.106; FEMA GRAS; Japan approved with restrictions; FDA approved for ophthalmics, orals, topicals; USP/NF, BP, Ph.Eur. compliance

Manuf./Distrib.: Aldrich; Allchem Ind.; Ashland; J.T. Baker; Baychem; Berje; BP Chems. Ltd; Brown; Chemcentral; Chisso Am.; Eastman; Fluka; Harcros; Hoechst Celanese; Hüls AG; Lonza Ltd.; Mallinckrodt; Monsanto; Penta Mfg.; Punda Mercantile; Samson; Sigma; Sunnyside; Union Carbide; Van Waters & Rogers

Trade names containing: Filmex® C; HC-913; Tecsol® 1; Tecsol® 3; Tecsol® C

Ethylacetoacetate

CAS 141-97-9; EINECS 205-516-1

Synonyms: EEA; 3-Oxobutanoic acid ethyl ester; Diacetyl ether; Acetoacetic acid ethyl ester; Acetoacetic ester; Ethyl 3-oxobutanoate

Classification: Aliphatic organic compd.

Empirical: $C_6H_{10}O_3$

Formula: $CH_3COCH_2COOC_2H_5$

Properties: Colorless liq., fruity odor; sol. in ≈ 35 parts water; misc. with common org. solvs.; m.w. 130.14; dens. 1.0213 (25/4 C); m.p. -45 C; b.p. 180.8 C (760 mm); flash pt. (CC) 184 F; ref. index 1.4180-1.4195

Precaution: DOT: Flamm. liq.; combustible liq. when exposed to heat or flame; can react with oxidizing materials

Toxicology: LD50 (oral, rat) 3.98 g/kg; mod. toxic by ingestion; mod. irritating to skin, mucous membranes, eyes; heated to decomp., emits acrid smoke and irritating fumes

Uses: Catalyst for esterification and olefin polymerization; resin crosslinking agent for automotive prods., coatings, elastomers, films/paints, graphic arts, plastics; flavoring agent

Regulatory: FDA 21CFR §172.515; FEMA GRAS; Japan approved as flavoring

Manuf./Distrib.: Aceto; Aldrich; Berje; Eastman; Fluka; Lonza; Penta Mfg.; Sigma; Spectrum Chem. Mfg.

Trade names: Tyzor DC

Ethyl acetone. *See* Methyl propyl ketone
Ethyl alcohol. *See* Alcohol
Ethyl alcohol, undenatured. *See* Alcohol
Ethyl aldehyde. *See* Acetaldehyde

Ethylbenzene

CAS 100-41-4; EINECS 202-849-4

Synonyms: Phenylethane; Ethylbenzol

Classification: Aromatic solvent

Empirical: C_8H_{10}

Formula: $C_6H_5C_2H_5$

Properties: Colorless liq., aromatic odor; sol. in alcohol, benzene, CCl_4, ether; almost insol. in water; m.w. 106.17; dens. 0.867 (20 C); f.p. -95 C; b.p. 136.187 C; flash pt. 15 C; ref. index 1.495 (20 C)

Precaution: Highly flamm.; dangerous fire risk; keep away from ignition sources

Toxicology: LD50 (oral, rat) 3500 mg/kg, (dermal, rabbit) 17,800 mg/kg; moderately toxic by ingestion, intraperitoneal route; mildly toxic by inhalation and skin contact; TLV:TWA 100 ppm; STEL 125 ppm

Uses: Solvent

Manuf./Distrib.: Aldrich; Ashland; Fluka; Phibro Energy USA; Sigma

Trade names containing: Acryloid® AU-1003; Acryloid® AU-1033

Ethylbenzol. *See* Ethylbenzene

7-Ethyl bicyclooxazolidine

CAS 7747-35-5; EINECS 231-810-4

Synonyms: 7A-Ethyldihydro-1H,3H-5H-oxazolo (3,4-C) oxazole

Uses: Antibacterial for cosmetics and toiletries; biocide, in-can preservative for latex paints, adhesives, slurries, inks, and resin emulsion systems; does not increase VOC content

Regulatory: USA not restricted; Europe provisional list 3000 ppm max.

Manuf./Distrib.: Aldrich; Fluka

Trade names: Canguard® 442

Ethyl carbinol. *See* n-Propyl alcohol

Ethylcellulose
CAS 9004-57-3
Synonyms: EC; Cellulose ethyl ether; Ethocel
Definition: Ethyl ether of cellulose
Properties: Wh. to lt. tan powd.; hygroscopic; sol. in most org. liqs.; insol. in water, glycerol, propylene glycol; dens. 1.07-1.18; ref. index 1.47
Precaution: Combustible
Toxicology: No known toxicity; irritant; heated to decomp., emits acrid smoke and irritating fumes
Uses: Hot-melt adhesives and coatings for cables, etc.; extrusion wire insulation, protective coatings, pigment-grinding bases; food additive; protective coating, tablet binder in pharmaceutical vitamin/mineral preps., color diluent, flavor fixative
Regulatory: FDA 21CFR §73.1, 172.868, 175.300, 182.90, 573.420, GRAS; FDA approved for orals, topicals; USP/NF, BP, Ph.Eur. compliance
Manuf./Distrib.: Aldrich; Aqualon; Ashland; Chemcentral; Colorcon; Fluka; FMC; Hercules/Aqualon; Hoechst Celanese; Punda Mercantile; Sigma; Van Waters & Rogers
Trade names: Hercules® Ethylcellulose Series; Hercules® K; Hercules® N; Hercules® T

Ethyl citrate. *See* Triethyl citrate
Ethyl 2-cyano-3,3-diphenylacrylate. *See* Etocrylene
Ethyl 2-cyano-3,3-diphenyl-2-propenoate. *See* Etocrylene
7A-Ethyldihydro-1H,3H-5H-oxazolo (3,4-C) oxazole. *See* 7-Ethyl bicyclooxazolidine

Ethyl 4-(dimethylamino) benzoate
CAS 10287-53-3; EINECS 233-634-3
Synonyms: Ethyl-p-(dimethylamino) benzoate
Classification: Amine
Empirical: $C_{11}H_{15}NO_2$
Properties: Sol. in acrylates; m.w. 193.25; m.p. 61-64 C
Uses: UV photoinitiator, synergist for uv-cured coatings, inks
Manuf./Distrib.: Aldrich; First Chem.; Fluka; Hampford Research; Nobel Chems.
Trade names: Esacure® EDB; Firstcure™ EDAB

Ethyl-p-(dimethylamino) benzoate. *See* Ethyl 4-(dimethylamino) benzoate

Ethylene/acrylic acid copolymer
CAS 9010-77-9
Definition: Copolymer of ethylene and acrylic acid monomers
Properties: Dens. 0.960
Uses: Binder for nonwoven fibers; lubricant and processing aid for plastics; pigment dispersant
Regulatory: FDA 21CFR §177.1310, 178.1005
Manuf./Distrib.: Aldrich
Trade names: A-C® 540; A-C® 540A; A-C® 580; A-C® 5120; A-C® 5180; Esi-Cryl 2540N; Luwax™ EAS 1; Luwax™ 9656; Michem® Emulsion 00925; Michem® Emulsion 02125; Michem® Emulsion 02925; Michem® Prime 4983; Michem® Prime 4983-45; Michem® Prime 4983HS; Michem® Prime 4990

Ethylene/acrylic acid/vinyl acetate copolymer
Uses: Used in adhesives, floor finishes, general coatings
Trade names: Esi-Cryl 1450N

Ethylene alcohol. *See* Glycol

Ethylenebis dibromonorbornane dicarboximide
CAS 52907-07-0; EINECS 258-250-3
Synonyms: Ethylenebis (5,6-dibromonorbornane-2,3-dicarboximide)
Uses: Flame retardant for PP, polyesters, polyamides, polyurethane elastomers, castables, coatings
Manuf./Distrib.: Albemarle
Trade names: Saytex® BN-451

Ethylenebis (5,6-dibromonorbornane-2,3-dicarboximide). *See* Ethylenebis dibromonorbornane dicarboximide
Ethylenebis (dithiocarbamate) disodium salt. *See* Nabam

N,N′-Ethylenebis 12-hydroxystearamide
Uses: Internal lubricant, mold release, slip additive for PVC; antiblocking agent for textile coatings; slip agent for varnishes, lacquers; also in elec. potting compds., plug valve lubricants, caryons, wax blends, high-temp. greases
Trade names: Paricin® 285

N,N′-Ethylene bis-ricinoleamide
Uses: Lubricant/antistat for plastics, metals; mold release, antiblocking agent for textile coatings; slip agent for varnishes and lacquers; also for elec. potting compds., crayons, wax blends, high-temp. greases

N,N′-Ethylene bisstearamide. *See* Ethylene distearamide

Ethylene butyl acrylate copolymer
Definition: Copolymer of ethylene and butyl acrylate monomers
Uses: For extrusion coating, adhesives; for coatings and laminations with oriented PP, PET, papers, clay-coated board and nylon substrates for specialty pkg.
Trade names: Enathene® EA 705-009; Enathene® EA 719-009; Enathene® EA 720-009

Ethylene/calcium acrylate copolymer
CAS 26445-96-5
Uses: Processing and performance additive; improves dispersion of additives in plastics; adhesion to variety of substrates; encapsulant for personal care prods.; low m.w. ionomer flatting agent that reduces gloss in polyester powder coatings
Regulatory: FDA 21CFR §175.105
Trade names: AClyn® 201; AClyn® 201A

Ethylene carboxylic acid. *See* Acrylic acid
Ethylene chloride. *See* Ethylene dichloride

Ethylene-chlorotrifluoroethylene copolymer
Synonyms: ECTFE
Uses: Melt processable fluoropolymer for extrusion (wire coating, tubing, film), inj. molding, blow molding, compr. molding, rotomolding, electrostatic coating
Trade names: Halar® 300; Halar® 500; Halar® 558; Halar® 6013; Halar® 6014; Halar® 6613; Halar® 6614

Ethylene diacetate. *See* Ethylene glycol diacetate

Ethylenediamine
CAS 107-15-3 (anhyd.); 6780-13-8 (monohydrate); EINECS 203-468-6
Synonyms: EDA; 1,2-Diaminoethane; 1,2-Ethanediamine; Dimethylenediamine
Classification: Aliphatic organic compd.; alkaline compd.
Empirical: $C_2H_8N_2$
Formula: $NH_2CH_2CH_2NH_2$
Properties: Colorless to sl. yel. clear volatile liq., ammonia-like odor; hygroscopic; misc. with water, alcohol; m.w. 60.12; dens. 0.8994 (20/4 C); m.p. 8.5 C; b.p. 117.2 C; flash pt. (CC) 110 F; ref. index 1.4565
Precaution: DOT: Corrosive material; flamm. exposed to heat, flame, oxidizers; can react violently with acetic acid, acetic anhydride, acrylic acid, epichlorohydrin, many others
Toxicology: TLV:TWA 10 ppm; LD50 (oral, rat) 500 mg/kg; human irritant poison by inh.; mod. toxic by ingestion, skin contact; corrosive; severe skin and eye irritant; allergen, sensitizer; mutagenic data; heated to decomp., emits toxic fumes of NO_x and NH_3
Uses: Fungicide, mfg. of chelating agents, dimethylolethylene-urea resins, chemical intermediate, solvent, emulsifier, textile lubricants, antifreeze inhibitor, agric. chems., corrosion inhibitors, surfactants, ion-exchange resins, rubber chems.; thermoplastics lubricants; stabilizer for dyes; solv. for shellac
Regulatory: FDA 21CFR §173.315, 173.320 (1 ppm max.), 178.3120, 556.270, 181.30; FDA approved for IV, injectables, orals, rectals, topicals; USP, BP, Ph.Eur., JP compliance
Manuf./Distrib.: Aldrich; Allchem Ind.; BASF; BASF AG; Coyne; Diamine & Chems. Ltd; Fluka; Nova Molecular Tech.; Sigma; Spectrum Chem. Mfg.; Texaco; Tosoh; Union Carbide
Trade names containing: Cardolite® NC-541

trans-1,2-Ethylenedicarboxylic acid. *See* Fumaric acid

Ethylene dichloride
CAS 107-06-2; EINECS 203-458-1

Synonyms: 1,2-DCE; Dichloroethylene; Dutch oil; 1,2-Dichloroethane; Ethylene chloride
Classification: Halogenated aliphatic hydrocarbon
Empirical: $C_2H_4Cl_2$
Formula: $Cl \cdot CH_2CH_2 \cdot Cl$
Properties: Colorless oily liq., chloroform-like odor, sweet taste; misc. with most common solvs.; sl. sol. in water; m.w. 98.96; dens. 1.2554 (20/4 C); b.p. 83.5 C; f.p. -35.5 C; flash pt. 56 C; ref. index 1.445
Precaution: Fire hazard exposed to heat, flame, or oxidizers; violent reaction with Al, NH_3
Toxicology: TLV:TWA 10 ppm; LD50 (oral, rat) 670 mg/kg; human poison by ingestion; mod. toxic by inh., intraperitoneal route; experimental carcinogen; strong narcotic; skin/severe eye irritant; heated to dec., emits highly toxic fumes of Cl^-, phosgene
Uses: Production of vinyl chloride, trichloroethylene, vinylidene chloride, trichloroethane; lead scavenger in gasoline; paint, varnish remover; metal degreasing; soaps; wetting/penetrating agents; organic synthesis; ore flotation; solvent; fumigant; extract solvent for foods, flume wash water additive, pesticide; intermdiate for paints
Regulatory: FDA 21CFR §172.560, 172.710, 173.230, 173.315, 175.105, 573.440
Manuf./Distrib.: Albemarle; Albright & Wilson Am.; Aldrich; Ashland; BASF; BP Chems. Ltd; Fluka; Georgia Gulf; Norsk Hydro AS; Occidental; PPG Ind.; Sigma; Vulcan

Ethylene dimethacrylate. *See* Ethylene glycol dimethacrylate
Ethylenedinitrilotetra-2-propanol. *See* Tetrahydroxypropyl ethylenediamine

Ethylene dioleamide
CAS 110-31-6; EINECS 203-756-1
Synonyms: N,N′-1,2-Ethanediylbis-9-octadecenamide; 9-Octadecenamide, N,N′-1,2-ethanediylbis-; N,N′-Dioleoylethylenediamine
Classification: Diamide
Empirical: $C_{38}H_{72}N_2O_2$
Uses: Syn. wax used as plastics processing lubricant and release agent, antistat, antiblock, slip agent, m.p. modifier for waxes, industrial asphalts and tar, pigment dispersing agent for resin systems; polyamide-paraffin coupling agent; used in adhesive tapes, coatings, food pkg. materials
Manuf./Distrib.: Aldrich
Trade names: Advawax® 240

Ethylene distearamide
CAS 110-30-5; 68955-45-3; EINECS 203-755-6, 273-277-0
Synonyms: N,N′-Ethylene bisstearamide; N,N′-1,2-Ethanediylbisoctadecanamide; N,N-Bis-stearoylethylenediamide
Classification: Diamide
Empirical: $C_{39}H_{76}O_2$
Formula: $CCH_3(CH_2)_{16}CONH(CH_2)_2NHCO(CH_2)_{16}CH_3$
Uses: Lubricant, processing aid for PVC, PS, ABS, nylon; peptizing and dispersing agent for tech. NR articles; paint raw material
Manuf./Distrib.: Aldrich; Aquatec Quimica SA; Chemax; Lonza
Trade names: Acrawax® C; Advawax® 280; Advawax® 290; Hoechst Wax C; Ross Wax #140

Ethylene glycol. *See* Glycol
Ethylene glycol butyl ether. *See* Butoxyethanol
Ethylene glycol butyl ether acetate. *See* Butoxyethanol acetate

Ethylene glycol diacetate
CAS 111-55-7; EINECS 203-881-1
Synonyms: Glycol diacetate; Ethylene diacetate; 1,2-Diacetoxyethane
Classification: Ester
Empirical: $C_6H_{10}O_4$
Formula: $CH_3COOCH_2CH_2OOCCH_3$
Properties: Colorless liq., faint odor; sol. in alcohol, ether, benzene; sl. sol. in water (10%); m.w. 146.14; dens. 1.1063 (20/20 C); b.p. 190.5 C; f.p. -31 C; flash pt. 96 C; ref. index 1.416
Precaution: Combustible
Toxicology: LD50 (oral, rat) 6850 mg/kg, (skin, rabbit) 8480 mg/kg; moderately toxic by intraperitoneal route; mildly toxic by ingestion and skin contact; eye irritant; heated to decomp., emits acrid smoke and fumes
Uses: Extraction solv., foundry resins, perfume fixative; solv. for coatings, lacquers; plasticizer for cellulosics, PVAc; printing inks
Manuf./Distrib.: Aceto; Aldrich; Ashland; BP Chems. Inc; Chemoxy Int'l. plc; CPS; Eastman; Fluka; Sigma

Ethylene glycol diethyl ether
CAS 629-14-1; EINECS 211-076-1
Synonyms: 1,2-Diethoxyethane; Ethyl glyme
Empirical: $C_6H_{14}O_2$
Formula: $C_2H_5OCH_2CH_2OC_2H_5$
Properties: Colorless liq., sl. odor; immiscible in water; m.w. 118.20; dens. 0.8417 (20/20 C); f.p. -74 C; b.p. 121.4 C; flash pt. 35 C; ref. index 1.393 (20 C)
Precaution: Flamm.; mod. fire risk; can react with oxidizing materials
Toxicology: LD50 (oral, rat) 4390 mg/kg; moderately toxic by ingestion; mildly toxic by inhalation; eye irritant
Uses: Organic synthesis (reaction medium); solvent and diluent for detergents; solv. for paints
Manuf./Distrib.: Aldrich; Brand-Nu Labs; Eastern Chem.; Ferro/Grant; Fluka; Sigma

Ethylene glycol dilaurate. *See* Glycol dilaurate

Ethylene glycol dimethacrylate
CAS 97-90-5; EINECS 202-617-2
Synonyms: 1,2-Ethanediol dimethacrylate; 1,2-Bis(methacryloyoxy) ethane; Ethylene dimethacrylate
Empirical: $C_{10}H_{14}O_4$
Formula: $CH_2:C(CH_3)COOCH_2CH_2OCOC(CH_3):CH_2$
Properties: M.w. 198.1; dens. 1.053 (20/4 C); b.p. 66-68 C; flash pt. 73 C; ref. index 1.545 (20 C)
Precaution: Photosensitive
Toxicology: Moderately toxic by ingestion, intraperitoneal routes; irritating to eyes, respiratory system
Uses: Crosslinker and modifier for ABS, acrylic sheet and rods, PVC, ion exchange resins, glaze coatings, dental polymers, paper processing aids, rubber modifier, adhesives, optical polymers, leather finishing, moisture barrier films
Manuf./Distrib.: Akzo Nobel; Aldrich; CPS; Fluka; Hampford Research; Monomer-Polymer & Dajac Labs; Rohm Tech; Sartomer; Sigma
Trade names containing: Ageflex EGDMA; SR-206

Ethylene glycol dimethyl ether
CAS 110-71-4; EINECS 203-794-9
Synonyms: DME; 1,2-Dimethoxyethane; Dimethylglycol; Monoglyme; Glycol dimethyl ether
Empirical: $C_4H_{10}O_2$
Formula: $CH_3OCH_2CH_2OCH_3$
Properties: Water-wh. liq., sharp ethereal odor; misc. with water, alcohol; sol. in hydrocarbons; m.w. 90.12; dens. 0.86285 (20/4 C); m.p. -58 C; b.p. 82-83 (760 mm); flash pt. 4.5 C; ref. index 1.3813 (20 C)
Precaution: DOT: Flamm. liq.; mod. fire hazard; may form explosive peroxides
Toxicology: Experimental reproductive effects
Uses: Solvent for paints
Regulatory: BP compliance
Manuf./Distrib.: Aldrich; Brand-Nu Labs; Ferro/Grant; Fluka; Sigma; Spectrum Chem. Mfg.

Ethylene glycol distearate. *See* Glycol distearate
Ethylene glycol ethyl ether. *See* Ethoxyethanol
Ethylene glycol ethyl ether acetate. *See* Ethoxyethanol acetate

Ethylene glycol 2-ethylhexyl ether
CAS 1559-35-9
Synonyms: Ethylene glycol mono-2-ethylhexyl ether
Trade names containing: Eastman® EEH

Ethylene glycol hexyl ether
CAS 112-25-4; EINECS 203-951-1
Synonyms: Ethylene glycol monohexyl ether; Hexylglycol; Ethylene glycol n-hexyl ether; 2-(Hexyloxy)ethanol; n-Hexyl 'Cellosolve'
Classification: Glycol ether
Empirical: $C_8H_{18}O_2$
Formula: $C_6H_{13}OC_2H_4OH$
Properties: Water-wh. liq.; m.w. 148.20; sp.gr. 0.887 (20/20 C); f.p. -58 F; b.p. 208.1 C; flash pt. 90.5 C; ref. index 1.4290 (20 C)
Precaution: Combustible
Uses: High-boiling solv. for coatings

Manuf./Distrib.: Ashland; Eastman; Fluka; Union Carbide
Trade names: Hexyl Cellosolve®

Ethylene glycol n-hexyl ether. *See* Ethylene glycol hexyl ether
Ethylene glycol methacrylate. *See* 2-Hydroxyethyl methacrylate
Ethylene glycol methyl ether. *See* Methoxyethanol
Ethylene glycol methyl ether acetate. *See* Methoxyethanol acetate
Ethylene glycol monobutyl ether. *See* Butoxyethanol
Ethylene glycol monobutyl ether acetate. *See* Butoxyethanol acetate
Ethylene glycol monoethyl ether. *See* Ethoxyethanol
Ethylene glycol monoethyl ether acetate. *See* Ethoxyethanol acetate
Ethylene glycol mono-2-ethylhexyl ether. *See* Ethylene glycol 2-ethylhexyl ether
Ethylene glycol monohexyl ether. *See* Ethylene glycol hexyl ether
Ethylene glycol monohydroxystearate. *See* Glycol hydroxystearate
Ethylene glycol monomethacrylate. *See* 2-Hydroxyethyl methacrylate
Ethylene glycol monomethyl ether. *See* Methoxyethanol
Ethylene glycol monomethyl ether acetate. *See* Methoxyethanol acetate
Ethylene glycol monophenyl ether. *See* Phenoxyethanol
Ethylene glycol monopropyl ether. *See* Ethylene glycol propyl ether
Ethylene glycol monostearate. *See* Glycol stearate
Ethylene glycol nonyl phenyl ether. *See* Nonoxynol-1
Ethylene glycol octyl phenyl ether. *See* Octoxynol-1
Ethylene glycol phenyl ether. *See* Phenoxyethanol

Ethylene glycol propyl ether
CAS 2807-30-9; EINECS 220-548-6
Synonyms: Propyl 'Cellosolve'; 2-Propoxyethanol; Ethanol, 2-propoxy; Ethylene glycol monopropyl ether
Empirical: $C_5H_{12}O_2$
Formula: $CH_2OHCH_2OCH_2CH_2CH_3$
Properties: Liq.; sol. in water; m.w. 104.17; dens. 0.913 (20/20 C); b.p. 149.5 C (760 mm); f.p. -90 C; flash pt. (TCC) 49 C
Precaution: Combustible
Toxicology: Moderately toxic by ingestion and skin contact; mild toxicity by inhalation; skin/eye irritant; heated to dec., emits acrid smoke, irritating fumes
Uses: Solvent; Solvent for coatings, cosmetics, resins; coupling solv. for resin/water systems
Manuf./Distrib.: Aldrich; Ashland; Eastman
Trade names: Eastman® EP; Propyl Cellosolve®

Ethylene glycol stearate. *See* Glycol stearate
Ethylene homopolymer. *See* Polyethylene
Ethylene/MA copolymer (INCI). *See* Ethylene-maleic anhydride copolymer

Ethylene/magnesium acrylate copolymer
Synonyms: 2-Propenoic acid, polymer with ethene, magnesium salt
Definition: Copolymer of ethylene and magnesium acrylate monomers
Formula: $(C_3H_4O_2 \cdot C_2H_4)_x \cdot xMg$
Uses: Processing and performance additive; improves dispersion of additives in plastics; adhesion to variety of substrates; encapsulant for personal care prods.; low m.w. ionomer flatting agent that reduces gloss in polyester powder coatings
Trade names: AClyn® 246

Ethylene-maleic anhydride copolymer
CAS 9006-26-2
Synonyms: Ethylene/MA copolymer (INCI); 2,5-Furandione, polymer with ethene
Definition: Polymer of ethylene and maleic anhydride monomers
Uses: Bonding/compatibilizing agent for substrates, polymers, and polymer blends for paper coatings, wood prods., adhesives/sealants, plastics (olefin blends and filled olefins)
Regulatory: FDA 21CFR §175.105
Manuf./Distrib.: Aldrich; Chemical; Monomer-Polymer & Dajac Labs; Sigma
Trade names: Polyace™ 573

Ethylene methacrylic acid copolymer
Trade names: Nucrel® Resins

Ethylene-methyl acrylate. *See* Ethylene-methyl acrylate copolymer

Ethylene-methyl acrylate copolymer
Synonyms: Ethylene-methyl acrylate
Trade names: EMAC® SP2205; EMAC® SP2207; EMAC® SP2255; EMAC® SP2257; EMAC® SP2260; EMAC® SP2261T; EMAC® SP2268T; EMAC+™ SP2305T; Lotryl® 20 MA 08; PE 2207; PE 2260; SP 2207; SP 2260
Trade names containing: Ampacet 11187

Ethylene monochloride. *See* Vinyl chloride
Ethylene polymers. *See* Polyethylene
Ethylene/propylene copolymer (INCI). *See* EPM rubber
Ethylene-propylene-diene terpolymer. *See* EPDM rubber
Ethylene propylene terpolymer. *See* EPDM rubber

Ethylene/sodium acrylate copolymer
CAS 25750-82-7
Synonyms: 2-Propenoic acid, polymer with ethene, sodium salt
Definition: Copolymer of ethylene and sodium acrylate monomers
Formula: $(C_3H_4O_2 \cdot C_2H_4)_x \cdot xNa$
Regulatory: FDA 21CFR §175.105
Manuf./Distrib.: Aldrich
Trade names: AClyn® 285

Ethylenesuccinic acid. *See* Succinic acid
Ethylene tetrachloride. *See* Perchloroethylene

Ethylene/tetrafluoroethylene copolymer
Synonyms: ETFE
Uses: Melt processable high performance thermoplastic for melt extrusion, coating of fine wire and inj. molding of slender, thin-walled, or intricate shapes; in elec. applics., mechanical applics., and in chemical service
Trade names: Tefzel® 210

Ethylene thioglycol. *See* 2-Mercaptoethanol

Ethylene VA/acid terpolymer
Trade names: Elvax® 4260

Ethylene/VA copolymer
CAS 24937-78-8
Synonyms: EVA; EVA copolymer; Ethylene vinyl acetate; Ethylene/vinyl acetate copolymer; Acetic acid, ethenyl ester, polymer with ethene
Definition: Polymer of ethylene and vinyl acetate monomers
Formula: $(C_4H_6O_2 \cdot C_2H_4)_x$
Properties: Dens. 0.930
Uses: Syn. rubber for tech. moldings and extrudates, lamp seals, cable sheathings and insulations, cellular rubber goods, footwear soles, waterproof sheeting, hot-melt and pressure-sensitive hot-melt adhesives, coatings, inks, lacquers, sealants; impact modifier for PVC
Regulatory: FDA 21CFR §175.300, 177.1200, 177.1210, 177.1350, 178.1005; FDA approved for ophthalmics, otics
Manuf./Distrib.: Aldrich; Allchem Ind.; AlliedSignal; Bayer; Chemcentral; Focus; Monomer-Polymer & Dajac Labs; Nat'l. Casein; Tamms Ind.; Union Carbide
Trade names: A-C® 400; A-C® 400A; A-C® 405M; A-C® 405S; A-C® 405T; A-C® 430; AT 1806M; AT 4030M; Elvax® 40-W; Elvax® 150; Elvax® 150-W; Elvax® 240; Elvax® 250; Elvax® 260; Elvax® 265; Elvax® 310; Elvax® 350; Elvax® 360; Elvax® 410; Elvax® 420; Elvax® 450; Elvax® 460; Elvax® 470; Fusabond® MC-189D; Fusabond® MC-190D; Fusabond® MC-197D; Luwax™ EVA 1; Luwax™ EVA 2; Ultrathene® UE 612-04; Ultrathene® UE 613-00; Ultrathene® UE 631-04; Ultrathene® UE 634-04; Ultrathene® UE 635-004; Ultrathene® UE 636-04; Ultrathene® UE 639-35; Ultrathene® UE 639-67; Ultrathene® UE 645-04; Ultrathene® UE 646-04; Ultrathene® UE 649-04; Ultrathene® UE 653-67; Ultrathene® UE 654-67

Ethylene vinyl acetate. *See* Ethylene/VA copolymer
Ethylene/vinyl acetate copolymer. *See* Ethylene/VA copolymer

Ethylene vinyl alcohol copolymer
Synonyms: EVOH copolymer
Uses: Thermoplastic barrier resin used in pkg. materials, cast and blown film, gas fill pkg., pkg. for oily foods, edible oils, min. oils, agric. pesticides, org. solvs.
Trade names: Eval® E105; Eval® F104

Ethylene/vinyl chloride copolymer
Uses: Crosslinkable flexible binder and saturant for paper and paperboard applics.; nonwoven binder for fiberfill and high loft stock requiring flame retardancy; imparts flexibility and water resist. to caulks, mastics, barrier coats in building applics.; high build flexible coatings, low MVTR coatings
Trade names: Airflex® 4500; Airflex® 4514; Airflex® 4530

Ethylene/zinc acrylate copolymer
Synonyms: 2-Propenoic acid, polymer with ethene, zinc salt
Definition: Copolymer of ethylene and zinc acrylate monomers
Formula: $(C_3H_4O_2 \cdot C_2H_4)_x \cdot xZn$
Uses: Low m.w. ionomer; encapsulant for personal care prods.; flatting agent for powd. coatings
Regulatory: FDA 21CFR §175.105
Trade names: AClyn® 291A; AClyn® 293A; AClyn® 295; AClyn® 295A

Ethyl 3-epoxypropionate
Trade names: Eastman® EEP

Ethyl ester of PVM/MA copolymer (INCI). *See* PVM/MA copolymer, ethyl ester
N-Ethylethanamine. *See* Diethylamine
Ethyl ethanoate. *See* Ethyl acetate

Ethyl ether
CAS 60-29-7; EINECS 200-467-2
Synonyms: Diethyl ether; Ether; Ethyl oxide; Sulfuric ether
Empirical: $C_4H_{10}O$
Formula: $CH_3CH_2OCH_2CH_3$
Properties: Clear volatile liq., sweet pungent odor; hygroscopic; sol. in water, chloroform, benzene; misc. with alcohol, ether; m.w. 74.12; dens. 0.7135 (20/4 C); m.p. -116.2 C; b.p. 34.6 C; flash pt. -49 F
Precaution: DOT: Flamm. liq.; severe fire and explosion hazard when exposed to heat or flame; forms explosive peroxides
Toxicology: LD50 (oral, rat) 1215 mg/kg; TLV:TWA 400 ppm (air); moderately toxic by ingestion, intraperitoneal, intravenous routes; poison by subcutaneous route; vapor inh. produces CNS effects, possible respiratory failure; irritating and defatting to skin
Uses: Organic synthesis; smokeless powders; industrial solv. for paints; analytical chemistry; anesthetic; extractant; solv. for waxes, fats, and oils
Regulatory: FDA approved for transmucosal pharmaceuticals
Manuf./Distrib.: Aldrich; J.T. Baker; Exxon; Fluka; Hüls AG; Mallinckrodt; Quantum; Sigma; Spectrum Chem. Mfg.; Van Waters & Rogers

2-Ethyl, 2´-ethoxy-oxalanilide
CAS 23946-66-8
Synonyms: Ethanediamine, N-(2-ethoxyphenyl)-N´-(2-ethylphenyl)-; N-(2-Ethoxyphenyl)-N´-(2-ethylphenyl) ethanediamide; Ethanediamide, n-(2-ethoxyphenyl)-n´-(4-ethylphenyl)
Classification: Oxalic anilide
Empirical: $C_{18}H_{20}N_2O_3$
Formula: $OC_2H_5NHCOCOHNC_2H_5$
Properties: Wh. powd., mild odor; insol. in water; sp.gr. 1.209; m.p. 126 C; flash pt. > 200 F
Precaution: Dusts may be explosive with spark or flame initiation; incompat. with strong oxidizing agents; thermal decomp. may produce CO_x and NO_x
Toxicology: LD50 (oral, rat) > 5000 mg/kg; virtually nontoxic; nonirritating to skin and eyes
Uses: UV absorber for acrylates, alkyd/melamines, PU, thermosetting polyester, coatings
Trade names: Sanduvor® VSU

Ethyl 3-ethoxypropionate
CAS 763-69-9
Synonyms: EEP; Ethyl-β-ethoxypropionate; Ethoxypropionic acid ethyl ester; 3-Ethoxypropionic acid ethyl ester

Ethyl-β-ethoxypropionate

Empirical: $C_7H_{14}O_3$
Formula: $C_2H_5OC_3H_5O_2C_2H_5$
Properties: Liq.; sl. sol. in water; m.w. 146.19; dens. 0.9496 (20/20 C); b.p. 165-172 C; f.p. < -50 C; flash pt. (Seta) 58 C; ref. index 1.4074 (20 C)
Precaution: Flamm. when exposed to heat or flame; can react with oxidizing materials
Toxicology: LD50 (oral, rat) 5000 mg/kg, (skin, rabbit) 10 g/kg; mildly toxic by ingestion and skin contact; skin and eye irritant; heated to decomp., emits acrid smoke and irritating fumes
Uses: Solvent for polymerization, esp. high-solids acrylics; retarder solv. for coatings; let-down solv. for thinning synthetic coating resins, esp. moisture-sensitive resins for PU coatings; air-dry and baking finishes; intermediate for vitamin B_1
Manuf./Distrib.: Aldrich; Ashland; Eastman; Union Carbide
Trade names: Ektapro® EEP Solvent; Ucar® Ester EEP

Ethyl-β-ethoxypropionate. *See* Ethyl 3-ethoxypropionate
Ethyl glycol. *See* Ethoxyethanol
Ethyl glyme. *See* Ethylene glycol diethyl ether
4-Ethyl-2-(8-heptadecenyl)-4,5-dihydro-4-oxazolemethanol. *See* Ethyl hydroxymethyl oleyl oxazoline
2-Ethylhexanoic acid vinyl ester. *See* Vinyl 2-ethylhexanoate

2-Ethylhexanol
CAS 104-76-7; EINECS 203-234-3
Synonyms: 2-EH; 2-Ethylhexyl alcohol; 2-Ethyl-1-hexanol; Octyl alcohol; Alcohol C_8
Empirical: $C_8H_{18}O$
Formula: $CH_3(CH_2)_3CHC_2H_5CH_2OH$
Properties: Colorless liq., mild oily sweet sl. rose fragrance; misc. with most org. solvs.; sl. sol. in water; m.w. 130.26; dens. 0.83 (20 C); b.p. 183.5 C; f.p. -76 C; flash pt. 81.1 C; ref. index 1.4300 (20 C)
Precaution: Dangerous fire hazard when exposed to heat or flame; can react vigorously with oxidizing materials
Toxicology: LD50 (rat, oral) 12.46 ml/kg; moderately toxic by ingestion, skin contact, IP, subcut., parenteral routes; severe eye irritant; moderate skin irritant; heated to decomp., emits acrid smoke and fumes
Uses: Plasticizer for PVC resins; defoaming agent, wetting agent, organic synthesis, solvent mix for nitrocellulose; solv. for paints, waterborne coatings; penetrant for plasticizing inks, etc.
Regulatory: FEMA GRAS
Manuf./Distrib.: Aldrich; Aristech; Ashland; BASF; BP Chems. Ltd; Coyne; Eastman; Fluka; Hoechst AG; Penta Mfg.; Shell
Trade names containing: Surfynol® 104A

2-Ethyl-1-hexanol. *See* 2-Ethylhexanol; Isooctyl alcohol
2-Ethyl-1-hexanol ester with diphenyl phosphate. *See* Diphenyl octyl phosphate
2-Ethylhexanyl acetate. *See* 2-Ethylhexyl acetate

2-Ethylhexoic acid
CAS 149-57-5
Synonyms: Butylethylacetic acid
Formula: $C_4H_9CH(C_2H_5)COOH$
Properties: Liq., mild odor; sl. sol. in water; dens. 0.9077 (20/20 C); f.p. -83 C; b.p. 226.9 C; flash pt. 126 C
Precaution: Combustible
Uses: Paint and varnish driers (metallic salts); esters as plasticizers
Manuf./Distrib.: Aldrich; Ashland; BASF; Coyne; Eastman; Fluka; Neste UK; Sigma; Union Carbide

2-Ethylhexoic acid vinyl ester. *See* Vinyl 2-ethylhexanoate

2-Ethylhexyl acetate
CAS 103-09-0
Synonyms: Acetic acid 2-ethylhexyl ester; 2-Ethylhexanyl acetate; 2-Ethylhexyl ethanoate
Empirical: $C_{10}H_{20}O_2$
Formula: $CH_3COOCH_2CHC_2H_5C_4H_9$
Properties: Water-wh. stable liq.; very sl. sol. in water; m.w. 172.27; sp.gr. 0.873 (20/20 C); f.p. -135 F; b.p. 199-205 C; flash pt.(TCC) 160 F; ref. index 1.4201 (20 C)
Precaution: Combustible
Toxicology: LD50 (oral, rat) 3000 mg/kg; mod. toxic by ingestion; heated to decomp., emits acrid smoke and fumes
Uses: Solvent for nitrocellulose, resins, lacquers, baking finishes

Manuf./Distrib.: Aceto; Allan; Ashland; Eastman; Hüls AG; MTM Spec. Chems.; Penta Mfg.; Union Carbide; Van Waters & Rogers

2-Ethylhexyl acrylate. *See* Octyl acrylate
2-Ethylhexyl alcohol. *See* 2-Ethylhexanol
2-Ethylhexyl 2-cyano-3,3-diphenylacrylate. *See* Octocrylene
2-Ethylhexyl 2-cyano-3,3-diphenyl-2-propenoate. *See* Octocrylene
2-Ethylhexyl-4(dimethylamino) benzoate. *See* Octyl dimethyl PABA
2-Ethylhexyl p-dimethylaminobenzoate. *See* Octyl dimethyl PABA
2-Ethylhexyl diphenyl ester phosphoric acid. *See* Diphenyl octyl phosphate
2-Ethylhexyl diphenyl phosphate. *See* Diphenyl octyl phosphate

Ethylhexyl diphenyl phosphite
CAS 15647-08-2
Formula: $C_4H_9CHC_2H_5CH_2OP(OC_6H_5)_2$
Properties: Liq.; m.w. 346; dens. 1.040-1.047; b.p. 185 C (5 mm); ref. index 1.5200-1.5250; flash pt. (PMCC) 182 C
Uses: Color and processing stabilizer for ABS, PC, polyurethane, coatings, PET fiber; sec. stabilizer for PVC

2-Ethylhexyl ethanoate. *See* 2-Ethylhexyl acetate
2-Ethylhexyl fumarate. *See* Dioctyl fumarate

2-Ethylhexyl glycidyl ether
CAS 2461-15-6; EINECS 219-553-6
Synonyms: EHGE; Oxirane, [[2-ethylhexyl)oxy]methyl]-
Empirical: $C_{11}H_{22}O_2$
Properties: Colorless to ylsh. liq.; sol. in most org. solvs.; m.w. 186.30; dens. 0.891; b.p. 60-61.5 C (0.3 mm); flash pt. 96 C
Precaution: Vapors can form explosive mixts. with air; incompat. with acids, alkalies, oxidizing agents, metals; protect from exposure to air/oxygen; hazardous decomp. prods.: CO, gaseous hydrocarbons
Toxicology: LD50 (oral, rat) > 5000 mg/kg, (dermal, rat) > 3500 mg/kg; irritating to eyes, respiratory system, skin; harmful by ingestion
Uses: Reactive epoxy diluent for exposed aggregates, potting, flooring, casting, tooling, laminates, solv.-free coating systems, fiber-reinforced composites; stabilizer for chlorinated hydrocarbons
Manuf./Distrib.: Aldrich; Raschig

2-Ethylhexyl hydrogenated tallowalkyl methosulfate
Uses: Surfactant, dispersant for protective coatings, pigments, inks, textiles, agric., acid pickling baths, marine applics, metalworking, electroplating, fuel treatment, emulsion/plastic mfg., waste water treatment, min. processing, paper
Regulatory: EPA listed
Trade names: Arquad® HTL8(W) MS-85

2-Ethylhexyl methacrylate
CAS 688-84-6; EINECS 211-708-6
Synonyms: 2-Ethyl-1-hexyl methacrylate
Empirical: $C_{12}H_{22}O_2$
Properties: M.w. 198.34
Toxicology: LD50 (IP, mouse) 2614 mg/kg; mod. toxic by intraperitoneal route; heated to decomp., emits acrid smoke and irritating fumes
Manuf./Distrib.: Allchem Ind.; Ashland; BASF; Hoechst Celanese; Monomer-Polymer & Dajac Labs; Rohm Tech; Union Carbide
Trade names containing: Empicryl® 6068B

2-Ethyl-1-hexyl methacrylate. *See* 2-Ethylhexyl methacrylate

2-Ethylhexyl phosphate
CAS 12645-31-7
Uses: Detergent in dry-cleaning, emulsifier in formulation of metal cleaners, paints in emulsion polymerization; pesticide and cosmetic preparations
Trade names: Rhodafac® PEH

2-Ethylhexyl-2-propenoate. *See* Octyl acrylate

2-Ethylhexyl sulfosuccinate sodium. *See* Dioctyl sodium sulfosuccinate

Ethyl hydroxyethyl cellulose
CAS 9004-58-4
Synonyms: EHEC
Properties: Wh. gran. solid; sol. in water; sol. in mixt.of aliphatic hydrocarbons contg. alcohol
Precaution: Combustible
Uses: Stabilizer, thickener, binder, film-former in silk screen and gravure printing inks, protective coatings, aq., aq.-organic, and organic solv. systems

N-Ethyl-N-hydroxyethyl-m-toluidine
CAS 91-88-3
Synonyms: 2-(N-Ethyl-m-toluidino) ethanol
Formula: $CH_3C_6H_4N(C_2H_5)CH_2CH_2OH$
Properties: M.w. 179.26; dens. 1.019; b.p. 114-115 C (1 mm); flash pt. > 110 C; ref. index 1.5550 (20 C)
Uses: Coupling agent for disperse dyes for syn. fibers; photosensitive chemical for paper coatings; accelerator for polyester resins
Manuf./Distrib.: Aldrich; First

Ethyl hydroxymethyl oleyl oxazoline
CAS 68140-98-7; 88543-32-2; EINECS 268-820-3
Synonyms: 4-Ethyl-2-(8-heptadecenyl)-4,5-dihydro-4-oxazolemethanol
Classification: Substituted heterocyclic compd.
Empirical: $C_{23}H_{43}NO_2$
Uses: Detergent, emulsifier, wetting agent, antifoamer, antioxidant; used in salt, soap, paper, textiles, and metal cleaners; emulsion stabilizer; acid acceptor; pigment grinding and dispersion
Trade names: Alkaterge®-E

2-Ethyl-2-(hydroxymethyl)-1,2,3-propanediol. *See* 1,1,1-Trimethylolpropane triacrylate
2-Ethyl-2-hydroxymethyl-1,3-propanediol. *See* Trimethylolpropane
Ethyl 2-hydroxypropionate. *See* Ethyl lactate
Ethyl α-hydroxypropionate. *See* Ethyl lactate
Ethyl-S(-)-2-hydroxypropionate. *See* Ethyl lactate
Ethylidenelactic acid. *See* Lactic acid

Ethyl lactate
CAS 97-64-3; EINECS 202-598-0
Synonyms: Lactic acid ethyl ester; 2-Hydroxypropionic acid ethyl ester; Ethyl α-hydroxypropionate; Ethyl 2-hydroxypropionate; Ethyl-S(-)-2-hydroxypropionate
Definition: Ethyl ester of lactic acid
Empirical: $C_5H_{10}O_3$
Formula: $CH_3CHOHCOOC_2H_5$
Properties: Colorless liq., mild odor; misc. with water, alcohol, ketones, esters, hydrocarbons, oil; m.w. 118.13; dens. 1.020-1.036 (20/20 C); m.p. -26 C; b.p. 154 C; flash pt. 46.1 C; ref. index 1.410-1.420
Precaution: Flamm. or combustible; can react with oxidizers; sl. explosion hazard in vapor form exposed to flame
Toxicology: LD50 (oral, mouse) 2500 mg/kg; moderately toxic by ingestion, intraperitoneal, subcutaneous, intravenous routes; heated to decomp., emits acrid smoke and irritating fumes
Uses: Dipped latex prods., coatings, specialty finishes, inks, diluents, adhesives, intermediates, photoresist solvs., screen printing of electronic parts, flavors and fragrances, solvent resins, safety glass
Regulatory: FDA 21CFR §172.515; FEMA GRAS
Manuf./Distrib.: Aldrich; CPS; Farleyway Chem. Ltd; Int'l. Chem. Inc.; Penta Mfg.; Purac Am.
Trade names: Purasolv® EL; Purasolv® ELECT; Purasolv® ELS

Ethylmethyl carbinol. *See* 2-Butanol
Ethyl methyl ketone. *See* Methyl ethyl ketone
Ethyl methyl ketone peroxide. *See* Methyl ethyl ketone peroxide

3-Ethyl-2-methyl-2-(3-methylbutyl)-1,3-oxazolidine
CAS 143860-04-2
Uses: Moisture scavenger for use in urethane coatings, sealants and elastomers; eliminates moisture from solvents, polyols and pigment grinds without adversely affecting film props.

Manuf./Distrib.: Aldrich
Trade names: Zoldine® MS-Plus

Ethyl octynol
Classification: Sec. acetylenic alcohol
Uses: Corrosion inhibitor in oil well acidizing, mild steel pickling, mild steel cleaning in acid systems; as electroplating additive; curative in paints
Manuf./Distrib.: Air Prods.
Trade names: Polamine® 650; Polamine® 1000; Polamine® 2000

Ethylolamine. *See* Ethanolamine
Ethyl orthosilicate. *See* Ethyl silicate
Ethyl oxide. *See* Ethyl ether
Ethyl 3-oxobutanoate. *See* Ethylacetoacetate

Ethyl phenyl ethanolamine
CAS 92-50-2
Formula: $C_6H_5N(C_2H_5)CH_2CH_2OH$
Properties: M.w. 165.24; m.p. 36-38 C; b.p. 268 C
Toxicology: Irritant
Uses: Coupling agent for dyes for syn. fibers; cure promoter for polyester resins; photosensitive chemical for paper coatings; intermediate for oil-sol. dyestuffs
Manuf./Distrib.: Aldrich

Ethyl polysilicate
Uses: Binder for use in two-pkg. inorg. zinc primers
Manuf./Distrib.: Hüls AG
Trade names: Silbond® 40; Silbond® H-6; Silbond® H-12A

Ethyl propionate
CAS 105-37-3; EINECS 203-291-4
Synonyms: Propanoic acid ethyl ester
Definition: Ester of propionic acid and ethyl alcohol
Empirical: $C_5H_{10}O_2$
Formula: $CH_3CH_2COOC_2H_5$
Properties: Colorless liq., fruity odor; sol. in $\approx$60 parts water; misc. with alcohol, ether; m.w. 102.14; dens. 0.890 (20/4 C); m.p. -73 C; b.p. 96-99 C; flash pt. (CC) 12 C; ref. index 1.384 (20 C)
Precaution: Highly flamm. liq.; dangerous fire risk
Toxicology: LD50 (oral, rabbit) 3500 mg/kg; mod. toxic by ingestion and IP routes; skin irritant; reacts with oxidizing materials; heated to decomp., emits acrid smoke and fumes
Uses: Solv. for paints, various natural and synthetic resins; flavoring agent, fruit syrups
Regulatory: FDA 21CFR §172.515; FEMA GRAS; Japan approved as flavoring
Manuf./Distrib.: Aldrich; Ashland; Berje; Fluka; Northwestern Flavors; Penta Mfg.; Sigma; Union Carbide

Ethyl silicate
CAS 78-10-4; EINECS 201-083-8
Synonyms: Ethyl orthosilicate; Tetraethyl orthosilicate; Tetraethoxysilane
Empirical: $C_8H_{20}O_4Si$
Formula: $(C_2H_5)_4SiO_4$
Properties: Colorless liq., faint odor; pract. insol. in water; misc. with alcohol; m.w. 208.33; dens. 0.933 (20/4 C); m.p. -77 C; b.p. 165-166 C; flash pt. (OC) 52 C; ref.index 1.383 (20 C)
Precaution: Flamm.
Toxicology: Poison by intravenous route; moderately toxic by other routes; TLV 10 ppm in air; strong irritant to eyes, nose, throat
Uses: Intermediate for mfg. of ethyl silicate prods.; produces binders; chemical-and heat-resistant paints, cements, weatherproofing; protective coatings
Manuf./Distrib.: Akzo Nobel; Aldrich; Fluka; R.W. Greeff; Hüls Am.; PCR; Sigma; Wacker Silicones
Trade names: Silbond® Condensed; Silbond® H-28A; Silbond® H-28D
Trade names containing: Silbond® H-6C

Ethyl toluenesulfonamide
CAS 80-39-7; 1077-56-1; EINECS 214-073-3; 201-275-1
Synonyms: Ethyl tosylamide

2-(N-Ethyl-m-toluidino) ethanol

Classification: Mixture of isomers of aromatic amides
Empirical: $C_9H_{13}NO_2S$
Formula: $C_2H_5NHSO_2C_6H_4CH_3$
Properties: Colorless cryst.; m.w. 199.2; m.p. 64 C sol. in alc.; flash pt. 126 C (316 F)
Precaution: Combustible
Uses: Plasticizer for resins, coatings, adhesives, inks, electroplating sol'ns., thermoplastics and thermosets, nitrocellulose lacquers, shellac
Regulatory: FDA 21CFR §175.105, 175.300
Manuf./Distrib.: Aldrich; BASF; Charkit; ICI Spec. Chems.; Rit-Chem; Seal Sands Chems. Ltd; Spectrum Chem. Mfg.
Trade names: Ketjenflex® 8; Rit-Cizer #8; Uniplex 108

2-(N-Ethyl-m-toluidino) ethanol. *See* N-Ethyl-N-hydroxyethyl-m-toluidine
Ethyl tosylamide. *See* Ethyl toluenesulfonamide

Etocrylene
CAS 5232-99-5; EINECS 226-029-0
Synonyms: Ethyl 2-cyano-3,3-diphenylacrylate; Ethyl 2-cyano-3,3-diphenyl-2-propenoate; UV Absorber-2
Classification: Organic ester
Empirical: $C_{18}H_{15}NO_2$
Properties: Powd.; m.p. 96 C
Uses: UV absorber/stabilizer for NC lacquers, PVC, VDC, urea-formaldehyde, epoxyamine, and in cosmetics
Manuf./Distrib.: Aldrich
Trade names: Uvinul® 3035

EtOH. *See* Alcohol
EVA. *See* Ethylene/VA copolymer
EVA copolymer. *See* Ethylene/VA copolymer
EVOH copolymer. *See* Ethylene vinyl alcohol copolymer

Fatty acids, C18, unsaturated, trimers. *See* Trilinoleic acid
Fatty acids, soya. *See* Soy acid
Fatty acids, tall oil. *See* Tall oil acid

Feldspar
EINECS 270-666-7
Synonyms: Potassium aluminosilicate
Precaution: Noncombustible
Toxicology: Toxic as fine-ground powd.
Uses: Extender for paints; pottery, enamel, and ceramic ware; glass; abrasive bond for abrasive wheel; poultry grit
Manuf./Distrib.: KMG Mins.; Process, Materials & Eng.; H.M. Royal; Unimin Spec. Mins.
Trade names: Unispar 40

FEP. *See* Fluorinated ethylene-propylene
Ferric octoate. *See* Iron octoate
Ferric orthophosphate. *See* Ferric phosphate

Ferric oxide
CAS 1309-37-1 (anhyd.); EINECS 215-168-2
Synonyms: Ferric oxide red; Iron (III) oxide; Yellow ferric oxide; Red iron trioxide; Red iron oxide; Ferrosoferric oxide
Empirical: Fe_2O_3
Properties: Red-brn. to blk. cryst.; sol. in acids; insol. in water, alcohol, ether; m.w. 159.69; dens. 5.240; m.p. 1538 C (dec.)
Precaution: Catalyzes the potentially explosive polymerization of ethylene oxide
Toxicology: TLV:TWA 5 mg (Fe)/m^3 (vapor, dust); LD50 (IP, rat) 5500 mg/kg; irritant; poison by subcutaneous route; suspected human carcinogen; experimental tumorigen
Uses: Metallurgy, gas purification, in thermite, polishing compounds, mordant, laboratory reagent, catalyst, feed additive, electronic pigments for TV, permanent magnets, memory cores for computers, magnetic tapes; pigment in paints and coatings, latex dipped goods, latex foam, latex thread, fabric coatings and

saturants, and cements/concrete
Regulatory: FDA 21CFR §73.200 (limitation 0.25%), 186.1300, 186.1374, GRAS as indirect food additive; FDA approved for orals; USP/NF compliance
Manuf./Distrib.: Advance Research Chems.; Aldrich; BASF; Bayer/Fibers, Orgs., Rubber; Crompton & Knowles; Fluka; Kerr-McGee; Spectrum Chem. Mfg.
Trade names: Octotint 103; Octotint 825; OSO® 440; OSO® 1905; OSO® NR 830 M; OSO® NR 950

Ferric oxide red. *See* Ferric oxide

Ferric phosphate
CAS 10045-86-0 (anhyd.); 14940-41-1 (tetrahydrate); EINECS 233-149-7; 239-018-0 (tetrahydrate)
Synonyms: Ferric orthophosphate; Ferric (III) phosphate; Ferriphosphate; Iron phosphate
Empirical: $FePO_4 \cdot xH_2O$, x = 1-4
Properties: Ylsh.-wh. to buff-colored powd.; sol. in HCl; slowly sol. in HNO_3; pract. insol. in water; m.w. 150.83; dens. 2.87 (dihydrate)
Toxicology: Heated to decomp., emits toxic fumes of PO_x
Uses: Rust-inhibitive pigment for paints; fertilizers; feed and food additive
Regulatory: FDA 21CFR §182.5301, 184.1301, GRAS
Manuf./Distrib.: Fluka; Lohmann; Nat'l. Chem.; Spectrum Chem. Mfg.

Ferric (III) phosphate. *See* Ferric phosphate
Ferriphosphate. *See* Ferric phosphate

Ferro-aluminum silicate
CAS 12178-41-5
Uses: Filler for abrasion-resistant plastic systems; extender pigment for primers and other coatings
Trade names: Ferrosil™ 14

Ferrosoferric oxide. *See* Ferric oxide

Ferrous sulfate heptahydrate
CAS 7782-63-0; EINECS 231-753-5
Synonyms: Iron (II) sulfate heptahydrate; Ferrous sulfate-7-hydrate; Copperas; Green vitriol; Iron vitriol
Empirical: $O_4S \cdot Fe \cdot 7H_2O$
Formula: $FeSO_4 \cdot 7H_2O$
Properties: Pale bluish-grn. cryst. or gran., odorless; sol. in water; pract. insol. in alcohol; m.w. 278.05; dens. 2.99-3.08
Precaution: Incompat. with alkalies, sol. carbonates, Au and Ag salts, lime water, K and Na tartrate, tannin, etc.
Toxicology: TLV:TWA 1 mg(Fe)/m^3; LD50 (oral, mouse) 1.52 g/kg; poison by IV, IP, subcut. routes; mod. toxic by ingestion and rectal routes; human G.I. disturbances; mutagenic data; heated to decomp., emits toxic fumes of SO_x
Uses: Flocculant and coagulant in water treatment; fertilizer, animal feed supplement, paint pigment; food additive
Regulatory: FDA 21CFR §182.5315, 182.8315, 184.1315, GRAS; BATF 27CFR §240.1051
Manuf./Distrib.: Aldrich; Crown Tech.; Fluka; Penta Mfg.; Sigma; Spectrum Chem. Mfg.

Ferrous sulfate-7-hydrate. *See* Ferrous sulfate heptahydrate
Fischer-Tropsch wax. *See* Synthetic wax
Fischer-Tropsch wax, oxidized. *See* Synthetic wax

Fish glycerides
CAS 100085-40-3
Definition: Mixture of mono, di and triglycerides expressed or extracted from menhaden, hake or similar oil-bearing fish
Trade names containing: Ross Japan Wax Substitute 525

Flaxseed oil. *See* Linseed oil
Flowers of antimony. *See* Antimony trioxide
Flowers of zinc. *See* Zinc oxide

Fluorinated ethylene-propylene
CAS 9002-84-0

Fluorite

Synonyms: FEP; Tetrafluoroethylene/hexafluoropropylene copolymer; Polyethylene tetrafluoride
Uses: Lubricant powd.; thickening and EP agent in hydrocarbon and fluorosilicone base oils and greases; wire and cable insulation, inj. molded parts, high performance film; in antistick, antifriction, corrosion resistant coatings and impregnations
Manuf./Distrib.: Aldrich; Fluka
Trade names: Teflon® 120; Teflon® 160N

Fluorite. *See* Calcium fluoride
Fluorotributylstannane. *See* Tributyltin fluoride
Fluorspar. *See* Calcium fluoride
Fluxing lime. *See* Calcium oxide

Formaldehyde
CAS 50-00-0; EINECS 200-001-8
Synonyms: Oxymethylene; Formalin; Formic aldehyde; Aldehyde C-1; Methanal
Empirical: CH_2O
Formula: HCHO
Properties: Colorless gas, strong pungent odor; avail. commercially as aq. sol'ns. (37-50% in methanol); sol. in water, alcohol; m.w. 30.03; dens. 1.083; b.p. -19 C; f.p. -118 C; flash pt. 56 C; ref. index 1.3765
Precaution: Combustible when exposed to heat, or flame; mod. fire risk; explosive limits 7-73%; can react vigorously with oxidizers
Toxicology: TLV 1 ppm in air; LD50 (oral, rat) 800 mg/kg; human poison by ing., systemic effects, skin/eye irritant, mutagenic data; toxic by inh.; vapor intensely irritating to mucous membranes; suspected human carcinogen; heated to decomp., emits acrid smoke
Uses: Urea and melamine resins, polyacetal resins, phenolic resins, fertilizers, preservatives, reducing agent, corrosive inhibitor; intermediate for paints
Regulatory: FDA 21CFR §173.340 (1% of dimethicone content), 175.105, 175.210, 176.170, 1876.180, 176.200, 176.210, 177.2410, 177.2800, 178.3120, 573.460; FDA approved for topicals; BP compliance (sol'n.)
Manuf./Distrib.: Aqualon; Ashland; J.T. Baker; DuPont; Farleyway Chem. Ltd; Fluka; Georgia-Pacific Resins; C.P. Hall; Hoechst Celanese; Mallinckrodt; Mitsubishi Gas; Monomer-Polymer & Dajac Labs; Monsanto; Sigma; Spectrum Chem. Mfg.; Van Waters & Rogers; Veckridge; Wright
Trade names: Hercules® 37M6-8

Formaldehyde/toluenesulfonamide polymer
Uses: Modifier, adhesion promoter for coatings, lacquers, printing inks, adhesives, oil and gasoline resist. coatings; compat. with alkyds, acrylics, urethanes, nitrocellulose and vinyls
Trade names: Rit-O-Lite MHP-S

Formalin. *See* Formaldehyde

Formic acid
CAS 64-18-6; EINECS 200-579-1
Synonyms: Hydrogen carboxylic acid; Methanoic acid; Aminic acid
Classification: Organic acid
Empirical: CH_2O_2
Formula: HCOOH
Properties: Colorless fuming liq., penetrating odor; sol. in water, alcohol, ether; m.w. 46.03; dens. 1.22 (20/4 C); m.p. 8.3 C; b.p. 100.8 C; flash pt. (OC) 69 C; ref. index 1.3714
Precaution: Flamm.; DOT: corrosive material; can react vigorously with oxidizers; explosive with furfuryl alcohol
Toxicology: LD50 (oral, rat) 1100 mg/kg; poison by IV route; mod. toxic by ingestion, intraperitoneal routes; skin and severe eye irritant; TLV 5 ppm in air; heated to decomp., emits acrid smoke and irritating fumes
Uses: Reducing agent, dyeing and finishing of textiles, leather treatment, chemicals, mfg. of fumigants, insecticides, refrigerants, solvs. for perfumes, lacquers; electroplating; silvering glass; ore flotation; flavoring adjunct; paper mfg. aid
Regulatory: FDA 21CFR §172.515, 186.1316, GRAS as indirect additive, 573.480; FEMA GRAS; Europe listed; prohibited in UK
Manuf./Distrib.: Aldrich; Ashland; BASF; BASF AG; BP Chems. Ltd; Coyne; Fluka; Hoechst Celanese; Mallinckrodt; Norsk Hydro AS; Sigma; Spectrum Chem. Mfg.

Formic acid isobutyl ester. *See* Isobutyl formate
Formic aldehyde. *See* Formaldehyde

Fossil wax. *See* Ozokerite
French chalk. *See* Talc
β-D-Fructofuranosyl-α-D-glucopyranoside benzoate. *See* Sucrose benzoate
Fuller's earth. *See* Attapulgite

Fumaric acid
CAS 110-17-8; EINECS 203-743-0
Synonyms: Allomaleic acid; Boletic acid; trans-1,2-Ethylenedicarboxylic acid; 2-Butenedioic acid; trans-Butenedioic acid; Lichenic acid
Classification: Dicarboxylic acid
Empirical: $C_4H_4O_4$
Formula: HOOCCH:CHCOOH
Properties: Wh. cryst., odorless, acidic taste; sol. in alcohol; sl. sol. in water, ether; very sl. sol. in chloroform; m.w. 116.08; dens. 1.635 (20/4 C); m.p. 287 C; b.p. 290 C; flash pt. (COC) 282 C
Precaution: Combustible when exposed to heat or flame; can react vigorously with oxidizers
Toxicology: LD50 (oral, rat) 10,700 mg/kg, (dermal, rabbit) 20,000 mg/kg; poison by intraperitoneal route; mildly toxic by ingestion and skin contact; skin and eye irritant; heated to decomp., emits acrid smoke and irritating fumes
Uses: Modifier for polyester, alkyd and phenolic resins; paper sizing resins; plasticizers, rosin esters and adducts, alkyd resin coatings, upgrading natural drying oils, mordant, organic synthesis, printing inks; food additive, acidulant, flavoring agent, leavening agent
Regulatory: FDA 21CFR §172.350; USDA 9CFR §318.7, 381.147; BATF 27CFR §240.1051, limitation 25 lb/1000 gal wine; FEMA GRAS; Japan approved; Europe listed; UK approved; FDA approved for orals; USP/NF, BP compliance
Manuf./Distrib.: Aceto; Aldrich; Allchem Ind.; Am. Biorganics; Ashland; Balchem; Bayer/Fibers, Orgs., Rubber; Browning; Chemical; Chemie Linz UK; Fluka; Haarmann & Reimer; Hüls Am.; Lonza; Mitsubishi Gas; Monomer-Polymer & Dajac Labs; Monsanto; Penta Mfg.; Schweizerhall; Sigma; Spectrum Chem. Mfg.; United Min. & Chem.; Van Waters & Rogers

Fumaric acid, dibutyl ester. *See* Dibutyl fumarate
Fumed silica. *See* Silica, fumed
Fural. *See* Furfural
2-Furaldehyde. *See* Furfural
Furale. *See* Furfural
Furan. *See* Furan polymer
2-Furanaldehyde. *See* Furfural
2-Furancarbinol. *See* Furfuryl alcohol
2-Furancarbonal. *See* Furfural
2-Furancarboxaldehyde. *See* Furfural
2,5-Furandione. *See* Maleic anhydride
2,5-Furandione, polymer with ethene. *See* Ethylene-maleic anhydride copolymer
2,5-Furandione, polymer with ethenylbenzene. *See* Styrene/MA copolymer
2,5-Furandione, polymer with methoxyethylene. *See* PVM/MA copolymer
2,5-Furandione, polymer with 2-methyl-1-propene. *See* Isobutylene/MA copolymer
Furanidine. *See* Tetrahydrofuran
2-Furanmethanol. *See* Furfuryl alcohol

Furan polymer
CAS 110-00-9; EINECS 203-727-3
Synonyms: Furan; Furan resin; Furfuran; Divinylene oxide
Empirical: C_4H_4O
Properties: M.w. 68.08; dens. 0.936 (20/4 C); b.p. 31-33 C; flash pt. -36 C; ref. index 1.421 (20 C)
Precaution: Extremely flamm.; keep away from ignition sources; photosensitive; store refrigerated
Uses: Chemical intermediate in the mfg. of herbicides, pharmaceuticals, plastics, and fine chemicals; coating asphaltic pavements, foundry sand cores
Manuf./Distrib.: Aldrich; Atlas Mins. & Chems.; Delta Resins & Refractories; Fluka; Great Lakes; QO; Sigma; Spectrum Chem. Mfg.
Trade names: Hetron® 800

Furan resin. *See* Furan polymer

Furfural
CAS 98-01-1; EINECS 202-627-7

2-Furfural

Synonyms: 2-Furaldehyde; Fural; Furale; 2-Furanaldehyde; 2-Furancarbonal; 2-Furfural; Furfuraldehyde; Furfurol; Furfurole; 2-Furancarboxaldehyde; Furole; α-Furole; 2-Furyl methanal; Pyromucic aldehyde; Artificial oil of ants; Artificial ant oil

Classification: Cyclic aldehyde

Empirical: $C_5H_4O_2$

Formula: C_4H_3OCHO

Properties: Colorless liq. (pure); reddish-brown (on exposure to air and light); almond-like odor; sol. in alcohol, ether, benzene, 8.3% in water; m.w. 96.08; dens. 1.1598 (20/4 C); f.p. -36.5 C; b.p. 161.7 C; flash pt. 60 C; ref. index 1.5260 (20 C)

Precaution: Flamm. or combustible; can react with oxidizing materials; mod. explosion hazard exposed to heat or flame or by chem. reaction; incompat. with strong min. acids or alkalies

Toxicology: LD50 (oral, rat) 65 mg/kg; poison by ingestion, intraperitoneal, subcutaneous, IV, IM routes; moderately toxic by inhalation; TLV 2 ppm in air; irritates mucous membranes, acts on CNS; heated to decomp., emits acrid smoke and irritating fumes

Uses: Chemical intermediate for mfg. of derivs. (furan, THF), paints; solvent for petrol. lube, nitrocellulose; wetting agent; in mfg. of furfural-phenol plastics; vulcanization accelerator; insecticide, fungicide, germicide; reagent in analytical chemistry; flavoring agent

Regulatory: FDA 21CFR §175.105; FEMA GRAS; Japan approved as flavoring

Manuf./Distrib.: Aldrich; Allchem Ind.; Fluka; Great Lakes; Penta Mfg.; QO; Sigma; Spectrum Chem. Mfg.; Van Waters & Rogers

2-Furfural. *See* Furfural
Furfuralcohol. *See* Furfuryl alcohol
Furfuraldehyde. *See* Furfural
Furfuran. *See* Furan polymer
Furfurol. *See* Furfural
Furfurole. *See* Furfural

Furfuryl alcohol

CAS 98-00-0; EINECS 202-626-1

Synonyms: 2-Furancarbinol; 2-Furanmethanol; 2-Hydroxymethylfuran; α-Furylcarbinol; 2-Furylcarbinol; (2-Furyl) methanol; Furfuralcohol; Furyl alcohol

Empirical: $C_5H_6O_2$

Formula: $C_4H_3OCH_2OH$

Properties: Colorless mobile liq., brn.-dk. red (air/lt. exposed), low odor, cooked sugar taste; sol. in alc., chloroform, benzene; misc. with water but unstable; m.w. 98.10; dens. 1.1285 (20/4 C); m.p. -29 C; b.p. 170 C; flash pt. (OC) 75 C; ref. index 1.485

Precaution: Combustible; can react with oxidizing materials; mod. explosion hazard when exposed to heat or flame; may react explosively with mineral and some organic acids

Toxicology: TLV 10 ppm in air; TLV:TWA 2 ppm (skin); LD50 (oral, rat) 275 mg/kg, (dermal, rabbit) 450 mg/kg; poison by ingestion, skin contact, subcut. route; mod. toxic by inh, IP routes; heated to deocmp., emits acrid smoke and fumes

Uses: Wetting agent, furan polymers, foundry sand binders, corrosion-resist. resins; intermediate for esterification and etherification, paints; plasticizer for phenolic resins; solvent for dyes and resins; flavoring; visc. reducer, cure promoter, and carrier in amine-cured epoxy resins

Regulatory: FEMA GRAS

Manuf./Distrib.: AC Ind.; Aldrich; Allchem Ind.; Ashland; Browning; Fluka; Great Lakes; Penta Mfg.; QO; Schweizerhall; Spectrum Chem. Mfg.; Van Waters & Rogers

Furnace black. *See* Carbon black
Furole. *See* Furfural
α-Furole. *See* Furfural
Furyl alcohol. *See* Furfuryl alcohol
2-Furylcarbinol. *See* Furfuryl alcohol
α-Furylcarbinol. *See* Furfuryl alcohol
2-Furyl methanal. *See* Furfural
(2-Furyl) methanol. *See* Furfuryl alcohol
Fused boric acid. *See* Boron oxide

GAE. *See* Glyceryl-1-allyl ether

Gelatin
CAS 9000-70-8; EINECS 232-554-6
Synonyms: Gelatine; White gelatin
Definition: Obtained from partial hydrolysis of collagen derived from animal skin, connective tissues, and bones; Type A is derived from acid-treated precursor, Type B from alkali-treated precursor
Properties: Faint yel. or amber flake or powd., sl.char. bouillon-like odor in sol'n., tasteless; sol. in warm water, glycerol, acetic acid; insol. in org. solvs., alcohol, chloroform, ether, fixed and volatile oils; amphoteric
Precaution: Stable in air when dry, but subject to microbic decomp. when moist or in sol'n.
Toxicology: LD50 (rat, oral) 5 g/kg; may cause anaphylactoid reactions
Uses: Photographic film, sizing, textile and paper adhesives, cements, capsules for medications, matches; clarifying agent; protective colloid in ice cream; stabilizer, thickener, texturizer in food; dietary supplements; chewing gum base; raw material in paints
Regulatory: FDA 21CFR §133.133, 133.134, 133.162, 133.178.133.179, 182.70; GRAS; Japan approved; FDA approved for dentals, inhalants, intramuscular injectables, intravenous, IV (infusion), orals, topicals; USP/NF, BP, Ph.Eur. compliance
Manuf./Distrib.: Aldrich; Am. Gelatin; Ashland; Atlantic Gelatin; Croda; DynaGel; G. Fiske; GMI Prods.; Hormel; Int'l. Chem. Inc.; Nitta Gelatin; Norland Prods.; Sigma; Spectrum Chem. Mfg.; Triple Crown Am.; Vyse Gelatin
Trade names: Technical Grade Gelatin

Gelatine. *See* Gelatin

Gilsonite
CAS 12002-43-6
Classification: An asphaltic material or solidified hydrocarbon
Definition: A blk. solid hydrocarbon deriving from petroleum millions of years ago
Properties: Blk. solid
Precaution: Flamm. when exposed to heat or open flame
Toxicology: Irritant, skin sensitizer; heated to decomp., emits acrid fumes
Uses: Acid, alkali and waterproof coatings; black varnishes, lacquers, baking enamels; wire-insulation compounds; linoleum, floor tile; paving; insulation; diluent in low-grade rubber compounds; possible source of gasoline, fuel oil
Manuf./Distrib.: Am. Gilsonite; R.E. Carroll; Lexco; Superior Materials; Ziegler Chem. & Min.; Zophar Mills

Glicerol. *See* Glycerin
Glicidyl phenyl ether. *See* Phenyl glycidyl ether
1,4-D-Glucan glucanohydrolase. *See* Amylase
D-Glucitol. *See* Sorbitol
D-Gluconic acid monosodium salt. *See* Sodium gluconate
Gluconic acid sodium salt. *See* Sodium gluconate
Glutaronitrile, 2-bromo-2-(bromomethyl). *See* Methyldibromo glutaronitrile
Glycerides, hydrogenated tallow mono-. *See* Hydrogenated tallow glyceride
Glycerides, palm kernel oil mono-, di-, and tri-, acetates. *See* Acetylated palm kernel glycerides
Glycerides, tallow mono-, di- and tri-. *See* Tallow glycerides
Glycerides, tallow mono-, di- and tri-, hydrogenated. *See* Hydrogenated tallow glycerides

Glycerin
CAS 56-81-5; EINECS 200-289-5
Synonyms: Glicerol; Glycerol; Glycyl alcohol; 1,2,3-Propanetriol; Propane-1,2,3-triol; Glycerine; Trihydroxypropane glycerol
Classification: Polyhydric alcohol
Empirical: $C_3H_8O_3$
Formula: $HOCH_2COHHCH_2OH$
Properties: Clear colorless syrupy liq., odorless, sweet taste; hygroscopic; sol. in water, alcohol; insol. in ether, benzene, chloroform; m.w. 92.09; dens. 1.26201 (25/25 C); m.p. 17.8 C; b.p. 290 C (dec.); flash pt. (OC) 176 C; ref. index 1.4730 (25 C)
Precaution: Combustible liq. exposed to heat, flame, strong oxidizers; highly explosive with hydrogen peroxide
Toxicology: LD50 (oral, rat) > 20 ml/kg, (IV, rat) 4.4 ml/kg; poison by subcutaneous route; mildly toxic by ingestion; human systemic and GI effects by ingestion; human mutagenic data; skin and eye irritant; nuisance dust; heated to dec., emits acrid smoke
Uses: Alkyd resins, paints, dynamite, ester gums, pharmaceuticals, cosmetics, perfumery, lubricants, softener, bacteriostat, penetrant, emollient; antifreeze; in prod. of antibiotics in OTC drugs; solvent, humectant, plasticizer, bodying agent for baked goods, candy, marshmallows, pkg. materials

Glycerin-1-allyl ether

Usage level: 0.2-65.7% (topicals), 1-50% (liq. orals), 50% (parenterals), 7-10% (dentifrices), 0.5-3.0% (ophthalmics)
Regulatory: FDA 21CFR §169.175, 169.176, 169.177, 169.178, 169.180, 169.181, 172.866, 175.300, 178.3500, 182.90, 182.1320, GRAS; FEMA GRAS; FDA approved for orals, parenterals, ophthalmics, dentifrices, intramuscular injectables, rectals; USP/NF compliance; Japan approved; Europe listed; UK approved; BP compliance
Manuf./Distrib.: Akzo Nobel; Alba Int'l.; Aldrich; Alnor Oil; Asahi Denka Kogyo; Ashland; Baychem; Boehle; Chemcentral; Dial; Farleyway Chem. Ltd; Fina Chems.; Fluka; C.P. Hall; Henkel/Emery; Lonza; Procter & Gamble; Samson; Sigma; Unichema Int'l.; United Min. & Chem.; Van Waters & Rogers; Witco

Glycerin-1-allyl ether. *See* Glyceryl-1-allyl ether
Glycerine. *See* Glycerin
Glycerol. *See* Glycerin
Glycerol monoricinoleate. *See* Glyceryl ricinoleate
Glycerol monostearate. *See* Glyceryl stearate
Glycerol ricinoleate. *See* Glyceryl ricinoleate
Glycerol stearate. *See* Glyceryl stearate

Glyceryl-1-allyl ether
CAS 123-34-2; EINECS 204-620-4
Synonyms: GAE; 1,2-Propanediol, 3-(2-propenyloxy)-; 1-Allyloxy-2,3-propanediol; Allyl-(2,3-dihydroxypropane) ether; Glycerin-1-allyl ether
Empirical: $C_6H_{12}O_3$
Formula: $CH_2=CHCH_2OCH_2CHOHCH_2OH$
Properties: Colorless to ylsh. visc. liq.; sol. in water, ethanol, and many org. solvs.; m.w. 132.16; dens. 1.07 g/cc (20 C); m.p. < -90 C; b.p. 245 C
Precaution: Incompat. with acids, alkalies, oxidizing agents, peroxides; hazardous decomp. prods.: CO, gaseous hydrocarbons
Toxicology: LD50 (oral, rat) 5400 mg/kg, (dermal, rat) 4300 mg/kg; irritating to eyes, respiratory system, skin
Uses: Can be polymerized into polyesters, PU, polyacetals, epoxy resins, acrylates, methacrylates, or styrene; for prod. of printed circuit boards, prod. and coating of PU rubber and foams, unsat. polyesters for radiation-resist. coatings
Manuf./Distrib.: Aldrich; Fluka
Trade names: GAE

Glyceryl hydrogenated rosinate
Synonyms: Rosin, hydrogenated, glycerol ester
Definition: Monoester of glycerin and hydrogenated mixed long chain acids derived from rosin
Uses: Thermoplastic syn. resin; as softener/plasticizer for the masticatory agent in chewing gum bases; as resin modifier for film-formers, elastomers, and waxes in adhesive and protective coating compositions; as tackifier in adhesives, coatings
Trade names: Foral® 85; Staybelite® Ester 10

Glyceryl hydroxystearate
CAS 1323-42-8; EINECS 215-355-9
Synonyms: Glyceryl 12-hydroxystearate; Hydroxystearic acid, monoester with glycerol
Definition: Monoester of glycerin and hydroxystearic acid
Empirical: $C_{21}H_{42}O_5$
Uses: Emulsifier, emollient, opacifier, bodying and thickening agent for cosmetics, pharmaceuticals, household prods.; beeswax substitute; pigment wetting in paints, inks; paper industry; lubricant and mold release in plastics; textile and leather processing; detergency and cleaning prods.
Regulatory: FDA 21CFR §175.105, 176.170, 176.200, 177.1210, 177.2800

Glyceryl 12-hydroxystearate. *See* Glyceryl hydroxystearate

Glyceryl mono/dioleate
CAS 25496-72-4; 68424-61-3
Properties: Yel. oil or soft solid; dens. 0.95; m.p. 14-19 C
Precaution: Combustible
Uses: Emulsifier and antifoam for foods; emulsifier, solubilizer for cosmetic, pharmaceutical, and industrial applics.; lubricant, softener for syn. fibers; rust preventive additive in oils; lubricant, antistat, antifog for PVC film processing
Trade names: Alkamuls® GMR-55LG

Glyceryl monooleate. *See* Glyceryl oleate
Glyceryl monoricinoleate. *See* Glyceryl ricinoleate
Glyceryl monorosinate. *See* Glyceryl rosinate
Glyceryl monostearate. *See* Glyceryl stearate

Glyceryl oleate
CAS 111-03-5; 25496-72-4; 37220-82-9; EINECS 203-827-7; 253-407-2
Synonyms: Glyceryl monooleate; Monoolein; 9-Octadecenoic acid, monoester with 1,2,3-propanetriol
Definition: Monoester of glycerin and oleic acid
Empirical: $C_{21}H_{40}O_4$
Formula: $CH_3(CH_2)_7CH=CH(CH_2)_7COOCH_2CCH_2OHHOH$
Properties: Yel. oil or soft solid; insol. in water; somewhat sol. in alcohol, most org. solvs.; sp.gr. 0.940-0.960; m.p. 149-19 C; acid no. 3 max.; iodine no. 65-80; sapon. no. 166-174
Precaution: Combustible
Toxicology: Heated to decomp., emits acrid smoke and irritating fumes
Uses: Emulsifier, coemulsifier, stabilizer, wetting agent, lubricant, and antistat; used in cosmetic, pharmaceutical, industrial, food applics.; plasticizer; internal antistat for PE, PP, PVC; antifog, cling agent, lubricant for PVC film; flavor adjuvant, solv., vehicle, defoamer, dispersant, emulsifier, plasticizer for food use (coffee whiteners, baking mixes, beverages, chewing gum, meat prods., pkg. materials, veg. oil); odorless paints; rust-preventive oils
Regulatory: FDA 21CFR §175.105, 175.300, 176.210, 177.2800, 181.22, 181.27, 182.4505, 184.1323, GRAS; FEMA GRAS; FDA approved for orals
Manuf./Distrib.: ABITEC; Am. Ingreds.; Aquatec Quimica SA; Calgene; Croda Surf. Ltd.; Ferro/Keil; Fluka; Grindsted Prods.; Henkel/Emery; ICI Surf. Am.; Inolex; Lonza; Mona Ind.; Sigma; Spectrum Chem. Mfg.; Stepan; Unichema N. Am.; Witco/Oleo-Surf.
Trade names: Dur-Em® GMO; Witco® 942

Glyceryl ricinoleate
CAS 141-08-2; EINECS 205-455-0
Synonyms: 12-Hydroxy-9-octadecenoic acid, monoester with 1,2,3-propanetriol; Monoricinolein; Glycerol ricinoleate; Glycerol monoricinoleate; Glyceryl monoricinoleate
Definition: Monoester of glycerin and ricinoleic acid
Formula: $C_3H_5(OOCC_{16}H_{32}OH)_3$
Properties: Yel. liq.; sp.gr. 0.981; m.p. < -50 C; flash pt. (COC) 265 C; ref. index 1.4770
Uses: Emulsifying agent; plasticizer for PVB, cellulosics; emulsifier, solubilizer for cosmetic, pharmaceutical, household prods., coatings, and printing inks
Regulatory: FDA 21CFR §175.105, 176.170, 176.210, 178.3130; FDA approved for orals, topicals
Manuf./Distrib.: Calgene; CasChem; Lonza; Witco/Oleo-Surf.
Trade names: Aldo® MR

Glyceryl rosinate
Synonyms: Glyceryl monorosinate; Rosin, glyceryl ester
Definition: Monoester of glycerin and mixed long chain acids derived from rosin
Properties: Drop soften. pt. 88-96 C
Toxicology: Heated to decomp., emits acrid smoke and irritating fumes
Uses: Thermoplastic resin gum; clouding agent in beverages; in chewing gums, adhesives, inks, coatings; tackifier, softener, plasticizer, modifier for adhesives, coatings; process aid for rubber
Regulatory: FDA 21CFR §172.615, 172.735, limitation 100 ppm in finished beverage, 175.105, 175.300, 178.3120, 178.3800, 178.3870
Trade names: Lewisol® 28; Poly-Pale® Ester 10; Uni-Tac® R85; Uni-Tac® R85-Light

Glyceryl stearate
CAS 123-94-4; 11099-07-3; 31566-31-1; 85666-92-8; 85251-77-0; EINECS 250-705-4; 234-325-6; 204-664-4; 286-490-9
Synonyms: GMS; Monostearin; 1,2,3-Propanetriol octadecanoate; Octadecanoic acid, monoester with 1,2,3,-propanetriol; Glyceryl monostearate; Glycerol stearate; Glycerol monostearate; 2,3-Dihydroxypropyl octadecanoate
Definition: Monoester of glycerin and stearic acid
Empirical: $C_{21}H_{42}O_4$
Formula: $CH_3(CH_2)_{16}COOCH_2COHHCH_2OH$
Properties: Wh. to cream wax-like flakes, sl. fatty odor and taste; sol. in hot org. solvs.; insol. in water, ethanol, glycerin, propylene glycol; disp. in min. oil; m.p. 56-59 C; acid no. 6 max.; iodine no. 3 max.; sapon. no. 162-175; hyd. no. 300-330

Glyceryl triacetate

Precaution: Combustible; affected by light
Toxicology: LD50 (IP, mouse) 200 mg/kg; poison by IP route; heated to decomp., emits acrid smoke and irritating fumes
Uses: Nonionic sec. o/w emulsifier for creams and lotions; emulsifier for oils, waxes, solvs.; thickener, visc. booster for emulsions; plasticizer for cellulose nitrate; antistat, antifog, lubricant, processing aid for plastics; protective coating for hygroscopic powds.; cosmetics, pharmaceuticals; opacifier; detackifier
Regulatory: FDA 21CFR §139.110, 139.115, 139.117, 139.120, 139.121, 139.122, 139.125, 139.135, 139.138, 139.140, 139.150, 139.155, 139.160, 139.165, 139.180, 175.105, 175.210, 175.300, 176.200, 176.210, 177.2800, 184.1324, GRAS; FEMA GRAS; Europe listed; FDA approved for orals, ophthalmics, otics, rectals, topicals; USP/NF, BP, Ph.Eur. compliance
Manuf./Distrib.: ABITEC; Am. Ingreds.; Aquatec Quimica SA; Calgene; Croda Surf. Ltd.; Eastman; Goldschmidt; Grindsted Prods.; Hart Prods.; Henkel/Emery; ICI Surf. Am.; Inolex; ISP Van Dyk; Koster Keunen; Lanaetex Prods.; Lipo; Lonza; Protameen; Rhone-Poulenc Surf. & Spec.; Spectrum Chem. Mfg.; Stepan; Witco/Oleo-Surf.
Trade names: Aldo® MSLG FG; Petrac® GMS

Glyceryl triacetate. *See* Triacetin

Glyceryl (triacetoxystearate)
CAS 139-43-5; EINECS 295-625-0
Synonyms: Castor oil, acetylated and hydrogenated; Glyceryl tri(12-acetoxystearate)
Formula: $C_3H_5(OOCC_{17}H_{34}OCOCH_3)_3$
Properties: Clear pale yel. oily liq., mild odor; sol. in most org. solvs.; insol. in water; dens. 0.967 (25/25 C)
Precaution: Combustible
Toxicology: Heated to decomp, emits acrid smoke and irritating fumes
Uses: Plasticizer for nitrocellulose, ethylcellulose, and PVC, food pkg.; lubricants; protective coatings
Regulatory: FDA 21CFR §178.3505, limitation with $CaCO_3$ 1% of total mixt.
Manuf./Distrib.: CasChem

Glyceryl tri(12-acetoxystearate). *See* Glyceryl (triacetoxystearate)
Glyceryl tri-(12-acetylricinolate). *See* Glyceryl triacetyl ricinoleate

Glyceryl triacetyl ricinoleate
CAS 101-34-8; EINECS 202-935-1
Synonyms: 9-Octadecenoic acid, 12-(acetyloxy)-, 1,2,3-propanetriol ester; Castor oil, acetylated; 1,2,3-Propanetriyl 12-(acetyloxy)-9-octadecenoate; Glyceryl tri-(12-acetylricinolate)
Definition: Triester of glycerin and acetyl ricinoleic acid
Formula: $C_3H_5(OOCC_{17}H_{32}OCOCH_3)_3$
Properties: Clear pale yel. oily liq., mild odor; sol. in most org. liqs.; insol. in water; dens. 0.967 (25/25 C); solidifies @ -40 C; iodine no. 76; sapon. no. 300
Precaution: Combustible
Uses: Plasticizer for nitrocellulose, ethylcellulose, and PVC; lubricants; protective coatings
Manuf./Distrib.: CasChem
Trade names: Flexricin® P-8

Glyceryl tribenzoate
CAS 614-33-5
Synonyms: GTB; Tribenzoin
Definition: Benzoic acid triester with glycerin
Empirical: $C_{24}H_{20}O_6$
Properties: Colorless liq.; insol. in water; sol. in alcohol, ether; m.w. 404.44; dens. 1.032; m.p. < -75 C; b.p. 305-309 C
Toxicology: LD50 (oral, rat) 11,700 mg/kg; mildly toxic by ingestion; heated to decomp., emits acrid smoke and irritating fumes
Uses: Process aid, modifier for thermoplastics, hot-melt adhesives, lacquers; plasticizer in vinyl and acrylic systems
Regulatory: FEMA GRAS
Manuf./Distrib.: Aldrich; Unitex; Velsicol
Trade names: Benzoflex® S-404; Uniplex 260

Glyceryl tri (12-hydroxystearate). *See* Glyceryl tris-12-hydroxystearate; Trihydroxystearin

Glyceryl tris-12-hydroxystearate
Synonyms: Glyceryl tri (12-hydroxystearate); Castor oil, hydrogenated
Definition: Glyceryl triricinoleate where hydrogen has saturated the ricinoleic groups
Formula: $C_3H_5(OOCC_{17}H_{34}OH)_3$
Properties: Ylsh. to milky wh. hard brittle wax-like solid; dens. 0.899 (100/25 C); m.p. 86-88 C
Uses: Stabilizer, thickener, thixotrope, suspending agent; lubricants; heavy-metal soaps; waxes; plasticizers; cosmetics; chemical intermediate; suspending agent for solv.-borne systems, polyester patch and fillers, powd. coatings
Trade names: Thixcin® E

3-Glycidoxypropyltrimethoxysilane
CAS 2530-83-8; EINECS 219-784-2
Synonyms: γ-Glycidoxypropyltrimethoxysilane; 3-Glycidyloxypropyltrimethoxysilane; 3-(2,3-Epoxypropyl)-propyltrimethoxysilane
Empirical: $C_9H_{20}O_5Si$
Formula: $OCH_2CHCH_2O(CH_2)_3Si(OCH_3)_3$
Properties: Liq.; sol. in acetone, benzene, ether; reacts with water; m.w. 236.34; dens. 1.070 (25 C); f.p. > 110 C; b.p. ≈ 120 C (2 mm); flash pt. 79 C; ref. index 1.429 (20 C)
Toxicology: LD50 (oral, rat) 23 g/kg, (dermal, rat) 3970 mg/kg; mod. toxic by skin contact; mildly toxic by ingestion; irritating to eyes, skin, respiratory system
Uses: Coupling agent, filler for glass- and mineral-filled plastics, epoxies, urethanes, acrylics, PBT, polysulfides; adhesion promoter for solv., water-based, solventless, and powd. coating systems; pigment treatment
Manuf./Distrib.: Aldrich; Fluka; Sigma
Trade names: Dow Corning® Z-6040; Silquest® A-187
Trade names containing: Aktisil® EM

γ-Glycidoxypropyltrimethoxysilane. *See* 3-Glycidoxypropyltrimethoxysilane

Glycidyl acrylate
CAS 106-90-1
Synonyms: Acrylic acid 2,3-epoxypropyl ester; Glycidyl propenate; 2-Propenoic acid oxiranylmethyl ester; 2,3-Epoxy-1-propanol acrylate; 2,3-Epoxypropyl ester acrylic acid
Empirical: $C_6H_8O_3$
Formula: $H_2C{:}CHCOOCH_2CHCH_2O$
Properties: Liq.; insol. in water; m.w. 128.13; dens. 1.1074 (20/20 C); b.p. 57 C (2 mm); f.p. -41.5 C; flash pt. (TOC) 60.5 C
Precaution: Flamm. when exposed to heat or flame; can react vigorously with oxidizers
Toxicology: LD50 (oral, rat) 210 mg/kg, (dermal, rabbit) 400 mg/kg; toxic by ingestion, inh., skin contact; irritant to skin and eyes; mutagenic data; heated to decomp., emits acrid smoke and fumes
Uses: Polyfunctional monomer; in coatings to improve adhesion to substrate and solv. resistance
Manuf./Distrib.: Estron; Monomer-Polymer & Dajac Labs
Trade names: SR-378

Glycidyl butyl ether. *See* Butyl glycidyl ether

Glycidyl methacrylate
CAS 106-91-2; EINECS 203-441-9
Synonyms: GMA; 2,3-Epoxypropyl methacrylate; Methacrylic acid 2,3-epoxypropyl ester
Empirical: $C_7H_{10}O_3$
Properties: M.w. 142.15; dens. 1.042; b.p. 189 C; flash pt. 76 C; ref. index 1.450 (20 C)
Toxicology: LD50 (oral, rat) 597 mg/kg, (skin, rabbit) 469 mg/kg; mod. toxic by skin contact and IP rout; poison by ingestion; skin and eye irritant; heated to decomp., emits acrid smoke and fumes
Uses: Polyfunctional monomer; in hydrogels for contact lenses and membranes, molding and casting compds., impregnating paper, concrete, wood, coatings, printing inks, adhesives, sealants, elastomers, paints
Manuf./Distrib.: Aceto; Aldrich; Estron; Fluka; Mitsubishi Gas; Monomer-Polymer & Dajac Labs; Polysciences; Richman; San Esters; Sartomer; Sigma; Spectrum Chem. Mfg.
Trade names containing: SR-379

Glycidyl 2-methylphenyl ether. *See* o-Cresyl glycidyl ether

Glycidyl naphthenate
Uses: Raw material for paints
Manuf./Distrib.: Merichem

Glycidyl neodecanoate
CAS 26761-45-5
Synonyms: Neodecanoic acid glycidyl ester
Uses: Reactive diluent for epoxy resins; resin modifier; for coatings applics.; produces resins with reduced VOC content, exc. color and gloss retention, high hardness, exc. exterior durability
Manuf./Distrib.: Aldrich
Trade names: Glydexx® N-10; Glydexx® ND-101

3-Glycidyloxypropyltrimethoxysilane. *See* 3-Glycidoxypropyltrimethoxysilane
Glycidyl phenyl ether. *See* Phenyl glycidyl ether
Glycidyl propenate. *See* Glycidyl acrylate
Glycinol. *See* Ethanolamine

Glycogen
CAS 9005-79-2; EINECS 232-683-8
Synonyms: Animal starch; Liver starch
Definition: Glycose polysaccharide
Formula: $(C_6H_{10}O_5)_n$
Properties: Wh. powd., sweet taste; hygroscopic; insol. in alcohol, sol. in water; m.w. $(162.07)_n$; m.p. 240 C
Precaution: Store refrigerated
Uses: Biochemical research; in lacquers and varnishes as a plasticizer
Manuf./Distrib.: Aldrich; Fluka; Sigma; Spectrum Chem. Mfg.

Glycol
CAS 107-21-1; EINECS 203-473-3
Synonyms: Ethylene glycol; Glycol alcohol; 1,2-Ethanediol; Ethane-1,2-diol; Ethylene alcohol
Classification: Aliphatic diol
Empirical: $C_2H_6O_2$
Formula: $HOCH_2CH_2OH$
Properties: Clear liq., sweet taste (poisonous); very hygroscopic; misc. with water, lower aliphatic alcohols, glycerol; m.w. 62.07; dens. 1.1135 (20/4 C); visc. 17.3 cps (25 C); m.p. -13 C; b.p. 197.6 C (760 mm); flash pt. (OC) 115 C; ref. index 1.43063
Precaution: Combustible
Toxicology: LD50 (oral, rat) 4700 mg/kg; irritant; toxic by ingestion and inhalation; lethal human dose 1.4 ml/kg; TLV 50 ppm (vapor ceiling); heated to decomp., emits acrid smoke and fumes
Uses: Antifreeze in cooling and heating systems; in hydraulic brake fluids; industrial humectant; solv. in paints, plastics, printing inks; softening agent for cellophane; stabilizer; in explosives, alkyd resins, elastomers, syn. fibers and waxes; asphalt; wood stains; raw material for prod. of syn. polyester fibers, latex paints, polyester and alkyd resins; plasticizer; humectant; heat transfer agent; low pressure laminates; textile processing; foam stabilizer
Regulatory: FDA 21CFR §175.105, 176.300; FDA approved for topicals; BP compliance
Manuf./Distrib.: Aldrich; Ashland; J.T. Baker; BASF; Baychem; CasChem; Chemcentral; Coyne; Eastman; Fluka; C.P. Hall; Hoechst Celanese; Huntsman; Inspec Group plc; Mitsui Petrochem. Ind.; Mitsui Toatsu; Mobil; Occidental; Olin; Oxiteno; Shell; Sigma; Spectrum Chem. Mfg.; Sunnyside; Texaco; Union Carbide; Van Waters & Rogers
Trade names containing: Nacure® 2558; Surfynol® 104E; Surfynol® 104H; Surfynol® TG

Glycol alcohol. *See* Glycol
Glycol butyl ether. *See* Butoxyethanol
Glycol diacetate. *See* Ethylene glycol diacetate

Glycol dilaurate
CAS 624-04-4; EINECS 210-827-0
Synonyms: Ethylene glycol dilaurate; Lauric acid, 1,2-ethanediyl ester; Dodecanoic acid 1,2-ethanediyl ester
Definition: Diester of ethylene glycol and lauric acid
Empirical: $C_{26}H_{50}O_4$
Formula: $CH_3(CH_2)_{10}COOCH_2CH_2OCO(CH_2)_{10}CH_3$
Properties: Colorless amorphous mass; insol. in alcohol, ether; m.w. 426.66; m.p. 50-52 C; b.p. 188 C (20 mm)
Uses: Plasticizer in lacquers and varnishes; emulsifier, dispersant, antistat for textile, paper processing,

cutting oils, polishes, emulsion cleaners, rubber latex, wool lubricants

Glycol dimethyl ether. *See* Ethylene glycol dimethyl ether

Glycol distearate
CAS 627-83-8; EINECS 211-014-3
Synonyms: EGDS; Ethylene glycol distearate; Octadecanoic acid, 1,2-ethanediyl ester
Definition: Diester of ethylene glycol and stearic acid
Formula: $CH_3(CH_2)_{16}COOCH_2CH_2OCO(CH_2)_{16}CH_3$
Properties: Sp.gr. 0.97; m.p. 60 C; flash pt. (COC) 171 C
Uses: Pearlescent and opacifier; thickener, intermediate, lubricant, emulsifier, emollient; for emulsion shampoos and foam baths; plasticizer, lubricant, antistat for plastics and rubber; surfactant paints
Regulatory: FDA 21CFR §73.1, 176.210
Manuf./Distrib.: Ashland; Cedar Concepts; Inolex; Lonza; PPG Ind.; Rhone-Poulenc Surf. & Spec.; Witco/Oleo-Surf.
Trade names: Calgene EGDS

Glycol ether EM acetate. *See* Methoxyethanol acetate
Glycol ethylene ether. *See* 1,4-Dioxane

Glycol hydroxystearate
CAS 33907-46-9; EINECS 251-732-4
Synonyms: Ethylene glycol monohydroxystearate; Glycol monohydroxystearate; Hydroxyoctadecanoic acid, 2-hydroxyethyl ester
Definition: Ester of ethylene glycol and hydroxystearic acid
Empirical: $C_{20}H_{40}O_4$
Formula: $CH_3(CH_2)_5COHH(CH_2)_{10}COOCH_2CH_2OH$
Uses: Wax modifier, firming agent in pharmaceuticals and cosmetics, slip additives in varnishes

Glycol methacrylate. *See* 2-Hydroxyethyl methacrylate
Glycol monohydroxystearate. *See* Glycol hydroxystearate
Glycol monomethacrylate. *See* 2-Hydroxyethyl methacrylate
Glycol monomethyl ether acetate. *See* Methoxyethanol acetate
Glycol monostearate. *See* Glycol stearate

Glycol stearate
CAS 111-60-4; 97281-23-7; EINECS 203-886-9; 306-522-8
Synonyms: EGMS; Ethylene glycol monostearate; Ethylene glycol stearate; PEG-1 stearate; Glycol monostearate; POE (1) stearic acid; 2-Hydoxyethyl octadecanoate
Definition: Ester of ethylene glycol and stearic acid
Empirical: $C_{20}H_{40}O_3$
Formula: $CH_3(CH_2)_{16}COOCH_2CH_2OH$
Properties: Yel. waxy solid; sol. in alcohol, hot ether, acetone; insol. in water; m.w. 328.60; dens. 0.96 (25 C); m.p. 57-60 C
Precaution: Combustible
Toxicology: Poison by intraperitoneal route; skin irritant; heated to decomp., emits acrid smoke and fumes
Uses: Opacifier and pearling agent for cream shampoos, other cosmetics; plasticizer for cellulose nitrate; lubricant for plasticized PVC; surfactant for industrial and cosmetic applics., pulp/paper, pharmaceuticals, metalworking, lubricants, textilesi; agric., paints, adhesives
Regulatory: FDA 21CFR §176.210; FDA approved for topicals
Manuf./Distrib.: Ashland; C.P. Hall; Inolex; ISP Van Dyk; PPG Ind.; Rhone-Poulenc Surf. & Spec.; Stepan; Witco/Oleo-Surf.
Trade names: Calgene EGS

Glycyl alcohol. *See* Glycerin
GMA. *See* Glycidyl methacrylate
GMS. *See* Glyceryl stearate
Gold bronze. *See* Copper

Graphite
CAS 7782-42-5; EINECS 231-955-3
Synonyms: Black lead; Plumbago; Mineral carbon
Empirical: C

Green cinnabar

Properties: Steel gray to black powd., flake, cryst., rods, plates, or fibers; soft greasy feel, metallic sheen; dens. 2.0-2.25; resistant to oxidation and thermal shock
Precaution: Fire risk (powd., natural)
Toxicology: TLV 2.5 mg/m^3 respirable dust; mildly irritating to lungs
Uses: Reinforcing agent for plastics, carbon brushes, batteries, electrochemistry, pencils, hard metals, lubricants, catalysts, prepregging, filament winding; lubricant additive for greases, engine oils, etc.; paints, coatings, self-lubricating bearings
Manuf./Distrib.: Aldrich; Alfa Aesar Johnson Matthey; Cerac; Fluka; Johnson Matthey; Lonza; Premier Services; San Yuan; Shamokin Filler; Sigri GmbH; Ucar Carbon
Trade names: Dylon Grade AA Ultra Graphite Coat
Trade names containing: Kynar® 320; Kynar® 370

Green cinnabar. *See* Chromium oxide (ic)
Green vitriol. *See* Ferrous sulfate heptahydrate
Griffith's zinc white. *See* Lithopone

Groundnut acid
Uses: Used in mfg. of alkyd resins for stoving enamels, acid curing lacquers and NC lacquers; in wood and metal varnishes
Trade names: Prifac 7912

Groundnut oil. *See* Peanut oil
GTB. *See* Glyceryl tribenzoate
Guar flour. *See* Guar gum

Guar gum
CAS 9000-30-0; EINECS 232-536-8
Synonyms: Guar flour; Jaguar gum; Gum cyamopsis; Cyamopsis gum
Definition: Natural material derived from the ground endosperms of *Cyamopsis tetragonolobus*; consists of high m.w. hydrocolloidal polysaccaride composed of galactomannan units
Properties: Ylsh.-wh. free-flowing powd.; aq. sol'ns. tasteless, odorless; sol. in hot or cold water; insol. in oil, greases, hydrocarbons, ketones, esters; m.w. $\approx$ 220,000
Toxicology: Mildly toxic by ingestion; may cause contact dermatitis; heated to decomp., emits acrid smoke and irritating fumes
Uses: Paper coating, cosmetics, pharmaceuticals; as protective colloid, stabilizer; binding agent in tablets; flocculant in mining industry; coagulant aid in water treatment; stabilizer, thickener, emulsifier for foods; 5-8 times the thickening power of starch
Regulatory: FDA 21CFR §133.124, 133.133, 133.134, 133.162, 133.178, 133.179, 150.141, 150.161, 184.1339, GRAS; FEMA GRAS; FDA approved for buccals, orals; USP/NF compliance; Japan approved; Europe listed; UK approved; ADI not specified (WHO)
Manuf./Distrib.: Aldrich; Alfa; Aqualon; Fabrichem; Folexco; Gumix Int'l.; Hercules; Int'l. Chem. Inc.; Multi-Kem; Nat'l. Starch & Chem.; Penta Mfg.; Rhone-Poulenc; Sigma; Stan Chem Int'l. Ltd; TIC Gums

Gum camphor. *See* Camphor
Gum cyamopsis. *See* Guar gum
Gum rosin. *See* Rosin
Guncotton. *See* Nitrocellulose
Gypsum. *See* Calcium sulfate
Gypsum. *See* Calcium sulfate dihydrate

Hard paraffin. *See* Paraffin
HBCD. *See* Hexabromocyclododecane
HBD. *See* Tributyltin oxide
H.E. cellulose. *See* Hydroxyethylcellulose
HDDA. *See* 1,6-Hexanediol diacrylate
HDI. *See* Hexamethylene diisocyanate
HDO. *See* Hexamethylene glycol
HDODA. *See* 1,6-Hexanediol diacrylate
HDPE. *See* Polyethylene, high-density
Heavy mineral oil. *See* Mineral oil
Heavy sparr. *See* Barium sulfate

HEC. *See* Hydroxyethylcellulose
Hector clay. *See* Hectorite

Hectorite
CAS 12173-47-6; EINECS 235-340-0
Synonyms: Hector clay
Definition: One of the montmorillonite minerals that are the principal constituent of bentonite clay
Formula: ≈ $Na_0.67(Mg,Li)_6Si_8O_{20}(OH,F)_4$
Toxicology: No known toxicity to skin; dust can be irritating to respiratory tract
Uses: Emulsifier, extender; VOC-free thickener, sag control and suspension agent for latex paints, construction materials (grouts, tile adhesives, plasters, joint compds., textured coatings, traffic paints, underbody coatings, asphalt emulsions, mortars; in chillproofing of beer
Trade names: Benaqua™ 4000; Benaqua™ 6000; Bentone® AD; Bentone® CT; Bentone® EW; Bentone® HC; Bentone® MA; Optigel SH
Trade names containing: Bentone® LT; Hectabrite® LT

HEMA. *See* 2-Hydroxyethyl methacrylate
Hemellitol. *See* 1,2,3-Trimethylbenzene
Hemimellitene. *See* 1,2,3-Trimethylbenzene
2-(8-Heptadecenyl)-4,5-dihydro-1H-imidazole-1-ethanol. *See* Oleyl hydroxyethyl imidazoline

Heptadecenyl hydroxyethyl imidazoline
CAS 95-38-5
Synonyms: 1-(2-Hydroxyethyl)-2-heptadecenyl-2-imidazoline; 1-(2-Hydroxyethyl)-2-heptadecenylglyoxalidine; 2-(8-Heptadecenyl)-2-imidazoline-1-ethanol; 1-(2-Hydroxyethyl)-2-N-heptadecenyl-2-imidazoline; 1-Hydroxyethyl-2-heptadecenylglyoxalidine
Empirical: $C_{22}H_{42}N_2O$
Properties: Liq.; m.w. 350.66; dens. 0.9300 (20/20 C); b.p. 235 C (1 mm); flash pt. (OC) 465 F
Precaution: Combustible; can react with oxidizing materials
Toxicology: LD50 (oral, rat) 3130 mg/kg; mod. toxic by ingestion; heated to decomp., emits toxic fumes of NO_x
Uses: Corrosion inhibitor, emulsifier, penetrant, wetting agent, coupling agent, antistat; raw material for quat. reactions; paint industry; leather; metalworking
Trade names: Rewomine IM-CA; Rewomine IM-OA

2-(8-Heptadecenyl)-2-imidazoline-1-ethanol. *See* Heptadecenyl hydroxyethyl imidazoline
2-Heptadecyl-4,5-dihydro-1H-imidazole. *See* Stearyl hydroxyethyl imidazoline

Heptane
CAS 142-82-5; 64742-89-8; EINECS 205-563-8
Synonyms: n-Heptane; Alkane C_7; Dipropyl methane; Heptyl hydride
Classification: Aliphatic hydrocarbon
Empirical: C_7H_{16}
Formula: $CH_3(CH_2)_5CH_3$
Properties: Volatile colorless liq.; sol. in alcohol, ether, chloroform; insol. in water; m.w. 100.21; dens. 0.68368 (20 C); m.p. -91 C; b.p. 98.428 C; flash pt. (CC) -3.89 C; ref. index 1.38764 (20 C)
Precaution: Flamm. when exposed to heat or flame; dangerous fire risk; can react vigorously with oxidizing materials; mod. explosive exposed to heat or flame; violent reaction with phosphorus + chlorine
Toxicology: LD50 (IV, mouse) 222 mg/kg; TLV 400 ppm in air (ACGIH); STEL 500 ppm; lethal conc. for mice 15,900 ppm in air; irritating to respiratory tract; narcotic in high concs.; poison by IV route; heated to decomp., emits acrid smoke and fumes
Uses: Standard for octane rating determinations; anesthetic; solvent for rubber compounding, cements/ sealants, extraction of oils and fats; organic synthesis
Regulatory: FDA 21CFR §175.105, 177.1580, 27CFR §21.111
Manuf./Distrib.: Aldrich; Ashland; Burdick & Jackson; Coyne; Exxon; Fluka; Great Western; Humphrey; Phibro Energy USA; Phillips; Sigma; Texaco
Trade names containing: GP-197 Resin Sol'n.; GPRI™ CKU-2266; Tecsol® 1; Tecsol® 3

n-Heptane. *See* Heptane
1,7-Heptanedicarboxylic acid. *See* Azelaic acid
2-Heptanone. *See* Methyl n-amyl ketone
3,6,9,12,15,18,21-Heptaoxatricosane-1,2,3-diol. *See* PEG-8
Heptyl carbinol. *See* Caprylic alcohol
Heptyl hydride. *See* Heptane

Hexabromocyclododecane
CAS 3194-55-6; 25637-99-4; EINECS 221-695-9
Synonyms: HBCD; 1,2,5,6,9,10-Hexabromocyclododecane
Empirical: $C_{12}H_{18}Br_6$
Properties: Hygroscopic; m.w. 641.73; m.p. 188-191
Uses: Flame retardant for expandable PS, foamed PS, crystal and high-impact PS, SAN, adhesives, coatings
Manuf./Distrib.: Albemarle; Aldrich; Ameribrom; Fluka; Great Lakes
Trade names: FR-1206; Great Lakes CD-75P™; Great Lakes SP-75™; Saytex® HBCD-LM

1,2,5,6,9,10-Hexabromocyclododecane. *See* Hexabromocyclododecane
Hexabutyldistannoxane. *See* Tributyltin oxide
Hexadecafluoroheptane. *See* Perfluoroheptane
Hexadecanoic acid. *See* Palmitic acid
Hexadecanoic acid, hexadecyl ester. *See* Cetyl palmitate
1-Hexadecanol. *See* Cetyl alcohol
1-Hexadecanol, hydrogen sulfate, sodium salt. *See* Sodium cetyl sulfate

Hexadecyl acrylate
Definition: Ester of acylic acid and C14-C16 alcohols
Uses: Contributes to performance characteristics of all radiation curable formulations
Manuf./Distrib.: Monomer-Polymer & Dajac Labs
Trade names: Photomer® 4816

Hexadecyl alcohol. *See* Cetyl alcohol
Hexadecylic acid. *See* Palmitic acid
Hexaethylene glycol. *See* PEG-6

Hexafluoropropylene/vinylidene fluoride copolymer
Uses: Fluoroelastomer for molded goods, calendered goods, profiles; exc. mold release for transfer and inj. molding of complex shapes, e.g., gaskets, extrusions, sol'n. coatings (fabric, tanks or chem. containers)
Trade names: Viton® A-100; Viton® A-200

Hexafluoropropylene/vinylidene fluoride/tetrafluoroethylene terpolymer
Uses: Fluoroelastomer for molded goods, e.g., shaft seals, calendered goods, e.g., flue duct expansion joints; sol'n. coatings (fabric, tanks or chem. containers)
Trade names: Viton® B-200

Hexaglycerol. *See* Trimethylolpropane
Hexahydrobenzene. *See* Cyclohexane
Hexahydrophenol. *See* Cyclohexanol

Hexahydrophthalic acid
Uses: Intermediate for paints
Manuf./Distrib.: Hüls Am.

Hexahydrophthalic anhydride
CAS 85-42-7
Synonyms: HHPA; 1,2-Cyclohexanedicarboxylic anhydride
Empirical: $C_8H_{10}O_3$
Formula: $C_6H_{10}(CO)_2O$
Properties: Clear colorless visc. liq., glassy solid; misc. with benzene, toluene, acetone, CCl_4, chloroform, ethanol, ethyl acetate; sl. sol. in petrol. ether; m.w. 154; dens. 1.18 (40 C); b.p. 158 C (17 mm)
Toxicology: Toxic by inhalation; strong irritant to eyes and skin
Uses: Epoxy curing agent; intermediate for paints, alkyds, plasticizers, insect repellents, rust inhibitors, hardener in epoxy resins
Manuf./Distrib.: Allchem Ind.; Anhydrides & Chems.; Buffalo Color; Cambridge Ind. Co. of Am.; Dixie; GCA; Hüls AG; Milliken; Punda Mercantile
Trade names: Milldride® HHPA

Hexahydro terephthalic acid. *See* 1,4-Cyclohexanedimethanol dibenzoate

Hexahydro-1,3,5-triethyl-s-triazine
CAS 7779-27-3

Empirical: $C_9H_{21}N_3$
Properties: Lt. yel. liq.; sol. in water; m.w. 171.33; dens. 0.89 (25 C)
Toxicology: LD50 (oral, rat) 315 mg/kg; poison by ingestion; moderately toxic by skin contact; severe eye irritant; heated to decomp., emits acrid smoke and fumes
Uses: Fungicide, industrial preservative for latex, adhesives, cutting fluids, marine lubricants
Manuf./Distrib.: Aldrich
Trade names: Vancide® TH

Hexa(hydroxymethyl) melamine. *See* Hexamethylol melamine resin
Hexakis(hydroxymethyl) melamine. *See* Hexamethylol melamine resin
Hexakis(hydroxymethyl)1,3,5-triazine-2,4,6-triamine. *See* Hexamethylol melamine resin
Hexalin. *See* Cyclohexanol

Hexamethoxymethylmelamine
CAS 3089-11-0
Synonyms: HMMM
Empirical: $C_{15}H_{30}N_6O_6$
Formula: M.w. 390.51
Uses: Crosslinking agent in melamine resin coating systems, general industrial finishes, appliance finishes; also with alkyd, polyester, thermosetting acrylic, epoxy, and cellulose resins; condensation agent for resoricnol-type bonding systems
Trade names: Cymel 303; Resimene® 745; Resimene® 747

Hexamethyldisilazane
CAS 999-97-3; EINECS 213-668-5
Synonyms: HMDS; OAP; Bis (trimethylsilyl) amine; 1,1,1-Trimethyl-N-(trimethylsilyl) silanamine; Hexamethylsilazane
Empirical: $C_6H_{19}NSi_2$
Formula: $(CH_3)_3SiNHSi(CH_3)_3$
Properties: Liq.; sol. in acetone, benzene, ethyl ether, heptane, perchloroethylene; m.w. 161.44; dens. 0.77; b.p. 125 C; flash pt. 25 C; ref. index 1.4057
Precaution: Flamm.; dangerous fire hazard exposed to heat or flame; can react vigorously with oxidizers
Toxicology: LDLo (IP, mouse) 650 mg/kg; moderately toxic by intraperitoneal route; experimental tumorigen; PEL 6 mg/m^3; heated to decomp., emits toxic fumes of NO_x
Uses: Chemical intermediate, release agent, lubricant, coupling agent, chromatographic packings, silylating agent
Regulatory: BP compliance
Manuf./Distrib.: Aldrich; Austin; Dow Corning; FAR Research; Fluka; Gelest; Great Western; Hüls Am.; Hüls AG; Janssen Chimica; PCR; Schweizerhall; Sigma; Wacker-Chemie GmbH
Trade names containing: Cab-O-Sil® TS-530

Hexamethylenamine. *See* Hexamethylenetetramine
Hexamethylene. *See* Cyclohexane
Hexamethyleneamine. *See* Hexamethylenetetramine

Hexamethylene diisocyanate
CAS 822-06-0; EINECS 212-485-8
Synonyms: HDI; 1,6-Diisocyanatohexane; Isocyanic acid, hexamethylene ester; Hexamethylene 1,6-diisocyanate
Empirical: $C_8H_{12}N_2O_2$
Formula: $OCN(CH_2)_6NCO$
Properties: Liq.; sol. in hexane, benzene, chloroform; m.w. 168.20; dens. 1.04 (25/15.5 C); b.p. 61 C (0.08 mm); flash pt. 140 C; ref. index 1.453 (20 C)
Toxicology: LD50 (oral, rat) 738 mg/kg, (dermal, rabbit) 593 mg/kg; poison by inhalation, intravenous routes; moderately toxic by ingestion, skin contact
Uses: Chemical intermediate; paints; adhesives
Manuf./Distrib.: Aldrich; Fluka; Sigma
Trade names: Luxate® HB3000; Luxate® HB9000; Luxate® HD100; Luxate® HM; Luxate® HT2000
Trade names containing: Luxate® HB9075; Luxate® HT2090

Hexamethylene 1,6-diisocyanate. *See* Hexamethylene diisocyanate

Hexamethylene glycol
CAS 629-11-8; EINECS 211-074-0
Synonyms: HDO; 1,6-Hexanediol; 1,6-Dihydroxyhexane
Empirical: $C_6H_{14}O_2$
Formula: $HO(CH_2)_6OH$
Properties: Clear colorless crystals; sol. in water, alcohol; sparingly sol. in hot ether; m.w. 118.17; dens. 0.953 (50 C); m.p. 42.8 C; b.p. 208 C (760 mm); flash pt. 147 C; ref. index 1.4579 (25 C)
Precaution: Combustible when exposed to heat or flame; can react with oxidizers
Toxicology: LD50 (rat, oral) 3.73 g/kg; mod. toxic by ingestion; eye irritant; heated to decomp., emits acrid smoke and fumes
Uses: Intermediate in the prod. of nylon; mfg. of hexamethylenediamine, polyesters, polyurethanes, paints; in gasoline refining; as plasticizer
Manuf./Distrib.: Aldrich; Allchem Ind.; BASF; Eastern Chem.; Fluka; Penta Mfg.

Hexamethylenetetraamine. *See* Hexamethylenetetramine

Hexamethylenetetramine
CAS 100-97-0; EINECS 202-905-8
Synonyms: HMT; HMTA; Methenamine; Ammonioformaldehyde; Hexamethylenamine; Hexamethyleneamine; Hexamethylenetetraamine; Aminoform; Urotropine; Methamin; 1,3,5,7-Tetraazaadamantane; Hexamine
Empirical: $C_6H_{12}N_4$
Formula: $(CH_2)_6N_4$
Properties: Wh. cryst. powd. or colorless lustrous crystals, pract. odorless; hygroscopic; sol. in water, alcohol, chloroform; insol. in ether; m.w. 140.22; dens. 1.27 (25 C)
Precaution: DOT: Flamm. solid
Toxicology: Skin irritant; poison by subcutaneous route; moderately toxic by ingestion, intraperitoneal routes
Uses: Curing of phenolformaldehyde and resorcinolformaldehyde resins, adhesives, fungicide, antibacterial; intermediate for paints
Regulatory: FDA 21CFR §181.30; Europe listed; UK approved
Manuf./Distrib.: Aldrich; Allchem Ind.; Browning; Chemie Linz N. Am.; Fluka; R.W. Greeff; Harcros; Hüls Am.; Kowa Am.; Mitsubishi Gas; Monomer-Polymer & Dajac Labs; Occidental/Durez; San Yuan; Sigma; Spectrum Chem. Mfg.; Stanchem; Van Waters & Rogers; Vanguard Chem. Int'l.; Wright

Hexamethylolmelamine. *See* Hexamethylol melamine resin

Hexamethylol melamine resin
CAS 531-18-0
Synonyms: Hexa(hydroxymethyl) melamine; Hexakis(hydroxymethyl) melamine; Hexakis(hydroxymethyl) 1,3,5-triazine-2,4,6-triamine; 2,4,6-Tris(bis(hydroxymethyl)amino)-s-triazine; 1,3,5-Triazine-2,4,6-triyltrinitrilo)hexakis methanol; Hexamethylolmelamine
Empirical: $C_9H_{18}N_6O_6$
Properties: M.w. 306.33
Toxicology: LD50 (IV, mouse) 180 mg/kg; poison by IV route; skin and eye irritant; heated to decomp., emits toxic fumes of NO_x
Uses: Thermosetting resin producing durable press finishes on cellulosic fibers; stiffener for syn. fabrics
Trade names: Glazamine M

Hexamethylsilazane. *See* Hexamethyldisilazane
Hexamine. *See* Hexamethylenetetramine
6-Hexanalactone, 2-oxepanone. *See* ε-Caprolactone monomer
Hexanaphthene. *See* Cyclohexane

Hexane
CAS 110-54-3; 64742-49-0; EINECS 203-777-6
Synonyms: n-Hexane; Alkane C-6; Hexyl hydride
Classification: Aliphatic compd.
Empirical: C_6H_{14}
Formula: $CH_3(CH_2)_4CH_3$
Properties: Colorless volatile liq., faint odor; sol. in alcohol, acetone, ether; insol. in water; m.w. 86.20; dens. 0.65937 (20/4 C); m.p. -95 C; b.p. 68.742 C; flash pt. -22.7 C; ref. index 1.37486 (20 C)
Precaution: Flamm.; very dangerous fire/explosion hazard exposed to heat or flame; can react vigorously with oxidizers

Toxicology: TLV 50 ppm in air; LD50 (oral, rat) 28.710 mg/kg; sl. toxic by ingestion, inh.; human systemic effects by inh.; mutagenic data; eye irritant; irritating to respiratory tract; narcotic in high concs.; heated to decomp., emits acrid smoke and fumes
Uses: Solvent for veg. oil and pharmaceutical extraction, compounding rubber cements, alcohol denaturant, paint diluent, polymerization reaction medium; filling for thermometers
Regulatory: FDA 21CFR §173.270; Japan approved with restrictions
Manuf./Distrib.: Aldrich; Ashland; BP Chems. Ltd; Burdick & Jackson; Coyne; Exxon; Fluka; Great Western; Humphrey; Mitsui Petrochem. Ind.; Phibro Energy USA; Phillips; Shell; Sigma; Spectrum Chem. Mfg.; Texaco

n-Hexane. *See* Hexane
1,6-Hexanediamine, trimethyl. *See* Trimethylhexamethylene diamine
Hexanedioic acid. *See* Adipic acid
1,6-Hexanedioic acid. *See* Adipic acid
Hexanedioic acid, bis (2-ethylhexyl) ester. *See* Dioctyl adipate
Hexanedioic acid diisobutyl ester. *See* Diisobutyl adipate
Hexanedioic acid, diisodecyl ester. *See* Diisodecyl adipate
1,6-Hexanediol. *See* Hexamethylene glycol

1,6-Hexanediol diacrylate
CAS 13048-33-4; EINECS 235-921-9
Synonyms: HDDA; HDODA; 2-Propenoic acid 1,6-hexanediyl ester
Formula: $[H_2C{=}CHCO_2(CH_2)_3]_2$
Properties: Sol. in org. solvs.; difficulty sol.; m.w. 226.28; dens. 1.010; flash pt. > 110 C; ref. index 1.4560
Precaution: Hygroscopic
Uses: Crosslinking agent used in inks, coatings, adhesives, textile prod. modifiers, photoresists, modifiers for castings, polyesters, fiberglass, or radiation cured prods.
Manuf./Distrib.: Aldrich; CPS; Monomer-Polymer & Dajac Labs; UCB Radcure
Trade names: Photomer® 4017
Trade names containing: Ageflex HDDA; CN 104 B80; CN 120 B80; CN 945 B85; CN 960 B85; CN 961 B85; CN 962 B85; CN 963 B80; CN 963 B85; CN 964 B85; CN 965 B85; CN 966 B85; CN 970 B75; CN 971 B75; CN 972 B85; CN 981 B88; Craynor 104 B80; Ebecryl® 244; Ebecryl® 264; Ebecryl® 284; Ebecryl® 3700-20H; Ebecryl® 4849; SR-238

1,6-Hexanediol dimethacrylate. *See* 1,6-Hexanediol methacrylate

1,6-Hexanediol methacrylate
CAS 6606-59-3
Synonyms: 1,6-Hexanediol dimethacrylate; 1,6-Hexanediyl 2-methyl-2-propenoate
Empirical: $C_{14}H_{22}O_4$
Properties: M.w. 254.33
Uses: Crosslinking agent used in casting compds., glass fiber-reinforced plastics, adhesives, coatings, ion-exchange resins, textile prods., plastisols, dental polymers, rubber compding.
Trade names containing: SR-239

1,6-Hexanediyl 2-methyl-2-propenoate. *See* 1,6-Hexanediol methacrylate
1,2,3,4,5,6-Hexanehexol. *See* Sorbitol
Hexanoic acid. *See* Caproic acid
Hexanoic acid, 2-ethyl-, zinc salt. *See* Zinc 2-ethylhexanoate
1-Hexanol. *See* Hexyl alcohol
n-Hexanol. *See* Hexyl alcohol
Hexoic acid. *See* Caproic acid
Hexone. *See* Methyl isobutyl ketone
n-Hexoxyethoxyethanol. *See* Diethylene glycol n-hexyl ether

Hexyl acetate
CAS 142-92-7; EINECS 205-572-7
Synonyms: Hexyl ethanoate; Acetic acid hexyl ester; n-Hexyl acetate; 1-Hexyl acetate; Hexyl alcohol, acetate
Empirical: $C_8H_{16}O_2$
Formula: $CH_3COO(CH_2)_5CH_3$
Properties: Colorless oily liq., pleasant fruity odor, bittersweet taste; sol. in alcohol, ether; insol. in water; m.w. 144.22; sp.gr. 0.873 (20/4 C); m.p. -81 C; b.p. 167-169 C; flash pt. 41 C; ref. index 1.409 (20 C)
Precaution: Flamm.

1-Hexyl acetate

Toxicology: LD50 (oral, rat) 42 g/kg; mildly toxic by ingestion; heated to decomp., emits acrid smoke and fumes
Uses: Solv. for cellulosic and acrylic coatings
Regulatory: FDA 21CFR §172.515; FEMA GRAS
Manuf./Distrib.: Aldrich; Berje; Fluka; Penta Mfg.; Sigma

1-Hexyl acetate. *See* Hexyl acetate
n-Hexyl acetate. *See* Hexyl acetate
s-Hexyl acetate. *See* Methyl amyl acetate

n-Hexyl acrylate
CAS 2499-95-8
Synonyms: Acrylic acid, hexyl ester
Classification: Monomer
Empirical: $C_9H_{16}O_2$
Formula: $CH_2=CHCOOC_6H_{13\text{-}n}$
Properties: M.w. 156.23
Toxicology: LD50 (oral, rat) 26 g/kg, (dermal, rabbit) 5660 mg/kg; mildly toxic by skin contact; heated to decomp., emits acrid smoke and fumes
Uses: Monomer for UV-cured inks and coatings, glass coating, visc. index improver for functional oils, polymer cements and sealants; polymer modifier
Manuf./Distrib.: Aldrich; CPS; Hampford Research; Monomer-Polymer & Dajac Labs
Trade names containing: Ageflex FA-6; SR-439

Hexyl alcohol
CAS 111-27-3; 68526-79-4; EINECS 203-852-3
Synonyms: n-Hexyl alcohol; 1-Hexanol; n-Hexanol; Alcohol C-6; 1-Hydroxyhexane; Pentylcarbinol; Amylcarbinol
Classification: Aliphatic alcohol
Empirical: $C_6H_{14}O$
Formula: $CH_3(CH_2)_4CH_2OH$
Properties: Colorless liq., fruity odor, aromatic flavor; sol. in alcohol and ether; sl. sol. in water; m.w. 102.20; dens. 0.8186; f.p. -51.6 C; b.p. 157.2 C; flash pt. (TOC) 65 C; ref. index 1.1469 (25 C)
Precaution: Flamm. or combustible liq.; reactive with oxidizing materials
Toxicology: LD50 (rat, oral) 4.59 g/kg, (skin, rabbit) 3100 mg/kg; poison by intravenous route; moderately toxic by ingestion, skin contact; skin and severe eye irritant
Uses: Solv. for waterborne coatings; pharmaceuticals (antiseptics, perfume esters); plasticizer; flavoring agent; intermediate; emulsion polymerization; intermediate for textile and leather finishing agents
Regulatory: FDA 21CFR §172.515, 172.864, 178.3480; FEMA GRAS
Manuf./Distrib.: Albemarle; Aldrich; Ashland; Fluka; Penta Mfg.; Sigma; Vista
Trade names: Exxal® 6
Trade names containing: Nansa® EVM70/B

n-Hexyl alcohol. *See* Hexyl alcohol
Hexyl alcohol, acetate. *See* Hexyl acetate
Hexyl Carbitol. *See* Diethylene glycol n-hexyl ether
n-Hexyl 'Cellosolve'. *See* Ethylene glycol hexyl ether
Hexyldiglycol. *See* Diethylene glycol n-hexyl ether

Hexylene glycol
CAS 107-41-5; EINECS 203-489-0
Synonyms: 2-Methyl-2,4-pentanediol; 4-Methyl-2,4-pentanediol; α,α,α'-Trimethyltrimethyleneglycol
Classification: Aliphatic alcohol
Empirical: $C_6H_{14}O_2$
Formula: $(CH_3)_2COHCH_2CHOHCH_3$
Properties: Colorless liq., nearly odorless; hygroscopic; misc. with water, hydrocarbons, fatty acids; m.w. 118.18; dens. 0.9216 (20/4 C); b.p. 198.3 C; flash pt. (OC) 93 C; ref. index 1.4276 (20 C)
Precaution: Combustible; can react with oxidizing materials
Toxicology: TLV:CI 25 ppm in air; LD50 (rat, oral) 4.70 g/kg; heated to decomp., emits acrid smoke and fumes
Uses: Hydraulic brake fluids, printing inks, coupling agent and penetrant for textiles, cosmetics; ice inhibitor in carburetors; fuel and lubricant additive; emulsifier
Regulatory: FDA 21CFR §175.105, 176.180, 176.200, 176.210, 177.1210, 177.2800; FDA approved for topicals; USP/NF compliance
Manuf./Distrib.: Aldrich; Allchem Ind.; Ashland; BP Chems. Ltd; Coyne; Elf Atochem SA; Fluka; Great Western;

Mitsui Petrochem. Ind.; Penta Mfg.; Shell; Sigma; Union Carbide
Trade names containing: Activ-8 in Hexylene Glycol; Monawet MT-70

Hexyl ethanoate. *See* Hexyl acetate
Hexylglycol. *See* Ethylene glycol hexyl ether
Hexyl hydride. *See* Hexane

Hexyl methacrylate
CAS 142-09-6
Synonyms: N-Hexyl methacrylate
Formula: $C_6H_{13}OOCC(CH_3):CH_2$
Properties: Liq.; dens. 0.88; b.p. 67-85 C (8 mm)
Uses: Monomer for plastics, molding powder, etc.; emulsions for textile, leather, and paper finishing
Manuf./Distrib.: Monomer-Polymer & Dajac Labs; Rohm Tech
Trade names containing: SR-211

N-Hexyl methacrylate. *See* Hexyl methacrylate
Hexyl methyl ketone. *See* Methyl hexyl ketone
2-(Hexyloxy)ethanol. *See* Ethylene glycol hexyl ether
2-(2-Hexyloxyethoxy)ethanol. *See* Diethylene glycol n-hexyl ether

Hexyl phosphate
CAS 3900-04-7
Uses: Surfactant for fiber lubricant finishes; antistat
Trade names: Findet NHP

HHPA. *See* Hexahydrophthalic anhydride
HMDS. *See* Hexamethyldisilazane
HMMM. *See* Hexamethoxymethylmelamine
HMT. *See* Hexamethylenetetramine
HMTA. *See* Hexamethylenetetramine
HPC. *See* Hydroxypropylcellulose
HSA. *See* Hydroxystearic acid
Huile de ricini. *See* Castor oil
2-Hydoxyethyl octadecanoate. *See* Glycol stearate
Hydrated alumina. *See* Alumina trihydrate; Aluminum hydroxide
Hydrated aluminum oxide. *See* Alumina trihydrate
Hydrated aluminum silicate. *See* Kaolin; Pyrophyllite

Hydroabietyl alcohol
CAS 1333-89-7; 26266-77-3; EINECS 247-574-0
Synonyms: Dihydroabietyl alcohol; Dodecahydro-1,4a-dimethyl-7-(1-methylethyl)-1-phenanthrenemethanol
Definition: Organic alcohol derived from wood rosin
Empirical: $C_{20}H_{34}O$
Formula: $C_{19}H_{31}CH_2OH$
Properties: Solid; insol. in water; dens. 1.007-1.008; m.p. 32-33 C; flash pt. (COC) 185 C; ref. index 1.526 (20 C)
Precaution: Combustible
Uses: Plasticizer for cellulose nitrate, ethylcelluose, PVC, paints
Regulatory: FDA 21CFR §175.105, 176.180
Manuf./Distrib.: Hercules
Trade names: Abitol®

Hydroabietyl phthalate
Properties: Sp.gr. 1.055 (20 C); m.p. 63 C; ref. index 1.513
Uses: Resin used in hot-melt and pressure-sensitive adhesives, specialty nitrocellulose lacquers, printing inks; plasticizer for PVC, cellulose nitrate
Manuf./Distrib.: Hercules
Trade names: Cellolyn® 21

Hydrocarbon solvent
CAS 64742-46-7
Uses: Solvent, foam control agent, in waterless hand cleaners, agric. sprays, polishes, fruit and veg. processing,

cleaning oils, paper, coatings
Manuf./Distrib.: Penreco
Trade names: Penreco 2257 Oil; Penreco 2259 Oil; Penreco 2260 Oil

Hydrogenated castor oil
CAS 8001-78-3; EINECS 232-292-2
Synonyms: Opalwax; Castorwax; Castor oil, hydrogenated
Definition: End prod. of controlled hydrogenation of castor oil, consisting mainly of the triglyceride of hydroxystearic acid
Properties: Wh. hard wax; very insol. in water and in the more common org. solvs.; m.w. ≈ 932; m.p. 86-88 C; iodine no. 5 max.; sapon. no. 176-182; hyd. no. 154-162
Toxicology: Ingestion of large amts. may cause pelvic congestion
Uses: In water-repellent coatings, candles, polishes, ointments, cosmetics; impregnant for paper, wood, cloth; as lubricant, mold release in mfg. of formed plastics and rubber goods
Regulatory: FDA 21CFR §175.105, 175.300, 176.170, 176.210, 177.1200, 177.1210, 177.2420, 177.2800, 178.3280; USP/NF compliance
Manuf./Distrib.: Akzo Nobel; Amber Syn.; Arista Ind.; Ashland; Hoechst AG; Southern Clay Prods.; Welch, Holme & Clark
Trade names: Castorwax® MP-70; Castorwax® MP-80; Cenwax® G; Ceroxin Special; Hydrogenated Castor Wax; Rilanit® HT Extra; Rilanit® HT-EZ; Rilanit® Special

Hydrogenated methyl ester of rosin. *See* Methyl hydrogenated rosinate

Hydrogenated rosin
Uses: Thermoplastic acidic resin; as tackifier or modifying resin in adhesives and hot-melt-applied decorative, pressure sensitive, and heat-sealable coatings
Trade names: Foral® AX

Hydrogenated soybean oil
CAS 8016-70-4; 68002-71-1; EINECS 232-410-2
Synonyms: Soybean oil hydrogenated
Definition: End prod. of controlled hydrogenation of soybean oil
Uses: Textile lubricant, pharmaceutical intermediate, emulsifier, mold release agent, buffing compd.
Regulatory: FDA 21CFR §175.105, 176.210, 177.2800, 182.70, 182.170
Manuf./Distrib.: Alnor Oil
Trade names containing: Ross Japan Wax Substitute 966

Hydrogenated tallow amide
CAS 61790-31-6; EINECS 263-123-0
Synonyms: Amides, tallow, hydrogenated; Tallow amides, hydrogenated
Classification: Amide
Formula: $RCONH_2$, RCO- represents the fatty acids derived from hydrog. tallow
Uses: Antiblock, lubricant, slip agent for plastics, coatings, films; foam booster, builder for syn. detergents; water repellent for textiles; intermediate for syn. waxes; pigment dispersant; rubber processing
Trade names: Armid® HT

Hydrogenated tallow glyceride
CAS 61789-09-1; EINECS 263-031-0
Synonyms: Hydrogenated tallow monoglyceride; Glycerides, hydrogenated tallow mono-
Definition: Monoglyceride of hydrogenated tallow
Uses: Emulsifier, stabilizer, dispersant, opacifier for cosmetics, foods and drugs
Regulatory: FDA 21CFR §176.210, 177.2800
Trade names containing: Ross Synthetic Candelilla Wax

Hydrogenated tallow glycerides
CAS 68308-54-3; EINECS 269-658-6
Synonyms: Hydrogenated tallow mono-, di- and tri- glycerides; Glycerides, tallow mono-, di- and tri-, hydrogenated
Definition: Mixture of mono, di and triglycerides of hydrogenated tallow acid
Uses: Emulsifier, stabilizer, dispersant, opacifier for cosmetics, foods, and drugs; lubricating greases, synthetic waxes, textile lubricants
Regulatory: FDA 21CFR §176.210, 177.2800
Trade names containing: Ross Beeswax Substitute No. 628/5

Hydrogenated tallow mono-, di- and tri- glycerides. *See* Hydrogenated tallow glycerides
Hydrogenated tallow monoglyceride. *See* Hydrogenated tallow glyceride
Hydrogen carboxylic acid. *See* Formic acid
Hydrogen dioxide. *See* Hydrogen peroxide

Hydrogen peroxide
CAS 7722-84-1; EINECS 231-765-0
Synonyms: Hydrogen dioxide; Hydroperoxide
Classification: Inorganic oxide
Empirical: H_2O_2
Formula: HOOH
Properties: Colorless visc. liq., cryst. solid at low temp., bitter taste; sol. in ether, alcohol; misc. with water; dec. by many org. solvs.; m.w. 34.02; dens. (liq.) 1.450 g/cc (20 C); f.p. -0.41 C; b.p. 150.2 C
Precaution: Dangerous fire hazard by chem. reaction with flamm. materials; explosion hazard; strong oxidizer
Toxicology: LD50 (oral, mouse) 2 g/kg, (skin, rat) 4060 mg/kg; mod. toxic by inh., ingestion, skin contact; corrosive irritant to skin, eyes, mucous membranes; tumorigenic; human mutagenic data; TLV: 1 ppm in air
Uses: Bleaching and deodorizing textiles, wood pulp, hair, fur, etc.; plasticizers; refining and cleaning metals; visc. control for starch and cellulose derivatives; intermediate for paints
Regulatory: FDA 21CFR §133.133, 160.105, 160.145, 160.185, 172.814, 172.892, 175.105, 178.1005, 178.1005 (35% sol'n. max.), 178.1010, 184.1366, GRAS; BATF 27CFR §240.1051 (3 ppm max. in wine), 240.1051a (200 ppm max. in distilling materials); Japan restricted; FDA approved for topicals; BP compliance (sol'n.)
Manuf./Distrib.: Aldrich; AlliedSignal; Ashland; J.T. Baker; Browning; Coyne; Degussa; DuPont; Eka Nobel; Elf Atochem N. Am.; Farleyway Chem. Ltd; Fluka; FMC; C.P. Hall; Harcros; Mallinckrodt; Mitsubishi Gas; Olin; Sigma; Spectrum Chem. Mfg.; Stanchem; Van Waters & Rogers; Veckridge

Hydroperoxide. *See* Hydrogen peroxide
2-Hydroperoxy-2-methylpropane. *See* t-Butyl hydroperoxide
Hydroquinol. *See* Hydroquinone

Hydroquinone
CAS 123-31-9; EINECS 204-617-8
Synonyms: 1,4-Benzenediol; 1,4-Dihydroxybenzene; p-Dihydroxybenzene; Hydroquinol
Classification: Aromatic organic compd.
Empirical: $C_6H_6O_2$
Formula: $C_6H_4(OH)_2$
Properties: Wh. crystals; sol. in water, alcohol, ether; m.w. 110.11; dens. 1.330; m.p.170 C; b.p. 285 C; flash pt. 165 C
Precaution: Combustible; light-sensitive
Toxicology: LD50 (rat, oral) 320 mg/kg; TLV 2 mg/m^3 of air; toxic by ingestion and inhalation; irritant to eyes, skin, respiratory tract; ingestion of large amts. has caused nausea, vomiting, ringing in ears, delirium, sense of suffocation, and collapse
Uses: Photographic developer (not for color film); dye intermediate; inhibitor; stabilizer in paints and varnishes; motor fuels and oils; antioxidant for fats and oils, syn. latexes, polyester resins
Regulatory: FDA 21CFR §175.105, 176.170, 177.2420; BP compliance
Manuf./Distrib.: Aldrich; Alfa; Allchem Ind.; Charkit; Eastman; Fluka; Goodyear; Kraeber GmbH; Penta Mfg.; San Yuan; Sigma; Spectrum Chem. Mfg.

Hydroquinone methyl ether. *See* Hydroquinone monomethyl ether

Hydroquinone monomethyl ether
CAS 150-76-5; EINECS 205-769-8
Synonyms: MEHQ; Hydroquinone methyl ether; 4-Methoxyphenol; p-Hydroxyanisole (INCI)
Classification: Substituted phenolic compd.
Definition: Monomethyl ether of hydroquinone
Empirical: $C_7H_8O_2$
Formula: $CH_3OC_6H_4OH$
Properties: Wh. waxy solid; hygroscopic; sl. sol. in water; sol. in benzene, acetone, ethyl acetate, alcohol; m.w. 124.14; dens. 1.55 (20/20 C); m.p. 52.5 C; b.p. 243 C
Precaution: Combustible
Toxicology: LD50 (oral, rat) 1600 mg/kg; mildly toxic by ingestion; skin irritant; heated to decomp., emits acrid smoke and fumes

Hydrous magnesium calcium silicate

Uses: Mfg. of antioxidants, pharmaceuticals, plasticizers, dyestuffs; stabilizer for chlorinated hydrocarbons and ethylcellulose; UV inhibitor; inhibitor for acrylic and vinyl monomers and acrylonitrile
Regulatory: FDA 21CFR §177.1010
Manuf./Distrib.: Aldrich; Alemark; Alfa; Allchem Ind.; Arenol; ChemDesign; Eastman; Fluka; Kincaid Enterprises; Monomer-Polymer & Dajac Labs; Penta Mfg.; Rhone-Poulenc N. Am.; Sigma; Specialty Chem Prods.; Spectrum Chem. Mfg.

Hydrous magnesium calcium silicate. *See* Talc
Hydrous magnesium silicate. *See* Talc
p-Hydroxyanisole (INCI). *See* Hydroquinone monomethyl ether
Hydroxybenzene. *See* Phenol
4-Hydroxybenzenesulfonic acid. *See* Phenol sulfonic acid

2-Hydroxybenzophenone
CAS 117-99-7
Synonyms: o-Benzoylphenol; Phenyl 2-hydroxyphenyl ketone
Empirical: $C_{13}H_{10}O_2$
Formula: $C_6H_5COC_6H_4OH$
Properties: Solid; insol. in water; sol. in alcohol; m.w. 198l.22; m.p. 41 C; b.p. 210 C (27 mm); flash pt. > 110 C
Toxicology: Irritant
Uses: UV absorber in plastics, coatings
Manuf./Distrib.: Aldrich
Trade names: Sanduvor® 3041 Disp.

2-Hydroxybutane. *See* 2-Butanol
Hydroxybutanedioic acid. *See* N-Hydroxysuccinic acid
4-Hydroxybutanoic acid lactone. *See* Butyrolactone
2-(2´-Hydroxy-3´-t-butyl-t´-methylphenyl)-5-chlorobenzotriazole. *See* Bumetrizole

2-(2´-Hydroxy-3,5´-di-t-amylphenyl) benzotriazole
CAS 25973-55-1
Synonyms: 2-(2H-Benzotriazol-2-yl)-4,6-bis(1,1-dimethylpropyl)phenol
Properties: Powd.; m.p. 81 C
Uses: UV absorber, light stabilizer for PVC, ABS, cellulosics, epoxy, PS, PE, PP, PS, VDC, coatings, automotive coatings, styrenics, polyolefins, acrylic, other substrates, food contact applics.
Manuf./Distrib.: Aldrich
Trade names: Tinuvin® 328

m-Hydroxy-N,N-dimethyl aniline. *See* Dimethylaminomethyl phenol
Hydroxydimethylbenzene. *See* Xylenol
1-Hydroxyethane 1-carboxylic acid. *See* Lactic acid
1-Hydroxy-1,2-ethanedicarboxylic acid. *See* N-Hydroxysuccinic acid
2-Hydroxy-1-ethanethiol. *See* 2-Mercaptoethanol
2-Hydroxyethylamine. *See* Ethanolamine
β-**Hydroxyethylamine.** *See* Ethanolamine

Hydroxyethylcellulose
CAS 9004-62-0
Synonyms: HEC; Cellulose, 2-hydroxyethyl ether; H.E. cellulose
Definition: Partially substituted poly(hydroxyethyl) ether of cellulose
Formula: $C_6H_7O_2(OH)_2OCH_2CH_2OH$
Properties: Wh. free-flowing powd., odorless, tasteless; nonionic; hygroscopic; insol. in org. solvs.; sol. in hot or cold water; grease and oil resistant; m.p. 288-290 C (dec.); ref. index 1.336; pH 6.0-8.5 (1%)
Precaution: Combustible
Uses: Thickener, suspending agent; stabilizer for vinyl polymerization; retards evaporation of water in mortars and cements; binder in ceramic glazes; used in paper and textile sizing; latex paints; protective colloid
Regulatory: FDA 21CFR §175.105, 175.300, 177.1200, 177.1400; FDA approved for ophthalmics, orals, otics, topicals; USP/NF, BP, Ph.Eur. compliance
Manuf./Distrib.: Aldrich; Allchem Ind.; Amerchol Europe; Aqualon; Spectrum Chem. Mfg.; Sumisho Plaschem; Union Carbide
Trade names: Cellosize® ER-15M; Cellosize® ER-30M; Cellosize® ER-52M; Cellosize® ER- 4400; Cellosize® QP-300; Cellosize® QP-300; Cellosize® QP-4400; Cellosize® QP-15,000; Cellosize® QP-

30,000; Cellosize® QP-52,000; Cellosize® QP-100,000; Natrosol® 250; Natrosol® Hydroxyethylcellulose; Natrosol® Plus; Natrosol® Plus HMHEC, Grade 330; Natrosol® Plus HMHEC, Grade 430
Trade names containing: Bentone® LT; Hectabrite® LT

2-Hydroxyethyl ester methacrylic acid. *See* 2-Hydroxyethyl methacrylate

2-Hydroxyethylethylene urea
CAS 3699-54-5
Synonyms: 1-(2-Hydroxyethyl)-2-imidazolidinone
Empirical: $C_5H_{10}N_2O_2$
Properties: M.w. 130.13; dens. 1.190; ref. index 1.4660
Toxicology: Irritant
Uses: Monomer used in coatings (metal, plastic, wood), inks, paints, sealants, adhesives, and chem. intermediates; reduces VOCs, improves gloss, hydrophilic
Manuf./Distrib.: Aldrich
Trade names: SR-511

1-Hydroxyethyl-2-heptadecenylglyoxalidine. *See* Heptadecenyl hydroxyethyl imidazoline
1-(2-Hydroxyethyl)-2-heptadecenylglyoxalidine. *See* Heptadecenyl hydroxyethyl imidazoline
1-(2-Hydroxyethyl)-2-heptadecenyl-2-imidazoline. *See* Heptadecenyl hydroxyethyl imidazoline
1-(2-Hydroxyethyl)-2-N-heptadecenyl-2-imidazoline. *See* Heptadecenyl hydroxyethyl imidazoline

N (2-Hydroxyethyl) 12-hydroxystearamide
Uses: Internal mold release agent, lubricant for polyolefins, PVC, styrenics; antiblocking agent for textile coatings; slip agent for varnishes and lacquers; elec. potting compds., plug valve lubricants, crayons, wax blends, high-temp. greases
Trade names: Paricin® 220

1-(2-Hydroxyethyl)-2-imidazolidinone. *See* 2-Hydroxyethylethylene urea
2-Hydroxyethylmercaptan. *See* 2-Mercaptoethanol
Hydroxyethyl methacrylate. *See* 2-Hydroxyethyl methacrylate

2-Hydroxyethyl methacrylate
CAS 868-77-9; EINECS 212-782-2
Synonyms: HEMA; Ethylene glycol methacrylate; Ethylene glycol monomethacrylate; Glycol methacrylate; Glycol monomethacrylate; 2-Hydroxyethyl ester methacrylic acid; Hydroxyethyl methacrylate; β-Hydroxyethylmethacrylate
Empirical: $C_6H_{10}O_3$
Formula: $CH_2:C(CH_3)COOCH_2CH_2OH$
Properties: Clear mobile liq.; sol. in common org. solvs.; m.w. 130.16; dens. 1.0644 (77/60 F); f.p. -2 C; b.p. 205-208 C; ref. index 1.4505
Precaution: Flamm.; mod. fire risk
Toxicology: LD50 (IP, rat) 1250 mg/kg, (oral, mouse) 5888 mg/kg; mod. toxic by IP route; mildy toxic by ingestion; heated to decomp., emits acrid smoke and irritating fumes
Uses: Monomer for creating and modifying wide range of polymers, acrylic resins, binder for nonwoven fabrics, enamels, adhesives; reactive thinner for radiation curing; rubber modifier
Manuf./Distrib.: Aldrich; Allchem Ind.; Ashland; BP Chems. Inc; Fluka; Mitsubishi Gas; Monomer-Polymer & Dajac Labs; Rohm & Haas; Rohm Tech; San Esters; Scientific Polymer Prods.; Sigma
Trade names: BM-903; Sipomer® HEM-D

β-Hydroxyethylmethacrylate. *See* 2-Hydroxyethyl methacrylate
Hydroxyethylmethylcellulose. *See* Methyl hydroxyethylcellulose
1-Hydroxyethyl-2-oleyl imidazoline. *See* Oleyl hydroxyethyl imidazoline

N(β-Hydroxyethyl) ricinoleamide
Classification: Hydroxyamide wax
Properties: Sp.gr. 1.00; m.p. 46 C
Uses: Lubricant, antistat, mold release; in plastics, metals, textile coatings; slip agent for varnishes and lacquers; elec. potting compds., crayons, wax blends, high-temp. greases

1-Hydroxyethyl-2-tall oil imidazoline. *See* Tall oil hydroxyethyl imidazoline
1-Hydroxyethyl 2-undecyl imidazoline. *See* Lauryl hydroxyethyl imidazoline
1-Hydroxyhexane. *See* Hexyl alcohol

α-Hydroxy-ω-hydroxy poly(oxy-1,2-ethanediyl). *See* Polyethylene glycol

α-Hydroxy ketone
Synonyms: Alpha hydroxy ketone
Trade names containing: Esacure® KIP 100F; Esacure® KT37

Hydroxylamine sulfate
CAS 10039-54-0; EINECS 233-118-8
Synonyms: Hydroxylammonium sulfate; Oxammonium sulfate; Bis (hydroxylamine) sulfate
Empirical: $H_6N_2O_2 \cdot H_2O_4S$
Formula: $(NH_2OH)_2 \cdot H_2SO_4$
Properties: Cryst.; m.w. 164.16; m.p. 177 C
Precaution: DOT: Corrosive material
Toxicology: LDLo (IP, mouse) 102 mg/kg; corrosive irritant to skin, eyes, mucous membranes; poison by IP route
Uses: Used in paints, coatings, metal extractants, personal care, pharmaceutical intermediates, agricultural intermediates, also in fragrance, photography, rubber and environmental markets
Manuf./Distrib.: Aldrich; AlliedSignal; Am. Int'l.; BASF; Charkit; Eastern Chem.; Fluka; Penta Mfg.; Sigma; Spectrum Chem. Mfg.

Hydroxylammonium sulfate. *See* Hydroxylamine sulfate

Hydroxylated lecithin
CAS 8029-76-3; EINECS 232-440-6
Synonyms: Lecithin, hydroxylated
Definition: Prod. obtained by the controlled hydroxylation of lecithin
Properties: Lt. yel. liq. to paste, char. odor; mod. sol. in water
Toxicology: Nontoxic; heated to decomp., emits acrid smoke and irritating fumes
Uses: Wetting agent, emulsifier for personal care products and pharmaceuticals; wetting and dispersing agent for air-drying and stoving paints; raw material for textile and leather compds.
Regulatory: FDA 21CFR §136.110, 136.115, 136.130, 136.160, 136.165, 136.180, 172.814, 173.340, 176.170, 176.200
Trade names: Lipotin H

2-Hydroxy-4-methoxybenzophenone. *See* Benzophenone-3
(2-Hydroxy-4-methoxyphenyl) phenylmethanone. *See* Benzophenone-3

2[(-Hydroxymethyl) amino] ethanol
CAS 34375-28-5; EINECS 251-974-0
Synonyms: Ethanol, 2-(hydroxymethylamino)-
Uses: Preservative for protection against bacterial spoilage in aq. systems, coatings
Trade names: Nuosept® 91; Troysan® 174
Trade names containing: Troysan® 364

2-[(-Hydroxymethyl) amino]-2-methylpropanol
CAS 52299-20-4
Synonyms: 2-((Hydroxymethyl)amino) 2-methyl-1-propanol; 1-Propanol, 2-((hydroxymethyl)amino)-2-methyl
Empirical: $C_5H_{13}NO_2$
Properties: M.w. 119.16
Uses: Preservative used in aq. systems, coatings
Trade names: Troysan® 192

2-((Hydroxymethyl)amino) 2-methyl-1-propanol. *See* 2-[(-Hydroxymethyl) amino]-2-methylpropanol

N-(Hydroxymethyl)-N-(1,3-dihydroxymethyl-2,5-dioxo-4-imidazolidinyl)-N-(hydroxymethyl) urea
Uses: Industrial biocide for household, industrial, institutional products, latex paint systems and in-can preservative, adhesives, water-based inks, polishes, waxes, pulp and paper, metalworking fluids
Trade names: Integra™ 22

1-(Hydroxymethyl)-5,5-dimethyl hydantoin. *See* MDM hydantoin
1-(Hydroxymethyl)-5,5-dimethyl-2,4-imidazolinedione. *See* MDM hydantoin

Hydroxymethyl dioxoazabicyclooctane
CAS 6542-37-6; EINECS 229-457-6
Synonyms: 7-Hydroxymethyl-1,5-dioxo-3-aza-bicyclooctane
Classification: Heterocyclic organic compd.
Empirical: $C_6H_{11}NO_3$
Uses: Cross-linking agent for resorcinol phenol-formaldehyde or protein-based resin systems; raw material for synthesis; for paints, coatings, adhesives, and inks
Manuf./Distrib.: Aldrich
Trade names: Zoldine® ZT-65

7-Hydroxymethyl-1,5-dioxo-3-aza-bicyclooctane. *See* Hydroxymethyl dioxoazabicyclooctane
2-Hydroxymethylfuran. *See* Furfuryl alcohol
4-Hydroxy-4-methyl-2-pentanone. *See* Diacetone alcohol
2-(2´-Hydroxy-5´-methylphenyl) benzotriazole. *See* Drometrizole

2-Hydroxy 2-methyl 1-phenyl 1-propanone
CAS 7473-98-5
Synonyms: 2-Hydroxy-2-methylpropiophenone; 1-Phenyl-2-hydroxy-2-methyl-propan-1-one
Empirical: $C_{10}H_{12}O_2$
Formula: $C_6H_5COC(CH_3)_2OH$
Properties: M.w. 164.2; dens. 1.077; b.p. 102-103 C (4 mm); flash pt. > 110 C; ref. index 1.5330
Precaution: Light sensitive
Toxicology: Irritant
Trade names containing: Esacure® KIP 100F

2-Hydroxy-2-methylpropiophenone. *See* 2-Hydroxy 2-methyl 1-phenyl 1-propanone
1-Hydroxynaphthalene. *See* 1-Naphthol
α-Hydroxynaphthalene. *See* 1-Naphthol
12-Hydroxyoctadecanoic acid. *See* Hydroxystearic acid
Hydroxyoctadecanoic acid, 2-hydroxyethyl ester. *See* Glycol hydroxystearate
12-Hydroxyoctadecanoic acid, methyl ester. *See* Methyl hydroxystearate
12-Hydroxyoctadecanoic acid, 1,2,3-propanetriyl ester. *See* Trihydroxystearin
12-Hydroxy-9-octadecenoic acid. *See* Ricinoleic acid
12-Hydroxy-cis-9-octadecenoic acid. *See* Ricinoleic acid
cis-12-Hydroxyoctadec-9-enoic acid. *See* Ricinoleic acid
12-Hydroxy-9-octadecenoic acid, monoester with 1,2,3-propanetriol. *See* Glyceryl ricinoleate
2-Hydroxy-4-n-octoxybenzophenone. *See* Benzophenone-12
2-Hydroxy-4-(octyloxy) benzophenone. *See* Benzophenone-12
[2-Hydroxy-4-(octyloxy)phenyl]phenylmethanone. *See* Benzophenone-12
2-(2´-Hydroxy-5´-t-octylphenyl)benzotriazole. *See* Octrizole
12-Hydroxyoleic acid. *See* Ricinoleic acid
d-12-Hydroxyoleic acid. *See* Ricinoleic acid
1-(4-Hydroxyphenyl)-2-bromoethanone. *See* 2-Bromo-4´-hydroxyacetophenone
2-Hydroxy-1,2,3-propanetricarboxylic acid. *See* Citric acid
2-Hydroxy-1,2,3-propanetricarboxylic acid, tributyl ester. *See* Tributyl citrate
2-Hydroxy-1,2,3-propanetricarboxylic acid, triethyl ester. *See* Triethyl citrate
2-Hydroxypropanoic acid. *See* Lactic acid
2-Hydroxypropionic acid. *See* Lactic acid
α-Hydroxypropionic acid. *See* Lactic acid
2-Hydroxypropionic acid ethyl ester. *See* Ethyl lactate
Hydroxypropyl alginate. *See* Propylene glycol alginate

N-(2-Hydroxypropyl) benzenesulfonamide
CAS 35325-02-1
Uses: Plasticizer for polyamide, polyurethane, polyacrylic, cellulose ester; antistat; for paints, lacquers; stabilizer for pigmented unsat. polyester resins
Trade names: Uniplex 225

Hydroxypropylcellulose
CAS 9004-64-2
Synonyms: HPC; Cellulose, 2-hydroxypropyl ether; Oxypropylated cellulose; Hyprolose
Definition: Partially substituted poly(hydroxypropyl) ether of cellulose
Properties: Wh. powd., odorless, tasteless; sol. in water, methanol, ethanol, other org. solvs.; insol. in water

> 37.7 C; thermoplastic; can be extruded and molded; softens at 130 C; visc.various; ref. index 1.337; pH 5-8.5 (1%)
Precaution: Combustible
Toxicology: LD50 (oral, rat) 10,200 mg/kg; sl. toxic by ingestion; heated to decomp., emits acrid smoke and fumes
Uses: Emulsifier, film-former, protective colloid, stabilizer, thickener, food additive, suspending agent; binder in ceramics; hair and cosmetic preps.; in blow-molded bottles; PVC polymerization; tablet coating aid in pharmaceuticals
Regulatory: FDA 21CFR §172.870, 177.1200; Europe listed; UK approved; FDA approved for orals, topicals; USP/NF, BP, Ph.Eur. compliance
Manuf./Distrib.: Aldrich; Aqualon; Nippon Soda; Shin-Etsu
Trade names: Klucel® E, G, H, J, L, M

Hydroxypropyl-β-cyclodextrin
CAS 94035-02-6
Properties: M.w. 1500 (avg.)
Uses: Complex hosting guest molecules; increases the sol. and bioavailability of other substances; masks flavor, odor, or coloration; stabilizes against light, oxidation, heat, and hydrolysis; turns liqs. or volatiles into stable solid powds.; for use in pharmaceuticals, cosmetics, toiletries, foods, tobacco, pesticides, textiles, paints, plastics, synthesis, polymers
Manuf./Distrib.: Aldrich

Hydroxypropyl-γ-cyclodextrin
CAS 99241-25-5
Properties: M.p. 250 C
Uses: complex hosting guest molecules; increases the sol. and bioavailability of other substances; masks flavor, odor, or coloration; stabilizes against light, oxidation, heat, and hydrolysis; turns liqs. or volatiles into stable solid powds.; for use in pharmaceuticals, cosmetics, toiletries, foods, tobacco, pesticides, textiles, paints, plastics, synthesis, polymers
Manuf./Distrib.: Aldrich
Trade names: Gamma W8 HP0.6

Hydroxypropyl methacrylate
CAS 27813-02-1; EINECS 248-666-3
Synonyms: 2-Propenoic acid-2-methyl-2-hydroxymethylethyl ester; 1,2-Propanediol-2-methyl monomethacrylate; Hydroxypropyl monomethacrylate
Empirical: $C_7H_{12}O_3$
Formula: $CH_3CHOHCH_2OOCC(CH_3):CH_2$
Properties: Clear mobile liq.; limited sol. in water; sol. in common org. solvs.; m.w. 144.17; sp.gr. 1.03; b.p. 205-209 C; flash pt. 96.6 C; ref. index 1.4447
Precaution: Combustible
Toxicology: LD50 (oral, rats) > 5000 mg/kg, (dermal, rabbits) > 5000 mg/kg; severe eye irritant, mod. skin irritation; nontoxic by inhalation
Uses: Monomer for acrylic resins, nonwoven fabric binders, detergent lube oil additives, coatings, adhesives, contact lenses
Manuf./Distrib.: Aldrich; Allchem Ind.; Ashland; BP Chems. Inc; Fluka; Monomer-Polymer & Dajac Labs; Rhone-Poulenc; Rohm & Haas; Rohm Tech; Sigma
Trade names: BM-951
Trade names containing: Sipomer® HPM

2-Hydroxypropyl methacrylate. *See* Propylene glycol methacrylate

Hydroxypropyl methylcellulose
CAS 9004-65-3
Synonyms: MHPC; Methyl hydroxypropyl cellulose; Cellulose hydroxypropyl methyl ether; Cellulose 2-hydroxypropyl methyl ether; Hypromellose
Definition: Propylene glycol ether of methyl cellulose
Properties: Wh. powd.; swells in water to produce a clear to opalescent visc. colloidal sol'n.; insol. in hot water; insol. in anhyd. alcohol, ether, chloroform; sol. in most polar solvs.; nonionic
Precaution: Combustible
Toxicology: LD50 (intraperitoneal, rat) 5200 mg/kg; mildly toxic by intraperitoneal route; heated to decomp., emits acrid smoke and fumes
Uses: Thickener, stabilizer, emulsifier in food prods.; thickener in paint stripping preps.; protective colloid,

suspending agent; tablet excipient; in adhesives, asphalt emulsions, caulks, cements, paints; visc. stabilizer for latex and emulsion paints; plasticizer for ceramics, refractory shapes

Regulatory: FDA 21CFR §172.874, 175.105, 175.300; Europe listed; FDA approved for ophthalmics, orals, topicals; USP/NF compliance

Manuf./Distrib.: Aceto; Aldrich; Ashland; Fluka; Sigma

Trade names: Methocel® E

Hydroxypropyl monomethacrylate. *See* Hydroxypropyl methacrylate

Hydroxystearic acid

CAS 106-14-9; EINECS 203-366-1

Synonyms: HSA; 12-Hydroxyoctadecanoic acid; 12-Hydroxystearic acid; Octadecanoic acid, 12-hydroxy-

Classification: Fatty acid

Definition: C-18 straight chain fatty acid

Empirical: $C_{18}H_{36}O_3$

Formula: $CH_3(CH_2)_5(CHOH)(CH_2)_{10}COOH$

Properties: Flakes; m.w. 300.49; sp.gr. 1.021; m.p.79-82 C

Precaution: Combustible

Toxicology: Experimental carcinogen; heated to decomp., emits acrid smoke and fumes

Uses: Major component in lithium greases; chemical intermediate; lubricant for PVC; in cosmetics, toiletries, wax blends, polishes, inks, hot-melt adhesives, paints

Regulatory: FDA 21CFR §175.105, 176.210, 178.3570

Manuf./Distrib.: Aldrich; Allchem Ind.; Alnor Oil; Amber Syn.; CasChem; Fluka; Penta Mfg.

12-Hydroxystearic acid. *See* Hydroxystearic acid
12-Hydroxystearic acid methyl ester. *See* Methyl hydroxystearate
Hydroxystearic acid, monoester with glycerol. *See* Glyceryl hydroxystearate

N-Hydroxysuccinic acid

CAS 6915-15-7 (±); 97-67-6 (L); 617-48-1 (DL); 636-61-3 (+); EINECS 202-601-5

Synonyms: Malic acid (INCI); Apple acid; 1-Hydroxy-1,2-ethanedicarboxylic acid; Hydroxybutanedioic acid

Empirical: $C_4H_6O_5$

Formula: $COOHCH_2CH(OH)COOH$

Properties: Wh. or colorless cryst. powd. or gran., strongly acid taste; dl, l, and d isomeric forms; very sol. in water, alcohol; sl. sol. in ether; m.w. 134.09; dens. 1.595 (20/40 C, d or l), 1.601 (dl); m.p. 100 C (d or l), 128 C (dl); b.p. 140 C (dec.)

Precaution: Combustible

Toxicology: Mod. toxic by ingestion; skin and severe eye irritant; dust and aq. sol'ns. may irritate skin, eyes, mucous membranes; heated to decomp., emits acrid smoke and irritating fumes

Uses: Manufacture of esters and salts, wines; chelating agent, food acidulant, flavoring; in pharmaceuticals; in paint mixing, metal cleaning/finishing, metal plating; dyeing acid for textiles; chem. intermediate; deodorizer

Regulatory: FDA 21CFR §146.113, 150, 150.161, 169.115, 169.140, 169.150, 184.1069, GRAS; USDA 9CFR §318.7, 0.01% max.; BATF 27CFR §240.1051, GRAS; not for use in baby foods; FEMA GRAS; Japan approved; Europe listed; UK approved; FDA approved for orals; USP/NF, BP compliance

Manuf./Distrib.: Aldrich; Allchem Ind.; Am. Biorganics; Bayer/Fibers, Orgs., Rubber; Brown; Chemical; Croda Colloids Ltd.; Eastern Chem.; Fluka; Haarmann & Reimer; Janssen Chimica; Kyowa Hakko USA; Schweizerhall; Sigma; Spectrum Chem. Mfg.

β-Hydroxytricarballylic acid. *See* Citric acid
Hydroxyxylene. *See* Xylenol
Hypnone. *See* Acetophenone
Hyprolose. *See* Hydroxypropylcellulose
Hypromellose. *See* Hydroxypropyl methylcellulose

IBOA. *See* Isobornyl acrylate
IIR. *See* Isobutylene/isoprene copolymer
1H-Imidazole-1-ethanol, 4,5-dihydro-2-docosanyl-. *See* Behenyl hydroxyethyl imidazoline
1H-Imidazole-1-ethanol, 4,5-dihydro-2-nonyl-. *See* Capryl hydroxyethyl imidazoline
1H-Imidazole-1-ethanol, 4,5-dihydro-2-norcocoyl-. *See* Cocoyl hydroxyethyl imidazoline
1H-Imidazole-1-ethanol, 4,5-dihydro-2-undecyl-. *See* Lauryl hydroxyethyl imidazoline

1H-Imidazolium, 1-(carboxymethyl)-4,5-dihydro-1-(2-hydroxyethyl)-2-undecyl-, hydroxide, sodium salt. *See* Sodium lauroamphoacetate
Imidodicarbonimidic diamide, N-(2-methylphenyl)-. *See* o-Tolyl biguanide
2,2´-Iminobisethanol. *See* Diethanolamine
1,1´-Iminobis(propan-2-ol). *See* Diisopropanolamine
1,1´-Iminobis-2-propenol. *See* Diisopropanolamine
2,2´-Iminodiethanol. *See* Diethanolamine
2,2´-Iminodiethylamine. *See* Diethylenetriamine
Inactive limonene. *See* dl-Limonene
Indian tragacanth. *See* Karaya gum
India tragacanth. *See* Karaya gum

Indium
CAS 7440-74-6; EINECS 231-180-0
Classification: Metallic element
Empirical: In
Properties: Shiny silver-wh. ductile metal; softer than lead; sol. in acids; insol. in alkalies; m.w. 114.82; dens. 7.31; m.p. 156 C; b.p. 2075 C; stable in air
Toxicology: Relatively nontoxic orally; highly toxic when administered subcutaneously or intravenously
Uses: Automobile bearings, electronic and semiconductor devices, brazing and soldering alloys, reactor control rods, electroplated coatings on aircraft bearings
Manuf./Distrib.: Aldrich; Alfa Aesar Johnson Matthey; Atlantic Equip. Engrs.; Atomergic Chemetals; Belmont Metals; Cerac; Fluka; Indium Corp. of Am.; Noah; Reade Advanced Materials; Spectrum Chem. Mfg.; United Min. & Chem.

Industrial talc. *See* Talc
Infusorial earth. *See* Diatomaceous earth

Iodopropynyl butylcarbamate
CAS 55406-53-6; EINECS 259-627-5
Synonyms: IPBC; 3-Iodo-2-propynyl butyl carbamate; Butyl-3-iodo-2-propynylcarbamate; Carbamic acid, butyl-3-iodo-2-propynyl ester
Classification: Organic compd.
Empirical: $C_8H_{12}INO_2$
Formula: $IC_2CH_2OCONH(CH_2)_3CH_3$
Properties: Wh. to off-wh. powd.; m.w. 281.1; m.p. 65-67 C
Uses: Fungicide, antimildew additive, wood preservative; used in oil-based and latex paints, wood prods., cutting oils, textiles, paper coatings, inks, plastics, adhesives
Regulatory: USA EPA registered; Japan approved; Europe listed
Trade names: Protectal™ FB1; Protectal™ FB45; Troysan® Polyphase® AF1; Troysan® Polyphase® EC17; Troysan® Polyphase® P-20T; Troysan® Polyphase® P-100; Troysan® Polyphase® WD17
Trade names containing: Idex™-400; Idex™-1000

3-Iodo-2-propynyl butyl carbamate. *See* Iodopropynyl butylcarbamate
IPA. *See* Isopropyl alcohol
IPBC. *See* Iodopropynyl butylcarbamate
IPDI. *See* Isophorone diisocyanate
IPGE. *See* Isopropyl glycidyl ether

Iron drier
Uses: Used in org. coatings, inks, polyesters

Iron linoleate
Toxicology: Heated to decomp., emits acrid smoke and irritating fumes
Uses: Drier for paints
Regulatory: FDA 21CFR §181.25

Iron naphthenate
Toxicology: Heated to decomp., emits acrid smoke and irritating fumes
Uses: Drier for paints
Regulatory: FDA 21CFR §181.25
Manuf./Distrib.: Akzo Nobel; Archway Sales; Baychem; Boehle; ; Dussek Campbell Ltd; Hüls Am.; OM Group
Trade names: Nuodex Napthenate® Iron 6%

Iron octoate
Synonyms: Ferric octoate
Uses: Drier for paints
Manuf./Distrib.: Archway Sales; Baychem; Hüls Am.; D.N. Lukens

Iron (III) oxide. *See* Ferric oxide
Iron phosphate. *See* Ferric phosphate
Iron (II) sulfate heptahydrate. *See* Ferrous sulfate heptahydrate

Iron tallate
Toxicology: Heated to decomp, emits acrid smoke and irritating fumes
Uses: Drier for paints
Regulatory: FDA 21CFR §181.25
Manuf./Distrib.: Baychem; Dussek Campbell Ltd

Iron vitriol. *See* Ferrous sulfate heptahydrate
Isoacetophorone. *See* Isophorone

Isoamyl alcohol
CAS 123-51-3; EINECS 204-633-5
Synonyms: 3-Methylbutanol; 3-Methyl-1-butanol; Isopentyl alcohol; Isobutyl carbinol
Classification: Aliphatic organic compd.
Empirical: $C_5H_{12}O$
Formula: $(CH_3)_2CHCH_2CH_2OH$
Properties: Colorless liq., pungent taste, disagreeable odor; sl. sol. in water; misc. with alcohol, ether; m.w. 88.15; dens. 0.813 (15/4 C); b.p. 132 C; f.p. -117.2 C; flash pt. (CC) 42.7 C; ref. index 1.407 (20 C)
Precaution: DOT: Flamm. liq.; explosive limits in air 1.2-9%; vapor is toxic and irritant
Toxicology: TLV 100 ppm in air; highly toxic; ing. has caused human deaths from respiratory failure; may cause heart, lung, kidney damage; CNS depressant; vapor exposure has caused marked irritation of eyes, nose, throat, and headache
Uses: Flavors and fragrances, fine chemicals and pharmaceuticals, extraction solv., herbicides, plasticizers, mining chemicals, solv. for gums, resins, and lacquers, oil additives, intermediates, photographic chemicals, organic synthesis, microscopy
Regulatory: FDA 21CFR §172.515; FEMA GRAS
Manuf./Distrib.: Aldrich; Allchem Ind.; AMRESCO; BASF; Coyne; CPS; Fluka; Hoechst Celanese; Oxiteno; Penta Mfg.; Sigma; Spectrum Chem. Mfg.

Isoamyl methyl ketone. *See* Methyl isoamyl ketone
1,3-Isobenzofurandione. *See* Phthalic anhydride
1,3-Isobenzofurandione, 4,5,6,7-tetrabromo-(9CI). *See* Tetrabromophthalic anhydride

Isobornyl acrylate
CAS 5888-33-5; EINECS 227-561-6
Synonyms: IBOA
Empirical: $C_{13}H_{20}O_2$
Properties: M.w. 208.3; dens. 0.986; m.p. -60 C; b.p. 104 C (5 mm); flash pt. 97 C; ref. index 1.4760 (20 C)
Precaution: Light sensitive
Toxicology: Irritant
Uses: Curing agent; diluent monomer (UV EB resin) and modifier; for inks, electronics applics., paints, textile finishes, adhesives, industrial and textile coatings; resists uv degradation
Manuf./Distrib.: Aldrich; Ashland; CPS; Monomer-Polymer & Dajac Labs; Osaka Org. Chem. Ind.; Rhone-Poulenc Surf. & Spec.; San Esters; UCB Radcure
Trade names: Sipomer® IBOA; Sipomer® IBOA HP
Trade names containing: Ageflex IBOA; CN 960 J75; CN 961 J75; CN 962 J75; CN 963 J75; CN 964 J75; CN 965 J75; CN 966 J75; CN 970 J75; CN 971 J75; CN 972 J75; CN 973 J75; SR-506

Isobornyl methacrylate
CAS 7534-94-3
Synonyms: Methacrylic acid isobornyl ester
Empirical: $C_{14}H_{22}O_2$
Properties: M.w. 222.33; dens. 0.983; b.p. 127-129 (15 mm); flash pt. 107 C; ref. index 1.4770 (20 C)

Toxicology: Irritant
Uses: Monomer for creating and modifying a wide range of polymers; for high-performance coatings, paints, textile finishes, adhesives, industrial and textile coatings
Manuf./Distrib.: Aldrich; Ashland; CPS; Monomer-Polymer & Dajac Labs; Rhone-Poulenc Surf. & Spec.; Rohm Tech; San Esters
Trade names: Sipomer® IBOMA HP
Trade names containing: Ageflex IBOMA; Sipomer® IBOMA; SR-423

Isobutane
CAS 75-28-5; EINECS 200-857-2
Synonyms: 2-Methylpropane; Trimethylmethane
Classification: Hydrocarbon gas
Definition: A constituent of natural gas and illuminating gas
Empirical: C_4H_{10}
Formula: $CH(CH_3)_3$
Properties: Colorless gas, odorless; easily liquefied under pressure at R.T.; insol. in water; m.w. 58.12; dens. 0.5572; vapor pressure 2950 mm Hg (31 psig, 21 C); f.p. -159 C; b.p. -11.73 C
Precaution: Highly flamm. gas; explosive; very dangerous fire and explosion hazard exposed to heat, flame, or oxidizers
Toxicology: Asphyxiant; narcotic at high concs.; heated to decomp., emits acrid smoke and irritating fumes
Uses: Aerosol propellant for paint aerosols
Regulatory: FDA 21CFR §184.1165; FDA approved for topicals; USP/NF compliance
Manuf./Distrib.: Air Prods.; Aldrich; Hüls Am.; Phillips
Trade names containing: Konform® AR 2000

Isobutanol. *See* Isobutyl alcohol
Isobutanolamine. *See* 2-Amino-2-methyl-1-propanol
Isobutanol-2 amine. *See* 2-Amino-2-methyl-1-propanol
Isobutenyl methyl ketone. *See* Mesityl oxide
2-Isobutoxy-2-phenyl-acetophenone. *See* 2-(2-Methylpropoxy)-1,2-diphenylethanone

Isobutyl acetate
CAS 110-19-0; EINECS 203-745-1
Synonyms: 2-Methylpropyl acetate; 2-Methyl-1-propyl acetate; β-Methylpropyl ethanoate; Acetic acid isobutyl ester; Acetic acid 2-methylpropyl ester
Definition: Ester of isobutyl alcohol and acetic acid
Empirical: $C_6H_{12}O_2$
Formula: $CH_3COOCH_2CH(CH_3)_2$
Properties: Colorless liq., fruit-like odor; very sol. in alcohol, fixed oils, propylene glycol; sl. sol. in water; m.w. 116.18; dens. 0.8685 (15 C); m.p. -98.9 C; b.p. 118 C; flash pt. (CC) 18 C; ref. index 1.389
Precaution: Highly flamm.; very dangerous fire and mod. explosion hazard on exposure to heat, flame, oxidizers
Toxicology: TLV 150 ppm; LD50 (oral, rat) 13,400 mg/kg; mildly toxic by ingestion and inh.; skin and eye irritant; heated to decomp., emits acrid smoke and fumes
Uses: Solvent for nitrocellulose; in thinners, sealants, topcoat lacquers; perfumery; flavoring agent
Regulatory: FDA 21CFR §172.515; FEMA GRAS; BP compliance
Manuf./Distrib.: Aldrich; Allchem Ind.; Ashland; BASF; Coyne; Eastman; Fluka; Hoechst Celanese; Janssen Chimica; Penta Mfg.; Union Carbide
Trade names containing: Acryloid® AT-70IBA

Isobutylacetone. *See* Methyl isoamyl ketone

Isobutyl alcohol
CAS 78-83-1; EINECS 201-148-0
Synonyms: Isobutanol; Isopropylcarbinol; 2-Methyl-1-propanol; 2-Methylpropanol
Classification: Alcohol
Empirical: $C_4H_{10}O$
Formula: $(CH_3)_2CHCH_2OH$
Properties: Colorless liq., sweet odor; partly sol. in water; sol. in alcohol, ether; m.w. 74.12; dens. 0.806 (15 C); m.p. -108 C; b.p. 106-109 C; f.p. -108 C; flash pt. (TCC) 29 C; ref. index 1.396
Precaution: Flamm.; dangerous fire hazard with heat, flame; mod. explosive as vapor with heat, flame, oxidizers
Toxicology: TLV 50 ppm in air; LD50 (oral, rat) 2460 mg/kg; poison by IV, intraperitoneal route; mod. toxic by

ingestion, skin contact; experimental carcinogen, tumorigen; severe skin/eye irritant; mutagenic data; heated to decomp., emits acrid smoke and fumes

Uses: Flavors and fragrances; organic synthesis; latent solvent; intermediate; solv. for paints; paint removers; fluorometric determinations; liq. chromatography

Regulatory: FDA 21CFR §172.515; FEMA GRAS

Manuf./Distrib.: Aldrich; Allchem Ind.; Ashland; BASF; CPS; Eastman; Fluka; Hoechst Celanese; Neste UK; Penta Mfg.; Shell; Sigma; Spectrum Chem. Mfg.; Union Carbide

Trade names containing: Beckamine 27-566; Beckamine 27-809; Filmex® B; Nacure® 155; Nacure® 3327; Nacure® 3525; Nacure® X49-110; Nansa® EVM70/E; Resimene® 872; Resimene® 933; Resimene® 970; Resimene® HM-2608; SILIKOFTAL® Nonstick 50; Silikophen® Non-stick 50; Silikophen® P 50/X; Silikophen® P 80/X; Solvenon® I; Uformite® 27-802; Uformite® 27-806

Isobutyl benzoin ether

CAS 22499-12-3

Empirical: $C_{18}H_{20}O_2$

Properties: Liq.; m.w. 268.34; insol. in water; sol. in styrene and unsat. acrylate monomers; dens. 1.04-1.06; ref. index 1.5450-1.5485; flash pt. (TOC) > 149 C

Uses: Photosensitizer for uv curable systems, e.g., coatings, inks, graphic arts

Manuf./Distrib.: Aldrich

Trade names: Vicure® 10

Isobutyl carbinol. *See* Isoamyl alcohol

Isobutylene/isoprene copolymer

CAS 9010-85-9

Synonyms: IIR; Butyl rubber; 3-Methyl-1,3-butadiene polymer with 2-methyl-1-propene; 1,3-Butadiene-2-methyl polymer with 2-methyl-1-propene

Definition: Copolymer of isobutylene and isoprene monomers

Formula: $(C_5H_8 \cdot C_4H_8)_x$

Toxicology: Heated to decomp., emits acrid smoke and irritating fumes

Uses: Butyl rubber used in tires, molded, extruded and calendered goods, mech. goods, dips, sealants, tapes, adhesives, membranes, dynamic parts, diaphragms, chewing gum base, pharmaceuticals, elec. wire insulation, paints

Regulatory: FDA 21CFR §172.615, 175.105, 177.1210, 177.2600

Manuf./Distrib.: Aldrich; Nat'l. Chem.

Trade names: Kalar® 5214; Kalar® 5245; Kalar® 5263; Kalar® 5264; Kalene® 800; Kalene® 800V70; Kalene® 1300; Kalene® 1300V70

Isobutylene/MA copolymer

CAS 26426-80-2

Synonyms: Isobutylene/maleic anhydride copolymer; 2-Methyl-1-propene, polymer with 2,5-furandione; 2,5-Furandione, polymer with 2-methyl-1-propene

Definition: Copolymer of isobutylene and maleic anhydride monomers

Empirical: $(C_4H_8 \cdot C_4H_2O_3)_x$

Uses: Dispersant for dyes, pigments; in cosmetics, latex paints, polymerization, leather tanning, water treatment

Trade names: Daxad® 31

Isobutylene/maleic anhydride copolymer. *See* Isobutylene/MA copolymer

Isobutyl formate

CAS 542-55-2

Synonyms: Tetryl formate; Formic acid isobutyl ester

Empirical: $C_5H_{10}O_2$

Formula: $HCOOCH_2CH(CH_3)_2$

Properties: Liq., fruity ether-like odor, rum-like taste; sol. in 100 parts water; misc. with alcohol, ether; m.w. 102.13; dens. 0.885 (20/4 C); m.p. -95 C; b.p. 98-99 C; flash pt. 50 F; ref. index 1.3858 (20 C)

Precaution: Flamm. liq.; very dangerous fire hazard

Toxicology: Mod. toxic by ingestion

Regulatory: FDA 21CFR §172.515; FEMA GRAS

Manuf./Distrib.: Aldrich

Trade names containing: Solvenon® I

Isobutyl heptyl ketone
CAS 123-18-2
Synonyms: 2,6,8-Trimethyl-4-nonanone
Empirical: $C_{12}H_{24}O$
Formula: $(CH_3)_2CHCH_2COCH_2CH(CH_3)CH_2CH(CH_3)_2$
Properties: Water-wh. liq., pleasant odor; insol. in water; m.w. 184.32; sp.gr. 0.822 (20/20 C); b.p. 224 C; flash pt. (CC) 189 F
Uses: Solv. for coatings (NC emulsions, vinyls); dispersant; intermediate; high solvating power for vinyl resins, cellulose esters and ethers
Manuf./Distrib.: Union Carbide

Isobutyl isobutyrate
CAS 97-85-8; EINECS 202-612-5
Synonyms: 2-Methylpropanoic acid 2-methylpropyl ester
Empirical: $C_8H_{16}O_2$
Formula: $(CH_3)_2CHCOOCH_2CH(CH_3)_2$
Properties: Liq., fruity odor; misc. with alcohol; insol. in water; m.w. 144.22; dens. 0.854 (20/4 C); m.p. -81 C; b.p. 149-151 C; flash pt. 40 C; ref. index 1.399 (20 C)
Precaution: Flamm.
Toxicology: LD50 (oral, rat) 12,800 mg/kg; mildly toxic by ingestion and inhalation; irritant
Uses: Solv. for paints
Regulatory: FDA 21CFR §172.515; FEMA GRAS
Manuf./Distrib.: Aldrich; Ashland; Eastman; Fluka; Penta Mfg.

Isobutyl methacrylate
CAS 97-86-9; EINECS 202-613-0
Synonyms: Isobutyl-α-methacrylate; 2-Methylpropyl methacrylate
Empirical: $C_8H_{14}O_2$
Formula: $CH_2{:}C(CH_3)COOCH_2CH(CH_3)$
Properties: Colorless transparent liq.; m.w. 142.22; dens. 0.886 (20/4 C); b.p. 155 C; flash pt. 46 C; ref. index 1.420 (20 C)
Precaution: Flamm.
Toxicology: Moderately toxic by intraperitoneal route; mildly toxic by ingestion; irritating to eyes, skin, respiratory system
Uses: Thermoplastic resin for plastic coatings, printing inks, overprint varnishes, refinish lacquers
Manuf./Distrib.: Aldrich; Ashland; Fluka; Monomer-Polymer & Dajac Labs; Rohm Tech
Trade names containing: Acryloid® B-67; Acryloid® B-67MT

Isobutyl-α-methacrylate. *See* Isobutyl methacrylate
Isobutyl methyl carbinol. *See* Methyl amyl alcohol
Isobutyl methyl ketone. *See* Methyl isobutyl ketone

Isobutyl stearate
CAS 646-13-9; 85865-69-6; EINECS 211-466-1; 288-668-1
Synonyms: 2-Methylpropyl octadecanoate; Stearic acid, 2-methylpropyl ester
Definition: Ester of isobutyl alcohol and stearic acid
Empirical: $C_{22}H_{44}O_2$
Formula: $CH_3(CH_2)_{16}COOCH_2CH(CH_3)_2$
Properties: Waxy cryst. solid; m.w. 340.57; m.p. 20 C
Toxicology: No known toxicity
Uses: Emollient for cosmetics, topical pharmaceuticals; inks, waterproof coatings, polishes, ointments, rubber mfg., dyes; plasticizer/lubricant for PVC, PS
Regulatory: FDA 21CFR §176.210, 177.2260, 177.2800, 178.3910

Isobutyltrimethoxysilane
CAS 18395-30-7
Synonyms: Silane trimethoxy (2-methylpropyl); Trimethoxy (2-methylpropyl) silane
Empirical: $C_7H_9O_3Si$
Properties: Liq.; m.w. 178.3; dens. 0.93; b.p. 154-157 C; ref. index 1.396; flash pt. 14 C
Uses: Coupling agent, release agent, lubricant, blocking agent, chemical intermediate; used in paints and coatings
Manuf./Distrib.: Aldrich; Fluka
Trade names: ASil-100

Isobutyl vinyl ether. *See* Vinyl isobutyl ether

Isobutyric acid
CAS 79-31-2; EINECS 201-195-7
Synonyms: Dimethylacetic acid; 2-Methylpropanoic acid; Isopropylformic acid
Classification: Organic acid
Empirical: $C_4H_8O_2$
Formula: $(CH_3)_2CHCOOH$
Properties: Colorless liq., pungent odor of rancid butter; sol. in 6 parts of water; misc. with alcohol, ether, chloroform; m.w. 88.11; dens. 0.946-0.950 (20/20 C); b.p. 154.4 C (760 mm); f.p. -47 C; flash pt. (TOC) 76.6 C; ref. index 1.393 (20 C)
Precaution: Flamm., corrosive; reactive with oxidizing materials
Toxicology: LD50 (oral, rat) 280 mg/kg; poison by ingestion; mod. toxic by skin contact; corrosive irritant to eyes and tissue; heated to decomp., emits acrid smoke and fumes
Uses: Mfg. of esters for solvents, flavors (pungent, butter-like, fruity on dilution), perfume bases, disinfecting agent, varnish, deliming hides, tanning agent
Regulatory: FDA 21CFR §172.515; FEMA GRAS
Manuf./Distrib.: Aldrich; BASF; Eastman; Fluka; Hoechst Celanese; Hüls Am.; Penta Mfg.; Sigma

Isobutyric acid, 1-isopropyl-2,2-dimethyltrimethylene ester. *See* Trimethyl-1,3-pentanediol, 2,2,4-diisobutyrate
3-Isocyanatomethyl-3,5,5-trimethyl cyclohexylisocyanate. *See* Isophorone diisocyanate

Isocyanatopropyltriethoxysilane
CAS 24801-88-5; EINECS 246-467-6
Empirical: $C_7H_{18}O_3Si$
Properties: M.w. 178.30; dens. 0.925 (20/4 C); b.p. 154 C; flash pt. 43 C; ref. index 1.397 (20 C)
Precaution: Flamm.; keep away from ignition sources
Toxicology: Corrosive lachrymator; eye, skin, and respiratory system irritant
Uses: Coupling agent, chem. intermediate, blocking agent, release agent, lubricant, primer, reducing agent; provides improved wet adhesion, corrosion resist., weatherability, pigment disp. to adhesives, coatings, inks, sealants; min./resin coupling for improved composite str. in rubber and elastomers
Manuf./Distrib.: Aldrich; Fluka
Trade names: Silquest® A-1310

Isocyanic acid, hexamethylene ester. *See* Hexamethylene diisocyanate

Isodecyl acrylate
CAS 1330-61-6
Synonyms: Isodecyl propenoate; Acrylic acid isodecyl ester; Isodecyl alcohol acrylate
Classification: Monomer
Empirical: $C_{13}H_{24}O_2$
Formula: $CH_2=CHCOOC_{10}H_{21}$
Properties: M.w. 212.33; b.p. 121 C (10 mm)
Precaution: Combustible
Toxicology: LD50 (oral, rat) 12 g/kg, (skin, rabbit) 3540 mg/kg; skin irritant
Uses: Monomer for adhesives, coatings, uv-curable reactive diluent in inks and coatings, visc. index improver
Manuf./Distrib.: Aldrich; CPS; Monomer-Polymer & Dajac Labs
Trade names containing: Ageflex FA-10; Sipomer® IDA; SR-395

Isodecyl alcohol acrylate. *See* Isodecyl acrylate

Isodecyl benzoate
CAS 131298-44-7
Properties: Sp.gr. 0.95; m.p. -70 C; flash pt. (COC) 174 C; ref. index 1.4878
Uses: Plasticizer for PVC, PVAc, plastisols, adhesives, sealants, caulks; latex paint coalescing aid and specialty plasticizer; suitable for interior flat, semi or high gloss latex paint, exterior low gloss latex
Manuf./Distrib.: Velsicol
Trade names: Velate® 262

Isodecyl diphenyl phosphate
CAS 29761-21-5
Synonyms: Phosphoric acid isodecyl diphenyl ester

Isodecyl methacrylate

Empirical: $C_{22}H_{31}O_4P$
Properties: Sp.gr. 1.069-1.079; flash pt. (COC) 241 C; ref. index 1.503-1.509
Uses: Flame-retardant plasticizer for PVC and copolymers, PVAc, acrylics; for finished film or coated fabric applics., vinyl plastisols; also for PVC adhesives, ethyl cellulose, NC, SBR and butyl rubbers
Manuf./Distrib.: Akzo Nobel; Ashland; Harwick; Monsanto
Trade names: Santicizer 148

Isodecyl methacrylate
CAS 29964-84-9
Classification: Monomer
Empirical: $C_{14}H_{26}O_2$
Formula: $CH_2=C(CH_3)COOC_{10}H_{21}$
Properties: M.w. 226.36
Precaution: Nonhazardous (DOT)
Toxicology: Moderately toxic by intraperitoneal route
Uses: Pressure-sensitive adhesives, coatings for leather, textiles, paper, nonwoven fiber, polymer modifier and stabilizer, visc. index improver, dispersion for plastics and rubber, floor waxes, potting compds., sealants
Manuf./Distrib.: Aldrich; CPS; Monomer-Polymer & Dajac Labs; Rohm & Haas; Sartomer
Trade names: Empicryl® 6315; Sipomer® IDM
Trade names containing: Ageflex FM-10; SR-242

Isodecyl oxypropyl amine acetate
Uses: Latex paint coalescing aid and specialty plasticizer; suitable for interior flat, semi or high gloss latex paint, exterior low gloss latex
Trade names: Tomah PA-14 Acetate

Isodecyl propenoate. *See* Isodecyl acrylate

Isoheptyl alcohol
CAS 70914-20-4
Uses: Used in coatings
Trade names: Exxal® 7

β-Isonitrosopropane. *See* Acetone oxime

Isononyl alcohol
CAS 2430-22-0; 68526-84-1; 68527-05-9; EINECS 222-376-7
Synonyms: 7-Methyl-1-octanol
Empirical: $C_7H_{20}O$
Formula: $C_6H_{17}CH_2OH$
Properties: Water-wh. liq., mild odor; insol. in water; m.w. 144.29; sp.gr. 0.838 (60/60 F); b.p. 260-212 F; m.p. -65 F; flash pt. (PMCC) 195 F
Precaution: Combustible
Toxicology: LD50 (oral, rat) 2980 mg/kg; severely irritating to eyes, moderately to severely irritating to skin; irritating to respiratory passages; low acute toxicity by ingestion; heated to decomp., emits acrid smoke and fumes
Uses: Basis of plasticizers such as diisononyl adipate
Trade names: Exxal® 9

Isooctanol. *See* Isooctyl alcohol

Isooctyl acid phosphate
CAS 12645-53-3
Synonyms: Isooctyl phosphate; Phosphoric acid isooctyl ester
Empirical: $C_8H_{19}O_4P$
Properties: M.w. 210.21
Uses: Catalyst used in solder fluxes; in coatings and inks (appliance, automotive, floor, furniture, paper, and other coatings); antistatic agents, corrosion inhibitors, lubricants, tanning chems.
Manuf./Distrib.: Akzo Nobel; Albright & Wilson Am.; Dexter
Trade names: Albrite® Isooctyl Acid Phosphate

Isooctyl acrylate
CAS 29590-42-9
Classification: Monomer
Properties: M.w. 184.0
Uses: Monomer for pressure-sensitive adhesives, coatings, caulks, sealants
Regulatory: FDA approved for topicals
Manuf./Distrib.: Aldrich; Ashland; CPS; Monomer-Polymer & Dajac Labs
Trade names containing: Ageflex FA-8; Macrobase 600; Sipomer® IOA; SR-440

Isooctyl alcohol
CAS 68526-83-0; 26952-21-6; EINECS 248-133-5
Synonyms: Isooctanol; 2-Ethyl-1-hexanol; Alcohol C_8
Empirical: $C_8H_{18}O$
Formula: $CH_3(CH_2)_3CH(C_2H_5)CH_2OH$
Properties: Clear liq.; m.w. 130.26; dens. 0.832 (20/20 C); b.p. 182-185 C; flash pt. (TOC) 180 F; ref. index 1.431
Precaution: Combustible
Toxicology: LD50 (oral, rat) 1480 mg/kg, (skin, rabbit) 2520 mg/kg; moderately toxic by skin absorption and ingestion; severe eye irritant; TLV 50 ppm
Uses: In plasticizers; intermediate for nonionic detergents and surfactants, hydraulic fluids; resin, solvent, emulsifier, antifoaming agent; in coatings
Trade names: Exxal® 8
Trade names containing: Ken-React® NZ 38J

Isooctyl phosphate. *See* Isooctyl acid phosphate
Isopentyl alcohol. *See* Isoamyl alcohol
Isopentyl methyl ketone. *See* Methyl isoamyl ketone

Isophorone
CAS 78-59-1; EINECS 201-126-0
Synonyms: 3,3,5-Trimethyl-2-cyclohexen-1-one; Isoacetophorone; 1,1,3-Trimethyl-3-cyclohexene-5-one
Empirical: $C_9H_{14}O$
Formula: $OCHC{:}C(CH_3)CH_2C(CH_3)_2CH_2$
Properties: Wh. liq.; m.w. 138.20; sp.gr. 0.922 (20/20 C); f.p. 17 F; b.p. 210-218 C; flash pt. (TCC) 179 F; ref. index 1.4781 (20 C)
Precaution: Combustible
Toxicology: LD50 (oral, rat) 2330 mg/kg; TLV 5 ppm; mildly toxic by inhalation; irritating to eyes, skin, respiratory system
Uses: In solvent mixtures for finishes, for polyvinyl and nitrocellulose resins, pesticides, stoving lacquers
Regulatory: FEMA GRAS
Manuf./Distrib.: Aceto; Aldrich; Allchem Ind.; Ashland; BP Chems. Inc; Chemcentral; Coyne; Elf Atochem N. Am.; Exxon; Fabrichem; Fluka; Great Western; Hüls AG; Sigma; Stanchem; Sunnyside; Union Carbide; Van Waters & Rogers

Isophorone diamine
CAS 2855-13-2; EINECS 220-666-8
Synonyms: 5-Amino-1,3,3-trimethylcyclohexylmethanamine
Classification: Cycloaliphatic diamine
Formula: $H_2NC_6H_7(CH_3)_3CH_2NH_2$
Properties: Colorless liq., amine odor; misc. with water, org. solvs.; m.w. 170.30; dens. 0.922; m.p. 10 C; b.p. 247 C; flash pt (PMCC) > 110 C; ref. index 1.4880 (20 C)
Precaution: Sl. hygroscopic; tends to form carbamates; avoid contact with air and moisture
Toxicology: LD50 (oral, rat) 1030 mg/kg; harmful in contact with skin and if swallowed; causes burns; risk of serious eye damage; may cause sensitization by skin contact
Uses: Epoxy curative for coatings, adhesives, castings and composites
Manuf./Distrib.: Aldrich; BASF; BASF AG; Degussa; Fluka; Sigma
Trade names: Degamin IPDA™; Vestamin® IPD

Isophorone diamine diisocyanate. *See* Isophorone diisocyanate

Isophorone diisocyanate
CAS 4098-71-9; EINECS 223-861-6
Synonyms: IPDI; 3-Isocyanatomethyl-3,5,5-trimethyl cyclohexylisocyanate; Isophorone diamine diisocyanate

Isophthalic acid

Empirical: $C_{12}H_{18}N_2O_2$
Properties: Colorless to sl. yel. liq.; misc. with esters, ketones, ethers, aromatic and aliphatic hydrocarbons; m.w. 222.29; dens. 1.056; m.p. -60 C; b.p.158 C (10 torr); flash pt. 163 C
Toxicology: LD50 (skin, rat) 1060 mg/kg; TLV 0.01 ppm; TLV:TWA 0.005 ppm; severe irritant; poison by inhalation; moderately toxic by skin contact; irritating to skin, eyes, respiratory system
Uses: Reactive building block in coatings for automotive, flooring, roofing, maintenance, and textile applics., cast elastomers, potting and encapsulation compds., optical prods., adhesives, sealants, as crosslinkers for powd. coatings; yields polyurethanes with high stability, resistance to light discoloration, chemical resistance
Manuf./Distrib.: Aldrich; Fluka; Sigma
Trade names: Desmodur® I; Luxate® IM

Isophthalic acid
CAS 121-91-5; EINECS 204-506-4
Synonyms: Benzene-1,3-dicarboxylic acid; m-Benzenedicarboxylic acid
Empirical: $C_8H_6O_4$
Formula: $C_6H_4(COOH)_2$
Properties: Colorless cryst.; sol. in alcohol, acetic acid; insol. in benzene, petrol. ether; sl. sol. in water; m.w. 166.14; m.p. > 300 C, sublimes
Precaution: Combustible
Toxicology: LD50 (oral, rat) 10,400 mg/kg, (IP, mouse) 4200 mg/kg
Uses: Intermediate for paints
Manuf./Distrib.: Aldrich; Allchem Ind.; Amoco; Chemical; Fluka; Sigma; Stanchem
Trade names: Amoco® PIA

Isopropanol. *See* Isopropyl alcohol
Isopropenylbenzene. *See* α-Methylstyrene monomer
2-Isopropoxypropane. *See* Isopropyl ether

1-Isopropoxy-2-propanol
Uses: Solvent for resins and dyes, surface coatings
Trade names containing: Solvenon® IPP

2-Isopropoxy-1-propanol
Uses: Solvent for resins and dyes, surface coatings
Trade names containing: Solvenon® IPP

Isopropyl acetate
CAS 108-21-4; EINECS 203-561-1
Synonyms: Acetic acid 1-methylethyl ester; 2-Propyl acetate; 1-Methylethyl acetate
Empirical: $C_5H_{10}O_2$
Formula: $CH_3COOCH(CH_3)2$
Properties: Colorless aromatic liq., fruity odor; sl. sol. in water; misc. with alcohol, ether, fixed oils; m.w. 102.15; dens. 0.874 (20/20 C); m.p. 073 C; f.p. -69.3 C; b.p. 88.4 C; flash pt. 40 F; ref. index 1.377
Precaution: Highly flamm.; dangerous fire hazard with heat, flame, oxidizers; mod. explosive with heat or flame
Toxicology: TLV 250 ppm; LD50 (oral, rat) 3000 mg/kg; mod. toxic by ingestion; mildly toxic by inh.; human systemic effects on inh.; narcotic in high conc.; chronic exposure can cause liver damage
Uses: Solv. for paint mfg.
Regulatory: FDA 21CFR §172.515, 175.105, 177.1200; FEMA GRAS
Manuf./Distrib.: Aldrich; Allchem Ind.; Ashland; J.T. Baker; BP Chems.; Chemcentral; Coyne; Eastman; Fluka; Harcros; Samson; Stanchem; Sunnyside; Union Carbide; Van Waters & Rogers

Isopropylacetone. *See* Methyl isobutyl ketone

Isopropyl alcohol
CAS 67-63-0; EINECS 200-661-7
Synonyms: IPA; Isopropanol; Petrohol; 2-Propanol; s-Propyl alcohol; Dimethyl carbinol
Classification: Aliphatic alcohol
Empirical: C_3H_8O
Formula: $(CH_3)_2CHOH$
Properties: Colorless volatile liq., pleasant odor, sl. bitter taste; sol. in water, alcohol, ether, chloroform; m.w. 60.11; dens. 0.7863 (20/20 C); f.p. -86 C; b.p. 82.4 C (760 mm); flash pt. (TOC) 11.7 C; ref. index 1.3756 (20 C)

Precaution: DOT: Flamm. liq.; very dangerous fire hazard with heat, flame, oxidizers; reacts with air to form dangerous peroxides; heated to decomp., emits acrid smoke and fumes
Toxicology: TLV:TWA 400 ppm; STEL 500 ppm; LD50 (oral, rat) 5045 mg/kg; poison by ingestion, subcutaneous routes; human systemic effects by ingestion/inhalation (headache, nausea, vomiting, narcosis); 100 ml can be fatal; experimental reproductive effects
Uses: Solv. for essential oils, alkaloids, gums, resins, cellulose derivs., coatings; deicing agent for liq. fuels; lacquers; extraction processes; dehydrating agent; denaturing ethyl alcohol; in cosmetics; mfg. of acetone, glycerol, isopropyl acetate
Regulatory: FDA 21CFR §73.1 (no residue), 73.1001, 172.515, 172.560, 172.712, 173.240 (limitation 50 ppm in spice oleoresins, 6 ppm in lemon oil, 2% in hops extract), 173.340; 175.105, 176.200, 176.210, 177.1200, 177.2800, 178.1010, 178.3910; 27CFR §21.112; use in bread is permitted in Ireland and Japan; FEMA GRAS; FDA approved for orals, topicals; USP/NF, BP compliance
Manuf./Distrib.: Aldrich; Allchem Ind.; ARCO; Ashland; J.T. Baker; Baychem; BP Chems.; R.E. Carroll; Chemcentral; Coyne; Eastman; Exxon; Fluka; General Chem.; C.P. Hall; Harcros; Hüls AG; Mallinckrodt; Mitsui Toatsu; Olin; Samson; Shell; Sigma; Spectrum Chem. Mfg.; Union Carbide; Van Waters & Rogers; Veckridge
Trade names containing: Acryloid® WR-97; Adogen® 432; Ancamide 100-IT-60; Ancamide 220-IPA-73; Arquad® 2C-70 Nitrite; Carboset® 514A; Cyastat® SN; Cyastat® SP; Dow Corning® 14; Ethoduoquad® T/15-50; Ethoquad® C/12 Nitrate; Ethoquad® CB/12; Ethoquad® T/13-50; EXP-38-X20 Epoxy Functional Silicone Sol'n.; Filmex® D-1; Gantrez® ES-335; Gantrez® ES-435; GP-197 Resin Sol'n.; GP-RA-159 Silicone Polish Additive; Joncryl® 56; K-Cure® 1040; Ken-React® KR 7; Ken-React® KR 9S; Ken-React® KR 12; Ken-React® KR 26S; Ken-React® KR 38S; Ken-React® KR 41B; Ken-React® KR 44; Ken-React® KR 46B; Ken-React® KR 55; Ken-React® KR 133DS; Ken-React® KR 134S; Ken-React® KR 138S; Ken-React® KR 158FS; Ken-React® KR 212; Ken-React® KR 238S; Ken-React® KR 262ES; Ken-React® KR OPP2; Ken-React® KR OPPR; Ken-React® KR TTS; Ken-React® NZ 01; Ken-React® NZ 09; Ken-React® NZ 12; Ken-React® NZ 33; Ken-React® NZ 38; Ken-React® NZ 39; Ken-React® NZ 44; Monawet MM-80; Nacure® 2500; Nacure® 2501; Nacure® 2530; Nacure® 3327; Nacure® 3525; Nacure® 5076; Nacure® 5225; Nacure® 5528; Nacure® 5925; Nacure® X49-110; Noramium M2C; PVP/VA I-335; PVP/VA I-535; PVP/VA I-735; Resimene® 730; Resimene® 735; Resimene® 741; Slip-Ayd® SL-508; Slip-Ayd® SL-511; Surfynol® 104PA; Tecsol® A; Tecsol® B; Tecsol® D; Tecsol® H; Uformite® 27-806; Uni-Rez® 2115-I75

Isopropyl alcohol, titanium (4+) salt. *See* Tetraisopropyl titanate

Isopropylamine dodecylbenzenesulfonate
CAS 26264-05-1; 68584-24-7; EINECS 247-556-2
Synonyms: Dodecylbenzenesulfonic acid, comp. with 2-propanamine (1:1)
Classification: Aromatic compd.
Definition: Salt of isopropylamine and doecylbenzene sulfonic acid
Empirical: $C_{21}H_{39}O_3NS$
Formula: $C_{18}H_{30}O_3S \cdot C_3H_9N$
Uses: Emulsifier, solubilizer, detergent, and wetting agent for oil-based systems; dispersant in oil and water-based systems; used in dry-cleaning surfactants; hydrotrope for liq. detergents; latex emulsifier; emulsion polymerization
Regulatory: FDA 21CFR §176.210
Trade names: Polystep® A-11; Rhodacal® 330; Rhodacal® IPAM

Isopropyl 4-aminobenzenesulfonyl di (dodecylbenzenesulfonyl) titanate
Uses: Coupling agent, adhesion promoter, antioxidant, antistat, antifoam, accelerator, blowing agent activator, catalyst, curative, corrosion inhibitor, dispersion aid, emulsifier, flame retardant, foaming agent, grinding and process aid
Trade names containing: Ken-React® KR 26S

Isopropylcarbinol. *See* Isobutyl alcohol

Isopropyl dimethacryl isostearoyl titanate
Uses: Coupling agent, adhesion promoter, antioxidant, antistat, antifoam, accelerator, blowing agent activator, catalyst, curative, corrosion inhibitor, dispersion aid, emulsifier, flame retardant, foaming agent, grinding and process aid
Trade names containing: Ken-React® KR 7

Isopropyl diphenyl. *See* 1,1´-Biphenyl (1-methylethyl)
Isopropyl ester of PVM/MA copolymer (INCI). *See* PVM/MA copolymer, isopropyl ester

Isopropyl ether
CAS 108-20-3; EINECS 203-560-6
Synonyms: Diisopropyl oxide; 2-Isopropoxypropane
Empirical: $C_6H_{14}O$
Formula: $(CH_3)_2CHOCH(CH_3)_2$
Properties: Colorless liq., ethereal odor; misc. with water, ethanol, ether; m.w. 102.20; dens. 0.719; m.p. -60 C; b.p. 68.5 C; flash pt. (CC) -18 F; ref. index 1.368 (20 C)
Precaution: DOT: Flamm. liq.; severe explosion hazard when exposed to heat, flame, sparks, or oxidizers; may form explosive peroxides; shock may cause explosion; violent reaction with chlorosulfonic acid, HNO_3
Toxicology: LD50 (oral, rat) 8470 mg/kg, (dermal, rabbit) 20 g/kg; TLV 250 ppm in air; mod. toxic by IP and other routes; mildly toxic by ingestion, inh., skin contact; skin irritant; heated to decomp., emits acrid smoke and fumes
Uses: Solv. for paints, animal, vegetable, and min. oils, waxes, and resins; extraction solv.; rubber cements
Manuf./Distrib.: Allchem Ind.; Ashland; J.T. Baker; Exxon; Fluka; Shell

Isopropylformic acid. *See* Isobutyric acid

Isopropyl glycidyl ether
CAS 4016-14-2; EINECS 223-672-9
Synonyms: IPGE; Oxirane, [(1-methylethoxy)methyl]-,
Empirical: $C_6H_{12}O_2$
Formula: $(H_3C)_2CHOCH_2CHOCH_2$
Properties: Colorless liq., ethereal odor; sol. 20% in water; misc. with ethanol, methanol, toluene; m.w. 116.16; dens. 0.92 g/cc (20 C); b.p. 137 C; flash pt. 35 C
Precaution: Flamm.; may form explosive peroxides; incompat. with acids, alkalies, amines, oxidizing agents; hazardous decomp. prods.: CO, hydrocarbons
Toxicology: LD50 (oral, rat) 4200 mg/kg, (dermal, rabbit) 9650 mg/kg; TLV 50 ppm; harmful by inhalation, ingestion; may cause sensitization by inh. and skin contact; skin, eye, and respiratory irritant
Uses: Reactive dilent for epoxy resins; for solv.-free coating systems, laminating resins, fiber-reinforced composites; stabilizer for chlorinated hydrocarbons; thickener for paints and varnishes
Manuf./Distrib.: Aldrich; Fluka; Raschig

Isopropyl-S-(-)-2-hydroxy propionate. *See* Isopropyl lactate
Isopropylideneacetone. *See* Mesityl oxide
4,4´-Isopropylidenebis (2,6-dibromophenol). *See* Tetrabromobisphenol A
p,p-Isopropylidenediphenol. *See* Bisphenol A
4,4´-Isopropylidenediphenol. *See* Bisphenol A

Isopropyl lactate
CAS 617-51-6
Synonyms: Isopropyl-S-(-)-2-hydroxy propionate
Definition: Isopropyl ester of lactic acid
Formula: $CH_3CH(OH)COOCH(CH_3)_2$
Uses: Solv. for paints and coatings, electronics, printed circuit boards, metal industry
Manuf./Distrib.: Purac Am.
Trade names: Purasolv® IPL

Isopropyl thioxanthone
CAS 5495-84-1 (2-isomer); 83846-86-0 (4-isomer); EINECS 226-827-9 (2-isomer); 281-065-4 (4-isomer)
Uses: UV photoinitiator for inks, adhesives, coatings, and photoresists
Manuf./Distrib.: Aceto; Aldrich; First Chem.; Triple Crown Am.
Trade names: Esacure® ITX; Firstcure™ ITX
Trade names containing: Esacure® X15

Isopropyl titanate. *See* Tetraisopropyl titanate

Isopropyl titanium triisostearate
CAS 61417-49-0; EINECS 262-774-8
Synonyms: Isopropyl triisostearoyl titanate; Tris(isooctadecanoato-O)(2-propanolato) titanium
Classification: Organic compd.
Empirical: $C_{57}H_{112}O_7Ti$
Formula: $CH_3CH_3CHOTi(OCOC_{17}H_{35})_3$
Trade names containing: Ken-React® KR TTS

Isopropyl tri (dioctylphosphato) titanate
Uses: Coupling agent, adhesion promoter, antioxidant, antistat, antifoam, accelerator, blowing agent activator, catalyst, curative, corrosion inhibitor, dispersion aid, emulsifier, flame retardant, foaming agent, grinding and process aid
Trade names containing: Ken-React® KR 12

Isopropyl tri (dioctylpyrophosphato) titanate
Uses: Coupling agent, adhesion promoter, antioxidant, antistat, antifoam, accelerator, blowing agent activator, catalyst, curative, corrosion inhibitor, dispersion aid, emulsifier, flame retardant, foaming agent, grinding and process aid
Trade names containing: Ken-React® KR 38S

Isopropyl tridodecylbenzenesulfonyl titanate
Trade names containing: Ken-React® KR 9S

Isopropyl tri (N ethylamino-ethylamino) titanate
Uses: Coupling agent, adhesion promoter, antioxidant, antistat, antifoam, accelerator, blowing agent activator, catalyst, curative, corrosion inhibitor, dispersion aid, emulsifier, flame retardant, foaming agent, grinding and process aid
Trade names containing: Ken-React® KR 44

Isopropyl triisostearoyl titanate. *See* Isopropyl titanium triisostearate
4-Isothiazolin-3-one, 5-chloro-2-methyl-. *See* Methylchloroisothiazolinone
3(2H)-Isothiazolone, 2-methyl-. *See* Methylisothiazolinone

Isotridecyl methacrylate
Uses: Monomer for mfg. of polymers used as oil additives, visc. index improvers, and pour pt. depressants; internal plasticizer for adhesives, uv-curable resins, and coatings
Trade names: Empicryl® 6325

Isourethane. *See* Polyurethane, thermoplastic
Isovalerone. *See* Diisobutyl ketone

Itaconic acid
CAS 97-65-4; EINECS 202-599-6
Synonyms: Methylenesuccinic acid; Propylenedicarboxylic acid
Empirical: $C_5H_6O_4$
Formula: $HOOCCH_2C(:CH_2)COOH$
Properties: Wh. cryst. powd., char. odor; hygroscopic; sol. (1 g/ml): 12 ml water, 5 ml alcohol; very sl. sol. in benzene, chloroform, ether, CS_2, petroleum ether; m.w. 130.10; dens. 1.63; m.p. 165-167 C (dec.)
Precaution: Keep well closed
Uses: Intermediate for acrylic latex and S/B latex for paper coatings, carpet backings, nonwoven textiles, adhesives, and paints; prep. of acrylic fibers; in acrylonitrile, acrylate, and methacrylate copolymers
Regulatory: FDA 21CFR §181.30; Japan approved (itaconic acid)
Manuf./Distrib.: Aldrich; Allchem Ind.; Ashland; Fluka; ICD Group; Rhone-Poulenc Surf. & Spec.; San Yuan; Sigma; United Min. & Chem.

IVE. *See* Vinyl isobutyl ether

Jaguar gum. *See* Guar gum
Japan tallow. *See* Japan wax

Japan wax
CAS 8001-39-6
Synonyms: Rhus succedanea wax; Japan tallow; Sumac wax
Definition: Fat expressed from the mesocarp of the fruit of *Rhus succedanea*, contg. 10-15% palmitin, stearin, olein, 1% japanic acid
Properties: Pale yel. solid, greasy feel, rancid odor and taste; sol. in benzene, CS_2, ether, hot alcohol, alkalies; insol. in water; dens. 0.97-0.98; m.p. 53.5-55 C; acid no. 22-23; iodine no. 10-15; sapon. no. 217-237
Precaution: Combustible

Kadaya gum

Uses: Substitute for beeswax in wax varnishes, candles; in plasters, ointments, floor waxes, furniture polish; plasticizer in dental impression compds.
Regulatory: FDA 21CFR §73.1, 175.105, 175.350, 176.170, 182.70
Manuf./Distrib.: Koster Keunen; Robeco
Trade names: Koster Keunen Synthetic Japan Wax

Kadaya gum. *See* Karaya gum

Kaolin
CAS 1332-58-7; EINECS 296-473-8
Synonyms: Bolus alba; China clay; Hydrated aluminum silicate
Definition: Native hydrated aluminum silicate
Formula: $\approx Al_2O_3 \cdot 2SiO_2 \cdot 2H_2O$
Properties: Wh. to yel. or grayish fine powd., clay-like odor when moist, earthy taste; insol. in water, dilute acids, alkali hydroxides; dens. 1.8-2.6; high lubricity
Precaution: Noncombustible
Toxicology: Nuisance dust; large doses may cause obstructions, perforations, or granuloma (tumor formation)
Uses: Filler and coatings for paper, rubber, refractories, ceramics; in anticaking preps., paint; adsorbent for clarification of liqs.
Regulatory: FDA 21CFR §178.3550, 182.2727, 182.2729, 186.1256, GRAS as indirect additive; BATF 27CFR §240.1051; Japan restricted (0.5% max. residual); Europe listed; UK approved; FDA approved for orals; USP/NF, BP compliance
Manuf./Distrib.: 20 Microns Ltd.; Aldrich; Burgess Pigment; CE Mins.; Dry Branch Kaolin; ECC Int'l.; Feldspar; J.M. Huber; Kaopolite; San Yuan; Sigma; Southeastern Clay; Spectrum Chem. Mfg.; Thiele Kaolin; United Catalysts; U.S. Silica; R.T. Vanderbilt; Whittaker, Clark & Daniels
Trade names: Bilt-Plates® 156; DB-BASE™; DB-GLAZE™; DB-KOTE™ 1; DB-KOTE™ 2; DB-PAQUE™; DB-PAQUE™ LB; DB-PLATE™; DB-SHEEN™; Dixie® Clay; Fiberfrax® EF-119; Huber 35; Huber 40C; Huber 65A; Huber 70C; Huber 80; Huber 80B; Huber 80C; Huber 90; Huber 90B; Huber 90C; Huber 95; Huber HG; Huber HG90; Hydrite 121-S; Hydrite Flat D; Hydrite PX; Hydrite R; Hydrite UF; Peerless® Coating Clay; Peerless® No. 1; Peerless® No. 2; Peerless® No. 3; Polyplate 90; Polyplate 852; Polyplate P; Polyplate P01

Kaolinite
CAS 1318-74-7
Definition: Clay mineral; main constituent of kaolin
Formula: $Al_2O_3 \cdot 2SiO_2 \cdot 2H_2O$
Uses: Extender pigment offering hiding power and suspension, flow control; used in interior emulsions and primer paints and polyester premix plastics
Trade names containing: Aktisil® EM; Aktisil® MAM; Aktisil® PF 224; Silfin® Z; Sillikolloid P 82; Sillikolloid P 87; Sillitin N 82; Sillitin N 85; Sillitin V 85; Sillitin V 88; Sillitin Z 86; Sillitin Z 89

Karaya gum
CAS 9000-36-6; EINECS 232-539-4
Synonyms: Sterculia gum; Sterculia urens gum; India tragacanth; Indian tragacanth; Kadaya gum
Definition: A hydrophilic polysaccharide from trunks of the genus *Sterculia*
Properties: Wh. fine powd., sl. acetic acid odor; insol. in alcohol; swells in water to a gel; produces highly stable emulsions, resist. to acids
Toxicology: Very mildly toxic by ingestion; mild allergen causing hay fever, dermatitis, gastrointestinal diseases, and asthma; may cause intolerance; laxative effect, may reduce nutrient intake
Uses: Pharmaceuticals, textile coatings, ice cream and other food products, adhesives, protective colloids, stabilizers, thickeners, emulsifiers; as denture adhesive
Regulatory: FDA 21CFR §133.133, 133.134, 133.162, 133.178, 133.179, 150.141, 150.161, 184.1349, GRAS; FEMA GRAS; Japan approved; Europe listed; UK approved; BP compliance
Manuf./Distrib.: Meer; Penta Mfg.; Rhone-Poulenc/Perf. Resins & Coatings; Sigma; TIC Gums

Katchung oil. *See* Peanut oil
Ketohexamethylene. *See* Cyclohexanone
Ketone C-7. *See* Methyl n-amyl ketone
Ketone methyl phenyl. *See* Acetophenone
β-Ketopropane. *See* Acetone

Kieselguhr. *See* Diatomaceous earth
KTPP. *See* Potassium tripolyphosphate
Kyanol. *See* Aniline

Lacca. *See* Shellac

Lactic acid

CAS 50-21-5; 598-82-3 (DL); 79-33-4 (L); 10326-41-7 (D); EINECS 200-018-0; 209-954-4 (DL); 201-296-2 (L); 233-713-2 (D)
Synonyms: 2-Hydroxypropanoic acid; 1-Hydroxyethane 1-carboxylic acid; 2-Hydroxypropionic acid; α-Hydroxypropionic acid; Milk acid; Acetonic acid; Ethylidenelactic acid
Classification: Organic acid
Definition: Prod. of the metabolism of glucose and glycogen
Empirical: $C_3H_6O_3$
Formula: $CH_3CHOHCOOH$
Properties: Colorless to yellowish cryst. or syrupy liq., nearly odorless; hygroscopic; misc. with water, alcohol, glycerol, furfural; insol. in chloroform; m.w. 90.09; dens. 1.249; m.p. 18 C; b.p. 122 C (15 mm); flash pt. > 230 F; ref. index 1.4251
Precaution: DOT: Corrosive material; mixts. with nitric acid + hydrofluoric acid may react vigorously
Toxicology: LD50 (oral, rat) 3730 mg/kg; mod. toxic by ingestion, rectal routes; mutagenic data; severe skin and eye irritant; heated to decomp., emits acrid smoke and irritating fumes
Uses: Cultured dairy prods., as acidulant, chemicals (salts, plasticizers, adhesives, pharmaceuticals), mordant in wool dyeing, food additive, mfg. of lactates; sequestrant in metal plating, cataphoresis paints; leather tanning delimer
Usage level: Up to 10%
Regulatory: FDA 21CFR §131.144, 133, 150.141, 150.161, 172.814, 184.1061, GRAS; USDA 9CFR §318.7, 381.147; BATF 27CFR §240.1051, GRAS; not for use in infant foods; FEMA GRAS; Japan approved; Europe listed; UK approved; FDA approved for injectables, parenterals, orals, topicals, vaginals; USP/NF, BP, Ph.Eur., JP compliance
Manuf./Distrib.: Aldrich; Am. Biorganics; Balchem; Chemical; Coyne; Fluka; Lohmann; Penta Mfg.; Pfanstiehl Labs; Purac Biochem; Research Organics; Rhone-Poulenc N. Am.; San Yuan; Sigma; Spectrum Chem. Mfg.; Wilke Int'l.
Trade names: Purac® HS; Purac® PH 88; Purac® USP 88

Lactic acid, ammonium salt. *See* Ammonium lactate
Lactic acid ethyl ester. *See* Ethyl lactate
Lactic acid methyl ester. *See* Methyl lactate
Lamp black. *See* Carbon black
Lanolic acids. *See* Lanolin acid

Lanolin

CAS 8006-54-0 (anhyd.); 8020-84-6 (hyd.); EINECS 232-348-6
Synonyms: Anhydrous lanolin; Adeps lanae; Wool wax; Wool fat
Definition: Deriv. of unctuous fatty sebaceous secretion of sheep, *Ovis aries*, consistg. of complex mixt. of esters of high m.w. aliphatic, steroid, or triterpenoid alcohol and fatty acids
Properties: Yel.-wh. semisolid; sol. in chloroform, ether; insol. in water; m.p. 38-42 C; iodine no. 18-36; flash pt. > 230 F
Precaution: Heated to decomp., emits acrid smoke and irritating fumes
Toxicology: Can cause allergic reactions, contact dermatitis
Uses: Ointments, soaps, face creams, facial tissues, hair set and sun-tan preps.; plasticizer for rubber; lubricant for textiles, metalworking compds.; EP and slip agent for metalworking/rust preventative coatings; in printing inks; waterproofing agent for leather
Regulatory: FDA 21CFR §172.615, 175.300, 176.170, 176.210, 177.1200, 177.2600, 178.3910; Japan approved; FDA approved for ophthalmics, topicals; USP/NF, BP, Ph.Eur. compliance
Manuf./Distrib.: Aldrich; Amerchol; Charkit; Croda; Fluka; Henkel/Emery; Integra; Lanaetex Prods.; Penta Mfg.; R.I.T.A.; Ruger; Sigma; Stevenson Cooper; Westbrook Lanolin
Trade names: Anhydrous Lanolin Technical; Anhydrous Lanolin USP Cosmetic; Anhydrous Lanolin USP Pharmaceutical; Anhydrous Lanolin USP Superfine; Anhydrous Lanolin USP Ultrafine
Trade names containing: Amerchol® C

Lanolin acid

CAS 68424-43-1; EINECS 270-302-7
Synonyms: Lanolic acids; Lanolin fatty acids; Acids, lanolin
Definition: Mixture of organic acids obtained from hydrolysis of lanolin
Uses: Emulsifier, stabilizer, superfatting agent, emollient for fatty acid systems, aerosol shave creams, cream shampoos, wax systems, household prods.; pigment dispersant; increases tack and plasticity of wax films; used in industrial leather treating, coatings, polishes, corrosion inhibitors, lubricants
Trade names: Fancor LFA

Lanolin alcohol

CAS 8027-33-6; EINECS 232-430-1
Synonyms: Alcohols, lanolin; Wool wax alcohol
Definition: Mixture of organic alcohols obtained from hydrolysis of lanolin
Properties: Amber waxy solid, char. odor; sol. in ether, chloroform; sl. sol. in alcohol; insol. in water; m.p. 56 C; acid no. 2 max.; sapon. no. 12 max.
Toxicology: Less likely to cause allergic reaction than lanolin
Uses: W/o emulsifier, stabilizer, softener, emollient, gelling agent, thickener, plasticizer, moisturizer for absorption bases, cosmetics, cleansing preps.
Regulatory: FDA approved for ophthalmics, topicals; USP/NF compliance
Manuf./Distrib.: Ashland; ChemMark; Croda; Henkel/Emery; Heterene
Trade names containing: Amerchol® C; Amerchol® CAB; Amerchol L-101®

Lanolin fatty acids. *See* Lanolin acid

(3-Lauramidopropyl) trimethyl ammonium methyl sulfate

CAS 10595-49-0
Empirical: $C_{19}H_{42}O_5N_2S$
Properties: Off-wh. to lt. tan powd.; m.w. 410
Toxicology: Relatively low toxicity, but moderately to severely irritating to skin and eyes; LD50 (rat, oral) 1.8 g/kg; LD50 (rabbit, dermal) 2.8 g/kg
Uses: Antistatic agent for PVC, PS, polyolefins, ABS for pkg., electronic and polymer deflashing applics., and surface coatings
Trade names: Cyastat® LS

Lauramine oxide

CAS 1643-20-5; 70592-80-2; EINECS 216-700-6
Synonyms: DDAO; LDAO; Lauryl dimethylamine oxide; Lauryldimethylamine-N-oxide; N,N-Dimethyl-1-dodecanamine-N-oxide; N,N-Dimethyldodecylamine-N-oxide
Classification: Tertiary amine oxide
Empirical: $C_{14}H_{31}NO$
Formula: $CH_3(CH_2)_{11}NO(CH_3)_2$
Properties: M.w. 229.41; dens. 0.966 (20/4 C); ref. index 1.379 (20 C)
Toxicology: Skin and eye irritant
Uses: Detergent, foam stabilizer, thickener, emollient in cosmetics, detergents, textile softeners, foam rubber, electroplating, paper coatings, bleach
Regulatory: FDA approved for topicals
Manuf./Distrib.: Aldrich; Fluka; Sigma
Trade names: Rhodamox® LO

Laureth-2

CAS 3055-93-4 (generic); 9002-92-0; 68002-97-1; 68439-50-9; EINECS 221-279-7
Synonyms: Diethylene glycol dodecyl ether; PEG-2 lauryl ether; 2-[2-(Dodecyloxy)ethoxy]ethanol
Definition: PEG ether of lauryl alcohol
Empirical: $C_{16}H_{34}O_3$
Formula: $CH_3(CH_2)_{10}CH_2(OCH_2CH_2)_nOH$, avg. n = 2
Uses: Emulsifier, foam booster, superfatting agent; solubilizer for solvents; bases for prod. of sulfates; raw material for dishwashing, cleansing agent and cold cleaners; emulsion polymerization; in coning and textile spin finishes
Manuf./Distrib.: Aldrich; Fluka; Sigma
Trade names: Prox-onic LA-1/02

Laureth-4

CAS 5274-68-0; 68002-97-1; 68439-50-9; EINECS 226-097-1

Synonyms: PEG-4 lauryl ether; PEG 200 lauryl ether; 3,6,9,12-Tetraoxatetracosan-1-ol
Definition: PEG ether of lauryl alcohol
Empirical: $C_{20}H_{42}O_5$
Formula: $CH_3(CH_2)_{10}CH_2(OCH_2CH_2)_nOH$, avg. n = 4
Uses: Emulsifier, solubilizer, lubricant, detergent for cosmetics, silicone polish, mold releases; bases for prod. of sulfates; raw material for dishwashing, cleansing agent and cold cleaners; antistat for PE, PS; surfactant for pulp/paper, pharmaceuticals, metalworking, lubricants, textiles, agric., paints, adhesives
Regulatory: FDA 21CFR §178.3520; FDA approved for topicals
Manuf./Distrib.: Aldrich; Fluka; Sigma
Trade names: Calgene Nonionic L-4; Prox-onic LA-1/04

Laureth-5
CAS 3055-95-6; EINECS 221-281-8
Synonyms: PEG-5 lauryl ether; POE (5) lauryl ether; 3,6,9,12,15-Pentaoxyheptacosan-1-ol
Definition: PEG ether of lauryl alcohol
Empirical: $C_{22}H_{46}O_6$
Formula: $CH_3(CH_2)_{10}CH_2(OCH_2CH_2)_nOH$, avg. n = 5
Uses: Dispersant, wetting agent, for washing, cleaning, soil suspending, and homogenizing applics.
Manuf./Distrib.: Fluka; Sigma
Trade names: Mulsifan RT 23

Laureth-8
CAS 3055-98-9; 9002-92-0 (generic)
Synonyms: PEG-8 lauryl ether; POE (8) lauryl ether; 3,6,9,12,15,18,21,24-Octaoxahexatriacontan-1-ol
Definition: PEG ether of lauryl alcohol
Empirical: $C_{28}H_{58}O_9$
Formula: $CH_3(CH_2)_{10}CH_2(OCH_2CH_2)_nOH$, avg. n = 8
Uses: Wetting agent, detergent, emulsifier, dispersant used for maintenance and institutional cleaners, in textile, paper, and paint industries
Regulatory: FDA 21CFR §177.2800
Manuf./Distrib.: Aldrich; Fluka; Sigma

Laureth-9
CAS 3055-99-0; 9002-92-0 (generic); 68439-50-9; EINECS 221-284-4
Synonyms: PEG-9 lauryl ether; POE (9) lauryl ether; 3,6,9,12,15,18,21,24,27-Nonaoxanonatriacontan-1-ol
Definition: PEG ether of lauryl alcohol
Empirical: $C_{30}H_{62}O_{10}$
Formula: $CH_3(CH_2)_{10}CH_2(OCH_2CH_2)_nOH$, avg. n = 9
Uses: Wetting agent, detergent, emulsifier, dispersant used for maintenance and institutional cleaners, in textile, paper, and paint industries; emulsion polymerization
Regulatory: FDA 21CFR §177.2800, 178.3130
Manuf./Distrib.: Aldrich; Fluka; Sigma
Trade names: Prox-onic LA-1/09

Laureth-12
CAS 3056-00-6; 9002-92-0 (generic); EINECS 221-286-5
Synonyms: PEG-12 lauryl ether; POE (12) lauryl ether; PEG 600 lauryl ether
Definition: PEG ether of lauryl alcohol
Empirical: $C_{36}H_{64}O_{13}$
Formula: $CH_3(CH_2)_{10}CH_2(OCH_2CH_2)_nOH$, avg. n = 12
Properties: M.w. 1199.57; m.p. 41-45 C; b.p. 100 C; flash pt. > 110 C
Uses: Emulsifier for cosmetic, pharmaceutical, paints, and industrial uses; emollient, thickener for shampoos; emulsion polymerization
Regulatory: FDA 21CFR §177.2800
Manuf./Distrib.: Aldrich; Fluka; Sigma
Trade names: Calgene Nonionic L-12; Prox-onic LA-1/012

Laureth-23
CAS 9002-92-0 (generic)
Synonyms: PEG-23 lauryl ether; POE (23) lauryl ether
Definition: PEG ether of lauryl alcohol
Formula: $CH_3(CH_2)_{10}CH_2(OCH_2CH_2)_nOH$, avg. n = 23
Uses: Emulsifier, stabilizer, solubilizer, surfactant, emollient, thickener, dispersant for cosmetics; post ad

stabilizer for syn. latexes
Regulatory: FDA 21CFR §177.2800; FDA approved for topicals
Manuf./Distrib.: Aldrich; Fluka; Sigma
Trade names: Calgene Nonionic L-23; Prox-onic LA-1/023; Rhodasurf® L-25

Lauric acid, 1,2-ethanediyl ester. *See* Glycol dilaurate
Lauroamphoacetate. *See* Sodium lauroamphoacetate
Lauroamphoglycinate. *See* Sodium lauroamphoacetate

Lauroyl lysine
CAS 52315-75-0; EINECS 257-843-4
Synonyms: Lauroyl-1-lysine; Lauroyl-L-lysine
Definition: Lauroyl deriv. of lysine
Empirical: $C_{18}H_{36}N_2O_3$
Formula: $CH_3(CH_2)_{10}CONH(CH_2)_4CHNH_2COOH$
Properties: Wh. cryst. powd.; insol. in almost all solvs. except strong acidic and alkaline sol'ns.; sol. in water @ pH < 1 and > 12
Toxicology: LD50 (mice, oral) > 5.0 g/kg; nonirritating and nonsensitizing to skin, nonirritating to eyes
Uses: Powder material for cosmetics and medicinals; filler for ink and paint; surface improver of inorganic powders; chelating agent
Trade names: Amihope LL

Lauroyl-1-lysine. *See* Lauroyl lysine
Lauroyl-L-lysine. *See* Lauroyl lysine

Lauryl acrylate
CAS 2156-97-0; EINECS 218-463-4
Synonyms: Dodecyl acrylate
Classification: Monomer
Empirical: $C_{15}H_{28}O_2$
Formula: $CH_2=CHCOOC_{12}H_{25}$
Properties: M.w. 240.39; b.p. 120 C; dens. 0.87 (20/4 C); flash pt. 83 C
Precaution: Combustible
Toxicology: Irritating to eyes, skin, respiratory system
Uses: UV-curable reactive diluent in inks and coatings, adhesives, visc. index improver, finishing aid for leather
Manuf./Distrib.: Aldrich; CPS; Fluka; Monomer-Polymer & Dajac Labs; Sartomer
Trade names: Photomer® 4812
Trade names containing: Ageflex FA-12; SR-335

Lauryl dimethylamine oxide. *See* Lauramine oxide
Lauryldimethylamine-N-oxide. *See* Lauramine oxide

Lauryl hydroxyethyl imidazoline
CAS 136-99-2; EINECS 205-271-0
Synonyms: Lauryl imidazoline; 1H-Imidazole-1-ethanol, 4,5-dihydro-2-undecyl-; 1-Hydroxyethyl 2-undecyl imidazoline
Classification: Heterocyclic compd.
Empirical: $C_{16}H_{32}N_2O$
Uses: Corrosive inhibitor, emulsifier, dispersant and fluidizing agent for pigments, emulsions, cosmetics, polishes, textile, leather, and agric. prods.; acid detergent for food and dairy prods.; intermediate for quat. ammonium compds.
Trade names: Calgene C-100-L

Lauryl imidazoline. *See* Lauryl hydroxyethyl imidazoline

Lauryl methacrylate
CAS 142-90-5; EINECS 205-570-6
Synonyms: Dodecyl methacrylate; Dodecyl 2-methyl-2-propenoate
Definition: Ester of lauryl alcohol and methacrylic acid
Empirical: $C_{16}H_{30}O_2$
Formula: $CH_2=C(CH_3)COOC_{12}H_{25}$
Properties: M.w. 254.43; bulk dens. 0.868 g/ml; b.p. 272-344 C; flash pt. (COC) 132 C

Precaution: Nonhazardous (DOT); combustible
Toxicology: Skin and eye irritant
Uses: Lube oil additives, coatings for nonwoven fiber, floor waxes, paints, adhesives, varnishes, sealants, caulks, stabilizers in nonaq. disp. and inks
Manuf./Distrib.: Albright & Wilson Am.; Aldrich; CPS; Elf Atochem N. Am.; Fluka; Monomer-Polymer & Dajac Labs; Rhone-Poulenc Spec. Chem.; Rohm Tech
Trade names: Photomer® 2812
Trade names containing: Ageflex FM-12; Ageflex FM-246; SR-313

Lauryl-myristyl methacrylate
Uses: Monomer for mfg. of polymers used as oil additives, visc. index improvers, pour pt. depressants; internal plasticizer for adhesives, uv-curable resins, coatings
Trade names containing: Empicryl® 6047

Lauryl pyrrolidone
CAS 2687-96-9
Synonyms: N-Dodecyl-2-pyrrolidone
Classification: Substituted heterocyclic compd.
Empirical: $C_{16}H_{31}NO$
Uses: Conditioner, foam stabilizer, wetting agent; specialty solv. for commercial cleaning, textile processing, lithographic printing prods., water-borne coatings, inks; VOC replacement
Manuf./Distrib.: Aldrich
Trade names: Surfadone® LP-300

LDAO. *See* Lauramine oxide
LDPE. *See* Polyethylene, low-density

Lead
CAS 7439-92-1; EINECS 231-100-4
Classification: Metallic element
Empirical: Pb
Properties: Metallic, heavy ductile soft gray solid; sol. in dilute nitric acid; insol. in water; at.wt. 207.2; dens. 11.35; m.p. 327.4 C; b.p. 1755 C
Precaution: Noncombustible; dust can be flamm. and moderately explosive when exposed to heat or flame
Toxicology: TLV 0.15 mg/m^3 of air; poison by ingestion; moderately toxic by intraperitoneal route; toxic by inhalation of dust or fume; cumulative poison; affects nervous, renal, reproductive, blood, and GI systems
Uses: Storage batteries, gasoline additive, radiation shielding, cable covering, ammunition, chemical reaction equip., solder and fusible alloys; in heat and light stabilizers for PVC; paints, shellac, caulking compds.; bearing metals and alloys
Manuf./Distrib.: Aldrich; Asarco; Belmont Metals; Cerac; Fluka; Fry's Metals Ltd; Noah; Sigma

Lead acetate
CAS 301-04-2; EINECS 206-104-4
Synonyms: Acetic acid, lead salt; Normal lead acetate; Sugar of lead; Salt of saturn
Classification: Inorganic salt
Empirical: $C_2H_4O_2 \cdot 1/2Pb$
Formula: $[CH_3COO]_2Pb$
Properties: Wh. cryst. or flakes, sweetish taste, sl. acetic odor; sol. in water, glycerin, alcohol; m.w. 325.30; dens. 3.251 (20 C); m.p. 280 C
Precaution: Combustible
Toxicology: Toxic by ingestion, inhalation, and skin absorption; use may be restricted
Uses: Dyeing of textiles, waterproofing varnishes, insecticides, lead driers, chrome pigments, hair dye, weighting silks
Regulatory: FDA 21CFR §73.2396
Manuf./Distrib.: Alemark; Allan; Amber Syn.; Am. Biorganics; Am. Int'l.; Barker Ind.; Cerac; Hoechst Celanese; Mallinckrodt; Noah; Poly Research; Spectrum Chem. Mfg.

Lead caprylate. *See* Lead octoate

Lead carbonate (basic)
CAS 598-63-0; 1319-46-6; EINECS 215-290-6
Synonyms: White lead; Dibasic lead carbonate; Lead subcarbonate
Formula: $2PbCO_3 \cdot Pb(OH)_2$

Lead drier

Properties: Wh. amorphous powd.; sol. in acids; insol. in water; m.w. 267.20; dens. 6.86; dec. 400 C
Precaution: Noncombustible
Toxicology: DOT: Poisonous material; toxic by inhalation; TLV (as Pb) 0.15 mg/m^3 of air; moderately toxic by ingestion
Uses: Heat and light stabilizer for PVC; exterior paint pigment, ceramic glazes
Manuf./Distrib.: Aldrich; EM Ind./Pigments; Halstab; D.N. Lukens; Nat'l. Chem.; Revelli; Sigma; Spectrum Chem. Mfg.

Lead drier
Uses: Used in org. coatings, inks, polyesters
Trade names: Troymax™ Drier Lead 24%; Troymax™ Drier Lead 36%

Lead linoleate
CAS 16996-51-3
Synonyms: Lead plaster; 9,12-Octadecadienoic acid lead salt
Empirical: $C_{18}H_{32}O_2 \cdot xPb$
Properties: Ylsh.-wh. paste; sol. in oils; insol. in water
Precaution: Combustible
Toxicology: Toxic material; absorbed by skin
Uses: Drier for paints; in medicine

Lead metasilicate. *See* Lead silicate
Lead monoxide. *See* Lead (II) oxide

Lead naphthenate
CAS 61790-14-5; EINECS 263-109-4
Synonyms: Naphthenic acid lead salt
Uses: Drier for paints; lubricant additive; flotation reagent; emulsifier; cutting oil additive
Manuf./Distrib.: Aceto; Akzo Nobel; Archway Sales; Baychem; Boehle; CasChem; Dussek Campbell Ltd; Hüls Am.; D.N. Lukens; OM Group; Shepherd
Trade names: Nuodex Napthenate® Lead 24%

Lead octanoate. *See* Lead octoate

Lead octoate
CAS 15696-43-2
Synonyms: Lead caprylate; Lead octanoate; Octanoic acid lead salt
Empirical: $C_8H_{16}O_2 \cdot xPb$
Uses: Drier for paints
Manuf./Distrib.: Archway Sales; Baychem; Boehle; CasChem; Dussek Campbell; Hüls Am.; D.N. Lukens; Shepherd
Trade names: Nuodex Octoate® Lead 24%

Lead orthoplumbate. *See* Lead oxide, red

Lead (II) oxide
CAS 1317-36-8; EINECS 215-267-0
Synonyms: Lead (II) oxide yellow; Lead oxide yellow; Plumbous oxide; Lead monoxide; Lead protoxide; Calcined litharge; Plumbous oxide; C.I. 77577; C.I. Pigment Yellow 46; Litharge
Empirical: PbO
Properties: Yel. cryst.; sol. in acids, alkalies; insol. in water, alcohol; dens. 9.53; m.p. 888 C
Precaution: Violent or explosive reaction with chlorinated rubber (> 200 C), fluoroelastomers (@ 200 C); incompat. with Cl, metals, nonmetals
Toxicology: DOT: Poisonous material; toxic by ing. and inh.; ACGIH TLV:TWA 0.15 mg (Pb)/m^3; mutagenic data; skin irritant; heated to decomp., emits toxic fumes of Pb
Uses: Activator, vulcanizing agent for rubber; mfg. of dry colors, greases, high-pressure lubricants, brake linings, ceramics, glass, piezoelectric devices, various chemical processes, storage batteries, paint
Manuf./Distrib.: Aarbor Int'l.; Aldrich; Asarco; Atlantic Equip. Engrs.; J.T. Baker; Fluka; Hammond Group; Nat'l. Chem.; Noah; O&C; Sigma; Spectrum Chem. Mfg.

Lead (II,III) oxide. *See* Lead oxide, red

Lead oxide, red
CAS 1314-41-6

Synonyms: Red lead; Red lead oxide; Minium; Mineral red; Lead (II,III) oxide; Plumboplumbic oxide; C.I. 77578; C.I. Pigment red 105; Lead orthoplumbate; Lead tetroxide; Trilead tetroxide
Empirical: Pb_3O_4
Properties: Red powd.; partly sol. in acids; insol. in water; m.w. 685.57; dens. 8.32-9.16; vapor pressure 1 mm @ 943 C; m.p. 890 C (dec.); b.p. 1472 C
Precaution: Oxidizer; combustible by reaction with reducing agents; explodes on contact with peroxyformic acid; incompat. with Al, Na, SO_3, Ti, Zr
Toxicology: DOT: Poisonous material; LD50 (IP, rat) 630 mg/kg; poison by IP route; mod. toxic by ing.; irritant; toxic as dust; TLV (as Pb) 0.15 mg/m^3 of air; heated to decomp., emits toxic fumes of Pb
Uses: Storage batteries, glass, pottery, enameling, varnish, purification of alcohol, packing pipe joings, metal protective paints, fluxes, ceramic glazes; processing aid for rubber
Manuf./Distrib.: Aldrich; Atlantic Equip. Engrs.; Hammond Group; O&C; Spectrum Chem. Mfg.

Lead oxide yellow. *See* Lead (II) oxide
Lead (II) oxide yellow. *See* Lead (II) oxide

Lead phosphite dibasic
CAS 1344-40-7
Formula: $2PbO \cdot PbHPO_3 \cdot {}^1/_2H_2O$
Properties: Wh. fine acicular crystals; insol. in water; dens. 6.94; ref. index 2.25
Toxicology: Toxic material; TLV (as Pb) 0.15 mg/m^3 of air
Uses: Heat and light (UV screening and antioxidizing) stabilizer for vinyl plastics and chlorinated paraffins; wh. rust-inhibitive pigment in paints and plastics
Manuf./Distrib.: D.N. Lukens; Nat'l. Chem.

Lead phthalate, basic
EINECS 290-588-7
Synonyms: 1,2-Benzenedicarboxylic acid, lead (2+) salt, basic
Formula: $2PbO \cdot PbHPO_3 \cdot {}^1/_2H_2O$
Properties: Wh. fine acicular cryst.; insol. in water; dens. 6.94
Toxicology: TLV (Pb) 0.15 mg/m^3 of air
Uses: Heat and light stabilizer, antioxidant for vinyl plastics and chlorinated paraffins; in paints and plastics

Lead plaster. *See* Lead linoleate
Lead protoxide. *See* Lead (II) oxide

Lead silicate
CAS 10099-76-0; 11120-22-2
Synonyms: Lead metasilicate; Basic lead white silicate
Empirical: $O_3Si \cdot Pb$
Formula: $PbSiO_3$
Properties: Wh. cryst. powd.; insol. in most solvs.; m.w. 283.28; dens. 6.49; m.p. 766 C
Precaution: Noncombusible
Toxicology: Toxic material; TLV (as Pb) 0.15 mg/m^3
Uses: Pigment, heat stabilizer, rust inhibitor; in ceramics, fireproofing fabrics, paints, plastics
Manuf./Distrib.: Eagle-Picher; Hammond Lead Prods.; D.N. Lukens; Nat'l. Chem.
Trade names: BSWL 202
Trade names containing: BSWL 201

Lead silicochromate
CAS 11113-70-5
Synonyms: Basic lead silica chromate
Toxicology: TLV: 0.15 (Pb)mg/m^3 of air
Uses: Corrosion inhibiting pigment for metal protective coatings
Manuf./Distrib.: Am. Disps.; D.N. Lukens; Nat'l. Chem.
Trade names: Oncor M-50

Lead silicosulfate, basic
Toxicology: Toxic material; TLV (as Pb) 0.15 mg/m^3 of air
Uses: Corrosive inhibitive pigment for metal protective coatings, primers, and finishes; industrial enamels requring a high gloss; heat stabilizer for PVC

Lead stearate
CAS 1072-35-1; 7428-48-0
Synonyms: Stearic acid, lead salt
Empirical: $C_{36}H_7O_4 \cdot Pb$
Formula: $Pb(C_{18}H_{35}O_2)_2$
Properties: Wh. powd.; sol. in hot alcohol; insol. in water; m.w. 774.25; dens. 1.4; m.p. 100-115 C
Precaution: Combustible
Toxicology: LDLo (oral, guinea pig) 6000 mg/kg; toxic material, absorbed by the skin; mildly toxic by ingestion; heated to decomp., emits toxic fumes of Pb
Uses: Varnish and lacquer drier; in extreme pressure lubricants; lubricant in extrusion processes; stabilizer for vinyl polymers; corrosion inhibitor for petroleum; component of greases, waxes, and paints
Manuf./Distrib.: AC Ind.; Cookson Spec. Additives; Ore & Chem.; Syn. Prods.; R.T. Vanderbilt

Lead subcarbonate. *See* Lead carbonate (basic)

Lead sulfate
CAS 7446-14-2; EINECS 231-198-9
Synonyms: Lead (II) sulfate (1:1); Lead (II) sulfate
Empirical: O_4PbS
Formula: $PbSO_4$
Properties: Wh. rhombic crystals; sl. sol. in hot water; insol. in alcohol, sodium hydroxide sol, conc. hydriodic acid; m.w. 303.25; dens. 6.12-6.39; m.p. 1170 C
Precaution: Noncombustible
Toxicology: DOT: Poisonous material; strong irritant to tissues; TLV (as Pb) 0.15 mg/m^3 of air; toxic by inhalation; poison by intraperitoneal route; moderately toxic by ingestion
Uses: Storage batteries, paint pigments; weighting fabrics; preparing rapidly drying oil varnishes
Manuf./Distrib.: Aldrich; Am. Biorganics; Fluka; Halstab; Spectrum Chem. Mfg.

Lead (II) sulfate. *See* Lead sulfate
Lead (II) sulfate (1:1). *See* Lead sulfate

Lead sulfate, basic
Synonyms: White lead, sublimed; White lead sulfate
Formula: $PbSO_4 \cdot PbO$
Properties: Wh. monoclinic crystals; sl. sol. in hot water or acids; dens. 6.92; m.p. 977 C
Precaution: Noncombustible
Toxicology: Toxic material; TLV (as Pb) 0.15 mg/m^3 of air
Uses: Heat stabilizer for PVC; wh. pigment for paints, ceramics
Manuf./Distrib.: Halstab; D.N. Lukens; Nat'l. Chem.; Spectrum Chem. Mfg.

Lead sulfate, blue basic
Synonyms: Sublimed blue lead; Blue lead
Definition: Mixture of lead sulfate (45% min.), lead oxide (30% min.), lead sulfide (12% max.), lead sulfite (5% max.), zinc oxide (5%), carbon, etc.
Properties: Bluish gray; insol. in water or alcohol; dens. 6.2
Precaution: Noncombustible
Toxicology: Toxic material; TLV (as Pb) 0.15 mg/m^3 of air
Uses: Corrosion-inhibiting pigment; component of structural-metal priming coat paints, lubricants, vinyl plastics, rubber prods.
Manuf./Distrib.: D.N. Lukens

Lead tallate
Toxicology: Toxic material; absorbed by skin
Uses: Drier for paints
Manuf./Distrib.: Dussek Campbell Ltd; OM Group

Lead tetroxide. *See* Lead oxide, red

Lecithin
CAS 8002-43-5; 8029-76-3; 8030-76-0; 97281-47-5; EINECS 232-307-2
Synonyms: Soya lecithin
Definition: Mixture of the diglycerides of stearic, palmitic and oleic acids linked to the choline ester of phosphoric acid; found in plants and animals

Formula: $C_8H_{17}O_5NRR'$, R and R′ are fatty acid groups
Properties: Nearly wh. to yel. or brn. waxy mass or thick fluid, nutlike odor, bland taste; insol. but swells in water and salt sol'ns.; sol. in chloroform, ether, petrol. ether, min. oils, fatty acids; dens. 1.0305 (24/4 C); sapon. no. 196
Toxicology: May cause bronchoconstriction in people with asthma; heated to decomp., emits acrid smoke and irritating fumes
Uses: Edible surfactant and emulsifier for food use, pharmaceuticals, cosmetics, leather treatment, textiles; release agent for silicone rubber molds; raw material for paints; dispersing agent, wetting agent; antioxidant
Regulatory: FDA 21CFR §133.169, 133.173, 133.179, 136.110, 136.115, 136.130, 137.160, 136.165, 136.180, 163.123, 163.130, 163.135, 163.140, 163.145, 163.150, 163.155, 166.40, 166.110, 169.115, 169.140, 169.150, 175.300, 184.1400, GRAS; USDA 9CFR §318.7, 0.5% max. in oleomargarine, 381.147; Japan approved; Europe listed; UK approved; FDA approved for orals, topicals; USP/NF compliance
Manuf./Distrib.: ADM Lecithin; Am. Lecithin; Central Soya; Fluka; W.A. Cleary; Cosan; Great Western; Int'l. Chem. Inc.; Landers-Segal Color; Lucas Meyer; Penta Mfg.; Reichhold; Sigma; Solvay Duphar BV; Spice King; U.S. Biochemical
Trade names: Alcolec® 439-C; Alcolec® 440-WD; Amisol™ F-100; Blendmax 322; Centrol® 2F SB; Centrol® 2F UB; Centrol® 3F SB; Centrol® 3F UB; Centrolex®; Centrolex® P; Clearate LV; Clearate Special Extra; Clearate WD; Kelecin 1081; Kelecin F; Lecithin W.D.; Lipotin 100, 100J, SB; R & R 551; R & R 552; Stablec IDC 30; Stablec IDC 50; Stablec IDC 52; Stablec IDC 85; TLV 68-SB; TLV 68-UB; TLV 70-SB; TLV 70-UB; Troykyd® Lecithin W.D.; Yelkin® DS; Yelkin® SS; Yelkin® T; Yelkin® TS
Trade names containing: Centromix® LP250; Centrophil® M

Lecithin, hydroxylated. *See* Hydroxylated lecithin
Lemon chrome. *See* Barium chromate
Lichenic acid. *See* Fumaric acid
Light aromatic petroleum naphtha. *See* Naphtha, light aromatic
Light mineral oil. *See* Mineral oil
Lignite wax. *See* Montan wax
Lignoceric acid, myristyl ether. *See* Myristyl lignocerate
Ligroin. *See* Mineral spirits
Lime. *See* Calcium oxide
Limed rosin. *See* Calcium resinate

Limestone
CAS 1317-65-3
Synonyms: Calcium carbonate, natural; Agricultural limestone; Lithographic stone
Empirical: $CaCO_3$
Properties: Solid, odorless, tasteless; insol. in water; sol. in dilute acids; m.w. 100.09; dens. 2.7-2.95; m.p. 825 C
Precaution: Noncombustible
Toxicology: TLV/TWA 10 mg/m^3 (total dust); severe eye and moderate skin irritant
Uses: Filler for rubber, plastics, building stone, caulks, cements, ceramics, coatings, metallurgy (flux), mfg. of lime; source of CO_2; Portland and natural cement; removal of sulfur dioxide from stack gases and sulfur from coal
Trade names: Dolocron® 15-16; Dolocron® 32-15; Dolocron® 40-13; Dolocron® 45-12; Franklin T-11; Franklin T-12; Franklin T-13; Franklin T-14; Franklin T-325; Marblewhite® 200; Marblewhite® 325; Marblewhite® A 200; Marblewhite® A 325; Marblewhite® A 4500; Vicron® 10-20; Vicron® 15-15; Vicron® 25-11; Vicron® 31-6; Vicron® 41-8; Vicron® 45-3
See also Calcium carbonate

(±)-Limonene. *See* dl-Limonene

dl-Limonene
CAS 138-86-3; EINECS 205-341-0
Synonyms: Dipentene (INCI); Cinene; Cajeputene; p-Mentha-1,8-diene; Inactive limonene; 1-Methyl-4-isopropenyl-1-cyclohexene; (±)-Limonene
Empirical: $C_{10}H_{16}$
Properties: Colorless liq., pleasant lemon-like odor; misc. with alcohol; pract. insol. in water; m.w. 136.23; dens. 0.847 (15.5/15.5 C); m.p. -96.9 C; b.p. 175-176 C (763 mm); ref. index 1.4744
Precaution: DOT: Flamm. liq.; can react vigorously with oxidizers
Toxicology: LD50 (oral, rat) 5000 mg/kg; skin irritant and sensitizer; heated to decomp., emits acrid smoke and irritating fumes
Uses: Solvent for oleoresinous prods., rosin, ester gum, etc.; rubber compding. and reclaiming; dispersant for

oils, resins and combinations, pigments, driers, paints, enamels, lacquers; general wetting; printing inks, perfumes, flavors, waxes, polishes

Regulatory: FDA 21CFR §175.105, 177.2600, 182.60, GRAS; FDA approved for topicals

Manuf./Distrib.: Aldrich; Arizona; Fluka; Hercules; Penta Mfg.; SCM Glidco Organics; Spectrum Chem. Mfg.; Veitsiluoto Oy

Limonene polymer. *See* Polydipentene

Linoleic acid

CAS 60-33-3; EINECS 200-470-9

Synonyms: 9,12-Octadecadienoic acid; (Z,Z)-9,12-Octadecadienoic acid; Linolic acid

Classification: Unsaturated essential fatty acid

Empirical: $C_{18}H_{32}O_2$

Formula: $CH_3(CH_2)_4$=$CHCH_2CH$=$CH(CH_2)_7COOH$

Properties: Colorless to pale yel. oil; freely sol. in ether; sol. in chloroform, abs. alcohol; misc. with dimethylformamide, fat solvs., oils; insol. in water; m.w. 280.44; dens. 0.9007 (22/4 C); m.p. -12 C; b.p. 230 C (16 mm); ref. index 1.4699 (20 C)

Precaution: Combustible; easily oxidized by air; store refrigerated

Toxicology: Human skin irritant; ingestion can cause nausea and vomiting; heated to decomp., emits acrid smoke and irritating fumes

Uses: Mfg. of paints, coatings, emulsifiers, vitamins

Regulatory: FDA 21CFR §175.105, 182.5065, 184.1065, GRAS

Manuf./Distrib.: Aldrich; Am. Biorganics; Arizona; CasChem; Fluka; Henkel/Emery; Hercules; Langley Smith Ltd; Penta Mfg.; Research Organics; Rhone-Poulenc Surf. & Spec.; Sigma; Spectrum Chem. Mfg.

Trade names: Pamolyn® 200; Pamolyn® 240

Linoleyl alcohol

Definition: Fatty alcohol derived from linoleic acid

Formula: $CH_3(CH_2)_4CH{:}CHCH_2CH{:}CH(CH_2)_7CH_2OH$

Uses: Paints, flotation, paper, surfactants, resins, leather

Linolic acid. *See* Linoleic acid

Linseed oil

CAS 8001-26-1; EINECS 232-278-6

Synonyms: Oils, linseed; Flaxseed oil; Linseed oil, raw; Raw linseed oil

Definition: Expressed oil from the dried ripe seed of *Linum usitatissimum*

Properties: Golden-yel., amber, or brown drying oil, peculiar odor, bland taste; sol. in ether, chloroform, carbon disulfide, turpentine; sl. sol. in alcohol; dens. 0.921-0.936; m.p. -19 C; b.p. 343 C; flash pt. 222 C

Precaution: Combustible liq. exposed to heat or flame; can react with oxidizers; subject to spontaneous heating; violent reaction with Cl_2

Toxicology: Allergen and skin irritant to humans

Uses: Drying oil for paints, varnishes, oil cloth, putty, printing inks, core oils, linings and packings, alkyd resins, soap, pharmaceuticals; hot melt adhesives

Regulatory: FDA 21CFR §175.105, 175.300, 176.200, 176.210, 181.22, 181.26; BP compliance

Manuf./Distrib.: ADM; Aldrich; Alnor Oil; Arista Ind.; Ashland; Chemcentral; Degen; Ferro/Bedford; C.P. Hall; Lomas Int'l.; Penta Mfg.; Reichhold; John L Seaton Ltd; Sigma; Spectrum Chem. Mfg.; Van Waters & Rogers; Welch, Holme & Clark; Jesse S. Young

Trade names: A-R Varnish & Grinding; Cykelin; Cykelin 70; Esskol Y-Z; Kellin U-V; K.V.O.; Neutral K.V.O.; S-70 OKO; Superflo Grinding; Superior; Superior C

Linseed oil, raw. *See* Linseed oil
Liquid paraffin. *See* Mineral oil
Liquid petrolatum. *See* Mineral oil
Liquid rosin. *See* Tall oil
Litharge. *See* Lead (II) oxide

Lithium molybdate

CAS 13568-40-6

Empirical: Li_2MoO_4

Properties: Wh. crystalline compd.; sol. in water; dens. 2.66; m.p. 705 C

Uses: Steel coating, petroleum cracking catalyst

Manuf./Distrib.: AAA Molybdenum Prods.; Aldrich; Atomergic Chemetals; Cerac; Chem-Met; FMC; Great Western

Lithographic stone. *See* Limestone

Lithopone
CAS 1345-05-7
Synonyms: Griffith's zinc white
Definition: Mixture of zinc sulfide (26-60%), barium sulfate, and zinc oxide
Properties: Wh. powd.
Toxicology: Poison liberating hydrogen sulfide upon dec. by heat, moisture, acids; heated to decomp., emits highly toxic fumes of SO_x, ZnO, H_2S
Uses: White pigment in plastics, paints, paper; provides thixotropy, improves gloss and flow
Manuf./Distrib.: Landers-Segal Color; D.N. Lukens; Ore & Chem.; Primachem; H.M. Royal; Sachtleben Chemie GmbH; San Yuan; Sino-Am. Pigment Systems; Van Waters & Rogers
Trade names: Lithopone 30% DS; Lithopone 30% L; Lithopone DS (Red Seal 30% ZnS)

Liver starch. *See* Glycogen
LLDPE. *See* Polyethylene, linear low density

Locust bean gum
CAS 9000-40-2; EINECS 232-541-5
Synonyms: Carob flour; Carob bean gum; Locust gum; St. John's bread; Algaroba
Classification: Polysaccharide plant mucilage
Definition: Ground seed of the ripe fruit of St. John's Bread (*Ceratonia siliqua*)
Properties: Yel.-grn. powd., odorless, tasteless; swells in cold water; insol. in org. solvs.; visc. increases when heated; m.w. $\approx$ 310,000
Precaution: Combustible
Toxicology: LD50 (oral, rat) 13 g/kg; mildly toxic by ingestion; heated to decomp., emits acrid smoke and irritating fumes
Uses: Stabilizer, thickener, emulsifier, suspending agent, water-binder in foods, feeds, cosmetics, pharmaceuticals; sizing and finishes for textiles; paints; fiber bonding agent in paper mfg.; drilling fluid; coffee, chocolate substitute; packaging material
Regulatory: FDA 21CFR §133.133, 133.134, 133.162, 133.178. 133.179, 150.141, 150.161, 182.20, 184.1343, 240.1051, GRAS; FEMA GRAS; Japan approved; Europe listed; UK approved; ADI not specified (JECFA)
Manuf./Distrib.: Fluka; Grindsted Prods.; Hercules; Rhone-Poulenc Food Ingreds.; Sigma

Locust gum. *See* Locust bean gum
Lye. *See* Sodium hydroxide

Macrogol 300. *See* PEG-6
Macrogol 400. *See* PEG-8
Macrogol 600. *See* PEG-12
Macrogol 1000. *See* PEG-20
Macrogol 1540. *See* PEG-32
Macrogol 6000. *See* PEG-150
Magnesia. *See* Magnesium oxide
Magnesia clinker. *See* Magnesium oxide
Magnesia magma. *See* Magnesium hydroxide
Magnesia usta. *See* Magnesium oxide

Magnesium aluminum silicate
CAS 1327-43-1; 12199-37-0; EINECS 235-374-6
Synonyms: Aluminum magnesium silicate; Aluminosilicic acid, magnesium salt
Definition: Complex silicate refined from naturally occurring minerals; blend of colloidal montmorillonite and saponite
Empirical: $Al_2MgO_8Si_2$
Properties: Fine powd. or sm. flakes, odorless, tasteless; swells in water or glycerin; insol. in water or alcohol; m.w. 262.45; visc. 100-2200 cps; pH 9-10 (5% susp.)
Toxicology: Not harmful at presently used levels; WHO recommends further studies because of kidney

damage in dogs that ingested it
Uses: Suspending agent, thickener; antacid
Regulatory: FDA approved for dentals, orals, rectals, topicals, vaginals; USP/NF, BP compliance
Manuf./Distrib.: Am. Colloid; Chemisphere; Dry Branch Kaolin; ECC Int'l.; Kaopolite; R.T. Vanderbilt; Volclay Ltd
Trade names: Gelwhite® MAS-H, MAS-L; Magnabrite® T; Van Gel® B; Veegum®; Veegum® T

Magnesium calcium carbonate hydroxide
Uses: Fire retardant, smoke suppressant for engineering thermoplastics, coatings
Trade names: Hydramax™ HM-C9, HM-C9-S

Magnesium dinonylnaphthalene sulfonate
Trade names containing: Nacorr® 1652

Magnesium fluosilicate. *See* Magnesium silicofluoride
Magnesium hexafluorosilicate. *See* Magnesium silicofluoride
Magnesium hydrate. *See* Magnesium hydroxide
Magnesium hydrogen metasilicate. *See* Talc

Magnesium hydroxide
CAS 1309-42-8 (anhyd.); EINECS 215-170-3
Synonyms: Magnesium hydrate; Milk of magnesia; Magnesia magma
Classification: Inorganic base
Empirical: H_2MgO_2
Formula: $Mg(OH)_2$
Properties: Wh. amorphous powd., odorless; sol. in sol'n. of ammonium salts and dilute acids; almost insol. in water and alcohol; m.w. 58.33; dens. 2.36; m.p. 350 C (dec.)
Precaution: Noncombustible; incompat. with maleic anhdyride
Toxicology: Variable toxicity; toxic when inhaled; harmless to skin
Uses: Intermediate for obtaining magnesium metal, sugar refining, medicine (antacid, laxative), residual fuel oil additive, sulfite pulp, uranium processing, dentifrices, in foods as drying agent, frozen desserts; plastics flame retardant and filler; extender pigment for flame retardant coatings
Regulatory: FDA 21CFR §184.1428, GRAS; Europe listed; UK approved; FDA approved for orals; BP, Ph.Eur. compliance
Manuf./Distrib.: Aldrich; Allchem Ind.; AluChem; Ameribrom; Am. Int'l.; Baymag; Chemisphere; Climax Perf. Materials; Coyne; Croxton & Garry; J.W.S. Delavau; EM Ind.; Fluka; Generichem; Giles; J.M. Huber; Lohmann; Mallinckrodt; Martin Marietta Magnesia Spec.; Morton Int'l.; Reheis; Sigma; Spectrum Chem. Mfg.; Tetra; Zinkan Enterprises
Trade names: Hydramax™ HM-B8, HM-B8-S; Magoh-S

Magnesium octadecanoate. *See* Magnesium stearate

Magnesium oxide
CAS 1309-48-4; EINECS 215-171-9
Synonyms: Magnesia; Periclase; Magnesia clinker; Deadburned magnesite; Calcined magnesia; White charcoal; Magnesia usta
Classification: Inorganic oxide
Definition: Inorganic salt of magnesium
Empirical: MgO
Properties: Wh. powd., odorless; sol. in acids, ammonium salt sol'ns.; sl. sol. in water; insol. in alcohol; m.w. 40.31; dens. 0.36; m.p. 2800 C; b.p. 3600 C
Precaution: Noncombustible; violent reaction or ignition with interhalogens; incandescent reaction with phosphorus pentachloride; moisture-sensitive
Toxicology: Toxic by inhalation of fume; experimental tumorigen; TLV (as magnesium) 10 mg/m^3 (fume)
Uses: Refractories, esp. for steel furnace linings, polycrystalline ceramic for aircraft windshields, elec. insulations, pharmaceuticals, cosmetics, inorg. rubber accelerator, paper mfg.; white color standard; reflector in optical instruments; thickener for polyester resins; extender for paints
Regulatory: FDA 21CFR §163.110, 175.300, 177.1460, 177.2400, 177.2600, 182.5431, 184.1431, GRAS; Japan restricted; Europe listed; FDA approved for orals; BP, Ph.Eur. compliance
Manuf./Distrib.: Aldrich; Alfa Aesar Johnson Matthey; Allan; AluChem; Am. Int'l.; Baymag; Cerac; Chemisphere; J.W.S. Delavau; EM Ind.; Fluka; Generichem; Harwick; Hüls Am.; K3; Lohmann; Magnesia GmbH; Mallinckrodt; Marine Magnesium; Martin Marietta Magnesia Spec.; Morton Int'l.; Nat'l. Magnesia;

Premier Services; Reade Advanced Materials; Sigma; Southeastern Mins.; Spectrum Chem. Mfg.; Tetra; Tomita Pharmaceutical
Trade names: Magotex™; Magox® Super Premium

Magnesium silicofluoride
CAS 18972-56-0
Synonyms: Magnesium hexafluorosilicate; Magnesium fluosilicate
Empirical: $F_6Si \cdot Mg$
Properties: M.w. 166.40
Toxicology: LD50 (oral, guinea pig) 200 mg/kg; OSHA PEL:TWA 2.5 mg (F)/m^3; poison by ingestion; mod. toxic by subcut. route; heated to decomp., emits toxic fumes of F^-
Manuf./Distrib.: Atomergic Chemetals; Browning; Peridot; Triple Crown Am.
Trade names containing: Sylysia 358; Sylysia 435; Sylysia 445

Magnesium stearate
CAS 557-04-0; EINECS 209-150-3
Synonyms: Magnesium octadecanoate; Octadecanoic acid, magnesium salt
Definition: Magnesium salt of stearic acid
Empirical: $C_{36}H_{70}MgO_4$
Formula: $[CH_3(CH_2)_{16}COO]_2Mg$
Properties: Wh. soft oily powd., tasteless, odorless; insol. in water, alcohol, ether; dec. by dilute acids; m.w. 591.27; dens. 1.028; m.p. 88.5 C (pure)
Precaution: Nonflamm.
Toxicology: No known toxicity; heated to decomp., emits acrid smoke and toxic fumes
Uses: Baby dusting powder; lubricants in making tablets; drier in paints and varnishes; release, stabilizer, and lubricant for plastics; emulsifying agent for cosmetics; surfactant
Regulatory: FDA 21CFR §172.863, 173.340; 175.105, 175.300, 175.320, 176.170, 176.200, 176.210, 177.1200, 177.2260, 178.3910, 179.45, 181.22, 181.29, 184.1440, GRAS; must conform to FDA specs for salts of fats or fatty acids derived from edible oils; Europe listed; UK approved; FDA approved for buccals, parenterals, orals, vaginals; USP/NF, BP, Ph.Eur., JP compliance
Manuf./Distrib.: Aldrich; Allan; Am. Int'l.; Avrachem AG; Chemisphere; Cometals; Cookson Spec. Additives; EM Ind.; Ferro/Grant; Lohmann; Magnesia GmbH; Mallinckrodt; Miljac; Norac; Ruger; San Yuan; Spectrum Chem. Mfg.; Syn. Prods.; Witco/Oleo-Surf.; Witco/Polymer Addit.

Maize oil. *See* Corn oil
Maize starch. *See* Corn starch
MAK. *See* Methyl n-amyl ketone

Maleic acid
CAS 110-16-7; EINECS 203-742-5
Synonyms: Maleinic acid; cis-Butenedioic acid; 2-Butenedioic acid; Toxilic acid
Classification: Cis unsaturated organic acid
Empirical: $C_4H_4O_4$
Formula: HOOCCH:CHCOOH
Properties: Wh. cryst. powd., odorless; sol. in water, alcohol, acetone; m.w. 116.07; dens. 1.590; m.p. 132-140 C; flash pt. 100 C
Precaution: DOT: Corrosive material; combustible when exposed to heat or flame
Toxicology: LD50 (oral, rat) 708 mg/kg, (dermal, rabbit) 1560 mg/kg; mod. toxic by ingestion and skin contact; skin and severe eye irritant; heated to decomp., emits acrid smoke and irritating fumes
Uses: Organic synthesis (malic, succinic, aspartic, tartaric, propionic, lactic, malonic, and acrylic acids); dyeing and finishing of cotton, wool, silk; preservative for oils and fats (rancidity retardant); intermediate for paints; pharmaceutical intermediate (antihistamines)
Regulatory: FDA 21CFR §175.105, 177.1200; FDA approved for intramuscular injectables, orals; BP, Ph.Eur. compliance
Manuf./Distrib.: AC Ind.; Aldrich; Am. Biorganics; Ashland; J.T. Baker; Baychem; Chemical; Chemie Linz N. Am.; Croda Colloids Ltd.; Fluka; General Chem.; Harcros; Hüls Am.; Penta Mfg.; Research Organics; Sigma; Spectrum Chem. Mfg.; Schweizerhall; Thor; United Min. & Chem.; Van Waters & Rogers

Maleic acid/acrylic acid copolymer
Uses: Dispersant and anti-incrustation agent; aux. for phosphate-reduced or phosphate-free detergents; for water treatment, laundry detergents, agric., paints/coatings, I&I cleaning formulations
Trade names: Sokalan® CP 12S

Maleic acid/acrylic acid copolymer, sodium salt
Uses: Dispersant and anti-incrustation agent; for water treatment, laundry detergents, agric., paints/coatings, I&I cleaning formulations
Trade names: Sokalan® CP 5; Sokalan® CP 5 Powd.

Maleic acid anhydride. *See* Maleic anhydride

Maleic acid/olefin copolymer, sodium salt
Uses: Dispersant for org. and inorg. solids; for water treatment, laundry detergents, agric., paints/coatings, I&I cleaning formulations
Trade names: Sokalan® CP 9

Maleic anhydride
CAS 108-31-6; EINECS 203-571-6
Synonyms: 2,5-Furandione; 2,5-Dihydrofuran-2,5-dione; Maleic acid anhydride; Toxilic anhydride; cis-Butenedioic anhydride
Empirical: $C_4H_2O_3$
Properties: Colorless needles; sol. in water, acetone, alcohol, dioxane; partly sol. in chloroform, benzene; m.w. 98.06; dens. 0.934 (20/4 C); m.p. 53 C; b.p. 200 C; flash pt. 218 F
Precaution: DOT: Corrosive material; combustible exposed to heat or flame; can react vigorously with oxidizing materials; explosive as vapor epxoed to heat or flame; violent reaction with bases
Toxicology: LD50 (oral, rat) 481 mg/kg, (IP, rat) 97 mg/kg, (skin, rabbit) 2620 mg/kg; poison by ingestion, IP routes; mod. toxic by skin contact; corrosive irritant to tissues, eyes, skin; TLV 0.25 ppm in air; heated to dec., emits acrid smoke, irritating fumes
Uses: Polyester resins, alkyd coating resins; fumaric and tartaric acid mfg.; a pesticide; preservative for oils and fats; pharmaceuticals
Regulatory: BP compliance
Manuf./Distrib.: Aldrich; Allchem Ind.; Amoco; Aristech; Ashland; J.T. Baker; Baychem; BP Chems. Inc; Brown; Browning; Chemical; Elf Atochem SA; Fluka; Hüls AG; Huntsman; Mitsui Toatsu; Monsanto; NOF; Occidental; Primachem; Punda Mercantile; Royale Pigments & Chems.; Schweizerhall; Sigma; Spectrum Chem. Mfg.; United Min. & Chem.; Van Waters & Rogers
Trade names: Amoco® MAN

Maleic anhydride/methyl vinyl ether copolymer, sodium salt
Uses: Dispersant, anti-incrustation agent; for water treatment, laundry detergents, agric., paints/coatings, I&I cleaning formulations
Trade names: Sokalan® CP 2; Sokalan® CP 2 Powd.

Maleic resin
Synonyms: Maleinic resin
Uses: Used in nitrocellulose lacquers, printing inks; promotes hardness, high gloss and adhesion
Manuf./Distrib.: Chemcentral; Chemical; Monomer-Polymer & Dajac Labs
Trade names: Uni-Rez® 7003

Maleinic acid. *See* Maleic acid
Maleinic resin. *See* Maleic resin
Malic acid (INCI). *See* N-Hydroxysuccinic acid

Manganese acetate
CAS 638-38-0; 6156-78-1; EINECS 211-334-3
Synonyms: Acetic acid manganese (2+) salt; Acetic acid manganese (II) salt (2:1); Manganese (II) acetate; Manganese diacetate; Manganous acetate; Diacetyl manganese; Manganese (2+) acetate
Empirical: $C_4H_6O_4 \cdot Mn$
Formula: $Mn(C_2H_3O_2)_2 \cdot 4H_2O$
Properties: Lt. red monoclinic system; sol. 64.5 g/100 g water (50 C); sol. in alcohol; m.w. 173.04 (anhyd.), 245.10 (tetrahydrate)
Toxicology: LD50 (oral, rat) 2940 mg/kg; ACGIH TLV: 5 mg(Mn)/m^3; mod. toxic by ingestion; mutagenic data; heated to decomp., emits acrid smoke and irritating fumes
Uses: Textile dyeing, oxidation catalyst, paint and varnish drier, fertilizer, food packaging, feed additive
Manuf./Distrib.: Aldrich; Alemark; Atomergic Chemetals; Barker Ind.; Hoechst Celanese; Lohmann; Min. R&D; Nihon Kagaku Sangyo; OM Group; Poly Research; Shepherd; Spectrum Chem. Mfg.; Verdugt BV

Manganese (2+) acetate. *See* Manganese acetate

Manganese (II) acetate. *See* Manganese acetate
Manganese ammonium pyrophosphate. *See* Manganese violet
Manganese diacetate. *See* Manganese acetate

Manganese drier
Uses: Used in org. coatings, inks, polyesters
Trade names: Troymax™ Drier Manganese 6%; Troymax™ Drier Manganese 9%; Troymax™ Drier Manganese 12%

Manganese linoleate
Formula: $Mn(C_{18}H_{31}O_2)_2$
Properties: Dk. brn. plaster-like mass; sol. in linseed oil
Precaution: Combustible
Toxicology: Heated to decomp., emits acrid smoke and irritating fumes
Uses: Drier for paints
Regulatory: FDA 21CFR §181.25
Manuf./Distrib.: Shepherd; Troy

Manganese naphthenate
Properties: Hard brn. resinous mass; sol. in min. spirits; m.p. 130-140 C
Precaution: Combustible; sol'n. is flamm.
Toxicology: Heated to decomp., emits acrid smoke and irritating fumes
Uses: Drier for paints
Regulatory: FDA 21CFR §181.25
Manuf./Distrib.: Akzo Nobel; Archway Sales; Baychem; Boehle; ; Dussek Campbell Ltd; Hüls Am.; Lomas Int'l.; D.N. Lukens; OM Group; Shepherd; Troy
Trade names: Nuodex Napthenate® Manganese 6%

Manganese octoate
Uses: Drier for paints
Manuf./Distrib.: Aceto; Archway Sales; Baychem; Boehle; Hüls Am.; Lomas Int'l.; D.N. Lukens; OM Group; Shepherd; Troy
Trade names: Nuodex Octoate® Manganese 6%

Manganese (II) sulfate (1:1). *See* Manganese sulfate (ous)

Manganese sulfate (ous)
CAS 7785-87-7 (anhyd.); 10034-96-5 (monohydrate); EINECS 232-089-9
Synonyms: Manganous sulfate; Manganese (II) sulfate (1:1); Sulfuric acid, manganese (2+) salt
Empirical: $MnSO_4$ (anhyd.), $MnSO_4 \cdot H_2O$ (monohydrate)
Properties: Anhyd.: Pink gran. powd., odorless; very sol. in water; insol. in alcohol; m.w. 151.00; dens. 3.25; m.p. 700 C; b.p. 850 C (dec.); Monohydrate: Pale red, sl. efflorescent cryst.; sol. in 1 part water; insol. in alcohol; m.w. 169.00
Toxicology: TLV 5 mg(Mn)/m^3; LD50 (IP, mouse) 332 mg/kg; poison by IP route; experimental neoplastigen; mutagenic data; heated to decomp., emits toxic fumes of SO_x and manganese
Uses: Fertilizers, feed additive, paints and varnishes, ceramics, textile dyes, medicine, fungicide, ore flotation, catalyst in viscose process, synthetic manganese dioxide
Regulatory: FDA 21CFR §182.5461, 184.1461, GRAS
Manuf./Distrib.: Aldrich; Chemetals; Fluka; Lohmann; Mallinckrodt; Nihon Kagaku Sangyo; Sigma

Manganese tallate
Definition: Manganese salts of tall oil fatty acids
Precaution: Combustible
Toxicology: Heated to decomp., emits acrid smoke and irritating fumes
Uses: Drier for paints
Regulatory: FDA 21CFR §181.25
Manuf./Distrib.: Akzo Nobel; Baychem; Dussek Campbell Ltd; OM Group; Shepherd; Troy

Manganese violet
CAS 10101-66-3; EINECS 233-257-4
Synonyms: Ammonium manganese pyrophosphate; Pigment violet 16; CI 77742; Diphosphoric acid, ammonium manganese (3+) salt (1:1:1); Manganese ammonium pyrophosphate
Classification: Inoganic salt

Manganous acetate

Empirical: $H_4O_7P_2 \cdot H_3N \cdot Mn$
Uses: Colorant for plastics, powd. coatings, artists' colors, and cosmetics
Regulatory: FDA 21CFR §73.2775
Manuf./Distrib.: Shepherd Color; Whittaker, Clark & Daniels
Trade names: Manganese Violet

Manganous acetate. *See* Manganese acetate
Manganous sulfate. *See* Manganese sulfate (ous)
MBT. *See* Methylenebis (thiocyanate)
MC. *See* Methylcellulose
MCC. *See* Microcrystalline cellulose
MDA. *See* 4,4´-Methylene dianiline

MDI
CAS 101-68-8
Synonyms: Methylene di-p-phenylene isocyanate; Bis(p-isocyanatophenyl) methane; Bis (1,4-isocyanatophenyl) methane; Bis (4-isocyanatophenyl) methane; Diphenylmethane-4,4´-diisocyanate; 4,4´-Diphenylmethane diisocyanate; p,p-Diphenylmethane diisocyanate; 4,4´-Diisocyanatodiphenylmethane; Methylene bisphenyl isocyanate
Empirical: $C_{15}H_{10}N_2O_2$
Formula: $CH_2(C_6H_4NCO)_2$
Properties: Lt. yel. fused solid; sol. in acetone, benzene, kerosene, nitrobenzene; m.w. 250.27; dens. 1.197 (70 C); m.p. 37.2 C; b.p. 194-199 C (5 mm)
Precaution: Combustible
Toxicology: LDLo (oral, rat) 31,690 mg/kg; TLV:CL 0.01 ppm in air; TLV 0.005 ppm; toxic by inhalation of fumes; mildly toxic by ingestion; strong irritant; human systemic effects; allergic sensitizer; heated to decomp., emits toxic fumes of NO_x, SO_x
Uses: Vulcanizing agent for rubbers; prep. of polyurethane resin and spandex fibers; bonding rubber to rayon and nylon; processing aid, intermediate for the prod. of cast, RIM, and thermoplastic PU elastomers, adhesives, binders, coatings, and sealants
Manuf./Distrib.: Aldrich; Allchem Ind.; BASF; Bayer; ICI Polyurethanes
Trade names: Isonate® 125M; Isonate® 143L; Isonate® 181; Isonate® 191; Isonate® 240; Isonate® 2125M; Isonate® 2143L; Isonate® 2181; Isonate® 2191; Isonate® 2240; Rubinate® 0044; Rubinate® 1000; Rubinate® 1209; Rubinate® 1234; Rubinate® 1680; Rubinate® 1790; Rubinate® 1920; Rubinate® 9009; Rubinate® 9016; Rubinate® 9040; Rubinate® 9041; Rubinate® 9210; Rubinate® 9225; Rubinate® 9234; Rubinate® 9236; Rubinate® 9257; Rubinate® 9258; Rubinate® 9259; Rubinate® 9271; Rubinate® 9272; Rubinate® 9415; Rubinate® LF-168; Rubinate® XI-244
Trade names containing: Rubinate® M

MDMH. *See* MDM hydantoin

MDM hydantoin
CAS 116-25-6; EINECS 204-132-1
Synonyms: MDMH; 1-(Hydroxymethyl)-5,5-dimethyl hydantoin; Monomethylol dimethyl hydantoin; 1-(Hydroxymethyl)-5,5-dimethyl-2,4-imidazolinedione
Classification: Organic compd.
Empirical: $C_6H_{10}N_2O_3$
Uses: Intermediate for cosmetics, other applics.; preservative; formaldehyde donor
Regulatory: USA not restricted; Japan not approved
Manuf./Distrib.: Int'l. Sourcing
Trade names containing: Biochek 240

MDPE. *See* Polyethylene, medium density
2-ME. *See* 2-Mercaptoethanol
MEA. *See* Ethanolamine

Meadowfoam seed oil
Definition: Oil extracted from the seeds of the meadowfoam plant, *Limnanthes alba*
Uses: {Skin/hair conditioner, color enhancer for cosmetics and pharmaceuticals}; color enhancer in paints, inks, fabric dyeing industry
Manuf./Distrib.: Charkit
Trade names: EmCon Limnanthes Alba

MEHQ. *See* Hydroquinone monomethyl ether
MEK (INCI). *See* Methyl ethyl ketone
MEK peroxide. *See* Methyl ethyl ketone peroxide

Melamine
CAS 108-78-1; EINECS 203-615-4
Synonyms: Cyanuramide; sym-Triaminotriazine; 1,3,5-Triazine-2,4,6-triamine; Cyanurotriamide; Cyanourotriamine; 2,4,6-Triamino-s-triazine; 2,4,6-Triamino-sym-triazine
Empirical: $C_3H_6N_6$
Properties: Colorless monoclinic prisms; sl. sol. in water; very sl. sol. in hot alcohol; insol. in ether; m.w. 126.15; dens. 1.573 (250 C); m.p. < 250 C; flash pt. > 300 C; b.p. sublimes
Toxicology: LD50 (oral, rat) 3161 mg/kg; mod. toxic by ingestion, IP routes; eye, skin, and mucous membrane irritant; causes dermatitis in humans; experimental carcinogen, tumorigen; mutagenic data; heated to decomp., emits toxic fumes of NO_x and CN^-
Uses: Forms synthetic resins with formaldehyde; intermediate for paints
Regulatory: FDA 21CFR §181.30
Manuf./Distrib.: Aldrich; Allchem Ind.; BASF; Browning; Charkit; Chemical; Chemie Linz N. Am.; Cytec Ind.; DSM; Fluka; Chemical; ICD Group; Int'l. Chem. Inc.; Melamine Chems; Monsanto; Punda Mercantile; Reichhold; San Yuan; Sigma

Melamine/formaldehyde resin
CAS 9003-08-1
Synonyms: Melamine resin; 1,3,5-Triazine,2,4,6-triamine, polymer with formaldehyde
Classification: Amino resin
Definition: Reaction prod. of melamine and formaldehyde
Formula: $(CH_3H_6N_6 \cdot CH_2O)_x$
Properties: Syrup or insol. wh. powd.; sol. in water
Uses: Thermosetting resin; crosslinking agent for coatings; fiber and paper processing
Regulatory: FDA 21CFR §175.105, 175.300, 175.320, 177.1200, 177.1460, 177.2260, 177.2470, 181.22, 181.30
Manuf./Distrib.: Akzo Nobel; Astro Industries; Bakelite GmbH; BASF; Chemical; Cytec Ind.; DSM Melamine Am.; Georgia-Pacific Resins; McWhorter; Monsanto; Nat'l. Casein; Reichhold; Rhone-Poulenc Water Treatment; Sybron
Trade names: Cymel 373; Cymel 380; Glazamine DP2; Luwipal®; Uformite® 27-801
Trade names containing: Beckamine 27-556; Beckamine 27-566; Hostaflam® AP 462; Uformite® 27-802; Uformite® 27-803; Uformite® 27-804

Melamine-formaldehyde resin, butylated
Trade names containing: Resimene® 750; Resimene® 872; Resimene® 881; Resimene® 1406; Resimene® 7512; Resimene® BM-5901

Melamine-formaldehyde resin, methylated
Uses: Crosslinker for thermosetting surf. coatings; used in aq. coatings incl. metal, paper, automotive, coil, and wood furniture and cabinet coatings, inks
Trade names: Resimene® AQ-7550
Trade names containing: Resimene® 717; Resimene® 730; Resimene® 735; Resimene® 741; Resimene® 797; Resimene® 2040; Resimene® HM-2608

Melamine-formaldehyde resin, methylated-butylated
Uses: Crosslinker for thermosetting surf. coatings; resist. to hydrolysis; for auto, can/container, metal, metal deco, appliance, and extrusion coatings, and inks
Trade names: Resimene® 751; Resimene® 755; Resimene® 757; Resimene® CE-6550
Trade names containing: Resimene® CE-4514; Resimene® CE-6517

Melamine phosphate
Uses: Flame retardant for polymers; flame retardant additive for latex intumescent coating formulations
Manuf./Distrib.: Akzo Nobel; Albright & Wilson Am.; Chemie Linz N. Am.; DSM Melamine Am.; Miljac
Trade names: Aeroguard® MP; Amgard® NH

Melamine pyrophosphate
Uses: Flame retardant for polymers; flame retardant additive for latex intumescent coating formulations
Manuf./Distrib.: Akzo Nobel; Anhydrides & Chems.; Miljac; StanChem
Trade names: Aeroguard® MPP

Melamine resin. *See* Melamine/formaldehyde resin

Menhaden oil

CAS 8002-50-4; EINECS 232-311-4

Synonyms: Oils, menhaden; Pogy oil; Mossbunker oil

Definition: Oil obtained from the small North Atlantic fish, *Brevoortia tyrannus*

Properties: Yellowish-brown or reddish-brown oil, characteristic fishy odor and taste; sol. in ether, benzene, petrol. ether, naphtha, kerosene, CS_2; dens. 0.925-0.933; m.p. 38.5-47.2 C; iodine no. 115-160; sapon. no. 191-200; ref. index 1.480 (20 C)

Precaution: Combustible

Toxicology: No known toxicity

Uses: Substitute for linseed oil; leather dressing; hydrog. fats for cooking and industrial use, printing inks, animal feed, lubricants, paint drier, cleansers, lipstick, as tallow substitute in soap mfg.

Regulatory: FDA 21CFR §175.300, 176.200, 176.210, 177.2800

Manuf./Distrib.: ABITEC; Arista Ind.; Croda; Werner G. Smith; Sigma

p-Mentha-1,8-diene. *See* dl-Limonene
2-Mercaptobenzothiazole sodium salt. *See* Sodium 2-mercaptobenzothiazole
2-Mercaptobenzothiazole zinc salt. *See* Zinc 2-mercaptobenzothiazole
Mercaptoethanol. *See* 2-Mercaptoethanol

2-Mercaptoethanol

CAS 60-24-2; EINECS 200-464-6

Synonyms: 2-ME; Mercaptoethanol; β-Mercaptoethanol; Thioethylene glycol; Ethylene thioglycol; 1-Ethanol-2-thiol; 2-Hydroxy-1-ethanethiol; Thioglycol; Monothioethyleneglycol; 2-Thioethanol; Thiomonoglycol; 2-Hydroxyethylmercaptan

Classification: Aliphatic organic compd.

Empirical: C_2H_6OS

Formula: $HSCH_2CH_2OH$

Properties: Water-wh. mobile liq., disagreeable odor; misc. with water; m.w. 78.13; dens. 1.114; b.p. 157 C; flash pt. 165 F; ref. index 1.5006

Precaution: Combustible; flamm. exposed to heat, flame, oxidizers

Toxicology: LD50 (oral, rat) 244 mg/kg, (skin, rabbit) 150 mg/kg; poison by ingestin, skin contact, IP routes; mod. toxic by IV route; irritating to eyes, skin, nose; human mutagenic data; heated to decomp., emits highly toxic fumes of SO_x

Uses: Solvent for dyestuffs, intermediate for producing dyestuffs, pharmaceuticals, rubber chemicals, flotation agents, insecticides, plasticizers, reducing agent, biochemical reagent, PVC staiblizers, agricultural chemicals, textile auxiliary; chain transfer agent in mfg. of acrylic resins for high-solids automotive topcoats; mfg. of polythioethers for high-solids elastomers, coatings, and adhesives

Manuf./Distrib.: Aldrich; BASF; Fluka; Morton Int'l.; Rhone-Poulenc Surf. & Spec.; Sigma; Spectrum Chem. Mfg.; Toray Thiokol

Trade names: Sipomer® 2ME

β-Mercaptoethanol. *See* 2-Mercaptoethanol
3-Mercapto-1,2-propanediol. *See* Thioglycerin

Mercaptopropyltrimethoxysilane

CAS 4420-74-0; EINECS 224-588-5

Synonyms: γ-Mercaptopropyltrimethoxy silane; Trimethoxysilylpropanethiol; 3-Trimethoxysilyl-1-propanethiol; (3-Mercaptopropyl)trimethoxysilane

Empirical: $C_6H_{16}O_3SSi$

Properties: Liq.; m.w. 196.37; dens. 1.057 (20/4 C); b.p. 219-220 C; flash pt. 93 C; ref. index 1.440

Toxicology: LD50 (oral, rat) 2940 mg/kg, (skin, rabbit) 5880 mg/kg; mod. toxic by ingestion, IP route; mildly toxic by skin contact; skin irritant; heated to decomp., emits toxic fumes of SO_x

Uses: Filler for sulfur and metal oxide-cured systems; coupling agent, chem. intermediate, blocking agent, release agent, lubricant; coupling agent, crosslinking agent, adhesion promoter for coatings

Manuf./Distrib.: Aldrich; Fluka; Sax TLC000; Sigma

Trade names: Silquest® A-189

(3-Mercaptopropyl)trimethoxysilane. *See* Mercaptopropyltrimethoxysilane
γ-Mercaptopropyltrimethoxy silane. *See* Mercaptopropyltrimethoxysilane
Mercuric oxide. *See* Mercury oxide (ic), red and yellow
Mercury (II) oxide. *See* Mercury oxide (ic), red and yellow

Mercury oxide (ic), red and yellow
CAS 21908-53-2; EINECS 244-654-7
Synonyms: Mercuric oxide; Yellow oxide of mercury; Red precipitate; Yellow precipitate; Mercury (II) oxide
Empirical: HgO
Properties: M.w. 216.59; m.p. 500 C (dec.)
Precaution: Strong oxidizer capable of igniting combustible materials; light-sensitive
Toxicology: DOT: Poisonous material; inh. may cause respiratory tract irritation, abdominal pain, vomiting, diarrhea, inflamm. of gums; chronic exposure results in anxiety, depression, insomnia, nervous system effects, kidney damage
Uses: Red: Chemicals, paint pigment, perfumery, cosmetics, pharmaceuticals, ceramics, dry batteries, polishes, analytical reagent, antifouling paints, fungicide, antiseptic; Yellow: Antiseptic, mercury compounds
Regulatory: BP compliance
Manuf./Distrib.: Aldrich; Alfa Aesar Johnson Matthey; Allan; BASF; Centerchem; Cerac; Cosan; Fluka; Noah; Sigma; Spectrum Chem. Mfg.; Thor UK

Meroxapol 105
CAS 9003-11-6 (generic)
Classification: Polyoxypropylene, polyoxyethylene block polymer
Formula: $HO(CHCH_3CH_2O)_x(CH_2CH_2O)_y(CH_2CHCH_3O)_zH$, avg. x=7, y=22, z=7
Properties: Nonionic liq.
Uses: Emulsifier, wetting agent, binder, stabilizer, plasticizer, lubricant, solubilizer, dispersant, visc. control agent, defoamer, intermediate for hard surf. detergents, rinse aids, automatic dishwashing, textile processing, paper processing, paints; latexes
Regulatory: FDA 21CFR §172.808, 173.340
Manuf./Distrib.: Aldrich; Fluka; Sigma
Trade names: Pluronic® 10R5

Meroxapol 171
CAS 9003-11-6 (generic)
Classification: Polyoxypropylene, polyoxyethylene block polymer
Formula: $HO(CHCH_3CH_2O)_x(CH_2CH_2O)_y(CH_2CHCH_3O)_zH$, avg. x=12, y=4, z=12
Uses: Emulsifier, wetting agent, binder, stabilizer, plasticizer, lubricant, solubilizer, dispersant, visc. control agent, defoamer, intermediate for hard surface detergents, rinse aids, automatic dishwashing, textile processing, paper processing; paints, adhesives
Manuf./Distrib.: Aldrich; Fluka; Sigma
Trade names: Calgene Nonionic 1017-R

Meroxapol 172
CAS 9003-11-6 (generic)
Classification: Polyoxypropylene, polyoxyethylene block polymer
Formula: $HO(CHCH_3CH_2O)_x(CH_2CH_2O)_y(CH_2CHCH_3O)_zH$, avg. x=12, y=9, z=12
Uses: Emulsifier, wetting agent, binder, stabilizer, plasticizer, lubricant, solubilizer, dispersant, visc. control agent, defoamer, intermediate for hard surface detergents, rinse aids, automatic dishwashing, textile processing, paper processing; paints, adhesives
Manuf./Distrib.: Aldrich; Fluka; Sigma
Trade names: Antarox® 17-R-2; Calgene Nonionic 2017-R; Pluronic® 17R2

Meroxapol 174
CAS 9003-11-6 (generic)
Classification: Polyoxypropylene, polyoxyethylene block polymer
Formula: $HO(CHCH_3CH_2O)_x(CH_2CH_2O)_y(CH_2CHCH_3O)_zH$, avg. x=12, y=23, z=12
Uses: Emulsifier, wetting agent, binder, stabilizer, plasticizer, lubricant, solubilizer, dispersant, visc. control agent, defoamer, intermediate for hard surface detergents, rinse aids, automatic dishwashing, textile processing, paper processing, paints
Manuf./Distrib.: Aldrich; Fluka; Sigma
Trade names: Antarox® 17-R-4; Pluronic® 17R4

Meroxapol 252
CAS 9003-11-6 (generic)
Classification: Polyoxypropylene, polyoxyethylene block polymer
Formula: $HO(CHCH_3CH_2O)_x(CH_2CH_2O)_y(CH_2CHCH_3O)_zH$, avg. x=18, y=14, z=18
Uses: Emulsifier, wetting agent, binder, stabilizer, plasticizer, lubricant, solubilizer, dispersant, visc. control

agent, defoamer, intermediate for hard surface detergents, rinse aids, automatic dishwashing, textile processing, paper processing; paints, latexes
Manuf./Distrib.: Aldrich; Fluka; Sigma
Trade names: Antarox® 25-R-2; Pluronic® 25R2

Meroxapol 254
CAS 9003-11-6 (generic)
Classification: Polyoxypropylene, polyoxyethylene block polymer
Formula: $HO(CHCH_3CH_2O)_x(CH_2CH_2O)_y(CH_2CHCH_3O)_zH$, avg. x=18, y=34, z=18
Uses: Emulsifier, wetting agent, binder, stabilizer, plasticizer, lubricant, solubilizer, dispersant, visc. control agent, defoamer, intermediate for hard surface detergents, rinse aids, automatic dishwashing, textile processing, paper processing; paints, latexes
Manuf./Distrib.: Aldrich; Fluka; Sigma
Trade names: Pluronic® 25R4

Meroxapol 258
CAS 9003-11-6 (generic)
Classification: Polyoxypropylene, polyoxyethylene block polymer
Formula: $HO(CHCH_3CH_2O)_x(CH_2CH_2O)_y(CH_2CHCH_3O)_zH$, avg. x=18, y=163, z=18
Uses: Emulsifier, wetting agent, binder, stabilizer, plasticizer, lubricant, solubilizer, dispersant, visc. control agent, defoamer, intermediate for hard surface detergents, rinse aids, automatic dishwashing, textile processing, paper processing; paints, latexes
Manuf./Distrib.: Aldrich; Fluka; Sigma
Trade names: Pluronic® 25R8

Meroxapol 311
CAS 9003-11-6 (generic)
Classification: Polyoxypropylene, polyoxyethylene block polymer
Formula: $HO(CHCH_3CH_2O)_x(CH_2CH_2O)_y(CH_2CHCH_3O)_zH$, avg. x=21, y=7, z=21
Properties: Nonionic liq.
Uses: Emulsifier, wetting agent, binder, stabilizer, plasticizer, lubricant, solubilizer, dispersant, visc. control agent, defoamer, intermediate for hard surface detergents, rinse aids, automatic dishwashing, textile processing, paper processing, paints
Manuf./Distrib.: Aldrich; Fluka; Sigma
Trade names: Antarox® 31-R-1; Pluronic® 31R1

Mesitylene. *See* 1,3,5-Trimethylbenzene

Mesityl oxide
CAS 141-79-7; EINECS 205-502-5
Synonyms: 4-Methyl-3-penten-2-one; 4-Methyl-3-pentene-2-one; 2-Methyl-2-penten-4-one; 4-Methyl-3-penten-2-one; Isobutenyl methyl ketone; Isopropylideneacetone; Methyl isobutenyl ketone
Empirical: $C_6H_{10}O$
Formula: $(CH_3)_2C{:}CHCOCH_3$
Properties: Colorless oily liq., strong odor; somewhat water-sol. @ 20 C; misc. with alcohol, ether, most org. liqs.; m.w. 98.15; dens. 0.854 (20/4 C); m.p. -59 C; b.p. 128-130 C; flash pt. 31 C; ref. index 1.446 (20 C)
Precaution: DOT: Flamm. liq.; dangerous fire hazard exposed to heat, sparks, flame; reactive with oxidizers; violent reactions possible; heated to decomp., emits acrid smoke, irritating fumes
Toxicology: LD50 (oral, rat) 1120 mg/kg, (skin, rabbit) 5150 mg/kg; ACGIH TLV:TWA 15 ppm; STEL 25 ppm; poison by IP route; mod. toxic by ing.; mildly toxic by inh., skin contact; human systemic effects; highly irritating to tissue; narcotic in high concs.
Uses: Solv. for paints; insect repellent
Manuf./Distrib.: Fluka; Penta Mfg.; Van Waters & Rogers

Metaphenylenediamine. *See* m-Phenylenediamine
Methacetone. *See* Diethyl ketone

Methacrylamide
CAS 79-39-0; EINECS 201-202-3
Synonyms: Methacrylic acid amide; Methacrylic amide; 2-Methyl-2-propaneamide; 2-Methylpropenamide; 2-Methacrylamide; α-Methyl acrylic amide
Empirical: C_4H_7NO
Formula: $CH_2{:}C(CH_3)CONH_2$

Properties: M.w. 85.11; m.p. 106-109 C
Toxicology: LD50 (oral, rat) 459 mg/kg, (IP, mouse) 200 mg/kg; poison by IP route; mod. toxic by ingestion; human systemic effects by inh.; heated to decomp., emits toxic fumes of NO_x
Uses: Used in self-crosslinking emulsions, heat-curing coatings, crosslinked acrylic sheets
Manuf./Distrib.: Aldrich; Fluka; Mitsui Toatsu; Monomer-Polymer & Dajac Labs; Polysciences; Rohm Tech
Trade names: BM-801

2-Methacrylamide. *See* Methacrylamide

Methacrylamidoethylethyleneurea
Uses: Monomer for adhesion promotion in paints
Manuf./Distrib.: Rhone-Poulenc
Trade names containing: Sipomer® WAM II

Methacrylate copolymer
Synonyms: Methacrylic acid copolymer
Uses: Visc. index improver, pour pt. depressant, low temp. sludge dispersant; in hydraulic fluids; lt. stabilizer for PP, ABS, PS, nylon, LDPE, LLDPE, HDPE, acrylics, PC, thermoplastic polyester, SAN, thermoplastic elastomers, surf. coatings
Manuf./Distrib.: Esschem; Monomer-Polymer & Dajac Labs
Trade names containing: Empicryl® 6059; Empicryl® 6070

Methacrylic acid
CAS 79-41-4; EINECS 201-204-4
Synonyms: α-Methylacrylic acid (monomer); 2-Methylpropenoic acid; 2-Methyl-2-propenoic acid; α-Methyl acrylic acid
Empirical: $C_4H_6O_2$
Formula: $H_2C{:}C(CH_3)COOH$
Properties: Colorless liq., acrid odor; sol. in warm water, alcohol, ether, most org. solvs.; m.w. 86.09; dens. 1.015 (20 C); m.p. 15-16 C; b.p. 163 C; flash pt. (OC) 76 C
Precaution: Combustible; corrosive; flamm. exposed to heat, flame, oxidizers; a storage hazard; spontaneous exothermic polymerization can occur
Toxicology: LD33 (oral, rat) 60 mg/kg; LD50 (skin, rabbit) 500 mg/kg; poison by ingestion, IP routes; mod. toxic by skin contact; strong irritant to skin, eyes, mucous membranes; mutagenic data; TLV 20 ppm; heated to dec., emits acrid smoke, irritating fumes
Uses: Monomer for large-volume resins and polymers, organic synthesis; many of the polymers are based on esters of the acid, as the methyl, butyl, or isobutyl esters; paints; adhesives; ion-exchange membrane; processing agent for paper and textile, leather treatment
Regulatory: FDA approved for ophthalmics
Manuf./Distrib.: Aldrich; Allchem Ind.; Ashland; Degussa; Elf Atochem N. Am.; Fluka; Kuraray; Mitsubishi Gas; Monomer-Polymer & Dajac Labs; Rohm & Haas; Rohm Tech; Sigma
Trade names containing: Baypren Latex 4 R; Sipomer® WAM II

β-Methacrylic acid. *See* Crotonic acid
Methacrylic acid amide. *See* Methacrylamide
Methacrylic acid copolymer. *See* Methacrylate copolymer
Methacrylic acid 2,3-epoxypropyl ester. *See* Glycidyl methacrylate
Methacrylic acid isobornyl ester. *See* Isobornyl methacrylate
Methacrylic acid methyl ester polymers. *See* Methyl methacrylate polymer
Methacrylic acid, tetrahydrofurfuryl ester. *See* Tetrahydrofurfuryl methacrylate
Methacrylic amide. *See* Methacrylamide

3-Methacryloxypropyltrimethoxysilane
CAS 2530-85-0; EINECS 219-785-8
Synonyms: γ-Methacryloxypropyltrimethoxysilane
Classification: Methacrylate silane
Empirical: $C_{10}H_{20}O_5Si$
Formula: $CH_2{:}CCH_3COO(CH_2)_3Si(OCH_3)_3$
Properties: Liq.; sol. in acetone, benzene, ether, methanol, hydrocarbons; m.w. 248.1; dens. 1.045; b.p. 80 (1 mm); ref. index 1.429 (25 C); flash pt. 135 F
Precaution: Combustible; moderate fire risk
Uses: Coupling agent to promote resin-to-glass, resin-to-metal, resin-to-resin bonds, unsaturated polyesters, acrylics, EVA, adhesives; adhesion promoter for solv., water-based, solventless, and powd. coating

systems; pigment treatment
Manuf./Distrib.: Aldrich; Fluka; Sigma
Trade names: Dow Corning® Z-6030; Silquest® A-174
Trade names containing: Aktisil® MAM

γ-Methacryloxypropyltrimethoxysilane. *See* 3-Methacryloxypropyltrimethoxysilane
2 (Methacryloyloxy) ethyl acetoacetate. *See* 2-(Acetoacetoxy) ethyl methacrylate
Methamin. *See* Hexamethylenetetramine
Methanal. *See* Formaldehyde
Methane dichloride. *See* Methylene chloride
Methane-, dichloro-. *See* Methylene chloride
Methane, oxybis-. *See* Dimethyl ether
Methanesulfinic acid, hydroxy-, monosodium salt. *See* Sodium formaldehyde sulfoxylate

Methanesulfonic acid
CAS 75-75-2; EINECS 200-898-6
Synonyms: Sulfomethane; Methylsulfonic acid
Classification: Aliphatic organic compd.
Empirical: CH_4O_3S
Formula: CH_3SO_3H
Properties: Colorless solid; sol. in water, alcohol, ether; m.w. 96.10; dens. 1.483 (20/4 C); m.p. 20 C; b.p. 167 C (10 mm); flash pt. > 230 F; ref. index 1.4300
Precaution: Corrosive to iron, steel, brass, copper, lead; explosive reaction with ethyl vinyl ether; incompat. with hydrogen fluoride
Toxicology: LDLo (oral, rat) 200 mg/kg, (IP, rat) 50 mg/kg; poison by ingestion, IP routes; corrosive to tissue, skin, eyes, mucous membranes; causes burns; heated to decomp., emits toxic fumes of SO_x
Uses: Pharmaceutical; polymerization catalyst; dehydrating agent; acid catalyst; curing accelerator for coatings; plating assistant; pesticides; anticorrosives; textile oil
Regulatory: FDA approved for parenterals; BP compliance
Manuf./Distrib.: Aldrich; Allchem Ind.; AMRESCO; Atotech USA; Elf Atochem N. Am.; Fluka; Sigma; Toyo Kasei Kogyo

Methane, sulfonylbis trichloro-. *See* Bis (trichloromethyl) sulfone
Methane tetramethylol. *See* Pentaerythritol
Methanoic acid. *See* Formic acid
Methanol. *See* Methyl alcohol
4,7-Methano-3a,4,7,7a-tetrahydroindene. *See* Dicyclopentadiene
Methenamine. *See* Hexamethylenetetramine
Methocel. *See* Methylcellulose

Methoxybutyl acetate
CAS 4435-53-4; EINECS 224-644-9
Synonyms: Methyl-1,3-butylene glycol acetate; Butoxyl; 3-Methoxybutyl acetate
Empirical: $C_7H_{14}O_3$
Properties: Colorless liq.; sol. in water, org. solvs.; m.w. 146.21; b.p. 171 C; flash pt. 62.5 C
Uses: Nitrocellulose, acryl, thinner, urethane coatings, epoxy resin, aminoalkyd resin, coating use solvent
Manuf./Distrib.: Daicel Chem. Ind.
Trade names containing: TEGO® Flow 300

3-Methoxybutyl acetate. *See* Methoxybutyl acetate

Methoxydiglycol
CAS 111-77-3; EINECS 203-906-6
Synonyms: Diethylene glycol methyl ether; Diethylene glycol monomethyl ether; Ethanol, 2-(2-methoxyethoxy)-; (2-β-Methyl 'Carbitol'), methoxyethoxy ethanol
Classification: Aliphatic ether alcohol
Empirical: $C_5H_{12}O_3$
Formula: $CH_3OCH_2CH_2OCH_2CH_2OH$
Properties: Water-wh. liq., hygroscopic; sol. in water; m.w. 120.17; dens. 1.0211 (20/4 C); b.p. 194 C; flash pt. 93.3 C; ref. index 1.4264 (27 C)
Precaution: Combustible
Toxicology: Moderately toxic by skin contact, intraperitoneal route; mildly toxic by ingestion
Uses: Solvent; brake fluid component; intermediate; solv. for coating formulations incl. nongrain-raising wood

stains, printing inks, dye pastes for textiles; coalescing aid for PVAc latex paints; solv. in stamp pad and stencil inks
Regulatory: FDA 21CFR §175.105
Manuf./Distrib.: Aldrich; Ashland; Brown; Eastman; Fluka; Great Western; Harcros; Hoechst AG; Occidental; Oxiteno; Union Carbide
Trade names: Eastman® DM; Ektasolve® DM; Hisolve DM; Methyl Carbitol®; Methyl Di-Icinol; Poly-Solv® DM

Methoxy dipropylene glycol. *See* PPG-2 methyl ether

Methoxyethanol
CAS 109-86-4; EINECS 203-713-7
Synonyms: Ethylene glycol methyl ether; Ethylene glycol monomethyl ether; 2-Methoxyethanol; Methyl Cellosolve®
Classification: Aliphatic ether alcohol
Empirical: $C_3H_8O_2$
Formula: $CH_3OCH_2CH_2OH$
Properties: Colorless liq., mild agreeable odor.; misc. with water, alcohol, ether, benzene, glycerol, acetone, dimethylformamide; m.w. 76.10; dens. 0.964 (20/4 C); f.p. -86.5 C; b.p. 123-124 C (760 mm); flash pt. (OC) 115 F; ref. index 1.4028 (20 C)
Precaution: Flamm. on exposure to heat and flame; mod. explosion hazard; can react with oxidizers to form explosive peroxides
Toxicology: TLV:TWA 5 ppm (skin); LD50 (oral, rat) 2460 mg/kg; mod. toxic to humans by ingestion; human systemic effects; experimental teratogen, reproductive effects; mutagenic data; skin and eye irritant; heated to dec., emits acrid smoke, irritating fumes
Uses: Solvent for cellulose acetate, natural resins, some synthetic resins, paints, inks, dyes, leather dyeing, nail polishes, varnishes, enamels, wood stains; plastics and plasticizers; in modified Karl Fischer reagent
Regulatory: FDA 21CFR §73.1 (no residue), 175.105; USDA 9CFR §381.147
Manuf./Distrib.: Aldrich; ARCO; Ashland; Brown; Fluka; Great Western; Harcros; Hoechst AG; Occidental; Oxiteno; Sigma; Union Carbide
Trade names: Hisolve MC; Methyl Cellosolve®; Methyl Icinol; Poly-Solv® EM

2-Methoxyethanol. *See* Methoxyethanol

Methoxyethanol acetate
CAS 110-49-6; EINECS 203-772-9
Synonyms: Ethylene glycol methyl ether acetate; Ethylene glycol monomethyl ether acetate; Glycol monomethyl ether acetate; 2-Methoxyethyl acetate; 2-Methoxyethanol acetate; Glycol ether EM acetate; 1-Acetoxy-2-methoxyethane; Methyl glycol monoaceetate; Methyl glycol acetate
Empirical: $C_5H_{10}O_3$
Formula: $CH_3COOC(CH_2)_2OCH_3$
Properties: Colorless liq.; m.w. 118.14; dens. 1.007 (20/20 C); f.p. -70 C; b.p. 143 C; flash pt. (CC) 111 F; ref. index 1.402 (20 C)
Precaution: DOT: Flamm. liq.; explosive limits 1.7-8.2%; mod. explosion hazard; can react with oxidizing materials; heated to decomp., emits acrid smoke and irritating fumes
Toxicology: LD50 (oral, rat) 3390 mg/kg, (dermal, rabbit) 5250 mg/kg; ACGIH TLV/TWA 5 ppm (skin); mod. toxic by ing., IP, SC routes; mildly toxic by inh., skin contact; human systemci effects; mutagenic data; eye irritant; inh. irritant
Uses: Solv. for coatings
Manuf./Distrib.: Aldrich; Fluka; Sigma; Spectrum Chem. Mfg.
Trade names: Methyl Cellosolve® Acetate

2-Methoxyethanol acetate. *See* Methoxyethanol acetate
Methoxyethene, homopolymer. *See* Polyvinyl methyl ether
Methoxy ether of propylene glycol. *See* Methoxyisopropanol
2-Methoxyethyl acetate. *See* Methoxyethanol acetate

Methoxyethyl acrylate
CAS 3121-61-7
Synonyms: Methyl 'Cellosolve' acrylate
Classification: Monomer
Empirical: $C_6H_{10}O_4$
Properties: Liq.; m.w. 130.15; dens. 1.1034 (20 C); b.p. 61 C (17 mm); flash pt. (OC) 180 F

2-Methoxyethyl ether

Precaution: Flamm. on exposure to heat, flame, or sparks
Toxicology: Poison by skin contact, moderately toxic by ingestion and inhalation
Uses: Solv.-resistant elastomers, polyacrylate rubbers, UV-curable reactive diluent, soft contact lenses, PVC impact modifier, fabric coatings, barrier coatings for polyethylene, textile coatings
Manuf./Distrib.: Aldrich; Ashland; CPS; Elf Atochem N. Am.; Monomer-Polymer & Dajac Labs
Trade names containing: Ageflex MEA

2-Methoxyethyl ether. *See* Diethylene glycol dimethyl ether

Methoxyisopropanol
CAS 107-98-2; EINECS 203-539-1
Synonyms: Monopropylene glycol methyl ether; Polypropylene glycol monomethyl ether; Methoxy ether of propylene glycol; Propylene glycol methyl ether; Propylene glycol monomethyl ether; 1-Methoxy-2-propanol; 2-Propanol, 1-methoxy
Classification: Aliphatic ether alcohol
Empirical: $C_4H_{10}O_2$
Formula: $CH_3OCH_2CH(OH)CH_3$
Properties: Colorless liq.; m.w. 90.14; dens. 0.919; m.p. -96.7 C; b.p. 120 C; flash pt. 102 F
Precaution: Combustible; dangerous fire hazard when exposed to heat or flame; can react with oxidizing materials; heated to decomp., emits acrid smoke and irritating fumes
Toxicology: ACGIH TLV:TWA 100 ppm; STEL 150 ppm; LD50 (oral, rat) 5660 mg/kg, (dermal, rabbit) 13 g/kg; moderately toxic by intravenous route; mildly toxic by ingestion, inhalation, skin contact; skin and eye irritant
Uses: Antifreeze and coolant for diesel engines; solvent for paints, cleaners, inks, electronics, textile dyes, agric. prods., cosmetics, floor polish, adhesives, fuel additives; solv. sealing of cellophane; oil field, mining, epoxy laminates; chem. intermediate applics.
Regulatory: FDA 21CFR §175.105, 181.22, 181.30
Manuf./Distrib.: Aldrich; Allchem Ind.; Ashland; Fluka; Harcros; Norman, Fox
Trade names: Arcosolv® PM; Dowanol® PM; Eastman® PM; Icinol PM; Methyl Propasol®; Poly-Solv® MPM; Solvenon® PM
Trade names containing: D.E.R. 667-PMT55; D.E.R. 671-PM75; D.E.R. 671-PMK75; D.E.R. 671-PMT70; D.E.R. 671-PMX75

(2-Methoxymethylethoxy) propanol. *See* PPG-2 methyl ether

Methoxy methyl melamine resin
Uses: Resin for crosslinking solv. and water-based polyols, e.g., alkyds, epoxies, oil-free polyesters, and thermosetting acrylic resins
Trade names containing: Uformite® 27-806

4-Methoxy-4-methyl-pentanone-2
CAS 107-70-0
Synonyms: 4-Methoxy-4-methylpentan-2-one
Empirical: $C_7H_{14}O_2$
Formula: $CH_3C(CH_3)(OCH_3)CH_2COCH_3$
Properties: Water-wh. liq.; b.p. 147-163 C; flash pt. 141 F
Precaution: Combustible; moderate fire risk
Toxicology: Irritant to skin and eyes; midly toxic by inhalation
Uses: Solvent for a variety of resin coatings
Trade names: PentoXone

4-Methoxy-4-methylpentan-2-one. *See* 4-Methoxy-4-methyl-pentanone-2
2-Methoxy-2-methylpropane. *See* Methyl t-butyl ether
4-Methoxyphenol. *See* Hydroquinone monomethyl ether
1-Methoxy-2-propanol. *See* Methoxyisopropanol

Methoxypropyl acetate
Trade names containing: Carboflow® 21S; Cellokyd 2829 PMA/T; EPON® Resin 876-QX-50; EPON® Resin 1004-QX-55; EPON® Resin 1007-Q-40; EPON® Resin 1007-Q-45; Epotuf Resin 38-509

2-Methoxypropyl acetate. *See* Propylene glycol methyl ether acetate

Methoxypropylamine
CAS 5332-73-0; EINECS 226-241-3
Synonyms: 3-MPA; 3-Methoxypropylamine; 1-Amino-3-methoxypropane
Empirical: $C_4H_{11}NO$
Formula: $CH_3OCH_2CH_2CH_2NH_2$
Properties: Colorless liq.; misc. with water, ethanol, toluene, acetone, CCl_4, hexane, ether; m.w. 89.16; dens. 0.873 (20/20 C); m.p. -75.7 C; b.p. 119 C; flash pt. (TCC) 80 F; ref. index 1.417 (20 C)
Precaution: Flamm.; moderate fire risk; can react with oxidizing materials
Toxicology: LD50 (IV, mouse) 180 mg/kg; toxic by ingestion, inhalation, IV route; irritating to skin, eyes, mucous membranes; causes burns; heated to decomp., emits toxic fumes of NO_x
Uses: Organic intermediate; emulsifier in anionic coatings and wax formulations
Manuf./Distrib.: Aldrich; BASF; Elf Atochem N. Am.; Fluka

3-Methoxypropylamine. *See* Methoxypropylamine

Methoxy tripropylene glycol acrylate
Synonyms: Monomethoxy tripropylene glycol monoacrylate
Uses: Reactive diluent with exc. solvating props.; exhibits good adhesion to aluminum, paper, paperboard, and selected plastics; useful in UV/EB adhesives
Trade names: Photomer® 8061

Methyl abietate
CAS 127-25-3
Synonyms: Abietic acid methyl ester; Methyl ester of wood rosin
Empirical: $C_{21}H_{32}O_2$
Properties: Colorless to yel. thick liq., odorless; misc. with alcohol, ether, common org. solvs., aliphatic hydrocarbons; insol. in water; m.w. 316.47; dens. 1.03; b.p. 360 C; flash pt. (COC) 180 C; ref. index 1.53 (20 C)
Precaution: Combustible exposed to heat or flame; can react with oxidizers
Toxicology: Skin irritant; sl. toxic
Uses: Plasticizer and solvent for vinyls, cellulosics, PS, paints, ester gums, adhesives
Manuf./Distrib.: Hercules

Methylacetaldehyde. *See* Propionaldehyde

Methyl acetate
CAS 79-20-9; EINECS 201-185-2
Synonyms: Acetic acid methyl ester
Definition: Ester of methyl alcohol and acetic acid
Empirical: $C_3H_6O_2$
Formula: CH_3COOCH_3
Properties: Colorless volatile liq., fragrant odor, sl. bitter flavor; misc. with common hydrocarbon solvs.; sol. in water; m.w. 74.09; dens. 0.92438; f.p. -98.05 C; b.p. 54.05 C; flash pt. (CC) -10 C; ref. index 1.3614 (20 C)
Precaution: Flamm.; dangerous fire and explosion risk; explosive limits 3-16% in air
Toxicology: Irritant to respiratory tract; narcotic in high concs.; TLV 200 ppm in air
Uses: Paint remover compds., lacquer solv., intermediate, synthetic flavoring; solv. for nitrocellulose, acetylcellulose; mfg. of artificial leather
Regulatory: FDA 21CFR §172.515, 175.105; FEMA GRAS
Manuf./Distrib.: Akzo Nobel; Aldrich; ARCO; Ashland; J.T. Baker; Eastern Chem.; Fluka; Hampshire; Hoechst Celanese; Monsanto; Penta Mfg.; Van Waters & Rogers; Wacker-Chemie GmbH

Methyl acetone. *See* Methyl ethyl ketone
Methyl 12-acetoxyoleate. *See* Methyl acetyl ricinoleate

Methyl acetyl ricinoleate
CAS 140-03-4; EINECS 205-392-9
Synonyms: Methyl 12-acetoxyoleate
Properties: Sol. in most organic solvents; insol. in water; pale-yellow liquid; mild odor; m.w. 354.59; sp.gr. 0.938; m.p. -15 C; flash pt. (COC) 196 C; ref. index 1.4545
Precaution: Combustible
Uses: Plasticizer, lubricant, protective coatings, synthetic rubbers, vinyl compds. and cellulosics
Regulatory: FDA 21CFR §175.105; Japan approved

Methyl acrylate (monomer)

Manuf./Distrib.: CasChem
Trade names: Flexricin® P-4

Methyl acrylate (monomer)
CAS 96-33-3; EINECS 202-500-6
Synonyms: 2-Propenoic acid methyl ester; Acrylic acid methyl ester
Classification: Aliphatic organic compd.
Definition: Ester of methyl alcohol and acrylic acid
Empirical: $C_4H_6O_2$
Formula: CH_2:CHCOOCH_3
Properties: Colorless volatile liq.; sl. sol. in water; m.w. 86.1; dens. 0.9574 (20/20 C); m.p. -76.5 C; b.p. 79-80 C; flash pt. (TOC) 25 F; ref. index 1.401 (20 C)
Precaution: Flamm.; dangerous fire and explosion risk
Toxicology: LD50 (rat, oral) 0.3 g/kg; TLV 10 ppm in air; toxic by inhalation, ingestion, and skin absorption; irritant to skin and eyes
Uses: Acrylic polymers; amphoteric surfactants; chemical intermediate; leather finish resins; textile and paper coatings; plastic films
Regulatory: FDA approved for orals
Manuf./Distrib.: Aldrich; Allchem Ind.; Asahi Chem. Ind.; Ashland; BASF; Elf Atochem N. Am.; Fluka; Hoechst Celanese; Mitsubishi Chem.; Monomer-Polymer & Dajac Labs; Showa Denko

α-Methyl acrylic acid. *See* Methacrylic acid
α-Methylacrylic acid (monomer). *See* Methacrylic acid
α-Methyl acrylic amide. *See* Methacrylamide
Methyl adipate. *See* Dimethyl adipate

Methyl alcohol
CAS 67-56-1; EINECS 200-659-6
Synonyms: Methanol; Wood alcohol; Wood naphtha; Wood spirit; Carbinol; Columbian spirits; Methyl hydroxide; Methylol
Empirical: CH_4O
Formula: CH_3OH
Properties: Clear colorless liq., alcoholic odor (pure), pungent odor (crude); highly polar; misc. with water, alcohol, ether, benzene, ketones; m.w. 32.05; dens. 0.7924; m.p. -97.8 C; b.p. 64.5 C (760 mm); flash pt.(OC) 54 F; ref. index 1.3292 (20 C)
Precaution: DOT: Flamm. liq.; dangerous fire risk; explosive limits 6.0-36.5% vol. in air; reacts vigorously with oxidizers; heated to decomp., emits acrid smoke and irritating fumes
Toxicology: LD50 (oral, rat) 5628 mg/kg; toxic (causes blindness); poisonous by ingestion, inhalation, or percutaneous absorption; experimental teratogen, reproductive effects; eye and skin irritant; narcotic; usual fatal dose 100-250 ml; TLV 200 ppm in air
Uses: Industrial solvent; mfg. of formaldehyde, acetic acid, dimethyl terephthalate; chemical synthesis; antifreeze; solv. for nitrocellulose, polyvinyl butyral, shellac, rosin, manila resin, dyes, paints, waterborne coatings; denaturant
Regulatory: FDA 21CFR §172.560, 173.250, 173.385, 175.105, 176.180, 176.200, 176.210, 177.2420, 177.2460, 27CFR §21.115; FDA approved for orals; USP/NF, BP compliance
Manuf./Distrib.: Air Prods.; Albright & Wilson Am.; Aldrich; Allchem Ind.; AlliedSignal; Ashland; J.T. Baker; Baychem; Brown; Chemcentral; Coyne; CPS; DuPont; Eastman; Fluka; General Chem.; C.P. Hall; Harcros; Hoechst Celanese; Mitsui Toatsu; Nissan Chem. Ind.; Norsk Hydro AS; Quantum; Samson; Sigma; Spectrum Chem. Mfg.; Sunnyside; Van Waters & Rogers; Veckridge
Trade names containing: Arquad® 2C-70 Nitrite; Dow Corning® Z-6032; Filmex® A-2; Filmex® B; Filmex® C; Filmex® D-1; Filmex® D-2; GP-197 Resin Sol'n.; Nacure® 2501; Nacure® 2530; Nacure® XP-357; Silikophen® P 50/X; Silikophen® P 80/X; Silquest® A-1160; Tecsol® A-2; Tecsol® B; Tecsol® D-2

Methyl amyl acetate
CAS 108-84-9
Synonyms: Methylisobutyl carbinol acetate; s-Hexyl acetate; 4-Methyl-2-pentanol acetate
Formula: $CH_3COOCH(CH_3)C_4H_9$
Properties: Colorless liq., mild odor; insol. in water; sol. in alcohol; m.w. 144.21; sp.gr. 0.858 (20/20 C); f.p. -64 C; b.p. 146-150 C; flash pt. (TCC) 96 F; ref. index 1.4008 (20 C)
Precaution: Combustible; moderate fire risk
Toxicology: Toxic by inhalation; TLV 50 ppm in air
Uses: Solvent for coatings, nitrocellulose, and other lacquers
Manuf./Distrib.: Eastman; Penta Mfg.

Methyl amyl alcohol
CAS 108-11-2; EINECS 203-551-7
Synonyms: MIBC; Methylisobutyl carbinol; Isobutyl methyl carbinol; 4-Methylpentanol-2; 4-Methyl-2-pentanol
Empirical: $C_6H_{14}O$
Formula: $(CH_3)_2CHCH_2CH(CH_3)OH$
Properties: Colorless liq.; misc. with most common org. solvs.; sl. sol. in water; m.w. 102.18; sp.gr. 0.8079 (20/20 C); b.p. 131.8 C; flash pt. (OC) 40.5 C; ref. index 1.4089
Precaution: Combustible; moderate fire risk; explosive limits in air 1-5.5%
Toxicology: TLV 25 ppm in air; irritating to respiratory system
Uses: Solvent for dyestuffs, oils, gums, resins, waxes, nitrocellulose, ethylcellulose; latent solv. for coatings; organic synthesis; froth flotation; brake fluids
Manuf./Distrib.: Aldrich; Allchem Ind.; Ashland; Fluka; O'Brien Ind.; Shell; Union Carbide; Zinkan Enterprises

Methyl n-amyl ketone
CAS 110-43-0; EINECS 203-767-1
Synonyms: MAK; 2-Heptanone; Ketone C-7; n-Amyl methyl ketone; Methyl pentyl ketone
Empirical: $C_7H_{14}O$
Formula: $CH_3CH_2CH_2CH_2CH_2COCH_3$
Properties: Water-wh. liq., penetrating fruity odor; almost insol. in water; misc. with org. solvs.; m.w. 114.18; dens. 0.8166 (20/20 C); b.p. 150.6 C; flash pt. 49 C; ref. index 1.4110 (20 C)
Precaution: Combustible; moderate fire risk; reactive with oxidizing materials
Toxicology: LD50 (oral, rat) 1670 mg/kg; mod. toxic by ingestion; mildly toxic by inhalation, skin contact; skin irritant; narcotic in high concs.; TLV 50 ppm in air; heated to decomp., emits acrid smoke and fumes
Uses: Industrial solvent for nitrocellulose lacquers, synthetic flavoring and perfumery
Regulatory: FDA 21CFR §172.515; FEMA GRAS
Manuf./Distrib.: Aldrich; Allchem Ind.; Ashland; Eastman; Fluka; Sigma; Union Carbide; Van Waters & Rogers
Trade names containing: Acryloid® AT-400; Acryloid® AT-410; Acryloid® AT-954; Acryloid® AU-946; Araldite® EPN 1138 MAK-80; Aroplaz 3667-Z-80; D.E.N. 438-MAK80; D.E.N. 444-MAK75; D.E.R. 671-MAK75; EPON® Resin 875-O-70; Joncryl® 500; SCD 16875; SCX™-901; SCX™-906

Methylbenzene. *See* Toluene
4-Methylbenzenesulfonic acid. *See* p-Toluene sulfonic acid
p-Methylbenzenesulfonic acid. *See* p-Toluene sulfonic acid

4-Methylbenzophenone
CAS 134-84-9
Synonyms: Phenyl p-tolyl ketone; 4-Methyl-p-benzophenone
Empirical: $C_{14}H_{12}O$
Formula: $CH_3C_6H_4COC_6H_5$
Properties: Colorless cryst.; sol. in alcohol, benzene; m.w. 196.26; b.p. 326 C
Toxicology: LD50 (IP, mouse) 250 mg/kg; poison by IP route; heated to decomp., emits acrid smoke and irritating fumes
Uses: Fixative in perfumes
Manuf./Distrib.: Aceto; Aldrich
Trade names containing: Esacure® KT37; Esacure® TZT

4-Methyl-p-benzophenone. *See* 4-Methylbenzophenone
3-Methyl-1,3-butadiene polymer with 2-methyl-1-propene. *See* Isobutylene/isoprene copolymer
2-Methyl-1-butanol. *See* Amyl alcohol
3-Methylbutanol. *See* Isoamyl alcohol
3-Methyl-1-butanol. *See* Isoamyl alcohol
Methyl-1,3-butylene glycol acetate. *See* Methoxybutyl acetate

Methyl t-butyl ether
CAS 1634-04-4; EINECS 216-653-1
Synonyms: MTBE; Methyl tertiary butyl ether; Methyl 1,1-dimethylethyl ether; Propane, 2-methoxy-2-methyl; t-Butyl methyl ether; 2-Methoxy-2-methylpropane
Classification: Aliphatic ether
Empirical: $C_5H_{12}O$
Formula: $CH_3OC(CH_3)_3$
Properties: Clear liq., terpene-like odor; sl. sol. in water; misc. with all gasoline-type hydrocarbons; m.w. 88.15; dens. 0.7335; b.p. 91.1 C; f.p. -75 C; flash pt. -25.6 C
Precaution: Flamm. exposed to heat or flame

Toxicology: Slight skin and eye irritant; heated to decomp., emits acrid smoke and irritating fumes
Uses: Octane booster in gasoline; intermediate; solv. for paints; extraction solv.; reaction medium in pharmaceuticals, for polymerizations, Grignard reactions
Manuf./Distrib.: Aldrich; Allchem Ind.; ARCO; Ashland; Fluka; Hüls Am.; Sigma; Texas Petrochem.

(2-β-Methyl 'Carbitol'), methoxyethoxy ethanol. *See* Methoxydiglycol
Methyl 'Cellosolve' . *See* Methoxyethanol
Methyl 'Cellosolve' acrylate. *See* Methoxyethyl acrylate

Methylcellulose
CAS 9004-67-5
Synonyms: MC; Cellulose methyl ether; Cologel; Methocel; Citrucel
Definition: Methyl ether of cellulose
Properties: Grayish-wh. fibrous powd., odorless, tasteless; aq. suspension swells in water to visc. colloidal sol'n.; sol. in cold water, glacial acetic acid, some org. solvs.; insol. in alcohol, ether, chloroform, warm water; m.w. 86,000-115,000
Precaution: Combustible
Toxicology: Nontoxic; nonallergenic; may cause immune responses; heated to decomp., emits acrid smoke and irritating fumes
Uses: Protective colloid in water-based paints to prevent flocculation of pigment; film and sheeting; binder in ceramic glazes; leather tanning; dispersing, thickening, and sizing agent; food additive; adhesive; paper greaseproofing; pharmaceuticals; visc. stabilizer for latex and emulsion paints; plasticizer for ceramic and refractory shapes
Regulatory: FDA 21CFR §150.141, 150.161, 175.105, 175.210, 175.300, 176.200, 182.1480, GRAS; USDA 9CFR §318.7, limitation 0.15% in meat and vegetable prods.; FEMA GRAS; Japan restricted (2% max.); Europe listed; UK approved; FDA approved for buccals, injectables, ophthalmics, orals, topicals, vaginals; USP/NF, BP, Ph.Eur. compliance
Manuf./Distrib.: Aceto; Aldrich; Allchem Ind.; Aqualon; Ashland; Chemcentral; Courtaulds Water Soluble Polymers; Fluka; Hoechst Celanese; Punda Mercantile; Shin-Etsu; Sigma; Spectrum Chem. Mfg.
Trade names: Methocel® A

Methylchloroform. *See* Trichloroethane

Methylchloroisothiazolinone
CAS 26172-55-4; EINECS 247-500-7
Synonyms: 5-Chloro-2-methyl-4-isothiazolin-3-one; 4-Isothiazolin-3-one, 5-chloro-2-methyl-; Chloromethylisothiazolinone
Classification: Heterocyclic organic compd.
Empirical: C_4H_4ClNOS
Uses: Antimicrobial, preservative for metalworking fluids, polymer emulsions, cooling tower water treatment, paints; slimicide for paper mills
Regulatory: FDA 21CFR §175.105, 176.170
Trade names containing: Amerstat® 251; Biochek 430

Methyl-α-cyclodextrin
Uses: Complex hosting guest molecules; increases the sol. and bioavailability of other substances; masks flavor, odor, or coloration; stabilizes against light, oxidation, heat, and hydrolysis; turns liqs. or volatiles into stable solid powds.; for use in pharmaceuticals, cosmetics, toiletries, foods, tobacco, pesticides, textiles, paints, plastics, synthesis, polymers

Methyl-β-cyclodextrin
CAS 128446-36-6
Uses: Complex hosting guest molecules; increases the sol. and bioavailability of other substances; masks flavor, odor, or coloration; stabilizes against light, oxidation, heat, and hydrolysis; turns liqs. or volatiles into stable solid powds.; for use in pharmaceuticals, cosmetics, toiletries, foods, tobacco, pesticides, textiles, paints, plastics, synthesis, polymers
Trade names: Beta W7 M1.8

Methyl-γ-cyclodextrin
Definition: Complex hosting guest molecule
Uses: Complex hosting guest molecules; increases the sol. and bioavailability of other substances; masks flavor, odor, or coloration; stabilizes against light, oxidation, heat, and hydrolysis; turns liqs. or volatiles into stable solid powds.; for use in pharmaceuticals, cosmetics, toiletries, foods, tobacco, pesticides, textiles,

paints, plastics, synthesis, polymers
Trade names: Gamma W8 M1.8

2-Methyl-1,5-diaminopentane. *See* 2-Methylpentamethylenediamine

Methyldibromo glutaronitrile
CAS 35691-65-7; EINECS 252-681-0
Synonyms: 2-Bromo-2-(bromomethyl) glutaronitrile; Glutaronitrile, 2-bromo-2-(bromomethyl); 2-Bromo-2-(bromomethyl) pentanedinitrile; 1,2-Dibromo-2,4-dicyanobutane
Definition: Brominated methylene glutaronitrile
Empirical: $C_6H_6Br_2N_2$
Formula: $NCCBrCH_2BrCH_2CH_2CN$
Properties: Mildly pungent odor; sol. in methanol, ethanol, ether; insol. in water; m.w. 265.96; m.p. 51.2-52.5 C
Uses: Antimicrobial for preserving cosmetics, personal care prods., aq. systems incl. paints, emulsions, adhesives, joint cements, metalworking fluids, inks, polishes, waxes, dispersed pigments
Regulatory: USA not restricted; Europe listed
Manuf./Distrib.: Calgon; Schülke & Mayr
Trade names: Tektamer® 38; Tektamer® 38 A.D.; Tektamer® 38 L.V.
Trade names containing: Biochek 410; Biochek 430

Methyl 1,1-dimethylethyl ether. *See* Methyl t-butyl ether
4-Methyl-1,3-dioxolan-2-one. *See* Propylene carbonate
4,4´-Methylenebis(aniline). *See* 4,4´-Methylene dianiline
2,2´-Methylenebis (6-t-butyl-p-cresol). *See* 2,2´-Methylenebis (6-t-butyl-4-methylphenol)

2,2´-Methylenebis (6-t-butyl-4-methylphenol)
CAS 119-47-1; EINECS 204-327-1
Synonyms: BKF; 2,2´-Bis(6-t-butyl-p-cresyl)methane; 2,2´-Methylenebis (6-t-butyl-p-cresol); Phenol, 2,2´-methylene-bis-6-[(1,1-dimethyl)-4-methyl-]; 2,2´-Methylenebis(4-methyl-6-t-butylphenol)
Empirical: $C_{23}H_{32}O_2$
Properties: Wh. to cream powd.; m.w. 340.55; dens. 1.08 mg/m^3; m.p. 125 C; flash pt. (COC) 190 C
Precaution: Avoid strong oxidizers
Toxicology: LD50 (rat, oral) 5000 mg/kg; mildlly toxic by ingestion; may cause eye irritation
Uses: Antioxidant, stabilizer for plastics, rubber, latex, sol'n. and emulsion polymers, fats, oils, and paraffin wax; polymerization inhibitor in chemical processes; antidegradant for polyolefins and coatings
Manuf./Distrib.: Aldrich; Great Lakes; Raschig
Trade names: Santowhite® PC

2,2´-Methylenebis(4-methyl-6-t-butylphenol). *See* 2,2´-Methylenebis (6-t-butyl-4-methylphenol)
Methylene bisphenyl isocyanate. *See* MDI

Methylenebis (thiocyanate)
CAS 6317-18-6
Synonyms: MBT; MTC
Empirical: $C_3H_2S_2N_2$
Formula: $CH_2(SCN)_2$
Properties: Yel. to lt. orange powd., irritating pungent odor; m.w. 130; m.p. 105-107 C
Uses: Biocide for water treatment, pulp and paper, antifoulant paint, leather, timber preservation
Manuf./Distrib.: Albright & Wilson Am.; Aldrich
Trade names: Amerstat® 282; Tolcide MBT
Trade names containing: Amerstat® 1910; Busan® 1009; Busan® 1025; Busan® 1105

Methylene chloride
CAS 75-09-2; EINECS 200-838-9
Synonyms: DCM; Dichloromethane; Methylene dichloride; Methane dichloride; Methane-, dichloro-
Classification: Halogenated organic compd.
Empirical: CH_2Cl_2
Properties: Colorless clear volatile liq., penetrating ether-like odor; sol. in alcohol, ether; misc. with fixed and volatile oils; sl. sol. in water; m.w. 84.93; dens. 1.335 (15/4 C); f.p. -97 C; b.p. 40.1 C; ref. index 1.4244 (20 C); KB value 136
Precaution: Nonflamm.; explosive as vapor exposed to heat or flame; contact with hot surfaces cause decomp., yielding toxic fumes; heated to decomp., emits highly toxic fumes of phosgene and Cl^-

Toxicology: LD50 (oral, rat) 2136 mg/kg; poison by IV route; mod. toxic by ingestion, subcutaneous, IP routes; experimental carcinogen; human systemic effects; eye and severe skin irritant; human mutagenic data; narcotic in high concs.; ACGIH TLV:TWA 50 ppm
Uses: Paint removers, solvent degreasing, plastics processing, blowing agent in foams, solvent extraction, solvent for cellulose acetate, propellant for paint aerosols, pharmaceutic aid
Regulatory: FDA 21CFR §73.1 (no residue), 173.255; FDA approved for orals; USP/NF compliance
Manuf./Distrib.: Aldrich; Allchem Ind.; Ashland; R.E. Carroll; Chemcentral; Coyne; Elf Atochem N. Am.; Farleyway Chem. Ltd; Fluka; C.P. Hall; Harcros; Hüls Am.; ICI Spec. Chems.; Mallinckrodt; Mitsui Toatsu; Occidental; Primachem; Samson; Sigma; Spectrum Chem. Mfg.; Stanchem; TR-AMC; Van Waters & Rogers; Vulcan

4,4′-Methylene dianiline
CAS 101-77-9; EINECS 202-974-4
Synonyms: MDA; p,p′-Diaminodiphenylmethane; 4,4′-Methylenebis(aniline); p,p′-Methylene dianiline
Empirical: $C_{13}H_{14}N_2$
Formula: $H_2NC_6H_4CH_2C_6H_4NH_2$
Properties: Tan flakes or lumps, faint amine-like odor; very sol. in alcohol, benzene, ether; sol. in cold water; m.w. 198.29; m.p. 90 C; b.p. 398-399 C; flash pt. 440 F
Precaution: Combustible
Toxicology: TLV:TWA 0.1 ppm (skin); LD50 (oral, rat) 347 mg/kg; poison by ingestion, subcutaneous, intraperitoneal routes; eye irritant
Uses: Curing agent; corrosion inhibitor; epoxy resin hardening agent; intermediate for paints
Manuf./Distrib.: Aldrich; BASF; Fluka; Hoechst Celanese; Sigma; Uniroyal

p,p′-Methylene dianiline. *See* 4,4′-Methylene dianiline
Methylene dichloride. *See* Methylene chloride
Methylene di-p-phenylene isocyanate. *See* MDI
Methylenesuccinic acid. *See* Itaconic acid
Methyl ester stearic acid. *See* Methyl stearate
Methyl ester of wood rosin. *See* Methyl abietate
Methyl ether. *See* Dimethyl ether
1-Methylethyl acetate. *See* Isopropyl acetate
(1-Methylethyl)benzene, monosulfo deriv., sodium salt. *See* Sodium cumenesulfonate
Methyl ethyl carbinol. *See* 2-Butanol
Methyl ethylene glycol. *See* Propylene glycol
4,4′-(1-Methylethylidene) bisphenol. *See* Bisphenol A

Methyl ethyl ketone
CAS 78-93-3; EINECS 201-159-0
Synonyms: MEK (INCI); Ethyl methyl ketone; 2-Butanone; 2-Oxobutane; Methyl acetone
Classification: Aliphatic ketone
Empirical: C_4H_8O
Formula: $CH_3COCH_2CH_3$
Properties: Colorless liq., acetone-like odor; sol. in 4 parts water, benzene, alcohol, ether; misc. with oils; m.w. 72.10; dens. 0.8255 (0/4 C); m.p. -86 C; b.p. 79.6 C; flash pt. (TOC) 24 F; visc. 0.40 cp (25 C); ref. index 1.3814 (15 C)
Precaution: DOT: Flamm. liq; dangerous fire risk; explosive limits in air 2-10%
Toxicology: LD50 (oral, rat) 2737 mg/kg; mod. toxic by ingestion, skin contact, IP routes; toxic by inhalation; experimental teratogen, reproductive effects; strong irritant; affects CNS; TLV 200 ppm in air; heated to decomp., emits acrid smoke and fumes
Uses: Solvent in nitrocellulose coatings and vinyl films, paint removers, cements, adhesives, organic synthesis; mfg. of smokeless powder; cleaning fluids; priming, catalyst carrier; acrylic coatings
Regulatory: FDA 21CFR §172.515, 175.105, 175.320, 177.1200
Manuf./Distrib.: Aldrich; Allchem Ind.; AlliedSignal; Ashland; J.T. Baker; Baychem; BP Chems. Ltd; R.E. Carroll; Chemcentral; Elf Atochem N. Am.; Exxon; Fluka; General Chem.; C.P. Hall; Harcros; Hoechst Celanese; Mallinckrodt; Olin; Oxiteno; Penta Mfg.; Primachem; Samson; Shell; Spectrum Chem. Mfg.; Sunnyside; Texaco; Union Carbide; Van Waters & Rogers
Trade names containing: Acryloid® A-21LV; Acryloid® A-101; Araldite® GZ 488 N-40; Araldite® GZ 7071 N-80; Araldite® GZ 7488 N-50; D.E.N. 438-EK85; D.E.N. 439-EK85; D.E.R. 671-PMK75; D.E.R. 684-EK40; Konform® AR 2000; Resin QR-1281

Methyl ethyl ketone oxime. *See* Methyl ethyl ketoxime

Methyl ethyl ketone peroxide
CAS 1338-23-4; EINECS 215-661-2
Synonyms: MEK peroxide; Ethyl methyl ketone peroxide; 2-Butanone peroxide
Empirical: $C_8H_{18}O_6$
Formula: $C_2H_5C(OOH)(CH_3)OOC(OOH)(CH_3)C_2H_5$
Properties: Colorless liq.; m.w. 210.23; dens. 1.053 (20/4 C) ; ref. index 1.442 (20 C)
Toxicology: TLV:CL 0.2 ppm; poison by intraperitoneal route; moderately toxic by ingestion, inhalation; skin and eye irritant
Uses: Polymerization initiator/catalyst for cure of unsaturated polyester resins; used in gel coats to eliminate or reduce porosity
Manuf./Distrib.: Akzo Nobel; Aldrich; Cook Composites & Polymers; Elf Atochem N. Am.; Fluka; Great Western; Hastings Plastics; Norac; Witco/Polymer Addit.
Trade names: Superox® 702; Superox® 732

Methyl ethyl ketoxime
CAS 96-29-7; EINECS 202-496-6
Synonyms: 2-Butanone oxime; Methyl ethyl ketone oxime
Empirical: C_4H_9NO
Formula: $CH_3C(:NOH)CH_2CH_3$
Properties: Colorless liq.; sol. in water, ether, ethanol; m.w. 87.14; b.p. 70-73 C; dens. 0.923 (20 C); flash pt. 60 C; ref. index 1.443 (20 C)
Toxicology: Poison by intraperitoneal route; moderately toxic by subcutaneous route; eye irritant
Uses: Antiskinning agent for paint industry and in environmental market
Manuf./Distrib.: Aceto; Akzo Nobel; Aldrich; Allchem Ind.; AlliedSignal; Fluka; KMZ Chem. Ltd; OM Group
Trade names: Troykyd® Anti-Skin B

Methyl glucoside
Uses: Intermediate for paints
Manuf./Distrib.: Grain Processing; Horizon Prods.

Methyl glycol. *See* Propylene glycol
Methyl glycol acetate. *See* Methoxyethanol acetate
Methyl glycol monoaceetate. *See* Methoxyethanol acetate
2-Methyl glyoxaline. *See* 2-Methyl imidazole
5-Methyl-2-hexanone. *See* Methyl isoamyl ketone

Methyl hexyl ketone
CAS 111-13-7; EINECS 203-837-1
Synonyms: 2-Octanone; Hexyl methyl ketone
Empirical: $C_8H_{16}O$
Formula: $CH_3(CH_2)_5COCH_3$
Properties: Colorless liq., apple odor, camphor taste; misc. with alcohol, ether, esters, hydrocarbons; insol. in water; m.w. 128.22; dens. 0.818 (20/4 C); m.p. -16 C; b.p. 170-172 C; flash pt. 56 C; ref. index 1.416 (20 C)
Precaution: Flamm. exposedto heat, flame, oxidizers
Toxicology: LD50 (mouse) 1600 mg/kg; mod. toxic; skin irritant; heated to decomp., emits acrid smoke and irritating fumes
Uses: Solv. for paints
Regulatory: FDA 21CFR §172.515; FEMA GRAS
Manuf./Distrib.: Aldrich; Fluka; Penta Mfg.; Union Camp; UOP

Methyl hydrogenated rosinate
CAS 8050-13-3
Synonyms: Hydrogenated methyl ester of rosin
Definition: Ester of methyl alcohol and the hydrogenated mixed long chain acids derived from rosin
Uses: Plasticizer and tackifier in lacquers, inks, adhesives, floor tiles, vinyl plastisols, artificial leather, and antifouling paints; fixative and carrier in perfumes and cosmetic preps.
Trade names: Hercolyn® D

Methyl hydroxide. *See* Methyl alcohol

Methyl hydroxyethylcellulose
CAS 9032-42-2

Methyl 12-hydroxyoctadecanoate

Synonyms: Hydroxyethylmethylcellulose
Definition: Methyl ether of hydroxyethylcellulose
Uses: Binder, thickener, pigment, foam, and filler stabilizer, dispersant, emulsifier, plasticizer, visc. control and sedimenting aid, and protective colloid used in coatings, paints, resins, mining, batteries, insecticidal prods.; rubber, textile, leather, ceramics, suspension polymerization, pharmaceuticals
Regulatory: FDA approved for orals; BP compliance
Manuf./Distrib.: Aldrich
Trade names: Tylose® MH Grades; Tylose® MHB

Methyl 12-hydroxyoctadecanoate. *See* Methyl hydroxystearate
Methyl-S-(-)-2-hydroxy propionate. *See* Methyl lactate
Methyl hydroxypropyl cellulose. *See* Hydroxypropyl methylcellulose

Methyl hydroxystearate
CAS 141-23-1; EINECS 205-471-8
Synonyms: Methyl 12-hydroxyoctadecanoate; 12-Hydroxystearic acid methyl ester; Methyl 12-hydroxystearate; 12-Hydroxyoctadecanoic acid, methyl ester
Definition: Ester of methyl alcohol and hydroxystearic acid
Empirical: $C_{18}H_{38}O_3$
Formula: $C_{16}H_{34}OHCOOCH_3$
Properties: Wh. waxy solid, flat rods; insol. in water; sl. sol. in org. solvs.; m.w. 314.57; m.p. 48 C
Precaution: Combustible
Toxicology: Experimental tumorigen; heated to decomp., emits acrid smoke and fumes
Uses: Lubricant, processing aid for butyl rubber, adhesives, inks, cosmetics, greases, paints
Regulatory: FDA 21CFR §176.210

Methyl 12-hydroxystearate. *See* Methyl hydroxystearate

2-Methyl imidazole
CAS 693-98-1; EINECS 211-765-7
Synonyms: 2MZ; 2-Methyl glyoxaline
Empirical: $C_4H_6N_2$
Formula: $CHCHNC(CH_3)NH$
Properties: Solid; m.w. 82.11; m.p. 142-143 C
Toxicology: Moderately toxic by ingestion and intraperitoneal routes
Uses: Dyeing auxiliary for acrylic fibers, plastic foams; curing agent for printed circuit board laminates, powd. coatings, adhesives, encapsulation; accelerator
Manuf./Distrib.: Aldrich; Allchem Ind.; BASF; Fabrichem; Fluka; Janssen Chimica; Sigma; SKW Chems.
Trade names: Imicure® AMI-2

Methyl isoamyl ketone
CAS 110-12-3; EINECS 203-737-8
Synonyms: MIAK; 5-Methyl-2-hexanone; Isoamyl methyl ketone; Methyl isopentyl ketone; Isobutylacetone; Isopentyl methyl ketone
Empirical: $C_7H_{14}O$
Formula: $CH_3COC_2H_4CH(CH_3)_2$
Properties: Colorless liq., pleasant odor; sl. sol. in water; misc. with most org. solvs.; m.w. 114.19; sp.gr. 0.8164; f.p. -101 F; b.p. 286 F; flash pt. (TCC) 96 F; ref. index 1.4078
Precaution: Flamm.; combustible exposed to heat, flame, oxidizers
Toxicology: LD50 (oral, rat) 4760 mg/kg, (skin, rabbit) 10 g/kg; TLV 50 ppm; mod. toxic by ingestion; mildly toxic by inh., skin contact; heated to decomp., emits acrid smoke and irritating fumes
Uses: Solv. for paints
Manuf./Distrib.: Aldrich; Allchem Ind.; Ashland; Burdick & Jackson; Eastman; Fluka; Sigma

Methyl isobutenyl ketone. *See* Mesityl oxide
Methylisobutyl carbinol. *See* Methyl amyl alcohol
Methylisobutyl carbinol acetate. *See* Methyl amyl acetate

Methyl isobutyl ketone
CAS 108-10-1; EINECS 203-550-1
Synonyms: MIBK(INCI); 4-Methyl-2-pentanone; Hexone; Isopropylacetone; Isobutyl methyl ketone
Classification: Aliphatic ketone
Empirical: $C_6H_{12}O$

Formula: $CH_3COCH_2CH(CH_3)_2$
Properties: Colorless volatile liq., faint ketonic/camphoraceous odor; sl. sol. in water; misc. with alcohol, ether, benzene, most org. solvs.; m.w. 100.18; dens. 0.8042 (20/20 C); f.p. -85 C; b.p. 115.8 C; flash pt. 13 C; ref. index 1.396
Precaution: DOT: Flamm. liq.; dangerous fire risk; may form explosive peroxides on exposure to air; can react vigorously with reducing materials; explosive limits 1.4-7.5% in air
Toxicology: LD50 (oral, rat) 2080 mg/kg; poison by IP route; mod. toxic by ingestion; mildly toxic by inh.; very irritating to skin, eyes, mucous membranes; narcotic in high conc.; TLV 50 ppm in air
Uses: Solvent for paints, varnishes, nitrocellulose, lacquers, mfg. of methyl amyl alcohol, extraction processes including extraction of uranium from fission prods., organic synthesis, denaturant for alcohol
Regulatory: FDA 21CFR §172.515; FEMA GRAS; USP/NF compliance
Manuf./Distrib.: Aldrich; Allchem Ind.; Ashland; J.T. Baker; Baychem; Chemcentral; Eastman; Elf Atochem SA; Exxon; Fluka; C.P. Hall; Harcros; Hüls Am.; Primachem; Samson; Shell; Sigma; Spectrum Chem. Mfg.; Union Carbide; Van Waters & Rogers
Trade names containing: Araldite® GZ 571 KX-75; Araldite® GZ 597 KT-55; Aroplaz 3667-Z-80; D.E.N. 438-MK75; D.E.R. 671-MK75; D.E.R. 671-XM75; EPON® Resin 1001-CX-75; Epotuf Resin 38-519; Filmex® A-2; Filmex® B; Filmex® C; Filmex® D-1; Filmex® D-2; Nacure® 1419; Tecsol® 3; Tecsol® A; Tecsol® A-2; Tecsol® B; Tecsol® C; Tecsol® D; Tecsol® D-2

Methyl isopentyl ketone. *See* Methyl isoamyl ketone
1-Methyl-4-isopropenyl-1-cyclohexene. *See* dl-Limonene

Methylisothiazolinone
CAS 2682-20-4; EINECS 220-239-6
Synonyms: 2-Methyl-4-isothiazolin-3-one; 3(2H)-Isothiazolone, 2-methyl-; 2-Methyl-3(2H)-isothiazolone
Classification: Heterocyclic organic compd.
Empirical: C_4H_5NOS
Uses: Preservative for paints, etc.; paper mill slimicide
Regulatory: FDA 21CFR §175.105, 176.170
Trade names containing: Amerstat® 251; Biochek 430

2-Methyl-4-isothiazolin-3-one. *See* Methylisothiazolinone
2-Methyl-3(2H)-isothiazolone. *See* Methylisothiazolinone

Methyl lactate
CAS 547-64-8; EINECS 208-930-0
Synonyms: Methyl-S-(-)-2-hydroxy propionate; Lactic acid methyl ester; Methyl DL-lactate
Definition: Methyl ester of lactic acid
Empirical: $C_4H_8O_3$
Formula: $CH_3CH(OH)COOCH_3$
Properties: M.w. 104.11; sp.gr. 1.085-1.098 (20 C); b.p. 140-150 C; flash pt. 51 C; ref. index 1.410-1.415
Precaution: Flamm.
Toxicology: Mild irritant
Uses: Solv. for paints and coatings, electronics, printed circuit boards, metal industry
Manuf./Distrib.: Fluka; Purac Am.; Rhone-Poulenc N. Am.
Trade names: Purasolv® ML

Methyl DL-lactate. *See* Methyl lactate

Methyl linoleate
CAS 112-63-0; EINECS 203-993-0
Synonyms: 9,12-Octadecadienoic acid methyl ester; Methyl cis,cis-9,12-octadecadienoate
Definition: Ester of methyl alcohol and linoleic acid
Empirical: $C_{19}H_{34}O_2$
Formula: $CH_3(CH_2)_4CH:CHCH_2CH:CH(CH_2)_7COOCH_3$
Properties: Colorless oil; misc. with dimethylformamide, fat solvs., oils; m.w. 294.48; dens. 0.887 (20/4 C); m.p. -35 C; b.p. 207-208 C (11 mm); iodine no. 172.4; ref. index 1.466
Precaution: Combustible
Uses: Plasticizer for paints
Regulatory: FDA 21CFR §172.225
Manuf./Distrib.: Air Prods.; Aldrich; Fluka; Hart Prods.; Penta Mfg.; Sigma

Methyl maleate. *See* Dimethyl maleate
Methyl methacrylate homopolymer. *See* Methyl methacrylate polymer

Methyl methacrylate monomer
CAS 80-62-6; EINECS 201-297-1
Definition: Methyl ester of methacrylic acid
Empirical: $C_5H_8O_2$
Formula: $CH_2:C(CH_3)COOCH_3$
Properties: Colorless volatile liq.; sol. in MEK, THF, esters, aromatic and chlorinated hydrocarbons; sl. sol. in water; m.w. 100.1; dens. 0.940 (25/25 C); b.p. 99-100 C; f.p. -48.2 C; flash pt. (OC) 10 C; ref. index 1.414 (20 C)
Precaution: Highly flamm.; dangerous fire risk; explosive limits 2.1-12.5% in air
Toxicology: TLV 100 ppm in air; LD50 (rat, oral) 8.4 g/kg; potent sensitizer; may cause contact dermatitis on handling, occupational asthma; irritating to skin, eyes, respiratory system
Uses: Monomer for polymethacrylate resins, paints, impregnation of concrete
Manuf./Distrib.: Aldrich; Allchem Ind.; Cyro Ind.; Degussa; Fluka; Mitsubishi Gas; Monomer-Polymer & Dajac Labs; Rohm & Haas; Sigma; Transol Chem. UK Ltd; Van Waters & Rogers

Methyl methacrylate polymer
CAS 9011-14-7
Synonyms: Polymethylmethacrylate; Methyl methacrylate homopolymer; 2-Methyl-2-propenoic acid methyl ester homopolymer; Methyl methacrylate resin; Methacrylic acid methyl ester polymers
Empirical: $(C_5H_8O_2)_n$
Toxicology: Experimental tumorigen by implant; heated to decomp., emits acrid smoke and irritating fumes
Uses: Thermoplastic acrylic resin used in acrylic sheet, molding, and extrusion powds., coatings, barrier coatings for PS, vinyl topcoats, product finishes, printing inks; processing aid for PVC
Manuf./Distrib.: Aldrich; Aristech; Coz; Cyro Ind.; Cytec Ind.; Degussa; Elf Atochem N. Am.; Esschem; Fluka; Monomer-Polymer & Dajac Labs; Sigma; StanChem; Sybron
Trade names: Acryloid® B-44; Resin QR-1225
Trade names containing: Acryloid® A-21; Acryloid® A-21LV; Acryloid® A-101; Acryloid® B-44S; Acryloid® B-48N; Acryloid® B-50S; Acryloid® B-82; Acryloid® B-99N

Methyl methacrylate resin. *See* Methyl methacrylate polymer
Methyl-2(methyl-2) oxybispropanol. *See* Dipropylene glycol
Methyl namate. *See* Sodium dimethyldithiocarbamate
Methyl cis,cis-9,12-octadecadienoate. *See* Methyl linoleate
Methyl octadecanoate. *See* Methyl stearate
Methyl 9-octadecenoate. *See* Methyl oleate
7-Methyl-1-octanol. *See* Isononyl alcohol
Methylol. *See* Methyl alcohol

Methyl oleate
CAS 112-62-9; 67762-38-3; EINECS 203-992-5; 267-015-4
Synonyms: Methyl 9-octadecenoate; 9-Octadecenoic acid, methyl ester
Definition: Ester of methyl alcohol and oleic acid
Empirical: $C_{19}H_{36}O_2$
Formula: $CH_3(CH_2)_7CH=CH(CH_2)_7COOCH_3$
Properties: Clear to amber liq., faint fatty odor; sol. in alcohols, most org. solvs.; insol. in water; dens. 0.8739 (20 C); f.p. -19.9 C; b.p. 218.5 C (20 mm); ref. index 1.4510 (26 C)
Precaution: Combustible
Toxicology: Low oral toxicity; mildly irritating to skin
Uses: Intermediate for detergents, emulsifiers, wetting agents, stabilizers, paints, textile treatment, plasticizers for PS, cellulosics, duplicating inks, rubbers, waxes, etc.; chromatographic reference standard
Regulatory: FDA 21CFR §172.225, 175.105, 176.200, 176.210, 177.2260, 177.2800; BP compliance
Manuf./Distrib.: Aldrich; Calgene; Ferro/Bedford; Fluka; Henkel/Emery; Norman, Fox; Sigma; Stepan; Unichema Int'l.; Union Camp; Witco/Oleo-Surf.

N-Methylol methacrylamide
Uses: Used in self-crosslinking emulsions, heat-curing coatings
Manuf./Distrib.: Monomer-Polymer & Dajac Labs; Rohm Tech
Trade names: BM-818

Methyl oximino silane
Uses: Used in sealants and coatings
Trade names: OS-1000

Methyl oxirane polymers. *See* Poloxamer 101
N-Methyl-N-(1-oxo-9-octadecenyl)glycine. *See* Oleoyl sarcosine
2-Methyl-2-((1-oxo-2-propenyl)-amino)-1-propanesulfonic acid, sodium salt. *See* Sodium 2-acrylamido-2-methylpropanesulfonate
Methylpentamethylenediamine. *See* 2-Methylpentamethylenediamine

2-Methylpentamethylenediamine
CAS 15520-10-2; EINECS 239-556-6
Synonyms: MPMD; 1,5-Pentanediamine, 2-methyl-; Methylpentamethylenediamine; 2-Methyl-1,5-diaminopentane
Empirical: $C_6H_{16}N_2$
Formula: $H_2NCH_2CH(CH_3)(CH_2)_3NH_2$
Properties: Colorless liq., weak ammonia, fishy odor; m.w. 116.2; sp.gr. 0.86; b.p. 193 C; f.p. -50 to -60 C; flash pt. (CC) 83 C
Precaution: Avoid strong oxidants; emits toxic fumes of nitrogen oxides on decomp.; combustible
Toxicology: Corrosive; can cause burns and ulceration of skin and eye tissue, nose, throat, and gastrointestinal irritation; LD50 (rat, oral) 1690 mg/kg
Uses: For mfg. of high m.w. polyamide polymers and copolymers, nonplastic copolyamide resins, coatings, adhesives, inks, corrosion inhibitors, emulsion breakers; epoxy curing agent
Manuf./Distrib.: Aldrich; Fluka
Trade names: Dytek® A

2-Methyl-2,4-pentanediol. *See* Hexylene glycol
4-Methyl-2,4-pentanediol. *See* Hexylene glycol
4-Methylpentanol-2. *See* Methyl amyl alcohol
4-Methyl-2-pentanol. *See* Methyl amyl alcohol
4-Methyl-2-pentanol acetate. *See* Methyl amyl acetate
4-Methyl-2-pentanone. *See* Methyl isobutyl ketone
4-Methyl-3-pentene-2-one. *See* Mesityl oxide
2-Methyl-2-penten-4-one. *See* Mesityl oxide
4-Methyl-3-penten-2-one. *See* Mesityl oxide
Methyl pentyl ketone. *See* Methyl n-amyl ketone

Methylphenyl ethanolamine
CAS 93-90-3
Uses: Coupling agent for disperse dyes for syn. fibers; photosensitive chemical for paper coatings

N-(2-Methylphenyl)imidodicarbonimidic diamide. *See* o-Tolyl biguanide
Methyl phenyl ketone. *See* Acetophenone
Methyl phenyl polysiloxane. *See* Phenyl trimethicone

Methyl phenyl siloxane
Uses: High solids resin imparting enhanced uv and temp. resist. to coatings
Trade names: Silres® MP 42 E; Silres® SY 201; Silres® SY 231; Silres® SY 550
Trade names containing: Silres® REN 50; Silres® REN 60; Silres® REN 80; Silres® SY 409

p-Methylphenylsulfonic acid. *See* p-Toluene sulfonic acid
Methyl phthalate. *See* Dimethyl phthalate

Methyl polysiloxane
Uses: Hammer finish additive for solv.-based paints
Trade names: TEGO® Hammer 300000; TEGO® Phobe SK 262; TEGO® Wet ZFS 454
Trade names containing: BYK®-3105; BYK®-A 525; BYK®-A 530

2-Methylpropane. *See* Isobutane
2-Methyl-2-propaneamide. *See* Methacrylamide

2-Methyl-1,3-propanediol
CAS 2163-42-0

2-Methylpropanoic acid

Empirical: $C_4H_{10}O_2$
Formula: $HOCH_2CH(CH_3)CH_2OH$
Properties: M.w. 90.12; dens. 1.015; m.p. -91 C; b.p. 123-125 C (20 mm); flash pt. > 110 C; ref. index 1.4450 (20 C)
Toxicology: Irritant
Uses: Intermediate used in prod. of solvs., urethanes, unsat. polyesters, gel coats, sat. polyester and alkyd coatings, polymeric plasticizers
Manuf./Distrib.: Aldrich; ARCO; Fluka
Trade names: MPDiol® Glycol

2-Methylpropanoic acid. *See* Isobutyric acid
2-Methylpropanoic acid 2-methylpropyl ester. *See* Isobutyl isobutyrate
1-Methyl propanol. *See* 2-Butanol
2-Methylpropanol. *See* Isobutyl alcohol
2-Methyl-1-propanol. *See* Isobutyl alcohol
2-Methyl-2-propanol. *See* t-Butyl alcohol
2-Methylpropenamide. *See* Methacrylamide
2-Methyl-1-propene, homopolymer. *See* Polyisobutene
2-Methyl-1-propene, polymer with 2,5-furandione. *See* Isobutylene/MA copolymer
2-Methylpropenoic acid. *See* Methacrylic acid
2-Methyl-2-propenoic acid. *See* Methacrylic acid
2-Methyl-2-propenoic acid methyl ester homopolymer. *See* Methyl methacrylate polymer

2-(2-Methylpropoxy)-1,2-diphenylethanone
CAS 22499-12-3
Synonyms: Benzoin isobutyl ether; 2-Isobutoxy-2-phenyl-acetophenone
Empirical: $C_{18}H_{20}O_2$
Formula: $C_6H_5CH[OCH_2CH(CH_3)_2]COC_6H_5$
Properties: M.w. 268.36; dens. 0.985; b.p. 133 C (0.5 mm); flash pt. 85 C; ref. index 1.5485
Manuf./Distrib.: Aldrich
Trade names containing: Esacure® EB3

2-Methylpropyl acetate. *See* Isobutyl acetate
2-Methyl-1-propyl acetate. *See* Isobutyl acetate

Methyl-n-propylene glycol acetate
Trade names containing: TEGO® Flow ZFS 460

β-Methylpropyl ethanoate. *See* Isobutyl acetate

Methyl propyl ketone
CAS 107-87-9; EINECS 203-528-1
Synonyms: MPK; 2-Pentanone; Ethyl acetone; Methyl n-propyl ketone
Empirical: $C_5H_{10}O$
Formula: $CH_3COCH_2CH_2CH_3$
Properties: Water-wh. liq., fruity ethereal odor; sl. sol. in water; misc. with alcohol, ether; m.w. 86.14; dens. 0.801-0.806; m.p. -78 C; b.p. 216 F; flash pt. 45 F
Precaution: Highly flamm.; very dangerous fire hazard exposed to heat or flame; reacts vigorously with oxidizers; explosive limits 1.5-8.2%
Toxicology: ACGIH TLV:TWA 200 ppm; LD50 (oral, rat) 3730 mg/kg; mod. toxic by ingestion, IP; mildly toxic by skin contact, inh.; human systemic effects; skin irritant; mutagenic data; heated to decomp., emits acrid smoke and irritating fumes
Uses: Solvent for paints, substitute for diethyl ketone, flavoring
Regulatory: FDA 21CFR §172.515; FEMA GRAS
Manuf./Distrib.: Aldrich; Ashland; Chisso Am.; Eastman; Fluka; Janssen Chimica; Penta Mfg.
Trade names containing: Cellokyd 2708

Methyl n-propyl ketone. *See* Methyl propyl ketone
2-Methylpropyl methacrylate. *See* Isobutyl methacrylate
2-Methylpropyl octadecanoate. *See* Isobutyl stearate
1-Methyl-2-pyrrolidinone. *See* N-Methyl-2-pyrrolidone
N-Methylpyrrolidinone. *See* N-Methyl-2-pyrrolidone
N-Methyl-2-pyrrolidinone. *See* N-Methyl-2-pyrrolidone

Methylpyrrolidone. *See* N-Methyl-2-pyrrolidone
1-Methyl-2-pyrrolidone. *See* N-Methyl-2-pyrrolidone

2-Methyl-2-pyrrolidone
Uses: Solv. and cosolv.; used as coatings solv., in stripping and cleaning of paints and varnishes, industrial cleaning, mold cleaning, petrochem. processing, agric. solv.; polymer solv. for PVAc, PVDF, PS, vinyl copolymers, nylon and aromatic polyamides and polyimides, polyesters, acrylics, PC, cellulose derivs., syn. elastomers, waxes
Trade names: NMP

N-Methylpyrrolidone. *See* N-Methyl-2-pyrrolidone

N-Methyl-2-pyrrolidone
CAS 872-50-4; EINECS 212-828-1
Synonyms: NMP; Methylpyrrolidone; N-Methylpyrrolidone; 1-Methyl-2-pyrrolidone; N-Methylpyrrolidinone; N-Methyl-2-pyrrolidinone; 1-Methyl-2-pyrrolidinone
Empirical: C_5H_9NO
Properties: Colorless liq., mild amine odor; misc. with water, org. solvs., castor oil; m.w. 99.13; dens. 1.032 (20/4 C); f.p. -24 C; b.p. 202 C; flash pt. 95 C; ref. index 1.470 (20 C)
Precaution: Combustible exposed to heat, open flame, powerful oxidizers; hygroscopic; photosensitive
Toxicology: LD50 (oral, rat) 7000 mg/kg; severely irritating to eyes, skin; mod. toxic by IP and IV routes; mildly toxic by ing., skin contact; experimental teratogen, reproductive effects; heated to decomp., emits toxic fumes of NO_x
Uses: Solv. for resins, paints, acetylene, industrial cleaning, mold cleaning; pigment dispersant; petrol. processing; spinning agent for PVC; intermediate; microelectronics industry plastic solv.
Manuf./Distrib.: Aldrich; Allchem Ind.; ARCO; Ashland; BASF; Coyne; Dynaloy; Fluka; ISP; Janssen Chimica; Olin; Sigma; Spectrum Chem. Mfg.
Trade names: M-Pyrol®; NMP-EL
Trade names containing: Eymyd® Resin L-20N; Eymyd® Resin L-30N

Methyl rosinate
CAS 68186-14-1; EINECS 269-035-9
Synonyms: Rosin acid, methyl ester
Definition: Methyl ester of acids recovered from rosin
Uses: Resin with surf.-wetting properties, visc., and tack used in lacquers, inks, paper coatings, varnishes, adhesives, sealing compds., plastics, wood preservatives, and perfumes
Regulatory: FDA 21CFR §172.615, 175.105, 175.300, 176.170, 176.200, 176.210, 177.1200, 177.2600, 178.3120, 178.3800, 178.3870
Trade names: Abalyn®

Methylsilanetriol sodium salt. *See* Sodium methyl siliconate

Methyl siloxane
Uses: High temp. resins for powd. coatings, elec., and over 800 F service
Trade names: Resin MK; Silres® HK 46; Silres® KX; Silres® M 50 E; Silres® MSE 100

Methyl stearate
CAS 112-61-8; 85586-21-6; EINECS 203-990-4; 287-824-6
Synonyms: Methyl octadecanoate; Stearic acid methyl ester; Octadecanoic acid, methyl ester; Methyl ester stearic acid
Definition: Ester of methyl alcohol and stearic acid
Empirical: $C_{19}H_{38}O_2$
Formula: $CH_3(CH_2)_{16}COOCH_3$
Properties: Wh. crystals; insol. in water; sol. in ether, alcohol; m.w. 298.57; m.p. 37.8 C; b.p. 234.5 C (30 mm); flash pt. 307 F
Precaution: Combustible exposed to heat or flame; can react with oxidizing materials
Toxicology: Experimental tumorigen; heated to decomp., emits acrid smoke and irritating fumes
Uses: Intermediate for stearic acid detergents, emulsifiers, wetting agents, stabilizers, resins, lubricants, plasticizers, paints
Regulatory: FDA 21CFR §172.225, 176.200, 176.210, 177.2260, 177.2800, 178.3910; FDA approved for topicals; BP compliance
Manuf./Distrib.: Aldrich; Ashland; Ferro/Bedford; Fluka; Henkel/Emery; Penta Mfg.; Sea-Land; Sigma; Union Camp; Witco/Oleo-Surf.

Methylstyrene. *See* Vinyltoluene monomer

α-Methylstyrene monomer
CAS 98-83-9; EINECS 202-705-0
Synonyms: Isopropenylbenzene; 2-Phenylpropene; 2-Phenylpropylene
Empirical: C_9H_{10}
Formula: $C_6H_5C(CH_3):CH_2$
Properties: M.w. 118.2; b.p. 164-165 C
Uses: Polymerization monomer, esp. for polyesters; intermediate, plasticizer in prod. of paints
Manuf./Distrib.: Aldrich; AlliedSignal; Aristech; Ashland; Chemcentral; Fluka; Honeywill & Stein; Mitsui Petrochem. Ind.; Mitsui Toatsu; Monomer-Polymer & Dajac Labs; Texaco

α-Methylstyrene/N-(2,2,6,6-tetramethyl piperidinyl-4) maleimide/N-stearyl maleimide terpolymer
EINECS 202-705-0, 241-467-2, 283-117-2 resp.
Toxicology: LD50 (rat) > 6000 mg/kg; nontoxic
Uses: Light stabilizer for polyolefins, ethylene copolymers, PU, surf. coatings
Trade names: Lowilite® 62

Methyl succinate. *See* Dimethyl succinate
Methylsulfonic acid. *See* Methanesulfonic acid
Methyl tertiary butyl ether. *See* Methyl t-butyl ether
Methyl toluene. *See* Xylene
o-Methyltoluene. *See* o-Xylene

1-Methyl-3,5,7-triaza-1-azoniatricyclo-[3.3.1.1]decane chloride
Uses: Bactericide for preservation of aq. industrial prods. incl. latex emulsions (acrylic, vinyl acetate, styrene, etc.), water-thinned paints, adhesives
Trade names: Busan® 1024

Methyltriethoxysilane
CAS 2031-67-6; EINECS 217-983-9
Synonyms: Triethoxymethylsilane
Empirical: $C_7H_{18}O_3Si$
Formula: $CH_3Si(OCH_2CH_3)_3$
Properties: Liq.; m.w. 178.34; dens. 0.90; b.p. 141-143 C; flash pt. 38 C; ref. index 1.383
Precaution: Flamm.
Toxicology: Mildly toxic by ingestion, inhalation; skin, eye, and respiratory tract irritant
Uses: Coupling agent, release agent, lubricant, blocking agent, chemical intermediate; agent providing durability, gloss, hiding power to coatings
Manuf./Distrib.: Aldrich; Fluka; Howard Hall; Spectrum Chem. Mfg.
Trade names: Silquest® A-162

Methyltriglycol. *See* Triglycol monomethyl ether

Methyltrimethoxysilane
CAS 1185-55-3; EINECS 214-685-0
Synonyms: Trimethoxymethylsilane
Empirical: $C_4H_{12}O_3Si$
Formula: $CH_3Si(OCH_3)_3$
Properties: Liq.; m.w. 136.25; dens. 0.955; b.p. 102-103 C; flash pt. 8 C; ref. index 1.3696
Precaution: Highly flamm.; keep away from ignition sources
Toxicology: LD50 (oral, rat) 12,500 mg/kg; mildly toxic by ingestion; skin and eye irritant; heated to decomp., emits acrid smoke and fumes
Uses: Coupling agent, release agent, lubricant, blocking agent, chemical intermediate; pigment and filler treatment in solv. and water-based systems; primer; reduces pigment separation and floating; improves disp.; also promotes adhesion
Manuf./Distrib.: Aldrich; Fluka; Sigma
Trade names: Dow Corning® Z-6070; Silquest® A-163

Methyltrimethylolmethane. *See* Trimethylolethane
Methyl vinyl ether/maleic anhydride copolymer. *See* PVM/MA copolymer
MHPC. *See* Hydroxypropyl methylcellulose
MIAK. *See* Methyl isoamyl ketone

MIBC. *See* Methyl amyl alcohol
MIBK(INCI). *See* Methyl isobutyl ketone

Mica

CAS 12001-26-2
Synonyms: Muscovite mica; CI 77019
Classification: Silicate minerals
Properties: Colorless to sl. red, brown to greenish-yel. soft, translucent solid; dens. 2.6-3.2; Mohs hardness 2.8-3.2; heat resistant to 600 C
Precaution: Noncombustible
Toxicology: TLV:TWA 3 mg/m^3 (respirable dust); irritant by inhalation; nontoxic to skin
Uses: Filler/extender for plastics, rubber, coatings, and pearlescent pigment applics.; binder and reinforcement in lipsticks
Regulatory: FDA 21CFR §73.1496, 73.2496, 175.300, 177.1460, 177.2410, 177.2600; permanently listed
Manuf./Distrib.: 20 Microns Ltd.; Feldspar; Franklin Ind. Mins.; C.P. Hall; J.M. Huber; ISP Van Dyk; KMG Mins.; Lomas Int'l.; D.N. Lukens; Mearl; Mykroy/Macalex Ceramics; Nyco Mins.; Punda Mercantile; H.M. Royal; Spectrum Chem. Mfg.; Tamms Ind.; Van Waters & Rogers; Whittaker, Clark & Daniels; Jesse S. Young
Trade names: AlbaFlex 25; AlbaFlex 50; AlbaFlex 100; AlbaFlex 200; AlbaShield 15; AlbaShield 25; AlbaShield 50; AlbaShield 1000; AlbaShield 2000; C-500; C-1000; C-3000; C-4000; Huber WG-1; Huber WG-2; Micawhite 200; PM-325; Polymica 200; Polymica 325; Polymica 400; Polymica 3105; WG-160; WG-325
Trade names containing: Afflair® Lustre Pigments; Mearlin®; Mearlin® Aztec Gold; Mearlin® Card Gold; Mearlin® Card Silver; Mearlin® Dynacolor® BG; Mearlin® Dynacolor® BP; Mearlin® Dynacolor® BY-B; Mearlin® Dynacolor® GB; Mearlin® Dynacolor® GP; Mearlin® Dynacolor® GY; Mearlin® Dynacolor® RB; Mearlin® Dynacolor® VP; Mearlin® Hi-Lite Super Blue; Mearlin® Hi-Lite Super Gold; Mearlin® Hi-Lite Super Green; Mearlin® Hi-Lite Super Orange; Mearlin® Hi-Lite Super Red; Mearlin® Hi-Lite Super Violet; Mearlin® Inca Gold; Mearlin® MagnaPearl 1000; Mearlin® MagnaPearl 2000; Mearlin® MagnaPearl 3000; Mearlin® MagnaPearl 3100; Mearlin® MagnaPearl 4000; Mearlin® MagnaPearl 5000; Mearlin® Majestic Gold; Mearlin® Mayan Gold; Mearlin® Nu-Antique Bronze; Mearlin® Nu-Antique Copper; Mearlin® Nu-Antique Gold; Mearlin® Nu-Antique Silver; Mearlin® Pearl White; Mearlin® Satin White; Mearlin® Silk White; Mearlin® Sparkle Gold; Mearlin® Sparkle; Mearlin® Sunset Gold; Mearlin® Super Blue-Russet; Mearlin® Super Brass; Mearlin® Super Bronze; Mearlin® Super Copper; Mearlin® Super Red-Russet; Mearlin® Super Russet; Mearlin® Supersparkle

Microcrystalline cellulose

CAS 9004-34-6
Synonyms: MCC; Cellulose gel
Definition: Isolated, colloidal crystalline portion of cellulose fibers; partially depolymerized acid hydrolysis prod. of purified wood cellulose
Properties: Wh. fine cryst. powd., odorless; insol. in water, dil. acids, and most org. solvs.; pH 5-7
Toxicology: LD50 (oral, rat) > 5 g/kg, no significant hazard; irritant by inhalation (dust); may be damaging to lungs
Uses: Electrical equip., vacuum tubes, incandescent lamps, dusting agent, lubricant, windows in high-temp. equip.; filler in exterior paints, cosmetics, roofing, rubber
Regulatory: FDA GRAS; Europe listed; UK approved; USP/NF, BP, Ph.Eur., JP compliance
Manuf./Distrib.: Aldrich; Alfa; Asahi Chem. Ind.; Ashland; Barrington; Chemisphere; Fabrichem; Fluka; FMC; Harrisons Trading; Howard Hall; Int'l. Chem. Inc.; Mendell; Schweizerhall; Sigma
See also Cellulose

Microcrystalline wax

CAS 8063-08-9; 63231-60-7; 64742-42-3; EINECS 264-038-1
Synonyms: Petroleum wax, microcrystalline; Waxes, microcrystalline
Definition: Wax derived from petroleum and char. by fineness of crystals; consists of high m.w. saturated aliphatic hydrocarbons
Properties: Wh. or cream-colored waxy solid, odorless; sol. in chloroform, ether, volatile oils, most warm fixed oils; insol. in water; very sl. sol. in dehydrated alcohol; m.p. 54-102 C
Toxicology: May be carcinogenic
Uses: Wax used in hot-melt coatings and adhesives, paper coatings, printing inks, plastic modification (as lubricant and processing aid), lacquers, paints, and varnishes, as binder in ceramics, for elec. potting
Regulatory: FDA 21CFR §172.886, 173.340, 175.105, 175.320, 176.170, 176.200, 177.2600; Europe listed; UK approved for restricted use; FDA approved for orals, topicals; USP/NF compliance
Manuf./Distrib.: Astor Corp.; Barco Chem. Prods.; Biwax; Dussek Campbell Inc; Ferro; IGI; Joy; Koster Keunen; Mobil; Sea-Land; Shamrock Tech.; Shell; Witco/Oleo-Surf.; Witco/Petroleum Spec.

Microcrystalline wax, oxidized

Trade names: Be Square® 175; Be Square® 185; Be Square® 195; Ecco Wax 010-S; Forbest MW 23; Fortex®; Mekon® White; Michem® Emulsion 01546; Michem® Emulsion 48040; Michem Lube® 124; Microlube™ C; Multiwax® 180-M; Paracol® 404C; Petrolite® C-700; Petrolite® C-1035; Ross No. 165; Ross No. 170; Ross No. 170 Amber; Ross No. 170 Black; Ross No. 190; Ross No. 214; Ross No. 669; Ross No. 863; Ross No. 916; Ross No. 1135/15W; Ross No. 1149/14; Ross No. 1160/14; Ross No. 1251/7; Ross No. 1275ML; Ross No. 1275W; Ross No. 1275WH; Ross No. 1329/1; Ross No. 1365; Ross No. 1385; Ross No. 2305; Starwax® 100

Trade names containing: Michem® Emulsion 01250; Michem® Emulsion 41540; Michem® Experimental Emulsion 70250; Michem Lube® 118; Michem Lube® 162; Michem Lube® 188; Michem Lube® 209; Petrolite® 29 Disp.; Petrolite® 37 Disp.; Ross Japan Wax Substitute 525; Shellwax

See also Petroleum wax

Microcrystalline wax, oxidized

Uses: Wax used in the formulation of emulsions, polishes, and coatings; modifier in solv. polish systems; carnauba substitute; lubricant, process aid, slip and antiblock agent in plastics

Trade names: Cardis® 314; Cardis® 319; Cardis® 320; Cardis® 370

Milk acid. *See* Lactic acid

Milk of magnesia. *See* Magnesium hydroxide

Milk protein

CAS 9000-71-9; EINECS 232-555-1

Synonyms: Casein

Definition: Mixture of proteins obtained from cow's milk

Properties: Light-yel. powd.

Uses: Cheesemaking, plastic items, paper coatings, water-dispersed paints, adhesives, textile sizing, foods and feeds, textile fibers, dietetic preparations, binder in foundry sands

Manuf./Distrib.: Aldrich; Fluka; Meggle Marketing GmbH; Nat'l. Casein; Sigma; U.S. Biochemical; Worthington Biochemical

Milk protein, casein. *See* Casein

Mineral carbon. *See* Graphite

Mineral oil

CAS 8012-95-1; 8020-83-5 (wh.); 8042-47-5; EINECS 232-384-2; 232-455-8

Synonyms: Heavy mineral oil; Light mineral oil; White mineral oil; Paraffin oil; Liquid paraffin; Petrolatum liquid; Liquid petrolatum

Definition: Liq. mixture of hydrocarbons obtained from petroleum

Properties: Colorless transparent oily liq., odorless, tasteless; insol. in water, alcohol; sol. in benzene, chloroform, ether, petrol. ether, volatile oils; dens. 0.83-0.86 (light), 0.875-0.905 (heavy); flash pt. (OC) 444 F; surf. tens. < 35 dynes/cm

Precaution: Combustible

Toxicology: Eye irritant; human carcinogen and teratogen by inhalation; heated to decomp., emits acrid smoke and fumes

Uses: Cathartic; laxative; protectant; lubricant; binder; carrier; mold release for foods, coating for fruits and vegetables, food pkg. materials; plasticizer, lubricant for plastics; in cosmetics, pharmaceuticals, plastics, agric., paper, textiles, etc.

Regulatory: FDA 21CFR §172.878, 173.340 (limitation 0.008% in wash water for sliced potatoes, 150 ppm in yeast), 175.105, 175.210, 175.230, 175.300, 176.170, 176.200, 176.210, 177.1200, 177.2260, 177.2600, 177.2800, 178.3570, 178.3620, 178.3740, 178.3910, 179.45; 573.680; ADI not specified (FAO/WHO); FDA approved for ophthalmics, orals, topicals; USP/NF, BP, Ph.Eur. compliance

Manuf./Distrib.: Air-Scent Int'l.; Aldrich; Amoco/Lubricants; Chemisphere; Coyne; Exxon; Fluka; Magie Bros. Oil; Mobil; Penreco; Penta Mfg.; Ruger; San Yuan; Sea-Land; Sigma; Spectrum Chem. Mfg.; Surco Prods.; Total Petrol.; Witco/Golden Bear; Witco/Petroleum Spec.

Trade names: TEGO® Foamex KS 6

Trade names containing: Amerchol L-101®; BSWL 201; Drewplus® L-108; Drewplus® L-123; Drewplus® L-131; Drewplus® L-139; Drewplus® L-175; Drewplus® L-191; Drewplus® L-198; Drewplus® L-435; Drewplus® L-467; Drewplus® L-468; Drewplus® L-475; Drewplus® L-477; Drewplus® L-493; Drewplus® L-496; Drewplus® L-593; Drewplus® L-790; Drewplus® Y-166; Drewplus® Y-250; Drewplus® Y-281; Drewplus® Y-601; Faktogel® Badenia T; Foamaster® TCX-Special; Forbest 2000C; TEGO® Foamex KS 10

Mineral oil sulfonic acids, sodium salts. *See* Sodium petroleum sulfonate

Mineral red. *See* Lead oxide, red
Mineral soap. *See* Bentonite

Mineral spirits
CAS 8032-32-4; 64475-85-0; EINECS 232-453-7
Synonyms: White spirits; Ligroin; Petroleum spirits
Definition: Mixture of hydrocarbons from petroleum with distillation range of 300-415 F; avail. in type I (reg.), II (high flash), III (odorless), IV (low dry pt.
Properties: Clear colorless, volatile, nonfluorescent liq.; dens. @ 15.6/15.6 C: 0.654-0.820 (I), 0.768-0.820 (II), 0.775 max. (III), 0.754-0.800 (IV); i.b.p. 149 C (I), 177 C (II), 149 C (III), 149 C (IV); flash pt. 38 C min. except 60 C (II)
Precaution: Combustible
Toxicology: TLV:TWA 300 ppm; STEL 400 ppm; mod. toxic in humans; mildly toxic by inh., IP routes; defatting on skin contact; vapor inh. can cause headache, dizziness, unconsciousness with high concs.; heated to dec., emits acrid smoke
Uses: Solvent; paint thinner; in coatings, drycleaning
Regulatory: FDA 21CFR §178.3800
Manuf./Distrib.: Aldrich; Ashland; Texaco
Trade names: Shell Mineral Spirits 135; Shell Mineral Spirits 145-EC; Texsolve S; Texsolve S-66; Texsolve S-LO
Trade names containing: Aroplaz 111-M-70; Aroplaz 6056-MX-90; Beckosol® 10-027; Beckosol® 10-539; Beckosol® 11-035; Blown Menhaden Fish Oil M&M; Blown Menhaden Fish Oil M80; Blown Menhaden Fish Oil Z-9; Dehydran® 1208; Disparlon® 1950; Disparlon® 1970; Disparlon® A670-20M; Disparlon® OX-70; Epi-Tex® 199; Flat-Ayd® FA-3B; Flat-Ayd® FA-HSP-2; Flat-Ayd® FA-NCO-6; Formulator; G-4067-M-50; G-4412-M-50; G-4412-M-50HV; G-4594-OMS-35; G-4617-M-35; G-4635-M-60; G-4689-HM-30; G-4696-M-50; G-4699-V-50; GP-210 Silicone Antifoam Emulsion; GP-RA-158 Silicone Polish Additive; GP-RA-159 Silicone Polish Additive; GPRI™ CKSB-2001; Kelpol 835-M-50; Keltrol 1001-M-60; Kelvar 3758-M-85; K-Sperse® 131; K-Sperse® 152MS; Monawet MO-70S; SACI® 100a; SACI® 300; SACI® 300a; SACI® 350; SACI® 350a; SACI® 500; SACI® 500a; SACI® 552a; SACI® 560; SACI® 560a; SACI® 570; SACI® 700; SACI® 760; SACI® 2452; SACI® 8100; SACI® 8200; Sobral AD-640; Sobral P470; Suspend-Ayd® 1; Tint-Ayd® AL Series; Uni-Rez® 1042; Wallkyd 11-024

Mineral wax. *See* Ceresin; Ozokerite
Minium. *See* Lead oxide, red

Molybdenum
CAS 7439-98-7; EINECS 231-107-2
Classification: Metallic element
Formula: Mo
Properties: Blk. powd. with metallic luster; m.w. 95.94; dens. 10.28; m.p. 2610 C; b.p. 4800 C; chemically inactive; acid resist.
Toxicology: Low toxicity
Uses: Alloying agent in steels and cast iron; pigments for printing inks, paints, ceramics; catalyst; solid lubricants; missile and aircraft parts; reactor vessels; cermets; die-casting copper-base alloys; special batteries
Manuf./Distrib.: AAA Molybdenum Prods.; Aldrich; Alfa Aesar Johnson Matthey; Atlantic Equip. Engrs.; Atomergic Chemetals; Cerac; Climax Molybdenum; Fluka; Noah; Powmet; Sigma

Molybdenum anhydride. *See* Molybdenum trioxide

Molybdenum trioxide
CAS 1313-27-5; EINECS 215-204-7
Synonyms: Molybdenum anhydride; Molybdic oxide; Molybdic acid anhydride; Molybdic acid hydride
Empirical: MoO_3
Properties: Wh. or yel. powd.; sl. sol. in water; sol. in conc. mixture of nitric acid and HCl; m.w. 143.94; dens. 4.69; m.p. 795 C; begins to sublime at 700 C; b.p. 1150 C
Toxicology: Toxic material; poison by ingestion; TLV (as Mo) 5 mg/m^3 of air
Uses: Source of Mo; reagent for analytical chemistry; agriculture; mfg. of metallic Mo; corrosive inhibitor; ceramic glazes; enamels; pigments; catalyst; smoke suppressant, flame retardant for plastics
Manuf./Distrib.: AAA Molybdenum Prods.; AC Ind.; Advance Research Chems.; Aldrich; All Chemie Ltd; Atlantic Equip. Engrs.; Atomergic Chemetals; Carbochem; Cerac; Climax Molybdenum; Fluka; Noah; Sigma; Spectrum Chem. Mfg.

Molybdic acid anhydride. *See* Molybdenum trioxide
Molybdic acid hydride. *See* Molybdenum trioxide
Molybdic oxide. *See* Molybdenum trioxide
Monochlorobenzene. *See* Chlorobenzene

Monochlorotoluene
CAS 25168-05-2
Uses: Solvent, extender, diluent for dye carrier, fuel oil additive, sludge solv., component of paint thinners and strippers, metal parts cleaners, adhesives
Trade names: Halso® 99
Trade names containing: Oxsol® 73; Oxsol® 253; Oxsol® 325; Oxsol® 550

Monocobalt oxide. *See* Cobalt oxide (ous)
Mono- and diglycerides citrates. *See* Citric acid esters of mono- and diglycerides of fatty acids
Monoethanolamine. *See* Ethanolamine

Monoethyl maleate
CAS 3990-03-2
Formula: $CH_3CH_3OOCCHCHCOOH$
Uses: Used in films and coatings with improved stiffness and adhesion
Trade names: Sipomer® MEM

Monoglyme. *See* Ethylene glycol dimethyl ether

Monoisopropanolamine
Uses: In soaps, shampoos, emulsifiers, textile specialties, agric. and polymer curing chems., adhesives, coatings, metalworking, petrol, rubber processing, gas conditioning chems.
Trade names: MIPA

Monomethoxy tripropylene glycol monoacrylate. *See* Methoxy tripropylene glycol acrylate
Monomethylol dimethyl hydantoin. *See* MDM hydantoin
Monoolein. *See* Glyceryl oleate
Monopentaerythritol. *See* Pentaerythritol
Monopropylene glycol methyl ether. *See* Methoxyisopropanol
Monoricinolein. *See* Glyceryl ricinoleate
Monosodium carbonate. *See* Sodium bicarbonate
Monosodium gluconate. *See* Sodium gluconate
Monosodium hydroxymethane sulfinate. *See* Sodium formaldehyde sulfoxylate
Monosodium D(-)-pentahydroxy capronate. *See* Sodium gluconate
Monostearin. *See* Glyceryl stearate
Monothioethyleneglycol. *See* 2-Mercaptoethanol
Monothioglycerol. *See* Thioglycerin
α-Monothioglycerol. *See* Thioglycerin

Montan wax
CAS 8002-53-7; EINECS 232-313-5
Synonyms: Lignite wax; Waxes, montan
Definition: Wax obtained by extraction of lignite
Properties: Dark brown lumps or white hard earth wax; sol. in CCl_4, benzene, chloroform, hot petrol. ether; insol. in water; m.p. 80-90 C; sapon. no. 88-112
Precaution: Combustible
Uses: Substitute for carnauba and beeswax; shoe and furniture polishes; waterproof and roofing paints; adhesive pastes; candles; paper sizing compds.; wire coatings; lubricant for plastics
Regulatory: FDA 21CFR §175.105, 176.210, 177.2600
Manuf./Distrib.: Hoechst Celanese; Frank B. Ross; Stevenson Cooper; Strahl & Pitsch

Montmorillonite
CAS 1318-93-0; EINECS 215-288-5
Classification: Complex silicate clay mineral
Formula: $Al_2O_5 \cdot 4SiO_2 \cdot 4H_2O$
Properties: Lt. yel. or green, cream, pink, gray to black; insol. in water and common org. solvs.
Toxicology: Poison by intravenous route
Uses: Major component of bentonite and Fuller's earth; thickener and rheological modifier in inks, paints,

sealants, cosmetics; thickener for lithographic, letterpress, metal deco and uv-cured inks
Manuf./Distrib.: Aldrich; Fluka
Trade names: Bentone® 500; Bentone® SD-1; Bentone® SD-2; Claytone® 40; Claytone® AF; Claytone® HY; Claytone® TG; Gelwhite® GP; Gelwhite® L; Mineral Colloid BP; Mineral Colloid MO; Perchem® 97; Suspengel 200; Suspengel 325; Suspengel Plus 200; Suspengel Plus 325

Morpholinium, 4-ethyl-4-hexadecyl, ethyl sulfate. *See* Cetethyl morpholinium ethosulfate
Mossbunker oil. *See* Menhaden oil
MPA. *See* Propylene glycol methyl ether acetate
3-MPA. *See* Methoxypropylamine
mPDA. *See* m-Phenylenediamine
MPK. *See* Methyl propyl ketone
MPMD. *See* 2-Methylpentamethylenediamine
MTBE. *See* Methyl t-butyl ether
MTC. *See* Methylenebis (thiocyanate)
Muscovite mica. *See* Mica
Mylase 100. *See* Amylase

Myristyl lignocerate
CAS 42233-51-2
Synonyms: Lignoceric acid, myristyl ether; Tetracosanoic acid, tetradecyl ester; Tetradecyl tetracosanoate
Definition: Ester of myristyl alcohol and lignoceric acid
Empirical: $C_{38}H_{76}O_2$
Formula: $CH_3(CH_2)_{22}COOCH_2(CH_2)_{12}CH_3$
Trade names containing: Ross Synthetic Candelilla Wax

2MZ. *See* 2-Methyl imidazole

Nabam
CAS 142-59-6
Synonyms: DSE; Disodium ethylene bisdithiocarbamate; 1,2-Ethanediylbiscarbamodithioic acid disodium salt; Ethylenebis (dithiocarbamate) disodium salt; Disodium ethylene-1,2-bisdithiocarbamate
Empirical: $C_4H_6N_2S_4 \cdot 2Na$
Formula: $NaSSCNHCH_2CH_2NHCSSNa$
Properties: Colorless crystals; sol. in water; m.w. 256.34
Toxicology: LD50 (oral, rat) 395 mg/kg; poison by ingestion; moderately toxic by IP route; skin irritant; experimental teratogen, reproductive effects; mutagenic data; heated to decomp., emits toxic fumes of Na_2O, NO_x, SO_x
Uses: Plant fungicide; ingredient for pesticides; industrial applics.
Regulatory: FDA 21CFR §173.320
Manuf./Distrib.: Alco

NaDBS. *See* Sodium dodecylbenzenesulfonate
NaMBT. *See* Sodium 2-mercaptobenzothiazole

Naphtha
CAS 8030-30-6; 68920-06-9; 64742-95-6 (lt. aromatic)
Synonyms: Coal tar naphtha; Benzin; Benzine; Petroleum naphtha; VM&P naphtha; Petroleum benzin; Petroleum ether; Petroleum spirit
Definition: Petroleum distillate
Properties: Dark straw-colored to colorless liq.; sol. in benzene, toluene, xylene; dens. 0.862-0.892; b.p. 149-216 C; flash pt. (CC) 107 F
Precaution: Flamm. when exposed to heat or flame; sl. explosion hazard; can react with oxidizing materials; keep containers tightly closed
Toxicology: Mildly toxic by inhalation; human poison and systemic effects by IV route; irritating to eyes, skin, respiratory tract; vapor inh. may cause nausea, dizziness, unconsiousness; common air contaminant
Uses: Solv., thinner in paint, drycleaning fluid, rubber compounding, sealants, chem. absorption; blending with natural gas
Regulatory: FDA 21CFR §73.1, 172.250
Manuf./Distrib.: Ashland; J.T. Baker; Baychem; R.E. Carroll; Chemcentral; Crowley Chem.; Exxon; Harcros;

Kerr-McGee; Mobil; Monsanto; Norsk Hydro AS; Phillips; Samson; Shell; Sunnyside; Texaco; Van Waters & Rogers

Trade names: Aromatic 150; Shell Sol B; Shell Super VM&P Naphtha EC; Shell Tolu-Sol 5

Trade names containing: Acryloid® B-67; Acryloid® NAD-10; Aroplaz 310-V-50; Disparlon® AP-20; Disparlon® AP-30; Disparlon® AQ-500; Disparlon® LC-900; Disparlon® LC-915; Disparlon® LC-955; G-4412-V-50; G-4699-V-50; G-4732-VBT-50; Sobral 1341; Styresol 13-040

Naphthalane. *See* Decahydronaphthalene
Naphthalenesulfonic acid, bis-(1-methylethyl)-, sodium salt. *See* Sodium isopropyl naphthalene sulfonate
Naphthalenesulfonic acid, polymer with formaldehyde, sodium salt. *See* Sodium polynaphthalene sulfonate
Naphthalene-1,2,3,4-tetrahydride. *See* Tetrahydronaphthalene
1-Naphthalenol. *See* 1-Naphthol

Naphtha, light aromatic

CAS 64742-95-6

Synonyms: Light aromatic petroleum naphtha

Trade names containing: BYK®-S 706; Ircogel® 905; Ircogel® 906; Ircogel® 907

Naphtha safety solvent. *See* Stoddard solvent

Naphthenic acid

CAS 1338-24-5; EINECS 215-662-8

Classification: Cyclopentane carboxylic acid

Properties: Cryst., odorless; sl. sol. in water; m.w. 235-280; dens. 1.034; m.p. 31 C; b.p. 233 C

Toxicology: LD50 (oral, rat) 3000 mg/kg, (IP, rat) 640 mg/kg; moderately toxic by ingestion, IP routes; heated to decomp., emits acrid smoke and irritating fumes

Uses: Paint dryers, fungicides, metal catalysts, corrosion inhibitors, lubricants, fracturing fluids, cellulose preservatives, solvents, detergents, rubber reclaiming agent

Manuf./Distrib.: Crowley Chem.; Esprit; Fabrichem; Fluka; OM Group; Orange Chems. Ltd

Naphthenic acid cobalt salt. *See* Cobalt naphthenate
Naphthenic acid copper salt. *See* Copper naphthenate
Naphthenic acid lead salt. *See* Lead naphthenate
Naphthenic acid zinc salt. *See* Zinc naphthenate

1-Naphthol

CAS 90-15-3; EINECS 201-969-4

Synonyms: α-Naphthol; 1-Naphthalenol; 1-Hydroxynaphthalene; α-Hydroxynaphthalene

Classification: Polycyclic phenol

Empirical: $C_{10}H_8O$

Formula: $C_{10}H_7OH$

Properties: Colorless or yel. prisms or powd., disagreeable taste, phenolic odor; sublimable; sol. in benzene, alcohol, ether; insol. in water; m.w. 144.16; dens. 1.224 (4 C); m.p. 96 C; b.p. 288 C; ref. index 1.6206 (98.7 C)

Precaution: Combustible exposed to heat or flame; light-sensitive; heated to decomp., emits acrid smoke and irritating fumes

Toxicology: LD50 (oral, rat) 2400 mg/kg, (skin, rabbit) 880 mg/kg; mod. toxic by ingestion and skin absorption; experimental teratogen, reproductive effects; severe eye and skin irritant; mutagenic data; ingestion of lg. amts. can be fatal

Uses: Mfg. of intermediates, paints, dyes, synthetic perfumes; also in microscopy

Manuf./Distrib.: Aceto; Aldrich; Fluka; Hoechst Celanese; Jarchem Ind.; Monomer-Polymer & Dajac Labs; San Yuan; Sigma; Spectrum Chem. Mfg.

α-Naphthol. *See* 1-Naphthol
Native calcium sulfate. *See* Calcium sulfate dihydrate

Natural rubber

CAS 9006-04-6

Synonyms: NR

Uses: Used for adhesive, dipping, coating, foam, and molded materials, latex applics., foamed rubber, textile, medical, cement, asphalt; processing aid for blending with other rubber

Manuf./Distrib.: Firestone Syn. Rubber; General Latex & Chem.; Hardman; A. Schulman

Trade names: Denflex 3060; Denflex 3062; Heveanol; NC 405; Unitex; Vultex

NBR. *See* Acrylonitrile-butadiene rubber

Neodecanoic acid
CAS 26896-20-8
Definition: A mixt. of isomeric 10-carbon saturated monocarboxylic acids
Empirical: $C_{10}H_{20}O_2$
Properties: Sol. 0.02 g/10 ml water; m.w. 172.30; sp.gr. 0.915 (20/20 C); b.p. 250-257 C; flash pt. (TCC) 221 F
Toxicology: LD50 (oral, mammal) 3400 mg/kg; may cause mod. to marked eye and skin irritation; mod. toxic by ingestion; heated to decomp., emits acrid smoke and irritating fumes
Uses: Intermediate for prod. of derivs. used in coatings, cosmetics, agric., pharmaceuticals, etc.
Manuf./Distrib.: Biddle Sawyer; Exxon

Neodecanoic acid glycidyl ester. *See* Glycidyl neodecanoate

Neoheptanoic acid
Properties: Negligible sol. in water; sp.gr. 0.930 (20/20 C); b.p. 207-210 C; flash pt. (TCC) 206 F
Toxicology: May cause mod. to marked eye and mod. skin irritation
Uses: Intermediate for prod. of derivs. used in coatings, cosmetics, agric., pharmaceuticals, etc.
Manuf./Distrib.: Exxon

Neopentanoic acid
CAS 75-98-9
Synonyms: Pivalic acid; 2,2-Dimethylpropanoic acid; α,α-Dimethylpropionic acid; 2,2-Dimethylpropionic acid; t-Pentanoic acid; Trimethylacetic acid; Propanoic acid
Properties: Cryst.; very sol. in alcohol, ether; sol. 2.10 g/10 ml water; m.w. 102.15; sp.gr. 0.913 (38/38 C); m.p. 35.5 C; b.p. 163-165 C; flash pt. (TCC) 145 F
Precaution: Combustible
Toxicology: LD50 (oral, rat) 900 mg/kg, (skin, rat) 190 mg/kg; mod. toxic by ingestion and skin contact; may cause mod. eye and severe skin irritation; experimental tumorigen; heated to decomp., emits acrid smoke, irritating fumes
Uses: Intermediate for prod. of derivs. used in coatings, cosmetics, agric., pharmaceuticals, etc.
Manuf./Distrib.: Exxon

Neopentyl (diallyl) oxy, triacryl zirconate
Trade names containing: Ken-React® NZ 39

Neopentyl (diallyl) oxy, tri (m-amino) phenyl zirconate
Uses: Coupling agent
Trade names containing: Ken-React® NZ 97

Neopentyl (diallyl) oxy, tri (dioctyl) phosphato zirconate
Uses: Coupling agent
Trade names containing: Ken-React® NZ 12

Neopentyl (diallyl) oxy, tri (dioctyl) pyrophosphato titanate
CAS 103432-54-8
Synonyms: Titanium IV neoalkanolato, tris(diisooctyl) pyrophosphato-o

Neopentyl (diallyl) oxy, tri (dioctyl) pyrophosphato zirconate
Uses: Coupling agent
Trade names containing: Ken-React® NZ 38

Neopentyl (diallyl) oxy, tri (dodecyl) benzene-sulfonyl zirconate
Uses: Coupling agent
Trade names containing: Ken-React® NZ 09

Neopentyl (diallyl) oxy, tri (N-ethylenediamino) ethyl zirconate
Uses: Coupling agent
Trade names containing: Ken-React® NZ 44

Neopentyl (diallyl) oxy, trimethacryl zirconate
Uses: Coupling agent

Neopentyl (diallyl) oxy, trineodecanoyl zirconate

Trade names containing: Ken-React® NZ 33

Neopentyl (diallyl) oxy, trineodecanoyl zirconate
Uses: Coupling agent
Trade names containing: Ken-React® NZ 01

Neopentylene glycol. *See* Neopentyl glycol

Neopentyl glycol
CAS 126-30-7; EINECS 204-781-0
Synonyms: NPG; 2,2-Dimethyl-1,3-propanediol; Neopentylene glycol; Dimethyl trimethylene glycol; Dimethylolpropane
Empirical: $C_5H_{12}O_2$
Formula: $HOCH_2C(CH_3)_2CH_2OH$
Properties: Wh. cryst.; misc. with alcohol and ether; partly sol. in water; m.w. 104.15; dens. 1.066 (25/4 C); m.p. 127 C; b.p. 210 C; flash pt. (TCC) 109 C
Precaution: Combustible exposed to heat or flame; can react with oxidizers
Toxicology: LDLo (oral, rat) 3200 mg/kg; moderately toxic by ingestion; heated to decomp., emits acrid smoke and irritating fumes
Uses: Resin intermediate; intermediate for paints; insect repellent; polyester-based plasticizer, lubricant
Manuf./Distrib.: Aldrich; Allchem Ind.; BASF; Browning; Chemical; Eastman; Fluka; Hüls Am.; Hüls AG; Int'l. Chem. Inc.; Mitsubishi Gas; Perstorp Polyols
Trade names: NPG® Glycol

Neopentyl glycol diacrylate
CAS 2223-82-7
Synonyms: 2,2-Dimethyltrimethylene acrylate; 2-Propenoic acid-2,2-dimethyl-1,3-propanediyl ester; Dimethylolpropane diacrylate; 2,2-Dimethyl-1,3-propanediol diacrylate; 2,2-Dimethyltrimethylene ester acrylic acid
Empirical: $C_{11}H_{16}O_4$
Properties: M.w. 212.27
Toxicology: LD50 (oral, rat) 6730 mg/kg, (skin, rabbit) 400 mg/kg; poison by skin contact; mildly toxic by ingestion; severe skin irritant; heated to decomp., emits acrid smoke and irritating fumes
Uses: Curing agent; diluent for UV-cured systems (coatings, printing inks, etc.)
Manuf./Distrib.: Aldrich; Monomer-Polymer & Dajac Labs
Trade names containing: SR-247

Neopentyl glycol dibenzoate
CAS 4196-89-8
Synonyms: 1,3-Propanediol-2,2-dimethyl dibenzoate; 2,2-Dimethyl-1,3-propanediol dibenzoate
Empirical: $C_{19}H_{20}O_4$
Properties: Liq.; color APHA 200; m.w. 312; m.p. 49 C
Uses: Process aid, modifier, plasticizer for thermoplastics, hot-melt adhesives, coatings
Manuf./Distrib.: Velsicol
Trade names: Benzoflex® 312; Benzoflex® S-312

Neopentyl glycol diglycidyl ether
CAS 17557-23-2
Synonyms: 1,3-Bis(2,3-epoxypropoxy)-2,2-dimethyl propane; 2,2´-((2,2-Dimethyl-1,3-propanediyl)bis(oxymethylene))bisoxirane; Diglycidyl ether of neopentyl glycol
Empirical: $C_{11}H_{20}O_4$
Properties: M.w. 216.31
Toxicology: Experimental tumorigen; heated to decomp., emits acrid smoke and irritating fumes
Uses: Reactive epoxy diluent for civil engineering applics.; increases impregnation of resin systems, and level of filler loading; used in elec., laminating, casting, tooling, flooring, and coatings
Manuf./Distrib.: Aldrich
Trade names: Heloxy® 68

Neopentyl glycol dimethacrylate
Manuf./Distrib.: Monomer-Polymer & Dajac Labs
Trade names containing: SR-248

Neoprene. *See* Polychloroprene

Nepheline syenite
CAS 37244-96-5
Synonyms: Nephylene syenite
Definition: Feldspathoid igneous rock primarily composed of the minerals microcline ($KAlSi_3O_8$), albite ($NaAlSi_3O_6$), and nepheline [$(Na,K)AlSiO_4$]
Properties: Wh. solid; sp.gr. 2.61; m.p. 1223 C
Toxicology: ACGIH TLV 10 mg/m^3; excessive/prolonged inh. of dust may harm respiratory system
Uses: Filler in paints, plastics, coatings, adhesives, caulks, and sealants, inks, rubber, friction prods.
Manuf./Distrib.: Hammill & Gillespie; D.N. Lukens; Unimin Canada; Jesse S. Young
Trade names: Minex® 2; Minex® 3; Minex® 4; Minex® 7; Minex® 10

Nephylene syenite. *See* Nepheline syenite
Nickel chloride. *See* Nickel chloride (ous)
Nickel (II) chloride. *See* Nickel chloride (ous)
Nickel (II) chloride (1:2). *See* Nickel chloride (ous)

Nickel chloride (ous)
CAS 7718-54-9; EINECS 231-743-0
Synonyms: Nickelous chloride; Nickel chloride; Nickel (II) chloride (1:2); Nickel (II) chloride; Nickel dichloride
Empirical: Cl_2Ni
Formula: $NiCl_2$
Properties: Grn. monoclinic cryst. powd.; m.w. 129.62; dens. 3.550
Toxicology: ACGIH TLV:TWA 0.1 mg (Ni)/m^3; LD50 (oral, rat) 105 mg/kg, (IP, mouse) 26 mg/kg; poison by ingestion, IV, IM, IP routes; cancer suspect agent; experimental reproductive effects; mutagenic data; heated to decomp., emits very toxic fumes of Cl^-
Uses: Electroplated nickel coatings, chemical reagent
Manuf./Distrib.: Aldrich; Ashland; Atomergic Chemetals; Elf Atochem N. Am.; Fluka; Mallinckrodt; Nihon Kagaku Sangyo

Nickel dichloride. *See* Nickel chloride (ous)

Nickel octoate
Uses: Drier for paints
Manuf./Distrib.: Archway Sales; Baychem; Boehle; OM Group; Shepherd

Nickelous chloride. *See* Nickel chloride (ous)

Nickel titanate
CAS 12035-39-1
Synonyms: Nickel titanium oxide; Titanium nickel oxide
Empirical: NiO_3Ti
Properties: M.w. 154.61
Toxicology: ACGIH TLV:TWA 0.1 mg(Ni)/m^3; experimental carcinogen and tumorigen
Uses: Pigment for paints
Manuf./Distrib.: Atomergic Chemetals; BASF; Bayer; Ciba-Geigy; Engelhard; Ferro/Color; Great Western; Heucotech Ltd; Ishihara; Landers-Segal Color; Shepherd Color

Nickel titanium oxide. *See* Nickel titanate
Nitrile elastomer. *See* Acrylonitrile-butadiene rubber
Nitrile rubber. *See* Acrylonitrile-butadiene rubber
2,2′,2′′-Nitrilotris(ethanol). *See* Triethanolamine
1,1′,1′′-Nitrilotris-2-propanol. *See* Triisopropanolamine
Nitrocarbol. *See* Nitromethane

Nitrocellulose
CAS 9004-70-0
Synonyms: Cellulose nitrate; Cellulose tetranitrate; Celluloid; Pyroxylin; Nitrocotton; Collodion cotton; Collodion wool; Collodion; Guncotton; Soluble guncotton
Classification: Cellulose deriv.
Empirical: $C_{12}H_{16}O_{18}N_4$
Formula: $C_{12}H_{16}(ONO_2)_4O_6$

Nitrocotton

Properties: Colorless liq. or wh. amorphous solid; sol. in acetone, glac. acetic acid; insol. in water, ether-alcohol mixt.; m.w. 504.3; dens. 1.66; flash pt. 55 F
Precaution: Flamm. solid; highly dangerous exposed to heat, flame, strong oxidizers; ignites easily; explodes
Toxicology: No known toxicity
Uses: Thermoplastic; in lacquers, high explosives, rocket propellant; printing ink base; leather finishing, molded prods., etc.
Regulatory: FDA 21CFR §175.105, 175.300, 176.170, 177.1200, 181.22, 181.30
Manuf./Distrib.: Aarbor Int'l.; Aldrich; Allchem Ind.; Aqualon; Asahi Chem. Ind.; Bayer; Daicel Chem. Ind.; Fluka; Hercules; Punda Mercantile; San Yuan; SNPE Chimie; Unicel; Vanguard Chem. Int'l.
Trade names: ParCell™ A; ParCell™ R; ParCell™ S; RS Nitrocellulose
Trade names containing: Mearlite® GPN; Mearlite® Ultra Bright USD; Mearlite® Ultra Bright USS

Nitrocotton. *See* Nitrocellulose

Nitroethane
CAS 79-24-3; EINECS 201-188-9
Classification: Aliphatic organic compd.; nitroparaffin
Empirical: $C_2H_5NO_2$
Formula: $CH_3CH_2NO_2$
Properties: Oily colorless liq., agreeable odor; sol. in water, acid, alkali; misc. with alcohol, chloroform, ether; m.w. 75.07; dens. 1.048 (20/4 C); m.p. -90 C; b.p. 112-116 C; flash pt. (OC) 106 F; ref. index 1.391 (20 C)
Precaution: DOT: Flamm. liq.; explodes when heated
Toxicology: Poison by intraperitoneal route; moderately toxic by ingestion; irritating to eyes, mucous membranes; TLV 100 ppm in air; LD50 (rat, oral) 1100 mg/kg
Uses: Intermediate for synthesis; stabilizer for chlorinated solvs.; fuel additive; specialty solv. for coating, adhesive, and ink formulations
Regulatory: BP compliance
Manuf./Distrib.: Aldrich; ANGUS; Fluka; Spectrum Chem. Mfg.
Trade names: NE™ 101
Trade names containing: NiPar 640™

2-Nitro-2-ethyl-1,3-propanediol
CAS 597-09-1
Synonyms: Nitro alcohol
Empirical: $C_4H_{13}O_4N$
Formula: $HOCH_2C(CH_2H_5)(NO_2)CH_2OH$
Properties: White crystals; sol. in water, org. solvs.; m.w. 149.14; m.p. 56-65 C
Uses: Chemical intermediate; formaldehyde donor; organic synthesis

Nitromethane
CAS 75-52-5; EINECS 200-876-6
Synonyms: Nitrocarbol
Classification: Nitroparaffin
Empirical: CH_3NO_2
Properties: Colorless oily liq., disagreeable odor; sol. in water, alcohol, ether; m.w. 61.04; dens. 1.139 (20/20 C); m.p. -29 C; b.p. 100-103 C; flash pt. 36 C (112 F); ref. index 1.382 (20 C)
Precaution: DOT: Flamm. liq.; dangerous fire and explosion risk; explosive limit 7.3% in air; reacts violently, may detonate under high temps. and pressures
Toxicology: LD50 (mice, oral) 1.44 g/kg; TLV 100 ppm in air; poison by ingestion, inhalation, intraperitoneal routes; vapor inh. may cause mild respiratory tract irritation; CNS depression occurs in animals
Uses: Stabilizer for chlorinated solvs.; chemical intermediate; solv. for cellulosic compds., polymers, waxes; rocket fuel, gasoline additive; in coatings industry
Regulatory: BP compliance
Manuf./Distrib.: Aldrich; Allchem Ind.; ANGUS; Fluka; Sigma; Spectrum Chem. Mfg.

Nitropropane
CAS 108-03-2; EINECS 203-544-9
Synonyms: 1-Nitropropane

Classification: Nitroparaffin
Empirical: $C_3H_7NO_2$
Formula: $CH_3CH_2CH_2NO_2$
Properties: Colorless liq.; very sl. sol. in water; misc. with alcohol, ether; m.w. 89.09; dens. 1.003 (20/20 C); m.p. -108 C; b.p. 129-133 C; flash pt. (TOC) 93 F; ref. index 1.4018 (20 C)
Precaution: Flamm. liq.; reacts violently
Toxicology: Poison by intraperitoneal route and ingestion; irritating to mucous membranes; TLV 25 ppm in air
Uses: Solv. for inks and coatings, cellulose acetate, vinyl resins, lacquers, synthetic rubbers, fats, oils, dyes; chemical intermediate; rocket propellant; gasoline additive
Manuf./Distrib.: Aldrich; ANGUS; Fluka
Trade names: NiPar S-10™
Trade names containing: NiPar 640™

1-Nitropropane. *See* Nitropropane

2-Nitropropane
CAS 79-46-9; EINECS 201-209-1
Synonyms: sec-Nitropropane
Classification: Nitroparaffin
Empirical: $C_3H_7NO_2$
Formula: $CH_3CH(NO_2)CH_3$
Properties: Colorless liq.; sl. sol. in water; misc. with many org. solvs.; m.w. 89.09; dens. 0.992 (20/20 C); m.p. -93 C; b.p. 119-122 C; flash pt. (TOC) 75 F; ref. index 1.394 (20 C)
Precaution: Flamm. when exposed to heat, open flame, oxidizers; may explode on heating
Toxicology: TLV:CL 25 ppm in air; LD50 (oral, rat) 725 mg/kg; poison by ingestion, inhalation, and intraperitoneal routes (nausea, diarrhea, anorexia); suspected carcinogen; mutagenic
Uses: Solv. for coatings and inks, cellulose acetate, vinyl resins, lacquers, synthetic rubbers, fats, oils, dyes; chemical intermediate; rocket propellant; gasoline additive
Manuf./Distrib.: Aldrich; ANGUS; Ashland; Fluka

sec-Nitropropane. *See* 2-Nitropropane
NMP. *See* N-Methyl-2-pyrrolidone
Nonanedoic acid. *See* Azelaic acid

Nonanoic acid
CAS 112-05-0; EINECS 203-931-2
Synonyms: Pelargonic acid (INCI); Carboxylic acid C_9; Nonylic acid; Nonoic acid
Classification: Acid
Empirical: $C_9H_{18}O_2$
Formula: $CH_3(CH_2)_7COOH$
Properties: Colorless oily liq., char. fatty odor; cryst. when cooled; sol. in alcohol, chloroform, ether; pract. insol. in water; m.w. 158.24; dens. 0.907 (20/4 C); m.p. 10-12 C; b.p. 252-253 C (756 mm); flash pt. 129 C; ref. index 1.433 (20 C)
Precaution: Corrosive
Toxicology: LD50 (IV, mouse) 224 ± 4.6 mg/kg; strong irritant to skin and eyes
Uses: Intermediate for paints; in prod. of hydrotropic salts; lacquers; plastics
Regulatory: FDA 21CFR §172.515, 173.315; FEMA GRAS
Manuf./Distrib.: Aldrich; Exxon; Fluka; Henkel/Emery; Hoechst Celanese; Penta Mfg.; Sigma

3,6,9,12,15,18,21,24,27-Nonaoxanonatriacontan-1-ol. *See* Laureth-9
Noncarbinol. *See* n-Decyl alcohol
Nonoic acid. *See* Nonanoic acid

Nonoxynol-1
CAS 26027-38-3 (generic); 37205-87-1 (generic); 27986-36-3; EINECS 248-762-5
Synonyms: Ethylene glycol nonyl phenyl ether; PEG-1 nonyl phenyl ether; 2-(Nonylphenoxy) ethanol
Classification: Ethoxylated alkyl phenol
Empirical: $C_{17}H_{28}O_2$
Formula: $C_9H_{19}C_6H_4OCH_2CH_2OH$

Nonoxynol-2

Properties: Yel. to almost colorless liq.; sol. in oil
Toxicology: Moderately toxic by ingestion, skin contact; severe eye and mild skin irritant in humans; heated to dec., emits acrid smoke and fumes
Uses: Nonionic surfactant; as detergent, emulsifier, wetting agent, dispersant, stabilizer, defoamer; intermediate in synthesis of anionic surfactants; emulsion polymerization; latex paints
Regulatory: FDA 21CFR §175.105, 176.180
Manuf./Distrib.: Aldrich
Trade names: Alkasurf® CO-210; Prox-onic NP-1.5; Surfonic® N-10; T-Det® N-1.5

Nonoxynol-2
CAS 26027-38-3 (generic); 37205-87-1 (generic); 27176-93-8 (generic); 9016-45-9 (generic); EINECS 248-291-5
Synonyms: PEG-2 nonyl phenyl ether; POE (2) nonyl phenyl ether; PEG 100 nonyl phenyl ether
Classification: Ethoxylated alkyl phenol
Empirical: $C_{19}H_{32}O_3$
Formula: $C_9H_{19}C_6H_4(OCH_2CH_2)_nOH$, avg. n = 2
Properties: Yel. to almost colorless liq.; sol. in oil
Toxicology: Moderately toxic by ingestion, skin contact; severe eye and mild skin irritant in humans; heated to dec., emits acrid smoke and fumes
Uses: Nonionic surfactant; as detergent, emulsifier, wetting agent, dispersant, stabilizer, defoamer; intermediate in synthesis of anionic surfactants; in metalworking, paints
Regulatory: FDA 21CFR §176.105, 176.180, 176.210
Manuf./Distrib.: Aldrich; Fluka; Sigma
Trade names: Chemax NP-1.5; Igepal® CO-210; Teric N2

Nonoxynol-3
CAS 27176-95-0 (generic); 84562-92-5 (generic); 51437-95-7 (generic); 9016-45-9 (generic)
Synonyms: PEG-3 nonyl phenyl ether; POE (3) nonyl phenyl ether; 2-[2-[2-(Nonylphenoxy)ethoxy]ethoxy]ethanol
Classification: Ethoxylated alkyl phenol
Empirical: $C_{21}H_{36}O_4$
Formula: $C_9H_{19}C_6H_4(OCH_2CH_2)_nOH$, avg. n = 3
Properties: Yel. to almost colorless liq.; sol. in oil
Toxicology: Moderately toxic by ingestion, skin contact; severe eye and mild skin irritant in humans; heated to dec., emits acrid smoke and fumes
Uses: Nonionic surfactant; as detergent, emulsifier, wetting agent, dispersant, stabilizer, defoamer; intermediate in synthesis of anionic surfactants; emulsion polymerization; latex paints
Regulatory: FDA 21CFR §175.105, 176.180, 176.210
Manuf./Distrib.: Fluka; Sigma
Trade names: Surfonic® N-31.5; Teric N3

Nonoxynol-4
CAS 7311-27-5; 9016-45-9 (generic); 26027-38-3 (generic); 37205-87-1 (generic); 27176-97-2; EINECS 230-770-5
Synonyms: PEG-4 nonyl phenyl ether; POE (4) nonyl phenyl ether; PEG 200 nonyl phenyl ether
Classification: Ethoxylated alkyl phenol
Empirical: $C_{23}H_{40}O_5$
Formula: $C_9H_{19}C_6H_4(OCH_2CH_2)_nOH$, avg. n = 4
Properties: Yel. to almost colorless liq.; sol. in oil
Toxicology: Moderately toxic by ingestion, skin contact; severe eye and mild skin irritant in humans; heated to dec., emits acrid smoke and fumes
Uses: Nonionic surfactant; as detergent, emulsifier, wetting agent, dispersant, stabilizer, defoamer; intermediate in synthesis of anionic surfactants; pharmaceutic aids; plasticizer, antistat for plastics; in cosmetics, fat liquoring, cutting oils, agric., petrol. oils
Regulatory: FDA 21CFR §175.105, 176.180, 176.210, 178.3400; FDA approved for ophthalmics, topicals, vaginals
Manuf./Distrib.: Aldrich; Fluka; Sigma
Trade names: Alkasurf® CO-430; Chemax NP-4; DeSonic™ 4N; Igepal® CO-430; Polystep® F-1; Prox-onic NP-04; Remcopal 334; Surfonic® N-40; Teric N4

Nonoxynol-5
CAS 9016-45-9 (generic); 26027-38-3 (generic); 37205-87-1 (generic); 26264-02-8; 20636-48-0; EINECS 247-555-7

Synonyms: PEG-5 nonyl phenyl ether; POE (5) nonyl phenyl ether; 14-(Nonylphenoxy)-3,6,9,12-tetraoxatetradecan-1-ol
Classification: Ethoxylated alkyl phenol
Empirical: $C_{25}H_{44}O_6$
Formula: $C_9H_{19}C_6H_4(OCH_2CH_2)_nOH$, avg. n = 5
Properties: Yel. to almost colorless liq.; sol. in oil
Toxicology: Moderately toxic by ingestion, skin contact; severe eye and mild skin irritant in humans; heated to dec., emits acrid smoke and fumes
Uses: Nonionic surfactant; as detergent, emulsifier, wetting agent, dispersant, stabilizer, defoamer; intermediate in synthesis of anionic surfactants; emulsion polymerization; paint
Regulatory: FDA 21CFR §175.105, 176.180, 176.210, 178.3400
Manuf./Distrib.: Aldrich; Fluka; Sigma
Trade names: Alkasurf® CO-520; Igepal® CO-520; Teric N5

Nonoxynol-6
CAS 9016-45-9 (generic); 26027-38-3 (generic); 37205-87-1 (generic); 27177-01-1; 27177-05-5
Synonyms: PEG-6 nonyl phenyl ether; POE (6) nonyl phenyl ether; PEG 300 nonyl phenyl ether
Classification: Ethoxylated alkyl phenol
Empirical: $C_{27}H_{48}O_7$
Formula: $C_9H_{19}C_6H_4(OCH_2CH_2)_nOH$, avg. n = 6
Properties: Yel. to almost colorless liq.
Toxicology: Moderately toxic by ingestion, skin contact; severe eye and mild skin irritant in humans; heated to dec., emits acrid smoke and fumes
Uses: Nonionic surfactant; as detergent, emulsifier, wetting agent, dispersant, stabilizer, defoamer; intermediate in synthesis of anionic surfactants; plasticizer, antistat for plastics; prep. of emulsified paints
Regulatory: FDA 21CFR §175.105, 176.180, 176.210, 178.3400
Manuf./Distrib.: Aldrich; Fluka; Sigma
Trade names: Alkasurf® CO-530; Alkasurf® NP-6; Chemax NP-6; Dehydrophen PNP 4; Dehydrophen PNP 6; Igepal® CO-530; Nissan Nonion NS-206; Polystep® F-2; Prox-onic NP-06; Surfonic® N-60

Nonoxynol-8
CAS 9016-45-9 (generic); 26027-38-3 (generic); 37205-87-1 (generic); 26571-11-9; 27177-05-5; EINECS 248-293-6; 247-816-5
Synonyms: PEG-8 nonyl phenyl ether; POE (8) nonyl phenyl ether; PEG 400 nonyl phenyl ether
Classification: Ethoxylated alkyl phenol
Empirical: $C_{31}H_{56}O_9$
Formula: $C_9H_{19}C_6H_4(OCH_2CH_2)_nOH$, avg. n = 8
Properties: Yel. to almost colorless liq.
Toxicology: Moderately toxic by ingestion, skin contact; severe eye and mild skin irritant in humans; heated to dec., emits acrid smoke and fumes
Uses: Nonionic surfactant; as detergent, emulsifier, wetting agent, dispersant, stabilizer, defoamer; intermediate in synthesis of anionic surfactants; pigment dispersant, emulsifier for polymerization (styrene acrylic, acrylic, vinyl acrylic, S/B, paints, coatings); contributes mech. and freeze/thaw stability to latexes
Regulatory: FDA 21CFR §175.105, 176.180, 176.210, 178.3400
Manuf./Distrib.: Aldrich; Fluka; Sigma
Trade names: Igepal® CO-610; Polystep® F-3; Rexol 25/8; Surfonic® N-85; T-Det® N-8; Teric N8

Nonoxynol-9
CAS 9016-45-9 (generic); 26027-38-3 (generic); 26571-11-9; 37205-87-1 (generic); 14409-72-4
Synonyms: PEG-9 nonyl phenyl ether; POE (9) nonyl phenyl ether; PEG 450 nonyl phenyl ether
Classification: Ethoxylated alkyl phenol
Empirical: $C_{33}H_{60}O_{10}$
Formula: $C_9H_{19}C_6H_4(OCH_2CH_2)_nOH$, avg. n = 9
Properties: Colorless to lt. yel. clear visc. liq.; sol. in water, ethanol, ethylene glycol, xylene, corn oil; m.w. 617; dens. 1.06 (25/4 C); solid. pt. 26 F; pour pt. 37 F; flash pt. 535-555 F; cloud pt. 126-133 F (1% aq.); visc. 175-250 cps
Toxicology: Moderately toxic by ingestion, skin contact; severe eye and mild skin irritant in humans; heated to dec., emits acrid smoke and fumes
Uses: Nonionic surfactant; as detergent, emulsifier, wetting agent, dispersant, stabilizer, defoamer; intermediate in synthesis of anionic surfactants; spermaticide; plastics antistat; emulsion and suspension polymerization; paints, textile and pesticide formulation

Nonoxynol-10

Regulatory: FDA 21CFR §175.105, 176.180, 176.210, 176.300, 178.3400; USP/NF compliance
Manuf./Distrib.: Aldrich; Fluka; Sigma
Trade names: Chemax NP-9; Empilan® NP9; Igepal® CO-630; Macol® NP-9.5; Prox-onic NP-09; Teric N9

Nonoxynol-10
CAS 9016-45-9 (generic); 26027-38-3 (generic); 27177-08-8; 37205-87-1 (generic); 27942-26-3; EINECS 248-294-1
Synonyms: PEG-10 nonyl phenyl ether; POE (10) nonyl phenyl ether; PEG 500 nonyl phenyl ether
Classification: Ethoxylated alkyl phenol
Empirical: $C_{35}H_{64}O_{11}$
Formula: $C_9H_{19}C_6H_4(OCH_2CH_2)_nOH$, avg. n = 10
Properties: Colorless to lt. amber visc. liq., aromatic odor; sol. in polar org.solvs., water; hyd. no. 81-97
Toxicology: Moderately toxic by ingestion, skin contact; severe eye and mild skin irritant in humans; heated to dec., emits acrid smoke and fumes
Uses: Nonionic surfactant; as detergent, emulsifier, wetting agent, dispersant, stabilizer, defoamer; intermediate in synthesis of anionic surfactants; emulsifier for emulsion polymerization (styrene acrylic, acrylic, vinyl acrylic, S/B, paints, coatings)
Regulatory: FDA 21CFR §175.105, 176.180, 176.210, 178.3400; USP/NF compliance
Manuf./Distrib.: Aldrich; Fluka; Sigma
Trade names: Chemax NP-10; Igepal® CO-660; Lutensol® AP 10; Polystep® F-4; Prox-onic NP-010; Rexol 25/10; Surfonic® N-95; Surfonic® N-100; Surfonic® N-102; T-Det® N-9.5; Teric N10
Trade names containing: Surfonic® HDL-1

Nonoxynol-11
CAS 9016-45-9 (generic); 26027-38-3 (generic); 37205-87-1 (generic)
Synonyms: PEG-11 nonyl phenyl ether; POE (11) nonyl phenyl ether
Classification: Ethoxylated alkyl phenol
Formula: $C_9H_{19}C_6H_4(OCH_2CH_2)_nOH$, avg. n = 11
Properties: Yel. to almost colorless liq.
Toxicology: Moderately toxic by ingestion, skin contact; severe eye and mild skin irritant in humans; heated to dec., emits acrid smoke and fumes
Uses: Nonionic surfactant; as detergent, emulsifier, wetting agent, dispersant, stabilizer, defoamer; intermediate in synthesis of anionic surfactants; spermaticide; used in concrete mfg., agric. sprays, solv. cleaners, paints; detergents
Regulatory: FDA 21CFR §175.105, 176.180, 176.210, 178.3400
Manuf./Distrib.: Aldrich; Fluka; Sigma
Trade names: Igepal® CO-710; T-Det® N-10.5; Teric N11

Nonoxynol-12
CAS 9016-45-9 (generic); 26027-38-3 (generic); 37205-87-1 (generic)
Synonyms: PEG-12 nonyl phenyl ether; POE (12) nonyl phenyl ether; PEG 600 nonyl phenyl ether
Classification: Ethoxylated alkyl phenol
Formula: $C_9H_{19}C_6H_4(OCH_2CH_2)_nOH$, avg. n = 12
Properties: Yel. to almost colorless liq.
Toxicology: Moderately toxic by ingestion, skin contact; severe eye and mild skin irritant in humans; heated to dec., emits acrid smoke and fumes
Uses: Nonionic surfactant; as detergent, emulsifier, wetting agent, dispersant, stabilizer, defoamer; intermediate in synthesis of anionic surfactants; emulsifier and stabilizer for emulsion polymerization (styrene acrylic, acrylic, vinyl acrylic, S/B); paints, coatings
Regulatory: FDA 21CFR §175.105, 176.180, 176.210, 178.3400
Manuf./Distrib.: Aldrich; Fluka; Sigma
Trade names: Igepal® CO-720; Nissan Nonion NS-212; Polystep® F-5; Surfonic® N-120; T-Det® N-12; Teric N12

Nonoxynol-13
CAS 9016-45-9 (generic); 26027-38-3 (generic); 37205-87-1 (generic)
Synonyms: PEG-13 nonyl phenyl ether; POE (13) nonyl phenyl ether
Classification: Ethoxylated alkyl phenol
Formula: $C_9H_{19}C_6H_4(OCH_2CH_2)_nOH$, avg. n = 13
Properties: Yel. to almost colorless liq.
Toxicology: Moderately toxic by ingestion, skin contact; severe eye and mild skin irritant in humans; heated to dec., emits acrid smoke and fumes

Uses: Nonionic surfactant; as detergent, emulsifier, wetting agent, dispersant, stabilizer, defoamer; intermediate in synthesis of anionic surfactants; emulsion polymerization; used in concrete mfg., agric. sprays, solv. cleaners, paints; detergents
Regulatory: FDA 21CFR §175.105, 176.180, 176.210, 178.3400
Manuf./Distrib.: Aldrich; Fluka; Sigma
Trade names: DeSonic™ 13N; Teric N13

Nonoxynol-14
CAS 9016-45-9 (generic); 26027-38-3 (generic); 37205-87-1 (generic)
Synonyms: PEG-14 nonyl phenyl ether; POE (14) nonyl phenyl ether
Classification: Ethoxylated alkyl phenol
Formula: $C_9H_{19}C_6H_4(OCH_2CH_2)_nOH$, avg. n = 14
Properties: Yel. to almost colorless liq.
Toxicology: Moderately toxic by ingestion, skin contact; severe eye and mild skin irritant in humans; heated to dec., emits acrid smoke and fumes
Uses: Nonionic surfactant; as detergent, emulsifier, wetting agent, dispersant, stabilizer, defoamer; intermediate in synthesis of anionic surfactants; emulsifier for emulsion polymerization (styrene acrylic, acrylic, vinyl acrylic, S/B, paints, coatings)
Regulatory: FDA 21CFR §175.105, 176.180, 176.210, 178.3400
Manuf./Distrib.: Aldrich; Fluka; Sigma
Trade names: Polystep® F-6

Nonoxynol-15
CAS 9106-45-9 (generic); 37205-87-1 (generic); 26027-38-3 (generic)
Classification: Ethoxylated alkyl phenol
Formula: $C_9H_{19}C_6H_4(OCH_2CH_2)_nOH$, avg. n = 15
Uses: Emulsifier, detergent, wetting agent, dispersant, solubilizer, coupling agent, defoamer for emulsion polymerization, latex carpet, textiles, metalworking, household, industrial, agric., paper, paint, and cosmetics industries
Regulatory: FDA 21CFR §175.105, 176.180, 176.210
Manuf./Distrib.: Aldrich
Trade names: Chemax NP-15; Igepal® CO-730; Prox-onic NP-015; Surfonic® N-150; Teric N15

Nonoxynol-20
CAS 9016-45-9 (generic); 26027-38-3 (generic); 37205-87-1 (generic); 64812-54-4
Synonyms: PEG-20 nonyl phenyl ether; POE (20) nonyl phenyl ether; PEG 1000 nonyl phenyl ether
Classification: Ethoxylated alkyl phenol
Formula: $C_9H_{19}C_6H_4(OCH_2CH_2)_nOH$, avg. n = 20
Properties: Pale yel. to off-white pastes or waxes
Toxicology: Moderately toxic by ingestion, skin contact; severe eye and mild skin irritant in humans; heated to dec., emits acrid smoke and fumes
Uses: Nonionic surfactant; as detergent, emulsifier, wetting agent, dispersant, stabilizer, defoamer; intermediate in synthesis of anionic surfactants; emulsion polymerization; surfactant, detergent, wetting and rewetting agent, emulsifier in cosmetics, textile, leather, paper, paint, and metal processing
Regulatory: FDA 21CFR §175.105, 176.180
Manuf./Distrib.: Aldrich; Fluka; Sigma
Trade names: Alkasurf® CO-850; Chemax NP-20; Igepal® CO-850; Lutensol® AP 20; Prox-onic NP-020; Surfonic® N-200; Teric N20

Nonoxynol-30
CAS 9016-45-9 (generic); 26027-38-3 (generic); 37205-87-1 (generic)
Synonyms: PEG-30 nonyl phenyl ether; POE (30) nonyl phenyl ether
Classification: Ethoxylated alkyl phenol
Formula: $C_9H_{19}C_6H_4(OCH_2CH_2)_nOH$, avg. n = 30
Properties: Pale yel. to off-white pastes or waxes
Toxicology: Moderately toxic by ingestion, skin contact; severe eye and mild skin irritant in humans; heated to dec., emits acrid smoke and fumes
Uses: Nonionic surfactant; as detergent, emulsifier, wetting agent, dispersant, stabilizer, defoamer, spreading agent; intermediate in synthesis of anionic surfactants; pharmaceutic aids; pigment wetting and grinding agent for aq. systems; emulsifier for vinyl acetate and acrylate emulsion polymerization; used in latex paints, floor finishes, paper coatings, textiles
Regulatory: FDA 21CFR §175.105, 176.180, 178.3400
Manuf./Distrib.: Aldrich; Fluka; Sigma

Nonoxynol-34

Trade names: Ablunol NP30; Ablunol NP30 70%; Alkasurf® CO-880; Alkasurf® CO-887; Chemax NP-30; Chemax NP-30/70; Iconol NP-30; Iconol NP-30-70%; Igepal® CO-880; Igepal® CO-887; Polystep® 9S; Polystep® F-9; Prox-onic NP-030; Prox-onic NP-030/70; Surfonic® N-300; Surfonic® NB-5; Teric N30; Witconol™ NP-300

Nonoxynol-34

Uses: Emulsifier for emulsion polymerization (acrylics and vinyl acetate), paints and coatings
Trade names: Polystep® F-95B

Nonoxynol-40

CAS 9016-45-9 (generic); 26027-38-3 (generic); 37205-87-1 (generic)
Synonyms: PEG-40 nonyl phenyl ether; POE (40) nonyl phenyl ether; PEG 2000 nonyl phenyl ether
Classification: Ethoxylated alkyl phenol
Formula: $C_9H_{19}C_6H_4(OCH_2CH_2)_nOH$, avg. n = 40
Properties: Pale yel. to off-white pastes or waxes
Toxicology: Moderately toxic by ingestion, skin contact; severe eye and mild skin irritant in humans; heated to dec., emits acrid smoke and fumes
Uses: Nonionic surfactant; as detergent, emulsifier, wetting agent, dispersant, stabilizer, defoamer; intermediate in synthesis of anionic surfactants; emulsion polymerization; used in latex paints, floor finishes, paper coatings, textiles
Regulatory: FDA 21CFR §175.105, 176.180, 178.3400
Manuf./Distrib.: Aldrich; Fluka; Sigma
Trade names: Ablunol NP40; Ablunol NP40 70%; Alkasurf® CO-890; Alkasurf® CO-897; Chemax NP-40; Chemax NP-40/70; Iconol NP-40; Iconol NP-40-70%; Igepal® CO-890; Igepal® CO-897; Polystep® F-10; Prox-onic NP-040; Prox-onic NP-040/70; Rexol 25/40; T-Det® N-407; Tergitol® NP-40; Tergitol® NP-40 (70% Aq.); Teric N40

Nonoxynol-50

CAS 9016-45-9 (generic); 26027-38-3 (generic); 37205-87-1 (generic)
Synonyms: PEG-50 nonyl phenyl ether; POE (50) nonyl phenyl ether
Classification: Ethoxylated alkyl phenol
Formula: $C_9H_{19}C_6H_4(OCH_2CH_2)_nOH$, avg. n = 50
Properties: Pale yel. to off-white pastes or waxes
Toxicology: Moderately toxic by ingestion, skin contact; severe eye and mild skin irritant in humans; heated to dec., emits acrid smoke and fumes
Uses: Nonionic surfactant; as detergent, emulsifier, wetting agent, dispersant, stabilizer, defoamer; intermediate in synthesis of anionic surfactants; emulsion polymerization; used in latex paints, floor finishes, paper coatings, textiles
Regulatory: FDA 21CFR §176.180, 178.3400
Manuf./Distrib.: Aldrich; Fluka; Sigma
Trade names: Ablunol NP50; Ablunol NP50 70%; Alkasurf® CO-970; Alkasurf® CO-977; Chemax NP-50; Chemax NP-50/70; Iconol NP-50; Iconol NP-50-70%; Igepal® CO-970; Igepal® CO-977; Prox-onic NP-050; Prox-onic NP-050/70; Rexol 25/50; Rexol 25/507; T-Det® N-507

Nonoxynol-70

Synonyms: PEG-70 nonyl phenyl ether; POE (70) nonyl phenyl ether
Classification: Ethoxylated alkyl phenol
Formula: $C_9H_{19}C_6H_4(OCH_2CH_2)_nOH$, avg. n = 70
Properties: Pale yel. to off-white pastes or waxes
Toxicology: Moderately toxic by ingestion, skin contact; severe eye and mild skin irritant in humans; heated to dec., emits acrid smoke and fumes
Uses: Nonionic surfactant; as detergent, emulsifier, wetting agent, dispersant, stabilizer, defoamer; intermediate in synthesis of anionic surfactants; emulsion polymerization; emulsifier/stabilizer for floor waxes and polishes, paints
Trade names: Alkasurf® CO-987; Iconol NP-70; Iconol NP-70-70%; Igepal® CO-980; Igepal® CO-987; T-Det® N-70; T-Det® N-705; T-Det® N-707

Nonoxynol-100

CAS 9016-45-9 (generic); 26027-38-3 (generic); 37205-87-1 (generic)
Synonyms: PEG-100 nonyl phenyl ether; POE (100) nonyl phenyl ether
Classification: Ethoxylated alkyl phenol
Formula: $C_9H_{19}C_6H_4(OCH_2CH_2)_nOH$, avg. n = 100
Properties: Pale yel. to off-white pastes or waxes

Toxicology: Moderately toxic by ingestion, skin contact; severe eye and mild skin irritant in humans; heated to dec., emits acrid smoke and fumes
Uses: Nonionic surfactant; as detergent, emulsifier, wetting agent, dispersant, stabilizer, defoamer; intermediate in synthesis of anionic surfactants; emulsion polymerization; stabilizer for syn. latexes
Regulatory: FDA 21CFR §176.180
Manuf./Distrib.: Aldrich; Fluka; Sigma
Trade names: Alkasurf® CO-990 FLK; Alkasurf® CO-997; Chemax NP-100; Chemax NP-100/70; Iconol NP-100; Iconol NP-100-70%; Igepal® CO-990 FLK; Igepal® CO-997; Prox-onic NP-0100; Prox-onic NP-0100/70; Rexol 25/100-70%; Teric N100

Nonoxynol-10 carboxylic acid
CAS 28212-44-4 (generic); 53610-02-9
Synonyms: PEG-10 nonyl phenyl ether carboxylic acid; PEG 500 nonyl phenyl ether carboxylic acid; POE (10) nonyl phenyl ether carboxylic acid
Classification: Organic acid
Formula: $C_9H_{19}C_6H_4(OCH_2CH_2)_nOCH_2COOH$, avg. n = 9
Uses: Detergent, wetting agent; dispersant for aq. coatings and inks
Trade names: Sandopan® MA-18

Nonoxynol-6 phosphate
CAS 51811-79-1; 68412-53-3; 29994-44-3; 51609-41-7 (generic); EINECS 249-992-9
Synonyms: PEG-6 nonyl phenyl ether phosphate; POE (6) nonyl phenyl ether phosphate; PEG 300 nonyl phenyl ether phosphate
Definition: Complex mixture of esters of phosphoric acid and nonoxynol-6
Uses: Lubricant, extreme pressure agent; in metalworking, cutting fluids; antisoil redeposition, antistat in drycleaning detergents; emulsifier for herbicides/insecticides, vinyl acetate, acrylates, SBR; emulsifier, dispersant for agric. formulations, paints; corrosion inhibitor
Regulatory: FDA 21CFR §175.105, 178.3400
Trade names: Rhodafac® PE-510

Nonoxynol-9 phosphate
CAS 51609-41-7 (generic); 51811-79-1; 66197-78-2; 68412-53-3; EINECS 266-231-6
Synonyms: PEG-9 nonyl phenyl ether phosphate; 26-(Nonylphenoxy)-3,6,9,12,15,18,21,24-octaoxahexacosan-1-ol, dihydrogenphosphate; POE (9) nonyl phenyl ether phosphate; PEG 450 nonyl phenyl ether phosphate
Definition: Complex mixture of esters of phosphoric acid and nonoxynol-9
Uses: Emulsifier, lubricant, antistat, detergent, corrosion inhibitor for agric., industrial use; antisoil redeposition for dry cleaning; emulsion polymerization; household and industrial detergents; fabric finishes; paints, textile wet processing, metals
Regulatory: FDA 21CFR §175.105, 178.3400
Trade names: Rhodafac® RE-610

Nonylcarbinol. *See* n-Decyl alcohol
2-Nonyl-4,5-dihydro-1H-imidazole-1-ethanol. *See* Capryl hydroxyethyl imidazoline
Nonylic acid. *See* Nonanoic acid

Nonyl nonoxynol-7
CAS 9014-93-1 (generic)
Uses: Emulsifier for agric., emulsion polymerization; leather fat liquoring; paints
Manuf./Distrib.: Aldrich
Trade names: Alkasurf® DM-430; Igepal® DM-430

Nonyl nonoxynol-8
CAS 9014-93-1 (generic)
Synonyms: PEG-8 dinonyl phenyl ether; POE (8) dinonyl phenyl ether
Classification: Ethoxylated alkyl phenol
Formula: $(C_9H_{19})_2C_6H_3(OCH_2CH_2)_nOH$, avg. n = 8
Uses: Emulsifier in textile dye carrier applics., insecticides, wax emulsions; foam control agent; spreading agent in pigment printing; post-stabilizer in emulsion polymerization; intermediate; solubilizer for paints, paper, agric., textiles, etc.
Manuf./Distrib.: Aldrich
Trade names: Chemax DNP-8; Prox-onic DNP-08

Nonyl nonoxynol-15

Nonyl nonoxynol-9
CAS 9014-93-1 (generic)
Synonyms: PEG-9 dinonyl phenyl ether; POE (9) dinonyl phenyl ether
Classification: Ethoxylated alkyl phenol
Formula: $(C_9H_{19})_2C_6H_3(OCH_2CH_2)_nOH$, avg. n = 9
Uses: Emulsifier for cutting oils, agric., textile finishing, dry cleaning soaps, inks, lacquers, paints, metalworking fluids, pesticides, cosmetics, emulsion polymerization
Manuf./Distrib.: Aldrich
Trade names: Alkasurf® DM-530; Igepal® DM-530

Nonyl nonoxynol-15
CAS 9014-93-1 (generic)
Synonyms: PEG-15 dinonyl phenyl ether; POE (15) dinonyl phenyl ether
Classification: Ethoxylated alkyl phenol
Formula: $(C_9H_{19})_2C_6H_3(OCH_2CH_2)_nOH$, avg. n = 15
Uses: Detergent, emulsifier, antistat for textiles, leather, metal cleaners, paper, latex, pesticides; coemulsifier for syn. latexes
Manuf./Distrib.: Aldrich
Trade names: Alkasurf® DM-710

Nonyl nonoxynol-18
Synonyms: PEG-18 dinonyl phenyl ether; POE (18) dinonyl phenyl ether
Classification: Ethoxylated alkyl phenol
Formula: $(C_9H_{19})_2C_6H_3(OCH_2CH_2)_nOH$, avg. n = 18
Uses: Emulsifier, detergent, solubilizer
Trade names: Chemax DNP-18

Nonyl nonoxynol-24
CAS 9014-93-1 (generic)
Synonyms: PEG-24 dinonyl phenyl ether; POE (24) dinonyl phenyl ether
Classification: Ethoxylated alkyl phenol
Formula: $(C_9H_{19})_2C_6H_3(OCH_2CH_2)_nOH$, avg. n = 24
Uses: Detergent, emulsifier for textiles, leather, metal cleaners, paper, latex, pesticides, emulsion polymerization, latex stabilization
Manuf./Distrib.: Aldrich
Trade names: Alkasurf® DM-730

Nonyl nonoxynol-49
CAS 9014-93-1 (generic)
Synonyms: PEG-49 dinonyl phenyl ether; POE (49) dinonyl phenyl ether
Classification: Ethoxylated alkyl phenol
Formula: $(C_9H_{19})_2C_6H_3(OCH_2CH_2)_nOH$, avg. n = 49
Uses: Emulsifier, solubilizer for emulsion polymerization, latex stabilization, pesticides, essential oils, cleaners
Manuf./Distrib.: Aldrich
Trade names: Alkasurf® DM-880

Nonyl nonoxynol-150
CAS 9014-93-1 (generic)
Synonyms: PEG-150 dinonyl phenyl ether; POE (150) dinonyl phenyl ether
Classification: Ethoxylated alkyl phenol
Formula: $(C_9H_{19})_2C_6H_3(OCH_2CH_2)_nOH$, avg. n = 150
Uses: Detergent, emulsifier, dispersant, wetter, stabilizer for laundry, household, textile, hard-surface detergents, cosmetics, insecticides, paper, petrol., paints
Manuf./Distrib.: Aldrich
Trade names: Chemax DNP-150; Chemax DNP-150/50; Igepal® DM-970 FLK; Prox-onic DNP-0150; Prox-onic DNP-0150/50

2-(Nonylphenoxy) ethanol. *See* Nonoxynol-1
2-[2-[2-(Nonylphenoxy)ethoxy]ethoxy]ethanol. *See* Nonoxynol-3
26-(Nonylphenoxy)-3,6,9,12,15,18,21,24-octaoxahexacosan-1-ol, dihydrogenphosphate. *See* Nonoxynol-9 phosphate
14-(Nonylphenoxy)-3,6,9,12-tetraoxatetradecan-1-ol. *See* Nonoxynol-5

Nonylphenyl phosphite (3:1). *See* Tris(nonylphenyl) phosphite
Normal lead acetate. *See* Lead acetate

Novolac resin
Synonyms: Novolak resin; Two-stage phenolic resin; Phenol-formaldehyde (novolak) resin
Classification: Thermoplastic phenol-formaldehyde type resin
Properties: Sol. in alcohol
Uses: Molding materials; bonding agent in brake linings, abrasive grinding wheels, electrical insulation; reinforcing agent and modifier for nitrile rubber; air-drying varnishes; bonding materials

Novolak resin. *See* Novolac resin; Phenolic resin
NPG. *See* Neopentyl glycol
NR. *See* Natural rubber
NVF. *See* N-Vinyl formamide

Nylon
CAS 63428-83-1
Definition: A family of polyamide polymers char. by presence of amide group -CONH
Empirical: $(C_6H_{11}NO)_n$
Properties: Cryst. solid; sol. in phenol, cresols, xylene, formic acid; insol. in alcohol, esters, ketones, hydrocarbons
Precaution: Reacts violently with F_2
Toxicology: Experimental tumorigen by implant; heated to decomp., emits toxic fumes of NO_x
Uses: Thermoplastic resin for tire cord, hosiery, wearing apparel, brush bristles, cordage, fish lines, tennis rackets, rugs, artificial turf, parachutes, composites, sails, film, gears/bearings, insulation, surgical sutures, metal coating, fuel tanks
Manuf./Distrib.: Allchem Ind.; Ashley Polymers; Bamberger Polymers; BASF; Bayer; Chemical; Coz; DSM; EMS-Am. Grilon; Ferro; Hoechst Celanese; Hüls Am.; ICI GmbH; LNP; Monsanto; A. Schulman
Trade names: Elvamide® 8023, 8061, 8063, 8066
See also Polyamide

Nylon-6
CAS 25038-54-4
Synonyms: Poly[imino(1-oxo-1,6-hexanediyl)]; Poly(iminocarbonylpentamethylene); Polyamide 6
Classification: Polyamide
Empirical: $(C_6H_{11}NO)_n$
Formula: $[NH(CH_2)_5CO]_x$
Properties: Dens. 1.14 (20/4 C); m.p. 223 C; resistant to most org. chem., dissolved by phenol, cresol, strong acids; immune to biological attack
Toxicology: Moderately toxic by ingestion; mildly toxic by inhalation
Uses: Tire cord; fishing lines; tow ropes; hose mfg.; woven fabrics; electronic and electric parts
Regulatory: FDA 21CFR §177.1500, 177.2260, 177.2470, 177.2480; BP compliance (sterile sutures)
Manuf./Distrib.: Aldrich; SNIA UK
Trade names: Nycoa® 471

Nylon 11
CAS 25035-04-5
Synonyms: Poly(undecanoamide); Poly[imino(1-oxo-1,11-undecanediyl)]
Classification: Polyamide
Empirical: $(C_{11}H_{21}NO)_n$
Formula: $[NH(CH_2)_{10}CO]_n$
Properties: Dens. 1.026
Uses: Thermoplastic resin for inj. molding and extrusion applics.; tire cord; fishing line; tow ropes; woven fabrics
Regulatory: FDA 21CFR §177.1500, 177.2260
Manuf./Distrib.: Aldrich
Trade names: Rilsan® RDP15-10 Natural

Nylon-12
CAS 25038-74-8; 24937-16-4
Synonyms: Azacyclotridecane-2-one polyamide; Poly(laurolactam); Azacyclotridecane-2-one, homopolymer
Classification: Polyamide
Definition: Polyamide derived from 12-aminododecanoic acid

Nylon-66

Empirical: $(C_{12}H_{23}NO)_n$
Properties: Dens. 1.010
Uses: Thermoplastic resin for inj. molding and extrusion applics.; for extrusion coatings for plastics films, metal foils, and paper
Regulatory: FDA 21CFR §177.1500, 177.2260
Manuf./Distrib.: Aldrich; Elf Atochem SA; Daicel-Hüls; Hüls AG
Trade names: Vestamid® Series; Vestamid® L 1700; Vestamid® L 1801; Vestosint®

Nylon-66
CAS 32131-17-2
Synonyms: Poly[imino (1,6-dioxo-1,6-hexanediyl) imino-1,6-hexanediyl; Poly(hexamethyleneadipamide); Poly(N,N´-hexamethyleneadipinediamide)
Classification: Polyamide
Definition: Polymeric amide formed by the reaction of adipic acid with hexylenediamine
Empirical: $(C_{12}H_{22}N_2O_2)_x$
Formula: $[NH(CH_2)_6NHCO(CH_2)_4CO]_n$
Properties: Crystalline solid; sol. in phenol, cresols, xylene, formic acid; insol. in alcohol, esters, ketones, hydrocarbons; dens. 1.090
Precaution: Reacts violently with F_2
Toxicology: Heated to dec., emits toxic fumes of NO_x
Uses: Thermoplastic resin for inj. molding, extrusion; used in inks, coatings, adhesives
Regulatory: FDA 21CFR §177.1200, 177.1500, 177.2260, 177.2470, 177.2480, 177.2600, 178.2010, 179.45
Manuf./Distrib.: Aldrich; Asahi Chem. Ind.; Fluka; SNIA UK
Trade names: Versamid® 930

OAP. *See* Hexamethyldisilazane
OBPA. *See* 10,10´-Oxybisphenoxyarsine
Oceol. *See* Oleyl alcohol
OCTA. *See* Octabromodiphenyl oxide
Octabenzone. *See* Benzophenone-12

Octabromodiphenyl oxide
CAS 32536-52-0; EINECS 251-087-9
Synonyms: OCTA
Empirical: $C_{12}H_2Br_8O$
Properties: Powd.; m.w. 801.37
Uses: Flame retardant for ABS, nylon, other engineering thermoplastics
Manuf./Distrib.: Albemarle; Allchem Ind.; Great Lakes
Trade names: Great Lakes DE-79™; Saytex® 111

9,12-Octadecadienoic acid. *See* Linoleic acid
(Z,Z)-9,12-Octadecadienoic acid. *See* Linoleic acid
(Z,Z)-9,12-Octadecadienoic acid, copper salt. *See* Copper linoleate
9,12-Octadecadienoic acid, dimer. *See* Dilinoleic acid
9,12-Octadecadienoic acid lead salt. *See* Lead linoleate
9,12-Octadecadienoic acid methyl ester. *See* Methyl linoleate
9,12-Octadecadienoic acid, trimer. *See* Trilinoleic acid
Octadecanamide. *See* Stearamide
n-Octadecanoic acid. *See* Stearic acid
Octadecanoic acid aluminum salt. *See* Aluminum stearate
Octadecanoic acid butyl ester. *See* Butyl stearate
Octadecanoic acid calcium salt. *See* Calcium stearate
Octadecanoic acid, 1,2-ethanediyl ester. *See* Glycol distearate
Octadecanoic acid, 12-hydroxy-. *See* Hydroxystearic acid
Octadecanoic acid, magnesium salt. *See* Magnesium stearate
Octadecanoic acid, methyl ester. *See* Methyl stearate
Octadecanoic acid, monoester with 1,2-propanediol. *See* Propylene glycol stearate
Octadecanoic acid, monoester with 1,2,3,-propanetriol. *See* Glyceryl stearate
Octadecanoic acid, 2-[(1-oxooctadecyl)amino]ethyl ester. *See* Stearamide MEA-stearate
Octadecanoic acid, tridecyl ester. *See* Tridecyl stearate

Octadecanoic acid zinc salt. *See* Zinc stearate
9-Octadecenamide. *See* Oleamide
9-Octadecenamide, N,N′-1,2-ethanediylbis-. *See* Ethylene dioleamide
9-Octadecenoic acid. *See* Oleic acid
cis-9-Octadecenoic acid. *See* Oleic acid
9-Octadecenoic acid, 12-(acetyloxy)-, 1,2,3-propanetriol ester. *See* Glyceryl triacetyl ricinoleate
9-Octadecenoic acid, 12-hydroxy-. *See* Ricinoleic acid
9-Octadecenoic acid, methyl ester. *See* Methyl oleate
9-Octadecenoic acid, monoester with 1,2,3-propanetriol. *See* Glyceryl oleate
9-Octadecenoic acid, tetraester with decaglycerol. *See* Polyglyceryl-10 tetraoleate
9-Octadecen-1-ol. *See* Oleyl alcohol
cis-9-Octadecen-1-ol. *See* Oleyl alcohol
(Z)-9-Octadecen-1-ol. *See* Oleyl alcohol
Octadecyl acrylate. *See* Stearyl acrylate
Octadecylamine acetate. *See* Stearamine acetate
Octadecyl 3,5-bis (1,1-dimethylethyl)-4-hydroxybenzene propanoate. *See* Octadecyl 3,5-di-t-butyl-4-hydroxyhydrocinnamate

Octadecyl 3,5-di-t-butyl-4-hydroxyhydrocinnamate
CAS 2082-79-3; EINECS 218-216-0
Synonyms: Octadecyl 3,5-bis (1,1-dimethylethyl)-4-hydroxybenzene propanoate; Octadecyl 3-(3,5-di-t-butyl-4-hydroxyphenyl) propionate; Stearyl-3-(3′,5′-di-t-butyl-4-hydroxyphenyl) propionate
Classification: Hindered phenol
Empirical: $C_{35}H_{62}O_3$
Formula: $[(CH_3)_3C]_2C_6H_2(OH)CH_2CH_2CO_2(CH_2)_{17}CH_3$
Properties: Wh. powd.; sol. in acetone, benzene, chloroform, ethyl acetate, hexane; m.w. 530.9; m.p. 50-52 C; f.p. > 230 F
Toxicology: Irritant
Uses: Antioxidant stabilizer for styrenics, polyolefins, PVC, urethane and acrylic coatings, adhesives, elastomers; replaces BHT in polyolefins
Manuf./Distrib.: Aldrich
Trade names: Akrochem® Antioxidant 1076; BNX® 1076, 1076G; Irganox® 1076; Ultranox® 276

Octadecyl 3-(3,5-di-t-butyl-4-hydroxyphenyl) propionate. *See* Octadecyl 3,5-di-t-butyl-4-hydroxyhydrocinnamate
Octadecyl methacrylate. *See* Stearyl methacrylate

Octamethylcyclotetrasiloxane
CAS 556-67-2; 69430-24-6; EINECS 209-136-7
Synonyms: Octamethyltetracyclosiloxane
Empirical: $C_8H_{24}O_4Si_4$
Formula: $[(CH_3)_2SiO]_4$
Properties: Smooth visc. oily liq.; m.w. 296.6; dens. 0.955 (20/4 C); m.p. 17.5 C; b.p. 175-176 C; flash pt. 60 C; ref. index 1.3968 (20 C)
Precaution: Flamm.
Uses: Silicone oils, methyl silicone oils, and related prods.; reagent
Manuf./Distrib.: Aldrich; Fluka; Hüls AG; Sigma
Trade names containing: EXP-28 Silicone Fluid

Octamethyltetracyclosiloxane. *See* Octamethylcyclotetrasiloxane
1,8-Octanedicarboxylic acid. *See* Sebacic acid
Octanoic acid lead salt. *See* Lead octoate
1-Octanol. *See* Caprylic alcohol
n-Octanol. *See* Caprylic alcohol
2-Octanone. *See* Methyl hexyl ketone
3,6,9,12,15,18,21,24-Octaoxahexatriacontan-1-ol. *See* Laureth-8
[2,3,4,6,7,8,9,10-Octapyrimidol[1,2-a]azepine]. *See* Diazabicycloundecene

Octocrylene
CAS 6197-30-4; EINECS 228-250-8
Synonyms: 2-Ethylhexyl 2-cyano-3,3-diphenylacrylate; 2-Ethylhexyl 2-cyano-3,3-diphenyl-2-propenoate; UV Absorber-3
Classification: Substituted acrylate

Octoic acid

Empirical: $C_{24}H_{27}NO_2$
Formula: $(C_6H_5)_2C{=}C(CN)CO_2CH_2CH(C_2H_5)(CH_2)_3CH_3$
Properties: Liq.; m.p. -10 C
Uses: Active ingred. in OTC drug prods.; uv-B absorber for cosmetics; uv stabilizer for plastics (flexible and semirigid PVC, VDC), coatings
Manuf./Distrib.: Aldrich
Trade names: Uvinul® 3039; Uvinul® N-539

Octoic acid
Uses: Intermediate for paints
Manuf./Distrib.: P & G Global Oleochems.

Octoxynol-1
CAS 9002-93-1 (generic); 9036-19-5 (generic); 9004-87-9 (generic); 2315-67-5; EINECS 264-520-1
Synonyms: Ethylene glycol octyl phenyl ether; PEG-1 octyl phenyl ether; 2-[p-(1,1,3,3-Tetramethylbutyl) phenoxy]ethanol
Classification: Ethoxylated alkyl phenol
Empirical: $C_{16}H_{26}O_2$
Formula: $C_8H_{17}C_6H_4OCH_2CH_2OH$
Toxicology: LD50 (oral, rat) 1800 mg/kg; mod. toxic by ingestion and IV routes; experimental reproductive effects; human mutagenic data; eye and human skin irritant; heated to decomp., emits toxic fumes of NO_x
Uses: Nonionic surfactant, detergent, emulsifier, dispersant; spermaticide; emulsion polymerization, industrial/household cleaners, agric., latex stabilizer
Regulatory: FDA 21CFR §172.710, 175.105, 176.180; FDA approved for topicals
Manuf./Distrib.: Aldrich; Fluka; Sigma
Trade names: Alkasurf® CA-210; Igepal® CA-210; Triton® X-15

Octoxynol-3
CAS 9002-93-1 (generic); 9004-87-9 (generic); 9036-19-5 (generic); 2315-62-0; 27176-94-9
Synonyms: 2-[2-[2-[p-(1,1,3,3-Tetramethylbutyl)phenoxy]ethoxy]ethoxy]ethanol; PEG-3 octyl phenyl ether; POE (3) octyl phenyl ether
Classification: Ethoxylated alkyl phenol
Empirical: $C_{20}H_{34}O_4$
Formula: $C_8H_{17}C_6H_4(OCH_2CH_2)_nOH$, avg. n = 3
Uses: Nonionic surfactant, detergent, emulsifier, dispersant; spermaticide; industrial/household cleaners, emulsion polymerization, agric., latex stabilizer
Regulatory: FDA 21CFR §175.105, 176.180, 176.210
Manuf./Distrib.: Aldrich; Fluka; Sigma
Trade names: Alkasurf® CA-420; Igepal® CA-420; Triton® X-35

Octoxynol-4
Classification: Ethoxylated alkyl phenol
Formula: $C_8H_{17}C_6H_4(OCH_2CH_2)_nOH$, avg. n = 4
Uses: Nonionic surfactant, detergent, emulsifier, dispersant; spermaticide; for agric., leather and metal processing, paint, drycleaning detergents, degreasing applics.
Trade names: T-Det® O-4

Octoxynol-5
CAS 9002-93-1 (generic); 9036-19-5 (generic); 9004-87-9 (generic); 2315-64-2; 27176-99-4
Synonyms: PEG-5 octyl phenyl ether; POE (5) octyl phenyl ether; 14-(Octylphenoxy)-3,6,9,12-tetraoxatetradecan-1-ol
Classification: Ethoxylated alkyl phenol
Empirical: $C_{24}H_{42}O_6$
Formula: $C_8H_{17}C_6H_4(OCH_2CH_2)_nOH$, avg. n = 5
Uses: Nonionic surfactant, detergent, emulsifier, dispersant; spermaticide; for paints; anti-icing additive for gasoline
Regulatory: FDA 21CFR §172.710, 175.105, 176.180, 176.210, 178.3400
Manuf./Distrib.: Aldrich; Fluka; Sigma
Trade names: Alkasurf® CA-520; Igepal® CA-520; Teric X5; Triton® X-45

Octoxynol-7
CAS 9002-93-1 (generic); 9004-87-9 (generic); 9036-19-5 (generic); 27177-02-2
Synonyms: PEG-7 octyl phenyl ether; POE (7) octyl phenyl ether; 20-(Octylphenoxy)-3,6,9,12,15,18-

hexaoxaeicosan-1-ol
Classification: Ethoxylated alkyl phenol
Empirical: $C_{28}H_{50}O_8$
Formula: $C_8H_{17}C_6H_4(OCH_2CH_2)_nOH$, avg. n = 7
Uses: Nonionic surfactant, detergent, emulsifier, dispersant; spermaticide; textile lubricants, agric., latex paint; post-polymerization stabilizer
Regulatory: FDA 21CFR §172.710, 175.105, 176.180, 176.210, 178.3400
Manuf./Distrib.: Aldrich; Fluka; Sigma
Trade names: Alkasurf® CA-620; Hyonic® OP-70; Igepal® CA-620

Octoxynol-8
CAS 9004-87-9 (generic); 9036-19-5 (generic); 9002-93-1 (generic)
Synonyms: PEG-8 octyl phenyl ether; PEG 400 octyl phenyl ether; POE (8) octyl phenyl ether
Classification: Ethoxylated alkyl phenol
Empirical: $C_{30}H_{54}O_9$
Formula: $C_8H_{17}C_6H_4(OCH_2CH_2)_nOH$, avg. n = 8
Uses: Surfactant for detergents, agric. formulations, oil well drilling, leather processing, paints
Regulatory: FDA 21CFR §172.710, 175.105, 176.180, 176.210, 178.3400
Manuf./Distrib.: Aldrich; Fluka; Sigma
Trade names: T-Det® O-8; Triton® X-114

Octoxynol-9
CAS 9002-93-1 (generic); 9004-87-9 (generic); 9010-43-9; 9036-19-5 (generic); 42173-90-0
Synonyms: PEG-9 octyl phenyl ether; POE (9) octyl phenyl ether; PEG 450 octyl phenyl ether
Classification: Ethoxylated alkyl phenol
Empirical: $C_{32}H_{58}O_{10}$
Formula: $C_8H_{17}C_6H_4(OCH_2CH_2)_nOH$, avg. n = 9
Properties: Pale yel. clear visc. liq., faint odor, bitter taste; sol. in benzene, toluene; misc. with water, alcohol, acetone; insol. in hexane; dens. 1.059-1.068; hyd. no. 85-101; cloud pt. 63-69 C; pH 6-8
Uses: Nonionic surfactant, detergent, emulsifier, dispersant; spermaticide; household/industrial cleaners; emulsion polymerization; paints and coatings
Regulatory: FDA 21CFR §175.105, 176.180, 176.210, 178.3400; FDA approved for topicals
Manuf./Distrib.: Aldrich; Fluka; Sigma
Trade names: Igepal® CA-630; Polystep® OP-9; Triton® X-100

Octoxynol-10
CAS 9002-93-1 (generic); 9004-87-9 (generic); 9036-19-5 (generic); 2315-66-4; 27177-07-7
Synonyms: PEG-10 octyl phenyl ether; POE (10) octyl phenyl ether; PEG 500 octyl phenyl ether
Classification: Ethoxylated alkyl phenol
Empirical: $C_{34}H_{62}O_{11}$
Formula: $C_8H_{17}C_6H_4(OCH_2CH_2)_nOH$, avg. n = 10
Properties: Pale yel. visc. liq.; misc. with water, alcohol, acetone; sol. in benzene, toluene; m.w. 647; dens. 1.0595 (25/4 C)
Uses: Nonionic surfactant, detergent, emulsifier, dispersant; spermaticide; industrial/household cleaners, emulsion polymerization, latex stabilizer, asphalt emulsions
Regulatory: FDA 21CFR §172.710, 175.105, 176.180, 176.210, 178.3400
Manuf./Distrib.: Aldrich; Fluka; Sigma
Trade names: Iconol OP-10; Macol® OP-10 SP; Rexol 45/10

Octoxynol-13
CAS 9002-93-1 (generic); 9004-87-9 (generic); 9036-19-5 (generic)
Synonyms: PEG-13 octyl phenyl ether; POE (13) octyl phenyl ether
Classification: Ethoxylated alkyl phenol
Formula: $C_8H_{17}C_6H_4(OCH_2CH_2)_nOH$, avg. n = 13
Uses: Nonionic surfactant, detergent, emulsifier, dispersant; spermaticide; foam stabilizer at high temps., in presence of electrolytes; metal cleaning, industrial, household liq. detergents and cleaners, sanitizers; solubilizer of anionic detergents; paints and coatings, paper and textile processing
Regulatory: FDA 21CFR §172.710, 175.105, 176.180, 176.210, 178.3400
Manuf./Distrib.: Aldrich; Fluka; Sigma
Trade names: Igepal® CA-720; Triton® X-102

Octoxynol-16
CAS 9004-87-9 (generic); 9036-19-5 (generic); 9002-93-1 (generic)

Octoxynol-25

Synonyms: PEG-16 octyl phenyl ether; POE (16) octyl phenyl ether
Classification: Ethoxylated alkyl phenol
Formula: $C_8H_{17}C_6H_4(OCH_2CH_2)_nOH$, avg. n = 16
Uses: Nonionic surfactant, detergent, emulsifier, dispersant; spermaticide; household/industrial cleaners, emulsion polymerization, cosmetics, pharmaceuticals
Regulatory: FDA 21CFR §175.105, 176.180
Manuf./Distrib.: Aldrich; Fluka; Sigma
Trade names: Triton® X-165-70%

Octoxynol-25
CAS 9002-93-1 (generic); 9036-19-5 (generic); 9004-87-9 (generic)
Synonyms: PEG-25 octyl phenyl ether; POE (25) octyl phenyl ether
Classification: Ethoxylated alkyl phenol
Formula: $C_8H_{17}C_6H_4(OCH_2CH_2)_nOH$, avg. n = 25
Uses: Emulsifier for emulsion polymerization; perfume solubilizer
Regulatory: FDA 21CFR §175.105, 176.180
Manuf./Distrib.: Aldrich; Fluka; Sigma
Trade names: Alkasurf® CA-877; Igepal® CA-877

Octoxynol-30
CAS 9004-87-9 (generic); 9036-19-5 (generic); 9002-93-1 (generic)
Synonyms: PEG-30 octyl phenyl ether; POE (30) octyl phenyl ether
Classification: Ethoxylated alkyl phenol
Formula: $C_8H_{17}C_6H_4(OCH_2CH_2)_nOH$, avg. n = 30
Uses: Nonionic surfactant, detergent, emulsifier, dispersant; spermaticide; industrial/household cleaners; emulsion polymerization; asphalt
Regulatory: FDA 21CFR §172.710, 175.105, 176.180, 178.3400
Manuf./Distrib.: Aldrich; Fluka; Sigma
Trade names: Alkasurf® CA-887; Chemax OP-30/70; Iconol OP-30; Iconol OP-30-70%; Igepal® CA-880; Igepal® CA-887; Polystep® OP-3070

Octoxynol-40
CAS 9002-93-1 (generic); 9004-87-9 (generic); 9036-19-5 (generic)
Synonyms: PEG-40 octyl phenyl ether; POE (40) octyl phenyl ether
Classification: Ethoxylated alkyl phenol
Formula: $C_8H_{17}C_6H_4(OCH_2CH_2)_nOH$, avg. n = 40
Uses: Nonionic surfactant, detergent, emulsifier, dispersant; spermaticide; emulsion polymerization; industrial/household cleaners; asphalt; latex stabilizer
Regulatory: FDA 21CFR §172.710, 175.105, 176.180, 178.3400
Manuf./Distrib.: Aldrich; Fluka; Sigma
Trade names: Alkasurf® CA-897; Chemax OP-40/70; Hyonic® GL 400; Iconol OP-40; Iconol OP-40-70%; Igepal® CA-890; Igepal® CA-897; Nissan Nonion HS-240; Polystep® OP-4070

Octoxynol-70
CAS 9004-87-9 (generic); 9036-19-5 (generic); 9002-93-1 (generic)
Synonyms: PEG-70 octyl phenyl ether; POE (70) octyl phenyl ether
Classification: Ethoxylated alkyl phenol
Formula: $C_8H_{17}C_6H_4(OCH_2CH_2)_nOH$, avg. n = 70
Uses: Nonionic surfactant, detergent, emulsifier, dispersant; spermaticide; household/industrial cleaners, textiles, paints, inks, agric., latex/emulsion polymerization
Regulatory: FDA 21CFR §172.710, 176.180
Manuf./Distrib.: Aldrich; Fluka; Sigma
Trade names: Nissan Nonion HS-270

Octrizole
CAS 3147-75-9; EINECS 221-573-5
Synonyms: 2-(2H-Benzotriazol-2-yl)-4-(1,1,3,3-tetramethylbutyl)phenol; 2-(2´-Hydroxy-5´-t-octylphenyl)benzotriazole; UV Absorber-5
Classification: Organic compd.
Empirical: $C_{20}H_{25}N_3O$
Properties: Powd.; m.w. 323.44; m.p. 106-108 C
Toxicology: LD50 (rat, oral) > 10 g/kg; LD50 (rabbit, dermal) > 5 g/kg; irritant
Uses: UV absorber, stabilizer for cosmetics, polyesters, PVC, styrenics, acrylics, PC, polyvinyl butyral

(molding, sheet, and glazing materials for window, lighting, sign, marine, and auto applics.); also in coatings, photo prods., sealants, and elastomeric materials
Manuf./Distrib.: Aldrich
Trade names: Cyasorb® UV 5411

Octyl acrylate
CAS 103-11-7; EINECS 203-080-7
Synonyms: Acrylic acid 2-ethylhexyl ester; 2-Ethylhexyl-2-propenoate; 2-Ethylhexyl acrylate; 2-Propenoic acid-2-ethylhexyl ester
Empirical: $C_{11}H_{20}O_2$
Properties: M.w. 184.31; dens. 0.8867 (20/20 C); f.p. -90 C; b.p. 130 C (50 mm); flash pt. (OC) 180 F
Precaution: Flamm.; fire hazard exposed to heat or flame
Toxicology: LD50 (oral, rat) 5660 mg/kg, (skin, rabbit) 8480 mg/kg; moderately toxic by inhalation and others; severe skin and eye irritant; experimental tumorigen; heated to decomp., emits acrid smoke
Uses: As comonomer for radical polymerization; modifier for flexibility
Manuf./Distrib.: Aldrich; BASF; Fluka; Hoechst Celanese; Monomer-Polymer & Dajac Labs; Sartomer; Sigma; Union Carbide
Trade names containing: SR-333

Octyl alcohol. *See* 2-Ethylhexanol
n-Octyl alcohol. *See* Caprylic alcohol

Octyl/decyl acrylate
Synonyms: ODA
Uses: Diluent for coating, ink, and adhesive applics.

Octyl dimethyl p-aminobenzoate. *See* Octyl dimethyl PABA
Octyl-p-(dimethylamino)benzoate. *See* Octyl dimethyl PABA

Octyl dimethyl PABA
CAS 21245-02-3; EINECS 244-289-3
Synonyms: Octyl dimethyl p-aminobenzoate; Octyl-p-(dimethylamino)benzoate; Padimate O; 2-Ethylhexyl-4(dimethylamino) benzoate; 2-Ethylhexyl p-dimethylaminobenzoate
Definition: Ester of 2-ethylhexyl alcohol and dimethyl p-aminobenzoic acid
Empirical: $C_{17}H_{27}NO_2$
Properties: M.w. 277.4; sp.gr. 0.99-1.00; ref. index 1.5390-1.5430
Toxicology: LD50 (rat, oral) 14.9 g/kg
Uses: Sunscreen agent; photoinitiator synergist for uv-cured coatings, inks, adhesives
Manuf./Distrib.: Aldrich; First
Trade names: Firstcure™ ODAB

2-n-Octyl-4-isothiazolin-3-one
CAS 26530-20-1
Synonyms: 2-Octyl-3(2H)-isothiazolone
Empirical: $C_{11}H_{19}NOS$
Properties: M.w. 213.37
Toxicology: LD50 (oral, rat) 550 mg/kg, (dermal, rabbit) 690 mg/kg; mod. toxic by ingestion and skin contact; severe eye irritant; skin irritant; heated to decomp., emits very toxic fumes of SO_x and NO_x
Uses: Mildewcide for paints, plastics
Trade names: Skane® M-8
Trade names containing: Vinyzene® IT-3000 DIDP

2-Octyl-3(2H)-isothiazolone. *See* 2-n-Octyl-4-isothiazolin-3-one
Octyl-octadecyl dimethyl ethylbenzyl ammonium chlorides. *See* Benzalkonium chloride
20-(Octylphenoxy)-3,6,9,12,15,18-hexaoxaeicosan-1-ol. *See* Octoxynol-7
14-(Octylphenoxy)-3,6,9,12-tetraoxatetradecan-1-ol. *See* Octoxynol-5

Octylphenyl dihydrogen phosphate bis (diethylamine) salt
CAS 64051-37-2
Manuf./Distrib.: Albright & Wilson
Trade names containing: Virco-Pet® 50

Octyl phosphate. *See* Trioctyl phosphate
N-Octyl-2-pyrrolidone. *See* Caprylyl pyrrolidone
Octyl sulfate sodium salt. *See* Sodium octyl sulfate

Octyltriethoxysilane
CAS 2943-75-1; EINECS 220-941-2
Empirical: $C_{14}H_{32}SiO_3$
Properties: Liq.; m.w. 276.5; dens. 0.88; b.p. 98-99 C (2 mm); flash pt. 100 C; ref. index 1.417 (20 C)
Precaution: Flamm.
Uses: Coupling agent, release agent, lubricant, blocking agent, chemical intermediate; crosslinking agent providing durability, gloss, hiding power to coatings
Manuf./Distrib.: Aldrich; Fluka
Trade names: Silquest® A-137

ODA. *See* Octyl/decyl acrylate
Oil of Palma Christi. *See* Castor oil
Oils, linseed. *See* Linseed oil
Oils, menhaden. *See* Menhaden oil
Oils, peanut, sulfated. *See* Sulfated peanut oil
Oils, vegetable. *See* Vegetable oil
Okenite. *See* Calcium silicate

Oleamide
CAS 301-02-0; EINECS 206-103-9
Synonyms: 9-Octadecenamide; Oleyl amide; Oleic acid amide
Classification: Aliphatic amide
Empirical: $C_{18}H_{35}NO$
Formula: $CH_3(CH_2)_7CH:CH(CH_2)_7CONH_2$
Properties: Ivory-colored powd.; dens. 0.94; m.p. 72 C
Precaution: Combustible
Toxicology: Heated to decomp., emits acrid smoke and irritating fumes
Uses: Slip/antiblock agent for extrusion of polyethylene; wax additive; ink additive; internal lubricant and slip agent for processed plastics, coatings, and films; builder, foam visc. stabilizer, and foam booster in syn. detergent formulations; water repellent for textiles
Regulatory: FDA 21CFR §175.105, 175.300, 178.3860, 178.3910, 179.45, 181.22, 181.28
Manuf./Distrib.: Akzo Nobel; Chemax; Chemron; Cookson Spec. Additives; Croda Universal Ltd.; Henkel/Emery; Mona Ind.; Pilot; Sigma; Syn. Prods.; Witco/Oleo-Surf.
Trade names: Armid® O; Petrac® Slip-Eze®

Olefin/acrylate copolymer
CAS 67892-91-5
Synonyms: Olefin/acrylic graft polymer
Uses: Emulsion used for floor finishes; results in higher gloss; better detergent resist.; improved scuff resist.; superior color and UV stability
Trade names: Syntran® PA-1465

Olefin/acrylic graft polymer. *See* Olefin/acrylate copolymer

Oleic acid
CAS 112-80-1; EINECS 204-007-1
Synonyms: cis-9-Octadecenoic acid; Red oil; Elainic acid; 9-Octadecenoic acid
Classification: Aliphatic organic compd.; unsaturated fatty acid
Empirical: $C_{18}H_{34}O_2$
Formula: $CH_3(CH_2)_7CH:CH(CH_2)_7COOH$
Properties: Colorless liq., odorless; darkens when exposed to oxygen; insol. in water; sol. in alcohol, ether, benzene, chloroform, fixed/volatile oils; m.w. 282.47; dens. 0.895 (25/25 C); m.p. 6 C; b.p. 286 C (100 mm); acid no. 196-204; flash pt. 100 C; ref. index 1.463 (18 C)
Precaution: Combustible when exposed to heat or flame; incompat. with Al and perchloric acid; light-sensitive
Toxicology: LD50 (oral, rat) 74 g/kg; poison by intravenous route; mildly toxic by ingestion; experimental tumorigen; irritant to skin, mucous membranes; heated to decomp., emits acrid smoke and irritating fumes
Uses: Soap base, mfg. of oleates, ointments, cosmetics, polishing compds., lubricants, food-grade additives, Turkey red oil, driers; intermediates for paints; waterproofing textiles; oiling wool; pharmaceutic aid (solv.); activator, plasticizer, softener, in rubbers
Regulatory: FDA 21CFR §172.210, 172.860, 172.862, 173.315 (0.1 ppm max. in wash water), 173.340, 175.105, 175.320, 176.170, 176.200, 176.210, 177.1010, 177.1200, 177.2260, 177.2600, 177.2800, 178.3570, 178.3910, 182.70, 182.90; FEMA GRAS; FDA approved for inhalants, orals, topicals; USP/NF, BP compliance

Manuf./Distrib.: Akzo Nobel; Aldrich; Arizona; Ashland; J.T. Baker; Baychem; Brown; Browning; R.E. Carroll; Climax Perf. Materials; Ferro/Keil; Fluka; C.P. Hall; Henkel/Emery; Hercules; Schweizerhall; Sigma; Unichema Int'l.; Union Derivan SA; Van Waters & Rogers; Witco/Oleo-Surf.

Oleic acid amide. *See* Oleamide
Oleol. *See* Oleyl alcohol
Oleostearin. *See* Oleostearine

Oleostearine
Synonyms: Oleostearin
Manuf./Distrib.: Anar; Norman, Fox; Geo. Pfau's Sons
Trade names containing: Ross Japan Wax Substitute 525

Oleoyl sarcosine
CAS 110-25-8; EINECS 203-749-3
Synonyms: N-Methyl-N-(1-oxo-9-octadecenyl)glycine; Oleyl methylaminoethanoic acid; Oleyl sarcosine
Definition: Condensation prod. of oleic acid with N-methylglycine
Empirical: $C_{21}H_{39}NO_3$
Formula: $CH_3(CH_2)_7CH:CH(CH_2)_7CON(CH_3)CH_2COOH$
Uses: Corrosion inhibitor, detergent, wetting agent, foamer/foam stabilizer, emulsifier; for cosmetics, household/industrial cleaners, biotechnology, agric., textiles, petrol. prods., metalworking, surf. coatings, emulsion polymerization; mold release for RIM; stabilizer for polyols; antifog and antistat in plastics
Regulatory: FDA 21CFR §178.3130
Manuf./Distrib.: Sigma
Trade names: Crodasinic O

Oleth-4
CAS 9004-98-2 (generic); 5353-26-4
Synonyms: PEG-4 oleyl ether; POE (4) oleyl ether; PEG 200 oleyl ether
Definition: PEG ether of oleyl alcohol
Formula: $CH_3(CH_2)_7CH{=}CH(CH_2)_7CH_2(OCH_2CH_2)_nOH$, avg. n = 4
Properties: M.w. 709.02; dens. 0.997; flash pt. > 110 C
Toxicology: Irritant
Uses: Coupling agent, solubilizer, emulsion stabilizer for cosmetics, hair care preps.; emulsion polymerization; textile finishes
Manuf./Distrib.: Aldrich; Fluka; Sigma
Trade names: Prox-onic OA-1/04

Oleth-5
CAS 9004-98-2 (generic); 5353-27-5
Synonyms: PEG-5 oleyl ether; POE (5) oleyl ether; 3,6,9,12,15-Pentaoxatriacont-24-en-1-ol
Definition: PEG ether of oleyl alcohol
Empirical: $C_{28}H_{56}O_6$
Formula: $CH_3(CH_2)_7CH{=}CH(CH_2)_7CH_2(OCH_2CH_2)_nOH$, avg. n = 5
Uses: Emulsifier, lubricant, solubilizer for pesticides, cosmetics, floor polishes; dispersant for paints
Manuf./Distrib.: Aldrich; Fluka; Sigma

Oleth-9
CAS 9004-98-2 (generic)
Synonyms: PEG-9 oleyl ether; POE (9) oleyl ether; PEG 450 oleyl ether
Definition: PEG ether of oleyl alcohol
Formula: $CH_3(CH_2)_7CH{=}CH(CH_2)_7CH_2(OCH_2CH_2)_nOH$, avg. n = 9
Uses: Coupling agent, solubilizer, emulsion stabilizer for cosmetics, hair care preps.; emulsion polymerization; textile spin finishes
Regulatory: FDA 21CFR §176.200, 177.2800
Manuf./Distrib.: Aldrich; Fluka; Sigma
Trade names: Prox-onic OA-1/09

Oleth-20
CAS 9004-98-2 (generic)
Synonyms: PEG-20 oleyl ether; POE (20) oleyl ether; PEG 1000 oleyl ether
Definition: PEG ether of oleyl alcohol
Formula: $CH_3(CH_2)_7CH{=}CH(CH_2)_7CH_2(OCH_2CH_2)_nOH$, avg. n = 20

Oleum Gossypii seminis

Uses: Coupling agent, solubilizer, emulsion stabilizer for cosmetics, pharmaceuticals, polishes, leather, plastics, dye dispersant, metalworking; stabilizer for rubber latex
Regulatory: FDA 21CFR §175.105, 176.180, 176.200, 177.1210, 177.2800; FDA approved for topicals
Manuf./Distrib.: Aldrich; Fluka; Sigma
Trade names: Rhodasurf® ON-870; Rhodasurf® ON-877

Oleum Gossypii seminis. *See* Cottonseed oil
Oleum ricini. *See* Castor oil

Oleyl alcohol
CAS 143-28-2; EINECS 205-597-3
Synonyms: 9-Octadecen-1-ol; (Z)-9-Octadecen-1-ol; cis-9-Octadecen-1-ol; Oceol; Oleol
Classification: Unsaturated fatty alcohol
Empirical: $C_{18}H_{36}O$
Formula: $CH_3(CH_2)_7CH=CH(CH_2)_8OH$
Properties: Colorless to pale yel. oily visc. liq., faint char. odor, bland taste; insol. in water; sol. in alcohol, ether; m.w. 268.49; dens. 0.84; m.p. 13-19 C; b.p. 207 C (13 mm); acid no. 1 max.; iodine no. 85-95; hyd. no. 205-215; cloud pt. ≤ 10 C; flash pt. > 110 C; ref. index 1.4582 (27.5 C)
Toxicology: Human skin irritant; heated to decomp., emits acrid smoke and irritating fumes
Uses: Plasticizer, emulsion stabilizer, antifoam, coupler, pigment dispersant, detergent in cutting oils, inks, textile finishing, petrochem., pulp/paper, paints, plastics, food applics., pharmaceuticals, cosmetics; mfg. of sulfuric esters used as surfactants
Regulatory: FDA 21CFR §176.170, 176.210, 177.1010, 177.1210, 177.2800, 178.3910; FDA approved for topicals; USP/NF compliance
Manuf./Distrib.: Aldrich; Croda; Fluka; R.W. Greeff; Lanaetex Prods.; M. Michel; Rhone-Poulenc Surf. & Spec.; Ronsheim & Moore; Sigma; Witco/Oleo-Surf.
Trade names: Fancol OA-95

Oleyl amide. *See* Oleamide

Oleyl hydroxyethyl imidazoline
CAS 95-38-5; 21652-27-7; 27136-73-8; EINECS 248-248-0; 244-501-4; 202-414-9
Synonyms: 2-(8-Heptadecenyl)-4,5-dihydro-1H-imidazole-1-ethanol; 1-Hydroxyethyl-2-oleyl imidazoline; Oleyl imidazoline
Classification: Heterocyclic compd.
Empirical: $C_{22}H_{42}N_2O$
Uses: Emulsifier, corrosion inhibitor, antistat, wetting and flocculating agent, lubricant used in car wax emulsions, cleaners, paint mfg., agric. applics., syn. coolants; dispersant for clay and pigments in solv. systems
Regulatory: FDA 21CFR §178.3570
Trade names: Calgene C-100-O; Crodazoline O; Miramine® O; Miramine® OC; Monazoline O; Textamine O-1; Varine O®

Oleyl imidazoline. *See* Oleyl hydroxyethyl imidazoline
Oleyl methylaminoethanoic acid. *See* Oleoyl sarcosine

Oleyl propanediamine
CAS 68037-97-8
Uses: Chemical intermediate; corrosion inhibitor, fuel oil additive, flotation agent; bitumen emulsifier; antisettling agent for paint formulations; used in metals, textiles, plastics, herbicides; epoxy curing agent
Trade names: Duomeen® O

Oleyl propylene diamine dioleate
CAS 34140-91-5
Uses: Rust inhibitor; paint additive
Trade names: Inipol OO2

Oleyl propylene diamine ditallate
Uses: Rust inhibitor; paint additive
Trade names: Inipol OT2

Oleyl sarcosine. *See* Oleoyl sarcosine
One-stage resin. *See* Phenolic resin

Opalwax. *See* Hydrogenated castor oil
OPT. *See* 1,10-Phenanthroline
Optal. *See* n-Propyl alcohol
Organosiloxane. *See* Silicone
Orthoboric acid. *See* Boric acid
Orthophenanthroline. *See* 1,10-Phenanthroline
Orthophosphoric acid. *See* Phosphoric acid
Orthotitanic acid tetrabutyl ester. *See* Tetrabutyl titanate
Oxammonium sulfate. *See* Hydroxylamine sulfate
Oxazolidine A. *See* Dimethyl oxazolidine

Oxidized linseed oil
Uses: Oil used for coatings
Trade names: Diamond 'K'; Miscible Diamond 'K.'

Oxidized menhaden oil
Properties: Amber liq., mild marine oil odor; sp.gr. 1.003 (60 F); flash pt. (COC) 530 F
Uses: Flow, gloss, and leveling aid in paint systems
Trade names: Blown Menhaden Fish Oil Z-3; Blown Menhaden Fish Oil Z-6
Trade names containing: Blown Menhaden Fish Oil M&M; Blown Menhaden Fish Oil M80; Blown Menhaden Fish Oil Z-9

Oxidized polyethylene. *See* Polyethylene, oxidized

Oxidized soybean oil
Properties: Amber liq.; sp.gr. 0.992 (60 F); flash pt. (COC) 400 F
Uses: Used in caulking compds., putties, paints, lacquer plasticizers
Trade names: Blown Soya Z2, Z4

Oxirane, [[2-ethylhexyl)oxy]methyl]-. *See* 2-Ethylhexyl glycidyl ether
Oxirane, [(1-methylethoxy)methyl]-,. *See* Isopropyl glycidyl ether
Oxirane, (phenoxymethyl)-. *See* Phenyl glycidyl ether
2-Oxobutane. *See* Methyl ethyl ketone
3-Oxobutanoic acid ethyl ester. *See* Ethylacetoacetate

Oxo-heptyl acetate
CAS 90438-79-2
Synonyms: C7 alkyl acetate
Properties: Sp.gr. 0.87; flash pt. (TCC) 150 F
Uses: Solv. for high solids, maintenance, marine, and industrial waterborne coatings, acrylic resin polymerization media, metalworking fluids, urethane foam and industrial cleaners and degreasers, paint strippers, pigment dispersions; pesticide emulsifiable concs., textile scouring aids, NC or screen printing inks, penetrating and lubricating oils, vinyl plastisols; good solvency, low surf. tens. and interfacial tens., good wetting, lubricity; visc. modifier; latex coalescer
Manuf./Distrib.: Chemcentral
Trade names: Exxate® 700
Trade names containing: Ircogel® 907

2-Oxohexamethylenimine. *See* Caprolactam

Oxo-hexyl acetate
CAS 88230-35-7
Synonyms: C6 alkyl acetate
Properties: Sp.gr. 0.87; flash pt. (TCC) 134 F
Uses: Solv. for high solids, maintenance, marine, and industrial waterborne coatings, acrylic resin polymerization media, metalworking fluids, urethane foam and industrial cleaners and degreasers, paint strippers, pigment dispersions; pesticide emulsifiable concs., textile scouring aids, NC or screen printing inks, penetrating and lubricating oils, vinyl plastisols; good solvency, low surf. tens. and interfacial tens., good wetting, lubricity; visc. modifier; latex coalescer
Trade names: Exxate® 600
Trade names containing: SCX™-514

2-[(1-Oxooctadecyl)amino]ethyl octadecanoate. *See* Stearamide MEA-stearate
5-Oxo-L-proline. *See* PCA
5-Oxo-DL-proline, sodium salt. *See* Sodium PCA
Oxybenzene. *See* Phenol
Oxybenzone. *See* Benzophenone-3
2,2´-[Oxybis(2,1-ethanediyloxy)]bisethanol. *See* PEG-4
2,2´-Oxybisethanol. *See* Diethylene glycol
Oxybismethane. *See* Dimethyl ether
1,1´-Oxybis (2,3,4,5,6-pentabromobenzene). *See* Decabromodiphenyl oxide

10,10´-Oxybisphenoxyarsine
CAS 58-36-6
Synonyms: OBPA; 10,10´-Oxydiphenoxarsine; Bis(phenoxarsin-10-yl)ether; Bis(10-phenoxyarsinyl)oxide
Empirical: $C_{24}H_{16}As_2O$
Properties: Colorless prisms; sol. in alcohol, chloroform, methylene chloride; pract. insol. in water (5 ppm @ 20 C); m.w. 502.23; dens. 1.40-1.42; m.p. 184-185 C; dec. 380 C
Toxicology: LD50 (male rat) 35-50 mg/kg
Uses: Fungicide and bactericide for protection of plastics
Trade names containing: Vinyzene® BP-5-2; Vinyzene® BP-5-2 DIDP; Vinyzene® BP-5-2 DOP; Vinyzene® BP-5-2 PG; Vinyzene® BP-5-2 S-160; Vinyzene® BP-5-2 U; Vinyzene® BP-5-5; Vinyzene® BP-5-5 160; Vinyzene® BP-5-5 DIDP; Vinyzene® BP-5-5 DOP; Vinyzene® BP-5-5 PG; Vinyzene® SB-1 U

1,1´-Oxybis-2-propanol. *See* Dipropylene glycol
Oxybis (tributyltin). *See* Tributyltin oxide
Oxydiethylene diacrylate. *See* Diethylene glycol diacrylate
Oxydiethylene methacrylate. *See* Diethylene glycol dimethacrylate
10,10´-Oxydiphenoxarsine. *See* 10,10´-Oxybisphenoxyarsine
1,1´-Oxydi-2-propanol. *See* Dipropylene glycol
3,3´-Oxydi-1-propanol dibenzoate. *See* Dipropylene glycol dibenzoate
3,3´-Oxydyl-1-propanol dibenzoate. *See* Dipropylene glycol dibenzoate
Oxymethylene. *See* Formaldehyde
Oxypropylated cellulose. *See* Hydroxypropylcellulose
Ozocerite. *See* Ozokerite

Ozokerite
CAS 8021-55-4; EINECS 265-134-6
Synonyms: Ozocerite; Ceresin; Mineral wax; Fossil wax
Classification: Hydrocarbon wax
Definition: Hydrocarbon wax derived from mineral or petroleum sources
Properties: Yel.-brown to black or green translucent (pure), noxious odor; sol. in lt. petrol. hydrocarbons, benzene, turpentine, kerosene, ether, carbon disulfide; sl. sol. in alcohol; insol. in water; dens. 0.85-0.95; m.p. 55-110 C (usually 70 C)
Precaution: Combustible
Toxicology: No known toxicity
Uses: Filler for electrical insulation, rubber prods., paints, leather prods., printing inks, floor and furniture polishes, cosmetics, ointments; substitute for carnauba and beeswax
Regulatory: Japan approved; FDA approved for orals
Manuf./Distrib.: Eastman; ISP; Koster Keunen; Frank B. Ross; Strahl & Pitsch
Trade names: Koster Keunen Ozokerite; Ross Ozokerite Wax

Padimate O. *See* Octyl dimethyl PABA
PAI. *See* Polyamide-imide

Palladium
CAS 7440-05-3; EINECS 231-115-6
Classification: Metallic element
Empirical: Pd
Properties: Silvery white metal; insol. in water; at.wt. 106.4; dens. 12.02 (20/4 C); m.p. 1555 C; b.p. 3167 C
Precaution: Combustible; dust can be fire and explosion hazard; violent reactions with Al, H_2, IPA, S_2
Toxicology: May be a skin sensitizer; can interfere with enzymatic functions when taken intravenously

Uses: Alloys for electrical relays and switching systems in telecommunications equip.; air craft spark plugs; protective coatings
Manuf./Distrib.: Aldrich; Atomergic Chemetals; Colonial Metals; Degussa; Fluka; Johnson Matthey; Noah; Sigma

Palladium chloride anhyd. *See* Palladium chloride (ous)
Palladium (II) chloride anhyd. *See* Palladium chloride (ous)

Palladium chloride (ous)
CAS 7657-10-1; EINECS 231-596-2
Synonyms: Palladium chloride anhyd.; Palladium (II) chloride anhyd.; Palladous chloride; Palladium dichloride
Empirical: Cl_2Pd
Formula: $PdCl_2$
Properties: Red to dk. brn. cryst.; hygroscopic; deliq.; sol. in water, alcohol, acetone, HCl; m.w. 177.31; dens. 4.0 (18 C); m.p. 680 C (decomp.)
Toxicology: LD50 (oral, rat) 25 mg/kg, (IP, rat) 70 mg/kg, (IV, rat) 5 mg/kg; toxic by ingestion, IP, IV routes; experimental carcinogen; human mutagenic data; skin irritant; heated to decomp., emits highly toxic fumes of Cl^-
Uses: Analytical chemistry, 'electroless' coatings for metals, photography, leak detection in gas lines, indelible inks, catalyst
Manuf./Distrib.: Aldrich; Alfa Aesar Johnson Matthey; Atlantic Equip. Engrs.; Atomergic Chemetals; Colonial Metals; Degussa; Fluka; Johnson Matthey; Monomer-Polymer & Dajac Labs; Noah; Sigma; Spectrum Chem. Mfg.; Technic

Palladium dichloride. *See* Palladium chloride (ous)
Palladous chloride. *See* Palladium chloride (ous)

Palmitic acid
CAS 57-10-3; EINECS 200-312-9
Synonyms: Hexadecanoic acid; Cetylic acid; Hexadecylic acid
Classification: Aliphatic organic compd.; saturated fatty acid
Definition: A mixt. of solid organic acids
Empirical: $C_{16}H_{32}O_2$
Formula: $CH_3(CH_2)_{14}COOH$
Properties: Wh. cryst. scales, sl. char. odor/taste; insol. in water; sl. sol. in cold alcohol, petrol. ether; sol. in hot alcohol, ether, propyl alcohol, chloroform; m.w. 256.42; dens. 0.853 (62/4 C); m.p. 63-64 C; b.p. 215 C (15 mm); ref. index 1.4273 (80 C
Precaution: Combustible
Toxicology: LD50 (IV, mouse) 57 mg/kg; acute poison by intravenous route; experimental neoplastigen; human skin irritant; heated to decomp., emits acrid smoke and fumes
Uses: Mfg. of metallic palmitates, soaps, lube oils, paints, waterproofing, food-grade additives; texturizer for shampoos, soaps, shaving creams
Regulatory: FDA 21CFR §172.210, 172.860, 173.340, 175.105, 175.320, 176.170, 176.200, 176.210, 177.1010, 177.1200, 177.2260, 177.2600, 177.2800, 178.3570, 178.3910; must conform to FDA specs for fats or fatty acids derived from edible oils; FEMA GRAS; BP compliance
Manuf./Distrib.: Akzo Nobel; Aldrich; Am. Biorganics; Ashland; Baychem; Brown; Browning; Condor; Fluka; Henkel/Emery; Penta Mfg.; Research Organics; Ruger; Sigma; Spectrum Chem. Mfg.; Unichema Int'l.; Witco/Oleo-Surf.

Palmitic acid, hexadecyl ester. *See* Cetyl palmitate
Palmitic/stearic acid glycerol monodiester. *See* Palmitic/stearic acid mono/diglycerides

Palmitic/stearic acid mono/diglycerides
Synonyms: Palmitic/stearic acid glycerol monodiester
Uses: Antistat for polyolefins; surface coating for EPS; fabric softener; coemulsifier
Trade names: Tegotens 4100; Tegotens BL 130

Palmityl alcohol. *See* Cetyl alcohol
Palygorskite. *See* Attapulgite
Parachlorobenzotrifluoride. *See* p-Chlorobenzotrifluoride
Parachlorometacresol. *See* p-Chloro-m-cresol
Parachlorometaxylenol. *See* Chloroxylenol

Paraffin

CAS 8002-74-2; EINECS 232-315-6

Synonyms: n-Paraffin; Paraffin wax; Hard paraffin; Petroleum wax, crystalline

Classification: Aliphatic organic compd.; hydrocarbon

Definition: Solid mixture of hydrocarbons obtained from petroleum; characterized by relatively large crystals

Empirical: C_nH_{2n+2}

Properties: Colorless to white cryst. solid, odorless, tasteless; insol. in water, alcohol; sol. in benzene, chloroform, ether, carbon disulfide, oils; misc. with fats; dens. ≈ 0.9; m.p. 50-57 C; flash pt. 340 F

Precaution: Dangerous fire hazard; wil burn above 198 C; when heated, produces irritating fumes

Toxicology: Anesthetic effect; ACGIH TLV:TWA 2 mg/m^3 (fume); chronic skin exposure can cause dermatitis, abnormal pigmentation, etc.; experimental tumorigens by implantation; many paraffin waxes contain carcinogens

Uses: Candles, paper coating, protective sealant for food prods.; plastics lubricants; hot-melt carpet backing, floor polishes, cosmetics, chewing gum base; raising m.p. of ointments

Regulatory: FDA 21CFR §133.181, 172.615, 173.3210, 175.105, 175.210, 175.250, 175.300, 175.320, 176.170, 176.200, 177.1200, 177.2420, 177.2600, 177.2800, 178.3710, 178.3800, 178.3910, 179.45; FEMA GRAS; Canada, Japan approved; FDA approved for implants, orals, topicals; USP/NF, BP compliance

Manuf./Distrib.: Aldrich; Astor Corp.; Barco Chem. Prods.; Dussek Campbell Inc; EM Ind.; Exxon; Humphrey; IGI; Jonk BV; Koster Keunen; Mobil; Penreco; Penta Mfg.; Phillips; Polygon; Frank B. Ross; Shell; Spectrum Chem. Mfg.; Stevenson Cooper; Texaco; Vista

Trade names: Hostalub® XL 165FR; Michem® Emulsion 34935; Michem® Emulsion 47950; Michem® Emulsion 62330; Michem® Emulsion 65935; Michem Lube® 135; Michem Lube® 511; Michem Lube® 723; Michem Lube® 743; Michem Lube® 981; Microlube™ N; Octowax 321; Paracol® 404G; Paracol® 802A; Paracol® 802G; Paracol® 802N; Paracol® 2370; Ross 118/125 AMP; Ross 125/130 AMP; Ross 130/135 AMP; Ross 140/145 AMP; Ross 150/155 AMP; Ross 160/165 AMP; Slip-Ayd® SL-145E

Trade names containing: Drewax™ E-9040; Jonwax 120; Michem® Emulsion 01250; Michem® Experimental Emulsion 70250; Michem® Experimental Emulsion 77030; Michem® Experimental Emulsion 87140; Michem Lube® 118; Michem Lube® 155; Michem Lube® 180; Michem Lube® 182; Michem Lube® 270R; Michem Lube® 368; Michem Lube® 693; Petrolite® 01 Disp.; Polyace™ 804; Ross Beeswax Substitute No. 628/5; Ross Japan Wax Substitute 966; Ross Synthetic Candelilla Wax; Shellwax; Surfynol® DF-62

n-Paraffin. *See* Paraffin

Paraffin, brominated

Uses: Flame retardant for polyurethane foam, pkg., rubber, textiles, adhesives, paints; visc. reducer

Trade names: Doverguard® 8410; Doverguard® 8426

Paraffin, bromochlorinated

Uses: Flame retardant for flexible, rigid polyurethane foam, laminated board, board stock, pkg., textiles, carpet backing, adhesives, paints

Trade names: Doverguard® 8207-A; Doverguard® 8208-A; Doverguard® 8307-A; Doverguard® 9119; Fyarestor® 100

Trade names containing: Doverguard® 9122

Paraffin, chlorinated

CAS63449-39-8

Synonyms: Chlorocosane; Chlorcosane; Chlorinated paraffin

Definition: Paraffin hydrocarbons treated with chlorine; contains about 50% Cl

Properties: Pale to amber, neutral, light viscous oils or soft waxes; insol. in water; sl. sol. in alcohol; sol. in benzene, chloroform; dens. 0.900-1.50

Precaution: Nonflamm.

Uses: In mfg. of textiles, draperies, etc.; solv. for dichloramine-T; high-pressure lubricants; flame retardants in plastics; as plasticizer for PVC in PE sealants; plasticizer in paints

Manuf./Distrib.: AC Ind.; Allchem Ind.; R.E. Carroll; Chemcentral; Dover; Ferro/Keil; C.P. Hall; Harwick; Hoechst AG; Morton Int'l.; Occidental; Punda Mercantile; Sea-Land; StanChem; Tri-Iso; Witco/Polymer Addit.

Trade names: Chlorez® 700; Chlorez® 700-DD; Chlorez® 700-DF; Chlorez® 700-S; Chlorez® 700-SS; Chlorez® 725-S; Chlorez® 760; Chloroflo® 40; Doverguard® 152; Doverguard® 170; Doverguard® 700; Doverguard® 700-S; Doverguard® 700-SS; Doverguard® 760; Doverguard® 5761; Paroil® 10; Paroil® 45; Paroil® 50; Paroil® 57-61; Paroil® 140; Paroil® 142-A; Paroil® 145-A; Paroil® 150-A; Paroil® 150-LVA; Paroil® 152; Paroil® 170; Paroil® 170-HV; Paroil® 170-LV; Paroil® 170-T; Paroil® 1057; Paroil® 1061; Paroil® 1650; Paroil® 5761; Rez-O-Sperse® 3; Rez-O-Sperse® A-1

Paraffin oil. *See* Mineral oil
Paraffin wax. *See* Paraffin
Pareth-15-7. *See* C11-15 pareth-7
Pareth-15-9. *See* C11-15 pareth-9
Pareth-15-12. *See* C11-15 pareth-12
Pareth-25-7. *See* C12-15 pareth-7
Pareth-25-9. *See* C12-15 pareth-9
PC. *See* Polycarbonate resin

PCA
CAS 98-79-3; EINECS 202-700-3
Synonyms: Pyrrolidonecarboxylic acid; (S)-2-Pyrrolidone-5-carboxylic acid; 5-Oxo-L-proline; L-Pyroglutamic acid
Classification: Cyclic organic compd.
Empirical: $C_5H_7NO_3$
Properties: Wh. crystal, odorless, sl. acidic taste, nonhygroscopic; sol. in water; m.w. 129.11; m.p. 181 C
Toxicology: Sl. irritating to skin; very irritating to eyes
Uses: Humectant for cosmetics for skin and hair care, tobacco, cellulose film, paper prods., fiber prods., paints; dyeing agent, softening agent, finishing agent, and antistatic agent; intermediate for synthesis
Manuf./Distrib.: Aldrich; Fluka; Sigma
Trade names: Ajidew A-100

PC/ABS
Uses: Thermoplastic blend for inj. molding, extrusion, blow molding, thermoforming, machining, welding, bonding, screwing, coating, printing, vacuum metallizing, and electroplating; for automotive, appliance industries
Manuf./Distrib.: Bayer

PCA-Na. *See* Sodium PCA
PCA Soda. *See* Sodium PCA
PCBTF. *See* p-Chlorobenzotrifluoride
PCMC. *See* p-Chloro-m-cresol
PCMX. *See* Chloroxylenol

PC/PET
Uses: Thermoplastic resin blend for inj. molding, paints and coatings
Manuf./Distrib.: Bayer

PC resin. *See* Polycarbonate resin
PDDP. *See* Phenyl diisodecyl phosphite
PDEA. *See* Phenyldiethanolamine
PDMS. *See* Dimethicone
PE. *See* Pentaerythritol; Polyethylene
PEA. *See* 2-Phenoxyethyl acrylate

Peanut oil
CAS 8002-03-7; EINECS 232-296-4
Synonyms: Arachis oil; Groundnut oil; Katchung oil; Earthnut oil; Pecan shell powder
Classification: Fixed oil
Definition: Refined fixed oil obtained from seed kernels of one or more cultivated varieties of *Arachis hypogaea*
Properties: Pale yel. liq., nutty odor, bland taste; sol. in benzene, alcohol, ether, chloroform; insol. in alkalies; dens. 0.916-0.922; solid. pt. -5 C; iodine no. 84-100; sapon. no. 185-195; flash pt. 540 F; ref. index 1.466-1.470
Precaution: Combustible exposed to heat or flame; can react with oxidizing materials; light-sensitive
Toxicology: Experimental tumorigen; human skin irritant, mild allergen; mutagenic data; heated to decomp., emits acrid smoke and irritating fumes
Uses: Edible oil; substitute for olive oil; vehicle for medicine; in mfg. of margarine, soaps, paints; heat transfer medium in laboratory
Regulatory: FDA 21CFR §175.105, 176.200, 176.210, 177.2800, 182.70; GRAS; FDA approved for injecables, orals, vaginals; USP/NF, BP compliance
Manuf./Distrib.: ABITEC; Alnor Oil; Arista Ind.; Charkit; Croda; Penta Mfg.; Sigma; Tri-K Ind.; Welch, Holme & Clark

Peanut oil, sulfated. *See* Sulfated peanut oil
Pearl white. *See* Bismuth oxychloride
Pear oil. *See* Amyl acetate
Pecan shell powder. *See* Peanut oil
PEEK. *See* Polyetheretherketone
PEG. *See* Polyethylene glycol

PEG-4
CAS 25322-68-3 (generic); 112-60-7; EINECS 203-989-9
Synonyms: PEG 200; POE (4); 2,2´-[Oxybis(2,1-ethanediyloxy)]bisethanol
Definition: Polymer of ethylene oxide
Empirical: $C_8H_{18}O_5$
Formula: $H(OCH_2CH_2)_nOH$, avg. n = 4
Properties: Visc. liq., sl. char. odor; hygroscopic; m.w. 190-210; dens. 1.127 (25/25 C); visc. 4.3 cSt (210 F); supercools on freezing
Precaution: Solvent action on some plastics
Toxicology: LD50 (oral, rat) 28,900 mg/kg; mildly toxic by ingestion; heated to decomp., emits acrid smoke and irritating fumes
Uses: Lubricant for rubber molds, textile fibers, metalworking; in food and food pkg.; in cosmetics and hair preps.; pharmaceutic aid; in gas chromatography; in paints, paper coatings, polishes, ceramics
Regulatory: FDA 21CFR §73.1, 172.210, 172.770, 172.820, 173.310, 173.340, 175.105, 175.300, 178.3750; FDA approved for topicals
Manuf./Distrib.: Aldrich; Ashland; Fluka; C.P. Hall; Harwick; Henkel; Sigma; Union Carbide
Trade names: Calgene PEG 200; Pluracol® E200; Poly-G® 200; Teric PEG 200

PEG-6
CAS 25322-68-3 (generic); 2615-15-8; EINECS 220-045-1
Synonyms: Polyethylene glycol 300; PEG 300; Hexaethylene glycol; Macrogol 300
Definition: Polymer of ethylene oxide
Empirical: $C_{12}H_{26}O_7$
Formula: $H(OCH_2CH_2)_nOH$, avg. n = 6
Properties: Sp.gr. 1.124-1.127; visc. 5.4-6.4 cSt (99 C); pour pt. -15 to -8 C; flash pt. (COC) 196 C; ref. index 1.463-1.4641 (20 C); pH 4.5-7.5 (5%)
Toxicology: LD50 (oral, rat) 27,500 mg/kg; mildly toxic by ingestion; heated to decomp., emits acrid smoke and irritating fumes
Uses: Lubricant for rubber molds, textile fibers, metalworking; in food and food pkg.; in cosmetics and hair preps.; pharmaceutic aid; in gas chromatography; in paints, paper coatings, polishes, ceramics; plasticizer for cellulose nitrate
Regulatory: FDA 21CFR §172.210, 172.770, 172.820, 173.310, 173.340, 175.105, 175.300, 178.3750, 178.3910; FDA approved for parenterals, ophthalmics, topicals; USP/NF, BP compliance
Manuf./Distrib.: Aldrich; Ashland; Fluka; Harwick; Henkel; Sigma; Union Carbide
Trade names: Calgene PEG 300; Carbowax® PEG 300; Poly-G® 300; Teric PEG 300
Trade names containing: Calgene PEG 540; Carbowax® PEG 540 Blend

PEG-8
CAS 25322-68-3 (generic); 5117-19-1; EINECS 225-856-4
Synonyms: Macrogol 400; Polyethylene glycol 400; PEG 400; POE (8); 3,6,9,12,15,18,21-Heptaoxatricosane-1,2,3-diol
Definition: Polymer of ethylene oxide
Empirical: $C_{16}H_{34}O_9$
Formula: $H(OCH_2CH_2)_nOH$, avg. n = 8
Properties: Visc. liq., sl. char. odor, minimal taste; sl. hygroscopic; m.w. 380-420; dens. 1.128 (25/25 C); visc. 7.3 cSt (210 F); m.p. 4-8 C; pH 4.5-7.5 (5%)
Toxicology: LD50 (rat, oral) 30 ml/kg; low toxicity by ingestion, IV, IP routes; heated to decomp., emits acrid smoke and irritating fumes
Uses: Lubricant for rubber molds, textile fibers, metalworking; in food and food pkg.; in cosmetics and hair preps.; pharmaceutic aid; in gas chromatography; in paints, paper coatings, polishes, ceramics; plasticizer for cellulose nitrate
Regulatory: FDA 21CFR §172.210, 172.770, 172.820, 173.310, 173.340, 175.105, 175.300, 178.3750, 178.3910, 181.22, 181.30; USP/NF, BP, Ph.Eur. compliance
Manuf./Distrib.: Aldrich; Ashland; Fluka; C.P. Hall; Harwick; Henkel; Sigma; Union Carbide
Trade names: Calgene PEG 400; Poly-G® 400; Teric PEG 400

PEG-9
CAS 25322-68-3 (generic); 3386-18-3; EINECS 222-206-1
Synonyms: PEG 450; POE (9)
Definition: Polymer of ethylene oxide
Empirical: $C_{18}H_{38}O_{10}$
Formula: $H(OCH_2CH_2)_nOH$, avg. n = 9
Uses: Lubricant for rubber molds, textile fibers, metalworking; in food and food pkg.; in cosmetics and hair preps.; pharmaceutic aid; in gas chromatography; in paints, paper coatings, polishes, ceramics
Regulatory: FDA 21CFR §172.210, 172.770, 172.820, 173.310, 173.340, 175.105, 175.300, 178.3750, 178.3910
Manuf./Distrib.: Aldrich; Fluka; Sigma

PEG-12
CAS 25322-68-3 (generic); 6790-09-6; EINECS 229-859-1
Synonyms: Polyethylene glycol 600; PEG 600; POE (12); Macrogol 600
Definition: Polymer of ethylene oxide
Empirical: $C_{24}H_{50}O_{13}$
Formula: $H(OCH_2CH_2)_nOH$, avg. n = 12
Properties: Visc. liq., char. odor; sl. hygroscopic; m.w. 570-630; dens. 1.128 (25/25 C); m.p. 20-25 C; visc. 10.5 cSt (210 F); pH 4.5-7.5 (5%)
Toxicology: LD50 (oral, rat) 38,100 mg/kg; low toxicity by ingestion; eye irritant; heated to decomp., emits acrid smoke and irritating fumes
Uses: Lubricant for rubber molds, textile fibers, metalworking; in food and food pkg.; in cosmetics and hair preps.; pharmaceutic aid; in gas chromatography; in paints, paper coatings, polishes, ceramics; plasticizer for cellulose nitrate
Regulatory: FDA 21CFR §172.210, 172.770, 172.820, 173.310, 173.340, 175.105, 175.300, 178.3750, 178.3910; USP/NF compliance
Manuf./Distrib.: Aldrich; Ashland; Fluka; C.P. Hall; Harwick; Sigma; Union Carbide
Trade names: Calgene PEG 600; Macol® E-600; Poly-G® 600

PEG-14
CAS 25322-68-3 (generic)
Synonyms: POE (14); PEG 700
Definition: Polymer of ethylene oxide
Empirical: $C_{28}H_{58}O_{15}$
Formula: $H(OCH_2CH_2)_nOH$, avg. n = 14
Properties: Visc. 11.5-13.0 cSt
Uses: Lubricant for rubber molds, textile fibers, metalworking; in food and food pkg.; in cosmetics and hair preps.; pharmaceutic aid; in gas chromatography; in paints, paper coatings, polishes, ceramics
Regulatory: FDA 21CFR §172.210, 172.770, 172.820, 173.310, 173.340, 175.105, 175.300, 178.3750, 178.3910
Manuf./Distrib.: Aldrich; Fluka; Sigma

PEG-16
CAS 25322-68-3 (generic)
Synonyms: PEG 800; POE (16)
Definition: Polymer of ethylene oxide
Empirical: $C_{32}H_{66}O_{17}$
Formula: $H(OCH_2CH_2)_nOH$, avg. n = 16
Properties: Visc. 12.5-14.5 cSt
Uses: Lubricant for rubber molds, textile fibers, metalworking; in food and food pkg.; in cosmetics and hair preps.; pharmaceutic aid; in gas chromatography; in paints, paper coatings, polishes, ceramics
Regulatory: FDA 21CFR §172.210, 172.770, 172.820, 173.310, 173.340, 175.105, 175.300, 178.3750, 178.3910
Manuf./Distrib.: Aldrich; Fluka; Sigma

PEG-20
CAS 25322-68-3 (generic)
Synonyms: Polyethylene glycol 1000; PEG 1000; Macrogol 1000; POE (20)
Definition: Polymer of ethylene oxide
Empirical: $C_{40}H_{82}O_{21}$
Formula: $H(OCH_2CH_2)_nOH$, avg. n = 20
Properties: Solid; sp.gr. 1.085; visc. 16-19 cSt (99 C); pour pt. 37-40 C; flash pt. (COC) 265 C; pH 4.5-7.5 (5%)

Toxicology: LD50 (oral, rat) 42 g/kg; mod. toxic by IP, IV routes; mildly toxic by ingestion; experimental tumorigen; heated to decomp., emits acrid smoke and irritating fumes
Uses: Lubricant for rubber molds, textile fibers, metalworking; in food and food pkg.; in cosmetics and hair preps.; pharmaceutic aid; in gas chromatography; in paints, paper coatings, polishes, ceramics; plasticizer for cellulose nitrate
Regulatory: FDA 21CFR §172.210, 172.770, 172.820, 173.310, 173.340, 175.105, 175.300, 178.3750, 178.3910; USP/NF, BP, Ph.Eur. compliance
Manuf./Distrib.: Aldrich; Ashland; Fluka; Harwick; Sigma; Union Carbide
Trade names: Calgene PEG 1000; Poly-G® 1000

PEG-32
CAS 25322-68-3 (generic)
Synonyms: PEG 1540; Macrogol 1540; POE (32)
Definition: Polymer of ethylene oxide
Empirical: $C_{64}H_{130}O_{33}$
Formula: $H(OCH_2CH_2)_nOH$, avg. n = 32
Properties: Wh. powd.
Toxicology: LD50 (oral, rat) 44,200 mg/kg; mildly toxic by ingestion; human skin irritant; heated to decomp., emits acrid smoke and irritating fumes
Uses: Lubricant for rubber molds, textile fibers, metalworking; in food and food pkg.; in cosmetics and hair preps.; pharmaceutic aid; in gas chromatography; in paints, paper coatings, polishes, ceramics
Regulatory: FDA 21CFR §172.210, 172.770, 172.820, 173.310, 173.340, 175.105, 175.300, 178.3750, 178.3910; FDA approved for dentals, orals, rectals, topicals; BP compliance
Manuf./Distrib.: Aldrich; Fluka; Sigma
Trade names: Calgene PEG 1450
Trade names containing: Calgene PEG 540; Carbowax® PEG 540 Blend

PEG-40
CAS 25322-68-3 (generic)
Synonyms: PEG 2000; POE (40)
Definition: Polymer of ethylene oxide
Empirical: $C_{80}H_{162}O_{41}$
Formula: $H(OCH_2CH_2)_nOH$, avg. n = 40
Properties: Solid; dens. 1.127; visc. 38-49 cSt; ref. index 1.4590 (20 C)
Uses: Lubricant for rubber molds, textile fibers, metalworking; in food and food pkg.; in cosmetics and hair preps.; pharmaceutic aid; in gas chromatography; in paints, paper coatings, polishes, ceramics
Regulatory: FDA 21CFR §172.210, 172.770, 172.820, 173.310, 173.340, 175.105, 175.300, 178.3750, 178.3910; FDA approved for parenterals
Manuf./Distrib.: Aldrich
Trade names: Poly-G® 2000

PEG-75
CAS 25322-68-3 (generic)
Synonyms: Polyethylene glycol 3350; PEG 3350; POE (75)
Definition: Polymer of ethylene oxide
Empirical: $C_{150}H_{302}O_{76}$
Formula: $H(OCH_2CH_2)_nOH$, avg. n = 75
Properties: Wh. powd. or creamy-white flakes; m.w. 3000-3700; dens. 1.212 (25/25 C); visc. 76-110 cSt (210 F); m.p. 54-58 C; pH 4.5-7.5 (5%)
Toxicology: LD50 (oral, rat) 50 g/kg; mildly toxic by ingestion; skin irritant; heated to decomp., emits acrid smoke and irritating fumes
Uses: Lubricant for rubber molds, textile fibers, metalworking; in food and food pkg.; in cosmetics and hair preps.; pharmaceutic aid; in gas chromatography; in paints, paper coatings, polishes, ceramics
Regulatory: FDA 21CFR §172.210, 172.770, 172.820, 173.310, 173.340, 175.105, 175.300, 178.3750, 178.3910; FDA approved for injectables, orals, rectals, topicals, vaginals; USP/NF compliance
Manuf./Distrib.: Aldrich; Fluka; Sigma
Trade names: Calgene PEG 3350

PEG-100
CAS 25322-68-3 (generic)
Synonyms: PEG (100); POE (100)
Definition: Polymer of ethylene oxide
Empirical: $C_{200}H_{402}O_{101}$

Formula: $H(OCH_2CH_2)_nOH$, avg. n = 100
Uses: Lubricant for rubber molds, textile fibers, metalworking; in food and food pkg.; in cosmetics and hair preps.; pharmaceutic aid; in gas chromatography; in paints, paper coatings, polishes, ceramics
Regulatory: FDA 21CFR §172.210, 172.770, 172.820, 173.310, 173.340, 175.105, 175.300, 178.3750, 178.3910
Manuf./Distrib.: Aldrich; Fluka; Sigma

PEG (100). *See* PEG-100

PEG-150
CAS 25322-68-3 (generic)
Synonyms: PEG 6000; Macrogol 6000; POE (150)
Definition: Polymer of ethylene oxide
Formula: $H(OCH_2CH_2)_nOH$, avg. n = 150
Properties: Powd. or creamy-white flakes; water-sol.; m.w. 7000-9000; dens. 1.21 (25/25 C); visc. 470-900 cSt (210 F); m.p. 56-63 C; flash pt. > 887 F
Precaution: Combustible exposed to heat or flame
Toxicology: LD50 (rat, oral) > 50 g/kg; mildly toxic by ingestion; mutagenic data; skin irritant; heated to decomp., emits acrid smoke and irritating fumes
Uses: Lubricant for rubber molds, textile fibers, metalworking; in food and food pkg.; in cosmetics and hair preps.; pharmaceutic aid; in gas chromatography; in paints, paper coatings, polishes, ceramics
Regulatory: FDA 21CFR §172.210, 172.770, 172.820, 173.310, 173.340, 175.300, 177.2420, 178.3750, 178.3910
Manuf./Distrib.: Aldrich; Fluka; Sigma
Trade names: Calgene PEG 8000

PEG-200
CAS 25322-68-3 (generic)
Synonyms: PEG 9000; POE (200)
Definition: Polymer of ethylene oxide
Formula: $H(OCH_2CH_2)_nOH$, avg. n = 200
Properties: Solid
Uses: Lubricant for rubber molds, textile fibers, metalworking; in food and food pkg.; in cosmetics and hair preps.; pharmaceutic aid; in gas chromatography; in paints, paper coatings, polishes, ceramics
Regulatory: FDA 21CFR §172.210, 172.770, 172.820, 173.310, 173.340, 175.300, 178.3750, 178.3910; FDA approved for intramuscular injectables, orals, topicals
Manuf./Distrib.: Aldrich; Fluka; Sigma

PEG 200. *See* PEG-4
PEG 300. *See* PEG-6

PEG-350
CAS 25322-68-3 (generic)
Synonyms: PEG 20000; POE (350)
Definition: Polymer of ethylene oxide
Formula: $H(OCH_2CH_2)_nOH$, avg. n = 350
Uses: Lubricant for rubber molds, textile fibers, metalworking; in food and food pkg.; in cosmetics and hair preps.; pharmaceutic aid; in gas chromatography; in paints, paper coatings, polishes, ceramics
Regulatory: FDA 21CFR §172.770, 173.310, 175.300, 178.3910
Manuf./Distrib.: Aldrich; Fluka; Sigma

PEG 400. *See* PEG-8
PEG 450. *See* PEG-9
PEG 600. *See* PEG-12
PEG 700. *See* PEG-14
PEG 800. *See* PEG-16
PEG 1000. *See* PEG-20
PEG 1540. *See* PEG-32
PEG 2000. *See* PEG-40
PEG 3350. *See* PEG-75
PEG-5000. *See* PEG-5M
PEG 6000. *See* PEG-150
PEG-7000. *See* PEG-7M

PEG 9000. *See* PEG-200
PEG-9000. *See* PEG-9M
PEG-14000. *See* PEG-14M
PEG 20000. *See* PEG-350
PEG-23000. *See* PEG-23M
PEG-45000. *See* PEG-45M
PEG-90000. *See* PEG-90M
PEG 300,000. *See* PEG-7M
PEG 600,000. *See* PEG-14M

PEG-5M
CAS 25322-68-3 (generic)
Synonyms: PEG-5000; POE (5000)
Definition: Polymer of ethylene oxide
Formula: $H(OCH_2CH_2)_nOH$, avg. n = 5000
Uses: Water-sol. thermoplastic resin, thickener, lubricant, binder, flocculant, wet adhesive; for agric., paper mfg., wastewater treatment, ceramic and glass treatment, adhesive and paint industries; dispersant for vinyl polymerization
Regulatory: FDA 21CFR §172.770, 173.310, 175.300, 178.3910
Manuf./Distrib.: Aldrich; Fluka; Sigma
Trade names: Polyox® WSR N-80; RITA PEO-1

PEG-7M
CAS 25322-68-3 (generic)
Synonyms: PEG-7000; POE (7000); PEG 300,000
Definition: Polymer of ethylene oxide
Formula: $H(OCH_2CH_2)_nOH$, avg. n = 7000
Uses: Water-sol. thermoplastic resin, thickener, lubricant, binder, flocculant, wet adhesive; for agric., elec., paper; emollient for cosmetics; dispersant for vinyl polymerization
Regulatory: FDA 21CFR §172.770, 173.310, 175.300, 178.3910
Manuf./Distrib.: Aldrich; Fluka; Sigma
Trade names: Polyox® WSR N-750

PEG-9M
CAS 25322-68-3 (generic)
Synonyms: PEG-9000; POE (9000)
Definition: Polymer of ethylene oxide
Formula: $H(OCH_2CH_2)_nOH$, avg. n = 9000
Uses: Water-sol. thermoplastic resin, thickener, lubricant, binder, flocculant, wet adhesive in papermaking, wastewater treatment, ceramic and glass treatment, adhesive and paint industries; dispersant for vinyl polymerization; cosmetic thickener, suspension agent, friction reducer, coagulant, skin slip agent; resist. to bacterial degradation
Regulatory: FDA 21CFR §172.770, 173.310, 175.300, 178.3910
Manuf./Distrib.: Aldrich; Fluka; Sigma
Trade names: RITA PEO-2

PEG-14M
CAS 25322-68-3 (generic)
Synonyms: PEG-14000; PEG 600,000; POE (14000)
Definition: Polymer of ethylene oxide
Formula: $H(OCH_2CH_2)_nOH$, avg. n = 14,000
Uses: Water-sol. thermoplastic resin, thickener, lubricant, binder, flocculant, wet adhesive; for ceramics, papermaking; dispersant for vinyl polymerization
Regulatory: FDA 21CFR §172.770, 173.310, 175.300, 178.3910
Manuf./Distrib.: Aldrich; Fluka; Sigma
Trade names: Polyox® WSR 205; Polyox® WSR N-3000

PEG-23M
CAS 25322-68-3 (generic)
Synonyms: PEG-23000; POE (23000)
Definition: Polymer of ethylene oxide
Formula: $H(OCH_2CH_2)_nOH$, avg. n = 23000
Uses: Water-sol. thermoplastic resin, thickener, lubricant, binder, flocculant, wet adhesive; paper mfg.,

wastewater treatment, ceramic and glass treatment, adhesive and paint industries; dispersant for vinyl polymerization
Regulatory: FDA 21CFR §172.770, 173.310, 175.300, 178.3910
Manuf./Distrib.: Aldrich; Fluka; Sigma
Trade names: RITA PEO-3

PEG-45M
CAS 25322-68-3 (generic)
Synonyms: PEG-45000; POE (45000)
Definition: Polymer of ethylene oxide
Formula: $H(OCH_2CH_2)_nOH$, avg. n = 45,000
Uses: Water-sol. thermoplastic resin, thickener, lubricant, binder, flocculant, wet adhesive, paper mfg., wastewater treatment, ceramic and glass treatment, adhesives, paints; lubricant/emollient in cosmetics; dispersant for vinyl polymerization
Regulatory: FDA 21CFR §172.770, 173.310, 175.300, 178.3910
Manuf./Distrib.: Aldrich; Fluka; Sigma
Trade names: RITA PEO-8

PEG-90M
CAS 25322-68-3 (generic)
Synonyms: PEG-90000; POE (90000)
Definition: Polymer of ethylene oxide
Formula: $H(OCH_2CH_2)_nOH$, avg. n = 90000
Uses: Water-sol. thermoplastic resin, thickener, lubricant, binder, flocculant, wet adhesive; dispersant for vinyl polymerization; agric., mining, construction, cosmetics, paper mfg., wastewater treatment, ceramic and glass treatment, adhesives, paints; foam stabilizer in malt beverages
Regulatory: FDA 21CFR §172.770, 173.310, 175.300, 178.3910
Manuf./Distrib.: Aldrich; Fluka; Sigma
Trade names: RITA PEO-18

PEG-160M
Definition: Polymer of ethylene oxide
Uses: Polymer providing lubricity on tack to many formulations; for paper mfg., wastewater treatment, ceramic and glass treatment, adhesives, paints; cosmetic thickener, suspension agent, friction reducer, coagulant, skin slip agent; resist. to bacterial degradation
Trade names: RITA PEO-27

PEG-6 bisphenol A
Uses: Monomer for polyester and urethane coatings
Trade names: Ethal BPA-6

PEG-4 bisphenol A dimethacrylate
Trade names containing: CD-540

PEG-6 bisphenol A dimethacrylate
Trade names containing: CD-541

PEG-2 butyl ether. *See* Butoxydiglycol
PEG-3 butyl ether. *See* Butoxytriglycol

PEG-4 butyl ether
Synonyms: Tetraethylene glycol butyl ether; Tetraethylene glycol monobutyl ether
Trade names containing: Icinol BE33

PEG-5 castor oil
CAS 61791-12-6 (generic)
Synonyms: POE (5) castor oil
Definition: PEG deriv. of castor oil with avg. 5 moles of ethylene oxide
Uses: Surfactant used as emulsifier, dispersant, solubilizer, visc. control agent for cosmetics, pharmaceuticals, and industrial applics.; emulsifier in lubricants for plastics, metals, textiles; paper, textile, leather; pigment dispersant in latex paints
Regulatory: FDA 21CFR §175.105, 175.300

PEG-12 castor oil

Manuf./Distrib.: Fluka; Sigma
Trade names: Chemax CO-5; Prox-onic HR-05; Surfactol® 318; Trylox® 5900

PEG-12 castor oil
CAS 61791-12-6 (generic)
Synonyms: POE (12) castor oil
Definition: PEG deriv. of castor oil with avg. 12 moles of ethylene oxide
Uses: Detergent, emulsifier, coemulsifier in sol. oils, solv. cleaners, coatings, industrial/institutional cleaners, metalworking; mold release in plastics; textile fiber lubricant
Manuf./Distrib.: Fluka; Sigma
Trade names: Teric C12

PEG-15 castor oil
CAS 61791-12-6 (generic)
Synonyms: POE (15) castor oil
Definition: PEG deriv. of castor oil with avg. 15 moles of ethylene oxide
Uses: Surfactant used as emulsifier, lubricant, dispersant, solubilizer, visc. control agent for cosmetics, pharmaceuticals, and industrial applics.; coemulsifier for herbicides, metalworking, hydraulic fluids, rayon delustrants, paint
Regulatory: FDA 21CFR §175.105, 175.300, 176.210, 177.2800
Manuf./Distrib.: Fluka; Sigma
Trade names: Alkamuls® CO-15

PEG-16 castor oil
CAS 61791-12-6 (generic)
Synonyms: POE (16) castor oil
Definition: PEG deriv. of castor oil with avg. 16 moles of ethylene oxide
Uses: Lubricant additive, emulsifier in lubricants for plastics, metals, textiles; clay and pigment dispersants, rewetting agent, softener, dyeing assistant for paint, paper, textile, and leather industries
Manuf./Distrib.: Fluka; Sigma
Trade names: Chemax CO-16; Ethox CO-16; Prox-onic HR-016

PEG-20 castor oil
CAS 61791-12-6 (generic)
Synonyms: POE (20) castor oil; PEG 1000 castor oil
Definition: PEG deriv. of castor oil with avg. 20 moles of ethylene oxide
Properties: Pale yel. oil; HLB 9.0
Uses: Emulsifier, solubilizer for leather, metal, textile processing, paints, paper, ink, rubber, polishes, and agric. applics
Regulatory: FDA 21CFR §175.105, 175.300, 176.210, 177.2800
Manuf./Distrib.: Fluka; Sigma
Trade names: T-Det® C-20

PEG-25 castor oil
CAS 61791-12-6 (generic)
Synonyms: POE (25) castor oil
Definition: PEG deriv. of castor oil with avg. 25 moles of ethylene oxide
Uses: Lubricant additive, emulsifier in lubricants for plastics, metals, textiles; clay and pigment dispersants, rewetting agent, softener, dyeing assistant for paint, paper, textile, and leather industries
Regulatory: FDA 21CFR §175.105, 175.300, 177.2800
Manuf./Distrib.: Fluka; Sigma
Trade names: Calgene Nonionic GR-25; Chemax CO-25; Prox-onic HR-025; Trylox® 5904

PEG-28 castor oil
CAS 61791-12-6 (generic)
Synonyms: POE (28) castor oil
Definition: PEG deriv. of castor oil with avg. 28 moles of ethylene oxide
Uses: Emulsifier; pigment dispersant in textiles, paint, paper, leather
Manuf./Distrib.: Fluka; Sigma
Trade names: Chemax CO-28

PEG-30 castor oil
CAS 61791-12-6 (generic)

Synonyms: POE (30) castor oil
Definition: PEG deriv. of castor oil with avg. 30 moles of ethylene oxide
Uses: Lubricant additive, emulsifier in lubricants for plastics, metals, textiles; clay and pigment dispersants, rewetting agent, softener, dyeing assistant for paint, paper, textile, and leather industries
Regulatory: FDA 21CFR §175.105, 175.300, 177.2800
Manuf./Distrib.: Fluka; Sigma
Trade names: Alkamuls® EL-620; Alkamuls® EL-620L; Chemax CO-30; Prox-onic HR-030; Trylox® 5906

PEG-32 castor oil
CAS 61791-12-6 (generic)
Synonyms: POE (32) castor oil
Definition: PEG deriv. of castor oil with avg. 32 moles of ethylene oxide
Uses: Surfactant, emulsifier, softener, rewetting agent, pigment dispersant, dye assistant, leveling agent for paint, textile, leather, plastics lubricants, metal lubricants
Manuf./Distrib.: Fluka; Sigma
Trade names: Berol 199

PEG-36 castor oil
CAS 61791-12-6 (generic)
Synonyms: POE (36) castor oil; PEG 1800 castor oil
Definition: PEG deriv. of castor oil with avg. 36 moles of ethylene oxide
Properties: Liq.; sol. in water, xylene
Uses: Emulsifier, softener, lubricant, rewetting agent, dye leveler/assistant, antistat, pigment dispersant; in textiles, leather, paper, polyurethane foams, paint; lubricant additive/emulsifier in lubricants for plastics, metals, textiles
Regulatory: FDA 21CFR §175.105, 175.300, 176.210, 177.2800
Manuf./Distrib.: Fluka; Sigma
Trade names: Chemax CO-36; Prox-onic HR-036

PEG-40 castor oil
CAS 61791-12-6 (generic)
Synonyms: POE (40) castor oil; PEG 2000 castor oil; Polyoxyl 40 castor oil
Definition: PEG deriv. of castor oil with avg. 40 moles of ethylene oxide
Uses: Lubricant additive, emulsifier in lubricants for plastics, metals, textiles; clay and pigment dispersants, rewetting agent, softener, dyeing assistant for paint, paper, textile, and leather industries
Regulatory: FDA 21CFR §175.105, 175.300, 176.170, 176.180, 176.210, 177.2800; FDA approved for parenterals
Manuf./Distrib.: Fluka; Sigma
Trade names: Alkamuls® EL-719; Berol 108; Calgene Nonionic GR-40; Chemax CO-40; Prox-onic HR-040; Surfactol® 365; T-Det® C-40

PEG-52 castor oil
CAS 61791-12-6 (generic)
Synonyms: POE (52) castor oil
Definition: PEG deriv. of castor oil with avg. 52 moles of ethylene oxide
Uses: Surfactant for industrial and cosmetic applics., pulp/paper, pharmaceuticals, metalworking, lubricants, textiles, agric., paints, adhesives
Manuf./Distrib.: Fluka; Sigma
Trade names: Calgene Nonionic GR-52

PEG-75 castor oil
CAS 61791-12-6 (generic)
Synonyms: POE (75) castor oil; PEG (75) castor oil
Definition: PEG deriv. of castor oil with avg. 75 moles of ethylene oxide
Uses: Surfactant, emulsifier, softener, rewetting agent, pigment dispersant, dye assistant, leveling agent for paints, textiles, leather, plastics/metal lubricants
Trade names: Berol 190

PEG (75) castor oil. *See* PEG-75 castor oil

PEG-80 castor oil
CAS 61791-12-6 (generic)
Synonyms: POE (80) castor oil

PEG-160 castor oil

Definition: PEG deriv. of castor oil with avg. 80 moles of ethylene oxide
Uses: O/w emulsifier and solubilizer; emulsifier for industrial lubricants, fiber finish and textile lubricants; pigment dispersant in textiles, paint, paper, leather
Manuf./Distrib.: Fluka; Sigma
Trade names: Chemax CO-80; Prox-onic HR-080

PEG-160 castor oil
CAS 61791-12-6 (generic)
Uses: Surfactant, emulsifier, softener, rewetting agent, pigment dispersant, dye assistant, leveling agent for paints, textiles, leather, plastics/metal lubricants
Manuf./Distrib.: Fluka; Sigma
Trade names: Berol 198

PEG-200 castor oil
CAS 61791-12-6 (generic)
Synonyms: POE (200) castor oil; PEG (200) castor oil
Definition: PEG deriv. of castor oil with avg. 200 moles of ethylene oxide
Uses: Lubricant additive, emulsifier in lubricants for plastics, metals, textiles; clay and pigment dispersants, rewetting agent, softener, dyeing assistant for paint, paper, textile, and leather industries
Regulatory: FDA 21CFR §175.300
Manuf./Distrib.: Fluka; Sigma
Trade names: Berol 191; Chemax CO-200/50; Prox-onic HR-0200; Prox-onic HR-0200/50

PEG (200) castor oil. *See* PEG-200 castor oil
PEG 1000 castor oil. *See* PEG-20 castor oil
PEG 1800 castor oil. *See* PEG-36 castor oil
PEG 2000 castor oil. *See* PEG-40 castor oil
PEG-6 cetyl/stearyl ether. *See* Ceteareth-6
PEG-8 cetyl/stearyl ether. *See* Ceteareth-8
PEG-10 cetyl/stearyl ether. *See* Ceteareth-10
PEG-11 cetyl/stearyl ether. *See* Ceteareth-11
PEG-12 cetyl/stearyl ether. *See* Ceteareth-12
PEG-14 cetyl/stearyl ether. *See* Ceteareth-14
PEG-18 cetyl/stearyl ether. *See* Ceteareth-18
PEG-27 cetyl/stearyl ether. *See* Ceteareth-27
PEG-30 cetyl/stearyl ether. *See* Ceteareth-30
PEG-55 cetyl/stearyl ether. *See* Ceteareth-55
PEG 400 cetyl/stearyl ether. *See* Ceteareth-8
PEG-7 C12-15 fatty alcohol ether. *See* C12-15 pareth-7

PEG-10 cocamine
CAS 61791-14-8 (generic)
Synonyms: POE (10) coconut amine; PEG 500 coconut amine
Definition: PEG deriv. of cocamine
Formula: $R-N(CH_2CH_2O)_xH(CH_2CH_2O)_yH$, R rep. alkyl groups from coconut oil, avg. (x+y)=10
Uses: Coemulsifier, antistat for textiles and plastics; dispersant; emulsifier in industrial lubricants, agric., inks, cosmetics; wetting agent for acid or alkaline metal cleaners, stripping of surf. coatings; oil emulsifier with anticorrosive properties; antistat for syn. fibers with PS
Trade names: Ethylan® TN-10

PEG-15 cocamine
CAS 8051-52-3 (generic); 61791-14-8 (generic)
Synonyms: POE (15) coconut amine
Definition: PEG deriv. of cocamine
Formula: $R-N(CH_2CH_2O)_xH(CH_2CH_2O)_yH$, R rep. alkyl groups from coconut oil, avg. (x+y)=15
Uses: Hydrophilic emulsifier, textile dyeing agent, dye leveler, antiprecipitant, stripping agent; agric.; plastics/fiber antistat; intermediate for quats.; wetting agent for metal cleaning, stripping of surf. coatings, textiles, paints, agric., polishes
Trade names: Ethylan® TC

PEG-2 coco-benzonium chloride
CAS 61789-68-2
Synonyms: PEG-2 cocobenzyl ammonium chloride; PEG 100 coco-benzonium chloride; POE (2) coco-

benzonium chloride
Classification: Quaternary ammonium salt
Trade names containing: Ethoquad® CB/12

PEG 100 coco-benzonium chloride. *See* PEG-2 coco-benzonium chloride
PEG-2 cocobenzyl ammonium chloride. *See* PEG-2 coco-benzonium chloride
PEG-2 cocomethyl ammonium nitrate. *See* PEG-2 cocomonium nitrate

PEG-2 cocomonium nitrate
CAS 71487-00-8
Synonyms: PEG-2 cocomethyl ammonium nitrate
Trade names containing: Ethoquad® C/12 Nitrate

PEG 500 coconut amine. *See* PEG-10 cocamine

PEG-3 diacetate
Synonyms: TEGDA; Triethylene glycol diacetate; POE (3) diacetate
Uses: Plasticizer for films and coatings; lubricants; cosmetic emollients
Manuf./Distrib.: Bayer
Trade names: SR-322

PEG-3 diacrylate
CAS 1680-21-3
Synonyms: Triethylene glycol diacrylate
Properties: M.w. 258.3; m.p. 125 (2 mm)
Uses: Curing agent for flexible substrates, photopolymeric printing plates, and no-wax vinyl tile coatings
Manuf./Distrib.: CPS; Monomer-Polymer & Dajac Labs
Trade names containing: SR-272

PEG-4 diacrylate
CAS 17831-71-9
Synonyms: Tetraethylene glycol diacrylate; PEG 200 diacrylate
Properties: M.w. 302.3; b.p. > 120 (0.3 mm)
Uses: Crosslinking agent for radiation-cured coatings, inks, adhesives, textile prods.
Manuf./Distrib.: Aldrich; Monomer-Polymer & Dajac
Trade names: Photomer® 4050
Trade names containing: Ageflex T4EGDA; SR-259; SR-268

PEG-8 diacrylate
Uses: Curing agent
Trade names containing: CN 960 K75; CN 961 K75; CN 962 K75; CN 963 K75; CN 964 K75; CN 965 K75; CN 966 K75; CN 970 K60; CN 971 K75; CN 972 K75; SR-344

PEG-12 diacrylate
Uses: Mod. reactive monomer; flexible film former; for flexible substrates and photopolymeric printing plates
Trade names: Photomer® 4056

PEG 200 diacrylate. *See* PEG-4 diacrylate
PEG-2 dibenzoate. *See* Diethylene glycol dibenzoate

PEG-4 dibenzoate
Synonyms: PEG 200 dibenzoate
Properties: Sp.gr. 1.158; m.p. -40 C; flash pt. (COC) 248 C; ref. index 1.5252
Uses: Plasticizer for PVAc adhesive formulations, phenol-formaldehyde resins, alkyd-modified phenol-formaldehyde varnishes
Manuf./Distrib.: Kalama; Velsicol
Trade names: Benzoflex® P-200

PEG 100 dibenzoate. *See* Diethylene glycol dibenzoate
PEG 200 dibenzoate. *See* PEG-4 dibenzoate

PEG-4 di-2-ethylhexoate
Synonyms: Tetraethylene glycol di-2-ethylhexoate

PEG-8 dilaurate

Properties: Sp.gr. 0.984; m.p. -65 C; flash pt. (COC) 204 C; ref. index 1.4445
Uses: Lubricant for aluminum can industry; plasticizer for PVC, PVAc, PS, cellulosics, rubber
Manuf./Distrib.: C.P. Hall; Inolex
Trade names: TegMeR® 804

PEG-8 dilaurate
CAS 9005-02-1 (generic)
Synonyms: POE (8) dilaurate; PEG 400 dilaurate
Definition: PEG diester of lauric acid
Empirical: $C_{40}H_{78}O_{11}$
Formula: $CH_3(CH_2)_{10}CO(OCH_2CH_2)_nOCO(CH_2)_{10}CH_3$, avg. n = 8
Properties: Sp.gr. 1.030; m.p. 15 C; flash pt. (COC) 249 C; ref. index 1.459
Uses: Emulsifier, solubilizer, dispersing agent, wetting agent, lubricant, softener, release agent, coupling agent used in personal care prods. and industrial applics., agric. chemical sprays, industrial and textile lubricants; plasticizer
Regulatory: FDA 21CFR §175.105, 175.300, 176.210, 177.1210, 177.2260, 177.2800, 178.3520
Manuf./Distrib.: C.P. Hall; Inolex; Velsicol
Trade names: Emerest® 2652; Pegosperse® 400 DL

PEG 400 dilaurate. *See* PEG-8 dilaurate

PEG-3 dimethacrylate
CAS 109-16-0
Synonyms: Triethylene glycol dimethacrylate
Formula: $[H_2C{=}C(CH_3)CO_2CH_2CH_2OCH_2]_2$
Properties: M.w. 286.2; b.p. 162 C (1.2 mm)
Uses: Crosslinking monomeric ester; in vinyl plastisols reduces initial visc. and oil extractability, and improves ultimate hardness, heat distort., hot tear strength, and stain resistance
Manuf./Distrib.: Aldrich; CPS; Fluka; Monomer-Polymer & Dajac Labs; Rohm Tech; Sigma
Trade names containing: SR-205

PEG-4 dimethacrylate
CAS 25852-47-5
Synonyms: Tetraethylene glycol dimethacrylate; PEG 200 dimethacrylate
Formula: $H_2C{=}C(CH_3)CO(OCH_2CH_2)_nO_2CC(CH_3)$
Properties: M.w. 330.4; b.p. 220 (1.0 mm)
Precaution: Light-sensitive
Uses: Crosslinking agent used in castings, plastisols, coatings, fibers, papers, etc.
Manuf./Distrib.: Aldrich; Monomer-Polymer & Dajac
Trade names containing: SR-209

PEG-12 dimethacrylate
Trade names containing: SR-252

PEG 200 dimethacrylate. *See* PEG-4 dimethacrylate
PEG-8 dinonyl phenyl ether. *See* Nonyl nonoxynol-8
PEG-9 dinonyl phenyl ether. *See* Nonyl nonoxynol-9
PEG-15 dinonyl phenyl ether. *See* Nonyl nonoxynol-15
PEG-18 dinonyl phenyl ether. *See* Nonyl nonoxynol-18
PEG-24 dinonyl phenyl ether. *See* Nonyl nonoxynol-24
PEG-49 dinonyl phenyl ether. *See* Nonyl nonoxynol-49
PEG-150 dinonyl phenyl ether. *See* Nonyl nonoxynol-150

PEG-9 dioleate
Synonyms: POE (9) dioleate
Definition: PEG diester of oleic acid
Uses: Lipophilic emulsifier and solubilizer for min. oils, fats and solvs.; emulsifier for kerosene in agric. and pesticides sprays; emulsification of latex paints, metalworking fluids, solvs., specialty and industrial lubricants

PEG-12 dioleate
CAS 9005-07-6 (generic); 52688-97-0 (generic); 85736-49-8; EINECS 288-459-5
Synonyms: POE (12) dioleate; PEG 600 dioleate

Definition: PEG diester of oleic acid
Uses: Emulsifier, dispersant; additive for cutting oils; component of paper defoamers, softeners, lubricants; cosmetics, pharmaceuticals, foods, agric., plastics, solv. and solventless coatings; lubricant for textile spin finishes
Regulatory: FDA 21CFR §173.340, 175.105, 175.300, 176.200, 176.210, 177.2260, 177.2800
Trade names: Pegosperse® 600 DO

PEG 600 dioleate. *See* PEG-12 dioleate

PEG-150 distearate
CAS 9005-08-7 (generic); 52668-97-0
Synonyms: POE (150) distearate; PEG 6000 distearate; Polyoxyl 150 distearate
Definition: PEG diester of stearic acid
Formula: $CH_3(CH_2)_{16}CO(OCH_2CH_2)_nOCO(CH_2)_{16}CH_3$, avg. n = 150
Properties: M.p. 35-37 C; flash pt. > 110 C
Uses: Thickener, dispersant, emollient, emulsifier in cosmetics, pharmaceuticals, food, agric., plastics, textiles, metalworking; water-based paints and lubricants, and protective coatings
Regulatory: FDA 21CFR §175.300; FDA approved for topicals
Manuf./Distrib.: Aldrich; Sigma
Trade names: G-1821; Witconol™ L3245

PEG 6000 distearate. *See* PEG-150 distearate

PEG-8 ditallate
CAS 61791-01-3 (generic)
Synonyms: POE (8) ditallate; PEG 400 ditallate
Definition: PEG diester of tall oil acid
Formula: $RCO-(OCH_2CH_2)_nOCOR$, RCO- rep. tall oil fatty radicals, avg. n = 8
Uses: Industrial detergent; solubilizer for min. oils, fats, solvs.; for latex paints, metalworking fluids, industrial lubricants, textile specialties
Regulatory: FDA 21CFR §175.105, 175.30, 176.210, 177.1210, 177.2800
Trade names: Laurel PEG 400 DT; Pegosperse® 400 DOT

PEG-12 ditallate
CAS 61791-01-3 (generic)
Synonyms: POE (12) ditallate; PEG 600 ditallate
Definition: PEG diester of tall oil acid
Formula: $RCO-(OCH_2CH_2)_nOCOR$, RCO- rep. tall oil fatty radicals, avg. n = 12
Uses: Surfactant, emulsifier and solubilizer for min. oils, fats, solvs.; for latex paints, metalworking fluids, industrial lubricants, textile specialties
Regulatory: FDA 21CFR §175.105, 175.300, 176.210, 177.2800
Trade names: Laurel PEG 600 DT; Pegosperse® 600 DOT

PEG 400 ditallate. *See* PEG-8 ditallate
PEG 600 ditallate. *See* PEG-12 ditallate
PEG-10 dodecyl phenyl ether. *See* Dodoxynol-10
PEG-3 ethyl ether. *See* Ethoxytriglycol

PEG-5 hydrogenated castor oil
CAS 61788-85-0 (generic)
Synonyms: POE (5) hydrogenated castor oil; PEG (5) hydrogenated castor oil
Definition: PEG deriv. of hydrogenated castor oil with avg. 5 moles of ethylene oxide
Uses: Lubricant additive, emulsifier in lubricants for plastics, metals, textiles; clay and pigment dispersants, rewetting agent, softener, dyeing assistant for paint, paper, textile, and leather industries
Trade names: Chemax HCO-5; Prox-onic HRH-05

PEG (5) hydrogenated castor oil. *See* PEG-5 hydrogenated castor oil

PEG-16 hydrogenated castor oil
CAS 61788-85-0 (generic)
Synonyms: POE (16) hydrogenated castor oil
Definition: PEG deriv. of hydrogenated castor oil with avg. 16 moles of ethylene oxide

PEG-25 hydrogenated castor oil

Uses: Lubricant additive, emulsifier in lubricants for plastics, metals, textiles; clay and pigment dispersants, rewetting agent, softener, dyeing assistant for paint, paper, textile, and leather industries
Regulatory: FDA 21CFR §177.2800
Trade names: Chemax HCO-16; Prox-onic HRH-016

PEG-25 hydrogenated castor oil
CAS 61788-85-0 (generic)
Synonyms: POE (25) hydrogenated castor oil
Definition: PEG deriv. of hydrogenated castor oil with avg. 25 moles of ethylene oxide
Uses: Lubricant additive, emulsifier in lubricants for plastics, metals, textiles; clay and pigment dispersants, rewetting agent, softener, dyeing assistant for paint, paper, textile, and leather industries
Regulatory: FDA 21CFR §177.2800
Trade names: Chemax HCO-25; Prox-onic HRH-025

PEG-200 hydrogenated castor oil
CAS 61788-85-0 (generic)
Synonyms: POE (200) hydrogenated castor oil; PEG (200) hydrogenated castor oil
Definition: PEG deriv. of hydrogenated castor oil with avg. 200 moles of ethylene oxide
Uses: Lubricant additive, emulsifier in lubricants for plastics, metals, textiles; clay and pigment dispersants, rewetting agent, softener, dyeing assistant for paint, paper, textile, and leather industries
Trade names: Chemax HCO-200/50; Prox-onic HRH-0200; Prox-onic HRH-0200/50

PEG (200) hydrogenated castor oil. *See* PEG-200 hydrogenated castor oil

PEG-8 isolauryl thioether
Synonyms: POE (8) isolauryl thioether; PEG 400 isolauryl thioether
Definition: PEG ether of a branched chain dodecyl mercaptan
Formula: $C_{12}H_{25}$-S-$(CH_2CH_2O)_nH$, avg. n = 8
Uses: Wetting agent, metal cleaning; heavy duty detergent; inhibitor in steel processing; scouring of textiles; agric. formulations; antiskinning agent for paints; hair prods.; wood and paper industry
Trade names: Alcodet® SK

PEG 400 isolauryl thioether. *See* PEG-8 isolauryl thioether

PEG-2 laurate
CAS 141-20-8; 9004-81-3; EINECS 205-468-1
Synonyms: Diethylene glycol laurate; Diglycol laurate; PEG 100 monolaurate
Definition: PEG ester of lauric acid
Formula: $CH_3(CH_2)_{10}CO(OCH_2CH_2)_nOH$, avg. n = 2
Uses: W/o emulsifier, dispersant, antistat, defoamer for textile, paper processing, cutting oils, polishes, emulsion cleaners, emulsion polymerization, rubber latex, wool lubricants, paints
Regulatory: FDA 21CFR §175.105, 175.300, 176.200, 176.210, 177.2800, 178.3910
Manuf./Distrib.: ABITEC; Henkel/Emery; Inolex; Lonza; Mona Ind.; Stepan; Witco
Trade names: Calgene DGL

PEG-2 laurate SE
CAS 141-20-8
Synonyms: Diethylene glycol monolaurate self-emulsifying; PEG 100 monolaurate self-emulsifying; POE (2) monolaurate self-emulsifying
Definition: Self-emulsifying grade of PEG-2 laurate containing some sodium and/or potassium laurate
Uses: Spreading agent, w/o emulsifier, dispersant, lubricant, opacifier, emulsion stabilizer, emollient, visc. builder , antistat used in cosmetics, textiles, paints, adhesives, paper processing, cutting oils, polishes, rubber latex; defoamer for process applics.
Trade names: Pegosperse® 100 L

PEG-4 laurate
CAS 9004-81-3 (generic); 10108-24-4
Synonyms: POE (4) monolaurate; PEG 200 monolaurate
Definition: PEG ester of lauric acid
Empirical: $C_{20}H_{40}O_6$
Formula: $CH_3(CH_2)_{10}CO(OCH_2CH_2)_nOH$, avg. n = 4
Uses: Emulsifier, lubricant, dispersing and leveling agent, coupling agent, solubilizer, wetting agent, thickener, defoamer used in cosmetic, pharmaceuticals, textile, plastics, paint and other industrial uses

Regulatory: FDA 21CFR §175.105, 175.300, 176.210, 178.3910
Trade names: Ablunol 200ML; Chemax E-200 ML; Pegosperse® 200 ML

PEG-5 laurate
CAS 9004-81-3 (generic)
Synonyms: POE (5) monolaurate
Definition: PEG ester of lauric acid
Formula: $CH_3(CH_2)_{10}CO(OCH_2CH_2)_nOH$, avg. n = 5
Uses: Emulsifier and coupling agent; defoamer in water base coatings; visc. depressant in vinyl plastic sols.; control additive in hair rinse formulations; paper softener

PEG-8 laurate
CAS 9004-81-3 (generic); 35179-86-3; 37318-14-2; EINECS 253-458-0
Synonyms: POE (8) monolaurate; PEG 400 monolaurate
Definition: PEG ester of lauric acid
Empirical: $C_{28}H_{56}O_{10}$
Formula: $CH_3(CH_2)_{10}CO(OCH_2CH_2)_nOH$, avg. n = 8
Properties: Insol. in water
Uses: Emulsifier, lubricant, dispersing and leveling agent, solubilizer, visc. control agent, defoamer used in cosmetic, pharmaceutical, textile, paint and other industrial uses; antiblock in vinyls; plasticizer
Regulatory: FDA 21CFR §175.105, 175.300, 176.170, 176.210, 177.1200, 177.1210, 177.2260, 177.2800, 178.3520, 178.3760, 178.3910
Trade names: Ablunol 400ML; Alkamuls® PE/400; Chemax E-400 ML; Karapeg 400-ML; Pegosperse® 400 ML

PEG-9 laurate
CAS 106-08-1; 9004-81-3 (generic); EINECS 203-359-3
Synonyms: POE (9) monolaurate
Definition: PEG ester of lauric acid
Empirical: $C_{30}H_{60}O_{11}$
Formula: $CH_3(CH_2)_{10}CO(OCH_2CH_2)_nOH$, avg. n = 9
Uses: Nonionic surfactant, emulsifier, lubricant, dispersant, softener for cosmetics, cleaners, dyeing, metal cleaning, leather processing; defoamer, leveling agent for latex paints; dispersant for dyes and pigments
Regulatory: FDA 21CFR §175.105, 175.300, 177.2260, 177.2800, 178.3910
Trade names: Alkamuls® L-9

PEG-12 laurate
CAS 9004-81-3 (generic)
Synonyms: POE (12) monolaurate; PEG 600 monolaurate
Definition: PEG ester of lauric acid
Formula: $CH_3(CH_2)_{10}CO(OCH_2CH_2)_nOH$, avg. n = 12
Uses: Emulsifier, lubricant, dispersing and leveling agent, solubilizer used in cosmetic, textile, paint and other industrial uses; plastics antistat
Regulatory: FDA 21CFR §175.105, 175.300, 176.170, 176.210, 177.1200, 177.2260, 177.2800, 178.3910
Trade names: Ablunol 600ML; Karapeg 600-ML; Pegosperse® 600 ML

PEG-2 lauryl ether. *See* Laureth-2
PEG-4 lauryl ether. *See* Laureth-4
PEG-5 lauryl ether. *See* Laureth-5
PEG-8 lauryl ether. *See* Laureth-8
PEG-9 lauryl ether. *See* Laureth-9
PEG-12 lauryl ether. *See* Laureth-12
PEG-23 lauryl ether. *See* Laureth-23
PEG 200 lauryl ether. *See* Laureth-4
PEG 600 lauryl ether. *See* Laureth-12
PEG (1-4) lauryl ether sulfate, sodium salt. *See* Sodium laureth sulfate
PEG-3 methyl ether. *See* Triglycol monomethyl ether
PEG 100 monolaurate. *See* PEG-2 laurate
PEG 200 monolaurate. *See* PEG-4 laurate
PEG 400 monolaurate. *See* PEG-8 laurate
PEG 600 monolaurate. *See* PEG-12 laurate
PEG 100 monolaurate self-emulsifying. *See* PEG-2 laurate SE
PEG 200 monooleate. *See* PEG-4 oleate
PEG 400 monooleate. *See* PEG-8 oleate

PEG 600 monooleate. *See* PEG-12 oleate
PEG 1000 monooleate. *See* PEG-20 oleate
PEG 100 monooleate self-emulsifying. *See* PEG-2 oleate SE
PEG 100 monostearate. *See* PEG-2 stearate
PEG 200 monostearate. *See* PEG-4 stearate
PEG 400 monostearate. *See* PEG-8 stearate
PEG 600 monostearate. *See* PEG-12 stearate
PEG 1000 monostearate. *See* PEG-20 stearate
PEG 100 monostearate self-emulsifying. *See* PEG-2 stearate SE

PEG-4 nonyl phenol acrylate
Trade names: CD-504

PEG-1 nonyl phenyl ether. *See* Nonoxynol-1
PEG-2 nonyl phenyl ether. *See* Nonoxynol-2
PEG-3 nonyl phenyl ether. *See* Nonoxynol-3
PEG-4 nonyl phenyl ether. *See* Nonoxynol-4
PEG-5 nonyl phenyl ether. *See* Nonoxynol-5
PEG-6 nonyl phenyl ether. *See* Nonoxynol-6
PEG-8 nonyl phenyl ether. *See* Nonoxynol-8
PEG-9 nonyl phenyl ether. *See* Nonoxynol-9
PEG-10 nonyl phenyl ether. *See* Nonoxynol-10
PEG-11 nonyl phenyl ether. *See* Nonoxynol-11
PEG-12 nonyl phenyl ether. *See* Nonoxynol-12
PEG-13 nonyl phenyl ether. *See* Nonoxynol-13
PEG-14 nonyl phenyl ether. *See* Nonoxynol-14
PEG-20 nonyl phenyl ether. *See* Nonoxynol-20
PEG-30 nonyl phenyl ether. *See* Nonoxynol-30
PEG-40 nonyl phenyl ether. *See* Nonoxynol-40
PEG-50 nonyl phenyl ether. *See* Nonoxynol-50
PEG-70 nonyl phenyl ether. *See* Nonoxynol-70
PEG 100 nonyl phenyl ether. *See* Nonoxynol-2
PEG-100 nonyl phenyl ether. *See* Nonoxynol-100
PEG 200 nonyl phenyl ether. *See* Nonoxynol-4
PEG 300 nonyl phenyl ether. *See* Nonoxynol-6
PEG 400 nonyl phenyl ether. *See* Nonoxynol-8
PEG 450 nonyl phenyl ether. *See* Nonoxynol-9
PEG 500 nonyl phenyl ether. *See* Nonoxynol-10
PEG 600 nonyl phenyl ether. *See* Nonoxynol-12
PEG 1000 nonyl phenyl ether. *See* Nonoxynol-20
PEG 2000 nonyl phenyl ether. *See* Nonoxynol-40
PEG-10 nonyl phenyl ether carboxylic acid. *See* Nonoxynol-10 carboxylic acid
PEG 500 nonyl phenyl ether carboxylic acid. *See* Nonoxynol-10 carboxylic acid
PEG-6 nonyl phenyl ether phosphate. *See* Nonoxynol-6 phosphate
PEG-9 nonyl phenyl ether phosphate. *See* Nonoxynol-9 phosphate
PEG 300 nonyl phenyl ether phosphate. *See* Nonoxynol-6 phosphate
PEG 450 nonyl phenyl ether phosphate. *See* Nonoxynol-9 phosphate
PEG 300 nonyl phenyl ether phosphate, sodium salt. *See* Sodium nonoxynol-6 phosphate
PEG-4 nonyl phenyl ether sulfate, sodium salt. *See* Sodium nonoxynol-4 sulfate
PEG 200 nonyl phenyl ether sulfate, sodium salt. *See* Sodium nonoxynol-4 sulfate
PEG-1 octyl phenyl ether. *See* Octoxynol-1
PEG-3 octyl phenyl ether. *See* Octoxynol-3
PEG-5 octyl phenyl ether. *See* Octoxynol-5
PEG-7 octyl phenyl ether. *See* Octoxynol-7
PEG-8 octyl phenyl ether. *See* Octoxynol-8
PEG-9 octyl phenyl ether. *See* Octoxynol-9
PEG-10 octyl phenyl ether. *See* Octoxynol-10
PEG-13 octyl phenyl ether. *See* Octoxynol-13
PEG-16 octyl phenyl ether. *See* Octoxynol-16
PEG-25 octyl phenyl ether. *See* Octoxynol-25
PEG-30 octyl phenyl ether. *See* Octoxynol-30
PEG-40 octyl phenyl ether. *See* Octoxynol-40

PEG-70 octyl phenyl ether. *See* Octoxynol-70
PEG 400 octyl phenyl ether. *See* Octoxynol-8
PEG 450 octyl phenyl ether. *See* Octoxynol-9
PEG 500 octyl phenyl ether. *See* Octoxynol-10

PEG-2 oleate
CAS 106-12-7; EINECS 203-364-0
Synonyms: Diethylene glycol monooleate; Diglycol oleate; POE (2) monooleate
Definition: PEG ester of oleic acid
Empirical: $C_{22}H_{42}O_4$
Formula: $CH_3(CH_2)_7CHCH(CH_2)_7CO(OCH_2CH_2)_nOH$, avg. n = 2
Uses: Emulsifier, dispersant, antistat for cosmetic, textile, paper processing, cutting oils, polishes, emulsion cleaners, rubber latex, wool lubricants; leather softener; lubricant for paints, adhesives
Regulatory: FDA 21CFR §175.105, 175.300, 176.210, 177.2800
Manuf./Distrib.: ABITEC; Henkel/Emery; Inolex; Lipo; Lonza; Mona Ind.; Witco
Trade names: Calgene DGO

PEG-2 oleate SE
CAS 106-12-7
Synonyms: Diethylene glycol monooleate self-emulsifying; PEG 100 monooleate self-emulsifying; POE (2) monooleate self-emulsifying
Definition: Self-emulsifying grade of PEG-2 oleate containing some sodium and/or potassium oleate
Uses: Emulsifier, dispersant, antistat for textile, paper processing, cutting oils, polishes, emulsion cleaners, rubber latex, wool lubricants; w/o and aux. o/w emulsifier for latex and aq. paints
Trade names: Pegosperse® 100 O

PEG-4 oleate
CAS 9004-96-0 (generic); 10108-25-5; EINECS 233-293-0
Synonyms: POE (4) monooleate; PEG 200 monooleate
Definition: PEG ester of oleic acid
Empirical: $C_{26}H_{50}O_6$
Formula: $CH_3(CH_2)_7CHCH(CH_2)_7CO(OCH_2CH_2)_nOH$, avg. n = 4
Uses: Wetting agent, penetrant, spreading agent, defoamer, detergent, emulsifier, solubilizer, thickening agent, dispersant, textile aux., softener, lubricant for textiles, cosmetics, metalworking, food, agric., plastics
Regulatory: FDA 21CFR §175.105, 175.300, 176.210
Trade names: Ablunol 200MO

PEG-8 oleate
CAS 9004-96-0 (generic)
Synonyms: POE (8) monooleate; PEG 400 monooleate
Definition: PEG ester of oleic acid
Formula: $CH_3(CH_2)_7CHCH(CH_2)_7CO(OCH_2CH_2)_nOH$, avg. n = 8
Properties: Dk. red oil; sol. in alcohol; disp. in water; misc. with cottonseed oil
Uses: Emulsifier, dispersant, lubricant, chemical intermediate, solubilizer, visc. control agent; for cosmetics, pharmaceuticals, food, agric., plastics, coatings
Regulatory: FDA 21CFR §175.105, 175.300, 176.170, 176.200, 177.1200, 177.1210, 177.2260, 177.2800
Trade names: Ablunol 400MO; Nopalcol 4-O; Pegosperse® 400 MO; Witconol™ H31A; Witflow™ 916

PEG-12 oleate
CAS 9004-96-0 (generic)
Synonyms: POE (12) monooleate; PEG 600 monooleate
Definition: PEG ester of oleic acid
Formula: $CH_3(CH_2)_7CHCH(CH_2)_7CO(OCH_2CH_2)_nOH$, avg. n = 12
Uses: Dispersant, emulsifier, solubilizer, detergent, dye leveling agent for cosmetic and industrial applics.; plastics antistat
Regulatory: FDA 21CFR §175.105, 175.300, 176.170, 176.200, 177.1200, 177.2260, 177.2800
Trade names: Ablunol 600MO

PEG-14 oleate
CAS 9004-96-0 (generic)
Synonyms: POE (14) monooleate
Definition: PEG ester of oleic acid
Formula: $CH_3(CH_2)_7CHCH(CH_2)_7CO(OCH_2CH_2)_nOH$, avg. n = 14

PEG-20 oleate

Uses: Surfactant in cutting oils, degreasing solvs., metal cleaners; emulsifier; dyeing assistant; cosmetics, textile processing; defoamer and leveling agent in latex paints; dye leveler
Regulatory: FDA 21CFR §175.105, 175.300, 176.200, 177.2260, 177.2800
Trade names: Alkamuls® O-14

PEG-20 oleate
CAS 9004-96-0 (generic)
Synonyms: POE (20) monooleate; PEG 1000 monooleate
Definition: PEG ester of oleic acid
Formula: $CH_3(CH_2)_7CHCH(CH_2)_7CO(OCH_2CH_2)_nOH$, avg. n = 20
Toxicology: LD50 (IV, mouse) 500 mg/kg; mod. toxic by IV route; skin and eye irritant; heated to decomp., emits acrid smoke and irritating fumes
Uses: Raw material for finishing agents in the syn. fiber industry; nonionic surfactant, emulsifier, thickener, solubilizer, emollient, opacifier, wetting agent, dispersant for cosmetics, pharmaceuticals, food, agric., plastics, paints, etc.
Regulatory: FDA 21CFR §175.300, 176.200, 177.2260, 177.2800
Trade names: Ablunol 1000MO

PEG-4 oleyl ether. *See* Oleth-4
PEG-5 oleyl ether. *See* Oleth-5
PEG-9 oleyl ether. *See* Oleth-9
PEG-20 oleyl ether. *See* Oleth-20
PEG 200 oleyl ether. *See* Oleth-4
PEG 450 oleyl ether. *See* Oleth-9
PEG 1000 oleyl ether. *See* Oleth-20

PEG-8 pelargonate
Synonyms: POE (8) pelargonate; PEG 400 pelargonate
Uses: Surfactant, base lubricant for syn. fiber spin finishes, textile processing; coemulsifier
Trade names: Emerest® 2654

PEG-9 pelargonate
CAS 31621-91-7
Synonyms: POE (9) pelargonate
Uses: Surfactant as a base lubricant for syn. fiber finishes; coemulsifier and coupling agent

PEG 400 pelargonate. *See* PEG-8 pelargonate
PEG-20 sorbitan laurate. *See* Polysorbate 20

PEG-80 sorbitan laurate
CAS 9005-64-5 (generic)
Synonyms: POE (80) sorbitan monolaurate
Definition: Ethoxylated sorbitan monoester of lauric acid with avg. 80 moles ethylene oxide
Uses: Wetting agent, dispersant, mild cleanser for shampoos, baby prods., pharmaceuticals, agric., paints, pulp/paper, etc.; counterirritant; foamer
Regulatory: FDA 21CFR §175.300
Manuf./Distrib.: Aldrich; Fluka; Sigma
Trade names: Calgene PSML-80

PEG-5 sorbitan oleate. *See* Polysorbate 81
PEG-20 sorbitan oleate. *See* Polysorbate 80
PEG-4 sorbitan stearate. *See* Polysorbate 61
PEG-20 sorbitan stearate. *See* Polysorbate 60

PEG-17 sorbitan trioleate
CAS 9005-70-3
Definition: Triester of oleic acid and a PEG ether of sorbitol, avg. 17 moles ethylene oxide
Uses: O/w and w/o emulsifier for min. oils, veg. oils, train oils, waxes, etc.; for cattle feed, textiles, biocides, paints, varnishes, plastics, leather, fur, tech. applics., cosmetics, pharmaceuticals
Manuf./Distrib.: Aldrich; Fluka; Sigma
Trade names: TO-55-EL

PEG-18 sorbitan trioleate
CAS 9005-70-3
Definition: Triester of oleic acid and a PEG ether of sorbitol, avg. 18 moles ethylene oxide
Properties: M.w. 1838.60; dens. 1.028; flash pt. > 110 C; ref. index 1.4680 (20 C)
Uses: O/w and w/o emulsifier for min. oils, veg. oils, train oils, waxes, etc.; for cattle feed, textiles, biocides, paints, varnishes, plastics, leather, fur, tech. applics., cosmetics, pharmaceuticals
Manuf./Distrib.: Aldrich; Fluka; Sigma
Trade names: TO-55-E

PEG-20 sorbitan trioleate. *See* Polysorbate 85
PEG-20 sorbitan tristearate. *See* Polysorbate 65
PEG-1 stearate. *See* Glycol stearate

PEG-2 stearate
CAS 106-11-6; 9004-99-3 (generic); 85116-97-8; EINECS 203-363-5; 285-550-1
Synonyms: Diethylene glycol stearate; Diglycol stearate; Polyoxyl 2 stearate; PEG 100 monostearate
Definition: PEG ester of stearic acid
Empirical: $C_{22}H_{44}O_4$
Formula: $CH_3(CH_2)_{16}CO(OCH_2CH_2)_nOH$, avg. n = 2
Properties: Wh. wax-like solid, faint fatty odor; sol. in hot alcohol, oils
Precaution: Combustible
Toxicology: Poison by intravenous, intraperitoneal route; mildly toxic by ingestion
Uses: Emulsifier, plasticizer, lubricant, wetting agent, binding and thickening agent, dispersant, antistat, opacifier, pearlescent, stabilizer used in cosmetics, pharmaceuticals, dry cleaning, leather, textile industries, paper processing, rubber; protective coating for hygroscopic materials (tablets)
Regulatory: FDA 21CFR §175.300, 176.200, 176.210, 177.2800; FDA approved for topicals
Manuf./Distrib.: ABITEC; Henkel/Emery; Inolex; Lipo; Lonza; Sigma; Stepan; Witco
Trade names: Calgene DGS; Calgene DGS-N

PEG-2 stearate SE
CAS 106-11-6; 9004-99-3
Synonyms: POE (2) monostearate self-emulsifying; Diethylene glycol monostearate self-emulsifying; PEG 100 monostearate self-emulsifying
Definition: Self-emulsifying grade of PEG-2 stearate
Uses: Emulsifier, dispersant, antistat for textile, paper processing, cutting oils, polishes, emulsion cleaners, rubber latex, wool lubricants; protective coating for hygroscopic materials (tablets)
Trade names: Calgene DGS-C

PEG-4 stearate
CAS 106-07-0; 9004-99-3 (generic); EINECS 203-358-8
Synonyms: POE (4) stearate; PEG 200 monostearate
Definition: PEG ester of stearic acid
Empirical: $C_{26}H_{52}O_6$
Formula: $CH_3(CH_2)_{16}CO(OCH_2CH_2)_nOH$, avg. n = 4
Toxicology: Poison by intravenous, intraperitoneal route; mildly toxic by ingestion
Uses: Emulsifier, lubricant, dispersing and leveling agent, solubilizer, thickener, softener used in cosmetic, textile, paint, agric., plastics, food
Regulatory: FDA 21CFR §175.105, 175.300, 176.210
Manuf./Distrib.: Sigma
Trade names: Ablunol 200MS

PEG-8 stearate
CAS 9004-99-3 (generic); 70802-40-3
Synonyms: POE (8) stearate; PEG 400 monostearate; Polyoxyl 8 stearate
Definition: PEG ester of stearic acid
Empirical: $C_{34}H_{68}O_{10}$
Formula: $CH_3(CH_2)_{16}CO(OCH_2CH_2)_nOH$, avg. n = 8
Toxicology: Poison by intravenous, intraperitoneal route; mildly toxic by ingestion
Uses: Emulsifier, lubricant, dispersing and leveling agent used in cosmetic, textile, paint and other industrial uses; plastics antistat
Regulatory: FDA 21CFR §175.105, 175.300, 176.170, 176.200, 176.210, 177.1200, 177.1210, 177.2260, 177.2800, 178.3910; Europe listed; UK approved; FDA approved for topicals
Manuf./Distrib.: Sigma
Trade names: Ablunol 400MS; Alkamuls® S-8; Emerest® 2640; Pegosperse® 400 MS

PEG-12 stearate
CAS 9004-99-3 (generic)
Synonyms: POE (12) stearate; PEG 600 monostearate
Definition: PEG ester of stearic acid
Formula: $CH_3(CH_2)_{16}CO(OCH_2CH_2)_nOH$, avg. n = 12
Toxicology: Poison by intravenous, intraperitoneal route; mildly toxic by ingestion
Uses: Emulsifier, lubricant, dispersing and leveling agent, defoamer, leveling agent, visc. modifier used in cosmetic, textile, paints, food, agric., plastics, pharmaceuticals
Regulatory: FDA 21CFR §175.105, 175.300, 176.170, 176.210, 177.1200, 177.2260, 177.2800
Manuf./Distrib.: Sigma
Trade names: Ablunol 600MS

PEG-20 stearate
CAS 9004-99-3 (generic)
Synonyms: POE (20) stearate; PEG 1000 monostearate; Polyoxyl 20 stearate
Definition: PEG ester of stearic acid
Formula: $CH_3(CH_2)_{16}CO(OCH_2CH_2)_nOH$, avg. n = 20
Properties: Sol. in ethanol; partly sol. in propylene glycol; disp. in glycerin; insol. in water; m.p. 39.5-42.5 C; sapon. no. 40-50
Toxicology: Poison by intravenous, intraperitoneal route; mildly toxic by ingestion
Uses: Emulsifier, thickener, solubilizer, emollient, opacifier, wetting agent, dispersant for cosmetics, pharmaceuticals, food, agric., plastics, paints
Regulatory: FDA 21CFR §175.300, 176.210, 177.2260, 177.2800; FDA approved for orals
Manuf./Distrib.: Sigma
Trade names: Ablunol 1000MS

PEG-2 stearyl ether. *See* Steareth-2
PEG-10 stearyl ether. *See* Steareth-10
PEG-20 stearyl ether. *See* Steareth-20
PEG 100 stearyl ether. *See* Steareth-2
PEG 500 stearyl ether. *See* Steareth-10
PEG 1000 stearyl ether. *See* Steareth-20

PEG-2 tallowamine
CAS 61791-44-4
Synonyms: Bis (2-hydroxyethyl) tallow amine
Uses: Hydrophilic emulsifier, wetting agent, antistat, anticorrosive for agric., leather, textiles, metalworking, plastics; intermediate for quats.; textile dyeing agent, dye leveler, antiprecipitant, stripping agent; lubricant, softener, scouring aid, dye leveler and antistat for textiles; in syn. latex paints; emulsifier for latex, dyes, and oils
Trade names: Varonic® T-202

PEG-5 tallowamine
CAS 61791-44-4
Uses: Hydrophilic emulsifier, wetting agent, antistat, anticorrosive for agric., leather, textiles, metalworking, plastics; intermediate for quats.; textile dyeing agent, dye leveler, antiprecipitant, stripping agent
Trade names: Ethox TAM-5

PEG-15 tallowamine
Uses: Hydrophilic emulsifier, wetting agent, antistat, anticorrosive for agric., leather, textiles, metalworking, plastics; intermediate for quats.; textile dyeing agent, dye leveler, antiprecipitant, stripping agent
Trade names: Ethylan® TT-15

PEG-20 tallowamine
Uses: Hydrophilic emulsifier, wetting agent, antistat, anticorrosive for agric., leather, textiles, metalworking, plastics; intermediate for quats.; textile dyeing agent, dye leveler, antiprecipitant, stripping agent
Trade names: Varonic® T-220

PEG-10 tallow aminopropylamine
CAS 61790-85-0 (generic)
Synonyms: POE (10) tallow aminopropylamine; PEG 500 tallow aminopropylamine; PEG-10 N-tallow-1,3-diaminopropane

Classification: Ethoxylated diamine
Formula: $RN(CH_2CH_2O)_zH(CH_2)_3N(CH_2CH_2O)_xH(CH_2CH_2O)_yH$, R = alkyl groups fr tallow, (x+y+z)=10
Uses: Wetting agent used in coating preparation on paperboard
Manuf./Distrib.: Sigma
Trade names: Ethoduomeen® T/20

PEG 500 tallow aminopropylamine. *See* PEG-10 tallow aminopropylamine

PEG-3 tallow ammonium acetate
Trade names containing: Ethoquad® T/13-50

PEG-10 N-tallow-1,3-diaminopropane. *See* PEG-10 tallow aminopropylamine

PEG-2 tallowmonium chloride
CAS 67784-77-4
Trade names containing: Ethoquad® T/12

PEG-3 tallow propylene diamine
Uses: Dispersant, wetting agent, emulsifier, corrosion inhibitor, industrial detergent used in paint, agric., chemical, and textile industries
Trade names: Dinoramox S3

PEG-7 tallow propylene diamine
Uses: Dispersant, wetting agent, emulsifier, corrosion inhibitor, industrial detergent used in paint, agric., chemical, and textile industries
Trade names: Dinoramox S7

PEG-12 tallow propylene diamine
Uses: Dispersant, wetting agent, emulsifier, corrosion inhibitor, industrial detergent used in paint, agric., chemical, and textile industries
Trade names: Dinoramox S12

PEG-1.3 tetramethyl decynediol
Uses: Wetting agent, defoamer, dispersant for aq. coatings, inks, adhesives, agric., electroplating, oilfield chems., paper coatings
Trade names: Surfynol® 420

PEG-3.5 tetramethyl decynediol
CAS 9014-85-1
Uses: Defoamer, rewetting, and leveling agent for paperboard coatings, agric. formulations, water-based industrial finishes; metal cleaning and plating bath additive
Trade names: Surfynol® 440

PEG-10 tetramethyl decynediol
CAS 9014-85-1
Uses: Wetting agent, defoamer for aq. coatings, inks, adhesives; surfactant for emulsion polymerization; electroplating additive
Trade names: Surfynol® 465

PEG-30 tetramethyl decynediol
CAS 9014-85-1
Uses: Wetting agent, defoamer for aq. coatings, inks, adhesives, agric., electroplating, oilfield chems., paper coatings, emulsion polymerization
Trade names: Surfynol® 485

PEG-6 tridecyl ether. *See* Trideceth-6
PEG-12 tridecyl ether. *See* Trideceth-12
PEG-14 tridecyl ether. *See* Trideceth-14
PEG-15 tridecyl ether. *See* Trideceth-15
PEG 300 tridecyl ether. *See* Trideceth-6
PEG 600 tridecyl ether. *See* Trideceth-12

PEG-66 trihydroxystearin
CAS 61788-85-0
Synonyms: POE (66) trihydroxystearin
Definition: PEG deriv. of trihydroxystearin with avg. 66 moles ethylene oxide
Uses: Emulsifier for oils, waxes; solubilizer for fragrances; emollient; used in textiles, paints, household, cosmetics, dyeing, tanning, finishing, sizing, insecticides, herbicides, fungicides, kier boiling, making of cutting and sol. oils

PEG-6 trimethylolpropane
CAS 50586-59-9
Uses: Reactive diluent for coatings; enhances wetting, imparts good flow, mar resist., and impact str., extends pot life, improves adhesion to substrate, increases flexibility; for low VOC applics
Manuf./Distrib.: Aldrich
Trade names: Macol® RD 306 EM

PEG-6 trimethylolpropane triacrylate
Trade names containing: SR-499

PEG-9 trimethylolpropane triacrylate
Trade names containing: SR-502

Pelargonic acid (INCI). *See* Nonanoic acid
1,4,7,10,13-Pentaazatridecane. *See* Tetraethylenepentamine

Pentabromodiphenyl oxide
CAS 32534-81-9
Uses: Flame retardant for plastics; used for unsat. polyester, rigid and flexible urethane foams, epoxies, PU elastomer wire insulation, laminates, adhesives, coatings, and textiles
Manuf./Distrib.: Albemarle; Great Lakes
Trade names: Great Lakes DE-71™
Trade names containing: Great Lakes DE-60F™ Special; Great Lakes DE-61™; Great Lakes DE-62™

Pentabromoethylbenzene
CAS 85-22-3
Uses: Flame retardant for thermoset polyester resins, textiles, adhesives, coatings, PU
Manuf./Distrib.: Albemarle; Great Lakes

Pentabromophenyl ether. *See* Decabromodiphenyl oxide
Pentaerythrite. *See* Pentaerythritol

Pentaerythritol
CAS 115-77-5; EINECS 204-104-9
Synonyms: PE; 2,2-Bis(hydroxymethyl)-1,3-propanediol; Monopentaerythritol; Methane tetramethylol; Pentaerythrite; Tetrahydroxymethylmethane; Tetrakis(hydroxymethyl) methane; Tetramethylolmethane
Empirical: $C_5H_{12}O_4$
Formula: $C(CH_2OH)_4$
Properties: Clear, colorless crystals; hygroscopic; sl. sol. in alc.; insol. in benzene CCl_4, ether, and petroleum ether; sol. 1 g/18 ml of water; m.w. 136.15; dens. 1.38 (25/4 C); m.p. 260 C
Precaution: Flamm.; avoid heat, flame, oxidizers; mixts. with thiophosphoryl chloride react when heated to form explosive prod.
Toxicology: LD50 (oral, mouse) 25,500 mg/kg; ACGIH TLV:TWA 10 mg/m^3 (total dust); mildly toxic by ingestion; nuisance dust; heated to decomp., emits acrid smoke and irritating fumes
Uses: In synthetic resins, paints, varnishes, stabilizers, explosives, drugs, insecticides, lubricants; flame retardant for plastics; vinyl chloride plasticizer; surfactant
Manuf./Distrib.: Aldrich; Allchem Ind.; Aqualon; Browning; CasChem; Degussa; Fluka; Chemical; Degussa; Hoechst Celanese; ICD Group; Mitsubishi Gas; Mitsui Toatsu; Penta Mfg.; Perstorp Polyols; Sigma; Spectrum Chem. Mfg.; United Min. & Chem.; U.S. Petrochem. Ind.
Trade names: Hercules® Improved Tech. PE; Hercules® Mono-PE; Hercules® PE-200

Pentaerythritol ester of rosin. *See* Pentaerythrityl rosinate
Pentaerythritol hydrogenated rosinate. *See* Pentaerythrityl hydrogenated rosinate
Pentaerythritol rosinate. *See* Pentaerythrityl rosinate

Pentaerythritol tetrakis (3,5-di-t-butyl-4-hydroxyhydrocinnamate). *See* Tetrakis [methylene (3,5-di-t-butyl-4-hydroxyhydrocinnamate)] methane
Pentaerythritol tetrastearate. *See* Pentaerythrityl tetrastearate
Pentaerythritol triacrylate. *See* Pentaerythrityl triacrylate
Pentaerythritol triallyl ether. *See* Pentaerythrityl triallyl ether

Pentaerythrityl hydrogenated rosinate
Synonyms: Pentaerythritol hydrogenated rosinate
Definition: Ester of pentaerythritol and hydrogenated acids from rosin
Uses: Thermoplastic resin as tackifier for adhesives; in protective and barrier-type coatings
Regulatory: FDA 21CFR §176.210, 178.3120, 178.3800, 178.3870
Trade names: Foral® 105

Pentaerythrityl rosinate
CAS 8050-26-8; EINECS 232-479-9
Synonyms: Pentaerythritol rosinate; Pentaerythritol ester of rosin; Rosin pentaerythritol ester
Definition: Ester of rosin acids with the polyol, pentaerythritol
Properties: Amber hard solid; sol. in acetone, benzene; insol. in water
Toxicology: Heated to decomp., emits acrid smoke and irritating fumes
Uses: Thermoplastic resin as tackifier for rubbers, lacquers, ink vehicles, varnishes, adhesives
Regulatory: FDA 21CFR §172.615, 175.105, 175.300, 176.170, 176.210, 176.2600, 178.3120, 178.3800, 178.3870
Trade names: Cellolyn® 102; Pentalyn® 830; Pentalyn® 856; Pentalyn® A; Pentalyn® C; Pentalyn® G; Pentalyn® X; Uni-Tac® R99; Uni-Tac® R100; Uni-Tac® R100-Light; Uni-Tac® R101; Uni-Tac® R102; Uni-Tac® R112

Pentaerythrityl tetraacrylate
CAS 4986-89-4
Properties: M.w. 352.4; dens. 1.190; m.p. 18 C; flash pt. > 110 C; ref. index 1.4870 (20 C)
Uses: Crosslinking agent in adhesives, coatings, inks, textile prods., photoresists, castings
Manuf./Distrib.: Aldrich
Trade names containing: SR-295

Pentaerythrityl tetrabenzoate
CAS 4196-86-5
Formula: $(C_6H_5CO_2CH_2)_4C$
Properties: M.w. 552.59; sp.gr. 1.2801 (30 C); m.p. 99 C; flash pt. (COC) 600 C
Uses: Plasticizer/extender for PVC, CAB, ethyl cellulose, coatings, modifier for hot-melt adhesives, aq. adhesives, delayed tack adhesives, process aid for thermoplastics
Manuf./Distrib.: Aldrich; Unitex; Velsicol
Trade names: Benzoflex® S-552

Pentaerythrityl tetrastearate
CAS 115-83-3; 91050-82-7; EINECS 204-110-1; 293-029-5
Synonyms: Pentaerythritol tetrastearate
Definition: Tetraester of pentaerythritol and stearic acid
Empirical: $C_{77}H_{148}O_8$
Properties: Ivory-colored hard, high melting wax
Uses: Processing aid for rubbers; polishes, coatings, textile finishes
Regulatory: FDA 21CFR §175.105, 176.170, 176.210, 177.1200, 177.1580, 178.2010
Manuf./Distrib.: Calgene; Hercules; Lipo; Lonza

Pentaerythrityl triacrylate
CAS 3524-68-3; EINECS 222-540-8
Synonyms: 2-Propenoic acid-2-(hydroxymethyl)-2-(((1-oxo-2-propenyl) oxy) methyl)-1,3-propanediyl ester; Pentaerythritol triacrylate
Empirical: $C_{14}H_{18}O_7$
Formula: $(H_2C{:}CHCO_2CH_2)_3CCH_2OH$
Properties: M.w. 298.30; dens. 1.167 (20C); ref. index 1.143 (20C); flash pt.> 110C
Precaution: Hygroscopic
Toxicology: Irritant
Uses: Crosslinking agent used in adhesives, coatings, inks, textile prods., photoresists, castings, modifiers for polyester, fiberglass, or polymers

Pentaerythrityl triallyl ether

Manuf./Distrib.: Aldrich; Fluka; Monomer-Polymer & Dajac Labs; Sartomer; UCB Radcure
Trade names containing: PETA-K; PETA-LQ; SR-444

Pentaerythrityl triallyl ether
CAS 1471-17-6; EINECS 216-008-4
Synonyms: PETAE; 3-Allyloxy-2,2-bis-(allyloxymethyl)-1-propanol; 1-Propanol, 3-[(2-propenyloxy)-2,2-bis[(2-propenyloxy)methyl]; Pentaerythritol triallyl ether
Formula: $HOCH_2C(-CH_2O-CH_2CH=CH_2)_3$
Properties: Colorless to ylsh. liq.; sol. in ethanol, many org. solvs.; partly sol. in water; m.w. 256.34; dens. 0.98 g/cc (20 C); m.p. < -20 C; b.p. 268 C; flash pt. 70 C
Precaution: May form explosive peroxides; hazardous decomp. prods.: CO, hydrocarbons
Toxicology: LD50 (oral, rat) 3920 mg/kg, (dermal, rat) > 2000 mg/kg; irritating to eyes, respiratory system
Uses: Intermediate for coatings, varnishes, paints, and photoresists; can be polymerized into polyether, PU, epoxy resins
Manuf./Distrib.: Aldrich; Raschig
Trade names: Penta

Pentaethylene glycol methyl ether
Synonyms: Pentaethylene glycol monomethyl ether
Trade names containing: Icinol ME45

Pentaethylene glycol monomethyl ether. *See* Pentaethylene glycol methyl ether
Pentaglycerine. *See* Trimethylolethane

N,N,N′,N′,N′-Penta(2-hydroxyethyl)-N-tallowalkyl-1,3-propane diammonium diacetate
Uses: EPA listed
Trade names containing: Ethoduoquad® T/15-50

1,3-Pentanediamine
CAS 589-37-7
Synonyms: DAMP; 1,3-Diaminopentane
Formula: $H_2NCH_2CH_2CH(C_2H_5)NH_2$
Properties: Liq., sl. odor; misc. with water; m.w. 102; dens. 0.855 g/ml; visc. 1.89 cp; vapor pressure 1 mm Hg (20 C); b.p. 164 C; flash pt. (CC) 59 C; surf. tens. 32.2 dyn/cm
Precaution: Corrosive, flamm.; incompat. with strong oxidants; thermal decomp. may emit NO_x
Uses: Epoxy curative; solv. for gas treatment; surfactant (asphalt emulsifiers, textiles); scale/corrosion inhibitor; additive for fuel, oil, plastic, petrol. and chem. processing; extender, catalyst for PU; in coatings, adhesives, sealants, elastomers
Manuf./Distrib.: Aldrich
Trade names: Dytek® EP

1,5-Pentanediamine, 2-methyl-. *See* 2-Methylpentamethylenediamine
1,3-Pentanediol, 2,2,4-trimethyl-, dibenzoate. *See* 2,2,4-Trimethyl-1,3-pentanediol dibenzoate
2,4-Pentanedione. *See* Acetylacetone
Pentanedione-2,4. *See* Acetylacetone
Pentanoic acid. *See* n-Valeric acid
n-Pentanoic acid. *See* n-Valeric acid
t-Pentanoic acid. *See* Neopentanoic acid
1-Pentanol (n-). *See* Amyl alcohol
2-Pentanol (t-). *See* Amyl alcohol
2-Pentanone. *See* Methyl propyl ketone
3-Pentanone. *See* Diethyl ketone
3,6,9,12,15-Pentaoxatriacont-24-en-1-ol. *See* Oleth-5
3,6,9,12,15-Pentaoxyheptacosan-1-ol. *See* Laureth-5
Pentapotassium triphosphate. *See* Potassium tripolyphosphate
Pentapotassium tripolyphosphate. *See* Potassium tripolyphosphate
Pentyl acetate. *See* Amyl acetate
Pentyl alcohol. *See* Amyl alcohol
Pentylcarbinol. *See* Hexyl alcohol
Pentylformic acid. *See* Caproic acid
Pentyl propanoate. *See* n-Pentyl propionate

n-Pentyl propionate
CAS 624-54-4
Synonyms: Propanoic acid, pentyl ester; Pentyl propanoate
Empirical: $C_8H_{16}O_2$
Formula: $CH_3(CH_2)_4OCOCH_2CH_3$
Properties: Insol. in water; m.w. 144.12; sp.gr. 0.874 (20/20 C); vapor pressure 1.5 mm Hg (20 C); b.p. 164.9 C; flash pt. (CC) 135 F
Uses: Solv. for high solids coatings, cellulosic and urethane paints
Manuf./Distrib.: Aldrich; Penta Mfg.; Union Carbide
Trade names: Ucar® n-Pentyl Propionate

Perc. *See* Perchloroethylene

Perchloroethylene
CAS 127-18-4; EINECS 204-825-9
Synonyms: Perc; Tetrachloroethylene; Tetrachloroethene; Ethylene tetrachloride
Empirical: C_2Cl_4
Formula: $Cl_2C=CCl_2$
Properties: Colorless liq., chloroform-like odor; misc. with alcohol, ether, chloroform, benzene, oil; sol. in 10,000 vol water; m.w. 165.82; dens. 1.6311 (15/4 C); m.p. -23 C; b.p. 121 C; ref. index 1.5055 (20 C); flash pt. none; KB value 90
Precaution: Nonflamm.
Toxicology: LD50 (mice, oral) 8.85 g/kg; TLV:TWA 50 ppm (skin, air); poison by IV route; mod. toxic by inh.; narcotic in high concs.; vapor inh. affects CNS; severe acute/chronic exposure causes liver damage; suspected carcinogen
Uses: Dry-cleaning solvent; vermifuge; drying agent; degreasing metals
Manuf./Distrib.: Aldrich; Allchem Ind.; Asahi-Penn; Ashland; Coyne; Elf Atochem SA; Fluka; General Chem.; ICI Am.; Occidental; PPG Ind.; Primachem; Sigma; Vulcan
Trade names containing: Oxsol® 73; Oxsol® 253; Oxsol® 325

Perfluoroheptane
CAS 335-57-9; EINECS 206-392-1
Synonyms: Hexadecafluoroheptane; Perfluoro-n-heptane
Empirical: C_7F_{16}
Formula: $CF_3(CF_2)_5CF_3$
Properties: M.w. 388.07; dens. 1.745 (20/4 C); b.p. 82-84 C
Toxicology: LDLo (IP, mouse) 100 mg/kg; poison by IP route; heated to decomp., emits very toxic fumes of F^-
Uses: Dispersant for lubricants, mold release agents, protective coatings; reaction media for polymerizations, purification, separation processes; solv. for removal of halogenated lubricants, oils, greases; carrier for halogenated material; zero ozone depletion potential
Manuf./Distrib.: Aldrich; Fluka
Trade names: PF-5070

Perfluoro-n-heptane. *See* Perfluoroheptane

Perfluorohexane
CAS 355-42-0; EINECS 206-585-0
Synonyms: Tetradecafluorohexane
Empirical: C_6F_{14}
Formula: $CF_3(CF_2)_4CF_3$
Properties: M.w. 338.04; dens. 1.684 (20/4 C); b.p. 54-58 C; ref. index 1.2520
Uses: Dispersant for lubricants, mold release agents, protective coatings; reaction media for polymerizations, purification, separation processes; solv. for removal of halogenated lubricants, oils, greases; carrier for halogenated material; zero ozone depletion potential
Regulatory: FDA 21CFR §173.342
Manuf./Distrib.: Aldrich; Fluka
Trade names: PF-5060

Perfluoro-N-methylmorpholine
Uses: Dispersant for lubricants, mold release agents, protective coatings; reaction media for polymerizations, purification, separation processes; solv. for removal of halogenated lubricants, oils, greases; carrier for halogenated material; zero ozone depletion potential
Trade names: PF-5052

Perfluoropentane
CAS 678-26-2
Synonyms: Dodecafluoropentane
Empirical: C_3F_{12}
Properties: M.w. 288.04
Uses: Dispersant for lubricants, mold release agents, protective coatings; reaction media for polymerizations, purification, separation processes; solv. for removal of halogenated lubricants, oils, greases; carrier for halogenated material; zero ozone depletion potential
Trade names: PF-5050

Perhydronaphthalene. *See* Decahydronaphthalene
Periclase. *See* Magnesium oxide
PET. *See* Polyethylene terephthalate
PETAE. *See* Pentaerythrityl triallyl ether
Petrohol. *See* Isopropyl alcohol

Petrolatum
CAS 8009-03-8 (NF); 8027-32-5 (USP); EINECS 232-373-2
Synonyms: Petroleum jelly; White soft paraffin; Petrolatum amber; Vaseline; Petrolatum white; White petrolatum
Classification: Petroleum hydrocarbons
Definition: Semisolid mixture of hydrocarbons obtained from petroleum
Properties: Ylsh. to lt. amber or white semisolid, unctuous mass; pract. odorless and tasteless; sol. in benzene, chloroform, ether, petrol. ether, oils; pract. insol. in water; dens. 0.820-0.865 (60/25 C); m.p. 38-54 C; ref. index 1.460-1.474 (60 C)
Toxicology: Can cause allergic skin reactions in hypersensitive persons; generally nontoxic; heated to decomp., emits acrid smoke and irritating fumes
Uses: Laxative; dispersant; as ointment base in pharmaceuticals and cosmetics; leather grease; shoe polish; rust preventives; plastics, rubber, textiles lubricant
Regulatory: FDA 21CFR §172.880, 172.884, 173.340, 175.105, 175.125, 175.176, 175.300, 176.170, 176.200, 176.210, 177.2600, 177.2800, 1787.3570, 178.3700, 178.3910, 573.720; FDA approved for ophthalmics, orals, topicals, otics; USP/NF, BP compliance
Manuf./Distrib.: Barco Chem. Prods.; Exxon; Harcros; Magie Bros. Oil; Mobil; Penreco; Sea-Land; Stevenson Cooper; Witco/Oleo-Surf.; Witco/Petroleum Spec.
Trade names: Michem Lube® 343
Trade names containing: Amerchol® C; Amerchol® CAB

Petrolatum amber. *See* Petrolatum
Petrolatum liquid. *See* Mineral oil
Petrolatum white. *See* Petrolatum
Petroleum benzin. *See* Naphtha

Petroleum distillates
CAS 8002-05-9; 64742-14-9; 64742-47-8; EINECS 232-298-5
Classification: Petroleum hydrocarbons
Definition: Mixture of volatile hydrocarbons obtained from petroleum
Precaution: Combustible or flamm. liqs.
Toxicology: Human systemic effects; skin contact may cause defatting; vapor inh. can cause CNS depression, headache, unconsciousness; heated to decomp., emits acrid smoke and irritating fumes
Uses: Hydrocarbon processing solvent, foam control agent, in waterless hand cleaners, agric. sprays, polishes, fruit and veg. processing, cleaning oils, paper, coatings; vehicle for pesticides
Trade names: Penreco 2251 Oil; Penreco 2263 Oil; Shell Sol 71

Petroleum ether. *See* Naphtha
Petroleum jelly. *See* Petrolatum
Petroleum naphtha. *See* Naphtha
Petroleum spirit. *See* Naphtha
Petroleum spirits. *See* Mineral spirits
Petroleum sulfonic acid, monosodium salt. *See* Sodium petroleum sulfonate

Petroleum wax
CAS 8002-74-2; EINECS 232-315-6
Synonyms: Microcrystalline wax; Petroleum wax, synthetic; Refined petroleum wax
Classification: Petroleum hydrocarbon

Definition: Hydrocarbon derived from petroleum
Properties: Translucent wax, odorless, tasteless; very sl. sol. in org. solvs.; insol. in water; m.p. 48-93 C
Toxicology: Heated to decomp., emits acrid smoke and irritating fumes
Uses: Lubricant for formulating PVC and elec. wire and cable compds.; antisunchecking protectant for elastomers
Regulatory: FDA 21CFR §172.230, 172.615, 172.886 (1050 ppm max. of poly(alkylacrylate) as an antioxidant), 172.888, 173.340, 178.3710, 178.3720
Manuf./Distrib.: Aldrich; Astor Corp.; Exxon; Koster Keunen; Mobil; Penreco; Shell; Witco/Petroleum Spec.
Trade names containing: Flatting Agent OK412; Flatting Agent OK500

Petroleum wax, crystalline. *See* Paraffin
Petroleum wax, microcrystalline. *See* Microcrystalline wax
Petroleum wax, synthetic. *See* Petroleum wax
PF. *See* Phenolic resin
PGE. *See* Phenyl glycidyl ether

Phenalkamine
Uses: Epoxy resin hardener; curing agent; for coatings, floorings; permits VOC compliance
Trade names: Cardolite® NC-540; Cardolite® NC-541LV; Cardolite® NC-554; Cardolite® NC-556X80; Cardolite® NC-558; Cardolite® NC-559; Cardolite® NC-560
Trade names containing: Cardolite® NC-541; Cardolite® NC-541X90

1,10-Phenanthroline
CAS 66-71-7; EINECS 200-629-2
Synonyms: OPT; 4,5-Diazaphenanthrene; o-Phenanthroline; β-Phenanthroline; Orthophenanthroline; 1,10-o-Phenanthroline; 4,5-Phenanthroline
Classification: Heterotricyclic compd.
Empirical: $C_{12}H_8N_2$
Properties: Wh. cryst. powd.; sol. in 300 parts water, 70 parts benzene; sol. in alcohol, ether, acetone; m.w. 180.20; m.p. 93-94 C, 117 C (anhyd.)
Toxicology: LD50 (oral, rat) 132 mg/kg, (IP, mouse) 75 mg/kg; poison by ingestion and IP routes; mutagenic data; heated to decomp., emits toxic fumes of NO_x
Uses: Indicator; drier accelerator and stabilizer in coatings cured by oxidative polymerization
Manuf./Distrib.: Aldrich; AMRESCO; GFS; Spectrum Chem. Mfg.
Trade names: Activ-8
Trade names containing: Activ-8 in Hexylene Glycol

1,10-o-Phenanthroline. *See* 1,10-Phenanthroline
4,5-Phenanthroline. *See* 1,10-Phenanthroline
β-Phenanthroline. *See* 1,10-Phenanthroline
o-Phenanthroline. *See* 1,10-Phenanthroline
Phenethylene. *See* Styrene
Phenic acid. *See* Phenol

Phenol
CAS 108-95-2; EINECS 203-632-7
Synonyms: Benzenol; Oxybenzene; Phenic acid; Phenyl hydroxide; Phenyl hydrate; Phenylic acid; 'Carbolic acid'; Hydroxybenzene
Classification: Aromatic organic compd.; carbolic acid
Empirical: C_6H_6O
Formula: C_6H_5OH
Properties: Colorless to lt. pink needle-shaped cryst., char. odor of coal tar and wood; sol. in alcohol, glycerin, chloroform, ether, water; sl. sol. in min. oil; m.w. 94.11; dens. 1.07; m.p. 40-42 C; b.p. 182 C; flash pt. 175 F; darkens on exposure to lt., air
Precaution: Combustible; vapor is flamm.; avoid contact with skin; DOT: poisonous material; moisture- and light-sensitive
Toxicology: Ingestion of even sm. amts. may cause vomiting, circulatory collapse, paralysis, convulsions, coma, grnsh. urine, necrosis of mouth and gastrointestinal tract; death results from respiratory failure; may cause serious skin burns; possible carcinogen
Uses: Phenolic adhesives and resins, epoxy reins, pesticide intermediates, paints, lubricant additives, chem. deriv. intermediates
Usage level: 0.5% (topicals), 0.2-5% (parenterals)
Regulatory: FEMA GRAS; FDA approved for injectables, parenterals, topicals; USP/NF, BP, Ph.Eur.

compliance

Manuf./Distrib.: Aldrich; Allchem Ind.; AlliedSignal; AMRESCO; Aristech; Ashland; J.T. Baker; Barker Ind.; Chemical; Fluka; R.W. Greeff; Integra; Int'l. Chem. Inc.; Kalama; Penta Mfg.; PMC Specialties; Research Organics; Rhone-Poulenc N. Am.; Royale Pigments & Chems.; Shell; Sigma; Spectrum Chem. Mfg.; Texaco; Van Waters & Rogers

Phenol, 2-(2H-benzotriazole-2-yl)-4-methyl-6-dodecyl
CAS 23328-53-2
Properties: M.w. 393.6
Uses: UV absorber for plastics, latexes
Trade names: Tinuvin® 571

Phenol, 2-(5-chloro-2H-benzo-triazol-2-yl)-4,6-bis (1,1-dimethylethyl)-. *See* 2-(3′,5′-Di-t-butyl-2′-hydroxyphenyl)-5-chlorobenzotriazole
Phenol-formaldehyde. *See* Phenolic resin
Phenol-formaldehyde (novolak) resin. *See* Novolac resin

Phenolic resin
CAS 9003-35-4
Synonyms: PF; Phenol-formaldehyde; One step: A-stage resin; One-stage resin; Resole; Two step: Novolac resin; Novolak resin; Two-stage resin
Definition: Thermosetting resin from condensation of phenol or substituted phenol with aldehydes such as formaldehyde, acetaldehyde, and furfural
Properties: Brn. colored thermosets
Uses: Thermosetting resin for applics. requiring heat resistance, exhaust duct systems, wiring devices, switch gears, ovens, toasters, pot handles, elec. devices, coil bobbins, coatings, laminates; tackifier, plasticizer, hardening and reinforcing agent for rubber
Manuf./Distrib.: 3M; Akzo Nobel; Arakawa USA; Arizona; Asahi Yukizai Kogyo; Bakelite GmbH; Borden; BP Chems. Ltd; Ciba-Geigy; Focus; Georgia-Pacific Resins; Hüls AG; Int'l. Chem. Inc.; McWhorter; PMC Specialties; QO; Raschig; Seegott; Union Camp
Trade names: GPRI™ BKR-2620; GPRI™ CK-0036; GPRI™ CK-2103; GPRI™ CK-2400; GPRI™ CK-2432; GPRI™ CK-2500; Heresite L-66; Heresite RO-88; RIX 80482
Trade names containing: GPRI™ 4000; GPRI™ 7550; GPRI™ 7557; GPRI™ 7570; GPRI™ 7590; GPRI™ 7597; GPRI™ BKS-2600; GPRI™ BKS-2640; GPRI™ BKS-2700; GPRI™ BKS-2900; GPRI™ BKS-2901; GPRI™ BKUA-2370; GPRI™ CKS-3892; GPRI™ CKSB-2001; GPRI™ CKU-2266
See also Novolac resin

Phenol, 2,2′-methylene-bis-6-[(1,1-dimethyl)-4-methyl-]. *See* 2,2′-Methylenebis (6-t-butyl-4-methylphenol)

Phenol sulfonic acid
CAS 98-67-9; 1333-39-7; 74665-14-8; EINECS 202-691-6; 215-587-0; 277-962-5
Synonyms: p-Phenolsulfonic acid; Phenol-4-sulfonic acid; 4-Hydroxybenzenesulfonic acid; Sulfocarbolic acid
Classification: Aromatic organic compd.
Empirical: $C_6H_6O_4S$
Formula: $HOC_6H_4SO_3H$
Properties: Ylsh. liq. (brown in air); sol. in water, alcohol; m.w. 174.18; dens. 1.34 (20/4 C, 65% aq.)
Precaution: DOT: Corrosive material
Toxicology: Irritant to skin and tissues; causes burns
Uses: Water analysis, lab reagent, electroplated tin coating baths; mfg. of intermediates and dyes, pharmaceuticals
Manuf./Distrib.: Aldrich; Fluka; AJ & JO Pilar; Sloss Ind.; Spectrum Chem. Mfg.

Phenol-4-sulfonic acid. *See* Phenol sulfonic acid
p-Phenolsulfonic acid. *See* Phenol sulfonic acid
Phenol, 2,4,6-tris(dimethylaminomethyl). *See* 2,4,6-Tris (dimethylaminomethyl) phenol

Phenoxydiglycol
Uses: Solv. for resins, in metal cleaners, paint strippers, cleaning compds., as ink vehicle
Trade names containing: Igepal® OD-410

Phenoxyethanol
CAS 122-99-6; EINECS 204-589-7
Synonyms: 2-Phenoxyethanol; Phenylglycol; Ethylene glycol monophenyl ether; Ethylene glycol phenyl ether;

Phenoxytol
Classification: Aromatic ether alcohol
Definition: Phenol polyglycol ether
Empirical: $C_8H_{10}O_2$
Formula: $C_6H_5OCH_2CH_2OH$
Properties: Clear liq., faint aromatic odor, burning taste; sl. sol. in water; sol. in alcohol, ether, NaOH sol'ns.; m.w. 138.18; dens. 1.1094 (20/20 C); m.p. 14 C; b.p. 242 C; flash pt. 121 C; ref. index 1.534 (20 C)
Toxicology: LD50 (rat, oral) 1.26 g/kg; moderately toxic by ingestion and skin contact; skin and severe eye irritant
Uses: Solvent for resins, dyes, inks, in organic synthesis; perfume fixative, bactericidal agent, plasticizer, germicide, as insect repellent; coalescent in adhesives, water-based architectural and industrial coatings; plasticizer for cellulosics
Regulatory: FDA 21CFR §175.105; CIR approved, EPA reg.; JSCI listed 1.0% max.; Europe listed 1% max.; BP, Ph.Eur. compliance
Manuf./Distrib.: Aldrich; Amber Syn.; Fluka; Jan Dekker BV; Hüls AG; Penta Mfg.; Sigma; Spectrum Chem. Mfg.; Tri-K Ind.
Trade names: Dowanol® EPh; Ethylan® HB1-TG
Trade names containing: Igepal® OD-410; Ken-React® NZ 66A

2-Phenoxyethanol. *See* Phenoxyethanol

2-Phenoxyethyl acrylate
CAS 48165-04-6
Synonyms: PEA
Classification: Monomer
Properties: M.w. 192.2; b.p. 103-104 (0.6 mm)
Uses: Reactive diluent for radiation-hardenable systems, inks, coatings, adhesives; visc. index improver
Manuf./Distrib.: CPS; Monomer-Polymer & Dajac Labs
Trade names: Photomer® 4035
Trade names containing: Ageflex PEA; CN 980 M50; SR-339

2-Phenoxyethyl methacrylate
CAS 10595-06-9
Properties: M.w. 206.2; b.p. 130-132 C (8 mm)
Uses: Curing agent
Manuf./Distrib.: CPS; Monomer-Polymer & Dajac Labs; PPG Ind.
Trade names containing: SR-340

Phenoxyglycol
Trade names containing: Igepal® OD-410

1-Phenoxypropanol-2
Trade names containing: Solvenon® PP

2-Phenoxypropanol. *See* Propylene glycol phenyl ether

2-Phenoxypropanol-1
Uses: Solvent
Trade names containing: Solvenon® PP

2-Phenoxypropyl alcohol. *See* Propylene glycol phenyl ether

Phenoxy resin
CAS 25068-38-6
Definition: High molecular weight plastics copolymer of Bisphenol A and epichlorohydrin
Formula: $—[OC_6H_4C(CH_3)_2C_6H_4OCH_2CH(OH)CH_2]_n—$
Properties: Sol. in MEK
Uses: Coatings and adhesives; blow-molded containers, pipe, ventilating ducts, and other molded parts
Manuf./Distrib.: Aldrich; Matteson-Ridolfi; Union Carbide

Phenoxytol. *See* Phenoxyethanol

Phenyl acid phosphate
Uses: Catalyst used in coatings, chem. reactions, coupling agents and solubilizers in heavy-duty liq. detergents, curing agents in solder fluxes, binder for foundry prods
Manuf./Distrib.: Albright & Wilson Am.
Trade names: Albrite® Phenyl Acid Phosphate
Trade names containing: Albrite® PA-75

2,2′-(Phenylamino)diethanol. *See* Phenyldiethanolamine
Phenyl 'Carbitol'. *See* Diethylene glycol phenyl ether
Phenylcarboxylic acid. *See* Benzoic acid
Phenyl chloride. *See* Chlorobenzene
Phenyl cyanide. *See* Benzonitrile

Phenyldiethanolamine
CAS 120-07-0; EINECS 204-368-5
Synonyms: PDEA; N,N-Bis(2-hydroxyethyl)aniline; Diethanolaminobenzene; Diethanolaniline; N,N-Diethanolaniline; Dihydroxyethylaniline; N,N-Di(β-hydroxyethyl)aniline; N,N-Di(2-hydroxyethyl)aniline; N,N-Dioxyethylaniline; 2,2′-(Phenylamino)diethanol; N-Phenyldiethanolamine; 2,2′-(Phenylimino) diethanol
Empirical: $C_{10}H_{15}NO_2$
Formula: $(HOCH_2CH_2)_2NC_6H_5$
Properties: Colorless liq.; sl. sol. in water; sol. in ethanol, acetone; m.w. 181.26; dens. 1.1203 (60/20 C); m.p. 56-58 C; b.p. 190 C; flash pt. (OC) 190 C
Toxicology: LD50 (oral, rat) 1430 mg/kg; moderately toxic by ingestion; severe eye and mild skin irritant; heated to decomp., emits toxic fumes of NO_x
Uses: Organic synthesis, laboratory reagent; polyester cure promoter; curing agent for urethane elastomers; coupling agent, intermediate for dyes; in hair dyes, detergents, paint strippers
Manuf./Distrib.: Aldrich; Fluka; Hüls AG; Milliken

N-Phenyldiethanolamine. *See* Phenyldiethanolamine

Phenyl diisodecyl phosphite
CAS 25550-98-5; EINECS 247-098-3
Synonyms: PDDP
Empirical: $C_{26}H_{47}O_3P$
Formula: $(C_{10}H_{21}O)_2POC_6H_5$
Properties: Liq.; m.w. 438; dens. 0.938-0.947; b.p. 190 C (5 mm); ref. index 1.4780-1.4810; flash pt. (PMCC) 160 C
Uses: Color and processing stabilizer for ABS, PC, polyurethane, coatings, PET fiber; sec. stabilizer for PVC
Manuf./Distrib.: Aldrich; Dover
Trade names: Weston® PDDP

1-Phenyldodecane. *See* Dodecylbenzene
1,3-Phenylenediamine. *See* m-Phenylenediamine

m-Phenylenediamine
CAS 108-45-2; EINECS 203-584-7
Synonyms: mPDA; m-Benzenediamine; 1,3-Benzenediamine; 1,3-Diaminobenzene; m-Diaminobenzene; 1,3-Phenylenediamine; Metaphenylenediamine; 3-Aminoaniline; m-Aminoaniline
Classification: Aromatic amine
Empirical: $C_6H_8N_2$
Formula: $C_6H_4(NH_2)_2$
Properties: Wh. cryst.; sol. in water, methanol, ethanol, MEK, dioxane; sl. sol. in ether, m.w. 108.14; CCl_4, IPA; dens. 1.139; m.p. 62.8 C; b.p. 284-287 C; flash pt. > 110C
Precaution: Combustible exposed to heat or flame; photosensitive; keep well sealed and protected from light
Toxicology: LD50 (oral, rat) 650 mg/kg, (IP, rat) 283 mg/kg; irritant; poison by ingestion, IV, IP, subcut. routes; mildly toxic by skin contact; experimental tumorigen, teratogen; mutagenic data; heated to decomp., emits toxic fumes of NO_x
Uses: Mfg. of paints, dyes, hair dyes; rubber curing agent; in photography; as reagent for gold and bromine
Regulatory: FDA 21CFR §177.2280
Manuf./Distrib.: Aldrich; Bayer/Fibers, Orgs., Rubber; DuPont; Fabrichem; First Chem.; Fluka; Hoechst Celanese; Sigma

Phenylethane. *See* Ethylbenzene
1-Phenylethanone. *See* Acetophenone
Phenylethene. *See* Styrene
Phenylethylene. *See* Styrene
Phenylformic acid. *See* Benzoic acid

Phenyl glycidyl ether
CAS 122-60-1; EINECS 204-557-2
Synonyms: PGE; 1,2-Epoxy-3-phenoxypropane; Glicidyl phenyl ether; Glycidyl phenyl ether; Oxirane, (phenoxymethyl)-; 2,3-Epoxypropyl phenyl ether
Classification: Aromatic organic compd.
Empirical: $C_9H_{10}O_2$
Formula: $H_2COCHCH_2OC_6H_5$
Properties: Colorless liq.; sol. in ethanol; m.w. 150.19; dens. 1.113 (20/4 C); m.p. 3.5 C; b.p. 245 C; flash pt. 114 C
Precaution: Incompat. with acids, alkalies, amines, oxidizing agents; hazardous decomp. prods.: CO, hydrocarbons
Toxicology: LD50 (oral, rat) 2150-3850 mg/kg, (dermal, rabbit) 1500 mg/kg; moderately toxic by ingestion, skin contact, subcutaneous routes; severe eye and skin irritant; TLV 1 ppm in air; may cause sensitization by skin contact; may cause cancer
Uses: Reactive epoxy diluent for solv.-free coating systems, laminating resins, fiber-reinforced composites, elec. applics.; stabilizer for chlorinated hydrocarbons; chem. intermediate for pharmaceuticals
Manuf./Distrib.: Aldrich; Fluka; Monomer-Polymer & Dajac Labs; Raschig; Rhone-Poulenc Surf. & Spec.; Richman

Phenylglycol. *See* Phenoxyethanol

Phenyl glycol ether
Uses: Plasticizer; diluent and flexibilizer for epoxies
Manuf./Distrib.: Rhone-Poulenc Surf. & Spec.
Trade names containing: Ken-React® NZ 97

Phenylglyoxal 2-diethyl acetal. *See* 2,2-Diethoxyacetophenone
Phenyl hydrate. *See* Phenol
Phenyl hydroxide. *See* Phenol
1-Phenyl-2-hydroxy-2-methyl-propan-1-one. *See* 2-Hydroxy 2-methyl 1-phenyl 1-propanone
Phenyl 2-hydroxyphenyl ketone. *See* 2-Hydroxybenzophenone
Phenylic acid. *See* Phenol
2,2′-(Phenylimino)diethanol. *See* Phenyldiethanolamine

Phenylmercuric acetate
CAS 62-38-2; EINECS 200-532-5
Synonyms: PMAC; Phenylmercury acetate; (Acetato)phenylmercury; Acetoxyphenylmercury
Classification: Metallo-organic compd.
Empirical: $C_8H_8HgO_2$
Formula: $C_6H_5HgOCOCH_3$
Properties: Wh. to cream prisms, odorless; sol. in 600 parts water; sol. in alcohol, benzene, acetone, glacial acetic acid; m.w. 336.75; m.p. 148-150 C
Precaution: Inactivated by sulfides, thioglycollates; light-sensitive
Toxicology: Toxic by ingestion, inhalation, skin absorption; strong irritant; LD50 (rat, oral) 22 mg/kg
Uses: Fungicide, herbicide, mildewcide for paints, cosmetics; slimicide in paper mills
Usage level: 0.0065%; 0.001% (parenterals); 0.002-0.004% (ophthalmics)
Regulatory: USA EPA reg., limited to eye cosmetics/pharmaceuticals; Europe listed; USP/NF compliance
Manuf./Distrib.: Allchem Ind.; Aldrich; Atomergic Chemetals; W.A. Cleary; Cosan; EM Ind.; Noah; Spectrum Chem. Mfg.

Phenylmercuric oleate
CAS 104-60-9
Synonyms: Phenylmercury oleate
Empirical: $C_{24}H_{38}O_2Hg$
Formula: $C_6H_5HgOOC(CH_2)_7CH{=}CHC_8H_{17}$
Properties: Wh. cryst. powd.; insol. in water; sol. in org. solvs. and some oils; m.p. 45 C
Toxicology: Toxic by ingestion, inhalation, skin absorption

Phenylmercury acetate

Uses: Mildewproofing agent for paints; fungicide, germicide
Manuf./Distrib.: Cosan; Noah; Thor

Phenylmercury acetate. *See* Phenylmercuric acetate

Phenylmercury ammonium propionate
Uses: Mildewcide for outdoor textiles, water-based paints

Phenylmercury oleate. *See* Phenylmercuric oleate
Phenylmethane. *See* Toluene
Phenylmethyl acetate. *See* Benzyl acetate
N-(Phenylmethyl) dimethylamine. *See* N-Benzyldimethylamine

Phenylmethyl polysiloxane
CAS 68083-14-7
Uses: Binder for high temp. anticorrosion paints resist. to 650 C; used in industrial equip., chimneys, exhaust tubes
Manuf./Distrib.: Hüls AG; Sigma
Trade names: Silikophen® 300 Compd.; Silikophen® P 50/300; Silikophen® P 75/X; Silikophen® PA 80/20
Trade names containing: Silikophen® P 40/W; Silikophen® P 50/X; Silikophen® P 80/X
See also Phenyl trimethicone

2-Phenylpropene. *See* α-Methylstyrene monomer
2-Phenylpropylene. *See* α-Methylstyrene monomer

Phenylpropyl siloxane
Trade names: Silres® 4408 VP

Phenyl siloxane
Uses: Imparts enhanced uv and temp. resist. to coatings
Trade names: Silres® SY 430

Phenyl p-tolyl ketone. *See* 4-Methylbenzophenone
6-Phenyl-1,3,5-triazine-2,4-diamine. *See* Benzoguanamine

Phenyl trimethicone
CAS 2116-84-9; EINECS 218-320-6
Synonyms: Methyl phenyl polysiloxane; Polyphenylmethyl siloxane; Phenyl tristrimethyl siloxysilane
Classification: Siloxane polymer
Empirical: $C_{15}H_{32}O_3Si_4$
Properties: M.w. 372.8; dens. 0.97; m.p. < -60 C; b.p. 264-6 C; ref. index 1.459; flash pt. 127 C
Uses: Oil-phase ingred. in alcoholic milky-white lotions; hydropobing agent; co-binder for primers, esp. for facade coatings
Regulatory: FDA 21CFR §175.105, 175.300
Manuf./Distrib.: Fluka
Trade names: TEGO® Phobe 1020; TEGO® Phobe WF
Trade names containing: SILIKOFTAL® Nonstick 50; Silikophen® Non-stick 50

Phenyltrimethoxysilane
CAS 2996-92-1; EINECS 221-066-9
Synonyms: (Trimethoxysilyl)benzene
Empirical: $C_9H_{14}O_3Si$
Formula: $C_6H_5Si(OCH_3)_3$
Properties: M.w. 198.30; dens. 1.062 (20/4 C); ref. index 1.472 (20 C)
Toxicology: Skin, eyes, respiratory tract irritant
Uses: Coupling agent, release agent, lubricant, blocking agent, chemical intermediate; surf. treatment for paints, inks, adhesives
Manuf./Distrib.: Aldrich; Fluka; Gelest; Hüls Am.; Janssen Chimica; PCR
Trade names: Dow Corning® Z-6124

Phenyl tristrimethyl siloxysilane. *See* Phenyl trimethicone

Phosphoric acid
CAS 7664-38-2; EINECS 231-633-2
Synonyms: Orthophosphoric acid
Classification: Inorganic acid
Empirical: H_3O_4P
Formula: H_3PO_4
Properties: Colorless liq. or rhombic crystals, odorless; sol. in water, alcohol; m.w. 98.00; dens. 1.70 (20/4 C); m.p. 42.4 C; b.p. 158 C
Precaution: DOT: Corrosive material; mixts. with nitromethane are explosive; incompat. with alkalis; corrosive to many metals
Toxicology: LD50 (oral, rat) 1530 mg/kg; mod. toxic by ingestion and skin contact; conc. sol'ns. irritating to skin, mucous membranes; corrosive irritant to eyes; TLV:TWA 1 mg/m^3 of air; heated to decomp., emits toxic fumes of PO_x
Uses: In mfg. of inorganic phosphates, fertilizers, detergents, chemical polishing, paints, priming metals, petroleum refining; acid catalyst; as acidulant and flavor, antioxidant and sequestrant in food; pharmaceutic acid; in dental cements; in analytical chem.
Regulatory: FDA 21CFR §131.144, 133, 175.300,177.2260, 178.3520, 182.1073, GRAS; USDA 9CFR §318.7, 381.147 (0.01% max. in lard, shortening, poultry fat); FEMA GRAS; Japan approved; Europe listed; UK approved; FDA approved for parenterals, intramuscular injectables, orals, topicals, vaginals; USP/NF, BP, Ph.Eur. compliance
Manuf./Distrib.: Albright & Wilson UK; Aldrich; Ashland; Coyne; Farleyway Chem. Ltd; Fluka; FMC; General Chem.; Mallinckrodt; Mitsui Toatsu; Monsanto; Occidental; Olin; Rasa Ind.; Rhone-Poulenc N. Am.; Sigma; Spectrum Chem. Mfg.

Phosphoric acid, aluminum salt (1:1). *See* Aluminum orthophosphate

Phosphoric acid, bis (2-hexyloxyethyl) ester, diethylamine salt
CAS 64051-24-7
Manuf./Distrib.: Albright & Wilson Am.
Trade names containing: Virco-Pet® 40

Phosphoric acid calcium salt (1:1). *See* Calcium phosphate dibasic
Phosphoric acid isodecyl diphenyl ester. *See* Isodecyl diphenyl phosphate
Phosphoric acid isooctyl ester. *See* Isooctyl acid phosphate
Phosphoric acid methylphenyl diphenyl ester (9CI). *See* Diphenylcresyl phosphate

Phosphoric acid, mono (2-hexyloxyethyl) ester, bis (diethylamine) salt
CAS 64051-25-8
Manuf./Distrib.: Albright & Wilson Am.
Trade names containing: Virco-Pet® 40

Phosphoric acid, triphenyl ester. *See* Triphenyl phosphate
Phosphoric acid, tris(methylphenyl) ester. *See* Tricresyl phosphate
Phosphoric acid tritolyl ester. *See* Tricresyl phosphate
Phosphorodithioic acid, mixed O,O-bis(iso-bu and pentyl) esters, zinc salt. *See* Zinc dialkyl dithiophosphate
Phosphorous acid, triisooctyl ester. *See* Triisooctyl phosphite
Phosphorous acid triphenyl ester. *See* Triphenyl phosphite

Phosphorus
CAS 7723-14-0; EINECS 231-768-7
Empirical: P
Properties: Reddish-brown powd.; at.wt. 30.97; dens. 2.34; m.p. 590 C (43 atm); b.p. 280 C
Toxicology: Human poison by ingestion; external contact may cause severe burns; chronic poisoning
Uses: Mfg. of phosphoric acid; additive to semiconductors, fertilizers, safety matches; antioxiant and color suppresant for powd. coatings
Manuf./Distrib.: Aldrich; Fluka; Rhone-Poulenc N. Am.; Sigma
Trade names containing: Doverguard® 9122

Phosphorus chloride. *See* Phosphorus trichloride

Phosphorus oxychloride
CAS 10025-87-3; EINECS 233-046-7
Synonyms: Phosphoryl chloride

Phosphorus trichloride

Empirical: Cl_3OP
Formula: $POCl_3$
Properties: Colorless fuming liq., pungent odor; dec. by water and alcohol with evolution of heat; m.w. 153.35; dens. 1.675 (20/20 C); m.p. 1.2 C; b.p. 107.2 C; dens. 1.675 (20C); ref. index 1.461 (20C)
Precaution: DOT: Corrosive material; moisture-sensitive
Toxicology: Toxic by ingestion and inhalation; strong irritant to skin and tissue
Uses: Mfg. of insecticides, phosphate esters, paints, pharmaceuticals; gasoline additives; dopant for semiconductor grade silicon, tricresyl phosphate, and fire-retarding agents
Manuf./Distrib.: Albright & Wilson Am.; Aldrich; Cerac; Fluka; FMC; Hoechst Celanese; Rhone-Poulenc Basic; Spectrum Chem. Mfg.

Phosphorus trichloride
CAS 7719-12-2; EINECS 231-749-3
Synonyms: Phosphorus chloride
Empirical: Cl_3P
Formula: PCl_3
Properties: Clear colorless fuming liq.; sol. in ether, benzene, carbon disulfide, CCl_4; m.w. 137.33; dens. 2.852 (15 C); dec. in moist air; b.p. 175 C; f.p. -40 C
Toxicology: Corrosive to skin and tissue; poison by inhalation; moderately toxic by ingestion
Uses: Source of phosphorus in mfg. of phosphite and phosphonate esters; chlorinating agent in mfg. of organic acid chlorides; analysis; catalyst; intermediate for paints
Manuf./Distrib.: Advance Research Chems.; Albright & Wilson Am.; Aldrich; Atomergic Chemetals; Fluka; FMC; Rhone-Poulenc Basic; Schumacher; Spectrum Chem. Mfg.

Phosphorus trichloride, reaction prods. with 1,1´-biphenyl and 2,4-bis (1,1-dimethylethyl) phenol
CAS 119345-01-6
Classification: Aryl phosphonite
Properties: Wh./off-wh. powd., sl. odor; insol. in water; sp.gr. 1.04; m.p. 185-203 F; flash pt. 350 C
Precaution: Incompat. with strong oxidizing agents; thermal decomp. may produce oxides of carbon
Toxicology: LD50 (oral, rat) > 5000 mg/kg (dermal, rat) > 5000 mg/kg; nonirritating to eyes; mildly irritating to skin
Uses: Stabilizer, sec. antioxidant for polymers; antioxidant and color suppresant for powd. coatings
Regulatory: FDA 21CFR §178.2010
Manuf./Distrib.: Sandoz
Trade names: Sandostab P-EPQ

Phosphoryl chloride. *See* Phosphorus oxychloride
Phthalic acid, butyl octyl ester. *See* Butyl octyl phthalate
Phthalic acid diisobutyl ester. *See* Diisobutyl phthalate
Phthalic acid dimethyl ester. *See* Dimethyl phthalate
Phthalic acid, tetrabromo-bis (2-ethylhexyl) ester. *See* Tetrabromobis (2-ethylhexyl) phthalate

Phthalic anhydride
CAS 85-44-9; EINECS 201-607-5
Synonyms: 1,3-Isobenzofurandione
Empirical: $C_8H_4O_3$
Formula: $C_6H_4(CO)_2O$
Properties: Wh. cryst. needles, mild odor; very sl. sol. in water; sol. in alcohol; sl. sol. in ether; m.w. 148.12; dens. 1.527 (4 C); m.p. 130.8 C; b.p. 295 C; sublimes; flash pt. (CC) 151 C
Precaution: DOT: Corrosive material; moisture-sensitive
Toxicology: TLV:TWA 1 ppm (air); poison by ingestion
Uses: Retarding agent in rubber compoudning; in alkyd resins, paints, plasticizers, hardener for resins, polyester, insecticides, laboratory reagent; mfg. of phthaleins, phthalates, benzoic acid
Manuf./Distrib.: Aldrich; Allchem Ind.; Aristech; BASF; Browning; Chemical; Coyne; Elf Atochem SA; Exxon; Fluka; Harcros; Int'l. Chem. Inc.; Koppers Ind.; Mitsubishi Gas; Mitsui Toatsu; Monomer-Polymer & Dajac Labs; Occidental; Primachem; Punda Mercantile; Schweizerhall; Sigma; Spectrum Chem. Mfg.; Stepan; UCB SA; United Min. & Chem.; Van Waters & Rogers

Phthalic anhydride, tetrachloro. *See* Tetrachlorophthalic anhydride
Phynlamine. *See* Aniline
PIB. *See* Polybutene; Polyisobutene
Pigment blue 29. *See* Ultramarine blue

Pigment red 202
Uses: Colorant for plastics, powd. coatings, finishes, inks, artists' colors
Trade names: Quindo Magenta RV-6863

Pigment violet 15. *See* Ultramarine violet
Pigment violet 16. *See* Manganese violet
Pigment white 4. *See* Zinc oxide
Pigment white 6. *See* Titanium dioxide
Pigment white 14. *See* Bismuth oxychloride
Pigment white 26. *See* Talc

Pigment yellow 14
Uses: Colorant
Trade names containing: Octotint 150

Pigment yellow 53
CAS 8007-18-9
Synonyms: CI 77788
Empirical: NiSbTi
Properties: Sp.gr. 4.4
Uses: Colorant for plastics and industrial finishes
Trade names: Ferro V-9415

Pimelic ketone. *See* Cyclohexanone
2-Pinene. *See* α-Pinene

α-Pinene
CAS 80-56-8; 7785-26-4 (-); 7785-70-8 (+); EINECS 232-077-3 (-); 232-087-8 (+)
Synonyms: 2-Pinene; 2,6,6-Trimethylbicyclo(3.1.1)-2-hept-2-ene; 2,6,6-Trimethylbicyclo(3.1.1)-2-heptene
Classification: Terpene hydrocarbon
Empirical: $C_{10}H_{16}$
Properties: Colorless liq., turpentine odor; insol. in water; sol. in alcohol, chloroform, ether, glacial acetic acid; m.w. 136.26; dens. 0.8592 (20/4 C); m.p. -55 C; b.p. 155 C; flash pt. 91 F; ref. index 1.464-1.468
Precaution: DOT: Flamm. liq.; dangerous fire hazard exposed to heat, flame, oxidizers; explodes on contact with nitrosyl perchlorate
Toxicology: LD50 (oral, rat) 3700mg/kg; deadly poison by inh.; mod. toxic by ingestion; eye, mucous membrane, and severe skin irritant; toxic effects similar to turpentine
Uses: Solvent for protective coating, synthesis of camphene, pine oil, odorant, lube oil additives, flavoring, insecticides
Regulatory: FDA 21CFR §172.515; FEMA GRAS
Manuf./Distrib.: Aldrich; Arizona; Fluka; Hercules; Penta Mfg.; SCM Glidco Organics; Sigma; Spectrum Chem. Mfg.; Veitsiluoto Oy

Pine tar
CAS 8011-48-1; EINECS 232-374-8
Definition: Prod. obtained by destructive distillation of the wood of various species of pine, *Pinaceae*
Uses: Plasticizer, tackifier, process aid for rubber; raw material for paints
Regulatory: FDA 21CFR §177.2600
Manuf./Distrib.: H.M. Royal

2-Piperazinoethylamione. *See* Aminoethylpiperazine
Pivalic acid. *See* Neopentanoic acid
Plaster of Paris. *See* Calcium sulfate
Plastic sponge. *See* Polyurethane, thermoplastic
Platy talc. *See* Talc
Plumbago. *See* Graphite
Plumboplumbic oxide. *See* Lead oxide, red
Plumbous oxide. *See* Lead (II) oxide
PMAC. *See* Phenylmercuric acetate
PM acetate. *See* Propylene glycol methyl ether acetate
PMDA. *See* Pyromellitic dianhydride
POE (4). *See* PEG-4
POE (8). *See* PEG-8

POE (9). *See* PEG-9
POE (12). *See* PEG-12
POE (14). *See* PEG-14
POE (16). *See* PEG-16
POE (20). *See* PEG-20
POE (32). *See* PEG-32
POE (40). *See* PEG-40
POE (75). *See* PEG-75
POE (100). *See* PEG-100
POE (150). *See* PEG-150
POE (200). *See* PEG-200
POE (350). *See* PEG-350
POE (5000). *See* PEG-5M
POE (7000). *See* PEG-7M
POE (9000). *See* PEG-9M
POE (14000). *See* PEG-14M
POE (23000). *See* PEG-23M
POE (45000). *See* PEG-45M
POE (90000). *See* PEG-90M
POE (5) castor oil. *See* PEG-5 castor oil
POE (12) castor oil. *See* PEG-12 castor oil
POE (15) castor oil. *See* PEG-15 castor oil
POE (16) castor oil. *See* PEG-16 castor oil
POE (20) castor oil. *See* PEG-20 castor oil
POE (25) castor oil. *See* PEG-25 castor oil
POE (28) castor oil. *See* PEG-28 castor oil
POE (30) castor oil. *See* PEG-30 castor oil
POE (32) castor oil. *See* PEG-32 castor oil
POE (36) castor oil. *See* PEG-36 castor oil
POE (40) castor oil. *See* PEG-40 castor oil
POE (52) castor oil. *See* PEG-52 castor oil
POE (75) castor oil. *See* PEG-75 castor oil
POE (80) castor oil. *See* PEG-80 castor oil
POE (200) castor oil. *See* PEG-200 castor oil
POE (6) cetyl/stearyl ether. *See* Ceteareth-6
POE (8) cetyl/stearyl ether. *See* Ceteareth-8
POE (10) cetyl/stearyl ether. *See* Ceteareth-10
POE (11) cetyl/stearyl ether. *See* Ceteareth-11
POE (12) cetyl/stearyl ether. *See* Ceteareth-12
POE (14) cetyl/stearyl ether. *See* Ceteareth-14
POE (18) cetyl/stearyl ether. *See* Ceteareth-18
POE (27) cetyl/stearyl ether. *See* Ceteareth-27
POE (30) cetyl/stearyl ether. *See* Ceteareth-30
POE (55) cetyl/stearyl ether. *See* Ceteareth-55
POE (2) coco-benzonium chloride. *See* PEG-2 coco-benzonium chloride
POE (10) coconut amine. *See* PEG-10 cocamine
POE (15) coconut amine. *See* PEG-15 cocamine
POE (3) diacetate. *See* PEG-3 diacetate
POE (2) dibenzoate. *See* Diethylene glycol dibenzoate
POE (8) dilaurate. *See* PEG-8 dilaurate
POE (8) dinonyl phenyl ether. *See* Nonyl nonoxynol-8
POE (9) dinonyl phenyl ether. *See* Nonyl nonoxynol-9
POE (15) dinonyl phenyl ether. *See* Nonyl nonoxynol-15
POE (18) dinonyl phenyl ether. *See* Nonyl nonoxynol-18
POE (24) dinonyl phenyl ether. *See* Nonyl nonoxynol-24
POE (49) dinonyl phenyl ether. *See* Nonyl nonoxynol-49
POE (150) dinonyl phenyl ether. *See* Nonyl nonoxynol-150
POE (9) dioleate. *See* PEG-9 dioleate
POE (12) dioleate. *See* PEG-12 dioleate
POE (150) distearate. *See* PEG-150 distearate
POE (8) ditallate. *See* PEG-8 ditallate
POE (12) ditallate. *See* PEG-12 ditallate

POE (10) dodecyl phenyl ether. *See* Dodoxynol-10
POE (5) hydrogenated castor oil. *See* PEG-5 hydrogenated castor oil
POE (16) hydrogenated castor oil. *See* PEG-16 hydrogenated castor oil
POE (25) hydrogenated castor oil. *See* PEG-25 hydrogenated castor oil
POE (200) hydrogenated castor oil. *See* PEG-200 hydrogenated castor oil
POE (8) isolauryl thioether. *See* PEG-8 isolauryl thioether
POE (5) lauryl ether. *See* Laureth-5
POE (8) lauryl ether. *See* Laureth-8
POE (9) lauryl ether. *See* Laureth-9
POE (12) lauryl ether. *See* Laureth-12
POE (23) lauryl ether. *See* Laureth-23
POE (4) monolaurate. *See* PEG-4 laurate
POE (5) monolaurate. *See* PEG-5 laurate
POE (8) monolaurate. *See* PEG-8 laurate
POE (9) monolaurate. *See* PEG-9 laurate
POE (12) monolaurate. *See* PEG-12 laurate
POE (2) monolaurate self-emulsifying. *See* PEG-2 laurate SE
POE (2) monooleate. *See* PEG-2 oleate
POE (4) monooleate. *See* PEG-4 oleate
POE (8) monooleate. *See* PEG-8 oleate
POE (12) monooleate. *See* PEG-12 oleate
POE (14) monooleate. *See* PEG-14 oleate
POE (20) monooleate. *See* PEG-20 oleate
POE (2) monooleate self-emulsifying. *See* PEG-2 oleate SE
POE (2) monostearate self-emulsifying. *See* PEG-2 stearate SE
POE (2) nonyl phenyl ether. *See* Nonoxynol-2
POE (3) nonyl phenyl ether. *See* Nonoxynol-3
POE (4) nonyl phenyl ether. *See* Nonoxynol-4
POE (5) nonyl phenyl ether. *See* Nonoxynol-5
POE (6) nonyl phenyl ether. *See* Nonoxynol-6
POE (8) nonyl phenyl ether. *See* Nonoxynol-8
POE (9) nonyl phenyl ether. *See* Nonoxynol-9
POE (10) nonyl phenyl ether. *See* Nonoxynol-10
POE (11) nonyl phenyl ether. *See* Nonoxynol-11
POE (12) nonyl phenyl ether. *See* Nonoxynol-12
POE (13) nonyl phenyl ether. *See* Nonoxynol-13
POE (14) nonyl phenyl ether. *See* Nonoxynol-14
POE (20) nonyl phenyl ether. *See* Nonoxynol-20
POE (30) nonyl phenyl ether. *See* Nonoxynol-30
POE (40) nonyl phenyl ether. *See* Nonoxynol-40
POE (50) nonyl phenyl ether. *See* Nonoxynol-50
POE (70) nonyl phenyl ether. *See* Nonoxynol-70
POE (100) nonyl phenyl ether. *See* Nonoxynol-100
POE (10) nonyl phenyl ether carboxylic acid. *See* Nonoxynol-10 carboxylic acid
POE (6) nonyl phenyl ether phosphate. *See* Nonoxynol-6 phosphate
POE (9) nonyl phenyl ether phosphate. *See* Nonoxynol-9 phosphate
POE (6) nonyl phenylether phosphate, sodium salt. *See* Sodium nonoxynol-6 phosphate
POE (3) octyl phenyl ether. *See* Octoxynol-3
POE (5) octyl phenyl ether. *See* Octoxynol-5
POE (7) octyl phenyl ether. *See* Octoxynol-7
POE (8) octyl phenyl ether. *See* Octoxynol-8
POE (9) octyl phenyl ether. *See* Octoxynol-9
POE (10) octyl phenyl ether. *See* Octoxynol-10
POE (13) octyl phenyl ether. *See* Octoxynol-13
POE (16) octyl phenyl ether. *See* Octoxynol-16
POE (25) octyl phenyl ether. *See* Octoxynol-25
POE (30) octyl phenyl ether. *See* Octoxynol-30
POE (40) octyl phenyl ether. *See* Octoxynol-40
POE (70) octyl phenyl ether. *See* Octoxynol-70
POE (4) oleyl ether. *See* Oleth-4
POE (5) oleyl ether. *See* Oleth-5
POE (9) oleyl ether. *See* Oleth-9

POE (20) oleyl ether. *See* Oleth-20
POE (8) pelargonate. *See* PEG-8 pelargonate
POE (9) pelargonate. *See* PEG-9 pelargonate
POE (27) POP (24) monobutyl ether. *See* PPG-24-buteth-27
POE (20) sorbitan monolaurate. *See* Polysorbate 20
POE (80) sorbitan monolaurate. *See* PEG-80 sorbitan laurate
POE (5) sorbitan monooleate. *See* Polysorbate 81
POE (20) sorbitan monooleate. *See* Polysorbate 80
POE (20) sorbitan monopalmitate. *See* Polysorbate 40
POE (4) sorbitan monostearate. *See* Polysorbate 61
POE (20) sorbitan monostearate. *See* Polysorbate 60
POE (20) sorbitan trioleate. *See* Polysorbate 85
POE (20) sorbitan tristearate. *See* Polysorbate 65
POE (4) stearate. *See* PEG-4 stearate
POE (8) stearate. *See* PEG-8 stearate
POE (12) stearate. *See* PEG-12 stearate
POE (20) stearate. *See* PEG-20 stearate
POE (1) stearic acid. *See* Glycol stearate
POE (2) stearyl ether. *See* Steareth-2
POE (10) stearyl ether. *See* Steareth-10
POE (20) stearyl ether. *See* Steareth-20
POE (10) tallow aminopropylamine. *See* PEG-10 tallow aminopropylamine
POE (6) tridecyl ether. *See* Trideceth-6
POE (12) tridecyl ether. *See* Trideceth-12
POE (14) tridecyl ether. *See* Trideceth-14
POE (15) tridecyl ether. *See* Trideceth-15
POE (66) trihydroxystearin. *See* PEG-66 trihydroxystearin
Pogy oil. *See* Menhaden oil
Poloxalene. *See* Poloxamer 188

Poloxamer 101
CAS 9003-11-6 (generic)
Synonyms: Methyl oxirane polymers; Polyethylenepolypropylene glycols, polymers
Classification: Polyoxyethylene, polyoxypropylene block polymer
Formula: $HO(CH_2CH_2O)_x(CH(CH_3)CH_2O)_y(CH_2CH_2O)_zH$, avg. x=2, y=16, z=2
Toxicology: Moderately toxic by ingestion, intraperitoneal route
Uses: Nonionic surfactant; as food additives, defoamers, antistats, demulsifiers, detergents, wetting agents, gellants, emulsifiers, dispersants, dye levelers; water treatment, paints, latexes
Regulatory: FDA 21CFR §176.210; USP/NF compliance
Manuf./Distrib.: Aldrich; Fluka; Sigma
Trade names: Pluronic® L31

Poloxamer 105
CAS 9003-11-6 (generic)
Classification: Polyoxyethylene, polyoxypropylene block polymer
Formula: $HO(CH_2CH_2O)_x(CCH_3HCH_2O)_y(CH_2CH_2O)_zH$, avg. x=11, y=16, z=11
Toxicology: Moderately toxic by ingestion, intraperitoneal route
Uses: Nonionic surfactant; as food additives, defoamers, antistats, demulsifiers, detergents, wetting agents, gellants, emulsifiers, dispersants, dye levelers; water treatment, paints, latexes
Regulatory: FDA 21CFR §172.808, 173.340, 175.105, 176.180, 176.200, 176.210, 177.1200
Manuf./Distrib.: Aldrich; Fluka; Sigma
Trade names: Calgene Nonionic 1035-L; Pluronic® L35

Poloxamer 108
CAS 9003-11-6 (generic)
Classification: Polyoxyethylene, polyoxypropylene block polymer
Formula: $HO(CH_2CH_2O)_x(CCH_3HCH_2O)_y(CH_2CH_2O)_zH$, avg. x=46, y=16, z=46
Toxicology: Moderately toxic by ingestion, intraperitoneal route
Uses: Nonionic surfactant; as food additives, defoamers, antistats, demulsifiers, detergents, wetting agents, gellants, emulsifiers, dispersants, dye levelers; water treatment, paints
Regulatory: FDA 21CFR §172.808, 173.340, 175.105, 176.180, 176.200, 176.210, 177.1200, 177.1210
Manuf./Distrib.: Aldrich; Fluka; Sigma
Trade names: Pluronic® F38; Pluronic® F68LF; Pluronic® L62D; Pluronic® L62LF

Poloxamer 123

CAS 9003-11-6 (generic)
Classification: Polyoxyethylene, polyoxypropylene block polymer
Formula: $HO(CH_2CH_2O)_x(CCH_3HCH_2O)_y(CH_2CH_2O)_zH$, avg. x=7, y=21, z=7
Toxicology: Moderately toxic by ingestion, intraperitoneal route
Uses: Nonionic surfactant; as food additives, defoamers, antistats, demulsifiers, detergents, wetting agents, gellants, emulsifiers, dispersants, dye levelers; water treatment, paints, latexes
Regulatory: FDA 21CFR §172.808, 173.340, 175.105, 176.180, 176.200, 176.210, 177.1200
Manuf./Distrib.: Aldrich; Fluka; Sigma
Trade names: Pluronic® L43

Poloxamer 124

CAS 9003-11-6 (generic)
Classification: Polyoxyethylene, polyoxypropylene block polymer
Formula: $HO(CH_2CH_2O)_x(CCH_3HCH_2O)_y(CH_2CH_2O)_zH$, avg. x=11, y=21, z=11
Properties: Colorless liq., mild odor; sol. in water, alcohol, IPA, propylene glycol, xylene; m.w. 2090-2360; m.p. 16 C; pH 5.0-7.5 (1 in 40)
Toxicology: Moderately toxic by ingestion, intraperitoneal route
Uses: Nonionic surfactant; as food additives, defoamers, antistats, demulsifiers, detergents, wetting agents, gellants, emulsifiers, dispersants, dye levelers; paints, paper, pharmaceuticals, metalworking, agric., etc
Regulatory: FDA 21CFR §172.808, 173.340, 175.105, 176.180, 176.200, 176.210, 177.1200
Manuf./Distrib.: Aldrich; Fluka; Sigma
Trade names: Calgene Nonionic 1044-L; Pluronic® L44

Poloxamer 181

CAS 9003-11-6 (generic); 53637-25-5
Classification: Polyoxyethylene, polyoxypropylene block polymer
Formula: $HO(CH_2CH_2O)_x(CCH_3HCH_2O)_y(CH_2CH_2O)_zH$, avg. x=3, y=30, z=3
Toxicology: Moderately toxic by ingestion, intraperitoneal route
Uses: Nonionic surfactant; as food additives, defoamers, antistats, demulsifiers, detergents, wetting agents, gellants, emulsifiers, dispersants, dye levelers; paints, paper, agric., etc
Regulatory: FDA 21CFR §172.808, 173.340, 175.105, 176.180, 176.200, 176.210, 177.1200
Manuf./Distrib.: Aldrich; Fluka; Sigma
Trade names: Calgene Nonionic 1061-L; Pluronic® L61

Poloxamer 182

CAS 9003-11-6 (generic)
Classification: Polyoxyethylene, polyoxypropylene block polymer
Formula: $HO(CH_2CH_2O)_x(CCH_3HCH_2O)_y(CH_2CH_2O)_zH$, avg. x=8, y=30, z=8
Properties: Dens. 1.018; flash pt. > 110 C
Toxicology: Moderately toxic by ingestion, intraperitoneal route
Uses: Nonionic surfactant; as food additives, defoamers, antistats, demulsifiers, detergents, wetting agents, gellants, emulsifiers, dispersants, dye levelers; emulsion polymerization; lubricant for agric., cosmetics, metal cleaning, pulp/paper; textile scouring, water treatment, paints, latexes
Regulatory: FDA 21CFR §172.808, 173.340, 175.105, 176.180, 176.200, 176.210, 177.1200
Manuf./Distrib.: Aldrich; Fluka; Sigma
Trade names: Calgene Nonionic 1062-L; Pluronic® L62

Poloxamer 184

CAS 9003-11-6 (generic)
Classification: Polyoxyethylene, polyoxypropylene block polymer
Formula: $HO(CH_2CH_2O)_x(CCH_3HCH_2O)_y(CH_2CH_2O)_zH$, avg. x=13, y=30, z=13
Properties: Dens. 1.018; flash pt. > 110 C
Toxicology: Moderately toxic by ingestion, intraperitoneal route
Uses: Nonionic surfactant; as food additives, defoamers, antistats, demulsifiers, detergents, wetting agents, gellants, emulsifiers, dispersants, dye levelers; emulsion polymerization; resin plasticizer; defoamer in paper coatings
Regulatory: FDA 21CFR §172.808, 173.340, 175.105, 176.180, 176.200, 176.210, 177.1200, 177.1210
Manuf./Distrib.: Aldrich; Fluka; Sigma
Trade names: Antarox® L-64; Calgene Nonionic 1064-L; Pluronic® L64

Poloxamer 185

CAS 9003-11-6 (generic)

Poloxamer 188

Classification: Polyoxyethylene, polyoxypropylene block polymer
Formula: $HO(CH_2CH_2O)_x(CCH_3HCH_2O)_y(CH_2CH_2O)_zH$, avg. x=19, y=30, z=19
Toxicology: Moderately toxic by ingestion, intraperitoneal route
Uses: Nonionic surfactant; as food additives, defoamers, antistats, demulsifiers, detergents, wetting agents, gellants, emulsifiers, dispersants, dye levelers; lubricant for agric., cosmetics, metal cleaning, pulp/paper, textile scouring, water treatment, paints, latexes
Regulatory: FDA 21CFR §172.808, 173.340, 175.105, 176.180, 176.200, 176.210, 177.1200, 177.1210
Manuf./Distrib.: Aldrich; Fluka; Sigma
Trade names: Pluronic® P65

Poloxamer 188
CAS 9003-11-6 (generic)
Synonyms: Poloxalene
Classification: Polyoxyethylene, polyoxypropylene block polymer
Formula: $HO(CH_2CH_2O)_x(CCH_3HCH_2O)_y(CH_2CH_2O)_zH$, avg. x=75, y=30, z=75
Properties: Wh. flakeable solid, nearly odorless; sol. in water, alcohol; m.w. 7680-9510; m.p. 50 C min.; cloud pt. > 100 C (10% aq.); pH 5.0-7.5 (1 in 40)
Toxicology: Moderately toxic by ingestion, intraperitoneal route
Uses: Nonionic surfactant; as food additives, defoamers, antistats, demulsifiers, detergents, wetting agents, gellants, emulsifiers, dispersants, dye levelers; latex stabilizer; lubricant for agric., cosmetics, metal cleaning, pulp/paper, textile scouring, water treatment, paints
Regulatory: FDA 21CFR §172.808, 173.340, 175.105, 176.180, 176.200, 176.210, 177.1200, 177.1210; FDA approved for intravenous, orals; BP compliance
Manuf./Distrib.: Aldrich; Fluka; Sigma
Trade names: Calgene Nonionic 1068-F; Pluronic® F68

Poloxamer 217
CAS 9003-11-6 (generic)
Classification: Polyoxyethylene, polyoxypropylene block polymer
Formula: $HO(CH_2CH_2O)_x(CCH_3HCH_2O)_y(CH_2CH_2O)_zH$, avg. x=52, y=35, z=52
Toxicology: Moderately toxic by ingestion, intraperitoneal route
Uses: Nonionic surfactant; as food additives, defoamers, antistats, demulsifiers, detergents, wetting agents, gellants, emulsifiers, dispersants, dye levelers; in textiles, agric., paints, adhesives
Regulatory: FDA 21CFR §172.808, 173.340, 175.105, 176.180, 176.200, 176.210, 177.1200, 177.1210
Manuf./Distrib.: Aldrich; Fluka; Sigma
Trade names: Calgene Nonionic 1077-F; Pluronic® F77

Poloxamer 231
CAS 9003-11-6 (generic)
Classification: Polyoxyethylene, polyoxypropylene block polymer
Formula: $HO(CH_2CH_2O)_x(CCH_3HCH_2O)_y(CH_2CH_2O)_zH$, avg. x=6, y=39, z=6
Toxicology: Moderately toxic by ingestion, intraperitoneal route
Uses: Nonionic surfactant; as food additives, defoamers, antistats, demulsifiers, detergents, wetting agents, gellants, emulsifiers, dispersants, dye levelers; lubricant for agric., cosmetics, metal cleaning, pulp/paper, textile scouring, water treatment; paints, latexes
Regulatory: FDA 21CFR §172.808, 173.340, 175.105, 176.180, 176.200, 176.210, 177.1200, 177.1210
Manuf./Distrib.: Aldrich; Fluka; Sigma
Trade names: Pluronic® L81

Poloxamer 234
CAS 9003-11-6 (generic)
Classification: Polyoxyethylene, polyoxypropylene block polymer
Formula: $HO(CH_2CH_2O)_x(CCH_3HCH_2O)_y(CH_2CH_2O)_zH$, avg. x=22, y=39, z=22
Toxicology: Moderately toxic by ingestion, intraperitoneal route
Uses: Nonionic surfactant; as food additives, defoamers, antistats, demulsifiers, detergents, wetting agents, gellants, emulsifiers, dispersants, dye levelers; lubricant for agric., cosmetics, metal cleaning, pulp/paper, textile scouring, water treatment; paints, latexes
Regulatory: FDA 21CFR §172.808, 173.340, 175.105, 176.180, 176.200, 176.210, 177.1200, 177.1210
Manuf./Distrib.: Aldrich; Fluka; Sigma
Trade names: Pluronic® P84

Poloxamer 235
CAS 9003-11-6 (generic)

Classification: Polyoxyethylene, polyoxypropylene block polymer
Formula: $HO(CH_2CH_2O)_x(CCH_3HCH_2O)_y(CH_2CH_2O)_zH$, avg. x=27, y=39, z=27
Toxicology: Moderately toxic by ingestion, intraperitoneal route
Uses: Nonionic surfactant; as food additives, defoamers, antistats, demulsifiers, detergents, wetting agents, gellants, emulsifiers, dispersants, dye levelers; lubricant for agric., cosmetics, metal cleaning, pulp/paper, textile scouring, water treatment; paints, latexes
Regulatory: FDA 21CFR §172.808, 173.340, 175.105, 176.180, 176.200, 1876.210, 177.1200, 177.1210
Manuf./Distrib.: Aldrich; Fluka; Sigma
Trade names: Pluronic® P85

Poloxamer 237
CAS 9003-11-6 (generic)
Classification: Polyoxyethylene, polyoxypropylene block polymer
Formula: $HO(CH_2CH_2O)_x(CCH_3HCH_2O)_y(CH_2CH_2O)_zH$, avg. x=62, y=39, z=62
Properties: Wh. prilled or cast solid, odorless to mild odor; sol. in water, alcohol; sparingly sol. in IPA, xylene; m.w. 6840-8830; m.p. 49 C; pH 5.0-7.5 (1 in 40)
Toxicology: Moderately toxic by ingestion, intraperitoneal route
Uses: Nonionic surfactant; as food additives, defoamers, antistats, demulsifiers, detergents, wetting agents, gellants, emulsifiers, dispersants, dye levelers; surfactant for personal care applics., pulp/paper, pharmaceuticals, metalworking, lubricants, textiles, agric., paints, adhesives
Regulatory: FDA 21CFR §172.808, 173.340, 175.105, 176.180, 176.200, 176.210, 177.1200, 177.1210
Manuf./Distrib.: Aldrich; Fluka; Sigma
Trade names: Calgene Nonionic 1087-F; Pluronic® F87

Poloxamer 238
CAS 9003-11-6 (generic)
Classification: Polyoxyethylene, polyoxypropylene block polymer
Formula: $HO(CH_2CH_2O)_x(CCH_3HCH_2O)_y(CH_2CH_2O)_zH$, avg. x=97, y=39, z=97
Properties: Dens. 1.018; flash pt. > 110 C
Toxicology: Moderately toxic by ingestion, intraperitoneal route
Uses: Nonionic surfactant; as food additives, defoamers, antistats, demulsifiers, detergents, wetting agents, gellants, emulsifiers, dispersants, dye levelers; latex stabilizer; lubricant for agric., cosmetics, metal cleaning, pulp/paper, textile scouring, water treatment, paints, latexes
Regulatory: FDA 21CFR §172.808, 173.340, 175.105, 176.180, 176.200, 176.210, 177.1200, 177.1210
Manuf./Distrib.: Aldrich; Fluka; Sigma
Trade names: Antarox® F88 FLK; Calgene Nonionic 1088-F; Pluronic® F88

Poloxamer 282
CAS 9003-11-6 (generic)
Classification: Polyoxyethylene, polyoxypropylene block polymer
Formula: $HO(CH_2CH_2O)_x(CCH_3HCH_2O)_y(CH_2CH_2O)_zH$, avg. x=10, y=47, z=10
Toxicology: Moderately toxic by ingestion, intraperitoneal route
Uses: Nonionic surfactant; as food additives, defoamers, antistats, demulsifiers, detergents, wetting agents, gellants, emulsifiers, dispersants, dye levelers; lubricant for agric., cosmetics, metal cleaning, pulp/paper, textile scouring, water treatment, paints, latexes
Regulatory: FDA 21CFR §172.808, 173.340, 175.105, 176.180, 176.200, 176.210, 177.1200, 177.1210
Manuf./Distrib.: Aldrich; Fluka; Sigma
Trade names: Pluronic® L92

Poloxamer 288
CAS 9003-11-6 (generic)
Classification: Polyoxyethylene, polyoxypropylene block polymer
Formula: $HO(CH_2CH_2O)_x(CCH_3HCH_2O)_y(CH_2CH_2O)_zH$, avg. x=122, y=47, z=122
Toxicology: Moderately toxic by ingestion, intraperitoneal route
Uses: Nonionic surfactant; as food additives, defoamers, antistats, demulsifiers, detergents, wetting agents, gellants, emulsifiers, dispersants, dye levelers; lubricant for agric., cosmetics, metal cleaning, pulp/paper, textile scouring, water treatment, paints, latexes
Regulatory: FDA 21CFR §172.808, 173.340, 175.105, 176.180, 176.200, 176.210, 177.1200, 177.1210
Manuf./Distrib.: Aldrich; Fluka; Sigma
Trade names: Pluronic® F98

Poloxamer 331
CAS 9003-11-6 (generic)

Poloxamer 333

Classification: Polyoxyethylene, polyoxypropylene block polymer
Formula: $HO(CH_2CH_2O)_x(CCH_3HCH_2O)_y(CH_2CH_2O)_zH$, avg. x=7, y=54, z=7
Properties: Colorless liq.; sol. in alcohol; very sl. sol. in water; m.w. 3800; dens. 1.018 (25/25 C); visc. 756 cp; cloud pt. 11 C (10% aq.)
Toxicology: Moderately toxic by ingestion, intraperitoneal route; heated to decomp., emits acrid smoke and irritating fumes
Uses: Nonionic surfactant; as food additives, defoamers, antistats, demulsifiers, detergents, wetting agents, gellants, emulsifiers, dispersants, dye levelers; lubricant for agric., cosmetics, metal cleaning, pulp/paper, textile scouring, water treatment, paints, latexes
Regulatory: FDA 21CFR §172.808, 173.340, 175.105, 176.180, 176.200, 176.210, 177.1200, 177.1210; 9CFR §381.147; FDA approved for orals
Manuf./Distrib.: Aldrich; Fluka; Sigma
Trade names: Pluronic® L101

Poloxamer 333
CAS 9003-11-6 (generic)
Classification: Polyoxyethylene, polyoxypropylene block polymer
Formula: $HO(CH_2CH_2O)_x(CCH_3HCH_2O)_y(CH_2CH_2O)_zH$, avg. x=20, y=54, z=20
Toxicology: Moderately toxic by ingestion, intraperitoneal route
Uses: Nonionic surfactant; as food additives, defoamers, antistats, demulsifiers, detergents, wetting agents, gellants, emulsifiers, dispersants, dye levelers; lubricant for agric., cosmetics, metal cleaning, pulp/paper, textile scouring, water treatment, paints, latexes
Regulatory: FDA 21CFR §172.808, 173.340, 175.105, 176.180, 176.200, 176.210, 177.1200, 177.1210
Manuf./Distrib.: Aldrich; Fluka; Sigma
Trade names: Pluronic® P103

Poloxamer 334
CAS 9003-11-6 (generic)
Classification: Polyoxyethylene, polyoxypropylene block polymer
Formula: $HO(CH_2CH_2O)_x(CCH_3HCH_2O)_y(CH_2CH_2O)_zH$, avg. x=31, y=54, z=31
Toxicology: Moderately toxic by ingestion, intraperitoneal route
Uses: Nonionic surfactant; as food additives, defoamers, antistats, demulsifiers, detergents, wetting agents, gellants, emulsifiers, dispersants, dye levelers; lubricant for agric., cosmetics, metal cleaning, pulp/paper, textile scouring, water treatment, paints, latexes
Regulatory: FDA 21CFR §172.808, 173.340, 175.105, 176.180, 176.200, 176.210, 177.1200, 177.1210
Manuf./Distrib.: Aldrich; Fluka; Sigma
Trade names: Pluronic® P104

Poloxamer 335
CAS 9003-11-6 (generic)
Classification: Polyoxyethylene, polyoxypropylene block polymer
Formula: $HO(CH_2CH_2O)_x(CCH_3HCH_2O)_y(CH_2CH_2O)_zH$, avg. x=38, y=54, z=38
Toxicology: Moderately toxic by ingestion, intraperitoneal route
Uses: Nonionic surfactant; as food additives, defoamers, antistats, demulsifiers, detergents, wetting agents, gellants, emulsifiers, dispersants, dye levelers; lubricant for agric., cosmetics, metal cleaning, pulp/paper, textile scouring, water treatment, paints, latexes
Regulatory: FDA 21CFR §172.808, 173.340, 175.105, 176.180, 176.200, 176.210, 177.1200, 177.1210
Manuf./Distrib.: Aldrich; Fluka; Sigma
Trade names: Pluronic® P105

Poloxamer 338
CAS 9003-11-6 (generic)
Classification: Polyoxyethylene, polyoxypropylene block polymer
Formula: $HO(CH_2CH_2O)_x(CCH_3HCH_2O)_y(CH_2CH_2O)_zH$, avg. x=128, y=54, z=128
Properties: Wh. prilled or cast solid, odorless to mild odor; sol. in water, alcohol; sparingly sol. in propylene glycol; m.w. 12,700-17,400; m.p. 57 C; pH 5.0-7.5 (1 in 40)
Toxicology: Moderately toxic by ingestion, intraperitoneal route
Uses: Nonionic surfactant; as food additives, defoamers, antistats, demulsifiers, detergents, wetting agents, gellants, emulsifiers, dispersants, dye levelers; lubricant for agric., cosmetics, metal cleaning, pulp/paper, textile scouring, water treatment, paints, latexes
Regulatory: FDA 21CFR §172.808, 173.340, 175.105, 176.180, 176.200, 176.210, 177.1200, 177.1210
Manuf./Distrib.: Aldrich; Fluka; Sigma
Trade names: Calgene Nonionic 1108-F; Pluronic® F108

Poloxamer 401

CAS 9003-11-6 (generic)
Classification: Polyoxyethylene, polyoxypropylene block polymer
Formula: $HO(CH_2CH_2O)_x(CCH_3HCH_2O)_y(CH_2CH_2O)_zH$, avg. x=6, y=67, z=6
Toxicology: Moderately toxic by ingestion, intraperitoneal route
Uses: Nonionic surfactant; as food additives, defoamers, antistats, demulsifiers, detergents, wetting agents, gellants, emulsifiers, dispersants, dye levelers; lubricant for agric., cosmetics, metal cleaning, pulp/paper, textile scouring, water treatment, paints, latexes
Regulatory: FDA 21CFR §172.808, 173.340, 175.105, 176.180, 176.200, 176.210, 177.1200, 177.1210
Manuf./Distrib.: Aldrich; Fluka; Sigma
Trade names: Pluronic® L121

Poloxamer 402

CAS 9003-11-6 (generic)
Classification: Polyoxyethylene, polyoxypropylene block polymer
Formula: $HO(CH_2CH_2O)_x(CCH_3HCH_2O)_y(CH_2CH_2O)_zH$, avg. x=13, y=67, z=13
Toxicology: Moderately toxic by ingestion, intraperitoneal route
Uses: Nonionic surfactant; as food additives, defoamers, antistats, demulsifiers, detergents, wetting agents, gellants, emulsifiers, dispersants, dye levelers; lubricant for agric., cosmetics, metal cleaning, pulp/paper, textile scouring, water treatment, paints, latexes
Regulatory: FDA 21CFR §172.808, 173.340, 175.105, 176.180, 176.200, 176.210, 177.1200, 177.1210
Manuf./Distrib.: Aldrich; Fluka; Sigma
Trade names: Pluronic® L122

Poloxamer 403

CAS 9003-11-6 (generic)
Classification: Polyoxyethylene, polyoxypropylene block polymer
Formula: $HO(CH_2CH_2O)_x(CCH_3HCH_2O)_y(CH_2CH_2O)_zH$, avg. x=21, y=67, z=21
Toxicology: Moderately toxic by ingestion, intraperitoneal route
Uses: Nonionic surfactant; as food additives, defoamers, antistats, demulsifiers, detergents, wetting agents, gellants, emulsifiers, dispersants, dye levelers; lubricant for agric., cosmetics, metal cleaning, pulp/paper, textile scouring, water treatment, paints, latexes
Regulatory: FDA 21CFR §172.808, 173.340, 175.105, 176.180, 176.200, 176.210, 177.1200, 177.1210
Manuf./Distrib.: Aldrich; Fluka; Sigma
Trade names: Pluronic® P123

Poloxamer 407

CAS 9003-11-6 (generic)
Classification: Polyoxyethylene, polyoxypropylene block polymer
Formula: $HO(CH_2CH_2O)_x(CCH_3HCH_2O)_y(CH_2CH_2O)_zH$, avg. x=98, y=67, z=98
Properties: Wh. prilled or cast solid, odorless to mild odor; sol. in water and alcohol; m.w. 9840-14,600; m.p. ≈ 56 C; pH 5.0-7.5 (1 in 40)
Toxicology: Moderately toxic by ingestion, intraperitoneal route
Uses: Nonionic surfactant; as food additives, defoamers, antistats, demulsifiers, detergents, wetting agents, gellants, emulsifiers, dispersants, dye levelers; for paints
Regulatory: FDA 21CFR §172.808, 173.340, 175.105, 176.180, 176.200, 176.210, 177.1200, 177.1210
Manuf./Distrib.: Aldrich; Fluka; Sigma
Trade names: Calgene Nonionic 1127-F; Pluronic® F127

Poloxamine 304

CAS 11111-34-5 (generic)
Definition: Polyoxyethylene, polyoxypropylene block polymer of ethylene diamine
Uses: Emulsifier, thickener, wetting agent, dispersant, solubilizer, stabilizer for cosmetics and pharmaceuticals; demulsifier in petrol. industry; detergent ingred.; antistat for polyethylene and resin molding powds.; polymerization; used in latex-based paints, aq.-based syn. cutting fluids and vulcanization of rubber
Manuf./Distrib.: Aldrich
Trade names: Tetronic® 304

Poloxamine 701

CAS 11111-34-5 (generic)
Definition: Polyoxyethylene, polyoxypropylene block polymer of ethylene diamine

Poloxamine 704

Uses: Emulsifier, thickener, wetting agent, dispersant, solubilizer, stabilizer for cosmetics and pharmaceuticals; demulsifier in petrol. industry; detergent ingred.; antistat for polyethylene and resin molding powds.; polymerization; used in latex-based paints, aq.-based syn. cutting fluids and vulcanization of rubber
Manuf./Distrib.: Aldrich
Trade names: Tetronic® 701

Poloxamine 704
CAS 11111-34-5 (generic)
Definition: Polyoxyethylene, polyoxypropylene block polymer of ethylene diamine
Uses: Emulsifier, thickener, wetting agent, dispersant, solubilizer, stabilizer for cosmetics and pharmaceuticals; demulsifier in petrol. industry; detergent ingred.; antistat for polyethylene and resin molding powds.; polymerization; used in latex-based paints, aq.-based syn. cutting fluids and vulcanization of rubber
Manuf./Distrib.: Aldrich
Trade names: Tetronic® 704

Poloxamine 901
CAS 11111-34-5 (generic)
Definition: Polyoxyethylene, polyoxypropylene block polymer of ethylene diamine
Uses: Emulsifier, thickener, wetting agent, dispersant, solubilizer, stabilizer for cosmetics and pharmaceuticals; demulsifier in petrol. industry; detergent ingred.; antistat for polyethylene and resin molding powds.; polymerization; used in latex-based paints, aq.-based syn. cutting fluids and vulcanization of rubber
Manuf./Distrib.: Aldrich
Trade names: Tetronic® 901

Poloxamine 904
CAS 11111-34-5 (generic)
Definition: Polyoxyethylene, polyoxypropylene block polymer of ethylene diamine
Uses: Emulsifier, thickener, wetting agent, dispersant, solubilizer, stabilizer for cosmetics and pharmaceuticals; demulsifier in petrol. industry; detergent ingred.; antistat for polyethylene and resin molding powds.; polymerization; used in latex-based paints, aq.-based syn. cutting fluids and vulcanization of rubber
Manuf./Distrib.: Aldrich
Trade names: Tetronic® 904

Poloxamine 908
CAS 11111-34-5 (generic)
Definition: Polyoxyethylene, polyoxypropylene block polymer of ethylene diamine
Uses: Emulsifier, thickener, wetting agent, dispersant, solubilizer, stabilizer for cosmetics and pharmaceuticals; demulsifier in petrol. industry; detergent ingred.; antistat for polyethylene and resin molding powds.; polymerization; used in latex-based paints, aq.-based syn. cutting fluids and vulcanization of rubber
Manuf./Distrib.: Aldrich
Trade names: Tetronic® 908

Poloxamine 1107
CAS 11111-34-5 (generic)
Definition: Polyoxyethylene, polyoxypropylene block polymer of ethylene diamine
Uses: Emulsifier, thickener, wetting agent, dispersant, solubilizer, stabilizer for cosmetics and pharmaceuticals; demulsifier in petrol. industry; detergent ingred.; antistat for polyethylene and resin molding powds.; polymerization; used in latex-based paints, aq.-based syn. cutting fluids and vulcanization of rubber
Manuf./Distrib.: Aldrich
Trade names: Tetronic® 1107

Poloxamine 1307
CAS 11111-34-5 (generic)
Definition: Polyoxyethylene, polyoxypropylene block polymer of ethylene diamine
Uses: Emulsifier, thickener, wetting agent, dispersant, solubilizer, stabilizer for cosmetics and pharmaceuticals; demulsifier in petrol. industry; detergent ingred.; antistat for polyethylene and resin molding powds.; polymerization; used in latex-based paints, aq.-based syn. cutting fluids and vulcanization of rubber
Manuf./Distrib.: Aldrich
Trade names: Tetronic® 1307

Polyacrylamide
CAS 9003-05-8
Synonyms: 2-Propenamide, homopolymer
Definition: Polyamide of acrylic monomers
Empirical: $(C_3H_5NO)_x$
Formula: $[CH_2CHCONH_2]_x$
Properties: Wh. solid; water-sol. high polymer; m.w. 10,000-18,000,000; dens. 1.302
Toxicology: LD50 (mouse, IP) 170 mg/kg (monomer); heated to decomp., emits acrid smoke and irritating fumes
Uses: Thickener, dispersant, antiprecipitant, solubilizer, binder, sizing, flocculating, suspending, crosslinking agent, filtering aid, lubricant; used in adhesives, agric., cement, coatings, cosmetics, detergents, latex mfg., plaster, printing ink
Regulatory: FDA 21CFR §172.255, 173.10, 173.315 (10 ppm in wash water), 175.105, 176.180; BP compliance (gel)
Manuf./Distrib.: Aldrich; Allchem Ind.; Allied Colloids; AMRESCO; Calgon; Callaway; Complex Quimica SA; Cytec Ind.; Fluka; Monomer-Polymer & Dajac Labs; Quimicel; Rhone-Poulenc Water Treatment
Trade names: Cyanamer A-370; Cyanamer N300LMW

Polyacrylate. *See* Acrylic resin; Polyacrylic acid

Polyacrylic acid
CAS 9003-01-4
Synonyms: 2-Propenoic acid, homopolymer; Acrylic acid homopolymer; Acrylic acid resin; Acrylic polymer; Acrylic resin; Acrylic acid polymer; Polyacrylate
Definition: Polymer of acrylic acid
Empirical: $(C_3H_4O_2)_4$
Formula: $[CH_2CHCOOH]_4$
Properties: M.w. 168.06
Toxicology: Possible carcinogen; heated to decomp., emits acrid smoke and fumes
Uses: Emulsifier, thickener, stabilizer, suspending agent, coupler, moisturizer, emollient, dispersant, sequestrant, thickener, gellant; used for latexes, emulsion paints, drilling muds, photosensitive emulsions, water treatment, cosmetics, adhesives
Regulatory: FDA 21CFR §175.105, 175.300, 175.320, 176.180
Manuf./Distrib.: Alco; Aldrich; Anedco; CPS; BFGoodrich; Rhone-Poulenc; Sigma; Synthron
Trade names: Acumer 1510; Alcogum L-15; Alcosperse® 404; Daxad® 37L Acid; Dislon® L-1980; Dislon® L-1982; Dislon® L-1983; Dislon® L-1984; Esi-Cryl 724; Larodur®; Luprenal®; Perenol® F3; Sokalan® CP 10S; Sokalan® CP 13S; Sokalan® PA 80 S; Sokalan® PA 110 S
Trade names containing: BYK®-S 706; Dislon® L-1985-50
See also Acrylic resin

Poly(acrylic acid), ammonium salt. *See* Ammonium polyacrylate
Polyacrylic acid, potassium salt. *See* Potassium polyacrylate
Polyacrylic acid, sodium salt. *See* Sodium polyacrylate
Polyacrylic rubber. *See* Acrylic elastomer

Polyamic acid
Trade names containing: Eymyd® Resin L-20N; Eymyd® Resin L-30N

Polyamide
CAS 25038-54-4
Synonyms: Nylon
Classification: Polymer
Definition: A high m.w. polymer in which amide linkages (-CONH) occur along the molecular chain; may be either natural or synthetic
Empirical: $(C_6H_{11}NO)_x$
Properties: M.w. 22,000-45,000; dens. 1.20; m.p. 283-319 C
Uses: Natural polyamides include casein, soybean and peanut proteins, zein; synthetic polyamides typified by various nylons; used for paints, plastics, textile fibers, adhesives; epoxy curing agent
Manuf./Distrib.: Aldrich; Arizona; Ashley Polymers; BASF; Bayer; Chemcentral; Ciba-Geigy; Cray Valley; Diamine & Chems. Ltd; DSM; Elf Atochem N. Am.; EMS-Am. Grilon; Georgia-Pacific Chem.; Hoechst Celanese; Hüls AG; Lawter Int'l.; Matteson-Ridolfi; McWhorter; Monomer-Polymer & Dajac Labs; Monsanto; Reichhold; Rit-Chem; Samson; SNIA UK; Union Camp; Zimmer AG
Trade names: Ancamide 220; Ancamide 400; Ancamide 2050; Casamid® 360; Casamid® 362W; Dislon®

6500; Dislon® 6900-20X; Dislon® A670-20M; Disparlon® 6500; Disparlon® 6600; Epi-Cure® 3125; Epi-Cure® 3140; Epi-Cure® 3141; Epi-Cure® 3175; Epi-Cure® 3192; Epotuf Hardener 37-600; Epotuf Hardener 37-612; Epotuf Hardener 37-618; Epotuf Hardener 37-621; Epotuf Hardener 37-625; Epotuf Hardener 37-640; Forbest PAM; HY 815; HY 825; HY 840; HY 9130; HZ 340; Micromid® 121RC; Micromid® 141L; Micromid® 142LTL; Micromid® 321RC; Micromid® 1022; Micromid® 1354; Micromid® 3022; Uni-Rez® 1514; Uni-Rez® 2100; Uni-Rez® 2100-X65; Uni-Rez® 2115; Uni-Rez® 2125; Uni-Rez® 2126; Uni-Rez® 2140; Uni-Rez® 2142; Uni-Rez® 2180-B75; Uni-Rez® 2355; Uni-Rez® 2810; Versamid® 100; Versamid® 100IT60; Versamid® 100PMX60; Versamid® 100X65; Versamid® 115; Versamid® 115I73; Versamid® 115X70; Versamid® 125; Versamid® 140; Versamid® 150; Versamid® 253; XU HY 0360

Trade names containing: Ancamide 100-IT-60; Ancamide 220-IPA-73; Ancamide 220-X-70; Ancamide 400-BX-60; Ancamide 700-B-75; Casamid® 350PM; Disparlon® 6900-20X; Disparlon® A650-20X; Disparlon® A670-20M; Disparlon® NS-30; HZ 815 X-70; Uni-Rez® 2115-I75; Uni-Rez® 2415

See also Nylon

Polyamide 6. *See* Nylon-6

Polyamide/epichlorohydrin polymer

Uses: Wool shrinkproofing, antistat finishing; printing pretreatment; also gives water resist. to starch-based adhesives and coatings

Trade names: Polycup® 172; Polycup® 1884; Polycup® 2002

Polyamide/epoxy

Uses: Adhesive, coatings

Trade names: Versamid® 230XB60; Versamid® 280B75

Polyamide-imide

Synonyms: PAI

Uses: High-strength resin for connectors, switches, bearing and wear applics., adhesives; used as protective varnish or formulated to give pigmented coatings; provides self-lubricating, high heat resist., and decorative props.

Trade names: Nolicoat FE 71008

Polybutadiene

CAS 9003-17-2; 69102-90-5

Synonyms: BR; Butadiene rubber; cis-Polybutadiene; Polybutadiene rubber

Classification: Polymer

Empirical: $(C_4H_6)_n$

Formula: $(CH_2CH{=}CHCH_2)_n$

Properties: M.w. 1000-233,000

Precaution: May explode > 337 C

Uses: Elastomer for tire industry, footwear, molded goods; blending ingred. in SBR; impact modifier, additive for plastics; coating resin in liq. form

Manuf./Distrib.: Aldrich; Ameripol Synpol; ARCO; Asahi Chem. Ind.; BASF; Elf Atochem N. Am.; Firestone Syn. Rubber; Goodyear; Monomer-Polymer & Dajac Labs; Phillips; Reichhold/Emulsion Polymers; Revertex Ltd; A. Schulman; Scientific Polymer Prods.

Trade names: Poly bd® R-20LM; Poly bd® R-45HT; Poly bd® R-45M; Polyoil Hüls 110

1,2-Polybutadiene

CAS 29406-96-0 (syndiotactic)

Uses: Thermosetting resin for wire coating, EPDM peroxide-cured modifier, coatings, processing aid

Trade names: Ricon 131; Ricon 142; Ricon 153; Ricon 156; Ricon 157

cis-Polybutadiene. *See* Polybutadiene

Polybutadiene diacrylate

Uses: Optical fibers, pottants and encapsulants, conversion coatings, pressure-sensitive adhesives

Trade names containing: CN 300

Polybutadiene maleic anhydride polymer

Uses: Resin for elec. applics. (low dielec. and moisture absorp. flexible epoxy formulations), R.T. cured elastomers for sealant and coating applics.

Trade names: Ricon 130MA8; Ricon 130MA13; Ricon 131MA5; Ricon 131MA10; Ricon 131MA12; Ricon 131MA17; Ricon 181MA10; Ricon 184MA6

Polybutadiene-polystyrene copolymer. *See* Polybutadiene-styrene copolymer
Polybutadiene rubber. *See* Polybutadiene

Polybutadiene-styrene copolymer
CAS 9003-55-8
Synonyms: Butadiene-styrene polymer; 1,3-Butadiene-styrene copolymer; Butadiene-styrene resin; Styrene-butadiene copolymer; Styrene-butadiene polymer; Polybutadiene-polystyrene copolymer; Ethenylbenzene polymer with 1,3-butadiene; Styrene-1,3-butadiene copolymer; Styrene polymer with 1,3-butadiene
Classification: Polymer
Empirical: $(C_8H_8 \cdot C_4H_6)_x$
Toxicology: Eye irritant
Uses: Thermosetting resin used in coatings, liq. rubber; rubber and ink additive
Trade names: Ricon 170; Ricon 181
Trade names containing: Ricon 104

Poly (1,4-butane adipate) polyester
Uses: Used in PU industry for mfg. of prepolymers, thermoplastic elastomers, coatings and adhesives, dispersing agents, microcellular shoe sole systems, millable gums, cast elastomers
Trade names: Formrez® 44

Poly (1,4-butanediol adipate)
Uses: Functional polyol for formulation of urethanes for solv.-based coatings and adhesives and for thermoplastic urethane
Trade names: Lexorez® 1150-110

Poly (1,4-butanediol-1,6-hexanediol adipate)
Uses: Functional polyol for formulation of very flexible thermoplastic urethane, urethane coatings, and urethane elastomers
Trade names: Lexorez® 1460-30

Poly (1,4-butanediol neopentyl glycol adipate)
Uses: Functional polyol for formulation of polyurethane for flexible coatings, adhesives, and cast elastomers
Trade names: Lexorez® 1640-35; Lexorez® 1640-55; Lexorez® 1640-150

Poly (1,3-butanediol phthalate adipate)
Uses: Polyol for urethanes, sol'n. coatings, elastomers, castings, and solv.-free systems

Polybutene
CAS 9003-28-5
Synonyms: PIB; Polybutylene; 1-Butene, homopolymer
Definition: Polymer formed by polymerization of a mixture of iso- and normal butenes
Empirical: $(C_4H_8)_x$
Formula: $[CH_2CH(C_2H_5)]_n$
Properties: Liq.; m.w. 500-75,000; dens. 0.910
Toxicology: May asphyxiate
Uses: Thermoplastic resin; used as tackifier, strengthener, and extender in adhesives, as plasticizer for rubber, as vehicle and fugitive binder for coatings, as cling additive for LLDPE stretch wrap films, as reactive intermediate for specialty chemicals; insulating agent; lubricant
Regulatory: FDA 21CFR §175.105, 175.125, 177.1570, 177.2600,178.3750
Manuf./Distrib.: Aldrich; Amoco; Ashland; BP Chems. Ltd; Chemcentral; Harcros; Monomer-Polymer & Dajac Labs; Nat'l. Chem.; NOF; Punda Mercantile; H.M. Royal; Shell; Stanchem; Van Waters & Rogers
Trade names: Amoco® H-15; Amoco® H-25; Amoco® H-35; Amoco® H-50; Amoco® H-100; Amoco® H-300; Amoco® H-1500; Amoco® H-1900; Amoco® L-14; Amoco® L-50; Amoco® L-100; Amoco® Polybutene

Polybutene, activated
Uses: In adhesives, sealants, coatings, unsat. polyesters, elec. compds., foams, and other applics.

Polybutylene. *See* Polybutene
Polycarbonate. *See* Polycarbonate resin

Polycarbonate resin
Synonyms: PC; Polycarbonate; PC resin
Uses: Thermoplastic resin for molded prods., sol'n. cast or extruded film, structural parts, tube and piping, prosthetic devices, optical parts, windows, computer and business equip., household appliances, compact disks, paints, food contact and medical; modifier for other polymers; reinforcing modifier for recycled polymers and alloys
Manuf./Distrib.: Aldrich; Allchem Ind.; Ashland; Bayer; Chemical; Coz; Dow Plastics; DSM Chem. N. Am.; Ferro; GE Plastics; LNP; Monomer-Polymer & Dajac Labs; Sanncor Ind.; Shuman Plastics; Stanchem; Westlake Plastics

Poly(2-chloro-1,3-butadiene). *See* Polychloroprene

Polychloroprene
CAS 126-99-8
Synonyms: CR; Neoprene; Poly(2-chloro-1,3-butadiene); Chloroprene; β-Chloroprene; 2-Chlorobutadiene-1,3; 2-Chloro-1,3-butadiene; Chlorobutadiene; Chloroprene rubber
Classification: Synthetic elastomer
Definition: Unsaturated polymer of chloroprene
Empirical: C_4H_5Cl
Formula: $(CH_3ClC:CHCH_3)_n$
Properties: Colorless liq.; sl. sol. in water; misc. with alcohol, ether; dens. 0.958 (20/20 C); b.p. 59.4 C; flash pt. -4 F; As solid, latex, or flexible foam; m.w. 100,000-300,000; dens. 1.23-1.35; brittle pt. -35 C; softens. ≈ 80 C; resistant to oils, oxygen, ozone, elec. current
Precaution: Flamm. liq.; dangerous fire hazard exposed to heat or flame; autooxidizes in air to form unstable peroxide; incompat. with F; heated to decomp., emits toxic fumes of Cl^-
Toxicology: LD50 (oral, mousse) 260 mg/kg; LDLo (IV, rabbit) 96 mg/kg; ACGIH TLV:TWA 10 ppm (skin); poison by ing., IV, subcut. routes; mod. toxic by inh.; experimental reproductive effects; may cause dermatitis, CNS depression, liver/kidney damage
Uses: Syn. elastomer for molding, extrusion, and calendering for adhesive compounding, construction, automotive, hose and cable jackets, conveyor belts, closed cell sponge, binder for rocket fuels, fibers; roof coatings; vulcanizing plasticizer for elastomers
Trade names: Baypren Latex B; Baypren Latex GK; Baypren Latex L 200A; Baypren Latex L 345; Baypren Latex MKB; Baypren Latex SK; Baypren Latex T; Heveasyn Polychloroprene Latex; Neoprene AH; Neoprene Latex 115; Neoprene Latex 622; Neoprene Latex 654; Neoprene Latex 671A; Neoprene Latex 842-A; Neoprene Series
Trade names containing: Baypren Latex 4 R

Poly(dibromostyrene)
Uses: Flame retardant for thermoplastics and S/B latex coatings
Manuf./Distrib.: Great Lakes
Trade names: Great Lakes PDBS-80™

Poly (diethylene adipate) polyester
Uses: Used in PU industry for mfg. of prepolymers, thermoplastic elastomers, coatings and adhesives, dispersing agents, microcellular shoe sole systems, millable gums, cast elastomers
Trade names: Formrez® 11

Poly (diethylene glycol adipate)
CAS 26760-54-3; 28183-09-7
Uses: Polyol for adhesives, flexible coatings, and castable elastomers
Trade names: Lexorez® 1100-35; Lexorez® 1100-45; Lexorez® 1100-110; Lexorez® 1100-220; Lexorez® 1101-50A; Lexorez® 1101-60A; Lexorez® 1102-50A; Lexorez® 1102-60A; Lexorez® 1821-50; Lexorez® 1842-90

Poly (diethylene glycol, neopentyl glycol, 1,6-hexanediol adipate)
CAS 78492-71-1
Uses: Functional polyol for coatings for fabrics and flexible foams
Trade names: Lexorez® 1721-65P

Poly-(1,2-dihydro-2,2,4-trimethylquinoline). *See* 2,2,4-Trimethyl-1,2-dihydroquinoline polymer

Polydimethylsiloxane. *See* Dimethicone
Polydimethylsiloxane rubber. *See* Silicone elastomer

Polydipentene
CAS 9003-73-0
Synonyms: Cyclohexene; Polylimonene; Dipentene polymer; Limonene polymer
Definition: Prod. formed by polymerization of terpene hydrocarbons
Empirical: $(C_{10}H_{16})_x$
Uses: Thermoplastic resin used as tackifier in adhesives, modifier resin in rubber compds., in hot-melt coatings, laminations, wax modification, as masticatory agents in chewing gums
Trade names: Piccolyte® C115; Zonarez® 7085

Poly (dipropylene glycol adipate)
Uses: Polyol used for cast systems and sol'n. coatings

Poly (dipropyleneglycol) phenyl phosphite
CAS 80584-86-7; EINECS 279-499-4
Properties: Liq.; m.w. 2102; dens. 1.168-1.180; b.p. will not distill; flash pt. (PMCC) 179 C; ref. index 1.5340-1.5380
Uses: Stabilizer for PVC
Manuf./Distrib.: Dover; GE

Poly (dipropylene glycol phthalate adipate)
Uses: Polyol used for cast systems and sol'n. coatings, structural RIM applics.
Trade names: Lexorez® 5901-300

Polyester acrylate
Uses: Diluting oligomer for offset inks, overprint varnishes, coatings (paper, plastics, cardboard), paper upgrading; improves hardness and solv. resist.
Trade names: Ebecryl® 40; Ebecryl® 80; Ebecryl® 81; Ebecryl® 82; Ebecryl® 140; Ebecryl® 810; Ebecryl® 830; Ebecryl® 870; Ebecryl® 1810; Ebecryl® 2047; IRR 210; Photomer® 5018

Polyester adipate
Uses: Plasticizer for PVC, NC, CAP, chlorinated rubber; pigment grinding medium; nonmigratory in paper coatings
Manuf./Distrib.: C.P. Hall; Hüls
Trade names: Admex® 515; Admex® 525; Admex® 6187; Admex® 6996

Polyester glutarate
Uses: Plasticizer for PVC applics.; flexibilizing, permanent, nonmigrating; adhesive for film backing and varieties of tape; modifier for polymer coatings providing exc. weathering
Manuf./Distrib.: C.P. Hall
Trade names: Plasthall® P-550; Plasthall® P-7046; Plasthall® P-7092

Polyester hexaacrylate
Uses: Used in fast curing litho inks, flexo inks, and clear varnishes with good abrasion, hardness, and solv. resist.; good pigment wetting
Trade names: Ebecryl® 450

Polyester polyol
Uses: Produces crosslinked elastomeric polyurethanes, sol'n. laminating adhesives, sol'n. coatings, prepolymers, and one-shot castables
Manuf./Distrib.: Bayer; Chemical; Etna Prods.; C.P. Hall; ICI Polyurethanes; Inolex; King Ind.; Polyurethane Corp. of Am.; Polyurethane Spec.; Quimica Pumex SA; Stepan; U.S. Polymers; Unichema Int'l.; Witco/Oleo-Surf.; Witco/Resins
Trade names: Lexorez® 1100-220LG; Lexorez® 1140-190; Lexorez® 1405-065; Rucoflex® F-203; Rucoflex® F-207; Rucoflex® F-2014; Rucoflex® F-2016-185; Rucoflex® S-101 Series; Rucoflex® S-105 Series; Rucoflex® S-1011 Series; Rucoflex® S-1015 Series; Rucoflex® S-1017 Series; Rucoflex® S-1019 Series; Rucoflex® S-1028 Series; Rucoflex® S-1035 Series; Rucoflex® S-1037 Series; Rucoflex® S-1040 Series
Trade names containing: Nyacol® AB40

Polyester resin, saturated. *See* Polyester resin, thermoplastic

Polyester resin, thermoplastic
Synonyms: Polyester resin, saturated; Saturated polyester resin
Uses: Raw material for high solids industrial coatings with high weather and chem. resist.
Manuf./Distrib.: Akzo Nobel; BASF; Bayer; Bostik; Cambridge Ind. Co. of Am.; Chemical; Coz; DSM; Eastman; Estron; Etna Prods.; GCA; GE Plastics; Hoechst Celanese; Inolex; King Ind.; LNP; Loctite; Polyurethane Corp. of Am.; Ranbar Tech.; A. Schulman; U.S. Polymers; Witco/Resins; Zimmer AG
Trade names: Dynapol® HS 706-21; Dynapol® L 205; Dynapol® L 490; Dynapol® L 651; Dynapol® L 912; Dynapol® LH 818-05; Dynapol® LH 830-02; Dynapol® LH 910-01; Dynapol® LS 415-10; Dynapol® LS 436-12; Dynapol® LS 615
Trade names containing: SCD 1060; SCD 1092; SCD 16875; SCD 18674; SCD 18687; SCD 18764

Polyester resin, thermosetting
Synonyms: Polyester, unsaturated
Definition: Unsaturated polyester resin
Uses: Thermoset resin for reinforced plastics, automotive parts, protective coatings, ducts, housings, flues, laminates, pipes
Manuf./Distrib.: Advance Coatings; Aristech; Ashland; Bayer; CCP Polymers; Chemical; DSM Chem. N. Am.; Lawter Int'l.; Owens-Corning; Ranbar Tech.; Reichhold; Synray
Trade names: Aropol™ 7131; Aropol™ 8422; Atlac® 382-05A; Crystic Fireguard 75PA; Grilesta® P 7401; Grilesta® V 72-13; Grilesta® V 72-14; Grilesta® V 73-72; Hetron® 700; Ludopal®; Polylite® 32-737; Polylite® 32-738; Polylite® 33-773; Texicote® 1000, 1050
Trade names containing: Ebecryl® 505
See also Alkyd resin

Polyester tetraacrylate
Uses: Used in wet lithographic inks and coatings on paper, metal, and plastic; exc. pigment wetting
Trade names: Ebecryl® 657; Ebecryl® 1657

Polyester, unsaturated. *See* Polyester resin, thermosetting

Polyether acrylate
Uses: Oligomer for use in UV and EB curable prods.; used for water-dilutable screen and flexographic inks, spray-applied coatings for wood, monomer-free formulations; good surf. cures
Trade names: Ebecryl® 11

Polyetheretherketone
Synonyms: PEEK
Uses: Thermoplastic engineering resin with high thermal stability; also suitable for coating; compds. feature exc. flex. and tens. str. and elec. props. at high temps., toughness and abrasion resist.
Trade names: Xytrex® 450
Trade names containing: Xytrex® 451; Xytrex® 452; Xytrex® 455

Polyether glycol. *See* Polyethylene glycol

Polyether-polyester polyol
Uses: Combined with modified MDI for use in two-part chem. resist. coatings with high hardness and fast development of mech. props.; exc. hydrolytic stability
Trade names: Sovermol® 750

Polyether polyol
Uses: Functional polyol for polyurethane industry for industrial/consumer RIM and structural polymers, dynamic elastomers, adhesives, binders, coatings, and sealants; reactive diluent, dispersant for solv. coatings, etc.
Manuf./Distrib.: Allchem Ind.; BASF; Bayer; ICI Polyurethanes; Witco/Resins
Trade names: Isonol 93; Isonol 100; Syn Fac® 334-13; Syn Fac® 8009; Syn Fac® 8017; Syn Fac® 8018; Syn Fac® 8024; Syn Fac® 8025; Syn Fac® 8026; Syn Fac® 8027; Syn Fac® 8029; Syn Fac® 8031; Syn Fac® 8033; Syn Fac® 8100; Syn Fac® 8118; Syn Fac® 8120; Syn Fac® 8385; Voranol® 220-028, 220-037, 220-056, 220-056N, 220-094, 220-110, 220-260, 220-530; Voranol® 222-056; Voranol® 230-027, 230-056, 230-112, 230-238, 230-660; Voranol® 231-027, 231-057; Voranol® 232-023, 232-028, 232-034; Voranol® 234-630; Voranol® 235-044, 235-048, 235-056; Voranol® 237-061; Voranol® 240-360, 240-446, 240-490, 240-615, 240-770, 240-800; Voranol® 250-473; Voranol® 270-370

Polyethylene

CAS 9002-88-4; EINECS 200-815-3

Synonyms: PE; Ethene, homopolymer; Ethene polymer; Ethylene homopolymer; Ethylene polymers

Definition: Polymer of ethylene monomers

Empirical: $(C_2H_4)_x$

Formula: $[CH_2CH_2]_x$

Properties: Wh. translucent partially cryst./partially amorphous plastic solid, odorless; sol. in hot benzene; insol. in water; m.w. 1500-100,000; dens. 0.92 (20/4 C); m.p. 85-110 C

Precaution: Combustible; store in well closed containers; reacts violently with F_2

Toxicology: No known skin toxicity; ingestion of lg. oral doses has produced kidney and liver damage; suspected carcinogen and tumorigen by implants; heated to decomp., emits acrid smoke and irritating fumes

Uses: Laboratory tubing; prostheses; elec. insulation; pkg. materials; kitchenware; tank and pipe linings; textile stiffeners; additive in inks, adhesives, paper coatings, floor finishes, cosmetics, plastics, rubber, textiles, latex emulsions

Regulatory: FDA 21CFR §172.260, 172.615 (m.w. 2000-21,000), 173.20, 175.105, 175.300, 176.180, 176.200, 176.210, 177.1200, 177.1520, 177.2600, 178.3570, 178.3850; FDA approved for dentals, ophthalmics, orals, topicals, vaginals

Manuf./Distrib.: Aldrich; Asahi Chem. Ind.; Ashland; Chevron; Chisso Am.; Coz; DSM; Eastman; Elf Atochem SA; Exxon; Fabrichem; Fluka; Henkel; LNP; Mitsubishi Petrochem.; Monomer-Polymer & Dajac Labs; Quantum; Rohm & Haas; A. Schulman; Solvay Polymers; Union Carbide; Westlake Plastics; Zinchem

Trade names: A-C® 6; A-C® 6A; A-C® 8; A-C® 8A; A-C® 9, 9A, 9F; A-C® 15; A-C® 16; A-C® 617, 617A; A-C® 712; A-C® 715; A-C® 725; A-C® 735; A-C® 1702; ACumist® A-6; ACumist® A-12; ACumist® A-18; ACumist® A-45; ACumist® C-5; ACumist® C-9; ACumist® C-12; ACumist® C-18; ACumist® C-30; ACumist® D-9; Epolene® C-18; Epolene® N-10; Esi-Cryl 1E10N; Esi-Cryl 54N40; Esi-Cryl 392A; Esi-Cryl 625N; Esi-Cryl RT-239; Forbest 410; Michem® Emulsion 73635; MPP-123; MPP-230F; MPP-230VF; MPP-480F; MPP-480VF; MPP-611; MPP-611XF; MPP-620F; MPP-620VF; MPP-620XF; MPP-635F; MPP-635G; MPP-635VF; Polywax® 400; Polywax® 500; Polywax® 725; Polywax® 850; Polywax® 3000; Schercoat P-110; Shamrock S-395 N2; Shamrock S-395 N5; Shamrock S-395 SP5; Shamrock Taber Tiger 5512; Shamrock Versaflow Base; Tenite® Polyethylene

Trade names containing: Colloids PE 48/61/30; Drewax™ E-0146; Drewax™ E-9040; Lanco-Wax™ PP 1340F; Lanco-Wax™ PP 1350F; Lanco-Wax™ TF 1780; Michem® Emulsion 41540; Michem® Experimental Emulsion 77030; Michem Lube® 270R; Michem Lube® 368; Michem Lube® 693; Microspersion™ 19; Microspersion™ 411; Microspersion™ 411-50; Microspersion™ 523; Petrolite® 07 Disp.; Polyace™ 804; Polysilk 14; Polysilk 600; Polysilk 750; Shamrock Super Taber 5509 N1; Shamrock Super Taber 5509 SP5; Slip-Ayd® SL-018; Slip-Ayd® SL-031; Slip-Ayd® SL-050; Slip-Ayd® SL-078; Slip-Ayd® SL-177; Slip-Ayd® SL-280; Slip-Ayd® SL-295A; Slip-Ayd® SL-300; Slip-Ayd® SL-530; Slip-Ayd® SL-551; Slip-Ayd® SL-800; Slip-Ayd® SL-810; Synfluo 100XF; Synfluo 178VF; Synfluo 178XF; Synfluo 180VF; Synfluo 180XF

Poly (ethylene adipate) polyester

Uses: Used in PU industry for mfg. of prepolymers, thermoplastic elastomers, coatings and adhesives, dispersing agents, microcellular shoe sole systems, millable gums, cast elastomers

Trade names: Formrez® 22

Poly (ethylene/1,4-butane adipate) polyester

Uses: Used in PU industry for mfg. of prepolymers, thermoplastic elastomers, coatings and adhesives, dispersing agents, microcellular shoe sole systems, millable gums, cast elastomers

Trade names: Formrez® 24

Polyethylene elastomer, chlorinated

Synonyms: CM

Uses: Elastomer for molding and extrusion, cable and wire insulation, hose, tubing, tech. moldings; used in sol'n. coatings to give flexible films; blendable with other grades to reduce visc. and aid processability

Trade names: Paraclor 213

Polyethylene elastomer, chlorosulfonated

CAS 68037-39-8

Synonyms: CSM

Uses: Elastomer with high ozone resistance; for wire jackets, hose, tubing, sheet, footwear, life boats, windbreakers; protective and decorative coatings

Manuf./Distrib.: Aldrich

Trade names: Hypalon® Series; Hypalon® 20; Hypalon® 30; Hypalon® 40; Hypalon® 40S; Hypalon® 45; Hypalon® 48; Hypalon® 610; Hypalon® 623; Hypalon® 4085; Hypalon® LD-999; Toso-CSM® TS-220;

Polyethylene glycol

Toso-CSM® TS-320 Hypalon 45; Toso-CSM® TS-340 Hypalon 30; Toso-CSM® TS-430 Hypalon 40S; Toso-CSM® TS-530 Hypalon 40; Toso-CSM® TS-930 Hypalon 610-4085

Polyethylene glycol
CAS 25322-68-3; EINECS 203-473-3
Synonyms: PEG; α-Hydroxy-ω-hydroxy poly(oxy-1,2-ethanediyl); Polyglycol; Poly(ethylene oxide); Polyether glycol
Classification: Aliphatic organic compd.
Definition: Condensation polymer of ethylene glycol
Formula: $H(OC_2H_4)_nOH$, $n \geq 4$
Properties: Clear liq. or wh. solid; sol. in org. solvs., aromatic hydrocarbons; dens. 1.110-1.140 (20 C); m.p. 4-10 C; flash pt. 471 F; ref. index 1.4590
Precaution: Combustible liq.
Toxicology: LD50 (oral, rat) 33,750 mg/kg; sl. toxic by ingestion; skin and eye irritant; contact dermatitis sensitization; may cause hives; heated to decomp., emits acrid smoke and irritating fumes
Uses: Chemical intermediates, plasticizers, softeners, humectants, ointments, polishes, paper coating, mold lubricants, bases for cosmetics and pharmaceuticals, solvents, binders, metal and rubber processing, food additives, laboratory reagent; antistat; paints; textile fibers; food pkg.
Regulatory: FDA 21CFR §172.210, 172.820, 173.340; FDA approved for orals, topicals; USP/NF compliance
Manuf./Distrib.: Aldrich; BASF; BP Chems. Ltd; Calgene; Dow; DuPont; Fluka; Harcros; Henkel; Hüls; Inolex; Olin; Rhone-Poulenc Surf. & Spec.; Sigma; Texaco; Union Carbide
See also PEG...

Polyethylene glycol 300. *See* PEG-6
Polyethylene glycol 400. *See* PEG-8
Polyethylene glycol 600. *See* PEG-12
Polyethylene glycol 1000. *See* PEG-20
Polyethylene glycol 3350. *See* PEG-75

Poly (ethylene glycol adipate)
Uses: Functional polyol for formulating polyurethane for solÆn. coatings, adhesives, and castable elastomers
Manuf./Distrib.: Werner G. Smith

Poly (ethylene glycol 1,4-butanediol adipate)
Uses: Functional polyol for microcellular cast elastomer applics. such as shoe soles, for sol'n. coatings, adhesives, etc.
Trade names: Lexorez® 1600-55P

Polyethylene, high-density
CAS 9002-88-4; EINECS 200-815-3
Synonyms: HDPE
Uses: Blow- and inj.-molded goods, film and sheet, piping, fibers, gasoline and oil containers; additive for inks; lubricant in plastics processing; flatting/antisettling agent in paints; dispersant in color concs.; improves hardness in waxes
Manuf./Distrib.: AlliedSignal; Ampacet; BP Chems. Ltd; Chevron; Chisso Am.; Coz; DuPont; Exxon; Hüls Am.; Monomer-Polymer & Dajac Labs; Occidental; Quantum; A. Schulman; Solvay Polymers; Union Carbide; Westlake Plastics
Trade names: ACumist® B-6; ACumist® B-9; ACumist® B-12; ACumist® B-18; BASF Wax AH 6; Drewax™ E-3030; Drewax™ E-3035; Luwax™ AH 3; Luwax™ AH 6; Michem® Emulsion 46940; Octowax 518; Polywax® 1000; Polywax® 2000; Vestolen® A5016 F

Polyethylene, linear low density
Synonyms: LLDPE
Uses: Compatibilizer in blends and alloys, polymeric coupling agents in reinforced or recycled PE or PP, adhesives, sealants; extrusion coating resin
Manuf./Distrib.: BP Chems. Ltd; Neste AB
Trade names: Dowlex 3010; Fusabond® MB-110D; Fusabond® MB-226D; Nipolon-L; Novatec® AP 420L; Novatec® AP 706L

Polyethylene, low-density
EINECS 200-815-3
Synonyms: LDPE
Uses: Packaging film, food packaging, paper coating, liners for drums, wire and cable coating, toys, cordage,

waste bags, chewing gum base, squeeze bottles, electrical insulation; processing aid in rubbers
Manuf./Distrib.: AlliedSignal; Ampacet; Chevron; Coz; Eastman; Exxon; Hüls Am.; Monomer-Polymer & Dajac Labs; Neste AB; Quantum; A. Schulman; Union Carbide; Westlake Plastics
Trade names: LDPE 722; LDPE 4005, 4012; LDPE 5004 I; Luwax™ AL 3; Luwax™ AL 61; Luwax™ AM 3; Luwax™ AM 6; Nipolon; PE 1017; PE 1018; PE 1019; PE 1028; PE 4517; Petrothene® NA 204-000; Petrothene® NA 206-000; Petrothene® NA 216-000; Petrothene® NA 217-000; Petrothene® NA 593-00; Petrothene® NA 594-00; Petrothene® NA 596-00; Petrothene® NA 597-00; Petrothene® NA 598-00; Petrothene® NA 601-00/04; Rexene® PE 5000 Series
Trade names containing: MDI EC-940

Polyethylene, medium density
Synonyms: MDPE
Uses: Extrusion coating resin
Manuf./Distrib.: Quantum; Union Carbide
Trade names: Petrothene® NL 405-000; Petrothene® NL 409-000

Poly(ethylene oxide). *See* Polyethylene glycol

Polyethylene, oxidized
CAS 68441-17-8
Synonyms: Oxidized polyethylene; Ethene, homopolymer, oxidized
Definition: Reaction prod. of polyethylene and oxygen
Properties: Dens. 0.930
Uses: Wax for polishes, finishes, adhesives, and emulsions; processing lubricant, mold release aid, textile lubricant; PVC lubricant
Regulatory: FDA 21CFR §172.260, 175.105, 175.125, 176.170, 176.200, 176.210, 177.1200, 177.1620, 177.2800
Manuf./Distrib.: Aldrich
Trade names: A-C® 307, 307A; A-C® 316, 316A; A-C® 325; A-C® 330; A-C® 392; A-C® 395, 395A; A-C® 629; A-C® 629A; A-C® 655; A-C® 656; A-C® 680; A-C® 6702; Epolene® E-10; Epolene® E-15; Epolene® E-20; Esi-Cryl 316N; Esi-Cryl 325A; Esi-Cryl 325N; Esi-Cryl 392N; Esi-Cryl 629N38; Esi-Cryl E10N; Luwax™ OA; Luwax™ OA 2; Luwax™ OA 3; Luwax™ OA 5; Michem® Emulsion 18325; Michem® Emulsion 32535; Michem® Emulsion 39235; Michem® Emulsion 41040; Michem® Emulsion 61335; Michem® Emulsion 66630; Michem® Emulsion 66930; Michem® Emulsion 68725; Michem® Emulsion 74040; Michem® Emulsion 80325; Michem® Emulsion 93135; Michem® Experimental Emulsion 07430; Michem Lube® 103; Michem Lube® 126; Michem Lube® 190; Petrolite® C-8500
Trade names containing: Disparlon® 4200-10; Disparlon® 4200-20; Disparlon® NS-30; Michem Lube® 110

Polyethylenepolypropylene glycols, polymers. *See* Poloxamer 101

Polyethylene propylene adipate polyester
Uses: Used in PU industry for mfg. of prepolymers, thermoplastic elastomers, coatings and adhesives, dispersing agents, microcellular shoe sole systems, millable gums, cast elastomers
Trade names: Formrez® 23

Polyethylene resin. *See* Polyethylene wax

Polyethylene terephthalate
CAS 25038-59-9
Synonyms: PET; Poly(oxy-1,2-ethanediyloxycarbonyl-1,4-phenylenecarbonyl)
Classification: Organic compd.
Definition: Fiber-forming polyesters prepared from terephthalic acid or its esters and ethylene glycol
Empirical: $(C_{10}H_8O_4)_n$
Properties: Solid; sol. in hot m-cresol, trifluoroacetic acid, o-chlorophenol; sp.gr. 1.38; dec. ≈ 250 C
Uses: In fabric mfg.; as films; as base for magnetic coatings; surgical aid
Regulatory: FDA 21CFR §177.1630, 177.2260, 177.2800
Manuf./Distrib.: Aldrich; Chemical

Polyethylene tetrafluoride. *See* Fluorinated ethylene-propylene

Polyethylene, ultrahigh m.w.
CAS 9002-88-4; EINECS 200-815-3

Polyethylene wax

Uses: Surface modified particles producing cast elastomer composites with high abrasion resist. and low coeff. of friction; imparts improved abrasion, corrosion and chem. resist. in coatings
Trade names: Primax UH-1060; Primax UH-1080; Primax UH-1250; Vistamer™ UH-1060; Vistamer™ UH-1080; Vistamer™ UH-1250; Vistamer™ UH-1500; Vistamer™ UH-1700

Polyethylene wax
CAS 9002-88-4; EINECS 200-815-3
Synonyms: Polyethylene resin
Formula: $(-CH_2CH_2-)_n$
Uses: Wax for polishes, plastics and rubber processing, printing inks, pigment master batches, hot melts; PVC lubricant; mold release; thickener for solv.-based polishes, inks, paints, and lacquers
Manuf./Distrib.: Aldrich; AlliedSignal; Astor Corp.; Barco Chem. Prods.; BASF; Eastman; Fluka; Hoechst Celanese; Hüls AG; IGI; Koster Keunen; Micro Powders; Royce Assoc.; Sartomer; Shamrock Tech.; Stevenson Cooper; Syn. Prods.
Trade names: A-C® 7; Drewax™ D-1545; Drewax™ E-1235; Drewax™ E-4242; Drewax™ E-6532; Dymsol® MS-40; Epolene® C-10; Epolene® C-13; Epolene® C-14; Epolene® C-15; Epolene® C-15P; Epolene® C-16; Epolene® C-17; Epolene® N-11; Esi-Cryl 371N; Hoechst Wax 371 FP; Hoechst Wax PE 520; Jonwax 26; Lanco-Wax™ PE 1500F; Lanco-Wax™ PE 1500SF; Lanco-Wax™ PEW 1555; Luwax® A; Luwax™ AF; Luwax™ AF 30; Luwax™ AF 31; Luwax™ AF 32; Microspersion™ 230; Microspersion™ 250-50; Shamrock Flexoslip II; Shamrock Neptune I N1; Shamrock Neptune I SP5; Shamrock S-368 N5T; Shamrock S-379 H; Shamrock S-379 N; Shamrock S-379 N3; Shamrock S-381 N1; Shamrock S-381 N5; Shamrock S-394 N1; Shamrock S-394 N5; Shamrock S-394 SP5; Shamrock S-483; Slip-Ayd® SL-100; Slip-Ayd® SL-330E; Slip-Ayd® SL-340E; Slip-Ayd® SL-1606; Slip-Ayd® SL-1612; Slip-Ayd® SL-1618; Slip-Ayd® SL-1645; Slip-Ayd® SL-2206; Slip-Ayd® SL-2209; Slip-Ayd® SL-2212; Slip-Ayd® SL-2218; Slip-Ayd® SL-3505; Slip-Ayd® SL-3509; Slip-Ayd® SL-3512; Slip-Ayd® SL-3518; Slip-Ayd® SL-3530; Slip-Ayd® SL-4709; Vestowax A-217; Vestowax A-227; Vestowax A-235; Vestowax A-415; Vestowax A-616; Vestowax AO-1539
Trade names containing: Jonwax 120; Lanco-Matt™ MC; Lanco-Wax™ TF 1778; Lanco-Wax™ TF 1780EF; Lanco-Wax™ TF 1830; Lanco-Wax™ TFW 1765; Microspersion™ 250; Polyfluo 120; Polyfluo 150; Polyfluo 190; Polyfluo 200; Polyfluo 302; Polyfluo 400; Polyfluo 523XF; Polyfluo 540

Polyethylene wax, oxidized
Uses: Plastic lubricant; used for films, profiles, inj. molding, in dry-bright wax emulsions for floor maintenance and leather finishes
Trade names: Vestowax AS-1550

Polyfoam. *See* Polyurethane, thermoplastic

Polyglyceryl-10 tetraoleate
CAS 34424-98-1; EINECS 252-011-7
Synonyms: Decaglycerol tetraoleate; Decaglyceryl tetraoleate; 9-Octadecenoic acid, tetraester with decaglycerol
Definition: Tetraester of oleic acid and a glycerin polymer containing an avg. 10 glycerin units
Empirical: $C_{102}H_{190}O_{25}$
Toxicology: No known toxicity
Uses: Emulsifier, antistat, lubricant, visc. control agent, stabilizer, and coupler for textile finishes, cosmetics, lubricants, foods, plasticizers for syn. fabrics and plastics; solubilizer and carrier
Regulatory: FDA 21CFR §172.854
Trade names: Caplube 8440

Polyglycol. *See* Polyethylene glycol
Poly(hexamethyleneadipamide). *See* Nylon-66
Poly(N,N′-hexamethyleneadipinediamide). *See* Nylon-66

Poly (1,6-hexane/adipate/isophthalate) polyester
Uses: Used in PU industry for mfg. of prepolymers, thermoplastic elastomers, coatings and adhesives, dispersing agents, microcellular shoe sole systems, millable gums, cast elastomers
Trade names: Formrez® 8005-72; Formrez® 8008-120

Poly (1,6-hexane adipate) polyester
Uses: Used in PU industry for mfg. of prepolymers, thermoplastic elastomers, coatings and adhesives, dispersing agents, microcellular shoe sole systems, millable gums, cast elastomers
Trade names: Formrez® 66

Poly (1,6-hexane/diethylene adipate) polyester
Uses: Used in PU industry for mfg. of prepolymers, thermoplastic elastomers, coatings and adhesives, dispersing agents, microcellular shoe sole systems, millable gums, cast elastomers
Trade names: Formrez® 61

Poly (1,6-hexanediol adipate)
Uses: Functional polyol; raw material for formulation of urethane prepolymers, coatings, and thermoplastic urethanes
Trade names: Lexorez® 1130-30P

Poly (1,6-hexanediol adipate isophthalate)
Uses: Functional polyol for the formulation of PU for adhesives, coatings, and cast elastomers
Trade names: Lexorez® 3130-35; Lexorez® 3130-120

Poly (1,6-hexanediol adipate terephthalate)
Uses: Functional polyol for the formulation of PU for adhesives, cast elastomers, and coatings

Poly (1,6-hexanediol neopentyl glycol adipate)
Uses: Used in polyurethanes, semiflexible coatings, sol'n. adhesives, prepolymer systems
Trade names: Lexorez® 1400-35; Lexorez® 1400-120

Poly (1,6-hexanediol neopentyl glycol phthalate adipate)
Uses: Functional polyol for polyurethanes, high-performance flexible coatings, adhesives, and cast elastomers
Trade names: Lexorez® 3500-140

Poly (1,6-hexane/neopentyl adipate/isophthalate) polyester
Uses: Used in PU industry for mfg. of prepolymers, thermoplastic elastomers, coatings and adhesives, dispersing agents, microcellular shoe sole systems, millable gums, cast elastomers
Trade names: Formrez® 8009-93; Formrez® 8009-146

Poly (1,6-hexane/neopentyl adipate) polyester
Uses: Used in PU industry for mfg. of prepolymers, thermoplastic elastomers, coatings and adhesives, dispersing agents, microcellular shoe sole systems, millable gums, cast elastomers
Trade names: Formrez® 65

Polyimide, thermoplastic
Uses: Thermoplastic resin for compression molding, inj. molding and film casting; structural composites, adhesives, film, insulation, coatings
Trade names: Aurum Series

Poly(iminocarbonylpentamethylene). *See* Nylon-6
Poly[imino (1,6-dioxo-1,6-hexanediyl) imino-1,6-hexanediyl. *See* Nylon-66
Poly[imino(1-oxo-1,6-hexanediyl)]. *See* Nylon-6
Poly[imino(1-oxo-1,11-undecanediyl)]. *See* Nylon 11

Polyisobutene
CAS 9003-27-4
Synonyms: PIB; Polyisobutylene; 2-Methyl-1-propene, homopolymer
Definition: Homopolymer of isobutylene
Empirical: $(C_4H_8)_x$
Formula: $[CH_2C(CH_3)HCH_2]_x$
Uses: Syn. rubber; rubber additive; thickener for lubricating oils; for the adhesive and sealant industry, paints, elec. insulating oils, bases for chewing gums, for prod. of damp-proof courses containing fillers in construction industry
Regulatory: FDA 21CFR §172.615 (min. m.w. 37,000), 175.105, 175.125, 175.300, 176.180, 177.1200, 177.1210, 177.1420, 178.3570, 178.3740, 178.3910; Japan approved; FDA approved for injectables
Manuf./Distrib.: Aldrich; BASF; Monomer-Polymer & Dajac Labs; Rit-Chem

Polyisobutylene. *See* Polyisobutene

Polyketone
Uses: High performance thermoplastic for high temp. applics. in chemical processing, aviation/aerospace

composites, advanced elec./electronic uses, oil drilling, self-lubricating sleeve bearings, antifriction parts; wire and cable coatings, films
Trade names: Kadel® E-1000; Kadel® E-1130; Kadel® E-1140; Kadel® EP-3140

Poly(laurolactam). *See* Nylon-12
Polylimonene. *See* Polydipentene
Polymer of diphenylmethane 4,4′-diisocyanate. *See* Polymethylene polyphenyl isocyanate

Polymeric MDI
Uses: For polyurethane industry for adhesives, binders, coatings, and sealants, molded elastomers, structural foam molding, appliances
Trade names: Papi 901; Papi 2020; Papi 2027; Papi 2094; Papi 2580; Papi 2901

Polymethylene polyaniline
Uses: Amine curing agent for PU; intermediate in mfg. of polyamides, polyimides, coatings, and plastics

Polymethylene polyphenyl isocyanate
CAS 9016-87-9
Synonyms: Polymer of diphenylmethane 4,4′-diisocyanate
Empirical: $C_{15}H_{10}N_2O_2$
Properties: Crystals or yel. fused solid; m.w. 250.27; dens. 1.19 (50 C); m.p. 37.2 C; b.p. 194-199 (5 mm)
Toxicology: Poison by inhalation; mild poison by ingestion; skin and eye irritant; TLV 0.005 ppm
Uses: Crosslinking agent for adhesives
Manuf./Distrib.: Aldrich
Trade names containing: Bonding Agent 2001; Rubinate® M

Polymethylmethacrylate. *See* Methyl methacrylate polymer

Poly-α-methylstyrene
CAS 25014-31-7
Formula: $[CH_2C(CH_3)(C_8H_5)]_n$
Properties: M.w. 685-960; dens.1.075 (15.6 C); m.p. 118-141 C; ref. index 1.61 (20 C)
Uses: Plasticizer, extrusion and molding process aid in ABS, PVC, CPVC, and semirigid vinyl, thermoplastic urethanes, molded rubbers, and thermoplastic elastomers; modifier and reinforcer in adhesives, thermoplastic powd. coatings, hot-melt coatings
Manuf./Distrib.: Aldrich; Hercules; Monomer-Polymer & Dajac Labs

Poly(methylvinyl ether). *See* Polyvinyl methyl ether
Poly(methylvinyl ether/maleic acid) butyl ester. *See* PVM/MA copolymer, butyl ester
Poly(methylvinyl ether/maleic acid) isopropyl ester. *See* PVM/MA copolymer, isopropyl ester
Poly(methyl vinyl ether/maleic anhydride). *See* PVM/MA copolymer

Poly (neopentyl adipate) polyester
Uses: Used in PU industry for mfg. of prepolymers, thermoplastic elastomers, coatings and adhesives, dispersing agents, microcellular shoe sole systems, millable gums, cast elastomers
Trade names: Formrez® 55

Poly (neopentyl/1,6-hexane adipate) polyester
Uses: Used in PU industry for mfg. of prepolymers, thermoplastic elastomers, coatings and adhesives, dispersing agents, microcellular shoe sole systems, millable gums, cast elastomers
Trade names: Formrez® 56

Polynoxylin. *See* Urea-formaldehyde resin
Poly (oxy-1,4-butanediyl)-α-hydro-ω-hydroxy. *See* Polytetramethylene ether glycol
Poly[oxy(dimethylsilylene)], α-hydro-ω-hydroxy-. *See* Dimethiconol
Poly(oxy-1,2-ethanediyloxycarbonyl-1,4-phenylenecarbonyl). *See* Polyethylene terephthalate
Polyoxyl 40 castor oil. *See* PEG-40 castor oil
Polyoxyl 150 distearate. *See* PEG-150 distearate
Polyoxyl 2 stearate. *See* PEG-2 stearate
Polyoxyl 8 stearate. *See* PEG-8 stearate
Polyoxyl 20 stearate. *See* PEG-20 stearate
Polyoxymethylene. *See* Acetal copolymer
Polyoxymethylene urea (INCI). *See* Urea-formaldehyde resin
Polyoxypropylene (9). *See* PPG-9

Polyoxypropylene (12). *See* PPG-12
Polyoxypropylene (17). *See* PPG-17
Polyoxypropylene (20). *See* PPG-20
Polyoxypropylene (26). *See* PPG-26
Polyoxypropylene (30). *See* PPG-30
Polyoxypropylene (3) methyl ether. *See* PPG-3 methyl ether

Polyphenylene sulfide
CAS 9016-75-5
Synonyms: PPS
Uses: High-performance engineering thermoplastic for inj. molding; elec. connectors, automotive components, avionic housings, business machine enclosures, appliances, industrial parts; for compr. molding, extrusion, or coating applics
Manuf./Distrib.: Ashland; Hoechst Celanese
Trade names: Fortron® 0103B5; Ryton® V-1; Xytrex® 646
Trade names containing: Xytrex® 641; Xytrex® 642; Xytrex® 645

Polyphenylmethyl siloxane. *See* Phenyl trimethicone
Polypropene. *See* Polypropylene

Polypropylene
CAS 9003-07-0; 9010-79-1 (nucleated)
Synonyms: PP; 1-Propene, homopolymer; Propene polymer; Propylene polymer; Polypropene
Definition: Polymer of propylene monomers; three forms: isotactic (fiber-forming), syndiotactic, atactic (amorphous)
Empirical: $(C_3H_6)_x$
Formula: $[CH_2(CH_3)CH]_x$
Properties: Wh. translucent solid; m.w. > 40,000; dens. 0.90; m.p. 168-171 c; Isotactic: solid; pract. insol. in cold org. solvs.; sol. in hot decalin, hot tetralin, boiling tetrachloroethane; dens. 0.090-0.92; m.p. 165 C
Precaution: Combustible
Uses: Mfg. of paints; isotactic: fishing gear, ropes, filter cloths, laundry bags, protective clothing, blankets, fabrics, carpets, yarns; amorphous: plastics lubricant
Regulatory: FDA 21CFR §175.105, 175.300, 177.1200, 177.1520, 179.45; FDA approved for injectables; BP compliance (stockinette)
Manuf./Distrib.: Aldrich; Amoco; Ampacet; Aristech; Ashland; BASF; Chisso Am.; Cray Valley; DSM; Eastman; Exxon; Fina Chems.; GE Plastics; Hoechst AG; Hüls Am.; Huntsman; LNP; Mitsubishi Petrochem.; Mitsui Toatsu; Monomer-Polymer & Dajac Labs; Montell N. Am.; Nat'l. Chem.; Neste UK; Quantum; Rexene Prods.; A. Schulman; Shell; Solvay Polymers; Stanchem; Sumitomo Chem.; Tokuyama Soda; Westlake Plastics
Trade names: Esi-Cryl 43N40; Fusabond® MZ-109D; Fusabond® MZ-203D; Hercoflat® Texturing Pigments and Flatting Agent; Lanco-Wax™ CP 1481F; Lanco-Wax™ CP 1481SF; Lanco-Wax™ PP 1362D; Michem® Emulsion 43040; Micropro 400; Micropro 440W; Micropro 500; Micropro 600; Micropro 600VF; Microspersion™ 31-40; Pro-fax® PF-611; Propylmatte 31; PropylTex 20; PropylTex 50; PropylTex 100S; PropylTex 140S; PropylTex 200S; PropylTex 200SF; PropylTex 325S; Unite™ MP-320
Trade names containing: Lanco-Wax™ PP 1340F; Lanco-Wax™ PP 1350F; Microspersion™ 440W; OPPalyte® 250 ASW, 350 ASW; Shamrock S-363

Polypropylene, chlorinated
CAS 68442-33-1
Manuf./Distrib.: Advanced Polymer; Aldrich
Trade names: Hypalon® CP 826

Polypropylene glycol
CAS 25322-69-4; EINECS 200-338-0
Synonyms: PPG
Classification: Aliphatic organic compd.
Empirical: $(C_3H_8O_2)_n$
Formula: $HO(C_3H_6O)_nH$
Properties: Colorless clear visc. liq., sl. bitter taste; sol. in water, aliphatic ketones, alcohol; insol. in ether, aliphatic hydrocarbons; m.w. 400-2000; dens. 1.001-1.007; m.p. does not cryst.; flash pt. > 390 F
Precaution: Combustible exposed to heat or flame; reactive with oxidizers
Toxicology: LD50 (oral, rat) 4190 mg/kg; mildly toxic by ingestion; skin and eye irritant; linked to sensitive reactions; heated to decomp., emits acrid smoke and irritating fumes

Polypropylene glycol (9)

Uses: Hydraulic fluids, rubber lubricants, antifoam agents, intermediates for urethane foams, adhesives, coatings, elastomers, plasticizers, paint formulations, lab reagent
Regulatory: FDA 21CFR §173.340; FDA approved for ophthalmics, orals, topicals
Manuf./Distrib.: Aldrich; ARCO; Ashland; BASF; Bayer; BP Chems. Ltd; Calgene; Dow; Fluka; Harcros; Hüls AG; Olin; PPG Ind.; Rhone-Poulenc Surf. & Spec.; Texaco; Witco/Oleo-Surf.
Trade names containing: BYK®-024; Vinyzene® BP-5-5 PG
See also PPG...

Polypropylene glycol (9). *See* PPG-9
Polypropylene glycol (12). *See* PPG-12
Polypropylene glycol (12). *See* PPG-17
Polypropylene glycol (20). *See* PPG-20
Polypropylene glycol (26). *See* PPG-26
Polypropylene glycol (30). *See* PPG-30

Polypropylene glycol dibenzoate
CAS 72245-46-6
Uses: Plasticizer for PU and polysulfide sealants, acrylic coatings, caulks, PVC, adhesives
Trade names: Benzoflex® 400

Polypropylene glycol monomethyl ether. *See* Methoxyisopropanol

Polyquaternium-16
CAS 29297-55-0
Synonyms: Vinylpyrrolidone/vinyl imidazolinium methochloride copolymer
Classification: Polymeric quaternary ammonium salt
Definition: From methylvinylimidazolium chloride and vinylpyrrolidone
Uses: Substantive cationic polymer used as conditioner in prods. for hair and skin care; film former; foam stabilizing and lubricating effects; coatings; bactericidal effects
Trade names: Luviquat® FC 905

Polyquaternium-31
CAS 136505-02-7
Definition: Polymeric quaternary ammonium salt prepared by the reaction of DMAPA acrylates/acrylic acid/ acryonitrogens copolymer and diethyl sulfate
Uses: Hydrogel polymer substantive to skin and hair; thickener; carrier for active substances; provides protective coating; primary emulsifier; for cosmetic and personal care prods.
Trade names: Hypan® QT100

Polysiloxane
CAS 9011-19-2
Uses: Industrial foam control agent for high-solids coatings and water-reducible coatings, polyester/epoxy coatings, primers, and varnishes, floor coatings, top coats, solv.- and water-based printing inks
Trade names: BYK®-019; Drewplus® L-404; Drewplus® S-688; Drewplus® S-689; TEGO® Glide A 115; TEGO® Phobe 1010; TEGO® Phobe 1040; Troysol™ MS2; Troysol™ Q148; Troysol™ S366; W 090L; W 290
Trade names containing: Agitan 731; BYK®-022; BYK®-065; BYK®-066; BYK®-080; BYK®-141; Dehydran® 1208; Drewplus® S-684; Efka®-31; Efka®-34; Efka®-36; Efka®-39; Efka®-83; Efka®-88; Efka®-LP 3033; Efka®-LP 7022; Forbest G23; Perenol® S4; Perenol® S5; Perenol® S500; Solupret Brands; Surfynol® DF-62; TEGO® Airex 900; TEGO® Airex 960; TEGO® Airex 970; Tegosivin® HL 040; Tegosivin® HL 100; Tegosivin® HL 250; W 280; W 285

Polysiloxane polyether copolymer
CAS 68937-54-2
Synonyms: Dimethicone copolyol
Uses: Defoamer for emulsion paints, adhesives, polymer dispersions
Trade names: TEGO® Flow 425; TEGO® Foamex 800; TEGO® Foamex 805; TEGO® Foamex 810; TEGO® Foamex 815; TEGO® Foamex 825; TEGO® Foamex L 808; TEGO® Foamex L 822; TEGO® Glide 435; TEGO® Glide 440; TEGO® Glide 445; TEGO® Glide 450; TEGO® Wet 250; TEGO® Wet 260
Trade names containing: TEGO® Foamex 1495; TEGO® Foamex KS 10

Polysodium vinyl sulfonate
Synonyms: PSVS

Classification: Polymer
Definition: Polymer of sodium vinyl sulfonate
Properties: M.w. 10,000
Uses: As dispersant and deflocculating agent for calcium carbonate and other pigment slurries used in paints, coatings, inks, and paper coatings
Manuf./Distrib.: Air Prods.

Polysorbate 20
CAS 9005-64-5 (generic)
Synonyms: Sorbitan, monodecanoate, poly(oxy-1,2-ethanediyl) derivs.; POE (20) sorbitan monolaurate; PEG-20 sorbitan laurate; Sorbimacrogol laurate 300
Definition: Mixture of laurate esters of sorbitol and sorbitol anhydrides, with ≈ 20 moles ethylene oxide
Empirical: $C_{58}H_{114}O_{26}$
Properties: Lemon to amber liq., char. odor, bitter taste; sol. in water, alcohol, ethyl acetate, methanol, dioxane; insol. in min. oil, min. spirits; m.w. 1227.72; acid no. 2.2 max.; sapon. no. 40-50; hyd. no. 96-108
Toxicology: LD50 (oral, rat) 37 g/kg; mod. toxic by intraperitoneal, intravenous routes; mildly toxic by ingestion; human skin irritant; heated to decomp., emits acrid smoke and irritating fumes
Uses: O/w emulsifier, solubilizer; used in agric., cosmetics, pharmaceuticals, leather, metalworking, textiles, paints, emulsion polymerization; antistat for rigid PVC
Usage level: 1-15% (emulsifier for pharmaceuticals)
Regulatory: FDA 21CFR §172.515, 175.105, 175.300, 178.3400; FEMA GRAS; Europe listed; FDA approved for intravenous, parenterals, ophthalmics, orals, topicals, vaginals; USP/NF, BP, Ph.Eur. compliance
Manuf./Distrib.: Aldrich; Fluka; Sigma
Trade names: Calgene PSML-20; Crillet 1; Emsorb® 6915; Witflow™ 990
Trade names containing: Aldosperse® L L-20

Polysorbate 40
CAS 9005-66-7
Synonyms: POE (20) sorbitan monopalmitate; Sorbitan, monohexadecanoate, poly(oxy-1,2-ethaneidyl) derivs.; Sorbimacrogol palmitate 300
Definition: Mixture of palmitate esters of sorbitol and sorbitol anhydrides, with ≈ 20 moles of ethylene oxide
Empirical: $C_{62}H_{122}O_{26}$
Properties: Yel. liq., faint char. odor; sol. in water, alcohol; insol. in min. and veg. oils; m.w. 1283.84; acid no. 2.2 max.; sapon. no. 41-52; hyd. no. 89-105
Toxicology: Moderately toxic by intravenous route
Uses: O/w emulsifier, solubilizer; used in agric., cosmetics, pharmaceuticals, leather, metalworking, textiles, paints, polymerization; polymer additives
Regulatory: FDA 21CFR §175.105, 175.300, 178.3400; Europe listed; UK approved; FDA approved for parenterals, intramuscular injectables, orals, topicals; USP/NF compliance
Manuf./Distrib.: Aldrich; Fluka; Sigma
Trade names: Calgene PSMP-20

Polysorbate 60
CAS 9005-67-8 (generic)
Synonyms: POE (20) sorbitan monostearate; PEG-20 sorbitan stearate; Sorbimacrogol stearate 300
Definition: Mixture of stearate esters of sorbitol and sorbitol anhydrides, with ≈ 20 moles ethylene oxide
Empirical: $C_{64}H_{126}O_{26}$
Properties: Lemon to orange oily liq., faint char. odor, bitter taste; sol. in water, aniline, ethyl acetate, toluene; insol. in min. and veg. oils; m.w. 1311.90; acid no. 2 max.; sapon. no. 45-55; hyd. no. 81-96
Toxicology: LD50 (IV, rat) 1220 mg/kg; moderately toxic by intravenous route; experimental tumorigen, reproductive effects; heated to decomp., emits acrid smoke and irritating fumes
Uses: Emulsifier, solv. for polymerization, cosmetics, pharmaceuticals, industrial chemicals, agric., textiles, plastics additive, paints, pulp/paper, veterinary drug; solubilizer; foaming agent in beverage mixes
Regulatory: FDA 21CFR §73.1001, 172.515, 172.836, 172.878, 172.886, 173.340, 175.105, 175.300, 178.3400; USDA CFR9 §318.7, 381.147 (limitation 1% max., 1% total combined with polysorbate 80, 0.0175% in scald water); FEMA GRAS; Europe listed; UK approved; FDA approved for orals, rectals, topicals, vaginals; USP/NF, BP, Ph.Eur. compliance
Manuf./Distrib.: Aldrich; Fluka; Sigma
Trade names: Alkamuls® PSMS-20; Calgene PSMS-20

Polysorbate 61
CAS 9005-67-8 (generic)
Synonyms: POE (4) sorbitan monostearate; PEG-4 sorbitan stearate
Definition: Mixture of stearate esters of sorbitol and sorbitol anhydrides, with ≈ 4 moles ethylene oxide

Polysorbate 65

Empirical: $C_{64}H_{126}O_{26}$
Properties: M.w. 1311.70; dens. 1.044; flash pt. > 110 C
Toxicology: Moderately toxic by intravenous route
Uses: Emulsifier, solubilizer, lubricant for textile use, household formulations, suppositories in pharmaceuticals, emulsion polymerization, hydraulic fluids, metal treatment, paints; color dispersants for plastics, cosmetics, leather
Regulatory: FDA 21CFR §175.300
Manuf./Distrib.: Aldrich; Fluka; Sigma

Polysorbate 65
CAS 9005-71-4
Synonyms: POE (20) sorbitan tristearate; PEG-20 sorbitan tristearate; Sorbimacrogol tristearate 300
Definition: Mixture of stearate esters of sorbitol and sorbitol anhydrides, with ≈ 20 moles ethylene oxide
Properties: Tan waxy solid, faint odor, bitter taste; sol. in min. and veg. oils, min. spirits, acetone, ether, dioxane, alcohol, methanol; disp. in water, CCl_4; acid no. 2 max.; sapon. no. 88-98; hyd. no. 44-60
Toxicology: Heated to decomp., emits acrid smoke and irritating fumes
Uses: O/w emulsifier, solubilizer; used in agric., cosmetics, leather, metalworking, textiles, polymerization, paints, pulp/paper; emulsifier for food processing
Regulatory: FDA 21CFR §73.1001, 172.838, 173.340, 175.300, 178.3400; Europe listed; UK approved
Manuf./Distrib.: Fluka; Sigma
Trade names: Calgene PSTS-20

Polysorbate 80
CAS 9005-65-6 (generic); 37200-49-0; 61790-86-1; EINECS 215-665-4; 200-849-9
Synonyms: POE (20) sorbitan monooleate; PEG-20 sorbitan oleate; Sorbimacrogol oleate 300
Definition: Mixture of oleate esters of sorbitol and sorbitol anhydrides, with ≈ 20 moles ethylene oxide
Properties: Amber visc. liq.; nonionic; very sol. in water; sol. in alcohol, cottonseed oil, corn oil, ethyl acetate, methanol, toluene; insol. in min. oil; dens. 1.06-1.10; visc. 270-430 cSt; acid no. 2 max.; sapon. no. 45-55; hyd. no. 65-80; pH 5-7 (5% aq.)
Toxicology: LD50 (oral, mouse) 25 g/kg; mod. toxic by intravenous route; mildly toxic by ingestion; eye irritant; experimental tumorigen, reproductive effects; mutagenic data; heated to decomp., emits acrid smoke and irritating fumes
Uses: Pharmaceutic aid (surfactant); as emulsifier and dispersant in medicinal prods.; as defoamer and emulsifier in foods; adjuvant in herbicides; polymerization, paints
Regulatory: FDA 21CFR §73.1, 73.1001, 172.515, 172.840, 173.340, 175.105, 175.300, 178.3400; USDA 9CFR §318.7, 381.147 (limitation 1% alone, 1% total combined with polysorbate 60); FEMA GRAS; Europe listed; UK approved; FDA approved for buccals, intramuscular injectables, intravenous, parenterals, ophthalmics, orals, otics, rectals, topicals, vaginals; USP/NF, BP, Ph.Eur. compliance
Manuf./Distrib.: Aldrich; Fluka; Sigma
Trade names: Calgene PSMO-20; Emsorb® 6900; Witflow™ 991

Polysorbate 81
CAS 9005-65-5 (generic)
Synonyms: POE (5) sorbitan monooleate; PEG-5 sorbitan oleate
Definition: Mixture of oleate esters of sorbitol and sorbitol anhydrides, with ≈ 5 moles ethylene oxide
Uses: O/w emulsifier, solubilizer, emollient; used in agric., food, cosmetics, leather, textiles, emulsion polymerization, metalworking, metal treatment, paint, insecticides, herbicides, fungicides, cutting oils; color dispersant in plastics
Regulatory: FDA 21CFR §175.300
Trade names: Alkamuls® PSMO-5; Calgene PSMO-5; Emsorb® 6901

Polysorbate 85
CAS 9005-70-3
Synonyms: POE (20) sorbitan trioleate; PEG-20 sorbitan trioleate; Sorbimacrogol trioleate 300
Definition: Mixture of oleate esters of sorbitol and sorbitol anhydrides, with ≈ 20 moles ethylene oxide
Toxicology: Skin irritant
Uses: Surfactant, emulsifier for polymerization, cosmetics, agric., textiles, paints, pulp/paper, metalworking; plastic additives; solubilizer for perfume, flavors, essential oils
Regulatory: FDA 21CFR §175.300, 178.3400
Manuf./Distrib.: Aldrich; Fluka; Sigma; Spectrum Chem. Mfg.
Trade names: Calgene PSTO-20

Polystyrene
CAS 9003-53-6; EINECS 202-851-5

Synonyms: PS; Styrene polymer; Ethenylbenzene, homopolymer; Benzene, ethenyl-, homopolymer
Classification: Polymer
Definition: Grades: crystal, impact, expandable
Empirical: $(C_8H_8)_x$
Properties: Colorless to ylsh. oily liq., penetrating odor; sol. in alcohol; sl. sol. in water; m.w. 2500-250,000
Toxicology: May cause irritation to eyes and mucous membranes; can be narcotic in high concs.; experimental tumorigen by implant
Uses: Thermoplastic resin for inj. molding, extrusion of egg carton foam, pill bottles, pkg., appliances, electronics, toys, recreation, and construction; expandable PS for insulation, protective pkg.; modifier for latexes
Regulatory: FDA 21CFR §175.105, 175.125, 175.300, 175.320, 176.180, 177.1200, 177.1640, 177.2600
Manuf./Distrib.: Aldrich; Amoco; Ampacet; ARCO; Asahi Chem. Ind.; Ashland; BASF; Chevron; Coz; Dow Plastics; Elf Atochem SA; Fina Chems.; BFGoodrich; Fluka; Hüls AG; Huntsman; LNP; Mitsubishi Petrochem.; Mitsui Toatsu; Monomer-Polymer & Dajac Labs; Reichhold; Royce Assoc.; A. Schulman; Scott Bader; Sigma; Westlake Plastics; Zinchem
Trade names: Macromer® 13K-PSMA
Trade names containing: Macrobase 600; Macrobase 610; Macrobase 620

Polysulfide
CAS 9080-49-3
Classification: Synthetic polymer
Properties: Solid or liq.; exc. low-temp. flexibility; poor tensile strength and abrasion resistance
Uses: Elastomer for use in sealants, adhesives, potting and encapsulating compds., gasoline and oil-loading hose, casting of molds, barrier coatings, binder in solid rocket fuel; epoxy modifier
Manuf./Distrib.: Morton Int'l.; Phillips
Trade names: LP®-2; LP®-3; LP®-32; LP®-33; Thiokol® MC-2027; Thiokol® R-2100; Thiokol® RLP-2078
Trade names containing: Thiokol® FEC-2232

Polyterpene resin
Uses: Thermoplastic polymer imparting tack and adhesion to many elastomeric and polymeric materials, adhesives, coatings, rubber prods., concrete-curing compds., caulks
Manuf./Distrib.: Allchem Ind.; Arizona; Monomer-Polymer & Dajac
Trade names: Nevtac® 10°; Nevtac® 80; Nevtac® 100; Nevtac® 115; Piccolyte® A115; Piccolyte® A125; Piccolyte® A135; Piccolyte® AO; Piccolyte® C125; Piccolyte® C135; Piccolyte® S25; Piccolyte® S85; Piccolyte® S115; Piccolyte® S125; Piccolyte® S135; Super Nevtac® 99; Zonarez® 7115; Zonarez® 7115 LITE; Zonarez® 7125; Zonarez® 7125 LITE; Zonarez® Alpha 25; Zonarez® B-10; Zonarez® B-115; Zonarez® B-125; Zonarez® M-1115

Polytetrafluoroethylene
CAS 9002-84-0; EINECS 204-126-9
Synonyms: PTFE; TFE; Tetrafluoroethene homopolymer; Tetrafluoroethylene polymer; Polytetrafluoroethylene resin
Classification: Thermoplastic homopolymer
Formula: $[CF_2CF_2]_x$, $x \approx 20{,}000$
Properties: Wh. translucent to opaque solid; dens. 2.2; useful temp. range cryogenic to 260 C; melts to visc. gel @ 327 C; Shore hardness 55-56; tens. str. 3500-4500 psi
Precaution: Nonflamm.
Toxicology: Polymer fume fever reported under conditions of inadequate ventilation; inert as finished compd.
Uses: As tubing or sheeting for chemical laboratory and process work; gaskets and pump packings; as elec. insulators esp. in high frequency applics.; filtration fabrics; protective clothing; prosthetic aid; plastics lubricant, release agent; paint raw material
Manuf./Distrib.: Aldrich; Fluka; Janssens NV; Shamrock Tech.
Trade names: Algoflon® L; Ceridust® 3615; Ceridust® 3620; Ceridust® 3715; Ceridust® 3910; Ceridust® 9202 F; Ceridust® 9205 F; Ceridust® 9610 F; Ceridust® 9612 A; Ceridust® 9615 A; Ceridust® 9630 F; Fluo 300; Fluo 625F; Fluo HT; Fluo HTG; Fluo HTG-LS; Fluo HT-LS; Fluon® AD1; Fluon® L169; Fluoroglide® FL 1700; Hostaflon® TF 5032; Hostaflon® TF 5515; Hostaflon® TF 5537; Hostaflon® TF 9202; Hostaflon® TF 9203; McLube 1775; McLube 1777-1; PeFlu 727; Polymist® 284; Polymist® F-5; Polymist® F-5A; Polymist® F-5A EX; Polymist® F-510; Polymist® XPH-284; Shamrock Fluoro 45; Shamrock Fluoro-Tex 55; Shamrock Fluoro-Tex 75; Shamrock Fluoro-Tex 85; Shamrock Fluoro-Tex 100; Shamrock Neptune 5031; Shamrock Neptune SST-2 SP5; Shamrock SST-2 N5; Shamrock SST-2 SP5; Shamrock SST-3; Shamrock SST-3H; Shamrock SST-3K; Shamrock SST-3P; Teflon® 6C; Teflon® 6CJ; Teflon® 6CN; Teflon® 30; Teflon® 30B; Teflon® 30J; Teflon® 30N; Teflon® B; Teflon® BJ; Teflon® MP 1000; Teflon® MP 1100; Teflon® MP 1200; Teflon® MP 1300; Teflon® MP 1400; Teflon® MP 1500; Teflon® MP 1600;

Polytetrafluoroethylene resin

TL-102; Ultralon®; Whitcon® TL-010; Whitcon® TL-171

Trade names containing: Drewax™ E-0146; Lanco-Wax™ TF 1778; Lanco-Wax™ TF 1780; Lanco-Wax™ TF 1780EF; Lanco-Wax™ TF 1830; Lanco-Wax™ TFW 1765; Microspersion™ 19; Microspersion™ 411; Microspersion™ 411-50; Microspersion™ 523; Polyfluo 120; Polyfluo 150; Polyfluo 190; Polyfluo 200; Polyfluo 302; Polyfluo 400; Polyfluo 523XF; Polyfluo 540; Polysilk 14; Polysilk 600; Polysilk 750; Shamrock Super Taber 5509 N1; Shamrock Super Taber 5509 SP5; Slip-Ayd® SL-800; Slip-Ayd® SL-810; Synfluo 100XF; Synfluo 178VF; Synfluo 178XF; Synfluo 180VF; Synfluo 180XF; Teflon® TE-3667N

Polytetrafluoroethylene resin. *See* Polytetrafluoroethylene

Polytetramethylene ether glycol

CAS 24979-97-3; 25190-06-1

Synonyms: PTMEG; PTMG; Poly (oxy-1,4-butanediyl)-α-hydro-ω-hydroxy

Classification: Polyether glycol

Formula: $HO[(CH_2)_4O]_nH$

Properties: Wh. waxy solid melting to clear visc. liq. near R.T.; m.w. 650-2900; sp.gr. 1.0; flash pt. (TOC) 163 C

Precaution: Avoid strong oxidizers; may release very flammable THF, CO, etc.

Toxicology: May cause mild skin and eye irritation; LD50 (rat, oral) > 11,000 mg/kg; moderate aquatic toxicity

Uses: For polyurethane formulation for automotive hose and gaskets, tires, industrial belts, tank and pipe liners, floor and roof coatings, medical devices

Manuf./Distrib.: Aldrich; Allchem Ind.; BASF; Great Lakes; Monomer-Polymer & Dajac Labs; QO

Trade names: QO® Polymeg® 650; QO® Polymeg® 1000; QO® Polymeg® 2000; Terathane® 650; Terathane® 1000; Terathane® 1400; Terathane® 2000; Terathane® 2900

Polytetramethylene ether glycol diamine

Synonyms: PTMEG diamine

Classification: Oligomeric diamine

Uses: Curative for elastomers; coatings, adhesives, sealants, and spray systems; epoxy flexibilizer

Trade names: Polamine® 250

Poly(undecanoamide). *See* Nylon 11

Polyurethane-acrylate resin

Uses: Curing agent used for anaerobic adhesives, tile coatings, flexible overprint varnishes, outdoor applics., fiber optic coatings

Trade names: CN 954; CN 955; CN 960; CN 961; CN 963; CN 966; CN 970; CN 971; CN 973; Flexthane® 610; Flexthane® 620; Flexthane® 630

Trade names containing: CN 945 A60; CN 945 B85; CN 953; CN 960 A80; CN 960 B85; CN 960 C75; CN 960 E75; CN 960 H90; CN 960 I80; CN 960 J75; CN 960 K75; CN 960 L90; CN 961 A80; CN 961 B85; CN 961 C75; CN 961 E75; CN 961 H81; CN 961 H90; CN 961 I80; CN 961 J75; CN 961 K75; CN 961 L90; CN 962; CN 962 A80; CN 962 B85; CN 962 C75; CN 962 E75; CN 962 H90; CN 962 I80; CN 962 J75; CN 962 K75; CN 962 L90; CN 963 A80; CN 963 B80; CN 963 B85; CN 963 C75; CN 963 E75; CN 963 H90; CN 963 I80; CN 963 J75; CN 963 K75; CN 963 L90; CN 964; CN 964 A80; CN 964 B85; CN 964 C75; CN 964 E75; CN 964 H90; CN 964 I80; CN 964 J75; CN 964 K75; CN 964 L90; CN 965; CN 965 A80; CN 965 B85; CN 965 C75; CN 965 E75; CN 965 H90; CN 965 I 80; CN 965 J75; CN 965 K75; CN 965 L90; CN 966 A80; CN 966 B85; CN 966 C75; CN 966 E75; CN 966 H90; CN 966 I80; CN 966 J75; CN 966 K75; CN 966 L90; CN 970 A60; CN 970 B75; CN 970 C60; CN 970 E60; CN 970 H75; CN 970 I75; CN 970 J75; CN 970 K60; CN 970 L75; CN 971 A80; CN 971 B75; CN 971 C75; CN 971 E75; CN 971 H90; CN 971 I80; CN 971 J75; CN 971 K75; CN 971 L90; CN 972; CN 972 A80; CN 972 B85; CN 972 C75; CN 972 E75; CN 972 H90; CN 972 I80; CN 972 J75; CN 972 K75; CN 972 L90; CN 973 A80; CN 973 H85; CN 973 J75; CN 980; CN 980 M50; CN 981; CN 981 B88

Polyurethane elastomer or rubber

Uses: Elastomer for extrusion, inj. molding, and calendering; for sealants, caulks, adhesives, film and sheet, shoes, encapsulation of electronic components, automotive parts, flexible and rigid casting shapes; produces tough cured prods. in thin sections, e.g., cast films and coatings, solv.-resistant adhesives

Manuf./Distrib.: Air Prods.; BASF; Bayer; CasChem; Crowley Chem.; DSM UK; Ferro/Bedford; BFGoodrich; Hardman; Morton Int'l.; Polygenex Int'l.; Polyurethane Corp. of Am.; Soluol; UCB SA

Trade names: Adiprene® BL-16; Kollerdur® L 90

Polyurethane methacrylate; High flexibility; hydrophilic; used for adhesives, paper, wood, ink, metal coatings

Trade names: CN 974

Polyurethane, polycaprolactone
Uses: Features very high rate of crystallization; additive for adhesives for wood and automotive industry and plastics; plasticizer for EVA, PVC, ASA hot-melts for impregnation and textile coating
Trade names: Pearlstick 45-05/40; Pearlstick 65-05

Polyurethane, polyester
Uses: Adhesion promoter in metal/plastic composite structures, textile and leather coatings, primers for rigid surface coatings
Trade names: Bayhydrol 140 AQ; Morthane PS-62
Trade names containing: Vinyzene® SB-1 U

Polyurethane prepolymer
Uses: Dispersant, emulsifier for latex polymerization, waste water treatment
Manuf./Distrib.: Air Prods.; Bayer; CasChem; Conap; BFGoodrich; Hampshire; ICI Polyurethanes; Monomer-Polymer & Dajac Labs; Polyurethane Corp. of Am.; Polyurethane Spec.; Soluol; Uniroyal; Witco/Oleo-Surf.; Zeneca Resins
Trade names: Hypol® 2000; Hypol® 2002; Hypol® X6100; PCA 500, 501; Vibrathane® 6005, 6007, 8011

Polyurethane sponge. *See* Polyurethane, thermoplastic

Polyurethane, thermoplastic
CAS 9009-54-5
Synonyms: Polyfoam; Plastic sponge; Polyurethane sponge; Isourethane
Toxicology: Experimental carcinogen and tumorigen
Uses: Thermoplastic polymer for inj. molding, extrusion, vacuum forming, calendering, and blow molding; used in adhesives, coated fabrics and films, automotive parts, cable jacketing, scuba equip., shoe soles, ski boots, etc.
Trade names: Estane® 5714 F1
Trade names containing: Estane® 5707 F1; Estane® 5715

Polyurethane, thermoset
Uses: Thermoset polymer for casting and potting systems, conformal coatings, adhesives, insulating coatings, cable molding, automotive fascia, etc.; flame retardant
Trade names: Elastocoat®; Styrothane 5329/5330; Vesticoat® UB 909-06; Vesticoat® UB 1256-06; Witcobond® W-160; Witcobond® W-213; Witcobond® W-232; Witcobond® W-240; Witcobond® W-290H

Polyvidone. *See* PVP
Polyvidonum. *See* PVP
Polyvinyl acetate chloride. *See* Polyvinyl chloride acetate

Polyvinyl acetate (homopolymer)
CAS 9003-20-7
Synonyms: PVAc; Acetic acid, ethenyl ester, homopolymer; Acetic acid vinyl ester polymers; Ethenyl acetate, homopolymer
Classification: Homopolymer
Definition: Homopolymer of vinyl acetate
Empirical: $(C_4H_6O_2)_x$
Formula: $[CH_2CHOOCOCH_3]_x$
Properties: Water-wh. clear solid resin; sol. in benzene, acetone; insol. in water
Toxicology: Heated to decomp., emits acrid smoke and irritating fumes
Uses: Resin with weathering resistance; used for paints; adhesives for food pkg., paper, wood, glass, and metals; primer sealers; dry wall cement; intermediate for conversion to polyvinyl alcohol and acetals; paper coating; component of lacquers, inks; binder, emulsion stabilizer, film-former for oral pharmaceuticals
Regulatory: FDA 21CFR §73.1, 172.615 (m.w. 2000 min.), 175.105, 175.300, 175.320, 176.170, 176.180, 177.1200, 177.2800, 181.22, 181.30; Japan approved; FDA approved for orals
Manuf./Distrib.: Air Prods.; Aldrich; Ashland; H.B. Fuller; General Latex & Chem.; Hampshire; Lenape; Monomer-Polymer & Dajac Labs; Monsanto; Nat'l. Starch & Chem.; Reichhold; Rhone-Poulenc; Rohm & Haas; StanChem; Union Carbide; VYN-AC; Wacker-Chemie GmbH
Trade names: Daratak® 55L; Daratak® 56L; Daratak® SP1011; Daratak® SP1065; Daratak® SP1066; Everflex® BG; Everflex® GT; Resolvyl 610; Resolvyl BC; Resyn® 28-3307; Rovace 571; Vinac® 885; Vultacet

Polyvinyl alcohol
CAS 9002-89-5 (super and fully hydrolyzed); EINECS 209-183-3

Polyvinyl alcohol, partially hydrolyzed

Synonyms: PVA; PVAL; Ethenol homopolymer; PVOH; Poval; Vinyl alcohol polymer
Classification: Water-sol. synthetic polymer; aliphatic organic compd.
Empirical: $(C_2H_4O)_x$, avg. x = 500-5000
Formula: $[CH_2CHOH]_x$
Properties: Wh. to cream amorphous powd. or gran., odorless; sol. in water; insol. in petrol. solvs.; m.w. $(44.05)_x$, avg. 120,000; dens. 1.329; softens at 200 C with dec.; flash pt. (OC) 175 F; ref. index 1.49-1.53; pH 5-8 (4%)
Precaution: Flamm. exposed to heat or flame; reactive with oxidizers; dust exposed to flame presents sl. explosion hazard
Toxicology: Experimental carcinogen and tumorigen; heated to decomp., emits acrid smoke and irritating fumes
Uses: In plastics industry in molding compds., surface coatings, films resistant to gasoline, textile sizes; for elastomers (artificial sponges, fuel hoses); printing inks; pharmaceutical finishing; cosmetics; film and sheeting; ophthalmic lubricant; emulsion polymerization; release agent
Regulatory: FDA 21CFR §73.1, 175.105, 175.300, 175.320, 176.170, 176.180, 177.1200, 177.1670, 177.2260, 177.2800, 178.3910, 181.22, 181.30; FDA approved for ophthalmics, orals, injectables, topicals, vaginals; USP/NF compliance
Manuf./Distrib.: Air Prods.; Aldrich; Allchem Ind.; British Traders & Shippers; Fluka; GCA; Hoechst Celanese; Honeywill & Stein; Hunt; Itochu Spec.; Monomer-Polymer & Dajac Labs; Polysciences; Rhone-Poulenc; San Yuan; Shin-Etsu; Sigma; Spectrum Chem. Mfg.; Wacker-Chemie GmbH
Trade names: Elvanol® 71-30; Elvanol® 90-50; Polysol J

Polyvinyl alcohol, partially hydrolyzed
CAS 25213-24-5
Uses: Binder, carrier, compounding agent, dispersant, stabilizer, protective colloid in polymerizations; for textiles, paper, adhesives, cement/plaster additive, peelable caulks, ceramics, strippable coatings, mold release, nonwovens
Trade names: Airvol® 205

Polyvinyl butyral
CAS 63148-65-2; 9003-62-7
Synonyms: PVB; Vinyl acetal polymers, butyrals; Vinyl acetyl polymers, butyrates; Polyvinyl butyral resin
Classification: Polymer
Definition: Polymer produced by condensation of polyvinyl alcohol and butyraldehyde
Formula: $H_2 \cdot (C_8H_{14}O_2)_n$
Properties: M.w. 38,000-270,000
Toxicology: Severe eye irritant
Uses: Thermoplastic for extrusion, molding, coating, and casting processes; for adhesives, paints, lacquers, films, as sheet interlayer in safety glass and shatter-resistant protection in aircraft
Regulatory: FDA 21CFR §175.105, 175.300, 176.170
Manuf./Distrib.: Cairn Chems. Ltd; Denki Kagaku Kogyo; Hoechst Celanese; Monsanto; Sigma; StanChem; Union Carbide; Wacker-Chemie GmbH

Polyvinyl butyral resin. *See* Polyvinyl butyral
Poly (n-vinylbutyrolactam). *See* PVP

Polyvinyl chloride
CAS 9002-86-2; EINECS 208-750-2
Synonyms: PVC; Chloroethene homopolymer; Chloroethylene polymer; Poly(vinyl chloride)
Classification: Synthetic thermoplastic high polymer
Empirical: $(C_2H_3Cl)_n$
Formula: $[CH_2CHClCH_2CHCl]_n$
Properties: Wh. powd. or colorless gran.; m.w. 60,000-150,000; dens. 1.406; ref. index 1.54
Toxicology: Suspected tumorigen by ingestion and implantation; chronic inhalation health problems; may cause necrotizing or contact dermatitis
Uses: Rubber substitutes; elec. wire and cable coatings; pliable thin sheeting; film finishes for textiles; nonflamm. upholstery; raincoats; tubing; belting; gaskets; shoe soles; copolymers for latex emulsions; calendered film; compounding
Regulatory: FDA approved for parenterals
Manuf./Distrib.: Air Prods.; Aldrich; Asahi Glass; Ashland; Chisso; Colorite Plastics; E-A-R; Elf Atochem SA; Fluka; Georgia Gulf; BFGoodrich; Goodyear; Hüls Am.; Mitsui Toatsu; Monomer-Polymer & Dajac Labs; Nat'l. Starch & Chem.; Nippon Zeon; Norsk Hydro AS; Occidental; A. Schulman; Shin-Etsu; Sigma; Teknor Apex; Vista; Wacker-Chemie GmbH
Trade names: BCP 70; CP 1720, 1730; CP 1742; CP 1755; CP 1757; Geon® 140X31; Geon® 140X32; Georgia Gulf B7-44; Georgia Gulf EH-71; Georgia Gulf EH-71AH; Georgia Gulf EH-71L; Georgia Gulf EH-753;

Heveasyn PVC Latex; Lutofan®; Oxy 68GP; Oxy 68HC, 75HC, 80HC; Oxy 74GP; Oxy 160; Oxy 530; Oxy 567; Oxy 575; Oxy 605; Oxy 654; Oxy 654H; Oxy 6337; Oxy 6338; Ryuron Paste; VC-260SSF; VC-410M; VC-411; VC-438; VC-440; VC-440X2; VC-1069; VC-1070; Vestolit® B 7021; Vestolit® E 7001, E 8001; Vestolit® P 1330 K; Vestolit® P 2004 K

Poly(vinyl chloride). *See* Polyvinyl chloride

Polyvinyl chloride acetate
CAS 34149-92-3
Synonyms: Vinyl chloride-acetate copolymer; Polyvinyl acetate chloride
Classification: Polymer
Formula: $(C_2H_4O \cdot C_2H_3Cl)_x$
Uses: Mfg. of paints
Manuf./Distrib.: Hoechst Canada; Union Carbide

Polyvinyl ether. *See* Polyvinyl ethyl ether, Polyvinyl isobutyl ether

Polyvinyl ethyl ether
Synonyms: PVE; Polyvinyl ether
Classification: Polymer
Uses: Nonyel., very resilient, soft resin for odorless and taste-free finishes in the paint industry, for gravure and flexographic inks, for producing pressure-sensitive, building adhesives, and as tackifying soft resins
Manuf./Distrib.: BASF; ISP
Trade names: Lutonal® A

Polyvinyl formal
Uses: Mfg. of paints
Manuf./Distrib.: Chisso Am.; Monomer-Polymer & Dajac Labs; Monsanto

Polyvinylidene chloride
Synonyms: PVDC; Saran
Uses: Thermoplastic polymer for extrusion or inj. or blow molding; in adhesion coatings for paper, food pkg., pipes for chemical processing, upholstery, fibers, bristles, latex coatings
Manuf./Distrib.: Dow Plastics; BFGoodrich; Hampshire; Matteson-Ridolfi; Rhone-Poulenc; Scott Bader; Solvay Polymers; Van Leer Flexibles
Trade names: Lucidene™ 777; Lucidene™ 5527
Trade names containing: Dualite® M6001AE; OPPalyte® 250 ASW, 350 ASW
See also Vinylidene chloride monomer

Polyvinylidene fluoride
CAS 24937-79-9
Synonyms: PVDF; 1,1-Difluoroethene homopolymer
Classification: Polymer
Empirical: $C_7H_{12}O_3$
Properties: M.w. 144.17
Precaution: Distillation residue explodes violently when heated to 130 C
Uses: Thermoplastic polymer used as insulation for high-temp. wire, tank linings, chemical tanks and tubing, paints and coatings with high resistance to weathering and UV light; processing aid for polyolefins
Manuf./Distrib.: Aldrich; Ausimont USA; Elf Atochem N. Am.; Monomer-Polymer & Dajac Labs; Solvay Polymers; Westlake Plastics; Zeus Industrial Prods.
Trade names: Dykor 204; Hylar 5000™; KF Polymer® C-1000; KF Polymer® C-1500; KF Polymer® U-1000; Kynar® 301 F; Kynar® 460; Kynar® 700 Series; Kynar® 2800; Kynar® 2850
Trade names containing: Kynar® 320; Kynar® 370

Polyvinyl isobutyl ether
Synonyms: PVI; Polyvinyl ether
Uses: Nonyel., very resilient, soft resin for odorless and taste-free finishes in the paint industry, for gravure and flexographic inks, for producing pressure-sensitive, building adhesives, and as tackifying soft resins
Trade names: Lutonal® I; Perenol® E1

Poly(vinyl isobutyl ether). *See* Vinyl isobutyl ether

Polyvinyl isobutyl ether hexane
Uses: Tacky polymer with excellent adhesion to plastic, metal, and coated surfs.; plasticizer and leveling agent for surf. coatings

Polyvinyl methyl ether
CAS 9003-09-2
Synonyms: PVM; Methoxyethene, homopolymer; Poly(methylvinyl ether)
Empirical: $(C_3H_6O)_x$
Formula: $[CH_2CHOCH_3]_x$
Uses: Tackifier, binder, plasticizer for inks, textile sizes and finishes, latex modification, paints
Regulatory: FDA 21CFR §175.105, 177.1680
Manuf./Distrib.: Aldrich; BASF; ISP
Trade names: Lutonal® M

Polyvinyl octadecyl carbamate
CAS 36671-85-9; 70892-21-6
Synonyms: Carbamic acid, homopolymer (9C1), octadecyl-ethenyl ester
Classification: Polymer
Empirical: $(C_{21}H_{41}NO_2)_x$
Uses: Release coating for films and tapes
Manuf./Distrib.: Polyad
Trade names: Mayzo RA-95H; Mayzo RA-95HS

Polyvinylpyrrolidone. *See* PVP
Polyvinylpyrrolidone/vinyl acetate copolymer. *See* PVP/VA copolymer
POM. *See* Acetal copolymer
POP (2) methyl ether. *See* PPG-2 methyl ether
POP (36) monooleate. *See* PPG-36 oleate
POP (24) POE (27) monobutyl ether. *See* PPG-24-buteth-27
Potassium aluminosilicate. *See* Feldspar

Potassium dodecylbenzene sulfonate
CAS 27177-77-1; EINECS 248-296-2
Synonyms: Dodecylbenzenesulfonic acid, potassium salt
Classification: Substituted aromatic compd.
Empirical: $C_{18}H_{30}O_3S{\bullet}K$
Uses: Surfactant for S/B, vinyl chloride, VDC latexes
Regulatory: FDA 21CFR §178.3400
Trade names: Polystep® A-15-30K

Potassium metasilicate. *See* Potassium silicate

Potassium methyl siliconate
Uses: Additive for coatings; imparts water repellancy; water repellant for masonry surf.; additive to silicate coatings, pipe insulation, and wallboard; no VOCs, water-dilutable
Manuf./Distrib.: Hüls AG
Trade names: Dow Corning® 777; TEGO® Phobe 1310

Potassium naphthalene-formaldehyde sulfonate
Uses: Dispersant for dyes, dyestuffs, inks, latex paints, wax emulsions, wallboard coating, ore flotation
Trade names: Daxad® 11KLS

Potassium octoate
Uses: Catalyst for polyurethane; drier for coatings
Manuf./Distrib.: OM Group; Pelron; Shepherd
Trade names: Nuodex Octoate Potassium® 15%

Potassium polyacrylate
CAS 25608-12-2
Synonyms: Polyacrylic acid, potassium salt
Definition: Potassium salt of polyacrylic acid
Formula: $(C_3H_4O_2)_x \bullet xK$
Uses: Dispersant for latex paints and coatings, pigments
Manuf./Distrib.: Aldrich
Trade names: Daxad® 37LK9

Potassium polysilicate. *See* Potassium silicate

Potassium pyrophosphate. *See* Tetrapotassium pyrophosphate
Potassium pyrophosphate, normal. *See* Tetrapotassium pyrophosphate

Potassium silicate
CAS 1312-76-1
Synonyms: Silicic acid, potassium salt; Potassium metasilicate; Potassium polysilicate
Properties: Colorless anhyd. lump; sol. in water @ high temp. and pressure; insol. in alcohol
Uses: Mfg. of glass/refractory materials; binder in carbon arc-light electrodes; binder improving rheology of paints and coatings for cementious surfs.; outdoor and indoor bldg. paints, plasters, primers, industrial contruction, insulation board coatings
Manuf./Distrib.: Ashland; Crosfield; PQ; U.S. Synthetics; Welch, Holme & Clark; Zaclon
Trade names: Trasol® KA-L; Trasol® KA-N; Trasol® KC-K; Trasol® KD-K

Potassium sulfopropyl methacrylate
CAS 31098-21-2
Synonyms: 2-Propenoic acid, 2-methyl, 3-sulfopropyl ester, potassium salt; Sulfopropyl methacrylate, K salt
Formula: $H_2C{=}CCH_3COOCH_2CH_2CH_2SO_3K$
Properties: Wh. cryst. powd.; sol. in water, methanol; m.w. 246.3; m.p. > 300 C (dec.)
Uses: Functional monomer for applics. such as ion exchange resins, flocculants, thickeners, emulsions; emulsion polymerization
Manuf./Distrib.: Aldrich
Trade names: SPM

Potassium triphosphate. *See* Potassium tripolyphosphate

Potassium tripolyphosphate
CAS 13845-36-8; EINECS 237-574-9
Synonyms: KTPP; Pentapotassium tripolyphosphate; Potassium triphosphate; Pentapotassium triphosphate; Triphosphoric acid pentapotassium salt
Classification: Inorganic salt
Empirical: $K_5O_{10}P_3$
Formula: $K_5P_3O_{10}$
Properties: Wh. cryst. solid; hygroscopic; sol. in water; m.w. 448.4; dens. 2.54; m.p. 620-640 C
Toxicology: Nuisance dust
Uses: Detergents, paints, cleaners, specialty fertilizers, sequestrant
Regulatory: FDA 21CFR §173.310, 175.105, 182.1810, GRAS; USDA 9CFR §318.7 (with limitation), 381.147 (limitation 0.5% of total poultry prod.); Europe listed
Manuf./Distrib.: Albright & Wilson Am.; Allchem Ind.; Ashland; Coyne; FMC; Harcros; Int'l. Chem. Inc.

Poval. *See* Polyvinyl alcohol
Povidone. *See* PVP
Powdered cellulose. *See* Cellulose
PP. *See* Polypropylene
PPG. *See* Polypropylene glycol

PPG-9
CAS 25322-69-4 (generic)
Synonyms: Polyoxypropylene (9); Polypropylene glycol (9); PPG 400
Definition: Polymer of propylene oxide
Formula: $H(OCH_2CHCH_3)_nOH$, avg. n = 9
Uses: Chemical intermediate, antifoam agent in fermentation and in paint formulations, antiblooming agent for pentachlorophenol-treated wood; binder and lubricant for ceramics; plasticizer of resin-treated papers; mold release applics.
Regulatory: FDA 21CFR §173.310, 175.105, 176.200, 176.210
Manuf./Distrib.: Aldrich; Fluka
Trade names: Calgene PPG-400; Pluracol® P-410

PPG-12
CAS 25322-69-4 (generic)
Synonyms: Polyoxypropylene (12); Polypropylene glycol (12)
Definition: Polymer of propylene oxide
Formula: $H(OCH_2CHCH_3)_nOH$, avg. n = 12
Uses: Chemical intermediate, antifoam agent in fermentation and in paint formulations, antiblooming agent

for pentachlorophenol-treated wood; binder and lubricant for ceramics; plasticizer of resin-treated papers
Regulatory: FDA 21CFR §173.310, 175.105, 176.200, 176.210
Manuf./Distrib.: Aldrich; Fluka
Trade names: Pluracol® P-710

PPG-17
CAS 25322-69-4 (generic)
Synonyms: Polyoxypropylene (17); Polypropylene glycol (12)
Definition: Polymer of propylene oxide
Formula: $H(OCH_2CHCH_3)_nOH$, avg. n = 17
Uses: Chemical intermediate, antifoam agent in fermentation and in paint formulations, antiblooming agent for pentachlorophenol-treated wood; binder and lubricant for ceramics; plasticizer of resin-treated papers; defoamer for food-grade coatings and paper prods.
Regulatory: FDA 21CFR §173.310, 175.105, 176.170, 176.200, 176.210
Manuf./Distrib.: Aldrich; Fluka
Trade names: Dow P1000TB; Pluracol® P-1010

PPG-20
CAS 25322-69-4 (generic)
Synonyms: Polyoxypropylene (20); Polypropylene glycol (20); PPG 1200
Definition: Polymer of propylene oxide
Formula: $H(OCH_2CHCH_3)_nOH$, avg. n = 20
Uses: Defoamer, mold release applics., chemical intermediates for fatty acid esters, components for urethane resins; textile ubricants, metalworking compds., cosmetics, paints, urethane foams, hydraulic fluids, plasticizers, release agents
Regulatory: FDA 21CFR §173.310, 173.340, 175.105, 176.170, 176.200, 176.210, 178.3740
Manuf./Distrib.: Aldrich; Fluka
Trade names: Calgene PPG-1200; Dow P1200

PPG-26
CAS 25322-69-4 (generic)
Synonyms: Polyoxypropylene (26); Polypropylene glycol (26); PPG 2000
Definition: Polymer of propylene oxide
Formula: $H(OCH_2CHCH_3)_nOH$, avg. n = 26
Uses: Defoamer, mold release applics., chemical intermediates for fatty acid esters, components for urethane resins; boiler water additive for prep. of steam that will contact food; defoamer for beet sugar and yeast processing; defoamer for food-grade coatings and paper prods.
Regulatory: FDA 21CFR §173.310, 173.340, 175.105, 176.170, 176.200, 176.210, 178.3740
Manuf./Distrib.: Aldrich; Fluka
Trade names: Calgene PPG-2000; Dow P2000; Macol® P-2000; Pluracol® P-2010

PPG-30
CAS 25322-69-4 (generic)
Synonyms: Polyoxypropylene (30); Polypropylene glycol (30); PPG 4000
Definition: Polymer of propylene oxide
Formula: $H(OCH_2CHCH_3)_nOH$, avg. n = 30
Uses: Defoamer, mold release applics., chemical intermediates for fatty acid esters, components for urethane resins; antifoam agent in fermentation and in paint formulations, antiblooming agent for pentachlorophenol-treated wood; binder and lubricant for ceramics; plasticizer of resin-treated papers
Regulatory: FDA 21CFR §173.310, 173.340, 175.105, 176.170, 176.200, 178.3740
Manuf./Distrib.: Aldrich; Fluka
Trade names: Calgene PPG-4000; Dow P4000; Pluracol® P-4010

PPG (400)
Uses: Intermediate for surfactants, ethers, esters; metalworking lubricants; solv. for paints/varnishes, printing inks, veg. oils; textile lubricants

PPG 400. *See* PPG-9
PPG 1200. *See* PPG-20
PPG 2000. *See* PPG-26
PPG 4000. *See* PPG-30

PPG-24-buteth-27
CAS 9038-95-3 (generic); 9065-63-8 (generic)
Synonyms: POE (27) POP (24) monobutyl ether; POP (24) POE (27) monobutyl ether
Definition: Polyoxypropylene, polyoxyethylene ether of butyl alcohol
Empirical: $(C_7H_{14}O_2 \cdot C_6H_{12}O_2)_x$
Formula: $C_4H_9(OCH_3CHCH_2)_x(OCH_2CH_2)_yOH$, avg. x = 24, avg. y = 27
Uses: Nonionic surfactant, emulsifier, dispersant for agric. concs., latex polymerization, mfg. of iodophors for germicidal cleaners
Regulatory: FDA 21CFR §173.310, 175.105, 176.210, 178.3570
Manuf./Distrib.: Aldrich; Fluka; Sigma
Trade names: Tergitol® XD

PPG-6 C12-18 pareth-11
Definition: Polyoxyethylene, polyoxypropylene ether of a mixture of syn. alcohols
Formula: $R(OCH_3CHCH_2)_x(OCH_2CH_2)_yOH$, R rep. C12-18 alcohols, avg. x = 6, avg. y = 11
Uses: Detergent for commercial and home cleaning formulations; in latexes
Trade names: Plurafac® D-25

PPG-3 diacrylate
CAS 68901-05-3; 42978-66-5
Synonyms: TPGDA; TRPGDA; Propenoic acid, (1-methyl-1,2-ethanediyl)bis(oxy(methyl-2,1-ethanediyl)) ester; Acrylic acid, propylenebis-(oxypropylene) ester; Tripropylene glycol diacrylate
Empirical: $C_{15}H_{24}O_6$
Formula: $H_2C{=}CHCO(OC_3H_6)_3O_2CCH{=}CH_2$
Properties: M.w. 300.39; b.p. > 120 C (1 mm); flash pt. 110 C
Precaution: Hygroscopic
Toxicology: Irritant
Uses: Reactive thinner for radiation-curing systems, inks, coatings, floor tiles, wood coatings and fillers, adhesives, textile finishes, rubber compds.
Manuf./Distrib.: Aldrich; CPS
Trade names: Photomer® 4060; Photomer® 4061
Trade names containing: Ageflex TPGDA; CN 104 A80; CN 114 A80; CN 960 A80; CN 961 A80; CN 962 A80; CN 963 A80; CN 964 A80; CN 965 A80; CN 966 A80; CN 970 A60; CN 971 A80; CN 972 A80; Ebecryl® 265; Ebecryl® 285; Ebecryl® 505; Ebecryl® 3603; Ebecryl® 3604; Ebecryl® 3700-25R; Ebecryl® 4866; Ebecryl® 4883; Ebecryl® 8800-20R; Macrobase 610; Macrobase 620; Photomer® 6184; SR-306

PPG-4 diacrylate
Synonyms: TTEGDA; Tetraethylene glycol diacrylate
Uses: Difunctional monomer contributing flexibility and rapid-cure response in coatings and inks
Trade names containing: Ebecryl® 4830; Ebecryl® 4881

PPG-2 methyl ether
CAS 13429-07-7; 34590-94-8; EINECS 236-547-9; 252-104-2
Synonyms: DPM; Dipropylene glycol methyl ether; Dipropylene glycol monomethyl ether; Methoxy dipropylene glycol; (2-Methoxymethylethoxy) propanol; POP (2) methyl ether
Definition: PPG ether of methyl alcohol
Empirical: $C_7H_{16}O_3$
Formula: $CH_3(OCH_3CHCH_2)_nOH$, avg. n = 2
Properties: Liq.; sol. in water; m.w. 148.2; dens. 0.951; f.p. -112 F; b.p. 190 C; flash pt. 167 C; ref. index 1.4205
Precaution: Flamm. when exposed to heat or flame; can react with oxidizing materials
Toxicology: LD50 (oral, rat) 5660 mg/kg; ACGIH TLV:TWA 100 ppm; STEL 150 ppm (skin); mildly toxic by ing. and skin contact; skin and eye irritant; mild allergen; heated to decomp., emits acrid smoke, irritating fumes
Uses: Solv. for paints, cleaners, inks, cosmetics, agric., epoxy laminates, adhesives, floor polish, fuel additives, oilfield, mining, and electronic chems., chem. intermediate applics.
Regulatory: FDA 21CFR §175.105, 181.22, 181.30
Manuf./Distrib.: Aldrich; Ashland; Fluka; Hoechst AG; Oxiteno; Van Waters & Rogers
Trade names: Arcosolv® DPM; Dowanol® DPM; Icinol DPM; Methyl Dipropasol; Poly-Solv® DPM
Trade names containing: BYK®-LP W 6246; Canguard® 409-40; SCD 18687; SCD 18764; Surfynol® 104DPM

PPG-3 methyl ether
CAS 10213-77-1; 37286-64-9 (generic); 25498-49-1
Synonyms: Tripropylene glycol monomethyl ether; Tripropylene glycol methyl ether; Polyoxypropylene (3) methyl ether; PPG (3) methyl ether

PPG (3) methyl ether

Definition: PPG ether of methyl alcohol
Empirical: $C_{10}H_{22}O_4$
Formula: $CH_3(OCH_3CHCH_2)_nOH$, avg. n = 3
Properties: M.w. 206.32; dens. 0.967 (25/25 C); b.p. 243 C; flash pt. 250 F
Toxicology: Moderately toxic by ingestion; skin irritant
Uses: Solv. for paints, cleaners, inks, functional fluids, agric., cosmetics, epoxy laminates, adhesives, floor polish, fuel additives, oilfield, mining, and electronic chems.; chemical intermediate applics.
Regulatory: FDA 21CFR §181.22, 181.30
Manuf./Distrib.: Aldrich
Trade names: Arcosolv® TPM; Poly-Solv® TPM

PPG (3) methyl ether. *See* PPG-3 methyl ether

PPG-2 methyl ether acetate
CAS 88917-22-0
Synonyms: Dipropylene glycol methyl ether acetate
Formula: $CH_3(OC_3H_6)_2OAc$
Properties: M.w. 190.2
Uses: Solv. for paints, inks, cleaners, epoxy laminates, agric., adhesives, floor polish, oilfield, mining, and electronic chems.
Manuf./Distrib.: Aldrich; Ashland
Trade names: Arcosolv® DPMA; Dowanol® DPMA

PPGMM. *See* Propylene glycol methacrylate
PPG (36) monooleate. *See* PPG-36 oleate

PPG-36 oleate
CAS 31394-71-5 (generic)
Synonyms: POP (36) monooleate; PPG (36) monooleate
Definition: PPG ester of oleic acid
Formula: $CH_3(CH_2)_7CH=CH(CH_2)_7CO(OCH_3CHCH_2)_nOH$, avg. n = 36
Uses: Emollient, dispersant, spreading agent in cosmetics, industrial lubricants, hydraulic fluids; oil-based coatings; visc. depressant for PVC
Regulatory: FDA 21CFR §175.300, 176.210

PPS. *See* Polyphenylene sulfide
Precipitated barium sulfate. *See* Barium sulfate
Precipitated calcium carbonate. *See* Calcium carbonate
Precipitated calcium sulfate. *See* Calcium sulfate dihydrate
Precipitated chalk. *See* Calcium carbonate
Precipitated silica. *See* Silica, hydrated
pri-Octyl alcohhol. *See* Caprylic alcohol
Propanal. *See* Propionaldehyde

Propane
CAS 74-98-6; EINECS 200-827-9
Synonyms: Dimethylmethane; Propyl hydride
Classification: Hydrocarbon
Empirical: C_3H_8
Formula: $CH_3CH_2CH_3$
Properties: Colorless gas, nat. gas odor; easily liquefied under pressure at R.T.; noncorrosive; sol. in ether, alcohol; sl. sol. in water; m.w. 44.09; dens. 0.513 (0 C, as liq.), 1.56 (0 C, as vapor); m.p. -188 C; b.p. -42.5 C; f.p. -189.9 C; flash pt. -156 C
Precaution: Flamm.; autoignit. temp. 467 C; dangerous fire risk; explosive limits in air 2.4-9.5%; reactive with oxidizers
Toxicology: Asphyxiant; narcotic in high concs.; TWA 1000 ppm; heated to decomp., emits acrid smoke and irritating fumes
Uses: Hydrocarbon propellant; paint aerosols; organic synthesis, household and industrial fuel, mfg. of ethylene, extractant, solvent, refrigerant, gas enricher
Regulatory: FDA 21CFR §173.350, 184.1655, GRAS; FDA approved for topicals; USP/NF compliance
Manuf./Distrib.: Air Prods.; Aldrich; Exxon; Fina Chems.; Marathon Oil; Phillips; Stanchem
Trade names containing: Konform® AR 2000

Propanedioic acid, ((3,5-bis-(1,1-dimethylethyl)-4-hydroxyphenyl)methyl)-butyl, bis (1,2,2,6,6-pentamethyl-4-piperidinyl) ester. *See* Bis (1,2,2,6,6-pentamethyl-4-piperidinyl) (3,5-di-t-butyl-4-hydroxybenzyl) butyl propanedioate
1,2-Propanediol. *See* Propylene glycol
Propane-1,2-diol. *See* Propylene glycol
Propane-1,2-diol alginate. *See* Propylene glycol alginate
1,3-Propanediol, 2-bromo-2-nitro. *See* 2-Bromo-2-nitropropane-1,3-diol
1,3-Propanediol-2,2-dimethyl dibenzoate. *See* Neopentyl glycol dibenzoate
1,2-Propanediol-2-methyl monomethacrylate. *See* Hydroxypropyl methacrylate
1,2-Propanediol monomethyl ether acetate. *See* Propylene glycol methyl ether acetate
1,2-Propanediol monostearate. *See* Propylene glycol stearate
1,2-Propanediol, 3-(2-propenyloxy)-. *See* Glyceryl-1-allyl ether
Propane, 2-methoxy-2-methyl. *See* Methyl t-butyl ether
1,2,3-Propanetriol. *See* Glycerin
Propane-1,2,3-triol. *See* Glycerin
1,2,3-Propanetriol octadecanoate. *See* Glyceryl stearate
1,2,3-Propanetriol triacetate. *See* Triacetin
1,2,3-Propanetriol tri(12-hydroxystearate). *See* Trihydroxystearin
1,2,3-Propanetriyl 12-(acetyloxy)-9-octadecenoate. *See* Glyceryl triacetyl ricinoleate
Propanoic acid. *See* Neopentanoic acid
Propanoic acid butyl ester. *See* n-Butyl propionate
Propanoic acid ethyl ester. *See* Ethyl propionate
Propanoic acid, pentyl ester. *See* n-Pentyl propionate
Propanoic acid propyl ester. *See* Propyl propionate
1-Propanol. *See* n-Propyl alcohol
Propan-1-ol. *See* n-Propyl alcohol
2-Propanol. *See* Isopropyl alcohol
n-Propanol. *See* n-Propyl alcohol
1-Propanol, 2-((hydroxymethyl)amino)-2-methyl. *See* 2-[(-Hydroxymethyl) amino]-2-methylpropanol
2-Propanol, 1-methoxy. *See* Methoxyisopropanol
2-Propanol, 2-methyl-. *See* t-Butyl alcohol
1-Propanol, 3-[(2-propenyloxy)-2,2-bis[(2-propenyloxy)methyl]. *See* Pentaerythrityl triallyl ether
2-Propanone. *See* Acetone
2-Propanone oxime. *See* Acetone oxime
2-Propenamide, homopolymer. *See* Polyacrylamide
Propene acid. *See* Acrylic acid
1-Propene, homopolymer. *See* Polypropylene
Propenenitrile. *See* Acrylonitrile
2-Propenenitrile. *See* Acrylonitrile
Propenenitrile copolymer. *See* Acrylonitrile copolymer
2-Propenenitrile, polymer with 1,3-butadiene. *See* Butadiene/acrylonitrile copolymer
Propene polymer. *See* Polypropylene
2-Propenoic acid. *See* Acrylic acid
2-Propenoic acid, 2-carboxyethyl ester. *See* β-Carboxyethyl acrylate
2-Propenoic acid-2,2-dimethyl-1,3-propanediyl ester. *See* Neopentyl glycol diacrylate
2-Propenoic acid-2-ethylhexyl ester. *See* Octyl acrylate
2-Propenoic acid 1,6-hexanediyl ester. *See* 1,6-Hexanediol diacrylate
2-Propenoic acid homopolymer. *See* Acrylic resin
2-Propenoic acid, homopolymer. *See* Polyacrylic acid
2-Propenoic acid, homopolymer, ammonium salt. *See* Ammonium polyacrylate
2-Propenoic acid-2-(hydroxymethyl)-2-(((1-oxo-2-propenyl) oxy) methyl)-1,3-propanediyl ester. *See* Pentaerythrityl triacrylate
2-Propenoic acid methyl ester. *See* Methyl acrylate (monomer)
Propenoic acid, (1-methyl-1,2-ethanediyl)bis(oxy(methyl-2,1-ethanediyl)) ester. *See* PPG-3 diacrylate
2-Propenoic acid, 2-methyl-, homopolymer, sodium salt. *See* Sodium polymethacrylate
2-Propenoic acid-2-methyl-2-hydroxymethylethyl ester. *See* Hydroxypropyl methacrylate
2-Propenoic acid-1-methyl-13-propanediyl ester. *See* 1,3-Butylene glycol diacrylate
2-Propenoic acid, 2-methyl, 3-sulfopropyl ester, potassium salt. *See* Potassium sulfopropyl methacrylate
2-Propenoic acid oxiranylmethyl ester. *See* Glycidyl acrylate
2-Propenoic acid, polymer with ethene, magnesium salt. *See* Ethylene/magnesium acrylate copolymer
2-Propenoic acid, polymer with ethene, sodium salt. *See* Ethylene/sodium acrylate copolymer
2-Propenoic acid, polymer with ethene, zinc salt. *See* Ethylene/zinc acrylate copolymer

[(2-Propenyloxy)methyl]oxirane. *See* Allyl glycidyl ether

Propionaldehyde
CAS 123-38-6; EINECS 204-623-0
Synonyms: Aldehyde C-3; Propanal; Propionic aldehyde; Methylacetaldehyde; Propylaldehyde
Empirical: C_3H_6O
Formula: CH_3CH_2CHO
Properties: Colorless mobile liq., suffocating odor; sol. in 5 vols water; misc. with alcohol, ether; m.w. 58.08; dens. 0.807 (20/4 C); m.p. -81 C; b.p. 47-49 C; flash pt. -40 C; ref. index 1.362 (20 C)
Precaution: Highly flamm.; dangerous fire hazard exposed to heat or flame; reacts vigorously with oxidizers; explosive limits 2.9-17%; store refrigerated
Toxicology: LD50 (oral, rat) 1.4 g/kg; mod. toxic by skin contact, ingestion, subcut. routes; mildly toxic by inh.; skin, severe eye irritant; heated to decomp., emits acrid smoke and irritating fumes
Uses: Intermediate for paints
Regulatory: FDA 21CFR §172.515; FEMA GRAS
Manuf./Distrib.: Aldrich; BASF AG; Eastman; Fluka; Penta Mfg.; Sigma; Spectrum Chem. Mfg.; Union Carbide

Propione. *See* Diethyl ketone
Propionic acid, 2-methyl-, monoester with 2,2,4-trimethyl-1,3-pentanediol. *See* 2,2,4-Trimethyl-1,3-pentanediol monoisobutyrate
Propionic aldehyde. *See* Propionaldehyde
2-Propoxyethanol. *See* Ethylene glycol propyl ether
2-(2-Propoxyethoxy) ethanol. *See* Diethylene glycol propyl ether

Propoxypropanol
CAS 30136-13-1
Synonyms: n-Propoxypropanol
Empirical: $C_6H_{14}O_2$
Properties: M.w. 118.18
Trade names containing: G-4714-P-75; Kelsol 3922-G-80

n-Propoxypropanol. *See* Propoxypropanol

Propyl acetate
CAS 109-60-4; EINECS 203-686-1
Synonyms: Acetic acid n-propyl ester; n-Propyl acetate
Definition: Ester of propyl alcohol and acetic acid
Empirical: $C_5H_{10}O_2$
Formula: $CH_3COOCH_2CH_2CH_3$
Properties: Liq., pear-like odor; misc. with alcohol, ether; m.w. 102.14; dens. 0.887 (20/20 C); m.p. -92 C; b.p. 99-102 C; flash pt. (CC) 14 C; ref. index 1.384 (20 C)
Precaution: Highly flamm.
Toxicology: LD50 (oral, rat) 9370 mg/kg; may be irritating to skin, mucous membranes; narcotic in high concs.
Uses: Flavoring agent, perfumery, solvent for nitrocellulose and other cellulose derivatives, natural and synthetic resins, lacquers, plastics, organic synthesis, lab reagent
Regulatory: FDA 21CFR §172.515, 177.1200; FEMA GRAS
Manuf./Distrib.: Aldrich; BASF; Berje; BP Chems. Ltd; Coyne; Eastman; Fluka; Hoechst Celanese; Penta Mfg.; Union Carbide
Trade names containing: D.E.R. 660-PA80; Konform® AR 2000

2-Propyl acetate. *See* Isopropyl acetate
n-Propyl acetate. *See* Propyl acetate
Propylacetic acid. *See* n-Valeric acid
Propyl alcohol. *See* n-Propyl alcohol

n-Propyl alcohol
CAS 71-23-8; EINECS 200-746-9
Synonyms: 1-Propanol; Propan-1-ol; n-Propanol; Propylic alcohol; Albacol; Optal; Propyl alcohol; Ethyl carbinol; Alcohol C-3
Classification: Aliphatic alcohol
Empirical: C_3H_8O
Formula: $CH_3CH_2CH_2OH$
Properties: Colorless liq., alcoholic and sl. stupefying odor; misc. with water, alcohol, ether; dissolves fat; m.w.

60.10; dens. 0.804 (20/4 C); m.p. -127 C; b.p. 97-98 C; flash pt. (TCC) 23 C; ref. index 1.385 (20 C)
Precaution: Highly flamm.; dangerous fire risk; explosive limits in air 2-13%
Toxicology: LD50 (oral, rat) 1.87 g/kg; TLV 200 ppm in air; toxic by skin absorption; drying effect on skin may lead to cracking, fissuring, and infections; mildly irritating to eyes, mucous membranes; depressant action
Uses: Solv. for natural and synthetic resins, cellulose esters, paints, waxes, veg. oil; organic synthesis; chemical intermediate
Regulatory: FDA 21CFR §172.515, 175.105, 177.1200, 573.880; FEMA GRAS; FDA approved for topicals; BP compliance
Manuf./Distrib.: Aldrich; Allchem Ind.; ARCO; Ashland; J.T. Baker; Baychem; Burdick & Jackson; Chemcentral; Coyne; Eastman; Fluka; Hoechst Celanese; Mallinckrodt; Penta Mfg.; Samson; Sigma; Spectrum Chem. Mfg.; Sunnyside; Union Carbide; Van Waters & Rogers
Trade names containing: GPRI™ CKSB-2001; Ken-React® NZ 38J; Sobral 587; Surfynol® 104NP

s-Propyl alcohol. *See* Isopropyl alcohol
Propylaldehyde. *See* Propionaldehyde
Propyl carbinol. *See* Butyl alcohol
Propyl 'Cellosolve'. *See* Ethylene glycol propyl ether

Propylene carbonate
CAS 108-32-7; EINECS 203-572-1
Synonyms: 1,3-Dioxolan-2-one, 4-methyl; 1,3-Carbonyl dioxypropane; 4-Methyl-1,3-dioxolan-2-one; Carbonic acid, 1,2-propylene glycol ester
Classification: Organic compd.
Empirical: $C_4H_6O_3$
Properties: Clear liq.; m.w. 102.10; dens. 1.2069 (20/20 C); m.p. -48.8 C; b.p. 242.1 C; flash pt. (OC) 275 F; ref. index 1.422 (20 C)
Toxicology: Mildly toxic by ingestion; human skin and eye irritant
Uses: Solvent for pigments, dyes, paints; extracting agent; intermediate for organic syntheses; reactive diluent for urethane foams and coatings, foundry sand binders, textiles, natural gas treating; cosmetics lubricant
Regulatory: FDA 21CFR §175.105; FDA approved for topicals
Manuf./Distrib.: Aldrich; Allchem Ind.; ARCO; Ashland; BASF AG; Fluka; Great Western; Hüls AG; Sigma; Spectrum Chem. Mfg.
Trade names: Arconate® 1000; Arconate® HP; Arconate® PC; Jeffsol™ PC; Texacar® PC
Trade names containing: SarCat™ CD-1010; SarCat™ CD-1011

Propylenedicarboxylic acid. *See* Itaconic acid

Propylene glycol
CAS 57-55-6; 4254-15-3 (+); 4254-14-2 (-); 4254-16-4 (±); EINECS 200-338-0
Synonyms: 1,2-Propanediol; Propane-1,2-diol; 1,2-Dihydroxypropane; Methyl glycol; Methyl ethylene glycol
Classification: Aliphatic alcohol
Empirical: $C_3H_8O_2$
Formula: $CH_3CHOHCH_2OH$
Properties: Colorless clear visc. liq., odorless, sl. acrid taste; hygroscopic; sol. in essential oils; misc. with water, acetone, chloroform; m.w. 76.11; dens. 1.0362; b.p. 188.2 C; flash pt. (OC) 210 F
Precaution: Combustible exposed to heat or flame; reactive with oxidizers; explosive limits 2.6-12.6%
Toxicology: LD50 (oral, rat) 25 ml/kg; eye/human skin irritant; sl. toxic by ingestion, IP, IV, subcutaneous routes; human systemic effects; experimental teratogenic, reproductive effects; mutagenic data; heated to decomp., emits acrid smoke and irritating fumes
Uses: Solvent, emulsifier, production paints, polyester and alkyd resins, foods, drugs, antifreeze in breweries and dairies; substitute for ethylene glycol and glycerol; mold growth and fermentation inhibitor; plasticizer for adhesives, cork, paper prods.; raw material for resinous plasticizers
Usage level: 10-25% (oral sol'ns.), 10-80% (parenterals), 5-80% (topicals)
Regulatory: FDA 21CFR §169.175, 169.176, 169.177, 169.178, 169.180, 169.181, 175.300, 177.2600, 178.3300, 184.1666, 582.4666, GRAS; USDA 9CFR §318.7, 381.147; BATF 27CFR §240.1051; EPA reg., approved for some drugs; Japan approved with limitations; Europe listed; FEMA GRAS; FDA approved for orals, parenterals, topicals; USP/NF, BP, Ph.Eur., JP compliance
Manuf./Distrib.: Aldrich; ARCO; Asahi Denka Kogyo; Ashland; J.T. Baker; Baychem; BP Chems. Ltd; Chemcentral; Eastman; Fluka; Focus; C.P. Hall; Harcros; Hüls AG; Olin; Primachem; Samson; Seeler Ind.; Sigma; Sunnyside; Texaco; Union Carbide; Van Waters & Rogers; Veckridge; Westco
Trade names containing: Centrophil® M; Flat-Ayd® FA-W-34; Geropon® 99; Geropon® SS-O-70PG; Joncryl® 62; Monawet MO-70R; Monawet MO-84R2W; Slip-Ayd® SL-300; Surfynol® 104PG; Surfynol® 104PG-50; Surfynol® PG-50

Propylene glycol alginate
CAS 9005-37-2
Synonyms: Hydroxypropyl alginate; Alginic acid, ester with 1,2-propanediol; Propane-1,2-diol alginate
Definition: Mixture of propylene glycol esters of alginic acid
Empirical: $(C_9H_{14}O_7)_8$
Properties: Wh. to ylsh. fibrous or gran. powd., pract. odorless and tasteless; sol. in water, dil. organic acids, hydroalcoholic mixts.; m.w. 1873.6
Toxicology: LD50 (oral, rat) 7200 mg/kg; mildly toxic by ingestion; heated to decomp., emits acrid smoke and irritating fumes
Uses: Food additive (human); gellant, film-former, emulsifier, suspending aid; reactive with milk; for foods (dairy prods., extruded foods, beverages, salad dressings, dessert gels, sauces, frozen desserts, bakery prods.); cosmetics/pharmaceuticals (lotions, vitamin suspension), industrial (paper coatings, adhesives, textile printing, dyeing, explosives)
Regulatory: FDA 21CFR §133.133, 133.134, 133.162, 133.178. 133.179, 172.210, 172.820, 172.858, 173.340, 176.170, GRAS; FEMA GRAS; Japan approved (1% max.); Europe listed; UK approved; FDA approved for orals; USP/NF compliance
Manuf./Distrib.: Ashland; Kelco Int'l.; Meer; Spectrum Chem. Mfg.
Trade names: Colloid 602

Propylene glycol butyl ether. *See* Butoxypropanol
Propylene glycol 1-t-butyl ether. *See* Propylene glycol t-butyl ether
Propylene glycol n-butyl ether. *See* Butoxypropanol

Propylene glycol t-butyl ether
CAS 57018-52-7
Synonyms: PTB; 1-t-Butoxy-2-propanol; Propylene glycol 1-t-butyl ether
Empirical: $C_7H_{16}O_2$
Formula: $C_4H_9OCH_2CHOHCH_3$
Properties: Partly sol. in water; m.w. 132.2; dens. 0.874 (20/4 C); f.p. -69 F; b.p. 151 C; flash pt. (TCC) 113 F; ref. index 1.4116
Precaution: Flamm.
Toxicology: Skin and eye irritant
Uses: Solvent for water-based paints, cleaners, inks, cutting fluids; in polyester and alkyd resin prod.
Manuf./Distrib.: Aldrich; Eastman; Fluka
Trade names: Arcosolv® PTB; Butyl Propasol

Propylene glycol dibenzoate
CAS 19224-26-1
Properties: Sp.gr. 1.146; flash pt. (COC) 199 C; ref. index 1.544
Uses: Solvating plasticizer for PVC applic., latex caulk formulations, PVAc adhesives, and castable PU; used in latex paints as coalescing agent
Regulatory: FEMA GRAS
Manuf./Distrib.: Unitex; Velsicol
Trade names: Benzoflex® 284; Velate® 300

Propylene glycol ethyl ether
CAS 1569-02-4; 52125-53-8
Synonyms: 1-Ethoxy-2-propanol; Propylene glycol monoethyl ether
Empirical: $C_5H_{12}O_2$
Properties: Mw. 104.17
Toxicology: Mildly toxic by ingestion and skin contact; heated to dec., emits acrid smoke and irritating fumes; Fast evaporating solv. for coatings, cleaners, cosmetics, agric., electronics, ink, textile and adhesive prods.
Trade names: Arcosolv® PE
Trade names containing: Araldite® GZ 7071 PM-75

Propylene glycol ethyl ether acetate
CAS 98516-30-4
Synonyms: Propylene glycol monoethyl ether acetate; Highly misc. solv. with broad solvency for coatings, cleaners, agric., cosmetics, electronics, ink, textile and adhesive prods.
Manuf./Distrib.: Ashland; Harcros
Trade names: Arcosolv® PEA

Propylene glycol laurate
CAS 142-55-2; 27194-74-7; EINECS 205-542-3
Synonyms: Dodecanoic acid, 2-hydroxypropyl ester; Dodecanoic acid, monoester with 1,2-propanediol; Propylene glycol monolaurate
Definition: Ester of propylene glycol and lauric acid
Empirical: $C_{15}H_{30}O_3$
Formula: $CH_3(CH_2)_{10}COOCH_2CHCH_3OH$
Properties: Sp.gr. 0.911; m.p. 0-12 C; flash pt. (COC) 188 C
Toxicology: Nontoxic but can cause allergic reactions in hypersensitive persons
Uses: Surfactant, emulsifier, coemulsifier, stabilizer, wetting agent, lubricant, plasticizer, emollient, solvent, and antistat; used in cosmetic, pharmaceutical, industrial, food applics., pulp/paper, metalworking, lubricants, textiles, agric., paints; adhesives
Regulatory: FDA 21CFR §172.856, 173.340, 175.105, 175.300, 176.170, 176.210, 177.2800
Manuf./Distrib.: Calgene; Inolex; Stepan; Velsicol
Trade names: Calgene PGML

Propylene glycol methacrylate
Synonyms: PPGMM; Propylene glycol monomethacrylate; 2-Hydroxypropyl methacrylate
Properties: Clear liq.; insol. in water; sol. in most org. solvs.; sp.gr. 1.01; b.p. 290-295 C; flash pt. > 300 F
Uses: Monomer improving adhesion in thermoset acrylic and urethane coatings, fibers, syn. resins, radiation-curable inks
Manuf./Distrib.: Monomer-Polymer & Dajac Labs; Rhone-Poulenc N. Am.
Trade names containing: Sipomer® PPGMM

Propylene glycol methyl ether. *See* Methoxyisopropanol

Propylene glycol methyl ether acetate
CAS 108-65-6; EINECS 203-603-9
Synonyms: MPA; PM acetate; 2-Methoxypropyl acetate; Propylene glycol monomethyl ether acetate; 1,2-Propanediol monomethyl ether acetate
Empirical: $C_6H_{12}O_3$
Formula: $CH_3COOCHCH_3CH_2OCH_3$
Properties: Liq.; sol. 20% in water; m.w. 132.2; dens. 0.97 (20/20 C); b.p. 140 C (760 mm); flash pt. (Seta) 45 C; ref. index 1.402 (20 C)
Precaution: Flamm.
Uses: Solvent for paints, electronics, inks
Manuf./Distrib.: Aldrich; Allchem Ind.; Ashland; Fluka; Harcros; Oxiteno; Van Waters & Rogers
Trade names: Arcosolv® PMA; Dowanol® PMA; Methyl Propasol® Acetate
Trade names containing: Acryloid® AU-608S; Acryloid® AU-1003; Acryloid® AU-1004; Araldite® GZ 488 PMA-32; Araldite® GZ 7488 PMA-40; D.E.R. 671-PMA75; Disperse-Ayd 15; Konform® AR 2000; Tint-Ayd® ST Series

Propylene glycol monoethyl ether. *See* Propylene glycol ethyl ether
Propylene glycol monoethyl ether acetate. *See* Propylene glycol ethyl ether acetate
Propylene glycol monolaurate. *See* Propylene glycol laurate
Propylene glycol monomethacrylate. *See* Propylene glycol methacrylate
Propylene glycol monomethyl ether. *See* Methoxyisopropanol
Propylene glycol monomethyl ether acetate. *See* Propylene glycol methyl ether acetate
Propylene glycol monomyristate. *See* Propylene glycol myristate
Propylene glycol monostearate. *See* Propylene glycol stearate

Propylene glycol myristate
CAS 29059-24-3; EINECS 249-395-3
Synonyms: Propylene glycol monomyristate; Tetradecanoic acid, monoester with 1,2-propanediol
Definition: Ester of propylene glycol and myristic acid
Empirical: $C_{17}H_{34}O_3$
Formula: $CH_3(CH_2)_{12}COOCH_2CHCH_3OH$
Uses: Wetting aid, lubricant, opacifier, antistat, dispersant, w/o emulgent, scouring and detergent aid, defoamer, plasticizer, rust inhibitor; cosmetics, pharmaceuticals, lubricating and cutting oils, pigment grinding; paints, inks, plastics, waxes, insecticides
Regulatory: FDA 21CFR §172.856, 173.340, 175.300, 176.170, 176.210, 177.2800

Propylene glycol octadecanoate. *See* Propylene glycol stearate

Propylene glycol palmitate

Definition: Ester of propylene glycol and palmitic acid

Uses: Surfactant for industrial and cosmetic applics., pulp/paper, pharmaceuticals, metalworking, lubricants, textiles, agric., paints, adhesives; food emulsifier

Trade names: Calgene PGMP

Propylene glycol phenyl ether

CAS 4169-04-4; 770-35-4

Synonyms: 2-Phenoxypropanol; 2-Phenoxypropyl alcohol

Empirical: $C_9H_{12}O_2$

Formula: $C_6H_5OC_3H_6OH$

Properties: Colorless liq.; m.w. 152.20; sp.gr. 1.063; f.p. 55 F; b.p. 242.7 C; flash pt. (TCC) 240 F

Precaution: Combustible

Uses: Solvent for coatings formulation; bactericide agent; fixative for soaps and perfumes; intermediate for plasticizers

Trade names: Dowanol® PPh

Propylene glycol propyl ether. *See* Propylene glycol n-propyl ether

Propylene glycol n-propyl ether

CAS 1569-01-3

Synonyms: Propylene glycol propyl ether

Formula: $C_3H_7OCH_2CH(CH_3)OH$

Properties: Sol. in water; m.w. 118.18; sp.gr. 0.886 (20/20 C); f.p. -112 F; b.p. 149.8 C; flash pt. (TCC) 119 F; ref. index 1.4121

Precaution: Flamm.

Uses: Solv. for coatings, cleaners, cosmetics, agric., electronics, ink, textiles, adhesives

Manuf./Distrib.: Aldrich

Trade names: Arcosolv® PNP; Dowanol® PnP

Trade names containing: Dislon® AQ-261; Disparlon® AQ-261

Propylene glycol stearate

CAS 1323-39-3; EINECS 215-354-3

Synonyms: Propylene glycol monostearate; Propylene glycol octadecanoate; Octadecanoic acid, monoester with 1,2-propanediol; 1,2-Propanediol monostearate

Definition: Ester of propylene glycol and stearic acid

Empirical: $C_{21}H_{42}O_3$

Formula: $CH_3(CH_2)_{16}COOCH_2CHCH_3OH$

Properties: Wh. to cream flakes, bland typ. fatty odor and taste; sol. in min. oil, IPM, oleyl alcohol; insol. in water, glycerin, propylene glycol; m.p. 35-38 C; acid no. 4 max.; iodine no. 3 max.; sapon. no. 155-165; hyd. no. 160-175

Toxicology: Poison by intraperitoneal route

Uses: Emulsifier for lotions, soft creams, makeup, paints, adhesives; plasticizer for cellulose nitrate

Regulatory: FDA 21CFR §172.856, 172.860, 172.862, 173.340, 175.105, 175.300, 176.170, 176.210, 177.2800; USDA 9CFR §318.7, 381.147; FEMA GRAS; FDA approved for rectals, vaginals; USP/NF compliance

Manuf./Distrib.: ABITEC; Aquatec Quimica SA; Calgene; Eastman; Grindsted Prods.; Inolex; ISP Van Dyk; Lipo; Lonza; Stepan; Witco/Oleo-Surf.; Witco/Polymer Addit.

Trade names: Calgene PGMS

Propylene polymer. *See* Polypropylene
Propyl hydride. *See* Propane
Propylic alcohol. *See* n-Propyl alcohol

Propyl propionate

CAS 106-36-5; EINECS 203-389-7

Synonyms: Propanoic acid propyl ester; n-Propyl propionate

Empirical: $C_6H_{12}O_2$

Formula: $CH_3CH_2COOCH_2CH_2CH_3$

Properties: Liq., complex fruity odor; sol. in 200 parts water; misc. with alcohol, ether; m.w. 116.16; dens. 0.881 (20/4 C); m.p. -76 C; b.p. 120-122 C; flash pt. 22 C; ref. index 1.3935 (20 C)

Precaution: Flamm.

Uses: Solv. for paints

Regulatory: FDA 21CFR §172.515; FEMA GRAS
Manuf./Distrib.: Aldrich; Ashland; Fluka; Penta Mfg.; Union Carbide

n-Propyl propionate. *See* Propyl propionate
PS. *See* Polystyrene
Pseudocumene. *See* 1,2,4-Trimethylbenzene
Pseudocumol. *See* 1,2,4-Trimethylbenzene
PSVS. *See* Polysodium vinyl sulfonate
PTB. *See* Propylene glycol t-butyl ether
PTBP. *See* 4-t-Butylphenol
PTFE. *See* Polytetrafluoroethylene
PTMEG. *See* Polytetramethylene ether glycol
PTMEG diamine. *See* Polytetramethylene ether glycol diamine
PTMG. *See* Polytetramethylene ether glycol
PTSA. *See* p-Toluenesulfonamide
Purified gum spirits. *See* Turpentine oil
PVA. *See* Polyvinyl alcohol
PVAc. *See* Polyvinyl acetate (homopolymer)
PVAL. *See* Polyvinyl alcohol
PVB. *See* Polyvinyl butyral
PVC. *See* Polyvinyl chloride

PVC/PVAc copolymer
Uses: Resin for extruded and calendered rigid sheet, sol'n. coatings, phono records, floor tiles
Trade names: VC-172

PVC/VA copolymer
CAS 9003-22-9
Uses: For paste processing of floor coverings, carpet backings, adhesives/laminating coatings, spray coating
Manuf./Distrib.: Aldrich; Fluka; Sigma
Trade names: Vestolit® B 7090

PVDC. *See* Polyvinylidene chloride
PVDF. *See* Polyvinylidene fluoride
PVE. *See* Polyvinyl ethyl ether
PVI. *See* Polyvinyl isobutyl ether
PVM. *See* Polyvinyl methyl ether

PVM/MA copolymer
CAS 9011-16-9; 52229-50-2
Synonyms: Methyl vinyl ether/maleic anhydride copolymer; Poly(methyl vinyl ether/maleic anhydride); 2,5-Furandione, polymer with methoxyethylene
Definition: Copolymer of methyl vinyl ether and maleic anhydride
Empirical: $(C_4H_2O_3 \cdot C_3H_6O)_x$
Uses: Dispersant, coupling, stabilizer, thickener, emulsifier, solubilizer, corrosion inhibitor, film former, antistat, used in agric., paper and textile industries, chemical processing, industrial products, detergents, cosmetics, emulsion polymerization
Manuf./Distrib.: Aldrich; Sigma
Trade names: Gantrez® AN-119; Gantrez® AN-139; Gantrez® AN-149; Gantrez® AN-169; Gantrez® AN-179

PVM/MA copolymer, butyl ester
CAS 54018-18-7; 54578-91-5; 53200-28-5
Synonyms: Poly(methylvinyl ether/maleic acid) butyl ester; Butyl ester of PVM/MA copolymer (INCI); 2-Butenedioic acid, polymer with methoxyethene, butyl ester
Definition: Polymer consisting of partial butyl ester of the polycarboxylic acid formed from vinyl methyl ether and maleic anhydride
Empirical: $(C_{11}H_{18}O_5)_n$
Formula: $-[CH_2CHOCH_3CHOCOHCHCOOC_4H_9]_{n-}$
Uses: Film-forming binders for hair sprays and setting lotions
Trade names containing: Gantrez® ES-425; Gantrez® ES-435

PVM/MA copolymer, ethyl ester
CAS 50935-57-4; 67724-93-0; 54578-90-4

PVM/MA copolymer, isopropyl ester

Synonyms: Ethyl ester of PVM/MA copolymer (INCI)
Uses: Film-forming binders for hair sprays and setting lotions
Trade names containing: Gantrez® ES-225

PVM/MA copolymer, isopropyl ester
CAS 54578-88-0; 54077-45-1; 56091-51-1
Synonyms: Poly(methylvinyl ether/maleic acid) isopropyl ester; 2-Butenedioic acid, polymer with methoxyethene, 1-methylethyl ester; Isopropyl ester of PVM/MA copolymer (INCI)
Uses: Copolymer forming clear, glossy films with substantivity and moisture resistance; used in hairsprays, mousses, gels and lotions, coatings, polishes; emulsion stabilizer in creams and lotions
Trade names containing: Gantrez® ES-335

PVOH. *See* Polyvinyl alcohol

PVP
CAS 9003-39-8; EINECS 201-800-4
Synonyms: Polyvinylpyrrolidone; Poly (n-vinylbutyrolactam); Polyvidonum; Polyvidone; Povidone; 1-Vinyl-2-pyrrolidinone polymer; 1-Ethenyl-2-pyrrolidinone homopolymer
Classification: Synthetic linear polymer
Definition: Polymer of 1-vinyl-2-pyrrolidone monomers
Empirical: $(C_6H_9NO)_x$
Properties: Wh. amorphous powd., odorless; hygroscopic; sol. in water, chlorinated hydrocarbons, alcohol, amines, nitroparaffins, lower m.w. fatty acids; m.w. ≈ 10,000, ≈ 24,000, ≈ 40,000 (food use), ≈ 160,000, ≈ 360,000 (beer); dens. 1.23-1.29; pH 3.0-7.0 (5%)
Toxicology: LD50 (IP, mouse) 12 g/kg; mildly toxic by IP and IV routes; potent histamine-releasing agent in animals; heated to decomp., emits toxic fumes of NO_x
Uses: Film-forming agent, hair fixative, thickener, protective colloid, suspending agent, and dispersant for cosmetics industry, tech. applics., paints; drug vehicle and retardant; tablet binder, pharmaceutical excipient; in adhesives; in detergents
Usage level: 3-5%; 0.03-0.9% (injectable drug prods.); 1-50 mg (oral dosage forms)
Regulatory: FDA 21CFR §73.1, 73.1001, 172.210, 173.50, 173.55, 175.105, 175.300, 176.170, 176.180, 176.210; BATF 27CFR §240.1051; FDA approved for parenterals, intramuscular injectables, orals, topicals; USP/NF, BP, Ph.Eur. compliance
Manuf./Distrib.: Aldrich; Allchem Ind.; Ashland; BASF; Fluka; Great Western; Hickson Danchem; ISP; Jarchem Ind.; Monomer-Polymer & Dajac Labs; Research Organics; Sigma
Trade names: Luviskol® K17; Luviskol® K30; Luviskol® K60; Luviskol® K80; Luviskol® K90; PVP K-15; PVP K-15 Sol'n.; PVP K-30; PVP K-60 Sol'n.; PVP K-90; PVP K-90 Sol'n.; PVP K-120

PVP/styrene copolymer. *See* Styrene/PVP copolymer

PVP/VA copolymer
CAS 25086-89-9
Synonyms: 1-Ethenyl-2-pyrrolidinone, polymer with acetic acid ethenyl ester; Polyvinylpyrrolidone/vinyl acetate copolymer; Vinylpyrrolidone/vinyl acetate copolymer
Definition: Copolymer of vinyl acetate and vinylpyrrolidone monomers
Empirical: $(C_6H_9NO \cdot C_4H_6O_2)_x$
Uses: Film-former used in hairsprays, gels, hair thickeners, tints, and dyes; suspending agent, dispersant, thickener, stabilizer, adhesion promoter, coatings
Manuf./Distrib.: Aldrich
Trade names: Luviskol® VA37E; Luviskol® VA37I; Luviskol® VA55E; Luviskol® VA55I; Luviskol® VA64E; Luviskol® VA64I; Luviskol® VA64 Powd.; Luviskol® VA64W; Luviskol® VA73E; Luviskol® VA73W
Trade names containing: PVP/VA I-335; PVP/VA I-535; PVP/VA I-735

Pyridine
CAS 110-86-1; EINECS 203-809-9
Empirical: C_5H_5N
Properties: Colorless liq., char. disagreeable odor, sharp taste; misc. with water, alcohol, ether, petroleum ether, oils, other org. liqs.; m.w. 79.10; dens. 0.98272 (20/4 C); m.p. -41.6 C; b.p. 115-116 C; flash pt. (CC) 20 C; ref. index 1.510 (20 C)
Precaution: DOT: Flamm. liq.; highly flamm.; volatile with steam
Toxicology: LD50 (oral, rat) 1.58 g/kg; may cause human CNS depression, irritation of skin and respiratory tract; large does may produce GI disturbances, kidney and liver damage
Uses: Chem. intermediate, acid scavenger; in agrochemicals, pharmaceuticals, sedative, surfactants,

photographic materials, coatings, curing agents, rubber chemicals, vulcanization accelerator, plastics, antidandruff shampoos, textiles, dyestuffs; denaturant for alcohols
Regulatory: FDA 21CFR §172.515; FEMA GRAS; BP compliance
Manuf./Distrib.: AC Ind.; Aldrich; Allchem Ind.; Berje; Daicel Chem. Ind.; Fabrichem; Fluka; Koei Chem.; Nepera; Penta Mfg.; Raschig; Reilly Ind.; Schweizerhall; Sigma; Spectrum Chem. Mfg.
Trade names: Pyridine 1°

2-Pyridinethiol, 1-oxide, sodium salt. *See* Sodium pyrithione
Pyrithione zinc. *See* Zinc pyrithione
Pyroacetic ether. *See* Acetone
L-Pyroglutamic acid. *See* PCA

Pyromellitic acid
CAS 89-05-4; EINECS 201-879-5
Synonyms: 1,2,3,4,5-Benzenetetracarboxylic acid; Benzene-1,2,4,5-tetracarboxylic acid
Empirical: $C_{10}H_6O_8$
Properties: M.w. 254.16
Uses: Intermediate for paints
Manuf./Distrib.: Allco; Amoco; Fluka; Hüls Am.

Pyromellitic anhydride
Uses: Intermediate for paints
Manuf./Distrib.: Allco; Amoco; Hüls Am.

Pyromellitic dianhydride
CAS 89-32-7; EINECS 201-898-9
Synonyms: PMDA; Benzene-1,2,4,5-tetracarboxylic dianhydride; 1,2,4,5-Benzenetetracarboxylic dianhydride; 1H,3H-Benzo(1,2-c:4,5-c′)difuran-1,3,5,7-tetrone; 1,2,4,5-Benzenetetracarboxylic 1,2:4,5 dianhydride
Empirical: $C_{10}H_2O_6$
Formula: $C_6H_2(C_2O_3)_2$
Properties: Wh. powd.; sol. 0.4 g/100 g in water, reacts to give tetra-acid; m.w. 218.12; dens. 1.680 (20 C); vapor pressure 63 mm Hg (290 C); m.p. 284-286 C; b.p. 380-400 C; pH 2.2 (3.9 g/L)
Precaution: Incompat. with alkali metal hydroxides, amines, ammonia, alcohols, thiols, strong oxidizing/reducing agents
Toxicology: LD50 (oral, rat) 2250-2595 mg/kg; skin and eye irritant, but nonsensitizing; may cause respiratory irritation/allergic response on inhalation of dust or fumes
Uses: Intermediate for pyromellitic acid, polyimides, polyester resins, alkyd and PU resins, polypyrrones; epoxy curing agent; crosslinking agent for epoxy plasticizers in vinyls, alkyd resins; end-uses incl. electronic transfer molding compds., wire coatings, aerospace adhesives, advanced structural composites, uv-cured inks, insulating coatings and foams, flame retardant prods., release agents, molded high-performance engineering parts
Manuf./Distrib.: Aldrich; Allco; Fluka; Hüls AG; Monomer-Polymer & Dajac Labs
Trade names: Allco PMDA

Pyromucic aldehyde. *See* Furfural

Pyrophyllite
CAS 12269-78-2
Synonyms: Hydrated aluminum silicate; Agalmatolite
Definition: Naturally occurring mineral substance consisting predominantly of hydrous aluminum silicate
Empirical: $Al_2O_3 \cdot 4SiO_2 \cdot H_2O$
Formula: $Al_2Si_4O_{10}(OH)$
Properties: Colorless, white, green, gray, brown; dens. 2.8-2.9
Uses: Filler, extender, diluent, carrier for rubber, plastics, paints, ceramics, insecticides, slate pencil, feed additive; color additive for drugs and cosmetics
Regulatory: FDA 21CFR §73.1400, 73.2400; permanently listed
Manuf./Distrib.: Chem-Materials; D.N. Lukens; R.T. Vanderbilt; Whittaker, Clark & Daniels
Trade names: Pyrax® ABB; Pyrax® B; Pyrax® WA; Veecote®
See also Aluminum silicate

Pyroxylin. *See* Nitrocellulose

2-Pyrrolidone
CAS 616-45-5; EINECS 204-648-7
Synonyms: Butyrolactam; Pyrrolidone-2
Empirical: C_4H_7NO
Formula: $CH_2CH_2CH_2C(O)NH$
Properties: Lt. yel. liq.; sol. in water, ethanol, ethyl ether, chloroform, benzene, ethyl acetate, carbon disulfide; m.w. 85.12; dens. 1.1; b.p. 245 C; flash pt. 265 F
Toxicology: Mildly toxic by ingestion, subcutaneous route
Uses: Plasticizer and coalescing agent for acrylic latices; solv. for veterinary medicine
Manuf./Distrib.: Aldrich; Allchem Ind.; BASF; Fluka; ISP; UCB SA; Unitex
Trade names: 2-Pyrol®

Pyrrolidone-2. *See* 2-Pyrrolidone
Pyrrolidonecarboxylic acid. *See* PCA
(S)-2-Pyrrolidone-5-carboxylic acid. *See* PCA

Quantril. *See* Benzoguanamine

Quartz
CAS 14808-60-7; EINECS 238-878-4
Synonyms: Silicon dioxide; Silica; Silica glass; Quartz glass
Definition: Crystallized silicon dioxide
Empirical: O_2Si
Formula: SiO_2
Properties: Wh. to reddish; insol. in acids except HF; m.w. 60.08 m.p. 1713 C
Precaution: Noncombustible; avoid inhalation of fine particles
Toxicology: TLV (for resp. dust) 10 mg/m^3
Uses: Extender for paints; electronic components; TV components; optical fiber materials; format for IC
Manuf./Distrib.: Aldrich; Archway Sales; Asahi Glass; Engelhard; Fluka; Hammill & Gillespie; Lomas Int'l.; D.N. Lukens; Malvern Mins.; Mitsubishi Materials; Sigma; Unimin; U.S. Silica; Van Waters & Rogers; Westo Ind. Prods. Ltd
Trade names: Novacite® 200; Novacite® 325; Novacite® 1250; Novacite® Daper; Novacite® L-207A; Novacite® L-337; Novacite® S-325
Trade names containing: Aktisil® EM; Aktisil® MAM; Aktisil® PF 224; Silfin® Z; Sillikolloid P 82; Sillikolloid P 87; Sillitin N 82; Sillitin N 85; Sillitin V 85; Sillitin V 88; Sillitin Z 86; Sillitin Z 89
See also Silica

Quartz glass. *See* Quartz
Quaternary ammonium compounds, bis(hydrog. tallow alkyl) dimethyl, chlorides, reaction products with bentonite. *See* Quaternium-18 bentonite
Quaternary ammonium compounds, bis(hydrog. tallow alkyl) dimethyl, chlorides, reaction products with hectorite. *See* Quaternium-18 hectorite
Quaternium-5. *See* Distearyldimonium chloride

Quaternium-15
CAS 51229-78-8; 4080-31-3; EINECS 223-805-0
Synonyms: 1-(3-Chloroallyl)-3,5,7-triaza-1-azoniaadamantane chloride; N-(3-Chloroallyl)hexaminium chloride; Chlorallyl methenamine chloride
Classification: Quaternary ammonium salt
Empirical: $C_9H_{16}ClN_4 \cdot Cl$
Formula: $C_6H_{12}N_4(CH_2CHCHCl)Cl$
Uses: Preservative for adhesives, latex emulsions, paints, cutting fluids, pharmaceuticals
Usage level: 0.02-03%; 0.02% (topicals)
Regulatory: FDA 21CFR §175.105, 176.170; CIR approved; Europe listed; Japan not approved; FDA approved for topicals
Manuf./Distrib.: Sigma
Trade names containing: Dowicil® 75

Quaternium-18
CAS 61789-80-8; 68002-59-5; EINECS 263-090-2
Synonyms: Dimethyl di(hydrogenated tallow)ammonium chloride; Dihydrogenated tallow dimethyl ammo-

nium chloride; Ditallowalkonium chloride
Classification: Quaternary ammonium salt
Formula: $R(CH_3)_2NR]^+Cl^-$, R rep. hydrogenated tallow fatty radicals
Uses: Surfactant, fabric antistat, conditioner for household, industrial, textile, hair applics.; antistatic coating for cellulose acetate, polyacetal, PE, PP
Trade names: Adogen® 442

Quaternium-25. *See* Cetethyl morpholinium ethosulfate
Quaternium 31. *See* Dialkyl dimethyl ammonium chloride
Quaternium-34. *See* Dicocodimonium chloride

Quaternium-18 bentonite
CAS 68953-58-2; EINECS 273-219-4
Synonyms: Quaternary ammonium compounds, bis(hydrog. tallow alkyl) dimethyl, chlorides, reaction products with bentonite
Definition: Reaction prod. of quaternium-18 and bentonite
Uses: Thixotrope, gellant, thickener for solv.-based coatings; lip and eye care prods., antiperspirant creams and lotions
Regulatory: FDA 21CFR §175.300, 178.3570
Trade names: Bentone® 34; Tixogel® VP
Trade names containing: Suspend-Ayd® 1

Quaternium-18/benzalkonium bentonite
Definition: Reaction prod. of bentonite, quaternium-18, and benzalkonium chloride
Trade names: Claytone® HT

Quaternium-18 hectorite
CAS 12001-31-9; 71011-27-3; EINECS 234-406-6
Synonyms: Quaternary ammonium compounds, bis(hydrog. tallow alkyl) dimethyl, chlorides, reaction products with hectorite
Definition: Reaction prod. of hectorite and quaternium-18
Uses: Thixotrope, gellant, thickener for solv.-based coatings; lip and eye care prods.; suspending agent
Trade names: Bentone® 38
Trade names containing: M-P-A® 14; Suspend-Ayd® 2

Quicklime. *See* Calcium oxide

Rapeseed oil
CAS 8002-13-9; EINECS 232-299-0
Synonyms: Brassica campestris oil; Colza oil; Canola oil (low erucic acid rapeseed oil)
Definition: Vegetable oil expressed from seeds of *Brassica campestris*
Properties: Brn. viscous liq., yel. when refined, noxious odor; sol. in chloroform, ether, CS_2; dens. 0.913-0.916; m.p. 17-22 C; solidifies at 0 C; flash pt. 325 F; iodine no. 97-105; sapon. no. 170-177 C; ref. index 1.4720-1.4752
Precaution: Subject to spontaneous heating
Toxicology: Toxic-allergic potential; can cause acne-like skin eruptions
Uses: Lubricant and slip agent for plastics, polymers, rubber, metal lubricants; edible oil for salad dressings, margarine; soft soaps; drying oil for paints
Regulatory: FDA 21CFR §175.105, 176.210, 177.1200, 177.2800, 184.1555; Japan approved (extract); FDA approved for orals; BP compliance
Manuf./Distrib.: ABITEC; Alnor Oil; Arista Ind.; Calgene; Climax Perf. Materials; Degen; Jarchem Ind.; Penta Mfg.; Reilly-Whiteman; Werner G. Smith; Welch, Holme & Clark; Witco/Oleo-Surf.

Rapeseed oil, vulcanized
Uses: Extender, processing aid for rubber goods
Trade names containing: Faktogel® Badenia T

Raw linseed oil. *See* Linseed oil
Red copper oxide. *See* Copper oxide (ous)
Red cuprous oxide. *See* Copper oxide (ous)

Red iron oxide. *See* Ferric oxide
Red iron trioxide. *See* Ferric oxide
Red lead. *See* Lead oxide, red
Red lead oxide. *See* Lead oxide, red
Red oil. *See* Oleic acid
Red precipitate. *See* Mercury oxide (ic), red and yellow
Refined petroleum wax. *See* Petroleum wax
Resin. *See* Rosin
Resole. *See* Phenolic resin
Resorcin. *See* Resorcinol

Resorcinol
CAS 108-46-3; EINECS 203-585-2
Synonyms: 1,3-Benzenediol; m-Dihydroxybenzene; Resorcin
Classification: Phenol
Empirical: $C_6H_6O_2$
Formula: $C_6H_4(OH)_2$
Properties: Wh. cryst., unpleasant sweet taste; very sol. in alcohol, ether, glycerol; sl. sol. in chloroform; sol. in water; m.w. 110.12; dens. 1.285 (15 C); m.p. 110 C; b.p. 280.5 C; flash pt. (CC) 261 F
Precaution: Protect from light
Toxicology: TLV:TWA 10 ppm; moderately toxic by skin contact and intravenous route; poison by ingestion; skin and severe eye irritant
Uses: Topical antiseptic; keratolytic agent; in tanning; in mfg. of resins, adhesives, hexylresorcinol, p-aminosalicylic acid, explosives, dyes, paints; in cosmetics; dyeing and printing textiles
Regulatory: FDA 21CFR §177.1210; FEMA GRAS
Manuf./Distrib.: Aldrich; Allchem Ind.; J.T. Baker; Cardolite; Fairmount; R.W. Greeff; Fluka; Hoechst Celanese; Indspec; Janssen Chimica; Napp Tech.; Penta Mfg.; Richman; Sigma

Resorcinol-formaldehyde resin
CAS 65876-95-1; 24969-11-7
Uses: Bonding agent for rubber compds., latex dips, adhesives in wood gluing
Trade names: Heveagrip

Rhodium
CAS 7440-16-6; EINECS 231-125-0
Classification: Metallic element
Empirical: Rh
Properties: At.wt. 102.91; dens. 2.41; m.p. 1966 C; b.p. 3727 C
Uses: Alloy with platinum for high temp. thermocouples, furnace windings, laborabory crucibles, spinnerets for rayon, electrical contacts, jewelry, catalyst, optical instrument mirrors, electrodeposited metal coatings, vacuum-deposited glass coatings
Manuf./Distrib.: Aldrich; Alfa Aesar Johnson Matthey; Atlantic Equip. Engrs.; Atomergic Chemetals; Colonial Metals; Degussa; Fluka; Johnson Matthey; Noah; Powmet; Reade Advanced Materials; Technic

Rhus succedanea wax. *See* Japan wax
Ricinic acid. *See* Ricinoleic acid
Ricini oleum. *See* Castor oil

Ricinoleic acid
CAS 141-22-0; EINECS 205-470-2
Synonyms: Castor oil acid; 12-Hydroxy-9-octadecenoic acid; cis-12-Hydroxyoctadec-9-enoic acid; 12-Hydroxy-cis-9-octadecenoic acid; 9-Octadecenoic acid, 12-hydroxy-; Ricinic acid; Ricinolic acid; d-12-Hydroxyoleic acid; 12-Hydroxyoleic acid
Classification: Unsaturated fatty acid
Empirical: $C_{18}H_{34}O_3$
Formula: $CH_3(CH_2)_5CH(OH)CH_2CH=CH(CH_2)_7COOH$
Properties: Colorless to yel. visc. liq.; sol. in alcohol, acetone, ether, chloroform; insol. in water; m.w. 298.45; dens. 0.940 (27.4/4 C); m.p. 5.5 C; b.p. 245 C (10 mm); ref. index 1.4716 (20 C)
Precaution: Combustible
Toxicology: Experimental tumorigen; heated to decomp., emits acrid smoke and irritating fumes
Uses: Chem. intermediate; lubricant, rustproofing in cutting oils; in textile finishing, resin plasticizers, inks, coatings, cosmetics; modifier for coatings and adhesive polymers; sometimes added to Turkey red oil, drycleaning soaps

Manuf./Distrib.: Alnor Oil; Amber; CasChem; Fluka; Sigma
Trade names: P®-10 Acid

Ricinolic acid. *See* Ricinoleic acid
Ricinus oil. *See* Castor oil

Rosin
CAS 8050-09-7; 8052-10-6; EINECS 232-475-7
Synonyms: Colophony; Colophane; Colophonium; Gum rosin; Yellow pine rosin; Rosin gum; Resin
Definition: Residue from distilling off the volatile oil from the oleoresin obtained from *Pinus palustris* and other species of *Pinaceae*
Properties: Pale yel. to amber translucent, sl. turpentine odor and taste; insol. in water; sol. in alcohol, benzene, ether, glacial acetic acid, oils, carbon disulfide; dens. 1.07-1.09; m.p. 100-150 C; flash pt. 187 C
Precaution: Combustible
Toxicology: May cause contact dermatitis
Uses: Mfg. of varnishes, paint driers, printing inks, cements, soap, sealing wax, wood polishes, paper, plastics, fireworks, sizes, rosin oil; waterproofing paper, walls; pharmaceutic aid; emulsifier for SBR polymerization; tackifier resin in adhesives, sealants
Regulatory: FDA 21CFR §73.1, 172.210, 172.510, 172.615, 175.105, 175.125, 175.300, 176.170, 176.200, 176.210, 177.1200, 177.1210, 177.2600, 178.3120, 178.3800, 178.3870; Japan approved; FDA approved for orals
Manuf./Distrib.: Akzo Nobel; Aldrich; Arakawa USA; Arizona; Browning; Chemcentral; Cytec Ind.; Focus; Georgia-Pacific Chem.; Hercules BV; Meer; Natrochem; Punda Mercantile; Sigma; Spectrum Chem. Mfg.; Veitsiluoto Oy; Westvaco
Trade names: Acintol® R Type S; Acintol® R Type SB; Poly-Pale® Resin; Resin 731D; Resin NC-11; Staybelite®; Sylvaros® 20; Sylvaros® 80
Trade names containing: Uni-Rez® 1042

Rosin acid, methyl ester. *See* Methyl rosinate
Rosin, glyceryl ester. *See* Glyceryl rosinate
Rosin gum. *See* Rosin
Rosin, hydrogenated, glycerol ester. *See* Glyceryl hydrogenated rosinate
Rosin pentaerythritol ester. *See* Pentaerythrityl rosinate

Rosin, polymerized
Uses: Mfg. of paints; tackifier for adhesives
Manuf./Distrib.: Arizona; Chemcentral; Matteson-Ridolfi; S & S Chem.; T&R Chem.; U.S. Polymers

Rubber, chlorinated
CAS 9006-03-5
Definition: Rubber deriv. mfg. by passing chlorine into sol'n. of rubber in chloroform, CCl_4, etc.
Uses: For printing ink, overprint varnish, paint applics.
Manuf./Distrib.: Bayer; John K. Bice; Chemcentral; Revelli; Rit-Chem; Van Waters & Rogers; Zeneca Resins
Trade names: Adeka Chlorinated Rubber CR-5

Safflower oil
CAS 8001-23-8; EINECS 232-276-5
Synonyms: Carthanus tinctorious oil
Definition: Oily liq. obtained from seeds of *Carthanus tinctorius* consisting principally of triglycerides of linoleic acid
Properties: Lt. yel. oil; sol. in oil and fat solvs.; insol. in water; misc. with ether, chloroform; dens. 0.9211-0.9215 (25/25 C); iodine no. 135-150; sapon. no. 188-194; ref. index 1.472-1.475
Precaution: Light-sensitive; becomes rancid on exposure to air
Toxicology: Human skin irritant; ingestion in large volumes produces vomiting; heated to decomp., emits acrid smoke and irritating fumes
Uses: Alkyd resins; drying oil for paints, varnishes, medicine; salad oil blends
Regulatory: FDA 21CFR §175.105, 175.300, 176.200, 176.210, GRAS; Japan approved (safflower); USP/NF compliance
Manuf./Distrib.: ABITEC; Aldrich; Alnor Oil; Arista Ind.; Calgene; Charkit; Croda; Lipo; Penta Mfg.; Sigma; Traco Labs; Tri-K Ind.; Welch, Holme & Clark

SAIB. *See* Sucrose acetate isobutyrate
Salicyloyloxytributylstannane. *See* Tributyltin salicylate
Salt of saturn. *See* Lead acetate
Saran. *See* Polyvinylidene chloride
Saturated polyester resin. *See* Polyester resin, thermoplastic
S/B. *See* Styrene-butadiene polymer
SBA. *See* 2-Butanol
SBH. *See* Sodium borohydride
SBR. *See* Styrene-butadiene rubber
Schardinger α-dextrin. *See* Cyclodextrin
Schardinger β-dextrin. *See* β-Cyclodextrin

SD alcohol 1
Definition: Ethyl alcohol denatured with methyl alcohol and one of the following: denatonium benzoate, MIBK, mixed isomers of nitropropane, or methyl n-butyl ketone
Regulatory: FDA 27CFR §20.11, 21.32
Trade names containing: Tecsol® 1; Tecsol® 3

SD alcohol 3A
Definition: Ethyl alcohol denatured with methyl alcohol
Regulatory: FDA 27CFR §20.11, 21.35
Manuf./Distrib.: Eastman
Trade names containing: Tecsol® A; Tecsol® A-2; Tecsol® B; Tecsol® C; Tecsol® D; Tecsol® D-2; Tecsol® H

SDBS. *See* Sodium dodecylbenzenesulfonate
SDDC. *See* Sodium dimethyldithiocarbamate
SDS. *See* Sodium lauryl sulfate

Sebacic acid
CAS 111-20-6; EINECS 203-845-5
Synonyms: Decanedioic acid; 1,8-Octanedicarboxylic acid
Classification: Organic dicarboxylic acid
Empirical: $C_{10}H_{18}O_4$
Formula: $HOOC(CH_2)_8COOH$
Properties: Monoclinic prismatic tablets, white powd., fatty acid odor; very sol. in alcohols, esters, ketones; sl. sol. in hydrocarbons, chlorohydrocarbons; m.w. 202.28; dens. 1.207 (20/4 C); m.p. 134.5 C; b.p. 294.5 C (100 mm)
Toxicology: Moderately toxic by ingestion, intraperitoneal route
Uses: Stabilizer; raw material in mfg. of paints, alkyd resins, maleic and other polyesters, plasticizers, polyester rubbers, synthetic polyamide fibers
Regulatory: FDA 21CFR §175.105
Manuf./Distrib.: Aldrich; Browning; Fluka; Janssen Chimica; C.P. Hall; ICD Group; Kowa Am.; Merrand Int'l.; Penta Mfg.; Primachem; Punda Mercantile; San Yuan; Sigma; Spectrum Chem. Mfg.; Union Camp; United Min. & Chem.; Van Waters & Rogers

S-EB-S thermoplastic elastomer. *See* Styrene-ethylene/butylene-styrene block copolymer
Secondary butyl alcohol. *See* 2-Butanol

Shellac
CAS 9000-59-3; EINECS 232-549-9
Synonyms: White shellac; Bleached shellac; Button lac; Lacca
Classification: Fatty acid
Definition: Waxy amorphous protein which is the resinous secretion of the insect *Laccifer (Tachardia) lacca*
Properties: Off-wh. amorphous gran. solid, very little odor; slowly sol. in alcohol; sl. sol. in acetone, ether, benzene, petrol. ether; insol. in water; acid no. 73-89 (reg.), 75-91 (refined)
Toxicology: Nonallergenic; may cause contact dermatitis; heated to decomp., emits acrid smoke and irritating fumes
Uses: Resin for paints
Regulatory: FDA 21CFR §73.1, 175.105, 175.300, 175.380, 175.390, 182.99; 27CFR §21.126, 212.61, 212.90; 40CFR §180.1001; Japan approved; Europe listed; UK approved; FDA approved for orals; USP/NF compliance; JP compliance (purified shellac and white shellac)
Manuf./Distrib.: Colony Ind.; Mantrose-Haeuser; Punda Mercantile; Wm. Zinsser

Silane trimethoxy (2-methylpropyl). *See* Isobutyltrimethoxysilane

Silica

CAS 7631-86-9; EINECS 231-545-4

Synonyms: Silicon dioxide; Silicic anhydride

Classification: Inorganic oxide

Definition: Occurs in nature as agate, amethyst, chalcedony, cristobalite, flint, quartz, sand, tridymite, diatomite. The designation silica (silicon dioxide) incl. cryst., amorphous forms which are hydrated or hydroxylated

Empirical: O_2Si

Formula: SiO_2

Properties: Transparent crystals or amorphous very fine powd.; hygroscopic; pract. insol. in water, alc., and acids except hydrofluoric; m.w. 60.09; dens. 2.2 (amorphous), 2.65 (quartz, 0 C); lowest coeff. of heat expansion; melts to a glass; pH 3.5-4.4

Toxicology: LD50 (oral, rat) 3160 mg/kg; poison by intraperitoneal, intravenous, intratracheal routes; moderately toxic by ingestion; prolonged inhalation of dust can cause silicosis; suspected human carcinogen

Uses: Mfg. of glass, water glass, refractories, abrasives, ceramics, enamels, petrol. prods.; filler in paints, cosmetics; rubber reinforcing agent; as anticaking and defoaming agent; abrasive; thickener

Regulatory: FDA 21CFR §73.1, 172.230, 172.480 (limitation 2%), 173.340, 175.105, 175.300, 176.200, 176.210, 177.1200, 177.1460, 177.2420, 177.2600, 182.90, 182.1711, GRAS; USDA 9CFR §318.7; Japan approved (2% max. as anticaking), other restrictions; Europe listed; UK approved; FDA approved for orals, rectals, vaginals; USP/NF, BP, Ph.Eur. compliance

Manuf./Distrib.: 20 Microns Ltd.; Akzo Nobel; Byk-Chemie GmbH; Cabot Carbon Ltd; Catalysts & Chems. Ind.; Chisso Am.; Degussa; DuPont; Fluka; Geltech; J.M. Huber; Nippon Silica Ind.; Nissan Chem. Ind.; PPG Ind.; PQ; Spectrum Chem. Mfg.; Unimin

Trade names: A.F.S.; Flatting Agent HK125; Flatting Agent HK188; Flatting Agent HK400; Flatting Agent TK900; Flatting Agent TS100; Gasil® 200DF; Gasil® GM2; Gasil® HP 039; Gasil® HP 250; Gasil® HP 260; Gasil® HP 270; Gasil® HP 285; Gasil® IJ 1; Gasil® IJ 24; Gasil® IJ 35; Gasil® IJ 45; Imsil® 1240; Imsil® A-8; Imsil® A-30; Imsil® A-75; Min-U-Sil® 5; Min-U-Sil® 10; Min-U-Sil® 15; Min-U-Sil® 30; Min-U-Sil® 40; Nalco® 2302; Nalco® 2303; Nalco® 2309; Nalco® 2311; Nalco® 2321; Patcote® 801; Patcote® 802; Patcote® 803; Patcote® 806; Patcote® 811; Patcote® 812; Patcote® 813; Patcote® 814; Patcote® 841; Patcote® 849; Patcote® 868; Patcote® 8060; Rexfoam 150-A; Sil-Co-Sil® 40; Sil-Co-Sil® 45; Sil-Co-Sil® 47; Sil-Co-Sil® 49; Sil-Co-Sil® 51; Sil-Co-Sil® 52; Sil-Co-Sil® 53; Sil-Co-Sil® 63; Sil-Co-Sil® 75; Sil-Co-Sil® 90; Sil-Co-Sil® 106; Sil-Co-Sil® 125; Sil-Co-Sil® 250; Siltex™; Siltex™ 44; Siltex™ 44C; Silver Bond 30; Silver Bond 45; Silver Bond B; Tamsil 8; Tamsil 10; Tamsil 15; Tamsil 25; Tamsil 30; Tamsil 45; Tamsil 75; Tamsil 150; Tamsil Gold Bond; TS 100

Trade names containing: Advantage® 70DYX; Advantage® 91WW; Bubble Breaker® 776, 3017-A; Bubble Breaker® 3017-A; Bubble Breaker® 3056A; BYK®-2615; Cab-O-Sil® TS-530; Cab-O-Sil® TS-610; Cab-O-Sil® TS-720; DB-1 Defoamer; DB-12 Defoamer; Drewfax® S-600; Drewplus® HG-6020; Drewplus® L-108; Drewplus® L-123; Drewplus® L-131; Drewplus® L-139; Drewplus® L-175; Drewplus® L-191; Drewplus® L-198; Drewplus® L-405; Drewplus® L-410; Drewplus® L-418; Drewplus® L-419; Drewplus® L-421; Drewplus® L-424; Drewplus® L-435; Drewplus® L-464; Drewplus® L-468; Drewplus® L-474; Drewplus® L-475; Drewplus® L-475C; Drewplus® L-477; Drewplus® L-493; Drewplus® L-496; Drewplus® L-768; Drewplus® L-790; Drewplus® Y-166; Drewplus® Y-250; Drewplus® Y-281; Flat-Ayd® FA-3B; Flat-Ayd® FA-14C; Flat-Ayd® FA-HSP-2; Flat-Ayd® FA-NCO-6; Flat-Ayd® FA-W-34; Flatting Agent OK412; Flatting Agent OK500; Foamaster® DS; Gasil® HP 210; Gasil® HP 220; Gasil® HP 240J; Gasil® IJ 37; Hercules® 831 Defoamer; Hercules® 1512 Defoamer; Hercules® Eff-101 Defoamer; Lanco-Matt™ MC; Modarez MFP Powd.; Nalco® 1090; Nalco® 2300; Nalco® 2305; Nalco® 2314; Novakup®; Rhenocure® TP/S; Silikup®; Surfynol® 82S; Surfynol® 104S; SWS-290; SWS-295; SWS-299; TEGO® Airex 900; TEGO® Foamex 835; TEGO® Foamex 1495; TEGO® Foamex N; Tronox® CR-822

See also Diatomaceous earth; Quartz; Silica, amorphous; Silica, colloidal; Silica, fumed; Silica gel; Silica, hydrated

Silica aerogel. *See* Silica gel

Silica-alumina

Uses: Catalyst support; filler and flame retardant for plastics, epoxies; mastics, coatings (tin films, industrial topcoats, semigloss architectural paints), films, and other applics

Trade names: Zeeospheres® 200; Zeeospheres® 400; Zeeospheres® 600; Zeeospheres® 800; Zeeospheres® 850

Silica, amorphous

Synonyms: Amorphous silica

Properties: Sl. sol. in water

Silica, colloidal

Toxicology: Transient dermatitis which causes skin dehydration and loss of skin oils
Uses: Filler and reinforcing material in rubber and plastics; to improve ink retention on paper; as pigment and filler in paints and coatings; as an abrasive, adsorbent, dessicant and catalyst base
Manuf./Distrib.: Akzo Nobel; Archway Sales; Cabot/Cab-O-Sil; Chem-Materials; CR Mins.; Crosfield; Degussa; J.M. Huber/Chems.; Lenape Ind.; Lomas Int'l.; D.N. Lukens; Magnesium Elektron; Nalco; PPG Ind.; PQ; H.M. Royal; SCM; Seegott; Tamms Ind.; Tulco; Van Waters & Rogers; Whittaker, Clark & Daniels; Jesse S. Young
Trade names: Hi-Sil® T-600; Imsil® A-10; Imsil® A-15; Imsil® A-25; Lo-Vel® 27; Lo-Vel® 28; Lo-Vel® 29; Lo-Vel® 39A; Lo-Vel® 275; PumaSil; Zeomatt® 1855; Zeomatt® 1860
Trade names containing: Lo-Vel® 66; Lo-Vel® HSF; Zeomatt® 1869
See also Silica

Silica, colloidal
CAS 7631-86-9
Synonyms: Colloidal silica; Silica sol
Definition: A stable dispersion of discrete, colloid-size particles of amorphous silica in aq. sol'n.
Uses: Antislipping agent for paper and fiber binder for metal casting and refractory materials, catalyst carrier, reinforcing material for adhesives and paints
Manuf./Distrib.: Aldrich; Sigma
Trade names: Adelite; PST-1; PST-3; PST-5; Snowtex 20L; Snowtex 40; Snowtex 50; Snowtex C; Snowtex O; Snowtex OL; Snowtex S; Snowtex UP; Snowtex XL; Snowtex XS; Snowtex YL; Snowtex ZL
Trade names containing: Snowtex N
See also Silica; Silica, amorphous

Silica dimethyl silylate
CAS 60842-32-2
Uses: Anticaking agent, thickener, thixotrope, rheology aid for plastisol coating compds.; reinforcement for rubber sealants, plastics, adhesives
Trade names: Aerosil® R974

Silica, fumed
CAS 112945-52-5
Synonyms: Fumed silica
Definition: High surface area aggregate particles of silica
Empirical: SiO_2
Properties: Wh. dusty powd.; dens. ≈ 2 lb/ft^3 (unpacked); hydrophilic
Uses: Thickener, thixotropic and reinforcing agent in inks, resins, paints, and cosmetics; base material for high-temp. mortars
Manuf./Distrib.: Aldrich; Fluka; Sigma
Trade names: Aerosil® 200; Aerosil® 300; Aerosil® 380; Aerosil® OX50; Aerosil® R202; Aerosil® R805; Aerosil® R812; Aerosil® R812S; Aerosil® R972; Aerosil® TT600; Cab-O-Sil® EH-5; Cab-O-Sil® H-5; Cab-O-Sil® HS-5; Cab-O-Sil® L-90; Cab-O-Sil® LM-130; Cab-O-Sil® LM-150; Cab-O-Sil® M-5; Cab-O-Sil® M-7D; Cab-O-Sil® MS-55; Cab-O-Sil® PTG; Wacker HDK® H15; Wacker HDK® H18; Wacker HDK® H20; Wacker HDK® H30; Wacker HDK® N20; Wacker HDK® T30; Wacker HDK® T40
Trade names containing: Aerosil® COK 84
See also Silica

Silica gel
CAS 63231-67-4; EINECS 231-545-4
Synonyms: Silica aerogel
Definition: A regenerative adsorbent consisting of coherent, continuous three-dimensional network of spherical particles of colloidal silica
Empirical: O_2Si
Properties: Odorless; insol. in water; m.w. 60.09; sp.gr. 2.1
Toxicology: LD50 (mice) 8000 mg/kg; PEL 6 mg/m^3; nontoxic; avoid prolonged breathing of dust
Uses: Dehumidifying and dehydrating agent; used in drying compressed air and other gases, and liquids; anticaking agent in cosmetics and pharmaceuticals; in waxes to prevent slipping; in dietary supplements
Regulatory: FDA 21CFR §182.1711, GRAS; BATF 27CFR §240.1051; FDA approved for buccals, dentals, orals; BP compliance
Manuf./Distrib.: Brown; Crosfield; Magnesium Elektron; PQ; SCM; Sigma; Spectrum Chem. Mfg.; Zeochem
See also Silica; Silica, colloidal

Silica glass. *See* Quartz

Silica hydrate. *See* Silica, hydrated

Silica, hydrated
CAS 1343-98-2; EINECS 215-683-2
Synonyms: Silicic acid; Silicic acid hydrate; Silicic acid hydrated; Silica hydrate; Precipitated silica
Classification: Inorganic oxide
Definition: Occurs in nature as opal; jelly-like precipitate obtained when sodium silicate sol'n. is acidified
Formula: $SiO_2 \cdot xH_2O$, x varies with method of precipitation and extent of drying
Properties: Wh. amorphous powd. or lumps; insol. in water or acids except hydrofluoric; m.w. 60.08 + water
Toxicology: Eye irritant; poison by intravenous route; TLV:TWA 10 mg/m^3 (total dust)
Uses: Insecticide; for absorption of vapors; laboratory reagent; filler and reinforcing agent in rubber and plastics; prevents ink migration in paper coatings
Regulatory: FDA 21CFR §73.1, 160.105, 160.185, 172.480, 173.340, 175.105, 175.300, 176.170, 176.180, 176.200, 176.210, 177.1200, 177.2420, 177.2600, 182.90, 573.940
Manuf./Distrib.: Crosfield; Fluka; J.M. Huber; Magnesium Elektron; Osram Sylvania; PPG Ind.; PQ; Spectrum Chem. Mfg.
Trade names: FK 500LS; Sipernat® 22LS; Sylophobic; Sylysia 250; Sylysia 250N; Sylysia 310; Sylysia 320; Sylysia 350; Sylysia 370; Sylysia 430; Sylysia 440; Sylysia 450; Sylysia 470; Sylysia 530; Sylysia 540; Sylysia 550; Tixosil 311; Tixosil 321; Tixosil 331; Tixosil 333, 343; Tixosil 375; Tullanox HM-250; Zeothix® 95; Zeothix® 177; Zeothix® 265
Trade names containing: Sylysia 256; Sylysia 256N; Sylysia 358; Sylysia 435; Sylysia 436; Sylysia 445; Sylysia 446; Sylysia 456

Silica sol. *See* Silica, colloidal
Siliceous earth. *See* Diatomaceous earth
Silicic acid. *See* Silica, hydrated
Silicic acid, aluminum sodium salt, sulfurized. *See* Sodium alumino sulfo silicate
Silicic acid, calcium salt. *See* Calcium silicate
Silicic acid, disodium salt. *See* Sodium metasilicate
Silicic acid hydrate. *See* Silica, hydrated
Silicic acid hydrated. *See* Silica, hydrated
Silicic acid, potassium salt. *See* Potassium silicate
Silicic acid, sodium salt. *See* Sodium silicate
Silicic anhydride. *See* Silica

Silicon carbide
CAS 409-21-2; EINECS 206-991-8
Synonyms: Carbon silicide; Carborundeum; Carborundum; Silicon monocarbide; Silundum
Empirical: CSi
Formula: SiC
Properties: Bluish-blk. iridescent cryst.; m.w. 40.10; dens. 3.17; m.p. 2600 C; b.p. subl. > 2000 C; dec. @ 2210 C
Toxicology: ACGIH TLV:TWA 10 mg/m^3 (total dust); nuisance particulate; inh. may cause mild respiratory and eye irritation; fibers may cause changes in lungs similar to those caused by silica
Uses: Abrasive for cutting and grinding metals, refractory in nonferrous metallurgy, ceramic industry, boiler furnaces, composite tubes for steam reforming operations; fibrous form in filament-wound structures and heat-resistant, high-strength composites; raw material in paint mfg.
Manuf./Distrib.: Advanced Refractory Tech.; Aldrich; Atlantic Equip. Engrs.; Atomergic Chemetals; Carborundum; Dow Corning; Electro Abrasives; Fluka; Fujimi; Lonza; Mitsui Toatsu; New Metals & Chems.; Noah; Reade Advanced Materials; Showa Denko; Spectrum Chem. Mfg.

Silicon dioxide. *See* Quartz; Silica

Silicone
Synonyms: Organosiloxane
Classification: Siloxane polymers
Properties: Liq., semisolid, or solid; cis 1->1,000,000 cs; water repellant; sol. in most organic solvents
Precaution: Unhalogenated types combustible
Toxicology: Silicone-related diseases can occur, e.g., silicone-induced synovitis and lymphadenopathy, acute and chronic pneumonitis from 'bleeding' from ruptured bag-gel breast implants, pulmonary lesions, granulomatous reactions
Uses: Thermosetting siloxane polymers used as mold release for plastics and rubber, defoamers for mining, latex, ink, soaps, agric., food processing, as lubricants, conditioners, and emollients in personal care

prods.; molding compds., encapsulants

Manuf./Distrib.: Ashland; Bayer; Chemcentral; Clariant; Crucible; Dow Corning; GE Silicones; Genesee Polymers; Goldschmidt; Harcros; Hüls Am.; D.N. Lukens; Reichhold; Rhone-Poulenc; Seegott; Shin-Etsu; Tego GmbH

Trade names containing: Agitan 217; Agitan 218; Agitan 301; Agitan 700; Agitan P 801; Agitan VP 725; Agitan VP P 804; Calgene Antifoam CO-350; Calgene Antifoam F-2; Carboflow® 32W; Carboflow® 35S; Chemax DF-10, DF-10A; Chemax DF-30; Chemax DF-100; CMI 115-08; CoatOSil™ 1300; CoatOSil™ 1301; CoatOSil™ 1378; Colloid™ 1010; DB-1 Defoamer; DB-12 Defoamer; DB-19 Antifoam Compd.; Dislon® AQ-261; Disparlon® AQ-261; Disparlon® LC-915; Disparlon® LC-955; Dow Corning® 1-0418; Dow Corning® 1-0543; Dow Corning® 236 Dispersion; Dow Corning® Antifoam 1400; Drewfax® 412; Drewfax® 842; Drewfax® S-600; Drewplus® HG-6020; Drewplus® HG-6022; Drewplus® HG-6025; Drewplus® L-405; Drewplus® L-410; Drewplus® L-418; Drewplus® L-419; Drewplus® L-768; Drewplus® S-681; Edaplan LA 410; Edaplan LA 411; Edaplan LA 412; Edaplan LA 413; EXP-36-X20 Epoxy Functional Silicone Sol'n.; EXP-38-X20 Epoxy Functional Silicone Sol'n.; EXP-58 Silicone Wax; Foam Blast 10; Foam Blast 20; Foam Blast 28; Foam Blast 30; Foam Blast 187; Foam Blast 191; Foam Blast 240; Foamkill® 8R; Foamkill® 618; Foamkill® 639J; Foamkill® 639JOH; Foamkill® 639Q; Forbest 680; G623; GP-6 Silicone Fluid; GP-74 Paintable Silicone Fluid; GP-179 Silicone Resin; GP-197 Resin Sol'n.; GP-300-I Antifoam Compd.; GP-7200 Silicone Fluid; GP-RA-158 Silicone Polish Additive; GP-RA-159 Silicone Polish Additive; Hercules® 831 Defoamer; Hercules® 1512 Defoamer; Hercules® Eff-101 Defoamer; Konform® SR 2000; Masil® 260; Mazu® DF 200SX; Mazu® DF 200SXSP; Nalco® 1090; Nalco® 2300; Nalco® 2301; Nalco® 2305; Nalco® 2314; Patcote® 500; Patcote® 512; Patcote® 513; Patcote® 519; Patcote® 520; Patcote® 525; Patcote® 531; Patcote® 532; Patcote® 550; Patcote® 555; Patcote® 577; Patcote® 597; Patcote® 598; Rhodorsil® 1505; Rhodorsil® 10336; S 370; S 670; S 860; S 882; Sag 5440; Serdas GE 4050; Surfynol® DF-58; Surfynol® DF-60; Surfynol® DF-574; Surfynol® SM-740; Surfynol® SM-745; TEGO® Airex 975; Troysol™ 307; Union Carbide® R272

See also Dimethicone; Polysiloxane; Simethicone

Silicone acrylate

Uses: For release coatings to be cured by electron beam or uv-radiation in the paper and film industries

Trade names: TEGO® Silicone Acrylate RC

Silicone elastomer

CAS 63394-02-5

Synonyms: Polydimethylsiloxane rubber

Definition: Room-temp. vulcanizing silicone rubber

Properties: Sol. in org. solvs.; water repellant; low surf. tension; dens. 1.12 (20C)

Toxicology: Skin and eye irritant

Uses: Encapsulation of electronic parts, elec. insulation, gaskets, surgical membranes and implants; automotive engine components; miscellaneous mechanical parts

Manuf./Distrib.: Ambersil Ltd; Dow Corning; Dow Corning France; Fluka; GE Silicones Europe; Hardman; Hüls AG; Laur Silicone Rubber Compding.; Reliance Silicones; Wacker Silicones

Trade names: Laur 202

Silicone emulsions

Classification: Organosiloxane

Toxicology: No known toxicity

Uses: Lubricant, release agent for plastics and rubber; defoamer for paints, adhesives, polymerization processes; water repellent

Regulatory: FDA approved for orals, topicals; USP/NF compliance

Manuf./Distrib.: Atlas Refinery; Calgene; Clariant; CNC Int'l.; Crucible; Dow Corning; Genesee Polymers; Great Western; Harwick; Hickson Spec.; Hüls Am.; OSI Specialties; PPG Ind.; Ross Chem.; Soluol; Taylor; Wacker Silicones

Trade names: Agitan E 255; Agitan E 256; BYK®-023; BYK®-026; Calgene Antifoam FD-62; Calgene Antifoam FD-82; CNC Antifoam 30-FG; CNC Antifoam 495, 495-M; Dow Corning® 1-0468; Dow Corning® 1-0469; Dow Corning® 24 Emulsion; Dow Corning® 84; Dow Corning® 85; Dow Corning® 290 Paintable Release; Dow Corning® 1430 Emulsion; Dow Corning® 2210 Emulsion; Dow Corning® B Emulsion; Dow Corning® HV-490 Emulsion; Dow Corning® Antifoam 1410; Dow Corning® Antifoam 1430; Dow Corning® Antifoam 1520-US; Dow Corning® Antifoam FG-10; Dow Corning® Antifoam H-10; Drewplus® L-407; Drewplus® L-813; Drewplus® L-833; EXP-24-LS Silicone Wax Emulsion; Foam Blast 106; Foamex AD-100; Foamex AD-300; Foamkill® 400A; Foamkill® 679; Foamkill® 810; Foamkill® 830; Foamkill® 836A; Foamkill® 836B; Foamkill® CPD; Foamkill® MS-1; Foamkill® MS Conc.; GP-85-AE Silicone Emulsion; GP-86-AE Silicone Emulsion; GP-165 Silicone Resin Emulsion; GP-310-I Antifoam Emulsion; Kilfoam; Masil® EM 501; Mazu® DF 210SX; Mazu® DF 210SX Mod 1; Mazu® DF 210SXSP; Mazu® DF 230SX;

Mazu® DF 230SXSP; Mazu® DF 243; Michem® Emulsion 03230; Se 21; Se 23; Se 26; SM 2079; SRE; Surfynol® DF-695; TEGO® Foamex 1435; TEGO® Phobe 1050; TEGO® Phobe 1200; TEGO® Phobe 1300; TEGO® Phobe 1400; Veoceal® 1311; Veoceal® 2100
Trade names containing: Advantage® 70DYX; Advantage® 91WW; GP-210 Silicone Antifoam Emulsion; TEGO® Phobe 1000; TEGO® Phobe 1000 S
See also Dimethicone; Polysiloxane; Simethicone

Silicone glycol copolymer
CAS 68937-55-3; 68938-54-5
Uses: Surfactant for use in MDI-based polyurethane foam systems; mar resist. additive used for water-based, solv.-based, high solids and solventless org. paints, inks, and coatings; slip aid; enhances flow-out, wetting, and leveling; aids pigment disp.
Trade names: Dow Corning® 28; Dow Corning® 29; Dow Corning® 54; Silwet® L-7200; Silwet® L-7210; Silwet® L-7230; Silwet® L-7622
Trade names containing: Dow Corning® 11; Dow Corning® 14

Silicone hexaacrylate
Uses: Slip, wetting agent, flow improver for coatings on paper, plastics, and metal
Trade names: Ebecryl® 1360

Silicone-polyester
Uses: Resin for mfg. of paints
Manuf./Distrib.: Tego GmbH

Silicon monocarbide. *See* Silicon carbide
Silundum. *See* Silicon carbide

Simethicone
CAS 8050-81-5
Synonyms: α-(Trimethylsilyl)-ω-methylpoly[oxy(dimethylsilylene)], mixt. with silicon dioxide
Definition: Mixture of dimethicone with an avg. chain length of 200-350 dimethylsiloxane units and silica gel
Formula: $(CH_3)_3SiO[SI(CH_3)_2O]_nSi(CH_3)_3$, n = 200-350
Properties: Hazy translucent visc. fluid; sol. in chloroform, ether; insol. in water and alcohol
Toxicology: No known toxicity
Uses: Foam control agent, processing aid for foods, cosmetics, pharmaceuticals, fermentation, pesticide/fertilizer, paper/printing, textile, adhesives/coatings industries, in mfg. of plastics in contact with food, for degassing of monomers
Regulatory: Europe listed; UK approved; FDA approved for orals, rectals, topicals; USP/NF compliance
Manuf./Distrib.: PPG Ind.
Trade names: Calgene Antifoam F-1; Dow Corning® Antifoam C Emulsion

SLS. *See* Sodium lauryl sulfate
SMA. *See* Styrene/MA copolymer

Smectite
Uses: Antisettling agent for solv.-based coatings; for trade sales and industrials
Trade names: Bentone® 14; Claytone® 38H; Claytone® 50; Claytone® 2000; Claytone® ED; Claytone® LMW; Claytone® PS-3; Claytone® SO

SMO. *See* Sorbitan oleate
SMS. *See* Sorbitan stearate
Sn-II-ethylhexanoate. *See* Stannous octoate
Soap clay. *See* Bentonite
Soda lye. *See* Sodium hydroxide
Soda niter. *See* Sodium nitrate

Sodium 2-acrylamido-2-methylpropanesulfonate
CAS 5165-97-9
Synonyms: 2-Acrylamido-2-methylpropane sulfonic acid sodium salt; 2-Methyl-2-((1-oxo-2-propenyl)-amino)-1-propanesulfonic acid, sodium salt
Empirical: $C_7H_{13}NO_4S \cdot Na$
Properties: M.w. 230.23

Sodium alginate

Uses: Used in monomer polymerization; as a co-monomer where polarity or water sensitivity is desirable
Trade names: AMPS® 2403; AMPS® 2405; Lubrizol® 2405 Monomer

Sodium alginate. *See* Algin

Sodium allyloxy hydroxypropyl sulfonate
Synonyms: Sodium 1-allyloxy-2-hydroxypropyl sulfonate
Formula: CH_2=$CHCH_2OCH_2CHOHCH_2SO_3Na$
Properties: M.w. 218
Uses: Monomer for emulsion polymerization; copolymerizable surfactant/stabilizer; improves latex props.; for paints, adhesives, paper and textile coatings
Manuf./Distrib.: Rhone-Poulenc N. Am.
Trade names: Sipomer® COPS 1

Sodium 1-allyloxy-2-hydroxypropyl sulfonate. *See* Sodium allyloxy hydroxypropyl sulfonate
Sodium aluminosilicate. *See* Sodium silicoaluminate

Sodium alumino sulfo silicate
CAS 101357-30-6; EINECS 309-928-3
Synonyms: Silicic acid, aluminum sodium salt, sulfurized
Manuf./Distrib.: Holliday Pigments
Trade names containing: Ultramarine Blue

Sodium aluminum silicate. *See* Sodium silicoaluminate

Sodium bentonite
Uses: Suspending agent, gellant, binder for household prods., automotive prods., aerosols, paints, enamels, personal care prods.
Trade names: Optigel CK; Optigel LX; Optigel WA; Optigel WX; Organotrol; Organotrol 1665; Organotrol 2200; Organotrol 3330; Organotrol 4540; Organotrol 9080; Organotrol CF; Organotrol RJ; Volclay® HPM-20; Volclay® MPS-1

Sodium benzoate
CAS 532-32-1; EINECS 208-534-8
Synonyms: Benzoic acid sodium salt; Benzoate of soda; Benzoate sodium
Definition: Sodium salt of benzoic acid
Empirical: $C_7H_5NaO_2$
Formula: C_6H_5COONa
Properties: Wh. gran. or cryst. powd., odorless, sweetish astringent taste; hygroscopic; sol. in water; sparingly sol. in alcohol; m.w. 144.11; pH $\approx$ 8; stable in air
Precaution: Combustible when exposed to heat or flame; incompat. with acids, ferric salts; heated to decomp., emits toxic fumes of Na_2O
Toxicology: LD50 (oral, rat) 4.07 g/kg; poison by subcutaneous, IV routes; mod. toxic by ingestion, IP routes; may cause human intolerance reaction, asthma, rashes, hyperactivity; experimental teratogen, reproductive effects; mutagenic data
Uses: Fungicide; preservative in pharmaceuticals and foods, esp. in sl. acidic media; clinical reagent (bilirubin assay); nucleating agent for plastics
Usage level: 0.1% (parenterals), 0.08% (dentifrices), 4.75-5% (injectables), 0.02-0.5% (orals)
Regulatory: FDA 21CFR §146.152, 146.154, 150.141, 150.161, 166.40, 166.110, 181.22, 181.23, 184.1733; GRAS; USDA 9CFR §318.7; BATF 27CFR §240.1051; EPA reg.; FEMA GRAS; Japan approved with limitations; Europe listed 0.5% as acid; FDA approved for dentals, orals, rectals, topicals; USP/NF, BP, Ph.Eur. compliance
Manuf./Distrib.: Aceto; Aldrich; Allchem Ind.; Fluka; Jan Dekker BV; R.W. Greeff; Haarmann & Reimer; Int'l. Sourcing; Lohmann; Mallinckrodt; E. Merck; Pfizer Food Science; San Yuan; Sigma; Tri-K Ind.
Trade names containing: Monawet MO-85P

Sodium bicarbonate
CAS 144-55-8; EINECS 205-633-8
Synonyms: Baking soda; Sodium hydrogen carbonate; Bicarbonate of soda; Carbonic acid monosodium salt; Monosodium carbonate
Classification: Inorganic salt
Empirical: $CHNaO_3$
Formula: $NaHCO_3$

Properties: Wh. powd. or cryst. lumps, sl. alkaline taste; sol. in water; insol. in alcohol; m.w. 84.01; dens. 2.159; stable in dry air, slowly decomp. in moist air
Precaution: Moisture-sensitive
Toxicology: Harmless to the skin; leaves an alkaline residue that may cause irritation; nuisance dust; excessive ingestion of sodium may be detrimental to certain persons
Uses: Mfg. of effervescent salts and beverages, artificial mineral water; prevention of timber mold, cleaning preparations, lab reagent, antacid, mouthwash, OTC drug active; blowing and nucleating agent for thermoplastics
Regulatory: FDA 21CFR §137.180, 137.270, 163.110, 173.385, 182.1736, 184.1736, GRAS; USDA 9CFR §318.7, 381.147; Japan approved; Europe listed; FDA approved for injectables, parenterals, ophthalmics, orals; USP/NF, BP, Ph.Eur. compliance
Manuf./Distrib.: Aldrich; Allchem Ind.; Balchem; Captree; Church & Dwight; Coyne; EM Ind.; Farleyway Chem. Ltd; Fluka; FMC; ICI Spec. Chems.; Kraft Chem.; Lohmann; Rhone-Poulenc Basic; Sigma
Trade names containing: Dowicil® 75

Sodium 1,4-bis(2-ethylhexyl) sulfosuccinate. *See* Dioctyl sodium sulfosuccinate
Sodium bistridecyl sulfosuccinate. *See* Ditridecyl sodium sulfosuccinate

Sodium borate
CAS 1330-43-4 (anhyd.); EINECS 215-540-4
Synonyms: Sodium tetraborate anhyd.; Sodium pyroborate anhyd.; Borax, fused
Classification: Inorganic salt
Empirical: $B_4Na_2O_7$
Formula: $Na_2B_4O_7$
Properties: Colorless to wh. cryst. or powd., odorless; hygroscopic; sol. in water, glycerol; insol. in alcohol; m.w. 201.22; dens. 1.730
Toxicology: Irritant
Uses: Heat-resistant glass, porcelain enamel, detergents, herbicides, fertilizers, rust inhibitors, pharmaceuticals, leather, photography, bleaches, paint, boron compounds, flame-retardant fungicide for wood, soldering flux, cleaners, lab reagent
Regulatory: FDA 21CFR §175.105, 175.210, 176.180, 177.2800, 181.22, 181.30; FDA approved for ophthalmics; USP/NF, BP compliance
Manuf./Distrib.: Aldrich; Ashland; Coyne; Fluka; Sigma; Spectrum Chem. Mfg.; U.S. Borax

Sodium borohydride
CAS 16940-66-2; EINECS 241-004-4
Synonyms: SBH; Sodium tetrahydroborate; Sodium tetrahydridoborate
Empirical: BH_4Na
Formula: $NaBH_4$
Properties: Wh. cryst. powd.; sol. in water, ammonia, amines, pyridine, diethylene glycol dimethyl ether; insol. in other ethers, hydrocarbons, alkyl chlorides; hygroscopic; stable in dry air to 300 C; dec. in moist air; m.w. 37.83; m.p. 37 C
Precaution: Flammable; dangerous fire risk; store out of contact with water; do not store in glass containers
Toxicology: Corrosive, toxic; reacts with water to evolve hydrogen and sodium hydroxide
Uses: Source of H_2 and other borohydrides; bleaching wood pulp; blowing agent for plastics; process stream purification; color purification in epoxy resins, plasticizers, curing amines for coatings, laminates, adhesives, elec. coatings, tooling, casting, and molding resins; rec. for storage tank purification
Manuf./Distrib.: AC Ind.; Aldrich; Alfa Aesar Johnson Matthey; Atomergic Chemetals; Charkit; Fluka; Morton Int'l.; Sigma; Tennant Trading Ltd
Trade names: VenPure® AF Caplets; VenPure® AF Gran.; VenPure® Powd.
Trade names containing: VenPure® Sol'n.

Sodium borosilicate
Uses: Extender/filler for polymers and barrier coatings
Trade names: Eccosphere® IG-101; Eccosphere® IGD-101

Sodium capryl sulfate. *See* Sodium octyl sulfate
Sodium carboxymethylcellulose. *See* Carboxymethylcellulose sodium

Sodium cetyl sulfate
CAS 1120-01-0; EINECS 214-292-4
Synonyms: 1-Hexadecanol, hydrogen sulfate, sodium salt
Definition: Sodium salt of cetyl sulfate

Sodium chromate

Empirical: $C_{16}H_{34}O_4S•Na$
Formula: $CH_3(CH_2)_{14}CH_2OSO_3Na$
Uses: Emulsifier, detergent, flotation agent; collector in ore flotation; softener/lubricant for textiles; cosmetics and toiletries; paints
Trade names: Rhodapon® EC111

Sodium chromate
CAS 7775-11-3; EINECS 231-889-5
Synonyms: Sodium chromate (VI) (nonradioactive)
Empirical: $CrNa_2O_4$
Formula: Na_2CrO_4
Properties: M.w. 161.97
Toxicology: Irritating to eyes, skin, respiratory system; may cause cancer
Uses: Inks, dyeing, paint pigment, leather tanning, other chromates, protection of iron against corrosion, wood preservative
Manuf./Distrib.: Aldrich; Allan; Ashland; Elf Atochem N. Am.; Fluka; Occidental

Sodium chromate (VI) (nonradioactive). *See* Sodium chromate
Sodium CMC. *See* Carboxymethylcellulose sodium

Sodium C14-16 olefin sulfonate
CAS 68439-57-6; EINECS 270-407-8
Definition: Mixt. of long chain sulfonate salts from sulfonation of C14-16 alpha olefins; consists of sodium alkene sulfonates and sodium hydroxyalkane sulfonates
Uses: Detergent, emulsifier, visc. builder, foaming agent for personal care, commercial and industrial formulations; emulsion polymerization; in paints and coatings
Regulatory: FDA 21CFR §175.105, 178.3406
Trade names: Polystep® A-18; Witconate™ AOS-EP

Sodium cumenesulfonate
CAS 32073-22-6; EINECS 250-913-5; 248-938-7
Synonyms: (1-Methylethyl)benzene, monosulfo deriv., sodium salt
Classification: Substituted aromatic compd.
Empirical: $C_9H_{11}O_3S • Na$
Uses: Hydrotrope, solubilizer, coupling agent, sol'n. aid for liq. detergents and slurries
Manuf./Distrib.: Ashland; Hüls AG; Ruetgers-Nease
Trade names: Eltesol® SC 40; Eltesol® SC 93; Eltesol® SC Pellets

Sodium decyl sulfate
CAS 142-87-0; 84501-49-5
Empirical: $C_{10}H_{21}O_4S • Na$
Properties: M.w. 260.36
Toxicology: Poison by intravenous, intraperitoneal routes; moderately toxic by ingestion
Uses: Emulsifier, dispersant, detergent, and wetting agent for industrial and institutional cleansers, plastics, mfg. of pigments, alkaline cleansers; dust suppression
Trade names: Polystep® B-25

Sodium dibutyl naphthalene sulfonate
CAS 25417-20-3
Uses: Dispersant, wetting agent for pesticides, industrial cleaning, textiles, dyeing, leather, paper, dyes, rubber, latex polymerization
Trade names: Supragil® NK

Sodium di(2-ethylhexyl) sulfosuccinate. *See* Dioctyl sodium sulfosuccinate
Sodium dihexyl sulfosuccinate. *See* Dihexyl sodium sulfosuccinate
Sodium diisobutyl sulfosuccinate. *See* Diisobutyl sodium sulfosuccinate

Sodium diisopropyl naphthalene sulfonate
Uses: Wetting agent for pesticide wettable powds.; dispersant; latex stabilizer; prevents coagulation in SBR and other elastomers; rewetting agent; humectant
Trade names: Supragil® WP

Sodium dimethyldithiocarbamate
CAS 128-04-1 (hydrate); EINECS 204-876-7 (hydrate)
Synonyms: SDDC; Sodium N,N-dimethyl dithiocarbamate; Dimethyldithiocarbamic acid sodium salt; Methyl namate
Empirical: $C_3H_6NS_2$ • Na (anhyd.), $C_3H_6NNaS_2$ • aq. (hydrate)
Formula: $(CH_3)_2NCS_2Na$
Properties: Crystals; m.w. 143.19 + aq.; dens. 1.1 (20/20 C); m.p. 95 C
Toxicology: LD50 (oral, rat) 1000 mg/kg; moderately toxic by ingestion, intraperitoneal and subcutaneous routes; mutagenic data; heated to decomp., emits toxic fumes of NO_x, SO_x, and Na_2O
Uses: Pesticide, fungicide, corrosion inhibitor, rubber accelerator
Regulatory: FDA 21CFR §173.320 (3 ppm max.)
Manuf./Distrib.: Complex Quimica SA; Fluka; Uniroyal; R.T. Vanderbilt
Trade names containing: Amerstat® 274

Sodium N,N-dimethyl dithiocarbamate. *See* Sodium dimethyldithiocarbamate

Sodium dinaphthalene methane sulfonate
Uses: Dispersant, protective colloid, stabilizer of nat. and syn. elastomers; pigment grinding aid for paints
Trade names: Rhodacal® RM/77-D

Sodium dinonyl sulfosuccinate
CAS 63217-13-0; EINECS 221-820-7
Uses: Wetting and rewetting agent used in dyeing, drycleaning detergents, emulsion polymerization, firefighting, textile processing, latex paints
Trade names: Geropon® WS-25, WS-25-I

Sodium dioctyl sulfosuccinate. *See* Dioctyl sodium sulfosuccinate
Sodium dithionite. *See* Sodium hydrosulfite
Sodium ditridecyl sulfosuccinate. *See* Ditridecyl sodium sulfosuccinate

Sodium dodecylbenzenesulfonate
CAS 25155-30-0; 68081-81-2; 85117-50-6; EINECS 246-680-4
Synonyms: SDBS; NaDBS; Sodium lauryl benzene sulfonate; Dodecylbenzenesulfonic acid sodium salt; Dodecylbenzene sodium sulfonate
Classification: Substituted aromatic compd.
Empirical: $C_{18}H_{29}O_3S$ • Na
Properties: Wh. to lt. yel. flakes, granules, or powd.; m.w. 348.52
Precaution: Combustible
Toxicology: LD50 (oral, rat) 1260 mg/kg, (oral, mouse) 2 g/kg, (IV, mouse) 105 mg/kg; poison by intravenous route; moderately toxic by ingestion; skin and severe eye irritant; heated to decomp., emits toxic fumes of Na_2O
Uses: Anionic detergent, emulsifier, wetting agent for emulsion polymerization, industrial cleaning, cosmetics, textile and laundry, paints, etc.
Regulatory: FDA 21CFR §173.315, 175.105, 175.300, 175.320, 176.210, 177.1010, 177.1200, 177.1630, 177.2600, 177.2800, 178.3120, 178.3130, 178.3400; USDA 9CFR §318.7, 381.147; FDA approved for topicals
Manuf./Distrib.: Albright & Wilson Am.; Aldrich; Ashland; DuPont; Emkay; Fluka; Norman, Fox; Pilot; Rhone-Poulenc Spec. Chem.; Sigma; Spectrum Chem. Mfg.; Stepan; Tokyo Kasei Kogyo; Unger Fabrikker AS; Witco/Oleo-Surf.
Trade names: Calsoft F-90; Polystep® A-7; Polystep® A-15; Polystep® A-16; Polystep® A-16-22; Polystep® LAS-50; Rhodacal® DS-4; Rhodacal® DS-10; Rhodacal® LDS-22

Sodium dodecyl sulfate. *See* Sodium lauryl sulfate
Sodium 2-ethylhexyl sulfate. *See* Sodium octyl sulfate

Sodium fluoride
CAS 7681-49-4; EINECS 231-667-8
Synonyms: Sodium monofluoride; Sodium hydrofluoride
Empirical: FNa
Formula: NaF
Properties: Wh. powd. or colorless cryst., odorless; sol. 1 in 25 of water; pract. insol. in alcohol; m.w. 41.99; dens. 2 (41 C); m.p. 993 C; b.p. 1700 C
Precaution: Corrosive; heated to decomp., emits toxic fumes of F^- and Na_2O

Toxicology: LD50 (oral, rat) 52 mg/kg; ACGIH TLV:TWA 2.5 mg (F)/m^3; human poison by ingestion; human systemic effects by ing. and intradermal routes; corrosive irritant to skin, eyes, mucous membranes; mutagenic data; phytotoxic

Uses: Fluoridation of municipal water, degassing steel, wood preservative, insecticide, fungicide, rodenticide, chemical cleaning, electroplating, glass mfg., vitreous enamels, preservative for adhesives, toothpastes, disinfectant, dental prophylaxis

Regulatory: USP, BP, Ph.Eur. compliance

Manuf./Distrib.: Advance Research Chems.; Aldrich; Allchem Ind.; Am. Int'l.; AMRESCO; Cerac; Coyne; EM Ind.; Fluka; General Chem.; Hoechst Celanese; Sigma; Solvay GmbH; Spectrum Chem. Mfg.; Whiting, Peter Ltd

Sodium formaldehyde sulfoxylate

CAS 149-44-0; 6035-47-8 (dihydrate); EINECS 295-739-4

Synonyms: Methanesulfinic acid, hydroxy-, monosodium salt; Monosodium hydroxymethane sulfinate

Empirical: CH_3NaO_3S or $CH_3NaO_3S \cdot 2H_2O$

Formula: $HOCH_2SOONa$ or $HOCH_2SOONa \cdot 2H_2O$

Properties: Wh. cryst., garlic odor; sol. in water; sl. sol. in alcohol, ether, chloroform, benzene; m.w. 118.09 (anhyd.), 154.11 (dihydrate); pH 9.5-10.5 (2%)

Toxicology: Irritating to respiratory system

Uses: Catalyst/polymerization regulator for plastics; antioxidant in pharmaceuticals; paint raw material

Regulatory: FDA approved for parenterals, topicals; USP/NF compliance

Manuf./Distrib.: Aceto; Aldrich; Fluka; Henkel/Coatings & Inks; Passaic Color & Chem.; Phibrochem; Royce Assoc.

Sodium gluconate

CAS 527-07-1; EINECS 208-407-7

Synonyms: D-Gluconic acid monosodium salt; Monosodium D(-)-pentahydroxy capronate; Gluconic acid sodium salt; Monosodium gluconate; Sodium d-gluconate

Definition: Sodium salt of gluconic acid

Empirical: $C_6H_{12}NaO_7$

Formula: $HOCH_2CHOHCHOHCHOHCHOHCOONa$

Properties: Wh. to ylsh. cryst. powd., pleasant odor; sol. 59 g/100 ml water; sl. sol. in alcohol; insol. in ether; m.w. 219.17

Precaution: Avoid dust formation

Toxicology: Low toxicity by IV route; heated to decomp., emits acrid smoke and irritating fumes

Uses: As sequestering agent; in metal plating, tanning of hides, mordants for fabrics, paints; rust remover; mineral source in foods and pharmaceuticals

Regulatory: FDA 21CFR §182.6757, GRAS; Europe listed; UK approved; FDA approved for orals

Manuf./Distrib.: Akzo Nobel; Albright & Wilson Am.; Aldrich; Alfa; Am. Biorganics; Faesy & Besthoff; Fluka; Int'l. Chem. Inc.; Lohmann; Pfizer Int'l.; PMP Fermentation Prods.; Rit-Chem; Sigma

Sodium d-gluconate. *See* Sodium gluconate

Sodium hexyl diphenyloxide disulfonate

Uses: Solubilizer, surfactant for cleaning, emulsion polymerization, latex mfg., paints, adhesives, min. and metal processing, textiles

Trade names: Dowfax C6L; Dowfax XDS 8292.00

Sodium hydrate. *See* Sodium hydroxide
Sodium hydrofluoride. *See* Sodium fluoride
Sodium hydrogen carbonate. *See* Sodium bicarbonate

Sodium hydrosulfite

CAS 7775-14-6; 16721-80-5; EINECS 231-890-0

Synonyms: Disodium dithionite; Sodium dithionite; Sodium sulfoxylate; Sodium hyposulfite; Sodium hydrosulphite

Classification: Inorganic salt

Empirical: $Na_2O_4S_2$

Formula: NaO_2SSO_2Na

Properties: Wh. or yel.-wh. large transparent cryst., bitter taste; sol. in water; insol. in alcohol, conc. HCl; m.w. 174.10; dens. 2.189; dec. 267 C; m.p. 55C

Precaution: DOT: Flamm. solid; spontaneously combustible; ignites on contact with water or sodium chlorite; decomp. violently when heated to 190 C, emitting toxic fumes of SO_x and Na_2O

Toxicology: Toxic, irritant, allergen
Uses: Chemical reagent for the reduction of aldehydes and ketones to alcohols; vat dying of fibers and textiles; bleaching sugar, soap, oils; oxygen scavenger for synthetic rubbers; paint raw material
Regulatory: FDA 21CFR §177.2800, 182.90; Japan approved (0.03-5 g/kg residual as sulfur dioxide), not permitted in certain foods
Manuf./Distrib.: Aldrich; Allchem Ind.; Brown; Burlington Chem.; Coyne; Farleyway Chem. Ltd; Fluka; Henkel/ Coatings & Inks; Hoechst Celanese; Int'l. Chem. Inc.; LaRoche Ind.; Mitsubishi Gas; Monomer-Polymer & Dajac Labs; Morton Int'l.; Nissan Chem. Ind.; Olin; Sigma; Vulcan

Sodium hydrosulphite. *See* Sodium hydrosulfite

Sodium hydroxide
CAS 1310-73-2; EINECS 215-185-5
Synonyms: Caustic soda; Sodium hydrate; White caustic; Soda lye; Lye
Classification: Inorganic base
Empirical: HNaO
Formula: NaOH
Properties: Wh. deliq. solid beads or pellets; deliq.; absorbs water and CO_2 from air; sol. in water, alcohol, glycerin; m.w. 40; dens. 2.12 (20/4 C); m.p. 318 C; b.p. 1390 C
Precaution: DOT: Corrosive material; strong base; incompat. with acids and water; may ignite or react violently with many org. compds.; dangerous material to handle
Toxicology: TLV:Cl 2 mg/m^3 of air; LD50 (IP, mouse) 40 mg/kg; poison by IP route; mod. toxic by ingestion; mutagenic data; corrosive irritant to eye, skin, mucous membrane; mists and dusts cause small burns; heated to decomp., emits toxic fumes of Na_2O
Uses: Chemical mfg., rayon, cellophane, neutralizing agent in petrol. refining; textile processing; pulp, paper, soaps; vegetable oil refining; reclaiming rubber; food additive; etching and electroplating
Regulatory: FDA 21CFR §163.110, 172.560, 172.814, 172.892, 173.310, 184.1763, GRAS; USDA 9CFR §318.7, 381.147; BATF 27CFR §21.101, 240.1051a; Japan restricted; Europe listed; UK approved; FDA approved for injectables, dentals, parenterals, inhalants, ophthalmics, orals, rectals, topicals, vaginals; USP/NF, BP, Ph.Eur. compliance
Manuf./Distrib.: Akzo Nobel; Aldrich; Asahi Chem. Ind.; Asahi Denka Kogyo; Elf Atochem N. Am.; Fluka; Georgia Gulf; Georgia-Pacific Chem.; Hüls AG; ICI Spec. Chems.; Nissan Chem. Ind.; Norsk Hydro AS; Occidental; Olin; PPG Ind.; Rasa Ind.; Sigma; Spectrum Chem. Mfg.
Trade names containing: VenPure® Sol'n.

Sodium hydroxymethylaminoacetate. *See* Sodium hydroxymethylglycinate

Sodium hydroxymethylglycinate
CAS 70161-44-3; EINECS 274-357-8
Synonyms: Sodium hydroxymethylaminoacetate
Classification: Sodium salt of a substituted amino acid
Empirical: $C_3H_6NO_3Na$
Formula: $HOCH_2NHCH_2COONa$
Properties: Clear liq.; m.w. 127.10; dens. 1.28-1.30; pH 10.0-12.0
Uses: Antimicrobicide for water-based applics. (adhesives, coatings, liq. detergents, dishwashing liq., cleaning prods., water-based inks, etc.)
Regulatory: USA permitted; Europe listed; Japan not approved
Trade names: Integra™ 44

Sodium hyposulfite. *See* Sodium hydrosulfite

Sodium isodecyl sulfate
CAS 68299-17-2
Uses: Wetting agent, emulsifier, detergent, foamer, rinse aid; post-stabilizer in latex paints; metal treatment; textile and plywood mfg.; visc. control in plywood mfg.; fruit and veg. washing; hard surface cleaners; for emulsion polymerization
Trade names: Rhodapon® CAV

Sodium isopropyl naphthalene sulfonate
CAS 1322-93-6
Synonyms: Naphthalenesulfonic acid, bis-(1-methylethyl)-, sodium salt
Empirical: $C_{16}H_{20}O_3SNa$
Properties: M.w. 315.39

Sodium laureth sulfate

Uses: Emulsifier, dispersant and wetting agent; used in alkaline cleaning formulations; antigelling agents, automotive radiator cleaners, metal, cement, brick, and tile cleaners for crystal growth control; plastics, rubber; latex stabilization
Trade names: Rhodacal® BA-77

Sodium laureth sulfate
CAS 1335-72-4; 3088-31-1; 9004-82-4 (generic); 13150-00-0; 15826-16-1; 68891-38-3; 68585-34-2; EINECS 221-416-0
Synonyms: Sodium lauryl ether sulfate (n=1-4); PEG (1-4) lauryl ether sulfate, sodium salt
Definition: Sodium salt of sulfated ethoxylated lauryl alcohol
Formula: $CH_3(CH_2)_{10}CH_2(OCH_2CH_2)_nOSO_3Na$, avg. n = 1-4
Uses: Foam stabilizer, detergent, flash foamer, wetter for detergent systems, personal care prods.; emulsifier for emulsion polymerization (vinyl acetate, acrylates, methacrylates, styrene, butadiene); for paper coatings, textile coatings, paints, adhesives, industrial coatings; shampoo base
Regulatory: FDA approved for topicals
Manuf./Distrib.: Aquatec Quimica SA; Chemron; Clariant; Int'l. Chem. Inc.; Lonza; Norman, Fox; Pilot; Rhone-Poulenc Surf. & Spec.; Sea-Land; Stepan; Unger Fabrikker AS; U.S. Synthetics; Vista; Witco/Oleo-Surf.
Trade names: Abex® 23S; Polystep® B-12

Sodium lauroamphoacetate
CAS 14350-96-0; 68647-44-9; 26837-33-2; 68298-21-5; 68608-66-2; EINECS 271-949-8; 269-547-2; 238-305-8
Synonyms: 1H-Imidazolium, 1-(carboxymethyl)-4,5-dihydro-1-(2-hydroxyethyl)-2-undecyl-, hydroxide, sodium salt; Lauroamphoacetate; Lauroamphoglycinate
Classification: Amphoteric organic compd.
Empirical: $C_{18}H_{36}N_2O_4 \cdot Na$
Uses: Detergent, wetting and foaming agent, sequestrant, emulsifier, dispersant, germicidal; paint emulsifier
Trade names: Dehyton® PG; Miranol® HM Conc.

Sodium lauryl benzene sulfonate. *See* Sodium dodecylbenzenesulfonate
Sodium lauryl ether sulfate (n=1-4). *See* Sodium laureth sulfate

Sodium lauryl sulfate
CAS 151-21-3; 68585-47-7; 68955-19-1; EINECS 205-788-1; 271-557-7; 273-257-1
Synonyms: SDS; SLS; Sulfuric acid monododecyl ester sodium salt; Sodium monododecyl sulfate; Sodium dodecyl sulfate; Dodecyl sodium sulfate; Dodecylsulfate sodium salt
Definition: Sodium salt of lauryl sulfate
Empirical: $C_{12}H_{25}O_4S \cdot Na$
Formula: $CH_3(CH_2)_{10}CH_2OSO_3Na$
Properties: Wh. to cream crystals, flakes, or powd., faint fatty odor; sol. in water; m.w. 288.38; m.p. 204-207 C
Precaution: Burns above 93.3 C; heated to decomp., emits toxic fumes of SO_x and Na_2O
Toxicology: LD50 (oral, rat) 1288 mg/kg; poison by IV, IP routes; mod. toxic by ingestion; human skin irritant; experimental eye, severe skin irritant; mild allergen; mutagenic data; experimental teratogen, reproductive effects
Uses: Anionic detergent; surface tension depressant; emulsifier for fats; wetting agent; in textile industry, in toothpastes; emulsion polymerization (vinyl chloride, S/B, acrylic), paints, coatings
Usage level: 0.004-0.6 mg (oral solids); 0.01-0.02% (oral liqs.); 0.1-12.7% (topical pharmaceuticals)
Regulatory: FDA 21CFR §172.210, 172.822, 175.105, 175.300, 175.320, 176.170, 176.210, 177.1200, 177.1210, 177.1630, 177.2600, 177.2800, 178.1010, 178.3400; USDA 9CFR §318.7, 381.147; FDA approved for dentals, orals, topicals, vaginals; USP/NF, BP, Ph.Eur. compliance
Manuf./Distrib.: Accurate Chem. & Scientific; Albright & Wilson UK; Aldrich; Am. Biorganics; Chemron; Clariant; DuPont; Fluka; Henkel/Emery; Lonza; Monomer-Polymer & Dajac Labs; Norman, Fox; Pilot; Rhone-Poulenc Surf. & Spec.; Sea-Land; Sigma; Spectrum Chem. Mfg.; Stepan; Unger Fabrikker AS; U.S. Synthetics; Witco/Oleo-Surf.
Trade names: Polystep® B-3; Polystep® B-5; Polystep® B-24; Rhodapon® LCP; Rhodapon® LSB; Rhodapon® SM; Texapon® Z; Ufarol Na-30

Sodium magnesium aluminosilicate
Uses: Filler for the rubber industry; TiO_2 extender for paints
Trade names: Zeolex® 94HP

Sodium magnesium fluorosilicate
Trade names containing: Laponite® S

Sodium MBT. *See* Sodium 2-mercaptobenzothiazole

Sodium 2-mercaptobenzothiazole
CAS 2492-26-4
Synonyms: NaMBT; Sodium MBT; 2-Mercaptobenzothiazole sodium salt
Empirical: $C_7H_4NS_2 \cdot Na$
Properties: M.w. 189.23
Toxicology: LD50 (oral, rat) 3120 mg/kg; moderately toxic by ingestion; heated to decomp., emits very toxic fumes of NO_x, SO_x, and Na_2O
Uses: Preservative and mildew inhibitor, for paper industry, textile industry, food pkg. adhesives; chemical intermediate; corrosion inhibitor; metal deactivator; protects starch and casein coatings from bacterial or fungal activity
Regulatory: FDA 21CFR §181.30
Manuf./Distrib.: Allchem Ind.; Uniroyal; R.T. Vanderbilt
Trade names: Troysan® 28

Sodium metasilicate
CAS 6834-92-0; EINECS 229-912-9
Synonyms: Silicic acid, disodium salt; Sodium metasilicate anhydrous
Classification: Inorganic salt
Empirical: Na_2O_3Si
Formula: Na_2SiO_3 (anhyd.), $Na_2SiO_3 \cdot 5H_2O$ (pentahydrate), $Na_2SiO_3 \cdot 9H_2O$ (nonahydrate)
Properties: Anhyd.: White gran., dustless; sol. in water; insol. in alcohol, acids, salt sol'ns.; m.w. 122.07 (anhyd.); dens. 2.614; m.p. 1089 C; pH 12.6 (1%); ref. index 1.520 (glass, 25 C)
Precaution: Strongly alkaline; caustic material
Toxicology: LD50 (oral, rat) 1280 mg/kg; poison by ingestion, intraperitoneal routes; severe eye, skin, mucous membrane irritant; experimental reproductive effects; ingestion causes GI tract upset; heated to decomp., emits toxic fumes of Na_2O
Uses: Cosmetics, laundry, dairy, and metal cleaning as soap builder and detergent, bleaching agent; as flocculant and dispersant in metallurgy and mining; corrosion inhibitor protecting metal surfs. and org. finishes such as paints or ceramic glazes
Regulatory: FDA 21CFR §173.310, 184.1769a, GRAS; USDA 9CFR §318.7
Manuf./Distrib.: Aldrich; Am. Int'l.; Coyne; Crosfield; Eka Nobel AB; Fluka; Harcros; Occidental; PQ; Rhone-Poulenc Basic; Veckridge
Trade names: Metso Beads® 2048

Sodium metasilicate anhydrous. *See* Sodium metasilicate

Sodium metasilicate pentahydrate
CAS 10213-79-3; EINECS 229-912-9
Empirical: $Na_2O_3Si \cdot 5H_2O$
Formula: $Na_2OSiO_2 \cdot 5H_2O$
Properties: Wh. gran.; m.w. 212.14
Toxicology: Irritating to respiratory system
Uses: Ingredient in detergents for food processing equipment and general cleaning in dairies, bakeries, packing houses, in laundry, textile, paper, oil, metal industries; corrosion inhibitor protecting metal surfs. and org. finishes such as paints or ceramic glazes
Manuf./Distrib.: Ashland; Crosfield; Fluka; Occidental; PQ; Rhone-Poulenc N. Am.; Sigma; Spectrum Chem. Mfg.
Trade names: Metso Pentabead® 20

Sodium methane napthalene sulfonate. *See* Sodium methylnaphthalenesulfonate

Sodium methylnaphthalenesulfonate
CAS 26264-58-4; EINECS 247-561-6
Synonyms: Sodium methane napthalene sulfonate; Sodium polynapthalene methane sulfonate
Definition: Mixture of mono- and dimethyl substituted naphthalene sulfonates
Properties: M.w. 245-260
Uses: Dispersing/suspending agent for pesticide wettable powds., paints and coatings; fluidizing and plasticizing agent for concrete and mortar

Sodium methyl silanolate

Regulatory: FDA 21CFR §172.824, 173.315 (0.2% in wash water), 175.105, 176.170, 176.180, 176.210
Trade names: Supragil® MNS/90

Sodium methyl silanolate
Uses: Water repellent, cleaner for limestone, concrete and other masonry; paint additive
Trade names: Union Carbide® R20

Sodium methyl siliconate
CAS 16589-43-8
Synonyms: Methylsilanetriol sodium salt
Definition: Reaction prod. of aq. sodium hydroxide and resinous silicone
Empirical: $CH_6O_3Si \cdot xNa$
Uses: Water repellent for concrete and masonry; material additive or surf. treatment for coatings1
Trade names: Dow Corning® 772

Sodium molybdate
CAS 7631-95-0; 10102-40-6 (dihydrate); EINECS 231-551-7
Synonyms: Sodium molybdate(VI); Sodium molybdate dihydrate
Formula: $Na_2MoO_4 \cdot 2H_2O$
Properties: Wh. small lustrous crystalline plates; sol. in water; m.w. 205.92 (anhyd.), 241.95 (dihydrate); dens. 3.28 (18 C); m.p. 687 C
Toxicology: Irritant
Uses: Reagent in analytical chemistry, paint pigment, corrosion inhibitor, catalyst in dye and pigment prod., additive for fertilizers and feeds, micronutrient
Manuf./Distrib.: AAA Molybdenum Prods.; AC Ind.; Aldrich; Atlantic Equip. Engrs.; Barker Ind.; Cerac; Chem-Met; Climax Molybdenum; Fluka; Mallinckrodt; PMC Specialties; Powmet; Sigma; Spectrum Chem. Mfg.

Sodium molybdate(VI). *See* Sodium molybdate
Sodium molybdate dihydrate. *See* Sodium molybdate
Sodium monododecyl sulfate. *See* Sodium lauryl sulfate
Sodium monofluoride. *See* Sodium fluoride

Sodium myristyl sulfate
CAS 1191-50-0; 139-88-8; EINECS 214-737-2
Synonyms: Sodium tetradecyl sulfate; Sulfuric acid, monotetradecyl ester, sodium salt; 1-Tetradecanol, hydrogen sulfate, sodium salt
Definition: Sodium salt of myristyl sulfate
Empirical: $C_{14}H_{30}O_4S \cdot Na$
Formula: $CH_3(CH_2)_{12}CH_2OSO_3Na$
Properties: M.w. 316.48
Toxicology: Poison by intraperitoneal, intravenous routes
Uses: Anionic detergent; foaming, wetting, and emulsifying agent for cosmetics, household and industrial use; emulsion polymerization; metal processing; pharmaceuticals; leather; textiles; adhesives, sealants, coatings, photo chemicals
Regulatory: FDA 21CFR §175.105, 177.1210, 177.2800
Manuf./Distrib.: Aldrich; Sigma
Trade names: Niaproof® Anionic Surfactant 4

Sodium naphthalene-formaldehyde sulfonate. *See* Sodium polynaphthalene sulfonate

Sodium nitrate
CAS 7631-99-4; EINECS 231-554-3
Synonyms: Soda niter; Cubic niter; Chile saltpeter; Chile salpeter
Empirical: $NNaO_3$
Formula: $NaNO_3$
Properties: Colorless transparent crystals, odorless, saline sl. bitter taste; deliq.; sol. in water, glycerol; sl. sol. in alcohol; m.w. 85.01; dens. 2.267; m.p. 308 C; dec. @ 380 C; pH 5.5-8.3 (5%)
Precaution: Strong oxidizer; ignites with heat or friction; explodes @ 537 C
Toxicology: LD50 (oral, rabbit) 2680 mg/kg; poison by intavenous route; moderately toxic by ingestion; can produce nitrosamines associated with cancers; can reduce blood oxygen levels; human mutagenic data; heated to decomp., emits toxic fumes of NO_x and Na_2O
Uses: Oxidizing agent; solid rocket propellants; fertilizer, flux, glass mfg., pyrotechnics, dynamites, gunpowder; color fixative and preservative in cured meats, fish, enamel for pottery; modifying burning properties

of tobacco; prod. of salts (potassium nitrate), sodium arsenate, pharmaceuticals, heat treatment agents
Regulatory: FDA 21CFR §171.170, 172.170, 172.177, 173.310, 181.33; USDA 9CFR §318.7, 381.147; Japan approved with limitations; Europe listed; UK approved; FDA approved for ophthalmics, orals; BP compliance
Manuf./Distrib.: Advance Research Chems.; Aldrich; Alfa Aesar Johnson Matthey; Am. Int'l.; BASF; Chilean Nitrate; Coyne; Faesy & Besthoff; Farleyway Chem. Ltd; Fluka; GFS; Int'l. Chem. Inc.; Mallinckrodt; Mitsubishi Chem.; Nissan Chem. Ind.; Research Organics; San Yuan; Sigma; Spectrum Chem. Mfg.; Spice King; Sumitomo Chem.; Ube Ind.

Sodium nonoxynol-6 phosphate
CAS 12068-19-8; 37340-60-6; EINECS 235-093-9
Synonyms: Sodium POE (6) nonyl phenyl ether phosphate; PEG 300 nonyl phenyl ether phosphate, sodium salt; POE (6) nonyl phenylether phosphate, sodium salt
Definition: Sodium salt of a complex mixture of esters of phosphoric acid and nonoxynol-6
Uses: Detergent for household and industrial formulations; rust and corrosion retarder for detergent conc.; wax and floor finishes; antistat for acrylic carpet yarn and drycleaning detergents; emulsion polymerization, paints
Trade names: Rhodafac® LO-529

Sodium nonoxynol-4 sulfate
CAS 9014-90-8 (generic); 68891-39-4
Synonyms: PEG-4 nonyl phenyl ether sulfate, sodium salt; PEG 200 nonyl phenyl ether sulfate, sodium salt; Sodium PEG-4 nonyl phenyl ether sulfate
Definition: Sodium salt of the sulfate ester of nonoxynol-4
Formula: $C_9H_{19}C_6H_4(OCH_2CH_2)_nOSO_3Na$, avg. n = 4
Uses: Detergent, emulsifier, lime-soap dispersant, wetting agent, foamer; dishwashing formulations, scrub soaps, car washes, rug and hair shampoos, emulsion polymerization, paints, petrol. waxes, textile wet processing, cosmetics, pesticides
Regulatory: FDA 21CFR §175.105, 178.3400
Trade names: Polystep® B-27; Polystep® CM 4 S

Sodium octoxynol-3 sulfate
Uses: Emulsifier for vinyl acetate specialty copolymers; emulsion polymerization in paints and coatings
Trade names: Polystep® C-OP3S

Sodium octyl sulfate
CAS 126-92-1; 142-31-4; EINECS 204-812-8; 205-535-5
Synonyms: Sodium 2-ethylhexyl sulfate; Sodium capryl sulfate; Sulfuric acid, mono (2-ethylhexyl) ester sodium salt; Octyl sulfate sodium salt
Definition: Sodium salt of 2-ethylhexyl sulfate
Empirical: $C_8H_{17}NaO_4S$
Formula: $CH_3(CH_2)_7OSO_3Na$
Properties: M.w. 232.28; m.p. 195 C
Toxicology: LD50 (oral, mouse) 1550 mg/kg; poison by intraperitoneal route; moderately toxic by ingestion, skin contact; skin and eye irritant; heated to decomp., emits very toxic fumes of SO_x and Na_2O
Uses: Anionic detergent, wetting, dispersing, and emulsifying agent, stabilizer for plastics, rubber, adhesives, food contact paper, textiles, household/industrial cleaners; emulsion polymerization in paints and coatings
Regulatory: FDA 21CFR §173.315, 175.105, 176.170; USDA 9CFR §381.147; BP compliance
Manuf./Distrib.: Aldrich; Fluka; Sigma
Trade names: Polystep® B-29

Sodium α olefin sulfonate
CAS 68188-45-5; 68439-57-6
Uses: Detergent base for personal care prods.; emulsifier for emulsion polymerization; paints and coatings
Trade names: Polystep® A-18S

Sodium PCA
CAS 28874-51-3; 54571-67-4; EINECS 249-277-1
Synonyms: PCA-Na; PCA Soda; 5-Oxo-DL-proline, sodium salt; Sodium pyroglutamate
Definition: Sodium salt of pyroglutamic acid
Empirical: $C_5H_7NO_3 \cdot Na$
Properties: Colorless liq., odorless, sl. salty taste; extremely hygroscopic; m.w. 151.1
Toxicology: No known toxicity; nonirritant to skin, eye mucosa
Uses: Humectant, moisturizer for cosmetics, dermatological soap, medicinals, shampoo, nutritive creams and

lotions, hair balms, dentifrices, tobacco, cellulose film, paper, fibers, paints; dyeing agent, softening agent, finishing agent, antistat; intermediate for synthesis; thickener for shampoos
Trade names: Ajidew N-50

Sodium PEG-4 nonyl phenyl ether sulfate. *See* Sodium nonoxynol-4 sulfate

Sodium petroleum sulfonate
CAS 68608-26-4; 78330-12-8
Synonyms: Mineral oil sulfonic acids, sodium salts; Petroleum sulfonic acid, monosodium salt; Sulfonated petroleum, sodium salt
Uses: Surfactant, emulsifier, and corrosion inhibitor for metalworking fluids; dry cleaning solvs., leather processing, printing inks, oil well drilling fluids, water-based emulsions contg. oil-based ingreds.; for use in semi-finished coatings where wetting, rewetting, suspending, emulsifying, and corrosion resist. props. are desired
Trade names: Lubrizol® 5363

Sodium POE (6) nonyl phenyl ether phosphate. *See* Sodium nonoxynol-6 phosphate

Sodium polyacrylate
CAS 9003-04-7
Synonyms: Polyacrylic acid, sodium salt
Empirical: $(C_3H_4O_2)_x$•xNa
Properties: Hygroscopic; m.w. 2000-2300
Toxicology: Eye and skin irritant; heated to decomp., emits toxic fumes of Na_2O
Uses: Thickener, stabilizer, protective colloid for natural and syn. latexes for paints, films, coatings, and adhesives; dispersant; antiredeposition agents; antiscalant
Regulatory: FDA 21CFR §173.73, 173.310, 173.340; Japan approved (0.2% max.)
Manuf./Distrib.: Alco; Aldrich; Allchem Ind.; Arakawa USA; Dixie; Elf Atochem N. Am.; Fluka; Monomer-Polymer & Dajac Labs; Rhone-Poulenc Surf. & Spec.; Synthron; 3V
Trade names: Alcosperse® 102; Alcosperse® 107; Alcosperse® 149; Alcosperse® 602-N; BYK®-155; Colloid™ 112; Colloid™ 211-D; Daxad® 37LN7; Daxad® 37LN10-35; Nopcosperse® 44; Sokalan® CP 7; Sokalan® CP 10; Sokalan® CP 45; Sokalan® PA 15; Sokalan® PA 20; Sokalan® PA 20 PN; Sokalan® PA 25 PN; Sokalan® PA 30; Sokalan® PA 40; Sokalan® PA 40 Powd.; Sokalan® PA 50; Sokalan® PA 70 PN

Sodium polyisobutylene/maleic anhydride copolymer
Uses: Dispersant for latex paints and coatings, enamels, polymerization, leather tanning, and water treatment
Trade names: Daxad® 31S

Sodium polymannuronate. *See* Algin

Sodium polymethacrylate
CAS 25086-62-8; 54193-36-1
Synonyms: 2-Propenoic acid, 2-methyl-, homopolymer, sodium salt
Classification: Polymer
Empirical: $(C_4H_6O_2)_x$•xNa
Formula: $[CH_3CHCHCOONa]_x$
Uses: Dispersant for pigments, in paint formulations, emulsion polymerization, water treatment, agriculture, cosmetics, industrial cleaners, large particle suspensions
Regulatory: FDA 21CFR §173.310, 175.105
Manuf./Distrib.: Aldrich
Trade names: Alcosperse® 124; Alcosperse® 125; Colloid™ 225; Darvan® No. 7; Daxad® 30; Daxad® 30-30; Daxad® 30S; Daxad® 34N10; Tamol® 850

Sodium polynaphthalene sulfonate
CAS 1321-69-3; 9084-06-4
Synonyms: Sodium naphthalene-formaldehyde sulfonate; Naphthalenesulfonic acid, polymer with formaldehyde, sodium salt
Definition: Sodium salt of the prod. obtained by condensation polymerization of naphthalene sulfonic acid and formaldehyde
Empirical: $(C_{10}H_8O_3S \cdot CH_2O)_x \cdot$ xNa
Toxicology: Moderately toxic by ingestion
Uses: Dispersant for pigments, extenders, and fillers; used in dyeing syn. and natural fibers, ceramics,

emulsion polymerization, gypsum board, printing, rubber, food pkg. applics., agric., tanning; plasticizer for concrete; stabilizer
Regulatory: FDA 21CFR §175.105, 176.170, 176.180, 177.1200, 177.1210, 177.1550, 177.1650, 177.2600, 178.3910
Trade names: Darvan® No. 1; Daxad® 11G; Daxad® 13; Dehscofix 914/ASL; Dehscofix 915/ASL; Lomar® P62; Lomar® PW 5353; Rhodacal® N; Tamol® NH 7519

Sodium polynapthalene methane sulfonate. *See* Sodium methylnaphthalenesulfonate
Sodium 2-pyridinethiol-1-oxide. *See* Sodium pyrithione
Sodium (2-pyridylthio)-N-oxide. *See* Sodium pyrithione

Sodium pyrithione
CAS 3811-73-2; EINECS 223-296-5
Synonyms: Sodium 2-pyridinethiol-1-oxide; Sodium (2-pyridylthio)-N-oxide; 2-Pyridinethiol, 1-oxide, sodium salt
Classification: Organic compd.
Empirical: $C_5H_5NOS \cdot Na$
Uses: Industrial microbiostat; chelating agent; used in aq. metal coolant and cutting fluids, latex emulsion, inks, fiber lubricants
Manuf./Distrib.: Fluka; Olin; Sigma
Trade names: Sodium Omadine®, 40% Aq. Sol'n.

Sodium pyroborate anhyd. *See* Sodium borate
Sodium pyroglutamate. *See* Sodium PCA
Sodium pyrophosphate. *See* Tetrasodium pyrophosphate
n-Sodium pyrophosphate. *See* Tetrasodium pyrophosphate

Sodium silicate
CAS 1344-09-8; EINECS 215-687-4
Synonyms: Silicic acid, sodium salt; Water glass; Soluble glass
Definition: Sodium salt of silicic acid
Empirical: Na_2SiO_3 or $Na_6Si_2O_7$ or $Na_2Si_3O_7 \cdot$ aq.
Properties: Colorless to wh. or grayish-wh. cryst. lumps; very sl. sol. in cold water; m.w. 242.23 + water; dens. 1.390 (20C)
Toxicology: Corrosive; irritating and caustic to skin, mucous membranes; if swallowed, causes vomiting and diarrhea
Uses: Lining Bessemer converters, acid concentrators; mfg. of grindstones, abrasive wheels; as sol'n.: preserving eggs; fireproofing fabrics; detergent in soaps; as adhesive; waterproofing walls; extender in cements, paints
Regulatory: FDA 21CFR §173.310, 177.1200, 182.70, 182.90
Manuf./Distrib.: Aichi Silicate Chem Co Ltd; Aldrich; Am. Int'l.; Asahi Denka Kogyo; Coyne; Crosfield; Fluka; E&F King; Occidental; PQ; Sigma; Spectrum Chem. Mfg.; Van Waters & Rogers
Trade names: N® Sodium Silicate

Sodium silicoaluminate
CAS 1318-02-1; 1344-00-9; EINECS 215-684-8
Synonyms: Sodium aluminosilicate; Sodium aluminum silicate; Aluminum sodium silicate
Definition: Series of hydrated sodium aluminum silicates
Empirical: $Na_2O : Al_2O_3 : SiO_2$ with mole ratio ≈ 1:1:13.2
Properties: Wh. fine amorphous powd. or beads, odorless and tasteless; insol. in water, alcohol, org. solvs.; partly sol. in strong acids and alkali hydroxides @ 80-100 C; pH 6.5-10.5 (20% slurry)
Precaution: Noncombustible
Toxicology: Irritant to skin, eyes, mucous membranes; heated to decomp., emits toxic fumes of Na_2O
Uses: Anticaking agent in food preps.; reinforcing filler for rubbers; extender for paints
Regulatory: FDA 21CFR §133.146, 160.105, 160.185, 182.2727 (2% max.), 582.2727, GRAS; Europe listed; UK approved; FDA approved for orals
Manuf./Distrib.: Albemarle; Crosfield; Fluka; J.M. Huber/Chems.; PQ; Punda Mercantile; Sigma; Spectrum Chem. Mfg.
Trade names: EZA®; Ketjensil® SM 405; Zeolex® 70HP; Zeolex® 80
See also Zeolite

Sodium p-styrenesulfonate
CAS 2695-37-6; EINECS 220-266-3

p-Sodium styrenesulfonate

Synonyms: p-Sodium styrenesulfonate
Empirical: $C_8H_7SO_3Na$
Formula: $CH:CH_2C_6H_4SO_3Na$
Properties: Wh. to pale yel. cryst. powd., odorless; sol. in water; insol. in aromatics, high alcohols; m.w. 206.20; sp.gr. 5.5; m.p. 330 C
Uses: Emulsifier producing acrylic, vinyl acetate, and SBR latexes; in dyeing; flocculant, scale inhibitor; dispersant for cosmetics; hair fixing agent; for polyion complex; antistat; photochemical metal plating brightener
Manuf./Distrib.: Aceto; Aldrich; Fluka; Monomer-Polymer & Dajac Labs; Polysciences; Tosoh; Wako Pure Chem. Ind.
Trade names: Spinomar NaSS

p-Sodium styrenesulfonate. *See* Sodium p-styrenesulfonate
Sodium sulfoxylate. *See* Sodium hydrosulfite
Sodium tetraborate anhyd. *See* Sodium borate
Sodium tetradecyl sulfate. *See* Sodium myristyl sulfate
Sodium tetrahydridoborate. *See* Sodium borohydride
Sodium tetrahydroborate. *See* Sodium borohydride

Sodium vinyl sulfonate
Synonyms: SVS
Classification: Functional monomer
Uses: As organic intermediate in sulfonoethylation reactions, as functional monomer for stabilization of emulsion polymers
Manuf./Distrib.: Aceto; Air Prods.; Monomer-Polymer & Dajac Labs
Trade names: Hartomer™ 4900-25

Soluble glass. *See* Sodium silicate
Soluble guncotton. *See* Nitrocellulose
Sorbide dioleate. *See* Sorbitan dioleate
Sorbimacrogol laurate 300. *See* Polysorbate 20
Sorbimacrogol oleate 300. *See* Polysorbate 80
Sorbimacrogol palmitate 300. *See* Polysorbate 40
Sorbimacrogol stearate 300. *See* Polysorbate 60
Sorbimacrogol trioleate 300. *See* Polysorbate 85
Sorbimacrogol tristearate 300. *See* Polysorbate 65
Sorbit. *See* Sorbitol
Sorbitan, di-9-octadecenoate. *See* Sorbitan dioleate

Sorbitan dioleate
CAS 29116-98-1; EINECS 249-448-0
Synonyms: Sorbide dioleate; Sorbitan, di-9-octadecenoate; Anhydrosorbitol dioleate
Definition: Diester of oleic acid and hexitol anhydrides derived from sorbitol
Empirical: $C_{42}H_{76}O_7$
Uses: Surfactant, emulsifier for creams and ointments; pigment stabilizer for paint emulsions, min. oil emulsions, anticorrosion oils
Regulatory: FDA 21CFR §175.320
Trade names: DO-33-F

Sorbitan, esters, monododecanoate. *See* Sorbitan laurate
Sorbitan, esters, monohexadecanoate. *See* Sorbitan palmitate
Sorbitan, esters, monooctadecanoate. *See* Sorbitan stearate

Sorbitan laurate
CAS 1338-39-2; 5959-89-7; EINECS 215-663-3; 227-729-9
Synonyms: Sorbitan monolaurate; Sorbitan, esters, monododecanoate; Anhydrosorbitol monolaurate; Sorbitan monododecanoate
Definition: Monoester of lauric acid and hexitol anhydrides derived from sorbitol
Empirical: $C_{18}H_{34}O_6$
Properties: Yel. to amber oily liq., bland char. odor; sol. in methanol, alcohol, min. oil; sl. sol. in cottonseed oil, ethyl acetate; insol. in water; m.w. 346.47; dens. 1.032; acid no. 8 max.; sapon. no. 158-170; hyd. no. 330-358; flash pt. > 230 F; ref. index 1.4740
Toxicology: Experimental neoplastigen

Uses: Emulsifier for foods, cosmetics, household prods., industrial applics., coatings, household (polishes, cleaners), pharmaceutical uses; lubricant, antifog, antistat for PVC and other plastics; antifog for food pkg. films; polymerization emulsifier for thermoplastics; emollient for skin care prods.; lubricant, emulsifier in textile spin finishes
Regulatory: FDA 21CFR §175.320, 178.3400; Europe listed; UK approved; FDA approved for ophthalmics, orals; USP/NF, BP compliance
Manuf./Distrib.: Aldrich; Fluka; Sigma; Spectrum Chem. Mfg.
Trade names: Calgene SML; Crill 1; Disponil® SML 100 F1; Glycomul® L
Trade names containing: Aldosperse® L L-20

Sorbitan, monodecanoate, poly(oxy-1,2-ethanediyl) derivs. *See* Polysorbate 20
Sorbitan monododecanoate. *See* Sorbitan laurate
Sorbitan, monohexadecanoate, poly(oxy-1,2-ethaneidyl) derivs. *See* Polysorbate 40
Sorbitan monolaurate. *See* Sorbitan laurate
Sorbitan monooctadecanoate. *See* Sorbitan stearate
Sorbitan mono-9-octadecenoate. *See* Sorbitan oleate
Sorbitan monooleate. *See* Sorbitan oleate
Sorbitan monopalmitate. *See* Sorbitan palmitate
Sorbitan monostearate. *See* Sorbitan stearate

Sorbitan oleate
CAS 1338-43-8; 5938-38-5; EINECS 215-665-4
Synonyms: SMO; Sorbitan monooleate; Sorbitan mono-9-octadecenoate; Anhydrosorbitol monooleate
Definition: Monoester of oleic acid and hexitol anhydrides derived from sorbitol
Empirical: $C_{24}H_{44}O_6$
Properties: Yel. to amber visc. liq., char. bland odor; misc. with min. and veg. oils; sl. sol. in ether; insol. in water, acetone, propylene glycol; m.w. 428.62; dens. 0.986; acid no. 8 max.; iodine no. 62-76; sapon. no. 145-160; hyd. no. 193-210; flash pt. > 230 F; ref. index 1.4800
Toxicology: Heated to decomp., emits acrid smoke and irritating fumes
Uses: Emulsifier for foods, cosmetics, pharmaceuticals, household prods., industrial applics., polymerization, pulp/paper, metalworking, lubricants, textiles, agric., paints, adhesives; antifog, antistat, cling additive for plastic films; oil additive, corrosion inhibitor
Regulatory: FDA 21CFR §73.1001, 173.75, 175.105, 175.320, 178.3400; Europe listed; UK approved; FDA approved for orals, topicals; USP/NF, BP compliance
Manuf./Distrib.: ABITEC; Aldrich; Calgene; Fluka; Heterene; ICI Surf. Am.; Lonza; Norman, Fox; Rhone-Poulenc Surf. & Spec.; Sigma; Spectrum Chem. Mfg.; Witco/Oleo-Surf.
Trade names: Calgene SMO; Crill 4; Disponil® SMO 100 F-1; MO-33-F; Rheodol SP-O10

Sorbitan palmitate
CAS 26266-57-9; EINECS 247-568-8
Synonyms: Sorbitan monopalmitate; 1,4-Anhydro-D-glucitol, 6-hexadecanoate; Sorbitan, esters, monohexadecanoate
Definition: Partial ester of palmitic acid with sorbitol mono- and dianhydrides
Empirical: $C_{22}H_{42}O_6$
Properties: Tan gran. waxy solid, faint tallow-like odor; sol. in ethyl acetate, warm abs. alcohol; sol. hazy in warm peanut or min. oil; insol. in water; m.w. 1109.59; dens. 0.989; acid no. 8 max.; sapon. no. 140-150; hyd. no. 275-305; flash pt. > 230 F; ref. index 1.4780
Toxicology: No known toxicity; irritant
Uses: Nonionic surfactant, emulsifier for foods, cosmetics, pharmaceuticals, household prods., industrial applics., polymerization, pulp/paper, metalworking, lubricants, textiles, agric., paints, adhesives; oil additive, corrosion inhibitor
Regulatory: FDA 21CFR §175.320, 178.3400; Europe listed; UK approved; FDA approved for intramuscular injectables; USP/NF compliance
Manuf./Distrib.: Aldrich; Fluka; Sigma; Spectrum Chem. Mfg.
Trade names: Calgene SMP

Sorbitan stearate
CAS 1338-41-6; 69005-67-8; EINECS 215-664-9
Synonyms: SMS; Sorbitan monostearate; Sorbitan monooctadecanoate; Sorbitan, esters, monooctadecanoate; Anhydrosorbitol monostearate
Definition: Monoester of stearic acid and hexitol anhydrides derived from sorbitol
Empirical: $C_{24}H_{46}O_6$
Properties: Cream to tan beads, bland odor and taste; sol. in veg. and min. oils; insol. in water, acetone, alcohol,

propylene glycol; m.w. 430.70; acid no. 5-10; sapon. no. 147-157; hyd. no. 235-260
Toxicology: LD50 (oral, rat) 31 g/kg; very mildly toxic by ingestion; experimental reproductive effects; heated to decomp., emits acrid smoke and irritating fumes
Uses: Emulsifier, stabilizer, coupling agent, defoamer for foods, cosmetics, household prods., industrial applics., polymerization; used to prepare silicone defoamer emulsions for industrial applics., paraffin wax emulsions for processing paper coatings; textile process lubricant; internal PVC film lubricant
Regulatory: FDA 21CFR §73.1001, 163.123, 163.130, 163.135, 163.140, 163.145, 163.150, 163.153, 163.155, 172.515, 172.842, 173.340, 175.105, 175.320, 178.3400, 573.960; FEMA GRAS; Europe listed; UK approved; FDA approved for topicals, vaginals, orals; USP/NF, BP compliance
Manuf./Distrib.: Aldrich; Aquatec Quimica SA; Ashland; Calgene; Fluka; ICI Surf. Am.; Lonza; Norman, Fox; Rhone-Poulenc Surf. & Spec.; Sigma; Spectrum Chem. Mfg.; Witco/Oleo-Surf.
Trade names: Alkamuls® SMS; Calgene SMS; Rheodol SP-S10

Sorbitan tallate
Uses: Lipophilic emulsifier, fiber lubricant, softener; drier for paints
Manuf./Distrib.: Lonza

Sorbitan trioctadecanoate. *See* Sorbitan tristearate
Sorbitan tri-9-octadecenoate. *See* Sorbitan trioleate

Sorbitan trioleate
CAS 26266-58-0; 85186-88-5; EINECS 247-569-3; 286-074-7
Synonyms: STO; Anhydrosorbitol trioleate; Sorbitan tri-9-octadecenoate
Definition: Triester of oleic acid and hexitol anhydrides derived from sorbitol
Empirical: $C_{60}H_{108}O_8$
Properties: Yel. to amber oily liq.; sol. in alcohol, veg. oil, min. oil; insol. in water, ethylene glycol, propylene glycol; m.w. 957.52; dens. 0.956; acid no. 17 max.; iodine no. 77-85; sapon. no. 170-190; hyd. no. 50-75; flash pt. > 230 F; ref. index 1.4760
Toxicology: Irritant
Uses: Emulsifier for foods, cosmetics, household prods., industrial applics., polymerization; antifog, lubricant, cling additive for plastic film; emollient
Regulatory: FDA 21CFR §175.320, 178.3400; FDA approved for inhalants, orals, topicals
Manuf./Distrib.: Aldrich; Fluka; Sigma; Spectrum Chem. Mfg.
Trade names: Alkamuls® STO; Calgene STO; Rheodol SP-O30

Sorbitan tristearate
CAS 26658-19-5; 72869-62-6; EINECS 247-891-4; 276-951-2
Synonyms: STS; Anhydrosorbitol tristearate; Sorbitan trioctadecanoate
Definition: Triester of stearic acid and hexitol anhydrides derived from sorbitol
Empirical: $C_{60}H_{114}O_8$
Properties: Tan waxy beads; sol. in IPA; insol. in water; m.w. 963.56; acid no. 12-15; sapon. no. 176-188
Toxicology: No known toxicity
Uses: Emulsifier for foods, cosmetics, pharmaceuticals, household prods., industrial applics., polymerization, pulp/paper, metalworking, lubricants, textiles, agric., paints, adhesives
Regulatory: FDA 21CFR §175.320, 178.3400; Europe listed; UK approved
Manuf./Distrib.: Ashland; Calgene; Fluka; Henkel/Emery; ICI Surf. Am.; Lonza; Rhone-Poulenc Surf. & Spec.; Sigma; Spectrum Chem. Mfg.; Witco/Oleo-Surf.
Trade names: Calgene STS; Rheodol SP-S30

D-Sorbite. *See* Sorbitol

Sorbitol
CAS 50-70-4; EINECS 200-061-5
Synonyms: D-Glucitol; γ-Sorbitol; Sorbit; Sorbol; Sorbo; D-Sorbitol; D-Sorbite; 1,2,3,4,5,6-Hexanehexol
Classification: Hexahydric alcohol
Empirical: $C_6H_{14}O_6$
Formula: $CH_2OHHCOHHOCHHCOHHCOHCH_2OH$
Properties: Wh. cryst. powd., gran. or flakes, odorless, sweet taste; hygroscopic; sol. in water; sol. in hot alcohol, methanol, IPA, DMF, acetic acid, phenol, acetamide sol'ns.; m.w. 182.20; dens. 1.47 (-5 C); m.p. 93-97.5 C; b.p. 105 C; pH ≈ 7.0
Toxicology: LD50 (oral, rat) 17.5 mg/kg; mildly toxic by ingestion; excess consumption may have laxative effect; intolerance manifested by abdominal pain, bloating, diarrhea; no known toxicity if used externally
Uses: Nutrient and dietary supplement, food additive; bodying agent for paper, textile, liq. pharmaceuticals;

in mfg. of sorbose, ascorbic acid, propylene glycol, synthetic plasticizers, resins, paints; as humectant, sequestrant

Usage level: 70-72% (suspensions); 6-35% (sol'ns.); 5-25% (syrups); 5-20% (elixirs)

Regulatory: FDA 21CFR §175.300, 182.90, 184.1835, GRAS; FEMA GRAS; Japan approved; Europe listed; UK approved; FDA approved for dentals, intramuscular injectables, orals, rectals; USP/NF, BP, Ph.Eur. compliance

Manuf./Distrib.: Aldrich; Allchem Ind.; Am. Biorganics; Ashland; Baychem; Boehle; Brown; Browning; Cerestar UK; Coyne; EM Ind.; Fanning; Fluka; R.W. Greeff; Harcros; ICI Surf. Am.; Int'l. Chem. Inc.; Lipo; Lonza; Pfizer Food Science; Research Organics; Sigma; San Yuan; Van Waters & Rogers

Trade names: Liponic 70-NC; Liponic 76-NC

D-Sorbitol. *See* Sorbitol
γ-Sorbitol. *See* Sorbitol
Sorbo. *See* Sorbitol
Sorbol. *See* Sorbitol
Soya bean oil. *See* Soybean oil

Soy acid

CAS 68308-53-2; 67701-08-0; EINECS 269-657-0

Synonyms: Acids, soy; Fatty acids, soya

Definition: Mixture of fatty acids derived from soybean oil

Uses: Lubricant, release agent, binder, defoaming agent and intermediate for food additives; rubber compounding; alkyd resins; water repellents; polishes; soaps; cutting oils; candles; crayons; used as raw material for high quality long oil alkyds for consumer paints, for med. and short oil alkyd resins for air and oven-drying lacquers with high flexibility

Regulatory: FDA 21CFR §175.105, 177.2800, 178.3570

Trade names: Prifac 7951

Soya dimethyl ethyl ammonium ethyl sulfate. *See* Soyethyldimonium ethosulfate
Soyaethyldimonium ethosulfate. *See* Soyethyldimonium ethosulfate
Soya lecithin. *See* Lecithin
Soya oil. *See* Soybean oil

Soybean oil

CAS 8001-22-7; EINECS 232-274-4

Synonyms: Soya oil; Soya bean oil; Chinese bean oil

Classification: Fixed oil

Definition: Oil obtained from seeds of soya plant, *Glycine soja*, by extraction or expression; consists of triglycerides of oleic, linoleic, linolenic, and saturated acids

Properties: Pale yel. to brnsh. yel. oil, sl. char. odor and taste; sol. in alcohol, ether, chloroform, carbon disulfide; insol. in water; dens. 0.924-0.929; visc. 50.09 cps; m.p. 22-31 C; iodine no. 120-141; sapon. no. 189-195; flash pt. 540 F; ref. index 1.471

Precaution: Combustible; light-sensitive

Toxicology: May cause allergic reactions incl. hair damage and acne-like pimples; heated to decomp., emits acrid smoke and irritating fumes

Uses: Soap mfg., plasticizer (epoxidized), high-protein foods, drying oil for inks, paints, varnishes, cattle feeds

Regulatory: FDA 21CFR §175.105, 175.300, 176.200, 176.210, 177.2800, GRAS; FDA approved for orals, topicals; USP/NF, BP compliance

Manuf./Distrib.: ABITEC; ADM; Akzo Nobel; Alba Int'l.; Aldrich; Alnor Oil; Arista Ind.; Ashland; Central Soya; Charkit; Croda; Degen; Lomas Int'l.; Reichhold; Sigma; Werner G. Smith; Tri-K Ind.; Welch, Holme & Clark; Jesse S. Young

Trade names: Drisoy Y-Z; Drisoy Z2-Z3; Soy Solinox T; Soy Solinox Z

Soybean oil, epoxidized. *See* Epoxidized soybean oil
Soybean oil hydrogenated. *See* Hydrogenated soybean oil

Soyethyldimonium ethosulfate

CAS 68308-67-8

Synonyms: Soya dimethyl ethyl ammonium ethyl sulfate; Soyaethyldimonium ethosulfate

Classification: Quaternary ammonium compd.

Formula: $[RN(CH_3)_2CH_2CH_3]^+CH_3CH_2OSO_3^-$, R rep. alkyl groups from soy

Uses: Conditioner for hair conditioners and shampoos; antistat for cleaners, rug shampoos, plastics, fibers, fiberglass; mold release agent

Trade names: Larostat® 264 A

Spirit of Hartshorn. *See* Ammonium hydroxide

Stannic chloride

CAS 7646-78-8 (anhyd.); EINECS 231-588-9

Synonyms: Tin tetrachloride; Stannic chloride anhyd.; Tin chloride (ic); Tin perchloride; Tin (IV) chloride; Tin (IV) chloride anhydrous

Empirical: Cl_4Sn

Formula: $SnCl_4$

Properties: M.w. 260.50; dens. 2.217 (20/4 C)

Precaution: Moisture-sensitive

Toxicology: DOT: Corrosive material; causes burns; irritating to respiratory system

Uses: Electroconductive/electroluminescent coatings, textile dye mordant, perfume stabilizer, manufacture of fuchsin, blueprint paper, color lakes, ceramic coatings, bleaching agent for sugar, stabilizer for resins, tin salts, soap bactericide/fungicide

Manuf./Distrib.: Aldrich; Alfa Aesar Johnson Matthey; Atotech USA; Brandeis; Chemisphere Ltd; Elf Atochem N. Am.; Esprit; Fluka; M&T Harshaw; Mason; Nihon Kagaku Sangyo; Noah; Spectrum Chem. Mfg.; Witco/ Polymer Addit.

Stannic chloride anhyd. *See* Stannic chloride

Stannous chloride anhyd.

CAS 7772-99-8; EINECS 231-868-0

Synonyms: Tin (II) chloride anhydrous; Tin crystals; Tin (II) chloride (1:2); Tin salt; Tin dichloride; Tin protochloride

Definition: Chloride salt of metallic tin

Empirical: Cl_2Sn

Formula: $SnCl_2$

Properties: Colorless orthorhombic cryst. mass or flakes, fatty appearance; sol. in water, ethanol, acetone, ether, methyl acetate, MEK, isobutyl alcohol; pract. insol. in min. spirits, xylene; m.w. 189.60; dens. 3.95; m.p. 246.8 C; b.p. 652 C

Precaution: Potentially explosive reaction with metal nitrates; violent reactions with hydrogen peroxide, ethylene oxide, nitrates, K, Na; moisture-sensitive

Toxicology: TLV:TWA 2 mg(Sn)/m^3; LD50 (oral, rat) 700 mg/kg, (IP, mouse) 66 mg/kg; poison by ingestion, IP, IV, subcutaneous routes; experimental reproductive effects; human mutagenic data; heated to decomp., emits toxic fumes of Cl^-

Uses: Reducing agent for intermediates, dyes, polymers, phosphors; mfg. of lakes; textile dyeing and printing; tin galvanizing; analytical reagent; silvering mirrors; antisludge for lubricants; food preservative; perfume stabilizer; soldering flux; esterification catalyst for mfg. of plasticizers and polyesters; antioxidant for pharmaceuticals; approved as component in polymeric coatings in food pkg.

Regulatory: FDA 21CFR §155.200, 172.180, 175.300, 177.2600, 184.1845, GRAS; FDA approved for intravenous, parenterals

Manuf./Distrib.: Advance Research Chems.; Aldrich; Allchem Ind.; Atotech USA; Blythe, William Ltd; Brandeis; Cerac; Elf Atochem N. Am.; Fluka; Goldschmidt Ind. Chems.; Jacksonlea; Noah; Sigma; Spectrum Chem. Mfg.

Trade names: Fascat® 2005

Stannous-2-ethylhexoate. *See* Stannous octoate

Stannous octoate

CAS 301-10-0

Synonyms: Stannous-2-ethylhexoate; Tin octoate; Tin-(II)-octoate; Sn-II-ethylhexanoate

Formula: $Sn(C_8H_{15}O_2)_2$

Properties: Lt. yel. liq.; sol. in benzene, toluene, petroleum ether; insol. in water, methanol; m.w. 405.11; dens. 1.25; flash pt. > 230 F; ref. index 1.4930

Toxicology: Irritant

Uses: Catalyst for mfg. of PU foam, coatings, adhesives, sealants, elastomers

Manuf./Distrib.: Aceto; Aldrich; Sigma

Trade names: Metacure® T-9

Stannum. *See* Tin

Starch

CAS 9005-25-8; 9005-84-9; EINECS 232-686-4

Classification: Carbohydrate polymer
Definition: Complex polysaccharide composed of units of glucose consisting of about one quarter amylose and three quarters amylopectin; derived from corn, wheat, potatoes, or tapioca
Empirical: $(C_6H_{10}O_5)_n$
Properties: Wh. amorphous powd. or gran., odorless, sl. char. taste; insol. in cold water, alcohol; forms gels in hot water; m.w. 162.14_n
Toxicology: May cause contact dermatitis, peritonitis
Uses: Adhesives, machine-coated paper, textile filler and sizing agent, gelling agent for foods, filler in baking powd., urea-formaldehyde resin adhesive, chelating and sequestering agent in foods, paint raw material
Regulatory: FDA 21CFR §182.90, GRAS; FDA approved for buccals, parenterals, orals, rectals, topicals, vaginals; USP/NF, BP compliance
Manuf./Distrib.: ADM Corn Processing; Aldrich; Avebe Am.; Cerestar Int'l.; Eka Nobel; Fluka; Grain Processing; Nalco; Nat'l. Starch & Chem.; Penford Prods.; Penta Mfg.; Sigma; A.E. Staley Mfg.

Starch, corn. *See* Corn starch

Stearalkonium bentonite
Definition: Reaction prod. of bentonite and stearalkonium chloride
Uses: Thixotrope for paint systems contg. high amts. of aromatics, ketones, glycol ether, esters, and alcohols; exc. visc. build, sag control; enhanced suspension control
Trade names: Claytone® APA; Tixogel® VZ

Stearalkonium hectorite
CAS 94891-33-5; 12691-60-0
Definition: Reaction prod. of hectorite and stearalkonium chloride
Uses: Thixotrope, gellant, thickener for solv.-based coatings; lip and eye care prods.; pharmaceuticals
Regulatory: FDA 21CFR §175.300; FDA approved for topicals
Trade names: Bentone® 27

Stearamide
CAS 124-26-5; EINECS 204-693-2
Synonyms: Octadecanamide; Stearic acid amide; Amide C-18
Classification: Aliphatic amide
Empirical: $C_{18}H_{37}NO$
Formula: $CH_3(CH_2)_{16}CONH_2$
Properties: Colorless leaflets; sl. sol. in alcohol, ether; insol. in water; m.w. 283.56; m.p. 98-102 C; b.p. 250-251 C (12 mm)
Toxicology: Experimental tumorigen
Uses: Slip/antiblock agent for LDPE, HDPE, PP; mold release agent, internal lubricant for coatings, films; builder, foam visc. stabilizer, and foam booster in syn. detergent formulations, cosmetics; water repellent for textiles; improves dye solubility in printing inks, dyes, carbon paper coatings, and fusible coatings for glassware and ceramics; intermediate for syn. waxes; pigment dispersant; thickener for paint
Regulatory: FDA 21CFR §175.105, 177.1210, 178.3860, 178.3910, 179.45, 181.22, 181.28
Manuf./Distrib.: Akzo Nobel; Aldrich; Astor Corp.; Chemax; Cookson Spec. Additives; Croda Universal Ltd.; Fluka; Henkel/Emery; Syn. Prods.; Witco/Oleo-Surf.
Trade names: Armid® 18; Petrac® Vyn-Eze®

Stearamide MEA-stearate
CAS 14351-40-7; EINECS 238-310-5
Synonyms: Octadecanoic acid, 2-[(1-oxooctadecyl)amino]ethyl ester; Stearic monoethanolamide stearate; 2-[(1-Oxooctadecyl)amino]ethyl octadecanoate
Classification: Substituted ethanolamide
Properties: Flakes; insol. in water; gels in min. oil, IPM; m.p. 76-82 C; sapon. no. 97-107
Toxicology: No known toxicity
Uses: Thickener, opacifier, conditioner, lubricant, gellant for cosmetics, household, institutional liq. soaps; aux. emulsifier for hydrocarbon propellant gas in aq. aerosol systems; in polishes; coating agent for paper and textiles; mold release agent for industrial processing; additive for raising melting pts. of petrol. waxes or glyceride waxes and fats; ingred. of insulating coatings or barriers, and of water-repellent compds.
Trade names: Witcamide® MAS

Stearamidopropyl dimethyl-β-hydroxyethyl ammonium dihydrogen phosphate
Empirical: $C_{25}H_{53}O_6N_2P$
Properties: M.w. 510.8

Stearamidopropyl dimethyl-2-hydroxyethyl ammonium nitrate

Uses: Antistat for plastics, waxes, textiles, and glass; emulsifier, settling, dispersing, rewetting agent
Trade names containing: Cyastat® SP

Stearamidopropyl dimethyl-2-hydroxyethyl ammonium nitrate
CAS 2764-13-8
Synonyms: Stearamidopropyl dimethyl-β-hydroxyethylammonium nitrate
Empirical: $C_{25}H_{53}O_5N_3$
Properties: M.w. 475
Uses: Antistat for polymers, paper, glass, other materials
Trade names containing: Cyastat® SN

Stearamidopropyl dimethyl-β-hydroxyethylammonium nitrate. *See* Stearamidopropyl dimethyl-2-hydroxyethyl ammonium nitrate

Stearamine acetate
CAS 2190-04-7
Synonyms: Stearyl amine acetate; Octadecylamine acetate
Uses: Wetting agent, emulsifier, dispersant, corrosion inhibitor, flotation agent for pigment flushing, froth flotation of mins., flocculation; anticaking agent for fertilizers

Steareth-2
CAS 9005-00-9 (generic); 16057-43-5
Synonyms: PEG-2 stearyl ether; POE (2) stearyl ether; PEG 100 stearyl ether
Definition: PEG ether of stearyl alcohol
Empirical: $C_{22}H_{46}O_3$
Formula: $CH_3(CH_2)_{16}CH_2(OCH_2CH_2)_nOH$, avg. n = 2
Properties: Oily liq.
Toxicology: Moderately toxic by ingestion
Uses: Intermediate in the mfg. of high-foaming surfactants; emulsifier, detergent, dispersant, wetting agent; pulp and paper industry, textiles, paints, adhesives, corrosion inhibitors, petrol. oils
Regulatory: FDA approved for topicals
Manuf./Distrib.: Aldrich; Fluka; Sigma
Trade names: Calgene Nonionic S-2

Steareth-10
CAS 9005-00-9 (generic); 13149-86-5
Synonyms: PEG-10 stearyl ether; POE (10) stearyl ether; PEG 500 stearyl ether
Definition: PEG ether of stearyl alcohol
Empirical: $C_{38}H_{78}O_{11}$
Formula: $CH_3(CH_2)_{16}CH_2(OCH_2CH_2)_nOH$, avg. n = 10
Uses: Intermediate in the mfg. of high-foaming surfactants; emulsifier, detergent, dispersant, wetting agent; pulp and paper industry, textiles, paints, adhesives, corrosion inhibitors, petrol. oils
Regulatory: FDA approved for rectals, topicals
Manuf./Distrib.: Aldrich; Fluka; Sigma
Trade names: Calgene Nonionic S-10

Steareth-20
CAS 9005-00-9 (generic)
Synonyms: PEG-20 stearyl ether; POE (20) stearyl ether; PEG 1000 stearyl ether
Definition: PEG ether of stearyl alcohol
Empirical: $C_{58}H_{118}O_{21}$
Formula: $CH_3(CH_2)_{16}CH_2(OCH_2CH_2)_nOH$, avg. n = 20
Uses: Intermediate in the mfg. of high-foaming surfactants; emulsifier, detergent, dispersant, wetting agent; pulp and paper industry, textiles, paints, adhesives, corrosion inhibitors, petrol. oils
Regulatory: FDA 21CFR §177.2800
Manuf./Distrib.: Aldrich; Fluka; Sigma
Trade names: Calgene Nonionic S-20

Stearic acid
CAS 57-11-4; EINECS 200-313-4
Synonyms: n-Octadecanoic acid; Carboxylic acid C_{18}
Classification: Fatty acid
Empirical: $C_{18}H_{36}O_2$

Formula: $CH_3(CH_2)_{16}COOH_9$
Properties: Wh. to ylsh.-wh. amorphous solid, tallow-like odor and taste; very sl. sol. in water; sol. in alcohol, ether, acetone, CCl_4; m.w. 284.47; dens. 0.847 (70 C); m.p. 69.3 C; b.p. 383 C; acid no. 195-200; iodine no. 4 max.; flash pt. (CC) 385 F; ref. index 1.4299 (80 C)
Precaution: Combustible when exposed to heat or flame; heats spontaneously
Toxicology: LD50 (IV, rat) 21.5 ± 1.8 mg/kg; poison by intravenous route; experimental tumorigen; human skin irritant; possible sensitizer for allergic persons; heated to decomp., emits acrid smoke and irritating fumes
Uses: Cosmetics, chemicals, dispersant, softener in rubber compds., food packaging, suppositories, ointments; PVC lubricant; intermediate for paints
Regulatory: FDA 21CFR §172.210, 172.615, 172.860, 175.105, 175.300, 175.320, 176.170, 176.200, 176.210, 177.1010, 177.1200, 177.2260, 177.2600, 177.2800, 178.3570, 178.3910, 184.1090, GRAS; FEMA GRAS; Europe listed; FDA approved for buccals, implants, orals, topicals, vaginals; USP/NF, BP, JP compliance
Manuf./Distrib.: Akrochem; Akzo Nobel; Aldrich; Allchem Ind.; Ashland; J.T. Baker; Browning; R.E. Carroll; Condor; Cookson Spec. Additives; Fluka; Great Western; C.P. Hall; Henkel/Emery; Lonza; Penta Mfg.; Sigma; Syn. Prods.; TR-AMC; Unichema Int'l.; Union Camp; Van Waters & Rogers; Witco/Oleo-Surf.
Trade names containing: Ross Beeswax Substitute No. 628/5; Ross Synthetic Candelilla Wax

Stearic acid aluminum dihydroxide salt. *See* Aluminum stearate
Stearic acid amide. *See* Stearamide
Stearic acid calcium salt. *See* Calcium stearate
Stearic acid, lead salt. *See* Lead stearate
Stearic acid methyl ester. *See* Methyl stearate
Stearic acid, 2-methylpropyl ester. *See* Isobutyl stearate
Stearic monoethanolamide stearate. *See* Stearamide MEA-stearate

Stearyl acrylate
Synonyms: Octadecyl acrylate
Definition: Ester of acylic acid and C16-C18 alcohols
Properties: M.p. 31-32 C
Uses: Flexibilizing agent in coatings and inks; improves water resist. props.
Manuf./Distrib.: Monomer-Polymer & Dajac Labs
Trade names: Photomer® 4818
Trade names containing: SR-257

Stearyl amine acetate. *See* Stearamine acetate
Stearyl-3-(3′,5′-di-t-butyl-4-hydroxyphenyl) propionate. *See* Octadecyl 3,5-di-t-butyl-4-hydroxyhydrocinnamate

Stearyl hydroxyethyl imidazoline
CAS 95-19-2; EINECS 202-397-8
Synonyms: Stearyl imidazoline; 2-Heptadecyl-4,5-dihydro-1H-imidazole
Classification: Heterocyclic compd.
Empirical: $C_{22}H_{44}N_2O$
Uses: Emulsifier, dispersant, detergent, antistat; intermediate for quat. ammonium compds.; strongly absorbed on textiles, paper and many metal surfs.; for agric., asphalt, cleaners, corrosion inhibitors, demulsifiers, flotation, metalworking, paints; pigment grinding, inks, textiles, wax emulsions
Trade names: Calgene C-100-S; Crodazoline S

N-Stearyl 12-hydroxystearamide
Uses: Lubricant, antistat for plastics, metals; mold release; antiblocking agent for textile coatings; slip agent for coatings; elec. potting compds., crayons, wax blends

Stearyl imidazoline. *See* Stearyl hydroxyethyl imidazoline

Stearyl methacrylate
CAS 32360-05-7
Synonyms: Octadecyl methacrylate
Empirical: $C_{22}H_{42}O_2$
Formula: $CH_2C(CH_3)COOC_{18}H_{37}$
Properties: M.w. 338.58
Precaution: DOT: Nonhazardous
Uses: Lube oil additive, pour point depressant, paper coatings, textile finishes, paints, varnishes, pressure-sensitive adhesives

Manuf./Distrib.: Aldrich; CPS; Rohm & Haas; Rohm Tech; Sartomer
Trade names: Photomer® 2818
Trade names containing: Ageflex FM-68; SR-324

Sterculia gum. *See* Karaya gum
Sterculia urens gum. *See* Karaya gum
Stibic anhydride. *See* Antimony pentoxide
St. John's bread. *See* Locust bean gum
STO. *See* Sorbitan trioleate

Stoddard solvent
CAS 8052-41-3
Synonyms: White spirits; Naphtha safety solvent
Definition: Mixed isomer contg. 85% nonane and 15% trimethylbenzene
Properties: Colorless clear liq.; misc. with abs. alcohol, benzene, ether, chloroform, CCl_4, CS_2; insol. in water; dens. 1.0; b.p. 220-300 C; flash pt. 100 F
Precaution: Flamm. exposed to heat or flame; explosive in vapor form when exposed to heat or flame; heated to decomp., may explode; reactive with oxidizing materials
Toxicology: Mildly toxic by inhalation; human eye irritant; heated to decomp., emits acrid fumes
Uses: Drycleaning solv.; spot and stain removal
Manuf./Distrib.: Ashland; R.E. Carroll
Trade names containing: BYK®-A 525; Virco-Pet® 50

Strong ammonia solution. *See* Ammonium hydroxide

Strontium chromate
CAS 7789-06-2
Synonyms: Chromic acid strontium salt; Chromic acid strontium salt (1:1); Strontium chromate (1:1); CI Pigment yellow 32; Strontium yellow
Empirical: $CrO_4 \cdot Sr$
Properties: Yel. monoclinic cryst.; m.w. 203.62; dens. 3.89 (15 C)
Toxicology: LD50 (oral, rat) 3118 mg/kg; ACGIH TLV:TWA 0.05 mg (Cr)/m^3; mod. toxic by ingestion; human carcinogen; experimental tumorigen; mutagenic data
Uses: Rust-inhibitive pigment for paints
Manuf./Distrib.: Atlantic Equip. Engrs.; Barium & Chems.; BASF; Davis Colors; Kikuchi Color & Chem.; D.N. Lukens; Matteson-Ridolfi; Min. Pigments; Nat'l. Chem.; Noah; Revelli; Van Waters & Rogers; Wayne Pigment

Strontium chromate (1:1). *See* Strontium chromate

Strontium sulfate
Uses: Filler, extender for coatings, ceramics, caulks, and refractories
Manuf./Distrib.: Barium & Chems.; R.E. Carroll
Trade names: Celestite SR-90; Foamwate 200; Foamwate 325

Strontium yellow. *See* Strontium chromate
STS. *See* Sorbitan tristearate

Styrene
CAS 100-42-5; EINECS 202-851-5
Synonyms: Phenethylene; Phenylethene; Phenylethylene; Ethenylbenzene; Styrol; Styrole; Styrolene; Styron; Styropor; Vinylbenzene; Vinylbenzol; Styrene monomer; Cinnamene; Cinnamenol; Cinnamol
Empirical: C_8H_8
Formula: $C_6H_5CH:CH_2$
Properties: Colorless to ylsh. oily liq., penetrating odor; sol. in alcohol, ether, methanol, acetone, CS_2; sparingly sol. in water; m.w. 104.15; dens. 0.906 (20/4 C); m.p. -30.6 C; b.p. 145-146 C; flash pt. (CC) 31 C; ref. index 1.546 (20 C)
Precaution: Flamm.; dangerous fire hazard exposed to flame, heat, oxidizers; vapor explosive exposed to heat/flame; slowly undergoes polymerization and oxidation on exposure to light and air, yielding peroxides
Toxicology: LD50 (oral, rat) 5000 mg/kg, (IV, mouse) 90 mg/kg, (IP, mouse) 660 mg/kg; experimental poison by ing., inh., IV routes; skin irritant; may be irritating to eyes, mucous membranes; narcotic in high concs.; ACGIH TLV:TWA 50 ppm
Uses: Monomer for prod. of paints

Regulatory: FDA 21CFR §172.515; FEMA GRAS
Manuf./Distrib.: Aldrich; Allchem Ind.; Amoco; ARCO; Ashland; Chemcentral; Chemical; Chevron; Dow Plastics; Elf Atochem N. Am.; Fluka; Huntsman; Samson; Sigma; Van Waters & Rogers
Trade names: Amoco® Styrene Monomer
Trade names containing: CN 120 S80

Styrene/acrylate copolymer. *See* Styrene/acrylates copolymer

Styrene/acrylates copolymer
CAS 9003-54-7
Synonyms: Styrene/acrylate copolymer; Styrene-acrylic
Definition: Polymer of styrene and a monomer consisting of acrylic acid, methacrylic acid, or their simple esters
Empirical: $(C_6H_8 \cdot C_3H_5NO)_x$
Toxicology: Strong irritant
Uses: For inj. molding and extrusion applics., specialty medical, pharmaceutical, food, and cosmetic pkg., advertising displays, high-str. toys; binder for floor coverings, concrete roof tiles; in paints, mastics, thermal insulation, adhesives
Regulatory: FDA 21CFR §175.300, 175.320, 176.170, 177.1010, 177.1830
Manuf./Distrib.: Goodyear; Polysat; Reichhold; Seegott
Trade names: Arolon 820-W-49; Arolon 860-W-45; Carboset® 1161; Carboset® 1162; Carboset® CR750; Carboset® CR751; Carboset® CR760; Carboset® CR761; Carboset® CR763; Carboset® CR764; Carboset® GA1387; Carboset® GA1719; Fulatex® PD-600; Fulatex® PD-638; Fulatex® PD-3194-P; Fulatex® PD-3631-G; Fulatex® PD-3691-M; Morez™ 101 Resin; Morez™ 200 Resin; Mor-Flo™ 395; Mor-Flo™ 602; NeoCryl® A-633; NeoCryl® A-639; NeoCryl® A-640; NeoCryl® A-4042; NeoCryl® A-6037; Pliolite® AC; Pliolite® AC4; Pliolite® AC5G; Pliolite® AC-80; Pliolite® AC-L; Pliotec™ 7103; Pliotec™ 7104; Pliotec™ 7217; Synthemul® 40-422; Synthemul® 40-430; Synthemul® 40-431; Texicryl® 13-031; Texicryl® 13-034; Texicryl® 13-040; Texicryl® 13-105; Texicryl® 13-800; Texicryl® Ecobinder; Ucar® Latex 421; Ucar® Latex 422; Ucar® Latex 460; Ucar® Latex 470; Ucar® Latex 480; Ucar® Vehicle 442; Ucar® Vehicle 451; Ucar® Vehicle 458; Ucar® Vehicle 4579, 4580
Trade names containing: Carboset® GA1926; Carboset® GA1931

Styrene-acrylic. *See* Styrene/acrylates copolymer

Styrene allyl alcohol
CAS 25119-62-4
Definition: Copolymer of styrene and allyl alcohol
Uses: Hard thermoplastic material; additive to improve adhesion, heighten gloss, improve weatherability, increase water and chem. resist. in coatings
Trade names: Arcal™ SAA; SAA-100; SAA-101

Styrene-butadiene copolymer. *See* Polybutadiene-styrene copolymer
Styrene-1,3-butadiene copolymer. *See* Polybutadiene-styrene copolymer

Styrene-butadiene polymer
CAS 9003-55-8
Synonyms: S/B; Styrene polymer with 1,3 butadiene (INCI); Butadiene-styrene resin; 1,3-Butadiene-styrene copolymer
Formula: $(C_8H_8 \cdot C_4H_6)_x$
Properties: Dens. 0.965
Toxicology: Eye irritant; may cause irritation
Uses: Inj. and blow molding, extrusion, thermoforming resin for housings, blister packs, tubes, toys, containers, medical devices, bottles, tech. parts; reinforcement for rubber; latex emulsions
Manuf./Distrib.: Aldrich; Ashland; Dow Plastics; BFGoodrich; Goodyear; Hampshire; Polyad; Reichhold; Rhone-Poulenc; Van Waters & Rogers
Trade names: Darex® 515L; Darex® 537L; Darex® 643L; Good-rite® 2570X59; Pliolite® S-5A; Pliolite® S-5B; Pliolite® S-5D; Pliolite® S-5E; Ricon 500; Tylac® 97-422; Tylac® 68212-00; Tylac® 68225-00; Tylac® 97422-20; Tylac® 97757-02; Tylac® 97885-00; Tylac® 97902-00; Tylac® 97917-00; Tylac® 97920-00; Tylac® 97924-00; Tylac® 97936-00
See also Polybutadiene-styrene copolymer

Styrene-butadiene rubber
CAS 9003-55-8
Synonyms: SBR

Styrene-ethylene/butylene-styrene block copolymer

Uses: Latex as binder, for coatings, paper/paperboard, sealants, adhesives, carpet backing; SBR for tires, adhesives, chewing gums; processing aid, modifier
Manuf./Distrib.: Aldrich
Trade names: Goodyear LPR-6632; Heveasyn SBR Latex; Ricon 100; Stereon® 840A

Styrene-ethylene/butylene-styrene block copolymer
Synonyms: S-EB-S thermoplastic elastomer
Uses: Thermoplastic elastomer for extrusion, inj. and blow molding; medical devices, tubing, disp. coatings, seals, gaskets, automotive parts, sporting goods, containers, film, molded parts; impact modifier, processing aid; adhesives, sealants, coatings
Trade names: Kraton® G 1650; Kraton® G 1652; Kraton® GX 1657

Styrene/MA copolymer
CAS 9011-13-6
Synonyms: SMA; Styrene/maleic anhydride copolymer; 2,5-Furandione, polymer with ethenylbenzene; Styrene/maleic anhydride resin
Definition: Polymer of styrene and maleic anhydride monomers
Empirical: $(C_8H_8 \cdot C_4H_2O_3)_x$
Properties: Dens. 1.270
Toxicology: Irritant
Uses: Engineering thermoplastic for inj. molding (mech. and automotive parts, appliance and electronic housings); emulsifier, binder, visc. modifier, stabilizer, pigment dispersant, protective colloid, leveling agent, sizing, adhesives, coatings, polishes; emulsion polymerization
Regulatory: FDA 21CFR §176.170, 177.1200, 177.1210, 177.1820
Manuf./Distrib.: Aldrich; ARCO; Cargill; Chemical; Elf Atochem N. Am.; Hoechst Canada; Monomer-Polymer & Dajac Labs; Monsanto; Sigma
Trade names: Scripset® 520; SMA® 1000; SMA® 1440; SMA® 2000; SMA® 3000; Surfynol® CT-151
Trade names containing: SMA® 1440H

Styrene/maleic anhydride copolymer. *See* Styrene/MA copolymer
Styrene/maleic anhydride resin. *See* Styrene/MA copolymer
Styrene monomer. *See* Styrene
Styrene polymer. *See* Polystyrene
Styrene polymer with 1,3-butadiene. *See* Polybutadiene-styrene copolymer, Styrene-butadiene polymer

Styrene/PVP copolymer
CAS 25086-29-7
Synonyms: 1-Ethenyl-2-pyrroldinone, polymer with ethenylbenzene; Vinylpyrrolidone/styrene copolymer; PVP/styrene copolymer
Definition: Copolymer from vinylpyrrolidone and styrene monomers
Empirical: $(C_8H_8 \cdot C_6H_9NO)_x$
Uses: Binder and adhesive for wood, cotton, paper, glass fiber, flour, concrete; stabilizer and opacifier; laundry processing; stabilizer for detergents, textiles, paper coatings, latex rug backings, floor wax emulsions, and cosmetics
Manuf./Distrib.: Aldrich
Trade names: Polectron® 430

Styrol. *See* Styrene
Styrole. *See* Styrene
Styrolene. *See* Styrene
Styron. *See* Styrene
Styropor. *See* Styrene
Sublimed blue lead. *See* Lead sulfate, blue basic

Succinic acid
CAS 110-15-6; EINECS 203-740-4
Synonyms: 1,4-Butanedioic acid; Amber; Amber acid; 1,2-Ethanedicarboxylic acid; Ethylenesuccinic acid
Classification: Dicarboxylic acid
Empirical: $C_4H_6O_4$
Formula: $HOOCCH_2CH_2COOH$
Properties: Colorless monoclinic prisms, odorless, sour acid taste; very sol. in alcohol, ether, acetone, glycerin; sol. in water; m.w. 118.09; dens. 1.552; m.p. 185 C; b.p. 235 C
Precaution: Combustible

Toxicology: LD50 (oral, rat) 2260 mg/kg; moderately toxic by subcutaneous route; severe eye irritant; heated to decomp., emits acrid smoke and irritating fumes
Uses: Organic synthesis; mfg. of lacquers, dyes, esters for perfumes, photography, in foods as a sequestrant, buffer, neutralizing agent
Regulatory: FDA 21CFR §131.144, 184.1091, GRAS; Japan approved; Europe listed; UK approved; FDA approved for parenterals, orals
Manuf./Distrib.: AC Ind.; Aldrich; Allan; Allchem !nd.; Am. Biorganics; J.T. Baker; Chemical; Chemie Linz N. Am.; DuPont; Eastern Chem.; General Chem.; Fluka; Hüls AG; Hayashi Pure Chem. Ind.; Mallinckrodt; Mitsubishi Chem.; Nippon Shokubai; Penta Mfg.; Riken Vitamin Oil; Schweizerhall; Sigma; Spectrum Chem. Mfg.; Takeda Chem. Ind.; Ueno Fine Chems. Ind.

Succinic acid anhydride. *See* Succinic anhydride

Succinic anhydride
CAS 108-30-5; EINECS 203-570-0
Synonyms: Dihydro-2,5-furandione; Butanedioic anhydride; 2,5-Diketotetrahydrofuran; Succinic acid anhydride; Succinyl oxide
Empirical: $C_4H_4O_3$
Formula: $OCCH_2CH_2COO$
Properties: Orthorhombic prisms; sol. in chloforom, CCl_4, alcohol; very sl. sol. in ether, water; m.w. 100.08; dens. 1.503; m.p. 119-120 C; b.p. 261 C (760 mm); sublimes @ 115 C and 5 mm pressure
Precaution: Moisture-sensitive
Toxicology: Irritating to eyes, respiratory system
Uses: Manufacture of chemicals, pharmaceuticals, esters, paints; hardener for resins, starch modifier in foods
Regulatory: FDA
Manuf./Distrib.: AC Ind.; Aldrich; Allchem Ind.; Am. Biorganics; Anhydrides & Chems.; J.T. Baker; Buffalo Color; Cambridge Ind. Co. of Am.; Chemical; Chemie Linz N. Am.; Fluka; Hüls AG; Humphrey; Lubrizol; Monomer-Polymer & Dajac Labs; Penta Mfg.; Punda Mercantile; Schweizerhall; Sigma; Spectrum Chem. Mfg.

Succinyl oxide. *See* Succinic anhydride

Sucrose acetate isobutyrate
CAS 126-13-6; EINECS 204-771-6
Synonyms: SAIB
Classification: Sucrose derivative
Definition: Mixed ester of sucrose and acetic and isobutyric acids
Empirical: $C_{40}H_{62}O_{19}$
Formula: $(CH_3COO)_2C_{12}H_{14}O_3[OOCCH(CH_3)_2]_6$
Properties: Clear semisolid or sol'n.; m.w. 847.02; sp.gr. 1.146; flash pt. (COC) 260 C; ref. index 1.4540
Precaution: Combustible
Uses: Plasticizer for cellulosics, PS, PVAc; modifier for lacquers, hot-melt coating formulations, extrudable plastics
Regulatory: FDA 21CFR §175.105
Manuf./Distrib.: Ashland; Eastman; Tech. Chems. & Prods.

Sucrose benzoate
CAS 12738-64-6; EINECS 235-795-5
Synonyms: β-D-Fructofuranosyl-α-D-glucopyranoside benzoate
Classification: Disaccharide ester
Properties: Sp.gr. 1.25; m.p. 98 C; flash pt. (COC) 260 C; ref. index 1.577
Uses: Plasticizer for PVC, PVAc, VCA, PS, cellulosics; modifier for coatings
Regulatory: FDA 21CFR §175.105
Manuf./Distrib.: Velsicol
Trade names: Sucrose Benzoate, Alcohol Soluble; Sucrose Benzoate CG Grade

Sugar of lead. *See* Lead acetate

Sulfated castor oil
CAS 72-48-0; 8002-33-3; EINECS 232-306-7
Synonyms: Castor oil sulfated; Sulfonated castor oil; Turkey-red oil
Definition: Oil consisting primarily of sodium salt of the sulfated triglyceride of ricinoleic acid
Uses: Anionic wetting agent; emulsifier, plasticizer, lubricant, dispersant; in textile dyeing, adhesives;

cosmetics superfatting agent, shampoos; antisag/antisettling agent in paints; emulsifier for latex
Regulatory: FDA 21CFR §175.105, 176.170, 176.200, 177.1200
Manuf./Distrib.: Aldrich; Fluka; Graden; Sigma; Spectrum Chem. Mfg.
Trade names: Eureka 102; Haroil SCO-75; Laurel R-50; Laurel R-75; Monosulf; Rilanit® LD4; Troythix™ Anti-Sag4

Sulfated peanut oil
CAS 73138-79-1
Synonyms: Oils, peanut, sulfated; Peanut oil, sulfated
Definition: Prod. obtained by sulfation of peanut oil
Uses: Emulsifier, wetting agent, dispersant, antistat for nylon finishes
Regulatory: FDA 21CFR §175.105
Trade names: Standapol® 1610

Sulfated vegetable oil
Uses: Surfactant used in metalworking, textile, leather, printing inks, paints, detergents
Trade names: Hydrolene

Sulfobutanedioic acid, 1,4-ditridecyl ester, sodium salt. *See* Ditridecyl sodium sulfosuccinate
Sulfobutanedioic acid, nonoxynol-10 ester, disodium salt. *See* Disodium nonoxynol-10 sulfosuccinate
Sulfobutanedioic acid, tallow ester, disodium salt. *See* Disodium tallow sulfosuccinamate
Sulfocarbolic acid. *See* Phenol sulfonic acid

Sulfolane
CAS 126-33-0; EINECS 204-783-1
Synonyms: Tetrahydrothiophene 1,1-dioxide; Dapsone; Tetramethylene sulfone; Thiophan sulfone
Empirical: $C_4H_8O_2S$
Formula: $CH_2CH_2CH_2CH_2SO_2$
Properties: Wh. or creamy wh. cryst. powd., odorless, sl. bitter taste; very sl. sol. in water; sol. in alcohol, acetone, dilute min. acids; m.w. 120.16; dens. 1.27 (20 C); m.p. 25-28 C; b.p. 285 C; flash pt. 330 F; ref. index 1.485
Precaution: Combustible
Toxicology: LD50 (oral, rat) 1941 mg/kg; toxic by ingestion; irritant
Uses: Curing agent for epoxy resins; medicine (antibacterial); solv. for extraction of benzene, toluene, other aromatic hydrocarbons from oil refinery streams; used in Sulfinol process to remove acid gases; used for separation of low boiling alcohols, min. oils, tars; plasticizer; polymerization solv.; dielec. in elec. equipment; solv. in surf. coatings; component in hydraulic fluids
Manuf./Distrib.: Aldrich; Fluka; Phillips; Shell; Sigma; Syn. Chems. Ltd
Trade names: Sulfolane W

Sulfomethane. *See* Methanesulfonic acid
Sulfonated castor oil. *See* Sulfated castor oil
Sulfonated petroleum, sodium salt. *See* Sodium petroleum sulfonate

Sulfonic acid
CAS 72674-05-6
Classification: Organic compound containing one or more sulfo radicals
Uses: Dispersant, wetting agent, leveling agent, protective colloid for dyestuffs; intermediate for paints
Manuf./Distrib.: Lubrizol; Royale Pigments & Chems.; Stepan; Van Waters & Rogers; Vista; Yorkshire Chem. plc

Sulfonylbis (trichloromethane). *See* Bis (trichloromethyl) sulfone
4,4′-Sulfonyldianiline. *See* 4,4′-Diaminodiphenyl sulfone
Sulfopropyl methacrylate, K salt. *See* Potassium sulfopropyl methacrylate
Sulfosuccinic acid diisobutyl ester sodium salt. *See* Diisobutyl sodium sulfosuccinate
Sulfuric acid barium salt (1:1). *See* Barium sulfate
Sulfuric acid, manganese (2+) salt. *See* Manganese sulfate (ous)
Sulfuric acid, monododecyl ester, ammonium salt. *See* Ammonium lauryl sulfate
Sulfuric acid monododecyl ester sodium salt. *See* Sodium lauryl sulfate
Sulfuric acid, mono (2-ethylhexyl) ester sodium salt. *See* Sodium octyl sulfate
Sulfuric acid, monotetradecyl ester, sodium salt. *See* Sodium myristyl sulfate
Sulfuric ether. *See* Ethyl ether
Sumac wax. *See* Japan wax

Sunflower acid
Uses: For high quality med. to fast drying alkyd resins, suitable for nonyel. lacquer systems with exc. heat stability
Trade names: Prifac 7960

Sunflower oil. *See* Sunflower seed oil

Sunflower seed oil
CAS 8001-21-6; EINECS 232-273-9
Synonyms: Sunflower oil
Definition: Oil expressed from seeds of the sunflower, *Helianthus annuus*
Properties: Amber liq., pleasant odor, mild taste; sol. in alcohol, ether, chloroform, CS_2; dens. 0.924-0.926; iodine no. 125-140; sapon. no.188-194; ref. index 1.4611
Precaution: Combustible
Toxicology: No known toxicity; heated to decomp., emits acrid smoke and irritating fumes
Uses: Raw material in plastics, polymers, rubbers, pharmaceuticals; drying oil for paints; emollient for cosmetics; in prep. of margarine
Regulatory: FDA 21CFR §175.300, 176.200, GRAS; BP compliance
Manuf./Distrib.: ABITEC; ADM; Alnor Oil; Arista Ind.; Charkit; Floratech; Int'l. Chem. Inc.; Lipo; Penta Mfg.; Sigma; Werner G. Smith; Tri-K Ind.; Welch, Holme & Clark

Super cobalt. *See* Cobalt
SVS. *See* Sodium vinyl sulfonate
Sylvic acid. *See* Abietic acid
sym-Triaminotriazine. *See* Melamine
sym-Trimethylbenzene. *See* 1,3,5-Trimethylbenzene

Synthetic beeswax
CAS 71243-51-1; 97026-94-0; EINECS 275-286-5
Synonyms: Beeswax, synthetic
Definition: Synthetic wax with composition and properties generally indistinguishable from natural beeswax
Uses: Lipophilic emulsifier; suspending agent for anhyd. systems; w/o emulsifier, thickener; used where pure beeswax not required; for creams, lotions, lipstick, makeup; depilatories; ointments, salves; sustained release pharmaceuticals; furniture, wood, and leather polishes; leather, textile, wood, and paper finishes
Trade names: Koster Keunen Substitute Beeswax

Synthetic pearl. *See* Bismuth oxychloride
Synthetic spermaceti. *See* Cetyl esters
Synthetic spermaceti wax. *See* Cetyl esters

Synthetic wax
CAS 8002-74-2; 123237-14-9
Synonyms: Fischer-Tropsch wax; Fischer-Tropsch wax, oxidized
Definition: Hydrocarbon wax derived by Fischer-Tropsch or ethylene polymerization processes
Uses: As ingredient, finish or processing aid, release agent, lubricant in adhesives, ammunition, asphalt, explosives, paints, paper, pyrotechnics, lubricants, PVC, textile finishes, powd. metallurgy, floor wax, candles, hot melts, inks, asphalt
Regulatory: FDA 21CFR §172.615, 172.888, 173.340, 175.105, 175.250, 176.170, 177.1200, 178.3720
Manuf./Distrib.: Aldrich
Trade names: Aqua Poly 250; Aqua Polyfluo 411; Aqua Polysilk 19; Aquawax 114; Aquawax 214; Microspersion™ 114-50; MP-12; MP-22; MP-22C; MP-22VF; MP-22XF; MP-26; MP-26VF; MP-28C; Ross Wax #100; Ross Wax #160; Shamrock S-400 N1; Shamrock S-400 N5; Shamrock S-400 SP5; Vestowax FT-150; Vestowax FT-150P; Vestowax FT-200; Vestowax FT-300; Vybar® 103; Vybar® 253; Vybar® 260; Vybar® 825
Trade names containing: Microspersion™ 114; Slip-Ayd® SL-700; Synfluo 100XF; Synfluo 178VF; Synfluo 178XF; Synfluo 180VF; Synfluo 180XF

Tabular alumina. *See* Alumina

Talc
CAS 14807-96-6; EINECS 238-877-9

Talcum

Synonyms: Hydrous magnesium silicate; Magnesium hydrogen metasilicate; Hydrous magnesium calcium silicate; Industrial talc; Cosmetic talc; Platy talc; French chalk; Talcum; Pigment white 26; CI 77019
Definition: Native, hydrous magnesium silicate sometimes containing small portion of aluminum silicate
Formula: $Mg_3Si_4O_{10}(OH)_2$ or $3MgO \cdot 4SiO_2 \cdot HOH$
Properties: Wh., apple green, gray powd., pearly or greasy luster, greasy feel; insol. in water, cold acids or in alkalies; m.w. 379.29; dens. 2.7-2.8
Toxicology: TLV:TWA 2 mg/m^3, respirable dust; toxic by inhalation; talc with < 1% asbestos is nuisance dust; experimental tumorigen; human skin irritant; prolonged/repeated exposure can produce talc pneumoconiosis; talc-based powds. linked to ovarian cancer
Uses: Reinforcement, filler, and pigment in rubber, paints, plastics, soaps, ceramics, cosmetics, pharmaceuticals; dusting agent, lubricant, electrical insulation; antiblocking agent for plastic film
Usage level: 0.003-220.4 mg (oral solids)
Regulatory: FDA 21CFR §73.1550, 175.300, 175.380, 175.390, 176.170, 177.1210, 177.1350, 177.1460, 182.70, 182.90; GRAS; Japan restricted (5000 ppm); Europe listed; UK approved; FDA approved for orals, rectals, topicals; USP/NF, BP, Ph.Eur. compliance
Manuf./Distrib.: 20 Microns Ltd.; Aldrich; Archway Sales; Celite; Chem-Materials; Fluka; Landers-Segal Color; Lomas Int'l.; D.N. Lukens; Luzenac Am.; Mins. Tech.; Pfizer Int'l.; Punda Mercantile; H.M. Royal; L.A. Salomon; Seegott; Sigma; Specialty Mins.; Tamms Ind.; R.T. Vanderbilt; Van Waters & Rogers; Whittaker, Clark & Daniels
Trade names: Artic Mist; Beaverwhite 200; Cimpact 699; Cimpact 700; Cimpact 710; I.T. 3X; I.T. 5X; I.T. 325; I.T. FT; I.T. X; LVT-325; LVT-400; LVT-500; LVT-600; Microtalc® MP10-52; Microtalc® MP12-50; Microtalc® MP15-38; Microtalc® MP25-38; Microtalc® MP30-36; Nicron 325; Nicron 400; Nicron 665; Nicron JS 216; Nicron JS 322; Nicron JS 422; Nicron JS 426; Nicron JS 634; Nytal® 200; Nytal® 300; Nytal® 300H; Nytal® 400; Silverline 200; Silverline 400; Silverline 665; Talcron® MP 40-27; Talcron® MP 44-26; Talcron® MP 45-26; Ultratalc™ 609; Vantalc® 6H; Vantalc® D3509, D8013, D8114.; Vantalc® F2003; Vertal 710
Trade names containing: Mistron ZSC; Nipol® 1452X8

Talcum. *See* Talc

Tall oil
CAS 8002-26-4; EINECS 232-304-6
Synonyms: Liquid rosin; Tallol
Definition: Byprod. of wood pulp contg. rosin acids, oleic and linoleic acids, and long chain alcohols
Properties: Dk. brn. liq., acrid odor; dens. 0.95; acid no. 178 min.; flash pt. 360 F
Precaution: Combustible when exposed to heat or flame; can react with oxidizing materials
Toxicology: Mild allergen; heated to decomp., emits acrid smoke and irritating fumes
Uses: Drying oil for paints
Regulatory: FDA 21CFR §175.105, 175.300, 176.200, 176.210, 177.2600, 177.2800, 181.22, 181.26, 182.70, 186.1557, GRAS as indirect food additive; FDA approved for topicals
Manuf./Distrib.: Arakawa USA; Arizona; Chemcentral; Chemical; Climax Perf. Materials; Georgia-Pacific Resins; Harcros; Lomas Int'l.; Geo. Pfau's Sons; Spectrum Chem. Mfg.; Union Camp; Westvaco

Tall oil acid
CAS 61790-12-3; EINECS 263-107-3
Synonyms: Acids, tall oil; Fatty acids, tall oil; Tall oil fatty acids
Definition: Mixture of rosin acids and fatty acids recovered from the hydrolysis of tall oil
Properties: Liq.
Precaution: Combustible
Toxicology: Mildly irritating to skin; repeated skin contact may result in allergic reactions such as rash or dermatitis; inh. of mists can cause irritation
Uses: Polymerization emulsifier; mfg. of surfactants, soaps, amines, imidazolines; cosmetics, dyes, leather, coatings, petrol. industry
Regulatory: FDA 21CFR §175.105, 175.320, 176.200, 176.210, 177.2600, 177.2800, 178.3570, 178.3910
Trade names: Acintol® D25LR; Acintol® D30E; Acintol® D30LR; Acintol® D40LR; Acintol® D40T; Acintol® D60LR; Acintol® DFA; Acintol® EPG; Acintol® FA-1; Acintol® FA-1 Special; Acintol® FA-2; Acintol® FA-3

Tall oil aminoethyl imidazoline
Synonyms: Aminoethyl tall oil imidazoline
Uses: Intermediate for prod. of film-forming corrosion inhibitors; in automatic car wash rinse aids, oilfield applics.; pigment dispersant for paints
Trade names: Chemzoline T-11; Chemzoline T-33

Tall oil, copper salt. *See* Copper tallate
Tall oil fatty acids. *See* Tall oil acid

Tall oil hydroxyethyl imidazoline
CAS 61791-39-7; EINECS 263-171-2
Synonyms: 1-Hydroxyethyl-2-tall oil imidazoline; Tall oil imidazoline; 4,5-Dihydro-7-nortall oil-1H-imidazole-1-ethanol
Classification: Heterocyclic compd.
Uses: Emulsifier, corrosion inhibitor in oil burning systems, pickling bath operations, asphalt emulsions; dispersant for clay and pigments; in protective metal coatings, printing ink additive
Trade names: Calgene C-100-T; Monazoline T

Tall oil imidazoline. *See* Tall oil hydroxyethyl imidazoline

Tall oil rosin
Uses: Mfg. of paints
Manuf./Distrib.: Arakawa Chem. Ind.; Arizona; Chemical; Georgia-Pacific Resins; Hercules; Matteson-Ridolfi; Reichhold; TR-AMC; Union Camp
See also Rosin

Tallol. *See* Tall oil
N-Tallow alkyl trimethylene diamine. *See* Tallow propylene diamine
Tallow amides, hydrogenated. *See* Hydrogenated tallow amide
Tallow diamino propane. *See* Tallow propylene diamine
N-Tallow-1,3-diaminopropane dioleate. *See* Tallow propane diamine dioleate

Tallow dipropylene triamine
CAS 61791-57-9; 85632-63-9; EINECS 263-191-1, 288-048-0
Synonyms: N-3-Aminopropyl-N-tallow alkyl trimethylene diamine
Uses: Emulsifier for bitumen; antistripping agent for roadmaking; wetting and adhesion agent, flushing agent, pigment grinding and dispersion agent for pigment industry; curing agent for epoxy paints and coatings; corrosion inhibitor
Trade names: Lilamin LSP 33

Tallow glycerides
Synonyms: Tallow mono, di and tri glycerides; Glycerides, tallow mono-, di- and tri-
Definition: Mixture of mono, di and triglycerides derived from tallow
Uses: Emulsifier, stabilizer, dispersant, opacifier for cosmetics, foods and drugs
Regulatory: FDA 21CFR §175.105, 176.210
Trade names containing: Ross Japan Wax Substitute 525

Tallow mono, di and tri glycerides. *See* Tallow glycerides
N-Tallow-1,3-propanediamine. *See* Tallow propylene diamine

Tallow propane diamine dioleate
CAS 61791-53-5
Synonyms: N-Tallow-1,3-diaminopropane dioleate
Uses: Chemical intermediate; corrosion inhibitor, fuel oil additive, flotation agent, lubricant; used in metals, textiles, plastics, herbicides; epoxy curing agent; dispersant in paint industry
Trade names: Duomeen® TDO

Tallow propylene diamine
CAS 61791-55-7; EINECS 263-189-0
Synonyms: N-Tallow-1,3-propanediamine; Tallow diamino propane; N-Tallow alkyl trimethylene diamine
Uses: Asphalt emulsifier, corrosion inhibitor, wetting agent, dispersant used in water treatment, pigment flushing, ore flotation; gasoline, grease, fuel oil additive; antistripping agent for road construction; textile finishing agent; paints
Trade names: Duomeen® T

Tangantangan oil. *See* Castor oil
TATM. *See* Triallyl trimellitate
TBA. *See* t-Butyl alcohol

TBBA. *See* Tetrabromobisphenol A
TBC. *See* Tributyl citrate
TBEP. *See* Tributoxyethyl phosphate
TBHP. *See* t-Butyl hydroperoxide
TBT. *See* Tetrabutyl titanate
TBTO. *See* Tributyltin oxide
TCE. *See* Trichloroethane
TCMTB. *See* 2-Thiocyanomethylthiobenzothiazole
TCP. *See* Tricresyl phosphate
TDI. *See* Toluene diisocyanate
TDP. *See* Triisodecyl phosphite
TDQP. *See* 2,2,4-Trimethyl-1,2-dihydroquinoline polymer
TEA. *See* Triethanolamine

TEA-dodecylbenzenesulfonate
CAS 27323-41-7; 68411-31-4; 29381-93-9; EINECS 248-406-9
Synonyms: Dodecylbenzenesulfonic acid, compd. with 2,2′,2′′-nitrilotris[ethanol] (1:1); Triethanolamine dodecylbenzene sulfonate
Classification: Substituted aromatic compd.
Empirical: $C_{18}H_{30}O_3S \cdot C_6H_{15}NO_3$
Properties: M.w. 475.68
Uses: Detergent, wetting agent, flash foamer; liq. detergents, wool wash compds., cosmetics and shampoos, agric. emulsifiers, industrial cleaners, textile scouring, car wash compds.; pigment dispersant; emulsion polymerization
Trade names: Witconate™ TAB

TEC. *See* Triethyl citrate
TEDA. *See* Triethylene diamine
TEGDA. *See* PEG-3 diacetate
TEPA. *See* Tetraethylenepentamine

Terephthalic acid
CAS 100-21-0; EINECS 202-830-0
Synonyms: Benzene-1,4-dicarboxylic acid; 1,4-Benzenedicarboxylic acid; p-Benzenedicarboxylic acid
Empirical: $C_8H_6O_4$
Properties: Wh. cryst. or powd.; sol. in alkalies; sl. sol. in alcohol; insol. in water, chloroform, ether, acetic acid; m.w. 166.14; dens. 1.51; m.p. > 300 C; subl. @ 402 C
Precaution: Can explode during preparation
Toxicology: LD50 (oral, rat) 18,800 mg/kg, (IP, mouse) 1430 mg/kg; mod. toxic by IV and IP routes; mildly toxic by ingestion; eye irritant; heated to decomp., emits acrid smoke and irritating fumes
Uses: Intermediate for paints; forms polyesters with glycols; in analytical chemistry
Manuf./Distrib.: Aldrich; Allchem Ind.; Amoco; Eastman; Fluka; Monomer-Polymer & Dajac Labs; Sigma; Spectrum Chem. Mfg.; Stanchem
Trade names: Amoco® TA

Terpene resin
CAS 9003-74-1
Uses: Thermoplastic resin; tackifier in adhesives; modifier resin in rubber compounding, coatings, laminations, waxes; extender for paints
Manuf./Distrib.: Arizona; Cardolite; Chemcentral; Hercules; Monomer-Polymer & Dajac Labs; Stanchem
Trade names: Piccofyn® A135; Piccofyn® T125; Zonatac® 105; Zonatac® 105 LITE; Zonatac® 115 LITE

Terpene resin, natural
CAS 9003-74-1
Classification: Unsat. hydrocarbon
Toxicology: No known toxicity; heated to decomp., emits acrid smoke and irritating fumes
Uses: Thermoplastic resin for use in hot-melt adhesives and coatings, as masticatory agents in chewing gum; antiseptic in oral pharmaceuticals
Usage level: Limitation 0.07% (of wt. of capsule) 7% (of ascorbic acid and salts)
Regulatory: FDA 21CFR §73.1, 172.280, 172.615, 178.3930; FDA approved for orals
Manuf./Distrib.: Arizona; Cardolite; Hercules; Langley Smith Ltd
See also dl-Limonene; α-Pinene

TETA. *See* Triethylenetetramine
1,3,5,7-Tetraazaadamantane. *See* Hexamethylenetetramine

Tetrabromobis (2-ethylhexyl) phthalate
CAS 26040-51-7
Synonyms: Phthalic acid, tetrabromo-bis (2-ethylhexyl) ester
Empirical: $C_{24}H_{34}Br_4O_4$
Uses: Flame retardant for PVC, SBR, neoprene, EPDM; plasticizing and elec. props.; for fire retardant flexible PVC, wire and cable insulation, film and sheeting, carpet backing, coated fabrics, wall coverings, adhesives and coatings
Trade names: Uniplex FRP-45

Tetrabromobisphenol A
CAS 79-94-7; EINECS 201-236-9
Synonyms: TBBA; 4,4´-Isopropylidenebis (2,6-dibromophenol)
Empirical: $C_{15}H_{12}Br_4O_2$
Properties: Wh. powd.; sol. in acetone, benzene, alcohol; m.w. 543.9; m.p. 180 C; decomp. @ 240 C
Toxicology: Irritant
Uses: Flame retardant for epoxies, phenolics, ABS, unsaturated polyesters, PC, adhesives, coatings, textiles; intermediate for flame retardants
Manuf./Distrib.: Albemarle; Aldrich; Allchem Ind.; Ameribrom; Great Lakes
Trade names: Great Lakes BA-59P™; Saytex® RB-100

Tetrabromobisphenol A, bis (2-hydroxyethyl ether)
Uses: Flame retardant for unsat. polyester and epoxy thermoset resins, laminates for electronic circuit boards, adhesives, coatings
Manuf./Distrib.: Great Lakes
Trade names: Great Lakes BA-50™; Great Lakes BA-50P™

Tetrabromobisphenol A diacrylate
Uses: Fire retardant for automotive coatings, wire and cable coatings
Trade names containing: SR-640

Tetrabromophthalate diol
CAS 20566-35-2; EINECS 243-885-0
Uses: Reactive intermediate used to produce flame retardant rigid urethane foam, PU elastomer wire insulation, adhesives, coatings, and fibers; can replace chlorinated polyols
Trade names: Great Lakes PHT4-Diol™; Saytex® RB-79

Tetrabromophthalic anhydride
CAS 632-79-1; EINECS 211-185-4
Synonyms: 1,3-Isobenzofurandione, 4,5,6,7-tetrabromo-(9CI); Bromphthal
Empirical: $C_8B4_4O_3$
Formula: $C_6Br_4C_2O_3$
Properties: Pale yel. crystalline solid; m.w. 463.72; m.p. 280 C
Precaution: Moisture-sensitive
Toxicology: Irritant
Uses: Flame retardant for plastics (unsaturated polyesters), paper, textiles; reactive intermediate for the preparation of polyols, esters and imides, paints
Manuf./Distrib.: Aldrich; Albemarle; Great Lakes
Trade names: Great Lakes PHT4™

Tetrabutyl orthotitanate monomer. *See* Tetrabutyl titanate

Tetrabutyl titanate
CAS 5593-70-4; EINECS 227-006-8
Synonyms: TBT; Butyl titanate; Orthotitanic acid tetrabutyl ester; Titanium butylate; Titanium (IV) butoxide; Tetrabutyl orthotitanate monomer
Empirical: $C_{16}H_{36}O_4 \cdot Ti$
Formula: $Ti(OC_4H_9)_4$
Properties: Colorless to lt. yel. liq.; sol. in most org. solvs. except ketones; dec. in water; m.w. 340.42; dens. 0.996; b.p. 310-314 C; flash pt. 170 F; ref. index 1.4900
Precaution: Combustible; moisture-sensitive

Tetrachloroethene

Toxicology: Irritant
Uses: Ester exchanger reactions, heat-resistant paints; improving adhesion of paints, rubber, plastics to metal surfaces
Manuf./Distrib.: Aldrich; Fluka
Trade names: Tyzor TBT

Tetrachloroethene. *See* Perchloroethylene
Tetrachloroethylene. *See* Perchloroethylene

Tetrachloroisophthalonitrile
CAS 1897-45-6
Synonyms: Chlorothalonil; Chloroalonil; 1,3-Dicyanotetrachlorobenzene
Empirical: $C_8Cl_4N_2$
Properties: Colorless crystals; insol. in water; pract. insol. in org. solvs.; dens. 1.70; m.p. 245 C; b.p. 350 C
Uses: Bactericide, nematocide; film preservative for use in water- and solv.-based coatings
Trade names: Busan® 1192; Busan® 1192D; Nopcocide® N-40-D; Nopcocide® N-96; Nuocide® 404-D; Nuocide® 960

Tetrachlorophthalic acid
Uses: Intermediate for paints
Manuf./Distrib.: Monsanto; Punda Mercantile

Tetrachlorophthalic anhydride
CAS 117-08-8; EINECS 204-171-4
Synonyms: Phthalic anhydride, tetrachloro
Empirical: $C_8Cl_4O_3$
Formula: $C_6Cl_4(CO)_2O$
Properties: Wh. free-flowing powd., odorless; sl. sol. in water; m.w. 285.90; m.p. 254-255 C; b.p. 371 C; flash pt. 360 C
Precaution: Moisture-sensitive
Toxicology: Irritating to eyes, skin, respiratory system; suspected carcinogen
Uses: Flame retardant for plastics; intermediate in dyes, paints, pharmaceuticals, plasticizers, and other org. materials; reagent
Manuf./Distrib.: Aldrich; Fluka; Monsanto; Punda Mercantile

Tetracosanoic acid, tetradecyl ester. *See* Myristyl lignocerate

Tetradecabromodiphenoxy benzene
CAS 58965-66-5; EINECS 261-526-6
Uses: Flame retardant for nylon, PET, PBT, styrenics, adhesives, paints
Trade names: Saytex® 120

Tetradecafluorohexane. *See* Perfluorohexane
Tetradecanoic acid, monoester with 1,2-propanediol. *See* Propylene glycol myristate
1-Tetradecanol, hydrogen sulfate, sodium salt. *See* Sodium myristyl sulfate
Tetradecyl tetracosanoate. *See* Myristyl lignocerate

Tetra (2, diallyoxymethyl-1 butoxy titanium di (di-tridecyl) phosphite
Uses: Coupling agent, adhesion promoter, antioxidant, antistat, antifoam, accelerator, blowing agent activator, catalyst, curative, corrosion inhibitor, dispersion aid, emulsifier, flame retardant, foaming agent, grinding and process aid
Trade names containing: Ken-React® KR 55

Tetraethoxysilane. *See* Ethyl silicate
Tetraethylene glycol butyl ether. *See* PEG-4 butyl ether
Tetraethylene glycol diacrylate. *See* PEG-4 diacrylate
Tetraethylene glycol diacrylate. *See* PPG-4 diacrylate
Tetraethylene glycol di-2-ethylhexoate. *See* PEG-4 di-2-ethylhexoate
Tetraethylene glycol dimethacrylate. *See* PEG-4 dimethacrylate

Tetraethylene glycol ethyl ether
CAS 5650-20-4

Synonyms: Tetraethylene glycol monoethyl ether
Trade names containing: Icinol EE22

Tetraethylene glycol methyl ether
CAS 23783-42-8
Synonyms: Tetraethylene glycol monomethyl ether
Empirical: $C_9H_{20}O_5$
Trade names containing: Icinol ME33; Icinol ME45

Tetraethylene glycol monobutyl ether. *See* PEG-4 butyl ether
Tetraethylene glycol monoethyl ether. *See* Tetraethylene glycol ethyl ether
Tetraethylene glycol monomethyl ether. *See* Tetraethylene glycol methyl ether

Tetraethylenepentamine
CAS 112-57-2; EINECS 203-986-2
Synonyms: TEPA; 1,4,7,10,13-Pentaazatridecane
Empirical: $C_8H_{23}N_5$
Formula: $NH_2(CH_2CH_2NH)_3CH_2CH_2NH_2$
Properties: Viscous liq., hygroscopic; sol. in most org. solvs. and water; m.w. 189.31; dens. 0.9980 (20/20 C); m.p. -30 C; b.p. 333 C; flash pt. 325 F; ref. index 1.5055
Precaution: Combustible; corrosive
Toxicology: LD50 (oral, rat) 205 mg/kg, (dermal, rabbit) 660 mg/kg; strong irritant to skin and eyes; causes burns
Uses: Solvent for sulfur, acid gases, various resins and dyes; saponifying agent for acidic materials; mfg. of syn. rubber; dispersant in motor oils; intermediate for oil additives; ion-exchange resins; surfactants; R.T. curing agent often used in two-pkg. protective coating systems
Manuf./Distrib.: Aldrich; Allchem Ind.; Ashland; Coyne; Fluka; Great Western; Tosoh; Union Carbide
Trade names: D.E.H. 26; Epi-Cure® 3245; TEPA

Tetraethyl orthosilicate. *See* Ethyl silicate
Tetrafluoroethene homopolymer. *See* Polytetrafluoroethylene

Tetrafluoroethylene/hexafluoropropylene copolymer
Uses: Melt processable resin offering nonaging, chem. inertness, exc. dielec. props., heat resist., toughness, flexibility, low COF, nonstick chars., negligible moisture absorp., and weather resist.; high m.w. and stress crack resistance; for melt extrusion and transfer molding; used in the chemical industry for chemical linings, bellows, and valve components
Trade names: Teflon® 160

Tetrafluoroethylene/hexafluoropropylene copolymer. *See* Fluorinated ethylene-propylene
Tetrafluoroethylene polymer. *See* Polytetrafluoroethylene

Tetrafluoroethylene/propylene copolymer
Uses: Molding and extrusion resin; used for o-rings, shaft seals, gaskets, packings, diaphragms, elec. connectors, wire and cable insulation, flexible joints, fabric-reinforced parts, hose and tubings, profiles, linings, coatings, sheet
Trade names: Aflas FA 100H; Aflas FA 100S; Aflas FA 150E; Aflas FA 150L; Aflas FA 150P

Tetrahydro-3,5-dimethyl-2H-1,3,5-thiadiazine-2-thione. *See* 3,5-Dimethyl tetrahydro-2-H,1,3,5-thiadiazone-2-thione
Tetrahydro-1,4-dioxin. *See* 1,4-Dioxane

Tetrahydrofuran
CAS 109-99-9; EINECS 203-726-8
Synonyms: THF; Butylene oxide; Diethylene oxide; Furanidine
Empirical: C_4H_8O
Formula: $CH_2CH_2CH_2CH_2O$
Properties: Water-wh. liq., ethereal odor; sol. in water and org. solvs.; m.w. 72.11; dens. 0.888 (20 C); f.p. -65 C; b.p. 66 C; flash pt. (OC) -15 C; ref. index 1.4070
Precaution: Flamm. limits in air 2-11.8%
Toxicology: LDLo (oral, rat) 3000 mg/kg; LD50 (IP, rat) 2900 mg/kg; oxic by ingestion and inhalation; TLV 200 ppm in air
Uses: Solvent for Grignard reactions, reductions, polymerizations, paints; chemical intermediate and

monomer; pharmaceutical intermediate; extraction solv.; steroid hormone prod. for use in birth control pills
Regulatory: BP compliance
Manuf./Distrib.: Aldrich; Allchem Ind.; ARCO Europe; Ashland; BASF; J.T. Baker; Burdick & Jackson; Chemcentral; Coyne; DuPont; Great Lakes; Fluka; Harcros; Hüls UK; ISP; Janssen Chimica; QO; Richman; Sigma; Spectrum Chem. Mfg.; Stanchem; Van Waters & Rogers
Trade names: THF

Tetrahydro-2-furancarbinol. *See* Tetrahydrofurfuryl alcohol
Tetrahydro-2-furanmethanol. *See* Tetrahydrofurfuryl alcohol

Tetrahydrofurfuryl acrylate
CAS 2399-48-6; EINECS 219-268-7
Empirical: $C_8H_{12}O_3$
Properties: Colorless liq.; m.w. 156.18; dens. 1.061; b.p. 80 C; flash pt. 110 C; ref. index 1.46
Toxicology: Irritant
Uses: Polymerizes to hard, infusible, insol. thermoset resin; used in uv irradiated coatings; vulcanizing agent; rubber reforming
Manuf./Distrib.: Aldrich; Monomer-Polymer & Dajac Labs
Trade names containing: SR-285

Tetrahydrofurfuryl alcohol
CAS 97-99-4; EINECS 202-625-6
Synonyms: THFA; Tetrahydro-2-furanmethanol; Tetrahydro-2-furancarbinol; Tetrahydro-2-furylmethanol
Classification: Cyclic alcohol
Empirical: $C_5H_{10}O_2$
Formula: $C_4H_7OCH_2OH$
Properties: Liq., mild odor; hygroscopic; misc. with water, alcohol, ether, acetone, chloroform, benzene; m.w. 102.14; dens. 1.053 (20/4 C); visc. 6.24 cp (20 C); m.p. < -80 C; b.p. 173-177 C; flash pt. 75 C; ref. index 1.453 (20 C); surf. tens. 37 dyn/cm
Precaution: Explosive limit 1.5-9.7% by vol.
Toxicology: Eye irritant; moderately irritating to skin, mucous membranes
Uses: Solvent for vinyl resins, paints, dyes for leather, chlorinated rubber, cellulose esters, coupling agent, solvent-softener for nylon
Regulatory: FDA 21CFR §172.515, 175.105, 176.210; FEMA GRAS
Manuf./Distrib.: Aldrich; Allchem Ind.; Ashland; Fluka; Great Lakes; Great Western; Penta Mfg.; QO; Schweizerhall; Sigma; Van Waters & Rogers
Trade names: QO® Tetrahydrofurfuryl Alcohol (THFA®)

Tetrahydrofurfuryl methacrylate
CAS 2455-24-5
Synonyms: Methacrylic acid, tetrahydrofurfuryl ester
Empirical: $C_9H_{14}O_3$
Properties: M.w. 170.21; dens. 1.044; b.p. 52 C (0.4 mm); flash pt. 90 C; ref. index 1.4580
Uses: Anaerobic adhesives and sealants, printed circuit boards, artificial finger nails, modifier for hard rubber rolls, wire and cable coatings, screen printing inks, emulsion polymerization, plastic modifier, EB-curable coatings
Manuf./Distrib.: Aldrich; CPS; Monomer-Polymer & Dajac Labs; Polysciences; Rohm Tech; Sartomer

Tetrahydro-2-furylmethanol. *See* Tetrahydrofurfuryl alcohol

Tetrahydronaphthalene
CAS 119-64-2; EINECS 204-340-2
Synonyms: 1,2,3,4-Tetrahydronaphthalene; Tetralin; Naphthalene-1,2,3,4-tetrahydride
Empirical: $C_{10}H_{12}$
Properties: Colorless liq., pungent odor; misc. with most solvs.; insol. in water; m.w. 132.21; dens. 0.981 (13 C); m.p. -25 C; b.p. 206 C; flash pt. 160 F; ref. index 1.5410
Toxicology: LD50 (oral, rat) 2860 mg/kg, (dermal, rabbit) 17 g/kg; irritant to eyes, skin, respiratory system; narcotic in high concs.
Uses: Chemical intermediate, solvent for greases, fats, oils, waxes, rubber, asphalt, coatings; substitute for turpentine
Manuf./Distrib.: Aldrich; DuPont; Fluka; Hüls AG; Sigma; Spectrum Chem. Mfg.
Trade names: Perenol® F4HN

1,2,3,4-Tetrahydronaphthalene. *See* Tetrahydronaphthalene
Tetrahydrophthalic acid anhydride. *See* Tetrahydrophthalic anhydride

Tetrahydrophthalic anhydride
CAS 85-43-8
Synonyms: Tetrahydrophthalic acid anhydride
Empirical: $C_8H_8O_3$
Properties: M.w. 152.16; dens. 1.37; m.p. 102 C; flash pt. 315 F
Toxicology: LD50 (oral, rat) 5410 mg/kg; corrosive irritant to skin, eyes, mucous membranes
Uses: Intermediate for paints
Manuf./Distrib.: Allchem Ind.; Anhydrides & Chems.; Cambridge Ind. Co. of Am.; Dixie; GCA; Hüls Am.; Hüls AG; Punda Mercantile

Tetrahydrothiophene 1,1-dioxide. *See* Sulfolane
Tetrahydroxymethylmethane. *See* Pentaerythritol

Tetrahydroxypropyl ethylenediamine
CAS 102-60-3; EINECS 203-041-4
Synonyms: N,N,N′,N′-Tetrakis(2-hydroxypropyl) ethylenediamine; Ethylenedinitrilotetra-2-propanol
Classification: Substituted amine
Empirical: $C_{14}H_{32}N_2O_4$
Formula: $(HOC_3H_6)_2NCH_2CH_2N(C_3H_6OH)$
Properties: Water-wh. visc. liq.; sol. in ethanol, toluene, ethylene glycol; misc. with water; m.w. 292.42; dens. 1.013; b.p. 175-181 C (0.8 mm); flash pt. > 230 F; ref. index 1.4812
Precaution: Combustible
Toxicology: May be irritating to skin and mucous membranes; may cause skin sensitization
Uses: Chelating agent, intermediate, emulsifier, antistat, rewetting agent, grease additive, textile lubricant; for pharmaceuticals, herbicides, fungicides, insecticides, adhesives, latexes, resins, plasticizers, inks, cosmetics, resins
Manuf./Distrib.: Aldrich; Fluka; Sigma
Trade names: Quadrol®

Tetraisopropyl di (dioctylphosphito) titanate
CAS 74665-17-1
Properties: M.w. 253.16; dens. 1.087; flash pt. 16 C
Uses: Coupling agent, adhesion promoter, antioxidant, antistat, antifoam, accelerator, blowing agent activator, catalyst, curative, corrosion inhibitor, dispersion aid, emulsifier, flame retardant, foaming agent, grinding and process aid
Manuf./Distrib.: DuPont
Trade names containing: Ken-React® KR 41B

Tetraisopropyl orthotitanate. *See* Tetraisopropyl titanate

Tetraisopropyl titanate
CAS 546-68-9; EINECS 208-909-6
Synonyms: TPT; Titanium isopropylate; Tetraisopropyl orthotitanate; Titanium (IV) isopropoxide; Isopropyl titanate; Isopropyl alcohol, titanium (4+) salt
Empirical: $C_{12}H_{28}O_4Ti$
Formula: $Ti[OCH(CH_3)_2]_4$
Properties: Lt. yel. liq.; sol. in most org. solvs.; dec. rapidly in water; m.w. 284.26; dens. 0.954; m.p. 14.8 C; b.p. 102-104 C (10 mm); flash pt. 60 C; ref. index 1.468 (20 C)
Uses: Catalyst for esterification and olefin polymerization; adhesion of paints, rubber, and plastics to metals, condensation catalyst
Manuf./Distrib.: Aldrich; Fluka
Trade names: Tyzor TPT

Tetrakis (2-ethylhexyl) titanate
Uses: Catalyst for esterification and olefin polymerization; used in coatings; resin crosslinking agent for automotive prods., coatings, elastomers, films/paints, graphic arts, plastics
Trade names: Tyzor TOT

Tetrakis(hydroxymethyl) methane. *See* Pentaerythritol
N,N,N′,N′-Tetrakis(2-hydroxypropyl) ethylenediamine. *See* Tetrahydroxypropyl ethylenediamine

Tetrakis [methylene (3,5-di-t-butyl-4-hydroxyhydrocinnamate)] methane
CAS 6683-19-8; EINECS 229-722-6
Synonyms: Tetrakis [methylene-3(3′,5′-di-t-butyl-4-hydroxyphenyl) propionate] methane; Pentaerythritol tetrakis (3,5-di-t-butyl-4-hydroxyhydrocinnamate)
Classification: Hindered phenolic
Empirical: $C_{73}H_{108}O_{12}$
Properties: Wh. powd.; sol. in chloroform, benzene, acetone, ethyl acetate; insol. in water; m.w. 1178; dens. 1.05 (20 C); m.p. 110-125 C; flash pt. 299 C
Uses: Antioxidant for polyolefins, styrenics, elastomers, adhesives, lubricants, oils, latex, varnishes
Manuf./Distrib.: Aldrich
Trade names: Akrochem® Antioxidant 1010; BNX® 1010, 1010G; Ultranox® 210

Tetrakis [methylene-3(3′,5′-di-t-butyl-4-hydroxyphenyl) propionate] methane. *See* Tetrakis [methylene (3,5-di-t-butyl-4-hydroxyhydrocinnamate)] methane
Tetralin. *See* Tetrahydronaphthalene

Tetramethyl bisphenol F dicyanate monomer/prepolymer
Uses: For hot-melt processing of prepregs and adhesives; for electronic circuitry, laminates, powd. coatings, molding powders; co-reacts with and cures epoxy resins
Trade names: AroCy® M-50

2-[p-(1,1,3,3-Tetramethylbutyl)phenoxy]ethanol. *See* Octoxynol-1
2-[2-[2-[p-(1,1,3,3-Tetramethylbutyl)phenoxy]ethoxy]ethoxy]ethanol. *See* Octoxynol-3
2,4,7,9-Tetramethyl-5-decyn-4,7-diol. *See* Tetramethyl decynediol

Tetramethyl decynediol
CAS 126-86-3; EINECS 204-809-1
Synonyms: 2,4,7,9-Tetramethyl-5-decyn-4,7-diol; 2,4,7,9-Tetramethyl-5-decyne-4,7-diol; Acetylenic glycol
Classification: Unsaturated alcohol
Empirical: $C_{14}H_{26}O_2$
Formula: $(CH_3)_2CHCH_2CCH_3OHCCCOHCH_3CH_2CH(CH_3)_2$
Properties: M.w. 226.36; m.p. 42-44 C; b.p. 255 C; flash pt. > 230 F
Toxicology: Irritant
Uses: Defoamer and dye dispersant in paint and ink formulations, dyestuffs; surfactant in rinse aids; substrate pigment wetting agent for industrial coatings and adhesives; defoamer, antishock agent for paper coatings
Regulatory: FDA 21CFR §175.105, 175.300
Manuf./Distrib.: Aldrich
Trade names: Surfynol® 104; Surfynol® PC
Trade names containing: Surfynol® 104A; Surfynol® 104BC; Surfynol® 104DPM; Surfynol® 104E; Surfynol® 104H; Surfynol® 104NP; Surfynol® 104PA; Surfynol® 104PG; Surfynol® 104PG-50; Surfynol® 104S; Surfynol® DF-110D; Surfynol® DF-110L; Surfynol® PG-50; Surfynol® SE; Surfynol® TG

2,4,7,9-Tetramethyl-5-decyne-4,7-diol. *See* Tetramethyl decynediol
Tetramethylene glycol. *See* 1,4-Butanediol
Tetramethylene sulfone. *See* Sulfolane

α,α,4,4-Tetramethyl-2-(methylethyl)-N-(2-methylpropylidene)-3-oxazolidineethanamine
CAS 148348-13-4
Uses: Reactive diluent used in high-solids two-component polyurethane coatings; used as partial or total replacement for the polyol component, reduces VOC; lowers visc., improves weathering performance
Trade names: Zoldine® RD-04

Tetramethylolmethane. *See* Pentaerythritol

2,2,6,6-Tetramethylpiperidin-4-yl acrylate/methyl methacrylate copolymer
CAS 115340-81-3
Classification: HALS
Uses: Lt. stabilizer for plastics, surf. coatings; esp. effective in polyolefins for outstanding heat and processing stability
Trade names: UV-Chek® AM-806

Tetraoctyloxytitanium di (ditridecylphosphite)
Uses: Coupling agent, adhesion promoter, antioxidant, antistat, antifoam, accelerator, blowing agent

activator, catalyst, curative, corrosion inhibitor, dispersion aid, emulsifier, flame retardant, foaming agent, grinding and process aid
Trade names containing: Ken-React® KR 46B

3,6,9,12-Tetraoxatetracosan-1-ol. *See* Laureth-4

Tetra oximino silane
CAS 34206-40-1
Trade names containing: OS-3000

Tetrapotassium diphosphate. *See* Tetrapotassium pyrophosphate

Tetrapotassium pyrophosphate
CAS 7320-34-5; EINECS 230-785-7
Synonyms: TKPP; Diphosphoric acid tetrapotassium salt; Tetrapotassium diphosphate; Potassium pyrophosphate; Potassium pyrophosphate, normal
Empirical: $H_4O_7P_2 \cdot 4K$
Formula: $K_4P_2O_7 \cdot 3HOH$
Properties: Colorless crystals or white powd.; hygroscopic; sol. in water; insol. in alcohol; m.w. 330.4; dens. 2.33; m.p. 1090 C
Toxicology: Nuisance dust
Uses: Soap and detergent builder, sequestering agent, peptizing and dispersing agent; pigment dispersant and stabilizer in emulsion paints; clarifying agent in liq. soaps; mfg. of syn. rubber; boiler water treatment
Regulatory: FDA 21CFR §173.315; USDA 9CFR §318.7, 381.147; Japan approved; Europe listed; UK approved
Manuf./Distrib.: Albright & Wilson Am.; Aldrich; Elf Atochem N. Am.; FMC; Monsanto; Seeler Ind.; Sigma; Telechem Int'l.
Trade names: Empiphos 4KP

Tetrasodium dicarboxyethyl octadecyl sulfosuccinamate. *See* Tetrasodium dicarboxyethyl stearyl sulfosuccinamate

Tetrasodium dicarboxyethyl stearyl sulfosuccinamate
CAS 3401-73-8; 37767-39-8; 38916-42-6; EINECS 222-273-7
Synonyms: Tetrasodium dicarboxyethyl octadecyl sulfosuccinamate
Classification: Organic compd.
Empirical: $C_{26}H_{47}NO_{10}S \cdot 4Na$
Properties: M.w. 657.68
Uses: Emulsifier, dispersant, solubilizer, surfactant for textile, cosmetics, agric. applics.; flotation reagent; emulsion polymerization; polishing waxes; surf. tension depressant for inks; demulsifier
Regulatory: FDA 21CFR §175.105, 176.170, 176.180, 178.3400
Manuf./Distrib.: Sigma
Trade names: Monawet SNO-35

Tetrasodium diphosphate. *See* Tetrasodium pyrophosphate

Tetrasodium pyrophosphate
CAS 7722-88-5; EINECS 231-767-1
Synonyms: TSPP; Tetrasodium diphosphate; n-Sodium pyrophosphate; Diphosphoric acid tetrasodium salt; Sodium pyrophosphate
Classification: Inorganic salt
Empirical: $Na_4P_2O_7$
Properties: Wh. cryst. powd., gran.; sol. 8 g/100 g water; insol. in alcohol; m.w. 265.91; dens. 2.534; m.p. 988 C
Toxicology: TLV:TWA 5 mg/m^3; LD50 (oral, rat) 4000 mg/kg, (IP, rat) 59 mg/kg, (IV, rat) 100 mg/kg; poison by ingestion, IP, IV, subcutaneous routes; heated to decomp., emits toxic fumes of PO_x and Na_2O; not a cholinesterase inhibitor
Uses: Water softener, syn. detergent builder, dispersant, emulsifier, metal cleaner, boiler water treatment, viscosifier for drilling muds, deinking newsprint, synthetic rubber, textile dyeing, wool scouring, buffer, sequestrant, nutrient, food additive
Regulatory: FDA 21CFR §133.169, 133.173, 133.179, 173.310, 175.210, 175.300, 181.22, 181.29, 182.70, 182.6787, 182.6789, GRAS; Japan approved; Europe listed; UK approved; FDA approved for buccals, dentals

Tetryl formate

Manuf./Distrib.: Albright & Wilson Am.; Aldrich; Farleyway Chem. Ltd; Fluka; FMC; Lohmann; Mitsui Toatsu; Monsanto; Nippon Chem. Ind.; Rhone-Poulenc Food Ingreds.; Seeler Ind.; Sigma; Spectrum Chem. Mfg.; Veckridge; Yoneyama Chem. Ind.
Trade names containing: Laponite® S

Tetryl formate. *See* Isobutyl formate
TFE. *See* Polytetrafluoroethylene
THAM. *See* Tris (hydroxymethyl) aminomethane
Thermal black. *See* Carbon black
THF. *See* Tetrahydrofuran
THFA. *See* Tetrahydrofurfuryl alcohol
Thiaben. *See* Thiabendazole

Thiabendazole
CAS 148-79-8
Synonyms: Thiaben; 2-(Thiazol-4-yl) benzimidazole; 4-(2-Benzimidazolyl) thiazol; 2-(4-Thiazolyl)benzimidazole
Empirical: $C_{10}H_7N_3S$
Properties: Colorless cryst.; sol. in DMF, DMSO; sl. sol. in alcohol, esters, chlorinated hydrocarbons; m.w. 201.25; m.p. 300 C (subl.)
Toxicology: LD50 (oral, rat) 3100 mg/kg; toxic; experimental teratogen
Uses: Systemic fungicide; anthelmintic
Regulatory: Europe listed; UK approved
Manuf./Distrib.: Aldrich; Sigma
Trade names: Metasol TK-100®; Metasol TK-100 Disp. W

Thiabendazole hypophosphite salt
Uses: In adhesive films, interior/exterior paint films, paper and paperboard prods. (for nonfood use such as soap wrappers), hard surfaces, and natural and syn. fibers such as canvas textiles and nylon carpets
Trade names: Metasol TK-100 Liq. Conc.

2-(Thiazol-4-yl) benzimidazole. *See* Thiabendazole
2-(4-Thiazolyl)benzimidazole. *See* Thiabendazole

4,4´-Thiobis-6-(t-butyl-m-cresol)
CAS 96-69-5
Synonyms: 4,4´-Thiobis (2-t-butyl-5-methylphenol); Bis (4-hydroxy-5-t-butyl-2-methylphenyl) sulfide; 4,4´-Thiobis(6-t-butyl-3-methylphenol); 1,1´-Thiobis(2-methyl-4-hydroxy-5-t-butylbenzene); Bis (3-t-butyl-4-hydroxy-6-methylphenyl) sulfide
Empirical: $C_{22}H_{30}O_2S$
Formula: $[(CH_3)_2CC_6H_2(CH_3)OH]_2S$
Properties: Wh. to tan powd.; m.w. 358.58; dens. 1.10; m.p. 163-165 C; flash pt. (COC) 205 C
Toxicology: LD50 (IP, mouse) 50 mg/kg; poison by IP route and probably by ingestionand inhalation; irritant; TLV 10 mg/m^3 of air; mutagenic data; heated to decomp., emits highly toxic fumes of SO_x
Uses: Antioxidant for latex, adhesives, rubber, plastics, cross-linked polymers
Manuf./Distrib.: Aldrich; Great Lakes; James River
Trade names: Santonox®

4,4´-Thiobis (2-t-butyl-5-methylphenol). *See* 4,4´-Thiobis-6-(t-butyl-m-cresol)
4,4´-Thiobis(6-t-butyl-3-methylphenol). *See* 4,4´-Thiobis-6-(t-butyl-m-cresol)
1,1´-Thiobis(2-methyl-4-hydroxy-5-t-butylbenzene). *See* 4,4´-Thiobis-6-(t-butyl-m-cresol)
2,2´-Thiobis (4-t-octylphenolate) n-butylamine nickel II. *See* 2,2´-Thiobis (4-t-octylphenolato)-n-butylamine nickel

2,2´-Thiobis (4-t-octylphenolato)-n-butylamine nickel
CAS 14516-71-3; EINECS 238-523-3
Synonyms: 2,2´-Thiobis (4-t-octylphenolate) n-butylamine nickel II
Properties: Powd.; m.p. 258-281 C
Uses: UV stabilizer for polyolefins, ABS; protects syn. resins and paints, esp. in aerospace industry
Trade names: Modarez UV Ni-3

Thiocyanic acid. *See* 2-Thiocyanomethylthiobenzothiazole

2-Thiocyanomethylthiobenzothiazole
CAS 21564-17-0
Synonyms: TCMTB; Thiocyanic acid; 2-(Benzothiazolylthio) methyl ester
Empirical: $C_9H_6N_2S_3$
Properties: M.w. 238.35
Toxicology: LD50 (oral, rat) 1590 mg/kg, (skin, rabbit) 10 g/kg; mod. toxic by ingestion
Uses: Microbicide used in paints, adhesives and metal working
Trade names: Amerstat® 1930; Busan® 1030; Busan® 1118
Trade names containing: Amerstat® 1910; Busan® 1009; Busan® 1025; Busan® 1105

2-Thioethanol. *See* 2-Mercaptoethanol
Thioethylene glycol. *See* 2-Mercaptoethanol

Thioglycerin
CAS 96-27-5; EINECS 202-495-0
Synonyms: 3-Mercapto-1,2-propanediol; Monothioglycerol; α-Monothioglycerol; Thioglycerol; 1-Thioglycerol
Classification: Polyhydric alcohol
Empirical: $C_3H_8O_2S$
Formula: $HOCH_2CHOHCH_2SH$
Properties: Colorless to pale ylsh. visc. liq., sl. sulfidic odor; hygroscopic; sl. sol. in water; misc. with alcohol; insol. in ether; m.w. 108.16; dens. 1.295; b.p. 118 C (5 mm); flash pt. > 230 F; ref. index 1.5260; pH 3.5-7.0 (10%)
Toxicology: Irritant
Uses: Stabilizer for acrylonitrile polymers; crosslinking agent for coatings; accelerator for epoxy-amine reactions; reducing agent; in wound healing, hair waving, textiles, furs, pharmaceutical intermediate, surfactants, foam stabilizing additives; cosmetics
Regulatory: FDA approved for injectables, buccals, parenterals; USP/NF compliance
Manuf./Distrib.: Aldrich; Asahi Chem. Ind.; Fluka; Hampshire; NOF; Research Organics; SAF Bulk Chems.; Schweizerhall; Sigma
Trade names: Thiovanol®

Thioglycerol. *See* Thioglycerin
1-Thioglycerol. *See* Thioglycerin
Thioglycol. *See* 2-Mercaptoethanol
Thiomonoglycol. *See* 2-Mercaptoethanol
Thiophan sulfone. *See* Sulfolane

2,2′-(2,5-Thiophenediyl)bis[5-t-butylbenzoxazole]
CAS 7128-64-5; EINECS 230-426-4
Synonyms: BBOT; 2,5-Bis (5-t-butyl-2-benzoxazolyl) thiophene; 2,5-Bis (5-t-butyl-2-benzoxazol-2-yl) thiophene
Classification: Bis (benzoxazolyl) deriv.
Empirical: $C_{26}H_{26}N_2O_2S$
Properties: Yel. powd.; m.w. 431; m.p. 199-201 C
Uses: Fluorescent whitening agent for thermoplastics esp. PVC, coatings, printing inks, syn. fibers
Manuf./Distrib.: Aldrich; Fluka; Sigma
Trade names: Uvitex® OB

Tiff. *See* Barium sulfate

Tin
CAS 7440-31-5; EINECS 231-141-8
Synonyms: Stannum
Classification: Element
Empirical: Sn
Properties: Silver-white ductile solid; sol. in acids, hot KOH sol'ns.; insol. in water; at.wt. 118.69; dens. 7.29 (20 C); m.p. 232 C; b.p. 2260 C
Toxicology: TLV 2 mg/m^3 of air (inorg. compds.), TLV 0.1 mg/m^3 of air (org. compds.)
Uses: Tin plate, anodes, corrosion-resistant coatings, mfg. of chemicals
Regulatory: BP compliance
Manuf./Distrib.: Aldrich; Alfa Aesar Johnson Matthey; Atlantic Equip. Engrs.; Atomergic Chemetals; Atotech USA; Belmont Metals; Cerac; Fluka; Fry's Metals Ltd; Noah; Sigma
Tin (II) chloride (1:2). *See* Stannous chloride anhyd.

Tin (II) chloride (1:2). *See* Stannous chloride anhyd.
Tin (II) chloride anhydrous. *See* Stannous chloride anhyd.
Tin (IV) chloride. *See* Stannic chloride
Tin (IV) chloride anhydrous. *See* Stannic chloride
Tin chloride (ic). *See* Stannic chloride
Tin crystals. *See* Stannous chloride anhyd.
Tin dichloride. *See* Stannous chloride anhyd.
Tin octoate. *See* Stannous octoate
Tin-(II)-octoate. *See* Stannous octoate
Tin perchloride. *See* Stannic chloride
Tin protochloride. *See* Stannous chloride anhyd.
Tin salt. *See* Stannous chloride anhyd.
Tin tetrachloride. *See* Stannic chloride
TIPA. *See* Triisopropanolamine
Titanic acid anhydride. *See* Titanium dioxide
Titanic anhydride. *See* Titanium dioxide
Titanic earth. *See* Titanium dioxide
Titanic oxide. *See* Titanium dioxide

Titanium
CAS 7440-32-6; EINECS 231-142-3
Classification: Metallic element
Empirical: Ti
Properties: Dk. gray amorphous powd. or lustrous wh. metal; at.wt. 47.90; dens. 4.5 (20 C); m.p. 1677 C; b.p. 3277 C
Precaution: Highly flamm.
Uses: Alloys; as structural material in aircraft, jet engines, marine equipment, chemical equipment, surgical instruments, orthopedic appliances, food-handling equipment; x-ray tube targets; abrasives; cermets; electrodeposited and dipped coatings
Manuf./Distrib.: Aldrich; Belmont Metals; Cerac; Fluka; Inco Alloys Int'l.; New Metals & Chems.; Titanium Metals

Titanium acetylacetonate
CAS 14024-64-7
Synonyms: Titanylacetylacetonate; Bis (acetylacetonato) titanium oxide; Bis (2,4-pentanedionato) titanium oxide
Empirical: $C_{10}H_{14}O_5Ti$
Formula: $TiO[OC(CH_3):CHCOCH_3]_2$
Properties: Cryst. powd.; sl. sol. in water; m.w. 262.14
Toxicology: LD50 (IP, rat) 650 mg/kg; mod. toxic by IP route
Uses: Crosslinking agent for cellulosic lacquers
Manuf./Distrib.: Aldrich; Amspec; Hüls AG

Titanium (IV) butoxide. *See* Tetrabutyl titanate
Titanium butylate. *See* Tetrabutyl titanate

Titanium di (butyl, octyl pyrophosphate) di (dioctyl, hydrogen phosphite) oxyacetate
Uses: Coupling agent, adhesion promoter, antioxidant, antistat, antifoam, accelerator, blowing agent activator, catalyst, curative, corrosion inhibitor, dispersion aid, emulsifier, flame retardant, foaming agent, grinding and process aid
Trade names containing: Ken-React® KR 158FS

Titanium di (cumylphenylate) oxyacetate
Uses: Coupling agent, adhesion promoter, antioxidant, antistat, antifoam, accelerator, blowing agent activator, catalyst, curative, corrosion inhibitor, dispersion aid, emulsifier, flame retardant, foaming agent, grinding and process aid
Trade names containing: Ken-React® KR 134S

Titanium di (dioctylpyrophosphate) oxyacetate
Uses: Coupling agent, adhesion promoter, antioxidant, antistat, antifoam, accelerator, blowing agent activator, catalyst, curative, corrosion inhibitor, dispersion aid, emulsifier, flame retardant, foaming agent, grinding and process aid
Trade names containing: Ken-React® KR 138S

Titanium dimethacrylate oxyacetate

Uses: Coupling agent, adhesion promoter, antioxidant, antistat, antifoam, accelerator, blowing agent activator, catalyst, curative, corrosion inhibitor, dispersion aid, emulsifier, flame retardant, foaming agent, grinding and process aid

Trade names containing: Ken-React® KR 133DS

Titanium dioxide

CAS 1317-80-2; 13463-67-7; EINECS 236-675-5

Synonyms: Titanic anhydride; Titanic earth; Titanic oxide; Titanic acid anhydride; Titanium oxide; Titanium white; Pigment white 6; CI 77891

Classification: Inorganic oxide

Empirical: TiO_2

Properties: Wh. amorphous powd., odorless, tasteless; insol. in water, HCl, HNO_3, dil. H_2SO_4; sol. in HF, hot conc. H_2SO_4; m.w. 79.90; Anatase: dens. 3.90; Rutile: dens. 4.23

Precaution: Violent or incandescent reaction with metals (e.g., aluminum, calcium, magnesium, potassium, sodium, zinc, lithium)

Toxicology: TLV:TWA 10 mg/m^3 of total dust; experimental carcinogen, neoplastigen, tumorigen; human skin irritant; nuisance dust

Uses: White pigment, opacifying agent in paints, paper, rubber, pharmaceuticals (tableted drugs), cosmetics (lipsticks, nail enamels, face powds., eye makeup, rouge); radioactive decontamination of skin

Regulatory: FDA 21CFR §73.575, 73.1575, 73.2575, 175.105, 175.210, 175.300, 175.380, 175.390, 176.170, 177.1200, 177.1210, 177.1350, 177.1400, 177.1460, 177.1650, 177.2260, 177.2600, 177.2800, 181.22, 181.30; USDA 9CFR §318.7, 381.147; Japan restricted as colorant; Europe listed; UK approved; prohibited in Germany; FDA approved for ophthalmics, orals, topicals; USP/NF, BP, Ph.Eur. compliance

Manuf./Distrib.: Aceto; Aldrich; Ashland; J.T. Baker; Bayer NV; British Traders & Shippers; Chemcentral; Chemical; Degussa; DuPont; Engelhard; Ferro/Transelco; Fluka; Fuji Titanium Ind.; Furukawa; Harcros; Hilton Davis; Hitox; Kemira; Kerr-McGee; Landers-Segal Color; Lenape; D.N. Lukens; Ore & Chem.; Royale Pigments & Chems.; Sachtleben Chemie GmbH; SCM; Seegott; Tioxide Am.; Titan Kogyo; Tricon Colors; Van Waters & Rogers; Whittaker, Clark & Daniels

Trade names: AT-1; Dapro UVAL-40; Dapro UVCW-30; Dapro UVST-35; Hitox®; Kronos® 1000; Kronos® 1070; Kronos® 1072; Kronos® 1074; Kronos® 2020; Kronos® 2090; Kronos® 2101; Kronos® 2131; Kronos® 2160; Kronos® 2220; Kronos® 2310; Kronos® 3020; Kronos® 3025; Octotint 138; Octotint 601; RL-60; RL-68; Tiona® RCL-2; Tiona® RCL-3; Tiona® RCL-4; Tiona® RCL-6; Tiona® RCL-7; Tiona® RCL-9; Tiona® RCL-69; Tiona® RCL-535; Tiona® RCL-628; Tiona® RCS-2; Tiona® RCS-3; Tiona® RCS-9; Tiona® RCS-535; Ti-Pure® R-900; Titankup®; Tronox® CR-800PG; Tronox® CR-813; Tronox® CR-821; Tronox® CR-834; Tronox® CR-837

Trade names containing: Ampacet 11171; Ampacet 11171-101; Ampacet 11187; Ampacet 11338; Ampacet 11495-S; Ampacet 11919-S; Ampacet 110052; Mearlin®; Mearlin® Dynacolor® BG; Mearlin® Dynacolor® BP; Mearlin® Dynacolor® BY-B; Mearlin® Dynacolor® GB; Mearlin® Dynacolor® GP; Mearlin® Dynacolor® GY; Mearlin® Dynacolor® RB; Mearlin® Dynacolor® VP; Mearlin® MagnaPearl 1000; Mearlin® MagnaPearl 2000; Mearlin® MagnaPearl 3000; Mearlin® MagnaPearl 3100; Mearlin® MagnaPearl 4000; Mearlin® MagnaPearl 5000; Tronox® CR-800; Tronox® CR-822; Tronox® CR-828

Titanium (IV) isopropoxide. *See* Tetraisopropyl titanate
Titanium isopropylate. *See* Tetraisopropyl titanate
Titanium IV neoalkanolato, tris(diisooctyl) pyrophosphato-o. *See* Neopentyl (diallyl) oxy, tri (dioctyl) pyrophosphato titanate
Titanium nickel oxide. *See* Nickel titanate
Titanium oxide. *See* Titanium dioxide
Titanium white. *See* Titanium dioxide
Titanylacetylacetonate. *See* Titanium acetylacetonate
TKPP. *See* Tetrapotassium pyrophosphate
TMB. *See* 1,3,5-Trimethylbenzene
TME. *See* Trimethylolethane
TMPD. *See* 2,2,4-Trimethyl-1,3-pentanediol
TMPEOTA. *See* Trimethylolpropane ethoxy triacrylate
TMPTA. *See* 1,1,1-Trimethylolpropane triacrylate
TMPTMA. *See* 1,1,1-Trimethylolpropane trimethacrylate
TNPP. *See* Tris(nonylphenyl) phosphite

Toluene

CAS 108-88-3; EINECS 203-625-9

Toluene, p-chloro-α,α,α,-trifluoro

Synonyms: Methylbenzene; Phenylmethane; Toluol
Classification: Aromatic compd.
Empirical: C_7H_8
Formula: $C_6H_5CH_3$
Properties: Colorless liq., benzene odor; sol. in alcohol, benzene, ether; very sl. sol. in water; m.w. 92.13; dens. 0.866 (20/4 C); m.p. -94.5 C; b.p. 110.7 C; flash pt. (CC) 40 F; ref. index 1.4967 (20 C); KB value 105
Precaution: Highly flammable
Toxicology: Toxic by ingestion, inhalation, and skin absorption; irritant to eyes, skin, respiratory tract; narcotic in high concs. causing CNS depression, unconsciousness; possible liver/kidney damage; TLV 50 ppm in air (ACGIH); LD50 (rat, oral) 7.53 g/kg
Uses: Aviation gasoline additive; solvent for paint, gums, resins; diluent and thinner in nitrocellulose lacquers; adhesive solvent in plastic toys; mfg. of benzoic acid, benzaldehyde; in extraction of various principles from plants; starting material for TDI for PU resins, TNT for explosives, toluene sulfonates for detergents and dyestuffs
Regulatory: FDA 21CFR §175.105, 175.320, 176.180, 177.1010, 177.1200, 177.1440, 177.1580, 177.1650, 177.2460, 178.3010, 27CFR §21.131
Manuf./Distrib.: Aldrich; ARCO; Ashland; J.T. Baker; Baychem; BP Chems.; Burdick & Jackson; R.E. Carroll; Chemcentral; Chevron; Coyne; Exxon; Fina Chems.; Fluka; Harcros; Marathon Oil; Mitsubishi Petrochem.; Mitsui Petrochem. Ind.; Mobil; Olin; Phillips; Shell; Sigma; Spectrum Chem. Mfg.; Texaco; Triple Crown Am.; Van Waters & Rogers
Trade names: Shell Toluene
Trade names containing: Acryloid® A-21; Acryloid® A-21LV; Acryloid® AU-608S; Acryloid® B-44S; Acryloid® B-48N; Acryloid® B-50S; Acryloid® B-66; Acryloid® B-72; Acryloid® B-82; Acryloid® B-99N; Ancamide 100-IT-60; Araldite® GZ 571 T-75; Araldite® GZ 597 KT-55; Araldite® GZ 7071 T-65; Aroplaz 1247-T-70; Beckosol® 12-514; Cellokyd 2829 PMA/T; D.E.R. 667-PMT55; D.E.R. 671-PMT70; D.E.R. 671-T75; Dislon® L-1985-50; Disparlon® AP-20; Disparlon® AP-30; Disparlon® L-1985-50; Disparlon® LC-900; Disparlon® LC-915; Disparlon® LC-955; Disparlon® OX-70; Dow Corning® 11; EPON® Resin 1001-T-75; Epotuf Resin 38-519; G-4664-T-60; G-4732-VBT-50; OS-3000; Ricon 104

Toluene, p-chloro-α,α,α,-trifluoro. *See* p-Chlorobenzotrifluoride

Toluene diisocyanate
CAS 584-84-9; 26471-62-5; EINECS 209-544-5
Synonyms: TDI; Toluene 2,4-diisocyanate; 2,4-Tolylene diisocyanate; 2,4-Diisocyanatotoluene
Empirical: $C_9H_6N_2O_2$
Formula: $CH_3C_6H_3(NCO)_2$
Properties: Water-wh. to pale yel. liq., sharp pungent odor; sol. in ether, acetone, and other org. solvs.; m.w. 174.15; dens. 1.122 (25/15.5 C); m.p. 19.4-21.5 C; b.p. 251 C (760 mm); flash pt. (OC) 270 F; ref. index 1.5680
Precaution: Combustible
Toxicology: Toxic by ingestion, inhalation; strong irritant to skin and tissue; may be anticipated to be a carcinogen; TLV 0.005 ppm in air
Uses: In mfg. of polyurethane foams, elastomers, and coatings
Manuf./Distrib.: Aldrich; Allchem Ind.; Bayer Hispania Industrial SA; Fluka; ICI Polyurethanes; Nippon Polyurethane Ind.; Olin; Rhone-Poulenc N. Am.; Sigma; Tricon Trading Assoc.
Trade names: Rubinate® TDI 80/20

Toluene 2,4-diisocyanate. *See* Toluene diisocyanate
Toluene-4-sulfonamide. *See* p-Toluenesulfonamide

p-Toluenesulfonamide
CAS 70-55-3; EINECS 200-741-1
Synonyms: PTSA; Toluene-4-sulfonamide
Empirical: $C_7H_9NO_2S$
Formula: $CH_3C_6H_4SO_2NH_2$
Properties: Wh. leaflets; sol. in alcohol; very sl. sol. in water; m.w. 171.22 sp.gr. 1.353; m.p. 138-139 C; flash pt. (COC) 206 C
Precaution: Combustible
Uses: Organic synthesis; plasticizer for thermoplastic and thermoset resins; fungicide and mildewcide in paints and coatings
Manuf./Distrib.: Akzo Nobel; Aldrich; Allchem Ind.; Fluka; Focus; Howard Hall; Honeywill & Stein; ICI Spec. Chems.; Monsanto; Rit-Chem; Schweizerhall; Sigma; Unitex

Toluene sulfonamide formaldehyde resin
CAS 1338-51-8
Uses: Modifier and adhesion promoter for resins in adhesives and coatings; extender in polyamide resins
Trade names: Uniplex 600

o-Toluenesulfonate. *See* o-Toluene sulfonic acid
4-Toluenesulfonic acid. *See* p-Toluene sulfonic acid

o-Toluene sulfonic acid
Synonyms: o-Toluenesulfonate
Classification: Substituted aromatic acid
Formula: $C_6H_4(SO_3H)(CH_3)$
Properties: Colorless crystals; sol. in alcohol, water, ether; m.p. 67.5 C
Precaution: Combustible
Toxicology: Toxic by ingestion, inhalation; strong irritant to tissue
Uses: Dyes, organic synthesis, acid catalyst for resins in foundry cores, plastics, coatings
Manuf./Distrib.: Bayer/Fibers, Orgs. Rubber; Boliden Intertrade; Byk-Chemie GmbH; Eastman; Ferro/Grant; Nissan Chem. Ind.; PMC Specialties; Ruetgers-Nease; Witco/Oleo-Surf.

p-Toluene sulfonic acid
CAS 104-15-4; EINECS 203-180-0
Synonyms: 4-Methylbenzenesulfonic acid; 4-Toluenesulfonic acid; p-Methylbenzenesulfonic acid; Toluene-p-sulfonic acid; p-Methylphenylsulfonic acid; Tosic acid
Classification: Substituted aromatic acid
Empirical: $C_7H_8O_3S$
Formula: $C_6H_4(SO_3H)(CH_3)$
Properties: Colorless leaflets; sol. in alcohol, ether, water; m.w. 172.21; m.p. 107 C; b.p. 140 C (20 mm)
Precaution: DOT: Corrosive material; combustible; potentially explosive reaction with acetic anhydride + water; heated to decomp., emits toxic fumes of SO_x
Toxicology: LD50 (oral, rat) 2480 mg/kg; poison by ingestion; skin and mucous membrane irritant
Uses: Dyes, organic synthesis, acid catalyst for resins in foundry cores, plastics, coatings
Regulatory: FDA 21CFR §175.105; BP compliance
Manuf./Distrib.: Allchem Ind.; Bayer/Fibers, Orgs., Rubber; Boliden Intertrade; Byk-Chemie GmbH; Eastman; Ferro/Grant; Howard Hall; Nissan Chem. Ind.; PMC Specialties; Ruetgers-Nease; Sloss Ind.; Spectrum Chem. Mfg.; Witco/Oleo-Surf.
Trade names: Eltesol® TSX/A; K-Cure® 1040W; Manro PTSA/C; Nacure® 2547; Nacure® XC-2211; Nacure® XP-386
Trade names containing: K-Cure® 1040; Nacure® 2500; Nacure® 2501; Nacure® 2530; Nacure® 2558; Nacure® XP-357

Toluene-p-sulfonic acid. *See* p-Toluene sulfonic acid

p-Toluene sulfonyl isocyanate
CAS 4083-64-1; EINECS 223-810-8
Synonyms: Tosyl isocyanate
Empirical: $C_8H_7NO_3S$
Formula: $CH_3C_6H_4SO_2NCO$
Properties: Colorless liq., acrid odor; m.w. 197.2 sp.gr. 1.295 (20/4 C); vapor pressure 1.0 mm Hg (100 C); b.p. 275-278 C; flash pt. 145 C; ref. index 1.5340
Precaution: Prolonged exposure to temps. above 338 F could result in violent decomp.; keep dry; decomp. prods.: CO, NO_x, hydrogen cyanide, SO_x, hydrogen chloride
Toxicology: LD50 (oral, rat) > 500 mg/kg; severe eye irritant; harmful if inh., inhg., absorbed thru skin; severe irritant to respiratory tract, mucous membranes; may cause coughing, burning sensation, wheezing, headaches, nausea, vomiting; asthma (chronic)
Manuf./Distrib.: Aldrich; Amber Syn.; Fluka; Sigma
Trade names: PTSI

Toluol. *See* Toluene

o-Tolyl biguanide
CAS 93-69-6; EINECS 202-268-6
Synonyms: Imidodicarbonimidic diamide, N-(2-methylphenyl)-; N-(2-Methylphenyl)imidodicarbonimidic diamide

Tolyl diphenyl phosphate

Classification: Substituted aromatic compd.
Empirical: $C_9H_{13}N_5$
Formula: $NH_2(CNHNH)_2C_6H_4CH_3$
Properties: Wh. to off-wh. powd.; m.p. 138 C
Precaution: Combustible
Uses: Antioxidant for soaps produced from animal or vegetable oil; hardener for decorative and protective powd. coatings
Manuf./Distrib.: Aceto; Aldrich
Trade names: HT 2844

Tolyl diphenyl phosphate. *See* Diphenylcresyl phosphate
2,4-Tolylene diisocyanate. *See* Toluene diisocyanate
Tosic acid. *See* p-Toluene sulfonic acid
Tosyl isocyanate. *See* p-Toluene sulfonyl isocyanate
TOTM. *See* Trioctyl trimellitate
Toxilic acid. *See* Maleic acid
Toxilic anhydride. *See* Maleic anhydride
TPGDA. *See* PPG-3 diacrylate
TPP. *See* Triphenyl phosphate; Triphenyl phosphite
TPT. *See* Tetraisopropyl titanate

Triacetin
CAS 102-76-1; EINECS 203-051-9
Synonyms: Glyceryl triacetate; Acetin; Enzactin; 1,2,3-Propanetriol triacetate; Triacetyl glycerol; Triacetyl glycerin
Definition: Triester of glycerin and acetic acid
Empirical: $C_9H_{14}O_6$
Formula: $C_3H_5(OCOCH_3)_3$
Properties: Colorless oily liq., sl. fatty odor, bitter taste; sol. in water, alcohol, ether, other org. solvs.; sl. sol. in CS_2; m.w. 218.20; dens. 1.160 (20 C); m.p. -78 C; b.p. 258-260 C; flash pt. 300 F; ref. index 1.4307 (20 C)
Precaution: Combustible exposed to heat, flame, or powerful oxidizers
Toxicology: LD50 (oral, rat) 3000 mg/kg, (IV, mouse) 1600 ± 81 mg/kg; poison by ingestion; mod. toxic by IP, subcutaneous, IV routes; eye irritant; heated to decomp., emits acrid smoke and irritating fumes
Uses: Plasticizer for vinyl compds.; fixative in perfumery; mfg. of cosmetics; specialty solv.; mfg. of celluloid, photographic films; topical antifungal; adhesives, coatings, paper; food additive
Regulatory: FDA 21CFR §175.300, 175.320, 181.22, 181.27, 184.1901, GRAS; FEMA GRAS; FDA approved for orals; USP/NF, BP compliance
Manuf./Distrib.: Aldrich; Allan; Eastman; Fluka; K3; MTM Spec. Chems.; Penta Mfg.; Sigma; Spectrum Chem. Mfg.; Unichema Int'l.; Yuki Gosei Kogyo
Trade names: Eastman® Triacetin; Kodaflex® Triacetin

Triacetyl cellulose. *See* Cellulose triacetate
Triacetyl glycerin. *See* Triacetin
Triacetyl glycerol. *See* Triacetin

Triallylcyanurate
CAS 101-37-1; EINECS 202-936-7
Synonyms: 2,4,6-Triallyloxy-1,3,5-triazine
Empirical: $C_{12}H_{15}N_3O_3$
Formula: $(CH_2{:}CHCH_2OC)_3N_3$
Properties: Colorless liq. or solin.; misc. with acetone, benzene, chloroform, dioxane, ethyl acetate, ethanol, xylene; m.w. 249.27; dens. 1.1133 (30 C); m.p. 27 C; flash pt. > 176 F
Precaution: Combustible; moisture-sensitive
Toxicology: Toxic by ingestion and inhalation
Uses: Polymers as monomers and modifier; organic intermediate; coagent improving peroxide-induced crosslinking of rubber; intermediate for paints
Manuf./Distrib.: Akzo Nobel; Aldrich; Cytec Ind.; Degussa; Fluka; Monomer-Polymer & Dajac Labs; Nat'l. Starch & Chem.; Richman; Sigma

2,4,6-Triallyloxy-1,3,5-triazine. *See* Triallylcyanurate

Triallyl trimellitate
CAS 2694-54-4
Synonyms: TATM; Tri-2-propenyl 1,2,4-benzenetricarboxylate
Empirical: $C_{18}H_{18}O_6$
Properties: M.w. 330.33; b.p. 210 C (4.5 mm)
Manuf./Distrib.: C.P. Hall; Hardwicke; Monomer-Polymer & Dajac Labs; Natrochem; Nipa Hardwicke; Rhone-Poulenc Spec.
Trade names: Sipomer® TATM

2,4,6-Triamino-sym-triazine. *See* Melamine
2,4,6-Triamino-s-triazine. *See* Melamine

Triaryl phosphate
CAS 68937-41-7
Formula: $(CH_2{:}CHCH_2O)_3PO$
Properties: Water-wh. liq.; dens. 1.064 (25/15 C); f.p. -50 C; b.p. 80 C (0.5 mm); ref. index 1.448 (25 C)
Precaution: Combustible
Uses: Flame-retardant plasticizer for PVC, NC lacquers and coatings; compatibilizing agent; intermediate; catalyst carrier and pigment vehicle for polyurethane; processing aid in rubber belting and mech. goods
Manuf./Distrib.: Akzo Nobel; Albright & Wilson Am.; Ashland; FMC; C.P. Hall; Harwick; Monsanto; Velsicol
Trade names: Kronitex® 200; Kronitex® 1840; Kronitex® 1884; Kronitex® 1886; Reofos® 95

Triaryl sulfonium hexafluoroantimonate
Trade names containing: SarCat™ CD-1010

Triaryl sulfonium hexafluorophosphate
Trade names containing: SarCat™ CD-1011

Triazine formaldehyde
Uses: Thermosetting resin for wh., one-coat finishes for washing machines, refrigerators, stoves and cabinets, and in industrial baking enamels requiring max. whiteness and retention of color
Trade names: Uformite® 27-809; Uformite® 27-810

1,3,5-Triazine-2,4,6-triamine. *See* Melamine
1,3,5-Triazine,2,4,6-triamine, polymer with formaldehyde. *See* Melamine/formaldehyde resin
1,3,5-Triazine-2,4,6-triyltrinitrilo)hexakis methanol. *See* Hexamethylol melamine resin
Tribasic aluminum stearate. *See* Aluminum tristearate
Tribenzoin. *See* Glyceryl tribenzoate

Tribromophenyl allyl ether
CAS 3278-89-5
Synonyms: Allyl 2,4,6-tribromophenyl ether
Formula: $Br_3C_6H_2OCH_2CH{=}CH_2$
Properties: M.w. 370.88; m.p. 74-76 C
Uses: Flame retardant for expandable PS, adhesives, and coatings; synergist with hexabromocyclododecane
Manuf./Distrib.: Aldrich
Trade names: Great Lakes PHE-65™

Tributoxyethyl phosphate
CAS 78-51-3; EINECS 201-122-9
Synonyms: TBEP; Ethanol, 2-butoxy-, phosphate (3:1); Tris (2-butoxyethyl) phosphate; 2-Butoxyethanol phosphate
Empirical: $C_{18}H_{39}O_7$
Formula: $[CH_3(CH_2)_3O(CH_2)_2O]_3PO$
Properties: Sl. yel. oily iq.; insol. or limited sol. in glycerol, glycols, certain amines; sol. in most org. liqs.; m.w. 398; dens. 1.020 (20 C); m.p. -70 C; b.p. 215-228 C (4 mm); flash pt. 438 F; ref. index 1.4380
Precaution: Combustible
Toxicology: LD50 (oral, rat) 3000 mg/kg, (IV, mouse) 180 mg/kg; mod. toxic by ingestion; irritant
Uses: Primary plasticizer for most resins and elastomers; floor finishes and waxes; flame-retarding agent; latex paints; as defoamer
Manuf./Distrib.: Akzo Nobel; Albright & Wilson Am.; Aldrich; Ashland; FMC; C.P. Hall; Harwick; Hoechst AG; Rhone-Poulenc; Velsicol
Trade names: Amgard® TBEP; KP-140®; TBEP

Tributyl acetyl citrate. *See* Acetyl tributyl citrate

Tributyl citrate

CAS 77-94-1; EINECS 201-071-2
Synonyms: TBC; 2-Hydroxy-1,2,3-propanetricarboxylic acid, tributyl ester; Butyl citrate
Definition: Triester of butyl alcohol and citric acid
Empirical: $C_{18}H_{32}O_7$
Formula: $C_3H_5O(COOC_4H_9)_3$
Properties: Colorless or pale yel., odorless; insol. in water; m.w. 360.45; dens. 1.042 (25/25 C); m.p. -20 C; b.p. 233.5 C (22.5 mm); flash pt. (COC) 315 F; ref. index 1.4453
Precaution: Combustible
Uses: Plasticizer, antifoam agent, solvent for nitrocellulose
Regulatory: FDA 21CFR §175.105
Manuf./Distrib.: Morflex; Unitex
Trade names: Citroflex® 4

Tributyl phosphate

CAS 126-73-8; EINECS 204-800-2
Synonyms: Tri-n-butyl phosphate
Empirical: $C_{12}H_{26}O_4P$
Formula: $(C_4H_9)_3PO_4$
Properties: Stable colorless liq., odorless; misc. with most solvs. and diluents; sol. in water; m.w. 266.32; dens. 0.978 (20/20 C); m.p. < -80 C; b.p. 292 C; flash pt. (COC) 295 F; ref. index 1.4215 (25 C)
Precaution: Combustible
Toxicology: Toxic by ingestion, inhalation; irritant to skin, mucous membranes; TLV 2.5 mg/m^3 of air; LD50 (rat, oral) 3.0 g/kg
Uses: Heat-exchange medium; solv. extraction of metal ions; plasticizer for cellulose esters, lacquers, plastics, vinyl resins; antifoam agent; dielectric; flame retardant
Manuf./Distrib.: AC Ind.; Akzo Nobel; Albright & Wilson Am.; Aldrich; Ashland; Bayer; Chemron; EM Ind.; Fluka; FMC; C.P. Hall; Harwick; Sigma; Velsicol
Trade names: Albrite® TBP; Albrite® Tributyl Phosphate; Amgard® $TBPO_4$; Kronitex® TBP; Reomol® TBP; TBP

Tri-n-butyl phosphate. *See* Tributyl phosphate
Tri-n-butyl-stannane. *See* Tributyltin oxide

Tributyltin benzoate

CAS 4342-36-3
Synonyms: Benzoyloxytributylstannane
Empirical: $C_{19}H_{32}O_2Sn$
Properties: M.w. 411.2
Toxicology: LD50 (oral, rat) 132 mg/kg, (subcut., rat) 505 mg/kg, (IV, mouse) 178 mg/kg; poison by ingestion and IV routes; mod. toxic by subcut. route; ACGIH TLV:TWA 0.1 $mg(Sn)/m^3$ (skin); heated to decomp., emits acrid smoke and irritating fumes
Manuf./Distrib.: Syntech labs
Trade names containing: Fungitrol® 158

Tributyltin fluoride

CAS 1983-10-4; 27615-98-1; EINECS 217-847-9
Synonyms: Fluorotributylstannane
Empirical: $C_{12}H_{27}FSn$
Formula: $[CH_3(CH_2)_3]_3SnF$
Properties: M.w. 309.08; m.p. 267-271 C dec.; flash pt. ≈ 145 C
Toxicology: LDLo (oral, mouse) 320 mg/kg, (oral, rabbit) 50 mg/kg; poison by ingestion and inhalation; corrosive; causes burns; TLV:TWA 0.1 mg $(Sn)/m^3$ (skin)
Uses: Antifoulant for marine paints; reagent
Manuf./Distrib.: Aldrich; Fluka
Trade names: bioMeT TBTF

Tributyltin oxide

CAS 56-35-9; EINECS 200-268-0
Synonyms: HBD; TBTO; Bis(tributyltin) oxide; Bis (tri-n-butyltin) oxide; Tri-n-butyl-stannane; Hexabutyldistannoxane; Oxybis (tributyltin)

Empirical: $C_{24}H_{54}OSn_2$
Formula: $(CH_3CH_2CH_2CH_2)_3SnOSn(CH_2CH_2CH_2CH_3)_3$
Properties: Colorless to pale yel. liq.; sol. in many org. solvs.; insol. in water; m.w. 596.08; dens. 1.17; solidfies < -45 C; b.p. 180 C (2 mm)
Toxicology: LD50 (IP, rat) 7210 mg/kg, (dermal, rat) 11,700 mg/kg; poison by ingestion, IP, IV routes; mod. toxic by skin contact; TLV 0.1 mg/m^3 of air
Uses: Bactericide, fungicide, mildewcide in underwater and antifouling paints, pesticides, plastics; intermediate; PVAc latex paints
Manuf./Distrib.: Aceto; Akzo Nobel; Aldrich; Elf Atochem N. Am.; Fluka; KMZ Chem. Ltd
Trade names: bioMeT TBTO; Fungitrol® Tinox; Keycide® X-10
Trade names containing: Troysan® 364

Tributyltin salicylate
CAS 4342-30-7
Synonyms: Salicyloyloxytributylstannane; Tri-n-butyltin salicylate
Empirical: $C_{19}H_{32}O_3Sn$
Properties: M.w. 427.20
Toxicology: LD50 (oral, rat) 137 mg/kg, (IV, mouse) 8910 mg/kg; poison by ingestion and IV routes
Uses: In-can fungicide/preservative and dry-film fungicide for water- or solv.-based systems
Trade names: Fungitrol® 234

Tri-n-butyltin salicylate. *See* Tributyltin salicylate
Tricarballylic acid-β-acetoxytributyl ester. *See* Acetyl triethyl citrate

Tricetylmonium chloride
CAS 52467-63-7; 71060-72-5
Synonyms: Tricetyl monomethyl ammonium chloride
Classification: Quaternary ammonium salt
Empirical: $C_{49}H_{102}N{\bullet}Cl$
Formula: $[CH_3(CH_2)_{14}CH_2N[CH_2(CH_2)_{14}CH_3]_2CH_3]^+Cl^-$
Uses: Industrial surfactant for pigment dispersing, coatings, inks, paper processing
Trade names: Arquad® 316(W)

Tricetyl monomethyl ammonium chloride. *See* Tricetylmonium chloride
Trichlorethyl phosphate. *See* Trichloroethyl phosphate

Trichloroethane
CAS 71-55-6; EINECS 200-756-3; 201-166-9
Synonyms: TCE; 1,1,1-Trichloroethane; Chloroethene; Ethane, 1,1,1-trichloro-; Methylchloroform
Classification: Halogenated aliphatic hydrocarbon
Empirical: $C_2H_3Cl_3$
Formula: CH_3CCl_3
Properties: Colorless liq., strong sweet odor; insol. in water; sol. in alcohol, ether, acetone, benzene, CCl_4; m.w. 133.42; dens. 1.3376 (20/4 C); m.p. -32.5 C; b.p. 74.1 C (760 mm); flash pt. none; ref. index. 1.43838 (20 C); KB value 124
Precaution: Nonflamm., but vapor may burn in presence of oxidizers or an ignition source
Toxicology: LD50 (oral, rat) 10,300 mg/kg, (IP, rat) 3993 mg/kg; TLV 350 ppm in air; poison by IV; mod. toxic by ing., inh.; irritant to eyes, tissue, skin, respiratory tract; vapor inh. produces CNS effects; narcotic in high concs. causing unconsiousness, coma
Uses: Solvent for cleaning precision instruments, metal degreasing, pesticide, textile processing; diluent for paints; catalyst for PU foams; paint aerosols
Regulatory: FDA 21CFR §175.105, 177.1650
Manuf./Distrib.: Aldrich; Allchem Ind.; Asahi Chem. Ind.; Asahi-Penn; Ashland; Chemoxy Int'l. plc; Elf Atochem N. Am.; Fabrichem; Fluka; ICI Am.; Olin; PPG Ind.; Schumacher; Sigma; Stanchem; Vulcan

1,1,1-Trichloroethane. *See* Trichloroethane

Trichloroethyl phosphate
CAS 115-96-8; 306-52-5
Synonyms: 2-Chloroethanol phosphate; Trichlorethyl phosphate; Tri-β-chloroethyl phosphate; Tri (2-chloroethyl) phosphate; Tris (2-chloroethyl) ester phosphoric acid; Tris (β-chloroethyl) phosphate; Tris (2-chloroethyl) phosphate
Empirical: $C_6H_{12}Cl_3O_4P$

Tri (2-chloroethyl) phosphate

Properties: M.w. 285.5; dens. 1.425 (20/20 C); b.p. 210-220 (20 mm); flash pt. (COC) 421 F
Precaution: Combustible exposed to heat and flame
Toxicology: LD50 (oral, rat) 1230 mg/kg; LDLo (IP, mouse) 250 mg/kg; poison by IP route; mod. toxic by ing.; skin and eye irritant; experimental reproductive effects; heated to decomp., emits very toxic fumes of PO_x and Cl^-
Uses: Flame retardant for PVC and other plastics and coatings
Manuf./Distrib.: Akzo Nobel
Trade names: Disflamoll® TCA; Fyrol® CEF

Tri (2-chloroethyl) phosphate. *See* Trichloroethyl phosphate
Tri-β-chloroethyl phosphate. *See* Trichloroethyl phosphate
cis-N-[(Trichloromethyl) thio]-4-cyclohexene-1,2-dicarboximide. *See* Captan
N-Trichloromethylthio-4-cyclohexene-1,2-dicarboximide. *See* Captan
N-Trichloromethylthiotetrahydrophthalimide. *See* Captan
Tricresol. *See* Cresylic acid

Tricresyl phosphate
CAS 1330-78-5; 78-30-8; 68952-35-2; EINECS 215-548-8
Synonyms: TCP; Phosphoric acid, tris(methylphenyl) ester; Tris (tolyloxy) phosphine oxide; Tritolyl phosphate; Cresyl phosphate; Phosphoric acid tritolyl ester
Empirical: $C_{21}H_{21}O_4P$
Formula: $(CH_3C_6H_4O)_3PO$
Properties: Colorless liq., odorless; misc. with all common solvs. and thinners, veg. oil; insol. in water; m.w. 368.37; dens. 1.162 (25/25 C); b.p. 420 C; flash pt. 437 F; ref. index 1.5550
Toxicology: LD50 (oral, rat) 1160 mg/kg, (IP, rat) 2500 mg/kg; poison by ing., subcut, IM, IV, IP; mod. toxic by skin contact; human systemic effects by ing.; eye and skin irritant; experimental reproductive effects; heated to decomp., emits toxic fumes of PO_x
Uses: Plasticizer, fire retardant for plastics, air filter medium, waterproofing, additive to extreme pressure lubricants; plasticizer and coalescing agent for coatings
Manuf./Distrib.: Akzo Nobel; Aldrich; Chemron; Daihachi Chem. Ind.; Fluka; FMC; Great Western
Trade names: Antiblaze® TCP; Disflamoll® TKP; Kronitex® TCP

Tridecanol. *See* Tridecyl alcohol
1-Tridecanol. *See* Tridecyl alcohol
n-Tridecanol. *See* Tridecyl alcohol
Tridecanol stearate. *See* Tridecyl stearate

Trideceth-6
CAS 24938-91-8 (generic); 78330-21-9
Synonyms: PEG-6 tridecyl ether; PEG 300 tridecyl ether; POE (6) tridecyl ether
Definition: PEG ether of tridecyl alcohol
Formula: $C_{13}H_{26}(OCH_2CH_2)_nOH$, avg. n = 6
Uses: Intermediate in the mfg. of high-foaming surfactants; emulsifier, detergent, dispersant, wetting agent; pulp and paper industry, textile, corrosion inhibitors, petrol. oils, paints, adhesives
Regulatory: FDA 21CFR §175.105, 176.200, 176.210
Manuf./Distrib.: Sigma
Trade names: Calgene Nonionic TD-6; Renex® 36

Trideceth-12
CAS 24938-91-8 (generic)
Synonyms: PEG-12 tridecyl ether; PEG 600 tridecyl ether; POE (12) tridecyl ether
Definition: PEG ether of tridecyl alcohol
Empirical: $C_{37}H_{75}O_{13}$
Formula: $C_{13}H_{26}(OCH_2CH_2)_nOH$, avg. n = 12
Uses: Intermediate in the mfg. of high-foaming surfactants; emulsifier, detergent, dispersant, wetting agent; pulp and paper industry, textile, corrosion inhibitors, petrol. oils, paints, adhesives
Regulatory: FDA 21CFR §175.105, 176.200, 176.210
Manuf./Distrib.: Sigma
Trade names: Calgene Nonionic TD-12

Trideceth-14
CAS 24938-91-8 (generic); 78330-21-9
Synonyms: PEG-14 tridecyl ether; POE (14) tridecyl ether

Definition: PEG ether of tridecyl alcohol
Empirical: $C_{41}H_{83}O_{15}$
Formula: $C_{13}H_{26}(OCH_2CH_2)_nOH$, avg. n = 14
Uses: Emulsifier, stabilizer for syn. latexes, textile wetting; solubilizer for essential oils, solvs., fats, waxes
Manuf./Distrib.: Sigma
Trade names: Rhodasurf® BC-737

Trideceth-15
CAS 24938-91-8 (generic); 78330-21-9
Synonyms: PEG-15 tridecyl ether; POE (15) tridecyl ether
Definition: PEG ether of tridecyl alcohol
Empirical: $C_{43}H_{87}O_{16}$
Formula: $C_{13}H_{26}(OCH_2CH_2)_nOH$, avg. n = 15
Uses: Solubilizer, foam builder, and detergent; emulsifier and stabilizer for syn. latexes; leveling and scouring agent
Regulatory: FDA 21CFR §175.105, 176.200, 176.210
Manuf./Distrib.: Sigma
Trade names: Calgene Nonionic TD-15; Rhodasurf® BC-840

Tridecyl alcohol
CAS 112-70-9; 68526-86-3; EINECS 203-998-8
Synonyms: Tridecanol; 1-Tridecanol; n-Tridecanol; C13 linear primary alcohol
Classification: Synthetic alcohol; commercial mixt. of isomers
Empirical: $C_{13}H_{28}O$
Formula: $C_{12}H_{25}CH_2OH$
Properties: Low melting white solid, pleasant odor; m.w. 200.41; dens. 0.845 (20/20 C); b.p. 274 C; m.p. 31 C; flash pt. (TOC) 180 F
Precaution: Combustible
Toxicology: LD50 (oral, rat) 17,200 mg/kg, (skin, rabbit) 5600 mg/kg; mildly toxic by ingestion and skin contact; skin irritant
Uses: Esters for synthetic lubricants, detergents, antifoam agents, other tridecyl compds.; perfumery; solv. for paints
Regulatory: FDA 21CFR §175.105, 175.300, 176.210
Manuf./Distrib.: Aldrich; Allchem Ind.; Ashland; Browning; Chemcentral; Exxon; Fluka; Hoechst Celanese; Penta Mfg.; Sigma

Tridecyl methacrylate
CAS 2495-25-2
Formula: $H_2C{=}C(CH_3)CO_2C_{13}H_{27}$
Properties: M.w. 268.44; dens. 0.88; flash pt. > 110 C; ref. index 1.4500
Toxicology: Irritant
Manuf./Distrib.: Aldrich; CPS; Monomer-Polymer & Dajac Labs
Trade names containing: SR-493

Tridecyl stearate
CAS 31556-45-3; EINECS 250-696-7
Synonyms: Octadecanoic acid, tridecyl ester; Tridecanol stearate
Definition: Ester of tridecyl alcohol and stearic acid
Empirical: $C_{31}H_{62}O_2$
Formula: $CH_3(CH_2)_{16}COOC_{13}H_{27}$
Properties: M.w. 466.83
Uses: Emollient for creams and lotions, dye carrier, lubricant for textile/industrial filament yarns, fiber finishes, plastic extrusion, magnetic tapes
Trade names: Lexolube® B-109

2,4,6-Tri(dimethylaminomethyl) phenol. *See* 2,4,6-Tris (dimethylaminomethyl) phenol
Tri(dimethylphenyl)phosphite. *See* Trixylenyl phosphate
Trien. *See* Triethylenetetramine
Trientine. *See* Triethylenetetramine

Triethanolamine
CAS 102-71-6; EINECS 203-049-8
Synonyms: TEA; 2,2′,2′′-Nitrilotris(ethanol); Trolamine; Trihydroxytriethylamine; Tris (2-hydroxyethyl) amine

Triethanolamine dodecylbenzene sulfonate

Classification: Aliphatic organic compd.; alkanolamine
Empirical: $C_6H_{15}NO_3$
Formula: $N(CH_2CH_2OH)_3$
Properties: Colorless to pale yel. visc. liq., sl. ammoniacal odor, very hygroscopic; misc. with water, alcohol; sol. in chloroform; sl. sol. in benzene, ether; m.w. 149.19; dens. 1.126; m.p. 21.2 C; b.p. 335 C; flash pt. (OC) 375 F; ref. index 1.4835
Precaution: Combustible when exposed to heat or flame; can react vigorously with oxidizing materials; light-sensitive
Toxicology: LD50 (oral, rat) 8 g/kg; mod. toxic by IP; mildly toxic by ingestion; experimental carcinogen; liver and kidney damage in animals from chronic exposure; human skin irritant; eye irritant; heated to decomp., emits toxic fumes of NO_x and CN^-
Uses: Intermediate in mfg. of surfactants, textile specialties, waxes, polishes, toiletries, herbicides, cutting oils, fatty acid soaps (drycleaning), cosmetics, household detergents, emulsions; solvent for casein, shellac, dyes; water repellent; dispersant for dyes, casein, shellac, and rubber latex; sequestering agent; rubber chem. intermediate; emulsifier for pharmaceuticals
Regulatory: FDA 21CFR §173.315, 175.105, 175.300, 175.380, 175.390, 176.170, 176.180, 176.200, 176.210, 177.1210, 177.1680, 177.2260, 177.2600, 177.2800, 178.3120, 178.3910; FDA approved for orals, topicals, vaginals; USP/NF, BP compliance
Manuf./Distrib.: Aldrich; Allchem Ind.; BASF AG; Coyne; Fluka; Harwick; Hüls AG; Int'l. Chem. Inc.; Mitsui Toatsu; Nippon Shokubai; Occidental; Oxiteno; Research Organics; Schweizerhall; Sigma; Texaco; Union Carbide
Trade names: TEA 85 Low Freeze Grade
Trade names containing: Surfonic® HDL-1

Triethanolamine dodecylbenzene sulfonate. *See* TEA-dodecylbenzenesulfonate
Triethoxymethylsilane. *See* Methyltriethoxysilane
(Triethoxysilyl)ethylene. *See* Vinyltriethoxysilane
3-(Triethoxysilyl)-1-propanamine. *See* Aminopropyltriethoxysilane
Triethoxyvinylsilane. *See* Vinyltriethoxysilane
Triethoxyvinylsilicane. *See* Vinyltriethoxysilane
Triethyl acetylcitrate. *See* Acetyl triethyl citrate

Triethylamine
CAS 121-44-8; EINECS 204-469-4
Synonyms: N,N-Diethylethanamine; (Diethylamino) ethane
Empirical: $C_6H_{15}N$
Formula: $(C_2H_5)_3N$
Properties: Colorless liq., strong ammoniacal odor; sol. in water and alcohol; m.w. 101.22; dens. 0.7293 (20/20 C); b.p. 89.7 C; f.p. -115.3 C; flash pt. (OC) 10 F; ref. index 1.4000
Precaution: Flamm.; dangerous fire risk; explosive limits in air 1.2-8%
Toxicology: LCLo (oral, rat) 460 mg/kg, (inh., rat) 1000 ppm/4 h; moderately toxic by ingestion and inhalation; strong irritant to tissue; TLV 10 ppm in air
Uses: Propellant; corrosion inhibitor; catalytic solvent in chemical synthesis; accelerator activators for rubber; wetting, penetrating and waterproofing agents of quat. ammonium types; curing and hardening of polymers; solv. for paints
Manuf./Distrib.: Air Prods.; Aldrich; Allchem Ind.; Ashland; BASF; BASF AG; Coyne; Elf Atochem N. Am.; Fluka; Focus; Sigma; Union Carbide
Trade names containing: ECO-CRYL™ 9790

Triethyl citrate
CAS 77-93-0; EINECS 201-070-7
Synonyms: TEC; 2-Hydroxy-1,2,3-propanetricarboxylic acid, triethyl ester; Ethyl citrate
Definition: Triester of ethyl alchol and citric acid
Empirical: $C_{12}H_{20}O_7$
Formula: $C_3H_5O(COOC_2H_5)_3$
Properties: Colorless mobile oily liq., odorless, bitter taste; sol. 65 g/100 cc water; sol. 0.8g/100 cc oil; misc. with alcohol, ether; m.w. 276.32; dens. 1.136 (25 C); b.p. 294 C; flash pt. (COC) 303 F; ref. index 1.4420
Precaution: Combustible liq. when exposed to heat or flame
Toxicology: LD50 (oral, rat) 5900 mg/kg; mod. toxic by IP route; mildly toxic by ingestion, inh.; no known skin toxicity; heated to decomp., emits acrid smoke and irritating fumes
Uses: Solvent and plasticizer for nitrocellulose and natural resins; softener; paint removers; agglutinant; perfume base; food additive
Regulatory: FDA 21CFR §175.300, 175.320, 181.22, 181.27, 182.1911, GRAS; FEMA GRAS; FDA approved

for orals; USP/NF compliance
Manuf./Distrib.: Aldrich; Fluka; Great Western; Int'l. Chem. Inc.; Morflex; Penta Mfg.; Unitex
Trade names: Citroflex® 2

Triethylene diamine
CAS 280-57-9; EINECS 205-999-9
Synonyms: TEDA; 1,4-Diazobicyclo [2.2.2] octane; Diazobicyclooctane
Empirical: $C_6H_{12}N_2$
Formula: $N(CH_2CH_3)_3N$
Properties: Wh. cryst., ammoniacal odor, very hygroscopic; sublimes readily at R.T.; sol. (g/100 g): 45 g in water, 13 g in acetone, 51 g in benzene, 77 g in ethanol, 26.1 g in MEK; m.w. 112.17; m.p. 158 C; b.p. 174 C
Toxicology: LD50 (oral, rat) 1100 mg/kg; mod. toxic; skin and eye irritant
Uses: Catalyst for urethane foams, PU coatings, adhesives, sealants; dyeing assistant, crosslinking agent, stabilizer, surf. treatments, bond promoter, etching agent
Manuf./Distrib.: Aldrich; Fluka; Sigma
Trade names: Amicure® TEDA; Dabco® Crystalline
Trade names containing: Amicure® 33-LV; Dabco® 33-LV®

Triethylene glycol butyl ether. *See* Butoxytriglycol
Triethylene glycol diacetate. *See* PEG-3 diacetate
Triethylene glycol diacrylate. *See* PEG-3 diacrylate
Triethylene glycol dimethacrylate. *See* PEG-3 dimethacrylate
Triethylene glycol ethyl ether. *See* Ethoxytriglycol
Triethylene glycol monobutyl ether. *See* Butoxytriglycol
Triethylene glycol monoethyl ether. *See* Ethoxytriglycol
Triethylene glycol monomethyl ether. *See* Triglycol monomethyl ether

Triethylenetetramine
CAS 112-24-3; EINECS 203-950-6
Synonyms: TETA; N,N′-Bis(2-aminoethyl)-1,2-ethanediamine; Trien; Trientine
Empirical: $C_6H_{18}N_4$
Formula: $NH_2(C_2H_4NH)_2C_2H_4NH_2$
Properties: Moderately visc. ylsh. oily liq.; sol. in water, alcohol; m.w. 146.23; dens. 0.9818 (20/20 C); m.p. 12 C; b.p. 277.5 C; flash pt. (CC) 275 F; ref. index 1.4971 (20 C); pH 14
Precaution: DOT: Corrosive material; combustible
Toxicology: LD50 (oral, rat) 2.5 g/kg, (dermal, rabbit) 805 mg/kg; poison by IV route; strong irritant to tissue, mucous membranes; skin burns, eye damage
Uses: Detergents and softening agents, synthesis of dyestuffs, pharmaceuticals, rubber accelerator; as thermosetting resin; lubricating oil additive; analytical reagent for Cu, Ni; paper additive; surfactants; lubricating oil additive; curing agent for epoxy resins; used for civil engineering, protective coatings, adhesives
Manuf./Distrib.: Aldrich; Allchem Ind.; Ashland; Diamine & Chems. Ltd; Fluka; Great Western; Rit-Chem; Sigma; Sumitomo Seika Chems.; Texaco; Tosoh; Union Carbide
Trade names: D.E.H. 24; Epi-Cure® 3234; TETA

Tri (2-ethylhexyl) trimellitate. *See* Trioctyl trimellitate
Triflic acid. *See* Trifluoromethane sulfonic acid

Trifluoromethane sulfonic acid
CAS 1493-13-6; EINECS 216-087-5
Synonyms: Triflic acid
Classification: Aliphatic organic compd.
Empirical: CF_3HSO_3
Formula: CHF_3O_3S
Properties: Colorless to amber clear liq.; hygroscopic; sol. in water, alcohol; m.w. 150.08; dens. 1.708 (20/4 C); m.p. 40 C; b.p. 167-180 C; flash pt. none; ref. index 1.331 (20 C)
Precaution: Strong acid; violent reaction with acyl chlorides or aromatic hydrocarbons, evolving toxic hydrogen chloride gas
Toxicology: Corrosive irritant to skin, eyes, mucous membrane; heated to decomp., emits toxic fumes of F^- and SO_x
Uses: Chemical synthesis; catalyst or reactant; polymerization of epoxies, styrenes, and THF, alkylation and some acylation reactions; pharmaceuticals, explosives, dyes, paints, and intermediates; electrolytes;

formation of biaryls; polymerization reaction
Regulatory: FDA 21CFR §173.395
Manuf./Distrib.: Aldrich; Amber Syn.; 3M; Fluka; MTM Research; Rhone-Poulenc N. Am.; SAF Bulk Chems.; Schweizerhall; Sigma

Trifluoromethanesulfonic acid amine salt
Trade names containing: FC-520

Triglycol monoethyl ether. *See* Ethoxytriglycol

Triglycol monomethyl ether
CAS 112-35-6
Synonyms: PEG-3 methyl ether; Triethylene glycol monomethyl ether; Methyltriglycol
Empirical: $C_7H_{16}O_4$
Formula: $CH_3O(CH_2CH_2O)_2CH_2CH_2OH$
Properties: Misc. with water; m.w. 164.20; dens. 1.048 (20/4 C); b.p. 249 C; flash pt. 118 C; ref. index 1.439 (20 C)
Toxicology: LD50 (oral, rat) 11,300 mg/kg, (dermal, rabbit) 7100 mg/kg; mildly toxic by ingestion and skin contact; skin irritant
Uses: Solvent for brake fluids, hard-surf. cleaners, leather dyeing, paints, coatings, printing inks, textile vat dyeing and printing, adhesives, antifreeze, floor waxes/polishes, insect repellents; solubilizer for dyes; plasticizer
Manuf./Distrib.: Aldrich; Ashland; Fluka; Oxiteno; Sigma
Trade names: Poly-Solv® TM

Trihexyl 1,2,4,-benzenetricarboxylate. *See* Tri n-hexyl trimellitate
Trihexyl trimellitate. *See* Tri n-hexyl trimellitate

Tri n-hexyl trimellitate
CAS 1528-49-0
Synonyms: Trihexyl trimellitate; Trihexyl 1,2,4,-benzenetricarboxylate
Formula: $C_6H_3[CO_2(CH_2)_5CH_3]_3$
Properties: M.w. 462.63; sp.gr. 1.010; m.p. -45.5 C; b.p. 260-262 C (4 mm); flash pt. (COC) 260 C; ref. index 1.485
Precaution: Moisture-sensitive
Uses: Plasticizer for PVC; good permanence; used in automotive and furniture upholstery, wire coatings, and gasketing materials
Manuf./Distrib.: Aldrich; Morflex; Unitex
Trade names: Morflex® 560

Trihydrated alumina. *See* Aluminum hydroxide
Trihydroxypropane glycerol. *See* Glycerin

Trihydroxystearin
CAS 8001-78-3; 139-44-6
Synonyms: Glyceryl tri(12-hydroxystearate); 12-Hydroxyoctadecanoic acid, 1,2,3-propanetriyl ester; 1,2,3-Propanetriol tri(12-hydroxystearate)
Definition: Triester of glycerin and hydroxystearic acid
Empirical: $C_{57}H_{110}O_9$
Uses: Thixotrope, gellant for solv.-based coatings, aliphatic solv. systems; topical pharmaceuticals
Regulatory: FDA approved for topicals
Trade names: Flowtone® R; Rheocin; Thixcin® R

Trihydroxytriethylamine. *See* Triethanolamine

Triisodecyl phosphite
CAS 25448-25-3; EINECS 246-998-3
Synonyms: TDP
Formula: $(C_{10}H_{21}O)_3P$
Properties: Liq.; m.w. 502; dens. 0.884-0.904; b.p. 180 C (5 mm); ref. index 1.4530-1.4610; flash pt. (PMCC) 160 C
Uses: Antioxidant and extreme pressure additive in lubricant formulations; sec. heat stabilizer for ABS, PET, PVC, PU, coatings

Manuf./Distrib.: Aldrich; Ashland; Dover
Trade names: Doverphos® 6; Weston® TDP

Triisooctyl phosphite
CAS 25103-12-2; EINECS 246-614-4
Synonyms: Phosphorous acid, triisooctyl ester
Empirical: $C_{24}H_{54}O_3P$
Formula: $(C_8H_{17}O)_3P$
Properties: Colorless liq. with odor; misc. with most common org. solvs.; insol. in water; m.w. 418; dens. 0.891 (20/4 C); b.p. 161-164 C (0.3 mm); flash pt. (COC) 385 F
Precaution: Combustible
Toxicology: LD50 (oral, rat) 9200 mg/kg, (dermal, rabbit) 3970 mg/kg; skin irritant
Uses: Intermediate; insecticides; lubricant additive; specialty solvents; stabilizer for acrylics, nylon, unsat. polyester, PVC; improves antiwear and antifriction props.; stabilizer in coatings
Manuf./Distrib.: Albright & Wilson Am.; Ashland; GE Specialty
Trade names: Albrite® TIOP

Triisopropanolamine
CAS 122-20-3; EINECS 204-528-4
Synonyms: TIPA; 1,1′,1′′-Nitrilotris-2-propanol; Tris(2-hydroxypropyl)amine
Classification: Aliphatic amine
Empirical: $C_9H_{21}NO_3$
Formula: $N(CH_2CHOHCH_3)_3$
Properties: Wh. cryst. solid, hygroscopic; sol. in water; m.w. 191.27; dens. 0.9996 (50/20 C); m.p. 45 C; b.p. 305 C; flash pt. (OC) 320 F
Precaution: Combustible; corrosive
Toxicology: LD50 (oral, rat) 6500 mg/kg; LDLo (skin, rabbit) 10 g/kg; mod. toxic by ingestion; mildly toxic by skin contact; skin and severe eye irritant
Uses: Emulsifying agent; for soaps, cosmetics, pharmaceuticals, textile specialties, agric., polymer curing chems., adhesives, antistats, coatings, metalworking, rubber, gas conditioning chems.
Regulatory: FDA 21CFR §175.105, 176.200, 176.210
Manuf./Distrib.: Aldrich; Ashland; BASF AG; Fluka
Trade names: TIPA 99
Trade names containing: Doverphos® 10-HR; Doverphos® 10-HR Plus; Ultranox® 626; Ultranox® 626A; Ultranox® 627A; Weston® 399; Weston® 619F; Weston® EGTPP

Triisopropylphenyl phosphate
Properties: Liq.; sp.gr. 1.150-1.165; flash pt. (COC) 230 C; ref. index 1.552
Uses: Plasticizer, flame retardant for plastics incl. acrylics, cellulosics, epoxy, nitrile, phenolic, thermoset polyester, PP, PS, PVAc, PVC, flexible PU foam; plasticizer and coalescing agent for coatings
Manuf./Distrib.: Akzo Nobel; Albright & Wilson Am.; FMC
Trade names: Antiblaze® 519

Trilead tetroxide. *See* Lead oxide, red

Trilinoleic acid
CAS 7049-66-3; 68937-90-6
Synonyms: Trimer acid; Fatty acids, C18, unsaturated, trimers; 9,12-Octadecadienoic acid, trimer
Definition: 54-carbon tricarboxylic acid formed by the catalytic trimerization of linoleic acid
Uses: Corrosion inhibitor, lubricant; epoxy curing agent; polymer building block; used in polyamino-amides for PVC plastisols to be used as car underbody coatings and in water-sol. alkyd resins
Regulatory: FDA 21CFR §176.200, 178.3910
Manuf./Distrib.: Unichema N. Am.; Union Camp; Witco/Oleo-Surf.
Trade names: Pripol 1040

Trimellitic anhydride
CAS 552-30-7
Synonyms: Anhydrotrimellitic acid; 1,2,4-Benzenetricarboxylic acid anhydride
Empirical: $C_9H_4O_5$
Properties: Cryst.; sol. in acetone, ethyl acetate; m.w. 192.13; m.p. 162 C
Toxicology: LD50 (oral, rat) 5600 mg/kg; mildly toxic by ingestion; irritant to lungs and air passages; may be a powerful allergen
Uses: Intermediate for paints

Trimer acid

Manuf./Distrib.: Allchem Ind.; Amoco; Monomer-Polymer & Dajac Labs
Trade names: Amoco® TMA

Trimer acid. *See* Trilinoleic acid
Trimethoxy (2-methylpropyl) silane. *See* Isobutyltrimethoxysilane
Trimethoxymethylsilane. *See* Methyltrimethoxysilane
(Trimethoxysilyl)benzene. *See* Phenyltrimethoxysilane
Trimethoxysilylpropanethiol. *See* Mercaptopropyltrimethoxysilane
3-Trimethoxysilyl-1-propanethiol. *See* Mercaptopropyltrimethoxysilane
3-(Trimethoxysilyl)propylamine. *See* Aminopropyltrimethoxysilane
N-[3-(Trimethoxysilyl) propyl] ethylenediamine. *See* N-2-Aminoethyl-3-aminopropyl trimethoxysilane
Trimethoxyvinylsilane. *See* Vinyltrimethoxysilane
Trimethylacetic acid. *See* Neopentanoic acid
Trimethyl benzene. *See* 1,2,3-Trimethylbenzene

1,2,3-Trimethylbenzene
CAS 526-73-8; EINECS 208-394-8
Synonyms: Hemellitol; Hemimellitene; Trimethyl benzene
Empirical: C_9H_{12}
Formula: $C_6H_3(CH_3)_3$
Properties: M.w. 120.20; dens. 0.894; m.p. -25 C; flash pt. 53 C; ref. index 1.514
Precaution: Flamm.
Toxicology: LDLo (oral, rat) 5000 mg/kg; toxic by inhalation, ingestion, skin contact; irritant
Manuf./Distrib.: Aldrich; Fluka
Trade names containing: Ircogel® 905; Ircogel® 906; Ircogel® 907

1,2,4-Trimethylbenzene
CAS 95-63-6; EINECS 202-436-9
Synonyms: Pseudocumene; Pseudocumol; Asymmetrical trimethyl benzene; as-Trimethylbenzene
Empirical: C_9H_{12}
Formula: $C_6H_3(CH_3)_3$
Properties: Insol. in water; sol. in alcohol, benzene, ether; m.w. 120.21; dens. 0.889; m.p. -44 C; b.p. 168 C; flash pt. 48 C; ref. index 1.5040
Precaution: Moderate fire risk
Toxicology: LDLo (IP, rat) 1752 mg/kg; ACGIH TLV-TWA 123 mg/m^3, 25 ppm; mildly toxic by inh.; mod. toxic by IP route; irritant to mucous membranes; CNS depressant
Uses: Mfg. of trimellitic anhydride, dyes, pharmaceuticals, pseudocumidine
Manuf./Distrib.: Aldrich; Fluka; Sigma
Trade names containing: Ircogel® 905; Ircogel® 906; Ircogel® 907

1,3,5-Trimethylbenzene
CAS 108-67-8; EINECS 203-604-4
Synonyms: TMB; Trimethylbenzol; Mesitylene; sym-Trimethylbenzene
Empirical: C_9H_{12}
Formula: $C_6H_3(CH_3)_3$
Properties: Liq.; insol. in water; sol. in alcohol and ether; m.w. 120.21; dens. 0.863; f.p. -52.7 C; b.p. 164.6 C
Precaution: Moderate fire hazard; combustible
Toxicology: LC50 (inh., rat) 24 mg/m^3 (4 h); toxic by inhalation
Uses: Intermediate; uv oxidation stabilizer for plastics
Manuf./Distrib.: Aldrich; Fluka; Sigma
Trade names containing: Ircogel® 905; Ircogel® 906; Ircogel® 907

Trimethylbenzol. *See* 1,3,5-Trimethylbenzene

2,4,6-Trimethylbenzophenone
Trade names containing: Esacure® KT37; Esacure® TZT

1,7,7-Trimethylbicyclo[2.2.1] heptan-2-one. *See* Camphor
2,6,6-Trimethylbicyclo(3.1.1)-2-heptene. *See* α-Pinene
2,6,6-Trimethylbicyclo(3.1.1)-2-hept-2-ene. *See* α-Pinene
Trimethyl carbinol. *See* t-Butyl alcohol
1,1,3-Trimethyl-3-cyclohexene-5-one. *See* Isophorone
3,3,5-Trimethyl-2-cyclohexen-1-one. *See* Isophorone

Trimethyldihydroquinoline polymer. *See* 2,2,4-Trimethyl-1,2-dihydroquinoline polymer

2,2,4-Trimethyl-1,2-dihydroquinoline polymer
CAS 26780-96-1
Synonyms: TDQP; 1,2-Dihydro-2,2,4-trimethylquinoline homopolymer; 1,2-Dihydro-2,2,4-trimethylquinoline polymer; Poly-(1,2-dihydro-2,2,4-trimethylquinoline); Trimethyldihydroquinoline polymer
Formula: $(C_{11}H_{16}N)_n$
Properties: Amber pellets; insol. in water; misc. with ethanol, acetone, benzene, monochlorobenzene, isopropyl acetate, gasoline; dens. 1.08; soften. pt. 75 C; m.p. 95-100 C
Precaution: Avoid strong oxidizers; dust can present explosion hazard
Toxicology: Experimental neoplastigen; heated to decomp., emits toxic fumes of NO_x
Uses: Antioxidant for rubbers; stabilizer or polymerization inhibitor; staining antioxidant for stabilization of latex foams for carpet back coating
Manuf./Distrib.: Aldrich
Trade names: Ralox® TMQ-T

Trimethylhexamethylene diamine
CAS 25620-58-0
Synonyms: 1,6-Hexanediamine, trimethyl
Empirical: $C_9H_{22}N_2$
Properties: M.w. 158.29; dens. 0.867; b.p. 232 C; flash pt. 104 C; ref. index 1.4640
Precaution: Corrosive
Uses: Epoxy curing agent for coatings, adhesives, castings, composites
Manuf./Distrib.: Aldrich; Fluka
Trade names: Vestamin® TMD

Trimethylmethane. *See* Isobutane
2,6,8-Trimethyl-4-nonanone. *See* Isobutyl heptyl ketone

Trimethylolethane
CAS 77-85-0; EINECS 201-063-9
Synonyms: TME; Pentaglycerine; Methyltrimethylolmethane; 1,1,1-Tris(hydroxymethyl) ethane
Empirical: $C_5H_{12}O_3$
Formula: $CH_3C(CH_2OH)_3$
Properties: Colorless cryst.; hygroscopic; sol. in water, alcohol; m.w. 120; m.p. ≈ 190 C; flash pt. 160 C
Precaution: Combustible
Toxicology: Essentially nontoxic; mildly irritating to abraded skin; nonirritating to eyes
Uses: Raw material for alkyd and polyester resins, paints; conditioning agent; mfg. of varnishes, synthetic drying oils
Manuf./Distrib.: Aldrich; Fabrichem; Fluka; Great Western; Kowa Am.
Trade names: Trimet® Liquid; Trimet® Nitration Grade; Trimet® Pure; Trimet® Tech.

Trimethylolpropane
CAS 77-99-6; EINECS 201-074-9
Synonyms: 1,1,1-Tris(hydroxymethyl)propane; 2-Ethyl-2-hydroxymethyl-1,3-propanediol; Hexaglycerol
Empirical: $C_6H_{14}O_3$
Formula: $CH_3CH_2C(CH_2OH)_3$
Properties: Hygroscopic; m.w. 134.18; m.p. 56-58 C; b.p. 161 C; flash pt. 172 C
Uses: Conditioner, manufacture of varnishes, alkyd resins, synthetic drying oils, urethane foams and coatings, silicone lube oils, lactone plasticizers, textile finishes, surfactants, epoxidation products
Manuf./Distrib.: Aldrich; Allchem Ind.; Bayer; Browning; Chemical; Fluka; Hoechst Celanese; Mitsubishi Gas; Perstorp Polyols; Punda Mercantile

Trimethylolpropane adipate polyester
Uses: Used for two-part and moisture-cure coatings, chain extenders, cast PU elastomers, microcellular urethanes, potting compds., encapsulating agents, printing rolls, exterior high performance coatings, one-shot PU elastomers
Trade names: Formrez® 102; Formrez® 104; Formrez® 106; Formrez® 108; Formrez® 110R; Formrez® 118

Trimethylolpropane ethoxylate triacrylate
Formula: $C_2H_5C\text{-}[CH_2O(CH_2CH_2O)_nCOCH\text{-}CH_2]_3$
Uses: General purpose monomer for inks, metal deco, and plastics radiation-curing coatings; fast curing; forms homopolymers of high tens. and mod. elong.; exc. solvency and visc. reducing chars.
Trade names: Photomer® 4149

Trimethylolpropane ethoxy triacrylate
CAS 28961-43-5
Synonyms: TMPEOTA; Ethanol,2,2′,2′′-(propyldyne tris(methyleneoxy))tri-, triacrylate
Formula: $C_2H_5C[CH_2(OCH_2CH_2)_nOH]_3$
Properties: Dens. 1.102; flash pt. 110 C; ref. index 1.4710
Toxicology: Irritant
Uses: Trifunctional monomer for radiation curing of films; cured films retain hardness, abrasion resist., and high gloss props.
Manuf./Distrib.: Aldrich; UCB Radcure

Trimethylolpropane triacrylate. *See* 1,1,1-Trimethylolpropane triacrylate

1,1,1-Trimethylolpropane triacrylate
CAS 15625-89-5
Synonyms: TMPTA; Trimethylolpropane triacrylate; 2-Ethyl-2-(hydroxymethyl)-1,2,3-propanediol
Formula: $(H_2C{=}CHCO_2CH_2)_3CC_2H_5$
Properties: Hygroscopic; m.w. 296.3; dens. 1.100; flash pt. > 230 F; ref. index 1.4740
Toxicology: LD50 (oral, rat) 5190 mg/kg, (dermal, rabbit) 5170 mg/kg; human skin irritant
Uses: Crosslinker; uv-cured adhesives, wood fillers, coatings, inks, vinyl acrylic latex paint, highly crosslinked polybutadiene rubber
Manuf./Distrib.: Aldrich; CPS; Monomer-Polymer & Dajac Labs; Rhone-Poulenc Spec. Chem.; UCB Radcure
Trade names: Photomer® 4006
Trade names containing: Ageflex TMPTA; CN 104 C75; CN 112 C60; CN 120 C60; CN 120 C80; CN 960 C75; CN 961 C75; CN 962 C75; CN 963 C75; CN 964 C75; CN 965 C75; CN 966 C75; CN 970 C60; CN 971 C75; CN 972 C75; Craynor 104 C75; Craynor 112 C60; Craynor 114 E80; Ebecryl® 3700-20T; Ebecryl® 3701-20T; SR-351

1,1,1-Trimethylolpropane trimethacrylate
CAS 3290-92-4
Synonyms: TMPTMA
Formula: $[H_2C{=}C(CH_3)CO_2CH_2]_3CC_2H_5$
Properties: Hygroscopic; m.w. 338.4; dens. 1.060; b.p. 185 C (5 mm); flash pt. > 230 F; ref. index 1.4720
Toxicology: Irritant
Uses: Crosslinker, coagent for wire and cable, hard rubber rolls, polybutadiene, polyethylene, moisture barrier films and coatings, plastisols and vinyl acetate latexes, adhesives, molding compds., textile prods.
Manuf./Distrib.: Aldrich; CPS; Elf Atochem N. Am.; Monomer-Polymer & Dajac Labs; Rohm Tech; U.S. Chems.
Trade names: Sipomer® TMPTMA
Trade names containing: Ageflex TMPTMA; SR-350

2,2,4-Trimethyl-1,3-pentanediol
CAS 144-19-4; EINECS 205-619-1
Synonyms: TMPD
Empirical: $C_8H_{18}O_2$
Formula: $(CH_3)_2CHCH(OH)C(CH_3)_2CH_2OH$
Properties: Wh. cryst. solid; sl. sol. in water; sol. in alcohol, acetone, ether, benzene; m.w. 146.23; sp.gr. 0.93 (55/15 C); m.p. 50 C; b.p. 225 C; flash pt. (COC) 113 C
Precaution: Combustible
Toxicology: LDLo (oral, rat) 2000 mg/kg, (IP, rat) 800 mg/kg; poison by intravenous, moderately toxic by ingestion and intraperitoneal routes; skin irritant
Uses: Resin intermediate; paints; insect repellent
Manuf./Distrib.: Aldrich; Eastman; Fluka; Hüls Am.
Trade names: TMPD® Glycol

2,2,4-Trimethyl-1,3-pentanediol bis(2-methylpropanoate). *See* Trimethyl-1,3-pentanediol, 2,2,4-diisobutyrate

2,2,4-Trimethyl-1,3-pentanediol dibenzoate
CAS 68052-23-3
Synonyms: 1,3-Pentanediol, 2,2,4-trimethyl-, dibenzoate
Empirical: $C_{22}H_{26}O_4$
Properties: M.w. 354.45
Trade names: Benzoflex® 354

Trimethyl-1,3-pentanediol, 2,2,4-diisobutyrate
CAS 6846-50-0; EINECS 229-934-9
Synonyms: 2,2,4 Trimethylpentanediol-1,3-diisobutyrate; 2,2,4-Trimethyl-1,3-pentanediol bis(2-methylpropanoate); Isobutyric acid, 1-isopropyl-2,2-dimethyltrimethylene ester
Empirical: $C_{16}H_{30}O_4$
Formula: $(CH_3)_2CHCH[O_2CCH(CH_3)_2]C(CH_3)_2CH_2O_2CCH(CH_3)_2$
Properties: M.w. 286.4; sp.gr. 0.945; b.p. 280 C; f.p. -70 C; flash pt. (COC) 143 C; ref. index 1.4300
Toxicology: Skin irritant
Uses: Plasticizer for PVC, PS, cellulosics; primary plasticizer used in surf. coatings, vinyl floorings, moldings, and vinyl prods.; compatible with film-forming vehicles used in lacquers for wood, paper, and metals
Manuf./Distrib.: Aldrich; Chisso Am.; Eastman
Trade names: Eastman® TXIB

2,2,4 Trimethylpentanediol-1,3-diisobutyrate. *See* Trimethyl-1,3-pentanediol, 2,2,4-diisobutyrate

2,2,4-Trimethyl-1,3-pentanediol monoisobutyrate
CAS 25265-77-4; EINECS 246-771-9
Synonyms: Propionic acid, 2-methyl-, monoester with 2,2,4-trimethyl-1,3-pentanediol; 2,2,4-Trimethyl-1,3-pentanediol mono (2-methylpropanoate)
Classification: Ester alcohol
Empirical: $C_{12}H_{24}O_3$
Formula: $(CH_3)_2CHCH(OH)C(CH_3)_2CH_2OOCCH(CH_3)_2$
Properties: Liq.; insol. in water; sol. in benzene, alcohol, acetone, CCl_4; m.w. 216.3; dens. 0.945-0.955 (20/20 C); b.p. 180-182 C (125 mm); flash pt. (COC) 245 F; ref. index 1.4423
Precaution: Combustible
Toxicology: LDLo (oral, rat) 3200 mg/kg; mod. toxic by ingestion
Uses: Solv., intermediate; mfg. of plasticizers, surfactants, pesticides, resins, inks, latexes, drilling muds, lubricants, syn. detergents, herbicides; slow-evaporating solv. used as coalescing agent in latex finishes and water-base inks, PVAc latices, PVAc-acrylic latices, and acrylic, EVA, and B/S latices; used as solv. in electrodeposition coatings, lacquer coatings
Manuf./Distrib.: Aldrich; Ashland; Chisso Am.; Eastman
Trade names: Texanol® Ester-Alcohol; Ucar® Filmer IBT

2,2,4-Trimethyl-1,3-pentanediol mono (2-methylpropanoate). *See* 2,2,4-Trimethyl-1,3-pentanediol monoisobutyrate
α-(Trimethylsilyl)-ω-methylpoly[oxy(dimethylsilylene)], mixt. with silicon dioxide. *See* Simethicone
α,α,α´-Trimethyltrimethyleneglycol. *See* Hexylene glycol
1,1,1-Trimethyl-N-(trimethylsilyl) silanamine. *See* Hexamethyldisilazane

Trioctyl phosphate
CAS 78-42-2; EINECS 201-116-6
Synonyms: Tris(2-ethylhexyl) phosphate; Octyl phosphate
Empirical: $C_{24}H_{51}O_4P$
Formula: $[CH_3(CH_2)_3CH(C_2H_5)CH_2O]_3PO$
Properties: Liq.; sol. in alcohol, acetone, ether; m.w. 744.04; dens. 0.924 (26 C); b.p. 220-230 (8 mm); flash pt. > 230 F; ref. index 1.4440
Precaution: Combustible
Toxicology: LD50 (oral, rat) 37 g/kg, (dermal, rabbit) 20 g/kg; oxic by ingestion, inhalation; suspected carcinogen
Uses: Solvent, antifoaming agent for paints; plasticizer for vinyl resins, rubbers, nitrocellulose; flame retardant
Manuf./Distrib.: Akzo Nobel; Albright & Wilson Am.; Aldrich; Ashland; Bayer; Ferro/Keil; Fluka; FMC; Rhone-Poulenc; Sigma
Trade names: Amgard® TOF; Kronitex® TOF; Reomol® TOP

Trioctyl trimellitate
CAS 89-04-3; 3319-31-1
Synonyms: TOTM; Tri (2-ethylhexyl) trimellitate
Empirical: $C_{33}H_{54}O_6$
Formula: $C_6H_3(COOCH_2CH[C_2H_5[C_4H_9)_3$
Properties: Liq.; m.w. 547; dens. 0.989 (20/20 C); b.p. 414 C; f.p. -38 C; flash pt. (COC) 263 C
Precaution: Combustible
Uses: Monomeric plasticizer for PVC, cellulosics, paints, as vehicles for pigment dispersions
Manuf./Distrib.: Aldrich; Aristech; Ashland; BASF; BP Oil; Eastman; Fluka; C.P. Hall; Harwick; Hatco; Hüls Am.;

Tripentaerythritol

Unitex

Trade names: Diplast® TM; Diplast® TM8

Tripentaerythritol

CAS 78-24-0

Empirical: $C_{15}H_{35}O_8$

Properties: Wh.-ivory powd.; hygroscopic; m.w. 372.41; m.p. 225-240 C

Precaution: Combustible

Uses: Hard resins, varnishes, fast-drying tall oil vehicles

Manuf./Distrib.: Aldrich; Allchem Ind.

Triphenyl phosphate

CAS 115-86-6; EINECS 204-112-2

Synonyms: TPP; Phosphoric acid, triphenyl ester

Empirical: $C_{18}H_{15}O_4P$

Formula: $PO(OC_6H_5)_3$

Properties: Colorless cryst. powd. or needles, odorless; sol. in benzene, chloroform, ether, acetone, lacquers, solvent, thinners, oil; insol. in water; m.w. 326.28; dens. 1.268 (60 C); m.p. 50 C; b.p. 245 C (11 mm); flash pt. (CC) 220 C; ref. index 1.550

Precaution: Combustible

Toxicology: Toxic by inhalation; TLV 3 mg/m^3 of air

Uses: Fire-retarding agent; noncombustible substitute for camphor in celluloid; plasticizer for cellulose acetate and nitrocellulose; in lacquers and varnishes; impregnating roofing paper

Manuf./Distrib.: AC Ind.; Akzo Nobel; Aldrich; Ashland; Bayer; Fluka; FMC; Harwick; Monsanto

Trade names: Disflamoll® TP

Triphenyl phosphite

CAS 101-02-0; EINECS 202-908-4

Synonyms: TPP; Phosphorous acid triphenyl ester

Empirical: $C_{18}H_{15}O_3P$

Formula: $(C_6H_5O)_3P$

Properties: Water-wh. to pale yel. solid or oily liq., pleasant odor; m.w. 310.29; dens. 1.184 (25/25 C); m.p. 22-25 C; b.p. 155-160 C (0.1 mm); flash pt. (COC) 425 F; ref. index 1.589

Precaution: Combustible exposed to heat or flame

Toxicology: LD50 (oral, rat) 1600 mg/kg, (IP, mammal) 250 mg/kg; poison by IP and subcut. routes; mod. toxic by ingestion; eye and severe skin irritant; heated to decomp., emits toxic fumes of PO_x

Uses: Chemical intermediate, stabilizer systems for resins, metal scavenger, diluent for epoxy resins, antioxidant and antiwear agent in gear and transmission oils

Manuf./Distrib.: Albright & Wilson Am.; Aldrich; Ashland; Chemcentral; Dover; Fluka; GE Specialty; Spectrum Chem. Mfg.; Witco/Polymer Addit.

Trade names: Albrite® TPP; Doverphos® 10; Weston® TPP

Trade names containing: Doverphos® 10-HR; Doverphos® 10-HR Plus; Weston® EGTPP

Triphenyltin fluoride

CAS 379-52-2

Empirical: $C_{18}H_{15}FSn$

Properties: M.w. 369.02

Manuf./Distrib.: Aldrich

Trade names: bioMeT 204

Triphosphoric acid pentapotassium salt. *See* Potassium tripolyphosphate
Tri-2-propenyl 1,2,4-benzenetricarboxylate. *See* Triallyl trimellitate
Tripropylene glycol diacrylate. *See* PPG-3 diacrylate
Tripropylene glycol methyl ether. *See* PPG-3 methyl ether
Tripropylene glycol monomethyl ether. *See* PPG-3 methyl ether
Tris. *See* Tris (hydroxymethyl) aminomethane
Trisamine. *See* Tris (hydroxymethyl) aminomethane
2,4,6-Tris(bis(hydroxymethyl)amino)-s-triazine. *See* Hexamethylol melamine resin
Tris Buffer. *See* Tris (hydroxymethyl) aminomethane
Tris (2-butoxyethyl) phosphate. *See* Tributoxyethyl phosphate
Tris (2-chloroethyl) ester phosphoric acid. *See* Trichloroethyl phosphate
Tris (2-chloroethyl) phosphate. *See* Trichloroethyl phosphate
Tris (β-chloroethyl) phosphate. *See* Trichloroethyl phosphate

2,4,6-Tris (dimethylaminomethyl) phenol
CAS 90-72-2; EINECS 202-013-9
Synonyms: DMP; 2,4,6-Tri(dimethylaminomethyl) phenol; Phenol, 2,4,6-tris(dimethylaminomethyl)
Empirical: $C_{15}H_{27}N_3O$
Formula: $[(CH_3)_2NCH_2]_3C_6H_2OH$
Properties: M.w. 265.45; dens. 0.969; b.p. 130-135 C (1 mm); flash pt. > 230 F; ref. index 1.5160
Precaution: Corrosive
Toxicology: Moderately toxic by ingestion and skin contact; severe skin and eye irritant
Uses: Epoxy curing agent, activator; for coatings, adhesives, castings, potting, encapsulation
Manuf./Distrib.: Aldrich; Fluka; Sigma
Trade names: Ancamine® K54; Capcure® EH-30

2,4,6-Tris (dimethylaminomethyl) phenol, 2-ethylhexanoic acid salt
Uses: Epoxy curing agent; for small and med.-sized castings, potting and impregnation varnishes
Trade names: Ancamine® K61B

Tris(2-ethylhexyl) phosphate. *See* Trioctyl phosphate
Tris (2-hydroxyethyl) amine. *See* Triethanolamine

Tris (2-hydroxyethyl) isocyanurate triacrylate
Trade names containing: SR-368

Tris (2-hydroxyethyl) isocyanurate trimethacrylate
Trade names containing: SR-290

Tris (hydroxymethyl) aminomethane
CAS 77-86-1; EINECS 201-064-4
Synonyms: THAM; Tromethamine (INCI); Trometamol; Tris (hydroxymethyl) methylamine; Aminotrimethylolmethane; 2-Amino-2-(hydroxymethyl)-1,3-propanediol; Tris; Tris Buffer; Trisamine
Empirical: $C_4H_{11}NO_3$
Formula: $(CH_2OH)_3CNH_3$
Properties: Wh. cryst. gran. or powd., sl. char. odor; hygroscopic; sol. 80 g/100 cc water (20 C); sol. in low m.w. aliphatic alcohols; pract. insol. in chloroform, benzene, CCl_4; m.w. 121.14; m.p. 168-171 C; b.p. 219-220 C (10 mm); pH 10-11.5 (5%)
Precaution: Combustible
Toxicology: LD50 (oral, rat) 5900 mg/kg; mod. toxic by ingestion and IV routes; irritant to skin and eyes
Uses: Emulsifying agent for oils, fats, waxes; absorbent for acidic gases, medicine; chemical intermediate for paints, etc.
Regulatory: FDA approved for injectables, orals, topicals
Manuf./Distrib.: Aldrich; Am. Biorganics; Am. Int'l.; AMRESCO; ANGUS; Barker Ind.; Fabrichem; Fluka; GFS; Hampshire; Heico; Janssen Chimica; Monomer-Polymer & Dajac Labs; Research Organics; Schweizerhall; Sigma; Spectrum Chem. Mfg.; Van Waters & Rogers
Trade names: Tris Amino® Tech. Grade

1,1,1-Tris(hydroxymethyl) ethane. *See* Trimethylolethane
Tris (hydroxymethyl) methylamine. *See* Tris (hydroxymethyl) aminomethane
1,1,1-Tris(hydroxymethyl)propane. *See* Trimethylolpropane

1,1,1-Tris(hydroxyphenylethane)benzotriazole
Uses: UV light stabilizer for plastics and coatings
Trade names: THPE-BZT

Tris(2-hydroxypropyl)amine. *See* Triisopropanolamine
Tris(isooctadecanoato-O)(2-propanolato) titanium. *See* Isopropyl titanium triisostearate
Tris (methoxyethoxy) vinylsilane. *See* Vinyltris(2-methoxyethoxy) silane
Tris (2-methoxyethoxy)vinylsilane. *See* Vinyltris(2-methoxyethoxy) silane
(Tris (β-methoxyethoxy))vinylsilane. *See* Vinyltris(2-methoxyethoxy) silane

Tris(nonylphenyl) phosphite
CAS 26523-78-4; EINECS 247-759-6
Synonyms: TNPP; Nonylphenyl phosphite (3:1)
Definition: Phosphite triester of nonyl phenol
Empirical: $C_{45}H_{69}O_3P$

Tris (tolyloxy) phosphine oxide

Formula: $[C_9H_{19}C_6H_4]_3.OPO_2H$
Properties: Liq.; m.w. 688; dens. 0.980-0.992; b.p. will not distill; ref. index 1.5255-1.5280; flash pt. (PMCC) 207 C
Uses: Heat stabilizer for PVC, ABS, polyolefins, some rubber prods.; antioxidant in coatings
Manuf./Distrib.: Aldrich; Dover; Fabrichem
Trade names: Weston® TNPP
Trade names containing: Weston® 399

Tris (tolyloxy) phosphine oxide. *See* Tricresyl phosphate

Tris-[3-(trimethoxysilyl)propyl] isocyanurate
Properties: Sp.gr. 1.03; b.p. > 150 C; flash pt. 127 C
Trade names: Silquest® Y-11597

Tritolyl phosphate. *See* Tricresyl phosphate

Trixylenyl phosphate
CAS 25155-23-1
Synonyms: Tri(dimethylphenyl)phosphite; Xylyl phosphate
Empirical: $C_{24}H_{27}O_4P$
Formula: $[(CH_3)_2C_6H_3O]_3PO$
Properties: Liq.; m.w. 410.48; sp.gr. 1.130-1.155; pour pt. -35 C; b.p. 243-265 C (10 mm); flash pt. (COC) 235 C; ref. index 1.551-1.555
Precaution: Combustible
Uses: Flame retardant, plasticizer for PVC, PVAc, cellulosics, PS, ABS, PVC wire and cable insulation; lubricant additive; plasticizer and coalescing agent for coatings
Manuf./Distrib.: Akzo Nobel; Albright & Wilson Am.; Ashland; FMC; C.P. Hall; Harwick
Trade names: Antiblaze® TXP

Trolamine. *See* Triethanolamine
Trometamol. *See* Tris (hydroxymethyl) aminomethane
Tromethamine (INCI). *See* Tris (hydroxymethyl) aminomethane
TRPGDA. *See* PPG-3 diacrylate
TSPP. *See* Tetrasodium pyrophosphate
TTEGDA. *See* PPG-4 diacrylate
Tung nut oil. *See* Tung oil

Tung oil
CAS 8001-20-5
Synonyms: Chinawood oil; Tung nut oil; Chinese tung oil
Definition: Drying oil from seeds of *Aleurites cordata*
Properties: Pale yel. liq., char. disagreeable odor; sol. in chloroform, ether, carbon disulfide, oils; dens. 0.936-0.943; iodine no. 163-171; sapon. no. 190-197
Precaution: Combustible when exposed to heat or flame; can react with oxidizing materials
Toxicology: Toxic by ingestion; contact causes dermatitis; ingestion causes nausea, vomiting, cramps, diarrhea, dizziness, lethargy, disorientation; large doses can cause fever, tachycardia, respiratory effects
Uses: Mfg. of quick-drying wood varnishes, linoleum, floor cloth; in India rubber substitutes, insulating masses; for waterproofing paper, etc.
Regulatory: FDA 21CFR §181.26
Manuf./Distrib.: Aldrich; Alnor Oil; Degen; Lomas Int'l.; Sino-Am. Pigment Systems; Van Waters & Rogers; Welch, Holme & Clark

Turkey-red oil. *See* Sulfated castor oil

Turpentine oil
CAS 8006-64-2
Synonyms: Purified gum spirits; Turpentine, purified; Turpentine, rectified
Definition: Volatile essential oil obtained by distillation and rectification from turpentine, an oleoresin obtained from *Pinus* spp., contg. pinene and diterpene
Empirical: $C_{10}H_{16}$
Properties: Colorless liq., penetrating odor; immisc. with water; dens. 0.860-0.875 (15 C); flash pt. (CC) 32-46 C; ref. index 1.463-1.483 (20 C)
Precaution: DOT: Flamm. liq.; mod. fire risk; protect from light

Toxicology: TLV 100 ppm in air; toxic by ingestion; irritating to skin and mucous membranes; can cause allergic reactions; CNS depressant; death due to respiratory failure
Uses: Solvent; thinner for paints, lacquers; solv. and reclaiming agent for rubber; synthesis of camphor and menthol; polishes; medicines (liniments); perfumery
Regulatory: FDA approved for inhalants; BP compliance
Manuf./Distrib.: Berje; Penta Mfg.; Quest Int'l.; Spectrum Chem. Mfg.

Turpentine, purified. *See* Turpentine oil
Turpentine, rectified. *See* Turpentine oil
Two-stage phenolic resin. *See* Novolac resin
Two-stage resin. *See* Phenolic resin

Ultramarine blue
CAS 1317-97-1; 57455-37-5
Synonyms: Pigment blue 29; CI 77007
Classification: Inorganic pigment
Formula: $Na_6Al_6Si_6O_{24}S_4$
Properties: Blue powd.
Precaution: Noncombustible
Uses: Pigment for printing inks, thermoplastics, rubber compds., paints
Manuf./Distrib.: Aarbor Int'l.; Am. Colors; Browning; Cleveland Pigment & Color; DCS Color & Supply; Ferro; Hilton Davis; Landers-Segal Color; D.N. Lukens; PMC Specialties; Royale Pigments & Chems.; S.A.P. Atlantic; Tamms Ind.; Tiarco; Universal Color Disps.; Whittaker, Clark & Daniels
Trade names: Ferro BP-10; Ferro BP-91; Ferro CP-18; Ferro CP-50; Ferro CP-68; Ferro CP-78; Ferro DP-25; Ferro EP-37; Ferro EP-62; Ferro FP-40; Ferro FP-64; Ferro FT-64; Ferro RA-40; Ferro RB-30; Octotint 120; Octotint 572
Trade names containing: Ultramarine Blue

Ultramarine violet
CAS 12769-96-9
Synonyms: CI 77007; Pigment violet 15
Definition: Sodium aluminum sulfosilicate complex
Formula: $Na_4H_2Al_6Si_6O_{24}S_2$
Uses: Pigment for thermoplastics, rubbers, paints, printing inks
Manuf./Distrib.: Ferro/Color
Trade names: Ferro V-5; Ferro V-8

Ultramarine yellow. *See* Barium chromate
Unslaked lime. *See* Calcium oxide

Urea-formaldehyde resin
CAS 9011-05-6
Synonyms: Polyoxymethylene urea (INCI); Polynoxylin; Urea, polymer with formaldehyde
Classification: Amino resin
Definition: Reaction prod. of urea and formaldehyde
Empirical: $(CH_4N_2O \cdot CH_2O)_x$
Uses: Thermosetting resin; pigment-grinding medium; aids adhesion and toughness of coatings; wet strength resin in paper treatment; in automotive enamels and primers, metal decorating finishes; modifier for water-sol. polymers
Regulatory: FDA 21CFR §175.105, 175.300, 177.1200, 177.1650, 177.1900, 181.30
Manuf./Distrib.: Akzo Nobel; Bakelite GmbH; BASF; Cargill; Chemical; Cytec Ind.; DSM UK; Georgia-Pacific Chem.; Hercules; McWhorter; Monomer-Polymer & Dajac Labs; Monsanto; Reichhold; Sybron; TR-AMC
Trade names: Diamonine B; Plastopal®; Uformite® 21-805
Trade names containing: Beckamine 21-500; Beckamine 21-511; Uformite® 21-806; Uformite® 27-805

Urea-formaldehyde resin, butylated
Trade names containing: Resimene® 901; Resimene® 918; Resimene® 920

Urea-formaldehyde resin, methylated
Trade names: Resimene® 975; Resimene® 980
Trade names containing: Resimene® 933; Resimene® 970

Urea, polymer with formaldehyde. *See* Urea-formaldehyde resin

γ-Ureidopropyltriethoxysilane
CAS 116912-64-2; EINECS 245-876-7, 245-904-8
Properties: Dens. 0.91; flash pt. 58 F
Manuf./Distrib.: Aldrich
Trade names containing: Silquest® A-1160

γ-Ureidopropyltrimethoxysilane
CAS 23843-64-3
Uses: Adhesion promoter aiding bond between fillers and reinforcements (e.g., fiberglass, particulates, metals) and various polymers (phenolic, urea-melamine, epoxies, polyamide, PU); used in glass fiber sizes and finishes, wool insulation resin binders, primers, foundry sand binders, adhesives, sealants, and abrasive grinding wheel binders; contains no flamm. or combustible solv.; low VOC emissions
Trade names: Silquest® Y-11542

Urethane acrylate
Uses: For use in UV/EB-curasble films, water-dilutable screen inks and coatings, sprayable wood sealings, monomer-free formulations, water-strippable applics.; slow cure response
Trade names: Ebecryl® 2000; Ebecryl® 6700; IRR 213; IRR 221; IRR 222; RSX 92528
Trade names containing: Ebecryl® 220; Ebecryl® 1290; Ebecryl® 4830; Ebecryl® 4833; Ebecryl® 8301; Ebecryl® 8800

Urethane-acrylic
Uses: Copolymer used as additive to other prods. to reduce cost, add toughness and anti-skid props.; for use in architectural coatings such as floor and brushing varnish
Trade names: NeoPac® R-9000; NeoPac® R-9030; Witcobond® A-100

Urethane bis-oxazolidine
Uses: Drying agent and moisture scavenger for use in PU coatings, sealants, castings, and elastomers; reactive diluent to confer lower visc. in high solids systems; produces tough, elastic coatings without CO_2 gassing
Trade names: Incozol 2

Urethane diacrylate
Uses: Additive improving flexibility of epoxy and urethane formulations; film-former for inks, coatings for plastic and metal, laminating adhesives
Trade names: Ebecryl® 230; Ebecryl® 270; Ebecryl® 4827; Ebecryl® 8402; Ebecryl® 8803; Ebecryl® 8804; Photomer® 6010; Photomer® 6140; Photomer® 6210; Photomer® 6217; Photomer® 6230
Trade names containing: Ebecryl® 244; Ebecryl® 284; Ebecryl® 285; Ebecryl® 4834; Ebecryl® 4849; Ebecryl® 4881; Ebecryl® 4883; Ebecryl® 8800-20R; Photomer® 6788-20R

Urethane triacrylate
Uses: General purpose, fast curing oligomer; exc. film former exhibiting high tens., outstanding abrasion resist., good high temp. stability; for use in screen inks, solder masks, plate resists. encapsulants, and optical fiber coatings
Trade names: Photomer® 6008
Trade names containing: Ebecryl® 264; Ebecryl® 265; Ebecryl® 4866; Photomer® 6184

Urotropine. *See* Hexamethylenetetramine
UV Absorber-2. *See* Etocrylene
UV Absorber-3. *See* Octocrylene
UV Absorber-5. *See* Octrizole
UV Absorber 6. *See* Bumetrizole

VAE. *See* Vinyl acetate/ethylene copolymer
VA/ethylene copolymer. *See* Vinyl acetate/ethylene copolymer
Valerianic acid. *See* n-Valeric acid

n-Valeric acid
CAS 109-52-4; EINECS 203-677-2

Synonyms: Carboxylic acid C_5; Valerianic acid; Butanecarboxylic acid; 1-Butanecarboxylic acid; Pentanoic acid; n-Pentanoic acid; Propylacetic acid
Definition: Distilled from roots of *Valeriana officinalis*
Empirical: $C_5H_{10}O_2$
Formula: $CH_3(CH_2)_3COOH$
Properties: Colorless mobile liq., penetrating rancid odor; sol. in water; misc. with alcohol, ether; m.w. 102.14; dens. 0.940 (20/4 C); m.p. -34.5 C; b.p. 186 C; flash pt. 203 F; ref. index 1.405-1.14
Precaution: Combustible liq.; DOT: corrosive material
Toxicology: LD50 (oral, mouse) 600 mg/kg; mod. toxic by ingestion, IV, subcutaneous routes; mildly toxic by inh.; corrosive irritant to skin, eyes, mucous membranes; heated to decomp., emits acrid smoke and irritating fumes
Uses: Intermediate for flavors and perfumes, ester-type lubricants, plasticizers, pharmaceuticals, vinyl stabilizers, paints
Regulatory: FDA 21CFR 172.515; FEMA GRAS
Manuf./Distrib.: Aldrich; BASF; Berje; Exxon; Fluka; Hoechst Celanse; Penta Mfg.; Sigma; Spectrum Chem. Mfg.; Union Carbide

Vaseline. *See* Petrolatum
VC. *See* Vinyl chloride
VdC. *See* Vinylidene chloride copolymer
Vegetable carbon. *See* Carbon black

Vegetable oil
CAS 68956-68-3; 68938-35-2; EINECS 273-313-5
Synonyms: Oils, vegetable
Definition: Expressed oil of vegetable origin consisting primarily of triglycerides of fatty acids
Toxicology: May cause contact dermatitis
Uses: Paints as drying oils, shortening, salad dressings; rubber softeners; dietary supplements
Regulatory: FDA approved for orals, topicals
Manuf./Distrib.: ABITEC; Arista Ind.; Desert King Jojoba; Int'l. Flora Tech.; Koster Keunen; Lipo; Mendell; A.E. Staley Mfg.; Terry Labs; Welch, Holme & Clark

Vermiculite
CAS 1318-00-9
Definition: Hydrated magnesium-iron-aluminum silicate
Uses: Lightweight concrete aggregate, insulation, sound conditioning, fireproofing, plaster, soil conditioner, fertilizer additive, seed bed, refractory, lubricant, oil-well drilling mud, filler in plastics, rubber and paint; absorbent; packing; carrier
Manuf./Distrib.: Aldrich; Filter-Media; Whittemore

Vinegar naphtha. *See* Ethyl acetate
Vinyl acetal polymers, butyrals. *See* Polyvinyl butyral

Vinyl acetate
CAS 108-05-4; EINECS 203-545-4
Synonyms: Acetic acid, ethenyl ester; Ethenyl acetate; Acetic acid, vinyl ester
Classification: Unsaturated ester
Empirical: $C_4H_6O_2$
Formula: $CH_3COOCH=CH_2$
Properties: Liq.; polymerizes in light to a colorless transparent mass; sol. 1 g/50 ml water (20 C); misc. with alcohol, ether; m.w. 86.1; dens. 0.932 (20/4 C); m.p. -100 or -93 C; b.p. 71-73 C; flash pt. (CC) -8 C
Toxicology: LD50 (rat, oral) 2.92 g/kg
Uses: In polymerized form for plastic masses, films, and lacquers
Regulatory: FDA 21CFR §172.892
Manuf./Distrib.: Aldrich; Allchem Ind.; Ashland; BP Chems.; Exxon Belgium; Fluka; Hoechst Celanese; Quantum; Sigma; Union Carbide
Trade names: Daratak® SP1012; Synthemul® 40503-00; Synthemul® 97883-01; Vinnapas® M 50/300; Wallpol® 40100-20

Vinyl acetate/acrylate copolymer
Uses: Used in interior, exterior, and semigloss paints, paper saturating, wallpaper topcoats, textile adhesives; binder for fabrics, glass fiber
Trade names: Flexbond® 325; Texicote® 03-029; Wallpol® 40-160; Wallpol® 40-165

Vinyl acetate/ethylene copolymer
CAS 24937-78-8
Synonyms: VAE; VA/ethylene copolymer
Formula: $(C_4H_6O_2 \cdot C_2H_4)_x$
Uses: Coating binder and saturant for paper/paperboard, nonwovens, medical/surgical applics., fiber laminates, adhesives, carpet backings; vehicle base for paints, caulks, and mastics in the building industry; impact modifier for PVC
Manuf./Distrib.: Aldrich
Trade names: Airflex® 100 HS; Airflex® 125; Airflex® 140; Airflex® 199; Airflex® 500; Airflex® 510; Airflex® 525-BP; Airflex® RP-224; Airflex® RP-226; Airflex® RP-230; Airflex® RP-244; Airflex® RP-245; Elvace® 40-707; Vynathene® EY 902-30; Vynathene® EY 902-35

Vinyl acetate/ethylene/vinyl chloride terpolymer
Uses: Binder for emulsion paints, textured finishes, flame retardant fabrics, and thermal insulation systems; adhesive for flooring, walls, foam, tiles, paper/paperboard
Trade names: Airflex® 456; Airflex® 728; Airflex® 738

Vinyl acetate monomer
Uses: Monomer for prod. of paints
Manuf./Distrib.: Ashland; Union Carbide; Wacker-Chemie GmbH

Vinyl acetate/vinyl laurate/vinyl chloride terpolymer
Uses: Binder for emulsion paints, textured finishes, and thermal insulation systems

Vinyl acetyl polymers, butyrates. *See* Polyvinyl butyral

Vinyl acrylic copolymer
Uses: Binder, vehicle for paints, factory finishes, roof, paper, textile and industrial coatings, adhesives, caulks, sealants, tiles
Manuf./Distrib.: Air Prods.; Ashland; General Latex & Chem.; Hampshire; McWhorter; Nat'l. Starch & Chem.; Ohio Polychem.; Reichhold; Rhone-Poulenc; Rohm & Haas; StanChem; Thibaut & Walker; Tri-Iso; Union Carbide; VYN-AC; Zeneca Resins
Trade names: Everflex® MA; Flexbond® 381; Fulatex® PD-110; Fulatex® PD-124; Fulatex® PD-126; Fulatex® PD-231; Fulatex® PD-542; Parco® 58-C-55; Plioway® EC1; Plioway® EC-H; Plioway® EC-L; Plioway® EC-T; Rovace 9100; Synthemul® 40550-00; Synthemul® 40551-03; Synthemul® 40552-00; Ucar® Latex 347; Ucar® Latex 351; Ucar® Latex 354; Ucar® Latex 367; Ucar® Latex 369; Ucar® Latex 376; Ucar® Latex 379; Wallpol® 40-136; Wallpol® 40-143; Wave® 345

Vinyl alcohol polymer. *See* Polyvinyl alcohol
Vinylbenzene. *See* Styrene
Vinylbenzol. *See* Styrene

N-[2-(Vinylbenzylamino)-ethyl)-3-aminopropyltrimethoxysilane
Classification: Styrylamine cationic silane
Uses: Coupling agent for unsaturated polyesters, styrenics, epoxies, PP, PE
Trade names containing: Dow Corning® Z-6032

Vinyl bromide
CAS 593-60-2
Synonyms: Bromoethylene
Formula: CH_2CHBr
Properties: Gas; m.w. 106.96; dens. 1.51; m.p. -138 C; b.p. 15.6 C
Toxicology: LD50 (oral, rat) 500 mg/kg; carcinogen; TLV 5 ppm in air
Uses: Flame-retarding agent for acrylic fibers, PVAc, PVC; also used in modified acrylic fibers, textiles, adhesives, coatings, photographic plates
Manuf./Distrib.: Albemarle; Aldrich
Trade names: Saytex® VBR

Vinyl butadiene
Uses: Thermosetting resin for wire coating, coatings, processing aid; EPDM peroxide-cured modifier
Trade names: Ricon 154

Vinyl chloride
CAS 75-01-4; EINECS 200-831-0
Synonyms: VC; Chloroethene; Chloroethylene; Ethylene monochloride; Vinyl chloride monomer
Classification: Vinyl monomer; aliphatic organic compd.
Formula: CH_2:CHCl
Properties: Compressed gas, easily liquefied, ethereal odor; sol. in alcohol, ether; sl. sol. in water; dens. 0.9121 (liq., 20/20 C); f.p. -159.7 C; b.p. -13.9 C; flash pt. -77 C
Precaution: Explosive limits in air 4-22 by vol.
Toxicology: Extremely toxic; severe irritant to skin, eyes, mucous membranes; carcinogen; human mutagenic data; TLV:TWA 5 ppm in air; prohibited for use in aerosol sprays
Uses: Monomer for paints, polyvinyl chloride and copolymers, organic synthesis; adhesives for plastics
Regulatory: FDA approved for orals
Manuf./Distrib.: Aldrich; Occidental; PPG Ind.; Sigma; Vista

Vinyl chloride-acetate copolymer. *See* Polyvinyl chloride acetate

Vinyl chloride copolymer
Trade names: Laroflex®; Vilit®

Vinyl chloride/ethylene vinyl acetate copolymer. *See* Vinyl chloride/vinyl acetate/ethylene terpolymer
Vinyl chloride monomer. *See* Vinyl chloride

Vinyl chloride/vinyl acetate copolymer
CAS 9003-22-9
Uses: Binder for fabrics, glass fiber; textile auxiliary and coating; paper coatings; records; flooring
Manuf./Distrib.: Aldrich; Fluka; Sigma
Trade names: Ucar® VAGC; Ucar® VAGD; Ucar® VAGF; Ucar® VAGH; Ucar® VERR-40; Ucar® VMCA; Ucar® VMCC; Ucar® VMCH; Ucar® VROH; Ucar® VYES, VYES-4; Ucar® VYHD; Ucar® VYHH; Ucar® VYLF; Ucar® VYNC; Ucar® VYNS-3
See also PVC/VA copolymer

Vinyl chloride/vinyl acetate/ethylene terpolymer
Synonyms: Vinyl chloride/ethylene vinyl acetate copolymer
Uses: Binder for fabrics, glass fiber; textile auxiliary and coating; paper coatings

Vinyl compounds and polymers
Definition: A compound having the vinyl grouping (CH_2=CH-)
Uses: Basis for varieties of plastics including vinyl chloride, vinyl acetate, etc.
Trade names containing: Disparlon® 1950; Disparlon® AQ-500; Disparlon® LC-900; Disparlon® LC-915
See also Polyvinyl chloride; Vinyl chloride; Vinyl acetate

Vinyl cyanide. *See* Acrylonitrile
Vinyl cyanide copolymer. *See* Acrylonitrile copolymer

Vinyl ester monomer
Uses: Monomer for prod. of paints
Manuf./Distrib.: Union Carbide

Vinyl ester resin
Uses: Thermoset used in chem. processing industry, pulp and paper mills, corrosive-resist. applics., coatings
Manuf./Distrib.: Alpha/Owens-Corning; Ashland; Chemcentral; Cook Composites & Polymers; Monomer-Polymer & Dajac Labs; Union Carbide
Trade names: Hetron® 922; Hetron® 922L; Hetron® 922L-25; Hetron® 970/35; Hetron® 980; Hetron® 980/35; XUS 19039.00

Vinyl 2-ethylhexanoate
CAS 94-04-2
Synonyms: Vinyl-2-ethylhexoate; 2-Ethylhexanoic acid vinyl ester; 2-Ethylhexoic acid vinyl ester
Classification: Vinyl ester monomer
Definition: Vinyl ester of 2-ethylhexanoic acid
Formula: CH_2=CHOCOCHCH$_2$CH$_3$(CH$_2$)$_3$CH$_3$
Properties: Liq.; m.w. 170.25; dens. 0.875; m.p. -90 C; b.p. 128-130 C (20 mm); flash pt. 65 C; ref. index 1.4260
Toxicology: LD50 (oral, rat) 4290 mg/kg; mildly toxic by ingestion; eye and severe skin irritant

Vinyl-2-ethylhexoate

Uses: Reactive intermediate for polymerization for copolymer latexes, water-sol. polymers, thermosetting coatings, adhesives, inks, textile sizing
Manuf./Distrib.: Aldrich
Trade names: Vynate® 2-EH

Vinyl-2-ethylhexoate. *See* Vinyl 2-ethylhexanoate

N-Vinyl formamide
Synonyms: NVF
Formula: CH_2=CHNHCHO
Properties: Liq.; sp.gr. 1.080; b.p. 210 C; ref. index 1.492
Uses: Monomer for use in adhesives, inks, and coatings
Trade names: SR-497

Vinylformic acid. *See* Acrylic acid

Vinyl homopolymer
Uses: Blending resin for modifying plastisol compds.
Trade names: Pliovic® M-50; Pliovic® M-70; Pliovic® M-70SC; Pliovic® M-90

Vinylidene chloride acrylate copolymer
Uses: Used in latex paints for factory finishes, ceiling tile, wall board, in fire retardant coatings, and in textile treatments
Trade names: Permax™ 801; Permax™ 803

Vinylidene chloride copolymer
Synonyms: VdC
Uses: Binder for pigments, nonwovens, fire-retardant fabrics, textile back coating; base emulsion for paints, paper coatings, caulks, sealants, roof coatings; impregnant and adhesive for papers; film and sheet extrusion for medicine, foods, cosmetics
Regulatory: FDA 21CFR §181.30
Manuf./Distrib.: BFGoodrich; Hampshire; matteson-Ridolfi; Scott Bader; Zeneca Resins
Trade names: Daran® SL143; Diofan®; Polidene® 33-004; Polidene® 33-048; Polidene® 33-065; Polidene® 33-075; Polidene® 33-080; Saran F-239, F-278, F-310

Vinylidene chloride monomer
CAS 75-35-4; EINECS 200-864-0
Synonyms: 1,1-Dichloroethene; 1,1-Dichloroethylene; asym-Dichloroethylene
Empirical: $C_2H_2Cl_2$
Formula: CH_2=CCl_2
Properties: Liq., mild sweet odor resembling that of chloroform; pract. insol. in water; sol. in org. solvs.; m.w. 96.95; dens. 1.2129 (20/4 C); m.p. -122.5 C; b.p. 31.7 C (760 mm); ref. index 1.4249 (20 C); flash pt. -15 C
Precaution: Extremely flamm.; uncontrollable polymerization may lead to explosive reaction prods. with O_2 or ozone; keep away from ignition sources; keep in tightly closed containers; photosensitive
Toxicology: Irritant to skin, mucous membranes; narcotic in high concs.
Uses: Intermediate in prod. of vinylidene polymer plastics such as Saran, paints
Manuf./Distrib.: Aldrich; Fluka; PPG Ind.; Sigma

Vinyl isobutyl ether
CAS 109-53-5; EINECS 203-678-8
Synonyms: IVE; Isobutyl vinyl ether; Poly(vinyl isobutyl ether)
Empirical: $C_6H_{12}O$
Formula: CH_2=$CHOCH_2CH(CH_3)_2$
Properties: Colorless liq.; sl. sol. in water; sol. in alcohol, ether; m.w. 100.2; dens. 0.7706 (20/20 C); m.p. -132 C; b.p. 83.3 C; flash pt. (OC) 15 F; ref. index 1.396 (20 C)
Precaution: Highly flamm.; keep away from ignition sources
Toxicology: Irritant
Uses: Polymer and copolymers in surgical adhesives, coatings, and lacquers; plasticizer
Manuf./Distrib.: Aldrich; BASF AG; Fluka

Vinyl neodecanoate
CAS 45115-34-2

Classification: Vinyl ester monomer
Definition: Vinyl ester of neodecanoic acid
Formula: CH_2=$CHOCOCR_3$, where R is methyl or greater
Uses: Reactive intermediate for polymerization for copolymer latexes, water-sol. polymers, thermosetting coatings, adhesives, inks, textile sizing, terpolymers for hair sprays
Manuf./Distrib.: Exxon
Trade names containing: Vynate® Neo-10

Vinyl neononanoate
Classification: Vinyl ester monomer
Definition: Vinyl ester of neononanoic acid
Formula: CH_2=$CHOCOCR_3$, where R is methyl or greater
Uses: Reactive intermediate for polymerization for copolymer latexes, water-sol. polymers, thermosetting coatings, adhesives, inks, textile sizing
Trade names: Vynate® Neo-9

Vinyl neopentanoate
Classification: Vinyl ester monomer
Uses: Intermediate for latex coatings

Vinyl oximinosilane
CAS 2224-33-1
Uses: Used in sealants and coatings
Trade names: OS-2000

Vinyl pivalate
Classification: Vinyl ester monomer
Definition: Vinyl ester of neopentanoic acid
Formula: CH_2=$CHOCOC(CH_3)_3$
Uses: Reactive intermediate for polymerization for copolymer latexes, water-sol. polymers, thermosetting coatings, adhesives, inks, textile sizing
Trade names: Vynate® Neo-5

Vinyl propionate
CAS 105-38-4
Classification: Vinyl ester monomer
Definition: Vinyl ester of propionic acid
Formula: CH_2=$CHOCOCH_2CH_3$
Properties: Liq.; insol. in water; m.w. 100.12; dens. 0.919; m.p. -80 C; b.p. 94-95 C; flash pt. 6 C; ref. index 1.4030
Precaution: Flamm.
Toxicology: Irritant
Uses: Reactive intermediate for polymerization for copolymer latexes, water-sol. polymers, thermosetting coatings, adhesives, inks, textile sizing
Manuf./Distrib.: Aldrich; BASF; Monomer-Polymer & Dajac Labs
Trade names: Vynate® L-3

Vinyl pyridine
CAS 1337-81-1
Empirical: C_7H_7N
Formula: C_5H_4NCH=CH_2
Properties: Colorless liq.; sol. in dilute acids, hydrocarbons, alcohol, ketones, esters; m.w. 105.14; dens. 0.9746 (20 C); ref. index 1.5509 (20 C)
Precaution: Combustible
Toxicology: Irritant to skin and eyes; heated to dec., emits toxic fumes of NO_x
Uses: Used for adhesives, tire cord, and industrial goods dips, latex emulsions
Manuf./Distrib.: BFGoodrich; Goodyear; Penta Mfg.; Raschig; Reilly Ind.

1-Vinyl-2-pyrrolidinone polymer. *See* PVP
1-Vinyl-2-pyrrolidone. *See* N-Vinyl-2-pyrrolidone
N-Vinyl pyrrolidone. *See* N-Vinyl-2-pyrrolidone

N-Vinyl-2-pyrrolidone
CAS 88-12-0; EINECS 201-800-4
Synonyms: 1-Ethenyl-2-pyrrolidinone; N-Vinyl pyrrolidone; 1-Vinyl-2-pyrrolidone
Empirical: C_6H_9NO
Formula: CH_2=CHNCH$_2$CH$_2$CH$_2$CO
Properties: Colorless liq.; sol. in water; m.w. 111.16; dens. 1.04 (25 C); f.p. 13.5 C; b.p. 148 C (100 mm); flash pt. (COC) 209 F; ref. index 1.511
Precaution: Combustible
Toxicology: Moderately toxic by ingestion, inhalation, skin contact; severe eye irritant; narcotic; may cause cancer
Uses: Monomer for polyvinylpyrrolidone, organic synthesis, paints
Manuf./Distrib.: Albemarle; Aldrich; Allchem Ind.; BASF; Fluka; ISP; Monomer-Polymer & Dajac Labs; Polysciences
Trade names containing: CN 960 L90; CN 961 L90; CN 962 L90; CN 963 L90; CN 964 L90; CN 965 L90; CN 966 L90; CN 970 L75; CN 971 L90; CN 972 L90; Ken-React® NZ 37; V-Pyrol
See also PVP

Vinylpyrrolidone/styrene copolymer. *See* Styrene/PVP copolymer
Vinylpyrrolidone/vinyl acetate copolymer. *See* PVP/VA copolymer
Vinylpyrrolidone/vinyl imidazolinium methochloride copolymer. *See* Polyquaternium-16

Vinyl-toluene/acrylate copolymer
Trade names: Pliolite® AC-3; Pliolite® VTAC; Pliolite® VTAC-L

Vinyl-toluene/butadiene copolymer
Trade names: Pliolite® VT; Pliolite® VT-L

Vinyltoluene monomer
CAS 25013-15-4; EINECS 246-562-2
Synonyms: Methylstyrene
Empirical: C_9H_{10}
Formula: $CH_3C_6H_4CH$:CH_2
Properties: Liq.; m.w. 118.19; dens. 0.896 (20/4 C); b.p. 169-172 C; flash pt. 60 C; ref. index 1.542
Precaution: Flamm.; heated to dec., emits acrid smoke and fumes
Toxicology: LD50 (oral, rat) 4000 mg/kg; skin and eye irritant; moderately toxic by inhalation, ingestion; human systemic effects; TLV:TWA 50 ppm; heated to decomp., emits acrid smoke and irritating fumes
Uses: Monomer for prod. of paints
Manuf./Distrib.: Ashland; Chemcentral; Dow; Fluka; Monomer-Polymer & Dajac Labs; Van Waters & Rogers

Vinyltriethoxysilane
CAS 78-08-0; EINECS 201-081-7
Synonyms: (Triethoxysilyl)ethylene; Triethoxyvinylsilane; Triethoxyvinylsilicane
Empirical: $C_8H_{18}O_3Si$
Properties: Liq.; m.w. 190.31; dens. 0.911 (20/4 C); b.p. 160-161 C; flash pt. 34 C; ref. index 1.399 (20 C)
Precaution: Flamm.
Toxicology: LD50 (oral, rat) 23 g/kg, (skin, rabbit) 1 g/kg; mildly toxic by ingestion, inh., skin contact; heated to decomp., emits acrid smoke and irritating fumes
Uses: Intermediate, esp. when acidic by-products are undesirable; filler; coupling agent, release agent, lubricant; crosslinking agent; provides durability, gloss, hiding power to coatings
Manuf./Distrib.: Aldrich; Fluka; Hüls Am.; Monomer-Polymer & Dajac Labs; OSi Spec.; PCR
Trade names: Silquest® A-151

Vinyltrimethoxysilane
CAS 2768-02-7; EINECS 220-449-8
Synonyms: Trimethoxyvinylsilane
Empirical: $C_5H_{12}O_3Si$
Formula: H_2C=CHSi$(OCH_3)_3$
Properties: Liq.; m.w. 148.23; dens. 0.970 (20/4 C); b.p. 123 C; flash pt. 23 C; ref. index 1.3930 (20 C)
Precaution: Flamm.
Toxicology: LD50 (rat, oral) 11,300 mg/kg; irritating to skin, eyes, respiratory system
Uses: Crosslinking agent; in peroxide graft/moisture crosslinking of polyethylene; provides durability, gloss, hiding power to coatings

Manuf./Distrib.: Aldrich; Fluka; Sigma
Trade names: Silquest® A-171

Vinyltris(methoxyethoxy)silane. *See* Vinyltris(2-methoxyethoxy) silane

Vinyltris(2-methoxyethoxy) silane
CAS 1067-53-4; EINECS 213-934-0
Synonyms: Tris (methoxyethoxy) vinylsilane; Tris (2-methoxyethoxy)vinylsilane; (Tris (β-methoxyethoxy))vinylsilane; 6-Ethenyl-6-(methoxyethoxy)-2,5,7,10-tetraoxa-6-silaundecane; Vinyltris(methoxyethoxy)silane; Vinyltris (β-methoxyethoxy)silane; Vinyltris(2-methoxyethoxy)silane
Empirical: $C_{11}H_{24}O_6Si$
Properties: Liq.; m.w. 280.39; dens. 1.034 (25 C); b.p. 284-286 C; flash pt. 65 C; ref. index 1.427 (25 C)
Precaution: Flamm.
Toxicology: LD50 (rat, oral) 2960 mg/kg, (skin, rabbit) 1500 mg/kg; mod. toxic by ingestion and skin contact; skin irritant; heated to decomp., emits acrid smoke and irritating fumes
Uses: Filler for peroxide-cured systems; crosslinking agent; provides durability, gloss, hiding power to coatings
Manuf./Distrib.: Aldrich; Fluka
Trade names: Silquest® A-172

Vinyltris(2-methoxyethoxy)silane. *See* Vinyltris(2-methoxyethoxy) silane
Vinyltris (β-methoxyethoxy)silane. *See* Vinyltris(2-methoxyethoxy) silane
Vitamin L. *See* Anthranilic acid
VM&P naphtha. *See* Naphtha

Water glass. *See* Sodium silicate
Waxes, microcrystalline. *See* Microcrystalline wax
Waxes, montan. *See* Montan wax
White beeswax. *See* Beeswax
White caustic. *See* Sodium hydroxide
White charcoal. *See* Magnesium oxide
White gelatin. *See* Gelatin
White lead. *See* Lead carbonate (basic)
White lead, sublimed. *See* Lead sulfate, basic
White lead sulfate. *See* Lead sulfate, basic
White mineral oil. *See* Mineral oil
White ozokerite wax. *See* Ceresin
White petrolatum. *See* Petrolatum
White shellac. *See* Shellac
White soft paraffin. *See* Petrolatum
White spirits. *See* Mineral spirits; Stoddard solvent
White wax. *See* Beeswax
Wilkinite. *See* Bentonite

Wollastonite
CAS 13983-17-0
Synonyms: Calcium metasilicate
Definition: A calcium silicate mineral
Empirical: CaH_2O_3Si
Properties: M.w. 118.19
Toxicology: Experimental tumorigen; heated to decomp., emits acrid smoke and irritating fumes
Uses: Extender, filler, pigment for paints, plastics, rubber, friction, refractory, ceramic, construction, sealants, adhesives
Manuf./Distrib.: Am. Colloid; Nyco Mins.; Reade Advanced Materials; R.T. Vanderbilt
Trade names: G-RRIM™ WOLLASTOCOAT®; NYAD® 200; NYAD® 325; NYAD® 400; NYAD® 475; NYAD® 1250; NYAD® FP; NYAD® G; NYAD® G Special; NYAD G® WOLLASTOCOAT®; NYCOR® R; Vansil® W-10; Vansil® W-20; Vansil® W-30; 10 WOLLASTOCOAT®; 400 WOLLASTOCOAT®; Wollastokup®

Wood alcohol. *See* Methyl alcohol
Wood naphtha. *See* Methyl alcohol
Wood pulp, bleached. *See* Cellulose
Wood spirit. *See* Methyl alcohol

Wool fat. *See* Lanolin
Wool wax. *See* Lanolin
Wool wax alcohol. *See* Lanolin alcohol

Xanthan. *See* Xanthan gum

Xanthan gum
CAS 11138-66-2; EINECS 234-394-2
Synonyms: Corn sugar gum; Xanthan
Classification: Polysaccharide gum
Definition: High m.w. hetero polysaccharide gum produced by a pure-culture fermentation of a carbohydrate with *Xanthomonas campestris*; contains D-glucose, D-mannose, and D-glucuronic acid and is prepared as the sodium, potassium, or calcium salt
Properties: Wh. to cream-colored powd., sl. organic odor, tasteless; very hygroscopic; sol. in hot or cold water; insol. in oils, most org. solvs.; visc. 600 cps min.
Toxicology: No known toxicity; heated to decomp., emits acrid smoke and irritating fumes
Uses: Thickening and suspending agent, stabilizer, emulsifier; in drilling fluids, ore flotation, foods, cosmetics, pharmaceuticals; lubricant for industrial applics. incl. abrasives, adhesives, herbicides, fertilizers, ceramics, cleaners, emulsions, gels, mining, thixotropic paints, paper, petrol., pigments
Regulatory: FDA 21CFR §133.124, 133.133, 133.134, 133.162, 133.178, 133.179, 172.695, 176.170; USDA 9CFR §318.7 (limitation 8%), 381.147; Japan, JCID, Europe, UK approvals; FDA approved for orals, rectals, topicals; USP/NF compliance
Manuf./Distrib.: Aldrich; Colony Ind.; Fluka; Folexco; Gumix Int'l.; Kelco Int'l.; Meer; Rhone-Poulenc Surf. & Spec.; Sigma; TIC Gums
Trade names: KELZAN®; KELZAN® D; KELZAN® M; KELZAN® XC Polymer; Rhodopol® 23; Rhodopol® 50MD; Ticaxan® Regular

XSA. *See* Xylene sulfonic acid

Xylene
CAS 1330-20-7; EINECS 215-535-7
Synonyms: Methyl toluene; Dimethylbenzene; Xylol
Classification: Aromatic compd.
Definition: Commercial mixture of 3 isomers: o-, m-, and p-xylene (1,3-dimethylbenzene, 1,2-dimethylbenzene, 1,4-dimethylbenzene)
Empirical: C_8H_{10}
Formula: $C_6H_4(CH_3)_2$
Properties: Clear liq., sweet odor; pract. insol. in water; misc. with abs. alcohol, ether, many org. liqs.; m.w. 106.16; dens. 0.86; b.p. 137-144 C; flash pt. 29 C; ref. index 1.4970; KB value 98
Precaution: Highly flamm.; moderate fire risk
Toxicology: TLV 100 ppm in air; toxic by ing., inh.; irritant to eyes, skin, respiratory tract; may be narcotic in high concs. cuasing CNS depression, dizziness, nausea, unconsiousness; possible liver damage; repeated skin contact causes defatting, dermatitis
Uses: Solvent; raw material for prod. of benzoic acid, phthalic anhydride, dyes, paints, other organics; aviation gasoline; protective coating; solvent for alkyd resins, lacquers, enamels, rubber cements
Regulatory: FDA 21CFR §175.105, 176.180, 177.1010, 177.1650
Manuf./Distrib.: Aldrich; ARCO; Ashland; J.T. Baker; Baychem; BP Chems.; R.E. Carroll; Chemcentral; Coyne; Crowley Chem.; Exxon; Fina Chems.; Fluka; General Chem.; Harcros; Mallinckrodt; Marathon Oil; Mitsubishi Petrochem.; Mitsui Petrochem. Ind.; Mobil; Olin; Philli; Samson; Shell; Sigma; Spectrum Chem. Mfg.; Texaco; Van Waters & Rogers
Trade names: Shell Xylene
Trade names containing: Acryloid® AT-51; Acryloid® AT-63; Acryloid® AU-608X; Acryloid® AU-1004; Acryloid® B-99N; Ancamide 220-X-70; Ancamide 400-BX-60; Araldite® EPN 1138 X-85; Araldite® GZ 471 X-75; Araldite® GZ 540 X-90; Araldite® GZ 571 KX-75; Araldite® GZ 7071 OX-65; Araldite® GZ 9749 OX-65; Aroflint 303-X-90; Aroplaz 3667-Z-80; Aroplaz 6056-MX-90; Beckamine 21-500; Beckamine 27-566; Beckamine 27-809; Cardolite® NC-541X90; D.E.R. 337-X90; D.E.R. 671-PMX75; D.E.R. 671-X75; D.E.R. 671-XM75; Disparlon® 1970; Disparlon® 4200-10; Disparlon® 4200-20; Disparlon® 6900-20X; Disparlon® A650-20X; Disparlon® AP-20; Disparlon® LC-915; Disparlon® LC-955; Disparlon® NS-30; Disparlon® OX-60; ECO-CRYL™ 9790; EPON® Resin 834-X-90; EPON® Resin 1001-CX-75; EPON® Resin 1001-X-75; EPON® Resin 1004-QX-55; Epotuf Resin 38-515; EXP-36-X20 Epoxy Functional Silicone Sol'n.; EXP-38-X20 Epoxy Functional Silicone Sol'n.; Flat-Ayd® FA-14C; G-4070-X-70; G-4561-

X-70; GP-197 Resin Sol'n.; GPRI™ CKS-3892; HZ 815 X-70; Joncryl® 504; Joncryl®-510; Kelpol 3755-X-80; Luxate® HB9075; Multiflow® Resin; 8521 MX60; Nacure® 1323; Nacure® 1419; Nacure® 5414; Nacure® XP-383; Resimene® 881; Resimene® 901; Resimene® 920; Resimene® 933; Resimene® 1406; SCD 1092; SILIKOFTAL® Nonstick 50; Silikophen® Non-stick 50; Silikophen® P 40/W; Silikophen® P 50/X; Silikophen® P 80/X; Silres® REN 50; Silres® REN 60; Silres® REN 80; Silres® SY 409; Slip-Ayd® SL-031; Slip-Ayd® SL-050; Slip-Ayd® SL-078; Slip-Ayd® SL-177; Sobral 12-101; Sobral 71NX60; Sobral 81X; Sobral 91NX; Sobral 663X50; Sobral 1308; Sobral 1341; Sobral 2750; Sobral 9249; Sobral 9255; Sobral 9257; Sobral AA-536; Sobral AD-640; Sobral AN-001; Sobral AR-530; Sobral AS-531; Sobral EE-632; Styresol 13-031; Suspend-Ayd® 2; TEGO® Airex 960; TEGO® Airex 970; TEGO® Flow ZFS 460; TEGO® Phobe 1000; Tint-Ayd® EP Series; Uformite® 27-802; Uformite® 27-805; Uni-Rez® 2415

1,2-Xylene. *See* o-Xylene

o-Xylene
CAS 95-47-6; EINECS 202-422-2
Synonyms: o-Dimethylbenzene; 1,2-Dimethylbenzene; 1,2-Xylene; o-Xylol; o-Methyltoluene; Benzene, 1,2-dimethyl
Empirical: C_8H_{10}
Formula: $C_6H_4(CH_3)_2$
Properties: Colorless liq.; misc. with abs. alcohol, ether; insol. in water; m.w. 106.17; dens. 0.8802 (20 C); m.p. -25 C; b.p. 141.5-144.5 C; flash pt. 17 C
Precaution: DOT: Flamm. liq.; dangerous fire hazard exposed to heat or flame; explosive as vapor exposed to heat or flame
Toxicology: LDLo (oral, rat) 8800 mg/kg; LD50 (IP, mouse) 1364 mg/kg; OSHA PEL:TWA 100 ppm; mod. toxic by IP route; mildly toxic by ing., inh.; experimental teratogen; common air contaminant; heated to decomp., emits acrid smoke and irritating fumes
Uses: Solv. for paints
Manuf./Distrib.: Mobil; Shell; Spectrum Chem. Mfg.

Xylene sulfonic acid
CAS 25321-41-9; EINECS 246-839-8
Synonyms: XSA; Dimethylbenzenesulfonic acid; Benzenesulfonic acid, dimethyl-
Definition: Mixture of substituted aromatic acids
Empirical: $C_8H_{10}O_3S$
Properties: M.w. 186.24
Toxicology: LD50 (IP, mouse) 500 mg/kg; mod. toxic by IP route; heated to decomp., emits toxic fumes of SO_x
Uses: Intermediate, catalyst for coatings, in preparation of esters, hardening agent in plastics, activator for nicotine insecticides; curing agent for resins
Manuf./Distrib.: Sloss Ind.
Trade names: Eltesol® XA65; Eltesol® XA90

Xylenol
CAS 1300-71-6
Synonyms: Dimethylphenol; Hydroxydimethylbenzene; Hydroxyxylene; Dimethylhydroxybenzene
Empirical: $C_8H_{10}O$
Formula: $(CH_3)_2C_6H_3OH$
Properties: M.w. 122.17; Commercial mixt.: wh. cryst. solid; sol. in most org. solvs.; sl. sol. in water; dens. 1.02-1.03 (15 C); m.p. 20-76 C; b.p. 203-225 C
Precaution: Combustible
Toxicology: Toxic by ingestion, skin absorption; heated to decomp., emits acrid smoke and irritating fumes
Uses: Disinfectants, solvents, pharmaceuticals, insecticides, fungicides, herbicides, plasticizers, rubber chemicals, paints, additives to lubricants and gasoline, mfg. of polyphenylene oxide, wetting agents, dyestuffs, synthetic resin
Manuf./Distrib.: J.T. Baker; Coalite Chems. Div.; Crowley Chem.; Crowley Tar Prods.; Merichem

Xylol. *See* Xylene
o-Xylol. *See* o-Xylene
Xylyl phosphate. *See* Trixylenyl phosphate

Yellow beeswax. *See* Beeswax
Yellow cuprocide. *See* Copper oxide (ous)

Yellow ferric oxide. *See* Ferric oxide
Yellow oxide of mercury. *See* Mercury oxide (ic), red and yellow
Yellow pine rosin. *See* Rosin
Yellow precipitate. *See* Mercury oxide (ic), red and yellow
Yellow wax. *See* Beeswax

ZDA. *See* Zinc diacrylate
ZDMC. *See* Zinc dimethyldithiocarbamate
Zea mays oil. *See* Corn oil

Zeolite
CAS 1318-02-1; EINECS 215-283-8
Definition: Hydrated alkali aluminum silicate
Properties: Dens. 0.58-0.75; pH 5-12
Uses: Cryst. molecular sieve which selectively adsorbs molecules and exchange ions; detergent builder; in dessicant systems; to separate gases; wh. pigment in paper, paint, and plastics; carpet cleaners; dentifrices; flame retardants
Manuf./Distrib.: Fluka; Sigma; Zeochem
See also Sodium silicoaluminate

Zinc
CAS 7440-66-6; EINECS 231-175-3
Classification: Metallic element
Empirical: Zn
Properties: Shining wh. metal; sol. in acids, alkalies; insol. in water; at.wt. 65.38; dens. 7.14; m.p. 419 C; b.p. 907 C
Precaution: Flamm. solid; dust can ignite spontaneously in air; forms flamm. hydrogen gas in contact with alkali hydroxides, acids, or water; dangerous when wet
Toxicology: Human systemic effects by ing.; human skin irritant; heated to extreme temps. and inhaled causes 'metal fume fever'; heated to decomp., emits toxic fumes of ZnO
Uses: Alloys, galvanizing iron and other metals, fungicides, in PVC stabilizers/activators
Regulatory: BP compliance
Manuf./Distrib.: Aldrich; Alfa Aesar Johnson Matthey; Atlantic Equip. Engrs.; Belmont Metals; Cerac; Cuproquim; Eagle Zinc; Ferro/Bedford; Fluka; Noah; Pasminco Europe; Sigma; U.S. Zinc; Zinc Corp. of Am.

Zinc-2-benzothiazolethiolate. *See* Zinc 2-mercaptobenzothiazole
Zinc benzothiazolyl mercaptide. *See* Zinc 2-mercaptobenzothiazole
Zinc benzothiazol-2-ylthiolate. *See* Zinc 2-mercaptobenzothiazole
Zinc bis(dimethylthiocarbamoyl)disulfide. *See* Zinc dimethyldithiocarbamate

Zinc borate
CAS 1332-07-6 (anhyd.); 120007-67-9; 12767-90-7; 12447-61-9; 138265-88-0; 12513-27-8; EINECS 215-566-6; 233-471-8
Synonyms: Boric acid, zinc salt
Classification: Inorganic salt
Definition: Inorganic salt of indefinite composition, contg. zinc oxide and boric oxide in various ratios
Empirical: $B_4H_2O_7 \cdot Zn$
Formula: $xZnO \cdot yB_2O_3 \cdot zH_2O$
Properties: Wh. powd.; sol. in dilute acids; sl. sol. in water; dens. 3.64; m.p. 980 C
Precaution: Nonflamm.
Uses: Medicine, fireproofing textiles, fungistat, mildew inhibitor; afterglow suppressant and synergist with antimony trioxide for fire protection in plastics; rust-inhibitive pigment in paints
Manuf./Distrib.: Allchem Ind.; Anzon; Atomergic Chemetals; BA Chem. Ltd; BASF; R.E. Carroll; Climax Perf. Materials; Landers-Segal Color; Lohmann; Seegott; Sino-Am. Pigment Systems; Joseph Storey; U.S. Borax
Trade names: Borogard® ZB; Zb™-112; Zb™-237; Zb™-325; Zinc Borate 9506; Zinc Borate B9355

Zinc chromate
CAS 12018-19-8; 13530-65-9; 14018-95-2; EINECS 236-878-9; 237-843-0
Synonyms: Zinc yellow no. 2; Chromic acid zinc salt (1:1); Zinc chromite

Formula: $ZnCrO_4 \cdot 7H_2O$
Properties: Solid yel. pigment; m.w. 307.6
Uses: Yel. rust-inhibitive pigment for paints; catalyst used to synthesize methanol from carbon monoxide and hydrogen
Manuf./Distrib.: Am. Disps.; BASF; Browning; Colores Hispania SA; Cookson Pigments; Davis Colors; Landers-Segal Color; D.N. Lukens; Matteson-Ridolfi; Min. Pigments; Nat'l. Chem.; Punda Mercantile; Revelli; Royale Pigments & Chems.; S.A.P. Atlantic; Van Waters & Rogers; Wayne Pigment; Whittaker, Clark & Daniels

Zinc chromite. *See* Zinc chromate

Zinc diacrylate
CAS 14643-87-9
Synonyms: ZDA
Empirical: $C_6H_6O_4Zn$
Formula: $HCCH_2COOZnOCOHCCH_2$
Properties: M.w. 207.0
Uses: Crosslinker for molded polybutadiene compounds, conductive and protective coatings, coagent for SBR compds. and reactive pigments; activator for rubber compounding; scorch retarder
Manuf./Distrib.: Aceto; Aldrich; Monomer-Polymer & Dajac Labs

Zinc dialkyl dithiophosphate
CAS 68457-79-4
Synonyms: Phosphorodithioic acid, mixed O,O-bis(iso-bu and pentyl) esters, zinc salt
Trade names containing: Rhenocure® TP/G; Rhenocure® TP/S

Zinc dimethyldithiocarbamate
CAS 137-30-4; EINECS 205-288-3
Synonyms: ZDMC; Ziram; Bis(dimethylcarbamodithioato,S,S′) zinc; Zinc bis(dimethylthiocarbamoyl)disulfide
Empirical: $C_6H_{12}N_2S_4Zn$
Properties: Crystals; pract. insol. in water; sol. < 0.5 g/100 ml in acetone, benzene; sol. in dilute caustic sol'ns.; m.w. 305.82; m.p. 250 C
Precaution: Can form a flammable dust
Toxicology: LD50 (rat, oral) 1.4 g/kg; irritant to skin and mucous membranes; suspected carcinogen
Uses: Rubber vulcanization accelerator; agric. fungicide; preservative for starch and syn. latex adhesives, food pkg. adhesives; mold inhibitor for unmodified latex paints
Manuf./Distrib.: Aldrich; Fluka; Uniroyal
Trade names: Vancide® MZ-96
Trade names containing: Vancide® 51Z; Vancide® 51Z Disp.

Zinc drier
Trade names: Troymax™ Drier Zinc 8%; Troymax™ Drier Zinc 16%

Zinc dust
CAS 7440-66-6; EINECS 231-175-3
Synonyms: Zinc powder
Empirical: Zn
Properties: Gray powd.; m.w. 65.38; m.p. 419.5 C; b.p. 907 C
Uses: Zinc salts, reducing agent, precipitating agent, bleaches, catalyst; metallic rust-inhibitive paint pigment
Manuf./Distrib.: Aarbor Int'l.; Am. Chemet; Ashland; Atlantic Equip. Engrs.; Atomergic Chemetals; Belmont Metals; Boehle; R.E. Carroll; Eagle Zinc; Heucotech Ltd; Int'l. Chem. Inc.; D.N. Lukens; Punda Mercantile; Reade Advanced Materials; Revelli; San Yuan; Sino-Am. Pigment Systems; Van Waters & Rogers; Vanguard Chem. Int'l.; Zinc Corp. of Am.
See also Zinc

Zinc 2-ethylhexanoate
CAS 136-53-8
Synonyms: Hexanoic acid, 2-ethyl-, zinc salt
Empirical: $C_8H_{16}O_2 \cdot {}^1/_2Zn$
Properties: M.w. 176.90
Uses: Activator for natural rubber; drier for paints

Zinc 2-ethylhexoate
Synonyms: Zinc octoate

Zinc hydroxyphosphite

Formula: $Zn(OOCCH(C_2H_5)C_4H_9)_2$
Properties: Lt. straw-colored visc. liq.; sol. in hydrocarbon solvs.; insol. in water; dens. 1.16
Precaution: Combustible
Uses: Activator for natural and syn. rubbers; stabilizer for foam processing; catalyst; drier for coatings
Manuf./Distrib.: AC Ind.; Aceto; Alar Engineering; Archway Sales; Baychem; Boehle; CasChem; Dussek Campbell; Ferro/Bedford; Hüls Am.; Lomas Int'l.; D.N. Lukens; OM Group; Shepherd; Troy; R.T. Vanderbilt
Trade names: Nuodex Octoate Calcium® 5%; Nuodex Octoate Zinc® 8%

Zinc hydroxyphosphite
CAS 55799-16-1
Uses: Anticorrosive pigment for paints
Manuf./Distrib.: Rheox

Zinc hydroxystannate
CAS 12027-96-2
Formula: $ZnSn(OH)_6$
Properties: Wh. powd.; sol. in strong acids and bases; m.w. 286.12; sp.gr. 3.4; decomp. temp. 180 C
Toxicology: LD50 (oral, rat) > 5000 mg/kg, (dermal, rat) > 2466 mg/kg; nuisance dust-may cause respiratory tract irritation, drying of skin, eye irritation
Uses: Fire retardant and smoke suppressant for plastics, rubber, paints
Manuf./Distrib.: Joseph Storey

Zinc laurate
CAS 2452-01-9; EINECS 219-518-5
Synonyms: Dodecanoic acid, zinc salt
Formula: $Zn(C_{12}H_{23}O_2)_2$
Properties: Wh. powd.; sl. sol. in water and alcohol; m.w. 463.99; m.p. 128 C
Precaution: Combustible
Uses: Softener, activator for rubber compounding; heat stabilizer for rubber and PVC, paints, varnishes

Zinc mercaptobenzothiazolate. *See* Zinc 2-mercaptobenzothiazole

Zinc 2-mercaptobenzothiazole
CAS 155-04-4; EINECS 205-840-3
Synonyms: ZMBT; ZnMB; 2(3H)-Benzothiazolethione, zinc salt; Bis(2-benzothiazolylthio)zinc; Bis(mercaptobenzothiazolato)zinc; 2-Mercaptobenzothiazole zinc salt; Zinc mercaptobenzylthiazol; Zinc-2-benzothiazolethiolate; Zinc benzothiazolyl mercaptide; Zinc benzothiazol-2-ylthiolate; Zinc mercaptobenzothiazolate; Zinc mercaptobenzothiazole salt
Empirical: $C_{14}H_8N_2S_4Zn$
Formula: $Zn(C_7H_4NS_2)_2$
Properties: M.w. 397.85
Toxicology: LD50 (oral, rat) 540 mg/kg, (IP, mouse) 200 mg/kg; posion by IP route; mod. toxic by ing., subcut. routes; experimental carcinogen; heated to decomp., emitss very toxic fumes of SO_x, NO_x, and ZnO
Uses: Accelerator for rubber, latex foam curing systems; fungicide
Trade names containing: Vancide® 51Z; Vancide® 51Z Disp.

Zinc mercaptobenzothiazole salt. *See* Zinc 2-mercaptobenzothiazole
Zinc mercaptobenzylthiazol. *See* Zinc 2-mercaptobenzothiazole

Zinc molybdate
Formula: $ZnMoO_4 \cdot 2HOH$
Properties: Wh. powd.; insol. in water; dens. 3.3; m.p. 1650 C
Uses: Anticorrosion agent; corrosion inhibiting pigment for coatings
Manuf./Distrib.: AAA Molybdenum Prods.; Atomergic Chemetals; Noah
Trade names: Moly-White® 101; Moly-White® 331, ZNP

Zinc molybdenum phosphate
Uses: Anticorrosive pigment for coatings
Trade names: Heucophos™ ZMP

Zinc naphthenate
CAS 12001-85-3; EINECS 234-409-2
Synonyms: Naphthenic acid zinc salt

Formula: $Zn(C_6H_5COO)_2$
Properties: Solid
Precaution: Combustible exposed to heat or flame
Toxicology: LD50 (oral, rat) 4920 mg/kg; mildly toxic by ingestion; heated to decomp., emits toxic fumes of ZnO
Uses: Drier and wetting agent in paints, varnishes, resins; insecticide, fungicide, mildew preventive; wood preservative; waterproofing textiles; insulating materials
Manuf./Distrib.: Akzo Nobel; Archway Sales; Baychem; Boehle; CasChem; Dussek Campbell Ltd; Hüls Am.; King Ind.; Lomas Int'l.; D.N. Lukens; OM Group; Shepherd; Troy
Trade names: Nuodex Napthenate® Zinc 8%; Nuodex Napthenate® Zinc 10%

Zinc octadecanoate. *See* Zinc stearate
Zinc octoate. *See* Zinc 2-ethylhexoate
Zinc orthophosphate. *See* Zinc phosphate

Zinc oxide
CAS 1314-13-2; EINECS 215-222-5
Synonyms: Chinese white; Pigment white 4; CI 77947; Zinc white; Flowers of zinc
Classification: Inorganic oxide
Empirical: OZn
Formula: ZnO
Properties: White to gray amorphous powd. or crystals, odorless, bitter taste; sol. in dilute acetic or min. acids, alkalis; insol. in water, alcohol; m.w. 81.38; dens. 5.67; m.p. 1975 C; ref. index 2.0041-2.0203; pH 6.95 (Amer. process), 7.37 (French process)
Precaution: Heated to decomp., emits toxic fumes of ZnO
Toxicology: TLV/TWA 5 mg/m^3; nuisance particulate; LD50 (IP, rat) 240 mg/kg; poison by IP route; skin/eye irritant; fumes may cause metal fume fever with chills, fever, tightness in chest, cough, leukocytes; experimental teratogen; mutagenic data
Uses: UV absorber; accelerator activator for rubber; lithopone; pigment in white paints, cosmetics, driers, dental cements, pharmaceuticals; mold inhibitor in paints; in mfg. of opaque glass, enamels, tires, printing inks, porcelains; analytical reagent; as flame retardant
Regulatory: FDA 21CFR §73.1991, 73.2991, 175.300, 177.1460, 182.5991, 182.8991, 582.80, GRAS; FDA approved for parenterals, rectals; BP, Ph.Eur. compliance
Manuf./Distrib.: Aceto; Aldrich; Am. Chemet; Asarco; Ashland; Bayer; Browning; R.E. Carroll; Chemical; Eagle Zinc; Fluka; General Chem.; C.P. Hall; Harcros Durham; Landers-Segal Color; Lohmann; D.N. Lukens; Mallinckrodt; Nippon Chem. Ind.; Reade Advanced Materials; H.M. Royal; Royale Pigments & Chems.; Sachtleben Chemie GmbH; Sigma; Sino-Am. Pigment Systems; Tamms Ind.; Toho Zinc; Van Waters & Rogers; Zinc Corp. of Am.; ZOCHEM
Trade names: NAO 105; Ottalume 2100; USP-1
Trade names containing: SWS-299; Zinc Oxide #1

Zinc phosphate
CAS 7779-90-0
Synonyms: Zinc orthophosphate
Empirical: $O_8P_2Zn_3$
Formula: $Zn_3(PO_4)_2$
Properties: White powd., odorless; sol. in dilute min. acids, acetic acid, ammonium hydroxide, alkali hydroxide solns.; insol. in water, alcohol; m.w. 386.05; dens. 3.998 (15 C); m.p. 900 C
Uses: In dental cements, phosphors; flame retardant for plastics; rust-inhibitive pigment for paints
Manuf./Distrib.: BASF; Calgon; Colores Hispania SA; Elf Atochem N. Am.; Halox Pigments; Hammond Lead Prods.; Heucotech Ltd; Landers-Segal Color; Lohmann; McGean-Rohco; Min. Pigments; Nat'l. Chem.; Pasminco Europe; Sino-Am. Pigment Systems; Wayne Pigment; G. Whitfield Richards; Witco/Allied-Kelite
Trade names: Heucophos™ ZPO; Heucophos™ ZPZ; Zinc Phosphate ZP-10
Trade names containing: Alcophor® 827

Zinc powder. *See* Zinc dust
Zinc 2-pyridinethiol-1-oxide. *See* Zinc pyrithione

Zinc pyrithione
CAS 13463-41-7; EINECS 236-671-3
Synonyms: Bis[1-hydroxy-2(1H)-pyridinethinato-O,S]-(T-4) zinc; Zinc 2-pyridinethiol-1-oxide; Pyrithione zinc
Classification: Aromatic salt
Empirical: $C_{10}H_8N_2O_2S_2Zn$
Properties: M.w. 317.7

Zinc silicophosphate

Precaution: Do not store with strong oxidizing agents
Toxicology: LD50 (rat, oral) 260 mg/kg; LD50 (rat, dermal) > 2 g/kg; poison by ingestion, intraperitoneal; irritating to skin and extremely irritating to eyes
Uses: Antidandruff agent for shampoos; cosmetic preservative; antimicrobial for plastics, metalworking fluids, pharmaceuticals; metal coolant and cutting fluids, plastics (PVC, polyolefins, urethanes, nat. and syn. latex, SBR, EPDM), fabric coatings, paint; carpets, wire and cable insulation, adhesives, caulks, vinyl/urethane flooring
Regulatory: FDA approved for topicals
Manuf./Distrib.: Allchem Ind.; Olin; Pyrion-Chemie GmbH; Ruetgers-Nease; Sigma
Trade names: Zinc Omadine® 48% Fine Particle Disp.; Zinc Omadine® 48% Std. Disp.; Zinc Omadine® Powd.

Zinc silicophosphate
Uses: Anticorrosive pigment for coatings
Trade names: Heucophos™ ZBZ

Zinc soap. *See* Zinc stearate

Zinc stearate
CAS 557-05-1; EINECS 209-151-9
Synonyms: Zinc octadecanoate; Octadecanoic acid zinc salt; Zinc soap
Definition: Zinc salt of stearic acid
Empirical: $C_{36}H_{70}O_4Zn$
Formula: $Zn(C_{18}H_{35}O_2)_2$
Properties: Wh. powd., faint char. odor; sol. in acids, common solvs. (hot); insol. in water, alcohol, ether; dec. by dilute acids; m.w. 632.33; dens. 1.095; m.p. 130 C
Precaution: Combustible
Toxicology: No known toxicity to skin; inh. of powd. may cause lung problems and produce death in infants from pneumonitis, with lesions resembling those caused by talc but more severe
Uses: In cosmetics, pharmaceuticals, lacquers, ointments, tablet mfg.; mold release agent for plastic; filler, antifoamer; flatting agent in lacquers; as a drying lubricant and dusting agent for rubber; waterproofing agent for concrete, paper, textiles
Regulatory: FDA 21CFR §175.105, 175.300, 176.170, 176.180, 176.200, 176.210, 177.1200, 177.1460, 177.1900, 177.2410, 177.2600, 178.2010, 178.3910, 182.5994, 182.8994, GRAS; FDA approved for orals; USP/NF, BP, Ph.Eur. compliance
Manuf./Distrib.: AC Ind.; Aldrich; Allan; Allchem Ind.; Am. Int'l.; Chemisphere; Ferro/Grant; Harwick; Lohmann; Magnesia GmbH; Mallinckrodt; Norac; Spectrum Chem. Mfg.; Syn. Prods.; Witco/Polymer Addit.
Trade names: Akrochem® P 3100; Akrodip™ Z-50/50; Akrodip™ Z-200; Akrodip™ Z-250; Cecavon ZN 70; Cecavon ZN 71; Cecavon ZN 72; Cecavon ZN 73; Cecavon ZN 735; Coad 20; Hallcote® ZS; Hallcote® ZS 5050; Haro® Chem ZGN; Haro® Chem ZPR-2; Hydro Zinc™; Petrac® Zinc Stearate ZN-41; Petrac® Zinc Stearate ZW-45; Quikote™; Quikote™ M; Rubichem Zinc-Dip®; Rubichem Zinc-Dip® HS; Rubichem Zinc-Dip® W; Synpro® Zinc Stearate USP; Wet Zinc™; Witco® Zinc Stearate Regular; Zincote™; Zinc Stearate 42
Trade names containing: Mistron ZSC

Zinc sulfide
CAS 1314-98-3; EINECS 215-251-3
Classification: Inorganic salt
Empirical: SZn
Formula: ZnS
Properties: Ylsh.-wh. powd.; sol. in dilute min. acids; insol. in water, alkalies; m.w. 97.45 dens. 3.98
Precaution: Moisture-sensitive
Toxicology: Irritant
Uses: Pigment; in white and opaque glass, plastics, dyeing, paints, linoleum, leather, dental rubber; fungicide; anhydrous in x-ray screens, TV screens
Regulatory: FDA 21CFR §175.105, 177.2600, 178.3297, 178.3570
Manuf./Distrib.: Aceto; Aldrich; Cerac; Chemson Ltd; Eagle-Picher; Elf Atochem N. Am.; Fluka; D.N. Lukens; Nat'l. Chem.; Noah; Ore & Chem.; Reade Advanced Materials; H.M. Royal; Sachtleben Chemie GmbH; Sigma; Spectrum Chem. Mfg.; Ulysses

Zinc sulfonate
Trade names containing: K-Sperse® 152; K-Sperse® 152MS

Zinc tallate
Uses: Drier for paints
Manuf./Distrib.: Dussek Campbell Ltd; OM Group; Troy

Zinc white. *See* Zinc oxide
Zinc yellow no. 2. *See* Zinc chromate
Ziram. *See* Zinc dimethyldithiocarbamate
Zircat. *See* Zirconium
Zirconate (2-), bis[carbonate (2-)-0] dihydroxy-diammonium. *See* Ammonium zirconium carbonate

Zirconium
CAS 7440-67-7
Synonyms: Zircat
Classification: Element
Empirical: Zr
Properties: Hard cryst. scales or gray powd.; sol. in hot conc. acids; insol. in water, cold acids; at.wt. 91.22; dens. 6.4; m.p. 1850 C; b.p. 4377 C
Toxicology: TLV 5 mg/m^3 of air
Uses: Coating nuclear fuel rods, explosive primer, deoxidizer
Manuf./Distrib.: Aldrich; Fluka

Zirconium IV bis alkenolato, cyclo (pentathio) dialkenyl diphosphato-O,O
Formula: $C_{48}H_{84}O_2P_2S_5Zr$
Properties: Brn. liq., mild sulfur odor; sp.gr. 1.17; b.p. > 300 F; flash pt. (TCC) > 200 F
Precaution: Incompat. with strong acids, oxidizers; avoid excessive heat, ignition sources; hazardous decomp. prods.: CO_x, PO_x, SO_x; may emit toxic fumes on thermal decomp.
Toxicology: Potential skin, eye, respiratory tract irritant
Manuf./Distrib.: Kenrich Petrochems.

Zirconium IV bis alkenolato, cyclo (pentathio) dialkenyl phosphato-O,O
Trade names: Ken-React® NZ 69

Zirconium IV di-neoalkanolato, di(p-amino benzoato-o)
CAS 146955-66-0
Manuf./Distrib.: Kenrich Petrochems.
Trade names containing: Ken-React® NZ 37

Zirconium IV di neoalkanolato, di(3-mercapto) propionato-o
CAS 124319-53-5
Manuf./Distrib.: Kenrich Petrochems.
Trade names containing: Ken-React® NZ 66A

Zirconium drier
Uses: Drier used in org. coatings, inks, polyesters
Trade names: Troymax™ Drier Zirconium 6%; Troymax™ Drier Zirconium 12%; Troymax™ Drier Zirconium 18%; Troymax™ Drier Zirconium 24%

Zirconium IV neoalkanolato, tris (diisooctyl) pyrophosphato-o
CAS 113252-64-5
Manuf./Distrib.: Kenrich Petrochems.
Trade names containing: Ken-React® NZ 38J

Zirconium octoate
Uses: Drier for paints
Manuf./Distrib.: Aceto; Alar Engineering; Archway Sales; Baychem; Boehle; Dussek Campbell; Hüls Am.; Lomas Int'l.; D.N. Lukens; OM Group; Shepherd; Troy
Trade names: Nuodex Octoate Zirconium® 6%

ZMBT. *See* Zinc 2-mercaptobenzothiazole
ZnMB. *See* Zinc 2-mercaptobenzothiazole

Part II
Chemical Product Functional Cross-Reference

Chemical Product Functional Cross-Reference

Chemical additives and raw materials from the first part of this reference are grouped by broad functional categories derived from research and manufacturers' specifications.

Accelerators

Cobalt octoate

Diazabicycloundecene; Di (butyl, methyl pyrophosphato) ethylene titanate di (dioctyl, hydrogen phosphite); Dicyclo (dioctyl) pyrophosphato titanate; Di (dioctylphosphato) ethylene titanate; Di (dioctylpyrophosphato) ethylene titanate; N,N-Dimethyl-p-toluidine

Isopropyl 4-aminobenzenesulfonyl di (dodecylbenzenesulfonyl) titanate; Isopropyl dimethacryl isostearoyl titanate; Isopropyl tri (dioctylphosphato) titanate; Isopropyl tri (dioctylpyrophosphato) titanate; Isopropyl tri (N ethylamino-ethylamino) titanate

Methanesulfonic acid; 2-Methyl imidazole

1,10-Phenanthroline

Sodium dimethyldithiocarbamate

Tetra (2, diallyoxymethyl-1 butoxy titanium di (di-tridecyl) phosphite; Tetraisopropyl di (dioctylphosphito) titanate; Thioglycerin; Titanium di (butyl, octyl pyrophosphate) di (dioctyl, hydrogen phosphite) oxyacetate; Titanium di (cumylphenylate) oxyacetate; Titanium di (dioctylpyrophosphate) oxyacetate; Titanium dimethacrylate oxyacetate

Zinc dimethyldithiocarbamate; Zinc 2-mercaptobenzothiazole; Zinc oxide

Adhesion Promoters • Bonding Agents

Aminopropyltrimethoxysilane

Bis-(γ-trimethoxysilylpropyl) amine

Castor oil, polymerized

Di (butyl, methyl pyrophosphato) ethylene titanate di (dioctyl, hydrogen phosphite); Dicyclo (dioctyl) pyrophosphato titanate; Di (dioctylpyrophosphato) ethylene titanate; Dimethylaminoethyl acrylate

2-(3,4-Epoxycyclohexyl) ethyltrimethoxysilane; Ethylene glycol dimethacrylate

Formaldehyde/toluenesulfonamide polymer

3-Glycidoxypropyltrimethoxysilane; Glycidyl acrylate

Isopropyl dimethacryl isostearoyl titanate; Isopropyl tri (dioctylphosphato) titanate; Isopropyl tri (dioctylpyrophosphato) titanate; Isopropyl tri (N ethylamino-ethylamino) titanate

Mercaptopropyltrimethoxysilane; Methyltriethoxysilane; Methyltrimethoxysilane

PEG-6 trimethylolpropane; Polyterpene resin; Polyurethane, polyester; Propylene glycol methacrylate

Styrene allyl alcohol

Tallow dipropylene triamine; Tetrabutyl titanate; Tetraisopropyl di (dioctylphosphito) titanate; Tetraisopropyl titanate; Tetraoctyloxytitanium di (ditridecylphosphite); Titanium di (butyl, octyl pyrophosphate) di (dioctyl, hydrogen phosphite) oxyacetate; Titanium di (dioctylpyrophosphate) oxyacetate; Titanium dimethacrylate oxyacetate; Toluene sulfonamide formaldehyde resin

γ-Ureidopropyltrimethoxysilane

Antiblocking Agents

Dimethicone copolyol
2-(3,4-Epoxycyclohexyl) ethyltrimethoxysilane; Erucamide; N,N´-Ethylenebis 12-hydroxystearamide; N,N´-Ethylene bis-ricinoleamide; Ethylene dioleamide
Hydrogenated tallow amide; N (2-Hydroxyethyl) 12-hydroxystearamide
Isobutyltrimethoxysilane; Isocyanatopropyltriethoxysilane
Oleamide
N-Stearyl 12-hydroxystearamide

Anticaking Agents

Calcium silicate; Calcium sulfate; Carnauba
Dicocamine
Kaolin
Silica dimethyl silylate

Antifoams. *See* Defoamers

Antimicrobials • Bactericides • Fungicides • Mildewcides

Behenyl hydroxyethyl imidazoline; Benzalkonium chloride; 1,2-Benzisothiazolin-3-one; Benzoic acid; Bis (trichloromethyl) sulfone; Boric acid; 2-Bromo-4´-hydroxyacetophenone; 2-Bromo-2-nitropropane-1,3-diol
Capryl hydroxyethyl imidazoline; Captan; Cocoalkonium chloride; Copper naphthenate; Copper oxide (ous)
2,2-Dibromo-3-nitrilopropionamide; Diiodomethyl p-tolyl sulfone; Dimethyl oxazolidine; 3,5-Dimethyl tetrahydro-2-H,1,3,5-thiadiazone-2-thione
7-Ethyl bicyclooxazolidine; Ethylenediamine
Hexahydro-1,3,5-triethyl-s-triazine; N-(Hydroxymethyl)-N-(1,3-dihydroxymethyl-2,5-dioxo-4-imidazolidinyl)-N-(hydroxymethyl) urea
Iodopropynyl butylcarbamate
Methylchloroisothiazolinone; Methyldibromo glutaronitrile; Methylenebis (thiocyanate); 1-Methyl-3,5,7-triaza-1-azoniatricyclo-[3.3.1.1]decane chloride
Naphthenic acid
2-n-Octyl-4-isothiazolin-3-one; 10,10´-Oxybisphenoxyarsine
Phenoxyethanol; Phenylmercuric acetate; Phenylmercuric oleate; Phenylmercury ammonium propionate; Propylene glycol phenyl ether
Resorcinol
Sodium benzoate; Sodium dimethyldithiocarbamate; Sodium fluoride; Sodium hydroxymethylglycinate; Sodium 2-mercaptobenzothiazole; Sodium pyrithione
Tetrachloroisophthalonitrile; Tetrahydro-3,5-dimethyl-2H-1,3,5-thiadiazine-2-thione; Thiabendazole; 2-Thiocyanomethylthiobenzothiazole; p-Toluenesulfonamide; Tributyltin oxide; Tributyltin salicylate; N-Trichloromethylthio) phthalimide
Zinc borate; Zinc dimethyldithiocarbamate; Zinc naphthenate; Zinc pyrithione; Zinc sulfide

Antioxidants

BHA; Bis (2,4-di-t-butylphenyl) pentaerythritol diphosphite; 4,4´-Butylidenebis (6-t-butyl-m-cresol)
C20-40 alcohols; p-Cresol/dicyclopentadiene butylated reaction product
Di (butyl, methyl pyrophosphato) ethylene titanate di (dioctyl, hydrogen phosphite); Dicyclo (dioctyl) pyrophosphato titanate; Di (dioctylphosphato) ethylene titanate; Di (dioctylpyrophosphato) ethylene titanate; Diisobutyl nonyl phenol; Dimethylaminomethyl phenol
Ethyl hydroxymethyl oleyl oxazoline
Isopropyl 4-aminobenzenesulfonyl di (dodecylbenzenesulfonyl) titanate; Isopropyl dimethacryl isostearoyl titanate; Isopropyl tri (dioctylphosphato) titanate; Isopropyl tri (dioctylpyrophosphato) titanate; Isopropyl tri (N ethylamino-ethylamino) titanate

Antioxidants *(cont'd.)*

Lead phthalate, basic
2,2´-Methylenebis (6-t-butyl-4-methylphenol)
Octadecyl 3,5-di-t-butyl-4-hydroxyhydrocinnamate
Phosphorus; Phosphorus trichloride, reaction prods. with 1,1´-biphenyl and 2,4-bis (1,1-dimethylethyl) phenol
Tetra (2, diallyoxymethyl-1 butoxy titanium di (di-tridecyl) phosphite; Tetraisopropyl di (dioctylphosphito) titanate; Tetrakis [methylene (3,5-di-t-butyl-4-hydroxyhydrocinnamate)] methane; Tetraoctyloxytitanium di (ditridecylphosphite); 4,4´-Thiobis-6-(t-butyl-m-cresol); Titanium di (butyl, octyl pyrophosphate) di (dioctyl, hydrogen phosphite) oxyacetate; Titanium di (cumylphenylate) oxyacetate; Titanium di (dioctylpyrophosphate) oxyacetate; Titanium dimethacrylate oxyacetate; 2,2,4-Trimethyl-1,2-dihydroquinoline polymer; Tris(nonylphenyl) phosphite

Antisagging Agents

Cellulose
Dihydrogenated tallow benzylmonium hectorite; Disodium tallow sulfosuccinamate
Hectorite
Sulfated castor oil

Antisettling Agents

Oleyl propanediamine
PEG-15 cocamine; Polyacrylamide; Polyethylene, high-density
Smectite; Sulfated castor oil;

Antiskinning Agents

BHT; m-Butyraldehyde oxime
Methyl ethyl ketoxime
PEG-8 isolauryl thioether

Antislip Agents

Silica, colloidal

Antistats

t-Butylaminoethyl methacrylate
Capryl hydroxyethyl imidazoline; Cetethyl morpholinium ethosulfate; Cocoyl hydroxyethyl imidazoline
Di (butyl, methyl pyrophosphato) ethylene titanate di (dioctyl, hydrogen phosphite); Dicyclo (dioctyl) pyrophosphato titanate; Di (dioctylphosphato) ethylene titanate; Dimethyl diallyl ammonium chloride; Distearyldimonium chloride
N,N´-Ethylene bis-ricinoleamide
Glyceryl mono/dioleate; Glyceryl oleate; Glyceryl stearate
Heptadecenyl hydroxyethyl imidazoline; Hexyl phosphate; N(β-Hydroxyethyl) ricinoleamide; N-(2-Hydroxypropyl) benzenesulfonamide
Isopropyl 4-aminobenzenesulfonyl di (dodecylbenzenesulfonyl) titanate; Isopropyl dimethacryl isostearoyl titanate; Isopropyl tri (dioctylphosphato) titanate; Isopropyl tri (dioctylpyrophosphato) titanate; Isopropyl tri (N ethylamino-ethylamino) titanate
(3-Lauramidopropyl) trimethyl ammonium methyl sulfate
Nonyl nonoxynol-15
Oleyl hydroxyethyl imidazoline
Palmitic/stearic acid mono/diglycerides; PCA; PEG-36 castor oil; PEG-10 cocamine; PEG-2 laurate; PEG-2

Antistats *(cont'd.)*

tallowamine; PEG-5 tallowamine; PEG-15 tallowamine; PEG-20 tallowamine; Poloxamer 101; Poloxamer 108; Poloxamer 123; Poloxamer 124; Poloxamer 181; Poloxamer 182; Poloxamer 184; Poloxamer 185; Poloxamer 188; Poloxamer 217; Poloxamer 231; Poloxamer 234; Poloxamer 235; Poloxamer 237; Poloxamer 282; Poloxamer 288; Poloxamer 331; Poloxamer 333; Poloxamer 334; Poloxamer 335; Poloxamer 338; Poloxamer 401; Poloxamer 402; Poloxamer 403; Poloxamer 407; Poloxamine 304; Poloxamine 701; Poloxamine 704; Poloxamine 901; Poloxamine 904; Poloxamine 908; Poloxamine 1107; Poloxamine 1307; Polyamide/epichlorohydrin polymer; Polyglyceryl-10 tetraoleate; Propylene glycol laurate; Propylene glycol myristate; PVM/MA copolymer

Quaternium-18

Stearamidopropyl dimethyl-β-hydroxyethyl ammonium dihydrogen phosphate; Stearamidopropyl dimethyl-2-hydroxyethyl ammonium nitrate; Sulfated peanut oil

Tetra (2, diallyoxymethyl-1 butoxy titanium di (di-tridecyl) phosphite; Tetrahydroxypropyl ethylenediamine; Tetraisopropyl di (dioctylphosphito) titanate; Tetraoctyloxytitanium di (ditridecylphosphite); Titanium di (butyl, octyl pyrophosphate) di (dioctyl, hydrogen phosphite) oxyacetate; Titanium di (cumylphenylate) oxyacetate; Titanium di (dioctylpyrophosphate) oxyacetate; Titanium dimethacrylate oxyacetate

Antistripping Agents

Dicocamine

Tallow dipropylene triamine ; Tallow propylene diamine

Bactericides. *See* Antimicrobials

Binders

Acrylic/styrene; Aminopropyltrimethoxysilane; Ammonium zirconium carbonate

Ceteareth-18; Ceteareth-80; Cetyl palmitate; Coumarone-indene resin

Ethylene/vinyl chloride copolymer; Ethyl hydroxyethyl cellulose; Ethyl polysilicate

Meroxapol 105; Meroxapol 171; Meroxapol 172; Meroxapol 174; Meroxapol 252; Meroxapol 254; Meroxapol 258; Meroxapol 311; Methyl hydroxyethylcellulose; Mineral oil

Phenyl acid phosphate; Phenylmethyl polysiloxane; Phenyl trimethicone; Polyacrylamide; Polybutene; Polyvinyl alcohol, partially hydrolyzed; Polyvinyl methyl ether; Potassium silicate; PVM/MA copolymer, butyl ester; PVM/MA copolymer, ethyl ester

Sodium bentonite; Styrene/acrylates copolymer; Styrene-butadiene rubber; Styrene/MA copolymer; Styrene/PVP copolymer

Vinyl acetate/ethylene copolymer; Vinyl acetate/ethylene/vinyl chloride terpolymer; Vinyl acetate/vinyl laurate/vinyl chloride terpolymer; Vinyl acrylic copolymer; Vinyl chloride/vinyl acetate copolymer; Vinyl chloride/vinyl acetate/ethylene terpolymer; Vinylidene chloride copolymer

Bodying Agents. *See* Thickeners

Bonding Agents. *See* Adhesion Promoters

Carriers • Vehicles

Alkyd resin

C20-40 pareth-3; C20-40 pareth-40; C30-50 pareth-3; C30-50 pareth-10; C40-60 pareth-3; C40-60 pareth-10

Diisobutyl phthalate; Diisodecyl phthalate; Diisononyl phthalate; Dioctyl adipate

Glyceryl oleate

Carriers • Vehicles *(cont'd.)*

Peanut oil; Polybutene; Pyrophyllite
Trioctyl trimellitate
Vinyl acetate/ethylene copolymer; Vinyl acrylic copolymer

Catalysts

Acetylacetonate; Alumina; 2-Amino-2-methyl-1-propanol; Ammonia; Ammonium lactate; Antimony trioxide
Barium acetate; N-Benzyldimethylamine; Boron trifluoride; Butyl acid phosphate; t-Butyl hydroperoxide; Butyltin mercaptide
Cellulase; Chromium phosphate; Cobalt acetate (ous); Cobalt octoate; Cumene hydroperoxide
Diatomaceous earth, amorphous; Diazabicycloundecene; Di (butyl, methyl pyrophosphato) ethylene titanate di (dioctyl, hydrogen phosphite); Dibutyltin (bis) mercaptide; Dibutyltin disulfide; Dicyclo (dioctyl) pyrophosphato titanate; Di (dioctylphosphato) ethylene titanate; Dihydrogenated tallow methylamine; Dimethylaminoethyl acrylate; Dimethylaminomethyl phenol
Isooctyl acid phosphate; Isopropyl 4-aminobenzenesulfonyl di (dodecylbenzenesulfonyl) titanate; Isopropyl dimethacryl isostearoyl titanate; Isopropyl tri (dioctylphosphato) titanate; Isopropyl tri (N ethylamino-ethylamino) titanate
Methanesulfonic acid; Methyl ethyl ketone peroxide
Palladium chloride (ous); 1,3-Pentanediamine; Phenyl acid phosphate; Phosphorus trichloride
Stannous octoate
Tetra (2, diallyoxymethyl-1 butoxy titanium di (di-tridecyl) phosphite; Tetraisopropyl titanate; Tetrakis (2-ethylhexyl) titanate; Titanium di (butyl, octyl pyrophosphate) di (dioctyl, hydrogen phosphite) oxyacetate; Titanium di (cumylphenylate) oxyacetate; Titanium di (dioctylpyrophosphate) oxyacetate; Titanium dimethacrylate oxyacetate; o-Toluene sulfonic acid; p-Toluene sulfonic acid; Trichloroethane; Triethylene diamine
Xylene sulfonic acid
Zinc 2-ethylhexoate

Colorants. *See* Pigments

Corrosion Inhibitors

2-Amino-2-methyl-1-propanol
Barium chromate; Barium dinonylnaphthalene sulfonate; 1H-Benzotriazole; N-Benzyldimethylamine; Borosilicate glass; Butyl acid phosphate
Calcium chromate; Calcium dinonylnaphthalene sulfonate; Calcium sulfonate; Calcium-zinc-molybdenum complex; Capryl hydroxyethyl imidazoline; C12 dibasic acid; Chromium phosphate; Cocamine; Cocoyl hydroxyethyl imidazoline
Di (butyl, methyl pyrophosphato) ethylene titanate di (dioctyl, hydrogen phosphite); Dicocodimonium chloride; Di (dioctylphosphato) ethylene titanate; Di (dioctylpyrophosphato) ethylene titanate; Dihydrogenated tallow methylamine; Dilinoleic acid; Dimethyl caprylamide-capramide; Dodecenyl succinic anhydride
Ethyl octynol
Ferric phosphate
Heptadecenyl hydroxyethyl imidazoline
Isopropyl dimethacryl isostearoyl titanate; Isopropyl tri (dioctylphosphato) titanate; Isopropyl tri (dioctylpyrophosphato) titanate; Isopropyl tri (N ethylamino-ethylamino) titanate
Lauryl hydroxyethyl imidazoline; Lead silicate; Lead silicochromate; Lead silicosulfate, basic; Lead sulfate, blue basic
4,4′-Methylene dianiline
Naphthenic acid
Oleoyl sarcosine; Oleyl hydroxyethyl imidazoline; Oleyl propanediamine; Oleyl propylene diamine dioleate; Oleyl propylene diamine ditallate
PEG-2 tallowamine; PEG-5 tallowamine; PEG-15 tallowamine; PEG-20 tallowamine; PEG-3 tallow propylene diamine; PEG-7 tallow propylene diamine; PEG-12 tallow propylene diamine; 1,3-Pentanediamine; Phenylmethyl polysiloxane; PVM/MA copolymer
Sodium dimethyldithiocarbamate; Sodium gluconate; Sodium metasilicate pentahydrate; Sorbitan oleate;

Corrosion Inhibitors *(cont'd.)*

Sorbitan palmitate; Stearamine acetate; Stearyl hydroxyethyl imidazoline

Tall oil hydroxyethyl imidazoline; Tallow dipropylene triamine; Tallow propane diamine dioleate; Tallow propylene diamine; Tetra (2, diallyoxymethyl-1 butoxy titanium di (di-tridecyl) phosphite; Tetraisopropyl di (dioctylphosphito) titanate; Tin; Titanium di (butyl, octyl pyrophosphate) di (dioctyl, hydrogen phosphite) oxyacetate; Titanium di (cumylphenylate) oxyacetate; Titanium di (dioctylpyrophosphate) oxyacetate; Titanium dimethacrylate oxyacetate; Trilinoleic acid

Zinc borate; Zinc chromate; Zinc dust; Zinc hydroxyphosphite; Zinc molybdate; Zinc molybdenum phosphate; Zinc phosphate; Zinc silicophosphate

Coupling Agents

N-2-Aminoethyl-3-aminopropyl trimethoxysilane

Bis-(γ-trimethoxysilylpropyl) amine; t-Butyl alcohol

Di (butyl, methyl pyrophosphato) ethylene titanate di (dioctyl, hydrogen phosphite); Dicocodimonium chloride; Dicyandiamide; Dicyclo (dioctyl) pyrophosphato titanate; Di (dioctylphosphato) ethylene titanate; Di (dioctylpyrophosphato) ethylene titanate; Dimethyl hexynediol; Dioctyl phosphate

Epoxidized linseed oil acrylate; 2-(3,4-Epoxycyclohexyl) ethyltrimethoxysilane; Ethylene dioleamide; Ethylene glycol propyl ether; N-Ethyl-N-hydroxyethyl-m-toluidine; Ethyl phenyl ethanolamine

3-Glycidoxypropyltrimethoxysilane

Heptadecenyl hydroxyethyl imidazoline; Hexamethyldisilazane

Isocyanatopropyltriethoxysilane; Isopropyl 4-aminobenzenesulfonyl di (dodecylbenzenesulfonyl) titanate; Isopropyl dimethacryl isostearoyl titanate; Isopropyl tri (dioctylphosphato) titanate; Isopropyl tri (dioctylpyrophosphato) titanate; Isopropyl tri (N ethylamino-ethylamino) titanate

Mercaptopropyltrimethoxysilane; 3-Methacryloxypropyltrimethoxysilane; Methylphenyl ethanolamine; Methyltriethoxysilane

Neopentyl (diallyl) oxy, tri (m-amino) phenyl zirconate; Neopentyl (diallyl) oxy, tri (dioctyl) phosphato zirconate; Neopentyl (diallyl) oxy, tri (dioctyl) pyrophosphato zirconate; Neopentyl (diallyl) oxy, tri (dodecyl) benzenesulfonyl zirconate; Neopentyl (diallyl) oxy, tri (N-ethylenediamino) ethyl zirconate; Neopentyl (diallyl) oxy, trimethacryl zirconate; Neopentyl (diallyl) oxy, trineodecanoyl zirconate

Octyltriethoxysilane; Oleth-4

PEG-8 dilaurate; PEG-4 laurate; PEG-5 laurate; Phenyldiethanolamine; Phenyltrimethoxysilane; Polyacrylic acid; PVM/MA copolymer

Sodium cumenesulfonate

Tetra (2, diallyoxymethyl-1 butoxy titanium di (di-tridecyl) phosphite; Tetraoctyloxytitanium di (ditridecylphosphite); Titanium di (butyl, octyl pyrophosphate) di (dioctyl, hydrogen phosphite) oxyacetate; Titanium di (cumylphenylate) oxyacetate; Titanium di (dioctylpyrophosphate) oxyacetate; Titanium dimethacrylate oxyacetate

N-[2-(Vinylbenzylamino)-ethyl)-3-aminopropyltrimethoxysilane; Vinyltriethoxysilane

Crosslinking Agents

Acetylacetonate; Allyl methacrylate; Ammonium lactate

1,3-Butylene glycol dimethacrylate

DEDM hydantoin; Diallyl fumarate; Diethylene glycol dimethacrylate

Ethylacetoacetate; Ethylene glycol dimethacrylate

Hexamethoxymethylmelamine; 1,6-Hexanediol diacrylate; 1,6-Hexanediol methacrylate; Hydroxymethyl dioxoazabicyclooctane

Melamine/formaldehyde resin; Melamine-formaldehyde resin, methylated; Melamine-formaldehyde resin, methylated-butylated; Mercaptopropyltrimethoxysilane

Octyltriethoxysilane

PEG-4 diacrylate; PEG-3 dimethacrylate; PEG-4 dimethacrylate; Pentaerythrityl tetraacrylate; Pentaerythrityl triacrylate; Polyacrylamide; Polymethylene polyphenyl isocyanate

Tetrakis (2-ethylhexyl) titanate; Thioglycerin; Titanium acetylacetonate; Triethylene diamine; 1,1,1-Trimethylolpropane triacrylate; 1,1,1-Trimethylolpropane trimethacrylate

Vinyltriethoxysilane; Vinyltrimethoxysilane; Vinyltris(2-methoxyethoxy) silane

Zinc diacrylate

Curing Agents

Aminoethylpiperazine; Amodimethicone
Boron trifluoride-amine complex; 1,3-Butylene glycol diacrylate
Cumene hydroperoxide; Cyclohexyl acrylate
Diazabicycloundecene, 2-ethylhexanoic acid salt; Dicyandiamide; Diethylenetriamine; Diethyl toluene diamine; Ditrimethylolpropane tetraacrylate; Dodecenyl succinic anhydride
Epoxy acrylate; Ethyl octynol
Isobornyl acrylate; Isophorone diamine
4,4´-Methylene dianiline; 2-Methyl imidazole
Neopentyl glycol diacrylate
PEG-3 diacrylate; PEG-8 diacrylate; Phenalkamine; 2-Phenoxyethyl methacrylate; Phenyl acid phosphate; Phenyldiethanolamine; Polytetramethylene ether glycol diamine; Polyurethane-acrylate resin; Pyrophyllite
Sulfolane
Tallow dipropylene triamine; Triethylenetetramine; Trilinoleic acid; Trimethylhexamethylene diamine; 2,4,6-Tris (dimethylaminomethyl) phenol; 2,4,6-Tris (dimethylaminomethyl) phenol, 2-ethylhexanoic acid salt

Defoamers • Antifoams

Azodicarbonamide
Cyclodextrin; β-Cyclodextrin
Dicyclo (dioctyl) pyrophosphato titanate; Dimethicone; Dimethyl hexynediol; Disodium tallow sulfosuccinamate
Hydroxypropyl-β-cyclodextrin; Hydroxypropyl-γ-cyclodextrin
Isooctyl alcohol
Meroxapol 105; Meroxapol 171; Meroxapol 172; Meroxapol 252; Meroxapol 254; Meroxapol 258; Meroxapol 311
Nonoxynol-1; Nonoxynol-2; Nonoxynol-3; Nonoxynol-4; Nonoxynol-5; Nonoxynol-6; Nonoxynol-8; Nonoxynol-9; Nonoxynol-10; Nonoxynol-11; Nonoxynol-12; Nonoxynol-13; Nonoxynol-14; Nonoxynol-20; Nonoxynol-40; Nonoxynol-50; Nonoxynol-70; Nonoxynol-100
Oleyl alcohol
PEG-5 laurate; PEG-9 laurate; PEG-4 oleate; PEG-14 oleate; PEG-12 stearate; PEG-1.3 tetramethyl decynediol; PEG-3.5 tetramethyl decynediol; PEG-10 tetramethyl decynediol; PEG-30 tetramethyl decynediol; Poloxamer 101; Poloxamer 108; Poloxamer 123; Poloxamer 124; Poloxamer 181; Poloxamer 182; Poloxamer 184; Poloxamer 185; Poloxamer 188; Poloxamer 217; Poloxamer 231; Poloxamer 234; Poloxamer 235; Poloxamer 237; Poloxamer 282; Poloxamer 288; Poloxamer 331; Poloxamer 333; Poloxamer 334; Poloxamer 335; Poloxamer 338; Poloxamer 401; Poloxamer 402; Poloxamer 403; Poloxamer 407; Poloxamine 304; Polypropylene glycol; Polysiloxane; Polysiloxane polyether copolymer; PPG-9; PPG-12; PPG-17; PPG-20; PPG-26; PPG-30; Propylene glycol myristate
Silicone elastomer; Silicone emulsions; Simethicone
Tetraisopropyl di (dioctylphosphito) titanate; Tetramethyl decynediol; Tributoxyethyl phosphate; Tributyl citrate

Detergents

Capryl hydroxyethyl imidazoline; Ceteareth-8; Ceteareth-27; Ceteareth-30; C11-15 pareth-7; C12-15 pareth-7; C12-15 pareth-9
Ditridecyl sodium sulfosuccinate
Ethanolamine
Lauramine oxide; Laureth-8; Laureth-9
Naphthenic acid; Nonoxynol-1; Nonoxynol-2; Nonoxynol-3; Nonoxynol-4; Nonoxynol-5; Nonoxynol-6; Nonoxynol-8; Nonoxynol-9; Nonoxynol-10; Nonoxynol-11; Nonoxynol-12; Nonoxynol-13; Nonoxynol-14; Nonoxynol-15; Nonoxynol-20; Nonoxynol-40; Nonoxynol-50; Nonoxynol-70; Nonoxynol-100; Nonoxynol-10 carboxylic acid; Nonyl nonoxynol-15; Nonyl nonoxynol-18; Nonyl nonoxynol-24; Nonyl nonoxynol-150
Octoxynol-1; Octoxynol-3; Octoxynol-4; Octoxynol-5; Octoxynol-7; Octoxynol-9; Octoxynol-13; Octoxynol-16; Octoxynol-30; Octoxynol-70; Oleoyl sarcosine; Oleyl alcohol
PEG-12 castor oil; Poloxamer 101; Poloxamer 108; Poloxamer 123; Poloxamer 124; Poloxamer 181; Poloxamer 182; Poloxamer 184; Poloxamer 185; Poloxamer 188; Poloxamer 217; Poloxamer 231; Poloxamer 234; Poloxamer 235; Poloxamer 237; Poloxamer 282; Poloxamer 288; Poloxamer 331; Poloxamer 333; Poloxamer 334; Poloxamer 335; Poloxamer 338; Poloxamer 401; Poloxamer 402; Poloxamer 403; Poloxamer 407; Poloxamine 304; Poloxamine 701; Poloxamine 704; Poloxamine 901; Poloxamine 904; Poloxamine 1107; Poloxamine 1307; PPG-6 C12-18 pareth-11

Detergents *(cont'd.)*

Sodium cetyl sulfate; Sodium C14-16 olefin sulfonate; Sodium decyl sulfate; Sodium dodecylbenzenesulfonate; Sodium isodecyl sulfate; Sodium lauroamphoacetate; Sodium lauryl sulfate; Sodium nonoxynol-6 phosphate; Sodium nonoxynol-4 sulfate; Steareth-2; Steareth-10; Steareth-20

TEA-dodecylbenzenesulfonate; Trideceth-6; Trideceth-12

Diluents

Allyl glycidyl ether; Aromatic 100

Biscumylphenyl trimellitate; Butyl glycidyl ether; p-t-Butyl phenyl glycidyl ether

Calcium resinate; C10-11 isoparaffin; C11-12 isoparaffin; o-Cresyl glycidyl ether; Cumylphenyl acetate

2-Ethylhexyl glycidyl ether

Glycidyl neodecanoate

Hexane

Isobornyl acrylate; Isodecyl acrylate; Isopropyl glycidyl ether

Lauryl acrylate

Methoxy tripropylene glycol acrylate

Neopentyl glycol diacrylate; Neopentyl glycol diglycidyl ether

Octyl/decyl acrylate

PEG-6 trimethylolpropane; 2-Phenoxyethyl acrylate; Phenyl glycidyl ether; Phenyl glycol ether; Polyester acrylate; Propylene carbonate; Pyrophyllite

α,α,4,4-Tetramethyl-2-(methylethyl)-N-(2-methylpropylidene)-3-oxazolidineethanamine; Trichloroethane

Dispersants

Amine dodecylbenzene sulfonate; 2-Amino-2-methyl-1-propanol; Ammonium acrylate; Ammonium lauryl sulfate; Ammonium naphthalene sulfonate; Ammonium nonoxynol-4 sulfate; Ammonium polyacrylate; Ammonium polymethacrylate; Ammonium/sodium acrylate

Behenyl hydroxyethyl imidazoline; Butylated PVP

Calcium dodecylbenzene sulfonate; Calcium naphthalene sulfonate; Calcium sulfonate; Castor oil, polymerized; Ceteareth-18; Ceteareth-27; Ceteareth-30; Ceteareth-55; Ceteareth-80; Cetoleth-5; Cetyl alcohol; Cocaminobutyric acid

Di (butyl, methyl pyrophosphato) ethylene titanate di (dioctyl, hydrogen phosphite); Di (dioctylphosphato) ethylene titanate; Di (dioctylpyrophosphato) ethylene titanate; Diglycol/CHDM/isophthalates/SIP copolymer; Dihexyl sodium sulfosuccinate; Dimethicone copolyol; Dimethyl oleamide; Dioctyl sodium sulfosuccinate; Disodium maleic anhydride/diisobutylene copolymer; Disodium nonoxynol-10 sulfosuccinate; Ditridecyl sodium sulfosuccinate; Dodoxynol-10

Ethylene/acrylic acid copolymer; Ethylene dioleamide; Ethylene distearamide; 2-Ethylhexyl hydrogenated tallowalkyl methosulfate; Ethyl hydroxymethyl oleyl oxazoline

Hydrogenated tallow glycerides; Hydroxylated lecithin

Isobutylene/MA copolymer; Isocyanatopropyltriethoxysilane; Isopropyl 4-aminobenzenesulfonyl di (dodecylbenzenesulfonyl) titanate; Isopropyl dimethacryl isostearoyl titanate; Isopropyl tri (dioctylphosphato) titanate; Isopropyl tri (dioctylpyrophosphato) titanate; Isopropyl tri (N ethylamino-ethylamino) titanate

Lanolin acid; Laureth-5; Laureth-8; Laureth-9; Laureth-23; Lauryl hydroxyethyl imidazoline; dl-Limonene

Maleic acid/acrylic acid copolymer; Maleic acid/acrylic acid copolymer, sodium salt; Maleic acid/olefin copolymer, sodium salt; Maleic anhydride/methyl vinyl ether copolymer, sodium salt; Meroxapol 105; Meroxapol 171; Meroxapol 172; Meroxapol 174; Meroxapol 252; Meroxapol 254; Meroxapol 258; Meroxapol 311; Methacrylate copolymer; Methyl hydroxyethylcellulose; N-Methyl-2-pyrrolidone

Nonoxynol-1; Nonoxynol-2; Nonoxynol-3; Nonoxynol-4; Nonoxynol-5; Nonoxynol-6; Nonoxynol-8; Nonoxynol-9; Nonoxynol-10; Nonoxynol-11; Nonoxynol-12; Nonoxynol-13; Nonoxynol-14; Nonoxynol-20; Nonoxynol-40; Nonoxynol-50; Nonoxynol-70; Nonoxynol-100; Nonoxynol-10 carboxylic acid; Nonoxynol-6 phosphate; Nonyl nonoxynol-150

Octoxynol-1; Octoxynol-3; Octoxynol-4; Octoxynol-5; Octoxynol-7; Octoxynol-9; Octoxynol-13; Octoxynol-16; Octoxynol-30; Octoxynol-70; Oleth-5; Oleth-20; Oleyl alcohol

PEG-5M; PEG-7M; PEG-9M; PEG-14M; PEG-23M; PEG-45M; PEG-90M; PEG-5 castor oil; PEG-15 castor oil; PEG-16 castor oil; PEG-25 castor oil; PEG-28 castor oil; PEG-30 castor oil; PEG-32 castor oil; PEG-36 castor oil; PEG-40 castor oil; PEG-75 castor oil; PEG-80 castor oil; PEG-160 castor oil; PEG-200 castor oil; PEG-10

Dispersants *(cont'd.)*

cocamine; PEG-8 dilaurate; PEG-12 dioleate; PEG-150 distearate; PEG-16 hydrogenated castor oil; PEG-25 hydrogenated castor oil; PEG-200 hydrogenated castor oil; PEG-2 laurate; PEG-2 laurate SE; PEG-4 laurate; PEG-8 laurate; PEG-9 laurate; PEG-12 laurate; PEG-2 oleate; PEG-2 oleate SE; PEG-4 oleate; PEG-8 oleate; PEG-12 oleate; PEG-20 oleate; PEG-80 sorbitan laurate; PEG-2 stearate SE; PEG-4 stearate; PEG-8 stearate; PEG-12 stearate; PEG-20 stearate; PEG-3 tallow propylene diamine; PEG-7 tallow propylene diamine; PEG-12 tallow propylene diamine; PEG-1.3 tetramethyl decynediol; Perfluoroheptane; Perfluorohexane; Perfluoro-N-methylmorpholine; Perfluoropentane; Petrolatum; Poloxamer 101; Poloxamer 108; Poloxamer 123; Poloxamer 124; Poloxamer 181; Poloxamer 182; Poloxamer 184; Poloxamer 185; Poloxamer 188; Poloxamer 217; Poloxamer 231; Poloxamer 234; Poloxamer 235; Poloxamer 237; Poloxamer 282; Poloxamer 288; Poloxamer 331; Poloxamer 333; Poloxamer 334; Poloxamer 335; Poloxamer 338; Poloxamer 401; Poloxamer 402; Poloxamer 403; Poloxamer 407; Poloxamine 304; Poloxamine 701; Poloxamine 704; Poloxamine 901; Poloxamine 904; Poloxamine 908; Poloxamine 1107; Poloxamine 1307; Polyacrylamide; Polyacrylic acid; Poly (1,4-butane adipate) polyester; Poly (diethylene adipate) polyester; Polyethylene, high-density; Polyethylene propylene adipate polyester; Poly (1,6-hexane/adipate/isophthalate) polyester; Poly (1,6-hexane adipate) polyester; Poly (1,6-hexane/diethylene adipate) polyester; Poly (1,6-hexane/neopentyl adipate/isophthalate) polyester; Poly (neopentyl adipate) polyester; Poly (neopentyl/1,6-hexane adipate) polyester; Polysodium vinyl sulfonate; Polysorbate 80; Polysorbate 81; Polyurethane prepolymer; Polyvinyl alcohol, partially hydrolyzed; Potassium naphthalene-formaldehyde sulfonate; Potassium polyacrylate; PPG-24-buteth-27; PPG-36 oleate; Propylene glycol myristate; PVM/MA copolymer; PVP; PVP/VA copolymer

Sodium decyl sulfate; Sodium dibutyl naphthalene sulfonate; Sodium diisopropyl naphthalene sulfonate; Sodium isopropyl naphthalene sulfonate; Sodium lauroamphoacetate; Sodium lauryl sulfate; Sodium methylnaphthalenesulfonate; Sodium octyl sulfate; Sodium polyisobutylene/maleic anhydride copolymer; Sodium polymethacrylate; Sodium polynaphthalene sulfonate; Stearamine acetate; Steareth-2; Steareth-10; Steareth-20; Stearic acid; Styrene/MA copolymer; Sulfated peanut oil; Sulfonic acid

Tall oil aminoethyl imidazoline; Tall oil hydroxyethyl imidazoline; Tallow dipropylene triamine; Tallow propane diamine dioleate; Tallow propylene diamine; TEA-dodecylbenzenesulfonate; Tetra (2, diallyoxymethyl-1 butoxy titanium di (di-tridecyl) phosphite; Tetramethyl decynediol; Tetraoctyloxytitanium di (ditridecylphosphite); Tetrapotassium pyrophosphate; Tetrasodium dicarboxyethyl stearyl sulfosuccinamate; Tetrasodium pyrophosphate; Titanium di (butyl, octyl pyrophosphate) di (dioctyl, hydrogen phosphite) oxyacetate; Titanium di (cumylphenylate) oxyacetate; Titanium di (dioctylpyrophosphate) oxyacetate; Trideceth-6; Trideceth-12; Triethanolamine

Driers

Abietic acid; Aluminum naphthenate; Aluminum octoate; Aluminum stearate; Aluminum tristearate

Barium naphthenate; Bismuth octoate

Cadmium naphthenate; Calcium drier; Calcium linoleate; Calcium naphthenate; Calcium octoate; Calcium sulfate; Calcium tallate; Castor oil, dehydrated; Cerium naphthenate; Cobalt acetate (ous); Cobalt linoleate; Cobalt naphthenate; Cobalt octoate; Cobalt sulfate (ous); Cobalt tallate; Coconut oil; Copper linoleate; Copper naphthenate; Copper octoate; Copper tallate; Cottonseed oil

2-Ethylhexoic acid

Iron linoleate; Iron naphthenate; Iron octoate; Iron tallate

Lead linoleate; Lead naphthenate; Lead octoate; Lead stearate; Lead tallate; Linseed oil

Magnesium stearate; Manganese acetate; Manganese drier; Manganese linoleate; Manganese naphthenate; Manganese octoate; Manganese tallate; Menhaden oil

Naphthenic acid; Nickel octoate; Novolac resin

Perchloroethylene; 1,10-Phenanthroline; Potassium octoate

Sorbitan tallate; Soybean oil

Tall oil

Urethane bis-oxazolidine

Vegetable oil

Zinc 2-ethylhexanoate; Zinc 2-ethylhexoate; Zinc naphthenate; Zinc tallate; Zirconium drier; Zirconium octoate

Dyes. *See* Pigments

Emulsifiers

Aluminum hydroxide; Amine dodecylbenzene sulfonate; 2-Amino-2-methyl-1-propanol; Ammonium laureth sulfate; Ammonium laureth-12 sulfate; Ammonium laureth-30 sulfate; Ammonium lauryl sulfate; Ammonium nonoxynol-4 sulfate; Ammonium nonoxynol-9 sulfate

Behenyl hydroxyethyl imidazoline; Benzalkonium chloride; Butyl acetyl ricinoleate; Butyl methacrylate

C20-40 alcohols; C40-60 alcohols; Capryl hydroxyethyl imidazoline; Caprylic alcohol; Carbomer; Ceteareth-6; Ceteareth-8; Ceteareth-10; Ceteareth-11; Ceteareth-12; Ceteareth-14; Ceteareth-27; Ceteareth-30; Ceteareth-55; Cetoleth-5; Citric acid esters of mono- and diglycerides of fatty acids; Cocamine; Coconut oil; Cocoyl hydroxyethyl imidazoline; Corn oil; C11-15 pareth-7; C20-40 pareth-3; C20-40 pareth-40; C30-50 pareth-3; C30-50 pareth-10; C40-60 pareth-3; C40-60 pareth-10

Di (butyl, methyl pyrophosphato) ethylene titanate di (dioctyl, hydrogen phosphite); Dicocodimonium chloride; Di (dioctylphosphato) ethylene titanate; Di (dioctylpyrophosphato) ethylene titanate; Diethanolamine; Dihexyl sodium sulfosuccinate; Diisopropanolamine; Dimethicone copolyol; Disodium nonoxynol-10 sulfosuccinate; Ditridecyl sodium sulfosuccinate; Dodecylbenzene sulfonic acid; Dodoxynol-10

2-Ethylhexyl phosphate; Ethyl hydroxymethyl oleyl oxazoline

Glyceryl oleate; Glyceryl ricinoleate

Heptadecenyl hydroxyethyl imidazoline; Hexyl methacrylate; Hydrogenated soybean oil; Hydrogenated tallow glycerides; Hydroxypropylcellulose

Isodecyl oxypropyl amine acetate; Isooctyl alcohol; Isopropylamine dodecylbenzenesulfonate; Isopropyl 4-aminobenzenesulfonyl di (dodecylbenzenesulfonyl) titanate; Isopropyl dimethacryl isostearoyl titanate; Isopropyl tri (dioctylphosphato) titanate; Isopropyl tri (dioctylpyrophosphato) titanate; Isopropyl tri (N ethylamino-ethylamino) titanate

Lanolin acid; Lanolin alcohol; Laureth-2; Laureth-4; Laureth-8; Laureth-9; Laureth-12; Laureth-23; Lauryl hydroxyethyl imidazoline; Lead naphthenate; Lecithin; Locust bean gum

Meroxapol 105; Meroxapol 171; Meroxapol 172; Meroxapol 174; Meroxapol 252; Meroxapol 254; Meroxapol 258; Meroxapol 311; Methoxypropylamine; Methyl hydroxyethylcellulose

Nonoxynol-1; Nonoxynol-2; Nonoxynol-3; Nonoxynol-4; Nonoxynol-5; Nonoxynol-6; Nonoxynol-8; Nonoxynol-9; Nonoxynol-10; Nonoxynol-11; Nonoxynol-12; Nonoxynol-13; Nonoxynol-14; Nonoxynol-15; Nonoxynol-20; Nonoxynol-30; Nonoxynol-34; Nonoxynol-40; Nonoxynol-50; Nonoxynol-70; Nonoxynol-100; Nonoxynol-6 phosphate; Nonoxynol-9 phosphate; Nonyl nonoxynol-7; Nonyl nonoxynol-8; Nonyl nonoxynol-9; Nonyl nonoxynol-15; Nonyl nonoxynol-18; Nonyl nonoxynol-24; Nonyl nonoxynol-49; Nonyl nonoxynol-150

Octoxynol-1; Octoxynol-3; Octoxynol-4; Octoxynol-5; Octoxynol-7; Octoxynol-9; Octoxynol-13; Octoxynol-16; Octoxynol-25; Octoxynol-30; Octoxynol-70; Oleoyl sarcosine; Oleth-4; Oleth-9; Oleyl hydroxyethyl imidazoline

PEG-12 castor oil; PEG-15 castor oil; PEG-16 castor oil; PEG-20 castor oil; PEG-25 castor oil; PEG-28 castor oil; PEG-30 castor oil; PEG-32 castor oil; PEG-36 castor oil; PEG-40 castor oil; PEG-75 castor oil; PEG-80 castor oil; PEG-160 castor oil; PEG-200 castor oil; PEG-10 cocamine; PEG-15 cocamine; PEG-8 dilaurate; PEG-9 dioleate; PEG-12 dioleate; PEG-150 distearate; PEG-12 ditallate; PEG-5 hydrogenated castor oil; PEG-16 hydrogenated castor oil; PEG-25 hydrogenated castor oil; PEG-200 hydrogenated castor oil; PEG-2 laurate; PEG-2 laurate SE; PEG-4 laurate; PEG-5 laurate; PEG-8 laurate; PEG-9 laurate; PEG-12 laurate; PEG-2 oleate; PEG-2 oleate SE; PEG-4 oleate; PEG-8 oleate; PEG-12 oleate; PEG-20 oleate; PEG-8 pelargonate; PEG-9 pelargonate; PEG-17 sorbitan trioleate; PEG-18 sorbitan trioleate; PEG-2 stearate; PEG-2 stearate SE; PEG-4 stearate; PEG-8 stearate; PEG-12 stearate; PEG-20 stearate; PEG-2 tallowamine; PEG-5 tallowamine; PEG-15 tallowamine; PEG-20 tallowamine; PEG-3 tallow propylene diamine; PEG-7 tallow propylene diamine; PEG-12 tallow propylene diamine; PEG-66 trihydroxystearin; Poloxamer 101; Poloxamer 123; Poloxamer 124; Poloxamer 181; Poloxamer 182; Poloxamer 184; Poloxamer 185; Poloxamer 188; Poloxamer 217; Poloxamer 231; Poloxamer 234; Poloxamer 235; Poloxamer 237; Poloxamer 282; Poloxamer 288; Poloxamer 331; Poloxamer 333; Poloxamer 334; Poloxamer 335; Poloxamer 338; Poloxamer 401; Poloxamer 402; Poloxamer 403; Poloxamer 407; Poloxamine 304; Poloxamine 701; Poloxamine 704; Poloxamine 901; Poloxamine 904; Poloxamine 908; Poloxamine 1107; Poloxamine 1307; Polyacrylic acid; Polyglyceryl-10 tetraoleate; Polysorbate 20; Polysorbate 40; Polysorbate 60; Polysorbate 61; Polysorbate 65; Polysorbate 80; Polysorbate 81; Polysorbate 85; Polyurethane prepolymer; PPG-24-buteth-27; Propylene glycol; Propylene glycol alginate; Propylene glycol laurate; Propylene glycol stearate; PVM/MA copolymer

Sodium cetyl sulfate; Sodium C14-16 olefin sulfonate; Sodium decyl sulfate; Sodium dodecylbenzenesulfonate; Sodium isodecyl sulfate; Sodium isopropyl naphthalene sulfonate; Sodium laureth sulfate; Sodium lauroamphoacetate; Sodium lauryl sulfate; Sodium nonoxynol-6 phosphate; Sodium nonoxynol-4 sulfate; Sodium octoxynol-3 sulfate; Sodium octyl sulfate; Sodium p-styrenesulfonate; Sorbitan laurate; Sorbitan oleate; Sorbitan palmitate; Sorbitan stearate; Sorbitan trioleate; Sorbitan tristearate; Stearamine acetate; Steareth-2; Steareth-10; Steareth-20; Styrene/MA copolymer; Sulfated peanut oil; Synthetic beeswax

Tall oil acid; Tall oil hydroxyethyl imidazoline; Tetra (2, diallyoxymethyl-1 butoxy titanium di (di-tridecyl) phosphite; Tetrahydroxypropyl ethylenediamine; Tetraoctyloxytitanium di (ditridecylphosphite); Tetrasodium dicarboxyethyl stearyl sulfosuccinamate; Tetrasodium pyrophosphate; Titanium di (butyl, octyl pyrophosphate)

Emulsifiers *(cont'd.)*

di (dioctyl, hydrogen phosphite) oxyacetate; Titanium di (cumylphenylate) oxyacetate; Titanium di (dioctylpyrophosphate) oxyacetate; Trideceth-6; Trideceth-12; Trideceth-14; Trideceth-15; Triisopropanolamine; Tris (hydroxymethyl) aminomethane

Extenders. *See* Fillers

Fillers • Extenders

Alumina silicate; Alumina trihydrate; Aluminum nitride; Aluminum silicate; Aluminum silicate, calcined; Aluminum silicate, hydrous; Aminopropyltriethoxysilane
Barium metaborate; Barium sulfate; Bentonite
Calcium carbide; Calcium carbonate; Calcium metasilicate; Calcium sulfate dihydrate; Cellulose; Coumarone-indene resin
Diatomaceous earth; Diatomaceous earth, amorphous
Feldspar; Ferro-aluminum silicate
3-Glycidoxypropyltrimethoxysilane
Kaolin; Kaolinite
Lauroyl lysine; Limestone
Magnesium hydroxide; Magnesium oxide; Mercaptopropyltrimethoxysilane; Methyl hydroxyethylcellulose; Mica; Microcrystalline cellulose
Nepheline syenite
Ozokerite
Pentaerythrityl tetrabenzoate; Pyrophyllite
Quartz
Rapeseed oil, vulcanized
Silica; Silica-alumina; Silica, amorphous; Silica, colloidal; Silica dimethyl silylate; Silica, fumed; Silica, hydrated; Sodium borosilicate; Sodium magnesium aluminosilicate; Sodium silicoaluminate; Strontium sulfate; Styrene-butadiene polymer
Terpene resin
Vinyltriethoxysilane
Wollastonite

Fire Retardants. *See* Flame Retardants

Flame Retardants • Fire Retardants

Aluminum hydroxide; Ammonium polyphosphate sol'n.; Antimony pentoxide
Bisphenol A; Bis (tribromophenoxy) ethane; Boron oxide
Decabromodiphenyl oxide; Dibromoethyldibromocyclohexane; Dibromostyrene; Dimelamine phosphate; Diphenyl octyl phosphate; Disodium tetrabromophthalate
Epoxy resin, brominated; Ethylenebis dibromonorbornane dicarboximide
Hexabromocyclododecane
Isodecyl diphenyl phosphate
Magnesium calcium carbonate hydroxide; Melamine phosphate; Melamine pyrophosphate
Paraffin, brominated; Paraffin, bromochlorinated; Pentabromodiphenyl oxide; Poly(dibromostyrene)
Silica-alumina
Tetrabromobis (2-ethylhexyl) phthalate; Tetrabromobisphenol A; Tetrabromobisphenol A, bis (2-hydroxyethyl ether); Tetrabromobisphenol A diacrylate; Tetrabromophthalic anhydride; Tetrachlorophthalic anhydride; Tetradecabromodiphenoxy benzene; Tribromophenyl allyl ether; Tributoxyethyl phosphate; Trichloroethyl phosphate; Triisopropylphenyl phosphate; Triphenyl phosphate; Trixylenyl phosphate
Vinyl bromide
Zinc hydroxystannate; Zinc oxide; Zinc phosphate

Flatting Agents • Matte Agents

Attapulgite
Calcium stearate
Diatomaceous earth, amorphous
Ethylene/magnesium acrylate copolymer; Ethylene/zinc acrylate copolymer
Polyethylene, high-density
Zinc stearate

Flow Control Agents

Borosilicate glass
Cocamine
Oxidized menhaden oil
PEG-6 trimethylolpropane; Perfluoroheptane; Perfluorohexane; Perfluoro-N-methylmorpholine; Perfluoropentane
Silicone hexaacrylate

Fungicides. *See* Antimicrobials

Gelling Agents. *See* Thickeners

Gloss Aids

Cetyl esters
Oxidized menhaden oil
Trimethylolpropane ethoxy triacrylate
Vinyltriethoxysilane; Vinyltrimethoxysilane; Vinyltris(2-methoxyethoxy) silane

Grinding Aids

Di (butyl, methyl pyrophosphato) ethylene titanate di (dioctyl, hydrogen phosphite); Dicyclo (dioctyl) pyrophosphato titanate; Di (dioctylphosphato) ethylene titanate; Di (dioctylpyrophosphato) ethylene titanate
Ethylcellulose; Ethyl hydroxymethyl oleyl oxazoline
Isopropyl 4-aminobenzenesulfonyl di (dodecylbenzenesulfonyl) titanate; Isopropyl dimethacryl isostearoyl titanate; Isopropyl tri (dioctylphosphato) titanate; Isopropyl tri (dioctylpyrophosphato) titanate; Isopropyl tri (N ethylamino-ethylamino) titanate
Nonoxynol-30
Polyester adipate; Propylene glycol myristate
Sodium dinaphthalene methane sulfonate; Stearyl hydroxyethyl imidazoline
Tallow dipropylene triamine; Tetra (2, diallyoxymethyl-1 butoxy titanium di (di-tridecyl) phosphite; Tetraisopropyl di (dioctylphosphito) titanate; Tetraoctyloxytitanium di (ditridecylphosphite); Titanium di (butyl, octyl pyrophosphate) di (dioctyl, hydrogen phosphite) oxyacetate; Titanium di (cumylphenylate) oxyacetate; Titanium di (dioctylpyrophosphate) oxyacetate; Titanium dimethacrylate oxyacetate

Hardeners

Candelilla wax
4,4´-Diaminodiphenyl sulfone; 3,5-Diethyl-2,4-diaminotoluene; 3,5-Diethyl-2,6-diaminotoluene; Dihydrogenated tallow methylamine
Hexahydrophthalic anhydride

Hardeners *(cont'd.)*

Phenalkamine; Phthalic anhydride
Succinic anhydride
o-Tolyl biguanide

Intermediates

Acetaldehyde; Acetone oxime; Acetophenone; Acetylacetone; Acrylic monomer; Adipic acid; Allyl glycidyl ether; Aniline; Anthranilic acid

Benzaldehyde; Benzoguanamine; Benzonitrile; 3,3´,4,4´-Benzophenone tetracarboxylic dianhydride; N-Benzyldimethylamine; Bisphenol A; 1,4-Butanediol; Butanetriol; Butoxytriglycol; Butyl acrylate; t-Butylaminoethyl methacrylate; 1,3-Butylene glycol dimethacrylate; 4-t-Butylphenol; Butyl phosphate; n-Butyraldehyde; Butyrolactone

C30-50 alcohols; Caprolactam; ε-Caprolactone monomer; Caprylic alcohol; Ceteareth-8; Citric acid; C12-15 pareth-7; C12-15 pareth-9

Diallyl phthalate; Dicocamine; Dicyclopentadiene; Diethyl aniline; Dihydrogenated tallow methylamine; Dimer acid; n,n-Dimethylaniline; Dimethyl cocamine; Dimethyldichlorosilane; Dipentaerythritol; Dodecenyl succinic anhydride; Dodecylbenzene

Epichlorohydrin; Ethoxytriglycol; Ethyl phenyl ethanolamine; Ethyl silicate

Formaldehyde; Furfural; Furfuryl alcohol

Hexahydrophthalic acid; Hexahydrophthalic anhydride; Hexamethyldisilazane; Hexamethylene diisocyanate; Hexamethylene glycol; Hexamethylenetetramine; Hydrogen peroxide; Hydroquinone; Hydroquinone monomethyl ether; Hydroxystearic acid

Isobutyltrimethoxysilane; Isocyanatopropyltriethoxysilane; Isooctyl alcohol; Isophthalic acid; Itaconic acid

MDI; MDM hydantoin; Melamine; 2-Mercaptoethanol; Mercaptopropyltrimethoxysilane; Meroxapol 172; Meroxapol 252; Meroxapol 258; Meroxapol 311; Methoxypropylamine; 4,4´-Methylene dianiline; Methyl glucoside; Methyl oleate; 2-Methyl-1,3-propanediol; N-Methyl-2-pyrrolidone; Methyl stearate; α-Methylstyrene monomer; Methyltriethoxysilane; Methyltrimethoxysilane

Neodecanoic acid; Neoheptanoic acid; Neopentanoic acid; Neopentyl glycol; Nitroethane; Nitromethane; 2-Nitropropane; Nonanoic acid; Nonoxynol-1; Nonoxynol-2; Nonoxynol-3; Nonoxynol-4; Nonoxynol-5; Nonoxynol-6; Nonoxynol-8; Nonoxynol-9; Nonoxynol-10; Nonoxynol-11; Nonoxynol-12; Nonoxynol-13; Nonoxynol-14; Nonoxynol-15; Nonoxynol-20; Nonoxynol-40; Nonoxynol-50; Nonoxynol-70; Nonoxynol-100

Octoic acid; Octyltriethoxysilane; Oleic acid; Oleyl propanediamine

PCA; Pentaerythrityl triallyl ether; Phenol sulfonic acid; Phenyldiethanolamine; Phenyltrimethoxysilane; Phosphorus trichloride; Polyethylene glycol; Polymethylene polyaniline; Polypropylene glycol; PPG-9; PPG-12; PPG-17; PPG-20; PPG-26; PPG-30; Propionaldehyde; Propylene carbonate; Propylene glycol phenyl ether; Pyridine; Pyromellitic acid; Pyromellitic anhydride; Pyromellitic dianhydride

Ricinoleic acid

Steareth-10; Stearic acid; Stearyl hydroxyethyl imidazoline

Tall oil aminoethyl imidazoline; Tallow propane diamine dioleate; Terephthalic acid; Tetrabromobisphenol A; Tetrabromophthalate diol; Tetrachlorophthalic acid; Tetrahydrophthalic anhydride; Tetrahydroxypropyl ethylenediamine; Triallylcyanurate; Tributyltin oxide; Trideceth-6; Trideceth-12; Triethanolamine; Trimellitic anhydride; 1,3,5-Trimethylbenzene; Triphenyl phosphite; Tris (hydroxymethyl) aminomethane

n-Valeric acid; Vinyl 2-ethylhexanoate; Vinylidene chloride monomer; Vinyl neodecanoate; Vinyl neononanoate; Vinyl neopentanoate; Vinyl pivalate; Vinyl propionate; Vinyltriethoxysilane

Xylene sulfonic acid

Leveling Agents

C11-15 pareth-7
Dihydrogenated tallow benzylmonium hectorite
Oxidized menhaden oil
PEG-36 castor oil; PEG-75 castor oil; PEG-160 castor oil; PEG-15 cocamine; PEG-8 laurate; PEG-9 laurate; PEG-12 laurate; PEG-14 oleate; PEG-4 stearate; PEG-8 stearate; PEG-12 stearate; PEG-3.5 tetramethyl decynediol
Sulfonic acid

Lubricants

Acetylated palm kernel glycerides; Aminopropyltrimethoxysilane
Butyl acetyl ricinoleate; Butyl stearate
Ceteareth-6; Ceteareth-10; Ceteareth-14; Cetyl palmitate
Dilinoleic acid
Erucamide; Ethylene dioleamide
Fluorinated ethylene-propylene
Glyceryl mono/dioleate; Glyceryl oleate; Graphite
Hexamethyldisilazane; Hydrogenated tallow amide; N (2-Hydroxyethyl) 12-hydroxystearamide; N(β-Hydroxyethyl) ricinoleamide
Laureth-4; Lead naphthenate
Meroxapol 171; Meroxapol 172; Meroxapol 174; Meroxapol 252; Meroxapol 254; Meroxapol 258; Meroxapol 311; Methyl acetyl ricinoleate; Methyl hydroxystearate; Methyltriethoxysilane; Methyltrimethoxysilane; Mineral oil
Neopentyl glycol
PEG-4; PEG-6; PEG-8; PEG-9; PEG-12; PEG-14; PEG-16; PEG-20; PEG-32; PEG-40; PEG-75; PEG-100; PEG-150; PEG-200; PEG-350; PEG-160M; PEG-15 castor oil; PEG-16 castor oil; PEG-25 castor oil; PEG-30 castor oil; PEG-36 castor oil; PEG-40 castor oil; PEG-200 castor oil; PEG-8 dilaurate; PEG-12 dioleate; PEG-5 hydrogenated castor oil; PEG-16 hydrogenated castor oil; PEG-25 hydrogenated castor oil; PEG-12 laurate; PEG-2 oleate; PEG-2 stearate; PEG-4 stearate; PEG-8 stearate; PEG-12 stearate; Phenyltrimethoxysilane; Polyethylene, oxidized; Polyethylene wax, oxidized; Polyglyceryl-10 tetraoleate; Polysorbate 61; Propylene glycol myristate
Rapeseed oil
Silicone emulsions; Stearamide; N-Stearyl 12-hydroxystearamide; Stearyl methacrylate
Tallow propane diamine dioleate; Trilinoleic acid
Xanthan gum

Marproofing Agents. *See* Mar Resistance Aids

Mar Resistance Aids • Marproofing Agents

Diisodecyl adipate
Methyltriethoxysilane
PEG-6 trimethylolpropane
Trimethylolpropane ethoxy triacrylate
Vinyltrimethoxysilane; Vinyltris(2-methoxyethoxy) silane

Matte Agents. *See* Flatting Agents

Mildewcides. *See* Antimicrobials

Modifiers

Acrylonitrile-butadiene rubber; Allyl glycidyl ether; Allyl methacrylate
Bis (2,4-di-t-butylphenyl) pentaerythritol diphosphite
1,4-Cyclohexanedimethanol dibenzoate; Cyclohexyl methacrylate
Dimethyl caprylamide-capramide; Dimethyl maleate; Dimethyl oleamide
EPDM rubber; Ethylene glycol dimethacrylate
Formaldehyde/toluenesulfonamide polymer; Fumaric acid; Furan polymer
Glyceryl hydrogenated rosinate; Glyceryl rosinate; Glyceryl tribenzoate; Glycidyl neodecanoate
Methylcellulose
Neopentyl glycol dibenzoate
Polyester glutarate

Modifiers *(cont'd.)*

Ricinoleic acid
Sucrose acetate isobutyrate; Sucrose benzoate
Terpene resin; Toluene sulfonamide formaldehyde resin
Vinyl acetate/ethylene copolymer; Vinyl bromide

Opacifiers

Bismuth subcarbonate
PEG-2 laurate SE; PEG-20 oleate; PEG-20 stearate; Propylene glycol myristate
Titanium dioxide

Pigments • Colorants • Dyes

Aluminum silicate, calcined; Aluminum silicate, hydrous; Aluminum-zinc phosphate hydrate; Antimony trioxide; Antimony trisulfide; Arsenic pentasulfide
Barium chromate; Barium metaborate; Barium molybdate; Bismuth oxychloride; Bone black
Cadmium; Calcium chromate; Calcium fluoride; Calcium metaborate; Calcium phosphate dibasic; Calcium silicate; Calcium sulfate dihydrate; Calcium-zinc-molybdenum complex; Carbon black; Cashew nut shell oil; Chrome antimony titanate; Chromium oxide hydrated; Chromium oxide (ic); Chromium phosphate; CI Fluorescent Brightener 236; Cobalt; Cobalt oxide (ic); Cobalt oxide (ous); Copper nitrate (ic); Copper oxide (ous); Copper powder
Ferric oxide; Ferric phosphate; Ferro-aluminum silicate; Ferrous sulfate heptahydrate
Kaolinite
Lead carbonate (basic); Lead phosphite dibasic; Lead silicate; Lead silicochromate; Lead silicosulfate, basic; Lead sulfate; Lead sulfate, basic; Lead sulfate, blue basic; Lithopone
Magnesium hydroxide; Manganese violet; Meadowfoam seed oil; Mercury oxide (ic), red and yellow; Molybdenum; Molybdenum trioxide
Nickel titanate
Phenyl diisodecyl phosphite; Pigment red 202; Pigment yellow 14; Pigment yellow 53
Sodium chromate; Sodium molybdate; Strontium chromate
Titanium dioxide
Ultramarine blue; Ultramarine violet
Wollastonite
Zinc borate; Zinc chromate; Zinc dust; Zinc hydroxyphosphite; Zinc molybdate; Zinc molybdenum phosphate; Zinc silicophosphate; Zinc sulfide

Plasticizers

Acetylated palm kernel glycerides; Acetyl tributyl citrate; Acetyl triethyl citrate
Benzyl phthalate; Butoxytriglycol; Butyl acetyl ricinoleate; N,N-Butyl benzene sulfonamide; Butyl benzyl phthalate; Butyl cyclohexyl phthalate; Butyl octyl phthalate; 4-t-Butylphenol; n-Butyl phthalyl-n-butyl glycolate
Camphor; Castor oil; Castor oil, polymerized; Cetearyl methacrylate; Cumylphenyl acetate; 1,4-Cyclohexanedimethanol dibenzoate
Dialkyl dimethyl ammonium chloride; Diallyl phthalate; Dibutyl fumarate; Dibutyl maleate; Dibutyl phthalate; Dicyclohexyl phthalate; Diethylene glycol dibenzoate; Di-n-heptyl phthalate; Diisobutyl adipate; Diisobutyl phthalate; Diisodecyl adipate; Diisodecyl phthalate; Diisononyl phthalate; Diisotridecyl phthalate; Dimethyl phthalate; Dinonyl phthalate; Dioctyl adipate; Dioctyl dodecanedioate dioate; Dioctyl fumarate; Dioctyl maleate; Dioctyl phthalate; Dioctyl terephthalate; Diphenylcresyl phosphate
Epoxidized linseed oil; Epoxidized soybean oil; Epoxy resin; Ethyl toluenesulfonamide
Glyceryl hydrogenated rosinate; Glyceryl rosinate; Glyceryl stearate; Glyceryl (triacetoxystearate); Glyceryl triacetyl ricinoleate; Glyceryl tribenzoate; Glyceryl tris-12-hydroxystearate; Glycogen
Hydroabietyl alcohol; N-(2-Hydroxypropyl) benzenesulfonamide
Isobutyl stearate; Isodecyl benzoate; Isononyl alcohol; Isotridecyl methacrylate
Lanolin alcohol; Lauryl-myristyl methacrylate
Meroxapol 174; Methyl abietate; Methyl acetyl ricinoleate; Methylcellulose; Methyl hydrogenated rosinate; Methyl

Plasticizers *(cont'd.)*

hydroxyethylcellulose; Methyl linoleate
Neopentyl glycol dibenzoate
Oleyl alcohol
Paraffin, chlorinated; PEG-32 castor oil; PEG-75 castor oil; PEG-160 castor oil; PEG-3 diacetate; PEG-4 dibenzoate; PEG-4 di-2-ethylhexoate; PEG-5 hydrogenated castor oil; PEG-16 hydrogenated castor oil; PEG-25 hydrogenated castor oil; PEG-200 hydrogenated castor oil; PEG-2 stearate; Pentaerythrityl tetrabenzoate; Phenyl glycol ether; Poly-α-methylstyrene; Polypropylene glycol dibenzoate; Polyvinyl methyl ether; Propylene glycol dibenzoate; Propylene glycol myristate; Propylene glycol stearate; 2-Pyrrolidone
Sodium cetyl sulfate
p-Toluenesulfonamide; Triacetin; Triaryl phosphate; Tributoxyethyl phosphate; Tributyl citrate; Tributyl phosphate; Tricresyl phosphate; Triethyl citrate; Tri n-hexyl trimellitate; Triisopropylphenyl phosphate; Trimethyl-1,3-pentanediol, 2,2,4-diisobutyrate; Trioctyl phosphate; Trioctyl trimellitate; Trixylenyl phosphate
Urethane diacrylate
Xylenol
Zinc laurate

Preservatives

1,2-Benzisothiazolin-3-one; BHA
p-Chloro-m-cresol; Chloromethoxy propyl mercuric acetate; Chloroxylenol
2,2-Dibromo-3-nitrilopropionamide; Diiodomethyl p-tolyl sulfone; Dimethyl oxazolidine; Di (phenyl mercury) dodecenyl succinate
Hexahydro-1,3,5-triethyl-s-triazine; 2[(-Hydroxymethyl) amino] ethanol; 2-[(-Hydroxymethyl) amino]-2-methylpropanol; N-(Hydroxymethyl)-N-(1,3-dihydroxymethyl-2,5-dioxo-4-imidazolidinyl)-N-(hydroxymethyl) urea
Methylchloroisothiazolinone; Methyldibromo glutaronitrile; Methylisothiazolinone; 1-Methyl-3,5,7-triaza-1-azoniatricyclo-[3.3.1.1]decane chloride
Quaternium-15
Sodium benzoate; Sodium 2-mercaptobenzothiazole
Tetrachloroisophthalonitrile
Zinc dimethyldithiocarbamate

Processing Aids

t-Butyl alcohol
Castor oil, polymerized; C5 hydrocarbon resin
Dimethylethanolamine
Ethylene/calcium acrylate copolymer
Glyceryl tribenzoate
Methyl hydroxystearate
Neopentyl glycol dibenzoate
Pentaerythrityl tetrastearate
Rapeseed oil, vulcanized
Styrene-butadiene rubber; Synthetic wax

Resin Raw Materials

Acetal copolymer; Acrylic, thermosetting
Bisphenol A
Carboxymethylcellulose; Cashew nut shell oil
1,2-Diaminocyclohexane
Epoxidized 1,2-polybutadiene; Epoxy novolac; Ethylene vinyl alcohol copolymer
Furfuryl alcohol
Glycerin
Hexamethylol melamine resin; Hydroabietyl phthalate; Hydrogenated rosin

Resin Raw Materials *(cont'd.)*

Isobutyl methacrylate
Maleic anhydride; Methoxy methyl melamine resin; Methyl methacrylate polymer; Methyl rosinate; Methyl siloxane
Nylon; Nylon-66; Nylon 11; Nylon-12
Octabromodiphenyl oxide
PC/PET; PEG-5M; PEG-7M; PEG-9M; PEG-14M; PEG-23M; PEG-45M; PEG-90M; Pentaerythrityl hydrogenated rosinate; Pentaerythrityl rosinate; Phenolic resin; Polyamide-imide; 1,2-Polybutadiene; Polybutadiene maleic anhydride polymer; Polybutadiene-styrene copolymer; Polybutene; Polycarbonate resin; Polydipentene; Polyester resin, thermosetting; Polyetheretherketone; Polyethylene, medium density; Polyimide, thermoplastic; Polystyrene; Polyvinyl acetate (homopolymer); Polyvinyl ethyl ether; Polyvinyl isobutyl ether; PVC/PVAc copolymer
Safflower oil; Shellac; Silicone-polyester
Terpene resin, natural; Tetrafluoroethylene/hexafluoropropylene copolymer; Tetrafluoroethylene/propylene copolymer; Tetrahydrofurfuryl acrylate; Triazine formaldehyde; 2,2,4-Trimethyl-1,3-pentanediol
Urea-formaldehyde resin
Vinyl butadiene; Vinyl ester resin; Vinyl homopolymer

Rheology Control Agents. *See* Thickeners

Slip Agents

Butyl stearate
Cetyl esters
Dimethicone copolyol
Erucamide; N,N′-Ethylenebis 12-hydroxystearamide; N,N′-Ethylene bis-ricinoleamide
Glycol hydroxystearate
Hydrogenated tallow amide; N (2-Hydroxyethyl) 12-hydroxystearamide; N(β-Hydroxyethyl) ricinoleamide
Lanolin
Oleamide
Rapeseed oil
Silicone hexaacrylate; Stearamide; N-Stearyl 12-hydroxystearamide

Solubilizers

Amine dodecylbenzene sulfonate
Butylated PVP
Ceteareth-11
Dihexyl sodium sulfosuccinate; 2-Dimethylamino-2-methyl-1-propanol; Disodium nonoxynol-10 sulfosuccinate
Glyceryl ricinoleate
Isopropylamine dodecylbenzenesulfonate
Laureth-2; Laureth-4
Meroxapol 105; Meroxapol 171; Meroxapol 172; Meroxapol 174; Meroxapol 252; Meroxapol 254; Meroxapol 258; Meroxapol 311
Nonyl nonoxynol-8; Nonyl nonoxynol-18; Nonyl nonoxynol-49
PEG-15 castor oil; PEG-20 castor oil; PEG-80 castor oil; PEG-8 dilaurate; PEG-9 dioleate; PEG-8 ditallate; PEG-12 ditallate; PEG-8 laurate; PEG-12 laurate; PEG-4 oleate; PEG-8 oleate; PEG-12 oleate; PEG-20 oleate; PEG-20 stearate; PEG-66 trihydroxystearin; Poloxamine 304; Poloxamine 701; Poloxamine 704; Poloxamine 901; Poloxamine 904; Poloxamine 908; Poloxamine 1107; Poloxamine 1307; Polyacrylamide; Polysorbate 20; Polysorbate 40; Polysorbate 61; Polysorbate 65; PVM/MA copolymer
Sodium cumenesulfonate; Sodium hexyl diphenyloxide disulfonate
Tetrapotassium pyrophosphate; Tetrasodium dicarboxyethyl stearyl sulfosuccinamate; Trideceth-15

Solvents

Acetone; Acetophenone; Acetylacetone; Acetylated palm kernel glycerides; Alcohol; Alcohol denatured; Ammonium xylenesulfonate; Amyl acetate; Amyl alcohol; Aromatic 100; Aromatic 150

Solvents *(cont'd.)*

Benzene; Benzonitrile; Benzyl acetate; Bis(methylethyl)-1,1′-biphenyl; 2-Butanol; Butoxydiglycol; Butoxyethanol; Butoxyethanol acetate; Butoxypropanol; n-Butyl acetate; s-Butyl acetate; Butyl alcohol; t-Butyl alcohol; n-Butyl propionate; Butyl stearate

C8 alkyl acetate; C9 alkyl acetate; C10 alkyl acetate; C13 alkyl acetate; ε-Caprolactone monomer; Caprylic alcohol; Castor oil, polymerized; Chlorobenzene; p-Chlorobenzotrifluoride; C7-8 isoparaffin; C10-11 isoparaffin; C11-12 isoparaffin; C11-13 isoparaffin; C13-14 isoparaffin; C8-9 isoparaffin; Copper oxide (ic); Corn oil; Cyclohexane; Cyclohexanone; N-Cyclohexyl pyrrolidone

Decahydronaphthalene; Diacetone alcohol; Dibutyl ether; Diethylene glycol butyl ether acetate; Diethylene glycol dibutyl ether; Diethylene glycol diethyl ether; Diethylene glycol dimethyl ether; Diethylene glycol n-hexyl ether; Diethylene glycol propyl ether; Diethylenetriamine; Diethyl ketone; Diisoamyl ketone; Diisobutyl ketone; Dimethyl adipate; Dimethyl ether; Dimethyl formamide; Dimethyl glutarate; Dimethyl oleamide; Dimethyl phthalate; Dimethyl succinate; 1,4-Dioxane; Dipropylene glycol; Dipropylene glycol butyl ether; Dipropylene glycol t-butyl ether; Dipropylene glycol dimethyl ether; Dipropylene glycol n-propyl ether

Ethoxydiglycol; Ethoxydiglycol acetate; Ethoxyethanol; Ethoxyethanol acetate; Ethyl acetate; Ethylbenzene; Ethylenediamine; Ethylene dichloride; Ethylene glycol diacetate; Ethylene glycol diethyl ether; Ethylene glycol dimethyl ether; Ethylene glycol hexyl ether; Ethylene glycol propyl ether; Ethyl 3-epoxypropionate; Ethyl ether; Ethyl 3-ethoxypropionate; 2-Ethylhexanol; 2-Ethylhexyl acetate; Ethyl propionate

Formic acid

Glycol; Glycol dilaurate

Heptane; Hexyl acetate; Hexyl alcohol; Hydrocarbon solvent

Isoamyl alcohol; Isobutyl acetate; Isobutyl alcohol; Isobutyl heptyl ketone; Isobutyl isobutyrate; Isooctyl acid phosphate; 1-Isopropoxy-2-propanol; 2-Isopropoxy-1-propanol; Isopropyl acetate; Isopropyl alcohol; Isopropyl ether; Isopropyl lactate

Lauryl pyrrolidone; dl-Limonene

2-Mercaptoethanol; Mesityl oxide; Methoxybutyl acetate; Methoxydiglycol; Methoxyethanol; Methoxyethanol acetate; Methoxyisopropanol; 4-Methoxy-4-methyl-pentanone-2; Methyl acetate; Methyl alcohol; Methyl amyl acetate; Methyl amyl alcohol; Methyl n-amyl ketone; Methyl t-butyl ether; Methylene chloride; Methyl ethyl ketone; Methyl hexyl ketone; Methyl isoamyl ketone; Methyl isobutyl ketone; Methyl lactate; Methyl propyl ketone; N-Methyl-2-pyrrolidone; 2-Methyl-2-pyrrolidone; Mineral spirits; Monochlorotoluene

Naphtha; Naphthenic acid; Nitroethane; Nitromethane; Nitropropane; 2-Nitropropane

Oxo-heptyl acetate; Oxo-hexyl acetate

n-Pentyl propionate; Perchloroethylene; Petroleum distillates; Phenoxydiglycol; Phenoxyethanol; 2-Phenoxypropanol-1; α-Pinene; Polysorbate 60; PPG (400); PPG-2 methyl ether; PPG-3 methyl ether; PPG-2 methyl ether acetate; Propyl acetate; n-Propyl alcohol; Propylene carbonate; Propylene glycol; Propylene glycol t-butyl ether; Propylene glycol ethyl ether; Propylene glycol ethyl ether acetate; Propylene glycol methyl ether acetate; Propylene glycol phenyl ether; Propylene glycol n-propyl ether; Propyl propionate

Stoddard solvent

Tetraethylenepentamine; Tetrahydrofuran; Tetrahydrofurfuryl alcohol; Tetrahydronaphthalene; Toluene; Tributyl citrate; Tridecyl alcohol; Triethylamine; Triethyl citrate; Triglycol monomethyl ether; 2,2,4-Trimethyl-1,3-pentanediol monoisobutyrate; Trioctyl phosphate; Turpentine oil

Xylene; o-Xylene; Xylenol

Stabilizers

Acetone; Acetophenone; Acetylacetone; Acetylated palm kernel glycerides; Alcohol; Alcohol denatured; Ammonium xylenesulfonate; Amyl acetate; Amyl alcohol; Aromatic 100; Aromatic 150

Benzene; Benzonitrile; Benzyl acetate; Bis(methylethyl)-1,1′-biphenyl; 2-Butanol; Butoxydiglycol; Butoxyethanol; Butoxyethanol acetate; Butoxypropanol; n-Butyl acetate; s-Butyl acetate; Butyl alcohol; t-Butyl alcohol; n-Butyl propionate; Butyl stearate

C8 alkyl acetate; C9 alkyl acetate; C10 alkyl acetate; C13 alkyl acetate; ε-Caprolactone monomer; Caprylic alcohol; Castor oil, polymerized; Chlorobenzene; p-Chlorobenzotrifluoride; C7-8 isoparaffin; C10-11 isoparaffin; C11-12 isoparaffin; C11-13 isoparaffin; C13-14 isoparaffin; C8-9 isoparaffin; Copper oxide (ic); Corn oil; Cyclohexane; Cyclohexanone; N-Cyclohexyl pyrrolidone

Decahydronaphthalene; Diacetone alcohol; Dibutyl ether; Diethylene glycol butyl ether acetate; Diethylene glycol dibutyl ether; Diethylene glycol diethyl ether; Diethylene glycol dimethyl ether; Diethylene glycol n-hexyl ether; Diethylene glycol propyl ether; Diethylenetriamine; Diethyl ketone; Diisoamyl ketone; Diisobutyl ketone; Dimethyl adipate; Dimethyl ether; Dimethyl formamide; Dimethyl glutarate; Dimethyl oleamide; Dimethyl phthalate; Dimethyl succinate; 1,4-Dioxane; Dipropylene glycol; Dipropylene glycol butyl ether; Dipropylene glycol t-butyl ether; Dipropylene glycol dimethyl ether; Dipropylene glycol n-propyl ether

Stabilizers *(cont'd.)*

Ethoxydiglycol; Ethoxydiglycol acetate; Ethoxyethanol; Ethoxyethanol acetate; Ethyl acetate; Ethylbenzene; Ethylenediamine; Ethylene dichloride; Ethylene glycol diacetate; Ethylene glycol diethyl ether; Ethylene glycol dimethyl ether; Ethylene glycol hexyl ether; Ethylene glycol propyl ether; Ethyl 3-epoxypropionate; Ethyl ether; Ethyl 3-ethoxypropionate; 2-Ethylhexanol; 2-Ethylhexyl acetate; Ethyl propionate

Formic acid

Glycol; Glycol dilaurate

Heptane; Hexyl acetate; Hexyl alcohol; Hydrocarbon solvent

Isoamyl alcohol; Isobutyl acetate; Isobutyl alcohol; Isobutyl heptyl ketone; Isobutyl isobutyrate; Isooctyl acid phosphate; 1-Isopropoxy-2-propanol; 2-Isopropoxy-1-propanol; Isopropyl acetate; Isopropyl alcohol; Isopropyl ether; Isopropyl lactate

Lauryl pyrrolidone; dl-Limonene

2-Mercaptoethanol; Mesityl oxide; Methoxybutyl acetate; Methoxydiglycol; Methoxyethanol; Methoxyethanol acetate; Methoxyisopropanol; 4-Methoxy-4-methyl-pentanone-2; Methyl acetate; Methyl alcohol; Methyl amyl acetate; Methyl amyl alcohol; Methyl n-amyl ketone; Methyl t-butyl ether; Methylene chloride; Methyl ethyl ketone; Methyl hexyl ketone; Methyl isoamyl ketone; Methyl isobutyl ketone; Methyl lactate; Methyl propyl ketone; N-Methyl-2-pyrrolidone; 2-Methyl-2-pyrrolidone; Mineral spirits; Monochlorotoluene

Naphtha; Naphthenic acid; Nitroethane; Nitromethane; Nitropropane; 2-Nitropropane

Oxo-heptyl acetate; Oxo-hexyl acetate

n-Pentyl propionate; Perchloroethylene; Petroleum distillates; Phenoxydiglycol; Phenoxyethanol; 2-Phenoxypropanol-1; α-Pinene; Polysorbate 60; PPG (400); PPG-2 methyl ether; PPG-3 methyl ether; PPG-2 methyl ether acetate; Propyl acetate; n-Propyl alcohol; Propylene carbonate; Propylene glycol; Propylene glycol t-butyl ether; Propylene glycol ethyl ether; Propylene glycol ethyl ether acetate; Propylene glycol methyl ether acetate; Propylene glycol phenyl ether; Propylene glycol n-propyl ether; Propyl propionate

Stoddard solvent

Tetraethylenepentamine; Tetrahydrofuran; Tetrahydrofurfuryl alcohol; Tetrahydronaphthalene; Toluene; Tributyl citrate; Tridecyl alcohol; Triethylamine; Triethyl citrate; Triglycol monomethyl ether; 2,2,4-Trimethyl-1,3-pentanediol monoisobutyrate; Trioctyl phosphate; Turpentine oil

Xylene; o-Xylene; Xylenol

Surfactants

C30-50 alcohols; Citric acid esters of mono- and diglycerides of fatty acids; Cocamine acetate

Dimethyl hexynol; Dioctyl phosphate; Disodium nonoxynol-10 sulfosuccinate; Dodecylbenzene sulfonic acid

2-Ethylhexyl hydrogenated tallowalkyl methosulfate

Glycol distearate; Glycol stearate

Hexyl phosphate

Nonoxynol-1; Nonoxynol-2; Nonoxynol-3; Nonoxynol-4; Nonoxynol-5; Nonoxynol-6; Nonoxynol-8; Nonoxynol-9; Nonoxynol-11; Nonoxynol-13; Nonoxynol-14; Nonoxynol-15; Nonoxynol-20; Nonoxynol-40; Nonoxynol-50; Nonoxynol-70; Nonoxynol-100

Octoxynol-1; Octoxynol-3; Octoxynol-4; Octoxynol-5; Octoxynol-7; Octoxynol-8; Octoxynol-9; Octoxynol-13; Octoxynol-16; Octoxynol-30; Octoxynol-70

PEG-52 castor oil; PEG-75 castor oil; PEG-160 castor oil; PEG-9 laurate; PEG-8 pelargonate; 1,3-Pentanediamine; Poloxamer 101; Poloxamer 105; Poloxamer 108; Poloxamer 123; Poloxamer 124; Poloxamer 181; Poloxamer 182; Poloxamer 184; Poloxamer 185; Poloxamer 188; Poloxamer 217; Poloxamer 231; Poloxamer 234; Poloxamer 235; Poloxamer 237; Poloxamer 282; Poloxamer 288; Poloxamer 331; Poloxamer 333; Poloxamer 334; Poloxamer 335; Poloxamer 338; Poloxamer 401; Poloxamer 402; Poloxamer 403; Poloxamer 407; Poloxamine 304; Poloxamine 701; Poloxamine 704; Poloxamine 901; Poloxamine 904; Poloxamine 908; Poloxamine 1107; Poloxamine 1307; Polysorbate 85; Potassium dodecylbenzene sulfonate; PPG-24-buteth-27; Propylene glycol palmitate

Quaternium-18

Silicone glycol copolymer; Sodium allyloxy hydroxypropyl sulfonate; Sodium hexyl diphenyloxide disulfonate; Sodium α olefin sulfonate; Sulfated vegetable oil

Tetrasodium dicarboxyethyl stearyl sulfosuccinamate; Tricetylmonium chloride

Suspending Agents

Butylated PVP

Carbomer; Carboxymethylcellulose sodium

Suspending Agents *(cont'd.)*

Dihydrogenated tallow benzylmonium hectorite
Glyceryl tris-12-hydroxystearate
Hectorite
Locust bean gum
Magnesium aluminum silicate
Polyacrylic acid; Propylene glycol alginate; PVP; PVP/VA copolymer
Sodium bentonite; Sodium methylnaphthalenesulfonate
Xanthan gum

Tackifiers

2-(Acetoacetoxy) ethyl methacrylate
C5 hydrocarbon resin
Glyceryl rosinate
Methyl hydrogenated rosinate
Pentaerythrityl hydrogenated rosinate; Pentaerythrityl rosinate; Polyvinyl methyl ether

Thickeners • Bodying Agents • Gelling Agents • Rheology Control Agents • Viscosity Modifiers

Aluminum distearate; Aluminum 2-ethyl hexanate; Aluminum naphthenate; Ammonium naphthalene sulfonate; Ammonium polyacrylate; Ammonium polymethacrylate; Attapulgite
Butadiene/acrylonitrile copolymer
Calcium sulfonate; C20-40 alcohols; Candelilla synthetic; Carbomer; Cellulose; Cetyl esters
Diatomaceous earth, amorphous; Dihydrogenated tallow benzylmonium hectorite; Dimethylaminoethyl methacrylate
Ethyl hydroxyethyl cellulose
Fluorinated ethylene-propylene
Gelatin; Glyceryl stearate; Glyceryl tris-12-hydroxystearate
Hectorite; Hydroxyethylcellulose; Hydroxypropylcellulose; Hydroxypropyl methylcellulose
Isodecyl acrylate; Isodecyl methacrylate; Isotridecyl methacrylate
Lauryl acrylate; Lithopone; Locust bean gum
Magnesium aluminum silicate; Meroxapol 105; Meroxapol 171; Meroxapol 172; Meroxapol 174; Meroxapol 252; Meroxapol 254; Meroxapol 258; Meroxapol 311; Methacrylate copolymer; Methylcellulose; Methyl hydroxyethylcellulose; Methyl rosinate; Montmorillonite
Paraffin, brominated; PEG-15 castor oil; PEG-12 diacrylate; PEG-150 distearate; PEG-2 laurate SE; PEG-4 laurate; PEG-5 laurate; PEG-20 oleate; PEG-2 stearate; PEG-12 stearate; 2-Phenoxyethyl acrylate; Poloxamine 304; Poloxamine 701; Poloxamine 704; Poloxamine 901; Poloxamine 904; Poloxamine 908; Poloxamine 1107; Polyacrylamide; Polyacrylic acid; Polyethylene wax; Polyglyceryl-10 tetraoleate; Polyisobutene; Polyquaternium-31; Potassium silicate; Propylene glycol alginate; PVM/MA copolymer; PVP; PVP/VA copolymer
Quaternium-18 bentonite; Quaternium-18 hectorite
Silica dimethyl silylate; Silica, fumed; Sodium C14-16 olefin sulfonate; Sodium polyacrylate; Stearalkonium bentonite; Stearalkonium hectorite; Synthetic beeswax
Trihydroxystearin
Xanthan gum

UV Absorbers • UV Stabilizers

Barium-zinc complex; Benzil dimethyl ketal; Benzoin ether; Benzophenone; Benzophenone-1; Benzophenone-3; Benzophenone-6; Benzophenone-8; Benzophenone-9; Benzophenone-11; Benzophenone-12; Benzyldimethyl ketal; Bis-(1-octyloxy-2,2,6,6,tetramethyl-4-piperidinyl) sebacate; Bis (1,2,2,6,6-pentamethyl-4-piperidinyl) (3,5-di-t-butyl-4-hydroxybenzyl) butyl propanedioate; Bumetrizole
C20-40 alcohols; Carbon black; 2-Chlorothioxanthone; Cyclodextrin; β-Cyclodextrin
Diaryl iodonium hexafluoroantimonate; 2-(3′,5′-Di-t-butyl-2′-hydroxyphenyl)-5-chlorobenzotriazole; 2,2-

UV Absorbers *(cont'd.)*

Diethoxyacetophenone; 2,2-Dimethoxy-2-phenylacetophenone; Dimethyl succinate; Diphenyl isodecyl phosphite; Diphenyl isooctyl phosphite; Drometrizole
Epoxidized soybean oil acrylate; Ethane diamide, n-(2-ethoxyphenyl)-n′-(4-ethylphenyl); Ethane diamide, n-(2-ethoxyphenyl)-n′-(4-isododecyl phenyl); N-(p-Ethoxycarbonylphenyl)-N′-ethyl-N′-phenylformamidine; Ethyl 4-(dimethylamino) benzoate; 2-Ethyl, 2′-ethoxy-oxalanilide; Etocrylene
Hydroquinone; Hydroquinone monomethyl ether; 2-Hydroxybenzophenone; 2-(2′-Hydroxy-3,5′-di-t-amylphenyl) benzotriazole; Hydroxypropyl-β-cyclodextrin; Hydroxypropyl-γ-cyclodextrin
Isobornyl acrylate; Isobutyl benzoin ether; Isopropyl thioxanthone
Lead carbonate (basic); Lead phosphite dibasic; Lead phthalate, basic; Lead silicate
α-Methylstyrene/N-(2,2,6,6-tetramethyl piperidinyl-4) maleimide/N-stearyl maleimide terpolymer
Octocrylene; Octrizole; Octyl dimethyl PABA
Phenol, 2-(2H-benzotriazole-2-yl)-4-methyl-6-dodecyl
2,2,6,6-Tetramethylpiperidin-4-yl acrylate/methyl methacrylate copolymer; 2,2′-Thiobis (4-t-octylphenolato)-n-butylamine nickel; 1,3,5-Trimethylbenzene; 1,1,1-Trimethylolpropane triacrylate; 1,1,1-Tris(hydroxyphenylethane)benzotriazole
Zinc oxide

UV Stabilizers. *See* Thickeners

Vehicles. *See* Carriers

Viscosity Modifiers. *See* Thickeners

Water Repellents

Calcium resinate; Calcium stearate
Sodium methyl silanolate; Sodium methyl siliconate; Styrene allyl alcohol

Waxes

Ethylene dioleamide
Glycol hydroxystearate
Microcrystalline wax; Microcrystalline wax, oxidized
Polyethylene, oxidized; Polyethylene wax

Wetting Agents

Amine dodecylbenzene sulfonate
Behenyl hydroxyethyl imidazoline
Capryl hydroxyethyl imidazoline; Caprylic alcohol; Castor oil, polymerized; Ceteareth-12; Ceteareth-27; Ceteareth-30; Ceteareth-55; Cetoleth-5; Cetyl alcohol; Cocaminobutyric acid
Dihexyl sodium sulfosuccinate; Diisobutyl sodium sulfosuccinate; Dimethyl hexynol; Dipentaerythrityl pentaacrylate; Dipropylene glycol dibenzoate; Disodium nonoxynol-10 sulfosuccinate; Ditrimethylolpropane triacrylate
Epoxidized linseed oil acrylate; Epoxidized soybean oil acrylate
Glyceryl hydroxystearate; Glyceryl oleate
Heptadecenyl hydroxyethyl imidazoline; Hydroxylated lecithin
Isodecyl oxypropyl amine acetate; Isopropylamine dodecylbenzenesulfonate
Laureth-5; Laureth-8; Laureth-9

Wetting Agents *(cont'd.)*

Meroxapol 105; Meroxapol 171; Meroxapol 172; Meroxapol 174; Meroxapol 252; Meroxapol 254; Meroxapol 258; Meroxapol 311

Nonoxynol-1; Nonoxynol-2; Nonoxynol-3; Nonoxynol-4; Nonoxynol-5; Nonoxynol-6; Nonoxynol-8; Nonoxynol-9; Nonoxynol-10; Nonoxynol-11; Nonoxynol-12; Nonoxynol-13; Nonoxynol-14; Nonoxynol-15; Nonoxynol-20; Nonoxynol-30; Nonoxynol-40; Nonoxynol-50; Nonoxynol-70; Nonoxynol-100; Nonoxynol-10 carboxylic acid

Oleoyl sarcosine; Oleyl hydroxyethyl imidazoline

PEG-32 castor oil; PEG-36 castor oil; PEG-75 castor oil; PEG-160 castor oil; PEG-10 cocamine; PEG-8 dilaurate; PEG-5 hydrogenated castor oil; PEG-16 hydrogenated castor oil; PEG-25 hydrogenated castor oil; PEG-200 hydrogenated castor oil; PEG-8 isolauryl thioether; PEG-4 laurate; PEG-5 laurate; PEG-8 laurate; PEG-4 oleate; PEG-80 sorbitan laurate; PEG-2 stearate; PEG-20 stearate; PEG-2 tallowamine; PEG-5 tallowamine; PEG-15 tallowamine; PEG-20 tallowamine; PEG-10 tallow aminopropylamine; PEG-3 tallow propylene diamine; PEG-7 tallow propylene diamine; PEG-12 tallow propylene diamine; PEG-1.3 tetramethyl decynediol; PEG-3.5 tetramethyl decynediol; PEG-10 tetramethyl decynediol; PEG-30 tetramethyl decynediol; PEG-6 trimethylolpropane; Perfluoroheptane; Perfluorohexane; Perfluoro-N-methylmorpholine; Perfluoropentane; Poloxamer 101; Poloxamer 105; Poloxamer 108; Poloxamer 123; Poloxamer 124; Poloxamer 181; Poloxamer 182; Poloxamer 184; Poloxamer 185; Poloxamer 188; Poloxamer 217; Poloxamer 231; Poloxamer 234; Poloxamer 235; Poloxamer 237; Poloxamer 282; Poloxamer 288; Poloxamer 331; Poloxamer 333; Poloxamer 334; Poloxamer 335; Poloxamer 338; Poloxamer 401; Poloxamer 402; Poloxamer 403; Poloxamer 407; Poloxamine 304; Poloxamine 701; Poloxamine 704; Poloxamine 901; Poloxamine 904; Poloxamine 908; Poloxamine 1107; Poloxamine 1307; Polyester hexaacrylate; Polyester tetraacrylate; Propylene glycol laurate; Propylene glycol myristate

Silicone hexaacrylate; Sodium decyl sulfate; Sodium dibutyl naphthalene sulfonate; Sodium dinonyl sulfosuccinate; Sodium dodecylbenzenesulfonate; Sodium isodecyl sulfate; Sodium isopropyl naphthalene sulfonate; Sodium lauroamphoacetate; Sodium lauryl sulfate; Sodium nonoxynol-6 phosphate; Sodium nonoxynol-4 sulfate; Stearamine acetate; Steareth-2; Steareth-10; Steareth-20; Sulfonic acid

Tallow dipropylene triamine; Tallow propylene diamine; TEA-dodecylbenzenesulfonate; Tetrahydroxypropyl ethylenediamine; Tetramethyl decynediol; Trideceth-6; Trideceth-12

Zinc naphthenate

Part III
Manufacturers Directory

Manufacturers Directory

AAA Molybdenum Products, Inc.

7233 W. 116th Place, Broomfield, CO 88020 USA (Tel.: 303-460-0844, 800-443-6812; Telefax: 303-460-0851)

Aabbitt Adhesives Inc.

2403 N. Oakley Ave., Chicago, IL 60647 USA (Tel.: 312-227-2700, 800-222-2488; Telefax: 312-227-2103)

Aarbor International Corp.

9434 Maltby Rd., Brighton, MI 48116 USA (Tel.: 810-220-0080, 800-4-PIGMENTS; Telefax: 810-220-0088)

Aarhus

Aarhus Oliefabrik A/S, Postboks 50, Bruunsgade 27, DK-8100 Aarhus C Denmark (Tel.: 86-12 60 00; Telefax: 86-183839; Telex: 64341)

Aarhus Inc., 131 Marsh St., PO Box 4240, Newark, NJ 07114 USA (Tel.: 201-344-1300; Telefax: 201-344-9049)

ABITEC Corp., Subsid. of Associated British Foods (ABF)

525 W. First Ave., PO Box 569, Columbus, OH 43216-0569 USA (Tel.: 614-299-3131, 800-848-1340; Telefax: 614-299-8279; Telex: 245494 capctyprdcol)

Ablestik Laboratories

2021 Susanna Rd., Dominguez, CA 90221 USA (Tel.: 310-764-4600; Telefax: 310-764-2545)

Accurate Chemical & Scientific Corp.

300 Shames Dr., Westbury, NY 11590 USA (Tel.: 516-333-2221, 800-645-6264; Telefax: 516-997-4948; Telex: 4972582)

Aceto Corporation

1 Hollow Lane, Suite 201, Lake Success, NY 11042-1215 USA (Tel.: 516-627-6000; Telefax: 516-627-6093; Telex: 62662)

Acheson

Acheson Colloids Co., Div. of Acheson Industries, Inc., 1600 Washington Ave., PO Box 611747, Port Huron, MI 48061-1747 USA (Tel.: 810-984-5581, 800-255-1908; Telefax: 810-984-1446)

Acheson Colloids (Canada) Ltd., PO Box 665, Shaver St., Brantford, Ontario, N3T 5P9 Canada (Tel.: 519-752-5461)

Acheson do Brasil Ind. e Com. Ltda., Rua Howard A. Acheson Jr., 279, Cotia-SP CEP 06700 Brazil (Tel.: (011) 492-4000)

Acheson Industries (Europe) Ltd., Sun Life House, 85 Queens Rd., Reading, Berkshire, RG1 4PT UK (Tel.: 44 1734 588844; Telefax: 44 1734-574897)

Deutsche Acheson Colloids, Postfach 1165, 89001 Ulm Germany (Tel.: (0731) 966930)

Acheson Colloiden B.V., PO Box 1, 9679 ZG Scheemda The Netherlands (Tel.: 5979-1303)

Acheson Italiana s.r.l., Via G. Fruia 18, 20146 Milano MI Italy (Tel.: 02 4815285)

Acheson A.N.Z. Pty. Ltd., PO Box 98, Revesby, N.S.W., 2212 Australia (Tel.: (02) 755-3099)

Acheson (Japan) Ltd., PO Box 538, Kobe Port, Kobe, 651-01 Japan (Tel.: (078) 332-3601; Telefax: (078) 391-5903; Telex: 5623230)

AC Industries, Inc., Sattva Chemical Co. Div.

2 Landmark Sq., Suite 214, Stamford, CT 06901-2410 USA (Tel.: 203-348-8002, 800-736-7893; Telefax: 203-348-3666; Telex: 6819048 SATVA UW)

ADM

ADM Corn Processing, Div. Archer Daniels Midland Co., Box 1470, Decatur, IL 62525 USA (Tel.: 217-424-5200, 800-323-0735; Telefax: 217-424-5978)

ADM Lecithin, Div. Archer Daniels Midland Co., 4666 Faries Pkwy., Box 1470, Decatur, IL 62525 USA (Tel.: 217-424-5898, 800-637-5843; Telefax: 217-424-4119; Telex: 190021)

Advance Coatings, Inc.

Depot Rd., Westminster, MA 01473 USA (Tel.: 508-874-5921; Telefax: 508-874-2788)

Advanced Polymer, Inc.

654 Gotham Pkwy., Carlstadt, NJ 07601 USA (Tel.: 201-933-0600, 800-882-9940; Telefax: 201-933-0600)

Advanced Refractory Technologies, Inc.

699 Hertel Ave., Buffalo, NY 14207-2341 USA (Tel.: 716-875-4091; Telefax: 716-875-0106)

Advanced Web Products, Inc.

PO Box 117, Ingomar, PA 15127 USA (Tel.: 412-367-8820; Telefax: 412-367-8848)

Advance Research Chemicals Inc.

5005 N. Skiatook Rd., Catossa, OK 74015 USA (Tel.: 918-266-6789; Telefax: 918-266-6796; Telex: 74-8086)

AECI Aroma & Fine Chemicals (Pty) Ltd., Subsid. of AECI Ltd.

Private bag X2, Modderfontein 1620, Rep. of S. Africa (Tel.: 27 11 605 2188; Telefax: 27 11 608 3314)

Agri-Pharm Inc.

4740 La Rueda Dr., La Mesa, CA 91941 USA (Tel.: 619-441-9109; Telefax: 619-441-2739; Telex: 650-552-1658)

Aichi Silicate Chemical Co., Ltd.

92-2, Bogane-cho, Seto-shi, Aichi, 489 Japan (Tel.: (0561) 83-8711; Telefax: (0561) 83-0561)

Air Products and Chemicals

Air Products and Chemicals, Inc., 7201 Hamilton Blvd., Allentown, PA 18195-1501 USA (Tel.: 610-481-4911, 800-345-3148; Telefax: 610-481-5900; Telex: 847416)

Air Products and Chemicals, Inc./Performance Chemicals, 7201 Hamilton Blvd., Allentown, PA 18195-1501 USA (Tel.: 610-481-6799, 800-345-3148; Telefax: 610-481-2276; Telex: 847416)

Air Products and Chemicals, Inc./Polymer Chemicals Div., 7201 Hamilton Blvd., Allentown, PA 18195-1501 USA (Tel.: 610-481-6799, 800-345-3148; Telefax: 610-481-4381)

Air Products and Chemicals, Inc./Polyurethane Chemicals Div., 7201 Hamilton Blvd., Allentown, PA 18195-1501 USA (Tel.: 610-481-6799, 800-345-3148; Telefax: 610-481-4381)

Air Products, Valchem Div., 403 Carline Rd., Langley, SC 29834 USA (Tel.: 803-593-4466)

Air Products and Chemicals de Mexico, S.A. de C.V., Rio Guadiana 23, Piso 5, Colonia Cuauhtemoc, Mexico D.F., 06500 Mexico (Tel.: 525-591-0800; Telefax: 525-592-3018)

Air Products Nederland B.V., Kanaalweg 15, PO Box 3193, 3502 GD Utrecht The Netherlands (Tel.: 31-30-857100; Telefax: 31-30 857111)

Air Products and Chemicals PURA GmbH & Co., Postfach 5108, Robert-Koch-Strasse 27, D-2000 Norderstedt Germany (Tel.: 49 40 529009-0; Telefax: 49 40 52900999; Telex: 403975 PURA)

Air Products Asia, Inc., Room 1901, Peregrine Tower, Lippo Centre, 89 Queensway, Central Hong Kong (Tel.: (852)524-7110; Telefax: (852) 877-0094)

Air Products Japan, Inc., Shuwa No. 2 Kamiyacho Bldg., 3-18-19, Toranomon, Minato-ku, Tokyo, 105 Japan (Tel.: (81)(3)3432-7046; Telefax: (81)(3)3432-7048)

Air Products China, Inc., Rm. 951 Beijing New Century Hotel Office Tower, No. 6 Southern Rd. Capital Gym, Beijing, 100046, China (Tel.: 86 10 849 2301; Telefax: 86 10 849 2303)

Air-Scent Int'l.

PO Box 1000, 215 Eighth Ave., Braddock, PA 15104 USA (Tel.: 412-351-2100, 800-247-0770; Telefax: 412-351-7701)

Aiscondel SA

Aragon 182, E-08011 Barcelona Spain (Tel.: 343-323-1020; Telefax: 343-323-7921; Telex: 97887 ETIN E)

Ajinomoto

Ajinomoto Co., Inc., 15-1, Kyobashi 1-chome, Chuo-ku, Tokyo, 104 Japan (Tel.: (03) 5250-8111; Telex: J22690)

Ajinomoto USA, Inc., Glenpointe Centre West, 500 Frank W. Burr Blvd., Teaneck, NJ 07666-6894 USA (Tel.: 201-488-1212; Telefax: 201-488-6472; Telex: 275425 (AJNJ)

Akcros

Akcros Chemicals, Lankro House, PO Box 1, Eccles, Manchester, M30 0BH UK (Tel.: 44 161 789 7300; Telefax: 44 161 788 7886; Telex: 587135)

Akcros Chemicals France S.A., Zone Industrielle BP 40, 441220 St Laurent-Nouan France (Tel.: 33 54 87 73 89; Telefax: 33 54 87 79 40; Telex: 750679 Tinstab)

Akcros Chemicals GmbH & Co. KG, Chemiewerk Greiz-Dolau, Liebigstrasse 7, 07973 Greiz Germany (Tel.: 49 36 61 780; Telefax: 49 36 61 782 02; Telex: 331141)

Akcros Chemicals Iberica S.L., Autovia Castelldefels Km. 4,65, 08820 El Prat De Llobregat, Barcelona Spain (Tel.: 34 3 4785 755; Telefax: 34 3 4780 888; Telex: 16152 Akrcos dk)

Akcros Chemicals Italia Srl, Via Cristina Belgioioso 13/15, 20021 Baranzate DiBollate (MI) Italy (Tel.: 39 2 38200484; Telefax: 39 2 38200443)

Akcros Chemicals Nordic ApS, Naverland 2, 8. floor, DK-2600 Glostrup-Copenhagen Denmark (Tel.: 45 4344 4201; Telefax: 45 4344 4203; Telex: 16152 Akrcos dk)

Akcros Chemicals v.o.f., Haagen House, PO Box 44, 6040 AA Roermond The Netherlands (Tel.: 31 4750-91777; Telefax: 31 4750-17489; Telex: 58021 haro nl)

Akcros Chemicals America, 500 Jersey Ave., PO Box 638, New Brunswick, NJ 08903 USA (Tel.: 908-247-2202, 800-500-7890; Telefax: 908-247-2287)

Akcros Chemicals, Taiwan Branch, Room 1408, 145h Fl. No. 96, Chung Shan N. Rd., Sec. 2, Taipei 10449, Taiwan, R.O.C. (Tel.: 886 2 562 6922; Telefax: 886 2 536 5287; Telex: 15185)

Akcros Chemicals (Asia Pacific) Pte Ltd., 7500A Beach Rd., 15-309 The Plaza, Singapore 0719 (Tel.: 65 292 1966; Telefax: 65 292 9665; Telex: RS 23241 AKCROS)

Akrochem Chemical Co.

255 Fountain St., Akron, OH 44304 USA (Tel.: 216-535-2108, 800-321-2260; Telefax: 216-535-8947)

Akzo Nobel

Akzo Nobel België NV/SA, 13 Marnix Ave., 1050 Bruxelles Belgium (Tel.: 2-518 04 07; Telefax: 02 518 05 05; Telex: 62 664)

Akzo Nobel Chemicals Inc., 300 S. Riverside Plaza, Chicago, IL 60606 USA (Tel.: 312-906-7500, 800-227-7070; Telefax: 312-906-7680; Telex: 25-3233)

Akzo Nobel Coatings Inc., US Hwy. 341 E., PO Box 349, Baxley, GA 31513 USA (Tel.: 912-367-3616; Telefax: 912-367-5754; Telex: 810-788-5450)

Akzo Nobel Chemicals Ltd., 1 City Center Dr., Suite 320, Mississauga, Ontario, L53 IM2 Canada (Tel.: 905-273-5959; Telefax: 905-273-7339)

Akzo Nobel Chemicals Ltd., 1-5, Queens Road, Hersham, Walton-on-Thames, Surrey, KT12 5NL UK (Tel.: 44 1932 247891; Telefax: 44 1932-231204; Telex: 21997)

Akzo Nobel Chemicals bv, Postbus 247, Stationsstraat 48, 3800 AE Amersfoort The Netherlands (Tel.: 31-33-467 67 67; Telefax: 31-33-467 61 00; Telex: 79322)

Akzo Nobel Chemicals GmbH, Postfach 100132, Phillippstrasse 27, 52301 Düren Germany (Tel.: 2421-49201; Telefax: 2421-492487; Telex: 833911)

Akzo Nobel Chemicals Ltd., 6 Grand Ave., PO Box 80, Camellia, NSW, 2150 Australia (Tel.: (02) 6384555; Telefax: (02) 6384681; Telex: AA 21562)

Akzo Nobel K.K., Godo Kaikan Bldg., 3-27 Kioi-cho, Chiyoda-ku, Tokyo, 102 Japan (Tel.: (03)5275-6300; Telefax: (03)3222-0465; Telex: J24262)

Eka Nobel Ltd., Div. of Eka Nobel AB, Unit 304 Worle Pkwy., Summer Lane, Worle, Weston-Super-Mare, Avon, BS22 OWA UK (Tel.: 44 1934 522244; Telefax: 44 1934 522577; Telex: 444351)

Eka Nobel AB, Div. of Nobel Industrier Sverige AB, S-445 80 Bohus Sweden (Tel.: 46-31-58 70 00; Telefax: 46-31-98 17 74; Telex: 2435 EKAGBGS)

Alar Engineering Corp.

9651 W. 196th St., Mokena, IL 60448 USA (Tel.: 708-479-6100)

Alba International Inc.

508 Clearwater Dr., N. Aurora, IL 60542 USA (Tel.: 708-897-4200, 800-669-9333; Telefax: 708-377-5330)

Albemarle

Albemarle Corp., 451 Florida St., Baton Rouge, LA 70801 USA (Tel.: 504-388-7040, 800-535-3030; Telefax: 504-388-7686; Telex: 586441, 586431)

Albemarle SA, Div. of Albemarle Corp., 523 Ave. Louise, Box 19, B-1050 Brussels Belgium (Tel.: 32-2-642-4411; Telefax: 32-2-648-0560; Telex: 22549)

Albemarle Asia Pacific Co., #13-06 PUB Bldg., Devonshire Wing, 111 Somerset Rd., Singapore 0923 Singapore (Tel.: 65-732-6286; Telefax: 65-737-4123)

Albemarle Japan Corporation, Shiroyama Hills 19F, 4-3-1, Toranomon, Minato-ku, Tokyo, 105 Japan (Tel.: 81-3-5401-2901; Telefax: 81-3-5401-3368)

Albright & Wilson

Albright & Wilson Ltd., European & Corporate Hdqtrs., Box 3, 210-222 Hagley Rd. West, Oldbury, Warley, West Midlands, B68 0NN UK (Tel.: 44 121 429-4942; Telefax: 44 121-420-5151; Telex: 336291)

Albright & Wilson Americas Inc., Box 26229, Richmond, VA 23260-6229 USA (Tel.: 804-550-4300, 800-446-3700; Telefax: 804-550-4385)

Albright & Wilson Saint Mihiel SA, P. 19 Han-Sur-Meuse, 55300 Saint-Mihiel France (Tel.: 33 29 91 73 00; Telefax: 33 29 91 73 99; Telex: 961058)

Albright & Wilson Am. (Canada), Box 2220, St. John's, Newfoundland, A1C 6E6 Canada (Tel.: 709-753-0838; Telefax: 709-753-5353)

Albright & Wilson Castiglione Srl, a Cavour 50, Casella Postale No. 142, I-46043 Castiglione delle Stiviere, (Mantova) Italy (Tel.: 39-376-6371; Telefax: 39-376 637323; Telex: 300432)

Albright & Wilson GmbH, ty Center, Frankfurter Strasse 181, 63263 Ne-Isenburg Germany (Tel.: 49-6102 27051; Telefax: 49-6102 25286; Telex: 4032069)

Albright & Wilson (Australia) Limited, Box 20, Yarraville, Victoria, 3013 Australia (Tel.: 61 3-688-7777; Telefax: 61 3-688-7788)

Albright & Wilson Asia Pacific Pte Ltd., Jalan Besut, Jurong Industrial Estate, Singapore 2261 (Tel.: 011-65-261-2151; Telefax: 011-65-265-1941; Telex: 37007)

Albright & Wilson Asia Pacific Pte Ltd.,. 4051, Hua Ting Sheraton Hotel, 1200 Cao Xi Bei Lu, Shanghai, 200030, China (Tel.: 86 21 439100 x4051; Telefax: 86 21 4397856)

Albright & Wilson Ltd. Japan,. 2 Okamotoya Bldg. 6 Fl., 1-24, Toranomon 1-chome, Minato-ku, Tokyo, 105 Japan (Tel.: (03) 3508 9461; Telefax: (03) 3591 0733; Telex: 2226721)

Alcan

Alcan Chemicals, Div. of Alcan Aluminum Corp., 3690 Orange Place, Suite 400, Cleveland, OH 44122-4438 USA (Tel.: 216-765-2550, 800-321-3864; Telefax: 216-765-2570; Telex: 135069)

Alcan Chemicals Ltd., Div. of British Alcan Aluminum plc, Chalfont Park, Gerrards Cross, Buckinghamshire, SL9 0QB UK (Tel.: 44 1753-887373; Telefax: 44 1753 881556; Telex: 847343)

Alcoa

Alcoa Industrial Chemicals Div., 4701 Alcoa Rd., PO Box 300, Bauxite, AR, 72011 USA (Tel.: 501-776-4717, 800-643-8771; Telefax: 501-776-4904; Telex: 536447)

Alcoa Industrial Chemicals Europe, Im Atzeinest 3, D-5380 Bad Homburg Germany (Tel.: 49-6172-40680; Telefax: 49-6172-406836)

Alcoa Inter-America Inc., 396 Alhambra Circle, Suite 200, Coral Gables, FL 33114 USA (Tel.: 305-445-8544; Telefax: 305-444-8924)

Alcoa Kasei Limited, Toranomon 4-chome Mori BldgII 7F, 2-12 Toranomon 4-chome, Minato-ku, Tokyo, 105 Japan (Tel.: 813-5472-3201; Telefax: 813-5472-3209)

Alco Chemical Corp., Div. of National Starch & Chem.

909 Mueller Dr., PO Box 5401, Chattanooga, TN 37406-0401 USA (Tel.: 615-629-1405, 800-251-1080; Telefax: 615-698-8723; Telex: 755002)

Aldrich

Aldrich Chemical Co., Inc., 1001 W. St. Paul Ave., Milwaukee, WI 53233 USA (Tel.: 414-273-3850, 800-558-9160; Telefax: 414-273-4979; Telex: 26843 ALDRICH MI)

Sigma-Aldrich NV/SA, K. Cardijnplei 8, B-2880 Bornem Belgium (Tel.: 038991301; Telefax: 038991311)

SAF Bulk Chemicals, Kyodo Bldg. Shinkanda, 10-Kanda-Mikuracho, Chiyoda-Ku, Tokyo Japan (Tel.: 81 03 3 258 0155; Telefax: 81 03 3 258 0157)

Alemark Chemicals

1177 High Ridge Rd., Stamford, CT 06905 USA (Tel.: 203-966-7410; Telefax: 203-966-4276)

Alfa Aesar Johnson Matthey

30 Bond St., Ward Hill, MA 01835 USA (Tel.: 508-521-6300, 800-322-4757; Telefax: 508-521-6366)

Alfa Chem

1661 N. Spur Dr., Central Islip, NY 11722-4325 USA (Tel.: 516-277-7681, 800-375-6869; Telefax: 516-277-7681; Telex: 6504811992)

Allan Chemical Corp.

PO Box 1837, Fort Lee, NJ 07024-8337 USA (Tel.: 201-592-8122; Telefax: 201-592-9298; Telex: 170991)

All Chemie Ltd.

1429 John St., Fort Lee, NJ 07024 USA (Tel.: 201-947-7776; Telefax: 201-947-3343)

Allchem Industries Inc.

4001 Newberry Rd, Suite E-3, Gainesville, FL 32607 USA (Tel.: 904-378-9696; Telefax: 904-338-0400; Telex: 509540 ALLCHEM UD)

Allco

Allco Chemical Corp., PO Box 247, Galena, KS 66739 USA (Tel.: 316-783-1321, 800-437-1998; Telefax: 316-783-5253; Telex: (650) 173-6115)

Allco Chemical Corp., 17304 N. Preston Rd., Dallas, TX 75252 USA (Tel.: 214-733-6831; Telefax: 214-733-6846)

Allied Colloids

Allied Colloids Ltd., Div. of Allied Colloids Group plc, PO Box 38, Low Moor, Bradford, West Yorkshire, BD12 0JZ UK (Tel.: 44 1274 671267; Telefax: 44 1274-606499; Telex: 51646)

Allied Colloids Inc., 2301 Wilroy Rd., PO Box 820, Suffolk, VA 23439-0820 USA (Tel.: 804-538-3700; Telefax: 804-538-0204)

AlliedSignal

AlliedSignal Inc., PO Box 1053, 101 Columbia Rd., Morristown, NJ 07960 USA (Tel.: 201-455-2000, 800-526-0717; Telefax: 201-707-4555; Telex: 136410)

AlliedSignal Inc./Performance Additives, PO Box 1039, 101 Columbia Rd., Morristown, NJ 07962-1039 USA (Tel.: 201-455-2145, 800-222-0094; Telefax: 201-455-6154; Telex: 990433)

AlliedSignal Europe NV, Haasrode Research Park, Grauwmeer 1, B-3001 Heverlee (Leuven) Belgium (Tel.: 32-16-39 12 33; Telefax: 32-16-40 03 77)

AlliedSignal Europe NV/Performance Additives, Haasrode Research Park, Grauwmeer 1, B-3001 Heverlee (Leuven) Belgium (Tel.: 32-16/391 211; Telefax: 32-16/400 039)

AlliedSignal Int'l. NV-SA/A-C Performance Additives, Haasrode Research Park, Grauwmeer, B-3001 Heverlee (Leuven) Belgium (Tel.: 32-16-39 12 11; Telefax: 32-16-400-039)

Alnor Oil Company, Inc.

70 E. Sunrise Hwy., Suite 418, Valley Stream, NY 11581 USA (Tel.: 516-561-6146; Telefax: 516-561-6123; Telex: 221512 ALNOR UR)

Alox Corp.

3943 Buffalo Ave., PO Box 517, Niagara Falls, NY 14302 USA (Tel.: 716-282-1295; Telefax: 716-282-2289)

Alpha/Owens-Corning

950 Hwy. 57 East, Collierville, TN 38017 USA (Tel.: 901-854-2800; Telefax: 901-854-1183)

AluChem Inc.

One Landy Lane, Reading, OH 45215 USA (Tel.: 513-733-8519; Telefax: 513-733-0608; Telex: 298252 ALUC UR)

Alzo Inc.

6 Gulfstream Blvd., Matawan, NJ 07747 USA (Tel.: 908-254-1901; Telefax: 908-254-4423)

Amber Inc.

220 White Plains Rd., Tarrytown, NY 100591 USA (Tel.: 914-332-5855; Telefax: 914-332-1480)

Ambersil Ltd.

Wylds Road, Castlefield Industrial Estate, Bridgewater, Somerset, TA6 4DD UK (Tel.: 44 1278 424200; Telefax: 44 1278-425644; Telex: 46796)

Amber Synthetics, Amsyn Inc.

1177 High Ridge Rd., Stamford, CT 06905 USA (Tel.: 203-972-7401; Telefax: 203-966-4276)

Amerchol

Amerchol Corp., Div. of United-Guardian, Inc., PO Box 4051, 136 Talmadge Rd., Edison, NJ 08818-4051 USA (Tel.: 908-248-6000, 800-367-3534; Telefax: 908-287-4186; Telex: 833472)

Amerchol, D.F. Anstead Ltd., Victoria House, Radford Way, Billericay, Essex, CM12 0DE UK (Tel.: 44 1277 630063; Telefax: 44 1277 631356; Telex: 851-99410 ANSTED G)

Amerchol Europe, Havenstraat 86, B-1800 Vilvoorde Belgium (Tel.: 2-252-4012; Telefax: 2-252-4909; Telex: 846-69105 AMRCHL B)

AmeriBrom, Inc., Member of the Dead Sea Bromine Group

52 Vanderbilt Ave., New York, NY 10017 USA (Tel.: 212-286-4000, 800-280-2766; Telefax: 212-286-4475; Telex: RCA 220531)

American Biorganics, Inc.

2236 Liberty Dr, Niagara Falls, NY 14304 USA (Tel.: 716-283-1434, 800-648-6689; Telefax: 716-283-1570; Telex: 926074)

American Casein Co.

109 Elbow Lane, Burlington, NJ 08016 USA (Tel.: 609-387-3130; Telefax: 609-387-7204; Telex: 843368)

American Chemet Corp.

400 Lake Cook Rd., Deerfield, IL 60015 USA (Tel.: 708-948-0800; Telefax: 708-948-0811; Telex: 72-4301)

American Colloid Co.

1500 W. Shure Dr., Arlington Hts., IL 60004-1434 USA (Tel.: 708-394-8730; Telefax: 708-506-6199; Telex: 4330321)

American Colors

160 E. Market St., PO Box 397, Sandusky, OH 44870 USA (Tel.: 419-621-4000; Telefax: 419-625-3979)

American Cyanamid. *See* Cytec Industries

American Dispersions, Inc.

PO Box 11505, Louisville, KY 40251-0505 USA (Tel.: 502-776-9494; Telefax: 502-776-7525)

American Fillers & Abrasives

14 Industrial Park Dr., Bangor, MI 49013 USA (Tel.: 616-427-7955; Telefax: 616-427-5171)

American Gelatin Co., Div. of Health Processes Inc.

789 Condon Dr., Charleston, SC 29412 USA (Tel.: 908-382-2212, 800-206-6555; Telefax: 803-762-3937)

American Gilsonite Co.

136 E. South Temple, Ste. 1460, Salt Lake City, UT, 84111 USA (Tel.: 801-328-0311)

American Ingredients Co.

3947 Broadway, Kansas City, MO 64111 USA (Tel.: 816-561-9050, 800-669-4092; Telefax: 816-561-1132)

American International Chemical, Inc.

17 Strathmore Rd, Natick, MA 01760 USA (Tel.: 508-655-5805, 800-238-0001; Telefax: 508-655-0927; Telex: 948342)

American Lecithin. *See under* Rhone-Poulenc

American Maize Products Co./Amaizo

1100 Indianapolis Blvd., Hammond, IN 46320-1094 USA (Tel.: 219-659-2000, 800-348-9896; Telefax: 219-473-6601)

American Oil & Supply Co.

238 Wilson Ave., Newark, NJ 07105 USA (Tel.: 201-589-0250, 800-582-3267; Telefax: 201-465-5083)

Ameripol Synpol

Ameripol Synpol Corp., 146 S. High St., 7th floor, Akron, OH 44308-1493 USA (Tel.: 216-762-4422, 800-321-9001; Telefax: 216-762-2549)

Ameripol Synpol Corp., PO Box 667, 1215 Main St., Port Neches, TX 77651 USA (Tel.: 800-547-0622; Telefax: 704-547-8849)

Mallard Creek Polymers, Inc., Subsid. of Ameripol Synpol Corp., 146 S. High St., Akron, OH 44308-1493 USA (Tel.: 216-762-5973, 800-321-9001; Telefax: 216-762-2549)

Amoco

Amoco Chemical Co., 200 East Randolph Dr., Mail code 4106, Chicago, IL 60601 USA (Tel.: 312-856-3200, 800-621-4567; Telefax: 312-856-4151; Telex: 25-3731)

Amoco Chemical Co./Chem. Intermediates Business Group, 801 Warrenville Rd., MC 6018, Lisle, IL 60532 USA (Tel.: 800-621-4567)

Amoco Chemical (Europe) SA, Div. of Amoco Corp., 15, Rue Rothschild, 1211 Geneva 21 Switzerland (Tel.: 41-22-715-0701; Telex: 422787)

Amoco Oil Co., Lubricants Div., 2021 Spring Rd., Oak Brook, IL 60562-1857 USA (Tel.: 708-571-7100; Telefax: 708-571-7174)

Ampacet

Ampacet Corp., 660 White Plains Rd., Tarrytown, NY 10591-5130 USA (Tel.: 914-631-6600; Telefax: 914-631-7197; Telex: 62588)

Ampacet Europe S.A., Rue D'Ampacet 1, 6780 Messancy Belgium (Tel.: 32-63- 371490; Telefax: 32-63-371499; Telex: 42710)

AMRESCO

30175 Solon Ind. Pkwy., Solon, OH 44139 USA (Tel.: 216-349-1313, 800-829-2802; Telefax: 216-349-1182; Telex: 985582)

Amspec Chemical Corp.

Foot of Water St., Gloucester City, NJ 08030 USA (Tel.: 609-456-3930, 800-5AMSPEC; Telefax: 609-456-6704; Telex: 136714 GLCY)

Amstat Industries, Inc.

3012 N. Lake Terrace, Glenview, IL 60025-5794 USA (Tel.: 708-998-6210, 800-783-9999; Telefax: 708-998-6218)

Anar Chemical Co.

1765F Cortland Court, Addison, IL 60101 USA (Tel.: 708-953-1660, 800-344-1660; Telefax: 708-953-1698)

Andrulex Trading Ltd.

Unit 34, Saffron Court, Southfields Industrial Estate, Laindon, Basildon, Essex, SS15 6SS UK (Tel.: 44 1268 416441; Telefax: 44 1268 541639; Telex: 99339 RULEX G)

Anedco, Inc.

10429 Koenig Rd., Houston, TX 77034 USA (Tel.: 713-484-3900; Telefax: 713-484-3931)

ANGUS

ANGUS Chemical Co., 1500 E. Lake Cook Rd., Buffalo Grove, IL 60089-6556 USA (Tel.: 847-215-8600, 800-362-2580; Telefax: 847-215-8626; Telex: 275422 ANGUS UR)

ANGUS Chemie GmbH, Unit 7, Rotunda Business Centre, Thorncliffe Park Estate, Chapeltown, Sheffield, S30 4PH UK (Tel.: 44 114 2571322; Telefax: 44 114 2571336)

ANGUS Chemie GmbH, Huyssenallee 5, 45128 Essen Germany (Tel.: (49) 201-233531; Telefax: (49)201-238661; Telex: 8571563 ANGE D)

ANGUS Chemie GmbH, Le Bonaparte, Centre d'Affaires, Paris Nord, F-93153 Le Blanc Mesnil France (Tel.: (33)1-48-65-73-40; Telefax: (33)1-48-65-73-20; Telex: ANGUS 232089 F)

ANGUS Chemical (Singapore) Pte. Ltd., 150 Beach Rd., #17-01 Gateway West, Singapore 189720 (Tel.: (65) 293-1738; Telefax: (65) 293-3307)

Anhydrides & Chemicals Inc.

7-33 Amsterdam St., Newark, NJ 07105 USA (Tel.: 201-465-0077; Telefax: 201-465-7713)

Anzon Inc., Subsid. of Cookson Group plc

2545 Aramingo Ave., Philadelphia, PA 19125 USA (Tel.: 215-427-6920, 800-523-0882; Telefax: 215-427-6955)

Aqualon

Aqualon Co., A Hercules Inc. Co., 1313 North Market St., PO Box 8740, Wilmington, DE 19899-8740 USA (Tel.: 302-594-6600, 800-345-8104; Telefax: 302-594-6660; Telex: 4761123)

Aqualon France, 3 Rue Eugene & Armand, Peugeot, 92500 Rueil-Malmaison France (Tel.: 33 1 4751 2919; Telefax: 33 1 4777 0614; Telex: 6314244)

Aqualon UK Ltd., Genesis Centre, Garret & Field, Birchwood, Warnington, Cheshire, WA3 7BH UK (Tel.: 44 1925 830077; Telefax: 44 1925 830112; Telex: 626219)

Aquatec Quimica SA

Av. Paulista no. 37-12° andar, 01311-000 Sao Paulo-SP Brazil (Tel.: 55 11-284-4188; Telefax: 55 11-287 7841; Telex: 1121312)

Arakawa

Arakawa Chemical Industries Ltd., 3-7, Hiranomachi 1-chome, Chuo-ku, Osaka, 541 Japan (Tel.: (06) 209-8581; Telefax: (06) 209-8542; Telex: 5222296 ARKOSA J)

Arakawa Chemical (USA) Inc., 625 N. Michigan Ave., Suite 1700, Chicago, IL 60611 USA (Tel.: 312-642-1750; Telefax: 312-642-0089; Telex: 26-5514)

Archway Sales Inc.

4155 Manchester St., St. Louis, MO 63110 USA (Tel.: 314-533-4662; Telefax: 214-533-3386)

ARCO

ARCO Chemical/Headquarters, Research & Engineering Center, 3801 West Chester Pike, Newtown Sq., PA 19073-2387 USA (Tel.: 610-359-2000, 800-345-0252; Telefax: 610-359-2841)

ARCO Chemical Canada Inc., 100 Consilium Pl., Suite 306, Scarborough, Ontario, M1H 3E3 Canada (Tel.: 416-296-9864; Telefax: 416-296-9834)

ARCO Chemical Pan American, Inc., Paseo de la Reforma, 390 Decimo Piso, 06600 Mexico City Mexico (Tel.: 905-514-6833)

ARCO Chemical Europe Inc., Bridge Ave., Maidenhead, Berkshire, SL6 1YP UK (Tel.: 44-1628-77-5000; Telefax: 44-1628-77-5180; Telex: 847436)

ARCO Quimica do Brazil Ltda., Av. Roque Petroni jr., 999 cj 123, Sao Paulo SP 04708-000 Brazil (Tel.: 55-1-535-5673; Telefax: 55-11535-3321)

ARCO Chemical Japan, Inc., Hamacho Center Bldg., 2-31-1 Nihonbashi, Hama-cho, Chuo-ku, Tokyo, 103 Japan (Tel.: (03) 5641-4500; Telefax: (03) 5641-4550)

ARCO Chemical Asia/Pacific Ltd., Toranomon 37 Mori Bldg., 5th Floor, 5-1 Toranomon 3-Chome, Minato-Ku, Tokyo, 105 Japan

ARCO Chemical (Singapore) Pte. Ltd., 37th Floor, 152 Beach Rd., #37-00 The Gateway East, Singapore 0718 (Tel.: 65-291-2011; Telefax: 65-297-0517)

Arenol Chemical Corp.

189 Meister Ave., Somerville, NJ 08876 USA (Tel.: 908-526-5900; Telefax: 908-526-9688)

M. Argueso & Co., Inc.

441 Waverly Ave., PO Box E, Mamaroneck, NY 10543 USA (Tel.: 914-698-8500; Telefax: 914-698-0325)

Arista Industries, Inc.

1082 Post Rd., Darien, CT 06820 USA (Tel.: 203-655-0881, 800-255-6457; Telefax: 203-656-0328; Telex: 996493)

Aristech Chemical Corp.

600 Grant St., Room 1028, Pittsburgh, PA 15230-0250 USA (Tel.: 412-433-7873; Telefax: 412-433-1816; Telex: 6503608865)

Arizona Chemical Co., Div. of International Paper

1001 E. Business Hwy. 98, Panama City, FL 32401-3633 USA (Tel.: 904-785-6700, 800-526-5294; Telefax: 904-785-2203; Telex: 514411)

Arol Chemical Products Co.

649 Ferry St., Newark, NJ 07105 USA (Tel.: 201-344-1510; Telefax: 201-344-7127)

Asahi Chemical Industry Co., Ltd.

Hibiya Mitsui Bldg., 1-2, Yuraku-cho 1-chome, Chiyoda-ku, Tokyo, 100 Japan (Tel.: (03) 3507-2730; Telefax: (03) 3507-2495; Telex: 222-3518 BEMBRGJ)

Adeka Fine Chemical Co., Ltd., Subsid. of Asahi Denka Kogyo

Yoko Bldg., 4-5, Hongo 1-chome, Bunkyo-ku, Tokyo, 113 Japan (Tel.: (03) 5689-8681; Telefax: (03) 5689-8680)

Asahi Denka Kogyo K.K.

Furukawa Bldg. 2-8, Nihonbashi Muro-machi 2-chome Chuo-ku, Tokyo, 103 Japan (Tel.: (03) 5255-9002; Telefax: (03) 3270-2463; Telex: 222-2407 TOKADK)

Asahi-Penn Chemical Co. Ltd., Joint venture of Asahi Glass Co. Ltd./PPG Industries Inc.

13-3, Nihonbashi Kobuna-cho, Chuo-ku, Tokyo, 103 Japan (Tel.: (03) 3662-0520; Telefax: (03) 3662-0596)

Asahi Yukizai Kogyo Co., Ltd.

2-5955, Nakanose-cho, Nobeoka-shi, Miyazaki, 882 Japan (Tel.: (0982) 33-3311; Telefax: (0982) 21-8606)

Asarco Inc.

180 Maiden Lane, New York, NY 10038 USA (Tel.: 212-510-2000; Telex: ITT 420585)

Asarco Tech.

3422 S. 700 West, Salt Lake City, UT, 84119 USA (Tel.: 801-262-2459; Telefax: 801-261-2194)

Asbury Graphite Mills, Inc.

PO Box 144, Asbury, NJ 08802 USA (Tel.: 908-537-2155; Telefax: 908-537-2908)

Ash Grove Cement Co.

8900 Indian Creek Pkwy., Overland Park, KS 66225 USA (Tel.: 913-451-8900)

Ashland

Ashland Chemical Inc./Industrial Chemicals & Solvents, PO Box 2219, Columbus, OH 43216 USA (Tel.: 614-790-3333, 800-526-4032; Telefax: 614-889-3465)

Ashland Chemical Inc./Resin & Chemicals Division, Subsidiary of Ashland Oil Inc., 2463 Royal Windsor Drive, Mississauga, Ontario, L5J 4ET Canada (Tel.: 905-823-7975)

Drew Industrial Div., Div. of Ashland Chemical Co., One Drew Plaza, Boonton, NJ 07005 USA (Tel.: 201-263-7800, 800-526-1015 x7800; Telefax: 201-263-4483; Telex: DREWCHEMS BOON)

Ashley Polymers

5114 Fort Hamilton Parkway, Brooklyn, NY 11219 USA (Tel.: 718-851-8111; Telefax: 718-972-3256; Telex: 427884 ASHLEY)

Aspect Minerals Inc.

PO Box 277, Spruce Pine, NC 28777 USA (Tel.: 704-765-8022, 800-803-7979; Telefax: 704-765-7887)

H.M. Royal, Inc., Distrib. for Aspect Minerals, 689 Pennington Ave., PO Box 28, Trenton, NJ 08601 USA (Tel.: 609-396-9176; Telefax: 609-396-3185)

Aster, Inc.

160 Glaser St., Fairborn, OH 45324 USA (Tel.: 513-754-1400; Telefax: 513-754-1415)

Astor Corp.

200-E Piedmont Ct., Doraville, GA 30340 USA (Tel.: 770-448-8083; Telefax: 770-840-0954)

Astro Industries Inc., Subsid. of Borden Inc.

PO Box 2559, 114 Industrial Blvd., Morganton, NC 28655 USA (Tel.: 704-584-3800, 800-872-7876; Telefax: 704-584-3885)

Atlantic Equip. Engrs., Div. of Micron Metals Inc.

13 Foster St., PO Box 181, Bergenfield, NJ 07621 USA (Tel.: 201-384-5606, 800-486-2436; Telefax: 201-387-0291)

Atlantic Gelatin/Kraft General Foods

Hill St., Woburn, MA 01801 USA (Tel.: 617-933-2800; Telefax: 617-935-1566)

Atlas Minerals & Chemicals, Inc.

1227 Valley Rd., PO Box 38, Mertztown, PA 19539-0038 USA (Tel.: 610-682-7171, 800-523-8269; Telefax: 610-682-9200)

Atlas Refinery, Inc.

142 Lockwood St., Newark, NJ 07105 USA (Tel.: 201-589-2002; Telefax: 201-589-7377; Telex: 138-425 ATLASOIL)

Atomergic Chemetals Corp.

222 Sherwood Ave., Farmingdale, NY 11735-1718 USA (Tel.: 516-694-9000; Telefax: 516-694-9177; Telex: 6852289)

Atotech USA Inc.

PO Box 6768, Two Riverview Dr., Somerset, NJ 08875-6768 USA (Tel.: 908-302-3500, 800-321-6565; Telefax: 908-271-8960; Telex: 4332158)

AT Plastics Inc.

142 Kennedy Rd., Charlotte, NC 28203 USA (Tel.: 704-331-9161, 800-342-2990; Telefax: 704-343-0019)

Atramet, Inc.

222 Sherwood Ave., Farmingdale, NY 11735 USA (Tel.: 516-694-9000; Telefax: 516-694-9177)

Ausimont

Ausimont SpA, Via Principe Eugenio, 1/5, 20155 Milano Italy (Tel.: 2-6270-3718; Telefax: 2-6270-3948; Telex: 310679 Monted I)

Ausimont SpA, Via San Pietro 50, 20021 Bollate, Milano Italy (Tel.: 39-2-6270-6236; Telefax: 39-2-6270-6538; Telex: 310679 MONTED I)

Ausimont USA Inc., 44 Whippany Rd., PO Box 1838, Morristown, NJ 07962-1838 USA (Tel.: 201-292-6250, 800-323-AUSI; Telefax: 201-292-0886)

Ausimont USA Inc., Crown Pt. Rd. & Leonards Lane, PO Box 26, Thorofare, NJ 08086 USA (Tel.: 609-853-8119, 800-323-2874; Telefax: 609-853-6405)

Ausimont UK Ltd., 111 Upper Richmond Rd., Putney, London, SW15 2TJ UK (Tel.: 44 181 780-0399; Telefax: 44 181 780-2871)

Montedison Deutschland GmbH, Kolner Strasse 3A, Postfach 5648, D-6236 Eschborn/TS.1 Germany (Tel.: (06196) 92201; Telefax: (06196) 482389; Telex: 482389 MED D)

Montedison do Brazil Ltd., Avenida Brig. Faria, Lima 888-5½ andar, CEP 01452 Jardim Paulistano, Sao Paulo SP Brazil (Tel.: 11-2103325; Telefax: 11-8131700)

Ausimont K.K., 3rd Fl. Izumi Akaxaka Bldg., Minato-ku, Tokyo, 107 Japan (Tel.: 81-3-3224-7212; Telefax: 81-3-32247218)

Ausimont Singapore Pte. Ltd., 70 Shenton Way, #12-03 A/B/C Marina House, Singapore 0207 (Tel.: 65-223-3581; Telefax: 65-223-2127)

Austin Chemical Co. Inc.

1565 Barclay Blvd., Buffalo Grove, IL 60089 USA (Tel.: 708-520-9600; Telefax: 708-520-9160; Telex: 280342)

Avebe America Inc.

4 Independence Way, Princeton, NJ 08540 USA (Tel.: 609-520-1400; Telefax: 609-520-1473; Telex: 0820713)

Avrachem AG

PO Box 51, Gartenstrasse 12, CH-6340 Baar 1 Switzerland (Tel.: 41 42 318 355; Telefax: 41 42 310 250; Telex: 864-996)

BA Chemicals Ltd., Div. of Alcan

Chalfont Park, Gerrards Cross, Buckinghamshire, SL9 0QB UK (Tel.: 44 1753-887373; Telefax: 44 1753-889602; Telex: 847343)

Bacon Industries Inc.

192 Pleasant St., Watertown, MA 02172 USA (Tel.: 617-926-2550)

Bakelite GmbH, Div. of Rütgerswerke AG

Postfach 7154, Gennaer Strasse 2-4, W-5860 Iserlohn 7 Germany (Tel.: 49-2374-510; Telefax: 49-2374-51409; Telex: 827255 BGLH)

J.T. Baker, Inc.

600 North Broad St., Phillipsburg, NJ 08865 USA (Tel.: 908-859-2151)

Balchem Corp.

PO Box 175, Slate Hill, NY 10973 USA (Tel.: 914-355-2861; Telefax: 914-355-6314)

Bamberger Polymers, Inc.

1983 Marcus Ave., Lake Success, NY 11042 USA (Tel.: 516-328-2772, 800-888-8959; Telefax: 516-326-1005; Telex: 6852108)

Barco Chemical Products Inc.

PO Box 268448, 6900 N. Glenwood Ave., Chicago, IL 60626 USA (Tel.: 312-465-3827)

Barium & Chemicals Inc.

County Road 44, PO Box 218, Steubenville, OH 43952 USA (Tel.: 614-282-9776; Telefax: 614-282-9161)

Barker Industries, Inc.

2841 Old Steele Creek Rd., Charlotte, NC 28208 USA (Tel.: 704-391-1023; Telefax: 704-393-0464)

Bärlocher

Bärlocher GmbH, Riesstrasse 16, 80992 München 50 Germany (Tel.: 089 143730; Telefax: 089 14373 312; Telex: 5215701)

Baerlocher U.S.A., Joint venture between Bärlocher GmbH and ICC Industries, 3676 Davis Rd. N.W., PO Box 545, Dover, OH 44622 USA (Tel.: 216-364-6000, 800-342-6158; Telefax: 216-343-7025)

BASF

BASF AG, Carl-Bosch Str. 38, 67056 Ludwigshafen Germany (Tel.: 0621-60-99739; Telefax: 0621-60-93344; Telex: 62157120BASF)

BASF Corp., 3000 Continental Dr. North, Mount Olive, NJ 07828-1234 USA (Tel.: 201-426-2600, 800-367-9861)

BASF Corp./Performance Chemicals, 3000 Continental Dr. North, Mt. Olive, NJ 07828-1234 USA (Tel.: 201-426-2600, 800-669-BASF)

BASF plc, PO Box 4, Earl Road, Cheadle Hulme, Cheadle, Cheshire, SK8 6QG UK (Tel.: 44 161 485-6222; Telefax: 44 161-486-0891; Telex: 669211 BASFCH G)

BASF Belgium S.A., Ave. Hamoir-laan 14, B-1180 Brussels Belgium (Tel.: 32 2 373 21 11; Telefax: 32 2 375 10 42)

BASF Espanola S.A., Paseo de Gracia, 99 E-08008, Barcelona Spain (Tel.: 34 3488 1010; Telefax: 34 3488 2020; Telex: 97923 BASF B E)

BASF France & Co. Cie, 140, Rue Jules Guesde, BP 8T F-92303 Levallois-Perret, Cedex France (Tel.: 33 47 30 55 00; Telefax: 33 47 57 5104; Telex: 610132 BASF F)

BASF Japan Ltd., 3-3, Kioi-cho, Chiyoda-ku, Tokyo, 102 Japan (Tel.: (03) 3238-2300; Telex: 222-2130 BASFTK)

Baxenden Chemicals Ltd., Div. of Witco Chemicals Inc.

Union Lane, Droitwich, WR9 9BB UK (Tel.: 44-1905-794-795; Telefax: 44-1905-794002)

Baychem, Inc.

3200 Moon Station Rd., Kennesaw, GA 30144 USA (Tel.: 404-429-1405; Telefax: 404-425-9141)

Bayer

Bayer AG, Bayerwerk, D-5090 Leverkusen Germany (Tel.: 49 (214)30-1; Telefax: 49 (214)30 6 51 36; Telex: 85103-0 byd)

Bayer Antwerpen NV, Div. of Bayer AG, Haven 507, Scheldelaan 420, B-2040 Antwerp Belgium (Tel.: 32-3540 3011; Telefax: 32-3541 6936; Telex: 71175 BAYANT B)

Bayer Hispania Industrial SA, Pablo Claris 196, E-80037 Barcelona Spain (Tel.: 34-3-218 45 50; Telefax: 34-3-217 41 49; Telex: 54482 BAYIN E)

Bayer Inc./Fibers, Organics & Rubber, Bayer Rd., Pittsburgh, PA 15205-9741 USA (Tel.: 412-777-2000, 800-662-2927; Telefax: 412-777-7840; Telex: 1561261)

Bayer Inc./Fibers, Organics & Rubber, 2603 W. Market St., Akron, OH 44313 USA (Tel.: 216-836-0451, 800-321-0997; Telefax: 216-838-0200; Telex: 1561261)

Bayer Corp./Industrial Chemicals Div., Coatings, 100 Bayer Rd., Pittsburgh, PA 15205-9741 USA (Tel.: 412-777-2000, 800-662-2927)

Baymag

800, 10655 Southport Rd. SW, Calgary, Alberta, T2W 4Y1 Canada (Tel.: 403-271-9400; Telefax: 403-271-0010)

BDH. *See under* EM Industries

Belmont Metals Inc.

322 Belmont Ave., Brooklyn, NY 11207 USA (Tel.: 718-342-4900; Telefax: 718-342-0175)

Berje Inc.

5 Lawrence St., Bloomfield, NJ 07003 USA (Tel.: 201-748-8980; Telefax: 201-680-9618; Telex: 475-4165A)

Bernardy Chimie SA, Div. of Société des Produits Chimiques d'Harbonnieres

Thenioux, F-18100 Vierzon France (Tel.: 48 52 00 80; Telefax: 48 852 04 61; Telex: 760328)

Berol Nobel

Berol Nobel AB, S-44485 Stenungsund Sweden (Tel.: 46 303 85000; Telefax: 46 303 84659; Telex: 10513 benobl s)

Berol Nobel Ltd., Div. of Berol Nobel AB, 23 Grosvenor Road, St. Albans, Hertfordshire, AL1 3AW UK (Tel.: 44 1727 841421; Telefax: 44 1727-841529; Telex: 23242 BEROL G)

John K. Bice Co., Inc.

1319 Boyd St., Los Angeles, CA 90033 USA (Tel.: 213-264-5950; Telefax: 213-264-2285)

Biddle Sawyer Corp.

2 Penn Plaza, New York, NY 10121 USA (Tel.: 212-736-1580; Telefax: 212-239-1089; Telex: 427471BSCE)

Bio-Rad Laboratories

Bio-Rad Laboratories, Life Science Group, 2000 Alfred Nobel Dr., Hercules, CA 94547 USA (Tel.: 510-741-1000, 800-4-BIORAD; Telefax: 800-879-2289; Telex: 335358)

Bio-Rad Laboratories, 85A Marcus Dr., Melville, NY 11747 USA (Tel.: 516-756-2575, 800-4-BIORAD; Telefax: 516-756-2594; Telex: 71-3720184)

Biwax Corp.

45 E. Bradrock, Des Plaines, IL 60018 USA (Tel.: 708-824-0137; Telefax: 708-635-6675)

Blythe, William Ltd., Div. of Holliday Chemical Holdings plc

Holland Bank Works, Bridge St., Church, Accrington, Lancashire, BB5 4PD UK (Tel.: 44 1254 872872; Telefax: 44 1254 872000; Telex: 63142 BLYCO G)

Boehle Chemicals, Inc.

19306 W. 10 Mile Rd., PO Box 2001, Southfield, MI 48037 USA (Tel.: 313-255-2210; Telefax: 313-255-4474)

Boliden Intertrade Inc.

3379 Peachtree Rd. NE, Suite 300, Atlanta, GA 30326 USA (Tel.: 404-239-6700, 800-241-1912; Telefax: 404-239-6701; Telex: 981036)

The Boots Co. plc

D110 Main Office, Beeston, Nottingham, Nottinghamshire, NG2 3AA UK (Tel.: 44 1159 591648; Telefax: 44 1159 593715)

Borden

Borden Chemical Div., 180 E. Broad St., Columbus, OH 43215 USA (Tel.: 614-225-4000, 800-225-8044; Telefax: 614-225-3476)

Borden (UK) Ltd., Div. of Borden Inc., North Baddesley, Southampton, Hamsphire, SO52 9ZB UK (Tel.: 44 1703 732131; Telefax: 44 1703 738656; Telex: 47212)

Bostik Inc.

Boston St., Middleton, MA 01949 USA (Tel.: 508-777-0100, 800-726-7845; Telefax: 508-750-7212)

BP

BP Chemicals Ltd., Subsid. of The British Petroleum Co. plc, Britannic House, 6th Flr., 1 Finsbury Circus, London, EC2M 7BA UK (Tel.: 44 171 496-4867; Telefax: 44 171-496-4898; Telex: 266883 BPCLBHG)

BP Oil UK Ltd., Div. of The British Petroleum Co. plc, BP House, Breakspear Way, Hemel Hempstead, Hertsfordshire, HP2 4UL UK (Tel.: 44 1442 232323; Telefax: 44 1442-225225; Telex: 827711)

BP Chemicals Inc., 4440 Warrensville Center Rd., Warrensville Hts., OH 44128 USA (Tel.: 216-586-6455, 800-272-4367; Telefax: 216-586-5588; Telex: 6873120, 6873119)

BP Performance Polymers Inc., Phenolic Business, 60 Walnut Ave., Suite 100, Clark, NJ 07066 USA (Tel.: 908-815-7843; Telefax: 908-815-7844)

Brandeis, A Div. of Pechiney World Trade (USA) Inc.

475 Steamboat Rd., Greenwich, CT 06830 USA (Tel.: 203-625-9081; Telefax: 203-622-8669; Telex: 6819520 BRANDEIS)

Brand-Nu Laboratories, Inc.

PO Box 895, 30 Maynard St., Meriden, CT 06450 USA (Tel.: 203-235-7989, 800-243-3768; Telefax: 203-235-7163)

British Chrome & Chemicals Ltd., A Div. of Harcros Chemicals UK Ltd.

Urlay Nook, Eaglescliffe, Stockton-on-Tees, Cleveland, TS16 0QG UK (Tel.: 44 1642 787755/78193; Telefax: 44 1642-781935; Telex: 58363 Dicrom G)

British Traders & Shippers Ltd., Div. of Linton Park plc

6-7 Merrielands Crescent, Dagenham, Essex, RM9 6SL UK (Tel.: 44 181 595 4211; Telefax: 44 181-593 0933; Telex: 897438 SHIPEX G)

British Wax Refining Co. Ltd.

29 St John's Rd, Redhill, Surrey, RH1 6DT UK (Tel.: 44 1737 761242; Telefax: 44 1737 761472)

Bromine & Chemicals Ltd. *See under* Dead Sea Bromine

Brotherton Specialty Products Ltd.

Calder Vale Rd, Wakefield, West Yorkshire, WF1 5PH UK (Tel.: 44 1924 371919; Telefax: 44 1924 290408; Telex: 556320 BROKEM G)

Brown Chemical Co., Inc./Industrial and Fine Chems.

302 W. Oakland Ave., Oakland, NJ 07436-0785 USA (Tel.: 201-337-0900)

Browning Chemical Corp./Food Ingredients Div.

707 West Chester Ave., White Plains, NY 10604 USA (Tel.: 914-686-0300; Telefax: 914-686-0310)

Brunner Mond & Co. Ltd.

PO Box 4, Mond House, Northwich, Cheshire, CW8 4DT UK (Tel.: 44 1606 724582; Telefax: 44 1606 781353)

Buckman Laboratories

Buckman Laboratories Int'l., Inc., Bulab Holdings, Inc. Worldwide Headquarters, 1256 North McLean Blvd., Memphis, TN 38108-0305 USA (Tel.: 901-278-0330, 800-BUCKMAN; Telefax: 901-276-5343; Telex: 68-28020)

Buckman Laboratories of Canada, Ltd., 351 Joseph-Carrier Blvd., Vaudreuil, Quebec, J7V 5V5 Canada (Tel.: 514-424-4404; Telefax: 514-424-4294)

Buckman Laboratories, S.A., Wondelgemkaai, 159, B-9000 Ghent Belgium (Tel.: 32-9-257 92 11; Telefax: 32-9-253 62 95; Telex: 846-11516 Bulab-B)

Buckman Laboratories Ltd., Enterprise House, Manchester Science Pk., Lloyd St. North, Manchester, M15 6SE UK (Tel.: 44 (61) 226 12 27; Telefax: 44 (61) 227-9314)

Buckman Laboratories Pty. Ltd., East Bomen Rd., PO Box 1396, Wagga Wagga, NSW, 2650 Australia (Tel.: 61-69-213155, 61 (08) 25-7272; Telefax: 61-69-213677)

Buckman Laboratories, K.K., Ogawa Bldg. 4F, 9-16, Nihonbashi Kodenmacho, Chuo-ku, Tokyo, 103 Japan (Tel.: 81 (3) 3808-1199; Telefax: 81 (3) 3808-1590)

Buckton Scott Ltd.

Black Horse House, Bentalls, Pipps Hill Estate, Basildon, Essex, SS14 3BX UK (Tel.: 44 1268 531308; Telefax: 44 1268 531316; Telex: 995923)

Buffalo Color Corp.

959 Rte. 46 East, Suite 403, Parsippany, NJ 07054 USA (Tel.: 201-316-5600, 800-631-0171; Telefax: 201-316-5828; Telex: 7109885924)

Burdick & Jackson, Div. of Baxter Diagnostics Inc.

1953 S. Harvey St., Muskegon, MI 49442 USA (Tel.: 616-726-3171, 800-368-0050; Telefax: 616-728-8226)

Burgess Pigment Co.

PO Box 349, Sandersville, GA 31082 USA (Tel.: 912-552-2544, 800-841-8999; Telefax: 912-552-1772; Telex: 804523)

Burlington Chemical Co., Inc.

PO Box 111, Burlington, NC 27216 USA (Tel.: 910-584-0111, 800-672-5888; Telefax: 910-584-3548; Telex: 9102502503)

Byk-Chemie

Byk Chemie GmbH, Div. of Altana Industrie Aktien und Anlagen AG, Abelstrasse 14, PO Box 10 02 45, D-46463 Wesel Germany (Tel.: 011-49-281-6700; Telefax: 011-49-281-65735; Telex: 812772)

Byk-Chemie USA, 524 S. Cherry St., PO Box 5670, Wallingford, CT 06492-7656 USA (Tel.: 203-265-2086; Telefax: 203-284-9158; Telex: 643378)

Byk-Chemie France S.A.R.L., Le Bonaparte Centre d'Affaires, Paris Nord, F-93153 Le Blanc Mesnil France

Byk-Chemie Japan KK, Sunshine 6 Bldg., 2-20-12, Shiba, Minato-Ku, Tokyo, 105 Japan (Tel.: 81-3-3256-5409; Telefax: 81-3-3256-5420)

Cabot

Cabot Corp./Cab-O-Sil® Div., PO Box 188, Tuscola, IL 61953-0188 USA (Tel.: 217-253-3370, 800-222-6745; Telefax: 217-253-4334; Telex: 910-663-2542)

Cabot Corp./Special Blacks Div., 157 Concord Rd., PO Box 7001, Billerica, MA 01821-7001 USA (Tel.: 508-663-3455, 800-526-7591; Telefax: 508-670-7035; Telex: 6817525)

Cabot Brasil Industriai E Comercio Ltda., Rua Beira Rio 57, 12° andar, Sao Paulo, SP 04548-906 Brazil (Tel.: 55 11 820-2711; Telefax: 55 11 820-9193)

Cabot Carbon Ltd./Cab-O-Sil Div., Div. of Cabot Corp., Barry Site, Sully Moors Rd., Sully, S. Glamorgan, CF6 2XP UK (Tel.: 44 1446-736999; Telefax: 44 1446-737123)

Cabot Carbon Ltd./Special Blacks Div., Div. of Cabot Corp., Lees Lane, Stanlow-Ellesmere Port, South Wirral, Cheshire, L65 4HT UK (Tel.: 44 151 355 3677; Telefax: 44 151-356-0712; Telex: 629261 CABLAK G)

Cabot GmbH/Cab-O-Sil® Div., PO Box 1766, D-79607 Rheinfelden Germany (Tel.: 49 7623-9090-150; Telefax: 49 7623-90932; Telex: 773451)

Cabot Europe Ltd./Special Blacks Div., Le Nobel 4B, 2 rue Marcel Monge, F-92158 Suresnes Cedex France (Tel.: 33 1 46 97 58 00; Telefax: 33 1 47 72 66 47)

Cabot Australasia Ltd./Special Blacks Div., PO Box 19, 300 Millers Road, Altona, Victoria, 3018 Australia (Tel.: 61-3-391-1622; Telefax: 61-3-391-9370; Telex: AA30773 (Carbalt)

Cabot Pacific Asia Carbon Black Div. (PACBD), 6th Fl., RHB 1, 424 Jalan Tun Razak, 50400 Kuala Lumpur, Malaysia (Tel.: (60) 3-984-9730; Telefax: (60) 3-983-9749)

Cabot Specialty Chemicals, Inc./Special Blacks Div., 3-1-14 Shiba, Minato-ku, Tokyo, 105 Japan (Tel.: 81-3-3457-7561; Telefax: 81-3-3457-7658)

Cabot Plastics Ltd., Gate St., Dunkinfield, SK16 4RU UK (Tel.: 44 161 330-5051; Telefax: 44 161-308-2641; Telex: 667114 CABLAK G)

Cabot Plastics International, Interleuvenlaan 5, B-3001 Leuven Belgium (Tel.: 32-16390111; Telefax: 32-16401253)

Cairn Chemicals Ltd.

Cairn House, Elgiva Lane, Chesham, Buckinghamshire, HP5 2JD UK (Tel.: 44 1494 786066; Telefax: 44 1494 791816; Telex: 837075)

Calgene Chemical Inc.

7247 North Central Park Ave., Skokie, IL 60076-4093 USA (Tel.: 708-675-3950, 800-432-7187; Telefax: 708-675-3013; Telex: 72-4417)

Calgon Corp.

PO Box 1346, Pittsburgh, PA 15230 USA (Tel.: 412-494-8000, 800-648-9005; Telefax: 412-494-8927; Telex: 671183CCC)

Callaway Chemical Co.

6003 Hamilton Rd., PO Box 2335, Columbus, GA 31993-3599 USA (Tel.: 706-576-2000; Telefax: 706-576-6455)

Cambridge Industries Co. of America

7-33 Amsterdam St., Newark, NJ 07105 USA (Tel.: 201-465-4565; Telefax: 201-465-7713)

Canbro Inc.

29 E. Park St., Valleyfield, Quebec, J6S 1P8 Canada (Tel.: 514-866-8514)

Captree Chemical Corp.

32 B Nancy St., West Babylon, NY 11704 USA (Tel.: 516-491-7400, 800-899-2725; Telefax: 516-491-7130)

Carbochem, Inc.

551 Lancaster Ave., Haverford, PA 19041 USA (Tel.: 610-527-0808; Telefax: 610-527-6422; Telex: 272746)

The Carborundum Co.

PO Box 337, Niagara Falls, NY 14303-1597 USA (Tel.: 716-278-2798; Telefax: 716-278-2225)

Cardinal Stabilizers, Inc.

2010 S. Belt Line Blvd., Columbia, SC 29202 USA (Tel.: 803-799-7190; Telefax: 803-799-8742)

Cardolite Corp.

500 Doremus Ave., Newark, NJ 07105-4805 USA (Tel.: 201-344-5015, 800-322-7365; Telefax: 201-344-1197; Telex: 325446)

Cargill

Cargill, Inc., Box 5630, Minneapolis, MN 55440 USA (Tel.: 612-475-6478; Telex: CGL MPS 290625)

Cargill, Knowle Hill Park, Fairmile Lane, Cobham, Surrey, KT11 2PD UK (Tel.: 44 1932 861175; Telefax: 44 1932 861286)

R.E. Carroll, Inc.

1570 N. Olden Ave., PO Box 139, Trenton, NJ 08638 USA (Tel.: 609-695-6211, 800-257-9365; Telefax: 609-695-0102)

CasChem Inc., A Cambrex Co.

40 Ave. A, Bayonne, NJ 07002 USA (Tel.: 201-858-7900, 800-CASCHEM; Telefax: 201-437-2728; Telex: 710-729-4466)

Catalysts & Chemicals Industries Co., Ltd., Joint venture of Asahi Glass Co. Ltd./JGC Corp.

Nippon Bldg., 6-2, Ohte-machi 2-chome, Chiyoda-ku, Tokyo, 100 Japan (Tel.: (03) 3270-6086; Telefax: (03) 3246-0617; Telex: 2223480 PETCATJ)

CC Pollen Co.

3627 E. Indian School Rd., Suite 209, Phoenix, AZ 85018-5126 USA (Tel.: 602-957-0096, 800-875-0096; Telefax: 602-381-3130; Telex: 559834)

CCP Polymers Div.

217 Freeman Dr., PO Box 996, Port Washington, WI 53074-0996 USA (Tel.: 414-284-5541; Telefax: 414-284-7517)

CDC International Inc.

22 Portsmouth Rd., Amesbury, MA 01913 USA (Tel.: 508-388-2221)

Ceca SA. *See under* Elf Atochem

Cedar Concepts Corp.

4342 S. Wolcott Ave., Chicago, IL 60609 USA (Tel.: 312-890-5790; Telex: 297-220 CLTD)

Celite

Celite Corp., PO Box 519, Lompoc, CA 93438-0519 USA (Tel.: 805-735-7791; Telefax: 805-735-5699; Telex: 62776493 ESL UD)

Celite (UK) Ltd., Livingston Rd., Hessle, North Humberside, HU13 OEG UK (Tel.: 44 1482 64 52 65; Telefax: 44 1482 64 11 76; Telex: 592160)

Celite México S.A. de C.V., Alejandro Dumas No. 103, 3er, piso, Col. Polanco, C.P. 11560 Mexico (Tel.: 525-203-5611; Telefax: 525-255-1835)

Celite France, 9 rue du Colonel-de-Rochebrune, B.P. 240, 92504 Rueil-Malmaison Cedex France (Tel.: 47 49 05 60; Telefax: 47 08 30 25; Telex: Celite 631969 F)

Celite Italiana srl, Viale Pasubio No. 6, 20154 Milano Italy (Tel.: 39 2 654531; Telefax: 39 2 29005439; Telex: 311136 MANVII)

Celite Pacific, Suite 284, Sui On Centre, 8 Harbor Road Hong Kong (Tel.: 011 852 582 5609; Telefax: 011 852 827 9392)

Celtic Chemicals Ltd.

Gas Works Industrial Estate, Victoria Rd, Port Talbot, West Glamorgan, SA12 6DB, Wales UK (Tel.: 44 1639 886236; Telefax: 44 1639 893147; Telex: 94013871 CCPT G)

CE Minerals, Div. Combustion Engrg.

901 E. Eighth Ave., King of Prussia, PA 19406 USA (Tel.: 215-265-6880; Telefax: 215-337-7163)

Centerchem Inc.

225 High Ridge Rd, Stamford, CT 06905-3036 USA (Tel.: 203-975-9800; Telefax: 203-975-8777; Telex: 233322 UW)

Central Soya Co., Inc./Chemurgy Div.

PO Box 2507, 1946 West Cook Rd., Fort Wayne, IN 46801-2507 USA (Tel.: 219-425-5100, 800-348-0960; Telefax: 219-425-5301; Telex: 49609682)

Cerac, Inc.

PO Box 1178, 407 N. 13th St., Milwaukee, WI 53201 USA (Tel.: 414-289-9800; Telefax: 414-289-9805; Telex: RCA 286122)

Cerestar International Sales

Ave. Louise 149, Bte 13, B-1050 Brussels Belgium (Tel.: 32 2 535 1711; Telefax: 32 2 537 8554; Telex: 22648)

Cerestar UK Ltd.

Trafford Park Rd, Trafford Park, Manchester, M17 1PA UK (Tel.: 44 161 872 5959; Telefax: 44 161-848 9034; Telex: 667022)

Charkit Chemical Corp.

PO Box 1725, 330 Post Rd., Darien, CT 06820 USA (Tel.: 203-655-3400; Telefax: 203-655-8643; Telex: 6819184)

Chemax, Inc.

PO Box 6067, Greenville, SC 29606 USA (Tel.: 803-277-7000, 800-334-6234; Telefax: 803-277-7807; Telex: 570412 IPM15SC)

Chemcentral Corp.

7050 W. 71 St., PO Box 730, Bedford Park, IL 60499-0730 USA (Tel.: 708-594-7000, 800-331-6174; Telefax: 708-594-6328)

ChemDesign Corp., A Bayer Company

99 Development Rd., Fitchburg, MA 01420 USA (Tel.: 508-345-9999; Telefax: 508-342-9769)

Chemetals Inc.

610 Pittman Rd., Baltimore, MD 21226 USA (Tel.: 410-636-7100, 800-876-3464; Telefax: 410-636-7113)

Chemfax Inc.

3 Rivers Rd., PO Box 2389, Gulfport, MS, 39505 USA (Tel.: 601-863-6511; Telefax: 601-868-3669)

The Chemical Co.

PO Box 436, 19B Narragansett Ave, Jamestown, RI 02835 USA (Tel.: 401-423-3100; Telefax: 401-423-3102)

Chemie Linz

Chemie Linz UK Ltd., Div. of Chemie Linz GmbH, 12 The Green, Richmond, Surrey, TW9 1PX UK (Tel.: 44 181 948 6966; Telefax: 44 181-332 2516; Telex: 924941)

Chemie Linz North America, Inc., 65 Challenger Rd, Ridgefield Park, NJ 07660 USA (Tel.: 201-641-6410; Telefax: 201-641-2323; Telex: 853211 Chemie Linz)

Chemisphere Corp.

2101 Clifton Ave., St. Louis, MO 63139 USA (Tel.: 314-644-1300; Telefax: 314-644-1425)

Chemisphere Ltd.

38 King St., Chester, Cheshire, CH1 2AH UK (Tel.: 44 1244 320878; Telefax: 44 1244 320858; Telex: 61398 CHEMSPR G)

ChemMark Development Inc.

70 Tyler Place, South Plainfield, NJ 07080 USA (Tel.: 908-561-5200; Telefax: 908-561-9174)

Chem-Materials Co.

16600 Sprague Rd., Cleveland, OH 44130-6318 USA (Tel.: 216-243-5590; Telefax: 216-243-1940)

Chem-Met Co.

6419 Yochelson Pl., Clinton, MD 20735 USA (Tel.: 301-868-3355; Telefax: 301-868-8946)

Chemoxy International plc, Div. of Suter plc

All Saints Refinery, Cargo Fleet Rd, Middlesbrough, Cleveland, TS3 6AF UK (Tel.: 44 1642 248555; Telefax: 44 1642 244340; Telex: 587185)

Chemron Corp.

PO Box 2299, Paso Robles, CA 93447 USA (Tel.: 805-239-1550; Telefax: 805-239-8551; Telex: 501532)

Chemson Ltd.

Hayhole Works, Willington Quay, Wallsend, Tyne & Wear, NE28 0PB UK (Tel.: 44 191 258 5892; Telefax: 44 191 258 1549; Telex: 537726)

Chemtronics Inc.

8125 Cobb Center Dr., Kennesaw, GA 30144 USA (Tel.: 404-424-4888, 800-645-5244; Telefax: 404-423-0748; Telex: 968567)

Chevron

Chevron Chemical Co., 1301 McKinney, PO Box 3766, Houston, TX 77253-3766 USA (Tel.: 713-754-2000, 800-231-3260; Telex: 762799)

Chevron Chemical Co./Olefin & Derivs., PO Box 3766, Houston, TX 77253 USA (Tel.: 713-754-2000, 800-231-3260; Telex: 762799)

Chilean Nitrate Corp.

Town Point Center, 150 Boush St., Suite 701, Norfolk, VA 23510 USA (Tel.: 804-640-7270, 800-648-6827; Telefax: 804-640-7271; Telex: 823427)

Chisso

Chisso Corp., Tokyo Bldg., 7-3, Marunouchi 2-chome, Chiyoda-ku, Tokyo, 100 Japan (Tel.: (03) 3284-8411; Telefax: (03) 3284-8412; Telex: 02225212 CHISSO J)

Chisso America Inc., 1185 Ave of the Americas, New York, NY 10036 USA (Tel.: 212-302-0500; Telefax: 212-302-0643; Telex: WU 147029 CHISSO NYK)

Church & Dwight Co. Inc./Specialty Prods. Div.

Box CN5297, 469 N. Harrison St., Princeton, NJ 08543-5297 USA (Tel.: 609-497-7116, 800-221-0453; Telefax: 609-497-7176; Telex: 752226)

Ciba-Geigy

Ciba-Geigy Corp., 540 White Plains Rd., Tarrytown, NY 101591 USA (Tel.: 914-785-2000, 800-431-1874)

Ciba-Geigy Corp./Chemical Div., 410 Swing Rd., Greensboro, NC 27409-2080 USA (Tel.: 919-632-6000, 800-334-9481; Telefax: 919-632-7008)

Ciba-Geigy Corp./Additives Div., Seven Skyline Dr., Hawthorne, NY 10532-2188 USA (Tel.: 914-785-2000, 800-431-2360; Telefax: 914-347-5687)

Ciba-Geigy Corp./Polymers Div., 7 Skyline Dr., Hawthorne, NY 10532 USA (Tel.: 914-347-6600, 800-222-1906)

Ciba Polymers, Duxford, Cambridge, Cambridgeshire, CB2 4QA UK (Tel.: 44 1223 832121; Telefax: 44 1223-838404; Telex: 81101)

Cimbar Performance Minerals

25 Old River Rd. S.E., PO Box 250, Cartersville, GA 30120 USA (Tel.: 404-387-0319, 800-852-6868; Telefax: 404-386-6785)

Clariant

Clariant Corp., 4000 Monroe Rd., Charlotte, NC 28205 USA (Tel.: 704-331-7000, 800-631-8077; Telefax: 704-372-5787; Telex: 704-216-922)

Clariant UK, Calverley Lane, Horsforth, Leeds, West Yorkshire, LS18 4RP UK (Tel.: 44 1132 584646; Telefax: 44 1132 591232; Telex: 557114)

Clark Chemical Inc.

25 Trammel St., Marietta, GA 30064 USA (Tel.: 404-514-8909; Telefax: 404-514-8906)

W.A. Cleary Chemical Corp.

Southview Industrial Park, 178 Route #522 Suite A, Dayton, NJ 08810 USA (Tel.: 908-329-8399, 800-524-1662; Telefax: 908-274-0894; Telex: 710-991-0237)

Cleveland Pigment & Color Co.

1680 E. Market St., Akron, OH 44305 USA (Tel.: 216-794-9977; Telefax: 216-794-1510)

Climax

Climax Performance Materials Corp./Corporate Headquarters, PO Box 22015, Tempe, AZ 85285-2015 USA

Climax Fluids Additives, Div. of Climax Performance, 7666 West 63rd St., Summit, IL 60501 USA (Tel.: 708-458-8450; Telefax: 708-458-0286)

Climax Molybdenum Co., 1370 Washington Pike, Bridgeville, PA 15017 USA (Tel.: 412-257-5181; Telefax: 412-257-5191)

Climax Performance Materials Corp., 7666 West 63 St., Chicago, IL 60501 USA (Tel.: 708-458-8450, 800-323-3231; Telefax: 708-458-0286)

Climax Molybdenum BV, Postbus 1130, NL-3180 AC Rozenburg The Netherlands (Tel.: 31-15933; Telex: 23673 MOLY NL)

CM Chemical Products Inc.

50 Valley Rd., Berkeley Heights, NJ 07922 USA (Tel.: 908-665-8593, 800-654-8059; Telefax: 908-665-8543)

CNC International, Limited Partnership

PO Box 3000, 20 Privilege St., Woonsocket, RI 02895 USA (Tel.: 401-769-6100; Telefax: 401-769-4509)

Coalite Chemicals Div.

PO Box 152, Buttermilk Lane, Bolsover, Chesterfield, Derbyshire, S44 6AZ UK (Tel.: 44 1246 826816; Telefax: 44 1246 240309; Telex: 547624)

Colloids Ltd.

Dennis Road, Widnes, Cheshire, WA8 0SL UK (Tel.: 44 151 424 74247; Telefax: 44 151-423-3553; Telex: 629164 COLOID)

Colonial Metals, Inc.

Triumph Ind. Complex, PO Box 726, Elkton, MD 21921 USA (Tel.: 410-398-7200, 800-962-1537; Telefax: 410-398-2918)

Colony Industries, Inc.

226 Seventh St., Garden City, NY 11530 USA (Tel.: 516-746-2560, 800-433-4102; Telefax: 516-294-4575; Telex: 66382 UW/HEPLACK)

Colorcon, Div. of Berwind Pharmaceutical Services, Inc.

415 Moyer Blvd., West Point, PA 19486 USA (Tel.: 215-699-7733; Telefax: 215-661-2605)

Colores Hispania SA

Josep Pla 149, E-08019 Barcelona Spain (Tel.: 34-3 307 13 50; Telefax: 34-3 303 2505; Telex: 54116 COHI)

Colorite Plastics Co.

101 Railroad Ave., Ridgefield, NJ 07657 USA (Tel.: 201-941-2900, 800-631-1577; Telex: 134442)

Colorite Polymers

PO Box 116, Beverly Rd., Burlington, NJ 08016 USA (Tel.: 609-239-2200, 800-725-5155; Telefax: 609-387-4433)

Columbian Chemicals

Columbian Chemicals Co., Worldwide Headquarters, A Phelps Dodge Industries Co., 1600 Parkwood Circle, Suite 400, Atlanta, GA 30339 USA (Tel.: 770-951-5700, 800-235-4003; Telefax: 800-951-7554)

Columbian Chemicals Canada, Ltd., PO Box 3398, Station C, 755 Parkdale Ave. North, Hamilton, Ontario, L8H 7M2 Canada (Tel.: 905-544-3343; Telefax: 905-544-8641)

Columbian Chemicals, UK, Sevalco, Ltd., Severn Road, Avonmouth, Bristol, BS11 OYL UK (Tel.: 44 117-9235532; Telefax: 44 117-9235333)

Columbian Carbon International, 2 rue de la Couture, SILIC 229, 94528 Rungis Cedex France (Tel.: 33 146879241; Telefax: 33 146879062)

Columbian Chemicals Europea GmbH, Wiesenauer Strasse 11, D-30179 Hanover Germany (Tel.: 49 511-630-890; Telefax: 49 511-630-8912)

Columbian Carbon Japan, Ltd., No. 8-12 Horidomecho 1-Chome, Nihonbaski, Chuo-ku, Tokyo, 103 Japan (Tel.: 81 3-3663-2881; Telefax: 81 3-3667-1569)

Cometals Inc., Subsid. of Commercial Metals Co.

1 Penn Plaza, New York, NY 10119 USA (Tel.: 212-760-1200; Telefax: 212-564-7915; Telex: 424087)

Complex Quimica SA

PO Box 544, Monterrey, NL Mexico (64000 Mexico (Tel.: 011(528)336-2577; Telefax: 011 (528)336-3650)

Composite Particles, Inc.

2330 26th St. S.W., Allentown, PA 18103 USA (Tel.: 610-791-9900; Telefax: 610-791-2486)

Conap, Inc.

1405 Buffalo St., Olean, NY 14760 USA (Tel.: 716-372-9650; Telefax: 716-372-1594)

Condea. *See under* Vista

Condor Corp.

Executive Center, 560 Sylvan Ave., Englewood Cliffs, NJ 07632-3193 USA (Tel.: 201-567-3337, 800-321-3005; Telefax: 201-567-6489)

Cook Composites & Polymers

PO Box 419389, Kansas City, MO 64141-6389 USA (Tel.: 816-391-6000, 800-821-3590; Telefax: 816-391-6215; Telex: 26737)

Cookson Pigments Inc.

256 Vanderpool St., Newark, NJ 07114 USA (Tel.: 201-242-1800; Telefax: 201-242-7274)

Cookson Specialty Additives

1000 Wayside Rd., Cleveland, OH 44110 USA (Tel.: 216-531-6010, 800-321-4236; Telefax: 216-486-6638)

Cosan Chemical Corp., A Cambrex Co.

400 Fourteenth St., Carlstadt, NJ 07072 USA (Tel.: 201-460-9300; Telex: 642706)

Courtaulds

Courtaulds Chemicals, Nelson Acetate Works, Caton Rd., Lancaster, Lancashire, LA1 3PF UK (Tel.: 44 1524 66111; Telefax: 44 1524 846384)

Courtaulds Water Soluble Polymers, Div. of Courtaulds plc, P O Box 5, Spondon, Derbyshire, DE21 7BP UK (Tel.: 44 1332 661422; Telefax: 44 1332-661078; Telex: 32771)

Coyne Chemical

3015 State Rd., Croydon, PA 19021 USA (Tel.: 215-785-3000; Telefax: 215-785-1585)

Coz Corp., Allied Products Corp.

Providence Rd., Northbridge, MA 01534 USA (Tel.: 508-234-6151; Telefax: 508-234-4087)

CPS Chemical Co. Inc.

PO Box 162, Old Water Works Rd., Old Bridge, NJ 08857 USA (Tel.: 908-607-2700; Telefax: 908-607-2562; Telex: 844532-CPSOLDB)

Cray Valley

Cedex 101, F92970 Paris la Défense France (Tel.: 33 1 41 35 68 10; Telefax: 33 1 41 35 61 43; Telex: 615 289)

Creative Materials Inc., CMI

141 Middlesex Rd., Tyngsboro, MA 01879 USA (Tel.: 508-649-4700; Telefax: 508-649-2040)

Crescent Bronze Powder Co., Inc.

3400 N. Avondale Ave., Chicago, IL 60618-5432 USA (Tel.: 312-539-2441, 800-445-6810; Telefax: 312-539-1131)

CR Minerals Corp., Wholly owned subsid. of Canyon Resources Corp.

14142 Denver W. Pkwy., Suite 101, Golden, CO 80401 USA (Tel.: 303-278-1706, 800-527-7315; Telefax: 303-279-3772)

Croda

Croda Chemicals Ltd., Div. of Croda International plc, Cowick Hall, Snaith, Goole, North Humberside, DN14 9AA UK (Tel.: 44 1405 860551; Telefax: 44 1405 860205; Telex: 57601)

Croda Surfactants Ltd., Cowick Hall, Snaith, Goole, North Humberside, DN14 9AA UK (Tel.: 44 1405 860551; Telefax: 44 1405 860205; Telex: 57601)

Croda Universal Ltd., Div. of Croda International plc, Cowick Hall, Snaith, Goole, North Humberside, DN14 9AA UK (Tel.: 44 1405 860551; Telefax: 44 1405 860205; Telex: 57601)

Croda Colloids Ltd., Foundry Lane, Ditton Widnes, Cheshire, WA8 8UB UK (Tel.: 44 151 423 3441; Telefax: 44 151 423 3205; Telex: 629586)

Croda Food Products Ltd., Div. of Croda International plc, Cowick Hall, Snaith, Goole, North Humberside, DN14 9AA UK (Tel.: 44 1405 860551; Telefax: 44 1405 860205; Telex: 57601)

Croda Resins Ltd., Div. of Croda International plc, Crabtree Manorway South, Belvedere, Kent, DA17 6BA UK (Tel.: 44 181 311 9109; Telefax: 44 181-310 9878; Telex: 896384)

Croda Inc., 7 Century Dr., Parsippany, NJ 07054-4698 USA (Tel.: 201-644-4900; Telefax: 201-644-9222; Telex: 57601)

Croda Canada Ltd., 78 Tisdale Ave., Toronto, Ontario, M4A 1Y7 Canada (Tel.: 416-751-3571; Telefax: 416-751-9611)

Croda Japan KK, Aceman Bldg., 1-10, Tokui-cho 1-chome, Chuo-ku, Osaka, 540 Japan (Tel.: (06) 942-1791; Telefax: (06) 942-1790; Telex: 5233117)

Crompton & Knowles Corp./Dyes & Chems. Div.

PO Box 33188, Charlotte, NC 28233 USA (Tel.: 704-372-5890, 800-438-4122; Telefax: 704-372-1522)

Crosfield

Crosfield Company, Member of the Unilever Specialty Chemicals group, 101 Ingalls Ave., Joliet, IL 60435 USA (Tel.: 815-727-3651, 800-727-3651; Telefax: 815-727-5312)

Crosfield Brazil, Ave Maria Coehlo de Aguiar 215, Bloco C 3 andar, CEP 05804 Sao Paulo Brazil (Tel.: (55) 11 545 1481; Telefax: (55) 11 545 2226; Telex: 1155026 GELE BR)

Crosfield Group, Div. of Unilever plc, PO Box 26, Warrington, Cheshire, WA5 1AB UK (Tel.: 44 1925 416100; Telefax: 44 1925 59828; Telex: 627067)

Crosfield SpA, Via dei Cipressi, 10, 37033, Montorio Verona Italy (Tel.: (39) 45 557159; Telefax: (39) 45 884 0099)

Crosfield BV, Ir. Rocourstraat 28, 6245 ZG Eijsden The Netherlands (Tel.: 31 4409-9333; Telefax: 31 4409-3995; Telex: 56884 CROSF NL)

Crosfield Singapore, 320 Jalan Boon Lay, 2261 Singapore (Tel.: (65) 261 3490; Telefax: (65) 261 3254; Telex: LEVSIN RS 24757)

Crowley

Crowley Chemical Co., 261 Madison Ave., New York, NY 10016 USA (Tel.: 212-682-1200; Telefax: 212-953-3487; Telex: 12-7662)

Crowley Tar Products Co., Inc., 261 Madison Ave., New York, NY 10016 USA (Tel.: 212-682-1200)

Crown Metro Inc.

PO Box 5857, Greenville, SC 29606 USA (Tel.: 803-299-1331, 800-368-1331; Telefax: 803-299-1678; Telex: 805113)

Crown Technology, Inc.

7513 E. 96 St., PO Box 50426, Indianapolis, IN 46250-0426 USA (Tel.: 317-845-0045, 800-432-0045; Telefax: 317-845-9086)

Croxton & Garry Ltd.

Curtis Rd. Industrial Estate, Dorking, Surrey, RH4 1XA UK (Tel.: 44 1306 886688; Telefax: 44 1306-887780; Telex: 859567/8 cand g)

Crucible Chemical Co. Inc.

PO Box 6786, Donaldson Center, Greenville, SC 29606 USA (Tel.: 803-277-1284, 800-845-8873; Telefax: 803-299-1192)

Cuproquim Corp.

PO Box 171357, Memphis, TN 38187-1357 USA (Tel.: 901-537-7257; Telefax: 901-685-8372)

Custom Fibers

Custom Fibers International, PO Box 940, 2045 Lebec Rd., Lebec, CA 93243 USA (Tel.: 800-321-5324; Telefax: 805-248-1123)

Custom Fibers Europe, 13 Rassau Indust. Esatate, Ebbw Vale, Gwent, Wales UK (Tel.: 44 1495 350655)

CVC Specialty Chemicals, Inc.

600 Deer Rd., Cherry Hill, NJ 08034 USA (Tel.: 609-354-0040; Telefax: 609-354-6226)

Cyro Industries

100 Valley Rd., PO Box 950, Mt. Arlington, NJ 07856 USA (Tel.: 201-770-3000, 800-631-5384; Telefax: 201-770-6117)

Cytec Industries

Cytec Industries Inc., Five Garret Mountain Plaza, West Paterson, NJ 07424 USA (Tel.: 201-357-3100, 800-438-5615; Telefax: 201-357-3065; Telex: 130400)

Cytec de Argentina S.A., Charcas 5051, EP 1425, Buenos Aires, Argentina (Tel.: 541-772-4031; Telefax: 541-953-6619)

Cytec Industries UK Ltd., Bowling Park Dr., Bradford, West Yorkshire, BD4 7TT UK (Tel.: 44 1274 733891; Telefax: 44-1274 734770; Telex: 51295)

Cytec Industries BV, Coolsingel 139, PO Box 5195, 3197 ZH Botlek-Rotterdam The Netherlands (Tel.: 181 295400; Telefax: 181 216150; Telex: 23554)

Cytec Australia Ltd., 5 Gibbon Rd., Baulkham Hills, NSW, 2153 Australia (Tel.: 612-624-9223; Telefax: 612-838-9985; Telex: 70879 CYANAMI AA)

Cytec Japan Ltd., No. 30 Kowa Bldg., 4th Floor, 4-5 Roppongi 2-chome, Minato-ku, Tokyo, 106 Japan (Tel.: 813-3586-9716; Telefax: 813-3586-9710; Telex: 22439 CYANAMID J)

Daicel

Daicel Chemical Industries, Ltd., Toranomon Mitsui Bldg., 8-1 Kasumigaseki, 3-Chome, Chiyoda-ku, Tokyo, 100 Japan (Tel.: (03)507-3203; Telefax: (03)507-3198; Telex: 2224632 DAICEL J)

Daicel (USA) Inc., Subsid. of Diacel Chem. Industries, Ltd., One Parker Plaza, 400 Kelby St., Fort Lee, NJ 07024 USA (Tel.: 201-461-4466; Telefax: 201-461-2776)

Daicel (Europa) GmbH, Subsid. of Daicel Chem. Industries Ltd., Ost St. 22, 4000 Düsseldorf 1 Germany (Tel.: (211)369848; Telefax: (211)364429; Telex: (41)8588042 DCELD)

Daicel-Hüls Ltd., Joint venture of Daicel Chem. Industries, Ltd./Hüls AG

1-19-5, Toranomon, Minato-ku, Tokyo, 105 Japan (Tel.: (03) 3592-6333; Telefax: (03) 3592-6338; Telex: 222-7385 DAHU J)

Daihachi Chemical Industry Co., Ltd.

Sanyo Nissei Kawaramachi Bldg., 2-7, Kawaramachi 2-chome, Chuo-ku, Osaka, 541 Japan (Tel.: (06) 201-1455; Telefax: (06) 201-1458)

Dai-ichi Kogyo Seiyaku Co., Ltd.

New Kyoto Center Bldg., 614, Higashishiokoji-cho, Shimokyo-ku, Kyoto, 600 Japan (Tel.: (075) 343-1656; Telefax: (075) 343-5006)

Daniel Products Co., Inc.

400 Claremont Ave., Jersey City, NJ 07304 USA (Tel.: 201-432-0800; Telefax: 201-432-0266; Telex: 126-304)

Davis Colors, Subsid. of Rockwood Industries

3700 E. Olympic Blvd., PO Box 23100, Los Angeles, CA 90023 USA (Tel.: 213-269-7311, 800-356-4848; Telefax: 213-259-1053)

Day-Glo Color Corp., Subsid. of Nalco Chemical Co.

4515 St. Clair Ave., Cleveland, OH 44103 USA (Tel.: 216-391-7070; Telefax: 216-391-7751; Telex: 960687)

DCS Color & Supply Co., Inc.

2011 S. Allis St., Milwaukee, WI 53207 USA (Tel.: 414-769-2580; Telefax: 414-769-2598)

Dead Sea Bromine

Dead Sea Bromine Group, Flame Retardants Marketing Div., Member of ICL Group, Makleff House, PO Box 180, Beer-Sheva, 84101 Israel (Tel.: (972)57-297696; Telefax: (972)57-297846; Telex: 5335, 5343)

Bromine & Chemicals Ltd., Div. of Dead Sea Bromine, 6 Arlington Street, St James's, London, SW1A 1RE UK (Tel.: 44 171 493-9711; Telefax: 44 171-493-9714; Telex: 23845)

Eurobrom BV, PO Box 158, NL-2280 AD Rijswijk The Netherlands (Tel.: (31)70 340 84 08; Telefax: (31)70 399 90 35; Telex: 32137)

The Degen Co.

200 Kellogg St., PO Box 5240, Jersey City, NJ 07305 USA (Tel.: 201-432-1192; Telefax: 201-432-8483; Telex: 325999 DEGEN CO)

Degussa

Degussa AG, Postfach 110533, Frankfurt Germany (Tel.: +49 69-21801; Telefax: +49 69-2183218; Telex: 415200-25 dwd)

Degussa Ltd., Div. of Degussa AG, Earl Rd, Stanley Green, Handforth, Wilmslow, Cheshire, SK9 3RL UK (Tel.: 44 161 486 6211; Telefax: 44 161-485 6445; Telex: 51665053 DGMCHR G)

Degussa Corp., Wholly owned subsid. of Degussa AG, 65 Challenger Rd., Ridgefield Park, NJ 07660 USA (Tel.: 201-807-3224, 800-237-6745; Telefax: 201-807-3111; Telex: 221420 degus ur)

Jan Dekker BV

Postbus 10, NL-1520 AA Wormerveer The Netherlands (Tel.: 31-75-2782 78; Telefax: 31-75-21 38 83; Telex: 19273)

J.W.S. Delavau Co. Inc.

2140 Germantown Ave., Philadelphia, PA 19122 USA (Tel.: 215-235-1100; Telefax: 215-235-2202)

Delta Resins & Refractories Inc.

17350 Ryan Rd., Detroit, MI 48212 USA (Tel.: 313-368-7000)

Desert King Jojoba Corp.

1550 E. Missouri Ave., Suite 201, Phoenix, AZ 85014 USA (Tel.: 602-263-8350; Telefax: 602-263-8276)

Dexter

Dexter Chemical Corp., 845 Edgewater Rd., Bronx, NY 10474 USA (Tel.: 718-542-7700; Telefax: 718-991-7684; Telex: 127061)

Dexter Corp./Frekote Products, One Dexter Dr., Seabrook, NH 03874 USA (Tel.: 603-474-5541; Telefax: 603-474-5545; Telex: 6817306 HYSEA)

Dexter Distributor Programs, One Dexter Dr., Seabrook, NH 03874 USA (Tel.: 603-474-5541; Telefax: 603-474-5545)

The Dial Corp., A Greyhound Dial Co.

2000 Aucutt Rd., Montgomery, IL 60538 USA (Tel.: 708-892-4381, 800-323-5385)

Diamine and Chemicals Limited

1 National Chambers, 2nd Flr., Near Dena Bank, Ashram Rd., Ahmedabad, 380 009 India (Tel.: 91-79-409626; Telefax: 91-79-409552)

Diversified Compounders

5701 E. Union Pacific Ave., Los Angeles, CA 90022 USA (Tel.: 213-728-3000; Telefax: 213-725-7806)

Dixie Chem Co., Inc.

PO Box 130410, 300 Jackson Hill, Houston, TX 77219 USA (Tel.: 713-863-1947; Telefax: 713-863-8316; Telex: 91018815089)

D.J. Enterprises, Inc.

PO Box 31366, Cleveland, OH 44131-0366 USA (Tel.: 216-524-3879)

Dover Chemical Corp., Subsid. of ICC Industries Inc.

3676 Davis Rd. N.W., PO Box 40, Dover, OH 44622 USA (Tel.: 216-343-7711, 800-321-8805/6; Telefax: 216-364-1579; Telex: 983466)

Dow

Dow Chemical North America, 2040 Willard H. Dow Center, Midland, MI 48674 USA (Tel.: 517-636-1000, 800-441-4DOW; Telefax: 517-636-9752; Telex: 227455)

Dow Plastics, A Business Group of The Dow Chemical Co., 2040 W.H. Dow Center, Midland, MI 48674 USA (Tel.: 800-441-4DOW; Telefax: 517-638-9942)

Dow Chemical Canada Inc., PO Box 1012, 1086 Modeland Rd., Sarnia, Ontario, N7T 7K7 Canada (Tel.: 519-339-3131, 800-441-4369)

Dow Quimica Mexicana, S.A. de C.V., Av. Paseo de las Palmas 555-3, Lomas de Chapultepec Mexico (11000 D.F. Mexico (Tel.: 525-227-1900)

Dow Europe SA, Bachtobelstrasse 3, CH-8810 Horgen Switzerland (Tel.: 41-1-728-2095; Telefax: 41-1-728-3081; Telex: 826940)

Dow Chemical Co. Ltd., Div. of The Dow Chemical Co., Lakeside House, Stockley Park, Uxbridge, Middlesex, UB10 1BE UK (Tel.: 44 181 848-8688; Telefax: 44 181-848-5400; Telex: 934626)

Dow Corning

Dow Corning Corp., PO Box 0994, Midland, MI 48686-0994 USA (Tel.: 517-496-4000, 800-248-2481; Telefax: 517-496-4586; Telex: 227450)

Dow Corning Ltd., Div. of Dow Corning Corp., Kings Court, 185 Kings Rd, Reading, Berkshire, RG1 4EX UK (Tel.: 44 1734 507251; Telefax: 44 1734-575051)

Dow Corning France SA, Div. of Dow Corning Corp., BP 203, European Health Care Center, 300 route des Cretes, F-06904 Sophio Antipoles Cedex France (Tel.: 33 92 94 40 00; Telefax: 78 62 78 98; Telex: 300537)

Dow Corning Kabushiki Kaisha, Subsid. Dow Corning Corp., 507-1, Kishi Yamakita-cho, Ashigarakami-gun, Kanagawa, 258-01 Japan (Tel.: (0465) 76-3108; Telefax: (0465) 75-1064)

Dragoco Inc.

10 Gordon Dr., Totowa, NJ 07512 USA (Tel.: 201-256-3850; Telefax: 201-256-6420; Telex: 130449)

Drew Industrial Div. *See under* Ashland

Dry Branch Kaolin

Rt. 1, Box 468D, Dry Branch, GA 31020 USA (Tel.: 912-750-3500, 800-DBK-CLAY; Telefax: 912-746-0217)

DSM

DSM United Kingdom Ltd., Div. of DSM NV, Kingfisher House, Kingfisher Walk, Redditch, Worcestershire, B97 4EZ UK (Tel.: 44 1527 68254; Telefax: 44 1527-68-949; Telex: 339861)

DSM Resins UK Ltd., PO Box 8, Ellesmere Port, South Wirral, L65 0HB UK (Tel.: 44 151 355 6170; Telefax: 44 151 357 1282; Telex: 628213)

DSM Chemicals North America, Inc., 4751 Best Road, Ste 140, Atlanta, GA 30337 USA (Tel.: 404-766-3179, 800-825-4376; Telefax: 404-766-3540)

DSM Engineering Plastics, N. Am. Headquarters, PO Box 3333, 2267 West Mill Road, Evansville, IN 47732-3333 USA (Tel.: 812-435-7500, 800-333-4237; Telefax: 812-435-7702)

DSM Melamine America, Inc., 4751 Best Rd., Suite 140, Atlanta, GA 30337 USA (Tel.: 404-766-3179, 800-825-4376; Telefax: 404-766-3540)

DuPont

DuPont Chemicals, 1007 Market St., Wilmington, DE 19898 USA (Tel.: 800-441-7515)

DuPont Adhesive Polymers, 1007 Market St., Suite D-5070-3, Wilmington, DE 19898 USA (Tel.: 800-441-7111; Telefax: 302-773-2128)

DuPont Nylon, Barley Mill Plaza, PO Box 80025, Wilmington, DE 19880-0025 USA (Tel.: 302-999-3213, 800-231-0998; Telefax: 302-999-3441)

DuPont/Polymer Products Dept., Kirk Mill Bldg., Wilmington, DE 19898 USA (Tel.: 302-992-3010, 800-262-2745)

DuPont Elastomers, 505 Blue Ball Rd., PO Box 306, Elkton, MD 21922-0306 USA (Tel.: 410-392-2532, 800-452-1454; Telefax: 410-392-2540)

DuPont Canada Inc., PO Box 2200, Streetsville Postal Station, Mississauga, Ontario, L5M 2H3 Canada (Tel.: 416-821-3300, 800-6686942)

DuPont S.A. de C.V., Apartado Postal 5819, 06500 Mexico D.F. Mexico (Tel.: 52-5-250-9033; Telefax: 52-5-250-9033)

DuPont (UK) Ltd., Div. of E I Du Pont de Nemours & Co., Wedgewood Way, Stevenage, Hertsfordshire, SG1 4QN UK (Tel.: 44 1438 734026; Telefax: 44 1438 734379; Telex: 825591 DUPONT G)

DuPont France, 137 rue de L'Université, F-75334 Paris Cedex 07 France (Tel.: 33-45 50 63 32; Telefax: 33-47 53 09 65; Telex: 206772)

DuPont de Nemours International S.A., 2, chemin du Pavillon, PO Box 50, CH-1218 Le Grand-Saconnex Geneva Switzerland (Tel.: 41-22-717 51 11; Telex: 415 777 DUP CH)

DuPont (Australia) Ltd., Northside Gardens, 168 Walker St., PO Box 930, North Sydney, NSW, 2060 Australia (Tel.: 011-612 923-6111)

DuPont Far East Inc., Kowa Bldg. No. 2, 11-39 Akasaka 1-Chome, Minato-Ku, Tokyo, 107 Japan (Tel.: 585-5511)

DuPont Asia Pacific, Ltd., 1122 New World Office Bldg., East Wing, Salisbury Rd., Kowloon Hong Kong (Tel.: 852-734-5345; Telefax: 852-724-4458)

DuPont Singapore PTE, Ltd., Maritime Square, #07-01World Trade Center, Singapore 0409 (Tel.: 011-65-273-2244)

Dussek Campbell

Dussek Campbell Inc., National Wax Div., 3652 Touhy Ave., PO Box 549, Skokie, IL 60076 USA (Tel.: 708-679-6300, 800-628-9299; Telefax: 708-679-6312; Telex: 724434)

Dussek Campbell Ltd., Thames Road, Crayford, Kent, DA1 4QJ UK (Tel.: 44 1322 526966; Telefax: 44 1322 555364)

Dylon Industries, Inc.

7700 Clinton Rd., Cleveland, OH 44144-1045 USA (Tel.: 216-651-1300, 800-237-8246; Telefax: 216-651-1777)

Dymax Corporation, Div. of Dymax Engineering Adhesives

51 Greenwoods Road, Torrington, CT 06791 USA (Tel.: 203 482 1010; Telefax: 0203 496 0608; Telex: 179060 ACTC UT)

DynaGel Inc.

Wentworth Ave. & Plummer St., Calumet City, IL 60409 USA (Tel.: 708-891-8400; Telefax: 708-891-8432; Telex: 211666)

Dynaloy, Inc.

7 Great Meadow Lane, Hanover, NJ 07936 USA (Tel.: 201-887-9270; Telefax: 201-887-3678; Telex: 642033)

Eagle-Picher Industries, Inc./Chemicals Dept

PO Box 550, C & Porter Sts., Joplin, MO 64801 USA (Tel.: 417-623-8000; Telefax: 417-782-1923; Telex: 9102508335)

Eagle Zinc Co.

30 Rockefeller Plaza, New York, NY 10112 USA (Tel.: 212-582-0420; Telefax: 212-582-3412)

E-A-R Specialty Composites

7911 Zionsville Rd., Indianapolis, In, 46268 USA (Tel.: 317-692-1111; Telefax: 317-692-3111)

Eastern Chemical, Div. of Admiral Specialty Products, Inc.

PO Box 2500, Smithtown, NY 11787 USA (Tel.: 516-273-0900, 800-645-5566; Telefax: 516-273-0858; Telex: 4974275 GCC HAUP)

Eastern Color & Chemical Co.

35 Livingston St., PO Box 6161, Providence, RI 02904 USA (Tel.: 401-331-9000; Telefax: 401-331-2155)

Eastman

Eastman Chemical Products, Inc., PO Box 431, Kingsport, TN 37662-5280 USA (Tel.: 423-229-2000, 800-EASTMAN; Telefax: 423-229-1196; Telex: 6715569)

Kodak Ltd., Div. of Eastman Kodak Co., PO Box 66, Kodak House, Station Rd, Hemel Hempstead, Hertsfordshire, HP1 1TU UK (Tel.: 44 1442 61122; Telefax: 44 1442 40609; Telex: 825101)

Eastman Chemical International AG, Hertizentrum 6, 3263 Zug Switzerland (Tel.: 41 42 23 25 25; Telefax: 41 42 21 12 52; Telex: 868 824)

Ebonex Corp.

2380 S. Wabash St., PO Box 3247, Melvindale, MI 48122 USA (Tel.: 313-388-0060; Telefax: 313-388-6495)

ECC International

ECC International, 5775 Peachtree-Dunwoody Rd. NE, Suite 200G, Atlanta, GA 30342 USA (Tel.: 404-303-4415, 800-843-3222; Telefax: 404-303-4384; Telex: 6827225)

ECC International/Calcium Prods., 5775 Peachtree-Dunwoody Rd, Suite 200G, Atlanta, GA 30342 USA (Tel.: 404-843-1551, 800-251-6327; Telefax: 404-303-4384; Telex: 6827225)

ECC International Ltd., Div. of ECC Group plc, John Keay House, St. Austell, Cornwall, PL25 4DJ UK (Tel.: 44 1726 74482; Telefax: 44 1726 623019; Telex: 45526 ECCSAU G)

ECC International SA, Div. of ECC Group plc, 2 rue du Canal, B-4551 Lixhe Belgium (Tel.: 32-41 79 98 11; Telefax: 32-41 79 82 79)

ECC Japan Ltd., Div. of ECC International, 1-5-11, Shiba, Minato-ku, Tokyo, 105 Japan (Tel.: (03) 5443-3144; Telefax: (03) 5443-3137; Telex: J28915)

Eckart-Werke

Kaiserstrasse 30, D-90763 Fürth/Bay Germany (Tel.: (0911) 99 78-0; Telefax: (0911) 99 78-238; Telex: 623415)

Efka Chemicals B.V.

2182 GZ Hillegom The Netherlands (Tel.: 31-2520-22284; Telefax: 31-2520-16722; Telex: 71036 efka nl)

EGC Corp.

Box 16080, Houston, TX 77222 USA (Tel.: 713-447-8611, 800-342-7677; Telefax: 713-931-2201)

Eka Nobel. *See under* Akzo Nobel

Electro Abrasives Corp.

701 Willet Rd., Buffalo, NY 14218 USA (Tel.: 716-822-2500, 800-284-GRIT; Telefax: 716-822-2858)

Electrochem Ltd.

Unit 11, Newfield Industrial Estate, Tunstall, Stoke-on-Trent, Staffordshire, ST6 5PD UK (Tel.: 44 1782 822 058; Telefax: 44 1782 822 350; Telex: 848668 GASKEM G)

Elf Atochem

Elf Atochem S.A., 4, cours Michelet, La Défense 10, F-92091 Paris Cedex 42 France (Tel.: 49-00-8080; Telefax: 49-00-7447; Telex: 611922 ATO F)

Elf Atochem North America Inc., Headquarters, 2000 Market St., Philadelphia, PA 19103-3222 USA (Tel.: 215-419-7000, 800-225-7788; Telefax: 215-419-7591)

Elf Atochem North America Inc./Plastics Dept., 2000 Market St., Philadelphia, PA 19103 USA (Tel.: 215-587-7000, 800-328-2811; Telefax: 215-587-7497)

Elf Atochem North America Inc./Polymers & Plastic Additives, Two Appletree Sq., Suite 147, Bloomington, MN 55425 USA (Tel.: 612-854-0450; Telefax: 612-854-0669)

Elf Atochem North America Inc./Wire Mill Products Dept., 43 James St., Homer, NY 13077 USA (Tel.: 607-749-2652)

Elf Atochem Canada, PO Box 278, Oakville, Ontario, L6J 5A3 Canada (Tel.: 905-827-9841; Telefax: 905-827-7913)

Elf Atochem UK Ltd., Colthrop Lane, Thatcham, Newbury, Berkshire, RG13 4LW UK (Tel.: 44 1635 870000; Telefax: 44 1635-861212; Telex: 847689 ATOKEM G)

Ceca SA, Div. of Elf Atochem, 22, place de l'Iris, La Défense 2, Cedex 54, 92062 Paris-La Défense France (Tel.: 147-96-9090; Telefax: 147-96-9234; Telex: 611444 ckd)

Elkem Metals Co.

200 Deer Run Rd., Sewickley, PA 15143 USA (Tel.: 412-749-3900; Telefax: 412-749-3920)

Emco Services Inc.

PO Box 2191, Taunton, MA 02780 USA (Tel.: 508-823-8852; Telefax: 508-822-1931)

Emerson & Cuming Composite Materials Inc.

59 Walpole St., Canton, MA 02021-1838 USA (Tel.: 617-821-4250; Telefax: 617-828-0124)

EM Industries

EM Industries, Inc./Fine Chems. Div., An Associate of E. Merck, 5 Skyline Dr., Hawthorne, NY 10532 USA (Tel.: 914-592-4350; Telefax: 914-592-9469)

EM Industries, Inc./Pigments Div., 5 Skyline Dr., Hawthorne, NY 10532 USA (Tel.: 914-592-4350; Telefax: 914-592-9469)

BDH Inc., Pigments Div., 350 Evans Ave., Toronto, Ontario, M8Z 1K5 Canada (Tel.: 416-255-8521; Telex: 06-967678)

Emkay Chemical Co.

319-325 Second St., PO Box 42, Elizabeth, NJ 07206 USA (Tel.: 908-352-7053; Telefax: 908-352-6398)

EMS

EMS-Chemie AG, Selnaustrasse 16, CH-8039 Zurich Switzerland (Tel.: 1-284-1820; Telefax: 1-284-1859; Telex: 815312)

EMS-American Grilon Inc., Subsid. of EMS-Chemie AG, Switzerland, Corporate Way Road, PO Box 1717, Sumter, SC 29151-1717 USA (Tel.: 803-481-9173; Telefax: 803-481-6121; Telex: 805077)

Emulsion Systems Inc.

70 East Sunrise Hwy., Valley Stream, NY 11581-1233 USA (Tel.: 516-825-3232, 800-ESI-CRYL; Telefax: 516-825-3233)

Engelhard

Engelhard Corp., 101 Wood Ave. South, CN 770, Iselin, NJ 08830-0770 USA (Tel.: 908-205-5000, 800-631-9505; Telefax: 908-906-0337; Telex: 219984 ENGL UR)

Engelhard Canada Limited, 195 Riviera Dr., Markham, Ontario, L3R 5J6 Canada (Tel.: 905-940-4020; Telefax: 905-940-4470)

Engelhard Ltd., Chancery House, St. Nicholas Way, Sutton, Surrey, SM1 1JB UK (Tel.: 44-181-643-8080; Telefax: 44-181-643-6063)

Engelhard SA, 4 Rue de Beaubourg, 75004 Paris France (Tel.: 33-1-44-611000; Telefax: 33-1-44-611192)

Engelhard GmbH, Lise Meitner Strasse 7, 63303 Dreieich Germany (Tel.: 49-6103-9345-0; Telefax: 49-6103-34787)

Engelhard s.r.l., Via Ronchi, 17, 20134 Milan Italy (Tel.: 39-2-264-251; Telefax: 39-2-215-4602)

Engelhard Corp. (Japan), 9th Fl. Toranomon 3-Chome, Annex, 7-12 Toranomon 3-Chome, Minato-ku, Tokyo, 105 Japan

Engelhard (Hong Kong) Ltd., Block B2, 6/F, Eldex Industrial Bldg., 21 Ma Tau Wei Road, Hunghom, Kowloon Hong Kong (Tel.: 852-2365-0302; Telefax: 852-2765-6406)

Engelhard Australia Pty. Ltd., 10-12 Prospect St., Box Hill 3128, Victoria Australia (Tel.: 61-3-899-6330; Telefax: 61-3-899-6360)

EniChem America, Inc.

1211 Ave. of the Americas, New York, NY 10036 USA (Tel.: 212-382-6521; Telefax: 212-382-6584; Telex: 6801159 ENICHEM)

Ennar Latex, Inc.

PO Box 247, Middlebury, CT 06762 USA (Tel.: 203-597-9275; Telefax: 203-758-8654)

Enterprise Chemical Corp.

4 Parkview Dr., Dover, OH 44622 USA (Tel.: 216-343-8861, 800-875-8861; Telefax: 216-343-8853)

Enzyme Development Corp.

2 Penn Plaza, Ste. 2439, New York, NY 10121-0034 USA (Tel.: 212-736-1580; Telefax: 212-279-0056; Telex: 427471 BSCE)

Enzymol International, Inc.

2543 Westbelt Dr., Columbus, OH 43228 USA (Tel.: 614-529-7333; Telefax: 614-529-7342)

Epolin, Inc.

358-364 Adams St., Newark, NJ 07105 USA (Tel.: 201-465-9495; Telefax: 201-465-5353)

Esprit Chemical Co.

800 Hingham St., Suite 207S, Rockland, MA 02370 USA (Tel.: 617-878-5555, 800-2-ESPRIT; Telefax: 617-871-5431; Telex: 951346)

Esschem Co.

PO Box 56, Essington, PA 19029 USA (Tel.: 610-521-3800, 800-765-9637; Telefax: 800-765-9275)

Essex Specialty Products Inc., Subsid. of Dow Chemical Co.

1250 Harmon Rd., Auburn, MI 48326 USA (Tel.: 810-391-6300; Telefax: 810-391-6417; Telex: 62953879)

Estron Chemical, Inc.

P.O. Box 127, Hwy. 95, Calvert City, KY 42029 USA (Tel.: 502-395-4195; Telefax: 502-395-5070)

Ethox Chemicals, Inc.

PO Box 5094, Sta. B, Greenville, SC 29606 USA (Tel.: 803-277-1620; Telefax: 803-277-8981)

Etna Products Inc.

PO Box 630, Chagrin Falls, OH 44022-0630 USA (Tel.: 216-543-9845; Telefax: 216-543-1789; Telex: 980131 WDMR)

Eurobrom. *See under* Dead Sea Bromine

Eval Co. of America

1001 Warrenville Rd., Suite 201, Lisle, IL 60532 USA (Tel.: 708-719-4610, 800-423-9762; Telefax: 708-719-4622)

Evans Chemetics. *See* Hampshire Chemical Corp.

Expancel

Expancel, Div. of Nobel Industries, PO Box 13000, S-85013 Sundsvall Sweden (Tel.: 46 60-134000; Telefax: 46 60-569518; Telex: 71399 expancel s)

Expancel, Inc., Div. of Nobel Industries, 2150-H Northmont Pkwy., Duluth, GA 30136 USA (Tel.: 404-813-9126; Telefax: 404-813-8639)

Exxon

Exxon Chemical Co., PO Box 3272, Houston, TX 77253-3272 USA (Tel.: 713-870-6000, 800-526-0749; Telefax: 713-870-6661; Telex: 794588)

Exxon Chemical Co./Application Chemicals Div., Tomah Products, 1012 Terra Dr., PO Box 388, Milton, WI 53563 USA (Tel.: 608-868-6811, 800-441-0708; Telefax: 608-868-6810; Telex: 910-280-1401)

Exxon Chemical Geopolymers Ltd., Div. of Exxon Corp., PO Box 122, 4600 Parkway, Solent Business Park, Whiteley, Fareham, Hampshire, PO15 7AZ UK (Tel.: 44 1489 884400; Telefax: 44 1489 884403; Telex: 47437)

Exxon Chemical Mediterranea SpA, Div. of Exxon Corp., Via Paleocapa 7, I-20121 Milan Italy (Tel.: 39-2 88031; Telefax: 39-2 8803231; Telex: 311561 ESSOCH I)

Exxon Chemical Europe Inc., 280 Vorstlaan, Bld du Souverain, B-1160 Bruxelles Belgium (Tel.: 32-2-674-41 11; Telefax: 32-2-674 41 29)

Exxon Chemical Belgium, Div. of Exxon Corp., Boulevard du Souverain 280, B-1160 Brussels Belgium (Tel.: 32-2-674 41 11; Telefax: 32-2-674 41 29; Telex: 22364)

Exxon Chemical International Services Ltd., Div. of Exxon Corp., 33rd Floor, Shui on Centre, 8 Harbour Road, Wanchai Hong Kong (Tel.: 852 582-0888)

Fabrichem Inc.

211 Sigwin Dr., Fairfield, CT 0006430 USA (Tel.: 203-259-5512; Telefax: 203-254-7886)

Faesy & Besthoff, Inc.

143 River Rd., Edgewater, NJ 07020-0029 USA (Tel.: 201-945-6200; Telefax: 201-945-6145)

Fairmount Chemical Co., Inc.

117 Blanchard St., Newark, NJ 07105 USA (Tel.: 201-344-5790, 800-872-9999; Telefax: 201-690-5298; Telex: 138905)

The Fanning Corp.

2450 W. Hubbard St., Chicago, IL 60612-1408 USA (Tel.: 312-563-1234; Telefax: 312-563-0087)

Farleyway Chemicals Ltd.

Ham Lane,, Kingswinford, West Midlands, DY6 7JU UK (Tel.: 44 1384 400 222; Telefax: 44 1384 400 020; Telex: 339528 FAR G)

FAR Research, Inc.

2210 Wilhelmina Ct., N.E., Palm Bay, FL 32905 USA (Tel.: 407-723-6160; Telefax: 407-723-8753)

The Feldspar Corp., Subsid. of Zemex Corp.

1040 Crown Pointe Parkway, Suite 270, Atlanta, GA 30338 USA (Tel.: 404-392-8660; Telefax: 404-392-8670)

Ferro

Ferro Corp./World Headquarters, 1000 Lakeside Ave., Cleveland, OH 44114-1183 USA (Tel.: 216-641-8580; Telefax: 216-696-6958; Telex: 98-0165)

Ferro Corp./Bedford Chemical Div., 7050 Krick Rd., Bedford, OH 44146 USA (Tel.: 216-641-8580, 800-321-9946; Telefax: 216-439-7686; Telex: 98-165)

Ferro Corp./Color Div., 4150 E. 56th St., PO Box 6550, Cleveland, OH 44101 USA (Tel.: 216-641-8580; Telefax: 216-641-8831; Telex: 98-0165)

Ferro Corp./Grant Chemical Div., PO Box 263, Baton Rouge, LA 70821 USA (Tel.: 504-654-6801; Telefax: 504-654-3268; Telex: 980165)

Ferro Corp./Keil Chemical Div., 3000 Sheffield Ave., Hammond, IN 46320 USA (Tel.: 219-931-2630, 800-628-9079; Telefax: 219-931-0895; Telex: 725484)

Ferro Corp./Transelco Div., Box 217, Penn Yan, NY 14527 USA (Tel.: 315-536-3357; Telefax: 315-536-8091; Telex: 97 8373)

Filter-Media Inc.

3603 Westcenter Dr., Houston, TX 77224-9156 USA (Tel.: 713-780-9000; Telefax: 713-781-4320; Telex: 775144 FEMCO HOU)

Fina Chemicals, Div. of Petrofina SA

52 Rue de l'Industrie, B-1040 Brussels Belgium (Tel.: 32 2-288 9132; Telefax: 32-2-288-3322; Telex: 21 556 PFINA B)

Finetex Inc.

418 Falmouth Ave., PO Box 216, Elmwood Park, NJ 07407 USA (Tel.: 201-797-4686; Telefax: 201-797-6558; Telex: 710-988-2239)

Firestone Synthetic Rubber & Latex Co., Div. of Bridgestone/Firestone Inc.

PO Box 26611, Akron, OH 44319-0006 USA (Tel.: 216-379-7727, 800-282-0222; Telefax: 216-379-7875)

First Chemical Corp.

PO Box 1427, Pascagoula, MS, 39568-1427 USA (Tel.: 601-762-0870, 800-828-7940; Telefax: 601-762-5213; Telex: 510-990-3361)

G. Fiske & Co. Ltd.

64 Sheen Rd., Richmond, Surrey, TW9 1UF UK (Tel.: 44 181 948 5811; Telefax: 44 181-948 7059; Telex: 925878)

Flexible Products Co.

1007 Industrial Park Dr., Box 3190, Marietta, GA 30061 USA (Tel.: 770-428-2684; Telefax: 770-421-6495)

Floratech

2295 S. Coconino Dr., Apache Junction, AZ 85220 USA (Tel.: 602-983-7909; Telefax: 602-982-4183)

Floridin Co.

PO Box 510, 1101 N. Madison St., Quincy, FL 32351-0510 USA (Tel.: 904-627-7688, 800-228-1131; Telefax: 904-875-1757; Telex: 4931835 FLOQYUI)

Fluka Chemical Corp.

980 South Second St., Ronkonkoma, NY 11779 USA (Tel.: 516-467-0980, 800-FLUKA-US; Telefax: 800-441-8841; Telex: 96-7807)

Fluorocarbon Co. Ltd.

Caxton Hill, Hertford, Hertfordshire, SG13 7NH UK (Tel.: 44 1992 550731; Telefax: 44 1992 584697; Telex: 81435)

FMC

FMC Corp./Chemical Products Group, 1735 Market St., Philadelphia, PA 19103 USA (Tel.: 215-299-6000, 800-346-5101; Telefax: 215-299-5999; Telex: 685-1326)

FMC Corp./Process Additives Div., 1735 Market St., Philadelphia, PA 19103 USA (Tel.: 215-299-6121, 800-545-6532)

FMC Corp. (UK) Ltd./Process Additives Div., Tenax Rd., Trafford Park, Manchester, Lancaster, M17 1WT UK (Tel.: 44 161 872 2323; Telefax: 44 161 873 3177; Telex: 666177)

FMC Corp. N.V., Ave. Louise 480-B9, Brussels 1050 Belgium (Tel.: 322-645 5511; Telefax: 322-640 6350)

FMC Litex AS, Risingvej 1, D-2665 Vallensbaek Strand Denmark (Tel.: 45 42 73 1122; Telefax: 45 42 73 1225; Telex: 855-33194 DANGEL DK)

FMC International S.A., 4th Floor, Interbank Bldg., 111 Paseo de Roxas, Makati, Metro Manila Phillippines (Tel.: (632) 817 5546; Telefax: (632) 818 1485)

Focus Chemical

875 Greenland Rd., Orchard Park, Suite B9, Portsmouth, NH 03801 USA (Tel.: 603-430-9802)

Folexco Inc.

150 Domorah Dr., Montgomeryville, PA 18936 USA (Tel.: 215-628-8895; Telefax: 215-628-8651)

Franklin Industrial Minerals

Franklin Industrial Minerals, 612 Tenth Ave. North, Nashville, TN 37203 USA (Tel.: 615-259-4222, 800-872-3740; Telefax: 615-726-2693)

Franklin Industrial Minerals, 821 Tilton Bridge Rd., S.E., Dalton, GA 30721 USA (Tel.: 404-277-3740; Telefax: 404-277-9827)

Freedom Textile Chemicals Co.

801 Washington St., Conshohocken, PA 19428 USA (Tel.: 215-828-3800, 800-533-4514; Telefax: 215-834-7855; Telex: 5106608845)

Frinton Laboratories, Inc.

PO Box 2428, Vineland, NJ 08360 USA (Tel.: 609-692-6902; Telefax: 609-692-6922)

Fry Metals, Inc.

4100 Sixth Ave., Altoona, PA 16602 USA (Tel.: 814-946-1611, 800-289-3797; Telefax: 814-944-8094)

Fry's Metals Ltd., Div. of Cookson Group plc

Tandem House, Beddington Farm Rd, Croydon, Greater London, CR9 4BT UK (Tel.: 44 181 665 6666; Telefax: 44 181-665 6196; Telex: 265732)

Fujimi Corp.

747 Church Rd., F-3, Elmhurst, IL 60126 USA (Tel.: 708-941-1400; Telefax: 708-941-7060)

Fuji Silysia

Fuji Silysia Chemical Ltd., 2-1846, Kozoji-cho, Kasugai-shi, Aichi-ken, 487 Japan (Tel.: (81) 568-51-2511; Telefax: (81) 568-8557)

Fuji Silysia Chemical Ltd., Bank of Am. Financial Center, 121 SW Morrison St., Suite 865, Portland, OR 97204 USA (Tel.: 503-295-1933; Telefax: 503-295-1832)

Fuji Titanium Industry Co., Ltd.

3-6-32, Nakanoshima, Kita-ku, Osaka, 530 Japan (Tel.: (06) 441-6856; Telefax: (06) 441-6855)

H.B. Fuller Co.

3530 Lexington Ave. North, St. Paul, MN 55126-8076 USA (Tel.: 612-481-1816, 800-468-6358; Telefax: 612-481-1863)

Furukawa Co., Ltd.

6-1, Marunouchi 2-chome, Chiyoda-ku, Tokyo, 100 Japan (Tel.: (03) 3212-6561; Telefax: (03) 3287-0696; Telex: 02225614 FURUKOJ)

Futura Coatings

Futura Coatings, Inc., 9200 Latty Ave., Hazelwood, MO 63042 USA (Tel.: 314-521-4100; Telefax: 314-521-7255)

Futura Coatings Europe N.V., Route de Flandre 100, B-7780 Komen Belgium (Tel.: 32-56-55-44-22; Telefax: 32-56-55-45-00)

Gayson, Inc.

30 Second St. SW, Barberton, OH 44203 USA (Tel.: 216-848-9200, 800-442-9766)

GCA Chemical Corp.

916 West 13th St., Bradenton, FL 34205 USA (Tel.: 813-748-6090; Telefax: 813-748-0194)

GE. *See under* General Electric

Gelest Inc.

612 William Leigh Dr., Tullytown, PA 19007-6308 USA (Tel.: 215-547-1015; Telefax: 215-547-2484)

Geltech, Inc.

1 Progress Blvd., #8, Alachua, FL 32615 USA (Tel.: 904-462-2358, 800-323-6595; Telefax: 904-462-2993)

General Chemical Corp.

90 East Halsey Rd., Parsippany, NJ 07054-0373 USA (Tel.: 201-515-0900, 800-631-8050; Telefax: 201-515-2468; Telex: 139-450 GEN-CHEMPAPY)

General Chemical & Plastic Services Co.

88 Main St., Suite 432, Mendham, NJ 07945 USA (Tel.: 908-953-0070, 800-555-0239; Telefax: 908-953-0073)

General Electric

General Electric Co./Plastics Div., One Plastics Ave., Pittsfield, MA 01201 USA (Tel.: 413-448-7110, 800-845-0600; Telex: 926430)

General Electric Co./Silicone Products Div., 260 Hudson River Rd., Waterford, NY 12188 USA (Tel.: 518-237-3330, 800-255-8886; Telefax: 518-233-3931)

GE Specialty Chemicals, 501 Avery St., PO Box 1868, Parkersburg, WV 26102-1868 USA (Tel.: 304-424-5411, 800-872-0022; Telefax: 304-424-5871)

GE Silicones Europe, Postbus 117, Plasticslaan 1, NL-4600 AC Bergen op Zoom The Netherlands (Tel.: 31-1640-32291; Telefax: 31-1640-32708; Telex: 78421)

GE Plastics UK, Old Hall Rd., Sale, Manchester, M33 2HG UK (Tel.: 44 161 905 5000)

GE Silicones, Div. of GE Plastics Ltd., Old Hall Rd., Sale, Manchester, M33 2HG UK (Tel.: 44 161 905 5000; Telefax: 44 161-905 5022)

General Latex & Chemical Corp.

67 High St., N. Bellerica, MA 01862 USA (Tel.: 508-663-3485; Telefax: 508-663-3488)

General Polymers Corp.

2 Sherman St., Linden, NJ 07036 USA (Tel.: 908-925-6776)

Generichem Corp.

85 Main St., PO Box 369, Little Falls, NJ 07424 USA (Tel.: 201-256-9266; Telefax: 201-256-0069; Telex: 510-601-6431)

Genesee Polymers Corp.

G-5251 Fenton Rd., PO Box 7047, Flint, MI 48507-0047 USA (Tel.: 810-238-4966; Telefax: 810-767-3016)

Genstar Stone Products Co.

Executive Plaza IV, 11350 McCormick Rd., Hunt Valley, MD 21031 USA (Tel.: 410-527-4000; Telefax: 410-527-4535)

Geon Company

6100 Oak Tree Blvd., Cleveland, OH 44131 USA (Tel.: 216-447-6000, 800-438-4366)

Georgia Gulf

Georgia Gulf Corp., PO Box 105197, Atlanta, GA 30348 USA

Georgia Gulf Corp./PVC Div., PO Box 629, Plaquemine, LA 70765-0629 USA (Tel.: 504-685-1200, 800-241-2673; Telefax: 504-685-1270)

Georgia Kaolin Co. *See* ECC Int'l.

Georgia Marble Co.

1201 Roberts Blvd., Bldg. 100, Kennesaw, GA 30144-3619 USA (Tel.: 404-421-6500, FAX 404-421-6507)

Georgia-Pacific

Georgia-Pacific Chemical Div., 1754 Thorne Rd., Tacoma, WA 98421 USA (Tel.: 206-572-8181; Telefax: 206-572-4721)

Georgia-Pacific Resins Inc., Subsid. of Georgia-Pacific Corp., 2883 Miller Rd, Decatur, GA 30035 USA (Tel.: 770-593-6859, 800-765-7374; Telefax: 800-395-6885; Telex: 804600)

GFI, Inc.

111 Boston Post Rd., PO Box 777, Sudbury, MA 01776 USA (Tel.: 508-443-3674; Telefax: 508-443-5714)

GFS Chemicals, Inc.

PO Box 245, Powell, OH 43065 USA (Tel.: 614-881-5501, 800-858-9682; Telefax: 614-881-5989; Telex: 981282 GFS CHEM UD)

G + G International Inc.

PO Box 165, Basking Ridge, NJ 07920 USA (Tel.: 908-776-9820; Telefax: 908-953-9569)

Giles Chemical Corp.

PO Box 370, 214 Commerce St., Waynesville, NC 28786 USA (Tel.: 704-452-4784; Telefax: 704-452-4786)

Girindus Chemie GmbH & Co. KG

Buchenallee 20, PO Box 100 259, D-51402 Bergisch Germany (Tel.: 0-2204-6-30-52; Telefax: 0-2204-6-34-14; Telex: 8874555 GIR D)

Gist-brocades

Gist-brocades Food Ingredients, Inc., 2200 Renaissance Blvd., Suite 150, King of Prussia, PA 19406 USA (Tel.: 215-272-4040, 800-662-4478; Telefax: 215-272-5695; Telex: 216902)

Gist-brocades SpA, Via Milano 42, I-27045 Casteggio Italy (Tel.: 39-383-8931; Telefax: 39-383-805397; Telex: 321197 VINAL I)

Givaudan-Roure Corp.

100 Delawanna Ave., Clifton, NJ 07014 USA (Tel.: 201-365-8277; Telefax: 201-777-9304; Telex: 219259 givc ur)

GMI Products Inc.

2525 Davie Rd., Suite 330, Davie, FL 33317 USA (Tel.: 305-999-9373; Telefax: 305-474-0989)

Goldschmidt

Goldschmidt AG, Th., Goldschmidtstrasse 100, Postfach 101461, D-4300 Essen 1 Germany (Tel.: 49 201-173-01; Telefax: 49 201-224916; Telex: 857170)

Goldschmidt Ltd., Subsid. of Goldschmidt AG, Tego House, Victoria Road, Ruislip, Middlesex, HA4 0YL UK (Tel.: 44 181 422 7788; Telefax: 44 181-864 8159; Telex: 923146)

Goldschmidt Chemical Corp., 914 E. Randolph Rd., PO Box 1299, Hopewell, VA 23860 USA (Tel.: 804-541-8658, 800-446-1809; Telefax: 804-541-8689; Telex: 710-958-1350)

Goldschmidt Canada, 2150 Winston Park Dr., Unit 201, Oakville, Ontario, L6H 5V1 Canada (Tel.: 905-829-2233; Telefax: 905-829-2575)

Tego Chemie Service GmbH, Gerlingstr. 64, D-45139 Essen Germany (Tel.: 49-201-173-06; Telefax: 49-201-173-1939; Telex: 85717-20tgd)

Tego Chemie Service USA, Div. of Goldschmidt Chem. Corp., PO Box 1299, 914 E. Randolph Rd., Hopewell, VA 23860 USA (Tel.: 804-541-8658, 800-446-1809; Telefax: 804-541-2783)

Goldschmidt Industrial Chemicals Corp.

PO Box 279, McDonald, PA 15057 USA (Tel.: 412-796-1511, 800-426-7273; Telefax: 412-922-6657)

Goodrich

BFGoodrich Specialty Chemicals, 9911 Brecksville Rd., Brecksville, OH 44141-3247 USA (Tel.: 216-447-5000, 800-331-1144; Telefax: 216-447-5770; Telex: 4996831)

Goodrich Canada, 125 Northfield Dr. West, Waterloo, Ontario, N2L 6K4 Canada (Tel.: 519-888-3330; Telefax: 519-888-3337)

Goodrich Holding Corp., Subsid. of BFGoodrich, Av. La Paseo de Las Palmas No. 735-103, Col. Lomas de Barrilaco, CP 11010 Mexico, D.F. Mexico (Tel.: 52-5-520-3199; Telefax: 52-5-520-8542)

BFGoodrich Chemical (UK) Ltd., The Lawn, 100 Lampton Road, Hounslow, Middlesex, TW3 4EB UK (Tel.: 44 181 570 4700; Telefax: 44 181-570 0850)

BFGoodrich Chemical Europe, 742 Rue de Verdun, B-1130 Brussels Belgium (Tel.: 32-2-247-1911; Telefax: 32-2-247-1991)

BFGoodrich Chemical (Deutschland) GmbH, Goerlitzer Str. 1, D-41460 Neuss 1 Germany (Tel.: 49-2131-18050; Telefax: 49-2131-180530)

BFGoodrich Chemical (Italia) Srl, Viale Gian Galeazzo 25/27, 20136 Milano Italy (Tel.: 39-2-5830-6961; Telefax: 39-2-5831-0390)

BFGoodrich Chemical Ltd., 14 Queens Rd., Melbourne, Victoria, 3004 Australia (Tel.: 61-3-267-6488; Telefax: 61-3-820-1094)

BFGoodrich Chemical (Far East) Ltd., 2208 Fortress Tower, 250 Kings Rd. Hong Kong (Tel.: 852-508-1021; Telefax: 852-512-2241)

Goodyear

Goodyear Tire & Rubber Co./The Chemical Div., 1485 E. Archwood Ave., Akron, OH 44306-3299 USA (Tel.: 216-796-6400, 800-548-8107; Telefax: 216-796-2617; Telex: 640550 Gdyr)

Goodyear Chemicals Europe, 14 Ave. des Tropiques, Z.A. de Courtaboeuf 2, 91955 Les Ulis Cedex France (Tel.: 33-1-69-29-27-00; Telefax: 33-1-69-29-27-01; Telex: 602895F)

Graden Chemical Co., Inc.

426 Bryan St., Havertown, PA 19083 USA (Tel.: 215-449-3808)

Grain Processing Corp.

1600 Oregon St., Muscatine, IA 52761 USA (Tel.: 319-264-4265; Telefax: 319-264-4289; Telex: 46-8497)

Great Lakes

Great Lakes Chemical Corp., PO Box 2200, One Great Lakes Blvd., W. Lafayette, IN 47906-0200 USA (Tel.: 317-497-6100, 800-621-9521; Telefax: 317-497-6123)

Great Lakes Chemical Corp., Rua Itapaiuna 1800-Casa 56, Morumbi, Sao Paulo SP 05707-001 Brazil (Tel.: 55-11-844-6486; Telefax: 55-11-844-6787)

Great Lakes Chemical (Europe) Ltd., P O Box 44, Oil Sites Road, Ellesmere Port, South Wirral, L65 4GD UK (Tel.: 44 151 356 8489; Telefax: 44 151-356 8490)

Great Lakes-QO Chemicals, Inc., Industrieweg 12, Haven 391, B-2030 Antwerp Belgium (Tel.: 32-3-541-2165; Telefax: 32-3-541-6503)

Great Lakes Chemical Corp./Japan, LaVie Sakuragicho Bldg., 5-26-3, Sakuragicho, Nishi-Ku, Yokohama, 220 Japan (Tel.: 81-45-212-9541; Telefax: 81-45-212-9539)

Great Western Chemical Co.

808 SW 15th Ave., Portland, OR 97205 USA (Tel.: 503-228-2600; Telefax: 503-221-5752; Telex: 910-464-4733)

R.W. Greeff & Co. Inc.

777 West Putnam Ave, Greenwich, CT 06830 USA (Tel.: 203-532-2900; Telefax: 203-532-2980; Telex: 996609)

Grefco Inc.

3435 W. Lomita Blvd., Torrance, CA 90509 USA (Tel.: 310-517-0700; Telefax: 310-517-0794; Telex: 664266)

Gresco Mfg., Inc.

216 E. Hollyhill Rd., Thomasville, NC 27360 USA (Tel.: 919-475-8101; Telefax: 919-475-0100)

Grindsted

Grindsted Products A/S, Edwin Rahrs Vej 38, DK-8220 Brabrand Denmark (Tel.: 45 86-25-3366; Telefax: 45 86-25-1077; Telex: 64177 gvdan dk)

Grindsted Products, Ltd., Northern Way, Bury St Edmunds, Suffolk, IP32 6NP UK (Tel.: 44 1284 769631; Telefax: 44 1284 760839; Telex: 81203)

Grindsted Products GmbH, Roberts-Bosch Strasse 20-24, D-25451 Quickborn Germany (Tel.: 4106/70960; Telefax: 4106/709666; Telex: 2180684 gpd d)

Grindsted Products, Inc., 201 New Century Pkwy., PO Box 26, Industrial Airport, KS 66031 USA (Tel.: 913-764-8100, 800-255-6837; Telefax: 913-764-5407; Telex: 4-37295)

Grindsted Products, Inc., 10 Carlson Court, Suite 580, Rexdale, Ontario, M9W 6L2 Canada (Tel.: 416-674-7340; Telefax: 416-674-7378)

Grindsted de México, S.A. de C.V., Cerrada de las Granjas 623, Col. Jagüey, Delegación Azcapotzalco, 02300 México, D.F. Mexico (Tel.: (5) 352 9102; Telefax: (5) 561 3285)

Grindsted do Brazil, Indústria e Comércio Ltda., Rodovia Regisé Bittencourt, KM 275,5, 06818-900 Embú S.P. Brazil (Tel.: (11) 494-3899; Telefax: (11) 494-3823; Telex: 1171854 gpbr br)

Nippon Grindsted K.K., Daiichi Nishiwaki Bldg., 1-58-10 Yoyogi, Shibuya-ku, Tokyo, J-151 Japan (Tel.: 3-3375-3481; Telefax: 3-3375-3715)

Gumix International Inc.

2160 N. Central Rd., Fort Lee, NJ 07024-7552 USA (Tel.: 201-947-6300, 800-2GU-MIX2; Telefax: 201-947-9265; Telex: 134227)

Guthrie Latex, Inc.

7400 N. Oracle, Suite 330, Tucson, AZ 85712 USA (Tel.: 520-742-3087; Telefax: 520-575-0511; Telex: 187150 guth ut)

Haarmann & Reimer

Haarmann & Reimer GmbH, Postfach 1253, Rumohrtalstrasse 1, Holzminden Germany (Tel.: 49 5531 900; Telefax: 49 5531 901649; Telex: 965 330 HARM D)

Haarmann & Reimer Corp., PO Box 175, 70 Diamond Road, Springfield, NJ 07081 USA (Tel.: 201-4912-5707, 800-422-1559; Telefax: 201-912-0499; Telex: 219134 HAR UR)

The Hall Chemical Co.

PO Box 200, 28960 Lakeland Blvd., Wickliffe, OH 44092 USA (Tel.: 216-944-8500, 800-322-6666; Telefax: 216-944-1298; Telex: 980707)

C.P. Hall

C.P. Hall Co., 311 South Wacker Dr., Suite 4700, Chicago, IL 60606-6622 USA (Tel.: 312-554-7400, 800-881-5851; Telefax: 312-554-7499)

C.P. Hall Co., 7300 South Central Ave., Chicago, IL 60638-0428 USA (Tel.: 708-594-6000, 800-321-8242; Telefax: 708-458-0428)

Howard Hall, Div. of R.W. Greeff & Co. Inc.

777 West Putnam Ave., Greenwich, CT 06830 USA (Tel.: 203-532-2900; Telefax: 203-532-2980; Telex: 681 9012)

Halox Pigments

1326 Summer St., Hammond, IN 46320 USA (Tel.: 219-933-1560; Telefax: 219-933-1570)

Halstab. *See under* Hammond

Hammill & Gillespie, Inc.

154 S. Livingston Ave., PO Box 104, Livingston, NJ 07039 USA (Tel.: 201-994-3650; Telefax: 201-994-3847; Telex: 139114)

Hammond

Hammond Group Inc., PO Box 6408, 5231 Hohman Ave., Hammond, IN 46325-6408 USA (Tel.: 219-931-9360; Telefax: 219-931-2140)

Hammond Lead Products Inc., PO Box 6408, 5231 Hohman Ave., Hammond, IN 46325-6408 USA (Tel.: 219-931-9360; Telefax: 219-931-2140)

Halstab, Div. Hammond Lead Products, 3100 Michigan St., Hammond, IN 46323 USA (Tel.: 219-844-3980; Telefax: 219-844-7287; Telex: 72-5481)

Hampford Research Inc.

292 Longbrook Ave., PO Box 1073, Stratford, CT 06497 USA (Tel.: 203-375-1137; Telefax: 203-386-9754)

Hampshire Chemical Corp.

55 Hayden Ave., Lexington, MA 02173 USA (Tel.: 617-861-9700; Telefax: 617-863-8043; Telex: 200076 GRLX UR)

Harcros

Harcros Chemicals UK Ltd./Specialty Chemicals Div., Lankro House, PO Box 1, Eccles, Manchester, M30 0BH UK (Tel.: 44 161 789-7300; Telefax: 44 161-788-7886; Telex: 667725)

Harcros Durham Chemicals, Div. of Harcros Chemicals UK Ltd., Birtley, Chester-le-Street, Co. Durham, DH3 1QX UK (Tel.: 44 1914 102361; Telefax: 44 1914 106005; Telex: 53618 DURHAM G)

Harcros Chemicals Inc./Harcros Organics, PO Box 2930, 5200 Speaker Rd., Kansas City, KS 66110-2930 USA (Tel.: 913-621-7747; Telefax: 913-621-7746; Telex: 477266)

Hardman Inc., A Harcros Chemical Group Co.

600 Cortlandt St., Belleville, NJ 07109 USA (Tel.: 201-751-3000; Telefax: 201-751-8407; Telex: TWX: 710-995-4940)

Hardwicke Chemical Inc.

2114 Larry Jeffers Rd., Elgin, SC 29045 USA (Tel.: 803-438-3471; Telefax: 803-438-4497; Telex: 810-671-1814)

Harrisons Trading Co. Inc.

303 South Broadway, Tarrytown, NY 10591 USA (Tel.: 914-332-4600; Telefax: 914-332-8575; Telex: 427906)

Hart Products Corp.

173 Sussex St., Jersey City, NJ 07302 USA (Tel.: 201-433-6632; Telefax: 201-435-7268)

Harwick Chemical Corp.

60 S. Seiberling St., PO Box 9360, Akron, OH 44305-0360 USA (Tel.: 216-798-9300; Telefax: 216-798-0214; Telex: TWX: 810-431-2126)

Hastings Plastics Co.

1704 Colorado Ave., Santa Monica, CA 90404 USA (Tel.: 213-829-3449; Telefax: 213-328-6820)

Hatco Corp.

1020 King George Post Rd., Fords, NJ 08863-0601 USA (Tel.: 908-738-1000; Telefax: 908-738-9385; Telex: 84-4545)

Hayashi Pure Chemical Industries Co., Ltd.

2-14, Dosho-machi 2-chome, Chuo-ku, Osaka, 541 Japan (Tel.: (06) 231-0841; Telefax: (06) 222-1035)

Hefti Ltd. Chemical Products

PO Box 1623, CH-8048 Zurich Switzerland (Tel.: 01-432-1340; Telefax: 01-432-2940; Telex: 822225 hexa ch)

Heico Chemicals, Inc., A Cambrex Co.

Route 611, PO Box 160, Delaware Water Gap, PA 18327-0160 USA (Tel.: 717-420-3900, 800-34-HEICO; Telefax: 717-421-9012)

Henkel

Henkel KGaA, Henkelstrasse 67, D-40191 Düsseldorf Germany (Tel.: 49-211-797-3300; Telefax: 49-211-798-9638; Telex: 858170)

Henkel KGaA/Cospha, Postfach 101100, D-40191 Düsseldorf Germany (Tel.: 49-211-797-0; Telefax: 49-211-798-7696; Telex: 85817-0)

Henkel KGaA/COK-Coatings, D-40191 Düsseldorf Germany (Tel.: 49-211 797-2022)

Henkel Ltd., Div. of Henkel KG, Henkel House, 292-308 Southbury Road, Enfield, Middlesex, EN1 1TS UK (Tel.: 44 181 804 3343; Telefax: 44 181 443 2777; Telex: 922708 HENKEL G)

Henkel (Ireland) Ltd., Western Industrial Estate, Naas Road, Dublin, 12, Ireland (Tel.: 35 31 4 505 622; Telefax: 35 31 4 503 649)

Henkel France SA, Div. of Henkel KG, BP 309, 150 rue Gallieni, F-92102 Boulogne Billancourt France (Tel.: 33-46 84 90 00; Telefax: 33-46 84 90 90; Telex: 633177 HENKEL F)

Henkel Belgium SA, Div. of Henkel KG, 66 ave du Port, Havenlaan 65, 1210 Brussels Belgium (Tel.: 32-423 17 11; Telefax: 32-428 34 67; Telex: 21294 HENKEL B)

Henkel Chimica SpA, Via Scalabrini 24, 22073 Fino Mornasco (Co.) Italy (Tel.: 3931 88 42 01; Telefax: 3931 88 43 60)

Henkel Nopco A/S, Postboks 2040, Stromse, 3003 Drammen Norway (Tel.: 47 3220 2200; Telefax: 47 3288 0701)

Henkel South Africa, PO Box 3933, Johannesburg, 2000, South Africa (Tel.: 27 11 864-4950; Telefax: 27 11 864 7888; Telex: 4-29310)

Henkel Iberica s.a., Div. Pulcra, Pasaje Mariner no. 9, 08025 Barcelona Spain (Tel.: 34 3 290 47 63; Telefax: 34 3 290 48 79)

Henkel Iberica s.a., Div. Pulcra, Sector E C/42, Zona Franca, 08040 Barcelona Spain (Tel.: 34 3 290 48 50; Telefax: 34 3 290 48 78)

Henkel Corp./Process & Polymer Chemicals Div., 300 Brookside Ave., Ambler, PA 19002 USA (Tel.: 215-628-1456, 800-654-7588; Telefax: 215-628-1457)

Henkel Corp./Coatings & Inks Div., 300 Brookside Ave., Ambler, PA 19002-3498 USA (Tel.: 215-628-1000, 800-445-2207; Telex: 215-628-1111)

Henkel Corp./Cospha, 300 Brookside Ave., Ambler, PA 19002 USA (Tel.: 215-628-1476, 800-531-0815; Telefax: 215-628-1450; Telex: 125854)

Henkel Corp./Functional Products, 300 Brookside Ave., Ambler, PA 19002 USA (Tel.: 215-628-1583, 800-654-7588; Telefax: 215-628-1155; Telex: 215-685-1092)

Henkel Australia Pty. Ltd., 1 Clyde St., Silverwater, NSW, 2141 Australia (Tel.: 61 2 748 4355; Telefax: 61 2 748 3863; Telex: AA 25058)

Henkel Corp./Organic Products Div., 300 Brookside Ave., Ambler, PA 19002 USA (Tel.: 215-628-1441, 800-634-2436; Telefax: 215-628-1200; Telex: 6851092 amchm uw)

Henkel Corp./Textile Chemicals, 11709 Fruehauf Dr., Charlotte, NC 28241 USA (Tel.: 800-634-2436; Telefax: 704-587-3800)

Henkel Canada Ltd., 2290 Argentia Rd., Mississauga, Ontario, L5N 6H9 Canada (Tel.: 416-542-7588, 800-668-6023; Telefax: 416-542-7566)

Henkel S.A. Indústrias Químicas, Avenida das Nacoes, Unidas 10.989, CEP 04578 Sao Paulo, SP Brazil (Tel.: 55 11 828 2340; Telefax: 55 11 828 2326; Telex: 038-1138417 hbiq br)

Henkel Mexicana S.A. de C.V., Calz. de la Viga S/N, Fracc. Los Laureles en Tulpetlac, Ecatepec de Morelos, C.P. 55090 Mexico (Tel.: 525 787-1899; Telefax: 525 729-9804; Telex: 1762295 henk me)

Henkel Corp./Emery Group, 11501 Northlake Dr., Cincinnati, OH 45249 USA (Tel.: 513-530-7300, 800-543-7370; Telefax: 513-530-7581; Telex: 4333016)

Hercules

Hercules Inc., Hercules Plaza-6205SW, Wilmington, DE 19894-0001 USA (Tel.: 302-594-5000, 800-247-4372; Telefax: 302-594-5400; Telex: 835-479)

Hercules Inc./Paper Technology Div., 500 Hercules Rd., CSD Bldg. 8145, Wilmington, DE 18908-1599 USA (Tel.: 302-995-4584; Telefax: 302-995-4077)

Hercules Inc./Aqualon Div., 1313 North Market St., Wilmington, DE 19894-0001 USA (Tel.: 302-594-5000, 800-345-8104; Telefax: 302-594-6660)

Hercules Ltd., Div. of Hercules Inc., 31 London Road, Reigate, Surrey, RH2 9YA UK (Tel.: 44 1737 242434; Telefax: 44 1737-224288; Telex: 25803)

Hercules BV, 8 Veraartlaan, NL-2288GM Rijswijk The Netherlands (Tel.: 31-70-150-000; Telefax: 31-70-3989893; Telex: 31172)

Hercules BV/Aqualon Div., Postbus 5832, NL-2280 HV Rijswijk The Netherlands (Tel.: 31 70 315 0226; Telefax: 31 70 390 7560)

Heresite Protective Coatings Inc.

PO Box 249, 822 S. 14th St., Manitowoc, WI 54220 USA (Tel.: 414-684-6646)

Heterene Chemical Co., Inc.

PO Box 247, 295 Vreeland Ave., Paterson, NJ 07543 USA (Tel.: 201-278-2000; Telefax: 201-278-7512; Telex: 883358)

Heucotech Ltd., An Independent U.S. Co. within the Heubach Organization

99 Newbold Rd., Fairless Hills, PA 19030 USA (Tel.: 215-736-0712, 800-HEUBACH; Telefax: 215-736-2249)

Heveatex Corp.

106 Ferry St., PO Box 2573, Fall River, MA 02722 USA (Tel.: 508-675-0181)

Hickson Danchem Corp.

1975 Richmond Blvd., PO Box 400, Danville, VA 24540 USA (Tel.: 804-797-8105, 800-797-8100; Telefax: 804-799-2814; Telex: 940103 WU PUBTLXBSN)

Hickson Manro Ltd.

Bridge St., Stalybridge, Cheshire, SK15 1PH UK (Tel.: 44 161 338-5511; Telefax: 44 161-303-2991; Telex: 668442)

Hickson Specialties

2954 W. Hampton Ave., Milwaukee, WI 53209 USA (Tel.: 414-449-1204; Telefax: 414-449-4980)

Hilton Davis Chemical Co., A Freedom Chemical Co.

2235 Langdon Farm Rd., Cincinnati, OH 45237 USA (Tel.: 513-841-4000, 800-477-1022; Telefax: 800-477-4565)

Hitox Corp. of America

PO Box 2544, Corpus Christi, TX 78403 USA (Tel.: 512-882-5175; Telefax: 512-882-6948)

Hoechst

Hoechst AG, Postfach 800320, D-65926 Frankfurt am Main Germany (Tel.: 49 69 305-5753; Telefax: 49 69 316700; Telex: 41234-0 HOD)

Hoechst Celanese/Int'l. Headqtrs., Rt. 202-206 North, Somerville, NJ 08876 USA (Tel.: 201-635-2600, 800-235-2637; Telefax: 201-635-4330; Telex: 136346)

Hoechst Celanese/Advanced Materials Group, 90 Morris Ave., Summit, NJ 07901 USA (Tel.: 908-598-4000, 800-526-4960; Telefax: 908-598-4330; Telex: 136346)

Hoechst Celanese/Bulk Pharmaceuticals & Intermediates Div., 1601 West LBJ Freeway, PO Box 819005, Dallas, TX 75381-9005 USA (Tel.: 214-277-4783; Telefax: 214-277-3858)

Hoechst Celanese/Surfactants Dept., 5200 77 Center Dr., Charlotte, NC 28217 USA (Tel.: 704-599-4000, 800-255-6189; Telefax: 704-559-6323)

Hoechst Celanese/Engineering Plastics Div., 90 Morris Ave., Summit, NJ 07901 USA (Tel.: 201-635-2600; Telefax: 201-635-4300)

Hoechst Celanese/Specialty Chems. Group/Paper Chems., 5200 77 Center Dr., PO Box 1026, Charlotte, NC 28217 USA (Tel.: 704-559-6135, 800-365-2436; Telefax: 704-559-6342; Telex: 671-7746)

Hoechst Celanese/Polymer Additives, 5200 77 Center Dr., PO Box 1026, Charlotte, NC 28201-1026 USA (Tel.: 704-559-6038; Telefax: 704-559-6780)

Hoechst Celanese/Waxes, Lubricants & Polymers, Route 202-206, PO Box 2500, Somerville, NJ 08876-1258 USA (Tel.: 908-704-7043; Telefax: 908-704-7059)

Hoechst Canada Inc., Div. of Hoechst AG, 800 Blvd Rene Levesque O, Montreal, Quebec, PQH38121 Canada (Tel.: 514-871-5511)

Hoechst Japan Limited/Chemicals Dept., New Hoechst Bldg., 10-16, Akasaka 8-chome, Minato-ku, Tokyo, 107 Japan (Tel.: (03) 3479-5118; Telefax: (03) 3479-6715)

Hoffmann-LaRoche Inc.

340 Kingsland St., Nutley, NJ 07110 USA (Tel.: 201-909-8332, 800-526-0189; Telefax: 201-909-8414)

Hoffmann Mineral, Div. of Franz Hoffmann & Söhne KG

Münchenerstrasse 75, D-86633 Neuburg (Donau) Germany (Tel.: 49-8431/53-0; Telefax: 49-8431/53-330; Telex: 55223 hond-d)

Holliday Pigments

Morley Street, Kingston-upon-Hull, North Humberside, HU8 8DN UK (Tel.: 44 1482 329875; Telefax: 44 1482 223114; Telex: 597065)

Holtrachem, Inc.

159 Boden Ln., Natick, MA 01760 USA (Tel.: 508-655-2510, 800-343-6470; Telefax: 508-653-2682; Telex: 948456)

Honeywill & Stein Ltd., Div. of BP Chemicals Ltd.

Times House, Throwley Way, Sutton, Surrey, SM1 4AF UK (Tel.: 44 181 770 7090; Telefax: 44 181-770 7295; Telex: 946560 BPCLGH G)

Hoover Color Corp.

Rte. 693, PO Box 218, Hiwassee, VA 24347 USA (Tel.: 703-980-7233; Telefax: 703-980-8781)

Horizon Products

1600 Oregon St., PO Box 349, Muscatine, IA 52761 USA (Tel.: 319-264-4783)

Hormel Foods Corp.

1 Hormel Place, Austin, MN 55912-3680 USA (Tel.: 507-437-5676; Telefax: 507-437-5120; Telex: 49-0522)

J.M. Huber

J.M. Huber Corp./Chemicals Div., PO Box 310, 907 Revolution St., Havre de Grace, MD 21078 USA (Tel.: 410-939-3500; Telefax: 410-939-7313)

J.M. Huber Corp./ATH Div. *See* J.M. Huber Corp./Engineered Minerals

J.M. Huber Corp./Engineered Minerals, 1807 Park 270 Drive, Suite 210, St. Louis, MO 63146 USA (Tel.: 217-224-1100, 800-637-8176; Telefax: 217-224-7957)

J.M. Huber Corp./Engineered Minerals, One Huber Rd., Macon, GA 31298 USA (Tel.: 912-745-4751, 800-TRY-HUBER; Telefax: 912-745-1116; Telex: 544438)

J.M. Huber Corp./Engineered Minerals, 4940 Peachtree Industrial Blvd., Suite 340, Norcross, GA 30071 USA (Tel.: 404-441-1301; Telefax: 404-368-9908)

Hüls

Hüls AG, Postfach 1320, D-45764 Marl 1 Germany (Tel.: 49 2365-49-1; Telefax: 49-2365-49-2000; Telex: 829211 HSD)

Hüls (UK) Ltd., Featherstone Rd., Wolverton Mill South, Milton Keynes, Buckinghamshire, MK12 5TB UK (Tel.: 44 1908 226 444; Telefax: 44 1908 224950; Telex: 826500)

Hüls France SA, Div. of Hüls AG, 49-51 Quai de Dion Bouton, F-92815 Puteaux Cedex France (Tel.: 33 49 06 50 00; Telefax: 1 47 73 97 65; Telex: 611868 huels f)

Hüls Japan Ltd., Mita Kokusai Bldg., 4-28, Mita 1-cho, Minato-ku, Tokyo, 108 Japan (Tel.: (03) 3455 1981; Telefax: (03) 3453 3233; Telex: 2422288 huels jp j)

Hüls America Inc., PO Box 365, 80 Centennial Ave., Piscataway, NJ 08855-0456 USA (Tel.: 908-980-6800, 800-631-5275; Telefax: 908-980-6970; Telex: 4754585 Huls Ul)

Hüls America Inc., 220 Davidson Ave., Somerset, NJ 08873 USA (Tel.: 800-631-5275)

Hüls Canada, Inc., 235 Orenda Rd., Brampton, Ontario, L6T 1E6 Canada (Tel.: 905-451-3810; Telefax: 905-451-4469; Telex: 0697557)

Hüls de Mexico, S.A. de C.V., San Francisco 657 A, Desp. 8 B, Col. Del Valle, C.P. 03100 Mexico Mexico (Tel.: 011 (525) 523-4299; Telefax: 011(525)543-7257)

The Humphrey Chemical Co. Inc., A Cambrex Co.

45 Devine St., North Haven, CT 06473-0325 USA (Tel.: 203-281-0021, 800-652-3456; Telefax: 203-287-9197; Telex: 994487)

Hunt Chemicals Inc.

530 Permalume Pl. NW, Atlanta, GA 30318 USA (Tel.: 404-352-1418; Telefax: 404-352-0395)

Huntsman

Huntsman Corp., 3040 Post Oak Blvd., Houston, TX 77056 USA (Tel.: 713-235-6000, 800-826-0868; Telefax: 800-831-1782)

Huntsman Chemical Corp., 2000 Eagle Gate Tower, Salt Lake City, UT, 84111 USA (Tel.: 801-532-5200, 800-421-2411; Telefax: 801-536-1581)

Huntsman Corp. Canada Inc., 150 Research Lane, Suite 307, Guelph, Ontario, N1G 2T2 Canada (Tel.: 519-824-3280; Telefax: 519-824-4979)

Huntsman de Brasil Participacoes Ltda., R. Heloisa Pamplona, 628, Sao Caetano Do Sul, SP 09520, Sao Paulo Brazil (Tel.: 55 11-442-9264; Telefax: 55 11-441-7216)

Huntsman Corp. Belgium N.V., Woluwe Office Garden, Woluwedal 26, B-1932 Zaventem Belgium (Tel.: 32-2-716-0020; Telefax: 32-2-716-0111)

Huntsman International Trading Corp., 350 Orchard Rd., #11-07/10 Shaw House, Singapore 0922 (Tel.: 65 730-0288; Telefax: 65 730-0222)

ICC Industries Inc.

720 Fifth Ave., New York, NY 10019 USA (Tel.: 212-903-1732; Telefax: 212-903-1726; Telex: CCI 7607944)

ICD Group Inc.

1100 Valley Brook Ave., Lyndhurst, NJ 07071 USA (Tel.: 201-507-3300, 800-777-0505; Telefax: 201-507-1506; Telex: 6505113226)

ICI

ICI plc, Imperial Chemical House, 9 Millbank, London, SW11 3JS UK (Tel.: 44 171 834 4444; Telefax: 44 171 834 2040; Telex: 21324)

ICI plc/Chemicals & Polymers Group, PO Box 90, Wilton, Middlesbrough, Cleveland, TS6 8JE UK (Tel.: 44 1642 454144; Telefax: 44 1642 432444; Telex: 587 461)

ICI Surfactants Ltd. (UK), PO Box 90, Wilton, Middlesbrough, Cleveland, TS90 8JE UK (Tel.: 44 1642 454144; Telefax: 44 1642 437374; Telex: 587461)

ICI Fluoropolymers UK, Hillhouse International, PO Box 4, Thornton Cleveleys, Blackpool, Lancashire, FY5 4QD UK

ICI Americas, Inc., Subsid. of ICI plc, PO Box 15391, 3411 Silverside Rd., Wilmington, DE 19850 USA (Tel.: 302-887-3000, 800-441-7780; Telefax: 302-887-4320; Telex: 62032112)

ICI Fluoropolymers, 475 Creamery Way, Exton, PA 19341 USA (Tel.: 610-363-4741, 800-ICI-PTFE; Telefax: 610-363-4748)

ICI Polymer Additives, PO Box 751, Wilmington, DE 19897 USA (Tel.: 302-886-3564, 800-456-3669 x 3564; Telefax: 302-886-5267)

ICI Polyurethanes Group, 286 Mantua Grove Rd., West Deptford, NJ 08066-1732 USA (Tel.: 609-423-8300, 800-257-5547; Telefax: 609-423-8580; Telex: 4945649)

ICI Polyurethanes Group/Sterling Heights Div., A Business Unit of ICI Americas Inc., 6555 Fifteen Mile Rd., Sterling Heights, MI 48312 USA (Tel.: 810-826-7658, 800-553-8624; Telefax: 810-826-7733)

ICI Specialty Chemicals, Concord Pike & New Murphy Rd., Wilmington, DE 19897 USA (Tel.: 302-886-3000, 800-822-8215; Telefax: 302-886-2972)

ICI Surfactants Americas, Concord Plaza, 3411 Silverside Rd., PO Box 15391, Wilmington, DE 19850-5391 USA (Tel.: 302-887-3000, 800-822-8215; Telefax: 302-887-3525; Telex: 4945649)

ICI Australia Operations Pty. Ltd./ICI Surfactants Australia, ICI House, 1 Nicholson St., Melbourne, 300 Australia (Tel.: 3-665-7111; Telefax: 61-03-665-7009; Telex: 30192)

ICI Belgium SA, Div. of ICI plc, Everslaan 45, B-3078 Everberg Belgium (Tel.: 2-758 92 11; Telefax: 2-759 77 22; Telex: 21332 ICIEVB B)

ICI Surfactants, Everslaan 45, B-3078 Everberg Belgium (Tel.: 02-758-9361; Telefax: 02-758-9686; Telex: 26151 ICIEVB B)

ICI Deutsche GmbH, Postfach 500728, Emil-von-Behring-Strasse 2, W-6000 Frankfurt am Main Germany (Tel.: 49-69-5801-00; Telefax: 49-69-5801234; Telex: 416974 ICI D)

ICI Surfactants (Australia), Newscom St., Ascot Vale, Victoria, 3032 Australia (Tel.: (03) 2836411; Telefax: (03) 2725353)

ICI Surfactants Asia Pacific (ICI (China) Ltd.), PO Box 107, 1, Pacific Pl., 14th Floor, HK-88 Queensway Hong Kong (Tel.: 8434888; Telefax: 8685282; Telex: 73248)

Ideas Inc.

PO Box 262, Wood Dale, IL 60191 USA (Tel.: 708-766-2326; Telefax: 708-350-3121)

Idex International

1523 North Post Oak Rd., Houston, TX 77055 USA (Tel.: 713-686-9323; Telefax: 713-688-8273)

IGI

85 Old Eagle School Rd., Wayne, PA 19087 USA (Tel.: 215-687-9030, 800-852-6537)

IMC

IMC Group, Pitcairn Place, Suite 1500, 165 Township Line Rd., Jenkintown, PA 19046-3595 USA (Tel.: 215-333-6800, 800-220-6800; Telefax: 215-624-3420; Telex: 244417)

IMC/Americhem, 5129 Unruh Ave., Philadelphia, PA 19135-2990 USA (Tel.: 215-335-0990, 800-220-6800; Telefax: 215-624-3420; Telex: 244417)

Inchema, Inc.

213 Old Tappan Rd., Old Tappan, NJ 07675 USA (Tel.: 201-664-6035; Telefax: 201-664-5938; Telex: 494-6009)

In-Cide Technologies Inc.

50 N. 41 Ave., Phoenix, AZ 85009 USA (Tel.: 602-233-0756, 800-777-4569; Telefax: 602-272-5864)

Inco Alloys Int'l. Inc.

3200 Riverside Dr., Huntington, WV 25720-1771 USA (Tel.: 304-526-5100, 800-344-INCO; Telefax: 304-526-5643; Telex: 886413)

Indium Corp. of America

1676 Lincoln Ave., Utica, NY 13502 USA (Tel.: 315-853-4900, 800-4-INDIUM; Telefax: 800-221-5759; Telex: 937363)

Indspec Chemical Corp.

411 Seventh Ave., Suite 300, Pittsburgh, PA 15219 USA (Tel.: 412-765-1200; Telefax: 412-765-0439; Telex: 199187 Indspec)

Industrial Copolymers

Industrial Copolymers Limited, PO Box 347, Primrose Hill, London Rd., Preston, Lancashire, PR1 4LT UK (Tel.: 44 1772 201964; Telefax: 44 1772 883570)

Percy International Ltd., Agent for Industrial Copolymers Ltd., PO Box 1149, Averill Park, NY 12018 USA (Tel.: 518-674-0271; Telefax: 518-674-0271)

Industrial Fibers, Inc.

2889 N. Nagel Court, Lake Bluff, IL 60044 USA (Tel.: 708-295-0046; Telefax: 708-295-0520)

Industrias Quimicas del Valles SA

Avenida Rafael de Casanova 81, Mollet del Vallés, E-08100 Barcelona Spain (Tel.: 34-3-570 56 96; Telefax: 34-3-593 80 11; Telex: 52170)

Inolex Chemical Co.

Jackson & Swanson Sts., Philadelphia, PA 19148-3497 USA (Tel.: 215-271-0800, 800-521-9891; Telefax: 215-289-9065; Telex: 834617)

Inspec Group plc

Charleston Ind. Estate, Hythe, Southampton, SO45 3ZG UK (Tel.: 44-1703-894666; Telefax: 44-1703-243113)

Integra Chemical Co.

710 Thomas Ave. SW, Renton, WA 98055 USA (Tel.: 206-277-9244; Telefax: 206-277-9246)

Interface Research Corp.

100 Chastain Center Blvd., Suite 165, Kennesaw, GA 30144 USA (Tel.: 770-421-9555; Telefax: 770-424-1888)

Interfibe Corp.

6001 Cochran Rd. #A-202, Solon, OH 44139 USA (Tel.: 216-248-2266, 800-262-3771; Telefax: 216-248-2132)

International Casein

111 Great Neck Rd., Great Neck, NY 11021 USA (Tel.: 516-466-4363; Telefax: 516-466-4365; Telex: 221549)

International Chemical Inc.

One World Trade Center, Suite 8665, New York, NY 10048 USA (Tel.: 212-912-0655, 800-914-2436; Telefax: 212-912-0669; Telex: 420001 ITT UI)

International Flora Technologies, Inc./FloraTech Am.

2295 S. Coconino Dr., Apache Junction, AZ 85220 USA (Tel.: 602-983-7909; Telefax: 602-982-4183)

International Sourcing Inc.

121 Pleasant Ave., Upper Saddle River, NJ 07458 USA (Tel.: 201-934-8900, 800-772-7672; Telefax: 201-934-8291; Telex: 697-2957 INSOURC)

International Specialty Products. *See under* ISP

Interpolymer Corp.

200 Dan Rd., Canton, MA 02021 USA (Tel.: 617-828-7120, 800-262-1281; Telefax: 617-821-2485; Telex: 6974364)

Ishihara Chemical Co., Ltd.

5-16, Nishi Yanagihara-cho, Hyogo-ku, Kobe, 652 Japan (Tel.: (078) 681-4801; Telefax: (078) 651-6784; Telex: 5622-774)

ISK Biosciences Corp.

6075 Poplar Ave., Suite 306, Memphis, TN 38119 USA (Tel.: 901-683-9464, 800-238-2523; Telefax: 901-683-9475)

ISP

ISP, International Specialty Products, World Headquarters, 1361 Alps Rd., Wayne, NJ 07470-3688 USA (Tel.: 201-628-4000, 800-522-4423; Telefax: 201-628-4117; Telex: 219264)

ISP (Canada) Inc., 1075 The Queensway East, Box 1740, Station B, Mississauga, Ontario, L4Y 4C1 Canada (Tel.: 905-277-0381; Telefax: 905-272-0552; Telex: 06961186)

ISP Europe, 40 Alan Turing Rd., Surrey Research Park, Guildford, Surrey, GU2 5YF UK (Tel.: 44 1483 301757; Telefax: 44 1483 302175; Telex: 859142)

ISP Global Technologies Deutschland GmbH, Rudolf-Diesel-Strasse 25, Postfach 1380, 5020 Frechen Germany (Tel.: 02234 105-0; Telefax: 02234 105-211; Telex: 889931)

ISP (Österreich) GmbH, Belvederegasse 18/1, A-1040 Wien Vienna Austria (Tel.: 43 1 504-76-21; Telefax: 43 1 505-89-44; Telex: 133990)

ISP (Australasia) Pty. Ltd., 73-75 Derby St., Silverwater, N.S.W., 2141 Australia (Tel.: Sydney (02) 648-5177; Telefax: (02) 647-1608; Telex: 73711)

ISP Asia Pacific Pte. Ltd., 200 Cantonment Rd., Hex 06-07 Southpoint, 0208 Singapore (Tel.: (65) 2249406; Telefax: (65) 2260853; Telex: 25071)

ISP (Japan) Ltd., Shinkawa Iwade Bldg. 8F, 26-9, Shinkawa 1-Chome, Chuo-Ku, Tokyo, 104 Japan (Tel.: (03) 3555-1571; Telefax: (03) 3555-1660; Telex: J23568)

ISP Van Dyk, Inc., Member of the ISP Group, 11 William St., Belleville, NJ 07109 USA (Tel.: 201-450-7722; Telefax: 201-751-2047; Telex: 710-995-4928)

Itochu Specialty Chemicals Inc.

350 Fifth Ave., Suite 5822, New York, NY 10118 USA (Tel.: 212-629-2660; Telex: 12297 C ITOH NYK)

Jackson Lea, A Unit of Jason Inc.

Hwy. 70 East, PO Box 699, Conover, NC 28613 USA (Tel.: 704-464-1376; Telefax: 704-464-7094; Telex: 962426)

James River Corp.

Fourth & Adams St., Camas, WA 98607 USA (Tel.: 206-834-8134; Telefax: 206-834-8278; Telex: 152845)

Janssen Chimica, Div. of Janssen Pharmaceutica

Janssen Pharmaceuticalaan 3, B-2440 Geel Belgium (Tel.: 14-60 42 00; Telefax: 14-60 42 20; Telex: 34103)

Janssens NV

Bte 129, Europark-Oost 15, B-9100 Sint-Niklass Belgium (Tel.: 32-3-776 4747; Telefax: 32-3-776 80 02; Telex: 31046 JANSEN B)

Jarchem Industries Inc.

414 Wilson Ave., Newark, NJ 07105 USA (Tel.: 201-344-0600; Telefax: 201-344-5743; Telex: 362-660)

JLM Marketing, Inc.

8675 Hidden River Pkwy., Tampa, FL 33637 USA (Tel.: 813-632-3300; Telefax: 813-632-3301; Telex: 666886 JLM UW)

S.C. Johnson

S.C. Johnson Polymer, Div. of S.C. Johnson & Son, Inc., 1525 Howe St., Racine, WI 53403-5011 USA (Tel.: 414-631-2000, 800-231-7868; Telefax: 800-437-3266)

S.C. Johnson Polymer b.v., Groot Mijdrechstraat 81, 3641 RV Mijdrecht The Netherlands (Tel.: 31-2979-91100; Telefax: 31-2979-82051)

S.C. Johnson Polymer Corp., OSK Bldg., 25-14, Minami Ohi 6-chome, Shinagawa-ku, Tokyo, 140 Japan (Tel.: 81-3-3764-2231; Telefax: 81-3-3766-0711)

Johnson Matthey

Johnson Matthey plc, Orchard Rd, Royston, Hertfordshire, SG8 5HE UK (Tel.: 44 1763 253000; Telefax: 44 1763 253649)

Johnson Matthey Inc., 2003 Nolte Dr., West Deptford, NJ 08066 USA (Tel.: 800-444-8544)

Jonk BV, Div. of Witco

Postbus 5, Wezelstraat 12, NL-1540 AA Koog a/d Zaan The Netherlands (Tel.: 31-75-283 854; Telefax: 731-5-210 811; Telex: 19270)

Joy Chemical Co.

35 Livingston St., Providence, RI 02904 USA (Tel.: 401-331-9000; Telefax: 401-331-2155)

K3 Corp.

10115 Walker Lake Dr., Great Falls, VA 22066 USA (Tel.: 703-759-9731; Telefax: 703-759-3799)

Kalama Chemical, Inc.

1110 Bank of California Center, Seattle, WA 98164 USA (Tel.: 206-682-7890, 800-742-6147; Telefax: 206-682-1907; Telex: 910-444-2294)

Kao Corp. S.A.

Puig dels Tudons, 10, 08210 Barbera Del Valles, Barcelona Spain (Tel.: 3-729-0000; Telefax: 3-718-9829; Telex: 59749)

Kaopolite, Inc.

2444 Morris Ave., Union, NJ 07083 USA (Tel.: 908-789-0609; Telefax: 908-851-2974)

Kelco

Kelco, A Unit of Monsanto Co., 8355 Aero Drive, PO Box 23576, San Diego, CA 92123-1718 USA (Tel.: 619-292-4900, 800-535-2656; Telefax: 619-467-6520; Telex: 695454)

Kelco International Ltd., Westminster Tower, 3 Albert Embankment, London, SE1 7RZ UK (Tel.: 44 171 735-0333; Telefax: 44 171-735-1363; Telex: 23815 KAILIL G)

Kelco Oil Field Group, Inc., 3300 Bingle Rd., Houston, TX 77055 USA (Tel.: 713-895-7575; Telefax: 713-895-7586)

Kemira, Inc.

Box 368, Savannah, GA 31402 USA (Tel.: 912-236-6171, 800-4-KEMIRA; Telefax: 912-234-3584)

Kemira Kemi AB, Div. of Kemira Oy

Box 902, Industrigatan 83, S-251 09 Helsingborg Sweden (Tel.: 46-42-17 10 00; Telefax: 46-42-14 06 35; Telex: 72185 KEMWATS S)

Kenrich Petrochemicals, Inc.

140 E. 22nd St., PO Box 32, Bayonne, NJ 07002-0032 USA (Tel.: 201-823-9000, 800-LICA KPI; Telefax: 201-823-0691; Telex: 125023)

Kerr-McGee Chemical Corp.

Kerr-McGee Ctr., PO Box 25861, Oklahoma City, OK 73125 USA (Tel.: 405-270-1313, 800-654-3911; Telefax: 405-270-3123; Telex: 747-128)

Key Polymer Corp.

One Jacob's Way, Lawrence Ind. Park, Lawrence, MA 01842 USA (Tel.: 508-683-9411; Telefax: 508-686-7729; Telex: 940 103)

Kikuchi Color & Chemicals Corp.

5-1, Sugamo 3-chome, Toshima-ku, Tokyo, 170 Japan (Tel.: (03) 3918-6611; Telefax: (03) 3917-1912; Telex: 2722690 KSKTKOJ)

Kincaid Enterprises, Inc.

PO Box 549, Plant Rd., Nitro, WV 25143 USA (Tel.: 304-755-3377; Telefax: 304-755-4547; Telex: 3791803-KEINC)

E&F King & Co. Inc.

640 Pleasant St., Norwood, MA 02062 USA (Tel.: 617-762-3113; Telefax: 617-762-7743)

King Industries

King Industries, Inc., Science Rd., Norwalk, CT 06852 USA (Tel.: 203-866-5551, 800-431-7900; Telefax: 203-866-1268; Telex: 710-468-0247)

King Industries, Inc., Kattensingel 7, 2801 CA Gouda The Netherlands (Tel.: 1820-28577; Telefax: 1820-29249)

Kingston Technologies, Inc.

2235-B Route 130, Dayton, NJ 08810 USA (Tel.: 908-274-2288; Telefax: 908-274-2426)

KMG Minerals, Inc., Div. of Franklin Industrial Minerals

PO Box 729, 1469 S. Battleground Ave., Kings Mountain, NC 28086 USA (Tel.: 704-739-1321, 800-443-MICA; Telefax: 704-739-7888; Telex: 703063)

KMZ Chemicals Ltd.

48 Station Rd, Stoke D'Abernon, Cobham, Surrey, KT11 3BN UK (Tel.: 44 1932 866426; Telefax: 44 1932 867099; Telex: 929624)

Kodak Ltd. *See under* Eastman

Koei Chemical Co., Ltd., Subsid. of Sumitomo Chem. Co. Ltd.

Sumkia Fudosan Yokobori Bldg., 6-17, Koraibashi 4-chome, Chuo-ku, Osaka, 541 Japan (Tel.: (06) 204-1515; Telefax: (06) 204-1510; Telex: 522-7415 KOEI J)

Koppers Industries, Inc.

436 Seventh Ave, Room 1600, Koppers Bldg, Pittsburgh, PA 15219 USA (Tel.: 412-227-2001, 800-321-9876; Telefax: 412-227-2202)

Koster Keunen

Koster Keunen, Inc., 1021 Echo Lake Rd., PO Box 69, Watertown, CT 06795-0069 USA (Tel.: 860-945-3333; Telefax: 860-945-0330)

Koster Keunen Holland BV, Postbus 53, 5530 AB Bladel The Netherlands (Tel.: 4977-2929; Telex: 51422 KOKEU NL)

Kowa American Corp.

342 Madison Ave., Suite 710, New York, NY 10173 USA (Tel.: 212-972-4600, 800-221-2076; Telefax: 212-972-4625)

Koyo Chemical Co., Ltd.

Iidabashi, Hitown-Bldg., 2-28, Shimomiyabi-cho, Shinjuku-ku, Tokyo, 162 Japan (Tel.: (03) 3268-1717; Telefax: (03) 3268-1723)

Kraeber GmbH & Co.

Hochallee 80, W-2000 Hamburg 13 Germany (Tel.: 49-40-4450613; Telefax: 49-40-455163; Telex: 17403443)

Kraft Chemical Co.

61975 N. Hawthorne, Melrose Park, IL 60160 USA (Tel.: 708-345-5200; Telefax: 708-345-4005; Telex: 654268 KRAFT)

Kronos

Kronos, Inc., PO Box 4272, Houston, TX 77210 USA (Tel.: 713-987-6300, 800-866-5600; Telefax: 713-987-6358)

Kronos Ltd., Div. of Kronos Inc., Barons Court, Manchester Rd., Wilmslow, Cheshire, SK9 1BQ UK (Tel.: 44 1625 529511; Telefax: 44 1625 533123)

Kuraray Co. Ltd.

Shin Hankyu Bldg., 1-12-39, Umeda, Kita-ku, Osaka, 530 Japan (Tel.: (06) 348-2111; Telefax: (06) 348-2189; Telex: 222-2272 Kurart J)

Kureha Chemical Industry Co., Ltd.

9-11, Nihonbashi Horidome-cho 1-chome, Chuo-ku, Tokyo, 103 Japan (Tel.: (03) 3249-4666; Telefax: (03) 3661-1277; Telex: 252-2737 KUREHA J)

Kyowa Chemical Industry Co., Ltd.

305, Yashima-Nishimachi, Takamatsu-shi, Kagawa, 761-01 Japan (Tel.: 0877-47-2500; Telefax: 0877-47-4208; Telex: 5822220)

Kyowa Hakko USA Inc.

599 Lexington Ave., Suite 2780, New York, NY 10022 USA (Tel.: 212-319-5353; Telefax: 212-521-1283)

Lanaetex Products, Inc.

151-157 Third Ave., PO Box 52 Station A, Elizabeth, NJ 07206 USA (Tel.: 908-351-9700; Telefax: 908-351-8753; Telex: 3792268 TLAP1)

Landers-Segal Color Co. Inc. (LANSCO)

90 Dayton Ave., Passaic, NJ 07055 USA (Tel.: 201-779-5001; Telefax: 201-779-8948; Telex: 6971185)

Langley Smith & Co. Ltd.

36 Spital Square, London, E1 6DY UK (Tel.: 44 171 247 7473; Telefax: 44 171-375 1470; Telex: 883013)

Laporte

Laporte plc, Laporte House, Kingsway, Luton, Bedfordshire, LU4 8EW UK (Tel.: 44 1582 21212; Telefax: 44 1582-31818)

Laporte Absorbents, Subsid. of Laporte plc, PO Box 2, Moorfield Rd, Widnes, Cheshire, WA8 0JU UK (Tel.: 44 151 495-2222; Telefax: 44 151-420-4088)

Laporte Industries South East Asia, 171 Chin Swee Rd., #05-05 San Centre, Singapore 0316 (Tel.: 532-0676; Telefax: 532-0502)

Laporte Inc., Southern Clay Prods., 1212 Church St., PO Box 44, Gonzales, TX 78629 USA (Tel.: 210-672-2891, 800-324-2891; Telefax: 210-672-3650)

LaRoche Chemicals Inc.

PO Box 1031, Airline Hwy., Baton Rouge, LA 70821-1031 USA (Tel.: 504-356-8406, 800-548-6336; Telefax: 504-356-8405; Telex: 784581)

LaRoche Industries Inc.

1100 Johnson Ferry Rd. NE, Atlanta, GA 30342 USA (Tel.: 404-851-0300; Telefax: 404-851-0476)

Laurel Industries

30000 Chagrin Blvd., Cleveland, OH 44124-5794 USA (Tel.: 216-831-5747, 800-221-1304; Telefax: 216-831-8479)

Laur Silicone Rubber Compounding Inc.

4930 S. M-18, Box 509, Beaverton, MI 48612-0509 USA (Tel.: 517-435-7706; Telefax: 517-535-7707)

Lawter International, Inc.

990 Skokie Blvd., Northbrook, IL 60062 USA (Tel.: 708-498-4700; Telefax: 708-498-0066; Telex: RCA 210013)

Leepoxy Plastics, Inc.

3324 Ferguson Rd., Fort Wayne, IN 46809-3199 USA (Tel.: 219-747-7411; Telefax: 219-747-7413)

Lenape

Lenape Chemical, 210 E. High St., Bound Brook, NJ 08805 USA (Tel.: 908-469-7310; Telefax: 908-469-8672)

Lenape Industries, Inc., 210 E. High St., Bound Brook, NJ 08805 USA (Tel.: 908-469-7310; Telefax: 908-469-8672)

Lexco Inc.

PO Box 1198, Vernal, UT, 84078 USA (Tel.: 801-789-5361; Telefax: 801-789-5370)

LignoTech

LignoTech USA, Inc., 100 Highway 51 South, Rothschild, WI 54474-1198 USA (Tel.: 715-359-6544; Telefax: 715-355-3648)

LignoTech USA, Inc., 81 Holly Hill Lane, Greenwich, CT 06830 USA (Tel.: 203-625-0701; Telefax: 203-625-0864; Telex: 643994)

Lipo Chemicals, Inc.

207 19th Ave., Paterson, NJ 07504 USA (Tel.: 201-345-8600; Telefax: 201-345-8365; Telex: 130117)

LNP Engineering Plastics Inc.

475 Creamery Way, Exton, PA 19341 USA (Tel.: 610-363-4500, 800-854-8774; Telefax: 610-363-4749; Telex: 4973041)

Loctite Corp.

10001 Trout Brook Crossing, Rocky Hill, CT 06067 USA (Tel.: 203-571-5100, 800-562-0560; Telex: 275207)

Lohmann Chemicals, Dr. Paul Lohmann GmbH KG

PO Box 1220, D-31857 Emmerthal Germany (Tel.: 49 5155 63-0; Telefax: 49 5155 63-118; Telex: 92858 lohma d)

Lomas International

7705 N.E. Industrial Blvd., PO Box 10225, Macon, GA 31297 USA (Tel.: 912-784-1900, 800-652-9297; Telefax: 912-784-1745)

Lonza

Lonza Ltd./Fine Chemicals, A member of the A-L Alusuisse-Lonza Group, Münchensteinerstrasse 38, CH-4002 Basle Switzerland (Tel.: 41 61-316 81 11; Telefax: 41 61-316 83 01; Telex: 965960 lon ch)

Lonza SpA, Via Vittor Pisani, 31, I-20124 Milan Italy (Tel.: (02) 66 99 91; Telefax: (02) 66 98 76 30; Telex: 312431 ftalmi i)

Lonza UK Ltd., Imperial House, Lypiatt Road, Cheltenham, Gloucestershire, GL50 2QJ UK (Tel.: 44 1242 513211; Telefax: 44 1242 222294; Telex: 43152)

Lonza France SARL, Div. of Alusuisse, 55, rue Aristide Briand, F-92309 Levallois-Perret Cedex France (Tel.: 1/40 89 99 25; Telefax: 1/40 89 99 21; Telex: 613647)

Lonza Inc., 17-17 Route 208, Fair Lawn, NJ 07410 USA (Tel.: 201-794-2400, 800-777-1875 (tech.); Telefax: 201-794-2695; Telex: 4754539 LONZAF)

Lonza Japan Ltd., Kyowa Shinkawa Bldg., 8th Fl., 20-8, Shinkawa 2-chome, Chuo-ku, Tokyo, 104 Japan (Tel.: (03) 5566-0612; Telefax: (03) 5566-0619)

Lord Corp./Elastomer Prods. Div., Industrial Adhesives

2000 West Grandview Blvd., PO Box 10038, Erie, PA 16514-0038 USA (Tel.: 814-868-3611, 800-CHEMLOK; Telefax: 814-864-3452; Telex: 291935)

Lubrizol

The Lubrizol Corp., 29400 Lakeland Blvd., Wickliffe, OH 44092-2298 USA (Tel.: 216-943-4200; Telefax: 216-943-5337; Telex: 4332033)

Lubrizol France SA, 25 quai de France, F-76100 Rouen France (Tel.: 33-35 72 04 09; Telex: 180641 LUZOFRA F)

Lubrizol Japan Ltd., 3-5-1, Toranomon, Minato-ku, Tokyo, 105 Japan (Tel.: (03) 5041-4170; Telefax: (03) 5401-4178; Telex: 26814)

D.N. Lukens Inc.

15 Old Flanders Rd., Westboro, MA 01581 USA (Tel.: 508-366-1300, 800-3-LUKENS; Telefax: 508-366-0771)

Luzenac

Luzenac America, Inc., 9000 E. Nichols Ave., Suite 200, Englewood, CO 80112 USA (Tel.: 303-643-0400, 800-325-0299; Telefax: 303-643-0444)

Luzenac, Inc., 1075 North Service Rd. W., Suite 14, Oakville, Ontario, L6M 2G2 Canada (Tel.: 416-825-3930; Telefax: 416-825-3932)

MacDermid Inc.

245 Freight St., Waterbury, CT 06702 USA (Tel.: 203-575-5700)

Magie Bros. Oil, Div. Pennzoil Products Co.

9101 Fullerton Ave., Franklin Park, IL 60131 USA (Tel.: 708-455-4500, 800-MAGIE 47; Telefax: 708-455-0383)

Magnablend, Inc.

I-35E Sterrett Rd., Exit 406, Waxahachie, TX 75165 USA (Tel.: 214-223-2068; Telefax: 214-576-8721)

Magnesia GmbH

Postfach 2168, Kurt-Höbold-Strasse 6, W-2120 Lüneburg Germany (Tel.: 49-4131-52011-14; Telefax: 49-4131-53050; Telex: 2182159)

Magnesium Elektron

Magnesium Elektron, Inc., Corporate Headquarters, 500 Point Breeze Rd., Flemington, NJ 08822 USA (Tel.: 908-782-5800, 800-366-9596; Telefax: 908-782-7768)

MEL Chemicals, Lumns Lane, Swinton, Manchester UK (Tel.: 44-161-727-8100; Telefax: 44-161-727-8200)

Mallard Creek Polymers. *See under* Ameripol Synpol

Mallinckrodt

Mallinckrodt, Inc./Drug and Fine Chemicals Div., Mallinckrodt & 2nd Street, PO Box 5439, St Louis, MO 63147 USA (Tel.: 314-895-2000, 800-325-8888; Telefax: 314-539-1251)

Mallinckrodt Specialty Chemicals, 16305 Swingley Ridge Dr., Chesterfield, MO 63017 USA (Tel.: 314-895-2000, 800-325-7155; Telefax: 314-530-2562)

Malvern Minerals Co.

PO Box 1238, Hot Springs, AR, 71902-1238 USA (Tel.: 501-623-8893; Telefax: 501-623-5113; Telex: 536-120)

Mandoval Ltd.

Mark House, The Square, Lightwater, Surrey, GU18 5SS UK (Tel.: 44 1276 471617; Telefax: 44 1276 476910; Telex: 858094)

George Mann & Co., Inc.

PO Box 9066, Harborside Blvd., Providence, RI 02940-9066 USA (Tel.: 401-781-5600; Telefax: 401-941-0830)

Mantrose-Haeuser Co.

500 Post Road East, Westport, CT 06880 USA (Tel.: 203-454-1800, 800-344-4229; Telefax: 203-227-0558)

Marathon Oil Co.

539 S. Main St., Findlay, OH 45840 USA (Tel.: 419-421-2973; Telefax: 419-421-2264; Telex: TRT 196230)

Marine Magnesium Co.

995 Beaver Grade Rd., Coraopolis, PA 15108 USA (Tel.: 412-264-0200; Telefax: 412-264-9020)

Martin Marietta Magnesia Specialties

PO Box 15470, 2323 Eastern Blvd., Bldg. E, Baltimore, MD 21220-0470 USA (Tel.: 410-780-5500, 800-648-7400; Telefax: 410-780-5777; Telex: 710-862-2630)

Martinswerk GmbH, Member of Lonza Group

PO Box 1209, Kolner Strasse 110, 50102 Bergheim Germany (Tel.: (0 22 71) 90 2-0; Telefax: (02271) 902-557; Telex: 888 712)

Maruzen Fine Chemicals Inc., Div. of Maruzen Kasei Co. Ltd.

525 Yale Ave, Pitman, NJ 08071 USA (Tel.: 609-589-4042; Telefax: 609-582-8894; Telex: 333812 MARFINE)

Mason Chemical Co.

721 West Algonquin Rd., Arlington Heights, IL 60005 USA (Tel.: 708-290-1621, 800-362-1855; Telefax: 708-290-1625)

Master Bond Inc.

154 Hobart St., Hackensack, NJ 07601 USA (Tel.: 201-343-8983)

Matteson-Ridolfi, Inc.

14450 King Rd., PO Box 2129, Riverview, MI 48192 USA (Tel.: 313-479-4500)

Mayzo, Inc.

6577 Peachtree Industrial Blvd., Norcross, GA 30092 USA (Tel.: 404-449-9066; Telefax: 404-449-9070)

McGean-Rohco, Inc.

1250 Terminal Tower, Cleveland, OH 44113-2251 USA (Tel.: 216-441-4900 x4470, 800-932-7006; Telefax: 216-621-9697; Telex: 98-0403)

McIntyre Group Ltd.

24601 Governors Hwy., University Park, IL 60466 USA (Tel.: 708-534-6200; Telefax: 708-534-6216)

McLube, Div. of McGee Industries Inc.

9 Crozerville Rd., Aston, PA 19014 USA (Tel.: 215-459-1890, 800-2-MCLUBE; Telefax: 215-459-9538; Telex: 910- 3807571 MCLUBE)

McWhorter, Inc.

400 E. Cottage Pl., Carpentersville, IL 60110 USA (Tel.: 800-759-0490)

Mearl

Mearl Corp., PO Box 3030, 320 Old Briarcliff Rd., Briarcliff Manor, NY 10510 USA (Tel.: 914-923-8500; Telefax: 914-923-9594; Telex: 421841)

Mearl International BV, Emrikweg 18, 2031 BT Haarlem The Netherlands (Tel.: 31-23-318058; Telefax: 31-23-315365; Telex: 41492 MRLN)

Mearl Corp. Japan, Room No. 802, Mido-Suji Urban-Life Bldg., 4-3 Minami Semba 4-Chome Chuo-ku, Osaka, 542 Japan (Tel.: (06) 281-1560; Telefax: (06) 281 1291)

Meer Corp.

PO Box 9006, 9500 Railroad Ave., N. Bergen, NJ 07047-1206 USA (Tel.: 201-861-9500; Telefax: 201-861-9267; Telex: 219130)

Meggle GmbH

Postfach 40, Megglestrasse 6-12, W-8090 Wasserburg 2 Germany (Tel.: 49 8071-730; Telefax: 49 8071-73444; Telex: 525137)

MEI. *See under* Magnesium Elektron

Melamine Chems, Inc.

PO Box 748, Donaldsonville, LA 70346 USA (Tel.: 504-473-3121; Telefax: 504-473-0550; Telex: 8109526516)

Edward Mendell Co., Inc., A Penwest Company

2981 Rt. 22, Patterson, NY 12563-9970 USA (Tel.: 914-878-3414, 800-431-2457; Telefax: 914-878-3484; Telex: 4971034)

E. Merck

Postfach 4119, Frankfurter Strasse 250, D-6100 Darmstadt 1 Germany (Tel.: 06151-72-0; Telefax: 06151-72-2000; Telex: 419328-0 em d)

Merichem Co.

4800 Texas Commerce Tower, Houston, TX 77002 USA (Tel.: 713-224-3030, 800-231-3030; Telefax: 713-224-4403)

Merix Chemical Co.

2234 E. 75th St., Chicago, IL 60649 USA (Tel.: 312-221-8242; Telefax: 312-221-3047)

Merquinsa

Gran Vial, 17, E-08160 Montmeló (Barcelona) Spain (Tel.: 34 3 572 11 00; Telefax: 34 3 572 09 34; Telex: 94236)

Merrand Int'l. Corp.

187 Ballardvale St., Suite B-200, Wilmington, MA 01887 USA (Tel.: 508-694-9202; Telefax: 508-964-9404)

Metalspecialties Inc.

515 Commerce Dr., Fairfield, CT 06432 USA (Tel.: 203-384-0331; Telefax: 203-368-4082; Telex: 964289)

Lucas Meyer

Lucas Meyer GmbH & Co., Postfach 261665, D-2000 Hamburg 26 Germany (Tel.: 49-40-789-550; Telefax: 49-40-789-8329; Telex: 2163220 MYER D)

Lucas Meyer (UK) Ltd., Unit 46, Deeside Ind. Park, First Ave., Deeside, Clwyd, CH5 2NU, Wales UK (Tel.: 44 1244 281168; Telefax: 44 1244 281169)

Lucas Meyer Inc., 765 E. Pythian Ave., Decatur, IL 62526 USA (Tel.: 217-875-3660, 800-769-3660; Telefax: 217-877-5046)

M. Michel & Co., Inc.

90 Broad St., New York, NY 10004 USA (Tel.: 212-344-3878; Telefax: 212-344-3880; Telex: 421468)

Michelman

Michelman, Inc., 9080 Shell Rd., Cincinnati, OH 45236-1299 USA (Tel.: 513-793-7766, 800-333-1723; Telefax: 513-793-2504)

Michelman Int'l. & Co., SNC, Zoning Industriel, B-6790 Aubange Belgium (Tel.: 32 (63) 38 96 09; Telefax: 32 (63) 38 96 92)

Micro Powders, Inc.

580 White Plains Rd., Tarrytown, NY 10591 USA (Tel.: 914-793-4058; Telefax: 914-472-7098; Telex: 996584)

Midwest Grain Products Inc.

1300 Main St., PO Box 130, Atchison, KS 66002 USA (Tel.: 913-367-1480, 800-255-0302; Telefax: 913-367-0192)

Miljac Inc.

280 Elm St., New Canaan, CT 06840 USA (Tel.: 203-966-8777; Telefax: 203-966-3577)

Milliken

Milliken Chemical, A Div. of Milliken & Co., PO Box 1927, M-400, Spartanburg, SC 29304-1927 USA (Tel.: 803-503-2200, 800-345-0372; Telefax: 803-503-2430; Telex: 810-282-2580)

Milliken Europe N.V., 18 Ham, B-9000 Gent Belgium (Tel.: 32 9 265 1084; Telefax: 32 9 265 11 95)

Mineral Pigments Corp.

7011 Muirkirk Rd., Beltsville, MD 20705 USA (Tel.: 301-776-2400)

Mineral Research & Development Corp.

One Woodlawn Green, Charlotte, NC 28217 USA (Tel.: 704-525-2771, 800-334-0417; Telefax: 704-527-8232)

Minerals Technologies, Inc., Specialty Minerals Subsid.

405 Lexington Ave., New York, NY 10174 USA (Tel.: 212-878-1919; Telefax: 212-878-1903)

Mississippi Lime Management Co.

7 Alby St., PO Box 2247, Alton, IL 62002-2247 USA (Tel.: 618-465-7741, 800-437-5436; Telefax: 618-465-7786)

Mitsubishi

Mitsubishi Chemical Corp., Mitsubishi Bldg., 5-2, Marunouchi 2-chome, Chiyoda-ku, Tokyo, 100 Japan (Tel.: (03) 3283-6531; Telefax: (03) 3283-6658; Telex: BISICH J 24901)

Mitsubishi Chemical America, Inc., 81 Main St., Suite 401, White Plains, NY 10601 USA (Tel.: 914-286-3600; Telefax: 914-681-0995; Telex: 233570 MCI UR)

Mitsubishi Chemical Europe GmbH, Niederkasseler Lohweg 8, D-40547 Duesseldorf Germany (Tel.: (0211) 52392-0; Telefax: (0211) 591272; Telex: 8587716 MCI D)

Mitsubishi Gas Chemical Co., Inc.

Mitsubishi Bldg., 2-5-2, Marunouchi, Chiyoda-ku, Tokyo, 100 Japan (Tel.: (03) 3283-5000; Telefax: (03) 3287-0844; Telex: 222-2624 MGCHO J)

Mitsubishi Materials Corp.

1-5-1, Ohte-machi, Chiyoda-ku, Tokyo, 100 Japan (Tel.: (03) 5252-5200; Telefax: (03) 5252-5270/1)

Mitsubishi Petrochemical Co., Ltd.

Mitsubishi Bldg., 5-2, Marunouchi 2-chome, Chiyoda-ku, Tokyo, 100 Japan (Tel.: (03) 3283-5700; Telefax: (03) 3283-5472; Telex: 222-3172)

Mitsui Petrochemical Industries, Ltd.

Kasumigaseki Bldg., 2-5, Kasumigaseki 3-chome, Chiyoda-ku, Tokyo, 100 Japan (Tel.: (03) 3580-3616; Telefax: (03) 3593-0028; Telex: J22984 MIPECA)

Mitsui Toatsu Chemicals, Inc.

Kasumigaseki Bldg., 2-5, Kasumigaseki 3-chome, Chiyoda-ku, Tokyo, 100 Japan (Tel.: (03) 3592-4111; Telefax: (03) 3592-4267; Telex: 2223622 MTCHEM J)

Mobil Chemical Co.

PO Box 3029, Edison, NJ 08818-3029 USA (Tel.: 908-321-6000)

Modern Dispersions, Inc. (MDI)

22 Marguerite Ave., Leominster, MA 01453 USA (Tel.: 508-534-3370; Telefax: 508-537-6065)

Molycorp, Inc., A Unocal Co.

Exec. 46 Office Center, 710 Route 46 East, Fairfield, NJ 07004 USA (Tel.: 201-808-8880; Telefax: 201-808-9060)

Mona Industries Inc.

PO Box 425, 76 E. 24th St., Paterson, NJ 07544 USA (Tel.: 201-345-8220, 800-553-6662; Telefax: 201-345-3527; Telex: 130308)

Monomer-Polymer & Dajac Labs, Inc.

1675 Bustleton Pike, Feasterville, PA 19053 USA (Tel.: 215-364-1155; Telefax: 215-364-1583)

Monsanto

Monsanto Chemical Co., 800 N. Lindbergh Blvd., St. Louis, MO 63167 USA (Tel.: 314-694-1000, 800-824-7144; Telefax: 314-694-5468; Telex: 650 397 7820)

Monsanto Canada Inc., 2330 Argentia Rd., Box 787, Streetsville Postal Station, Mississauga, Ontario, L5M 2G4 Canada (Tel.: 905-826-9222; Telefax: 905-826-3119)

Monsanto Commercial SA de CV, Bosques de Durazno 61, 3R Piso, Bosques de las Lomas, Mexico City, DF 11700 Mexico (Tel.: 52-5-251-2715; Telefax: 52-5-251-7923)

Monsanto Europe SA, Ave. de Tervuren 270-272, PO Box 1, B-1150 Brussels Belgium (Tel.: 32-2-761-41-11; Telefax: 32-2-776-40-40; Telex: 62927 Mesab)

Monsanto plc, Monsanto House, Chineham Court, Great Binfield Rd., Basingstoke, Hampshire, RG24 0UL UK (Tel.: 44 1256 572-88; Telefax: 44 1256-54995; Telex: 858837)

Monsanto Japan Ltd., 31F, ARK Mori Bldg., 1-12-32, Akasaka, Minato-ku, Tokyo, 107 Japan (Tel.: 81-3 5562-2600; Telefax: 81-3 5562-2601; Telex: J22614)

Monsanto Australia Ltd., 600 St. Kilda Rd., 12th Floor, Melbourne, Victoria, 3004 Australia (Tel.: 61-3-522-7122; Telefax: 61-3-525-2253)

Montedipe SpA

Via Rosellini 15-17, I-20124 Milan Italy (Tel.: 39 2-63331; Telefax: 39 2-63338943; Telex: 310679)

Montedison. *See under* Ausimont

Montell

Montell North America, 2801 Centerville Rd., Wilmington, DE 19850-1609 USA (Tel.: 302-996-6000, 800-545-7719; Telefax: 302-996-5587; Telex: 198994 HIMONT-UT)

Montell Europe/International, Woluwe Garden, Woluwedal 24, 1932 Zaventem, Brussels Belgium (Tel.: 32 (0) 2 715 8000)

Morflex

Morflex, Inc., 2110 High Point Rd., Greensboro, NC 27403 USA (Tel.: 910-292-1781; Telefax: 910-854-4058; Telex: 910240 7846)

Reilly Chemicals S.A., Rue Defacqz 115, Bte 19, B-1050 Bruxelles Belgium (Tel.: 32-2-537-1299; Telefax: 32-2 537 1208)

Morre-Tec Industries, Inc.

1600 Rte. 22 East, Union, NJ 07083 USA (Tel.: 908-688-9009; Telefax: 908-688-9005; Telex: 131646 TUKA)

Morton International

Morton International Inc., 100 North Riverside Plaza, Chicago, IL 60606-1598 USA (Tel.: 312-807-2000; Telefax: 312-807-3150; Telex: 353-645)

Morton International, Inc./Plastics Additives, 150 Andover St., Danvers, MA 01923 USA (Tel.: 508-774-3100; Telefax: 508-750-9511)

Morton International, Inc./Performance Chemicals, 150 Andover St., Danvers, MA 01923 USA (Tel.: 508-750-9142; Telefax: 508-750-9512)

Morton International, Inc./Polymer Systems, 100 North Riverside Plaza, Chicago, IL 60606 USA (Tel.: 312-807-2000)

Morton International, Inc./Water-based Polymers, PO Box 3089, 130 Mountain Creek Church Rd., Greenville, SC 29602 USA (Tel.: 803-292-5700, 800-845-6810; Telefax: 803-292-5713)

Morton International, Inc./Specialty Chemicals Group, 2000 West St., Cincinnati, OH 45215 USA (Tel.: 513-733-2100; Telefax: 513-733-2133)

Morton International Ltd./Specialty Chem., Ind. Chem. & Addit., 7900-A Taschereau Blvd., Suite 106, Brossard, Quebec, J4X 1C2 Canada (Tel.: 514-466-7764; Telefax: 514-466-7771)

Morton International Ltd., Westward House, 155-157 Staines Rd., Hounslow, Middlesex, TW3 3JB UK (Tel.: 44 181 570 7766; Telefax: 44 181-570 6943; Telex: 262002)

Morton International SA, Chaussee de la Hulpe 130, Boite 5, B-1050 Brussels Belgium (Tel.: 32-2-790211; Telefax: 32-2-790250; Telex: 23708)

N.V. Morton International SA/Perf. Chems., Chantraie Plantsoen 1, Industriepark Guldendelle, B-3070 Kortenberg Belgium (Tel.: 755-02-11; Telefax: 755-02-50; Telex: 23708 Ventron B)

Mosselman NV

80 Boulevard Industriel, B-1070 Brussels Belgium (Tel.: 32-2-524 18 78; Telefax: 32-2-5200158; Telex: 23533 MOSS B)

Mountain Minerals Co. Ltd.

324 8th Ave. S.W., Suite 705, Calgary, Alberta, T2P 2Z2 Canada (Tel.: 403-261-3999; Telefax: 403-264-2959)

M&T Harshaw, Div. of Elf Atochem

Two Riverview Dr., PO Box 6768, Somerset, NJ 08875-6768 USA (Tel.: 908-302-3500; Telefax: 908-271-8960)

MTM Research Chemicals, Inc.

PO Box 1000, Windham, NH 03087 USA (Tel.: 603-889-3306, 800-238-2324; Telefax: 603-889-3326)

MTM Speciality Chemicals Ltd., Div. of MTM plc

Station Rd, Cheddleton, Leek, Staffordshire, ST13 7EF UK (Tel.: 44 1538 361302; Telefax: 44 1538 361330; Telex: 367443)

Multi-Kem Corp.

553 Broad Ave., PO Box 538, Ridgefield, NJ 07657-0538 USA (Tel.: 201-941-4520, 800-462-4425; Telefax: 201-941-5239; Telex: 134 611 multikem)

Münzing Chemie GmbH

Salzstrasse 174, Postfach 27 62, D-74076 Heilbronn Germany (Tel.: 07131/9 87-0; Telefax: 07131/9 87-1 25; Telex: 728614 munco d)

Mykroy/Mycalex Ceramics

125 Clifton Blvd., Clifton, NJ 07011 USA (Tel.: 201-779-8866; Telefax: 201-779-2013)

Nalco

Nalco Chemical Co., One Nalco Center, Naperville, IL 60563-1198 USA (Tel.: 708-305-1041, 800-527-7753; Telefax: 708-305-2998)

Nalco Ltd., Div. of Nalco Chemical Co., PO Box 11, Winnington Ave, Northwich, Cheshire, CW8 4DX UK (Tel.: 44 1606 74488; Telefax: 44 1606 79557; Telex: 668663)

Nalco Chemical BV, Div. of Nalco Chemical Co., Postbus 5131, NL-5004 EC Tilburg The Netherlands (Tel.: 13-63 55 55; Telefax: 13-67 45 45; Telex: 52532)

Nalco/Process Chems. Div., 6216 W. 66th Pl., Chicago, IL 60638-5299 USA (Tel.: 708-496-5041, 800-435-0861; Telefax: 708-496-5290)

Napp Technologies, Inc.

199 Main St., PO Box 900, Lodi, NJ 07644 USA (Tel.: 201-773-3900; Telefax: 201-773-2010; Telex: 13-4649)

National Casein Co.

601 W. 80 St., Chicago, IL 60620 USA (Tel.: 312-846-7300; Telefax: 312-487-5709; Telex: 910-221-2582)

National Chemical Co.

600 W. 52 St., Chicago, IL 60609 USA (Tel.: 312-924-3700, 800-525-3750; Telefax: 312-924-7760)

National Magnesia Chemicals, Div. Nat'l. Refractories & Minerals Corp.

Highway One, PO Box 30, Moss Landing, CA 95039 USA (Tel.: 408-633-2499, 800-622-2499; Telefax: 408-633-2904; Telex: 650-2730561 (MCI)

National Starch & Chemical

National Starch & Chemical Corp., Box 6500, 10 Finderne Ave., Bridgewater, NJ 08807-3300 USA (Tel.: 908-685-5000; Telefax: 908-685-5005; Telex: 710-480-9240)

National Starch & Chemical Ltd., Prestbury Court, Greencourts Business Park, 333 Styal Rd., Manchester, M22 5LW UK (Tel.: 44 161 435 3200; Telefax: 44 161 435 3300)

National Starch & Chemical (Holdings) Ltd., Canon House, 27 London End, Beaconsfield, Buckinghamshire, HP9 2HN UK (Tel.: 44 1494 677966; Telefax: 44 1494 673960; Telex: 848386 NATADH G)

Natrochem, Inc.

PO Box 1205, Exley Ave., Savannah, GA 31498 USA (Tel.: 912-236-4464; Telefax: 912-236-1919)

Nattermann Phospholipid. *See under* Rhone-Poulenc

Nayler Chemicals Ltd.

Old Rolling Mill, Edge Green Rd, Ashton in Makerfield, Wigan, Lancashire, WN4 8YA UK (Tel.: 44 1942 720988; Telefax: 44 1942 271251; Telex: 669367 BREEZE G)

NB Entrepreneurs

Uppalwadi, Kamptee Rd. nr. Industrial Estate, Nagpur, 440 026 India (Tel.: (0712) 640062; Telefax: (0712) 641163; Telex: 0715-379 SONS IN)

Nepera, Inc., A Cambrex Co.

Route 17, Harriman, NY 10926 USA (Tel.: 914-782-1202; Telefax: 914-782-2418; Telex: 510-249-4847)

Neste

Neste Polyeten AB, Div. of Neste Oy, Box 44, S-44486 Stenungsund Sweden (Tel.: 46-303-86000; Telefax: 46-303-86449; Telex: 2402)

Neste Chemicals UK Ltd., Div. of Neste Oy, Neste House, Water Lane, Wilmslow, Cheshire, SK9 5AR UK (Tel.: 44 1625 537390; Telefax: 44 1625 535218; Telex: 668398)

Neville Chemical Co.

2800 Neville Rd., Pittsburgh, PA 15225-1496 USA (Tel.: 412-331-4200; Telefax: 412-777-4234)

New Metals & Chemicals Ltd.

Abbey Chambers, Highbridge St, Waltham Abbey, Essex, EN9 1DF UK (Tel.: 44 1992 711111; Telefax: 44 1992 768393; Telex: 28816)

Niacet Corp.

PO Box 258, 400 47th St., Niagara Falls, NY 14304 USA (Tel.: 716-285-1474, 800-828-1207; Telefax: 716-285-1497; Telex: 6730170)

Nichia Chemical Industries, Ltd. (Nichia Kagaku Kogyo K.K.)

PO Box 6, Anan, Tokushima, 774 Japan (Tel.: (0884) 22-2311; Telefax: (0884) 23-1802; Telex: 5867790 NICHIA J)

Nihon Kagaku Sangyo Co., Ltd.

20-5, Shitaya 2-chome, Taito-ku, Tokyo, 110 Japan (Tel.: (03) 3876-3131; Telefax: (03) 3876-3278; Telex: J 28318 NIKKASAN)

Nipa Hardwicke Inc.

3411 Silverside Rd., 104 Hagley Bldg., Wilmington, DE 19810 USA (Tel.: 302-478-1522; Telefax: 302-478-4097; Telex: 905030)

Nippon Chemical Industrial Co., Ltd.

2-6-10, Iwamoto-cho, Chiyoda-ku, Tokyo, 101 Japan (Tel.: (03) 3862-4730; Telefax: (03) 3762-4705; Telex: 262-2161 JCIHOA)

Nippon Denko Co., Ltd.

11-8, Ginza 2-chome, Chuo-ku, Tokyo, 104 Japan (Tel.: (03) 3546-9331; Telefax: (03) 3542-3690; Telex: 2522209 NDKTOK J)

Nippon Nyukazai Co., Ltd., Subsid. of Sankyo Co., Ltd.

Yoshizawa Bldg., 9-19, Ginza 3-chome, Chuo-ku, Tokyo, 104 Japan (Tel.: (03) 3543-8571; Telefax: (03) 3546-3174)

Nippon Oils & Fats Co., Ltd. (NOF Corp.)

Yurakucho Bldg., 10-1, Yarakucho 1-chome Chiyoda-Ku, Tokyo, 100 Japan (Tel.: (03) 3283-7295; Telefax: (03) 3283-7178; Telex: 222-2041 NIPOIL J)

Nippon Polyurethane Industry Co., Ltd.

Kotohira Kaikan Bldg., 2-8, Toranomon 1-chome, Minato-ku, Tokyo, 105 Japan (Tel.: (03) 3508-0611; Telefax: (03) 3580-9304; Telex: 222-3066 NIPOLY J)

Nippon Shokubai Co., Ltd.

Kogin Bldg., 1-1, Koraibashi 4-chome, Chuo-ku, Osaka, 541 Japan (Tel.: (06) 223-9111; Telefax: (06) 201-3716; Telex: 522-5243 NSKKOJ)

Nippon Silica Industrial Co., Ltd.

Toso-Kyobashi Bldg., 2-4, Kyobashi 3-chome, Chuo-ku, Tokyo, 104 Japan (Tel.: (03) 3273-1641; Telefax: (03) 5255-6647)

Nippon Soda Co., Ltd.

Shin Ohtemachi Bldg., 2-1, Ohte-machi 2-chome, Chioyda-ku, Tokyo, 101 Japan (Tel.: (03) 3245-6054; Telefax: (03) 3242-2882)

Nippon Zeon Co., Ltd.

Furukawa Sogo Bldg., 2-6-1, Marunouchi, Chiyoda-ku, Tokyo, 100 Japan (Tel.: (03) 3216-2335; Telefax: (03) 3216-0501; Telex: 222-4001)

Nissan

Nissan Chemical Industries, Ltd., Kowa-Hitotsubashi Bldg., 3-7-1, Kanda-Nishiki-cho, Chiyoda-ku, Tokyo, 101 Japan (Tel.: (03) 3296-8065; Telefax: (03) 3296-8360; Telex: 222-3071)

Nissan Chemical America Corp., 303 South Broadway, Tarrytown, NY 10591 USA (Tel.: 914-332-4745; Telefax: 914-332-4808)

Nitta Gelatin Inc.

8-12, Honmachi 1-chome, Chuo-ku, Osaka, 541 Japan (Tel.: (06) 266-1911; Telefax: (06) 266-1900; Telex: 522-2882 NITTAO J)

Noah Chemical Div., Div. of Noah Technologies Corp.

7001 Fairgrounds Pkwy, San Antonio, TX 78238 USA (Tel.: 512-680-9000; Telefax: 512-521-3323)

Nobel Chemicals, Inc.

2 Gillon St. #A, Charleston, SC 29401-2106 USA

NOF Corp.

4-20-3, Ebisu, Sibuya-ku, Tokyo, 150 Japan (Tel.: (03) 5424-6600; Telefax: (03) 5424-6800; Telex: 222-2041 NIPOIL J)

Norac

Norac Co., Inc., 169 Kennedy Dr., Lodi, NJ 07644-0230 USA (Tel.: 201-779-4981)

Norac Co., Inc., PO Box 577, 405 S. Motor Ave., Azusa, CA 91702-0706 USA (Tel.: 818-334-2908; Telefax: 818-334-3512; Telex: 882552 NORAC CO AZSA)

Norland Products, Inc.

695 Joyce Kilmer Ave., PO Box 7145, New Brunswick, NJ 08902 USA (Tel.: 908-545-7828; Telefax: 908-545-9542)

Norman, Fox & Co.

5511 S. Boyle Ave., PO Box 58727, Vernon, CA 90058 USA (Tel.: 213-583-0016, 800-632-1777; Telefax: 213-583-9769)

Norsk Hydro AS

Bygdoyalle 2, N 0240 Oslo 2 Norway (Tel.: 472 243 2100; Telefax: 472 243 2725; Telex: 78350 HYDRO N)

Northwestern Flavors, Inc.

120 N. Aurora St., W. Chicago, IL 60185 USA (Tel.: 708-231-6111; Telefax: 708-231-1302)

Norton Chemical Process Products

PO Box 350, Akron, OH 44309 USA (Tel.: 216-677-7216; Telefax: 216-677-7245; Telex: 433-8012)

Nova Molecular Technologies, Inc.

1 Parker Place, Suite 495, Janesville, WI 53545 USA (Tel.: 608-754-NOVA; Telefax: 608-654-6878)

Novo Nordisk A/S, Bioindustrial Group

Novo Allé, DK-2880 Bagsvaerd Denmark (Tel.: 45 4444-8888; Telefax: 45 4444-5918; Telex: 37173)

Nuodex Espanola SA.

Gran Via 55, E-28013 Madrid Spain (Tel.: 1-241 35 03; Telefax: 1-247 53 79; Telex: 44818 NOUD E)

NuSil Technology

NuSil Technology, 1050 Cindy Lane, Carpinteria, CA 93013 USA (Tel.: 805-684-8780; Telefax: 805-684-2365)

NuSil Technology, Atlantic Parc-Les Pyramides No. 5, P.A. de Maignon, 64600 Anglet France (Tel.: 33-59-31-41-04; Telefax: 33-59-31-41-05)

Nyacol Products Inc.

Megunco Rd., PO Box 349, Ashland, MA 01721-0349 USA (Tel.: 508-881-2454, 800-GET-SOLS; Telefax: 508-881-1855)

Nyco Minerals

Nyco Minerals, Inc., A Canadian Pacific Ltd. Co., 124 Mountain View Dr., PO Box 368, Willsboro, NY 12996-0368 USA (Tel.: 518-963-4262; Telefax: 518-963-4187; Telex: 957014)

Nyco Minerals, Inc. Europe, Ordrupvej 24, PO Box 88, DK-2920 Charlottenlund Denmark (Tel.: 31 64 33 70; Telefax: 31 64 37 10)

Nyltech North America

333 Sundial Ave., Manchester, NH 03103 USA (Tel.: 603-627-5150, 800-851-2001; Telefax: 603-627-5154)

O'Brien Industries, Inc., A Zinkan Enterprises Co.

10574 Ravenna Rd., Twinsburg, OH 44087 USA (Tel.: 216-487-1500; Telefax: 216-425-8202)

Obron Atlantic Corp.

830 E. Erie St., Painesville, OH 44077 USA (Tel.: 216-354-0600, 800-556-1111; Telefax: 216-354-6224; Telex: 428904 Powders)

O&C Corp.

PO Box 681380, Indianapolis, IN 46251 USA (Tel.: 317-290-5000; Telefax: 317-290-5011)

Occidental

Occidental Chemical Corp., 5005 LBJ Freeway, Dallas, TX 75244 USA (Tel.: 214-404-3800, 800-752-5151; Telefax: 214-404-3669; Telex: 229835)

Occidental Chemical Corp., PO Box 344, Niagara Falls, NY 14302-0344 USA (Tel.: 800-733-1165; Telefax: 716-278-7297)

Occidental Chemical Corp./Durez Div., PO Box 809050, 5005 LBJ Freeway, Dallas, TX 75380-9050 USA (Tel.: 214-404-4932, 800-733-3339; Telefax: 214-404-4981)

Occidental Chemical Europe SA, Holidaystraat 5, B-1831 Diegem Belgium (Tel.: 32 725 44 50; Telefax: 32 725 46 76; Telex: 23046)

Ohio Polychemical Co.

1920 Leonard Ave., Columbus, OH 43236 USA (Tel.: 614-253-8511; Telefax: 614-253-6327)

Olin

Olin Corp., 120 Long Ridge Rd., PO Box 1355, Stamford, CT 06904 USA (Tel.: 203-356-2000, 800-462-6546; Telefax: 203-271-4060; Telex: 420202)

Olin Research Center, 350 Knotter Dr., PO Box 586, Cheshire, CT 06410-0586 USA (Tel.: 203-271-4000, 800-654-6763; Telefax: 203-271-4060)

Olin Quimica, S.A. de C.V., Campos Eliseos No. 385, Piso 9, Torre A, Col. Polanco, Delg. Miguel Hidalgo, 11560, Mexico D.F. (Tel.: (52-5)281-2045; Telefax: (52-5)281-2037)

Olin Brasil Ltda., Nacoes Unidas, 11.857, 12th Flr., Sao Paulo, 04578-000 Brazil (Tel.: 55-11-505-0382; Telefax: 55-11 505-1950; Telex: 11-25034)

Olin SA, ZAC Paris Nord II, 209 Ave. des Nations, B.P. 60019, F-95970 Roissy CDG Cedex France (Tel.: (33-1) 48 63 21 66; Telefax: (33-1) 48 63 22 25; Telex: 650769)

Olin Australia Ltd., PO Box 141, Level 2, Suite 1, 601 Pacific Hwy., St. Leonards, N.S.W., 2065 Australia (Tel.: 61-2-906-4455; Telefax: 61-2-439-4198; Telex: 26328)

Olin Far East Ltd., Korea, 80-6 Soosong-dong, Suktan Bldg., Chongro-ku, Seoul, 110-140, Korea (Tel.: (82-2)737-2840; Telefax: (82-2) 730-7387)

Olin Far East Ltd., Taiwan, 7F-2 No. 137, Fu Shing S. Road Sec. 1, Taipei, Taiwan/ROC (Tel.: (886-2) 752-4413; Telefax: (886-2) 741-2113)

Olin Industrial H.K. Ltd., 111 Peninsula Centre, 67 Mody Rd., Tsim Sha Tsui-East, Kowloon Hong Kong (Tel.: (852) 366-8303; Telefax: (852) 367-1309)

Olin Japan Inc., Shiozaki Bldg., 7-1 Hirakawa-Cho 2-Chome, Chiyoda-ku, Tokyo, 102 Japan (Tel.: (813) 3263-4615; Telex: (81-3) 3264-2750)

Olin Pte., Ltd., 501 Orchard Rd., 08-03 Lane Crawford Rd., Singapore 0923 (Tel.: (65) 735-1268; Telefax: (65) 735-1298)

Olin Pty. Ltd., PO Box 114, Bergvlei, 2012, Rep. of South Africa (Tel.: (27-11)444-2244; Telefax: (27-11)444-2241)

OM Group

2301 Scranton Rd., Cleveland, OH 44113 USA (Tel.: 216-781-8383, 800-321-9696; Telefax: 216-781-5919; Telex: 810-421-8322)

Orange Chemicals Ltd.

34 St Thomas Street, Winchester, Hampshire, SO23 9HJ UK (Tel.: 44 1962 842525; Telefax: 44 1962-841101; Telex: 477361 ORANGE G)

Ore & Chemical Corp., Div. of Chemetall

520 Madison Ave., New York, NY 10022 USA (Tel.: 212-715-5232; Telefax: 212-486-2742; Telex: ITT 422681)

Osaka Organic Chemical Ind. Co., Ltd.

Shin Toyama Bldg., 1-7-20, Azuchi-machi, Chuo-ku, Osaka, 541 Japan (Tel.: (06) 264-5071; Telefax: (06) 264-1675)

OSi Specialties

OSi Specialties, Inc., Div. of Witco Corp., 39 Old Ridgebury Rd., Danbury, CT 06810-5121 USA (Tel.: 203-794-4302, 800-523-2862; Telefax: 203-794-3088)

OSi Specialties, Inc./Customer Service Center, PO Box 38002, S. Charleston, WV 25303-3802 USA (Tel.: 800-523-5852; Telefax: 304-747-3397)

OSi Specialties Canada, Inc., 1210 Sheppard Ave. East, Suite 210, Box 38, Willowdale, Ontario, M2K 1E3 Canada (Tel.: 416-490-0466)

OSi Specialties do Brasil Ltda., Rua Dr. Eduardo De Souza, Aranha 153-11, Andar, Sao Paulo, 04530-904 Brazil (Tel.: 55-11-828-1104)

OSi Specialties S.A., 7 Rue de Pre-Bouvier, Meyrin, CH-1217 Geneva Switzerland (Tel.: 41-22-989-2111)

OSi Specialties Singapore PTE, Ltd., 12 Science Park Dr., #03-01 The Mendel, Singapore Science Park, 0511 Singapore (Tel.: 65-774-4800)

Osram Sylvania Inc.

100 Endicott St., Danvers, MA 01923 USA (Tel.: 508-777-1900)

Owens-Corning Fiberglas Corp.

Fiberglas Tower, Toledo, OH 43659 USA (Tel.: 419-248-7185, 800-462-3435; Telefax: 419-248-6712)

Oxiteno S/A Industria E Comercio

Av. Brigedeiro Luiz Antonio 1343, 7 andar, Sao Paulo SP, 01350-900 Brazil (Tel.: (55-11)283 6118; Telefax: (55-11)2893533; Telex: (55-11) TLX 31727)

Pantasote Polymers Inc.

411 Rte. 17 S., Hasbrouch Heights, NJ 07604 USA (Tel.: 201-398-0888, 800-253-8982; Telefax: 201-393-0808)

Pasminco Europe Ltd./ISC Alloys Div., Div. of Pasminco Ltd.

Alloys House, PO Box 36, Willenhall Lane, Bloxwich, Walsall, West Midlands, WS3 2XW UK (Tel.: 44 1922 408444; Telefax: 44 1922-710043; Telex: 338270)

Passaic Color & Chemical Co.

28-36 Paterson St., Paterson, NJ 07501 USA (Tel.: 201-279-0400; Telefax: 201-279-8561; Telex: 820907 PASDYE UD)

PCR, Inc.

PO Box 1466, Gainesville, FL 32602 USA (Tel.: 904-376-8246, 800-331-6313; Telefax: 904-371-6246)

Pelmor Laboratories, Inc.

401 Lafayette St., Newtown, PA 18940 USA (Tel.: 215-968-3334, 800-772-6969; Telefax: 215-968-6415)

Pelron Corp.

7847 W. 47 St., Lyons, IL 60534 USA (Tel.: 708-442-9100; Telefax: 708-442-0213)

Penford Products Co.

1001 First St. SW, Cedar Rapids, IA 52406 USA (Tel.: 319-398-3715, 800-553-7294; Telefax: 319-398-3797)

Penreco, Div. of Pennzoil Prods. Co.

138 Petrolia St., Karns City, PA 16041 USA (Tel.: 412-756-0110, 800-245-3952; Telefax: 412-756-1050; Telex: 1561596)

Penta Manufacturing Co.

50 Okner Parkway, Livingston, NJ 07039 USA (Tel.: 201-740-2300; Telefax: 201-740-1839; Telex: 219472 PENT UR)

Pentad Group

3401 N. Federal Hwy., Communication Plaza, Boca Raton, FL 33431 USA (Tel.: 407-362-8678; Telefax: 407-FOB-TRUK)

Pentagon Chemicals Ltd.

Northside, Workington, Cumbria, CA14 1JJ UK (Tel.: 44 1900 604371; Telefax: 44 1900 66943; Telex: 64353 PENTA G)

Peridot Chemicals (NJ) Inc.

1680 Rt. 23 North, Wayne, NJ 07470 USA (Tel.: 201-696-9000, 800-222-0121; Telefax: 201-696-2501)

The Permethyl Corp.

191 S. Keim St., #102, PO Box 643, Pottstown, PA 19464 USA (Tel.: 610-970-0251; Telefax: 610-970-9415)

Perstorp Polyols, Inc.

600 Matzinger Rd, Toledo, OH 43612 USA (Tel.: 419-729-5448, 800-537-0280; Telefax: 419-729-3291)

Petrolite

Petrolite Corp./Headquarters & Polymers Div., 6910 E. 14th St., Tulsa, OK 74112 USA (Tel.: 918-836-1601, 800-331-5516; Telefax: 918-834-9718)

Toyo-Petrolite Co. Ltd., TOIN Bldg., 12-15, 2-Chome Shinkawa, Chuo-ku, Tokyo, 104 Japan (Tel.: 81-33-555-1831; Telefax: 81-33-555-1836)

Petrolite Pacific Pte. Ltd., 2 Tanjong Penjuru Crescent, Jurong, Singapore 2260 (Tel.: 65-268-0517; Telefax: 65-262-0344)

Pfanstiehl Laboratories, Inc.

1219 Glen Rock Ave., PO Box 439, Waukegan, IL 60085 USA (Tel.: 708-623-0370, 800-383-0126; Telefax: 708-623-9173; Telex: 25 3672 PFANLAB)

Geo. Pfau's Sons., Co., Inc.

PO Box 7, Jeffersonville, IN 47131 USA (Tel.: 800-PFAU-OIL; Telefax: 812-283-0765; Telex: 20-4135)

Pfizer

Pfizer International, 235 East 42nd Street, New York, NY 10017 USA (Tel.: 212-573-2323)

Pfizer Food Science Group, 235 E. 42nd St., New York, NY 10017 USA (Tel.: 212-573-2323/2548, 800-TECK-SRV; Telefax: 212-573-1166; Telex: 420440)

P&G Global Oleochemicals

PO Box 599, Cincinnati, OH 45201 USA (Tel.: 513-983-2026; Telefax: 513-983-1436)

Phibrochem, Div. of Philipp Bros Chemicals Inc.

One Parker Plaza, Fort Lee, NJ 07090 USA (Tel.: 201-944-6020, 800-223-0434; Telefax: 201-944-6245)

Phibro Energy USA Inc.

500 Dallas, Suite 3200, Houston, TX 77002 USA (Tel.: 713-646-5042; Telefax: 713-646-5293; Telex: 3736083)

Phillips

Phillips Chemical Co., Div. of Phillips Petroleum Co., PO Box 968, Borger, TX 79008 USA (Tel.: 806-274-5236, 800-858-4327; Telefax: 806-274-5230)

Phillips Chemical Co., Div. of Phillips Petroleum Co., 101 ARB Plastics Tech. Center, Bartlesville, OK 74004 USA (Tel.: 918-661-9845; Telefax: 918-662-2929)

Phillips Petroleum Chemicals NV, Steenweg op Brussels 355, B-3090 Overijse Belgium (Tel.: 32-2-689 1211; Telefax: 32-2-689 1472; Telex: 22197)

Pierce & Stevens Corp., A Pratt & Lambert Co.

PO Box 1092, Buffalo, NY 14240-1092 USA (Tel.: 716-856-4910, 800-888-4910; Telefax: 716-856-7530)

AJ & JO Pilar, Inc.

145 Chapel St., Newark, NJ 07105-4198 USA (Tel.: 201-589-3808; Telefax: 201-589-0836)

Pilot Chemical Co.

11756 Burke St., Santa Fe Springs, CA 90670 USA (Tel.: 213-723-0036, 800-707-4568; Telefax: 213-945-1877; Telex: 4991200 PILOT)

Plasticolors, Inc.

2600 Michigan Ave., PO Box 816, Ashtabula, OH 44004 USA (Tel.: 216-997-5137; Telefax: 216-992-3613)

Plastics & Chemicals, Inc.

PO Box 306, Cedar Grove, NJ 07009 USA (Tel.: 908-221-0002; Telefax: 908-221-1097; Telex: 219744)

PMC

PMC Specialties Group, Inc., Div. of PMC, Inc., 20525 Center Ridge Road, Rocky River, OH 44116 USA (Tel.: 216-356-0700, 800-543-2466; Telefax: 216-356-2787; Telex: 4332035)

PMC Specialties Group, Inc., Div. of PMC, Inc., 501 Murray Rd., Cincinnati, OH 45217 USA (Tel.: 513-242-3300, 800-543-2466; Telefax: 513-482-7353; Telex: 5106000948)

PMC Specialities International Ltd., Div. of PMC Specialities Group, 65B Wigmore Street, London, W1H 9LG UK (Tel.: 44 171 935-4058; Telefax: 44 171-935 9895; Telex: 24358 PMCS G)

PMP Fermentation Products, Inc.

Columbia Center III, 9525 W. Bryn Mawr Ave., Suite 725, Rosemont, IL 60018 USA (Tel.: 708-928-0050, 800-558-1031; Telefax: 708-928-0065)

Polyad Co.

113 Rose Terrace, Barrington, IL 60010-1320 USA (Tel.: 708-526-3322; Telefax: 708-526-3326)

Polygenex International Inc.

115 Woodwinds Ind. Ct., Cary, NC 27511 USA (Tel.: 919-380-8100; Telefax: 919-380-8115)

Polygon Corp.

200 W. Second St., S. Boston, MA 02127 USA (Tel.: 617-268-4455; Telefax: 617-268-9636)

Polymer Extruded Products Inc.

297 Ferry St., Newark, NJ 07105 USA (Tel.: 201-344-2700)

Polymer Industries

PO Box 32, Henegar, AL 35978 USA (Tel.: 205-657-5197; Telefax: 205-657-5189)

Poly Research Corp.

125 Corporate Dr., Holtsville, NY 11742 USA (Tel.: 516-758-0460; Telefax: 516-758-0471)

Poly-Resyn, Inc.

534 Stevens Court, Dundee, IL 60118-1822 USA (Tel.: 708-428-4031, 800-233-4325; Telefax: 708-428-0305)

Polysat Inc.

7240 State Rd., Philadelphia, PA 19135 USA (Tel.: 215-332-7700, 800-858-2828; Telefax: 215-332-9997; Telex: 831869)

Polysciences Inc.

400 Valley Road, Warrington, PA 18976-2590 USA (Tel.: 215-343-6484, 800-523-2575; Telefax: 800-343-3291; Telex: 510-665-8542)

Polyurethane Corp. of America

624 Schuyler Ave., Lyndhurst, NJ 07071 USA (Tel.: 201-438-2325; Telefax: 201-507-1367)

Polyurethane Specialties Co. Inc.

624 Schuyler Ave., Lyndhurst, NJ 07071 USA (Tel.: 201-438-2325; Telefax: 201-507-1367)

Potters Industries Inc., Affiliate of The PQ Corp.

Southpoint Corporate Headquarters, PO Box 840, Valley Forge, PA 19482-0840 USA (Tel.: 610-651-4700; Telefax: 610-408-9723)

Powmet, Inc.

PO Box 5086, 2625 Sewell St., Rockford, IL 61125 USA (Tel.: 815-398-6900; Telefax: 815-398-6907)

PPG Industries

PPG Industries, Inc., One PPG Place, Pittsburgh, PA 15272 USA (Tel.: 412-434-3131, 800-CHEM-PPG; Telefax: 412-434-2891; Telex: 86 6570)

PPG Industries, Inc./Specialty Chemicals Business Unit, 3938 Porett Dr., Gurnee, IL 60031 USA (Tel.: 708-244-3410, 800-323-0856; Telefax: 708-244-9633; Telex: 25-3310)

PPG Canada Inc./Specialty Chem., 2 Robert Speck Pkwy., Suite 900, Mississauga, Ontario, L4Z 1H8 Canada (Tel.: 905-848-2500; Telefax: 905-848-2185; Telex: 38906960351canbizmis)

PPG Industrial do Brazil Ltda., Edificio Grande Avenida, Paulista Ave. 1754, Suite 153, 01310 Sao Paulo Brazil (Tel.: 55-011-284-0433; Telefax: 55-011-289-2105; Telex: 011-39104 PPGB BR)

PPG-Mazer Mexico, S.A. de C.V., Av. Presidente Juarex No. 1978, Tlalnepantla, Edo., C.P. 54090 Mexico (Tel.: 52-5-397-8222; Telefax: 52-5-398-5133)

PPG Industries (UK) Ltd./Specialty Chem., Carrington Business Park, Carrington,, Urmston, Manchester, M31 4DD UK (Tel.: 44 161 777-9203; Telefax: 44 161-777-9064; Telex: 851-94014896 mazu g)

PPG Ouvrie S.A., 64, rue Faldherbe, B.P. 127, 59811 Lesquin Cedex France (Tel.: 33-2087-0510; Telefax: 33-2087-5631; Telex: 131419F)

PPG Industries Taiwan, Ltd., Suite 601, Worldwide House, No. 131, Ming East Rd., Sec. 3, Taipei, 105, Taiwan, R.O.C. (Tel.: 886-2-514-8052; Telefax: 886-2-514-7957; Telex: 10985 PPGTWN)

PQ Corp.

PQ Corp., PO Box 840, Valley Forge, PA 19482-0840 USA (Tel.: 610-651-4200, 800-944-7411; Telefax: 610-251-9118; Telex: 476 1129 PQCO VAF)

PQ Corp., PO Box 2248, Chattanooga, TN 37409-0248 USA (Tel.: 615-629-7160, 800-252-0039; Telefax: 615-698-0614; Telex: 558442)

PQ Hollow Spheres Ltd., 211 Lange Leemstraat, B-2018 Antwerp Belgium (Tel.: Telex:)

Vitro-PQ S.A., Apdo. 382, Tlalnepantla, Edo. de Mexico Mexico (Tel.: 5-227-6800; Telefax: 5-227-6898)

Akzo-PQ Silica VOF, Amersfoort The Netherlands (Tel.: 033-676-777; Telefax: 033-676-169)

National Silicates Ltd., Subsid. of PQ Corp., PO Box 69, Toronto, Ontario, M8V 3S4 Canada (Tel.: 416-255-7771; Telefax: 416-255-0170)

PQ Australia Pty. Ltd., Melbourne, Victoria Australia (Tel.: 03-791-4066; Telefax: 03-791-6511)

PQ Asia Inc., Tokyo Japan (Tel.: 03-5275-7051; Telefax: 03-5275-6670)

Premier Services Corp.

7251 Engle Rd., Suite 415, Middleburg Hts., OH 44130 USA (Tel.: 216-234-4600, 800-227-4287; Telefax: 216-234-5772)

Primachem Inc.

12 Greenwoods Rd., Old Tappan, NJ 07675 USA (Tel.: 201-784-3434; Telefax: 201-784-7997; Telex: 175094 PRIMA)

Process, Materials & Eng., Inc.

1021 Sedgefield Dr., Sylacauga, AL 35150 USA (Tel.: 205-249-2724)

Procter & Gamble Co./Chemicals Div.

PO Box 599, Cincinnati, OH 45201 USA (Tel.: 513-983-2026, 800-543-1580; Telefax: 513-983-1436; Telex: 21-4185, P&GCIN)

Protameen Chemicals, Inc.

375 Minnisink Rd., PO Box 166, Totowa, NJ 07511 USA (Tel.: 201-256-4374; Telefax: 201-256-6764; Telex: 130125)

Protex

Protex SA, B.P. 177, 6 rue Barbès, 92305 Levallois-Paris France (Tel.: 33 1-41 34 14 00; Telefax: 33 1-41 34 14 16; Telex: 620987)

Protex Chemicals Ltd., Astley Lane Industrial Estate, Astley Way, Swillington, Leeds, L228 8XT UK (Tel.: (0113) 2876002; Telefax: (0113) 2875003)

Protex Nederland, Broekhovenseweg 130 R, 502 LJ Tilburg The Netherlands (Tel.: 13/36 74 77; Telefax: 13/36 65 92)

Protex Chemie Basel AG, Thannerstrasse 72, 4054 Basel (CH) Switzerland (Tel.: 061-302 0066; Telefax: 061-302 0076)

Protex Extrosa, Postfach 1415, 79504 Lorrach Germany (Tel.: 07621-84772; Telefax: 07621-12429)

Protex Korea Co. Ltd., PO Box 1193, Seoul, 1351010, Korea (Tel.: 02/548 69 93; Telefax: 02/548 65 64; Telex: Panwool K 28817)

Provençale s.a.

Av. F. Mistral, 83174 Brignoles Cedex France (Tel.: 33 1 94 69 03 42; Telefax: 33 1 94 59 04 55; Telex: 400 253)

Pulcra SA. *See* Henkel Iberica s.a., Div. Pulcra

Punda Mercantile Inc.

310 Victoria Ave., Montreal, Quebec, H3Z 2M9 Canada (Tel.: 514-489-7278)

Purac

Purac Biochem, Postbus 21, Arkelsedijk 46, NL-4200 AA Gorinchem The Netherlands (Tel.: 31 1830-41799; Telefax: 31 1830-22741; Telex: 23615 purac nl)

Purac Biochem (UK), 50-54 St. Paul's Square, Birmingham, West Midlands, B3 1QS UK (Tel.: 44 121 236 1828; Telefax: 44 121 236 1401)

Purac America, Inc., 111 Barclay Blvd., Suite 280, Lincolnshire Corporate Center, Lincolnshire, IL 60069 USA (Tel.: 708-634-6330; Telefax: 708-634-1992; Telex: 280231 PURACINC ARHT)

Purac Far East PTE Ltd., 09-02 Cecil Court, 138 Cecil St., S-0106 Singapore (Tel.: 65 220 6022; Telefax: 65 222 1707; Telex: 23463)

Pyrion-Chemie

Pyrion-Chemie GmbH, Geulenstr. 94, Neuss Germany (Tel.: 49-2101-591025; Telefax: 49-2101-593624; Telex: 8517459)

Ruetgers-Nease Corp., U.S. representative, 201 Struble Rd., State College, PA 16801 USA (Tel.: 814-231-9261, 800-458-3434; Telefax: 814-238-4235)

QO Chemicals

QO Chemicals, Inc., Subsid. of Great Lakes Chem. Corp., PO Box 2500, One Great Lakes Blvd., West Lafayette, IN 47906 USA (Tel.: 317-497-6100, 800-621-9521; Telefax: 317-497-7930; Telex: 446968 QOC UD)

QO Chemicals, Inc., Industrieweg 12, Haven 391, 2030 Antwerp Belgium (Tel.: 32-3-541-21-65; Telefax: 32-3541-65-03; Telex: 32338)

QO Chemicals GmbH, Hermannstrasse 54, D-63263 Neu-Isenburg Germany (Tel.: 49-6102-25098; Telefax: 49-6102-1620; Telex: 841-417678 qoche d)

QO Chemicals Europe, S.A.R.L., 8, rue Bellini, 75782 Paris Cedex 16 France (Tel.: 33-1-45031450; Telefax: 33-1-47043604)

Great Lakes-QO Chemicals Italy, s.r.l., Via Montecatini, 15, 20144 Milano Italy (Tel.: 39-2-423-9471; Telefax: 39-2-483-00036; Telex: 353503)

Great Lakes Chemical (Europe) Ltd./QO Chems. Div., Spring House, 231 Glossop Rd., Sheffield, S10 2GW UK (Tel.: 44-742-767842; Telefax: 44-742-758472)

QO Chemicals (Australia), Inc., 16 Princess St., Kew, Victoria, 3101 Australia (Tel.: 61-3-853-6464; Telefax: 61-3-853-5929)

Quantum

Quantum Chemical Co., 11500 Northlake Dr., PO Box 429550, Cincinnati, OH 45249 USA (Tel.: 513-530-6500, 800-323-4905; Telefax: 513-530-6119; Telex: 155116)

Quantum Chemical Co., 1 Pierce Place, Suite 250C, Itasca, IL 60143 USA (Tel.: 708-285-1111, 800-323-7659; Telefax: 708-285-3316)

Quantum Chemical Europe BV, Lange Bunder 7, 4854 MB Bavel The Netherlands (Tel.: 1613-6600; Telefax: 1613-3500)

Quest International

Quest International, Postbus 2, 1400 CA Bussum The Netherlands (Tel.: 31 2159-99111; Telefax: 31 2159-46067; Telex: 43050 QSTN NL)

Quest International, Lindtsedijk 8, 3336 le Zwijndrecht The Netherlands (Tel.: 78-128511; Telefax: 78-195279; Telex: 29477)

Quimica Pumex SA

Libramiento Noroeste KM 12.8, Villa Garcia, N.L. Mexico (Tel.: 528-369-0370; Telefax: 528-369-0355)

Quimicel

Mexico Nuevo #4, Atizapan de Zaragoza Mexico (Mexico (Tel.: 525-824-2685; Telefax: 525-822-3055)

Rad-Cure Corp.

9 Aldrich Place, Fairfield, NJ 07004 USA (Tel.: 201-808-1002; Telefax: 201-808-6886)

Ranbar Technology Inc.

1114 William Flinn Hwy., Glenshaw, PA 15116-2657 USA (Tel.: 412-486-1111; Telefax: 412-487-3313; Telex: 9102500271)

Rasa Industries, Ltd.

Yaesu Dai Bldg., 1-1, Kyobashi 1-chome, Chuo-ku, Tokyo, 104 Japan (Tel.: (03) 3278-3801; Telefax: (03) 3281-6697; Telex: 2225818 RASAKOJ)

Raschig

Raschig AG, Mundenheimer Strasse 100, D-67061 Ludwigshafen Germany (Tel.: 49 621 56180; Telefax: 49 621 582885; Telex: 464 877 ralu d)

Raschig Corp., 5000 Old Osborne Tpke., PO Box 7656, Richmond, VA 23231 USA (Tel.: 804-222-9516; Telefax: 804-226-1569)

Raschig UK Ltd., 124 Doveleys Rd., Salford, Lancashire, M6 8QW UK (Tel.: 44 161 745 8811; Telefax: 44 161 745 8371)

Raschig France S.A.R.L., 49, ave. de Versailles, F-75016 Paris France (Tel.: 33 1 452 40636; Telefax: 33 1 452 08408)

Reade Advanced Materials

PO Box 15039, Riverside, RI 02915-0039 USA (Tel.: 401-433-7000; Telefax: 401-433-7001)

Reed Corp.

233 W. Pkwy., Pompton Plains, NJ 07444 USA (Tel.: 201-831-0636; Telefax: 201-831-0791)

Reedy International Corp.

25 East Front St., Suite 200, Keyport, NJ 07735 USA (Tel.: 908-264-1777; Telefax: 908-264-1189)

Reheis

Reheis Inc., PO Box 609, 235 Snyder Ave., Berkeley Heights, NJ 07922 USA (Tel.: 908-464-1500; Telefax: 908-464-8094; Telex: 219463 RCCA UR)

Reheis Ireland, Div. of Reheis Inc., Kilbarrack Rd., Dublin, 5, Ireland (Tel.: 353-1-832261; Telefax: 353-1-8392205; Telex: 32532 REHI EI)

Reichhold

Reichhold Chemicals, Inc./Corporate Headquarters, 2400 Ellis Rd., Durham, NC 27703-5543 USA (Tel.: 919-990-7500, 800-448-3482; Telefax: 919-990-7711)

Reichhold Chemicals/Emulsion Polymers Div., PO Box 13582, Research Triangle Park, NC 27709-3582 USA (Tel.: 919-990-7500, 800-441-6461, 919-990-7746)

Reichhold Chemie AG, CH 5212 Hausen b. Brugg Switzerland (Tel.: 056 482222; Telefax: 056 482412; Telex: 825109 RCAG CH)

Reilly Industries Inc.

151 N. Delaware St., Suite 1510, Indianapolis, IN 46204 USA (Tel.: 317-638-7531; Telefax: 317-248-6413; Telex: 27 404)

Reilly-Whiteman Inc. *See* Freedom Textile Chemicals Co.

Reliance Silicones (I) Ltd.

A 12/13 MIDC Ind. Estate, TTC Thane Belapur Rd., Thane, 400 613 India (Tel.: 91-22-767 1156; Telefax: 91-22-763-2838)

Research Organics Inc.

4353 E. 49th St., Cleveland, OH 44125 USA (Tel.: 216-883-8025, 800-321-0570; Telefax: 216-883-1576)

Revelli Chemicals, Inc.

35 Mason St., Greenwich, CT 06830 USA (Tel.: 203-661-4040; Telefax: 203-661-4332)

Revertex Ltd.

Templefields, Harlow, Essex, CM20 2BH UK (Tel.: 44 1279 429555; Telefax: 44 1279-412984; Telex: 81318)

Rexene Products Co.

5005 LBJ Freeway, Occidental Tower, Dallas, TX 75244 USA (Tel.: 214-450-9000, 800-233-1159; Telefax: 214-450-9028)

Rheox

Rheox Inc., PO Box 700, Wyckoffs Mill Road, Hightstown, NJ 08520 USA (Tel.: 609-443-2500, 800-866-6800; Telefax: 609-443-2422; Telex: 642240)

Rheox Europe N.V./S.A., 31 rue de l'Hôpital, B-1000 Brussels Belgium (Tel.: (02)512 0048; Telefax: (02) 513 24 25; Telex: 24662)

Rho Chemical Company Inc.

Industry Ave., Box 55, Joliet, IL 60434 USA (Tel.: 815-727-4791; Telefax: 815-740-3238)

Rhone-Poulenc

Rhone-Poulenc SA, 25 quai Paul Doumer, 92408 Courbevoie Cedex France (Tel.: 33 1 47 68 12 34; Telex: 610500 F RHONE)

Rhone-Poulenc Chimie, Div. of Rhone-Poulenc SA, 25 Quai Paul Doumer, F-92408 Courbevoie Cedex France (Tel.: 47 68 1234; Telefax: 47 68 23 00; Telex: 610500)

Rhone-Poulenc Chimie (France)/Secteur Specialites Chimiques, Cedex 29, F-92097 Paris La Défense France (Tel.: 47-68 02 01; Telefax: 47-68 13 31)

Rhone-Poulenc, Inc., CN7500, Prospect Plain Rd., Cranbury, NJ 08512-7500 USA (Tel.: 609-860-4000, 800-922-2189; Telefax: 609-860-0269)

Rhone-Poulenc North America Inc./North American Sales, One Corporate Dr., Shelton, CT 06484 USA (Tel.: 203-925-8164; Telefax: 203-925-3670)

Rhone-Poulenc Basic Chemical Co., One Corporate Dr., Box 881, Shelton, CT 06484 USA (Tel.: 800-642-4200; Telefax: 203-925-3627)

Rhone-Poulenc Food Ingredients, CN 7500, Prospect Plains Rd., Cranbury, NJ 08512 USA (Tel.: 609-860-4600, 800-253-5052)

Rhone-Poulenc Rorer Pharmaceuticals Inc., 500 Arcola Road, Collegeville, PA 19426-0107 USA (Tel.: 610-454-8000)

Rhone-Poulenc Rubber Specialties, 1929 Trement St., Dover, OH 44627 USA (Tel.: 216-364-6002; Telefax: 216-364-5007)

Rhone-Poulenc Silicones, Cranbury, NJ 08512 (Tel.: 800-288-1175)

Rhone-Poulenc, Inc./Performance Resins & Coatings, 1525 Church St. Ext., Marietta, GA 30060 USA (Tel.: 404-422-1250; Telefax: 404-427-0874; Telex: 542-112)

Rhone-Poulenc, Inc./Specialty Chemicals, CN 7500, Prospect Plains Rd., Cranbury, NJ 08512-7500 USA (Tel.: 609-860-4000, 800-922-2189; Telefax: 609-860-0466)

Rhone-Poulenc, Inc./Surfactants & Specialty Chemicals, CN 7500, Prospect Plains Rd., Cranberry, NJ 08512-7500 USA (Tel.: 609-860-4000, 800-922-2189; Telefax: 609-860-0459)

Rhone-Poulenc, Inc./Water Soluble Polymers, CN 7500, Cranberry, NJ 08512-7500 USA (Tel.: 609-860-4000, 800-922-2189; Telefax: 609-860-0459)

Rhone-Poulenc, Inc./Water Treatment Chemicals, One Gatehall Dr., Parsippany, NJ 07054 USA (Tel.: 201-292-2900, 800-848-7659; Telefax: 201-292-5295)

Rhone-Poulenc Surfactants & Specialties Canada, 3265 Wolfdale Rd., Mississauga, Ontario, L5C 1V8 Canada (Tel.: 905-270-5534; Telefax: 905-270-5816)

Rhone-Poulenc Chemicals Ltd., Div. of Rhone-Poulenc SA, Staveley, Chesterfield, Derbyshire, S43 2PB UK (Tel.: 44 1246 277251; Telefax: 44 1246-280090; Telex: 577425 STAVEX G)

Rhone-Poulenc Rorer Ltd., Div. of Rhone-Poulenc SA, Rainham Rd South, Dagenham, Essex, RM10 7XS UK (Tel.: 44 181 592 3060; Telefax: 44 181-593 2140; Telex: 28691 MBDAGN G)

Nattermann Phospholipid GmbH, Div. of Rhone-Poulenc Rorer, PO Box 350120, Nattermannallee 1, D-5000 Cologne 30 Germany (Tel.: 49 221-509-2267; Telefax: 49 221-509-2816; Telex: 8882663)

American Lecithin Co., Div. of Rhone-Poulenc Rorer/Nattermann, 115 Hurley Rd., Unit 2B, Oxford, CT 06478 USA (Tel.: 203-262-7100; Telefax: 203-262-7101)

G. Whitfield Richards Co.

4202-10 Main St., Philadelphia, PA 19127 USA (Tel.: 215-487-1202; Telefax: 215-487-3090; Telex: 845-322 GWRCO)

Richman Chemical, Inc.

768 N. Bethlehem Pike, Lower Gwynedd, PA 19002 USA (Tel.: 215-628-2946; Telefax: 215-628-4262)

Ricon Resins, Inc.

569 24 1/4 Road, Grand Junction, CO 81505 USA (Tel.: 303-245-8148; Telefax: 303-245-4348)

Ridge Technologies, Inc.

117 Lyons Rd., Basking Ridge, NJ 07920 USA (Tel.: 908-766-1915; Telefax: 908-204-0312)

Riken Vitamin Oil Co., Ltd.

TDC Bldg., 9-18, Misaki-Cho 2-chome, Chiyoda-Ku, Tokyo, 101 Japan (Tel.: 81 (03) 5275 5130; Telefax: 81 (03) 5275 2905; Telex: 2322783)

R.I.T.A. Corp.

1725 Kilkenny Court, PO Box 585, Woodstock, IL 60098 USA (Tel.: 815-337-2500, 800-426-7759; Telefax: 815-337-2522; Telex: 72-2438)

Rit-Chem Co. Inc.

109 Wheeler Ave., PO Box 435, Pleasantville, NY 10570-0435 USA (Tel.: 914-769-9110; Telefax: 914-769-1408; Telex: 229 639 RTCH)

Robeco Chemicals Inc.

99 Park Ave., New York, NY 10016 USA (Tel.: 212-986-6410; Telefax: 212-986-6419; Telex: 23-3053 A(RCA)

Rogers Corp.

One Technology Drive, Rogers, CT 06263 USA (Tel.: 203-774-9605; Telefax: 203-774-9630)

Rohm and Haas

Rohm and Haas Co., 100 Independence Mall West, Philadelphia, PA 19106-2399 USA (Tel.: 215-592-3000, 800-323-4165; Telefax: 215-592-6909; Telex: 845-247)

Rohm and Haas Canada Inc., 2 Manse Rd., West Hill, Ontario, M1E 3T9 Canada (Tel.: 416-284-4711, 800-268-4201; Telefax: 416-284-2982)

Rohm and Haas Co./Industrial Coatings, 3724 Crescent Court West, PO Box E, Whitehall, PA 18052 USA (Tel.: 800-874-6917)

Rohm and Haas Europe, 185 rue de Bercy, 75579 Paris, Cedex 12 France (Tel.: 011-33-1 40 02 50 00; Telefax: 011-33-1 43 45 28 19)

Röhm GmbH Chemische Fabrik, Kirschenallee, Postfach 4242, D-6100 Darmstadt Germany (Tel.: 6151 18 01; Telefax: 6151 184007)

Rohm and Haas (UK) Ltd., Lennig House, 2 Mason's Avenue, Croydon, Greater London, CR9 3NB UK (Tel.: 44 181 686-8844; Telefax: 44 181-686 8329; Telex: 917266)

Rohm and Haas Deutschland GmbH, Postfach 94 03 22, 60461 Frankfurt/Main Germany (Tel.: 49-69-789-96-0; Telefax: 49-69-789-5356)

Rohm and Haas Pacific Region, 391B Orchard Road #16-05/07, 0 Ngee Ann City, 0923 Singapore (Tel.: 011-65-735-0855; Telefax: 011-65-735-0877)

Rohm Tech Inc., A Company of the Hüls Group

195 Canal St., Malden, MA 02148 USA (Tel.: 617-321-6984, 800-666-7646; Telefax: 617-322-0358; Telex: 200721 Rohm UR)

Ronsheim & Moore Ltd., Div. of Hickson & Welch Ltd.

Wheldon Road, Castleford, West Yorkshire, WF10 2JT UK (Tel.: 44 1977 556565; Telefax: 44 1977-518058; Telex: 55378)

Ross Chemical, Inc.

303 Dale Dr., PO Box 458, Fountain Inn, SC 29644 USA (Tel.: 803-862-4474, 800-521-8246; Telefax: 803-862-2912)

Frank B. Ross Co., Inc.

22 Halladay St., PO Box 4085, Jersey City, NJ 07304-0085 USA (Tel.: 201-433-4512; Telefax: 201-332-3555)

Rotuba Extruders, Inc.

1401 Park Ave. S., Linden, NJ 07036 USA (Tel.: 908-486-1000; Telefax: 908-486-0874)

H.M. Royal, Inc.

689 Pennington Ave., PO Box 28, Trenton, NJ 08601 USA (Tel.: 609-396-9176; Telefax: 609-396-3185)

Royale Pigments & Chems.

12 Rte. 17N, Suite 309, Paramus, NJ 07652 USA (Tel.: 201-845-4666; Telefax: 201-845-0719)

Royce Associates, ALP

207 Ave. L, Newark, NJ 07105 USA (Tel.: 201-465-3932; Telefax: 201-279-8561; Telex: 820-907 PASDYE US)

R S A Corp.

36 Old Sherman Tpke., Danbury, CT 06810 USA (Tel.: 203-790-8100; Telefax: 203-790-1709; Telex: 131148)

Ruco Polymer Corp.

New South Rd., Hicksville, NY 11802 USA (Tel.: 516-931-8100; Telefax: 516-931-8179; Telex: 5102220856)

Ruetgers-Nease Chemical Co., Inc., Subsid. of Rütgerswerke AG

201 Struble Rd., State College, PA 16801 USA (Tel.: 814-238-2424; Telefax: 814-238-1567)

Ruger Chemical Co. Inc.

85 Cordier St., Irvington, NJ 07111 USA (Tel.: 201-926-0331, 800-631-7844; Telefax: 201-926-4921)

Sachtleben Chemie

Sachtleben Chemie GmbH, Dr Rudolph-Sachtleben-Str. 4, 47198 Duisburg Germany (Tel.: 02066-22-2640; Telefax: 02066-22 2650; Telex: 855202 sc d)

Sachtleben Chemie GmbH, UK Sales Office, Huntingdon House, Princess St., Bolton, BL1 1EJ UK (Tel.: 44 1204 363634; Telefax: 44 1204 36 11 44)

Ore & Chemical Corp., Distrib. for Sachtleben Chemie GmbH, 520 Madison Ave., 27th floor, New York, NY 10022 USA (Tel.: 212-715-5200; Telefax: 212-486-2742)

SAF Bulk Chemicals

PO Box 14508, St. Louis, MO 63178 USA (Tel.: 800-336-9719; Telefax: 800-368-4661)

L.A. Salomon Inc.

150 River Rd., Suite A-4, Montville, NJ 07045 USA (Tel.: 201-335-8300; Telefax: 201-335-1236; Telex: 96-1470)

Samson Chem. Co.

8915 Sorenson Ave., PO Box 2163, Santa Fe Springs, CA 90670 USA (Tel.: 310-945-9188; Telefax: 310-698-7571)

Sandoz. *See* Clariant

San Esters Corp., Subsid. of Mitshbishi Rayon

342 Madison Ave., Suite 710, New York, NY 10173 USA (Tel.: 212-972-1122, 800-337-8377; Telefax: 212-972-4645)

Sanitized AG

Lyssachstrasse 95, Postfach 764, 3401 Burgdorf Switzerland (Tel.: (0041) 34 22 20 55; Telefax: (0041) 34 22 20 58)

Sanncor Industries, Inc.

300 Whitney St., PO Box 703, Leominster, MA 01453 USA (Tel.: 508-537-4748; Telefax: 508-537-8245; Telex: 701662)

Sanyo Chemical Industries, Ltd.

11-1, Ikkyo Nomoto-cho, Higashiyama-ku, Kyoto, 605 Japan (Tel.: (075) 541-4311; Telefax: (075) 551-2557; Telex: 05422110)

San Yuan Chemical Co., Ltd.

PO Box 26-1134, Taipei, Taiwan, R.O.C. (Tel.: 86286630343; Telefax: 86287770155)

S.A.P. Atlantic Inc.

12 Rte. 17 N., Suite 309, Paramus, NJ 07652 USA (Tel.: 201-845-4666; Telefax: 201-845-0719)

Sartomer

Sartomer Co., Oaklands Corp. Center, 502 Thomas Jones Way, Exton, PA 19341 USA (Tel.: 610-363-4100, 800-SARTOMER; Telefax: 610-363-4140; Telex: 173071)

Societe Cray Valley, Cedex 101, 92970 Paris La Defense France (Tel.: (33) 1 4135 68 21; Telefax: (33) 1 4135 62 70; Telex: 616309, 615289)

Sartomer International, Inc., 331 North Bridge Rd., #23-06 Odeon Towers, Singapore 188720 (Tel.: 65-334-2645/334-2646; Telefax: 65-334-2647)

Scheel Corp.

38 Franklin St., Brooklyn, NY 11222 USA (Tel.: 609-395-1100; Telefax: 609-395-5525)

Schenectady Chemicals, Inc.

PO Box 1046, Schenectady, NY 12301 USA (Tel.: 518-370-4200; Telefax: 518-346-3111; Telex: 145457)

Scher Chemicals, Inc.

Industrial West & Styertowne Rd., PO Box 4317, Clifton, NJ 07012 USA (Tel.: 201-471-1300; Telefax: 201-471-3783; Telex: 642643 Scherclif)

Schülke & Mayr GmbH

Robert-Koch-Strasse 2, D-22840 Norderstedt Germany (Tel.: 040/521-00-0; Telefax: 040/521-00-577; Telex: 215-486 sagro d)

A. Schulman Inc.

3550 West Market St., PO Box 1710, Akron, OH 44313 USA (Tel.: 216-666-3751; Telefax: 216-666-6311)

Schumacher

1969 Palomar Oaks Way, Carlsbad, CA 92009 USA (Tel.: 619-931-9555, 800-545-9241; Telefax: 619-931-7819; Telex: 910 322 1382)

Schweizerhall Inc.

10 Corporate Place South, Piscataway, NJ 08854 USA (Tel.: 908-981-8200, 800-243-6564; Telefax: 908-981-8282; Telex: 4754581 SUSA)

Scientific Adsorbents

PO Box 80998, Atlanta, GA 30366-0998 USA (Tel.: 404-455-1140, 800-4-ADSORB; Telefax: 404-455-7502)

Scientific Polymer Products, Inc.

6265 Dean Parkway, Ontario, NY 14519 USA (Tel.: 716-265-0413; Telefax: 716-265-1390)

SCM

SCM Chemicals, 7 St. Paul St., Suite 1010, Baltimore, MD 21202 USA (Tel.: 410-783-1120, 800-638-3234; Telefax: 410-783-1087/9)

SCM Chemicals Ltd., PO Box 26, Grimsby, South Humberside, DN37 8DP UK (Tel.: 44 1469 571000; Telefax: 44 1469 571234; Telex: 52595)

SCM Chemicals Ltd., PO Box 465, 11-13 Byrne St., Auburn, N.S.W., 2144 Australia (Tel.: 61-2-647-2566)

SCM Glidco Organics, PO Box 389, Jacksonville, FL 32201 USA (Tel.: 904-768-5800, 800-231-6728; Telefax: 904-768-2200; Telex: 441763)

SCM Glidco Organics, 210 Summit Ave., Montvale, NJ 07645-1526 USA (Tel.: 201-391-3040; Telefax: 201-391-3284)

SCM Metal Products, Inc., 2601 Weck Dr., PO Box 12166, Research Triangle Park, NC 27709-2166 USA (Tel.: 919-544-8090; Telefax: 919-544-7996; Telex: 196 072)

Scott Bader

Scott Bader Co. Ltd., Wollaston, Wellingborough, Northamptonshire, NN8 7RL UK (Tel.: 44 1933 663100; Telefax: 44 1933-665650; Telex: 31387 S BADER G)

Scott Bader SA, Div. of Scott Bader Co. Ltd., 65 rue Sully, F-80044 Amiens France (Tel.: 33-22 44 42 00; Telefax: 33-22 44 06 24; Telex: 150425 SCOTBAD F)

Sea-Land Chemical Co.

795 Sharon Dr., Westlake, OH 44145 USA (Tel.: 216-871-7887; Telefax: 216-871-7949)

Seal Sands

Seal Sands Chemicals, Ltd., A Cambrex Co., Seal Sands Rd., Seal Sands, Middlesbrough, Cleveland, TS2 1UB UK (Tel.: 44-642 546546; Telefax: 44-642 546068)

Seal Sands Chemicals, Ltd., A Cambrex Co., 40 Ave. A., Bayonne, NJ 07002 USA (Tel.: 201-858-7900; Telefax: 201-858-0308)

John L Seaton & Co. Ltd.

Bankside, Hull, North Humberside, HU5 1RR UK (Tel.: 44 1482 41345; Telefax: 44 1482 447157; Telex: 592207)

Seefast (Europe) Ltd.

59 Lampton Rd., Hounslaw, TW3 4DH UK (Tel.: 081-5770033; Telefax: 081-5706376)

Seegott Inc.

5400 Naiman Pkwy., Solon, OH 44139 USA (Tel.: 216-248-5400, 800-321-2865; Telefax: 216-248-3451)

Seeler Industries Inc.

2000 N. Broadway, Joliet, IL 60435 USA (Tel.: 815-740-2640, 800-336-2422; Telefax: 815-740-6469)

Seimi Chemical Co., Ltd.

2-10, Chigasaki 3-chome, Chigasaki-shi, Kanagawa, 253 Japan (Tel.: (0467) 82-4131; Telefax: (0467) 86-2767)

Sekisui Plastics Co., Ltd.

Shinjuku Mitsui Bldg., 2-1-1 Nishi Shinjuku, Shinjuku-ku, Tokyo, 163-04 Japan (Tel.: 03-3347-9689; Telefax: 03-3344-2335; Telex: 02324755 SEKIPLJ)

Seppic

Seppic, Div. of L'Air Liquide, 75 Quai d'Orsay, F-75321 Paris Cedex 07 France (Tel.: 40 62 55 55; Telefax: 40 62 52 53; Telex: 202901 SEPPI F)

Seppic Inc., Subsid. of Seppic France, 30 Two Bridges Rd., Suite 225, Fairfield, NJ 07004 USA (Tel.: 201-882-5597; Telefax: 201-882-5178)

Servo Delden B.V.

Postbus 1, NL-7490 AA Delden The Netherlands (Tel.: 5407-63535; Telefax: 31-5407-75025; Telex: 44347)

Shamokin Filler Co., Inc.

Venn Access Rd., Shamokin, PA 17872 USA (Tel.: 717-644-0437; Telefax: 717-648-4094)

Shamrock Technologies Inc.

Foot of Pacific St., Newark, NJ 07114 USA (Tel.: 201-242-2999; Telefax: 201-242-8074)

Shance Chemical Corp.

1 Depot Plaza, Mamaroneck, NY 10543 USA (Tel.: 914-698-1493; Telefax: 914-698-1520)

Sheffield Bronze Paint Corp.

17814 Waterloo Rd., Cleveland, OH 44119 USA (Tel.: 216-481-8330)

Shell

Shell Chemical Co., A Div. of Shell Oil Co., One Shell Plaza, PO Box 2463, Houston, TX 77252 USA (Tel.: 713-241-6161, 800-872-7435; Telefax: 713-241-4043; Telex: 762248)

Shell Chemical Co., 3200 Southwest Freeway, Suite 1230, Houston, TX 77027 USA (Tel.: 713-241-8101; Telefax: 713-241-8107)

Pecten Chemicals, Inc., representing for Int'l. sales, PO Box 4407, Houston, TX 77210 USA (Tel.: 713-241-6161; Telefax: 713-241-5194)

Shell Chemicals UK Ltd., Heronbridge House, Chester Business Park, Wrexham Rd, Chester, Cheshire, CH4 9QA UK (Tel.: 44 1244 685678; Telefax: 44 1244-685010; Telex: 21795)

Shell Chemicals Ireland Ltd., Gratton House, 68-72 Lower Mount Street, Dublin, Irish Republic (Tel.: 1-785177; Telefax: 1-767489; Telex: 93450)

Shell Chimie SA, BP 319, 23-25 ave de Republique, 7539 Paris Cedex 08, F-92506 Rueil-Malmaison France (Tel.: 47 52 27 00; Telefax: 47 52 28 02; Telex: 632051 SHELL F)

The Shepherd Chemical Co.

4900 Beech St., Cincinnati, OH 45212 USA (Tel.: 513-731-1110; Telefax: 513-731-1532; Telex: 21-4647)

The Shepherd Color Co.

4539 Dues Dr., PO Box 465627, Cincinnati, OH 45246 USA (Tel.: 513-874-0714; Telefax: 513-874-5061; Telex: 24-1659)

Sherex Polymers, Inc.

2525 S. Combee Rd., Lakeland, FL 33801 USA (Tel.: 800-872-7435)

Sherwin-Williams

Sherwin-Williams Co., PO Box 1028, Coffeyville, KS 67337 USA (Tel.: 316-251-7200)

Sherwin-Williams Co./Chemical Coatings Div., 11541 S. Champlain St., Chicago, IL 60628 USA (Tel.: 312-821-3182)

Shin-Etsu Chemical Co., Ltd.

Asahi Tokai Bldg., 6-1, Ohtemachi 2-chome, Chiyoda-ku, Tokyo, 100 Japan (Tel.: 81 (03) 3246 5011; Telefax: 81 (03) 3246 5350; Telex: SHINCHEM J-24790)

Showa Denko K.K.

13-9, Shiba-Daimon 1-chome, Minato-ku, Tokyo, 105 Japan (Tel.: (03) 5470-3533; Telefax: (03) 3436-2625; Telex: J26232)

Shuman Plastics

35 Neoga St., Depew, NY 14043 USA (Tel.: 716-685-2121; Telefax: 716-685-3236; Telex: 91 9121)

Sidobre-Sinnova S.A.

185 Ave. de Fontainebleau, 77981 St. Fargeau-Ponthierry France (Tel.: 33 1 60-65-21 00; Telefax: 33 1 60-65-21 05)

Sigma-Aldrich. *See under* Aldrich

Sigri GmbH

Postfach 1160, Werner von Siemens Strasse 18, W-8901 Meitingen Germany (Tel.: 49-8271-830; Telefax: 49-8271-833127; Telex: 53823 SIGRI D)

Silberline Manufacturing Co., Inc.

Lincoln Drive, PO Box B, Tamaqua, PA 18252-0420 USA (Tel.: 717-668-6050, 800-348-4824; Telefax: 717-668-0197)

Silbond Corp.

9901 Sand Creek Hwy., PO Box 200, Weston, MI 49289 USA (Tel.: 517-436-3171, 800-462-6082; Telefax: 517-436-3156)

Silbrico Corp.

6300 River Rd., Hodgkins, IL 60525-4257 USA (Tel.: 708-354-3350; Telefax: 708-354-6698; Telex: 317360)

Sino-American Pigment Systems Inc.

5801 Christie Ave., Suite 575, Emeryville, CA 94608-1933 USA (Tel.: 510-653-9931, 800-536-9932; Telefax: 510-653-7501)

SKW Chemicals Inc.

1509 Johnson Ferry Rd., Marietta, GA 30062 USA (Tel.: 404-971-1317; Telefax: 404-971-4306)

Sloss Industries Corp.

PO Box 5327, Birmingham, AL 35207 USA (Tel.: 205-808-7909; Telefax: 205-808-7885)

Smith Lime Flour Co.

60-70 Central Ave., S. Kearny, NJ 07032 USA (Tel.: 201-344-1700; Telefax: 201-690-5936)

Werner G. Smith, Inc.

1730 Train Ave., Cleveland, OH 44113 USA (Tel.: 216-861-3676, 800-535-8343; Telefax: 216-861-3680)

SNIA (UK) Ltd.

36 Broadway, St. Jame's, London, SW1H 0BH UK (Tel.: 44 171 222 8696; Telefax: 44 171 222 8705; Telex: 23377)

SNPE Chimie

12, quai Henri-IV, 75181 Paris Cedex 4 France (Tel.: 48 04 66 66; Telefax: 48 04 69 89; Telex: 220 380 SNPE F)

Soluol Chemical Co.

Green Hill & Market Sts., PO Box 112, W. Warwick, RI 02893 USA (Tel.: 401-821-8100; Telefax: 401-823-6673)

Solvay

Solvay SA, 33 rue du Prince Albert, B-1050 Brussels Belgium (Tel.: 32-509-6111; Telefax: 32-509-6617; Telex: 21337 SOLV B)

Solvay Deutschland GmbH, Postfach 220, W-3000 Hannover-1 Germany (Tel.: 511-857-0; Telefax: 511-282126; Telex: 922-755)

Solvay Duphar BV, Postbus 900, Bldg. WWO, Room B1066, 1380 DA Weesp The Netherlands (Tel.: 31 2940-77000; Telefax: 31 2940-15401; Telex: 14232)

Solvay Enzymes, Inc., PO Box 4859, 1230 Randolph St., Elkhart, IN 46514-0859 USA (Tel.: 219-523-3700, 800-342-2097; Telefax: 219-523-3800)

Solvay Polymers Inc., Subsid. of Solvay America Inc., 3333 Richmond Ave., PO Box 27328, Houston, TX 77227-7328 USA (Tel.: 713-525-4000, 800-231-6313; Telefax: 713-522-7890; Telex: 166307)

Southeastern Clay Co.
PO Box 1055, Alken, SC 29802 USA (Tel.: 803-648-3248; Telefax: 803-649-5701)

Southeastern Minerals, Inc.
PO Box 1866, 1100 Dothan Rd., Bainbridge, GA 31717 USA (Tel.: 912-246-3396; Telefax: 212-246-7309; Telex: 804620 SEM BBRG (WU)

Southern Clay Products, Inc.
1212 Church St., Gonzales, TX 78629 USA (Tel.: 210-672-2891, 800-324-2891; Telefax: 210-672-3081)

Sovereign Chemicals Co.
341 White Pond Dr., Akron, OH 44320-1145 USA

Spartan Color Corp.
5803 Northdale St., Houston, TX 77087 USA (Tel.: 713-644-1964, 800-231-4040)

Specialty Chem Products Corp.
Two Stanton St., Marinette, WI 54143 USA (Tel.: 715-735-9033; Telefax: 715-735-5304; Telex: 887445)

Specialty Minerals Inc., Subsid. of Minerals Technologies Inc.
640 N. 13th St., Easton, PA 18042 USA (Tel.: 610-250-3000; Telefax: 610-250-3344)

Specialty Products Co.
75 Montgomery St., PO Box 306, Jersey City, NJ 07303-0306 USA (Tel.: 201-434-4700, 800-321-8506; Telefax: 201-434-6052)

Spectrum Chemical Mfg. Corp.
14422 S. San Pedro St., Gardena, CA 90248 USA (Tel.: 310-516-8000, 800-772-8786; Telefax: 800-525-2299; Telex: 182395)

Spice King Corp.
6009 Washington Blvd., Culver City, CA 90232-7488 USA (Tel.: 213-836-7770; Telefax: 213-836-6454; Telex: 664350)

S & S Chemical Co., Inc.
333 Jericho Tpke., Jericho, NY 11753 USA (Tel.: 516-931-3333; Telefax: 516-931-2387)

A.E. Staley Manufacturing Co., Subsid. of Tate & Lyle PLC
2200 E. Eldorado St., PO Box 151, Decatur, IL 62525 USA (Tel.: 217-423-4411, 800-258-7536; Telefax: 217-421-2881)

StanChem, Inc.
401 Berlin St., East Berlin, CT 06023 USA (Tel.: 203-828-0571; Telefax: 203-828-3297)

Stan Chem International Ltd.
4 Kings Rd, Reading, Berkshire, RG1 3AA UK (Tel.: 44 1734 580247; Telefax: 44 1734 589580; Telex: 847746)

Stepan

Stepan Co., 22 West Frontage Rd., Northfield, IL 60093 USA (Tel.: 708-446-7500, 800-745-7837; Telefax: 708-501-2443; Telex: 910-992-1437)

Stepan Canada, PO Box 307, Orillia, Ontario, L3V 6J6 Canada (Tel.: 705-326-7329; Telefax: 705-326-4523)

Stepan Europe, BP127, 38340 Voreppe France (Tel.: 33-76-50-51-00; Telefax: 33-7656-7165; Telex: 320511 F)

Stevenson Cooper

PO Box 38349, 1039 West Venango St., Philadelphia, PA 19140 USA (Tel.: 215-223-2600; Telefax: 215-223-3597)

Stockhausen, Inc., A Company of the Hüls Group

2401 Doyle St., Greensboro, NC 27406 USA (Tel.: 910-333-3500; Telefax: 910-333-3545; Telex: 574405)

Joseph Storey & Co. Ltd.

Heron Chemical Works, Moor Lane, Lancaster, LA1 1QQ UK (Tel.: 44 1524 63252; Telefax: 44 1524 381805; Telex: 669755)

Strahl & Pitsch, Inc.

PO Box 1098, 230 Great E. Neck Rd., W. Babylon, NY 11704 USA (Tel.: 516-587-9000; Telefax: 516-587-9120; Telex: 221636 STRALUR)

Struktol Co.

201 E. Steels Corner Rd., PO Box 1649, Stow, OH 44224-0649 USA (Tel.: 216-928-5188, 800-327-8649; Telefax: 216-928-8726)

Süd-Chemie

Süd-Chemie AG, Postfach 200154, Munich 2 Germany (Tel.: 49 89-5110 0; Telefax: 49 89 51 10 375; Telex: 523872 SCMU D)

Süd-Chemie Rheologicals, Lenbachplatz 6, D-80333 München Germany (Tel.: 49 89-5110 0; Telefax: 49 89 51 10-412)

Sumisho Plaschem Co., Ltd.

1-1-8, Toranomon, Minato-ku, Tokyo, 105 Japan (Tel.: (03) 3502-1350; Telefax: (03) 3595-2924)

Sumitomo Chemical Co., Ltd.

New Sumitomo Bldg., 5-33, Kitahama 4-chome, Chuo-ku, Osaka, 541 Japan (Tel.: (06) 220-3272; Telefax: (06) 220-3345; Telex: 63823 SUMIKAJ)

Sumitomo Seika Chemicals Co., Ltd.

Sumitomo Bldg., No. 2, 4-7-28, Kitahama, Chuo-ku, Osaka, 541 Japan (Tel.: (06) 220-8508; Telefax: (06) 220-8541)

Summer Laboratories

PO Box 162, Ft. Washington, PA 19034 USA (Tel.: 215-646-1477, 800-523-5874; Telefax: 215-646-8931)

Sun Chemical

Sun Chemical Corp. Colors Group, Corporate Headquarters, 222 Bridge Plaza South, PO Box 1302, Fort Lee, NJ 07024 USA (Tel.: 201-224-4600; Telefax: 201-224-4392)

Sun Chemical Corp. Colors Group/Dispersions Div., 3922 Bach-Buxton Rd., Amelia, OH 45102 USA (Tel.: 513-753-9550, 800-321-3946; Telefax: 513-753-8374)

Sun Chemical Corp. Colors Group/Pigments Div., 5020 Spring Grove Ave., Cincinnati, OH 45232-1988 USA (Tel.: 513-681-5950, 800-543-2323; Telefax: 513-681-5505)

Sun Chemical Corp. Colors Group, Canadian Sales, 1260 Lake Shore Rd. East, Mississauga, Ontario, L5E 3B8 Canada (Tel.: 905-274-0037, 800-237-6325; Telefax: 905-274-6151)

Sun Chemical Corp./European Headquarters, 1 Rue Larmoyer, 1301 Bierge, Wavre Belgium (Tel.: 32 10 41 13 03; Telefax: 32 10 41 17 12; Telex: 59570)

Sun Chemical Limited, 2 Northfield Dr., Northfield, Milton Keynes, MK15 0DQ UK (Tel.: 44 908-234040; Telefax: 44-908-230775)

Sun Chemical Corp., Benelux & Switzerland, Tilburgeweg 2A, 5081 XJ Hilvarenbeek The Netherlands (Tel.: 31 4255-3347; Telefax: 31 4255-4446)

Sunnyside

225 Carpenter Ave., Wheeling, IL 60090 USA (Tel.: 312-541-5700, 800-323-8611; Telefax: 708-541-9043)

Superior Materials Inc.

585 Stewart Ave., Garden City, NY 11530 USA (Tel.: 516-222-1010; Telefax: 516-222-1332)

Surco Products, Inc.

PO Box 777, Eighth & Pine Aves., Braddock, PA 15104 USA (Tel.: 412-351-7700, 800-556-0111; Telefax: 412-351-7701)

Sutton Laboratories, Inc., Member of the ISP Inc. Group

116 Summit Ave., PO Box 837, Chatham, NJ 07928-0837 USA (Tel.: 201-635-1551; Telefax: 201-635-4964; Telex: 710-999-5607)

Sybron

Sybron Chemicals Inc., PO Box 125, Hwy. 29, Wellford, SC 29385 USA (Tel.: 803-439-6333, 800-677-3500; Telefax: 803-439-1612)

Sybron Chemie Nederland BV, Postbus 46, NL-6710 BA Ede The Netherlands (Tel.: 31-8380-70911; Telefax: 31-8380-30236; Telex: 37249)

Synray Corp.

209 N. Michigan Ave., Kenilworth, NJ 07033 USA (Tel.: 908-245-2600)

Syntech Labs Inc.

100 Jersey Ave., New Brunswick, NJ 08901 USA (Tel.: 908-545-8380; Telefax: 908-545-0120)

Synthetic Chemicals Ltd., Div. of Shell Chemicals UK Ltd.

Four Ashes, Wolverhampton, West Midlands, WV10 7BP UK (Tel.: 44 1902 794000; Telefax: 44 1902 794300; Telex: 337306)

Synthetic Products Co., Subsid. of Cookson America Inc.

1000 Wayside Rd., Cleveland, OH 44110 USA (Tel.: 216-531-6010, 800-321-4236; Telefax: 216-486-6638)

Synthron Inc.

305 Amherst Rd., Morganton, NC 28655-9362 USA (Tel.: 704-437-8611; Telefax: 704-437-4126; Telex: 752 629)

Taiwan Surfactant Corp.

No. 106, 8-1 Floor, Sec. 2, Changan East Rd., Taipei, Taiwan, R.O.C. (Tel.: 886-2-507-9155; Telefax: 886-2-507-7011; Telex: 27568)

Takeda Chemical Industries, Ltd.

1-1, Dosho-machi 4-chome, Chuo-ku, Osaka, 541 Japan (Tel.: (06) 204-2111; Telefax: (06) 204-2880; Telex: TAKEDA J63404)

Tamms Industries Co.

Rt. 72 West, Kirkland, IL 60146 USA (Tel.: 815-522-3394)

Taylor Chemical Co.

PO Box 768, Lawrenceville, GA 30246 USA (Tel.: 404-339-4460, 800-822-4460; Telefax: 404-339-4464)

Technic, Inc.

1 Spectacle St., Cranston, RI 02910 USA (Tel.: 401-781-6100; Telefax: 401-781-2890; Telex: ESYLINK 927 514)

Technical Chemicals & Products, Inc.

3341 SW 15th St., Pompano Beach, FL 33069 USA (Tel.: 305-979-0400; Telefax: 305-979-0009)

Tego Chemie. *See under* Goldschmidt

Teknor Apex Co.

505 Central Ave., Pawtucket, RI 02861 USA (Tel.: 401-725-8000, 800-554-9893; Telefax: 401-724-6250; Telex: 927530)

Telechem International, Inc.

12 South First St., Suite 817, San Jose, CA 95113 USA (Tel.: 408-977-0160; Telefax: 408-977-0164)

Tennant Trading (Chemicals) Ltd., Div. of Charles Tennant (London) Ltd.

Denny Avenue, Waltham Abbey, Essex, EN9 1NS UK (Tel.: 44 1992 715777; Telefax: 44 1992 700449; Telex: 24329)

Terry Laboratories, Inc.

390 Wickham Rd. N., Suite F, Melbourne, FL 32935-8647 USA (Tel.: 407-259-1630, 800-367-2563; Telefax: 407-242-0625)

Tetra Chemicals

25025 I-45 South, The Woodlands, TX 77380 USA (Tel.: 713-364-2233, 800-327-7817; Telefax: 713-367-6471; Telex: 214 874 TETRA UR)

Texaco

Texaco Chemical Co., PO Box 27707, Houston, TX 77227-7707 USA (Tel.: 713-961-3711, 800-231-3107; Telefax: 713-235-6437; Telex: 227-031 TEX UR)

Texaco Ltd., Div. of Texaco Chemical Europe, 195 Knightsbridge Green, London, SW7 1RU UK (Tel.: 44 171 581 5500; Telefax: 44 171-581 9163; Telex: 8956681 TEXACO G)

Texaco France S.A., 5, rue Bellini, Tour Arago, F-92806 Puteaux Cedex France (Tel.: 33-1-47-17 26 02; Telefax: 33-1-47 76 30 50)

Texaco Chemical Deutschland GmbH, Baumwall 5, 2000 Hamburg 11 Germany (Tel.: 49-40-37670147; Telefax: 49 40 37282 9)

Texaco Olie Matschappij BV, Weena 170, NL-3012 CR Rotterdam The Netherlands (Tel.: 31-614471; Telex: 31542)

Texas Petrochemicals Corp., 8707 Katy Frwy., Suite 300, Houston, TX 77024 USA (Tel.: 713-461-3322; Telefax: 713-461-1029; Telex: TWX 510-891-7831)

Thermoset Plastics Inc.

5101 East 65th St., PO Box 20902, Indianapolis, IN 46220 USA (Tel.: 317-259-4161; Telefax: 317-252-8402)

Thibaut & Walker Co.

PO Box 296, 49 Rutherford St., Newark, NJ 07101 USA (Tel.: 201-589-3331; Telefax: 201-589-7231)

Thiele Kaolin Co.

Box 1056, Sandersville, GA 31082 USA (Tel.: 912-552-3951; Telefax: 912-552-4131)

Thomas Swan & Co. Ltd.

Crookhall, Consett, Co. Durham, DH8 7ND UK (Tel.: 44 1207 505131; Telefax: 44 1207-590467; Telex: 53565)

Thor Chemicals

Thor Chemicals, Inc., Brook House, 37 North Ave., Norwalk, CT 06851 USA (Tel.: 203-846-8613; Telefax: 203-846-4810; Telex: 888630)

Thor Chemicals (UK) Ltd., Cowley House, Earl Road, Cheadle Hulme, Cheshire, SK8 6QP UK (Tel.: 44 161 486-1051; Telefax: 44 161-488-4125; Telex: 666679 THORUK G)

3M

3M Co./Industrial Chem. Prods. Div., 3M Center Bldg. 223-6S-04, St. Paul, MN 55144-1000 USA (Tel.: 612-736-1394, 800-541-6752)

3M Co./Specialty Chemicals Div., 3M Center Bldg. 223-6S-04, St. Paul, MN 55144-1000 USA (Tel.: 612-733-3064; Telefax: 612-737-7635)

3M Co./Electrical Specialty Corrosion Protection Prods. Div., 3M Center Bldg., St. Paul, MN 55144-1000 USA (Tel.: 800-722-6721)

3M Co./Specialty Fluoropolymers Dept., 3M Center Bldg. 220-10E-10, St. Paul, MN 55144-1000 USA (Tel.: 612-733-0419; Telefax: 612-737-7686)

3M Canada Inc., PO Box 5757 Terminal A, 1840 Oxford St. East, London, Ontario, N6A 4T1 Canada (Tel.: 519-451-2500)

3V

3V Inc., 1500 Harbor Blvd., Weehawken, NJ 07087 USA (Tel.: 201-865-3600, 800-441-5156; Telefax: 201-865-1892)

3V Inc., 9140 Arrowpoint Blvd., Suite 120, Charlotte, NC 28273-8120 USA (Tel.: 704-523-5252; Telefax: 704-522-1763)

Tiarco Chemical Div./Textile Rubber & Chemical Co.

1300 Tiarco Dr., Dalton, GA 30720 USA (Tel.: 706-277-1300; Telefax: 706-277-9039)

TIC Gums, Inc.

4609 Richlynn Dr., Belcamp, MD 21017-0369 USA (Tel.: 410-273-7300, 800-221-3953; Telefax: 410-273-6469; Telex: 221049)

Tioxide Americas Inc.

Esplanade at Locust Point, 2001 Butterfield Rd., Suite 501, Downers Grove, IL 60515 USA (Tel.: 708-663-4900; Telefax: 708-663-4902)

Titanium Metals Corp.

1999 Broadway, Suite 4300, Denver, CO 80202 USA (Tel.: 303-296-5600; Telefax: 303-296-5640; Telex: 825750)

Titan Kogyo Kabushiki Kaisha (Titan Kogyo K.K.)

1978, Kogushi, Ube-shi, Yamaguchi, 755 Japan (Tel.: (0836) 31-4156; Telefax: (0836) 31-5148)

Toho Chemical Industry Co., Ltd.

No. 1-2-5, Ningyo-cho, Nihonbashi, Chuo-ku, Tokyo, 103 Japan (Tel.: (81-3) 3668-2279; Telefax: (81-3) 3668-2278; Telex: 252-2332 TOHO K J)

Toho Zinc Co., Ltd.

Asahi Bldg., 12-2, Nihonbashi 3-chome, Chuo-ku, Tokyo, 103 Japan (Tel.: (03) 3272-5611; Telefax: (03) 3271-0070; Telex: 222-3725 TOHO ZN)

Tokuyama Soda Co., Ltd.

4-5, Nishi-Shimbashi 1-chome, Minato-ku, Tokyo, 105 Japan (Tel.: (03) 3597-5055; Telefax: (03) 3597-5180; Telex: 222-3258 TOKUSO)

Tokyo Kasei Kogyo Co., Ltd.

1-13, Nihonbashi-Honcho 3-chome, Chuo-ku, Tokyo, 103 Japan (Tel.: (03) 3808-2821; Telefax: (03) 3808-2827; Telex: 2223592 ASACEM J)

Tomita Pharmaceutical Co., Ltd.

85-1, Aza Maruyama, Akinokami, Seto-machi, Naruto-shi, Tokushima, 771-03 Japan (Tel.: (0886) 88-0511; Telefax: (0886) 88-0565)

Toray Thiokol Co., Ltd.

Toray Bldg., 1-8-1, Mihama, Urayasu-shi, Chiba, 279 Japan (Tel.: (0473) 50-6151; Telefax: (0473) 50-6091)

Tosoh

Tosoh Corp., 7-7 Akasaka 1-chome, Minato-ku, Tokyo, 107 Japan (Tel.: (03) 3585-9891; Telefax: (03) 3582-8120; Telex: J24475tosoh)

Tosoh USA Inc., 1100 Circle 75 Pkwy., Suite 600, Atlanta, GA 30339 USA (Tel.: 770-956-1100; Telefax: 770-956-7368; Telex: 542272 tosoh atl)

Total Petroleum Inc.

East Superior St., Alma, MI 48801 USA (Tel.: 517-463-9630, 800-292-9033; Telefax: 517-463-9623)

Toyo Kasei Kogyo Co., Ltd.

Shindai Bldg., 1-2-6, Dojimahama, Kita-ku, Osaka, 530 Japan (Tel.: (06) 346-6707; Telefax: (06) 341-6715)

Traco Labs, Inc.

205 S. Main St., Seymour, IL 61875 USA (Tel.: 217-687-2800, 800-798-7226; Telefax: 217-687-2700)

Tra-Con, Inc.

55 North St., PO Box 306, Medford, MA 02155 USA (Tel.: 617-391-5550, 800-872-2661; Telefax: 617-391-7380)

Tradig

PO Box 601, Bethlehem, PA 18016 USA (Tel.: 610-432-9466; Telefax: 610-433-1416; Telex: 5106512132)

TR-AMC Chemicals

PO Box 296, Hudson Ave., Ridgefield, NJ 07657 USA (Tel.: 201-941-7706; Telefax: 201-941-7702; Telex: 130594)

Trans-Chemco, Inc.

19235 84th St., Bristol, WI 53104-0009 USA (Tel.: 414-857-2363)

Transol Chemicals (UK) Ltd., Div. of Transol Chemicals Int'l. BV

Caledonian House, Tatton St, Knutsford, Cheshire, WA16 6AG UK (Tel.: 44 1565 650386; Telefax: 44 1565 653255)

T&R Chemicals, Inc.

700 Celum Rd., Box 330, Clint, TX 79836 USA (Tel.: 915-851-2761)

Tricon Colors Inc.

16 Leliarts Ln., Elmwood Park, NJ 07407-3291 USA (Tel.: 201-794-3800; Telefax: 201-797-4660; Telex: 4991537 TRICN)

Tricon Trading Associates

2487 Kaladar Ave., Suite 2117, Ottawa, Ontario, K1V 8B9 Canada (Tel.: 613-247-0944; Telefax: 613-247-0945)

Tri-Iso, Inc.

480 N. Indian Hill Blvd., Ste. 2A, Claremont, CA 91711 USA (Tel.: 909-621-4635; Telefax: 909-621-9119)

Tri-K Industries, Inc.

27 Bland St., PO Box 312, Emerson, NJ 07630 USA (Tel.: 201-261-2800, 800-526-0372; Telefax: 201-261-1432; Telex: 215085 TRIK UR)

Triple Crown America, Inc.

13 N. Seventh St., Perkasie, PA 18944 USA (Tel.: 215-453-2500; Telefax: 215-453-2508)

Troy

Troy Corp., PO Box 366, 72 Eagle Rock Ave., East Hanover, NJ 07936-0366 USA (Tel.: 201-884-4300; Telefax: 201-884-4317; Telex: 138930 Troychem Nwk)

Troy Chemical Co., Ltd., 157 Overture Rd., Scarborough, Ontario, M1E 2W5 Canada (Tel.: 416-287-9116; Telefax: 416-287-9779)

Troy Chemical Co. UK, Zenith House, Northolme Rd., Louth, Lincolnshire, LN11 0HQ UK (Tel.: 44 1507 609606; Telefax: 44 1507-607107)

Troy Chemical Co. BV, Uiverlaan 12e, 3145 XN Maassluis The Netherlands (Tel.: 31-18-992-7494; Telefax: 31-18-992-8877; Telex: 26473 Troy NL)

Troy Chemie GmbH, Uerdingerstrasse 541, 4800 Krefeld 1 Germany (Tel.: 49-2151-59-03-38; Telefax: 49-2151-59-81-45)

Troy Southeast Asia, 35-B Polaris St., Bel-Air, Makati, Metro-Manila, Philippines (Tel.: 63-2-812-4734; Telefax: 63-2-895-5551)

Trugman Nash Inc.

90 West St., New York, NY 10006 USA (Tel.: 212-964-9350; Telefax: 212-791-1863; Telex: 232098-RCA)

Tulco, Inc.

9 Bishop Rd., Ayer, MA 01432 USA (Tel.: 508-772-4412; Telefax: 508-772-1751)

20 Microns Limited

PO Box 4053, 307/308 Third Fl., Arundeep Complex, Race Course (S.), Baroda, 390 015 India (Tel.: (0265) 330714, fax (0265) 333755; Telex: 0175-6533 mcon in)

Ube Industries, Ltd.

UBE Bldg., 2-3-11, Higashi-shinagawa, Shinagawa-ku, Tokyo, 140 Japan (Tel.: (03) 5460-3311; Telefax: (03) 5460-3388; Telex: 2224645)

Ucar Carbon Co., Inc.

39 Old Ridgebury Rd.-J4, Danbury, CT 06817 USA (Tel.: 203-794-3684, 800-342-3698; Telefax: 203-794-3180; Telex: 126019 UCCHQDURY)

UCB

UCB SA/Chemical Sector, 326 avenue Louise, B-1050 Brussels Belgium (Tel.: 32(2)-641-1411; Telefax: 32(2)-640 9860; Telex: 21280)

UCB (Chem) Ltd., Div. of UCB SA, Star House, 69 Clarendon Road, Watford, Hertfordshire, WD1 1DJ UK (Tel.: 44 1923 248011; Telefax: 44 1923-250225; Telex: 23958)

UCB Chemicals Corp./Radcure Business Unit, Wholly owned subsid. of UCB Group, 2000 Lake Park Dr., Smyrna, GA 30080 USA (Tel.: 770-434-6188, 800-433-2873; Telefax: 770-434-8314)

Ueno Fine Chemicals Industry, Ltd.

2-4-8 Koraibashi, Chuo-ku, Osaka, 541 Japan (Tel.: (06) 203-0761; Telefax: (06) 222-2413; Telex: J63638 UENOFCI)

Ultra Additives, Inc.

460 Straight St., Paterson, NJ 07501 USA (Tel.: 201-279-1306, 800-524-0055; Telefax: 201-279-0602)

Ulysses Inc.

581 Boylston St., Ste. 802, Boston, MA 02116 USA (Tel.: 617-262-3232; Telefax: 617-424-0708)

Unger Fabrikker AS

PO Boks 254, N-1601 Fredrikstad Norway (Tel.: 47-9 32 00 20; Telefax: 47-9 32 37 75; Telex: 76382 UNGER N)

Unicel Corp.

1891 Main St., Mohegan Lake, NY 10547 USA (Tel.: 914-528-9142; Telefax: 914-528-9205)

Unichema

Unichema International, Part of the Unilever Speciality Chemicals Group, Postbus 2, 2800 AA Gouda The Netherlands (Tel.: 31-0-1820-42911; Telefax: 31-0-1820-42250; Telex: 20661)

Unichema Chemie BV, Postbus 2, 2800 AA Gouda The Netherlands (Tel.: 31 (0) 1820-42911; Telefax: 31 (0) 1820-42250; Telex: 20661)

Unichema Chemicals Ltd., Div. of Unichema International, Bebington, Wirral, Merseyside, L62 4UF UK (Tel.: 44 151 645-2020; Telefax: 44 151-645-9197; Telex: 629408)

Unichema Chemie GmbH, Postfach 100963, D-4240 Emmerich Germany (Tel.: 49 0 2822-720; Telefax: 49 0 2822-72276; Telex: 8125113)

Unichema France SA, 148 Boulevard Haussemann, 75008 Paris France (Tel.: 33 1 44 95 08 40; Telefax: 33 1 42563188; Telex: 643217)

Unichema North America, 4650 S. Racine Ave., Chicago, IL 60609 USA (Tel.: 312-376-9000, 800-833-2864; Telefax: 312-376-0095; Telex: 176068)

Unichema Japan, Sankei Bldg. 7F 708, 4-9, Umeda 2-chome, Kita-ku, Osaka, 530 Japan (Tel.: 81 6341-7221; Telefax: 81 6341-7725)

Unichema Australia Pty. Ltd., 164 Ingles St., Port Melbourne, Victoria, 3207 Australia (Tel.: 61-3 647-9311; Telefax: 61-3 645 3001; Telex: 30130)

Unimin

Unimin Corp., 258 Elm St., New Canaan, CT 06840 USA (Tel.: 203-966-8880, 800-243-9004; Telefax: 203-966-3453; Telex: 99-6355)

Unimin Specialty Minerals Inc., Subsid. of Unimin Corp., PO Box 33, Rt. 127, Elco, IL 62929 USA (Tel.: 618-747-2311, 800-743-7519; Telefax: 618-747-9318)

Unimin Canada Ltd., RR #4, PO Box 2000, Havelock, Ontario, K0L 1Z0 Canada (Tel.: 705-877-2210, 800-363-4140; Telefax: 705-877-3343)

Union Camp

Union Camp Corp./Chem. Prods. Div., 1600 Valley Rd., Wayne, NJ 07470 USA (Tel.: 201-628-2290, 800-733-1374; Telefax: 201-628-2840; Telex: 130735)

Union Camp Chemicals Ltd., Vigo Lane, Chester-le-Street, Co. Durham, DH3 2RB UK (Tel.: 44-91-410-2631; Telefax: 44-91-410-9391; Telex: 851 53163)

Union Camp Chemicals, Am Burgholz 17, D-5166 Kreuzau-Stockheim Germany (Tel.: 49-2421-59260; Telefax: 49-2421-592635)

Union Camp Chemicals, 11/F, Mappin House, 98, Texaco Rd., Tsuen Wan, N.T. Hong Kong (Tel.: 852-408-7170; Telefax: 852-407-6067; Telex: 780-57027 BBAHBHX)

Union Carbide

Union Carbide Corp., 39 Old Ridgebury Rd., Danbury, CT 06817-0001 USA (Tel.: 203-794-2000, 800-335-8550; Telefax: 203-794-3170)

Union Carbide Corp./Solvents & Coatings, 39 Old Ridgebury Road, Danbury, CT 06817-0001 USA (Tel.: 203-794-5300, 800-SOLVENT)

Union Carbide Canada Ltd., 7400 Blvd des Galleries, d'Anjou, Quebec, H1M 3M2 Canada (Tel.: 514-493-2610; Telefax: 514-493-2619)

Union Carbide (UK) Ltd./Chemicals & Plastics, 93-95 High Street, Rickmansworth, Hertfordshire, WD3 1RB UK (Tel.: 44 1923 720 366; Telefax: 44 1923-896721)

Union Carbide Chemicals & Plastics Europe S.A., 15 Chemin Louis-Dunant, CH-1211 Geneve 20 Switzerland (Tel.: 41-22-739-6111; Telefax: 41-22-739-6527; Telex: 419207 UNC CH)

Union Derivan SA

Av. Meridiana 133, Barcelona E-08026 Spain (Tel.: 343-2322113; Telefax: 343-2323951; Telex: 98204 UNDER E)

Uniroyal

Uniroyal Chemical Co. Inc./World Headquarters, Benson Road, Middlebury, CT 06749 USA (Tel.: 203-573-2000, 800-243-3024; Telefax: 203-573-2489; Telex: 6710383 uniroyal)

Uniroyal Chimica, SpA, Via delle Industrie 40, I-104013 Latina-Scalo Italy (Tel.: 39-773-43605; Telex: 680056)

United Catalysts Inc., Süd-Chemie Rheologicals Group

PO Box 32370, Louisville, KY 40232 USA (Tel.: 502-634-7200; Telefax: 502-637-3732; Telex: 204190, 204239)

United-Guardian, Inc.

230 Marcus Blvd, PO Box 2500, Smithtown, NY 11787 USA (Tel.: 516-273-0900, 800-645-5566; Telefax: 516-273-0858)

United Mineral & Chemical Corp., UMC

1100 Valley Brook Ave., Lyndhurst, NJ 07071-3608 USA (Tel.: 201-507-3300, 800-777-0505; Telefax: 201-507-1506; Telex: 6505113226)

United States Bronze Powders, Inc.

PO Box 31, Rte. 202, Flemington, NJ 08822 USA (Tel.: 908-782-5454; Telefax: 908-782-3489; Telex: 833488)

Unitex Chemical Corp.

PO Box 16344, 520 Broome Rd., Greenboro, NC 27406 USA (Tel.: 910-378-0965; Telefax: 910-272-4312)

Universal Preserv-A-Chem Inc./UPI

297 North 7th St., Brooklyn, NY 11211 USA (Tel.: 718-782-7429)

Universal Color Dispersions, Div. of Morton Int'l.

2701 E. 170 St., Lansing, IL 60438 USA (Tel.: 312-785-5500, 800-23-COLOR; Telefax: 708-868-7490)

UOP

UOP, 25 E. Algonquin Rd., Des Plaines, IL 60017-5017 USA (Tel.: 708-391-2395, 800-348-0832; Telex: 25-3285)

UOP Ltd., Liongate, Ladymeade, Guildford, Surrey, GU1 1AT UK (Tel.: 44 1483 304848; Telefax: 44 1483-304863; Telex: 858051 UOPINT G)

U.S. Aluminum Inc.

Route 202, PO Box 2190, Flemington, NJ 08822 USA (Tel.: 908-782-5454; Telefax: 908-782-3489; Telex: 833488)

U.S. Biochemical Corp.

PO Box 22400, Cleveland, OH 44122 USA (Tel.: 216-765-5000, 800-321-9322; Telefax: 216-464-5075; Telex: 980718)

U.S. Borax

U.S. Borax Inc., Member of the RTZ Corp. plc, 26877 Tourney Rd., Valencia, CA 91355-1847 USA (Tel.: 805-287-5400, 800-729-2672; Telefax: 805-287-5455; Telex: 371-6120)

Borax Limited, Member of the RTZ Corp. plc, Gorsey Lane, Widnes, Cheshire, WA8 0RP UK (Tel.: 44 151 420 5522; Telefax: 44 151 420 8815)

Borax Benelux, Member of the RTZ Corp. plc, 43, Rue de Namur, B-1000 Brussels Belgium (Tel.: 32 2-512 8858; Telefax: 32 2-514 0697)

Borax Argentina, Member of the RTZ Corp. plc, Crespo 2759, Buenos Aires, CP1437, Argentina (Tel.: 54 1-924 3960; Telefax: 54 1-924 1374)

Borax Japan Limited, c/o ICI Japan Ltd., Member of the RTZ Corp. plc, Palace Bldg., 1-1, 1-Chome, Marunouchi, Chiyoda-ku, Tokyo, 100 Japan (Tel.: 81 3-3211 3110; Telefax: 81 3-3211 7806)

U.S. Chemicals, Inc.

280 Elm St., New Canaan, CT 06840 USA (Tel.: 203-966-8777; Telefax: 203-966-3577)

U.S. Gypsum Co.

125 S. Franklin St., Chicago, IL 60606 USA (Tel.: 312-606-4018)

U.S. Petrochemical Industries, Inc.

675 Galleria Financial Ctr., 5075 Westheimer Rd., Houston, TX 77056 USA (Tel.: 713-871-1951; Telefax: 713-871-1963; Telex: 402814)

U.S. Polymers Inc.

300 E. Primm St., St. Louis, MO 63111 USA (Tel.: 314-638-1632; Telefax: 314-638-3100)

U.S. Silica Co.

PO Box 187, Rt. 522 North, Berkeley Springs, WV 25411 USA (Tel.: 304-258-2500, 800-243-7500; Telefax: 304-258-3500; Telex: 4942414)

U.S. Synthetics Co.

PO Box 2236, Danbury, CT 06813 USA (Tel.: 203-270-0187; Telefax: 203-790-6407; Telex: 4972481 mbe tam)

U.S. Zinc Corp.

6020 Esperson St., PO Box 611, Houston, TX 77001 USA (Tel.: 713-926-1705; Telefax: 713-924-4824; Telex: 3785919)

VanDeMark

VanDeMark Chemical Co., Inc., One N. Transit Rd., Lockport, NY 14094 USA (Tel.: 716-433-6764, 800-836-8253; Telefax: 716-433-2850; Telex: 990731)

Vanchem, Inc., Member of VanDeMark Group, One N. Transit Rd., Lockport, NY 14094 USA (Tel.: 716-434-2624, 800-336-8253; Telefax: 716-433-2850; Telex: 990731)

Vanchlor, Inc., Member of VanDeMark Group, One N. Transit Rd., Lockport, NY 14094 USA (Tel.: 716-434-2624; Telefax: 716-433-2850; Telex: 990731)

Van Den Bergh Foods Co.

2200 Cabot Dr., Lisle, IL 60532 USA (Tel.: 708-505-5300, 800-949-7344; Telefax: 708-955-5497)

R.T. Vanderbilt Co. Inc.

30 Winfield St, PO Box 5150, Norwalk, CT 06856 USA (Tel.: 203-853-1400, 800-243-6064; Telefax: 203-853-1452; Telex: 710-468-2941)

Vanguard Chem. Int'l.

Nationsbank Tower, 101 E. Park Blvd., Plano, TX 75074 USA (Tel.: 214-423-1120; Telefax: 214-423-1291; Telex: 49 75044)

Van Leer Flexibles

9505 Bamboo Rd., Houston, TX 77041 USA (Tel.: 713-462-6111, 800-VALERON; Telefax: 713-690-2746)

Van Waters & Rogers Inc., Subsid. of Univar Corp.

6100 Carillon Point, Kirkland, WA 98033 USA (Tel.: 206-889-3400, 800-234-4588; Telefax: 206-889-4133)

Veckridge Chemical Co. Inc.

60-70 Central Ave., Kearny, NJ 07032 USA (Tel.: 201-344-1818; Telefax: 201-690-5936)

Veitsiluoto Oy/Forest Chemicals Industry

PO Box 196, SF-90101 Oulu 10 Finland (Tel.: 358-81-316 3111; Telefax: 358-81-378 5755; Telex: 32125 oulpk sf)

Velsicol

Velsicol Chemical Corp., 10400 W. Higgins Road, Rosemont, IL 60018 USA (Tel.: 708-298-9000, 800-843-7759; Telefax: 708-298-9014; Telex: 3730755)

Velsicol Chemical Ltd., Worting House, Basingstoke, Hampshire, RG23 8PY UK (Tel.: 44 1256 817640; Telefax: 44 1256 817744; Telex: 9312131051 VC G)

Verdugt BV

Postbus 60, Papesteeg 91, NL-4000 AB Tiel The Netherlands (Tel.: 31-3440-15224; Telefax: 31-3440-11475; Telex: 47200)

Vista

Vista Chemical Company, 900 Threadneedle, PO Box 19029, Houston, TX 77224-9029 USA (Tel.: 713-588-3000, 800-231-8216; Telefax: 713-588-3236; Telex: 794557)

Vista Chemical Europe, Hilton Tower, Blvd. de Waterloo #39, B-21000 Brussels Belgium (Tel.: 32-2-513-7490; Telefax: 32-2-513-5449; Telex: 24727 VISTA B)

Condea Vista Japan Inc., PO Box 110, Kasumigaseki Bldg., 3-2-5 Kasamugaseki, Chiyoda-ku, Tokyo, 100 Japan (Tel.: 3593-0611; Telefax: 3593-0615; Telex: J 29368 VISTACHM)

Condea Vista Singapore, 152 Beach Rd. #02-08, Gateway East, 0718 Singapore (Tel.: Telex:)

Volclay Ltd., Div. of American Colloid Co.

Birkenhead Rd, Wallasey, Merseyside, L44 7BU UK (Tel.: 44 151 638 0967; Telefax: 44 151-630 2764; Telex: 627029)

Voltaix, Inc.

Box 5357, 197 Mesiter Ave., North Branch, NJ 08876 USA (Tel.: 908-231-9060, 800-VOLTAIX; Telefax: 908-231-9063)

Vulcan Chemicals, Div. of Vulcan Materials Co.

One Metroplex Dr., Birmingham, AL 35209 USA (Tel.: 205-877-3000, 800-633-8280; Telefax: 205-877-3448; Telex: 59-6108)

VYN-AC Inc.

PO Box 788, Ormond Beach, FL 32175-0788 USA (Tel.: 800-342-8475, #567)

Vyse Gelatin Co.

5010 N. Rose St., Schiller Park, IL 60176 USA (Tel.: 708-678-4780; Telefax: 708-628-0329)

Wacker

Wacker-Chemie GmbH, Hanns-Seidel-Platz 4, 81737 München Germany (Tel.: (089) 6279 01; Telefax: (089) 6279-1770; Telex: 5291210)

Wacker Chemicals Ltd., Div. of Wacker-Chemie GmbH, The Clock Tower, Mount Felix, Bridge Street, Walton-on-Thames, Surrey, KT12 1AS UK (Tel.: 44 1932 246111; Telefax: 44 1932-240141; Telex: 28 391 wacker g)

Wacker Chemie Danmark A/S, Hovedvejen 91, DK-2600 Glostrop Denmark (Tel.: (43) 43 03 00; Telefax: (43) 43 03 16)

Wacker Quimica Ibérica SA, Div. of Wacker-Chemie GmbH, Corcega, 303-2° 3a, E-08008 Barcelona Spain (Tel.: 34-3-217-5900; Telefax: 34-3-217-5766; Telex: 97801 wosa e)

Wacker Chemicals (USA) Inc., 535 Connecticut Ave., Norwalk, CT 06855 USA (Tel.: 203-866-9400, 800-634-6752; Telefax: 203-866-9427; Telex: 643 444)

Wacker Silicones Corp., Subsid. of Wacker-Chemie, 3301 Sutton Rd., Adrian, MI 49221-9397 USA (Tel.: 517-264-8500, 800-248-0063; Telefax: 517-264-8246; Telex: 510-450-2700 sadrnud)

Wacker Mexicana, S.A. de C.V., Apartado Postal 99-070, Av. Periferico Sur 3343/PH, San Jeronimo, 10200, Mexico, DF (Tel.: 525-595-7599; Telefax: 525-683-8434)

Wako Pure Chemical Industries Ltd.

1,2-Doshomachi 3-Chome, Chuo-ku, Osaka, 541 Japan (Tel.: (06) 203-3741; Telefax: (06) 222-1203; Telex: 65188 wakoos j)

Welch, Holme & Clark Co. Inc.

7 Ave. L, Newark, NJ 07105 USA (Tel.: 201-465-1200; Telefax: 201-465-7332)

Westbrook Lanolin Co., A Div. of Woolcombers (Holdings) Ltd.

Argonaut Works, Laisterdyke, Bradford, West Yorkshire, BD4 8AU UK (Tel.: 44 1274 663331; Telefax: 44 1274-667665; Telex: 51502)

Westco Chemicals, Inc.

11312 Hartland St., North Hollywood, CA 91605 USA (Tel.: 213-877-0077; Telefax: 818-766-7170; Telex: 673150)

Westlake Plastics Co.

PO Box 127, W. Lenni Rd., Lenni, PA 19052 USA (Tel.: 215-459-1000; Telefax: 215-459-1084; Telex: 83-5406)

Westo Industrial Products Ltd.

31 Pembridge Rd, London, W11 3HG UK (Tel.: 44 171 727 8700; Telefax: 44 171-792 0329; Telex: 888941 LCCI G)

Westvaco Corp., Chemical Div.

PO Box 70848, Charleston Hts., SC 29415-0848 USA (Tel.: 803-740-2300; Telefax: 803-740-2329; Telex: 4611159)

Whitfield Chemicals Ltd.

23 Albert St, Newcastle, Staffordshire, ST5 1JP UK (Tel.: 44 1782 711777; Telefax: 44 1782 717290; Telex: 367165)

Whitford Corp.

PO Box 2347, West Chester, PA 19380 USA (Tel.: 610-296-3200; Telefax: 610-647-4849; Telex: 83-5305)

Whiting, Peter (Chemicals) Ltd.

5 Lord Napier Place, Upper Mall, London, W6 9UB UK (Tel.: 44 181 741 4025; Telefax: 44 181-741 1737; Telex: 8814670 WHICHEM G)

Whittaker, Clark & Daniels

1000 Coolidge St., South Plainfield, NJ 07080 USA (Tel.: 800-732-0562; Telefax: 800-833-8139; Telex: 221478)

Whittemore Co. Inc.

30 Glenn St., Lawrence, MA 01843 USA (Tel.: 508-681-8833; Telefax: 508-682-3413)

Wiley Organics, Inc.

PO Box 670, 1245 South Sixth St., Coshocton, OH 43812 USA (Tel.: 614-622-0755; Telefax: 614-622-3231)

Wilke International Inc.

1375 N. Winchester, Olathe, KS 66061 USA (Tel.: 913-78-5544, 800-779-5545; Telefax: 913-780-5574)

Witco

Witco Corp., One American Lane, Greenwich, CT 06831-2559 USA (Tel.: 203-552-2000, 800-779-4826 x6400; Telefax: 203-552-2010)

Witco Corp./Allied-Kelite, 17050 Lathrop Ave., Harvey, IL 60426 USA (Tel.: 800-323-9784)

Witco Corp./Concarb, 10500 Richmond, Suite 116, PO Box 42817, Houston, TX 77242-2817 USA (Tel.: 713-978-5745, 800-231-4591; Telefax: 713-978-5728)

Witco Corp./Lubricants Group Golden Bear Prods., 10100 Santa Monica Blvd., Suite 1470, Los Angeles, CA 90067-4183 USA (Tel.: 310-277-4511, 800-429-4826; Telefax: 310-201-0383)

Witco Corp./New Markets, One American Lane, Greenwich, CT 06831-2559 USA (Tel.: 800-494-8673; Telefax: 203-552-2878)

Witco Corp./Oleochemicals and Surfactants Group, One American Lane, Greenwich, CT 06831-2559 USA (Tel.: 203-861-6278, 800-779-4826; Telefax: 203-552-2893)

Witco Corp./Organics Div., One American Lane, Greenwich, CT 06831-2559 USA (Tel.: 203-552-2000; Telefax: 203-552-2010)

Witco Corp./Petroleum Specialties Group, One American Lane, Greenwich, CT 06831-2559 USA (Tel.: 203-861-6280, 800-494-8673; Telefax: 203-552-2448)

Witco Corp./Polyester/Polyurethane Group, One American Lane, Greenwich, CT 06831-2559 USA (Tel.: 203-861-6277, 800-779-4826; Telefax: 203-861-6323)

Witco Corp./Polymer Additives Group, 8 Wright Way, Oakland, NJ 07436-3121 USA (Tel.: 201-337-2036; Telefax: 201-337-3753)

Witco Corp./Polymer Additives Group, One American Lane, Greenwich, CT 06831-2559 USA (Tel.: 203-861-6279, 800-494-8737; Telefax: 203-552-2449)

Witco Corp./Polymer Additives Group (PAG), Bussey Rd., PO Box 1439, Marshall, TX 75671-1439 USA (Tel.: 903-938-5141; Telefax: 903-938-2647)

Witco Corp./Resins Group, One American Lane, Greenwich, CT 06831-2559 USA (Tel.: 203-861-6277, 800-494-8287 x2500; Telefax: 203-552-2850)

Witco Canada Ltd., 2 Lansing Sq., Suite 1200, Willowdale, Ontario, M2J 4Z4 Canada (Tel.: 416-497-9991; Telefax: 416-497-7110)

Witco UK, Union Lane, Droitwich, Worcestershire, WR9 9BB UK (Tel.: 44 190 579 4795; Telefax: 44 190 579 4002)

Witco Chemical Ltd. (UK), Paragon Works, Baxenden, Near Accrington, Lancashire, BB5 2SL UK (Tel.: 44-125-439-8616; Telefax: 44-125-439-8586)

Witco SA, 20, rue de la Ville l'Evêque, 75008 Paris France (Tel.: 33 1 44 51 05 05; Telefax: 33 1 42 65 67 61)

Witco GmbH, Industriegebiet West, Max-Wolf Strasse 7, D-35396 Steinau an der Strasse Germany (Tel.: 49 66 63 54-0; Telefax: 49 66 63 54 10-5)

Witco BV, St. Canisiussingel 26, 6511 TJ Nijmegen The Netherlands (Tel.: 31-80-231534; Telefax: 31-80-224562)

Rewo Chemische Werke GmbH, Postfach 1160, 36392 Steinau an der Strasse, Max Wolf Strasse 7, Industriegebiet West, W-6497 Steinau Germany (Tel.: 06663-540; Telefax: 06663-54-129; Telex: 493589)

Witco Italy, Viale Lunigiana 40, 20125 Milano Italy (Tel.: 39 2 676 011; Telefax: 39 2 6698 1166)

Witco Ltd., PO Box 10245, 26112 Haifa Bay Israel (Tel.: 972 4-469-111; Telefax: 972 4-455778; Telex: 45198)

Worthen Industries, Inc.

3 East Spit Brook Rd., Nashua, NH 03060 USA (Tel.: 603-888-5443; Telefax: 603-888-7945)

Worthington Biochemical Corp.

Halls Mill Rd, Freehold, NJ 07728 USA (Tel.: 908-462-3838, 800-445-9603; Telefax: 800-368-3108; Telex: 3715614)

Wright Corp.

PO Box 9009, 102 Orange St., Wilmington, NC 28402 USA (Tel.: 910-251-8952; Telefax: 910-762-9223)

Wyo-Ben, Inc.

PO Box 1979, Billings, MT 59103 USA (Tel.: 406-652-6351, 800-548-7055; Telefax: 406-656-0748; Telex: WYOBEN)

Yoneyama Chemical Industries, Ltd.

Takahashi Bldg., Higashi-kan, 5-2-18, Nishitenma, Kita-ku, Osaka, 530 Japan (Tel.: (06) 363-0824; Telefax: (06) 365-9982)

Yorkshire Chemicals plc

Kirkstall Rd, Leeds, West Yorkshire, LS3 1LL UK (Tel.: 44 1132 443111; Telefax: 44 1132 421670; Telex: 55366)

Jesse S. Young Co., Inc.

520 Westfield Ave., Elizabeth, NJ 07208 USA (Tel.: 908-351-0140; Telefax: 908-351-8837)

Yuki Gosei Kogyo Co., Ltd.

Hirakawa-cho CH Bldg., 2-3-24, Hirakawa-cho, Chiyoda-ku, Tokyo, 102 Japan (Tel.: (03) 5275-5067; Telefax: (03) 5275-5079)

Zaclon Inc., Partner with Henkel KGaA

2981 Independence Rd., Cleveland, OH 44115 USA (Tel.: 216-271-1715, 800-356-7327; Telefax: 216-271-1911)

Zeeland Chemicals, Inc., A Cambrex Co.

215 N. Centennial St., Zeeland, MI 49464 USA (Tel.: 616-772-2193, 800-223-0453; Telefax: 616-772-7344; Telex: 226375)

Zeelan Industries Inc., Subsid. of 3M

3M Center, Bldg. 220-8E-04, St. Paul, MN 55144-1000 USA (Tel.: 617-737-1751; Telefax: 612-737-1764)

Zeneca

Zeneca Specialties, A Business Unit of Zeneca Inc., PO Box 751, Wilmington, DE 19897 USA

Zeneca Biocides, Part of Zeneca Specialties, A Business Unit of Zeneca Inc., 1800 Concord Pike, Wilmington, DE 19850 USA (Tel.: 800-456-3669; Telefax: 302-886-2972)

Zeneca Resins, A Business Unit of Zeneca Inc., 730 Main St., Wilmington, MA 01887-3386 USA (Tel.: 508-658-6600, 800-225-0947; Telefax: 508-657-7978)

Zeneca Biocides UK, PO Box 42, Hexagon House, Blackley, Manchester, M9 8ZS UK (Tel.: 44-161-740-1460; Telefax: 44-161-721-4123; Telex: 94028500 ICI G)

Zeneca Resins UK, PO Box 14, The Heath, Runcorn, Cheshire, WA7 4QG UK (Tel.: 44 1928 514444; Telefax: 44 1928 576675)

Zeneca Resins BV, Postbus 123, Sluisweg 12, 5140 AC Waalwijk The Netherlands (Tel.: 04160-8 99 11; Telefax: 04160-8 99 22; Telex: 35079 ZENR NL)

Zeneca Resins SA, Ctra. Nacional, 152 Km 23, PO Box 5, Poligono Llevant Industriel, 01850 Parets, Barcelona Spain (Tel.: (3) 5620047; Telefax: (3) 5620966)

Zeneca Resins Pte. Ltd., 135 Pioneer Rd., Tuas, Singapore 2263 (Tel.: (65) 861-5956; Telefax: (65) 861-4940)

Zeochem

1314 S. 12 St., PO Box 35940, Louisville, KY 40232 USA (Tel.: 502-634-7600; Telefax: 502-634-8133; Telex: 204190, 204239)

Zeon

Zeon Chemicals, Inc., 4111 Bells Lane, Louisville, KY 40211 USA (Tel.: 800-735-3388)

Zeon Europe GmbH, Am Seester 18 (Euro Center), D40547 Dusseldorf 11 Germany (Tel.: 49 211 52670; Telefax: 49 211 5267 160)

Zeus Industrial Products Inc.

PO Box 2167, Orangeburg, SC 29116 USA (Tel.: 803-531-2174, 800-526-3842; Telefax: 803-533-5694)

Ziegler Chemical & Mineral Corp.

100 Jericho Quad, Ste. 140, Jericho, NY 11753 USA (Tel.: 516-681-9600; Telefax: 516-681-9604; Telex: 96-1387)

Zimmer AG

Borsigalle 1, 60388 Frankfurt/Main Germany (Tel.: 69-4007-01; Telefax: 69-4007-546)

Zinc Corp. of America

300 Frankfort Rd., Monaca, PA 15061-2295 USA (Tel.: 412-774-1020, 800-962-7500; Telefax: 412-773-2217; Telex: 510-462-1899)

Zinchem, Inc., Subsid. of William Zinsser & Co., Inc.

173 Belmont Dr., Somerset, NJ 08875 USA (Tel.: 908-469-8100; Telefax: 908-469-4539)

Zinkan Enterprises, Inc.

10574 Ravenna Rd., Twinsbury, OH 44087 USA (Tel.: 216-487-1500; Telefax: 216-425-8202)

William Zinsser & Co., Inc.

173 Belmont Dr., Somerset, NJ 08875 USA (Tel.: 908-469-8100; Telefax: 908-563-9774)

Zircar Products Inc.

110 N. Main St., PO Box 458, Florida, NY 10921 USA (Tel.: 914-651-4481; Telefax: 914-651-3192; Telex: 996608)

ZOCHEM, Div. of Hudson Bay Mining & Smelting Co., Ltd.

1 Tilbury Court, PO Box 1120, Brampton, Ontario, L6V 2LB Canada (Tel.: 905-453-4100; Telefax: 905-453-2920)

Zohar Detergent Factory

PO Box 11 300, Tel-Aviv, 61 112 Israel (Tel.: 03-528-7236; Telefax: 03-5287239; Telex: 33557 zohar il)

Zophar Mills, Inc.

112 26 St., Brooklyn, NY 11232 USA (Tel.: 718-768-0907, 800-394-7890; Telefax: 718-768-0910; Telex: PHARMILLS, NY)

Zschimmer & Schwarz

Zschimmer & Schwarz GmbH & Co., Postfach 2179, Max-Schwarz-Str. 3-5, D-56112 Lahnstein Germany (Tel.: 49 2621 12-0; Telefax: 49 2621-12407; Telex: 17262193)

Zschimmer & Schwarz Italiana, Via Vercelli 81, 13038 Tricerro Italy (Tel.: 161-821421; Telex: 200313 ZSI I)

Zyvax, Inc.

PO Box 825, Boca Raton, FL 33429 USA (Tel.: 407-395-4405, 800-858-4111; Telefax: 407-395-5262)

Appendices

CAS Number to Chemical Cross-Reference

CAS	Chemical
50-00-0	Lactic acid
50-70-4	2-Bromo-2-nitropropane-1,3-diol
56-35-9	Tributyltin oxide
56-81-5	Glycerin
57-10-3	Palmitic acid
57-11-4	Stearic acid
57-55-6	Propylene glycol
58-36-6	10,10´-Oxybisphenoxyarsine
59-50-7	p-Chloro-m-cresol
60-24-2	2-Mercaptoethanol
60-29-7	Ethyl ether
60-33-3	Linoleic acid
62-38-2	Phenylmercuric acetate
62-53-3	Aniline
64-17-5	Alcohol
64-17-5	Alcohol denatured
64-18-6	Formic acid
65-85-0	Benzoic acid
66-71-7	1,10-Phenanthroline
67-56-1	Methyl alcohol
67-63-0	Isopropyl alcohol
67-64-1	Acetone
68-12-2	Dimethyl formamide
70-55-3	p-Toluenesulfonamide
71-23-8	n-Propyl alcohol
71-36-3	Butyl alcohol
71-41-0	n-Amyl alcohol
71-43-2	Benzene
71-55-6	Trichloroethane
72-48-0	Sulfated castor oil
74-98-6	Propane
75-01-4	Vinyl chloride
75-07-0	Acetaldehyde
75-09-2	Methylene chloride
75-20-7	Calcium carbide
75-28-5	Isobutane
75-35-4	Vinylidene chloride monomer
75-52-5	Nitromethane
75-65-0	t-Butyl alcohol
75-75-2	Methanesulfonic acid
75-78-5	Dimethyldichlorosilane
75-85-4	t-Amyl alcohol
75-91-2	t-Butyl hydroperoxide
75-98-9	Neopentanoic acid
76-22-2	DL-Camphor
77-58-7	Dibutyltin dilaurate
77-73-6	Dicyclopentadiene
77-85-0	Trimethylolethane
77-86-1	Tris (hydroxymethyl) aminomethane
77-89-4	Acetyl triethyl citrate
77-90-7	Acetyl tributyl citrate
77-92-9	Citric acid anhyd.
77-93-0	Triethyl citrate
77-94-1	Tributyl citrate
77-99-6	Trimethylolpropane
78-08-0	Vinyltriethoxysilane
78-10-4	Ethyl silicate
78-21-7	Cetethyl morpholinium ethosulfate
78-24-0	Tripentaerythritol
78-30-8	Tricresyl phosphate
78-42-2	Trioctyl phosphate
78-51-3	Tributoxyethyl phosphate
78-59-1	Isophorone
78-66-0	Dimethyl octynediol
78-83-1	Isobutyl alcohol
78-92-2	2-Butanol
78-93-3	Methyl ethyl ketone
79-06-1	Acrylamide monomer
79-10-7	Acrylic acid
79-20-9	Methyl acetate
79-24-3	Nitroethane
79-31-2	Isobutyric acid
79-33-4	L-Lactic acid
79-39-0	Methacrylamide
79-41-4	Methacrylic acid
79-46-9	2-Nitropropane
79-94-7	Tetrabromobisphenol A
80-05-7	Bisphenol A
80-08-0	4,4´-Diaminodiphenyl sulfone
80-15-9	Cumene hydroperoxide
80-39-7	Ethyl toluenesulfonamide
80-56-8	α-Pinene
80-62-6	Methyl methacrylate monomer
84-61-7	Dicyclohexyl phthalate
84-64-0	Butyl cyclohexyl phthalate
84-69-5	Diisobutyl phthalate
84-74-2	Dibutyl phthalate
84-78-6	Butyl octyl phthalate
85-22-3	Pentabromoethylbenzene
85-42-7	Hexahydrophthalic anhydride
85-43-8	Tetrahydrophthalic anhydride
85-44-9	Phthalic anhydride
85-60-9	4,4´-Butylidenebis (6-t-butyl-m-cresol)
85-68-7	Butyl benzyl phthalate
85-70-1	n-Butyl phthalyl-n-butyl glycolate
86-39-5	2-Chlorothioxanthone
88-04-0	Chloroxylenol
88-12-0	N-Vinyl-2-pyrrolidone
89-04-3	Trioctyl trimellitate
89-05-4	Pyromellitic acid
89-32-7	Pyromellitic dianhydride
90-15-3	1-Naphthol
90-72-2	2,4,6-Tris (dimethylaminomethyl) phenol
91-17-8	Decahydronaphthalene
91-66-7	Diethyl aniline
91-76-9	Benzoguanamine
91-88-3	N-Ethyl-N-hydroxyethyl-m-toluidine
92-50-2	Ethyl phenyl ethanolamine
93-69-6	o-Tolyl biguanide
93-90-3	Methylphenyl ethanolamine
94-04-2	Vinyl 2-ethylhexanoate
94-51-9	Dipropylene glycol dibenzoate
95-14-7	1H-Benzotriazole
95-19-2	Stearyl hydroxyethyl imidazoline
95-38-5	Heptadecenyl hydroxyethyl imidazoline
95-38-5	Oleyl hydroxyethyl imidazoline
95-47-6	o-Xylene
95-63-6	1,2,4-Trimethylbenzene
96-05-9	Allyl methacrylate
96-22-0	Diethyl ketone
96-27-5	Thioglycerin
96-29-7	Methyl ethyl ketoxime
96-33-3	Methyl acrylate (monomer)
96-48-0	Butyrolactone
96-69-5	4,4´-Thiobis-6-(t-butyl-m-cresol)
97-64-3	Ethyl lactate
97-65-4	Itaconic acid
97-67-6	L-N-Hydroxysuccinic acid
97-85-8	Isobutyl isobutyrate
97-86-9	Isobutyl methacrylate
97-88-1	Butyl methacrylate
97-90-5	Ethylene glycol dimethacrylate
97-99-4	Tetrahydrofurfuryl

CAS	Chemical
	alcohol
98-00-0	Furfuryl alcohol
98-01-1	Furfural
98-54-4	4-t-Butylphenol
98-56-6	p-Chlorobenzotrifluoride
98-67-9	Phenol sulfonic acid
98-79-3	PCA
98-83-9	α-Methylstyrene monomer
98-86-2	Acetophenone
99-97-8	N,N-Dimethyl-p-toluidine
100-21-0	Terephthalic acid
100-41-4	Ethylbenzene
100-42-5	Styrene
100-47-0	Benzonitrile
100-52-7	Benzaldehyde
100-97-0	Hexamethylenetetramine
101-02-0	Triphenyl phosphite
101-34-8	Glyceryl triacetyl ricinoleate
101-37-1	Triallylcyanurate
101-43-9	Cyclohexyl methacrylate
101-68-8	MDI
101-77-9	4,4′-Methylene dianiline
101-96-2	N,N′-Di-s-butyl-p-phenylenediamine
102-60-3	Tetrahydroxypropyl ethylenediamine
102-71-6	Triethanolamine
102-76-1	Triacetin
103-09-0	2-Ethylhexyl acetate
103-11-7	Octyl acrylate
103-23-1	Dioctyl adipate
103-83-3	N-Benzyldimethylamine
104-15-4	p-Toluene sulfonic acid
104-60-9	Phenylmercuric oleate
104-68-7	Diethylene glycol phenyl ether
104-76-7	2-Ethylhexanol
105-08-8	1,4-Cyclohexane-dimethanol
105-16-8	Diethylaminoethyl methacrylate
105-37-3	Ethyl propionate
105-38-4	Vinyl propionate
105-46-4	s-Butyl acetate
105-60-2	Caprolactam
105-76-0	Dibutyl maleate
105-76-9	Dibutyl fumarate
106-07-0	PEG-4 stearate
106-08-1	PEG-9 laurate
106-11-6	PEG-2 stearate
106-11-6	PEG-2 stearate SE
106-12-7	PEG-2 oleate
106-12-7	PEG-2 oleate SE
106-14-9	Hydroxystearic acid
106-36-5	Propyl propionate
106-65-0	Dimethyl succinate
106-89-8	Epichlorohydrin
106-90-1	Glycidyl acrylate
106-91-2	Glycidyl methacrylate
106-92-3	Allyl glycidyl ether
106-97-8	Butane
107-06-2	Ethylene dichloride
107-13-1	Acrylonitrile
107-15-3	Ethylenediamine anhyd.
107-21-1	Glycol
107-41-5	Hexylene glycol
107-54-0	Dimethyl hexynol
107-64-2	Distearyldimonium chloride
107-70-0	4-Methoxy-4-methyl-pentanone-2
107-87-9	Methyl propyl ketone
107-98-2	Methoxyisopropanol
108-01-0	Dimethylethanolamine
108-03-2	Nitropropane
108-05-4	Vinyl acetate
108-10-1	Methyl isobutyl ketone
108-11-2	Methyl amyl alcohol
108-20-3	Isopropyl ether
108-21-4	Isopropyl acetate
108-30-5	Succinic anhydride
108-31-6	Maleic anhydride
108-32-7	Propylene carbonate
108-45-2	m-Phenylenediamine
108-46-3	Resorcinol
108-65-6	Propylene glycol methyl ether acetate
108-67-8	1,3,5-Trimethylbenzene
108-78-1	Melamine
108-83-8	Diisobutyl ketone
108-84-9	Methyl amyl acetate
108-88-3	Toluene
108-90-7	Chlorobenzene
108-93-0	Cyclohexanol
108-94-1	Cyclohexanone
108-95-2	Phenol
109-16-0	PEG-3 dimethacrylate
109-52-4	n-Valeric acid
109-53-5	Vinyl isobutyl ether
109-60-4	Propyl acetate
109-86-4	Methoxyethanol
109-89-7	Diethylamine
109-99-9	Tetrahydrofuran
110-00-9	Furan polymer
110-05-4	Di-t-butyl peroxide
110-12-3	Methyl isoamyl ketone
110-15-6	Succinic acid
110-16-7	Maleic acid
110-17-8	Fumaric acid
110-19-0	Isobutyl acetate
110-25-8	Oleoyl sarcosine
110-30-5	Ethylene distearamide
110-31-6	Ethylene dioleamide
110-43-0	Methyl n-amyl ketone
110-49-6	Methoxyethanol acetate
110-54-3	Hexane
110-63-4	1,4-Butanediol
110-69-0	m-Butyraldehyde oxime
110-71-4	Ethylene glycol dimethyl ether
110-80-5	Ethoxyethanol
110-82-7	Cyclohexane
110-86-1	Pyridine
110-97-4	Diisopropanolamine
110-98-5	Dipropylene glycol
111-03-5	Glyceryl oleate
111-13-7	Methyl hexyl ketone
111-15-9	Ethoxyethanol acetate
111-20-6	Sebacic acid
111-27-3	Hexyl alcohol
111-40-0	Diethylenetriamine
111-42-2	Diethanolamine
111-46-6	Diethylene glycol
111-55-7	Ethylene glycol diacetate
111-60-4	Glycol stearate
111-76-2	Butoxyethanol
111-77-3	Methoxydiglycol
111-87-5	Caprylic alcohol
111-90-0	Ethoxydiglycol
111-96-6	Diethylene glycol dimethyl ether
112-05-0	Nonanoic acid
112-07-2	Butoxyethanol acetate
112-15-2	Ethoxydiglycol acetate
112-24-3	Triethylenetetramine
112-25-4	Ethylene glycol hexyl ether
112-30-1	n-Decyl alcohol
112-34-5	Butoxydiglycol
112-35-6	Triglycol monomethyl ether
112-36-7	Diethylene glycol diethyl ether
112-50-5	Ethoxytriglycol
112-57-2	Tetraethylenepentamine
112-59-4	Diethylene glycol n-hexyl ether
112-60-7	PEG-4
112-61-8	Methyl stearate
112-62-9	Methyl oleate
112-63-0	Methyl linoleate
112-70-9	Tridecyl alcohol
112-73-2	Diethylene glycol dibutyl ether
112-80-1	Oleic acid
112-84-5	Erucamide
115-10-6	Dimethyl ether
115-77-5	Pentaerythritol
115-83-3	Pentaerythrityl tetrastearate
115-86-6	Triphenyl phosphate
115-96-8	Trichloroethyl phosphate
116-25-6	MDM hydantoin
117-08-8	Tetrachlorophthalic anhydride
117-81-7	Dioctyl phthalate
117-84-0	Dioctyl phthalate
117-99-7	2-Hydroxybenzo-phenone
118-92-3	Anthranilic acid
119-47-1	2,2′-Methylenebis (6-t-butyl-4-methylphenol)
119-61-9	Benzophenone
119-64-2	Tetrahydronaphthalene
120-07-0	Phenyldiethanolamine
120-55-8	Diethylene glycol dibenzoate
121-44-8	Triethylamine
121-69-7	n,n-Dimethylaniline
121-91-5	Isophthalic acid
122-20-3	Triisopropanolamine
122-60-1	Phenyl glycidyl ether
122-99-6	Phenoxyethanol
123-01-3	Dodecylbenzene
123-18-2	Isobutyl heptyl ketone
123-31-9	Hydroquinone
123-34-2	Glyceryl-1-allyl ether
123-38-6	Propionaldehyde
123-42-2	Diacetone alcohol
123-51-3	Isoamyl alcohol
123-54-6	Acetylacetone
123-72-8	n-Butyraldehyde

CAS NUMBER TO CHEMICAL CROSS-REFERENCE

CAS	Chemical
123-77-3	Azodicarbonamide
123-86-4	n-Butyl acetate
123-91-1	1,4-Dioxane
123-94-4	Glyceryl stearate
123-95-5	Butyl stearate
123-99-9	Azelaic acid
124-04-9	Adipic acid
124-17-4	Diethylene glycol butyl ether acetate
124-26-5	Stearamide
124-29-8	Cetyl alcohol
124-68-5	2-Amino-2-methyl-1-propanol
126-13-6	Sucrose acetate isobutyrate
126-30-7	Neopentyl glycol
126-33-0	Sulfolane
126-58-9	Dipentaerythritol
126-73-8	Tributyl phosphate
126-86-3	Tetramethyl decynediol
126-92-1	Sodium octyl sulfate
126-99-8	Polychloroprene
127-06-0	Acetone oxime
127-18-4	Perchloroethylene
127-25-3	Methyl abietate
127-39-9	Diisobutyl sodium sulfosuccinate
128-04-1	Sodium dimethyldithiocarbamate (hydrate)
128-37-0	BHT
131-11-3	Dimethyl phthalate
131-17-9	Diallyl phthalate
131-53-3	Benzophenone-8
131-54-4	Benzophenone-6
131-56-6	Benzophenone-1
131-57-7	Benzophenone-3
133-06-2	Captan
133-07-3	N-Trichloromethylthio phthalimide
134-84-9	4-Methylbenzophenone
136-52-7	Cobalt octoate
136-53-8	Zinc 2-ethylhexanoate
136-99-2	Lauryl hydroxyethyl imidazoline
137-30-4	Zinc dimethyldithiocarbamate
138-22-7	Butyl lactate
138-86-3	dl-Limonene
139-43-5	Glyceryl (triacetoxystearate)
139-44-6	Trihydroxystearin
139-88-8	Sodium myristyl sulfate
140-03-4	Methyl acetyl ricinoleate
140-04-5	Butyl acetyl ricinoleate
140-11-4	Benzyl acetate
140-31-8	Aminoethylpiperazine
141-02-6	Dioctyl fumarate
141-04-8	Diisobutyl adipate
141-08-2	Glyceryl ricinoleate
141-20-8	PEG-2 laurate
141-20-8	PEG-2 laurate SE
141-22-0	Ricinoleic acid
141-23-1	Methyl hydroxystearate
141-32-2	Butyl acrylate
141-43-5	Ethanolamine
141-78-6	Ethyl acetate
141-79-7	Mesityl oxide
141-97-9	Ethylacetoacetate
142-09-6	Hexyl methacrylate
142-16-5	Dioctyl maleate
142-30-3	Dimethyl hexynediol
142-31-4	Sodium octyl sulfate
142-55-2	Propylene glycol laurate
142-59-6	Nabam
142-62-1	Caproic acid
142-82-5	Heptane
142-87-0	Sodium decyl sulfate
142-90-5	Lauryl methacrylate
142-92-7	Hexyl acetate
142-96-1	Dibutyl ether
143-22-6	Butoxytriglycol
143-28-2	Oleyl alcohol
144-19-4	2,2,4-Trimethyl-1,3-pentanediol
144-55-8	Sodium bicarbonate
148-79-8	Thiabendazole
149-44-0	Sodium formaldehyde sulfoxylate
149-57-5	2-Ethylhexoic acid
150-76-5	Hydroquinone monomethyl ether
151-21-3	Sodium lauryl sulfate
155-04-4	Zinc 2-mercaptobenzothiazole
280-57-9	Triethylene diamine
300-92-5	Aluminum distearate
301-02-0	Oleamide
301-04-2	Lead acetate
301-10-0	Stannous octoate
306-52-5	Trichloroethyl phosphate
328-84-7	3,4-Dichlorobenzotrifluoride
335-57-9	Perfluoroheptane
355-42-0	Perfluorohexane
379-52-2	Triphenyltin fluoride
409-21-2	Silicon carbide
422-86-2	Dioctyl terephthalate
461-58-5	Dicyandiamide
464-49-3	D-+-Camphor
471-34-1	Calcium carbonate
502-44-3	ε-Caprolactone monomer
514-10-3	Abietic acid
515-98-0	Ammonium lactate
526-73-8	1,2,3-Trimethylbenzene
527-07-1	Sodium gluconate
531-18-0	Hexamethylol melamine resin
532-32-1	Sodium benzoate
533-74-4	3,5-Dimethyl tetrahydro-2-H,1,3,5-thiadiazone-2-thione
540-10-3	Cetyl palmitate
542-55-2	Isobutyl formate
542-92-7	Cyclopentadiene
543-80-6	Barium acetate
546-68-9	Tetraisopropyl titanate
547-64-8	Methyl lactate
552-30-7	Trimellitic anhydride
556-67-2	Octamethylcyclotetrasiloxane
557-04-0	Magnesium stearate
557-05-1	Zinc stearate
577-11-7	Dioctyl sodium sulfosuccinate
584-84-9	Toluene diisocyanate
589-37-7	1,3-Pentanediamine
590-01-2	n-Butyl propionate
593-60-2	Vinyl bromide
598-63-0	Lead carbonate (basic)
598-82-3	DL-Lactic acid
614-33-5	Glyceryl tribenzoate
616-45-5	2-Pyrrolidone
617-48-1	DL-N-Hydroxysuccinic acid
617-51-6	Isopropyl lactate
624-04-4	Glycol dilaurate
624-48-6	Dimethyl maleate
624-54-4	n-Pentyl propionate
627-83-8	Glycol distearate
627-93-0	Dimethyl adipate
628-63-7	Amyl acetate
629-11-8	Hexamethylene glycol
629-14-1	Ethylene glycol diethyl ether
632-79-1	Tetrabromophthalic anhydride
636-61-3	+-N-Hydroxysuccinic acid
637-12-7	Aluminum tristearate
638-38-0	Manganese acetate
646-13-9	Isobutyl stearate
678-26-2	Perfluoropentane
688-84-6	2-Ethylhexyl methacrylate
693-23-2	C12 dibasic acid
693-98-1	2-Methyl imidazole
694-83-7	1,2-Diaminocyclohexane
763-69-9	Ethyl 3-ethoxypropionate
770-35-4	Propylene glycol phenyl ether
822-06-0	Hexamethylene diisocyanate
868-77-9	2-Hydroxyethyl methacrylate
872-50-4	N-Methyl-2-pyrrolidone
919-30-2	Aminopropyltriethoxysilane
999-97-3	Hexamethyldisilazane
1067-33-0	Dibutyltin diacetate
1067-53-4	Vinyltris(2-methoxyethoxy) silane
1072-35-1	Lead stearate
1076-97-7	1,4-Cyclohexanedicarboxylic acid
1077-56-1	Ethyl toluenesulfonamide
1119-40-0	Dimethyl glutarate
1120-01-0	Sodium cetyl sulfate
1163-19-5	Decabromodiphenyl oxide
1185-55-3	Methyltrimethoxysilane
1191-50-0	Sodium myristyl sulfate
1241-94-7	Diphenyl octyl phosphate
1300-71-6	Xylenol
1302-78-9	Bentonite
1303-86-2	Boron oxide
1304-85-4	Bismuth subnitrate
1305-78-8	Calcium oxide
1307-96-6	Cobalt oxide (ous)
1308-04-9	Cobalt oxide (ic)
1308-38-0	Chromium oxide (ic)

CAS	Chemical
1309-37-1	Ferric oxide anhyd.
1309-42-8	Magnesium hydroxide anhyd.
1309-48-4	Magnesium oxide
1309-64-4	Antimony trioxide
1310-73-2	Sodium hydroxide
1312-76-1	Potassium silicate
1313-27-5	Molybdenum trioxide
1314-13-2	Zinc oxide
1314-41-6	Lead oxide, red
1314-60-9	Antimony pentoxide
1314-98-3	Zinc sulfide
1317-36-8	Lead (II) oxide
1317-38-0	Copper oxide (ic)
1317-39-1	Copper oxide (ous)
1317-65-3	Calcium carbonate
1317-65-3	Limestone
1317-80-2	Titanium dioxide
1317-97-1	Ultramarine blue
1318-00-9	Vermiculite
1318-02-1	Sodium silicoaluminate
1318-02-1	Zeolite
1318-74-7	Kaolinite
1318-93-0	Montmorillonite
1319-46-6	Lead carbonate (basic)
1319-77-3	Cresylic acid
1321-69-3	Sodium polynaphthalene sulfonate
1321-87-5	Dimethyl octynediol
1322-93-6	Sodium isopropyl naphthalene sulfonate
1323-38-2	Castor oil
1323-39-3	Propylene glycol stearate
1323-42-8	Glyceryl hydroxy-stearate
1327-36-2	Aluminosilicate
1327-36-2	Aluminum silicate
1327-43-1	Magnesium aluminum silicate
1330-20-7	Xylene
1330-43-4	Sodium borate anhyd.
1330-61-6	Isodecyl acrylate
1330-78-5	Tricresyl phosphate
1332-07-6	Zinc borate anhyd.
1332-58-7	Kaolin
1333-39-7	Phenolsulfonic acid
1333-86-4	Carbon black
1333-89-7	Hydroabietyl alcohol
1335-72-4	Sodium laureth sulfate
1336-21-6	Ammonium hydroxide
1337-76-4	Attapulgite
1337-81-1	Vinyl pyridine
1338-02-9	Copper naphthenate
1338-23-4	Methyl ethyl ketone peroxide
1338-24-5	Naphthenic acid
1338-39-2	Sorbitan laurate
1338-41-6	Sorbitan stearate
1338-43-8	Sorbitan oleate
1338-51-8	Toluene sulfonamide formaldehyde resin
1341-54-4	Benzophenone-11
1343-98-2	Silica, hydrated
1344-00-9	Sodium silicoaluminate
1344-09-8	Sodium silicate
1344-28-1	Alumina
1344-40-7	Lead phosphite dibasic
1344-95-2	Calcium metasilicate
1344-95-2	Calcium silicate
1345-05-7	Lithopone
1345-44-6	Antimony trisulfide
1369-66-3	Dioctyl sodium sulfosuccinate
1471-17-6	Pentaerythrityl triallyl ether
1493-13-6	Trifluoromethane sulfonic acid
1528-49-0	Tri n-hexyl trimellitate
1559-35-9	Ethylene glycol 2-ethylhexyl ether
1559-36-0	Diethylene glycol 2-ethylhexyl ether
1569-01-3	Propylene glycol n-propyl ether
1569-02-4	Propylene glycol ethyl ether
1592-23-0	Calcium stearate
1634-04-4	Methyl t-butyl ether
1643-20-5	Lauramine oxide
1680-21-3	PEG-3 diacrylate
1760-24-3	N-2-Aminoethyl-3-aminopropyl trimethoxysilane
1843-05-6	Benzophenone-12
1897-45-6	Tetrachloroisophthalo-nitrile
1983-10-4	Tributyltin fluoride
2031-67-6	Methyltriethoxysilane
2082-79-3	Octadecyl 3,5-di-t-butyl-4-hydroxyhydro-cinnamate
2116-84-9	Phenyl trimethicone
2156-97-0	Lauryl acrylate
2163-42-0	2-Methyl-1,3-propanediol
2190-04-7	Stearamine acetate
2210-79-9	o-Cresyl glycidyl ether
2223-82-7	Neopentyl glycol diacrylate
2224-33-1	Vinyl oximinosilane
2235-54-3	Ammonium lauryl sulfate
2315-62-0	Octoxynol-3
2315-64-2	Octoxynol-5
2315-66-4	Octoxynol-10
2315-67-5	Octoxynol-1
2358-84-1	Diethylene glycol dimethacrylate
2373-38-8	Dihexyl sodium sulfosuccinate
2399-48-6	Tetrahydrofurfuryl acrylate
2421-28-5	3,3´,4,4´-Benzophe-none tetracarboxylic dianhydride
2426-08-6	Butyl glycidyl ether
2426-54-2	Diethylaminoethyl acrylate
2430-22-0	Isononyl alcohol
2439-35-2	Dimethylaminoethyl acrylate
2440-22-4	Drometrizole
2452-01-9	Zinc laurate
2455-24-5	Tetrahydrofurfuryl methacrylate
2461-15-6	2-Ethylhexyl glycidyl ether
2491-38-5	2-Bromo-4´-hydroxyacetophenone
2492-26-4	Sodium 2-mercapto-benzothiazole
2495-25-2	Tridecyl methacrylate
2499-95-8	n-Hexyl acrylate
2530-83-8	3-Glycidoxypropyltri-methoxysilane
2530-85-0	3-Methacryloxypropyltri-methoxysilane
2615-15-8	PEG-6
2634-33-5	1,2-Benzisothiazolin-3-one
2673-22-5	Ditridecyl sodium sulfosuccinate
2682-20-4	Methylisothiazolinone
2687-96-9	Lauryl pyrrolidone
2694-54-4	Triallyl trimellitate
2695-37-6	Sodium p-styrene-sulfonate
2764-13-8	Stearamidopropyl dimethyl-2-hydroxyethyl ammonium nitrate
2768-02-7	Vinyltrimethoxysilane
2807-30-9	Ethylene glycol propyl ether
2855-13-2	Isophorone diamine
2867-47-2	Dimethylaminoethyl methacrylate
2915-53-9	Dioctyl maleate
2943-75-1	Octyltriethoxysilane
2996-92-1	Phenyltrimethoxysilane
3006-15-3	Dihexyl sodium sulfosuccinate
3055-93-4	Laureth-2
3055-95-6	Laureth-5
3055-98-9	Laureth-8
3055-99-0	Laureth-9
3056-00-6	Laureth-12
3061-75-4	Behenamide
3064-70-8	Bis (trichloromethyl) sulfone
3066-71-5	Cyclohexyl acrylate
3072-84-2	Epoxy resin, brominated
3088-31-1	Sodium laureth sulfate
3089-11-0	Hexamethoxymethyl-melamine
3101-60-8	p-t-Butyl phenyl glycidyl ether
3121-60-6	Benzophenone-9
3121-61-7	Methoxyethyl acrylate
3147-75-9	Octrizole
3194-55-6	Hexabromocyclo-dodecane
3278-89-5	Tribromophenyl allyl ether
3290-92-4	1,1,1-Trimethylolpro-pane trimethacrylate
3319-31-1	Trioctyl trimellitate
3322-93-8	Dibromoethyldibromo-cyclohexane
3333-62-8	CI Fluorescent Brightener 236
3386-18-3	PEG-9
3388-04-3	2-(3,4-Epoxycyclo-hexyl) ethyltrimeth-oxysilane

CAS	Chemical
3401-73-8	Tetrasodium dicarboxyethyl stearyl sulfosuccinamate
3524-68-3	Pentaerythrityl triacrylate
3622-84-2	N,N-Butyl benzene sulfonamide
3648-21-3	Di-n-heptyl phthalate
3699-54-5	2-Hydroxyethylethylene urea
3724-65-0	Crotonic acid
3775-90-4	t-Butylaminoethyl methacrylate
3806-34-6	Distearyl pentaerythritol diphosphite
3811-73-2	Sodium pyrithione
3864-99-1	2-(3´,5´-Di-t-butyl-2´-hydroxyphenyl)-5-chlorobenzotriazole
3896-11-5	Bumetrizole
3900-04-7	Hexyl phosphate
3990-03-2	Monoethyl maleate
4016-14-2	Isopropyl glycidyl ether
4074-88-8	Diethylene glycol diacrylate
4080-31-3	Quaternium-15
4083-64-1	p-Toluene sulfonyl isocyanate
4098-71-9	Isophorone diisocyanate
4169-04-4	Propylene glycol phenyl ether
4196-86-5	Pentaerythrityl tetrabenzoate
4196-89-8	Neopentyl glycol dibenzoate
4254-14-2	(-)-Propylene glycol
4254-15-3	(+)-Propylene glycol
4254-16-4	(±)-Propylene glycol
4306-88-1	Diisobutyl nonyl phenol
4342-30-7	Tributyltin salicylate
4342-36-3	Tributyltin benzoate
4420-74-0	Mercaptopropyltrimethoxysilane
4435-53-4	Methoxybutyl acetate
4767-03-7	Dimethylolpropionic acid
4986-89-4	Pentaerythrityl tetraacrylate
5039-78-1	Dimethylaminoethyl methacrylate methyl chloride quat.
5117-19-1	PEG-8
5131-66-8	Butoxypropanol
5165-97-9	Sodium 2-acrylamido-2-methylpropanesulfonate
5205-93-6	Dimethylaminopropyl-methacrylamide
5232-99-5	Etocrylene
5274-68-0	Laureth-4
5332-73-0	Methoxypropylamine
5353-26-4	Oleth-4
5353-27-5	Oleth-5
5495-84-1	Isopropyl thioxanthone (2-isomer)
5593-70-4	Tetrabutyl titanate
5650-20-4	Tetraethylene glycol ethyl ether
5888-33-5	Isobornyl acrylate
5892-10-4	Bismuth subcarbonate
5938-38-5	Sorbitan oleate
5959-89-7	Sorbitan laurate
6001-97-4	Dihexyl sodium sulfosuccinate
6035-47-8	Sodium formaldehyde sulfoxylate (dihydrate)
6067-86-0	Diatomaceous earth
6144-28-1	Dilinoleic acid
6147-53-1	Cobalt acetate (ous)
6156-78-1	Manganese acetate
6175-45-7	2,2-Diethoxyacetophenone
6197-30-4	Octocrylene
6317-18-6	Methylenebis (thiocyanate)
6542-37-6	Hydroxymethyl dioxoazabicyclooctane
6606-59-3	1,6-Hexanediol methacrylate
6674-22-2	Diazabicycloundecene
6683-19-8	Tetrakis [methylene (3,5-di-t-butyl-4-hydroxyhydrocinnamate)] methane
6780-13-8	Ethylenediamine monohydrate
6790-09-6	PEG-12
6813-14-0	C11-15 pareth-12
6834-92-0	Sodium metasilicate
6837-24-7	N-Cyclohexyl pyrrolidone
6846-50-0	Trimethyl-1,3-pentanediol, 2,2,4-diisobutyrate
6881-94-3	Diethylene glycol propyl ether
6891-44-7	Dimethylaminoethyl methacrylate dimethyl sulfate quat.
6915-15-7	±-N-Hydroxysuccinic acid
7005-47-2	2-Dimethylamino-2-methyl-1-propanol
7047-84-9	Aluminum stearate
7049-66-3	Trilinoleic acid
7128-64-5	2,2´-(2,5-Thiophenediyl)bis[5-t-butyl-benzoxazole]
7311-27-5	Nonoxynol-4
7320-34-5	Tetrapotassium pyrophosphate
7398-69-8	Dimethyl diallyl ammonium chloride
7428-48-0	Lead stearate
7429-90-5	Aluminum
7439-92-1	Lead
7439-98-7	Molybdenum
7440-05-3	Palladium
7440-16-6	Rhodium
7440-31-5	Tin
7440-32-6	Titanium
7440-43-9	Cadmium
7440-44-0	Carbon
7440-47-3	Chromium
7440-48-4	Cobalt
7440-50-8	Copper
7440-50-8	Copper powder
7440-66-6	Zinc
7440-66-6	Zinc dust
7440-67-7	Zirconium
7440-74-6	Indium
7446-14-2	Lead sulfate
7473-98-5	2-Hydroxy 2-methyl 1-phenyl 1-propanone
7534-94-3	Isobornyl methacrylate
7585-39-9	β-Cyclodextrin
7631-86-9	Diatomaceous earth
7631-86-9	Silica
7631-86-9	Silica, colloidal
7631-95-0	Sodium molybdate
7631-99-4	Sodium nitrate
7637-07-2	Boron trifluoride
7646-78-8	Stannic chloride anhyd.
7657-10-1	Palladium chloride (ous)
7664-38-2	Phosphoric acid
7664-41-7	Ammonia
7681-49-4	Sodium fluoride
7718-54-9	Nickel chloride (ous)
7719-12-2	Phosphorus trichloride
7721-15-5	Copper linoleate
7722-84-1	Hydrogen peroxide
7722-88-5	Tetrasodium pyrophosphate
7723-14-0	Phosphorus
7727-43-7	Barium sulfate
7738-94-5	Chromic acid
7747-35-5	7-Ethyl bicyclooxazolidine
7757-93-9	Calcium phosphate dibasic
7758-98-7	Cupric sulfate anhyd.
7758-99-8	Cupric sulfate pentahydrate
7772-99-8	Stannous chloride anhyd.
7775-11-3	Sodium chromate
7775-14-6	Sodium hydrosulfite
7778-18-9	Calcium sulfate anhyd.
7779-27-3	Hexahydro-1,3,5-triethyl-s-triazine
7779-90-0	Zinc phosphate
7782-42-5	Graphite
7782-63-0	Ferrous sulfate heptahydrate
7784-30-7	Aluminum orthophosphate
7785-26-4	(-)-α-Pinene
7785-70-8	(+)-α-Pinene
7785-87-7	Manganese sulfate (ous) anhyd.
7787-59-9	Bismuth oxychloride
7789-04-0	Chromium phosphate
7789-06-2	Strontium chromate
7789-75-5	Calcium fluoride
8001-20-5	Tung oil
8001-21-6	Sunflower seed oil
8001-22-7	Soybean oil
8001-23-8	Safflower oil
8001-26-1	Linseed oil
8001-29-4	Cottonseed oil
8001-30-7	Corn oil
8001-31-8	Coconut oil
8001-39-6	Japan wax
8001-54-5	Benzalkonium chloride
8001-75-0	Ceresin
8001-78-3	Hydrogenated castor oil
8001-78-3	Trihydroxystearin

CAS	Chemical
8001-79-4	Castor oil
8002-03-7	Peanut oil
8002-05-9	Petroleum distillates
8002-13-9	Rapeseed oil
8002-23-1	Cetyl esters
8002-26-4	Tall oil
8002-33-3	Sulfated castor oil
8002-43-5	Lecithin
8002-50-4	Menhaden oil
8002-53-7	Montan wax
8002-74-2	Paraffin
8002-74-2	Petroleum wax
8002-74-2	Synthetic wax
8006-40-4	Beeswax, white
8006-44-8	Candelilla wax
8006-54-0	Lanolin anhyd.
8006-64-2	Turpentine oil
8007-18-9	Pigment yellow 53
8007-24-7	Cashew nut shell oil
8009-03-8	Petrolatum NF
8011-48-1	Pine tar
8012-89-3	Beeswax, yellow
8012-95-1	Mineral oil
8013-07-8	Epoxidized soybean oil
8015-86-9	Carnauba
8016-70-4	Hydrogenated soybean oil
8020-83-5	Mineral oil (white)
8020-84-6	Lanolin hydrous
8021-55-4	Ozokerite
8021-99-6	Bone black
8027-32-5	Petrolatum USP
8027-33-6	Lanolin alcohol
8029-76-3	Hydroxylated lecithin
8029-76-3	Lecithin
8030-30-6	Naphtha
8030-76-0	Lecithin
8032-32-4	Mineral spirits
8042-47-5	Mineral oil
8050-09-7	Rosin
8050-13-3	Methyl hydrogenated rosinate
8050-26-8	Pentaerythrityl rosinate
8050-81-5	Simethicone
8051-52-3	PEG-15 cocamine
8052-10-6	Rosin
8052-41-3	Stoddard solvent
8063-08-9	Microcrystalline wax
9000-11-7	Carboxymethylcellulose
9000-30-0	Guar gum
9000-36-6	Karaya gum
9000-40-2	Locust bean gum
9000-59-3	Shellac
9000-70-8	Gelatin
9000-71-9	Casein
9000-71-9	Milk protein
9000-92-4	Amylase
9002-84-0	Fluorinated ethylene-propylene
9002-84-0	Polytetrafluoroethylene
9002-86-2	Polyvinyl chloride
9002-88-4	Polyethylene
9002-88-4	Polyethylene, high-density
9002-88-4	Polyethylene, ultrahigh m.w.
9002-88-4	Polyethylene wax
9002-89-5	Polyvinyl alcohol (super and fully hydrolyzed)
9002-92-0	Laureth-2
9002-92-0	Laureth-8
9002-92-0	Laureth-9
9002-92-0	Laureth-12
9002-92-0	Laureth-23
9002-93-1	Octoxynol-1
9002-93-1	Octoxynol-3
9002-93-1	Octoxynol-5
9002-93-1	Octoxynol-7
9002-93-1	Octoxynol-8
9002-93-1	Octoxynol-9
9002-93-1	Octoxynol-10
9002-93-1	Octoxynol-13
9002-93-1	Octoxynol-16
9002-93-1	Octoxynol-25
9002-93-1	Octoxynol-30
9002-93-1	Octoxynol-40
9002-93-1	Octoxynol-70
9003-01-4	Acrylic resin
9003-01-4	Carbomer
9003-01-4	Polyacrylic acid
9003-03-6	Ammonium polyacrylate
9003-04-7	Sodium polyacrylate
9003-05-8	Polyacrylamide
9003-07-0	Polypropylene
9003-08-1	Melamine/formaldehyde resin
9003-09-2	Polyvinyl methyl ether
9003-11-6	Meroxapol 105
9003-11-6	Meroxapol 171
9003-11-6	Meroxapol 172
9003-11-6	Meroxapol 174
9003-11-6	Meroxapol 252
9003-11-6	Meroxapol 254
9003-11-6	Meroxapol 258
9003-11-6	Meroxapol 311
9003-11-6	Poloxamer 101
9003-11-6	Poloxamer 105
9003-11-6	Poloxamer 108
9003-11-6	Poloxamer 123
9003-11-6	Poloxamer 124
9003-11-6	Poloxamer 181
9003-11-6	Poloxamer 182
9003-11-6	Poloxamer 184
9003-11-6	Poloxamer 185
9003-11-6	Poloxamer 188
9003-11-6	Poloxamer 217
9003-11-6	Poloxamer 231
9003-11-6	Poloxamer 234
9003-11-6	Poloxamer 235
9003-11-6	Poloxamer 237
9003-11-6	Poloxamer 238
9003-11-6	Poloxamer 282
9003-11-6	Poloxamer 288
9003-11-6	Poloxamer 331
9003-11-6	Poloxamer 333
9003-11-6	Poloxamer 334
9003-11-6	Poloxamer 335
9003-11-6	Poloxamer 338
9003-11-6	Poloxamer 401
9003-11-6	Poloxamer 402
9003-11-6	Poloxamer 403
9003-11-6	Poloxamer 407
9003-17-2	Polybutadiene
9003-18-3	Butadiene/acrylonitrile copolymer
9003-20-7	Polyvinyl acetate (homopolymer)
9003-22-9	PVC/VA copolymer
9003-22-9	Vinyl chloride/vinyl acetate copolymer
9003-27-4	Polyisobutene
9003-28-5	Polybutene
9003-35-4	Phenolic resin
9003-39-8	PVP
9003-53-6	Polystyrene
9003-54-7	Styrene/acrylates copolymer
9003-55-8	Polybutadiene-styrene copolymer
9003-55-8	Styrene-butadiene polymer
9003-55-8	Styrene-butadiene rubber
9003-56-9	Acrylonitrile-butadiene-styrene
9003-62-7	Polyvinyl butyral
9003-73-0	Polydipentene
9003-74-1	Terpene resin
9003-74-1	Terpene resin, natural
9004-32-4	Carboxymethylcellulose sodium
9004-34-6	Cellulose
9004-34-6	Microcrystalline cellulose
9004-35-7	Cellulose acetate
9004-36-8	Cellulose acetate butyrate
9004-39-1	Cellulose acetate propionate
9004-57-3	Ethylcellulose
9004-58-4	Ethyl hydroxyethyl cellulose
9004-62-0	Hydroxyethylcellulose
9004-64-2	Hydroxypropylcellulose
9004-65-3	Hydroxypropyl methylcellulose
9004-67-5	Methylcellulose
9004-70-0	Nitrocellulose
9004-81-3	PEG-2 laurate
9004-81-3	PEG-4 laurate
9004-81-3	PEG-5 laurate
9004-81-3	PEG-8 laurate
9004-81-3	PEG-9 laurate
9004-81-3	PEG-12 laurate
9004-82-4	Sodium laureth sulfate
9004-87-9	Octoxynol-1
9004-87-9	Octoxynol-3
9004-87-9	Octoxynol-5
9004-87-9	Octoxynol-7
9004-87-9	Octoxynol-8
9004-87-9	Octoxynol-9
9004-87-9	Octoxynol-10
9004-87-9	Octoxynol-13
9004-87-9	Octoxynol-16
9004-87-9	Octoxynol-25
9004-87-9	Octoxynol-30
9004-87-9	Octoxynol-40
9004-87-9	Octoxynol-70
9004-96-0	PEG-4 oleate
9004-96-0	PEG-8 oleate
9004-96-0	PEG-12 oleate
9004-96-0	PEG-14 oleate
9004-96-0	PEG-20 oleate
9004-98-2	Oleth-4
9004-98-2	Oleth-5

CAS NUMBER TO CHEMICAL CROSS-REFERENCE

CAS	Chemical
9004-98-2	Oleth-9
9004-98-2	Oleth-20
9004-99-3	PEG-2 stearate
9004-99-3	PEG-2 stearate SE
9004-99-3	PEG-4 stearate
9004-99-3	PEG-8 stearate
9004-99-3	PEG-12 stearate
9004-99-3	PEG-20 stearate
9005-00-9	Steareth-2
9005-00-9	Steareth-10
9005-00-9	Steareth-20
9005-02-1	PEG-8 dilaurate
9005-07-6	PEG-12 dioleate
9005-08-7	PEG-150 distearate
9005-25-8	Corn starch
9005-25-8	Starch
9005-37-2	Propylene glycol alginate
9005-38-3	Algin
9005-64-5	PEG-80 sorbitan laurate
9005-64-5	Polysorbate 20
9005-65-5	Polysorbate 81
9005-65-6	Polysorbate 80
9005-66-7	Polysorbate 40
9005-67-8	Polysorbate 60
9005-67-8	Polysorbate 61
9005-70-3	PEG-17 sorbitan trioleate
9005-70-3	PEG-18 sorbitan trioleate
9005-70-3	Polysorbate 85
9005-71-4	Polysorbate 65
9005-79-2	Glycogen
9005-84-9	Starch
9006-03-5	Rubber, chlorinated
9006-04-6	Natural rubber
9006-26-2	Ethylene-maleic anhydride copolymer
9006-65-9	Dimethicone
9007-13-0	Calcium resinate
9007-16-3	Carbomer
9007-17-4	Carbomer
9007-20-9	Carbomer
9009-54-5	Polyurethane, thermoplastic
9010-43-9	Octoxynol-9
9010-77-9	Ethylene/acrylic acid copolymer
9010-79-1	EPM rubber
9010-79-1	Polypropylene (nucleated)
9010-85-9	Isobutylene/isoprene copolymer
9011-05-6	Urea-formaldehyde resin
9011-13-6	Styrene/MA copolymer
9011-14-7	Methyl methacrylate polymer
9011-16-9	PVM/MA copolymer
9011-19-2	Polysiloxane
9012-09-3	Cellulose triacetate
9012-54-8	Cellulase
9014-85-1	PEG-3.5 tetramethyl decynediol
9014-85-1	PEG-10 tetramethyl decynediol
9014-85-1	PEG-30 tetramethyl decynediol
9014-90-8	Sodium nonoxynol-4 sulfate
9014-92-0	Dodoxynol-10
9014-93-1	Nonyl nonoxynol-7
9014-93-1	Nonyl nonoxynol-8
9014-93-1	Nonyl nonoxynol-9
9014-93-1	Nonyl nonoxynol-15
9014-93-1	Nonyl nonoxynol-24
9014-93-1	Nonyl nonoxynol-49
9014-93-1	Nonyl nonoxynol-150
9016-00-6	Dimethicone
9016-45-9	Nonoxynol-2
9016-45-9	Nonoxynol-3
9016-45-9	Nonoxynol-4
9016-45-9	Nonoxynol-5
9016-45-9	Nonoxynol-6
9016-45-9	Nonoxynol-8
9016-45-9	Nonoxynol-9
9016-45-9	Nonoxynol-10
9016-45-9	Nonoxynol-11
9016-45-9	Nonoxynol-12
9016-45-9	Nonoxynol-13
9016-45-9	Nonoxynol-14
9016-45-9	Nonoxynol-20
9016-45-9	Nonoxynol-30
9016-45-9	Nonoxynol-40
9016-45-9	Nonoxynol-50
9016-45-9	Nonoxynol-100
9016-75-5	Polyphenylene sulfide
9016-87-9	Polymethylene polyphenyl isocyanate
9032-42-2	Methyl hydroxyethylcellulose
9036-19-5	Octoxynol-1
9036-19-5	Octoxynol-3
9036-19-5	Octoxynol-5
9036-19-5	Octoxynol-7
9036-19-5	Octoxynol-8
9036-19-5	Octoxynol-9
9036-19-5	Octoxynol-10
9036-19-5	Octoxynol-13
9036-19-5	Octoxynol-16
9036-19-5	Octoxynol-25
9036-19-5	Octoxynol-30
9036-19-5	Octoxynol-40
9036-19-5	Octoxynol-70
9038-95-3	PPG-24-buteth-27
9040-38-4	Disodium nonoxynol-10 sulfosuccinate
9051-57-4	Ammonium nonoxynol-4 sulfate
9065-63-8	PPG-24-buteth-27
9080-49-3	Polysulfide
9084-06-4	Sodium polynaphthalene sulfonate
9106-45-9	Nonoxynol-15
10016-20-3	α-Cyclodextrin
10025-87-3	Phosphorus oxychloride
10031-43-3	Copper nitrate (ic)
10034-96-5	Manganese sulfate (ous) monohydrate
10039-54-0	Hydroxylamine sulfate
10043-35-3	Boric acid
10045-86-0	Ferric phosphate anhyd.
10099-76-0	Lead silicate
10101-39-0	Calcium silicate
10101-41-4	Calcium sulfate dihydrate
10101-66-3	Manganese violet
10102-40-6	Sodium molybdate (dihydrate)
10108-24-4	PEG-4 laurate
10108-25-5	PEG-4 oleate
10124-43-3	Cobalt sulfate (ous)
10213-77-1	PPG-3 methyl ether
10213-79-3	Sodium metasilicate pentahydrate
10222-01-2	2,2-Dibromo-3-nitrilopropionamide
10287-53-3	Ethyl 4-(dimethylamino) benzoate
10294-40-3	Barium chromate
10326-41-7	D-Lactic acid
10361-46-3	Bismuth nitrate
10595-06-9	2-Phenoxyethyl methacrylate
10595-49-0	(3-Lauramidopropyl) trimethyl ammonium methyl sulfate
11099-07-3	Glyceryl stearate
11111-34-5	Poloxamine 304
11111-34-5	Poloxamine 701
11111-34-5	Poloxamine 704
11111-34-5	Poloxamine 901
11111-34-5	Poloxamine 904
11111-34-5	Poloxamine 908
11111-34-5	Poloxamine 1107
11111-34-5	Poloxamine 1307
11113-70-5	Lead silicochromate
11120-22-2	Lead silicate
11138-66-2	Xanthan gum
11890-98-8	1,3-Butylene glycol dimethacrylate
12001-26-2	Mica
12001-31-9	Quaternium-18 hectorite
12001-85-3	Zinc naphthenate
12002-43-6	Gilsonite
12018-19-8	Zinc chromate
12027-67-7	Ammonium molybdate
12027-96-2	Zinc hydroxystannate
12035-39-1	Nickel titanate
12068-19-8	Sodium nonoxynol-6 phosphate
12069-32-8	Boron carbide
12141-46-7	Aluminum silicate
12173-47-6	Hectorite
12178-41-5	Ferro-aluminum silicate
12199-37-0	Magnesium aluminum silicate
12269-78-2	Pyrophyllite
12447-61-9	Zinc borate
12513-27-8	Zinc borate
12542-30-2	Dicyclopentenyl acrylate
12645-31-7	2-Ethylhexyl phosphate
12645-53-3	Isooctyl acid phosphate
12691-60-0	Stearalkonium hectorite
12738-64-6	Sucrose benzoate
12767-90-7	Zinc borate
12769-96-9	Ultramarine violet
12788-93-1	Butyl acid phosphate
13048-33-4	1,6-Hexanediol diacrylate
13106-44-0	Dimethylaminoethyl acrylate dimethyl sulfate
13106-76-8	Ammonium molybdate
13149-86-5	Steareth-10
13150-00-0	Sodium laureth sulfate

CAS	Chemical
13429-07-7	PPG-2 methyl ether
13463-41-7	Zinc pyrithione
13463-67-7	Titanium dioxide
13530-65-9	Zinc chromate
13568-40-6	Lithium molybdate
13701-59-2	Barium metaborate
13822-56-5	Aminopropyltrimethoxy-silane
13845-36-8	Potassium tripolyphos-phate
13983-17-0	Wollastonite
14018-95-2	Zinc chromate
14024-64-7	Titanium acetylacetonate
14103-61-8	Diisononyl phthalate
14350-96-0	Sodium lauroampho-acetate
14351-40-7	Stearamide MEA-stearate
14409-72-4	Nonoxynol-9
14504-95-1	Aluminum silicate
14516-71-3	2,2′-Thiobis (4-t-octylphenolato)-n-butylamine nickel
14643-87-9	Zinc diacrylate
14728-39-3	Ammonium polyphos-phate sol'n.
14807-96-6	Talc
14808-60-7	Quartz
14940-41-1	Ferric phosphate tetrahydrate
15214-89-8	2-Acrylamido-2-methylpropanesulfonic acid
15520-10-2	2-Methylpentamethyl-enediamine
15625-89-5	1,1,1-Trimethylol-propane triacrylate
15647-08-2	Ethylhexyl diphenyl phosphite
15696-43-2	Lead octoate
15826-16-1	Sodium laureth sulfate
16057-43-5	Steareth-2
16589-43-8	Sodium methyl siliconate
16721-80-5	Sodium hydrosulfite
16940-66-2	Sodium borohydride
16996-51-3	Lead linoleate
17465-86-0	γ-Cyclodextrin
17557-23-2	Neopentyl glycol diglycidyl ether
17661-50-6	Cetyl esters
17831-71-9	PEG-4 diacrylate
18395-30-7	Isobutyltrimethoxysilane
18972-56-0	Magnesium silicofluoride
19224-26-1	Propylene glycol dibenzoate
19485-03-1	1,3-Butylene glycol diacrylate
20018-09-1	Diiodomethyl p-tolyl sulfone
20566-35-2	Tetrabromophthalate diol
20636-48-0	Nonoxynol-5
21245-02-3	Octyl dimethyl PABA
21282-96-2	2-(Acetoacetoxy) ethyl acrylate
21282-97-3	2-(Acetoacetoxy) ethyl methacrylate
21564-17-0	2-Thiocyanomethylthio-benzothiazole
21645-51-2	Alumina trihydrate
21645-51-2	Aluminum hydroxide
21652-27-7	Oleyl hydroxyethyl imidazoline
21810-39-9	Diethylaminoethyl acrylate dimethyl sulfate quat.
21908-53-2	Mercury oxide (ic), red and yellow
22499-12-3	Isobutyl benzoin ether
22499-12-3	2-(2-Methylpropoxy)-1,2-diphenylethanone
23328-53-2	Phenol, 2-(2H-benzotriazole-2-yl)-4-methyl-6-dodecyl
23783-42-8	Tetraethylene glycol methyl ether
23843-64-3	γ-Ureidopropyltrimeth-oxysilane
23844-24-8	Benzoguanamine
23946-66-8	2-Ethyl, 2′-ethoxy-oxalanilide
24304-00-5	Aluminum nitride
24615-84-7	β-Carboxyethyl acrylate
24650-42-8	Benzyldimethyl ketal
24650-42-8	2,2-Dimethoxy-2-phenylacetophenone
24801-88-5	Isocyanatopropyltri-ethoxysilane
24937-16-4	Nylon-12
24937-78-8	Ethylene/VA copolymer
24937-78-8	Vinyl acetate/ethylene copolymer
24937-79-9	Polyvinylidene fluoride
24938-91-8	Trideceth-6
24938-91-8	Trideceth-12
24938-91-8	Trideceth-14
24938-91-8	Trideceth-15
24969-11-7	Resorcinol-formalde-hyde resin
24979-97-3	Polytetramethylene ether glycol
25013-15-4	Vinyltoluene monomer
25013-16-5	BHA
25014-31-7	Poly-α-methylstyrene
25035-04-5	Nylon 11
25036-25-3	Epoxy, bisphenol A
25038-54-4	Nylon-6
25038-54-4	Polyamide
25038-59-9	Polyethylene terephthalate
25038-74-8	Nylon-12
25068-38-6	Epoxy, bisphenol A
25068-38-6	Phenoxy resin
25086-29-7	Styrene/PVP copolymer
25086-62-8	Sodium polymeth-acrylate
25086-89-9	PVP/VA copolymer
25103-12-2	Triisooctyl phosphite
25119-62-4	Styrene allyl alcohol
25155-23-1	Trixylenyl phosphate
25155-30-0	Sodium dodecyl-benzenesulfonate
25168-05-2	Monochlorotoluene
25190-06-1	Polytetramethylene ether glycol
25213-24-5	Polyvinyl alcohol, partially hydrolyzed
25265-71-8	Dipropylene glycol
25265-77-4	2,2,4-Trimethyl-1,3-pentanediol monoiso-butyrate
25321-41-9	Xylene sulfonic acid
25322-17-2	Dinonylnaphthalene sulfonic acid
25322-68-3	PEG-4
25322-68-3	PEG-6
25322-68-3	PEG-8
25322-68-3	PEG-9
25322-68-3	PEG-12
25322-68-3	PEG-14
25322-68-3	PEG-16
25322-68-3	PEG-20
25322-68-3	PEG-32
25322-68-3	PEG-40
25322-68-3	PEG-75
25322-68-3	PEG-100
25322-68-3	PEG-150
25322-68-3	PEG-200
25322-68-3	PEG-350
25322-68-3	PEG-5M
25322-68-3	PEG-7M
25322-68-3	PEG-9M
25322-68-3	PEG-14M
25322-68-3	PEG-23M
25322-68-3	PEG-45M
25322-68-3	PEG-90M
25322-68-3	Polyethylene glycol
25322-69-4	Polypropylene glycol
25322-69-4	PPG-9
25322-69-4	PPG-12
25322-69-4	PPG-17
25322-69-4	PPG-20
25322-69-4	PPG-26
25322-69-4	PPG-30
25338-55-0	Dimethylaminomethyl phenol
25377-73-5	Dodecenyl succinic anhydride
25417-20-3	Sodium dibutyl naphthalene sulfonate
25448-25-3	Triisodecyl phosphite
25496-72-4	Glyceryl mono/dioleate
25496-72-4	Glyceryl oleate
25498-49-1	PPG-3 methyl ether
25550-98-5	Phenyl diisodecyl phosphite
25608-12-2	Potassium polyacrylate
25619-56-1	Barium dinonylnaphtha-lene sulfonate
25620-58-0	Trimethylhexamethylene diamine
25637-99-4	Hexabromocyclo-dodecane
25640-78-2	1,1′-Biphenyl (1-methylethyl)
25750-82-7	Ethylene/sodium acrylate copolymer
25852-47-5	PEG-4 dimethacrylate
25928-94-3	Epoxy resin
25973-55-1	2-(2′-Hydroxy-3,5′-di-t-amylphenyl) benzotriazole
26027-38-3	Nonoxynol-1
26027-38-3	Nonoxynol-2
26027-38-3	Nonoxynol-4

CAS NUMBER TO CHEMICAL CROSS-REFERENCE

CAS	Chemical
26027-38-3	Nonoxynol-5
26027-38-3	Nonoxynol-6
26027-38-3	Nonoxynol-8
26027-38-3	Nonoxynol-9
26027-38-3	Nonoxynol-10
26027-38-3	Nonoxynol-11
26027-38-3	Nonoxynol-12
26027-38-3	Nonoxynol-13
26027-38-3	Nonoxynol-14
26027-38-3	Nonoxynol-15
26027-38-3	Nonoxynol-20
26027-38-3	Nonoxynol-30
26027-38-3	Nonoxynol-40
26027-38-3	Nonoxynol-50
26027-38-3	Nonoxynol-100
26040-51-7	Tetrabromobis (2-ethylhexyl) phthalate
26172-55-4	Methylchloroiso-thiazolinone
26264-02-8	Nonoxynol-5
26264-05-1	Isopropylamine dodecylbenzene-sulfonate
26264-06-2	Calcium dodecylben-zene sulfonate
26264-58-4	Sodium methylnaph-thalenesulfonate
26266-57-9	Sorbitan palmitate
26266-58-0	Sorbitan trioleate
26266-77-3	Hydroabietyl alcohol
26401-27-4	Diphenyl isooctyl phosphite
26401-47-8	Dodoxynol-10
26426-80-2	Isobutylene/MA copolymer
26444-49-5	Diphenylcresyl phosphate
26445-96-5	Ethylene/calcium acrylate copolymer
26447-10-9	Ammonium xylene-sulfonate
26471-62-5	Toluene diisocyanate
26523-78-4	Tris(nonylphenyl) phosphite
26530-20-1	2-n-Octyl-4-isothiazolin-3-one
26544-23-0	Diphenyl isodecyl phosphite
26544-38-7	Dodecenyl succinic anhydride
26571-11-9	Nonoxynol-8
26571-11-9	Nonoxynol-9
26658-19-5	Sorbitan tristearate
26741-53-7	Bis (2,4-di-t-butylphenyl) pentaerythritol diphosphite
26760-54-3	Poly (diethylene glycol adipate)
26761-40-0	Diisodecyl phthalate
26761-45-5	Glycidyl neodecanoate
26780-96-1	2,2,4-Trimethyl-1,2-dihydroquinoline polymer
26837-33-2	Sodium lauroampho-acetate
26850-24-8	DEDM hydantoin
26896-20-8	Neodecanoic acid
26952-21-6	Isooctyl alcohol
27136-73-8	Oleyl hydroxyethyl imidazoline
27138-31-4	Dipropylene glycol dibenzoate
27176-87-0	Dodecylbenzene sulfonic acid
27176-93-8	Nonoxynol-2
27176-94-9	Octoxynol-3
27176-95-0	Nonoxynol-3
27176-97-2	Nonoxynol-4
27176-99-4	Octoxynol-5
27177-01-1	Nonoxynol-6
27177-02-2	Octoxynol-7
27177-05-5	Nonoxynol-6
27177-05-5	Nonoxynol-8
27177-07-7	Octoxynol-10
27177-08-8	Nonoxynol-10
27177-77-1	Potassium dodecylben-zene sulfonate
27178-16-1	Diisodecyl adipate
27194-74-7	Propylene glycol laurate
27323-41-7	TEA-dodecylbenzene-sulfonate
27615-98-1	Tributyltin fluoride
27813-02-1	Hydroxypropyl methacrylate
27942-26-3	Nonoxynol-10
27986-36-3	Nonoxynol-1
28183-09-7	Poly (diethylene glycol adipate)
28212-44-4	Nonoxynol-10 carboxylic acid
28553-12-0	Dinonyl phthalate
28874-51-3	Sodium PCA
28961-43-5	Trimethylolpropane ethoxy triacrylate
29059-24-3	Propylene glycol myristate
29116-98-1	Sorbitan dioleate
29297-55-0	Polyquaternium-16
29381-93-9	TEA-dodecylbenzene-sulfonate
29406-96-0	1,2-Polybutadiene (syndiotactic)
29590-42-9	Isooctyl acrylate
29690-82-2	Epoxy cresol novolac
29761-21-5	Isodecyl diphenyl phosphate
29911-27-1	Dipropylene glycol n-propyl ether
29911-28-2	Dipropylene glycol butyl ether
29964-84-9	Isodecyl methacrylate
29994-44-3	Nonoxynol-6 phosphate
30136-13-1	Propoxypropanol
31098-21-2	Potassium sulfopropyl methacrylate
31394-71-5	PPG-36 oleate
31556-45-3	Tridecyl stearate
31566-31-1	Glyceryl stearate
31621-69-9	Dicyclopentenyl methacrylate
31621-91-7	PEG-9 pelargonate
31691-97-1	Ammonium nonoxynol-4 sulfate
31692-79-2	Dimethiconol
32073-22-6	Sodium cumenesulfonate
32131-17-2	Nylon-66
32360-05-7	Stearyl methacrylate
32534-81-9	Pentabromodiphenyl oxide
32536-52-0	Octabromodiphenyl oxide
32612-48-9	Ammonium laureth sulfate
32612-48-9	Ammonium laureth-12 sulfate
32612-48-9	Ammonium laureth-30 sulfate
33907-46-9	Glycol hydroxystearate
34140-91-5	Oleyl propylene diamine dioleate
34149-92-3	Polyvinyl chloride acetate
34206-40-1	Tetra oximino silane
34375-28-5	2[(-Hydroxymethyl) amino]ethanol
34424-98-1	Polyglyceryl-10 tetraoleate
34590-94-8	PPG-2 methyl ether
35179-86-3	PEG-8 laurate
35325-02-1	N-(2-Hydroxypropyl) benzenesulfonamide
35541-81-2	1,4-Cyclohexane-dimethanol dibenzoate
35691-65-7	Methyldibromo glutaronitrile
36653-82-4	Cetyl alcohol
36671-85-9	Polyvinyl octadecyl carbamate
37200-49-0	Polysorbate 80
37205-87-1	Nonoxynol-1
37205-87-1	Nonoxynol-2
37205-87-1	Nonoxynol-4
37205-87-1	Nonoxynol-5
37205-87-1	Nonoxynol-6
37205-87-1	Nonoxynol-8
37205-87-1	Nonoxynol-9
37205-87-1	Nonoxynol-10
37205-87-1	Nonoxynol-11
37205-87-1	Nonoxynol-12
37205-87-1	Nonoxynol-13
37205-87-1	Nonoxynol-14
37205-87-1	Nonoxynol-15
37205-87-1	Nonoxynol-20
37205-87-1	Nonoxynol-30
37205-87-1	Nonoxynol-40
37205-87-1	Nonoxynol-50
37205-87-1	Nonoxynol-100
37220-82-9	Glyceryl oleate
37244-96-5	Nepheline syenite
37286-64-9	PPG-3 methyl ether
37318-14-2	PEG-8 laurate
37340-60-6	Sodium nonoxynol-6 phosphate
37478-68-5	Capryl hydroxyethyl imidazoline
37767-39-8	Tetrasodium dicarboxyethyl stearyl sulfosuccinamate
38916-42-6	Tetrasodium dicarboxyethyl stearyl sulfosuccinamate
42173-90-0	Octoxynol-9
42233-51-2	Myristyl lignocerate
42978-66-5	PPG-3 diacrylate

CAS	Chemical
44992-01-0	Dimethylaminoethyl acrylate methyl chloride quat.
45115-34-2	Vinyl neodecanoate
48165-04-6	2-Phenoxyethyl acrylate
50586-59-9	PEG-6 trimethylol-propane
50935-57-4	PVM/MA copolymer, ethyl ester
50976-02-8	Dicyclopentenyl acrylate
51178-59-7	Dicyclopentenyl methacrylate
51200-87-4	Dimethyl oxazolidine
51229-78-8	Quaternium-15
51437-95-7	Nonoxynol-3
51609-41-7	Nonoxynol-6 phosphate
51609-41-7	Nonoxynol-9 phosphate
51811-79-1	Nonoxynol-6 phosphate
51811-79-1	Nonoxynol-9 phosphate
52125-53-8	Propylene glycol ethyl ether
52229-50-2	PVM/MA copolymer
52299-20-4	2-[(-Hydroxymethyl) amino]-2-methyl-propanol
52315-75-0	Lauroyl lysine
52467-63-7	Tricetylmonium chloride
52668-97-0	PEG-150 distearate
52688-97-0	PEG-12 dioleate
52907-07-0	Ethylenebis dibromonor-bornane dicarboximide
53200-28-5	PVM/MA copolymer, butyl ester
53610-02-9	Nonoxynol-10 carboxylic acid
53637-25-5	Poloxamer 181
54018-18-7	PVM/MA copolymer, butyl ester
54077-45-1	PVM/MA copolymer, isopropyl ester
54193-36-1	Sodium polymeth-acrylate
54571-67-4	Sodium PCA
54578-88-0	PVM/MA copolymer, isopropyl ester
54578-90-4	PVM/MA copolymer, ethyl ester
54578-91-5	PVM/MA copolymer, butyl ester
55406-53-6	Iodopropynyl butylcarbamate
55799-16-1	Zinc hydroxyphosphite
56091-51-1	PVM/MA copolymer, isopropyl ester
57018-52-7	Propylene glycol t-butyl ether
57455-37-5	Ultramarine blue
57855-77-3	Calcium dinonylnaph-thalene sulfonate
58965-66-5	Tetradecabromo-diphenoxy benzene
60842-32-2	Silica dimethyl silylate
61417-49-0	Isopropyl titanium triisostearate
61788-46-3	Cocamine
61788-63-4	Dihydrogenated tallow methylamine
61788-85-0	PEG-5 hydrogenated castor oil
61788-85-0	PEG-16 hydrogenated castor oil
61788-85-0	PEG-25 hydrogenated castor oil
61788-85-0	PEG-200 hydrogenated castor oil
61788-85-0	PEG-66 trihydroxy-stearin
61788-89-4	Dilinoleic acid
61788-93-0	Dimethyl cocamine
61789-09-1	Hydrogenated tallow glyceride
61789-22-8	Copper tallate
61789-51-3	Cobalt naphthenate
61789-68-2	PEG-2 coco-benzonium chloride
61789-71-7	Benzalkonium chloride
61789-76-2	Dicocamine
61789-77-3	Dicocodimonium chloride
61789-80-8	Quaternium-18
61789-86-4	Calcium sulfonate
61790-12-3	Tall oil acid
61790-14-5	Lead naphthenate
61790-31-6	Hydrogenated tallow amide
61790-53-2	Diatomaceous earth, amorphous
61790-57-6	Cocamine acetate
61790-85-0	PEG-10 tallow aminopropylamine
61790-86-1	Polysorbate 80
61791-01-3	PEG-8 ditallate
61791-01-3	PEG-12 ditallate
61791-12-6	PEG-5 castor oil
61791-12-6	PEG-12 castor oil
61791-12-6	PEG-15 castor oil
61791-12-6	PEG-16 castor oil
61791-12-6	PEG-20 castor oil
61791-12-6	PEG-25 castor oil
61791-12-6	PEG-28 castor oil
61791-12-6	PEG-30 castor oil
61791-12-6	PEG-32 castor oil
61791-12-6	PEG-36 castor oil
61791-12-6	PEG-40 castor oil
61791-12-6	PEG-52 castor oil
61791-12-6	PEG-75 castor oil
61791-12-6	PEG-80 castor oil
61791-12-6	PEG-160 castor oil
61791-12-6	PEG-200 castor oil
61791-14-8	PEG-10 cocamine
61791-14-8	PEG-15 cocamine
61791-38-6	Cocoyl hydroxyethyl imidazoline
61791-39-7	Tall oil hydroxyethyl imidazoline
61791-44-4	PEG-2 tallowamine
61791-44-4	PEG-5 tallowamine
61791-53-5	Tallow propane diamine dioleate
61791-55-7	Tallow propylene diamine
61791-57-9	Tallow dipropylene triamine
63148-55-0	Dimethicone copolyol
63148-62-9	Dimethicone
63148-65-2	Polyvinyl butyral
63148-69-6	Alkyd resin
63217-13-0	Sodium dinonyl sulfosuccinate
63231-60-7	Microcrystalline wax
63231-67-4	Silica gel
63351-73-5	Ammonium nonoxynol-4 sulfate
63394-02-5	Silicone elastomer
63428-83-1	Nylon
63449-39-8	Paraffin, chlorinated
63449-41-2	Benzalkonium chloride
63843-89-0	Bis (1,2,2,6,6-pentamethyl-4-piperidinyl) (3,5-di-t-butyl-4-hydroxybenzyl) butyl propanedioate
64051-24-7	Phosphoric acid, bis (2-hexyloxyethyl) ester, diethylamine salt
64051-25-8	Phosphoric acid, mono (2-hexyloxyethyl) ester, bis (diethylamine) salt
64051-35-0	Dioctylphenyl hydrogen phosphate diethylamine salt
64051-37-2	Octylphenyl dihydrogen phosphate bis (diethylamine) salt
64365-23-7	Dimethicone copolyol
64475-85-0	Mineral spirits
64742-14-9	Petroleum distillates
64742-42-3	Microcrystalline wax
64742-46-7	Hydrocarbon solvent
64742-47-8	Petroleum distillates
64742-48-9	C7-8 isoparaffin
64742-48-9	C8-9 isoparaffin
64742-48-9	C10-11 isoparaffin
64742-48-9	C11-12 isoparaffin
64742-48-9	C11-13 isoparaffin
64742-48-9	C13-14 isoparaffin
64742-49-0	Hexane
64742-89-8	Heptane
64742-95-6	Naphtha, light aromatic
64812-54-4	Nonoxynol-20
65816-20-8	N-(p-Ethoxycarbonyl-phenyl)-N′-ethyl-N′-phenylformamidine
65876-95-1	Resorcinol-formalde-hyde resin
65996-61-4	Cellulose
65997-17-3	Borosilicate glass
66197-78-2	Nonoxynol-9 phosphate
67700-99-6	Dihydrogenated tallow methylamine
67701-08-0	Soy acid
67724-93-0	PVM/MA copolymer, ethyl ester
67762-19-0	Ammonium laureth sulfate
67762-36-1	Caprylic/capric acid
67762-38-3	Methyl oleate
67762-96-3	Dimethicone copolyol
67784-77-4	PEG-2 tallowmonium chloride
67892-91-5	Olefin/acrylate copolymer
67999-57-9	Disodium nonoxynol-10 sulfosuccinate
68002-59-5	Quaternium-18

CAS	Chemical
68002-71-1	Hydrogenated soybean oil
68002-97-1	Laureth-2
68002-97-1	Laureth-4
68037-39-8	Polyethylene elastomer, chlorosulfonated
68037-74-1	Dimethicone (branched)
68037-97-8	Oleyl propanediamine
68052-23-3	2,2,4-Trimethyl-1,3-pentanediol dibenzoate
68081-81-2	Sodium dodecylbenzenesulfonate
68081-96-9	Ammonium lauryl sulfate
68083-14-7	Phenylmethyl polysiloxane
68131-39-5	C12-15 pareth-7
68131-39-5	C12-15 pareth-9
68131-40-8	C11-15 pareth-7
68131-40-8	C11-15 pareth-9
68140-98-7	Ethyl hydroxymethyl oleyl oxazoline
68155-09-9	Cocamidopropylamine oxide
68186-14-1	Methyl rosinate
68188-45-5	Sodium α olefin sulfonate
68298-21-5	Sodium lauroamphoacetate
68299-17-2	Sodium isodecyl sulfate
68308-53-2	Soy acid
68308-54-3	Hydrogenated tallow glycerides
68308-67-8	Soyethyldimonium ethosulfate
68309-95-5	Ammonium zirconium carbonate
68391-01-5	Benzalkonium chloride
68411-31-4	TEA-dodecylbenzenesulfonate
68411-32-5	Dodecylbenzene sulfonic acid
68412-53-3	Nonoxynol-6 phosphate
68412-53-3	Nonoxynol-9 phosphate
68424-43-1	Lanolin acid
68424-61-3	Glyceryl mono/dioleate
68424-85-1	Benzalkonium chloride
68439-49-6	Ceteareth-6
68439-49-6	Ceteareth-8
68439-49-6	Ceteareth-10
68439-49-6	Ceteareth-11
68439-49-6	Ceteareth-12
68439-49-6	Ceteareth-14
68439-49-6	Ceteareth-18
68439-49-6	Ceteareth-27
68439-49-6	Ceteareth-30
68439-49-6	Ceteareth-55
68439-50-9	Laureth-2
68439-50-9	Laureth-4
68439-50-9	Laureth-9
68439-57-6	Sodium C14-16 olefin sulfonate
68439-57-6	Sodium α olefin sulfonate
68441-17-8	Polyethylene, oxidized
68442-33-1	Polypropylene, chlorinated
68457-79-4	Zinc dialkyl dithiophosphate
68459-31-4	Alkyd resin
68479-98-1	Diethyltoluene diamine
68514-95-4	Dialkyl dimethyl ammonium chloride
68515-48-0	Diisononyl phthalate
68515-49-1	Diisodecyl phthalate
68526-79-4	Hexyl alcohol
68526-83-0	Isooctyl alcohol
68526-84-1	Isononyl alcohol
68526-85-2	n-Decyl alcohol
68526-86-3	Tridecyl alcohol
68527-05-9	Isononyl alcohol
68584-22-5	Dodecylbenzene sulfonic acid
68584-24-7	Isopropylamine dodecylbenzenesulfonate
68585-34-2	Sodium laureth sulfate
68585-47-7	Sodium lauryl sulfate
68608-26-4	Sodium petroleum sulfonate
68608-66-2	Sodium lauroamphoacetate
68608-88-8	Dodecylbenzene sulfonic acid
68610-51-5	p-Cresol/dicyclopentadiene butylated reaction product
68611-44-9	Dimethyldichlorosilane
68647-44-9	Sodium lauroamphoacetate
68649-05-8	Cocaminobutyric acid
68783-41-5	Dimer acid
68855-54-9	Diatomaceous earth
68891-38-3	Sodium laureth sulfate
68891-39-4	Sodium nonoxynol-4 sulfate
68901-05-3	PPG-3 diacrylate
68920-06-9	Naphtha
68937-41-7	Triaryl phosphate
68937-54-2	Polysiloxane polyether copolymer
68937-55-3	Silicone glycol copolymer
68937-90-6	Trilinoleic acid
68938-35-2	Vegetable oil
68938-54-5	Silicone glycol copolymer
68952-35-2	Tricresyl phosphate
68953-58-2	Quaternium-18 bentonite
68953-96-8	Calcium dodecylbenzene sulfonate
68955-19-1	Sodium lauryl sulfate
68955-45-3	Ethylene distearamide
68956-68-3	Vegetable oil
68989-00-4	Benzalkonium chloride
68990-05-6	Citric acid esters of mono- and diglycerides of fatty acids
69005-67-8	Sorbitan stearate
69009-90-1	Bis(methylethyl)-1,1´-biphenyl
69102-90-5	Polybutadiene
69430-24-6	Octamethylcyclotetrasiloxane
70161-44-3	Sodium hydroxymethylglycinate
70592-80-2	Lauramine oxide
70802-40-3	PEG-8 stearate
70892-21-6	Polyvinyl octadecyl carbamate
70914-20-4	Isoheptyl alcohol
71011-27-3	Quaternium-18 hectorite
71060-72-5	Tricetylmonium chloride
71243-51-1	Synthetic beeswax
71487-00-8	PEG-2 cocomonium nitrate
71487-01-9	Dicoco nitrite
72245-46-6	Polypropylene glycol dibenzoate
72674-05-6	Sulfonic acid
72869-62-6	Sorbitan tristearate
73138-79-1	Sulfated peanut oil
74665-14-8	Phenol sulfonic acid
74665-17-1	Tetraisopropyl di (dioctylphosphito) titanate
76050-42-5	Carbomer
78330-12-8	Sodium petroleum sulfonate
78330-21-9	Trideceth-6
78330-21-9	Trideceth-14
78330-21-9	Trideceth-15
78492-71-1	Poly (diethylene glycol, neopentyl glycol, 1,6-hexanediol adipate)
79185-77-6	1-Aza-3,7-dioxo-bicyclo-2,8-diisopropyl-5-ethyl(3.3.0)-octane
80584-86-7	Poly (dipropyleneglycol) phenyl phosphite
82985-35-1	Bis-(γ-trimethoxysilyl-propyl) amine
83846-86-0	Isopropyl thioxanthone (4-isomer)
84501-49-5	Sodium decyl sulfate
84562-92-5	Nonoxynol-3
85116-97-8	PEG-2 stearate
85117-50-6	Sodium dodecylbenzenesulfonate
85186-88-5	Sorbitan trioleate
85251-77-0	Glyceryl stearate
85409-22-9	Benzalkonium chloride
85536-14-7	Dodecylbenzene sulfonic acid
85586-21-6	Methyl stearate
85632-63-9	Tallow dipropylene triamine
85666-92-8	Glyceryl stearate
85736-49-8	PEG-12 dioleate
85865-69-6	Isobutyl stearate
88230-35-7	Oxo-hexyl acetate
88543-32-2	Ethyl hydroxymethyl oleyl oxazoline
88917-22-0	PPG-2 methyl ether acetate
90268-48-7	Disodium tallow sulfosuccinamate
90438-79-2	Oxo-heptyl acetate
91050-82-7	Pentaerythrityl tetrastearate
91744-38-6	Citric acid esters of mono- and diglycerides of fatty acids
94035-02-6	Hydroxypropyl-β-cyclodextrin

CAS	Chemical
94108-97-1	Ditrimethylolpropane tetraacrylate
94891-33-5	Stearalkonium hectorite
97026-94-0	Synthetic beeswax
97281-23-7	Glycol stearate
97281-47-5	Lecithin
97593-31-2	Citric acid esters of mono- and diglycerides of fatty acids
98516-30-4	Propylene glycol ethyl ether acetate
99241-25-5	Hydroxypropyl-γ-cyclodextrin
100085-40-3	Fish glycerides
101357-30-6	Sodium alumino sulfo silicate
103432-54-8	Neopentyl (diallyl) oxy, tri (dioctyl) pyrophosphato titanate
108419-32-5	C8 alkyl acetate
108419-33-6	C9 alkyl acetate
108419-34-7	C10 alkyl acetate
108419-35-8	C13 alkyl acetate
111109-77-4	Dipropylene glycol dimethyl ether
112945-52-5	Silica, fumed
113252-64-5	Zirconium IV neoalkanolato, tris (diisooctyl) pyrophosphato-o
115340-81-3	2,2,6,6-Tetramethylpiperidin-4-yl acrylate/ methyl methacrylate copolymer
116912-64-2	γ-Ureidopropyltriethoxysilane
119345-01-6	Phosphorus trichloride, reaction prods. with 1,1´-biphenyl and 2,4-bis (1,1-dimethylethyl) phenol
120007-67-9	Zinc borate
123237-14-9	Synthetic wax
124319-53-5	Zirconium IV di neoalkanolato, di(3-mercapto) propionato-o
128446-36-6	Methyl-β-cyclodextrin
129288-65-9	Epoxidized 1,2-polybutadiene
129757-67-1	Bis-(1-octyloxy-2,2,6,6,tetramethyl-4-piperidinyl) sebacate
131298-44-7	Isodecyl benzoate
132739-31-2	Dipropylene glycol t-butyl ether
136097-95-5	Candelilla synthetic
136097-96-6	Carnauba synthetic
136097-97-7	Cetyl esters
136505-02-7	Polyquaternium-31
138265-88-0	Zinc borate
143860-04-2	3-Ethyl-2-methyl-2-(3-methylbutyl)-1,3-oxazolidine
146955-66-0	Zirconium IV di-neoalkanolato, di(p-amino benzoato-o)
148348-13-4	α,α,4,4-Tetramethyl-2-(methylethyl)-N-(2-methylpropylidene)-3-oxazolidineethanamine

EINECS Number to Chemical Cross-Reference

EINECS	Chemical
200-001-8	Formaldehyde
200-018-0	Lactic acid
200-061-5	Sorbitol
200-143-0	2-Bromo-2-nitropropane-1,3-diol
200-268-0	Tributyltin oxide
200-289-5	Glycerin
200-312-9	Palmitic acid
200-313-4	Stearic acid
200-338-0	Polypropylene glycol
200-338-0	Propylene glycol
200-431-6	p-Chloro-m-cresol
200-464-6	2-Mercaptoethanol
200-467-2	Ethyl ether
200-470-9	Linoleic acid
200-532-5	Phenylmercuric acetate
200-539-3	Aniline
200-578-6	Alcohol
200-578-6	Alcohol denatured
200-579-1	Formic acid
200-618-2	Benzoic acid
200-629-2	1,10-Phenanthroline
200-659-6	Methyl alcohol
200-661-7	Isopropyl alcohol
200-662-2	Acetone
200-679-5	Dimethyl formamide
200-741-1	p-Toluenesulfonamide
200-746-9	n-Propyl alcohol
200-751-6	Butyl alcohol
200-752-1	Amyl alcohol
200-753-7	Benzene
200-755-8	Cobalt acetate (ous)
200-756-3	Trichloroethane
200-815-3	Polyethylene
200-815-3	Polyethylene, high-density
200-815-3	Polyethylene, low-density
200-815-3	Polyethylene, ultrahigh m.w.
200-815-3	Polyethylene wax
200-827-9	Propane
200-831-0	Vinyl chloride
200-836-8	Acetaldehyde
200-838-9	Methylene chloride
200-848-3	Calcium carbide
200-849-9	Polysorbate 80
200-857-2	Isobutane
200-864-0	Vinylidene chloride monomer
200-876-6	Nitromethane
200-889-7	t-Butyl alcohol

EINECS	Chemical
200-898-6	Methanesulfonic acid
200-901-0	Dimethyldichlorosilane
200-915-7	t-Butyl hydroperoxide
201-039-8	Dibutyltin dilaurate
201-052-9	Dicyclopentadiene
201-063-9	Trimethylolethane
201-064-4	Tris (hydroxymethyl) aminomethane
201-066-5	Acetyl triethyl citrate
201-067-0	Acetyl tributyl citrate
201-069-1	Citric acid
201-070-7	Triethyl citrate
201-071-2	Tributyl citrate
201-074-9	Trimethylolpropane
201-081-7	Vinyltriethoxysilane
201-083-8	Ethyl silicate
201-094-8	Cetethyl morpholinium ethosulfate
201-116-6	Trioctyl phosphate
201-122-9	Tributoxyethyl phosphate
201-126-0	Isophorone
201-131-8	Dimethyl octynediol
201-148-0	Isobutyl alcohol
201-159-0	Methyl ethyl ketone
201-166-9	Trichloroethane
201-177-9	Acrylic acid
201-178-4	Carboxymethylcellulose sodium
201-185-2	Methyl acetate
201-188-9	Nitroethane
201-195-7	Isobutyric acid
201-202-3	Methacrylamide
201-204-4	Methacrylic acid
201-209-1	2-Nitropropane
201-236-9	Tetrabromobisphenol A
201-245-8	Bisphenol A
201-248-4	4,4´-Diaminodiphenyl sulfone
201-275-1	Ethyl toluenesulfonamide
201-296-2	L-Lactic acid
201-297-1	Methyl methacrylate monomer
201-545-9	Dicyclohexyl phthalate
201-548-5	Butyl cyclohexyl phthalate
201-557-4	Dibutyl phthalate
201-562-1	Butyl octyl phthalate
201-607-5	Phthalic anhydride
201-622-7	Butyl benzyl phthalate
201-624-8	n-Butyl phthalyl-n-butyl glycolate
201-793-8	Chloroxylenol

EINECS	Chemical
201-800-4	PVP
201-800-4	N-Vinyl-2-pyrrolidone
201-879-5	Pyromellitic acid
201-898-9	Pyromellitic dianhydride
201-969-4	1-Naphthol
202-013-9	2,4,6-Tris (dimethylaminomethyl) phenol
202-046-9	Decahydronaphthalene
202-088-8	Diethyl aniline
202-268-6	o-Tolyl biguanide
202-340-7	Dipropylene glycol dibenzoate
202-394-1	1H-Benzotriazole
202-397-8	Stearyl hydroxyethyl imidazoline
202-414-9	Oleyl hydroxyethyl imidazoline
202-422-2	o-Xylene
202-436-9	1,2,4-Trimethylbenzene
202-473-0	Allyl methacrylate
202-490-3	Diethyl ketone
202-495-0	Thioglycerin
202-496-6	Methyl ethyl ketoxime
202-500-6	Methyl acrylate (monomer)
202-509-5	Butyrolactone
202-598-0	Ethyl lactate
202-599-6	Itaconic acid
202-601-5	N-Hydroxysuccinic acid
202-612-5	Isobutyl isobutyrate
202-613-0	Isobutyl methacrylate
202-615-1	Butyl methacrylate
202-617-2	Ethylene glycol dimethacrylate
202-625-6	Tetrahydrofurfuryl alcohol
202-626-1	Furfuryl alcohol
202-627-7	Furfural
202-679-0	4-t-Butylphenol
202-691-6	Phenol sulfonic acid
202-700-3	PCA
202-705-0	α-Methylstyrene monomer
202-708-7	Acetophenone
202-805-4	N,N-Dimethyl-p-toluidine
202-830-0	Terephthalic acid
202-849-4	Ethylbenzene
202-851-5	Polystyrene
202-851-5	Styrene
202-855-7	Benzonitrile
202-860-4	Benzaldehyde
202-905-8	Hexamethylenetetramine

EINECS	Chemical
202-908-4	Triphenyl phosphite
202-935-1	Glyceryl triacetyl ricinoleate
202-936-7	Triallylcyanurate
202-943-5	Cyclohexyl methacrylate
202-974-4	4,4´-Methylene dianiline
203-041-4	Tetrahydroxypropyl ethylenediamine
203-049-8	Triethanolamine
203-051-9	Triacetin
203-080-7	Octyl acrylate
203-090-1	Dioctyl adipate
203-149-1	N-Benzyldimethylamine
203-180-0	p-Toluene sulfonic acid
203-234-3	2-Ethylhexanol
203-268-9	1,4-Cyclohexanedi-methanol
203-291-4	Ethyl propionate
203-300-1	s-Butyl acetate
203-328-4	Dibutyl maleate
203-358-8	PEG-4 stearate
203-359-3	PEG-9 laurate
203-363-5	PEG-2 stearate
203-364-0	PEG-2 oleate
203-366-1	Hydroxystearic acid
203-389-7	Propyl propionate
203-419-9	Dimethyl succinate
203-439-8	Epichlorohydrin
203-441-9	Glycidyl methacrylate
203-442-4	Allyl glycidyl ether
203-448-7	Butane
203-458-1	Ethylene dichloride
203-466-5	Acrylonitrile
203-468-6	Ethylenediamine
203-473-3	Glycol
203-473-3	Polyethylene glycol
203-489-0	Hexylene glycol
203-500-9	Dimethyl hexynol
203-508-2	Distearyldimonium chloride
203-528-1	Methyl propyl ketone
203-539-1	Methoxyisopropanol
203-542-8	Dimethylethanolamine
203-544-9	Nitropropane
203-545-4	Vinyl acetate
203-550-1	Methyl isobutyl ketone
203-551-7	Methyl amyl alcohol
203-560-6	Isopropyl ether
203-561-1	Isopropyl acetate
203-570-0	Succinic anhydride
203-571-6	Maleic anhydride
203-572-1	Propylene carbonate
203-584-7	m-Phenylenediamine
203-585-2	Resorcinol
203-603-9	Propylene glycol methyl ether acetate
203-604-4	1,3,5-Trimethylbenzene
203-615-4	Melamine
203-620-1	Diisobutyl ketone
203-625-9	Toluene
203-628-5	Chlorobenzene
203-630-6	Cyclohexanol
203-631-1	Cyclohexanone
203-632-7	Phenol
203-677-2	n-Valeric acid
203-678-8	Vinyl isobutyl ether
203-686-1	Propyl acetate

EINECS	Chemical
203-713-7	Methoxyethanol
203-716-3	Diethylamine
203-726-8	Tetrahydrofuran
203-727-3	Furan polymer
203-733-6	Di-t-butyl peroxide
203-737-8	Methyl isoamyl ketone
203-740-4	Succinic acid
203-742-5	Maleic acid
203-743-0	Fumaric acid
203-745-1	Isobutyl acetate
203-749-3	Oleoyl sarcosine
203-755-6	Ethylene distearamide
203-756-1	Ethylene dioleamide
203-767-1	Methyl n-amyl ketone
203-772-9	Methoxyethanol acetate
203-777-6	Hexane
203-786-5	1,4-Butanediol
203-792-8	m-Butyraldehyde oxime
203-794-9	Ethylene glycol dimethyl ether
203-804-1	Ethoxyethanol
203-806-2	Cyclohexane
203-809-9	Pyridine
203-820-9	Diisopropanolamine
203-821-4	Dipropylene glycol
203-827-7	Glyceryl oleate
203-837-1	Methyl hexyl ketone
203-839-2	Ethoxyethanol acetate
203-845-5	Sebacic acid
203-852-3	Hexyl alcohol
203-865-4	Diethylenetriamine
203-868-0	Diethanolamine
203-872-2	Diethylene glycol
203-881-1	Ethylene glycol diacetate
203-886-9	Glycol stearate
203-905-0	Butoxyethanol
203-906-6	Methoxydiglycol
203-917-6	Caprylic alcohol
203-919-7	Ethoxydiglycol
203-924-4	Diethylene glycol dimethyl ether
203-931-2	Nonanoic acid
203-933-3	Butoxyethanol acetate
203-934-9	Butoxyethanol acetate
203-940-1	Ethoxydiglycol acetate
203-950-6	Triethylenetetramine
203-951-1	Ethylene glycol hexyl ether
203-956-9	n-Decyl alcohol
203-961-6	Butoxydiglycol
203-963-7	Diethylene glycol diethyl ether
203-978-9	Ethoxytriglycol
203-986-2	Tetraethylenepentamine
203-988-3	Diethylene glycol n-hexyl ether
203-989-9	PEG-4
203-990-4	Methyl stearate
203-992-5	Methyl oleate
203-993-0	Methyl linoleate
203-998-8	Tridecyl alcohol
204-001-9	Diethylene glycol dibutyl ether
204-007-1	Oleic acid
204-009-2	Erucamide
204-065-8	Dimethyl ether
204-104-9	Pentaerythritol

EINECS	Chemical
204-110-1	Pentaerythrityl tetrastearate
204-112-2	Triphenyl phosphate
204-126-9	Polytetrafluoroethylene
204-132-1	MDM hydantoin
204-171-4	Tetrachlorophthalic anhydride
204-211-0	Dioctyl phthalate
204-287-5	Anthranilic acid
204-327-1	2,2´-Methylenebis (6-t-butyl-4-methylphenol)
204-337-6	Benzophenone
204-340-2	Tetrahydronaphthalene
204-368-5	Phenyldiethanolamine
204-407-6	Diethylene glycol dibenzoate
204-442-7	BHA
204-469-4	Triethylamine
204-479-9	Benzalkonium chloride
204-493-5	n,n-Dimethylaniline
204-506-4	Isophthalic acid
204-528-4	Triisopropanolamine
204-557-2	Phenyl glycidyl ether
204-589-7	Phenoxyethanol
204-591-8	Dodecylbenzene
204-617-8	Hydroquinone
204-620-4	Glyceryl-1-allyl ether
204-623-0	Propionaldehyde
204-626-7	Diacetone alcohol
204-633-5	Isoamyl alcohol
204-634-0	Acetylacetone
204-646-6	n-Butyraldehyde
204-648-7	2-Pyrrolidone
204-650-8	Azodicarbonamide
204-658-1	n-Butyl acetate
204-661-8	1,4-Dioxane
204-664-4	Glyceryl stearate
204-666-5	Butyl stearate
204-669-1	Azelaic acid
204-673-3	Adipic acid
204-685-9	Diethylene glycol butyl ether acetate
204-693-2	Stearamide
204-709-8	2-Amino-2-methyl-1-propanol
204-771-6	Sucrose acetate isobutyrate
204-781-0	Neopentyl glycol
204-783-1	Sulfolane
204-800-2	Tributyl phosphate
204-809-1	Tetramethyl decynediol
204-812-8	Sodium octyl sulfate
204-820-1	Acetone oxime
204-825-9	Perchloroethylene
204-839-5	Diisobutyl sodium sulfosuccinate
204-876-7	Sodium dimethyldithio-carbamate hydrate
204-881-4	BHT
205-011-6	Dimethyl phthalate
205-016-3	Diallyl phthalate
205-026-8	Benzophenone-8
205-027-3	Benzophenone-6
205-029-4	Benzophenone-1
205-031-5	Benzophenone-3
205-087-0	Captan
205-271-0	Lauryl hydroxyethyl imidazoline

EINECS NUMBER TO CHEMICAL CROSS-REFERENCE

EINECS	Chemical
205-288-3	Zinc dimethyldithiocarbamate
205-341-0	dl-Limonene
205-392-9	Methyl acetyl ricinoleate
205-393-4	Butyl acetyl ricinoleate
205-399-7	Benzyl acetate
205-411-0	Aminoethylpiperazine
205-450-3	Diisobutyl adipate
205-455-0	Glyceryl ricinoleate
205-468-1	PEG-2 laurate
205-470-2	Ricinoleic acid
205-471-8	Methyl hydroxystearate
205-480-7	Butyl acrylate
205-483-3	Ethanolamine
205-500-4	Ethyl acetate
205-502-5	Mesityl oxide
205-516-1	Ethylacetoacetate
205-524-5	Dioctyl maleate
205-535-5	Sodium octyl sulfate
205-542-3	Propylene glycol laurate
205-550-7	Caproic acid
205-563-8	Heptane
205-570-6	Lauryl methacrylate
205-572-7	Hexyl acetate
205-592-6	Butoxytriglycol
205-597-3	Oleyl alcohol
205-619-1	2,2,4-Trimethyl-1,3-pentanediol
205-633-8	Sodium bicarbonate
205-769-8	Hydroquinone monomethyl ether
205-788-1	Sodium lauryl sulfate
205-840-3	Zinc 2-mercaptobenzothiazole
205-999-9	Triethylene diamine
206-101-8	Aluminum distearate
206-103-9	Oleamide
206-104-4	Lead acetate
206-392-1	Perfluoroheptane
206-585-0	Perfluorohexane
206-991-8	Silicon carbide
207-312-8	Dicyandiamide
207-355-2	Camphor
207-439-9	Calcium carbonate
207-938-1	ε-Caprolactone monomer
208-178-3	Abietic acid
208-394-8	1,2,3-Trimethylbenzene
208-407-7	Sodium gluconate
208-534-8	Sodium benzoate
208-736-6	Cetyl palmitate
208-750-2	Polyvinyl chloride
208-849-0	Barium acetate
208-909-6	Tetraisopropyl titanate
208-930-0	Methyl lactate
209-136-7	Octamethylcyclotetrasiloxane
209-150-3	Magnesium stearate
209-151-9	Zinc stearate
209-183-3	Polyvinyl alcohol
209-406-4	Dioctyl sodium sulfosuccinate
209-544-5	Toluene diisocyanate
209-954-4	DL-Lactic acid
210-827-0	Glycol dilaurate
210-848-5	Dimethyl maleate
211-014-3	Glycol distearate
211-020-6	Dimethyl adipate
211-047-3	Amyl acetate
211-074-0	Hexamethylene glycol
211-076-1	Ethylene glycol diethyl ether
211-185-4	Tetrabromophthalic anhydride
211-279-5	Aluminum tristearate
211-334-3	Manganese acetate
211-466-1	Isobutyl stearate
211-708-6	2-Ethylhexyl methacrylate
211-746-3	C12 dibasic acid
211-765-7	2-Methyl imidazole
211-776-7	1,2-Diaminocyclohexane
212-164-2	N-2-Aminoethyl-3-aminopropyl trimethoxysilane
212-485-8	Hexamethylene diisocyanate
212-782-2	2-Hydroxyethyl methacrylate
212-828-1	N-Methyl-2-pyrrolidone
213-048-4	Aminopropyltriethoxysilane
213-668-5	Hexamethyldisilazane
213-928-8	Dibutyltin diacetate
213-934-0	Vinyltris(2-methoxyethoxy) silane
214-073-3	Ethyl toluenesulfonamide
214-277-2	Dimethyl glutarate
214-292-4	Sodium cetyl sulfate
214-604-9	Decabromodiphenyl oxide
214-685-0	Methyltrimethoxysilane
214-737-2	Sodium myristyl sulfate
215-108-5	Bentonite
215-125-8	Boron oxide
215-136-8	Bismuth subnitrate
215-138-9	Calcium oxide
215-154-6	Cobalt oxide (ous)
215-156-7	Cobalt oxide (ic)
215-160-9	Chromium oxide (ic)
215-168-2	Ferric oxide
215-170-3	Magnesium hydroxide
215-171-9	Magnesium oxide
215-175-0	Antimony trioxide
215-185-5	Sodium hydroxide
215-204-7	Molybdenum trioxide
215-222-5	Zinc oxide
215-237-7	Antimony pentoxide
215-251-3	Zinc sulfide
215-267-0	Lead (II) oxide
215-269-1	Copper oxide (ic)
215-283-8	Zeolite
215-288-5	Montmorillonite
215-290-6	Lead carbonate (basic)
215-293-2	Cresylic acid
215-354-3	Propylene glycol stearate
215-355-9	Glyceryl hydroxystearate
215-475-1	Aluminum silicate
215-535-7	Xylene
215-540-4	Sodium borate
215-548-8	Tricresyl phosphate
215-566-6	Zinc borate
215-587-0	Phenol sulfonic acid
215-609-9	Carbon black
215-647-6	Ammonium hydroxide
215-661-2	Methyl ethyl ketone peroxide
215-662-8	Naphthenic acid
215-663-3	Sorbitan laurate
215-664-9	Sorbitan stearate
215-665-4	Polysorbate 80
215-665-4	Sorbitan oleate
215-683-2	Silica, hydrated
215-684-8	Sodium silicoaluminate
215-687-4	Sodium silicate
215-691-6	Alumina
215-710-8	Calcium silicate
215-713-4	Antimony trisulfide
216-008-4	Pentaerythrityl triallyl ether
216-087-5	Trifluoromethane sulfonic acid
216-472-8	Calcium stearate
216-653-1	Methyl t-butyl ether
216-700-6	Lauramine oxide
217-421-2	Benzophenone-12
217-847-9	Tributyltin fluoride
217-983-9	Methyltriethoxysilane
218-216-0	Octadecyl 3,5-di-t-butyl-4-hydroxyhydrocinnamate
218-320-6	Phenyl trimethicone
218-463-4	Lauryl acrylate
218-793-9	Ammonium lauryl sulfate
219-268-7	Tetrahydrofurfuryl acrylate
219-348-1	3,3´,4,4´-Benzophenone tetracarboxylic dianhydride
219-376-4	Butyl glycidyl ether
219-460-0	Dimethylaminoethyl acrylate
219-470-5	Drometrizole
219-518-5	Zinc laurate
219-553-6	2-Ethylhexyl glycidyl ether
219-784-2	3-Glycidoxypropyltrimethoxysilane
219-785-8	3-Methacryloxypropyltrimethoxysilane
220-045-1	PEG-6
220-120-9	1,2-Benzisothiazolin-3-one
220-219-7	Ditridecyl sodium sulfosuccinate
220-239-6	Methylisothiazolinone
220-266-3	Sodium p-styrenesulfonate
220-449-8	Vinyltrimethoxysilane
220-548-6	Ethylene glycol propyl ether
220-666-8	Isophorone diamine
220-688-8	Dimethylaminoethyl methacrylate
220-835-6	Dioctyl fumarate
220-941-2	Octyltriethoxysilane
221-066-9	Phenyltrimethoxysilane
221-109-1	Dihexyl sodium sulfosuccinate
221-279-7	Laureth-2
221-281-8	Laureth-5

EINECS	Chemical
221-284-4	Laureth-9
221-286-5	Laureth-12
221-304-1	Behenamide
221-416-0	Sodium laureth sulfate
221-498-8	Benzophenone-9
221-573-5	Octrizole
221-695-9	Hexabromocyclododecane
221-820-7	Sodium dinonyl sulfosuccinate
221-838-5	Copper nitrate (ic)
222-036-8	Dibromoethyldibromocyclohexane
222-206-1	PEG-9
222-217-1	2-(3,4-Epoxycyclohexyl) ethyltrimethoxysilane
222-273-7	Tetrasodium dicarboxyethyl stearyl sulfosuccinamate
222-376-7	Isononyl alcohol
222-540-8	Pentaerythrityl triacrylate
222-823-6	N,N-Butyl benzene sulfonamide
223-276-6	Distearyl pentaerythritol diphosphite
223-296-5	Sodium pyrithione
223-383-8	2-(3´,5´-Di-t-butyl-2´-hydroxyphenyl)-5-chlorobenzotriazole
223-445-4	Bumetrizole
223-672-9	Isopropyl glycidyl ether
223-805-0	Quaternium-15
223-810-8	p-Toluene sulfonyl isocyanate
223-861-6	Isophorone diisocyanate
224-588-5	Mercaptopropyltrimethoxysilane
224-644-9	Methoxybutyl acetate
225-091-6	Dioctyl terephthalate
225-306-3	Dimethylolpropionic acid
225-856-4	PEG-8
225-878-4	Butoxypropanol
226-002-3	Dimethylaminopropylmethacrylamide
226-029-0	Etocrylene
226-097-1	Laureth-4
226-241-3	Methoxypropylamine
226-827-9	Isopropyl thioxanthone (2-isomer)
227-006-8	Tetrabutyl titanate
227-561-6	Isobornyl acrylate
227-567-9	Bismuth subcarbonate
227-729-9	Sorbitan laurate
227-907-6	Amyl alcohol
228-220-4	2,2-Diethoxyacetophenone
228-250-8	Octocrylene
229-457-6	Hydroxymethyl dioxoazabicyclooctane
229-713-7	Diazabicycloundecene
229-722-6	Tetrakis [methylene (3,5-di-t-butyl-4-hydroxyhydrocinnamate)] methane
229-859-1	PEG-12
229-912-9	Sodium metasilicate
229-912-9	Sodium metasilicate pentahydrate
229-919-7	N-Cyclohexyl pyrrolidone
229-934-9	Trimethyl-1,3-pentanediol, 2,2,4-diisobutyrate
229-985-7	Diethylene glycol propyl ether
230-279-6	2-Dimethylamino-2-methyl-1-propanol
230-325-5	Aluminum stearate
230-426-4	2,2´-(2,5-Thiophenediyl)bis[5-t-butylbenzoxazole]
230-770-5	Nonoxynol-4
230-785-7	Tetrapotassium pyrophosphate
230-993-8	Dimethyl diallyl ammonium chloride
231-072-3	Aluminum
231-100-4	Lead
231-107-2	Molybdenum
231-115-6	Palladium
231-125-0	Rhodium
231-141-8	Tin
231-142-3	Titanium
231-152-8	Cadmium
231-153-3	Carbon
231-157-5	Chromium
231-158-0	Cobalt
231-159-6	Copper
231-159-6	Copper powder
231-175-3	Zinc
231-175-3	Zinc dust
231-180-0	Indium
231-198-9	Lead sulfate
231-493-2	β-Cyclodextrin
231-545-4	Diatomaceous earth
231-545-4	Silica
231-545-4	Silica gel
231-551-7	Sodium molybdate
231-554-3	Sodium nitrate
231-569-5	Boron trifluoride
231-588-9	Stannic chloride
231-596-2	Palladium chloride (ous)
231-633-2	Phosphoric acid
231-635-3	Ammonia
231-667-8	Sodium fluoride
231-743-0	Nickel chloride (ous)
231-749-3	Phosphorus trichloride
231-753-5	Ferrous sulfate heptahydrate
231-765-0	Hydrogen peroxide
231-767-1	Tetrasodium pyrophosphate
231-768-7	Phosphorus
231-784-4	Barium sulfate
231-801-5	Chromic acid
231-810-4	7-Ethyl bicyclooxazolidine
231-826-1	Calcium phosphate dibasic
231-847-6	Cupric sulfate anhyd.
231-847-6	Cupric sulfate pentahydrate
231-868-0	Stannous chloride anhyd.
231-889-5	Sodium chromate
231-890-0	Sodium hydrosulfite
231-900-3	Calcium sulfate
231-900-3	Calcium sulfate dihydrate
231-955-3	Graphite
232-077-3	(-)-α-Pinene
232-087-8	(+)-α-Pinene
232-089-9	Manganese sulfate (ous)
232-122-7	Bismuth oxychloride
232-188-7	Calcium fluoride
232-273-9	Sunflower seed oil
232-274-4	Soybean oil
232-276-5	Safflower oil
232-278-6	Linseed oil
232-280-7	Cottonseed oil
232-281-2	Corn oil
232-282-8	Coconut oil
232-290-1	Ceresin
232-292-2	Hydrogenated castor oil
232-293-8	Castor oil
232-296-4	Peanut oil
232-298-5	Petroleum distillates
232-299-0	Rapeseed oil
232-304-6	Tall oil
232-306-7	Sulfated castor oil
232-307-2	Lecithin
232-311-4	Menhaden oil
232-313-5	Montan wax
232-315-6	Paraffin
232-315-6	Petroleum wax
232-347-0	Candelilla wax
232-348-6	Lanolin
232-355-4	Cashew nut shell oil
232-373-2	Petrolatum
232-374-8	Pine tar
232-383-7	Beeswax
232-384-2	Mineral oil
232-391-0	Epoxidized soybean oil
232-399-4	Carnauba
232-410-2	Hydrogenated soybean oil
232-430-1	Lanolin alcohol
232-440-6	Hydroxylated lecithin
232-453-7	Mineral spirits
232-455-8	Mineral oil
232-475-7	Rosin
232-479-9	Pentaerythrityl rosinate
232-536-8	Guar gum
232-539-4	Karaya gum
232-541-5	Locust bean gum
232-549-9	Shellac
232-554-6	Gelatin
232-555-1	Casein
232-555-1	Milk protein
232-567-7	Amylase
232-674-9	Cellulose
232-679-6	Corn starch
232-683-8	Glycogen
232-686-4	Starch
232-734-4	Cellulase
233-007-4	α-Cyclodextrin
233-046-7	Phosphorus oxychloride
233-118-8	Hydroxylamine sulfate
233-139-2	Boric acid
233-149-7	Ferric phosphate
233-257-4	Manganese violet
233-293-0	PEG-4 oleate
233-471-8	Zinc borate

EINECS	Chemical
233-634-3	Ethyl 4-(dimethylamino) benzoate
233-660-5	Barium chromate
233-713-2	D-Lactic acid
233-792-3	Bismuth nitrate
234-325-6	Glyceryl stearate
234-394-2	Xanthan gum
234-406-6	Quaternium-18 hectorite
234-409-2	Zinc naphthenate
234-722-4	Ammonium molybdate
235-093-9	Sodium nonoxynol-6 phosphate
235-253-8	Aluminum silicate
235-340-0	Hectorite
235-374-6	Magnesium aluminum silicate
235-795-5	Sucrose benzoate
235-921-9	1,6-Hexanediol diacrylate
236-547-9	PPG-2 methyl ether
236-671-3	Zinc pyrithione
236-675-5	Titanium dioxide
236-878-9	Zinc chromate
237-511-5	Aminopropyltrimethoxysilane
237-574-9	Potassium tripolyphosphate
237-843-0	Zinc chromate
237-954-4	Dinonyl phthalate
238-305-8	Sodium lauroamphoacetate
238-310-5	Stearamide MEA-stearate
238-523-3	2,2´-Thiobis (4-t-octylphenolato)-n-butylamine nickel
238-877-9	Talc
238-878-4	Quartz
239-018-0	Ferric phosphate tetrahydrate
239-268-0	2-Acrylamido-2-methylpropanesulfonic acid
239-556-6	2-Methylpentamethylenediamine
240-029-8	2-Butanol
241-004-4	Sodium borohydride
241-467-2	N-(2,2,6,6-Tetramethyl piperidinyl-4) maleimide
241-482-4	γ-Cyclodextrin
241-640-2	Cetyl esters
243-468-3	Diiodomethyl p-tolyl sulfone
243-885-0	Tetrabromophthalate diol
244-289-3	Octyl dimethyl PABA
244-492-7	Aluminum hydroxide
244-501-4	Oleyl hydroxyethyl imidazoline
244-654-7	Mercury oxide (ic), red and yellow
245-876-7	γ-Ureidopropyltriethoxysilane
245-904-8	γ-Ureidopropyltriethoxysilane
246-140-8	Aluminum nitride
246-386-6	2,2-Dimethoxy-2-phenylacetophenone
246-467-6	Isocyanatopropyltriethoxysilane
246-562-2	Vinyltoluene monomer
246-563-8	BHA
246-614-4	Triisooctyl phosphite
246-680-4	Sodium dodecylbenzenesulfonate
246-770-3	Dipropylene glycol
246-771-9	2,2,4-Trimethyl-1,3-pentanediol monoisobutyrate
246-839-8	Xylene sulfonic acid
246-890-6	Disodium tetrabromophthalate
246-917-1	Dodecenyl succinic anhydride
246-998-3	Triisodecyl phosphite
247-098-3	Phenyl diisodecyl phosphite
247-500-7	Methylchloroisothiazolinone
247-555-7	Nonoxynol-5
247-556-2	Isopropylamine dodecylbenzenesulfonate
247-557-8	Calcium dodecylbenzene sulfonate
247-561-6	Sodium methylnaphthalenesulfonate
247-568-8	Sorbitan palmitate
247-569-3	Sorbitan trioleate
247-574-0	Hydroabietyl alcohol
247-658-7	Diphenyl isooctyl phosphite
247-710-9	Ammonium xylenesulfonate
247-759-6	Tris(nonylphenyl) phosphite
247-777-4	Diphenyl isodecyl phosphite
247-816-5	Nonoxynol-8
247-891-4	Sorbitan tristearate
247-952-5	Bis (2,4-di-t-butylphenyl) pentaerythritol diphosphite
247-977-1	Diisodecyl phthalate
248-052-5	DEDM hydantoin
248-133-5	Isooctyl alcohol
248-248-0	Oleyl hydroxyethyl imidazoline
248-289-4	Dodecylbenzene sulfonic acid
248-291-5	Nonoxynol-2
248-293-6	Nonoxynol-8
248-294-1	Nonoxynol-10
248-296-2	Potassium dodecylbenzene sulfonate
248-368-3	Diisotridecyl phthalate
248-406-9	TEA-dodecylbenzenesulfonate
248-666-3	Hydroxypropyl methacrylate
248-762-5	Nonoxynol-1
248-938-7	Sodium cumenesulfonate
249-079-5	Diisononyl phthalate
249-277-1	Sodium PCA
249-395-3	Propylene glycol myristate
249-448-0	Sorbitan dioleate
249-992-9	Nonoxynol-6 phosphate
250-696-7	Tridecyl stearate
250-705-4	Glyceryl stearate
250-913-5	Sodium cumenesulfonate
251-087-9	Octabromodiphenyl oxide
251-732-4	Glycol hydroxystearate
251-974-0	2[(-Hydroxymethyl) amino] ethanol
252-011-7	Polyglyceryl-10 tetraoleate
252-104-2	PPG-2 methyl ether
252-681-0	Methyldibromo glutaronitrile
253-149-0	Cetyl alcohol
253-407-2	Glyceryl oleate
253-458-0	PEG-8 laurate
253-521-2	Capryl hydroxyethyl imidazoline
257-048-2	Dimethyl oxazolidine
257-843-4	Lauroyl lysine
258-250-3	Ethylenebis dibromonorbornane dicarboximide
259-627-5	Iodopropynyl butylcarbamate
261-526-6	Tetradecabromodiphenoxy benzene
262-774-8	Isopropyl titanium triisostearate
262-977-1	Cocamine
262-991-8	Dihydrogenated tallow methylamine
263-020-0	Dimethyl cocamine
263-031-0	Hydrogenated tallow glyceride
263-080-8	Benzalkonium chloride
263-086-0	Dicocamine
263-087-6	Dicocodimonium chloride
263-090-2	Quaternium-18
263-107-3	Tall oil acid
263-109-4	Lead naphthenate
263-123-0	Hydrogenated tallow amide
263-147-1	Cocamine acetate
263-170-7	Cocoyl hydroxyethyl imidazoline
263-171-2	Tall oil hydroxyethyl imidazoline
263-189-0	Tallow propylene diamine
263-191-1	Tallow dipropylene triamine
264-038-1	Microcrystalline wax
264-151-6	Benzalkonium chloride
264-520-1	Octoxynol-1
264-705-7	Castor oil, dehydrated
265-134-6	Ozokerite
265-995-8	Carboxymethylcellulose sodium
266-231-6	Nonoxynol-9 phosphate
267-015-4	Methyl oleate
268-820-3	Ethyl hydroxymethyl oleyl oxazoline
268-938-5	Cocamidopropylamine oxide
269-035-9	Methyl rosinate
269-547-2	Sodium lauroampho-

EINECS	Chemical
	acetate
269-657-0	Soy acid
269-658-6	Hydrogenated tallow glycerides
269-919-4	Benzalkonium chloride
270-302-7	Lanolin acid
270-325-2	Benzalkonium chloride
270-407-8	Sodium C14-16 olefin sulfonate
270-666-7	Feldspar
271-557-7	Sodium lauryl sulfate
271-867-2	p-Cresol/dicyclopentadiene butylated reaction product
271-949-8	Sodium lauroamphoacetate
272-021-5	Cocaminobutyric acid
273-219-4	Quaternium-18 bentonite
273-257-1	Sodium lauryl sulfate
273-277-0	Ethylene distearamide
273-313-5	Vegetable oil
273-544-1	Benzalkonium chloride
274-273-1	C7-8 isoparaffin
274-357-8	Sodium hydroxymethylglycinate
275-286-5	Synthetic beeswax
276-951-2	Sorbitan tristearate
277-962-5	Phenol sulfonic acid
279-499-4	Poly (dipropyleneglycol) phenyl phosphite
280-084-5	Bis-(γ-trimethoxysilylpropyl) amine
281-065-4	Isopropyl thioxanthone (4-isomer)
283-117-2	N-Stearyl maleimide
285-550-1	PEG-2 stearate
286-074-7	Sorbitan trioleate
286-490-9	Glyceryl stearate
287-089-1	Benzalkonium chloride
287-824-6	Methyl stearate
288-048-0	Tallow dipropylene triamine
288-459-5	PEG-12 dioleate
288-668-1	Isobutyl stearate
290-588-7	Lead phthalate, basic
290-850-0	Disodium tallow sulfosuccinamate
293-029-5	Pentaerythrityl tetrastearate
295-625-0	Glyceryl (triacetoxystearate)
295-739-4	Sodium formaldehyde sulfoxylate
296-473-8	Kaolin
306-522-8	Glycol stearate
309-928-3	Sodium alumino sulfo silicate
404-640-5	Dipropylene glycol dimethyl ether

Glossary

absorption. The penetration of one into the inner structure of another ; it occurs between a liquid and a gas or vapor.

adsorption. Occurs when gases, vapors or materials which have been dissolved in liquid systems adhere to a solid material on contact.

agglomerate. Loose clusters of primary particles and/or aggregates which may be separated during dispersion.

antiblocking agent . A substance, such as a finely divided solid of mineral nature which is added to prevent adhesion of the surfaces of films made from plastic to each other or to other surfaces. These hard, infusible particles tend to roughen the surface and so maintain a small air space between adjacent layers of the film, thus preventing adhesion.

anticaking agent. An additive used to inhibit agglomeration.

antifloating agent. A substance that prevents the appearance of one pigment in a coating system 'floating' to the surface and creating a mottled appearance on the coated surface.

antiflooding agent. A substance which prevents the defect of flooding which appears soon after application, usually of gloss paints containing two or more pigments differing appreciably in particle size. The effect is gradual and uniform deepening of the color of the surface which dries to a deeper shade than the bulk of the paint.

antifoaming agent. *See* defoaming agent.

antifouling paint. Organic coating formulated especially for use on the hulls and bottoms of ships, boats, buoys, etc., to protect them from the attack by barnacles, teredos, and other marine organisms. The chief ingredient is metallic naphthenate as well as mercury compounds.

antifreeze agent. Any compound that lowers the freezing point of water.

antisag agent. A substance which prevents the downward movement of a paint film between the time of application and setting, resulting in an uneven coating having a thick lower edge.

antisettling agent. A substance that reduces the tendency of a pigment to settle in the paint upon storage.

antiskinning agent. A liquid antioxidant used in paints and varnishes to inhibit formation of an oxidized film on the exposed surface in cans, pails, or other open containers.

antistatic agent. The marked tendency of a coating to accumulate static charges which results in adherent particles, dust, and other foreign matter.

bactericide. Any agent that will kill bacteria.

bacteriostat. A substance which prevents or retards the growth of bacteria.

biocide. Material added to paints to obviate bacterial attack; a biocide functions by affecting the basic metabolism of bacteria and thereby interfering with enzyme synthesis.

binder. The nonvolatile part of a coating material excluding pigments and fillers, but including plasticizers, surface driers and other nonvolatile additives. The binder connects the individual pigment particles and the substrate, thus combining to make the finished coating.

bleeding. The diffusion of a coloring matter from a substrate and the discoloration resulting from such diffusion.

catalyst. A substance that promotes a reaction in which it does not participate; small amounts are needed.

CFR (Code of Federal Regulations). A codification of the general and permanent rules published in the Federal Register by the executive department and agencies of the federal government.

chalking. The condition of a paint surface which having lost most of its gloss, is coated with a white powder or chalk; it is the result of the photochemical breakdown of the surface layer of binder with the consequent release of pigment.

chelating agent. Substances used to remove ions from solutions, e.g., EDTA (ethylenediaminetetraacetic acid).

coating. Product in liquid or powder forms which form a covering layer on a substrate and has protective, decorative, or specific technical properties.

colorant. Any substance that imparts color to another material or mixture; broadly classified colors fall into three groups: natural pigments derived mostly from plant materials , inorganic pigments and lakes (combination of organic coloring materials with metals), and synthetic coal tar dyes.

corrosion inhibitor. The prevention of the electrochemical degradation of metals or alloys due to reaction with the environment, usually forming metal oxides.

cracking. A break in a paint film extending through the substrate.

crater. Small lipped depression in the coating caused when the wetting behavior of the film is disturbed.

critical pigment volume concentration (CPVC). The transition area to minimum gloss in the measurement of the pigment volume in the dry film. A number of properties of the paint film change rapidly at the CPVC, such as, decrease in gloss, which reaches a minimum; a rapid increase in permeability; and a decrease in scrubbability, cleanability, and flexibility. Hiding increases above the CPVC. Thus the volume of pigment is an important parameter in the formulation of paints.

crosslinking agent. A substance which aids in the attachment of two chains of polymers by bridges composed of either an element, a group or a compound which join certain carbon atoms of the chains by primary chemical bonds. Crosslinking can be effected by exposing a system to heat or by subjecting the polymer to high-energy radiation.

deaerator. A substance such as benzoin, which prevents the entrapment of bubbles of air or vapor beneath the surface of the film.

deflocculating aid. Substance which aids in the separation of primary pigment particles and aggregates to produce good flow behavior, high gloss, maximum hiding power, etc.

defoaming agent. A substance that reduces or inhibits foam formation due to proteins, gases, or nitrogenous materials which may interfere with processing; it coalesces the small bubbles before they rise to the surface of the liquid system.

ductility. The ability to withstand deformation without failure.

diluent. Addition of an organic liquid having no solvent power to a paint or lacquer to reduce viscosity and achieve suitable application properties.

dip coating. The complete immersion of an article to be coated, in a large tank containing a quantity of the paint to be applied. The article is then withdrawn and the coating is either allowed to air dry or is force dried; used for coating complex shaped objects.

dispersant. A surface active agent added to a suspending medium to promote uniform separation of extremely fine solid particles. often of colloidal size.

dispersion. A uniform distribution of pigments in binders or formulation components.

drier. A substance used to accelerate the drying of paints, varnishes, printing inks, etc., by catalyzing the oxidation of drying oils or synthetic resin varnishes.

drying agent. Substances with the ability to absorb moisture.

dye. Pigmented solid that is soluble in the paint medium. It is an organic material and more often used for decorative rather than protective coatings.

effective service. This period is the interval between the time of application of the paint system and the moment at which , through deterioration, it ceases to perform its required function.

emulsifier. A substance that prevents the separation of immiscible substances in an emulsion; helps to distribute evenly one substance in another; used to improve texture, homogeneity, consistency, and stability.

emulsion. A mixture of two immiscible liquids , one being dispersed in the other in the form of fine droplets, e.g., oil in water and water in oil. They are held in suspension by an emulsifier.

enamel. Opaque or semi-opaque vitreous composition applied by fusion to the surface of a metal at temperatures above 425 C. It protects against corrosion, decorates, and resists the attack of alkalies, acids, and other chemicals.

extender. Low-gravity material used in paints chiefly to reduce cost per volume by increasing bulk.

filler. An inert mineral powder with high specific gravity and used to provide a degree of stiffness and hardness as well as to decrease cost. Similar to extenders and diluents in their cost-reducing function. They have no reinforcing or coloring properties.

film formation. The transition of an applied coating material from the liquid to solid state. The film formation results from drying or curing. Both processes can take place simultaneously or one after the other.

fire/flame retardant. A substance applied to or incorporated into a combustible material to reduce or eliminate its tendency to ignite when exposed to a low energy flame such as a match or a cigarette.

flatting agent. A substance ground into minute particles of irregular shape and used in paints and varnishes to disperse incident light rays so that a dull or flat effect is produced. Some common flatting agents are: heavy metal soaps, finely divided silicas, and diatomaceous earth.

flocculation. The reversible aggregation of drops of particles in which interfacial forces allow the close approach or touching of individual units, but in which the separate identity of each unit is maintained.

foam. Stabilized gas in a liquid medium.

fungicide. Any substance which kills or inhibits the growth of fungi and is added to the coating at the formulation stage; especially used in paints and coatings for kitchens, food processing plants, dairies and breweries.

gel. The formation of a colloid in which the disperse phase has combined with the continuous phase to produce a viscous jelly-like product.

gel paint. A paint formulation which has a semi-solid or gel consistency when undisturbed, but which flows readily under the brush or when stirred or shaken. After removal of the stress, it becomes thick again and has little tendency to spill, drip, or run.

glaze. Transparent coating substance.

gloss aid. Additive which increases the sensory impression for the more or less directed reflection of beams of light on a surface.

grinding. Breaking up agglomerates into single particles or smaller agglomerates in the liquid.

hammer finish coating. Special effect coating which can be used to achieve intentional, visible and uniformly distributed irregularities on the surface of a coating film.

Hazardous Air Pollutants (HAPs). Airborne chemicals that cause serious health and environmental effects.

hydrophilic. Having an affinity to water; wettable by water.

hydrophobicity. Resistance of the interface between a given substrate and the air to being wetted by liquid water.

hygroscopic. Descriptive of a liquid or solid material that picks up atmospheric water vapor and thus acts as a drying agent, e.g., calcium chloride and silica gel.

inhibitive coating. Where pigments are used in the primer or in the coating to regulate corrosion. These types of coatings perform best in areas where the coating is subject to weathering,, atmospheric conditions, high humidities, or chemical fumes.

intermediate. A compound that acts as a stepping stone between a parent substance and an end product.

Lethal Dose Fifty (LD50). A calculated dose of a material which is expected to cause the death of 50% of an entire defined experimental animal population. It is determined from the exposure to the material by any other route than inhalation of a significant number from that population.

leveling agent. . Additives that allow the coating while it is still in liquid state to self-regulate the irregularities which occur during application.

lightfastness. The ability of pigments to maintain their color during exposure to sunlight .

mar resistance aid. Material added to a paint or coating that prevents scratching and blemishing of the applied surface.

matte agent. A compound which aids in the production of an uneven, uniformly rough surface.

mildewcide. A compound used to prevent the growth of parasitic fungi, usually stain producing, on organic materials, e.g., cresols, phenols, benzoic acid, formaldehyde.

Montreal Protocol. The terms 'Montreal Protocol' and 'the Protocol' refer to the Montreal Protocol on Substances that Deplete the Ozone Layer, a protocol to the Vienna Convention for the Protection of the Ozone Layer, including adjustments adopted by Parties thereto and amendments that have entered into force.

opacifier. Aids in hiding power by decreasing transparency. In paint, it refers to aiding in the concealment of another tint or shade, or the substrate itself, over which a coating is applied.

ozone-depletion potential (ODP). The term 'ozone-depletion potential' means a factor established by the Administrator to reflect the ozone-depletion potential of a substance, on a mass per kilogram basis, as compared to chlorofluorocarbon-11 (CFC-11). The factor is based on the substance's atmospheric lifetime, the molecular weight of bromine and chlorine, and the substance's ability to be photolytically disassociated, and on other factors determined to be an accurate measure of relative ozone-depletion potential.

paint. Collective term for a wide variety of coating substances based on organic film-forming substances.

pigment. An organic or inorganic chromatic or achromatic solid material, in the form of small discrete particles which is incorporated into but remains insoluble in the paint medium. It confers opacity and color to the paint and influences the degree of resistance to light of the film as well as modifying the flow properties of the liquid paint.

pigment to binder ratio. The weight ratio of the pigment and extender to the binder solids content and is used to classify paints according to their performance capabilities.

pigment volume concentration (PVC). The percent of the volume of the dried film that is occupied by the pigment.

plasticizer. An organic compound added to a high polymer both to facilitate processing and to increase the film flexibility and toughness of the final product by internal modification of the polymer molecule; some examples of plasticizers are dibutyl phthalate, dioctyl phthalate, trichloroethyl phosphate, and chlorinated paraffins.

polyurethane coating. Two-component systems which contain polyol and isocyanate components as characteristic film-forming substances.

powder coating. Paints in powder form which produce a coating on a substrate following application and melting.

primer. The first high-build pigmented coat applied to new surfaces or old cleaned surfaces prior to the application of other components of the finishing system.

propellant. A liquefied gas used to supply force to expel product.

refractive index. The ratio of the sine of the angle of incidence to the sine of the angle of refraction is the index of refraction of the second medium.

reinforcing agent. Fine powders, used in high percentages, to increase the strength, hardness, and abrasion resistance. The reinforcing effect is generally a function of the particle size of the powder. The finest is channel carbon black; other reinforcing agents are thermal and furnace blacks, magnesium carbonate, zinc oxide, kaolin, and hydrated silicas.

rheology control agent. An additive that affects the flow of the coating.

sealer. A low-viscosity material, not necessary pigmented, whose primary function is to penetrate and seal porous surfaces, such as masonry and wood.

sequestrant. A chemical compound that reacts with trace metal present in the environment to form a complex, so that the usual precipitation reaction of the trace metal is prevented.

shear rate. A measurement of the rate of fluid flow; for example, when two parallel surfaces are separated by a fluid film and are moving relative to one another, the shear rate is defined as the ratio of the relative velocity between the plates and the thickness of fluid film.

shear stress. A measurement of the force on a fluid which results in flow.

slip agent. A substance that increases the smoothness of a surface by providing a measure of hydrodynamic lubrication between sharp objects and the coating surface, thus making the coated surface more scratch resistant..

slushing agent. A nondrying oil, grease, or similar material used to coat metals to afford temporary protection from corrosion.

slurry. A thin watery suspension in which pigments are dispersed in the manufacture of paints.

solids content. The portion of a substance, given in percent by weight , which remains after the volatile components have been removed under set test conditions.

solvent. Volatile liquids added to paints and coatings to dissolve the resin component and to modify the viscosity of the coating.

stabilizer. Any substance which tends to keep a chemical compound mixture or solution from changing its chemical form or nature. Stabilizers may retard a reaction rate, preserve a chemical equilibrium, act as antioxidants, keep pigments and other components in emulsion form, or prevent the particles in a colloidal suspension from precipitating.

substrate. Any solid surface on which a coating of a different material is deposited.

surface active. Reducing the surface tension of a liquid.

surface tension. In any liquid, the attractive force exerted by the molecules below the surface upon those at the surface-air interface, resulting from the high molecular concentration of a liquid compared to the low molecular concentration of a gas. An inward pull or internal pressure, is thus created which tends to restrain the liquid from flowing.

surfactant. Any compound that reduces surface tension when dissolved in water or water solutions, or which reduces interfacial tension between two liquids or between a liquid and a solid; it is composed of molecules having both water soluble and water insoluble portions within the same molecule.

suspending agent. A substance that aids the formation of a suspension, i.e., a mixture of a solid dispersed in a liquid.

synergy. The chemical phenomenon in which the effect of two active components of a mixture is greater than the concentration of either component alone. A substance that induces this result when added to another substance is called a synergist.

tack-free. A coated surface which does not feel sticky to the touch.

tackifier. A substance that aids adhesion.

tensile strength. The maximum stress or load per unit cross-sectional area that it will withstand before failure. The tensile strengths of paints are usually determined by subjecting free films to tensile loading.

thickener. Any of a variety of hydrophilic substances used to increase the viscosity of liquid mixtures and solutions and to aid in maintaining stability by their emulsifying properties.

thixotropy. Time-dependent gel-sol-gel conversion which is reversible under isothermal conditions.

topcoat. The last coat applied to a substrate.

uv absorber. A substance which absorbs radiant energy in the ultraviolet wavelength. It is added to plastics, rubbers, and coatings to decrease light sensitivity and consequent discoloring and degradation.

vehicle. Term used in paint technology to indicate the liquid portion of a paint, composed of drying oil or resin, solvent and thinner in which the solid portions are dissolved or dispersed.

viscosity. The resistance to flow of a fluid; mathematically defined as the ratio of shear stress to shear rate.

volatile organic compounds (VOC). Chemical compounds that easily evaporate and contain one or more carbon atoms (organic). They include light alcohols, acetone, trichloroethylene, percholoroethylene, dichloroethylene, benzene, vinyl chloride, toluene, and methylene chloride. These potentially toxic chemicals are used as solvents, degreasers, paints, thinners, and fuels. Because of their volatile nature, they readily evaporate into the air, increasing the potential exposure to humans. Due to their low water solubility, environmental persistence, and widespread industrial use, they are commonly found in soil and ground water.

waterborne dispersion. A system containing two immiscible phases, water and a resin, in which one phase, (the resin) is dispersed as small particles in the continuous phase (the water).

water-dispersible system. A system consisting of 100% solids product that can be dispersed when it is mixed with water.

waterproofing agent. Any film-forming substance which coats a substrate with a water repellent layer, used in paint, rubber, or plastic films.

water-reducible system. A waterborne system that allows for the simple reduction of solids level by the addition of water.

water scavenger. A substance present in a mixture or added to a mixture to collect undesirable water in that mixture.

wax. Low-melting organic mixture or compound of high molecular weight, solid at room temperature and generally similar in composition to fats and oils except that it contains no glycerides. Properties associated with waxes are water repellency and thickening.

weathering. Term applied to the change, usually deterioration, that occurs from exposure while in service. Factors contributing to weathering are: sunlight, ambient temperature and humidity, atmospheric pollution, rainfall.

wetting agent. A surface-active agent which, when added to water causes it to penetrate more easily into, or to spread over the surface of another material by reducing the surface tension of the water.